S0-AVN-730

OLFACTION AND ELECTRONIC NOSE

OLFACTION AND ELECTRONIC NOSE

Proceedings of the 13th International Symposium on Olfaction and Electronic Nose

Brescia, Italy 15 – 17 April 2009

EDITORS

Matteo Pardo
SENSOR Laboratory,CNR-INFM, Brescia Italy

Giorgio Sberveglieri
SENSOR Laboratory,CNR-INFM, Brescia Italy
and University of Brescia, Italy

All papers have been peer-reviewed

SPONSORING ORGANIZATIONS
Alpha MOS
University of Brescia
CNR-INFM
ISOCS
IBEC
Fine Permeation Tubes
JLM Innovation
Labiotest
Osmo Tech
SACMI
SRA Instruments

Melville, New York, 2009
AIP CONFERENCE PROCEEDINGS ■ VOLUME 1137

Editors:

Matteo Pardo
Giorgio Sberveglieri

SENSOR Laboratory
CNR-INFM & University of Brescia
Via Valotti 9
I-25133 Brescia
Italy

E-mail: pardo@ing.unibs.it
sbervegl@ing.unibs.it

L.C. Catalog Card No. 2009903629
ISBN 978-0-7354-0674-2
ISSN 0094-243X
Printed in the United States of America

CONTENTS

KEYNOTE

SPECIAL SESSION: NEW MATERIALS, STRUCTURES AND DEVICE ARCHITECTURES

NANOSTRUCTURED SENSORS I

INSTRUMENTATION: SAMPLING I

SPECIAL SESSION: GAS SENSORS IN ROBOTICS

ACOUSTIC AND OPTICAL SENSING

DATA ANALYSIS

NANOSTRUCTURED SENSORS II

INSTRUMENTATION: SAMPLING II

SPECIAL SESSION: BIOLOGICALLY INSPIRED COMPUTATION FOR CHEMICAL SENSING

BIOLOGICAL AND BIOMETIC OLFACTION

NANOMEDICINE

INSTRUMENTATION: ELECTRONICS

APPLICATIONS

ANALYTICAL CHEMISTRY APPROACHES

APPLICATIONS: FOOD

POSTER SESSION I

BIOSENSORS

ELECTRONIC NOSES AND TONGUES INSTRUMENTATION: SOFTWARE / HARDWARE DESIGN

SAMPLING TECHNIQUES

SOLID STATE SENSOR TECHNOLOGY

ANALYTICAL CHEMISTRY TOPICS

POSTER SESSION II

APPLICATIONS

DATA ANALYSIS

Preface

The International Symposium on Olfaction and Electronic Nose (ISOEN), now at its 13th edition, addresses research in the field of gas sensors, artificial olfactory systems and natural olfaction.

The first symposium was held in 1994 in parallel to the EUROSENSORS conference in Toulouse (France) by the Alpha M.O.S. company. With a growing number of participants at the 1994 and 1995 meetings held in Europe, and based on the international attendance of the symposium, it was decided in 1996 to organise the symposium in Miami, USA. Since then, due to the success of the symposium on both continents, the location was alternated from Europe in Nice in 1997 to Baltimore in the USA in 1998.

With ISOEN 1999 in Tuebingen, the symposium organisation passed fromAlpha M.O.S. to an ad hoc instituted scientific committee. The symposium took then place predominantly in Europe. Until 2003 it had yearly frequence (2000 Brighton, 2001 Washington, 2002 Rome, 2003 Riga) and then biannual frequence (2005 Barcelona, 2007 St. Petersburg).

The 13th Edition of ISOEN organized by the Sensor Lab in Brescia, Italy, has a few new characteristics:

- ISOEN is organized in the framework of the recently instituted International Society for Olfaction & Chemical Sensing (ISOCS).
- Due to the unexpected number of submissions (more than 200) ISOEN has two parallel sessions with altogether 72 oral contributions.
- Referees' comments of papers submitted as orals were sent to authors and criteria have been explicitely stated for maximal transparency. All papers submitted as orals were sent for review to up to four referees. A mean of three reviews were received (at least two).
- Papers were corrected by authors following referees' comments. The proceedings are published online by the American Institute of Physics.

We wish to thank all reviewers and the technical program and local organizing committees for their efforts.

General Conference Chairman: Giorgio Sberveglieri
Technical Program Chairman: Matteo Pardo

Organizing Committee

General Conference Chair

Giorgio Sberveglieri — Brescia University, Italy

Local Organization Committee

Isabella Concina — CNR-INFM
Matteo Falasconi — Brescia University (Chair)
Thomas Heine — University of Tuebingen
Giselle Jimenez — Brescia University
Andrea Ponzoni — Brescia University
Silvia Todros — Brescia University

Technical Program Committee

Technical Program Chair

Matteo Pardo — CNR-INFM, Italy

TP Committee

Udo Weimar — University of Tuebingen, Germany
Julian Gardner — University of Warwick, UK
Santiago Marco — University of Barcelona, Spain
Peter Mombaerts — Max Planck Institute of Biophysics, Frankfurt am Main, Germany
Hiroshi Ishida — Tokyo University of Agriculture and Technology, Japan
Kenshi Hayashi — Kyushu University, Japan
Yongxiang Li — Shanghai Inst. of Ceramics, Chinese Academy of Sciences, China
Hitoshi Sakano — University of Tokyo, Japan
Andrei Kolmakov — Southern Illinois University, USA (Vice Chair)
Perena Gouma — State University of New York at Stony Brook, USA
Ken Suslick — University of Illinois at Urbana/Champaign, USA
Thomas Thundat — Oak Ridge National Laboratory, USA

ISOEN Steering Committee

Julian Gardner — Warwick University, UK (Chair)
Arnaldo D'Amico — Rome University, Italy
John Kauer — Tufts university, USA
Sandrine Isz — Alpha MOS, France
Andrey Legin — St. Petersburg University, Russia
Santiago Marco — University of Barcelona, Spain
Patrick Mielle — INRA, France
Krishna Persaud — Manchester University, UK
Jean-Pierre Rospars — INRA, France
Giorgio Sberveglieri — Brescia University, Italy
Joe Stetter — SRI International, USA
Kiyoshi Toko — Kyushu University, Japan
Udo Weimar — Tubingen University, Germany

Main Sponsor

During the 13th edition of the International Symposium on Olfaction and Electronic Nose, there will be for the first time an **Award for the best industrial presentation** on Electronic Nose or Electronic Tongue applications sponsored by Alpha MOS.

Sponsors

KEYNOTE

Progress of Biomimetic Artificial Nose and Tongue

Ping Wang* and Qingjun Liu

Biosensor National Special Laboratory, Key Laboratory of Biomedical Engineering of Education Ministry, Department of Biomedical Engineering, Zhejiang University, Hangzhou, 310027, P. R. China
State Key Laboratory of Transducer Technology, Chinese Academy of Sciences, Shanghai, 200050, P. R. China
*E-mail: cnpwang@zju.edu.cn *(P. Wang) Tel.: +86 571 87952832; fax: +86 571 87951676.*

Abstract. As two of the basic senses of human beings, olfaction and gustation play a very important role in daily life. These two types of chemical sensors are important for recognizing environmental conditions. Electronic nose and electronic tongue, which mimics animals' olfaction and gustation to detect odors and chemical components, have been carried out due to their potential commercial applications for biomedicine, food industry and environmental protection. In this report, the biomimetic artificial nose and tongue is presented. Firstly, the smell and taste sensors mimicking the mammalian olfaction and gustation was described, and then, some mimetic design of electronic nose and tongue for odorants and tastants detection are developed. Finally, olfactory and gustatory biosensors are presented as the developing trends of this field.

Keywords: Artificial nose; Artificial tongue; Biomimetic eNose; Biomimetic eTongue; Olfactory and gustatory cell sensors
PACS: 07.07.Df

INTRODUCTION

Olfactory and gustatory sensor systems are very useful in the biomedicine, food industry and environmental protection. At present, people have considerable amount of basic knowledge about vision and hearing. A robot possessing these senses has been put into practice already. However, the study of smell and taste is still at an early stage. Only a few kinds of olfactory and gustatory sensor systems are in commercial use. Therefore the artificial olfaction-electronic nose and artificial gustation-electronic tongue is very important for the future development.

We study the mechanisms of olfactory system using physiological quantitative technology based on physiological experiment of mammal. Thus provide basis for the study on the design of sense system, which simulate the sensing process of mammalian olfactory system. Otherwise, bionic modeling method and digital signal processing method are also used to accomplish the detection and recognition of mammalian olfactory information. During the study many bionic sensors and artificial neural network algorithms are developed in our laboratory.

And, based on the research about physiological mechanism of taste, we also designed different taste sensors, simulated and realized the apperception process of gustatory information by means of emulating and modeling. Besides, we realized the detection and differentiation of gustatory information utilizing the modern digital signal processing methods. Based on the research above all, we developed many kinds of apparatus systems to achieve sensing and detecting of taste intelligently.

In this report, the biomimetic artificial nose and tongue is presented. Firstly, the smell and taste sensors mimicking the mammalian olfaction and gustation was described, and then, some mimetic design of electronic nose and tongue for odorants and tastants detection are developed. Finally, olfactory and gustatory biosensors are presented as the developing trends of this field.

BIOMIMETIC ARTIFICIAL NOSE

Sensors of Artificial Nose

The recent advancement in micromachine techniques, such as chemical anisotropic etching and sacrificial layers adds a new dimension to the advancement of chemical sensor research. The sensor array can be constructed for a multichannel sensing system for identifying various gases and their concentrations. A thin oxide-based olfactory sensor

CP1137, *Olfaction and Electronic Nose: Proceedings of the 13th International Symposium*, edited by M. Pardo and G. Sberveglieri

and other chemical sensors will be used to illustrate the advantages of this novel sensor structure. The work includes the design of a novel integrated olfactory sensor array with heat insulation structure, microelectronic fabrication techniques, the design of independent gas sensing element and its control of operation temperature, the recognition methods of various gases and so on.

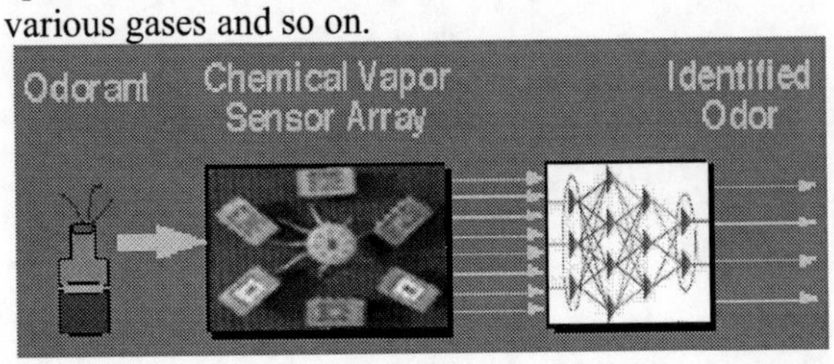

FIGURE 1. The structure and principle of artificial olfactory system with the sensor array and pattern recognition.

In our previous work, we used an odor or gas and taste sensor array combined with pattern recognition to implement the artificial olfactory and taste sensor system (Figure 1). Some integrated sensor arrays were produced. In order to optimize the system, a method that could adjust the structures of both the sensor array and the artificial neural network was introduced in the system [1]. Since the pattern recognition of odor and taste is carried out by the brain using neural networks and fuzzy logic, a pattern recognition method that combined artificial neural network and fuzzy recognition was applied to mimic human pattern recognition [2].

Applications of the Olfactory Sensor

From 1994, we began to study an on-invasive method to diagnose the disease based on an electronic nose, and in 1997 we published our first paper on the diagnosis of diabetes using a gas sensors array [3]. After that we tried to find an on-invasive and more convenient method to diagnose diabetes and lung cancer (Figure 2). In 2005, we published a paper on diagnosis of lung cancer based on a virtual SAW gas sensors array and imaging recognition [4]. And, these results and conclusions all depend on pathology statistics method to determine the biomarkers.

In our recent studies, an innovative method of lung cancer pathology analysis and early diagnosis at cell level has been introduced [5]. We used SPME-GC detection system to analyze the culture medium of several target cells including different kinds of lung cancer cells, normal lung cells, and other cells of cavy. Compared with the analysis results, volatile organic compounds can be seen in all lung cancer cells' culture medium, which are the metabolic product of lung cancer cells and can be viewed as biomarkers of lung cancer at cell level. Thus provide us a basis for electronic nose to monitor the healthy of our body and even diagnose non-invasively. Fortunately, in our recent studies, we have found the evidence, which prove that the breath diagnosis method is suitable for the lung cancer patients in early stage (Figure 3).

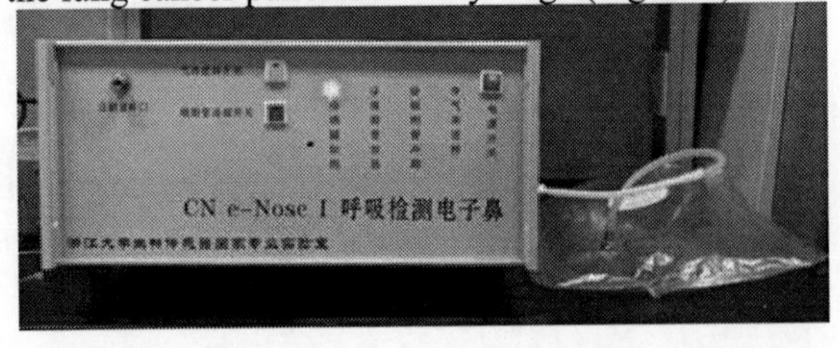

FIGURE 2. Electronic nose based on metal oxide sensors for diagnosis of diabetes and lung cancers from odors in breath of patients.

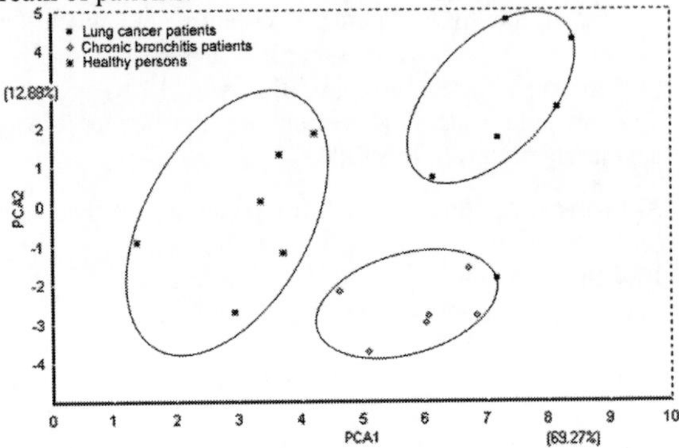

FIGURE 3. Principal component analysis (PCA) analysis results of lung cancer patients, chronic bronchitis patients and healthy persons.

BIOMIMETIC ARTIFICIAL TONGUE

Sensors of Artificial Tongue

Existing taste sensors can only detect some physics-chemical taste properties of the taste substance, and they can not simulate the taste sense of the real biological system. An effective method to solve this problem is to use a material which is very similar to the material existing in the real biological system as the sensitive membrane. Japanese researchers applied the biosensor mechanism and biological sensitive material to the realization of the taste sensor.

Recently, a novel integrated electronic tongue has been developed for the concentration determination of trace ions in aqueous media [6]. It combines an Au microelectrodes array (AuMEA) with a multiple light-addressable potentiometric sensor (LAPS) on a silicon substrate (Figure 4). AuMEA contains 30*30 individual 10 μm diameter Au disk microelectrodes, upon which thin mercury film is electrodeposited.

Based on our previous research of electronic tongue containing conventional electrodes and multiple LAPS, we improve our design by integrating AuMEA with multiple LAPS, and realize an inexpensive field-portable device for simultaneous determination of several heavy metals.

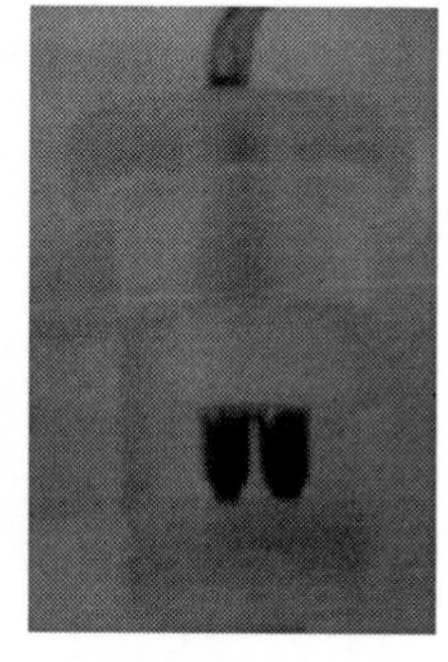

FIGURE 4. The taste sensor chip (left) and the field-portable device (right) of microelectrodes array with multiple light-addressable potentiometric sensor.

Applications of Artificial Tongue

In taste research, we introduced an experiment which uses a multichannel electrode taste sensor that is made up of bio-lipid material for taste sense. We adopted the dip method to make the lipid membrane. This method improved the characteristic of the membrane. Due to the fact that the thickness of the membrane is decreased by the dip method and the uniformity of membrane is improved by it. In another study we also improved this design by adopting the LAPS technique, in order to integrate many sensor devices on a single semiconductor substrate [7].

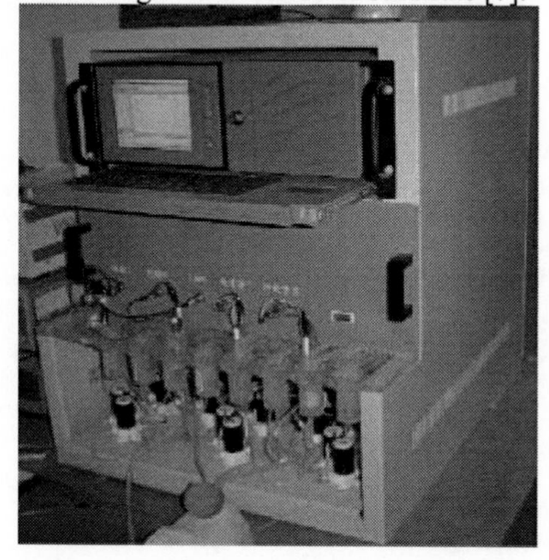

FIGURE 5. The taste sensor for detecting sea water.

There is an increasing need to measure rapidly heavy metals in various fields like the environmental detection or the industrial control. Convenient and applied instruments are rather keen to be realized. Stripping voltammetry method with the advances of high sensitivity and little sample, etc. has been proved feasible in marine trace metal measurements. It includes anodic stripping voltammetry (ASV) and adsorptive cathodic stripping voltammetry (CSV). We combined stripping voltammetry method with multiple-LAPS and completed an elementary design of an electronic tongue for heavy metal detection in sea water (Figure 5). Two kinds of chalcogenide glass thin films, respectively, sensitive to Fe(III) and Cr(VI) ions were prepared on the surface of the same LAPS. In the same measurement cell Mn was detected by ASV method, As by CSV method. In the other measurement cell Zn, Cd, Pb and Cu were detected by ASV method (Figure 6). Based on this device, we can effectively detect mixed solution with several kinds of heavy metals. This method can improve the accuracy and enlarge the detection kinds of heavy metals in waste water and sea water [6, 8].

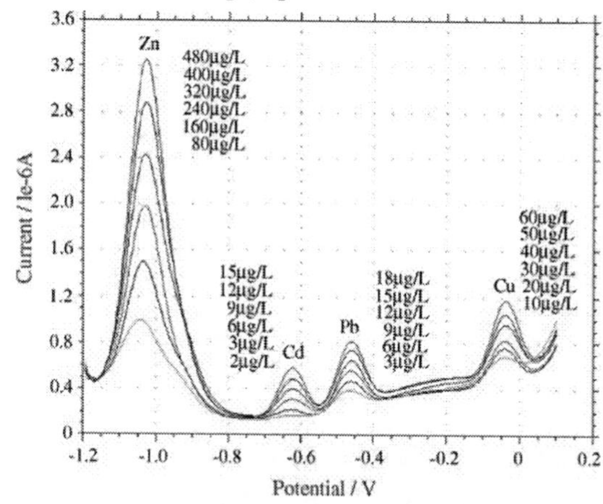

FIGURE 6. The taste sensor detecting results of Zn, Cd, Pb and Cu by anodic stripping voltammetry (ASV).

BIOMIMETIC ARTIFICIAL NOSE AND TONGUE USING BIOSENSORS

Biosensors for Artificial Nose and Tongue

Based on the olfactory and gustatory mechanisms, using biological functional components such as olfactory and gustatory organs, olfactory and gustatory receptor cells and even receptors themselves, many types of biomimetic olfactory and gustatory based biosensors have been fabricated by combining various secondary sensors, such as quartz crystal microbalance (QCM), field effect transistor (FET) microelectrode array (MEA), surface plasmon resonance (SPR). The great variety, exquisite specificity, high sensitivity and fast response of olfactory and gustatory biomaterials make them an ideal candidate for olfactory and gustatory biosensors.

Olfactory and gustatory biosensors have great potential commercial prospects and promising applications in such fields as biomedicine, environmental monitoring, pharmaceutical screening and the quality control of food and water. Meanwhile, olfactory biosensors also provide novel technologies for further studying olfactory and gustatory transduction, such as the identification of odor-receptor pairs and taste-receptor pairs, the characterization of moleculor and cellular physiology, consequently could promote the olfactory and gustatory studies greatly.

Olfactory and Gustatory Cell Sensor

Cells provide and express an array of naturally evolved receptor, ion-channels, and enzymes that may be the targets of biological or biologically active analytes. Cell-based biosensors that treat cells as biological sensing elements have the capacity to respond to analytes in a physiologically relevant manner. Our laboratory focuses on the theoretical design, fabrication technology and the design of detection system of cell-based biosensor following the principles of LAPS, MEA, and FET arrays [9]. We use those devices to monitor the extracellular microenvironmental ions (K^+, Ca^{2+}, H^+) and the action potentials of cultured cells (Figure 7).

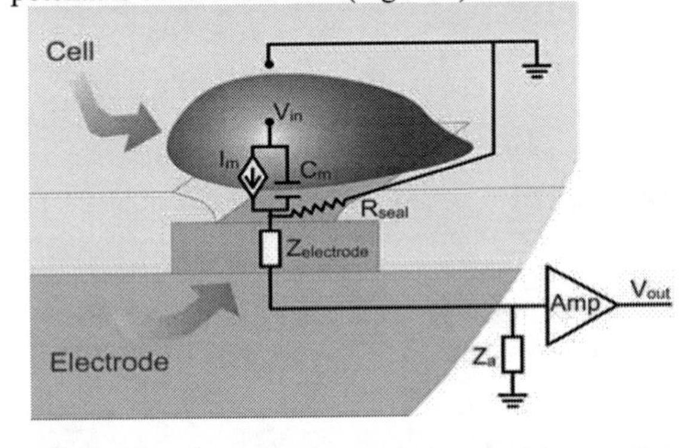

FIGURE 7. The principle of the cell sensor for extracellular potentials detection of the olfactory and taste cells.

Our laboratory have designed olfactory and gustatory cell-based biosensors to study the detecting progress of odors and taste in cell membrane, based on biomimetic olfactory biosensors and gustatory biosensors technology (Figure 8) [10].

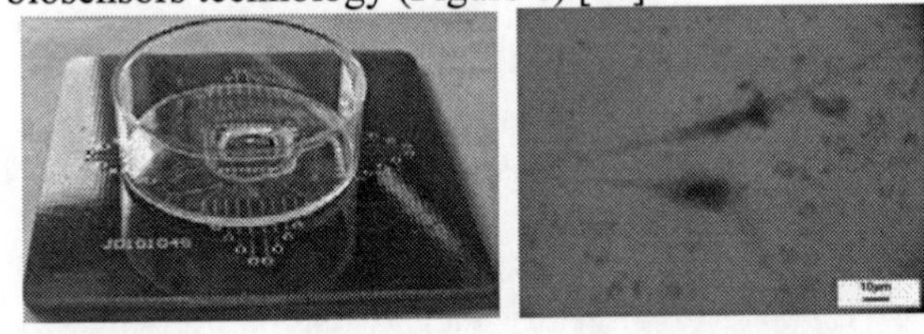

FIGURE 8. The cell culture device (left) and olfactory receptor cells cultured on the sensors (right).

At the same time, there is a growing interest in elaboration of biosensors based on the olfactory receptors expressed homologously and heterologously, employed in the membrane fraction or in the whole cell coupled to a solid transducer. We developed a QCM-based biosensor in which olfactory receptor expressed in a heterologous cell system was used as sensitive elements. An olfactory receptor protein of C. elegances, ODR-10, which was serve as a model of G-protein-coupled receptors (GPCRs), was expressed on the plasma membrane of human embryonic kidney (HEK)-293 cells. After cells containing ODR-10 was extracted and then coated on the electrode surface of QCM, the interactions between odorant molecules and olfactory receptors were monitored by QCM.

CONCLUSIONS

Though the history of study of biomimetic electronic nose and electronic tongue is not long, its progress has been rapid. Many methods have been applied in both the manufacture and signal processing of the sensor system, and some positive results applications have been obtained. On the basis existing research of the biomimetic artificial nose and tongue, it can be anticipated that some kind of sensor system with practical commercial use will appear in the next few years.

ACKNOWLEDGMENTS

This work was supported by the National Natural Science Foundation of China (Grant Nos. 30627002, 60725102, 30700167).

REFERENCES

1. P. Wang, L. Kong, X. Wang and J. Li, *Sensors Actuators B* **66**, 66-69 (2000).
2. P. Wang and J. Xie, *Sensors Actuators B* **37**, 169-174 (1997).
3. P. Wang, Y. Tan and R. Li, *Biosens. Bioelectron.* **12**, 1031-1036 (1997).
4. X. Chen, M. Cao, Y. Li, W. Hu, P. Wang, K. Ying and H. Pan, *Meas. Sci. Technol.* **16**, 1535-1546 (2005).
5. X. Chen, F. Xu, Y. Wang, Y. Pan, D. Lu, P. Wang and K. Ying, *Cancer.* **110**, 835-844 (2007).
6. H. Men, S. Zou, Y. Li, Y. Wang and P. Wang, *Sensors Actuators B*, **110**, 350-357 (2005).
7. H. He, G. Xu, X. Ye and P. Wang, *Meas. Sci. Technol.* **14**, 1040-1046 (2003).
8. W. Hu, H. Cai, J. Fu, P. Wang and G. Yang, *Sensors Actuators B,* **129**, 397-403 (2008).
9. P. Wang, G.X. Xu, L.F. Qin, Y. Xu, Y. Li and R. Li, *Sensors Actuators B,* **108**, 576-584 (2005).
10. Q.J. Liu, H. Cai, Y. Xu, Y. Li, R. Li and P Wang, *Biosens. Bioelectron.* **22**, 318-322 (2006).

SPECIAL SESSION:
NEW MATERIALS, STRUCTURES AND DEVICE ARCHITECTURES

Self-heated Nanowire Sensors: Opportunities, Optimization and Limitations

Evgheni Strelcov[1], Victor V. Sysoev[2], Serghei Dmitriev[1], Joshua Cothren[1] and Andrei Kolmakov[1*]

[1]*Department of Physics, Southern Illinois University, Carbondale, Illinois 62901, USA*
[2]*Saratov State Technical University, Saratov 410054, Russia*

Abstract. Semiconducting qusi-1D chemiresistors are nearly ideal substances for deployment of self-heating effect in gas sensorics. The latter is due to beneficial morphology of the nanowires, which hampers thermal losses to the environment, substrate and metal leads. Since the self-heated nanowire sensor device does not require an external heater, this effect can be used to fabricate nanoscopic sensors operable at room temperature with ultra-small power consumption. The analysis of the heat partitioning in self-heated nanowire device indicate the existence of its optimal architecture where power consumption of few μW is achievable in air.

Keywords: Oxide nanowire; self-heating; gas sensor; sensor array.
PACS: 07.07.Df; 68.47.Gh; 68.65._k; 73.63._b; 81.07._b; 85.85._j

INTRODUCTION

Many of the modern technologies and applications, which rely on gas sensors and E-nose systems, critically depend on the miniaturization of the active elements and minimization of the power consumption. On the other hand, the reversibility of the receptor function of the metal oxide sensing element requires an external heating to maintain its surface under elevated temperatures. Till recently, the compromise was addressed by using micro-machined micro-hot plates which reduced the overall dimensions of the sensor to sub millimeter size domain and power consumption to mW range [1]. The recent progress in implementation of metal oxide nanowires as chemiresistive sensing elements (see most recent reviews [2,3]) opens completely new opportunities in fabrication of nanosized and ultra-economic sensors and E-noses. Recently we demonstrated the possibility to use the Joule self-heating effect for operating metal oxide nanowire sensor at room temperature [2, 4]. This approach was successfully tested later in ref. [5].

The major advantage of the semiconducting nanowires with respect to self-heated thin film sensors [6] is their small thermal capacitance and drastic reduction of thermal losses to electrodes and gas environment. In this presentation, we elaborate on optimization of this kind of sensors.

EXPERIMENTAL AND METHODS

Synthesis of SnO_2 nanowires was conducted in a tube furnace via vapor-solid method. SnO precursor powder was heated at temperatures of 950 ^{0}C for 2 hours under 35 sccm Ar flow at pressure of 250 Torr. In order to eliminate the thermal losses into the substrate, the suspended geometry of the device is required. The planar suspended nanowire chemiresistor device was fabricated via dry transfer of the individual nanowires onto high-aspect-ratio walls of micro-machined glass (Fig. 1).

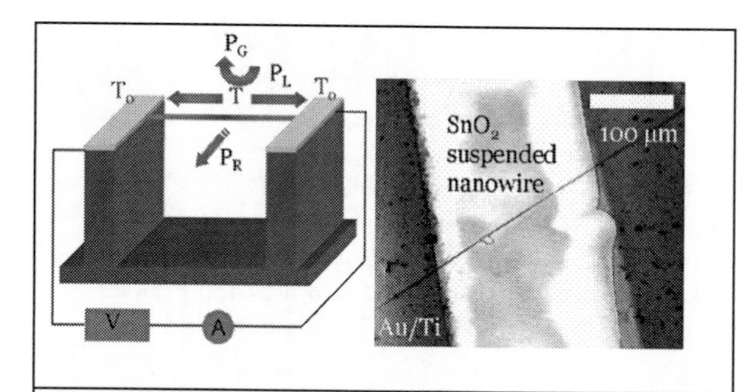

Figure 1. The sensor setup and principal thermal losses in the suspended nanowire heated by the Joule heat. SEM image of the suspended SnO_2 chemiresistor.

* Correspondence should be sent to akolmakov@physics.siu.edu

CP1137, *Olfaction and Electronic Nose: Proceedings of the 13th International Symposium,* edited by M. Pardo and G. Sberveglieri

Au/Ti electrodes were deposited on top of the nanowire using shadow mask deposition technique (see Fig.1).

In order to model redox reactions, the self-heated suspended nanowire was exposed to pulses of reducing gas (hydrogen) of $2\cdot10^{-4}$ Torr pressure in the background of oxygen of $1\cdot10^{-4}$ Torr pressure. Alternative device architecture was reported in [4] where the nanowire was suspended on Pt pads of the micro-hot plate.

RESULTS

The manifestation of the self-heating effect in the SnO_2 nanobelt can be seen as an increase of the current response (see insert in the Fig.2) to H_2 pulses (blue) in the background of oxygen with sequential increase of the voltage applied across the nanostructure. Fig. 2 demonstrates the increase of the sensor signal (reaction yield) and effective response rate (the rate of hydrogen oxidation reaction) concomitantly with increase of the Joule power released in the nanowire. As can be seen, the sensor is able to operate at 0.25 μW under these model conditions.

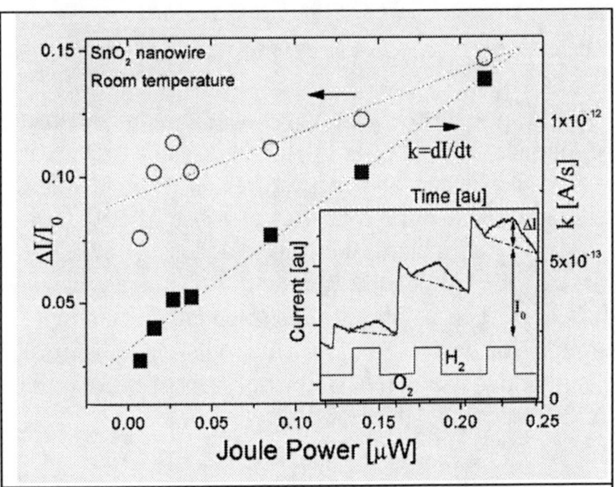

Figure 2. The increase of the sensor signal and decrease of the response time with Joule heat. NW and support are kept at room temperature.

To address the optimal parameters for suspended single-nanowire self-heated sensor, we have to consider a heat balance in the model device. As a reasonable assumption which simplify the analysis is to suggest a steady temperature gradient along the length of the nanowire from its center to metal leads.

The nanowire dissipates the Joule heat as: $Pj = I\cdot V \approx S\cdot n(T)\dfrac{e\mu V^2}{L}$ (where S is cross-sectional area of conducting channel in the nanowire, e is the elementary charge, μ, n are the mobility and concentration of electrons in the conduction band, L is the length of the nanowire).

For the case of suspended nanowire, this heat is transferred to the environment via three processes:

1. Losses to the contacts as $P_L \approx 4S_k\dfrac{(T-T_0)}{L}$.

2. Losses to the ambient gas as

$$P_G \sim \sqrt{\frac{R}{T_0}}\cdot\left(\frac{p_1}{\sqrt{M_1}}+\frac{p_2}{\sqrt{M_2}}\right)\cdot(T-T_0)\cdot S^*$$

(where S^* is the surface area of the nanowire; p_1, p_2 and M_1, M_2 are partial pressure and molar mass of the gases in the environment, respectively. In case of air and small concentrations of the analyte, p and M values can be taken for Nitrogen and Oxygen).

3. Radiation losses as $P_R \approx \varepsilon\cdot\sigma\cdot S\cdot\left(T^4-T_0^4\right)$.

At steady-state conditions, the equilibrium can be achieved at certain temperature T at the center of the nanowire. The results of the calculations are depicted in the Fig.3.

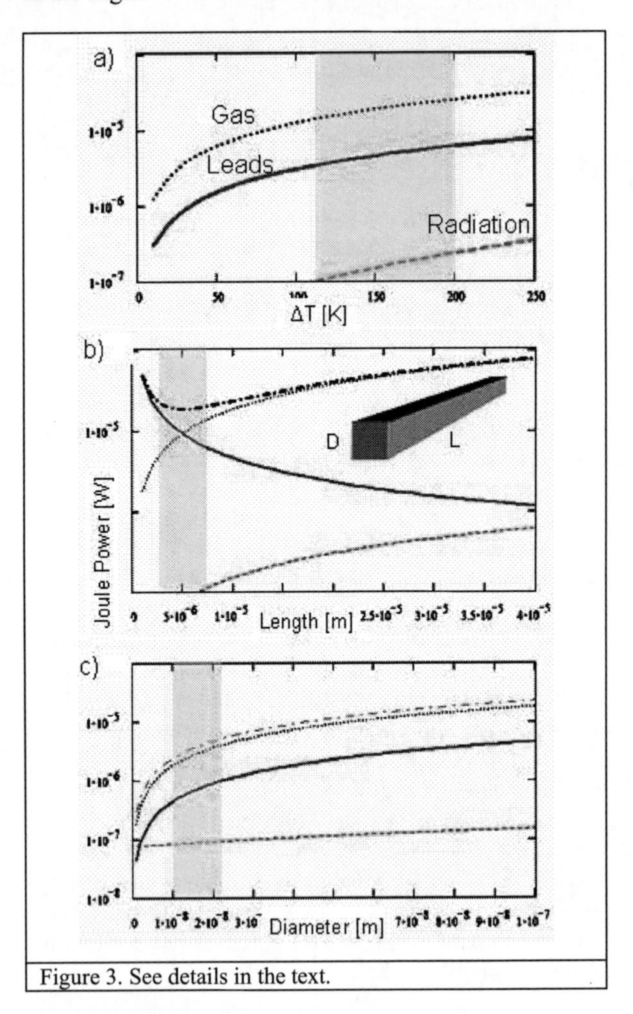

Figure 3. See details in the text.

As can be seen from the Fig.3, if we take a typical faceted nanowire resistor having a width of 100 nm and length of 10 microns, the thermal losses to the air would dominate over other channels in a practical temperature range. However, due to the fact that heat dissipation into the leads is decreased with the nanowire length, there exists an optimal nanowire length (depending on its diameter) when the cumulative thermal loss has a minimum (Fig. 3b). As can be seen from Fig. 3, in case of the typical nanowire this optimal length is ca 5 microns and only ca 2 μW is required to maintain the surface temperature of the nanowire center at 175 °C. Further reduction of the dissipated power can be achieved by reduction of the diameter of the nanowire (Fig. 3c).

DISCUSSION

It is apparent that the employment of self-heated nanowire chemiresistors will be beneficial for ultra-miniature sensors and E-noses. There are, however, some principal differences with respect to the sensors heated indirectly, which may impose some limitation on their application. There are two size domains in the self-heating approach depending on the ratio between nanowire diameter, D, and space charge width, W, at the nanowire surface:
(i) under $D >> W$, the redox reactions will not significantly affect the Joule heating of the nanowire;
(ii) while, under condition $D \approx W$, the adsorption of gases and redox reactions at the nanowire surface would "open" the conducting channel. The latter enhances the local temperature of the nanowire due to the Joule heat increase under constant bias condition. This positive feedback is favorable for sensitivity, though it eliminates the steadiness of the surface temperature and proportionality between the sensor response and the concentration of the analyte in environment. In addition, during the Joule heating, the steady-state temperature is not homogeneous along the nanowire length, what complicates the sensor signal and response time of self-heated chemiresistor.

ACKNOWLEDGMENTS

The research is partially supported through ACS PRF#45842-G5 and Caterpillar Inc. 21305 research grants. V.S. thanks Fulbright Scholarship for support.

REFERENCES

1. D. C. Meier, S. Semancik, B. Button, E. Strelcov and A. Kolmakov, *Appl. Phys. Letters* **91**, 063118 (2007).
2. A. Kolmakov, *International J. Nanotechnology* **5**, 250-274 (2008).
3. E. Comini, C. Baratto, G. Faglia, M. Ferroni, A. Vomiero and G. Sberveglieri, *Progress in Materials Science* **54**, 1-67 (2009).
4. E. Strelcov, S. Dmitriev, B. Button, J. Cothren, V. Sysoev and A. Kolmakov, *Nanotechnology* **19**, 355502 (2008).
5. J. D. Prades, R. Jimenez-Diaz, F. Hernandez-Ramirez, S. Barth, A. Cirera, A. Romano-Rodriguez, S. Mathur, and J. R. Morante, *Appl. Phys. Letters* **93**, 123110 (2008).
6. A. Salehi, *Sensors and Actuators B-Chemical* **96**, 88-93 (2003).

Metal Oxide Nanowires As Promising Materials For Miniaturised Electronic Noses

E. Comini, G. Faglia, M. Ferroni, A. Ponzoni, and G. Sberveglieri

CNR-INFM, SENSOR, Brescia University,
Via Valotti 9, 25133 Brescia, Italy

Abstract. Nanotechnology is in continuous evolution leading to production of quasi-one dimensional (Q1D) structures in a variety of morphologies: nanowires, nanotubes, nanobelts, nanorods, nanorings, hierarchical structures. In particular, metal oxides represent an appealing category of materials with properties from metals to semiconductors and covering practically all aspects of material science and physics in areas including superconductivity and magnetism. MOX nanowires are crystalline structures with well-defined surface terminations, chemical composition and almost dislocation and defect free. Due to their nanosized dimensions, they can exhibit properties significantly different from their coarse-grained polycrystalline counterpart. The increase in the specific surface causes an enhancement of the surface related properties, such as catalytic activity or surface adsorption, key properties for solid-state gas sensors development. The use of MOX nanowires as gas-sensing materials should reduce instabilities, suffered from their polycrystalline counterpart. The gas experiments confirm good sensing properties and the real integration in low power consumption transducers.

Keywords: Nanowires, metal oxide, chemical sensors.
PACS: 81.07.-b 07.07.Df

INTRODUCTION

Increasing concern on health hazard due to pollution [1] or terrorist attacks [2] encouraged an increasing research effort on gas sensing for real-time monitoring of all aspects of indoor and outdoor environments. Industrial requirements for a sensor are high sensitivity, high selectivity and good stability, together with low costs.

Among all possible technological approaches, conductometric gas sensors based on MOX semiconductors are the most promising for the development of low cost and reliable sensors. It has been confirmed that the sensitivity of MOX is improved as the crystallite size is decreased down to nanometer scale [3], thanks to the increased surface to volume ratio and the grains carriers' depletion. Unfortunately polycrystalline metal oxides suffer from grain coalescence induced by the high operating temperatures [4]. This affects sensor stability over long-term operation. On the contrary single crystalline nanosized MOX are stable materials and have the dimensions and surface to volume ratio required for high gas sensitivity.

The increasing scientific interest in Q1D systems such as NWs and nanorods has inspired their functional exploitation in new generation of electronic [5-7] and optoelectronic devices with improved performances [8,9].

The fabrication techniques of homogeneous 1-D nanostructures have pursued the control over shape, aspect ratio, and the crystalline arrangement to a considerable degree, and the improvement of the synthesis methods has recently achieved the fabrication of chemically non-homogeneous 1-D NWs, heterogeneous structures, and eventually the direct integration of functional nanostructures into nano-devices. [10]

After the first publication demonstrating the capability of NWs to detect gaseous species [15], huge research activity was carried out as demonstrated by the high number of publications and conferences on this topic. The challenge is the integration into sensing platforms necessary for the exploitation in the sensor market.

EXPERIMENTAL AND METHODS

Nanowires can be prepared by bottom up or top down approaches. The latter consist in reducing the lateral dimensions of a film to the nanoscale using

CP1137, *Olfaction and Electronic Nose: Proceedings of the 13th International Symposium,* edited by M. Pardo and G. Sberveglieri

standard micro-fabrication techniques. Top down approach uses the fully developed semiconductor industry technology and moreover the prepared NWs are on planar surface easing the manipulation and contacting issue. The drawbacks are a long preparation time and elevated costs. This technology can produce highly ordered NWs based structures [11-14], but does not accomplish the requirements for industrialization of NWs based devices. On the contrary bottom up approach consists in assembly of building blocks at molecular level or in synthesis by vapour phase transport, solution, electrochemical or template based techniques. The better crystallinity and purity together with smaller diameter, the possibility to easily form heterojunctions, the lower production costs make it the most promising method for NWs fabrication. The main disadvantage is the difficulty in integration in planar structures for nanodevices development.

For the fabrication of a conductometric sensing device the transduction of the signal produced by the sensing material is important. NWs are the sensing material doing a receptor function, thanks to the reactions between the gas and their surface, but the choice of the transducer and operation mode can affect the device response. Having in mind practical applications a proper design and choice of the transducers, electrical contacting, bonding and packaging should be made to accomplish each peculiar application requirements. In the case of NWs the contacting issue is much more difficult due to their peculiar morphology, they can be deposit after the NWs formation or NWs can be transferred to a transducer after removing from the growth substrate, dissolving in a solvent and using drop coating technique.

RESULTS

In 2002 Comini et al. [15] reported for the first time that use of nanowires network as conductometric gas sensor. Since then plenty of literature has been published on this topic. The electrical properties of MOX NWs were tested in a single NW or a multiple NWs configuration. We have focused our research on the second one since it does not require any nanomanipulation or expensive post processing procedures. Unfortunately some of the functional characterizations presented are not made in environmets reproducing real situations. Gas carrier, humidity and working temperature should be carefully selected and controlled. Humid air must be used as gas carrier, because oxygen and water vapor (present in all real environments) strongly interact with semiconducting surface and may change its interaction with the gas.

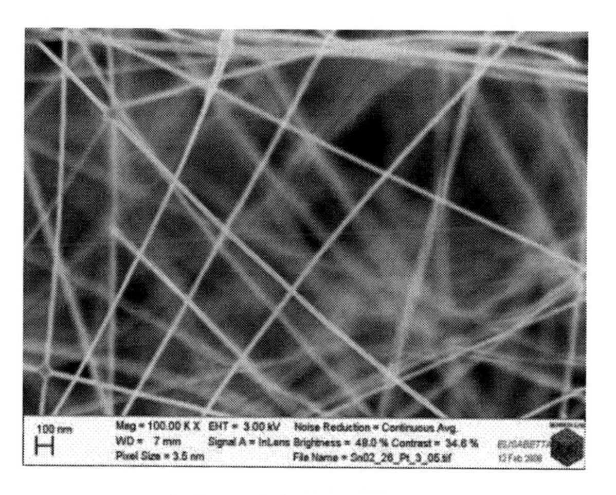

FIGURE 1. Tin oxide nanowires deposited with Pt catalyst at a deposition temperature of 390°C.

SnO_2 NWs resulted sensitive to gaseous polluting species like CO and NO_2, as well as to ethanol [15]. All measurements were done in humid air. Here the size of the NWs is large enough to exclude a full depletion. The conduction should be controlled by the barrier between different NWs.

After this first work, the research on MOX NWs based gas sensors continued with the exploitation of different MOX and with the deposition on low power consumption substrates for the integration of NWs into a miniutarised electronic nose thanks to their small dimensions.

Due to their high sensing performances, nanowire-based gas sensors have been tested in different fields, such as the detection of Chemical Warfare Agents (CWAs), where high sensitivity is mandatory. In this field, the threshold concentration is often expressed by means of the Immediately Dangerous for Life and Health value (IDLH).

It has been demonstrated that metal oxide nanowires can detect different CWAs simulants at concentrations close or even lower than the IDLH values of the respective real CWAs. As an example, the response of SnO_2 nanowires to ppm and sub-ppm concentrations of acetonitrile is shown in Figure 2. Acetonitrile is used in warfare applications as simulant for cyanide compounds, whose IDLH value ranges from few to few tens ppm, depending on the particular compound.

The capability of metal oxide nanowires to detect nerve agents simulants, as well as blistering compounds simulants, has also been demonstrated. Dimethyl methyl phosphonate (DMMP), a simulant for Sarin nerve agent, is detected at concentrations as low as 30 ppb [16], (Sarin IDLH is 50 ppb).

Dipropylene glycol methyl ether (DPGME), a simulant for blistering compounds, is detected at concentrations as low as 50 ppb [2], (IDLH for blistering agents varies from few ppm to few tens ppm depending on the blistering agent). These are only some of the possible use of MOX NWs for the detection of gaseous species; of course, since, like for all MOX based sensors, a real selectivity cannot be achieved, they must be used in an electronic nose configuration when operating in real environments with a proper data analysis.

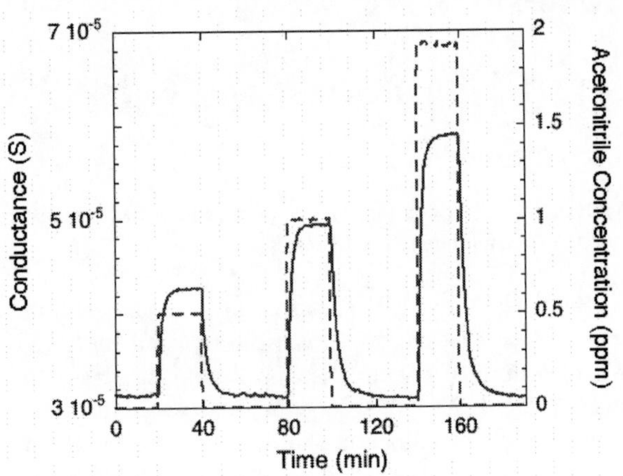

FIGURE 2. Dynamic response exhibited by SnO_2 nanowires to acetonitrile (simulant for cyanide-CWAs).

CONCLUSIONS

In the last year a great effort has been put into understanding and controlling the growth process for the preparation of high quality quasi one-dimensional nanostructures. Concerning gas sensing application MOX NWs demonstrate improved sensitivity to various gases. The greater surface to volume ratio, the better stoichiometry and greater crystallinity degree compared to polycrystalline oxides make these materials very promising for a better understanding of sensing principles and the development of a new generation of sensors. Selectivity still remains a concern for metal oxide based gas sensor. Possible ways to obtain different NWs response are surface coating with chemical selective membranes, modification by functional groups, or combining multi- component sensing modules coupled with signal processing functions, acting as an ''electronic nose'' to discriminate gas concentrations in a complex environment. The use of a single crystalline structure with lateral dimensions of less than hundreds nanometers allow the fabrication of an array of sensors in a chip with lateral dimensions of the order of microns.

Progress both in synthesis and manipulation of 1D nanostructures is developing fast and with it the range of materials prepared and foreseen applications. The future advances will increase quickly and in an unpredictable way.

ACKNOWLEDGMENTS

This work was partially supported, within the EU FP6, by the ERANET project "NanoSci-ERA: NanoScience in the European Research Area". This work was supported by the NATO project no. 982166 "Chemical threat detectors based on multisensor arrays and selective porous concentrators".

REFERENCES

1. Hoeck G, Bruneckreef B, Fischer P, Van Wijnen J(2001) Epidemiology 12:355-357
2. Ponzoni A, Baratto C, Bianchi S, Comini E, Ferroni M, Pardo M, Vezzoli M, Vomiero A, Faglia G, Sberveglieri G (2008) IEEE Sens. J. 8, 735
3. N. Yamazoe (1991) Sens. Actuators B 5-12
4. Guidi V, Carotta MC, Ferroni M, Martinelli G, Paglialonga L, Comini E, Sberveglieri G (1999) Sensors and Actuators B, 57: 197 – 200
5. Cui Y, Lieber CM (2006) Science 291:851
6. Cui Y, Duan X, Huang Y, Cui Y, Wang J, Lieber CM (2001) Nature 409:66, 291, 851
7. Huang, Y.; Duan, X.; Cui, Y.; Lieber, C.M. Nano Lett. 2002, 2, 101
8. Huang, Y, Duan X, Cui Y, Lieber CM (2002) Nano Lett 2:101,
9. Samuelson L (2003) Mater. Today 6, 22
10. Stellacci, F (2006) Adv. Funct. Mater. 16:15
11. Haghiri-Gosnet AM, Vieu C, Simon G, Mejıas M, Carcenac F, Launois H et al (1999) J Phys IV:92-133
12. Marrian CRK, Tennant DM et al (2003) Nanofabrication. J Vac Sci Tecnol A21:S207- S215
13. Candeloro P, Comini E, Baratto C, Faglia G, Sberveglieri G, Kumar R, Carpentiero A, Di Fabrizio E (2005) J Vac Sci Technol B 23: 2784-2788
14. Candeloro P, Carpentiero A, Cabrini S, Di Fabrizio E, Comini E, Baratto C, Faglia G., Sberveglieri G, Gerardino A. (2005) Micro Eng 178:78-79
15. E. Comini, G. Faglia, G. Sberveglieri, Z. Pan and Z. Wang, Applied Physics Letters 81, 1869 (2002)
16. Yu C, Hao Q, Saha S, Shi L, Kong XY, Wang ZL (2005) Appl. Phys. Lett. 86: 063101

Surface Ionization Gas Detection at SnO_2 Surfaces

A. Krenkow, C. Oberhüttinger, A. Habauzit, M. Kessler, J. Göbel, G. Müller

EADS Innovation Works, D-81663 München, Germany
angelika.krenkow@eads.net

Abstract. In surface ionization (SI) gas detection adsorbed analyte molecules are converted into ionic species at a heated solid surface and extracted into free space by an oppositely biased counter electrode. In the present work we consider the formation of positive and negative analyte gas ions at SnO_2 surfaces. We find that SI leads to positive ion formation only, with the SI efficiency scaling with the ionization energy of the analyte gas molecules. Aromatic and aliphatic hydrocarbons with amine functional groups exhibit particularly high SI efficiencies.

Keywords: SnO_2, metal oxide, surface ionization, gas detection
PACS: 82.45 Jn, 82.65 +r, 81.05 Hd, 82.50 Hp

INTRODUCTION

Resistive gas detection at metal oxide (MOX) surfaces is a widely investigated phenomenon. Commonly this phenomenon is explained in terms of a surface combustion process, in which atmospheric oxygen is ionosorbed at the MOX surface and in which the ionosorption is enabled by the transfer of electrons from the MOX conduction band to the adsorbed oxygen. In this way a reduced baseline conductivity is established. In case reducing analyte molecules such as hydrocarbons are co-adsorbed, these may undergo surface combustion events, emitting CO_2, CO and H_2O as neutral combustion products into the free space. Thereby some of the surface-trapped electrons are re-emitted into the conduction band again, resulting in an enhanced electrical conductivity which can serve as a sensor output signal.

The key essence of the resistive MOX gas detection process is that it relies on the exchange of electronic charge across the semiconductor surface. The principle idea in SI gas detection is that ionic species forming during ionosorption and surface combustion events may eventually be extracted from the MOX surface by applying a strong electrical field perpendicular to the MOX surface. Ion emission from MOX surfaces is a much less investigated phenomenon. Supporting evidence for the existence of SI comes from the work of Zandberg, Rasulev, and Nazarov [1, 2, 3] and from Morrison et. al. [4]. The first group studied SI gas detection on bulk noble metals and refractory MOX materials and the latter SI gas detection at heated Pt wires.

In the present work we have performed a screening study covering a wide range of possible analyte gases to assess their efficiency with regard to SI detection at SnO_2 surfaces. This study was performed in two steps: (i) optical investigations using UV and two-photon laser excitation [5] to map the energetic positions of molecular electronic states with regard to the electron receiving and electron donating states in the solid; (ii) actual SI detection experiments to assess the real SI efficiency of the investigated molecules. By a comparison of both data sets we try to elucidate the process of surface ionisation.

OPTICAL SCREENING OF ANALYTE MOLECULES

The basic principle of SI gas detection is displayed in **Figs.1a** and **b** with **a)** referring to positive and **b)** to negative ion emission on a SnO_2 surface. With SnO_2 being an n-type semiconductor the electron acceptor and donor states are predominantly those states lying close to the bottom of the conduction band.

Liberating an electron from the SnO_2 surface requires an energy input of $E = \Phi_{SnO2} \sim 3.8eV$. Ionizing a molecule M in free space requires a molecule-specific ionization energy $E_I = E_{vac} - E_{HOMO}$. Depending on the kind of molecule, E_I can range between 7 and 15 eV which means that the highest occupied molecular orbitals (HOMO) are always below the SnO_2 Fermi level. Assuming for the moment that HOMO and LUMO (lowest unoccupied molecular orbital) energies do not shift upon adsorption, one can

CP1137, *Olfaction and Electronic Nose: Proceedings of the 13th International Symposium*, edited by M. Pardo and G. Sberveglieri

expect that positive ion emission can occur, once it is possible to thermally excite a valence electron from the HOMO state of the adsorbate to the SnO_2 Fermi energy. This in any case requires a large energy input of several eV and therefore might only be expected at high temperatures and with the lowest-ionization-energy molecules. **Fig.1b** considers the case of negative ion emission. In this latter case, an electron from the conduction band needs to be excited from the SnO_2 Fermi level to the LUMO level of the adsorbed molecule M. Efficient negative ion emission in this case requires that the LUMO level is energetically close to the SnO_2 Fermi energy.

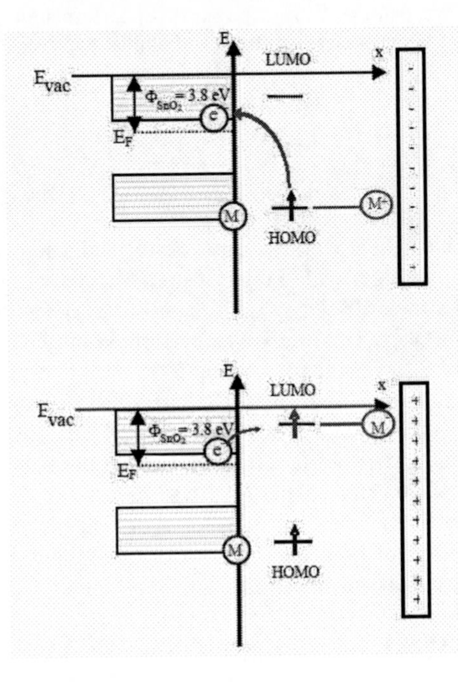

Figure 1. Electron transitions between adsorbed analytes and a SnO_2 surface leading to (a, top) positive and (b, bottom) negative surface ion emission.

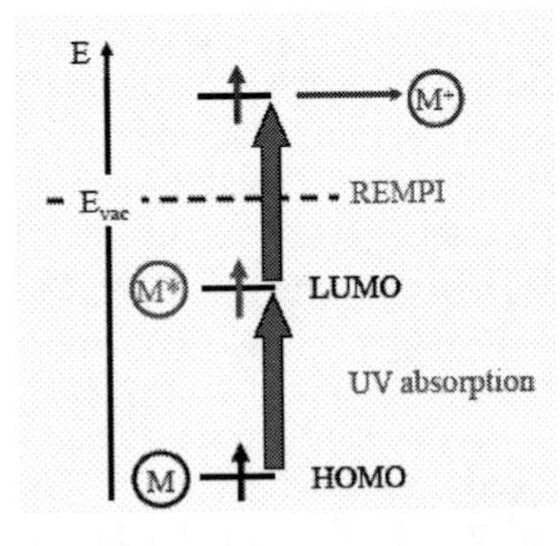

Figure 2. Identifying HOMO and LUMO energies in analyte molecules using single photon and two-photon UV absorption experiments.

In a search for analytes which might be good candidates for surface ionisation we have conducted both single-photon UV absorption as well as resonance enhanced two-photon laser ionisation (REMPI) experiments on a large number of molecules. As indicated in **Fig.2** such measurements yield the ionisation energy, i.e. the position of the HOMO energy relative to the vacuum energy E_{vac}, and the HOMO-LUMO splitting of the analyte molecules.

Both optical techniques yielded similar results for the HOMO-LUMO splittings. Quite generally also the LUMO states exhibited significant rotational and vibrational broadening. With the ionisation energies and HOMO-LUMO splittings having been determined, the LUMO energy positions can be compared to the SnO_2 conduction band edge. In this way, a map of potentially successful analytes can be drawn (**Fig.3**). Potentially successful candidates for positive ion emission are both aromatic (aniline, N,N-dimethyl-anline) and aliphatic hydrocarbons (dibutyl-amine) with amine functional groups. Such molecules derive from all kinds of organic decay reactions and are particularly odorous and often poisonous. Other hydrocarbon species and even air constituents such as O_2, H_2O and CO_2 are far less likely candidates.

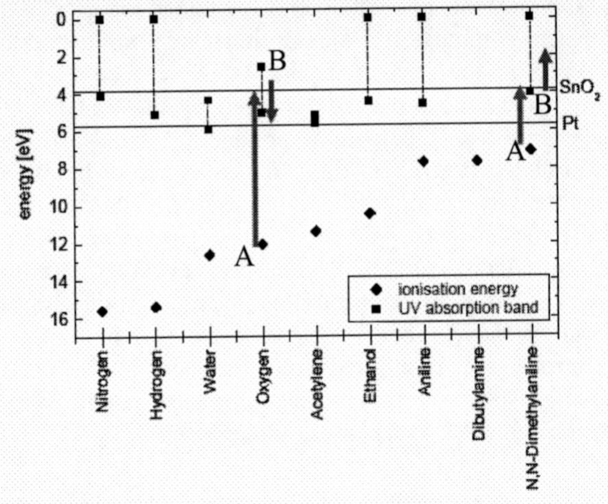

Figure 3. Energetic positions of HOMO and LUMO energies in a number of molecules as compared to the positions of the Pt Fermi energy and the SnO_2 conduction band. Arrows A indicate electron excitation processes leading to positive and arrows B ones to negative ion formation.

Concerning negative ion emission, **Fig.3** shows that many molecules exhibit LUMO bands in a broad range below E_{vac}. With the low energy wing of these bands often coinciding with the SnO_2 conduction band edge, it is suggested that conduction electrons may easily be transferred to the LUMO states of adsorbed molecules. We mention here that **Fig.3** predicts in particular that electronic charge should be easily transferred to O_2 and H_2O molecules adsorbed on SnO_2 surfaces. As is well known from resistive gas detection on SnO_2 surfaces, negative ionosorption is indeed observed upon O_2 and H_2O adsorption. Interestingly, a similar case of transfer is also suggested for a large number of other molecular species, in particular hydrocarbons.

ELECTRICAL DETECTION OF SURFACE ION EMISSION

For the actual detection of SI emission [6], SnO_2 films were deposited onto ceramic heater substrates. These substrates contained screen-printed Pt heater meanders and Pt thermometers at the back and Pt electrodes on the front side for contacting SnO_2 layers deposited onto this side. These electrodes firstly allowed resistive gas detection measurements and secondly SI currents to be detected that were extracted by an oppositely charged gold-plated counter electrode positioned about 1mm away from the SnO_2 surface [6]. Across this gap voltages up to +/- 1000V could be applied. The entire arrangement was contained in a stainless steel tube through which various gases and gas mixtures could be guided. Key results obtained with this arrangement are displayed in **Figs. 4**, **5** and **6**.

Fig.4 shows the variation of the ion currents with applied voltage using 1% ethene as an analyte gas. According to the electrode polarity the SI currents are positive. Secondly, the emitted ion currents steeply increase with increasing surface temperature. Qualitatively similar characteristics were observed with all kinds of analyte gases using both noble metal (Pt) and MOX (SnO_2, ZnO, CTO, $BaTiO_3$, BaO) materials as ion emitters.

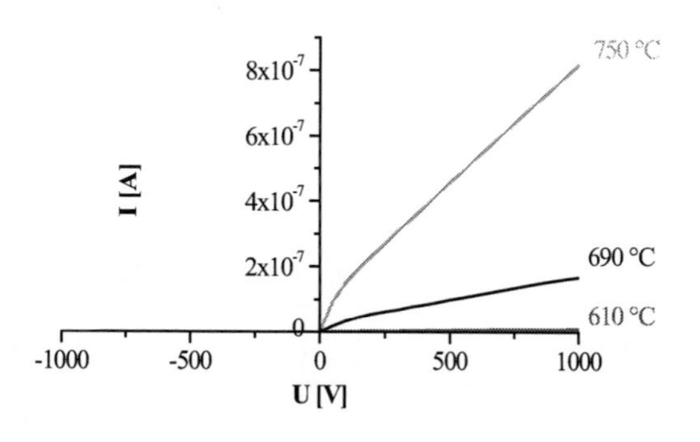

Figure 4. Ion current as a function of applied voltage as extracted from a screen-printed Pt surface operated at successively higher temperatures. The sample gas was 1% ethene diluted in synthetic air. Only positive ion emission is observed. The size of the Pt emitter area is 13 mm^2,

In the SI experiments that follow, the emitter material was biased with +1000V DC to observe positive ion emission. **Fig.5** shows that SI currents in response to toluene are not very sensitive to the kind of background gas as similar levels of toluene response were obtained both in N_2 and synthetic air (SA) backgrounds. This latter effect indicates that SI emission is unlikely to involve full or partial oxidation of the adsorbed toluene analytes.

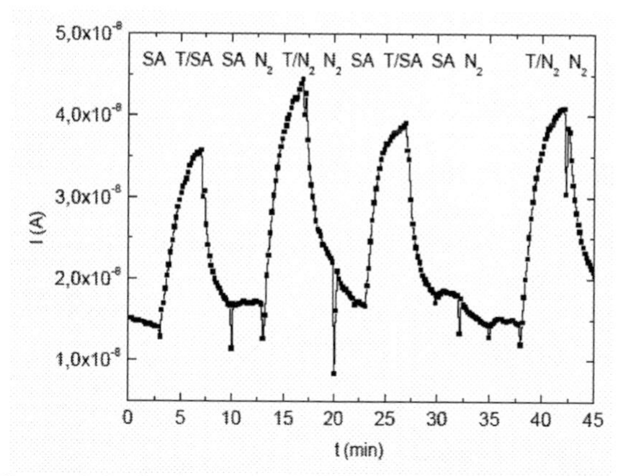

Figure 5. Ionization current induced by toluene (T) diluted in synthetic air (SA) and N_2, respectively (toluene concentration: 4.5%, flow rate: 450sccm, Pt Surface temperature: 750 °C).

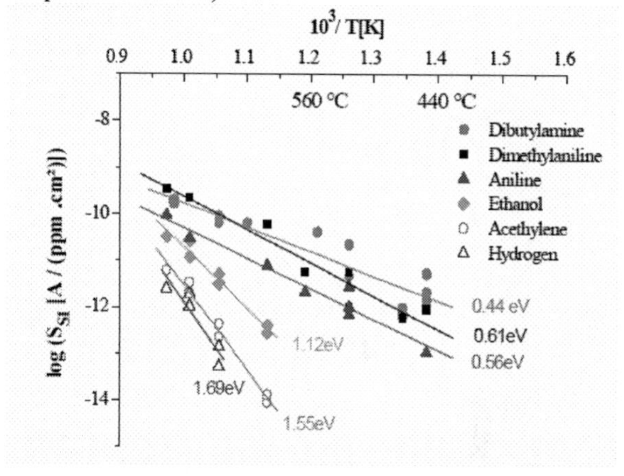

Figure 6. SI efficiency of aromatic and aliphatic hydrocarbons on SnO_2 films with a smooth surface morphology. Only amine-carrying molecules can be detected at relatively low temperatures (450°C to 550°C).

In order to assess the temperature dependence of the SI ion currents, data for different analyte gases were collected in Arrhenius plots as shown in **Fig.6**. For ease of comparison, the SI current values in this plot were normalized to the gas concentration applied and the heated emitter surface area. Whereas all analyte gases are detectable at the highest emitter operation temperatures (800 °C), the exponential drop-off of the SI efficiency towards lower temperatures leads to an increasingly clearer differentiation of the gas sensitivities. In this low temperature range only hydrocarbons with amine functional groups (dibutylamine, N,N-dimethylaniline, and aniline) could be detected in the ppm concentration range. All other gases, which do not contain such functional groups, cannot be detected, not even if their concentrations are raised to 1% or beyond in air.

Overall, all measured activation energies were much smaller than those extrapolated from **Fig.3**, which indicates that the free-space HOMO and LUMO energies suffer considerable shifts upon adsorption. The compilation of activation energy data in **Fig.7** indicates that the measured activation energies for positive ion emission amount in fact to only about 15% of the corresponding free-space HOMO-LUMO energy splittings. Such a strong shift indicates that ion emission occurs out of strongly chemisorbed rather than physisorbed molecular states.

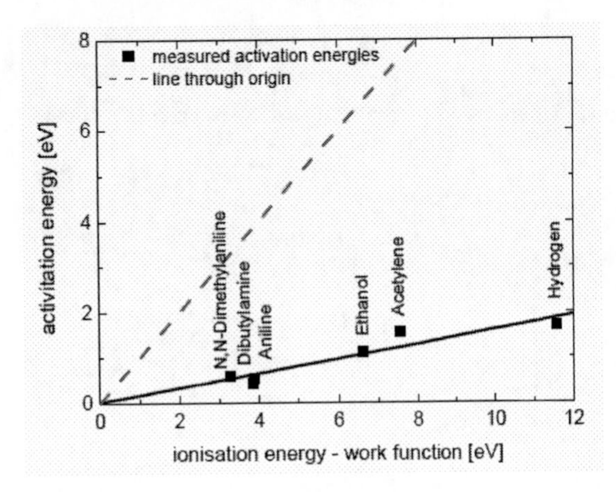

Figure 7. Measured (data points) and expected (dashed line) activation energies for SI emission from SnO_2 surfaces.

A scenario that can possibly explain the surprisingly low activation energies for positive ion formation is that all kinds of molecules, and in particular amines, do exhibit positive proton affinities. Considering for the sake of definiteness hydrocarbons with one or more amine functional groups, the energy expense for transferring of an electron from the adsorbed analyte to the adsorbent solid could be largely compensated by performing an intermolecular transfer of a proton from one of the hydrocarbon side groups (R) to the nitrogen lone-pair orbital of one of the amine functional groups. Such a reordering of chemical bonds inside an analyte will clearly have a large impact on the HOMO-LUMO states as actually observed in our experiments.

One final concern relates to the formation of negative analyte ions. The formation of O^- and NO_3^- ions is a well established fact in the field of resistive gas sensing. The evidence that has been produced in our experiments is that such ions cannot be extracted from a SnO_2 surface; neither could we get any compelling evidence for the extraction of other kinds of negative analyte ions. A possible explanation for such a failure could be the formation of strong covalent/ionic bonds between the adsorbates and lattice tin ions. Such strong bonds cannot be broken by the relatively weak external electric fields (~10^4V/cm) acting perpendicular to the emitter surface.

4. CONCLUSIONS

We have demonstrated positive ion emission from heated SnO_2 surfaces. Positive ion formation is a thermally activated process with the activation energy scaling with the difference of the molecular ionisation energy and the solid-state work function. Measured activation energies suggest that positive ion emission occurs out of strongly chemisorbed states. Positive analyte ion formation is unlikely to involve full or partial oxidation of the neutral analytes. It is likely to involve an electron transfer to the adsorbent solid and an additional inter- or intra-molecular transfer of a proton within the adsorbed analytes. Although the formation of negative surface ions appears to be energetically favourable, no compelling evidence for negative ion emission into the free space could be found so far.

ACKNOWLEDGMENTS

Part of this work was supported within the framework of the German national funded project "NACHOS" (CPNC03154705).

REFERENCES

1. U. Kh. Rasulev and E. Ya. Zandberg, Progress in Surface Science **28**, No. 3/4, 181-412 (1988).
2. U. Kh. Rasulev, U. Kasanov and V. V. Palitcin, Journal of Chromatography A **896**, No 1/2, 3-18 (2000).
3. U. Kh. Rasulev, E. G. Nazarov and G. B. Khudaeva, Journal of Chromatography A **704**, No. 2, 473-482 (1995).
4. W. M. Sears, V. A. Moen, B. K. Miremadi, R. F. Frindt and S. R. Morrison, Sensors and Actuators **11**, No. 3, 209-220 , (1987).
5. D. M. Lubman and M. N. Kronick, Anal Chem **55**, No. 6, 867 – 873 (1983)..
6. A. Krenkow, A. Habauzit, G. Müller, E. Comini, G. Faglia and G. Sberveglieri, IEEE Sensors, submitted 0/2008.

Nanotubes and Nanorods on CMOS Substrates for Gas Sensing

F. Udrea[1*], S. Santra[1], P. K. Guha[2], S. Z. Ali[1], J. A. Covington[2], W.I. Milne[1], J. W. Gardner[2], S. Maeng[3]

[1]Engineering Department, University of Cambridge, 9 J J Thomson Avenue, Cambridge CB3 0FA, UK
[2]School of Engineering, University of Warwick, Coventry CV4 7AL, UK
[3]Department of Electrical and Electronic Engineering, Woosuk University, Wanju, Jeonbuk, Rep. of Korea
**Corresponding author: email-Address: fu@eng.cam.ac.uk, tel: + 44 1223 748319, fax: +44 1223 748348*

Abstract. In this paper we discuss the combined use of integrated CMOS microhotplates employing nanomaterial sensing layers for intelligent, compact gas sensors with increased sensitivity, selecitivity and fast response times. We first review the status of nanomaterial-based gas sensors, their operating principles, discussing growth issues and their compatbility with CMOS substrates. We then describe Multiwall (MW) Carbon Nanotubes (CNTs) and ZnO Nanowires (NW) growth/deposition onto CMOS microhotplates. The paper continues by discussing the response of these nanomaterial sensing layers to vapours and gasses. Finally we discuss the future prospects of nanomaterial-based CMOS gas sensors, highlighting on one hand their future potential and on the other hand their present shortcomings and future challenges that need to be addressed before they can be released commercially.

Keywords: Gas Sensor, SOI CMOS, Carbon nanotube, Zinc oxide nanowire
PACS: 07.07.Df, 61.48.De

INTRODUCTION

Gas sensors are widely employed for a variety of applications, such as environmental monitoring (toxic and inflammable gases), automobiles, process control (e.g. petrochemical and food industry), personal safety, medical and even military scenarios. Commercial gas sensors are based on electrochemical, catalytic, optical or solid state technologies. These sensors suffer from poor compatibility with CMOS (complementary metal oxide semiconductor) processes and their power consumption is significant (hundreds of mW to 1W). Modern solid state (resistive) gas sensors are generally based on wide-bandgap semiconducting metal oxide sensing materials. The development of these gas sensors has been extensively studied in the literature [1-4]. They have attracted the attention of scientists and industrialists due to: relatively low cost, robustness, simplicity of their use, large number of detectable gases, possible application fields etc. However, they do suffer from selectivity problems, as the sensing material tends to be sensitive to more than one gas. In addition, batch-to-batch reproducibility of gas-sensitive thick films and the high power consumption have impeded the use of such sensors as accurate monitors of hazardous gases, while other sensor types (e.g. catalytic and electrochemical) are either too insensitive, power hungry or too expensive for mass markets.

In recent years, there has been increasing interest in integrating gas sensors with silicon CMOS technology. The aim is to increase the sensor's intelligence, reliability and ultimately reduce the physical dimensions of the system and its cost (by batch production). Nevertheless, the use of CMOS in sensors develops some constraints, e.g. any prolonged thermal treatment (which is often necessary for sensing material annealing) above 450°C of a CMOS substrate is not possible because of the low melting point of aluminium (Al, predominantly used in CMOS). In addition, most commercial gas sensor substrates use non-CMOS compatible materials to contact to the sensing film (such as Au or Pt). Hence the number of reported works [5-15] of CMOS gas sensors is still relatively small and there are few products available in the market. Last, but not least, CMOS is always marketed as a technology of low cost and high volume. In spite of its immense-potential, the gas sensor market is still considered today as being too small to justify the CMOS route.

CP1137, *Olfaction and Electronic Nose: Proceedings of the 13th International Symposium,* edited by M. Pardo and G. Sberveglieri

As already mentioned, CMOS comes with the 'power' of miniaturisation. The small size means more portability and lower cost, but conventional metal oxides (in the form of bulk material) are no longer effective in micro gas sensors, as the lower surface area and causes lower sensitivity. As a consequence, over the past couple of years, there has been a shift in sensor research towards more sensitive gas sensing layers. Nanomaterials are strong candidates for analytical detection, because of their reduced dimensions that create structures with exceptionally high surface area. Thus there is a potential increase in sensitivity even for a miniaturized sensor area. Researchers have already demonstrated growth or deposition of different nanomaterials and their promising performance in gas detection [16-20]. Thus, it is immensely beneficial to combine the two technologies (CMOS and nano-sensing material) to achieve a 'smart' low power, low cost, small size, reliable and reproducible sensor product, making them ideal for use in portable or wireless applications. The combination is also additionally useful as the sensor can be combined with on-chip circuitry for signal conditioning and to compensate for some of the short comings of the sensing material, i.e. drift, non-linearity, aging etc.

In this paper, we briefly discuss different nanomaterials used for gas sensor applications. We also review different processes for nanomaterials growth and discuss the challenges involved in integrating them with CMOS substrates. In the final part of the paper we review the work carried out at Cambridge and Warwick Universities, in the UK, to develop a CMOS gas sensor. Special emphasis is put on the design of CMOS micro-hotplates for low power and local growth of nanomaterials: carbon nanotubes (CNTs) and zinc oxide (ZnO) nanowires (NWs). Finally some gas testing results performed at Warwick University and ETRI (Korea) are included.

NANOMATERIAL GAS SENSOR

There have been reports of the use of different nanomaterials as gas sensing films. These materials show considerable resistance or temperature changes upon exposure to inorganic gases and volatile organics. Nanomaterials can be fabricated in the form of nanoparticles, nanoslabs, nanotubes, nanorods, nanowires etc.

Both multiwall (MW) and singlewall (SW) carbon nanotubes are strong contenders as nano gas sensing materials. Kong *et al.* [18] were among the first to demonstrate the response of Nanotube (NT) FET devices to NO_2 and NH_3 gases. After this initial report, in 2000, several groups [e.g. 21-24] have reported different Carbon Naotube (CNT) based chemical sensors. To date, researchers have already used CNTs to detect a variety of toxic gases including NH_3, NO_2, H_2, CH_4, CO, SO_2, H_2S and O_2 [25]. While CNTs show response to several gases, they seem to lack selectivity for a target gas. However, it is possible to increase the sensitivity and selectivity towards a particular gas by functionalizing the CNTs [e.g. 26-30]. This functionalisation can be done by metal and metal oxide decoration, polymer coating or atomic doping etc.

Another set of materials which have been widely studied in the literature are metal oxides. They are popular because of their better stability and their superior sensitivity towards a variety of gases. This is in contrast to CNTs that seem difficult to reproduce and often their drift in time and temperature is unacceptable. One idea that has been put forward is to create nano metal oxides taking advantage of both the nano geometry (with high surface area) while maintaining the stable sensing properties of metal oxides. Some of the materials already reported in literature for use as gas sensors are: Doped/ undoped ZnO nanowires, nanorods [31-35], tin oxide nanowires, nanoslab [36-39], titanium oxide nanoparticles, nanowires [20, 40], indium oxide [41, 42] and tungsten oxide nanoparticles [43].

However, all the above mentioned work is not on CMOS substrates. This is predominantly because of the constraints imposed by the CMOS process. In particular CMOS substrates are more sensitive to harsh environments (both thermal and chemical) that are often required for the synthesis and annealing of nanomaterials.

NANOTUBE AND NANOWIRE GROWTH OR DEPOSITION

There are many conventional ways to grow or deposit nanomaterials, however few are suitable for use on CMOS substrates. Two popular techniques are to grow nanomaterials by thermal or plasma enhanced (PE) chemical vapour deposition (CVD) and hydrothermal methods. Thermal CVD growth requires the substrates to be heated to a very high temperature (generally more than 500°C-depending on the specific recipe) which can be too high for the on-chip circuitry and internal metal layers. In PECVD the use of plasma can significantly damage the fragile micro electrical mechanical system (MEMS) structures. In addition, both the CVD methods are quite expensive. On the other hand, hydrothermal growth is simple, low cost and most importantly CMOS friendly. Though care needs to be taken to ensure that the chemicals used do not affect the passivation layers on chip. Apart from the above two methods, researchers have also been using commercially available nanomaterials to deposit onto the devices by different techniques such as drop coating, dip coating, inkjet printing, spray coating, spin coating etc.; so that they can avoid the potential harsh environments necessary for nanomaterial growth. Even though these methods are low cost and CMOS friendly, the dispersion is difficult due to strong Van der Waals force. Thus, deposition of agglomerated bunches of nanostructures is inevitable and this leads to the poor sensitivity and slow response. Also these methods have their own problems, e.g. in inkjet printing one needs to use a very dilute solution to avoid any nozzle clogging or in case of dip coating one needs to cover bond pads to avoid any chemical contaminations. It is also necessary to mention that some of the above methods (except inkjet printing and drop coating) require conventional lithography approach (i.e. steps like deposition of photoresist, exposure of ultra violet light, lift off etc.) or use of some masking process (e.g. shadow mask) to grow/deposit nanomaterials on predefined areas of the chip. Furthermore, the use of these techniques is strictly speaking restricted to low volumes and hence negates the use of CMOS technology.

Here we will discuss three methods of growing nanowires and nanotubes on a fully processed CMOS substrate. To start with, we will briefly describe the Cambridge-Warwick CMOS chemical sensors.

The starting device is a micro-hotplate structure – a resistive metal heater thermally isolated from the substrate by a membrane formed by a back-side etch. Such micro-hotplate structures are used in micro-gas sensors to provide a high temperature during operation for better sensitivity and faster chemical response time.

Our devices have been fabricated using a commercial SOI (silicon on insulator) CMOS process to form the heater as well as the interdigitated gas sensing electrodes using the top metal layer – that is exposed using the same process step that is used to etch the passivation layer above the bond pads. The electrodes are used to measure the resistance of the gas sensing material that is grown/deposited onto the device. The CMOS process step is followed by a back-side Deep Reactive Ion Etching (DRIE) step to form the membrane. Several devices have been fabricated – some with aluminium metallization, and others with tungsten (W). The tungsten metallization was used in the process to sustain high temperatures as conventional CMOS metal (Al) cannot achieve temperatures in excess of 300°C reliably.

The basic structure of the chemical sensor is shown in Fig. 1. The optical microscope picture of the fabricated device is shown in Fig. 2. SOI is particularly effective in this process, as the buried oxide acts as an etch stop during membrane formation. The membranes are very thin (~5 μm) yet very robust. The microhotplaes using such thin membranes formed by CMOS dielectric layers are particularly power efficient. At 600°C the power consumption is merely 45 mW for a membrane diameter of 500 μm and a heater diameter of 150 μm. The thermal mass of the membranes is very small and hence they have very fast transient time (10 ms to 600°C), which enables pulse mode operation, further reducing the power consumption. The details of the micro-heater design and characterisation were reported in [44]. The power versus temperature plots of the tungsten and aluminum micro-heaters is shown in Fig. 3 (without sensing film).

We were successfully able to grow/deposit ZnO nanowires and CNTs on our CMOS substrates. We give here, as examples, two different methods to grow/deposit ZnO nanowires, and one method to grow MWCNTs.

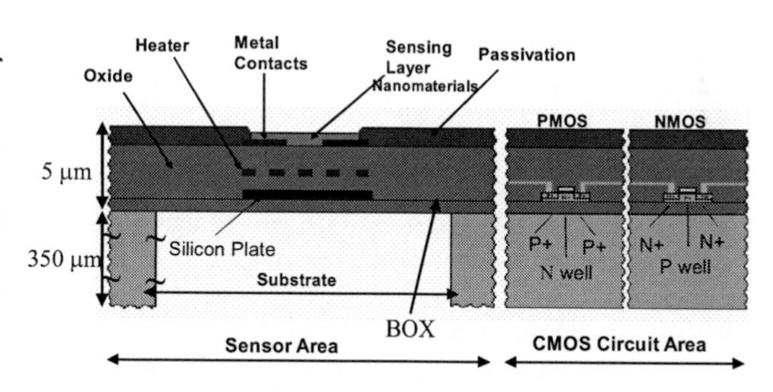

Fig. 1. Cross sectional view of the ultrathin (5 μm) SOI micro-hotplate and the CMOS electronic cells.

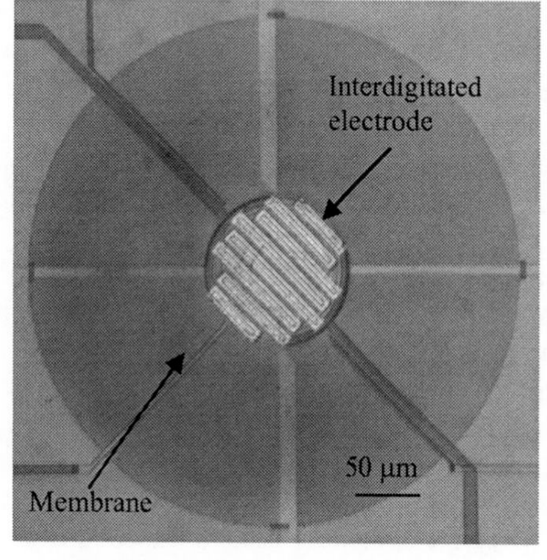

Fig. 2. An optical microscope picture of the fabricated micro-hotplate with interdigitated electrode.

(i) We have used hydrothermal methods to grow ZnO NWs on our above mentioned CMOS sensor device. For ZnO NWs growth, a thin ZnO seed layer was sputtered (~5 nm) on selective areas of our CMOS devices using a metal (shadow) mask. This was followed by dipping in an equimolar (25 mM) aqueous solution of zinc nitrate hexahydrate ($Zn(NO_3)_2.6H_2O$) and hexamethylenetramine (HMTA) at 90°C for two hours [45]. The devices were removed from the solution at the end of the growth, rinsed with de-ionised water and dried under nitrogen flow. The chips were then annealed at 300°C for one hour. The surface morphologies of the samples and size distribution of the nanowires were characterised using a field emission scanning electron microscope operated at 10 keV, as shown in Fig. 4. Typical Nanowire length was ~400 nm and their diameter was ~ 60 – 80 nm. These nanowires are touching each other and hence provide electrical paths between the pads of the interdigitated electrodes.

This deposition method is simple, relatively low cost and most importantly CMOS compatible (although in large volume the cost of it cannot be neglected compared to other costs). Nevertheless, ZnO NWs can be simultaneously grown on more than one micro-hotplate and hence this method can be extended to wafer level fabrication.

(ii) We also reported on the growth of ZnO NWs without using a ZnO seed layer [46]. Here conventional lithography was used to deposit on-chip NWs in predefined areas, although deposited NWs are non-uniform because of the absence of seed layer. The SEM picture is shown in Fig. 5.

(iii) Here we have taken a novel approach of growing CNTs on standard CMOS substrates. The technique is known as 'local growth' [47, 48], where, instead of heating the whole chip/wafer, local heating (using the tungsten micro-heaters) is used to grow CNTs on-chip (i.e. in-situ) over a single micro-heater region. This method allows unique control over the position (self aligned to the hotplate region) and time of growth.

For (iii) CNT growth, firstly the chips were covered (except the bond pads of the chips which were masked to avoid shorting them) with a 2 – 4 nm layer of iron (Fe) catalyst using sputtering. Then the devices were mounted onto a ceramic package and connected within a printed circuit board to a power supply. The chips were then transferred to a CVD chamber to grow the CNTs. The chamber was pumped down to 0.2 mbar using a rotary pump. The micro-heater was powered through a computer controlled external power supply so that the centre of the membrane (the hotspot) can reach a high temperature e.g. 700°C. When the temperature over the micro-heater region was 400°C, high purity ammonia (150 sccm) was introduced in the chamber using a mass flow controller (MFC). The micro-heater was kept at 725°C for 30 sec to form the small catalyst Fe islands. Then acetylene (75 sccm) was introduced through the separate MFC. The partial pressure was 4 mbar during growth process. The deposition time was typically 10 minutes, after which the gases were turned off and the devices were allowed to cool. It was found that the use of much higher temperatures (> 800°C) can give better quality CNTs, but can result in a rupture of the membrane. The nanotubes are formed due to the decomposition of acetylene, from which carbon dissolves and diffuses through the catalyst. The CNTs formed in this way are spaghetti like (and multiwall), but are useful for gas sensing applications and offer better conducting paths than vertically aligned nanotubes. The optical microscope picture of the gas sensor devices with MWCNTs and SEM pictures are shown in Fig. 6 and 7.

GAS SENSING PRINCIPLE

It is well known that the sensing mechanism in case of most semiconducting oxide gas sensors is a surface-controlled effect [17, 31]. An oxygen molecule adsorbs onto the surface of the metal oxide NWs when it exposed to air. As a result an O^-_2 ion forms by capturing an electron from the conduction band. When these sensors are exposed to reducing gas at high temperature, the gas reacts with the surface oxygen species, which decreases the surface concentration of O^-_2 ion and hence increases the electron concentration, meaning the conductivity of the metal oxide nanowires increases.

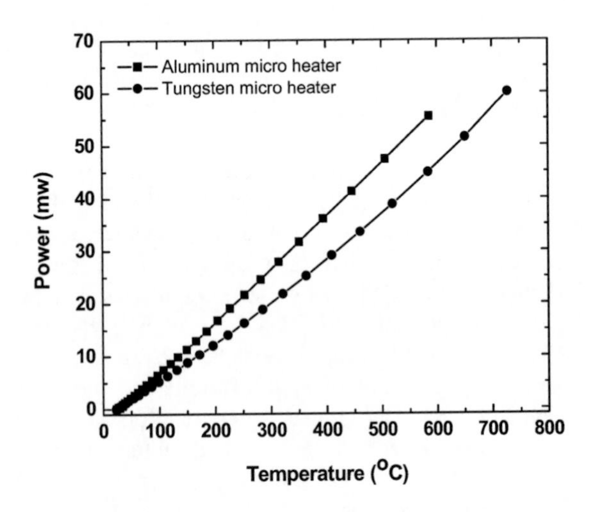

Fig. 3. Power versus temperature plots of aluminum and tungsten micro heaters.

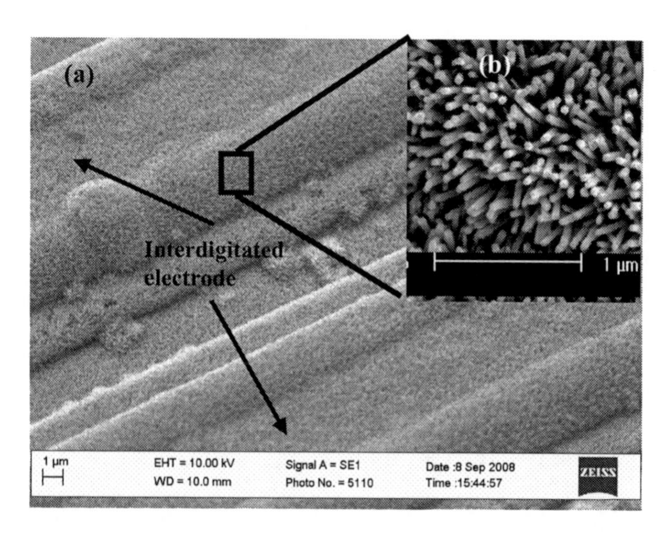

Fig. 4. (a) Top view of the SEM images of the ZnO NWs on interdigitated electrode, (b) zoom in view of NWs.

Fig. 5. SEM image of the deposited ZnO nanowires [46]

In the case of CNTs the sensing principles proposed by different groups are somewhat less rigorous and leave room for a more comprehensive understanding. The largest number of reported CNT based gas sensors target NO_2 and NH_3 gases. Several groups have attempted [49-53] to explain the sensing mechanism of Carbon Nanotubes (both SWCNT and MWCNT) when exposed to NO_2 and NH_3. Some theoreticians have used different tools (e.g. DFT calculation [50], self consistent field electronic structure- calculation [51], self consistent charge density functional based tight binding method [54] etc.) to figure out binding affinities between CNTs and NO_2/NH_3 and hence to devise a favorable mechanism that will be consistent with the experimental results. The popular two mechanisms are (i) physisorption and (ii) chemisorptions. In the case of physisorption physical adsorption (probably due to van der Waal forces) of gases (e.g. NO_2) takes place onto the sidewalls of CNTs leading to the formation of new states near the CNT Fermi levels that cause the change in resistance of the nanotubes. Whereas, in the second case, adsorbed molecules form bond structures with nanotube defect states and thus change the conductance of the CNTs.

GAS/VAPOUR TEST RESULTS

The response of the sensor is defined in this work as $[((R_b-R_a)/R_a)\times 100\%]$ where R_a is the baseline resistance of nanomaterial in presence of humid air/ dry nitrogen and R_b is the resistance in presence of target gas or vapour.

The response of the MWCNT-based sensors was investigated for NO_2 gas, which was balanced with dry N_2 carrier gas fixed at 1000 sccm [55]. The sensor was

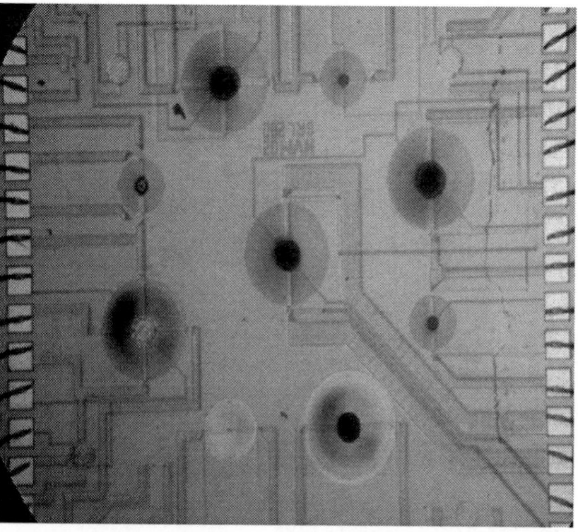

Fig. 6. Optical microscope picture of the locally grown MWCNTs on a fully processed CMOS chip.

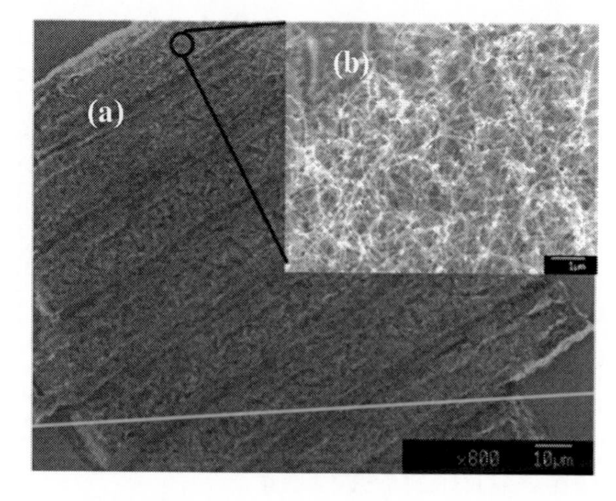

Fig. 7. (a) SEM pictures of the locally grown MWCNTS on interdigitated electrode, (b) zoom-in view of MWCNTS.

found to offer a response of 8 % to 100 ppb (parts per billion) of NO_2 and 20% at 20 ppm (parts per million) at room temperature before falling off at lower concentrations (Fig. 8). Following the removal of NO_2, the embedded micro-heater was used to speed up the recovery time (facilitate NO_2 desorption). It showed improved recovery time of the zero gas line at elevated temperatures (few seconds). The best response was seen at room temperature but a higher temperature was required to refresh quickly the baseline resistance. At higher temperatures the sensitivity is lower but no refreshing is required at an elevated temperature of 270°C for baseline recovery so there is a slight trade off between sensitivity and reversibility. To operate at elevated temperatures and/or refreshing the baseline at high temperatures, the micro-heater (placed underneath the sensing material region) was used for local heating with very low power consumption.

The sensing response of ZnO nanowire sensors were also investigated in the presence of NO_2 using the above mentioned setup [46]. Sensing and refreshing was performed at 18 and 25 mW, respectively. For this, the sensors were heated locally using the integrated tungsten micro-heaters. The measured gas response was shown in Fig. 9. In this case the response was found 40% at 100 ppb of NO_2 and the detection limit can be down to ppb level.

The ZnO NWs response at different ethanol concentrations (175 – 1484 ppm) were also measured, as shown in Fig. 10 [56]. These measurements were

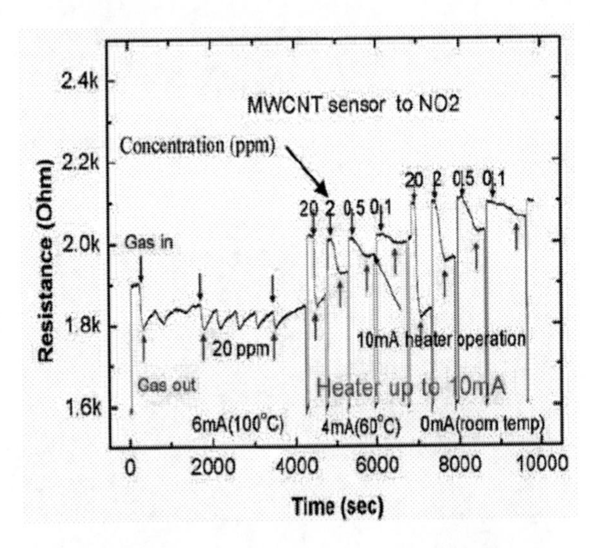

Fig.8. Response of CNTs to 20, 2, 0.5, 0.1 ppm of NO_2. The response time was in the order of minutes, and the of ethanol vapor from the system, recovery of the ZnO NWs was observed to take few tens of minutes before their resistance gets back to the base line value [55].

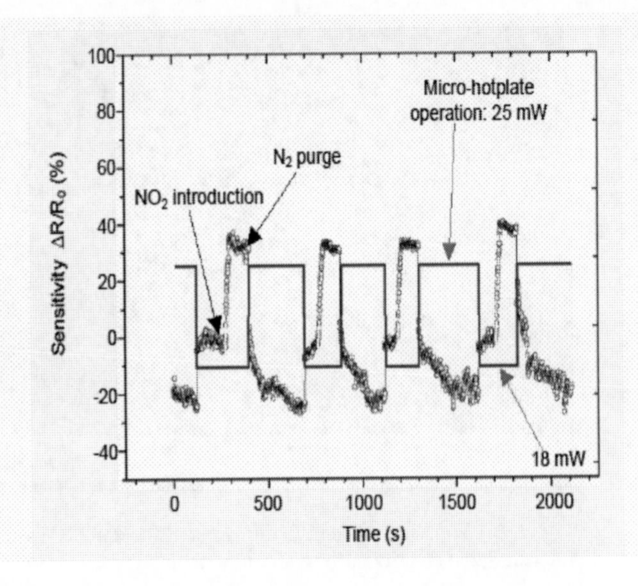

Fig. 9. Response of a ZnO nanowire-based sensor to NO_2 [46].

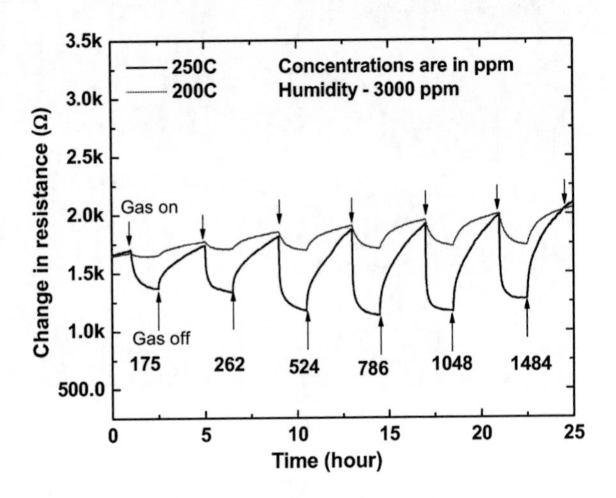

Fig.10. Change in resistance of the fabricated ZnO NWs ethanol gas sensor at different temperatures and different concentrations at a constant humidity of 3000 ppm [56].

carried out at five different concentrations of ethanol vapour in humid air. Humidity was kept constant at 3000 ppm throughout. This measurements were performed at 200 and 250°C (18 and 22 mW) using local micro-heater. We found that the response of NWs to ethanol vapour is significant and takes place within a few minutes.

DISCUSSION AND CONCLUSIONS

In this paper we presented the status of nanomaterials for gas sensing applications. We have paid particularly attention to issues related to growth compatibility, to CMOS processes and materials. We have given examples of the growth/deposition of ZnO nanowires and CNTs onto CMOS substrates. We covered briefly the gas sensing mechanisms reported in the literature and we have presented the response of ZNO NW and CNTs grown/deposited onto CMOS microhotplates to gases and vapours.

Nanomaterials with higher sensitivities (due to high surface to volume ratio) and higher selectivities (by functionalising or doping) are likely to form a new, future class of microsensors. Such nanomaterials need to be integrated with CMOS to get all the advantages of the conventional ICs (intelligence, cost, high volume, use of low power consumption microhotplates etc). However, the production of such sensors remains very challenging because of the requirements of respective fabrication steps of nanomaterials (which often need harsh environments) and the limitations imposed by the microelectronics technology. While the latter represents a hurdle that can be overcome at present, the former remains an issue to be addressed in the future. Such fabrication steps and recipies should deliver nanomaterials that are highly stable, selective and very importantly, reproducible.

Of the two materials discussed in this paper, ZnO nanowires have the advantage that they can be grown at lower temperatures using the hydrothermal methods. This growth method is more suitable for CMOS substrates compared to the local growth method for CNTs which requires high temperatures above 700°C, or for other metal oxide NWs that typically require high growth temperatures. CNTs, however, can instead be deposited by an inkjet printing or using a spray coating technique.

During operation, CNTs have the advantage of requiring lower operating temperatures, or even room temperature – compared to much higher temperatures required for ZnO NWs. This results in lower power consumption, making the sensors more suitable for portable and wireless applications.

However, these and other materials still need more study. For example, there are very few studies on the effects of dynamic environments with interfering species, long term stability and reproducibility (as already mentioned). Furthermore, most of the R&D works on chemical sensors using nanomaterials report excellent individual gas sensing performance, where the sensor measurements were carried out either in dry air, dry inert gas (e.g. Ar) or even in vacuum. Unfortunately this is far from real sensors' working conditions where humidity, presence of interfering gases, changes in ambient temperature etc. play a crucial role. Hence the effects of these factors also need to be carefully considered when evaluating sensing materials before these are deployed in real world applications.

The future of nanomaterial based sensor technology with CMOS integration looks promising, and we see continuous progress in this field. There are however numerous challenges that need to be overcome before such sensors can make it in the market.

ACKNOWLEDGMENTS

The work has been supported by Engineering and Physical Sciences Research Council (EPSRC) under the project no. EP/F004931/1. S. Maeng acknowledges the support of the RIC program of MKE in Woosuk University.

REFERENCES

1. G. Korotcenkov, *Sens. Actuators B* **107**, 287 – 304 (2005).
2. N. Barsan, D. Koziej, U. Weimar, *Sens. Actuators B* **121**, 18-35 (2007).
3. K. Ihokura and J.Watson, The Stannic Oxide Gas Sensor: Principle and Application, CRC Press inc., (1994).
4. D. E. Williams, *Sens. Actuators B* **57**, 1-16 (1999).
5. J. S. Suehle, R. E. Cavicchi, M. Gaitan, S. Semancik, *IEEE Electron Devices Letters* **14**, 118–120 (1993).
6. S. Semancik, R. E. Cavicchi, M. C. Wheeler, J. E. Tiffany, G. E. Poirier, R. M. Walton, J. S. Suehle, B. Panchapakesan, D. L. DeVoe, *Sens. Actuators B* **77**, 579–591 (2001).
7. D. Barrettino, M. Graf, H. S. Wan, K.-U. Kirstein, A. Hierlemann, H. Balte, IEEE *J. Solid-State Circuits* **39**, 1202–1207 (2004).
8. M. Y. Afridi, J. S. Suehle, M. E. Zaghloul, D. W. Berning, A. R. Hefner, R. E. Cavicchi, S. Semancik, C. B. Montgomery, C. J. Taylor, *IEEE Sensors J.* **2**, 644–655 (2002).
9. J. A. Covington, F. Udrea, J. W. Gardner, *Proc. IEEE Sensors Conf.* **2**, 1389–139 (2002).
10. M. Graf, S. K. Muller, D. Barrettino, A. Hierlemann, *IEEE Electron Device Lett*ers **26**, 295–297 (2005).
11. D. Barrettino, M. Graf, S. Taschini, S. Hafizovic, C. Hagleitner, A. Hierlemann, *IEEE Sensors Journal* **6**, 276 – 286 (2006).
12. Y. Li, C. Vancura, D. Barrettino, M. Graf, C. Hagleitner, A. Kummer, M. Zimmermann, K.-U. Kirstein, A. Hierlemann, *Sens. Actuators B* **126**, 431 – 440 (2007).
13. M. Graf, D. Barrettino, K.-U. Kirstein, A. Hierlemann, *Sens. Actuators B* **117**, 346 – 352 (2006).
14. F. Udrea, J. W. Gardner, D. Setiadi, J. A. Covington, T. Dogaru, C. C. Lu, W. I. Milne, *Sens. Actuators B* **78**, 180–190 (2001).

15. P.K. Guha, S.Z. Ali, C.C.C. Lee, F. Udrea, W.I. Milne, T. Iwaki, J.A. Covington, J.W. Gardner, *Sens. Actuators B* **127**, 260-266 (2007).
16. J. X. Wang, X. W. Sun, Y. Yang, H. Huang, Y. C. Lee, O. K. Tan, L. Vayssieres *Nanotechnology* **17**, 4995-4998 (2006).
17. Q. Wan, Q. H. Li, Y. J. Chen, T. H. Wang X. L. He, J. P .Li C. L. Lin, *Appl. Phys. Letters* **84**, 3654 – 3656 (2004).
18. J. Kong, N. R. Franklin, C. Zhou, M. G. Chapline, S. Peng, K. Cho, H. Dai, *Science* **287**, 622 – 625 (2000).
19. F Hernandez-Ramirez, S Barth, A Tarancon, O Casals, E Pellicer, J Rodriguez, A Romano-Rodriguez, J R Morante, S Mathur, *Nanotechnology* **18**, 424016 (6pp) (2007).
20. L. Francioso, A.M. Taurino, A. Forleo, P. Siciliano, *Sens. Actuators B* **130**, 70-76 (2008).
21. S. Santucci, S. Picozzi, F. di Gregorio, L. Lozzi, C. Cantalini, L. Valentini, J. M. Kenny, B. J. Delley, *Chem. Phys.* **119**, 10904 – 10910 (2003).
22. J. Kombarakkaran, C.F.M. Clewett, T. Pietra, *Chem. Phys. Lett*ers **441**, 282 – 285 (2007).
23. A. Goldoni, R. Larciprete, L. Petaccia, S. Lizzit, *J. Am. Chem. Soc.* **125**, 11329 – 11333 (2003).
24. Matranga C. and Bockrath B., *J. Phys. Chem. B*, **109**, 4853 –4864 (2005).
25. D. R. Kauffman and A. Star, *Angew. Chem. Int. Ed.* **47**, 6550 – 6570 (2008).
26. A. Star, V. Joshi, S. Skarupo, D. Thomas, J.-C. P. Gabriel, *J. Phys. Chem. B* **110**, 21014 – 21020 (2006).
27. L. Yijiang, C. Partride, M. Meyyappan, L. Jing, *J. Electroanal. Chem.* **593**, 105 – 110 (2006).
28. Y. X. Liang, Y. J. Chen, T. H. Wang, *Appl. Phys. Letters* **85**, 666 – 668 (2004).
29. P. Qi, O. Vermesh, M. Grecu, A. Javey, Q. Wang, H. Dai, *Nano Letters* **3**, 347 – 351 (2003).
30. T. Zhang, S. Mubeen, E. Bekyarova, B. Y. Yoo, R. C. Haddon, N. V. Myung, M. A. Deshusses, *Nanotechnology* **18**, 165504 (6pp) (2007).
31. T.-J. Hsueh, S.-J. Chang, C.-L. Hsu, Y.-R. Lin, I.-C. Chen, *J. of the Electrochemical Society* **155**, No. 9, K152 – K155 (2008).
32. A. Z. Sadek, S. Choopun, W. Wlodarski, S. J. Ippolito, K. Kalantar-zadeh, *IEEE Sensors Journal* **7**, No. 6, 919 – 924 (2007).
33. Y. Zhang, K. Yu, D. Jiang, Z. Zhu, H. Geng, L. Luo, *Appl. Surface Science* **242**, 212 – 217 (2005).
34. M.-W. Ahn, K.-S. Park, J.-H. Heo, J.-G. Park, D.-W. Kim, K. J. Choi, J.-H. Lee, S.-H. Hon, *Appl. Phy. Letters* **93**, 263103(3pp) (2008).
35. N. Hongsith, C. Viriyaworasakul, P. Mangkorntong, N. Mangkorntong and S. Choopun, *Ceramics International* **34**, 823–826 (2008).
36. G. An, N. Na, X. Zhang, Z. Miao, S. Miao, K. Ding, Z. Liu, *Nanotechonology* **18**, 435707 (6pp) (2007).
37. Y. J. Chen, X. Y. Xue, Y. G. Wang, T. H. Wang, *Appl. Phys. Lett*ers **87**, 233503 (3pp) (2005).
38. Y. J. Chen, L. Nie, X. Y. Xue, Y. G. Wang, T. H. Wang, *Appl. Phys. Letters* **88**, 083105 (3pp) (2006).
39. N. V. Hieu, H.-R. Kim, B.-K. Ju, J.-H. Lee, *Sens. Actuators B* **133**, 228-234 (2008).
40. E. Llobet, E. H. Espinosa, E. Sotter, R. Ionescu, X. Vilanova, J. Torres, A. Felten, J. J. Pireaux, X. Ke, G. Van Tendeloo, F. Renaux, Y. Paint, M. Hecq, C. Bittencourt, *Nanotechnology* **19**, 375501 (11pp) (2008)..
41. Ch.Y. Wang, M. Ali, Th. Kups, C.-C. Röhlig, V. Cimalla, Th. Stauden, O. Ambacher, *Sens. Actuators B* **130**, 589-593 (2008).
42. G. Neri, A. Bonavita, G. Micali, G. Rizzo, N. Pinna, M. Niederberger, J. Ba, *Sens. Actuators B* **130**, 222-230 (2008)..
43. T. Siciliano, A. Tepore, G. Micocci, A. Serra, D. Manno, E. Filippo, *Sens. Actuators B.* **133**, 321-326 (2008).
44. S. Z. Ali, F. Udrea, W. I. Milne, J. W. Gardner, *J. MEMS* **17**, 1408-1417 (2008).
45. H. E. Unalan, P. Hiralal, N. Rupesinghe, S. Dalal, W. I. Milne, G. A. J. Amaratunga, *Nanotechnology* **19**, 255608 (5pp) (2008).
46. S. Maeng, P. Guha, F. Udrea, S. Z. Ali, S. Santra, J. Gardner, J. Park, S.-H. Kim, S. E. Moon, K.-H. Park, J.-D. Kim, Y. Choi, W. I. Milne, *ETRI Journal* **30**, 516-525 (2008).
47. M. S. Haque, S. Z. Ali, P. K. Guha, S. P. Oei, J. Park, S. Maeng, K. B. K. Teo, F. Udrea, W. I. Milne, *J. Nanoscience Nanotechnology* **18**, 5667-5672 (2008).
48. M. S. Haque, K. B. K. Teo, N. L. Rupensinghe, S. Z. Ali, I. Haneef, S. Maeng, J. Park, F. Udrea, W. I. Milne, *Nanotechnology* **19**, 025607 (5pp) (2008).
49. S. Peng and K. Cho, *Nanotechnology* **11**, 57-60 (2000).
50. H. Chang, J. D. Lee, S. M. Lee, Y. H. Lee, *Appl. Phys. Lett*ers **79**, 3863-3865 (2001).
51. J. Zhao, A. Buldum, J. Han, J. P. Lu, *Nanotechnology* **13**, 195-200 (2002).
52. S. Peng, K. Cho, P. Qi, H. Dai, *Chem. Phys. Lett*ers **387**, 271 – 276 (2004).
53. M. Arab, F. Berger, F. Picaud, C. Ramseyer, J. Glory, M. Mayne-L'Hermite, *Chem. Phys. Lett*ers **433**, 175 – 181 (2006).
54. K. Seo, K. A. Park, C. Kim, S. Han, B. Kim, Y. H. Lee, *J. Am. Chem. Soc.* **127**, 15724 – 15729 (2005).
55. F. Udrea, S. Maeng, J.W. Gardner, J. Park, M. S. Haque, S. Z. Ali, Y. Choi, P. K. Guha, S. M. C. Vieira, H. Y. Kim, S. H. Kim, K. C. Kim, S. E. Moon, K. H. Park, W. I. Milne, S. Y. Oh, *IEEE International Electron Devices Meeting – IEDM*, 831-834 (2007).
56. S. Santra, S. Z. Ali, P. K. Guha, P. Hiralal, H. E. Unalan, S. H. Dalal, J. A. Covington, W.I. Milne, J. W. Gardner, F. Udrea, *International Symposium on Olfaction and Electronic Nose*, (2009), Accepted.

NANOSTRUCTURED SENSORS I

The Influence of Wide Range Humidity on Hydrogen Detection with Sensors Based on Nano-SnO_2 Materials

R.G. Pavelko[1]*, A.A. Vasiliev[1], E. Llobet[1], X. Vilanova[1], X. Correig[1], V.G. Sevastyanov[2]

[1]*University Rovira i Virgili, Electronic Department (DEEEA), 43007, Av. Paisos Catalans, 26, Tarragona, Spain*
[2]*N.S. Kurnakov Institute of General and Inorganic Chemistry RAS, 119991, Leninskiy prosp., 31, Moscow, Russia*

Abstract. The effect of environmental humidity on sensor signal to 1000 ppm of H_2 in air was investigated in a wide range of relative humidity (RH) and sensor operation temperature. To prepare sensing material we used highly dispersed nano-SnO_2 with 3-5 nm particle size and surface area ~ 150 m^2/g. Blank SnO_2, SnO_2 doped with Pd 1%, and SnO_2 doped with Pt 1% were used as materials for sensors. FTIR spectroscopy was employed to study water chemisorption on the surface of blank and doped SnO_2. Sensor signal measurements were performed in conditions close to equilibrium at 0, 20, 50 and 80% RH in the temperature range 30-600° C. The effect of humidity on H_2 detection mechanism is discussed.

Keywords: Gas Sensor, SnO_2, Humidity Effect, Hydrogen detection, Hydrogen Fuel Cell.
PACS: 85.30.De, 81.05.Dz, 81.07.Bc, 82.65.+r, 84.37.+q.

INTRODUCTION

It is well known that hydrogen is the most prospective energy carrier from various points of views: chemical energy per mass, availability in nature, ecological compatibility of energy production, etc. According to numerous investigations hydrogen fuel cells are close to commercial production for industry and personal use. However, apart from economic there are still technical challenges to make the H_2 fuel reality. The most crucial one is safe hydrogen storage [1].

Tin dioxide gas sensors are very promising to detect hydrogen in wide concentration range: from 100 ppb up to lower explosive limit (4 % for H_2). Humidity dependence of the sensor signal is the most important disadvantage of such detectors and this phenomenon still rises a lot of fundamental questions [2].

In this paper we report our recent results in the study of humidity effects on gas sensing characteristics of SnO_2 based gas sensors. We prepared sensing materials starting from nanocrystalline SnO_2 with high specific surface area. Two noble metals: Pd and Pt were used as catalytic additives to blank SnO_2. The signals to relatively high hydrogen concentration (1000 ppm) were measured in a wide range of relative humidity (RH) and operation temperature.

EXPERIMENTAL

Synthesis of nanocrystalline tin dioxide was performed from tin acetate complex dissolved in water free acetic acid. The precipitation of α-stannic acid occurred with dropwise addition of aqueous NH_3 solution. Solid phase was separated from the colloidal mixture by centrifugation. The product was dried at 300° C for 24 hours. To deposit Pd and Pt catalysts, tin dioxide powder was impregnated with water solutions of $Pd(NH_3)_4(NO_3)_2$ and $Pt(NH_3)_4(NO_3)_2$ respectively. The preset catalyst to SnO_2 ratio was equal for all sensing materials 1 *wt.* %. The powders were stirred and dried during 24 hours at 300° C. All heat treatments were carried out in ambient air at 50-60% RH.

Siemens D5000 diffractometer, EMAL-2 laser spark mass-spectrometer, JEM 1011 microscope, FTIR spectrometer JASCO 680 Plus and Micromeritics ASAP 2010 surface analyzer were used to characterize synthesized materials. FTIR spectroscopy was employed to study surface hydroxyl groups before and after annealing of as-prepared materials. Heat treatment was carried out at 700° C for 1 hour in ambient air. The samples were mixed with KBr powder and pressed into self-supporting disks. Spectra of dried blank and catalyst doped SnO_2 were recorded at room temperature in ambient air.

CP1137, *Olfaction and Electronic Nose: Proceedings of the 13th International Symposium,* edited by M. Pardo and G. Sberveglieri

Sensor fabrication was as follows. The powders were mixed with an organic vehicle to form printable ink and deposited on the heating platforms, which were made of screen printed alumina based material and platinum heaters. Deposited materials were annealed at 720° C. Before measurements the fabricated sensors was stabilized during 200 hours at 450° C.

Gas measurement set up consisted of DC system power supply Agilent 5751A, digital multimeter Agilent 34401A, gas mixing system Environics 4000 and gas chamber. Synthetic air was used as a carrier gas, and gas mixture of 1000 ppm of hydrogen in air was used to analyze sensor signals.

We characterized three different types of sensors on the basis of the following materials: pure SnO_2, SnO_2 with deposited 1 w. % of Pd and SnO_2 with deposited 1 w. % of Pt. To reach equilibrium between adsorbed species and gas phase before each measurement sensors were heated up to 300° C (under carrier gas or H_2 gas mixture) at certain RH during 90 min and then were left at room temperature for 10 hours. Resistance of sensing layer was measured separately in carrier gas and in 1000 ppm of hydrogen at preset relative humidity (RH): 0, 20, 50 and 80 % (experimental error was less than 10 %). For every measurement the heater temperature increased from RT to 600° C with the rate of 10°/min. Sensor signals were calculated as resistance ratio: R_{air}/R_{gas}, where R_{air} and R_{gas} – resistance in carrier and target gas (resp.).

RESULTS AND DISCUSSION

Material properties

The use of tin acetate complex as precursor for SnO_2 allowed us to obtain pure material with low impurity content. According to laser spark element analysis, the main impurity is Si – 0.05 *w*.%. The catalyst quantity was found equal to 1.2 and 0.9 *w*.% for Pd and Pt respectively.

The crystalline structure of all materials corresponds to rutile type. By means of XRD profile fitting software (TOPAS) the mean crystallite size of blank SnO_2, SnO_2-Pd and SnO_2-Pt (after drying at 300° C for 24 hours) was found to be 1.5, 2.5 and 2.1 nm respectively.

BET surface area of pure SnO_2 equals to 150 m^2/g, which corresponds to particle sizes of 3 nm (assuming spherical particle shape). The textural properties are mostly governed by the external surface area and the microporosity is very low. This means that catalytic sites are well-available for gas molecules and effect of steric hindrance is minimized. The low porosity of the material also allows one to reach homogeneous catalyst deposition due to small capillary effect.

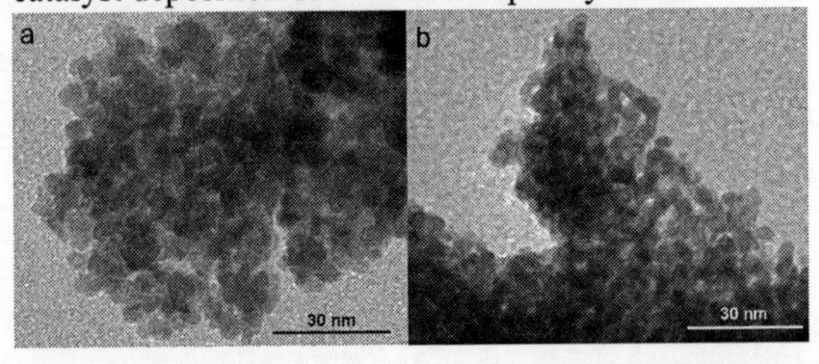

FIGURE 1. TEM pictures of blank (a) and Pd doped (b) SnO_2.

Figure 1 shows TEM picture of pure and Pd doped tin dioxide. The mean particle size for all materials is 3-5 nm. Due to small size, the deposited catalysts are difficult to see on the particle's surface. The particles of catalyst doped samples are slightly elongated in comparison with initial material, which consists of spherical ones.

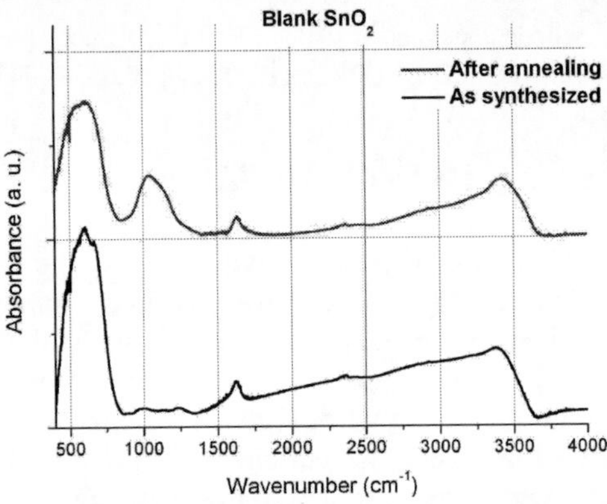

FIGURE 2. FTIR spectra of blank SnO_2 powder after drying of precipitate at 300° C for 24 hours (as synthesized) and after isothermal annealing at 700° C for 1 hour.

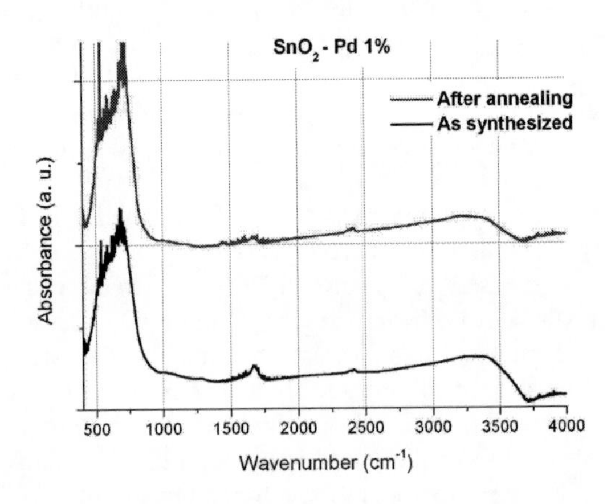

FIGURE 3. FTIR spectra of SnO_2 powder with deposited 1 w.% of Pd after drying at 300° C for 24 hours (as synthesized) and after isothermal annealing at 700° C for 1 hour.

Figures 2 and 3 depict room temperature FTIR spectra of the blank and doped SnO_2 after drying (as-synthesized) and after annealing at 700° C. Spectra of both materials before annealing demonstrate intense band between 400 and 800 cm^{-1}, which is assigned to different stretching vibration of Sn–O bond, small peak at 1620 cm^{-1}, corresponded to the bending mode of ligand-bonded molecular water, and broad band at 2000 - 3650 cm^{-1}, which conforms to hydroxyl stretching vibrations ν_{OH}(Sn–OH).

Heat treatment brings no new bands for doped SnO_2, causing only decrease of the peaks at 1620 cm^{-1} and 2000 - 3650 cm^{-1} (Fig. 3), which implies removing hydroxyl groups and water molecules from the surface. In contrary, spectrum of blank material changed significantly after annealing (Fig. 2). In a range of 900 – 1250 cm^{-1}, a broad intense band appeared together with evolution of the bands at 1620 cm^{-1} and 2000 - 3650 cm^{-1}.

Appearance of new band as well as evolution of the others could be tentatively attributed to the formation of surface OH groups multiply bonded to neighboring Sn atoms [3, 4]. Our studies revealed that this bonding is highly stable and is still present even after annealing at 700° C for 30 hours. This phenomenon is attributed to hydroxyl groups encapsulation inside the oxide lattice accordingly to [3, 5]. Apart from the formation of new Sn–OH bonds, there is an evidence of molecular water increase on the surface (band at 1620 cm^{-1}). Water accumulation after annealing occurs due to the adsorption from ambient air. It is well known that physisorbed water molecule bonds to several hydroxyls through hydrogen bonding [6]. Thus, the quantity of water molecules depends on the quantity of OH groups. An increase in OH groups and molecular water on the surface means that surface of blank SnO_2 is more polarized after calcination than before. This can be explained assuming oxygen vacancy formation during annealing [3].

Sensor characterization

The results of sensor characterization are represented as plots of sensor signals to 1000 ppm of H_2 in air against temperature at different RH (Fig. 4-6). To analyze the plots we used two main parameters: signal maximum (S_{max}) and temperature of the signal maximum (T_{Smax}).

In dry air S_{max}, is observed close to 300° C for all materials, and equals to 115 at 305° C for blank SnO_2, 130 at 300° C for Pd doped SnO_2, and 160 at 330° C for Pt doped SnO_2. As it can be seen, blank SnO_2 demonstrates quite high catalytic activity in comparison with doped ones which implies formation of numerous oxygen species on the surface (e.g. O^-, O_2^- and O^{2-}) even without catalysts. The explanation to this could be the fact that very low particle and crystallite size implies high density of surface defects. The last ones are known to participate in oxygen chemisorption.

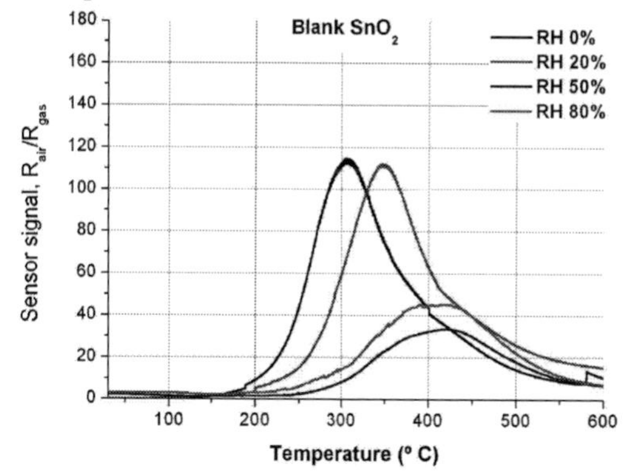

FIGURE 4. Signal of the sensor on the basis of blank SnO_2 against operation temperature at 0, 20, 50 and 80% RH.

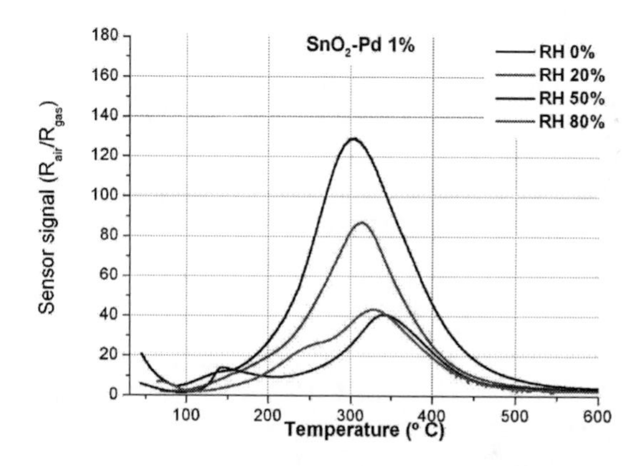

FIGURE 5. Signal of the sensor on the basis of Pd doped SnO_2 against operation temperature at 0, 20, 50 and 80% RH.

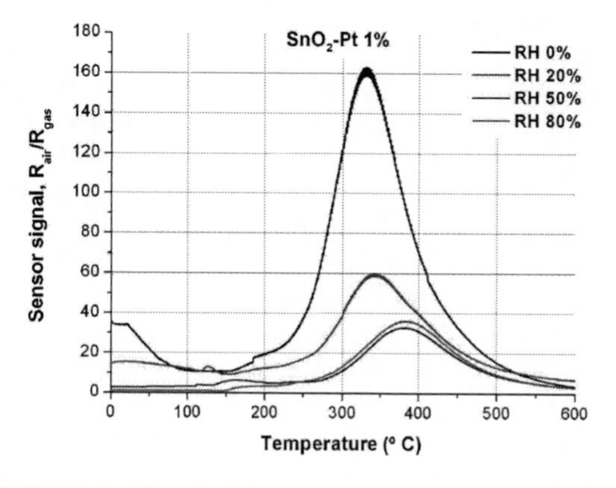

FIGURE 6. Signal of the sensor on the basis of Pt doped SnO_2 against operation temperature at 0, 20, 50 and 80% RH.

For all materials water vapors drastically change both S_{max} and T_{Smax} values. Let us consider first, how the RH affects the S_{max} value. An increase in relative humidity up to 20% leads to significant drop of S_{max} in the case of Pd and Pt doped SnO_2 (Fig. 5, 6). Blank SnO_2 demonstrates only slight decrease of this parameter from 115 to 112 (Fig. 4). In the range of 20-50% RH the S_{max} value still highly depends on the humidity. When relative humidity rises from 50 to 80% the change in S_{max} value becomes less, especially in the case Pt doped SnO_2. The results are summarized in Table 1.

TABLE 1. Signal maximum (S_{max}) and temperature of signal maximum (T_{Smax}) of the sensors.

Material	$S_{max}(a.u.) / T_{S\,max}(^{o}C)$			
	0% RH	**20% RH**	**50% RH**	**80% RH**
SnO_2	115/305	112/350	45/390-430	33/430
SnO_2-Pd 1%	130/300	87/315	43/330	41/340
SnO_2-Pt 1%	160/330	60/340	36/385	33/385

Apart from signal decreasing, humidity increase results in signal maximum shift towards higher temperatures. The most pronounced change was observed in the case of blank SnO_2. The total change of T_{Smax} in the 0-80 % range of RH sums up ~ 120°. The value does not change in the case of catalyst doped materials. The difference of T_{Smax} at 0% and 80% RH makes up 40° and 85° for Pd and Pt doped SnO_2 respectively (Tab. 1).

It is generally accepted, that a reducing agent molecule (e.g. H_2, CO, C_xH_y etc.) interacts on the surface with adsorbed oxygen species [7]:

$$H_2 + O_o \rightarrow H_2O + V_o^{n+} + ne^- \quad (n=1,2) \quad (1)$$

where O_o – adsorbed ionized oxygen (e.g. O^-, O_2^- and O^{2-} depending on the temperature [8]), V_o^{n+} – oxygen vacancy on the surface, e^- – electron. It is supposed that in dry air the active species are generated directly from surface oxygen. In the presence of water vapor the ionized oxygen formation mostly occurs due to dehydroxylation processes [9]:

$$2OH^- \rightarrow H_2O + O^- + e^- \quad (2)$$

Oxygen ion here represents highly reactive weakly bonded species, the formation of which defines the oxidation rate of hydrogen and, therefore, the quantity of injected electrons (i.e. signal).

However, hydrogen oxidation can also occur via interaction with hydroxyl groups, which were found to be stable on SnO_2 surface up to 500° C [10]. If we take into account that for materials with different surface polarity the stability could be even higher, this was suggested by our FTIR analysis of blank SnO_2, direct interaction between H_2 and OH groups should not be discarded [2]:

$$H_2 + 2OH^- \rightarrow [HO^- \cdots H{-}H \cdots {}^-OH] \rightarrow 2H_2O + 2e^- \quad (3)$$

Signal evolution in the case of blank SnO_2 speaks in favor of this assumption. Let us suppose that change in sensor signal (S_{max}) corresponds only to the change in oxidation kinetics, and change of the temperature of the highest oxidation rate (T_{Smax}) indicates the change in mechanism of H_2 oxidation. It will be evident then, that only for blank material there is a pronounced change in the mechanism from predominantly oxygen-type (1) at 0% RH, when the surface is covered with OH groups to a little degree, to hydroxyl-type (3) at 50-80% RH, when surface is highly hydroxylated even at high temperatures.

For catalyst doped SnO_2 the oxidation mechanism does not change much with humidity. Probably because the adsorption centers of SnO_2 are mostly covered with catalyst after wet deposition. It is known from the literature that dispersed Pd exists in oxidized form – PdO, which is stable up to ca 800° C. However, PtO_2 is highly unstable and deposited particles predominantly consist of Pt^0 [11], which suggest negligible affinity to water. But even in the case of PdO, water molecules are rebuilt through a disproportionation between two OH groups at low temperature – ca 250° C [12]. That is why FTIR spectra of doped SnO_2 show no evidence of strong Sn–OH interaction.

It is worth to note that for practical applications humidity effect on sensor signal to high H_2 concentrations could be minimized in the RH range 50-80% by employing SnO_2-Pt sensing layer. The operation temperature could be decreased to 300° C. At this temperature signal changes within ~18% depending on RH. The increase of operation temperature allows one to extend the operation limits up to 20-80% RH. In this range at 400° C the signal changes depending on RH within 30% in the case of both Pd and Pt doped SnO_2.

CONCLUSIONS

The humidity effect on sensor signals to hydrogen strongly depends on SnO_2 surface modification. It was found that synthesized blank SnO_2 being highly dispersive material with particle size 3-5 nm, is hydroxylated to high extent. High thermal stability of OH groups, which evidence was demonstrated by FTIR spectra of annealed material, results in remarkable change of H_2 oxidation mechanism depending on RH. We suppose that for this material at 50-80% RH the H_2 oxidation occurs mostly via interaction with OH groups, whereas at 0-20% the oxidation seems to occur with participation of ionized oxygen.

Deposition of Pd and Pt catalyst leads to significant decrease of hydroxyl groups on the SnO_2 surface even after annealing. The oxidation mechanism does not change much with humidity in this case. However, there is still strong kinetics dependency on humidity, which results in remarkable signal decrease for all sensing materials.

For indoor/outdoor application in the range 50-80% RH the signal change with humidity could be brought down to 18% in the case of SnO_2-Pt operating at 300° C. An increase of operation temperature leads to decrease on humidity dependency. For both doped materials signal changes depending on RH within 30% in the range 20-80% RH at 400° C.

REFERENCES

1. L. Schlapbach, and A. Zuttel, *Nature*, **414** (6861), 353-358 (2001).
2. N. Barsan, and U. Weimar, *Journal of Physics-Condensed Matter*, **15** (20), R813-R839 (2003).
3. N. Sergent, P. Gelin, L. Perier-Camby et al., *Sensors and Actuators B-Chemical*, **84** (2-3), 176-188 (2002).
4. D. Amalric-Popescu, and F. Bozon-Verduraz, *Catalysis Today*, **70** (1), 139-154 (2001).
5. E. W. Thornton, and P. G. Harrison, *J. Chem. Soc., Faraday Trans. 1*, **71** 461 - 472 (1975).
6. S. Kittaka, S. Kanemoto, and T. Morimoto, *J. Chem. Soc., Faraday Trans. 1*, **74** 676 - 685 (1978).
7. T. Sahm, A. Gurlo, N. Barsan et al., *Sensors and Actuators B-Chemical*, **118**, (1-2), 78-83 (2006).
8. A. Gurlo, *ChemPhysChem*, **7**, 2041-2052 (2006).
9. M. Caldararu, D. Sprinceana, V. T. Popa et al., *Sensors and Actuators B-Chemical*, **30** (1), 35-41 (1996).
10. M. Batzill, and U. Diebold, *Progress in Surface Science*, **79** (2-4), 47-154 (2005).
11. P. Gelin, and M. Primet, *Applied Catalysis B-Environmental*, **39** (1), 1-37 (2002).
12. J. M. Heras, G. Estiu, and L. Viscido, Applied Surface Science, **108** (4), 455-464 (1997).

Mixed-potential Behavior of Nanostructured RuO_2 Sensing Electrode of Water Quality Sensors in Strong Alkaline Solutions at a Temperature Range of 9-30°C

Dr. Serge Zhuiykov

Commonwealth Scientific Industrial Research Organization (CSIRO), Materials Science and Engineering Division, 37 Graham Rd., Highett, VIC. 3190, Australia
E-mail: serge.zhuiykov@csiro.au

Abstract. Mixed-potential behavior of the water quality monitoring sensors using nanostructured RuO_2 sensing electrode (SE) has been observed in strong alkaline solutions at dissolved oxygen (DO) measurements in the temperature range of 9-30°C. This behavior indicated that a Faradaic oxygen reduction reaction becomes not only a one-electron process, which is typical for DO measurements at a neutral pH, but rather multi-step process with superoxide oxygen ions (O_2^-), OH^- and RuO_4^{2-} ions involvement. The DO sensing characteristics were examined in the pH range of 2.0-13.0. The measured *emf* at strong alkaline solutions is a mixed potential from the reactions involved RuO_4^{2-} and OH^- ions and DO. Impedance spectroscopy was employed for confirmation the mixed-potential behavior of the sensor. It was also found during experiments that OH^- ions influence the response/recovery rate of the SE reactions as the pH of water increases.

Keywords: Mixed-potential, Water Quality Sensors, RuO_2, Nanostructures
PACS: 07.07.Df

INTRODUCTION

Oxygen is vital to all forms of aerobic life and as far as water quality is concern it is necessary to keep adequate DO level in order to sustain aerobic life. DO in water characterizes in the 0.5-8.0 ppm range (log[O_2], -4.82 to – 3.60) at a temperature range of 4-30°C. Our recent research focusing on the development inexpensive miniature integrated water quality monitoring sensors based on RuO_2-SE shown that the use of nanostructured RuO_2 and micro-fabrication technology provides a promising alternative to other DO instruments [1]. Using such nanostructured materials as RuO_2 allowing utilize several RuO_2-SEs onto one multi-sensor assembly capable to measure such parameters of water quality as pH and DO. Our previous DO measurements at the temperature range of 9-30°C at neutral pH revealed that although RuO_2 is not a good catalyst for molecular O_2 reduction, DO adsorption at the SE/water interface can be explained by the one electron redox reaction and straight linear dependence of the output sensor *emf* on log[O_2] is a consecutive evidence of that [2]. Hydrogen ions (H^+) diffusion in freshly-prepared RuO_2-SE can be interpreted by the following redox reaction

$$2RuO_2 + 2H^+_{aq} + 2e^- \leftrightarrow Ru_2O_3 + H_2O. \quad (1)$$

However, from the practical point of view, DO measurements in strong acid and/or alkaline solutions are also required. These measurements revealed that apart from DO, both OH^- and RuO_4^{2-} ions are also involved in the reactions contributing to the measuring potential. In case of simultaneous adsorption of the DO and other ions on the semiconductor SE, the charge transfer involves the binding energy and varies with the degree of coverage of the adsorbed species due to the strong electronic interactions between adsorbate and adsorbent [3]. As a result, the adsorption of ions and the appropriate potentials depend on the defect equilibrium which is attained via bulk DO diffusion through stoichiometric defects (vacancies) as well as kinetics of different electrochemical reactions. We therefore report the mixed-potential behavior at strong alkaline solutions for the integrated water quality monitoring sensors using nanostructured RuO_2-SE in this paper.

CP1137, *Olfaction and Electronic Nose: Proceedings of the 13th International Symposium,* edited by M. Pardo and G. Sberveglieri

EXPERIMENTAL

All reagents for preparation of nanostructured RuO_2-SEs are of high-purity analytical grade and were used as received: RuO_2 (ABCR GmbH & Co.) and Pt paste (Sigma-Adrich Australia Pty. Ltd.). Pt current conductors of 5 μm were deposited onto Al_2O_3 substrate and were sintered at 1000°C for 1 h in air prior to the RuO_2-SE deposition. Then, the RuO_2 nano-particles were applied onto platinized alumina substrate by using α–terpineol ($C_{10}H_{18}O$, 99.0%) and were sintered as described [4]. Ag/AgCl, Cl^- reference electrode (RE) was used for both pH and DO measurements. DO sensing properties of RuO_2-SEs were investigated on the conventional water-flow apparatus equipped with heating facilities. Since DO measurements are interconnected to the pH changes of the solution, various N_2/O_2 mixtures pumped through the solution in order to obtain the desired DO concentrations at selected pHs. The temperature of solution varied from 9 to 30°C. Commercial DO analyzer (HI-9142, HANNA Instruments) was used to measure DO concentrations against measured *emf* of the sensors with RuO_2-SEs. Analyzer was recalibrated before each measurement. Electrochemical measurements of complex impedance were performed using electrochemical analyzer AUTOLAB, PGSTAT, The Netherland. Impedance spectra were collected in the frequency range of 0.1 Hz – 1 MHz at amplitude 5 mV.

RESULTS

DO sensing performance of the multisensors based on nanostructured RuO_2-SE shown that, if pH of the solution is close to neutral (6.0-8.0 pH), a fast electrode reaction involves only one electron per O_2 molecule. Results of these measurements at 23°C and 8.0 pH are presented in Fig. 1. The standard *emf* deviations were found to be around ±5 mV within the whole DO measuring range. However, it was also discovered that although the gas flow controllers fixed pumping gas-mixture rate constant (100 cm^3/min) during all experiments, the deviation in DO measurement did occur, as shown in Fig. 1. This suggested that apart from the fast electrode reaction, other slower electrode reactions are possibly also involved into DO measurement even at neutral pH.

It is clear from Fig. 1 that DO changing in water is relatively slow process and it is required about ~50 min to drop DO concentration from 8.0 ppm down to about 3.5 ppm at 8.0 pH of the solution. Typical 90% response time T_{90} at 8.0 pH for the nanostructured RuO_2-SE was found to be about 8 min at 23°C for higher O_2 concentrations, while a much slow response was observed at the lowest measuring temperature of 9°C. T_{90} was found to be about 30 min at 9°C. The results obtained in strong alkaline solutions revealed that as pH of the solution increases to 11.0-13.0 pH the response/recovery rate becomes faster stabilizing more or less quickly depending upon the solution alkalinity. This behavior indicated that both OH^- and O_2^- ions act to accelerate the surface-forming process and possibly are directly involved through, for example, the heterogeneous dissolution of the nanostructured RuO_2–SE according to following equations

$$RuO_2 + 4OH^- \leftrightarrow RuO_4^{2-} + 2H_2O + 2e^- \quad (2)$$

$$RuO_2 + O_2^- \leftrightarrow RuO_4 + e^-, \quad (3)$$

whereby the reduced components are enriched at the surface. This explains the observed decrease in the degree of surface oxidation with increasing pH of the solution [5]. The decrease of the response time of the sensor with increasing the pH value of the solution is also shown a supporting influence of the OH^- ions on the kinetics of the O_2 adsorption reaction.

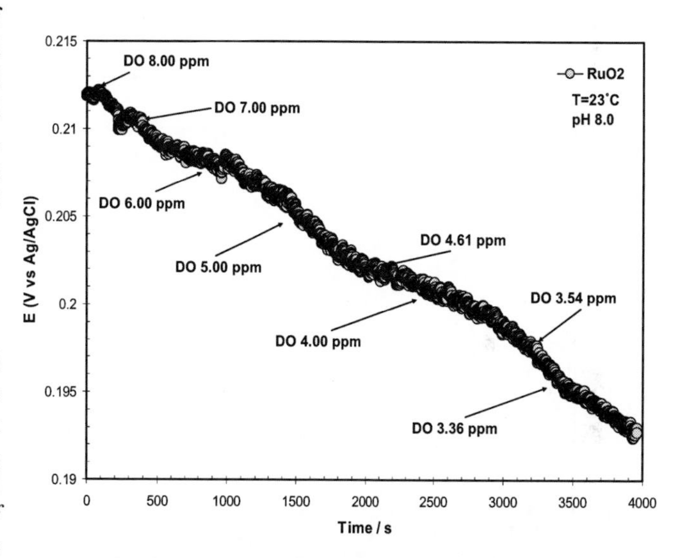

FIGURE 1. Changes of DO concentration in water measured by multisensor attached with nanostructured RuO_2-SE.

Nevertheless, both reactions (2) and (3) indicate that, as pH of the solution getting more alkaline (11.0-13.0 pH), a Faradaic oxygen reduction reaction becomes not only a one-electron process but rather mixed-potential multi-step process. This is in contrast to the first open-circuit potential measurements at pH close to neutral [1], which confirmed one-electron reaction. We hoped to get some answers to these questions from the impedance spectroscopy studies.

The impedance behavior is attributed to changes in charge-transfer kinetics at the substrate/SE interfaces in the situation when both SE and RE exposed to the same test environment. Contrary to the gaseous environment, the use of impedance spectroscopy for analysis of responses RuO_2–SE to DO changes in water, would allow to analyze possible changes in the SE/interface resistance due to reactions occurring at the interfaces. It should be also noted that interpretation of Nyquist plots is not always straightforward. The simplified equivalent circuit of the DO sensor is exhibiting both resistive and capacitive components (e.g., grain boundaries, interfaces, etc.). Unfortunately, in the real water sensor systems changes in resistive and capacitive behavior are not always easy quantified, which causes overlapping and convoluting in semicircular arcs, as can be seen in Fig. 2 for the nanostructured RuO_2-SE at 23°C. It is because the multiple phenomena on interfaces have similar time constants, which may lead to heterogeneity in the behavior at interfaces. The diameter of the semi-arcs reflects the charge-transfer resistance of the reactions (1)-(3) occurring on RuO_2-SE surface and on interfaces at changes pH of the solution at fixed DO (8.0 ppm).

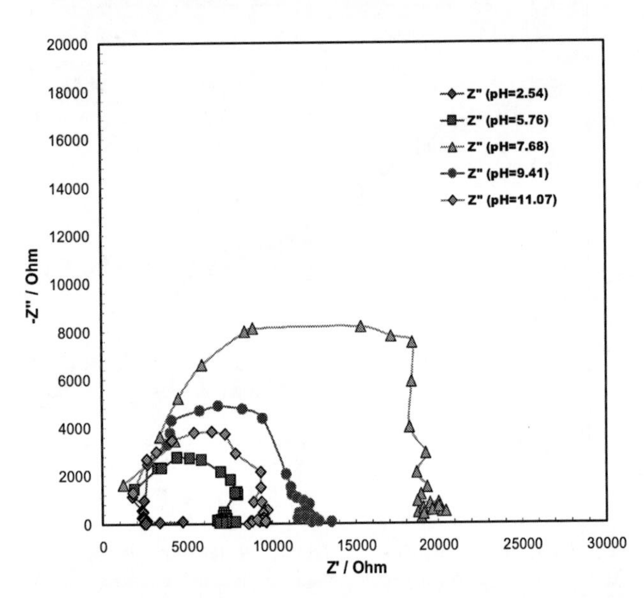

FIGURE 2. Nyquist plots for the multisensor attached with nanostructured RuO_2-SE at 23°C.

The summarized results of the complex impedance measurements for RuO_2–SE at changes pH of the test solution were presented in Fig. 3, which showed that the maximum complex impedance was found at the neutral pH (7.68 for tape water) and gradually decreases as the pH solution become more acid or, alternatively, more alkaline. The decrease in complex impedance for nanostructured RuO_2–SE at changes of the solution pH confirmed that apart form the fast super-oxide ions adsorption on RuO_2-SE, which was dominated at a neutral pH, slow adsorption of other ions (such as OH^-, RuO_4^{2-}) is also taken place and, consequently, the complex impedance is different at different pH of the solution. However, the kinetics of this adsorption in strong acid solution is different to the kinetics of adsorption in strong alkaline solution.

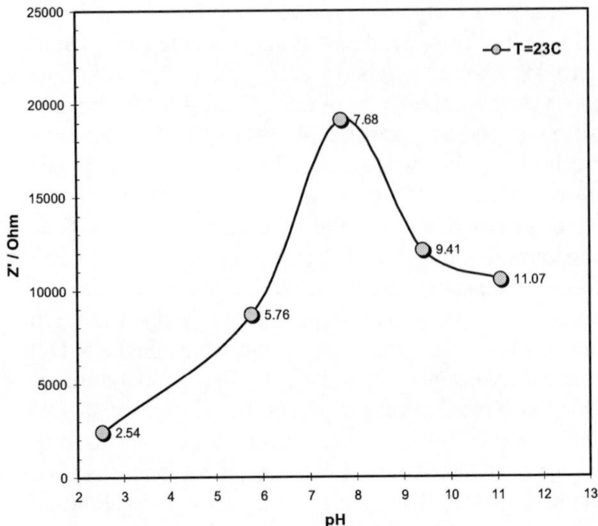

FIGURE 3. Summary of the complex impedance measurements for DO sensor attached with nanostructured RuO_2-SE.

The results obtained point out that the chemistry, structure modification and electronic band structure of the SE surface layer have considerable influence on the reactions contributing to the measuring potential, especially with DO.

The results can be summarized as follows:

- DO strongly influences the cathodic currents, i.e., the DO reaction on nanostructured RuO_2–SE is pure reduction reaction at a neutral pH.
- Kinetics of the oxygen reduction reaction enhanced by the presence of OH^- ions. This is in agreement with dependence of pH of the response time behavior of the *emf* to changes of oxygen concentration.
- The OH^- ions are involved in the anodic reactions as well.
- The RuO_4^- ions are involved in both the cathodic and anodic reactions.

The assumed steady-state equilibrium of antagonistic decomposing and surface layer developing reactions provides an explanation for an excellent reversibility of oxygen adsorption reaction $O_{2,aq} + e^- \leftrightarrow O_{2\ ,ads}^-$ observed on RuO_2-SE. The equilibrium determines the density of adsorbed superoxide ions in the wide range of ($O_{2,aq}$). However, we think that this statement should be verified by further measurements, for example, by the temperature-dependent X-ray photoelectron spectroscopy. Nevertheless, the results obtained so far are encouraging and they suggest that the use of the nanostructured RuO_2 as SE in the multisensor measuring such vital parameters of water quality as pH and DO is a promising concept. However, further investigation of reported behavior is necessary in order to make the final conclusions.

DISCUSSION

It can be postulated that the overall reactions observed on the nanostructured RuO_2–SEs at strong alkaline solutions is mixed-potential multi-step reactions. Only the first step activated fast, and therefore, is determining a potential at neutral pH. This means that if the water pH as well as (RuO_4^{2-}) are held constant, the sensor potential variations on ($O_{2,aq}$) are determined by this capacitive reaction only. The overall Faradaic current in acid and alkaline solutions may be provided not only by one-electrode reaction, but also by other much slower reactions (2) and (3).

We assume that the influence of the second step oxygen reduction reaction on RuO_2–SE potential is caused by the layer-forming reactions in (2) and (3) as well as the OH^- ions adsorption process. The latter may join the oxygen adsorption reaction (2) according to the following model:

$$O_{2,aq} + e^- \leftrightarrow O_{2}^{-}{}_{,ads}$$

(fast potential-determining reaction)

$$RuO_2 + O_2^- \leftrightarrow RuO_4 + e^-,$$

(slow surface oxidation reaction)

$$RuO_2 + 4OH^- \leftrightarrow RuO_4^{2-} + 2H_2O + 2e^-$$

(relatively fast reaction at high pH).

CONCLUSIONS

The electrochemical ceramic multisensor attached with nanostructured RuO_2-SE was designed for DO detection in water at a temperature range of 9-30°C. The use of nanostructured RuO_2–SE in DO sensors has been demonstrated to be a promising concept. Reversible dependence of the measuring *emf* on DO was obtained. However, it was also experimentally found that apart from O_2^- ions, OH^- and RuO_4^{2-} ions are also influenced the output potential indicating mixed-potential behavior at strong alkaline solutions.

ACKNOWLEDGMENTS

This work was partially supported by Research and Development Program of CSIRO Materials Science and Engineering Division and CSIRO ICT Centre, Australia – project "Integrated Water Quality Monitoring Sensors".

REFERENCES

1. S. Zhuiykov, *Ionics* (2009) – in press.
1. S. Zhuiykov, *Electrochem. Comm.* **10**, 839-843 (2008).
3. S. Zhuiykov, *Sens. Actuators B: Chem.* **136**, 248-256 (2009).
4. S. Zhuiykov. Australian provisional patent application 2008902285 (9 May 2008).
5. H. Kohler and W. Göpel, *Sens. Actuators B: Chem.* **4**, 345-354 (1991).

Carbon Nanomaterial Polymer Composite ChemFET and Chemoresistors For Vapour Sensing

Dr James A. Covington and Prof Julian W. Gardner

Sensors Research Laboratory
School of Engineering, University of Warwick, Coventry, CV4 7AL, UK
Tel: +44 (0) 2476 574494, Email: J.A.Covington@warwick.ac.uk

Abstract. Carbon nanotubes (CNTs) have been proposed for a broad spectrum of applications, including chemical sensing. Here we report on an investigation of multi-walled CNTs (MWCNTs) as conductive filler for composite polymer sensing films. Such materials combine conductive fillers with an insulating polymer to produce a chemically sensitive, electrically conducting material. These polymer composites offer several important advantages for chemical sensing, including room temperature operation (hence ultra low power), a broad range of selectivities (due to the wide choice of available polymers), and low manufacturing cost. Our approach is to compare the sensing qualities of these composite films, in resistive and field-effect configurations, with existing carbon black polymer composites. Their responses to propanol and toluene vapour in air show that the carbon black resistive sensors outperform CNT sensors by a factor of four in response magnitude. Thus we conclude that for these vapours and using this sensor fabrication method, carbon black polymer composite films are preferable for chemical sensing than MWCNT polymer composites.

Keywords: Chemical Sensor, Composite Polymer, CNTs.
PACS: 07.07.df

INTRODUCTION

The use of carbon black (CB) as a conductive filler for chemical sensing has been widely reported [1,2]. In such a scheme, carbon nanoparticles are dispersed in an insulating polymer film. The resulting combination produces an electrically conducting material that changes its properties when exposed to a range of vapours. This is traditionally explained by a solvation and swelling effect where the polymer (solvent) expands when exposed to a vapour (solute), altering the nanoparticle dispersion (i.e. increasing the resistance in a chemo-resistive configuration or changing the average work function for a chemFET).

In addition to the use of CB/polymer composites, there have been significant efforts in using carbon nano-tubes (CNTs) for chemical sensing. This is by either using the CNTs on their own, in a resistive or FET configuration, or by combining the material with an insulating polymer in a resistive configuration [3,4].

Here we report on an investigation to compare the use of multiwalled (MW) CNTs / polymer composite materials for both chemFETs and chemoresistive devices and compare it with carbon black composite materials. Our aim is to determine whether CNTs give any sensing advantage over the more traditional CB composites to test vapours.

EXPERIMENTAL METHODS

Here the two different types of chemical sensing devices are described, chemoresistive and chemFET.

Chemoresistive Device

The resistive device used in this study comprised of a silicon substrate (4.0 mm square), with a silicon nitride layer (200 nm thick), on to which a Ti/Au layer (10nm/250nm) had been deposited. The device was passivated with an addition 200 nm silicon nitride layer, with openings for bonding and sensor material deposition. The electrodes in the centre of the device were separated by a 50 μm gap and were 200 μm in length. The device was packaged onto a custom made PCB header for sensing material deposition and testing.

CP1137, *Olfaction and Electronic Nose: Proceedings of the 13th International Symposium,* edited by M. Pardo and G. Sberveglieri

CHEMFET DEVICE

The chemFET substrate was fabricated using a standard silicon process at Southampton University (UK). Here an enhanced *n*-channel MOSFET was created from a *p*-doped silicon substrate, using standard silicon processing techniques. After oxidation, the polysilicon gate was exposed through dry etching. The active area was then defined using lithography and a portion of the polysilicon gate was etched away to expose the gate oxide. Thus two polysilicon electrodes either side of the exposed gate oxide remained to provide electrical contact to the sensing layer.

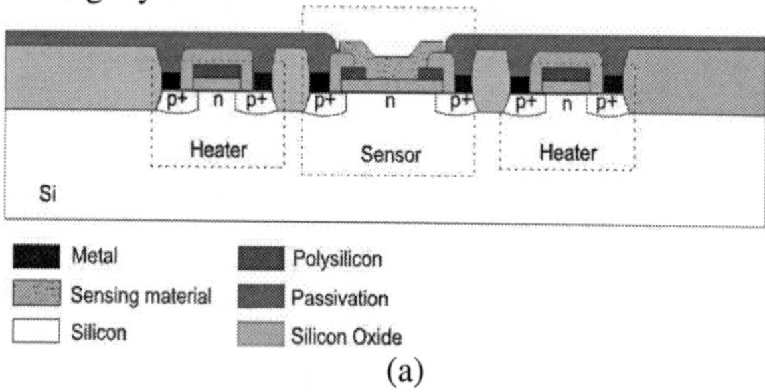

(a)

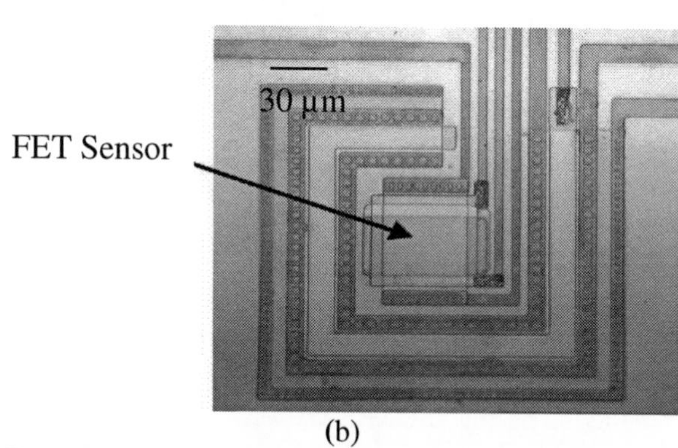

(b)

FIGURE 1. (a) Schematic and (b) photograph of the chemFET sensor.

Figure 1 shows (a) a schematic of the cross-structure of the FET device and (b) a microscope image of the active area of the FET before the sensing material was deposited.

Sensor Coatings

Three different recipes of MWCNT/Polymer and CB/Polymer were employed. Here the same weight of CNTs and CBs were used. Table 1 gives the details of these recipes. Note that 10 ml greater volume of toluene was used in depositing the CNT films. This is because the original recipe was too viscous for controlled air-brush deposition.

TABLE 1. Nano-material/polymer compositions.

Polymer (g)	CNT/CB (g)	Toluene (ml)
Poly(ethylene-co-vinyl acetate) / PEVA (1.2)	0.3	20 (CB) 30 (CNT)
Polycaprolactone / PCL - (1.2)	0.3	20 (CB) 30 (CNT)
Poly(Styrene-co-butadiene) / PSB - (1.2)	0.18	20 (CB) 30 (CNT)

The polymers were supplied by Sigma Aldrich (UK), the carbon black (Black Pearls 2000, Cabot Corporation, USA) and the CNTs from Nancyl (Belgium). The polymers were either in powder form or small crystals while the carbon black was supplied as nanospheres with diameters of typically 50 to 80 nm. The polymers were first dissolved with the aid of a magnetic stirrer in their respective solvent overnight at an elevated temperature (50 °C). Next, CB or CNT was added and the mixture sonicated for 10 min using a flask shaker (Griffin and George, UK). The mixture was then deposited onto the sensor electrodes using an airbrush (HP-BC Iwata, Japan) controlled by a micro-spraying system (RS precision liquid dispenser, UK). Before coating, the sensor substrate was mounted and wire-bonded on to a PCB header for monitoring the resistance whilst coating. This, in turn, was mounted onto an X-Y stage and aligned to an aluminium mask, containing a 1.0 mm hole. The device was coated multiple times until a resistance of around 5 kΩ was achieved. As stated above, Figure 2 shows the resistive sensor coated with a 3 µm ± 0.5 µm CB/PSB composite film.

FIGURE 2. Coated resistive sensor.

SENSOR TESTING

The sensors were tested to six different concentrations of 1-propanol and toluene vapour in air using an automated testing station. In addition, two different humidities were used of 3500 PPM (Parts Per Million) and secondly at 9800 PPM. 3500 PPM corresponds to 10% relative humidity in air at a temperature of 20 °C and 9800 PPM corresponds to

40% relative humidity in air at 20 °C. Before testing, the sensors were left for 24 h in a Dri-block™ heater at (30 ±1) °C to stabilize their baseline signals.

The sensors were driven at a constant current of 10 μA and the voltage drop across the sensors measured. In the case of the chemFET the gate and drain were shorted, thus the device was operated in the saturated region. In this case, with a constant current drive the change in output voltage is directly proportional to the change in threshold voltage. This is proportional to the change in the work function of the sensing material.

RESULTS

Figure 3 shows the transient responses of a CNT/PSB resistive, CB/PSB resistive and CNT/PSB chemFET sensors to six different concentrations of toluene vapour (1300, 1800, 2600, 5800, 8700 and 12400 PPM) in air (3500 PPM water).

Figures 4, 5 and 6 show the steady state isotherms of resistive and chemFET devices coated with CNTs and CB polymer composite films to toluene vapour in air. Here the water concentration was set at 3500 PPM. The data are displayed comparing the same polymer film, instead of device configuration for ease of comparison.

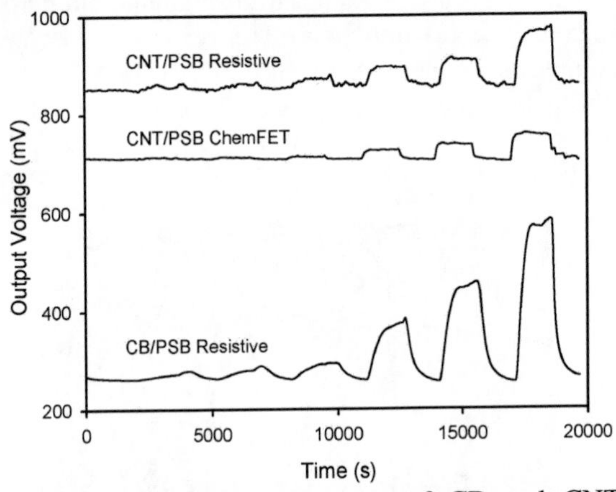

FIGURE 3. Transient responses of CB and CNT polymer composites to pulses of toluene vapour in air. Response times (t_{90}): CB/res - 375 s, CNT/Res - 35 s, CNT/FET – 260 s.

The data in figures 4 to 6 are plotted against the fractional response as defined by:

$$\text{Fractional Response} = (R_i - R_f)/R_i$$

Where R_i is the initial resistance, without vapour, R_f is the steady state resistance when exposed to a vapour. For the ChemFET sensors, the resistances R_i and R_f are replaced with the initial and steady state drain-source voltages. This pre-processing algorithm has been employed both to reduce signal noise and make the sensors more comparable.

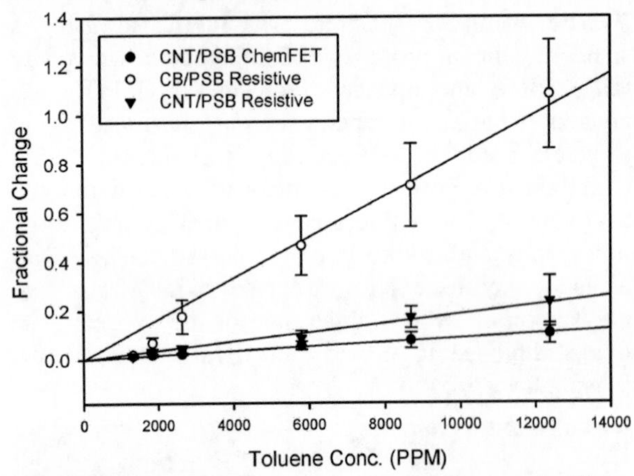

FIGURE 4. Response of sensors using PSB composite films to toluene vapour in air.

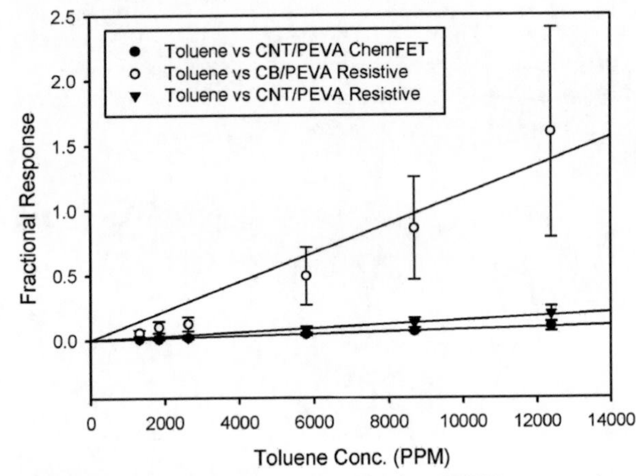

FIGURE 5. Response of sensors using PEVA composite films to toluene vapour in air.

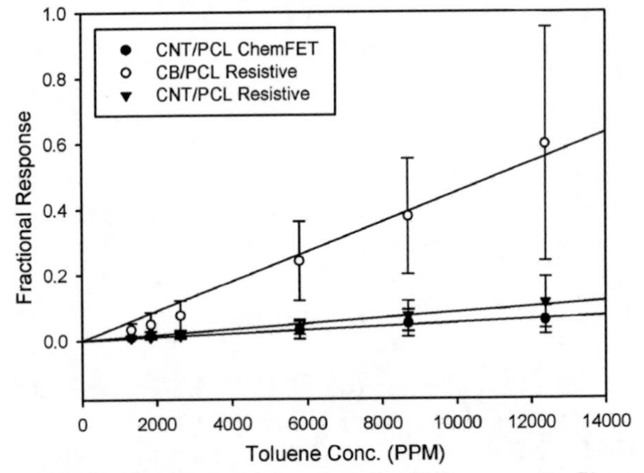

FIGURE 6. Response of sensors using PCL composite films to toluene vapour in air.

From this data and tests performed using 1-propanol and toluene vapour in air, we can define the sensitivity of each sensor type and sensing material. This is shown table 2 below.

TABLE 2. Sensitivity values in fractional response to toluene and 1-propanol vapour in air.

Sensor	Toluene $\times10^{-6}$/PPM	1-Propanol
CNT/PEVA FET	7.8 ± 2.8	2.4 ± 0.7
CNT/PCL FET	4.7 ± 3.5	3.5 ± 1.7
CNT/PSB FET	8.5 ± 3.4	3.3 ± 0.9
CB/PEVA Res	128.7 ± 65.4	11.6 ± 4.9
CB/PCL Res	48.1 ± 28.8	9.6 ± 3.9
CB/PSB Res	84.7 ± 18.0	4.2 ± 2.1
CNT/PEVA Res	15.7 ± 4.8	2.1 ± 0.3
CNT/PCL Res	9.0 ± 6.4	3.4 ± 1.8
CNT/PSB Res	19.1 ± 8.3	7.5 ± 7.3

In addition, to considering response to vapours, we have also considered the effect of water concentration on the magnitude of response. An example of this effect is shown in figure 7. Here the fractional response to different concentrations of toluene vapour is plotted with two different water concentrations. Results show that the magnitude of the response decreases with increasing water concentration. This was found for all the sensors and composite films tested.

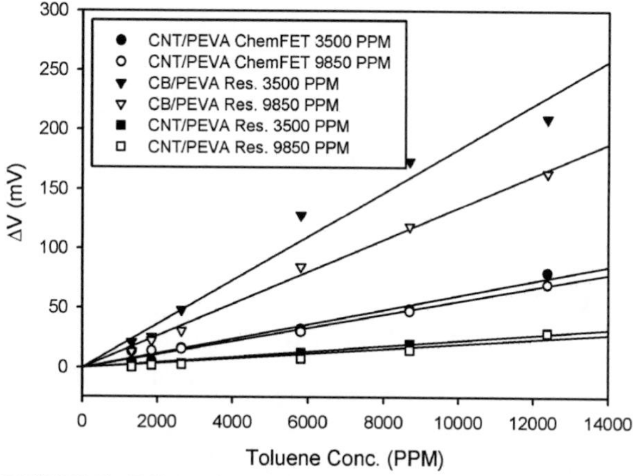

FIGURE 7. Effect of water concentration on the magnitude of the steady state response.

DISCUSSION

The results show that MWCNT FET and resistive sensors respond to simple organic vapours. We observed in these experiments that the carbon black resistive sensors had a larger response to these vapour concentrations than the CNT based resistive or chemFET devices; independent of the polymer used. We believe this reduced performance could possibly be due to nanoparticles clumping within the polymer; thus the swelling effect would not change the conduction path through these coagulated areas. Furthermore, multi-wall CNTs were used in our experiments. These are longer and wider than, for example, single wall CNTs. Thus it is possible that they are forming networks through the material (like a mesh), that restricts the expansion of the polymer. It might be possible to get a better response profile from these composites by using single walled CNTs. The carbon black is made of nanospheres, and so might not be as susceptible to these issues. However, it is possible and probably necessary to functionalise the CNTs to improve their sensing characteristics for main stream use.

Furthermore, we have observed a strong water dependence of these films. This is not surprising as similar effects have been seen with these polymers. In these cases we believe there is competition for absorption sites on the polymer between the water and the test analyte, though it is interesting to note that toluene, a non-polar molecule, suffers the same reduction in response magnitude as 1-propanol.

CONCLUSIONS

Here we report on a comparison between carbon black and MWCNTs as the conductive filler for polymer composite vapour sensors. We have fabricated both resistive and chemFET sensors with three different polymers combined with carbon black nanospheres and carbon nanotubes. These sensors have been tested to different concentrations of toluene vapour and 1-propanol vapour in air. Results show that the carbon black resistive sensors outperform the CNT and chemFET sensors by a considerable margin. We believe that the CNTs are restricting the swelling of the polymers. Furthermore, we have tested the water dependence of these sensors and we have found that increased water concentration reduces the steady state response. We believe that the low sensitivity of CNT sensors may be addressed by using single wall CNTs and functionalising the CNT to improve selectivity.

ACKNOWLEDGMENTS

We would like to thank Dr Danick Briand (Institute of Microtechnology, Neuchatel) for processing of the resistive device. Also, we would like to that Southampton University, UK, for processing the chemFET devices under an EPSRC funded project.

REFERENCES

1. J.A. Covington, J.W. Gardner, D. Briand, N.F. de Rooij, *Sensors and Actuators B* **77**, 1-2, 155-162 (2004).
2. E. Severin, B.J. Doleman, N.S. Lewis, *Anal. Chem.* **72**, 658-668 (2000).
3. S. Chopra, K. McGuire, N. Gothard, A.M. Rao, *Appl. Phys. Letters* **83**, 11, 2280-2291 (2003).
4. B. Philip, J.K. Abraham, A. Chandrasekhar, V.K. Varadan, *Smart Mater. Struct.* **12**, 935-939 (2003).

INSTRUMENTATION: SAMPLING I

Design Of A Sorbent/desorbent Unit For Sample Pre-treatment Optimized For QMB Gas Sensors

G. Pennazza[1],* , M. Santonico[2], E. Martinelli[2],
R. Paolesse[3], C. Di Natale[2], S. Cristina[1], A. D'Amico[2]

1 Faculty of Engineering, University 'Campus Bio-Medico di Roma'
Via Alvaro del Portillo, 21 - 00128Rome, Italy
2 Department of Electronic Engineering, University of Rome 'Tor Vergata'
Via del Politecnico, 1 – 00133 Rome, Italy
3 Department of technological and chemical Science, University of Rome 'Tor Vergata'
Via della ricerca scientifica, 1 - 00133 Rome, Italy
**Corresponding author: e-mail g.pennazza@unicampus.it*

Abstract. Sample pre-treatment is a typical procedure in analytical chemistry aimed at improving the performance of analytical systems. In case of gas sensors sample pre-treatment systems are devised to overcome sensors limitations in terms of selectivity and sensitivity. For this purpose, systems based on adsorption and desorption processes driven by temperature conditioning have been illustrated. The involvement of large temperature ranges may pose problems when QMB gas sensors are used. In this work a study of such influences on the overall sensing properties of QMB sensors are illustrated. The results allowed the design of a pret-reatment unit coupled with a QMB gas sensors array optimized to operate in a suitable temperatures range. The performance of the system are illustrated by the partially separation of water vapor in a gas mixture, and by substantial improvement of the signal to noise ratio.

Keywords: QMB, sample pre-treatment, preconcentration.
PACS: 07.07.Df

INTRODUCTION

In order to enrich the concentration of gases in a gaseous sample several techniques are currently available. In particular, analytical chemistry derived approaches. Among them the most used are Solid Phase Micro Extraction (SPME) [1], Adsorption/desorption and cryogenic.

All these methods require temperatures in the range o 150°C-250°C to be applied to the solid phase. As a consequence, hot gas streams are injected in the sensor cell and reach the sensors surface.

Hot gas flow are of limited problems for sensors working at high temperature, such as metal-oxide semiconductors, but may arise problems in sensors working at room temperature such as those based on polymeric or molecular sensing layers.

In case of QMB sensors, even the transduction mechanism, namely the resonating frequency of the electro-mechanical resonator is intrinsically sensitive to temperature changes. Even if some previous study claimed a compatibility between high temperature desorption pre-concentration stage and QMB sensors [2] in this paper a study of the temperature effects on QMB gas sensors is performed.

This study is aimed at determining a temperature range that can be conveniently used in pre-treatment stages without affecting the signal of QMB sensors.

The results are then utilized to design a gas pre-concentration unit for QMB sensors.

EXPERIMENTAL AND METHODS

The effects of temperature on sensors performance were studied considering the case of QMB coated by a molecular layer. A 20 MHz AT cut quartz crystal was utilized, the resonator was coated by a molecular layer of Manganese tetraPhenylPorhyrin (Mn-TPP) whose sensing properties have been largely studied in the past [3, 4].

The coated QMB was compared with the responses of an uncoated similar resonator. Fig. 1 shows the arrangement of the two sensors in a cell specially designed to be conditioned in temperature.

Cell Temperature was conditioned and controlled by a Peltier thermalized plate (MPA075-12 by Melcor) controlled by a MTTC-1410 by Melcor.

The designed pre-treatment unit consisted of a cell (8 cc volume) whose walls were coated with absorbing materials (thickness: 0.1 mm; surface: 8mm x 40mm).

CP1137, *Olfaction and Electronic Nose: Proceedings of the 13th International Symposium,* edited by M. Pardo and G. Sberveglieri

Two different coatings were tested: Mn-TPP and carbon black. In both cases a mixed membrane was formed dissolving the materials with polyvinylchloride.

The chamber was thermally connected to the temperature controller unit (plate and controller) above mentioned. The pre-concentrator was tested measuring the outflow gas with a GC-MS (Gas Chromatograph – Mass Spectrometer) and with a QMB based gas sensor array using metalloporphyrins as sensing material. Experiments were performed with butanol diluted in nitrogen and in humid synthetic air.

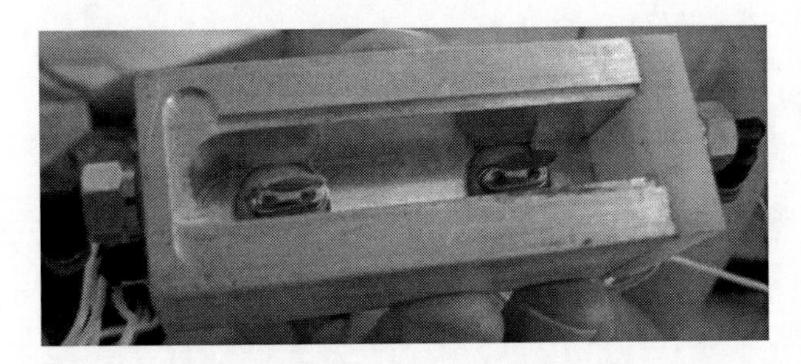

FIGURE 1. Measure test cell for temperature effects monitoring on two quartzes (one un-coated and the other covered with a Mn-TPP).

The experimental set-up is illustrated in figure 2: N_2 is used as reference and carrier gas, the gas flow is fixed by a mass flow controller, the low concentration of the sample gas is obtained by mean of a permeation tube.

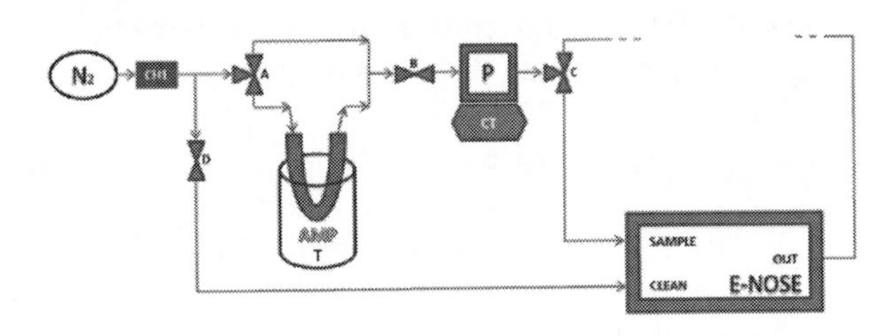

FIGURE 2. Experimental set-up, from left to right: nitrogen source, mass flow controller, permeation tube apparatus, preconcentrator, electronic nose chamber including the QMB sensors under test.

RESULTS

Typically, AT quartzes properties are granted in a temperature range from 10°C to 40°C.

From this study a temperature range has been selected (10°C-70°C); in this range the sensors responses showed to be influenced in magnitude by the temperature but this influence did not dramatically modified the sensors characteristic parameters (Sensitivity, Resolution and Limit Of Detection (LOD)).

On this basis we can say that no substantial interferences of the temperature with the sensors correct functioning has been observed in this temperature range.

Both the temperature of the sensor and of the gas are expected to influence the sensor signal. Here we investigated the response of sensors to heated gas streams.

In figure 3 the frequency shift versus the pre-concentrator temperature is plotted. Since the distance between pre-concentrator and sensors cell was rather short it can be assumed that the pre-concentrator temperature is roughly the same of the gas stream.

Frequency shift has a linear relationship with the temperature in the investigated range with a sensitivity of 1.4 Hz/°C.

Comparing these data with the increased signal due to the pre-concentration factor it is possible to conclude that the range of gas stream temperature between 10°C and 70°C obtains a good trade-off.

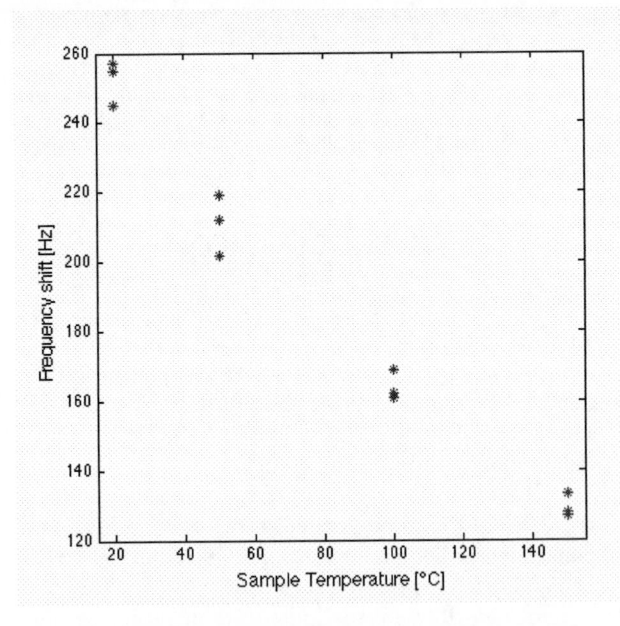

FIGURE 3. Frequency shift as a function of the gas stream temperature. Data are related to a QMB coated by Co-TPP.

The measurements performed with GC-MS and with the QMB sensors have shown a good repeatability.

From the QMB measurements it can be observed a good separation between n-butanol and water vapor (fig. 4). The second part of the signals shown in figure 4, is relative to the measurement of the pretreated sample.

The magnitude of the response appears amplified and two peaks are visible: the first can be attributed to water vapor, and the second as pertinent only to the n-butanol.

For sake of clarity only the signal of one sensor is reported even if all the sensors of the array displayed the same behavior.

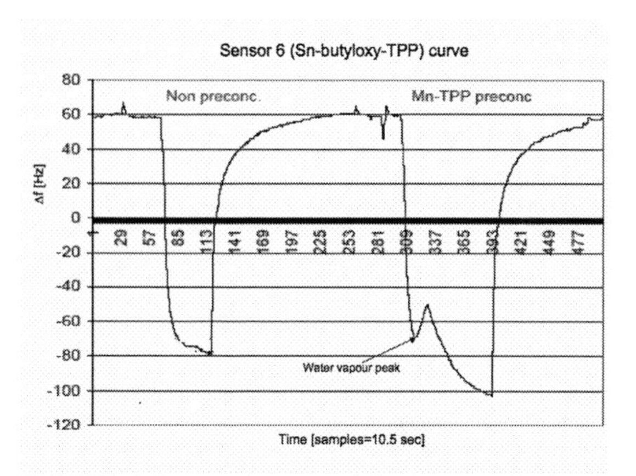

FIGURE 4. Response curves of the Sensor 6 (Sn-butyloxy-TPP) of the gas sensor array: the first on the left in case of no-pretreatment of the sample, the second on the right in case of sample pre-treated.

Figure 5 shows the gas-chromatographic peak of butanol recorded in samples pre-treated with different operating conditions. A maximum pre-concentrating factor of about 10 has been found. This factor has been calculated by comparing the area of the peaks.

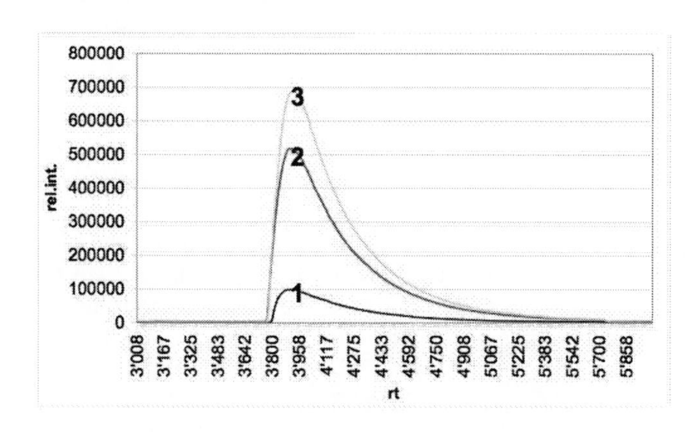

FIGURE 5. Detail on the n-butanol (270 ppm) peak in the three chromatograms relative to the measurement without pretreatment (1), with Mn-TPP (2) and with Mn-TPP+Carbon-black (3) as adsorbent materials. The area of peak 3 is ten times the peak 1 area.

The best result was obtained using an adsorbing material composed by a mixture of Mn-TPP and carbon-black.

It is worth to remark that the result reported in figure 4 about the separation of water vapor was observed in the QMB sensors responses, while the calculus of the pre-concentration factor reported in figure 5 was given by the GC-MS peaks, but it was also confirmed by the sensors responses.

CONCLUSIONS

In this work we illustrated the design of a device which allows the separation of mixtures such as water vapor and butanol, and an improvement of one order of magnitude in the concentration level. This device operates in a temperatures range that allows a safe operation of QMB sensors.

REFERENCES

1. C. Dietz, J. Sanz, C .Camara, *Journal of Chromatography A* **1103**, 183-192 (2006).
2. T. Hamacher, J. Niess, P. Schulze-Lammers, B. Diekmann, P. Bocker, *Sensors and Actuators B* **95**, 39-45. (2003).
3. C. Di Natale, R. Paolesse, A. D'Amico, *Sensors and Actuators B* **121,** 238-246 (2007).
4. C. Di Natale, R. Paolesse, A. Macagnano, S. Nardis, E. Martinelli, E. Dal Canale, M. Costa, A. D'Amico, *Journal of Material Chemistry* **14**, 1281-1287 (2004).

A New Hyphenated μTrap – GC – Surface Acoustic Wave (SAW) Based Electronic Nose For Monitoring Of Coffee Quality

Mauro Carvalho[1], Achim Voigt[2], Michael Rapp[2]

[1] *Instituto de Química, Universidade Federal do Rio de Janeiro, Rio de Janeiro, Brazil*
[2] *Forschungszentrum Karlsruhe, Institut für Mikrostrukturtechnik, Karlsruhe, Germany*
mauro@iq.ufrj.br

Abstract. An easy-to-use and versatile analytical method for complex matrix analisis like coffee was developed. The system consists of a microtrap sample preparation, a home made simplified gaschomatographic separation unit and an 8-fold surface acoustic wave based sensors (SAW) array detector. For the coffee quality analysis a successful discrimination of three coffee samples could be achieved. The system would be further developed into a fully automated, low cost version that can be broadly used by the coffee producers.

Keywords: headspace analysis, aroma, gas analysis, SAW sensors
PACS: 82.80.-d Chemical analysis and related physical methods of analysis

INTRODUCTION

The use of surface acoustic wave (SAW) sensor arrays coated with selective polymers is a very promising technique for analytical sensor systems for organic gas detection and electronic nose applications [1]. However, for complex matrices mostly arising from food headspaces such systems are often overcharged especially when simple marker profiles have to be extracted in order to evaluate the food quality [2].

In order to ease this problem, we present a combined method to achieve a remarkably higher analytical resolution, where two subsequent steps of a selective sample preparation is combined with a highly selective polymer based sensor array (electronic nose).

EXPERIMENTAL AND METHODS

In a first step this method consists of static head space enrichment by the use of an external miniaturized unit (μtrap) of the sample followed by a second step consisting of a sample separation by a self developed gas chromatographic unit. The third step is the detection with a miniaturized SAW sensor array with an extremely small sampling volume of 50μl and a low inner surface area minimizing response time and unwanted memory effects [3]. With the latter configuration very fast response times (1...2 seconds) can be achieved in order to resolve sharp peaks arising from short sample injection into the GC unit. The whole system is showed in the Figure 1. For this work we used a commercially available autosampler from the CTC/Gerstel Company with their commercially available "ITEX" option. We developed the temperature control as well sensor devices. The headspace sample was taken from thermally controlled sample vials and adsorbed in a μtrap directly attached to the sampling syringe. The trap was filled with tenax TA. After taking the sample the syringe with the μtrap is positioned over a GC inlet port and heated. The desorbed material is then injected into the GC-unit of the system where the separation occurs according to pre-programmed temperature profiles.

RESULTS

The separated components reach the SAW-sensor array at their specific retention times, and interact with each of the eight sensors. Due to the linear SAW sensor characteristic signals are proportional to the quantity of the substances in the headspace. An example of the resulting eight chromatograms obtained for each sample is showed in the Figure 2. The highest

CP1137, *Olfaction and Electronic Nose: Proceedings of the 13th International Symposium,* edited by M. Pardo and G. Sberveglieri

peak in the chromatogram of the figure 2 comes from water. The Figure 3a shows a scale expansion of the first part and the Figure 3b of the second part of the chromatogram of the figure 2. There can be observed the different profiles in between the sensors. For demonstration of the discrimination capability of this method we evaluated only the first five peaks in fig 3a. Treatment of these data was done by taking the same four peaks of the 8 chromatograms corresponding to the 8 sensors signals of the array. Then they were normalized to their maximum height, respectively. This procedure minimizes the variation due the sampling preparation and gives relative normalized maximum height values for each peak at each sensor for the different coffee samples. The results of the data analysis for three different samples of coffee – arranged in columns – are shown in Figure 4.

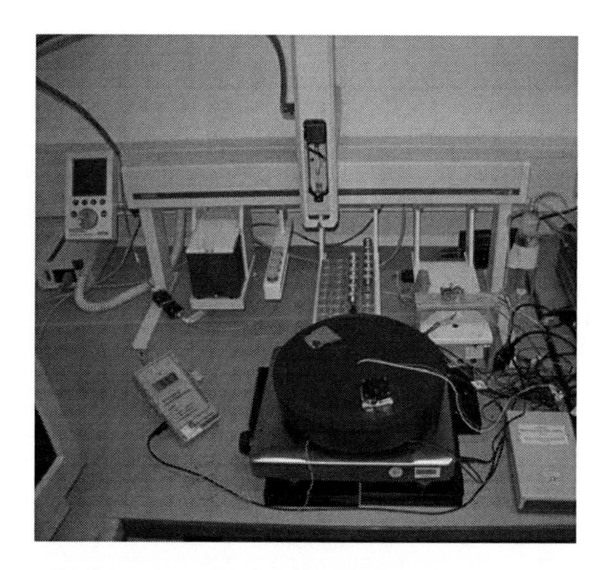

FIGURE 1. Headspace-GC-SAW system.

DISCUSSION

As it can be seen, the resulting profiles obtained for the sensor array permits a readily differentiation of the three coffee types. All implemented functionalities in the system can be controlled from a single PC with self made software enabling a maximum of flexibility for any kind of complex gas analyses. The greatest advantage of the method is the exclusion of typical interfering components at high concentrations, especially water. This can be done either by the choice of the absorber in the μtrap or by choosing suitable chromatographic parameters.

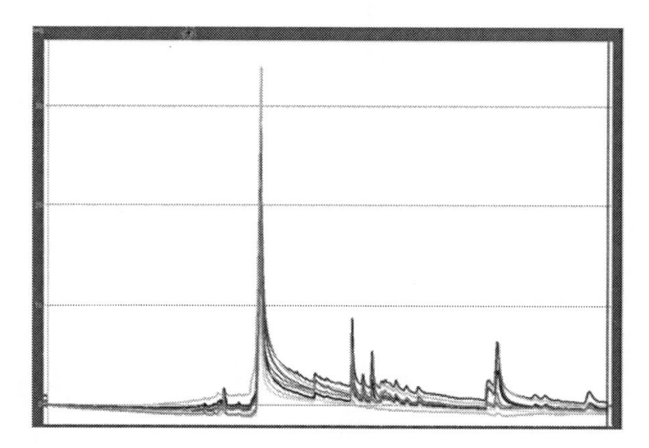

FIGURE 2. Total chromatogram of the eight sensors.

CONCLUSIONS

The system could easily differentiate the coffee samples and shows a great analytical versatility since it allows a free choice of the type of absorber material, the type of GC-column and the respective coating material for each of the eight sensors of the sensor array. Despite of its high analytical power, the setup is simple, easy to use and manipulate and can be fully automatized. With this system is possible a broad determination of volatile substances present in a headspace, especially in a complex food matrix. In a low cost version the proposed system will allow the easy detection of specific aroma markers in the coffee by an appropriate but fixed setting of evaluation time window and choice of GC parameters.

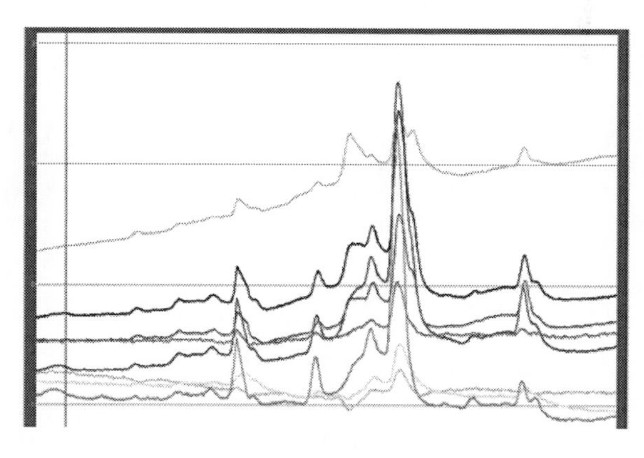

FIGURE 3A. Expansion of the chromatogram: First part.

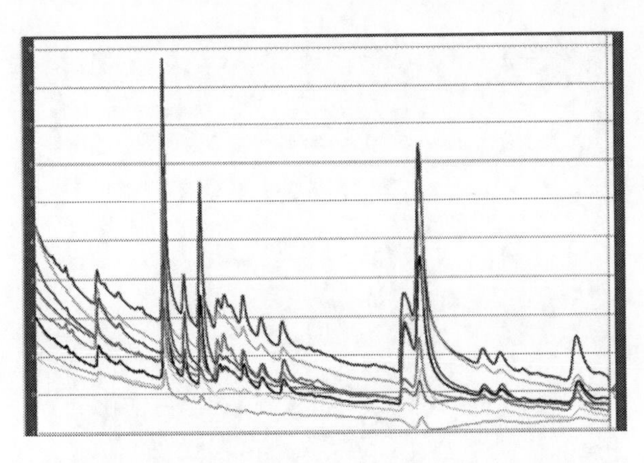

FIGURE 3B. Expansion of the chromatogram: Second part

REFERENCES

1. N. Barié and M. Rapp, "Discrimination of Alkanes Using Trap and GC enhanced SAW based Electronic Noses" in *Proceedings of IEEE Sensors 2004*, p. 1183-1186.
2. N. Barié, A. Voigt and M. Rapp, "Surface Acoustic Wave based Electronic Noses Coupled with SPME enhanced Headspace-Analysis for Food Quality Monitoring" at the *10. Int. Symp. on Olfaction and Electronic Noses - ISOEN 03, in Riga, 26. – 28. Juni 2003 Lettland*
3. F. Bender, K Länge, A. Voigt and M. Rapp, "Improvement of Surface Acoustic Wave Gas and Biosensor Response Characteristics Using a Capacitive Coupling Technique" in *Anal. Chem.*, vol. 76, No. 13 (2004), pp. 3837-3840.

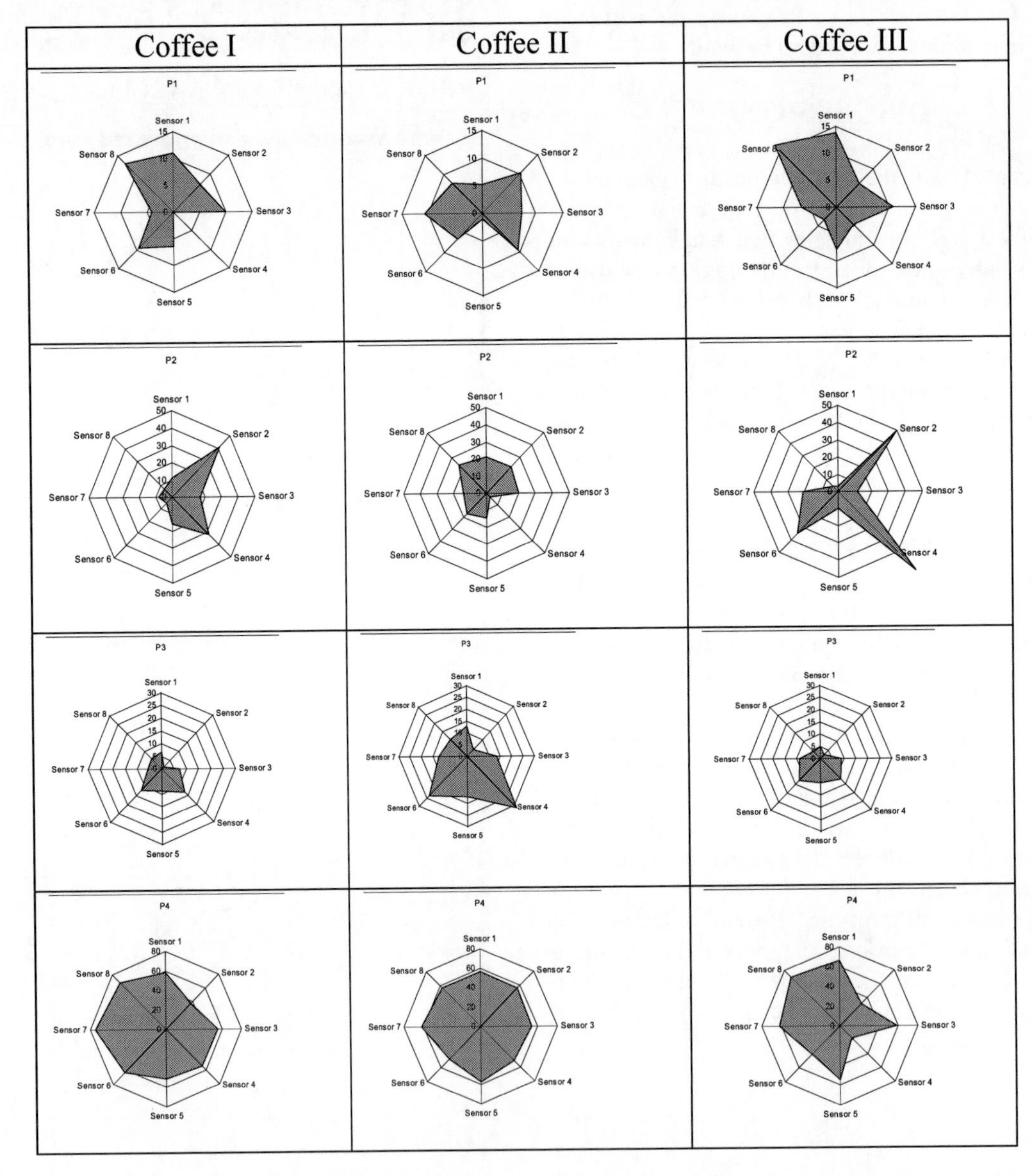

FIGURE 4. Relative normalized peak heights of three coffee samples (arranged in columns) with regard to four peaks (P1 – P4) arranged in rows) of the first part of their chromatogram of figure 3a.

Bringing Chromatography Back To Colour

F. Dini[1], R. Paolesse[2], D. Filippini[3], E. Martinelli[1], A. D'Amico[1], I. Lundström[3] and C. Di Natale[1]

[1]*Department of Electronic Engineering, University of Rome Tor Vergata, Via del Politecnico 1, 00133 Rome, Italy*
[2]*Department of Chemical Science and Technology, University of Rome Tor Vergata, Via della Ricerca Scientifica, 00133 Rome, Italy*
[3]*Department of Physics, Chemistry and Biology, Division of Applied Physics, Linköping University, S-58183 Linköping, Sweden*

Abstract. Biological systems enhance the odor discrimination capabilities by performing a small separation of complex mixtures. This principle may be successfully exploited in the development of new sensor platforms. The aim of the presented work is to implement the gas-chromatographic separation principle for chemical sensing. Herein we report a simple way of imaging gas diffusion through a channel by means of optical changes occurring in a track that contains a chemical indicator embedded in a polymeric matrix.

Keywords: Porphyrins, Polymers, Retention Time, Optical Chemical Sensors.
PACS: 07.07.Df

INTRODUCTION

In natural olfaction, the thin mucous layer, covering the receptors in the nasal epithelium, has partitioning properties. It is believed that this partitioning process contributes to the coding of olfactory information [1]. The mucous layer acts as a sorption layer where the adsorption and desorption of odorant molecules occur, in a way similar to the stationary phase of a gas-chromatographic (GC) column [2]. This idea has been recently exploited for the development of a microsystem that tries to mimic the olfactory mucosa. In this device, sensors have been distributed along a path covered with a polymeric stationary phase. This gives rise to a spatial and temporal distribution of signals, able to increase odor discrimination in the field of electronic noses [3].

Since polymers show different partitioning properties toward different analytes, we believe that this feature can be exploited to develop a sensor system that implements the gas-chromatographic (GC) separation principle. Polymers may act as a stationary phase, delaying in a peculiar way compounds flow along a given track. Combining the partitioning properties of polymers with the chemical sensitivities of dyes, the diffusion of the volatile compounds can be observed in real-time, through the optical changes occurring in a path functionalized with a colorimetric dye. These changes can be monitored by means of the computer screen photo-assisted technique (CSPT), which consists of a polychromatic light source, provided by a programmable computer screen and of a wide band detector, such as a web camera (Figure 1).

FIGURE 1. The CSPT platform consists of a computer screen as a light source and of a camera as a detector. The camera records the sample illuminated with a programmed color sequence.

The CSPT platform has already been demonstrated to be able to detect volatile compounds through optical changes in porphyrin layers [4]. Moreover, CSPT technology has been used to image a layer of

CP1137, *Olfaction and Electronic Nose: Proceedings of the 13th International Symposium,* edited by M. Pardo and G. Sberveglieri

distributed colour indicators that, exposed to an odour stimuli, responds with a characteristic spatio-temporal pattern, which mimics the response distribution of olfactory receptor neurons in the epithelium [5]. In the CSPT system, each pixel of the recorded image can be considered as an individual and independent sensor, allowing the imaging of thousand of sensors simultaneously. In this way, it is possible to observe the gas diffusion looking at the optical changes in the pixels along the path.

EXPERIMENTAL AND METHODS

Porphyrin tracks with a defined geometry were deposited onto a plastic substrate by means of microtransfer molding, a soft lithographic technique [6]. Microtransfer molding uses a patterned poly(dimethylsiloxane) (PDMS) mould that is covered with the membrane solution. The substrate is then placed in contact with the PDMS surface. The PDMS stamp is peeled away after solvent evaporation, leaving on the substrate a patterned porphyrin microstructure. The gas is delivered in a controlled way to the track by means of a diffusion path above it, defined by a patterned PDMS that provide a channel of 30 μm height and 100 μm wider than the printed track (400 μm wide). Figure 2 shows a printed linear track covered with its PDMS chamber and imaged with the CSPT setup.

FIGURE 2. Example of a printed track.

Measurements were carried out illuminating the sample with a three colours sequence: red (255, 0, 0), green (0, 255, 0) and blue (0, 0, 255), with the camera recording the intensity received in its three channels (red, green and blue). Vapors of various volatile compounds were injected directly into the channel through a gas line connection, using nitrogen flow as a carrier.

In the experiment described below Zinc(5,10,15,20-tetraphenylporphyrin) [ZnTPP] embedded in a polymeric matrix was used as a colorimetric indicator. Track printing was carried out using a polyvinylchloride (PVC) membrane, containing the porphyrin indicator, dissolved in tetrahydrofuran (membrane composition in weight: tetrapyrrole/PVC/bis(2-ethylhexyl)sebacate 1:33:66). The ZnTPP membrane was printed with a spiral geometry, since it reminds of the shape of a capillary column and because it represents a good and elegant way of exploiting the small area available.

RESULTS

Due to the different molecular weights, diffusion rate and chemical interactions with the polymeric matrix, different odor molecules react with the dye distributed along the path at different time and concentration, giving rise to a peculiar spatio-temporal pattern for each substance. First, we studied the diffusion of volatile compounds injected into the channel at a constant flow. For the acquired video, the mean intensity of pixels was calculated in selected areas of the spiral track, the so-called regions of interest (ROI) (Figure 3).

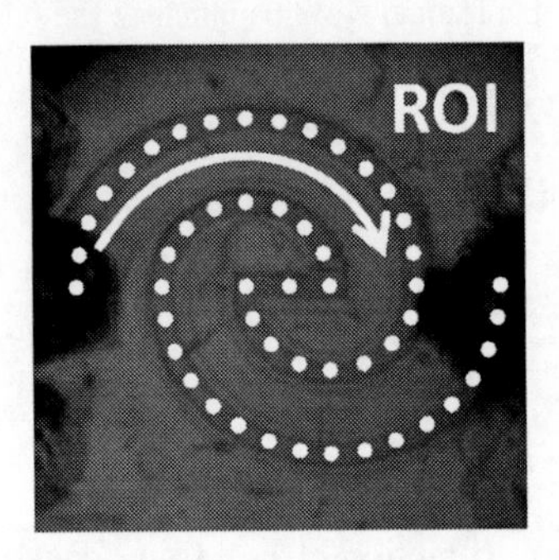

FIGURE 3. Printed track and selected regions of interest (ROI). Mean intensity of pixels belonging to the ROI was calculated and used for further analysis.

For the chosen color indicator, most of the information is retained in the green channel intensity of the video relative to the green illumination. The response of the porphyrin track is displayed as the difference between current frame and the first frame, taken in the reference gas. To visualize the time evolution of gas diffusion along the path, the change of light intensity in each ROI was displayed in a color-coded map scale versus time. Diffusion profile showed by butylamine is reported in Figure 4.

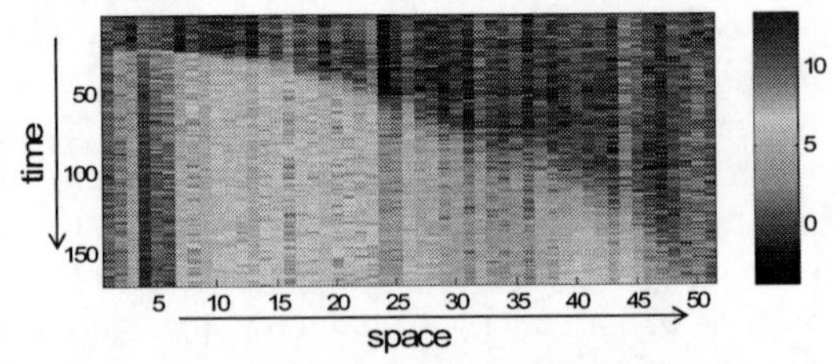

FIGURE 4. Example of the butylamine diffusion profile along the channel, displayed through the optical changes occurring in the selected ROIs of the porphyrin track.

As an example of the separation properties showed by our system, the results obtained with three amines, butylamine, trimethylamine and triethylamine, are shown. The time evolution of the sensor responses, at different locations along the spiral path, shows that different compounds take different time to permeate the whole channel. Figure 5a evidences that sensor signals, recorded under the exposure of butylamine, triethylamine and trimethylamine vapors in an area close to the gas entrance are almost completely overlapped. In Figure 5b, it is seen that the signals, concerning regions located at the end of the track for the same substances, are more separated, as a result of the delaying effect performed by the polymer.

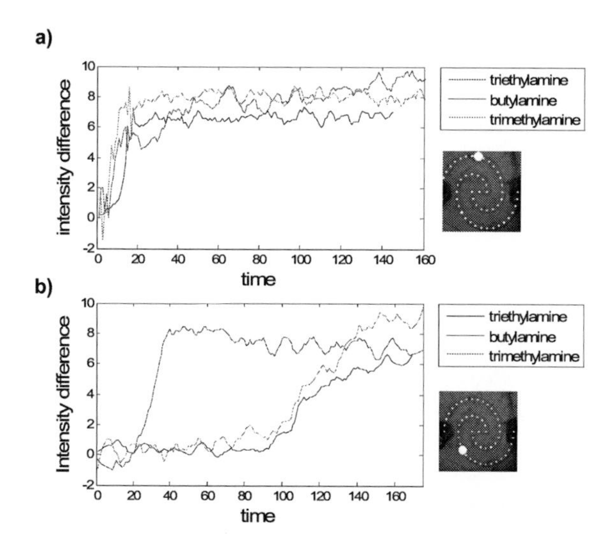

FIGURE 5. Time evolution of the response in ROI located in different region of the channel and displayed in the inset. At the end of the column, about 35 mm from the entrance, (b) the responses toward butylamine and triethylamine are more separated with respect to the signals calculated in the region 6 mm from the entrance (a).

Sorption processes of a vapor into a polymer are commonly described in the frame of the linear solvation energy relationship (LSER) as a combination of terms, which express the contribution of every kind of expected interaction to the overall sorption process [7]. These terms describes the contribution of polarity (rR), polarizability ($s\pi$), hydrogen-bond for an acid vapour ($a\alpha$), hydrogen-bond for a basic vapor ($b\beta$) and dispersion interactions ($l{\cdot}log\ L^{16}$). Parameters R, π, α, β and $log\ L^{16}$ describe solubility properties of the solute vapor, while r, s, a, b and l characterize the complementary properties of the polymer. It is interesting to analyze the relation between the time necessary to reach the maximum response and the LSER parameters of the ROI distributed along the path (Figure 6). Dispersion interactions are the only processes that are expected in the PDMS matrix, hence the relations existing with the other parameters may be likely attributed to the interactions with the dispersed porphyrins. Although the underlying phenomenon is very complex, these results suggest that porphyrins not only detect the molecules but they also have an active role in determining the diffusion properties of volatile molecules along the path.

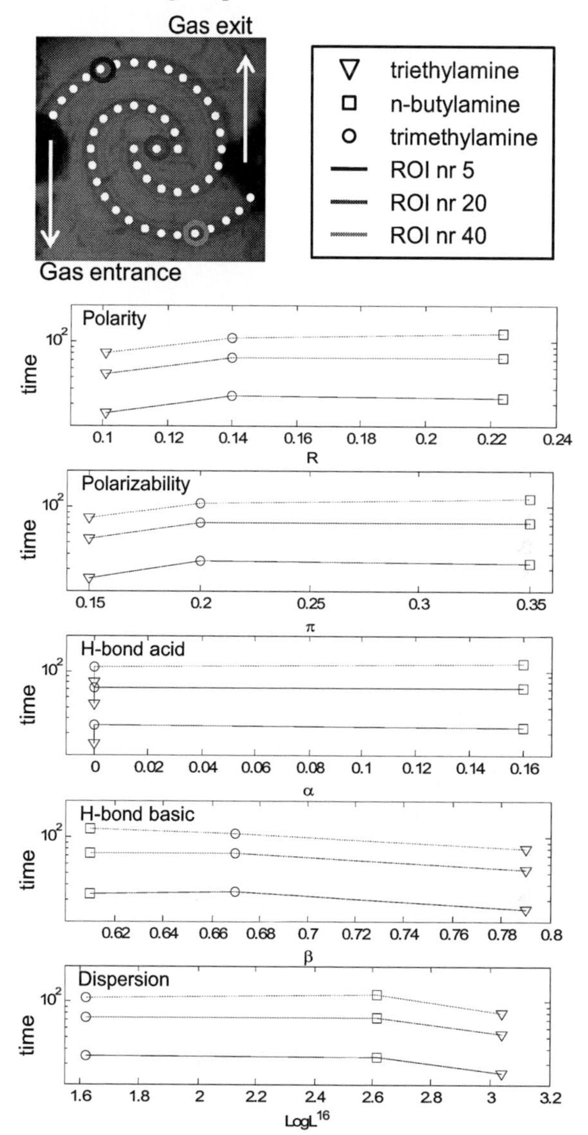

FIGURE 6. Time is plotted against values of LSER parameters for three selected ROIs, showing a relation between the intensity of interaction and the diffusion time.

The response to a concentration pulse was also analyzed. In this case, the substance package does not cover the entire path, because it is diluted by the subsequent nitrogen flow. However, different substances can still be discriminated on the basis of different flight time and, finally, looking at the length of the track interested by the reaction or at the depth of penetration of the substance through the channel. This effect is seen in Figure 7, where the intensity of the ROIs, recorded for the track subjected to a pulse of butylamine and triethylamine, are imaged versus time.

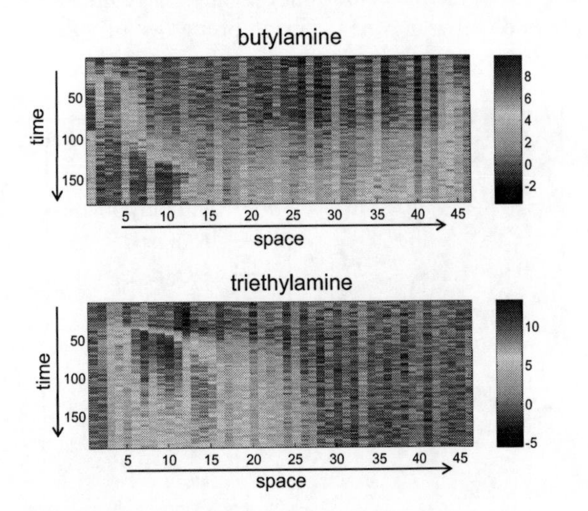

FIGURE 7. Image of the permeation depth of a 30 sec pulse of amines vapors.

CONCLUSIONS

A simple and efficient method of implementing the gas-chromatographic separation principles has been proposed in a sensor system. The polymeric matrix allows the separation of compounds, which are detected by means of colorimetric dye dispersed in the matrix. Modulation of the polymeric phase and of the chemical indicator offers the possibility to further increase the information retained in the spatio-temporal pattern, allowing a better sample discrimination and finally the separation of a complex mixture separation. Moreover, different geometries favoring compound separation may be thought of.

ACKNOWLEDGMENTS

This research project has been supported by a Marie Curie Early Stage Research Training Fellowship of the European Community Sixth Framework Programme under contract number MEST-CT-2004.

REFERENCES

1. P. Kent, M. Mozell, S. Youngetoub and S. Turco, *Brain Res.* **981**, 1-11 (2005).
2. D. E. Hornung and M. M. Mozell, *J. Gen. Physiol.* **69**, 343-361 (1977).
3. J.A. Covington, J.W. Gardner, A. Hamilton, T.C. Pearce and S.L. Tan, *IET Nanobiotechnol.* **1**, 15-21 (2007).
4. D. Filippini, A. Alimelli, C. Di Natale, R. Paolesse, A. D'Amico and I. Lundström, *Angew. Chemie* **118**, 3884 – 3887 (2006).
5. C. Di Natale, E. Martinelli, R. Paolesse, A. D'Amico, D. Filippini and I. Lundström, *PLoS ONE* **3(9)**, e3139 (2008)
6. Y. Xia and G. M. Whitesides, *Angew. Chem.* **37**, 550-575 (1998).
7. M. Abraham, *Chem. Soc. Rev.* **22**, 73–83 (1993).

Evaluation of Sample Recovery of Odorous VOCs and Semi-VOCs From Odor Bags, Sampling Canisters, Tenax TA Sorbent Tubes, and SPME

Jacek A. Koziel*[1], Jarett P. Spinhirne[2], Jenny D. Lloyd[2], David B. Parker[3], Donald W. Wright[4] and Fred W. Kuhrt[4]

[1]*Department of Agricultural & Biosystems Engineering, Iowa State University, Ames, IA 50011, USA; koziel@iastate.edu, formerly with Texas Agricultural Experiment Station, Amarillo, TX, USA*
[2] *Texas Agricultural Experiment Station, Amarillo, TX, USA*
[3]*West Texas A&M University, Killgore Research Center, Canyon, TX, USA.*
[4]*Microanalytics (a MOCON Company), Round Rock, TX, USA*

Abstract. Odor samples collected in field research are complex mixtures of hundreds if not thousands of compounds. Research is needed to know how best to sample and analyze these compounds. The main objective of this research was to compare recoveries of a standard gas mixture of 11 odorous compounds from the Carboxen/PDMS 75 μm SPME fibers, PVF (Tedlar), FEP (Teflon), foil, and PET (Melinex) air sampling bags, sorbent Tenax TA tubes and standard 6 L Stabilizer™ sampling canisters after sample storage for 0.5, 24, and 120 (for sorbent tubes only) hrs at room temperature. The standard gas mixture consisted of 7 volatile fatty acids (VFAs) from acetic to hexanoic, and 4 semi-VOCs including p-cresol, indole, 4-ethylphenol, and 2'-aminoacetophenone with concentrations ranging from 5.1 ppb for indole to 1,270 ppb for acetic acid. On average, SPME had the highest mean recovery for all 11 gases of 106.2%, and 98.3% for 0.5 and 24 hrs sample storage time, respectively. This was followed by the Tenax TA sorbent tubes (94.8% and 88.3%) for 24 and 120 hrs, respectively; PET bags (71.7% and 47.2%), FEP bags (75.4% and 39.4%), commercial Tedlar bags (67.6% and 22.7%), in-house-made Tedlar bags (47.3% and 37.4%), foil bags (16.4% and 4.3%), and canisters (4.2% and 0.5%), for 0.5 and 24 hrs, respectively. VFAs had higher recoveries than semi-VOCs for all bags and canisters. New FEP bags and new foil bags had the lowest and the highest amounts of chemical impurities, respectively. New commercial Tedlar bags had measurable concentrations of N,N-dimethyl acetamide and phenol. Foil bags had measurable concentrations of acetic, propionic, butyric, valeric and hexanoic acids.

Keywords: Odor, VOCs, sampling, sample recovery.
PACS: 01.30.Cc.

INTRODUCTION

Odorous gases encountered in field research such as livestock operations are very complex mixtures of hundreds if not thousands of compounds. The chemical characterization of individual compounds in these mixtures is extremely challenging. Many of these compounds are particularly susceptible to being adsorbed onto contact surfaces, with less than 100% recovery from sample containers such as air sampling bags. In addition, very low concentrations often preclude compound detection and identification with conventional GC-MS. These characteristics present a unique analytical challenge because special considerations are needed for air sample collection, preparation and analysis.

Scientists have recognized for some time that Tedlar is not the perfect material for gas sampling bags. Though it is relatively inert, there is evidence of water permeation, adsorption, and desorption of some chemical species to the Tedlar.[1,2] Keener et al. quantified recoveries of 19 odorous gases from Tedlar bags using sorbent tubes, and concluded that Tedlar bags emit acetic acid and phenol and greatly adsorb indole, skatole, p-cresol, 4-ethylphenol, nonanoic and octanoic acids.[3] Nagata (2003) used polyester odor bags and reported recovery rates from these bags for 35%, 40%, 39%, and 6.5% for isobutyric, butyric, isovaleric acids and indole, respectively. [4]

Wright et al. identified more than 60 odorous compounds in exhaust air from a swine barn, many of which were also present in air at and downwind from a

CP1137, *Olfaction and Electronic Nose: Proceedings of the 13th International Symposium*, edited by M. Pardo and G. Sberveglieri

beef cattle feedlot.[5] Wright et al. used SPME for air sample collection and for simultaneous chemical identification and olfactory analysis on a MDGC-MS-O system. The most preeminent compound for both the source and the distant locations was p-cresol. Other compounds including: 4-ethylphenol, 2'-aminoacetophenone, indole, and a suite of volatile fatty acids were also present in nearly every air sample collected with SPME. The highly polar and semivolatile compounds that appear to be odor-defining for livestock operations are potentially the most offensive odorants for swine and cattle feedlots.[5] However, no data exists on sample recoveries of these compounds from SPME and other sampling devices.

The primary objective of this study was to determine sample recoveries from a standard gas mixture of odorous gases from several popular air sampling devices including SPME, 5 types of air sampling bags, Tenax TA sorbent tubes, and stainless steel (SS) sampling canisters under typical sample storage conditions, i.e., room temperature and 24 hrs. A secondary objective was to identify impurities in new air sampling bags and sampling canisters that can potentially affect air samples.

EXPERIMENTAL AND METHODS

Standard Gases

Stable concentrations of 11 VOCs and semi-VOCs including acetic, propionic, isobutyric, butyric, valeric, isovaleric and hexanoic acids, p-cresol, 2'-aminoacetophone and 4-ethyl phenol were maintained using permeation sources. The 7 volatile fatty acids (VFAs) were generated with a standard gas generator based on permeation devices. The 4 semi-VOCs were generated in an in-house-made gas generator utilizing a permeation devices. In this research, we used SPME as an air sampler and also as a sampling and sample introduction technology used to evaluate sample recoveries from other devices except the sorbent tubes. Sampling with SPME was relatively short (5 min) and identical for all evaluations. Triplicate QA/QC samplings with a single Carboxen/PDMS 75 μm SPME fiber were used to confirm the stability of the standard gas daily..

The standard gas mixture was always sampled with SPME fiber for 5 min. This was followed by immediate insertion of the SPME fiber into GC injector for time = 0 hrs.

Sample Recovery from Air Sampling Bags

Commercial bags. All bags were 10 L capacity fitted with a single polypropylene septum valve fitting. The choice of bags was as follows: (a) Tedlar (polyvinyl fluoride, PVF), (b) in-house-made Tedlar, (c) FEP (Teflon), (d) foil, and (e) PET. Bags (a) (cat. #232-08), (c) (custom-made to 10 L capacity), and (d) (cat. #245-28) were purchased from SKC (Houston, TX).

In-house-made bags. The in-house-made Tedlar bags (b) were made at West Texas A&M University. The PVF film was ordered directly from DuPont. All in-house made bags were subjected to conditioning. Post-manufacturing conditioning was found to remove any residual odor.[3]

The PET (a.k.a. Melinex) bags were made from a 5000 m × 1.14 m wide roll of polyethylene terephthalate film and polypropylene fittings. PET bags meet the Australian and the European olfactometry standards where they are listed as "Nalophan" and are popular choice of air sampling bags for collection of livestock odor samples.

Standard gas sampling. All bags were filled with the standard gas using a special dispensing port located immediately downstream from the SPME sampling bulb.[6] Triplicate samples for 0.5 and 24 hrs holding/preservation time were collected in air sampling bags sequentially in 1 to 2 hrs intervals.

Interfering chemicals in new air sampling bags.

In addition, chemical backgrounds of all types of bags were studied using SPME extractions from new bags that came from the same batch as the bags used for sampling of standard gas.

Sample Recovery from Sampling Canisters

Sampling canisters are used in EPA methods TO-14 and TO-15 for sampling of VOCs in ambient air. Canisters are made from a low carbon 316L stainless steel and are subjected to electro-polishing that removes impurities from the inside surface while creating a passive layer enriched in chromium oxide.

Sample Recovery from Tenax TA Sorbent Tubes

Stainless steel desorption tubes (Supelco, Bellefonte, PA) filled with Tenax-TA adsorbent were used to collect standard gas mixture. Triplicate sorbent tubes were used to store samples for 24 hr and 120 hrs.

Analyses on GC/MS

The samples collected on Carboxen/PDMS fibers were analyzed with a Varian 3800/Saturn 2000 GC/MS system equipped with a 25 m × 0.25 mm × 0.2 μm film CP-WAX 58/FFAP capillary column (from Chrompack/Varian). The ion trap MS measured a wide mass range between 30 and 460 m/z to aid in identification of background compounds. MS responses to triplicate SPME samples of gas standard with 0 hr holding/preservation time were used as a reference for all samples with holding time of 0.5 and 24 hrs.

All desorption tubes were analyzed on a Turbomatrix automated thermal desorption (ATD) unit (Perkin-Elmer, Boston, MA) connected to the GC/MS system.

RESULTS

Interfering Chemicals in New Air Sampling Bags

Teflon bags had the lowest number and amount of interfering compounds. No compounds other than silanes and siloxanes were identified. The in-house made Tedlar bags had the second lowest background. The only interfering chemical from the in-house Tedlar bags was phenol. However, the amount of phenol was approximately 10 times lower than in commercial Tedlar bags. Large quantities of phenol in Tedlar bags were also reported by Keener et al.[3]

Commercial Tedlar bags also had a significant amount of DMAC. The foil bags had the greatest amounts of impurities of all bags tested. The largest impurities occurred in the low MW compound region of the chromatogram, where these types of bags are designed to preserve gases better than others. Foil bags also had significant impurities of acetic, propionic, butyric, valeric, and hexanoic acids, i.e., target gases in this study. The concentrations of VFAs in foil bags were on the order of those typical in livestock operations. Background impurities in PET bags and canisters were very low.

Sample Recovery from SPME

SPME fibers showed an excellent sample recovery. Average sample recovery for all 11 compounds sampled with 3 SPME fibers was 106% (±20.2%) for 0.5 hr storage time and 98% (±18.6%) for 24 hr storage time. This suggests that it could be expected to retain nearly all if not all target compounds in field samples if the sample is stored at room temperature and the storage time does not exceed 24 hrs. The sample recovery for p-cresol was 116.9% (±8.7%) and 92.5% (±10.3%) for 0.5 and 24 hrs sample storage times.

Sample Recovery from Air Sampling Bags

The sample recoveries for bags were generally less than those associated with SPME. There was also a greater variability in recoveries between target compounds. Sample recoveries after 0.5 hrs were generally greater than recoveries after 24 hrs. PET bags had the best mean recoveries for target compounds among all bags tested. The average sample recovery was 71.7% and 47.2% for 0.5 and 24 hrs sample storage time, respectively.

Teflon bags had the second best recoveries equal to 75.4% (±11.7%) and 39.4% (±9.5%) for 0.5 and 24 hrs sample storage time, respectively. Light MW VFAs had a better mean recovery equal to 94.2% (±7.2%) and 58.9% (±7.2%) for 0.5 and 24 hrs sample storage times compared to recoveries for semi-VOCs equal to 59.8% (±15.5%) and 23.2% (±11.5%) for 0.5 and 24 hrs sample storage times, respectively. Teflon bags had a low variability of less than 15% for the 3 bags tested and all target compounds except 2'-aminoacetophenone (up to 32.9%) and indole (up to 38.6 %). The sample loss was dependent on storage time. The sample recovery for p-cresol was 67.5% (±6.3%) and 29.0% (±3.0%) for 0.5 and 24 hrs sample storage times. The 24 hr sample recovery for acetic acid was 45.4%.

Tedlar bags made in-house had an average recovery of 11 target compounds equal to 47.3% (±14.9%) and 36.2% (±20.8%) for 0.5 and 24 hrs sample storage time, respectively. Sample recoveries had a greater variability compared to those of Teflon. Also, the recoveries of semi-VOCs were greater for 24 hrs storage compared to 0.5 hr storage. The recovery of p-cresol was 13.3% (±7.4%) and 22.7% (±34.3%) for 0.5 and 24 hrs sample storage time, respectively.

Commercial Tedlar bags had an average recovery of 11 target compounds equal to 67.6% (±7.6%) and 22.7% (±7.4%) for 0.5 and 24 hrs sample storage time, respectively. The initial losses at 0.5 hr storage time were smaller compared to the Tedlar bags made in-house. However, less standard gas was recovered from commercial Tedlar bags after 24 hrs. Sample recoveries were similar to Teflon bags and had a smaller variability compared to Tedlar bags made in-house. The 24 hr sample recovery for acetic acid was 20.4%, i.e., much smaller than those reported by Keener et al.[3] One possibility is a large amount of co-eluting and interfering DMAC found in Tedlar bags.

Foil bags had an average recovery of 11 target compounds equal to 57.1% (±5.2%) and 25.5% (±22.2%) for 0.5 and 24 hrs sample storage time, respectively. However, new foil bags have significant amounts of acetic, propionic, butyric, valeric, and hexanoic acids. When these impurities are subtracted, the recoveries are lowered to 16.4% (±15.3%) and 4.3% (±22.9%). The recovery of p-cresol was low and equal to 2.7% (±4.2%) and 3.7% (±3.4%).

Sample Recovery from Sampling Canisters

Average recoveries for canisters were equal to 4.2% (±7.3%) and 0.5% (±0.6%) for 0.5 and 24 hrs sample storage time, respectively. There was no recovery for all target compounds after 24 hrs with the exception of acetic and propionic acids (2.7%) and (±1.4%). This poor recovery could be caused by adsorption to the walls of canisters and/or reactions in the presence of chromium oxides which coat the inside surface of canisters.

Sample Recovery from Sorbent Tenax TA Tubes

Average recoveries from Tenax TA sorbent tubes were equal to 94.5% (±26.1%) and 88.3% (21.0%) for 24 and 120 hrs sample storage time, respectively. Recoveries of SVOCs (95.5% and 105.1%) were slightly better than recoveries for VFAs (93.9% and 76.4%). Recoveries for all compounds were excellent with the exception of acetic and propionic acids and 2'-aminoacetophenone.

DISCUSSIONS

Both the SPME Carboxen/PDMS 75 µm fibers and Tenax TA appear to be excellent samplers for the target gases used in this study with the average sample recovery equal to 98% (±18.6%) for SPME and the average sample recovery equal to 94.5% (±26.1%) for Tenax TA for 24 hrs storage time at room temperature. This information could be useful to researchers using SPME for field air sampling.

Sample recoveries for all other types of sampling media, i.e., bags and stainless steel (SS) canisters were significantly lower. Mean sample recoveries from all types of bags for 7 VFAs were always much higher than sample recoveries for 4 semi-VOCs. Mean recoveries for VFAs/semi-VOCs were 66.1%/2.4%, 47.4%/24.5%, 43.1%/24.1%, 31.0%/1.1%, 6.1%/1.1% for PET, Teflon, in-house Tedlar, Tedlar, and foil bags, respectively. This suggests that the bag material affects sample recoveries of chemical function groups of semi-VOCs from PET bags were poor. Teflon and in-house Tedlar had the best recoveries for semi-VOCs.

CONCLUSIONS

Several conclusions can be made:

(1) Tenax TA sorbent tubes and SPME Carboxen/PDMS 75 µm fibers had the highest average sample recovery equal to 94.5% (±26.1%) and 98% (±18.6%), respectively, for 24 hr storage time at room temperature for the 11 target gases.

(2) Sample recoveries for target gases from air sampling bags and sampling canisters were lower than SPME Carboxen/PDMS 75 µm. Sample recoveries were generally greater for 0.5 hr sample storage time compared to 24 hrs storage time. PET bags had the best recoveries for VFAs and in-house made Tedlar bags had the best sample recoveries for semi-VOCs. On average, PET bags had the best sample recoveries, followed by Teflon, commercial Tedlar, in-house made Tedlar, and foil bags.

(3) Sample recoveries from sampling canisters were lower than all others.

(4) New PET, Teflon bags, and sampling canisters had no residual interfering compounds. In-house made Tedlar bags had a small amount of phenol, however, the amount was 10 times less than phenol inside commercial Tedlar bags. These bags also had a measurable amount of DMAC. Foil bags had measurable amounts of acetic, propionic, butyric, valeric, and hexanoic acids.

(5) Further research is warranted to determine how recoveries from bags affect odor concentrations.

ACKNOWLEDGEMENTS

This work was published as reference #6.

REFERENCES

1. L. J. McGarvey and C.V. Shorten, *AIHA Journal*, **61**, 375-380(2000).
2. A. van Harreveld,. *JAWMA*, **53**, 51-60 (2003).
3. K.M. Keener, J. Zhang, R.W. Bottcher, R.D. Munilla, *Trans. ASAE*, **45**, 1579-1584 (2002).
4. Y. Nagata, Report to the Japan's Ministry of the Environment. Website http://www.env.go.jp/en/lar/odor_measure/02_3_2.pdf accessed on December 27, 2004.
5. D.W. Wright, D.K. Eaton, L.T. Nielsen, F.W. Kurth, J.A. Koziel, J.P. Spinhirne, and D.B. Parker, *ASAE Paper* #044128 (2004)
6. J.A. Koziel, J.P. Spinhirne, J. Lloyd, D.B. Parker, D.W. Wright, F.W. Kuhrt, *JAWMA*, **55**, 1147-1157 (2005).

SPECIAL SESSION:

GAS SENSORS IN ROBOTICS

Blimp Robot for Three-Dimensional Gas Distribution Mapping in Indoor Environment

Hiroshi Ishida

Department of Mechanical Systems Engineering, Tokyo University of Agriculture and Technology
2–24–16 Nakacho, Koganei, Tokyo 184–8588, JAPAN

Abstract. Mobile robots equipped with gas sensors can be used for automated measurement tasks including odor trail following, gas source localization, and gas distribution mapping. This article reports on the development of a blimp robot for mapping three-dimensional gas distribution in indoor environments. The blimp robot is programmed to fly randomly so that its trajectory covers everywhere in the given indoor environment. The blimp is equipped with gas sensors to measure gas concentrations and an ultrasonic sonar to measure the height from the floor. The measured data are transmitted to an external PC via a wireless communication module. At the same time, a camera placed on the floor takes a picture of the blimp, and its location is recorded with the gas sensor responses. The experimental results indicate that the blimp robot is effective in mapping three-dimensional gas concentration distribution in indoor environments.

Keywords: Metal-oxide gas sensors, aerial vehicle, mobile robot.
PACS: 87.19.lt, 07.07.Df, 45.40.Ln

INTRODUCTION

Olfaction has long been ignored in the robotics community because of the technical difficulties involved to realize artificial olfaction. However, chemical sensing capabilities allow mobile robots to have new functions that cannot be attained by other sensing modalities. Using mobile robots as a sensing platform is also beneficial for chemical sensors. Robots can work under harsh environments that are too dangerous for human workers. Moreover, laborious measurement can be automated.

Over the last fifteen years, the technological advances have been made in the field of gas sensors and electronic noses. Various commercial products are now available, and they have come to attract more interests from robotics researchers. Currently, challenges are being made to utilize gas-sensing capabilities in mobile robots. The tasks given to the gas-sensitive robots in the publications are classified into three categories, i.e., trail following, gas source localization, and gas distribution mapping [1].

In animal behavior, chemical substances are often used to mark trails or territories on the ground. A worker ant, for example, lays a pheromone trail on her way back home from a food source [2]. Other ants can locate the food source simply by tracking the smell of the pheromone. Robots can employ a similar technique to transport large quantities of material. In the scenario proposed in [3], an intelligent pathfinder robot first explores the environment. When an appropriate path is found, the pathfinder robot lays an odor trail. Much simpler trail-following robots can then be used to transport the material. An odor trail can be also used to bring the pathfinder robot back to its initial position after the exploration in the unknown environment [3]. Robots cleaning the floor can leave an odor trail to mark the already cleaned area. With the help of the odor trail, complete area coverage and multiple robot coordination can be easily attained [3, 4].

Many species of animals show olfactory search behavior in which they track aerial trails of invisible scents to find target objects. A famous example is male moths that track plumes of sexual pheromones to find conspecific females [5]. Robots equipped with gas sensors are being developed to achieve autonomous search for gas/odor sources by mimicking the olfactory search behavior of animals [6–10]. The gas concentration reaches its maximum at the location of the gas source. Therefore, one way to search for a gas source is to track the concentration gradient by comparing the responses of spatially separated gas sensors of the same kind [6, 8, 9]. However, the observed concentration gradient shows large fluctuations due to the turbulence of the airflow that carries the gas molecules. A male moth tracks a pheromone plume by turning in the upwind direction when the pheromone is detected. To mimic this

CP1137, *Olfaction and Electronic Nose: Proceedings of the 13th International Symposium,* edited by M. Pardo and G. Sberveglieri

behavioral mechanism, some robots are equipped with airflow sensors [7, 8, 10]. However, to search for a gas source in a real-life scenario is still a challenging task. Complicated turbulent airflow fields are created in offices and industrial facilities. The resultant gas distributions also have complicated structures. For this reason, most of the successful demonstrations on gas source localization robots are given in simplified environments with constant and uniform airflow fields [7, 8, 10].

In some applications, it is sufficient to know the distribution of the target gas in a given environment even though the exact location of the gas source cannot be tracked down [1]. In a rescue mission, it is necessary to determine where a highly concentrated hazardous gas is expected and where the rescue workers can safely do their jobs. The gas distribution can be easily measured if a sufficient number of gas sensors are deployed in advance. Even if such sensors are not available, a mobile robot equipped with gas sensors can scan the area of interest and map the gas distribution. One of the technical challenges involved in gas distribution mapping is how to obtain a reliable map with a sufficient resolution from fluctuating gas sensor responses measured at coarse grid points. Various interpolation/extrapolation algorithms are proposed to obtain reasonable maps from noisy gas sensor responses [11, 12].

Another issue in gas distribution mapping is the three-dimensional nature of the gas distribution. Diffusion of gas molecules into air is generally an extremely slow process. The diffusion length of typical gas molecules for one hour is only 20 cm. In outdoor environments and in most indoor environments, there usually exists airflow with much larger velocity. Therefore, the gas distribution is mainly determined by the airflow field in the given environment. There are various sources of airflow, e.g., ventilation, forced convection by air-conditioning systems, and natural convection. Most of them involve temperature variation, and therefore, the resultant airflow field is three dimensional in nature. The density mismatch between air and the gas released in the environment also causes three-dimensional gas distribution. In most of the experiments on gas-sensitive robots, a gas source was placed directly on the floor and a two-dimensional gas plume was formed along the floor. In reality, however, a gas leak may occur from a gas pipe placed on the ceiling.

To address these issues, here we propose to use a blimp as a robotic platform for three-dimensional gas distribution mapping. Among various types of aerial vehicles, a blimp was chosen because of its maneuverability in three-dimensional space and its flying mechanism. Unlike helicopters or airplanes, a blimp can hover without producing strong air currents.

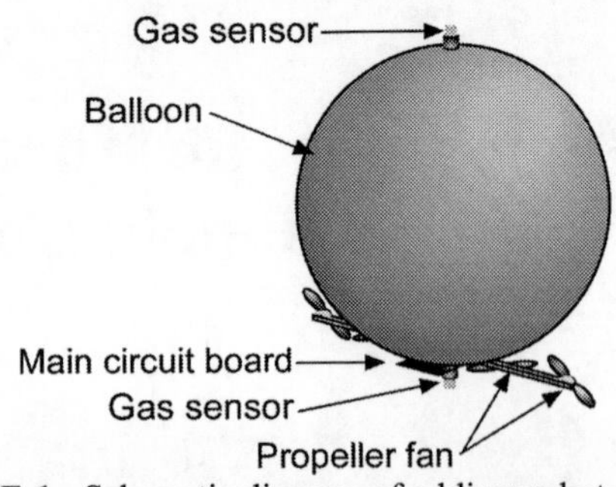

FIGURE 1. Schematic diagram of a blimp robot equipped with gas sensors.

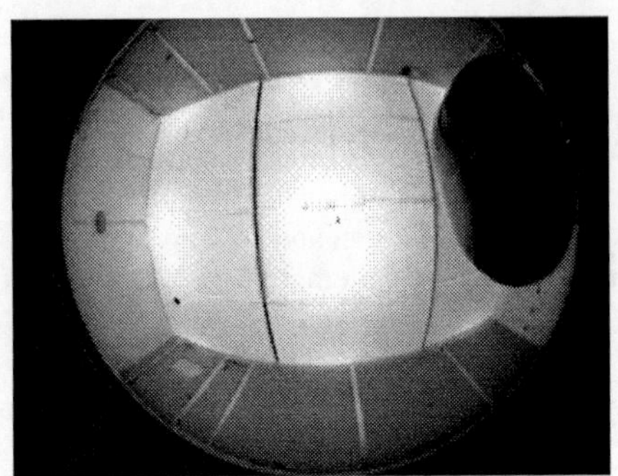

FIGURE 2. Image taken by a camera placed at the center of the floor.

Therefore, introduction of disturbance to the original gas distribution can be kept minimized.

GAS-SENSITIVE BLIMP ROBOT

The schematic diagram of the blimp robot is shown in Fig. 1. The robot is equipped with two metal-oxide gas sensors (TGS2620, Figaro Engineering), one on the top of the balloon and the other at the bottom. The balloon of 117 cm in diameter is filled with helium and produces lift of approximately 400 g. A small amount of weight is attached to the robot to achieve a neutrally buoyant condition. The altitude of the blimp is controlled by two propeller fans directed downward. The height from the floor is monitored using a sonar placed downward on the main circuit board. Thrust is produced by the left and right propeller fans directed backward. These four propeller fans can be turned on and off independently, and the directions of their rotation can be independently changed.

Ideally, the robot should be able to navigate itself to specified three-dimensional coordinates. However, it is difficult to achieve fully autonomous navigation and self-localization functions in a blimp robot of this size. Therefore, we chose to make the blimp robot fly randomly. The robot first measures the responses of the two gas sensors and obtain the reading of the sonar. A direction for the next step is then chosen randomly from three directions, i.e., the left front, the right front,

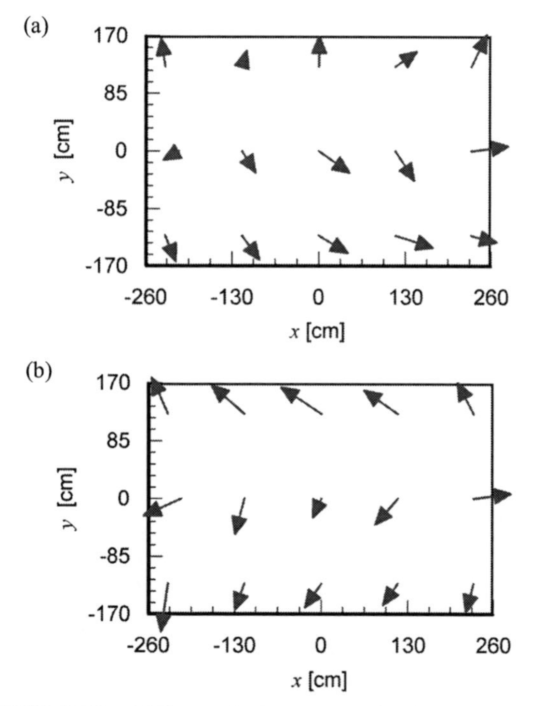

FIGURE 3. Airflow velocity measured using a stationary three-dimensional ultrasonic anemometer at (a) 195 cm and (b) 15 cm from the floor.

and the back. The two thruster propeller fans are controlled accordingly. The altitude of the blimp robot is modulated based on the reading of the sonar. If the altitude is higher than the half height of the room, the downward propellers are rotated to bring the blimp robot closer to the floor. If the altitude is shorter than the half height, the propellers are rotated reversely. This series of actions is repeated every five seconds. The trajectory of the blimp robot eventually covers the entire three-dimensional space in the given environment.

To map the gas distribution, the gas sensor responses must be recorded with the location where the measurement was conducted. This is done by an external system. A camera with a fish-eye lens is placed at the center of the floor. Every time the measurement of the sensor readings is conducted at a five-second interval, the robot transmits the data through a ZigBee wireless communication module to an external PC. When the transmission is received, an image from the camera is captured (Fig. 2). The three-dimensional coordinates of the blimp are calculated from the height of the blimp measured with the sonar and the location of the blimp in the two-dimensional camera image. Since the blimp robot has two gas sensors at the top and the bottom of the balloon, the gas concentrations at these two locations are thus recorded every five seconds together with the location of the blimp.

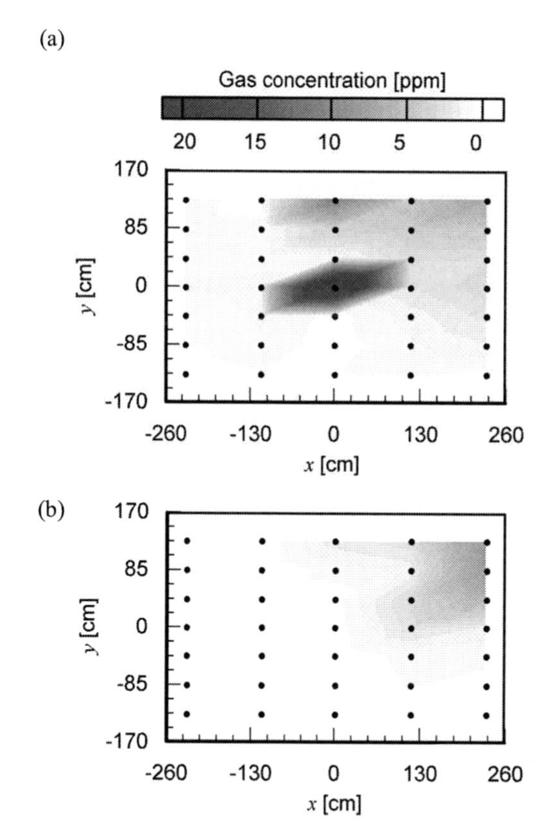

FIGURE 4. Gas concentration distribution measured with stationary gas sensors in the horizontal cross sections of the room at (a) 195 cm and (b) 15 cm from the floor.

EXPERIMENTAL

The experiments were conducted in a room (520 cm in width, 340 cm in depth, and 210 cm in height), which was prepared by setting partitions in a larger room. Although the room was thermally isolated from the outer walls of the building, weak natural convection was generated by the temperature variation in the room. Figure 3 shows the airflow velocity measured using a three-dimensional ultrasonic anemometer. Since the temperature of the right wall (16.8°C) was lower than that of the left wall (17.6°C), the airflow at the ceiling was pointing to the right whereas the airflow near the floor was pointing to the left. The time-averaged airflow velocity at each measurement point was in the rage of 0.01–0.9 cm/s.

Figure 4 shows the gas distribution measured using stationary gas sensors. A gas source releasing saturated ethanol vapor at 200 ml/min was placed at the center of the ceiling. To map the gas distribution, the time-averaged gas concentrations between nine and ten minutes from the start of the gas release were measured at 70 points indicated by dots in Fig. 4. Seven gas sensors were used in the measurement, and

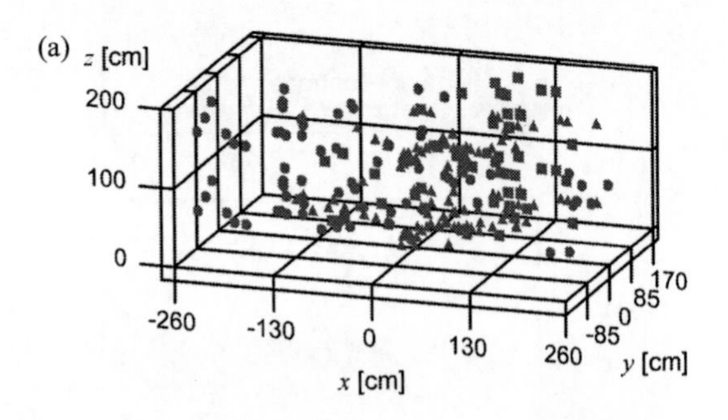

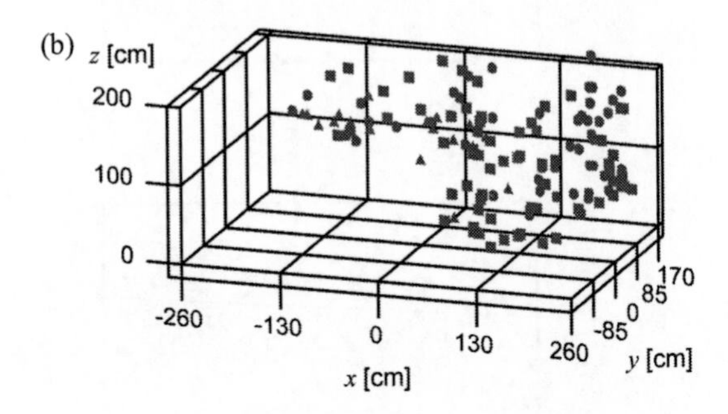

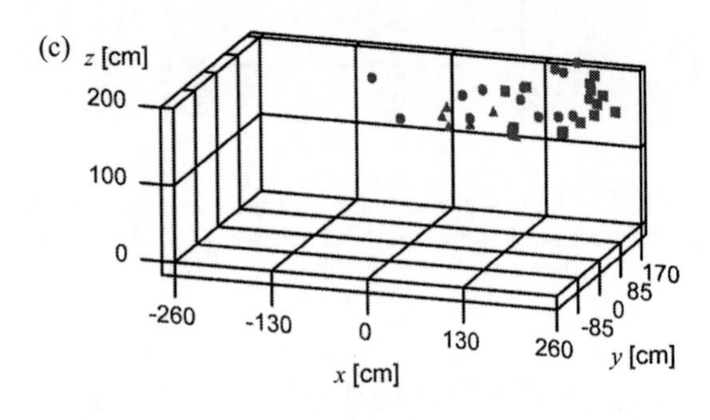

FIGURE 5. Gas distribution maps obtained by the blimp robot. Squares, circles, and triangles indicate the locations where the blimp robot measured the gas sensor responses in the first, second, and third trials, respectively. The locations are classified into three concentration ranges, i.e., (a) less than 1 ppm, (b) 1–10 ppm, and (c) over 10 ppm.

the concentrations at points in a single row are measured at a time. The density of ethanol vapor is slightly higher than that of air. Nevertheless, the released gas first trailed to the upper right corner of Fig. 4a since the airflow near the ceiling was pointing to the right. The gas plume then extended down along the right wall and spread to the left along the floor as shown in Fig. 4b. The results indicate that even a slight airflow has a significant impact on the resultant gas distribution.

Figure 5 shows the gas distribution maps obtained by the blimp robot. Three trials were made, and the blimp robot gathered the gas concentration data for 10 min in each trial. The locations where the blimp robot measured the gas concentrations are marked in Fig. 5. To show the concentration distribution, the marks are made in separate graphs according to the measured concentration values. The result coincides well with the gas distribution map measured using stationary gas sensors. Figure 5a shows that the entire space in the room was almost completely covered after the three trials. As shown in Fig. 5c, highly concentrated gas was detected only in the area ranging from the gas source location and the upper right corner of the room. Slightly lower concentrations were measured along the right wall as shown in Fig. 5b, suggesting that the plume was trailing down along the right wall.

CONCLUSIONS

A wheeled robot crawling on the floor cannot detect the presence of gas in the midair. The results of the experiments presented in this paper show that the proposed blimp robot can accomplish three-dimensional mapping of gas concentration distribution in an indoor environment. Due to the severe limitation in the payload, only the minimum set of sensors and intelligence are allowed on board the blimp. In the proposed system, simple reactive behavior was implemented to make the blimp robot fly randomly. The location of the robot was tracked by an external camera and a PC. Future work will be directed to attain better control of the location of the blimp robot for efficient data collection from the points of interest.

REFERENCES

1. A. J. Lilienthal, A. Loutfi, and T. Duckett, *Sensors* **6**, 1616–1678 (2006).
2. D. B. Dusenbery, *Sensory Ecology*, New York: W. H. Freeman, 1992.
3. R. A. Russell, *Odour Detection by Mobile Robots*, Singapore: World Scientific, 1999.
4. S. Larionova, N. Almeida, L. Marques, and A. T. de Almeida, *Auton. Robot.* **20**, 252–260 (2006).
5. N. J. Vickers, *Biol. Bull.* **198**, 203–212 (2000).
6. L. Marques, U. Nunes, and A. T. de Almeida, *Thin Solid Films* **418**, 51–58 (2002).
7. A. T. Hayes, A. Martinoli, and R. M. Goodman, *IEEE Sens. J.* **2**, 260–271 (2002).
8. R. A. Russell, A. Bab-Hadiashar, R. L. Shepherd, and G. G. Wallace, *Robot. Auton. Syst.* **45**, 83–97 (2003).
9. A. J. Lilienthal and T. Duckett, *Adv. Robotics* **18**, 817–834 (2004).
10. H. Ishida, G. Nakayama, T. Nakamoto, and T. Moriizumi, *IEEE Sens. J.* **5**, 537–545 (2005).
11. A. J. Lilienthal and T. Duckett, *Robot. Auton. Syst.* **48**, 3–16 (2004).
12. P. Pyk, S. Bermúdez i Badia, U. Bernardet, P. Knüsel, M. Carlsson, J. Gu, E. Chanie, B. S. Hansson, T. C. Pearce, and P. F. M. J. Verschure, *Auton. Robot.* **20**, 197–213 (2006).

Estimating Predictive Variance for Statistical Gas Distribution Modelling

Achim J. Lilienthal, Sahar Asadi and Matteo Reggente

AASS Research Center, Örebro University, Sweden

Abstract. Recent publications in statistical gas distribution modelling have proposed algorithms that model mean and variance of a distribution. This paper argues that estimating the predictive concentration variance entails not only a gradual improvement but is rather a significant step to advance the field. This is, first, since the models much better fit the particular structure of gas distributions, which exhibit strong fluctuations with considerable spatial variations as a result of the intermittent character of gas dispersal. Second, because estimating the predictive variance allows to evaluate the model quality in terms of the data likelihood. This offers a solution to the problem of ground truth evaluation, which has always been a critical issue for gas distribution modelling. It also enables solid comparisons of different modelling approaches, and provides the means to learn meta parameters of the model, to determine when the model should be updated or re-initialised, or to suggest new measurement locations based on the current model. We also point out directions of related ongoing or potential future research work.

Keywords: Gas distribution modelling; gas sensing; mobile robot olfaction; density estimation, model evaluation.
PACS: 01.30.Cc

1. INTRODUCTION

Gas distribution modelling (GDM) is the task of deriving a truthful representation of the observed gas distribution from a set of spatially and temporally distributed measurements. It is very challenging mainly since in many realistic scenarios gas is dispersed by turbulent advection, which creates a concentration field of fluctuating, intermittent patches of high concentration [1].

We can distinguish two types of GDM approaches: model-based and model-free. Model-based approaches infer the parameters of an analytical gas distribution model from the measurements. In principle, Computational Fluid Dynamics (CFD) models can be applied, which solve the governing equations numerically. They are, however, computationally very expensive, become intractable for higher resolutions in typical real world settings and depend sensitively on accurate knowledge of the environment state (boundary conditions), which is not available in practical situations. Many model-based approaches were developed for atmospheric dispersion [2]. Such models typically cannot efficiently incorporate sensor information on the fly and do not provide a sufficient level of detail. This is important since critical gas concentrations often have a local character in complex settings. Simpler analytical models such as [3] often rest on rather restrictive assumptions.

In this paper, we consider a class of model-free approaches, which create a statistical model of the observed gas distribution. In a pure form, these *statistical approaches to distribution modelling* treat input data as random variables and derive a statistical description without making strong assumptions on the functional form of the distribution. This includes that they do not assume certain environmental conditions (such as a uniform airflow, for example). Statistical approaches offer complementary strengths compared to model-based approaches: they do not rely on the validity of the underlying physical model, can provide a higher resolution, are computationally less expensive and generally less demanding in terms of the required knowledge about the state of the environment. Most of the available algorithms, discussed in Sec. 2, create a two dimensional spatial model that represents time-constant structures in the gas distribution in terms of the distribution mean. Three recently proposed approaches, introduced in Sec. 2.1, model not only the mean but also the variance of the distribution. In the following, we argue that this entails a significant improvement for statistical gas distribution modelling.

2. STATISTICAL GDM

This section reviews statistical GDM methods developed for mobile robots and tested at small scales.

A common approach to creating a representation of a time-averaged concentration field is to acquire measurements using a fixed grid of gas sensors over a prolonged period of time and to map average [3] or peak [4] concentrations obtained to the given grid approximation of the environment. Consecutive measurements with a single sensor were used in [5]. To make predictions at locations different from the measurement points, bi-cubic interpolation was applied in the case of equidistant measurements and triangle-based cubic interpolation in the

CP1137, *Olfaction and Electronic Nose: Proceedings of the 13th International Symposium,* edited by M. Pardo and G. Sberveglieri

case of non-equidistant measurements. A problem with such interpolation methods is that there is no means of "averaging out" instantaneous response fluctuations. Response values that were measured very close to each other appear independently in the gas distribution map and thus the representation tends to get more and more jagged while new measurements are added.

Histogram methods reflect the spatial correlation of concentration measurements to some degree by the quantization into histogram bins. The 2-d histogram proposed in [6] accumulates the number of "odor hits" received in an area assigned to the histogram bins. Odor hits are counted whenever the response of a gas sensor exceeds a defined threshold. Disadvantages of this method include the dependency on bin size and selected threshold, that a perfectly even coverage of the inspected area is required, and that only binary information is used and so useful information is discarded.

Kernel extrapolation distribution mapping (Kernel DM) is inspired by non-parametric estimation of density functions using a Parzen window and can be seen as an extension of histogram methods. The concentration field is represented in the form of a grid map. Spatial integration is carried out by convolving sensor readings and modelling the information content of the point measurements with a Gaussian kernel [7].

2.1. Gaussian Process Mixture GDM

None of the methods discussed so far models concentration fluctuations. The enhanced Kernel DM+V algorithm [8], detailed in Sec. 2.2, also estimates the observed distribution variance. Another method that predicts mean and concentration variance uses Gaussian process mixture (GPM) models [9]. It treats GDM as a regression problem. Two components of the GPM represent background signal and areas of high concentration. The components of the mixture model and a gating function, that decides to which component a data point belongs, are learned using Expectation Maximization. In contrast to Kernel DM+V, the model is represented directly using the training data. Because it requires the inversion of matrices that grow with the number of training samples n, the computational complexity of learning the GPM is $\mathcal{O}(n^3)$. This is addressed in [9] by adaptive sub-sampling of the observations to obtain a sparse training set. Similarly to Kernel DM+V, the dependency between nearby locations is modelled in the GPM approach by a radially symmetric, squared exponential covariance function.

2.2. Kernel DM+V

For the illustrating examples in Sec. 3 we use Kernel DM+V [8] to compute distribution models. The performance of this algorithm was found to be level with the GPM approach [9, 8] (see Sec. 3.2) but has the advantages of simplicity, lower computational complexity, and that it is more generally applicable.

Kernel DM+V uses a uni-variate Gaussian weighting function $\mathcal{N}$ to represent the importance of measurement r_i obtained at location $\mathbf{x}_i$ with respect to the measurement statistics over time at grid cell k. First, two temporary grid maps are computed – $\Omega^{(k)}$ by integrating importance weights and $R^{(k)}$ by integrating weighted readings:

$$\begin{aligned}\Omega^{(k)} &= \textstyle\sum_{i=1}^{n} \mathcal{N}(|\mathbf{x}_i - \mathbf{x}^{(k)}|, \sigma), \\ R^{(k)} &= \textstyle\sum_{i=1}^{n} \mathcal{N}(|\mathbf{x}_i - \mathbf{x}^{(k)}|, \sigma) \cdot r_i.\end{aligned} \tag{1}$$

Here, $\mathbf{x}^{(k)}$ denotes the center of cell k and the kernel width σ is a parameter of the algorithm. The integrated weights $\Omega^{(k)}$ are used for normalisation of the weighted readings $R^{(k)}$ (thus even coverage is not necessary) and to compute a further map $\alpha^{(k)}$, which estimates the confidence in the obtained estimates. The confidence map is used to compute the mean concentration estimate $r^{(k)}$ as

$$\begin{aligned}\alpha^{(k)} &= 1 - e^{-(\Omega^{(k)})^2/\sigma_\Omega^2} \\ r^{(k)} &= \alpha^{(k)} \tfrac{R^{(k)}}{\Omega^{(k)}} + \{1 - \alpha^{(k)}\} r_0\end{aligned} \tag{2}$$

where r_0 represents the mean concentration estimate for cells where we do not have sufficient information from nearby readings, indicated by a low value of $\alpha^{(k)}$. Currently, we set r_0 to be the average over all sensor readings. The scaling parameter σ_Ω^2 defines a soft margin for values of $\Omega^{(k)}$. Similarly to the distribution mean map, Eq. (2), the variance map $v^{(k)}$ is computed from *variance contributions* integrated in a further temporary map $V^{(k)}$

$$\begin{aligned}V^{(k)} &= \textstyle\sum_{i=1}^{n} \mathcal{N}(|\mathbf{x}_i - \mathbf{x}^{(k)}|, \sigma)(r_i - r^{(k(i))})^2, \\ v^{(k)} &= \alpha^{(k)} \tfrac{V^{(k)}}{\Omega^{(k)}} + \{1 - \alpha^{(k)}\} v_0\end{aligned} \tag{3}$$

where $k(i)$ is the cell closest to the measurement point $\mathbf{x}_i$, and v_0 is an estimate of the distribution variance in regions far from measurement points, computed here as the average over all variance contributions.

3. GDM EVALUATION

Ground truth evaluation has always been a critical issue for gas distribution modelling with mobile robots. The capability to identify hidden parameters, for example the location of the gas source, has been used to test gas distribution models. However, the distance of the distribution maximum to the gas source can only serve as a rough approach to validate the distribution model. Considering only fixed measurement points, a feasible experimental set-up would be to use a stationary grid of gas sensors and to compare the model derived from all but one or a few sensors with the measurements of the left out sensors. We apply a similar method here and create the model using a sub-set of measurements obtained with a mobile robot and compare the model predictions with unseen measurements also obtained with the robot.

In our experiments the robot followed a sweeping trajectory. It was driven at a maximum speed of 5 cm/s and periodically stopped at pre-defined points, constantly acquiring measurements at a rate of 0.8 Hz. The gas source was a small cup filled with ethanol. Apart from a SICK laser range scanner for pose correction, the robot was equipped with a Sensirion SHT11 digital humidity/temperature sensor and six Figaro gas sensors enclosed in an aluminum tube, actively ventilated through a fan. The tube was horizontally mounted at the front side of the robot at a height of 34 cm (see Fig. 1). We consider only the output of one TGS 2620 sensor here.

An obvious way to measure how well unseen measurements are predicted by the distribution model is to compute the average prediction error. Due to the large fluctuations of the instantaneous gas distribution, however, this measure of model quality is not particularly suitable for gas distribution modelling. A gas distribution model should represent the time-averaged concentration *and* the expected fluctuations. These properties are both captured by the negative log predictive density (NLPD), which is a standard criterion to evaluate distribution models. Under the assumption of a Gaussian posterior $p(r_i|\mathbf{x}_i)$, the NLPD of unseen measurements $\mathscr{D} = \{r_1, ..., r_n\}$ acquired at locations $\{x_1, ..., x_n\}$ is computed as

$$NLPD = \frac{\log(\pi)}{n} \sum_{i \in \mathscr{D}} \left\{ \log \hat{v}(x_i) + \frac{(r_i - \hat{r}(x_i))^2}{\hat{v}(x_i)} \right\}. \quad (4)$$

An estimate of the predictive variance is required to compute the NLPD. The importance of including the predictive variance into the criterion for the quality of gas distribution models can be seen in Fig. 1, which shows a comparison of a gas and a temperature model created from measurements recorded with a mobile robot along a sweeping path. The plots in the left part of the figure were created by computing the distribution model up to a certain time and comparing it to the true unseen values (red circles). Model predictions are indicated as predictive mean $\pm 3\times$predicted standard deviation. This comparison demonstrates that gas distribution models typically exhibit more pronounced spatial variance variations while the information is mainly located in the predictive mean in case of the temperature distribution model.

3.1. Learning meta parameters

As an example of an algorithm that provides an estimate of the predictive variance we use Kernel DM+V. It depends mainly on the meta-parameters kernel width σ and cell size c. Based on the NLPD defined in Eq. (4), we learn these meta parameters by dividing the samples $\mathscr{D}$ into disjoint sets $\mathscr{D}_{train}$ and $\mathscr{D}_{test}$ and determine optimal values of the model parameters by cross-validation on $\mathscr{D}_{train}$, keeping $\mathscr{D}_{test}$ for evaluation. Since we use Kernel DM+V we have $\hat{v}(x_i) = v^{(k(i))}$, $\hat{r}(x_i) = r^{(k(i))}$ in Eq. (4).

3.2. Comparison of GDM Approaches

In the same way, in which we evaluate a fixed distribution model depending on its meta parameters, we can compare different GDM approaches by comparing the respective NLPD for unseen measurements. Since the goal is to maximize the likelihood of unseen data, we will prefer models that minimise the NLPD. Tab. 1 shows a NLPD comparison of the GPM method (see Sec. 2.1) with Kernel DM+V (see Sec. 2.2) based on data sets from three different environments in which the robot carried out a sweeping movement consisting of two full sweeps (for details see [9]). The first sweep was used for training and the second sweep (in opposite direction) for testing. As a preliminary result from this investigation we find that GPM and Kernel DM+V offer a comparable performance for gas distribution modelling in the considered environments. This gives Kernel DM+V a slight edge because it scales better to larger numbers n of data samples (having complexity $\mathscr{O}[n \cdot (\frac{\sigma}{c})^2]$ compared to $\mathscr{O}[n^3]$) and the fact that the learning procedure is simpler.

Efficient approaches to GDM will probably apply some form of sub-sampling in a first stage. Again, the predictive variance via the NLPD allows for a meaningful comparison of different sub-sampling strategies.

3.3. Time-dependent GDM

A crucial assumption that we make when building stationary statistical models is that the statistical description is learned from measurements that are generated by a time-constant random process. It is clear that this assumption is not generally valid. We can address this issue in several ways. Regression approaches such as GPM could be extended by the dimension of time. The Kernel DM+V algorithm that is based on the idea of density estimation could be extended by recency weights or a method that detects when and how to change the time-window over which the distribution model is computed. A measure such as the NLPD does not only allow to compare different approaches to time-dependent GDM, it also provides a means to detect when the time-window

TABLE 1. Comparison of Kernel DM+V and GPM.

Dataset	NLPD, GPM	NPLD, Kernel DM+V
3-rooms	-1.54	-1.44
corridor	-1.60	-1.81
outdoor	-1.77	-1.75

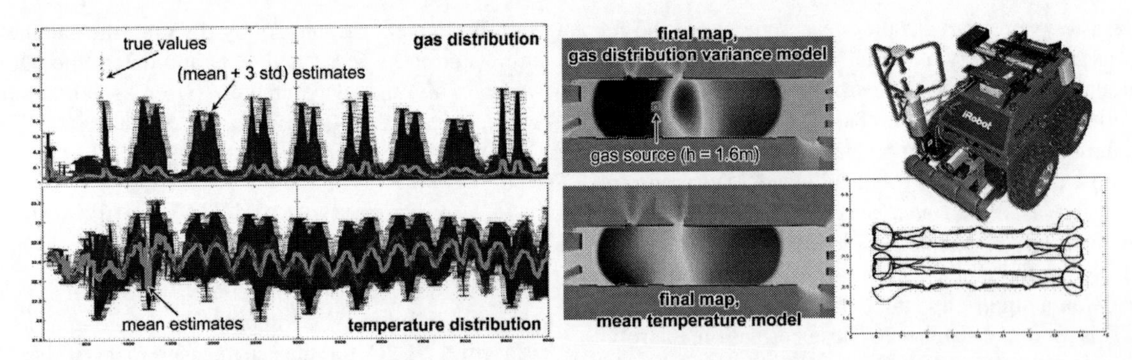

FIGURE 1. Comparison between gas and temperature model recorded with a robot along a sweeping path (shown on the right side). The models are visualised in terms of concatenated predictions of the next 60s of unseen measurements based on the model computed from all the measurements before (best viewed in color).

has to be modified and to decide how. An efficient solution might be a "lazy update" mechanism. The quality of the current distribution map could be continuously evaluated on new sensor readings and updated only if the data likelihood drops significantly. An update could also be compared to a model computed from fewer, more recent samples only. By eventually selecting this map, the representation could follow slow distribution changes over time. However, the main benefit of such an approach would be in efficiency of the algorithm since it would still not be possible to extrapolate on time-dependent trends.

3.4. Sensor Planning

A statistical distribution model can be considered good or *truthful* if it explains the measurements, which were used to build the model, and predicts new observations well. To obtain a truthful representation, we need to consider a *sufficient* number of measurements. To quantify this requirement, we can again use the NLPD to compute internal consistency (how well are training data explained) and predictive power of the model (how well are unseen measurements predicted).

A related question is at which locations the next measurements should be carried out in order to obtain a good model in minimum time. Again, the NLPD can be used to compare different sensing strategies regarding their suitability for gas distribution modelling. Further, the predictive variance is an important ingredient for techniques that suggest new measurement locations based on the current model (sensor planning). Appropriate sensor planning strategies need to be evaluated and it is to be expected that they will employ a cost function, which prioritises measurements in areas with high uncertainty, high concentration or high variance.

4. CONCLUSIONS

This paper argues that recently proposed gas distribution modelling algorithms (see Sections 2.1 and 2.2) entail more than a gradual improvement by providing an estimate of the predictive concentration variance. First, estimating the predictive variance captures the particular structure of gas distributions, which exhibit strong fluctuations with considerable spatial variations as a result of the intermittent character of gas dispersal. Accordingly, it is important to consider this feature of gas distributions for model evaluation. A second substantial step forwards is thus that the predictive variance allows to compute the negative log predictive density (NLPD) as a more meaningful measure to evaluate distribution models. The NLPD offers a solution to the problem of ground truth evaluation, which has always been a critical issue for gas distribution modelling. It not only enables solid comparisons of different modelling and sensor planning approaches, but also provides the means to learn meta parameters of the model, to determine when the model should be updated or re-initialised, or to suggest new measurement locations based on the current model.

REFERENCES

1. B. Shraiman, and E. Siggia, *Nature* **405**, 639–646 (2000).
2. H. Rørdam et al., Regulatory odour model development: Survey of modelling tools and datasets with focus on building effects, Tech. Rep. 541 (2005).
3. H. Ishida, T. Nakamoto, and T. Moriizumi, *Sensors and Actuators B* **49**, 52–57 (1998).
4. A. Purnamadjaja, and R. Russell, "Congregation Behaviour in a Robot Swarm Using Pheromone Communication," in *Proc. of the Australian Conf. on Robotics and Automation*, 2005.
5. P. Pyk et al., *Auton Robot* **20**, 197–213 (2006).
6. A. Hayes, A. Martinoli, and R. Goodman, *IEEE Sensors Journal, Special Issue on Electronic Nose Technologies* **2**, 260–273 (2002).
7. A. J. Lilienthal, and T. Duckett, *Robotics and Autonomous Systems* **48**, 3–16 (2004).
8. A. Lilienthal et al., "A Statistical Approach to Gas Distribution Modelling with Mobile Robots, The Kernel DM+V Algorithm," 2009, submitted to IROS.
9. C. Stachniss, C. Plagemann, and A. Lilienthal (2009), to appear.

Localizing an Odor Source and Avoiding Obstacles: Experiments in a Wind Tunnel using Real Robots

Thomas Lochmatter, Nicolas Heiniger and Alcherio Martinoli

Ecole Polytechnique Fédérale de Lausanne (EPFL)
Distributed Intelligent Systems and Algorithms Laboratory (DISAL)
{thomas.lochmatter, alcherio.martinoli}@epfl.ch

Abstract. We report on real-robot odor source localization experiments carried out in an environment with obstacles in the odor plume. The robot was equipped with an ethanol sensor and a wind direction sensor, and the experiments were carried out in a wind tunnel (controlled environment). An enhanced version of the surge-spiral algorithm was used, which was augmented with a dedicate behavior to manage obstacles (avoid them, or follow their contour). We compare the results in terms of distance overhead and success rate, and discuss the impact of obstacles on plume traversal.

Keywords: Odor source localization, mobile robots, chemical sensor, wind tunnel, obstacle.
PACS: 87.19.lr, 87.19.ls, 87.19.lt, 87.19.lu

INTRODUCTION

With the advances in robotics and chemicals sensor research in the last decade, odor sniffing robots have become an active research area. Notably the localization of odor sources would allow for very interesting robotic applications, such as search and rescue operations, safety and control operations on airports or industrial plants, and humanitarian demining [13, 2, 10, 4]. Many of these applications are time-critical, i.e. odor sources should be found as fast as possible. But as the structure of plumes in the air is intermittent in both time and space [14, 1], tracking plumes is a challenging problem.

In recent work, we carried out odor source localization experiments with bio-inspired algorithms both in simulation [7] and with real robots in a wind tunnel [8, 9], and showed that upwind surge algorithms (namely, *surge-spiral* and *surge-cast*) are more effective than pure casting. In all experiments, the wind flow was laminar (i.e. low Reynolds number) and the plume therefore almost straight. In this paper, we discuss results of real-robot experiments carried out in the same setup, but with obstacles. Two types of obstacles were considered:

1. Surface obstacles, which have a negligible impact on the wind flow, but put constraints to where the robot can move.
2. Obstacles which both change the wind flow and constrain the robot movements. In this case, the robot not only has to deal with the obstacle, but also with the turbulence induced by it.

Only very few researchers have so far considered obstacles when studying odor source localization algorithms. Jatmiko et al. [5] carried out simulation experiments with three PSO-based algorithms endowed with collision avoidance to avoid running into obstacles. The authors compared the algorithm in environments with no, 5 and 10 surface obstacles. While their basic algorithm (CPSO) suffered severely from putting obstacles, the algorithms using wind direction information (WU-I, WU-II) did not show statistically significant performance losses when adding obstacles.

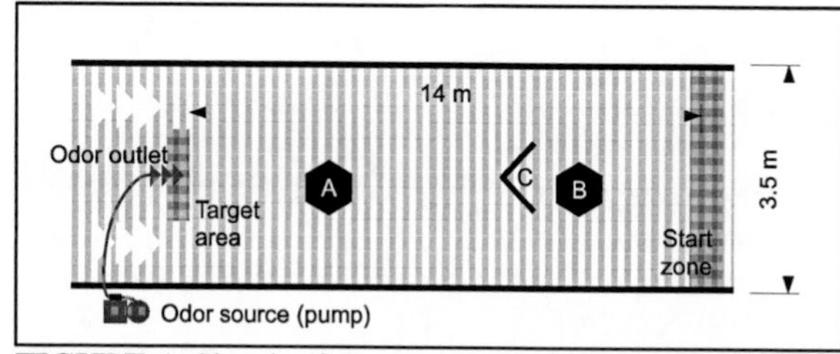

FIGURE 1. Sketch of the experimental setup with obstacles A, B, and C. Note that only one of the three obstacles was placed at a time.

Other experiments with obstacles are described by Russell et al. in [11, Chapter 7] and [12]. Instead of avoiding obstacles, the authors propose to follow the obstacle contour (wall) until the plume is reacquired. Their wall following algorithm is based on 15 rules

CP1137, *Olfaction and Electronic Nose: Proceedings of the 13th International Symposium*, edited by M. Pardo and G. Sberveglieri

and to simplify dead reckoning, the robot only moves along four orthogonal directions (north, south, west, east). Russell et al. also discuss obstacles in conjunction with source declaration, since moving around an obstacle without reacquiring the plume on the other side means that the obstacle is the source.

The remainder of this paper is organized as follows: The experimental setup is described in Section 2, and the algorithms are presented in Section 3. Section 4 discusses the results.

EXPERIMENTAL SETUP

The experimental setup was similar to the one used in our previous experiments without obstacles [8]. The experiments were thereby carried out in a 16 m long and 4 m wide wind tunnel with a wind flow of roughly 1 m/s speed. This wind flow is laminar unless obstacles are placed. An odor source releasing a constant amount of ethanol vapor was placed in proximity of the wind inlet.

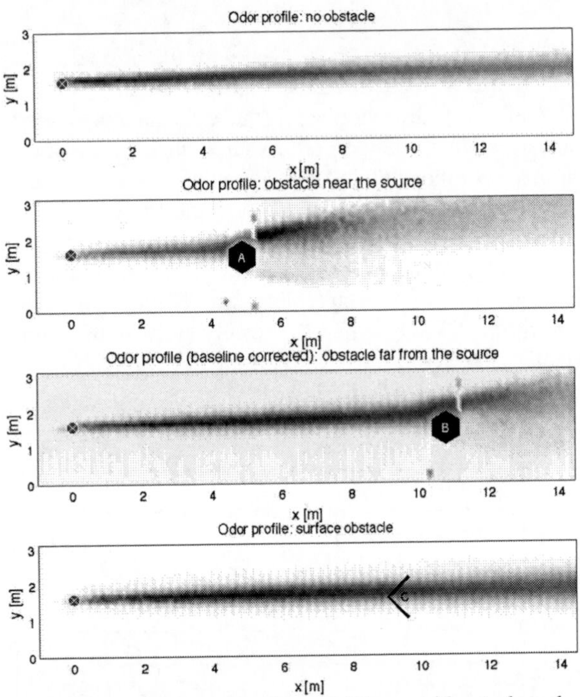

FIGURE 2. *From top to bottom:* Plume without obstacle, and with obstacles at positions A, B and C. Note that concentration levels are relative – only the shape of the plumes can be compared. The plume maps were recorded by systematically scanning the wind tunnel (traversing system) with a MiCS-5521 sensor.

We considered the three experimental configurations shown in Figure 1. Configuration (A) has a tall obstacle (changing wind flow) placed at 4 m downwind from the source. The shape of this obstacle can be described as a hexagon with an irregular border. Configuration (B) uses the same obstacle, but placed 10 m downwind from the source. As the wind flow behind these obstacles was turbulent, the plume got very diluted in these areas. The robot both had difficulties measuring the wind direction and discriminating between plume and fresh air behind the obstacles. Configuration (C) consists of a V-shaped surface obstacle (leaving the wind flow unchanged) placed at 9 m downwind. The resulting plumes are shown in Figure 2.

Our robotic platform is a Khepera III robot equipped with an odor sensor and a wind direction sensor board. The odor sensor board mainly consists of a MiCS-5521 volatile organic compound sensor with a fast response time (approx. 0.1 s). To take advantage of this response time, air was taken in and released with a small pump. The wind sensor board is based on 4 thermistors placed around a star-shape obstacle and can measure the wind direction with an accuracy of 10° in laminar flow. The sensor also provides us with a confidence value, which drops significantly in turbulent flow.

ALGORITHMS

The plume tracking strategy used here is based on the *surge-spiral* algorithm [8]. To deal with obstacles, we enhanced it with either a Braitenberg [6] obstacle avoidance or wall following algorithm using the 9 infrared sensors of the Khepera III robot. The left and right wheel speeds, s_l and s_r, are thereby calculated as a linear combination of the raw infrared proximity sensor values, p_i, i.e.

$$s_l = o_l + \sum_{i=1}^{9} w_i v_i \quad \text{and} \quad s_r = o_r - \sum_{i=1}^{9} w_i v_i$$

and the weights, w_i, and bias speeds, o_l and o_r, are chosen such that the resulting behavior is either obstacle avoidance or wall following.

With the obstacle avoidance algorithm, both the *surge-spiral* algorithm and the Braitenberg obstacle avoidance algorithm are running in parallel, and the output of the *surge-spiral* algorithm is simply the bias speed for the Braitenberg algorithm. As long as the robot is far away from any obstacle, the Braitenberg weights sum up to zero and leave the *surge-spiral* algorithm unmodified. When the robot is close to an obstacle, obstacle avoidance overrides *surge-spiral*.

With wall following, the *surge-spiral* algorithm is the only active algorithm in open space. When the robot approaches an obstacle, it switches to wall following and sticks to this mode until it has reached the other side of the obstacle. To find out when this has happened, the robot measures the wind direction in regular intervals and switches back to *surge-spiral* as soon as the wind is blowing towards the obstacle.

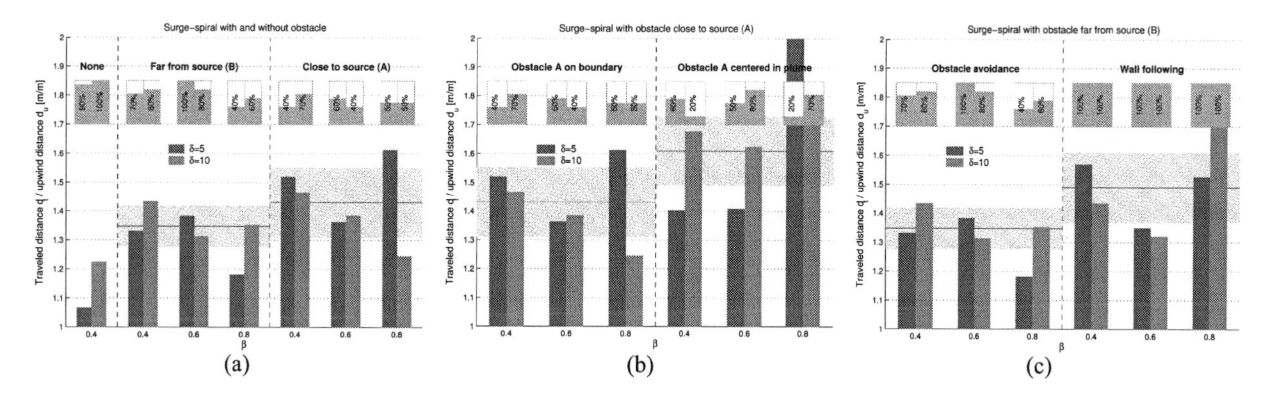

FIGURE 3. (a) Comparison of runs with and without obstacles. (b) Effect of the position of the obstacle within the plume. © Comparison between obstacle avoidance and wall following. In all diagrams, the blue bars show the distance overhead (lower is better) and the grey bars indicate the success rate. The blue horizontal line is the mean for the group with the 95% confidence interval.

To deal with turbulence and dilute plume, the sensory input was processed as follows:

1. If the wind direction sensor returned a low confidence value for a measurement, that measurement was repeated and the confidence threshold decreased.
2. The variable odor threshold, t_i, was dynamically adjusted using the following additive-increase-multiplicative-decrease scheme:

$$t_{i+1} = b_{i+1} + \delta$$

$$b_{i+1} = \begin{cases} b_i + \alpha & \text{if } v_i > b_i \\ b_i(1-\beta) + v_i\beta & \text{otherwise} \end{cases}$$

   with α=0.01 (experimentally found to be near-optimal) and varying β and δ. While β defines how fast the algorithm adapts to baseline changes, δ affects width of the plume, as perceived by the robot. v_i denotes the raw measurement, while b_i stands for the variable baseline.

Note that we only consider plume traversal and intentionally omit plume finding (i.e. randomized or systematic search until the plume is found) and source declaration (i.e. declaring that the source is in close vicinity), to prevent those two parts from interfering in the results. Hence, the robot starts in the plume, and source declaration is done by a supervisor (ideal source declaration). Experiments are considered successful when the robot has come in physical vicinity of the source, and unsuccessful if it bumped into an arena boundary.

We use the same two metrics as we did in our previous experiments [8]: the distance overhead, calculated as the traveled distance divided by the upwind distance, and the success rate, which is simply the fraction of successful runs.

RESULTS AND DISCUSSION

The different configurations we used to run our experiments and how they are related to each other are summarized in Figure 4.

No Obstacle vs. Obstacles A and B

In the first series of runs, we compared the impact of obstacle A and B on the performance. 10 runs each were carried out for different values of β and δ, and the results plotted in Figure 3 (a). The Braitenberg obstacle avoidance algorithm was used in all runs. In spite of the small number of runs carried out for each case, the results clearly show that the distance overhead increases and the success rate decreases when obstacles are present. Furthermore, there is evidence that the results are slightly worse in the case of obstacle A, which confirms our intuition that turbulent flow induces a performance penalty.

Even though the bars seem to suggest that higher β values (i.e., faster threshold adaptation) yield lower distance overheads, there is statistically not enough evidence to support this statement.

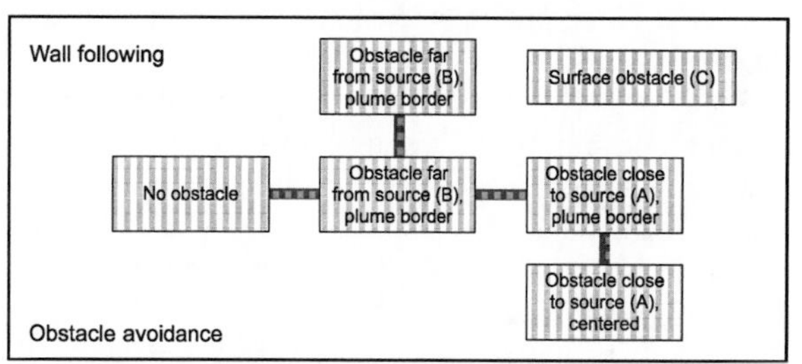

FIGURE 4. Obstacle configurations used in the experiments. Gray lines connect configurations for which the results can be directly compared.

Obstacle A on Plume Boundary vs. in Center

Figure 3 (b) shows the impact of moving obstacle A from the plume boundary to the plume center, with the effect that the plume splits into two almost equal lobes behind the obstacle. It turns out that this has a negative effect on the distance overhead, while keeping the success rates at similar levels.

Obstacle Avoidance vs. Wall Following

Finally, we carried out experiments with the wall following algorithm. As shown in Figure 3 (c), the success rate jumps to one, at the expense of a slightly higher distance overhead.

Surface Obstacles

Braitenberg obstacle avoidance can perform poorly in case of non-convex obstacles. As sketched in Figure 5, the obstacle avoidance version of our algorithm is likely to get trapped in obstacle C, while wall following is able to deal with it.

We carried out one set of 10 runs for β=0.4 and δ=10 with both variants of the algorithm. While none of the obstacle avoidance runs succeeded, all wall following runs were successful with a mean distance overhead of 1.59.

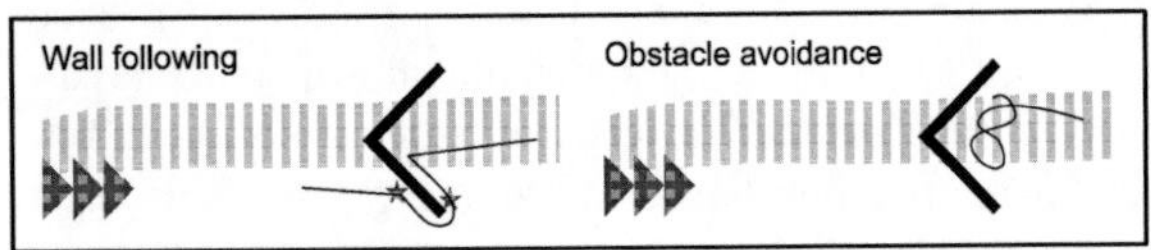

FIGURE 5. Wall following vs. Braitenberg obstacle avoidance with non-convex obstacles.

CONCLUSIONS

We presented results of 280 odor source localization runs with a real robot in scenarios with various obstacles. Three main conclusions can be drawn from the experiments.

First, obstacles induce a penalty both because of the path constraints and the turbulence downwind the obstacle. The latter causes the plume to get diluted and become more peaky, but also affects the wind direction sensor accuracy which we have previously shown to have a big impact on the performance [7].

Second, both Braitenberg obstacle avoidance and wall following are able to deal with convex obstacles. For non-convex obstacles, wall-following is the preferred technique.

Finally, the *surge-spiral* algorithm seems to be a good candidate algorithm for plume traversal in complex environments. A few initial runs (not systematically recorded) with the *casting* algorithm [8] did not provide satisfactory results in our scenarios.

Potential future research directions include studies with more than one obstacle, moving obstacles, or obstacles in immediate proximity of the source.

REFERENCES

1. J. A. Farrell, J. Murlis, X. Long, W. Li, and R. T. Cardé. *Filament-based atmospheric dispersion model to achieve short time-scale structure of odor plumes*, Environmental Fluid Mechanics, vol. 2, 2002, pp. 143ff.
2. D. W. Gage. *Many-Robot MCM Search Systems*, Proceedings of the Autonomous Vehicles in Mine Countermeasures Symposium, April 1995, pp. 9.56-9.64.
3. A. T. Hayes, A. Martinoli and R. M. Goodman. *Distributed Odor Source Localization*, IEEE Sensors Journal, vol. 2, no. 3, 2002, pp. 260-271.
4. H. Ishida, T. Nakamoto, T. Moriizumi, T. Kikas, and J. Janata. *Plume-Tracking Robots: A New Application of Chemical Sensors*, Biological Bulletin, no. 200, April 2001, pp. 222-226.
5. W. Jatmiko, K. Sekiyama, and T. Fukuda. *A PSO-Based Mobile Robot for Odor Source Localization in Dynamic Advection-Diffusion with Obstacles Environment*, IEEE Computational Intelligence Magazine, May 2007, pp. 37-51.
6. A. Lilienthal and T. Duckett. *Experimental Analysis of Gas-Sensitive Braitenberg Vehicles*, Advanced Robotics, vol. 18, no. 8, 2004, pp. 817-834.
7. T. Lochmatter and A. Martinoli. *Simulation Experiments with Bio-Inspired Algorithms for Odor Source Localization in Laminar Wind Flow*, Proceedings of ICMLA 2008, pp. 437-443.
8. T. Lochmatter and A. Martinoli. *Tracking Odor Plumes in a Laminar Wind Field with Bio-Inspired Algorithms*, Proceedings of ISER 2008, to appear.
9. T. Lochmatter, X. Raemy, L. Matthey, S. Indra, and A. Martinoli. *A Comparison of Casting and Spiraling Algorithms for Odor Source Localization in Laminar Flow*, Proceedings of ICRA 2008, pp. 1138-1143.
10. M. Long, A. Gage, R. Murphy, and K. Valavanis. *Application of the Distributed Field Robot Architecture to a Simulated Demining Task*, Proceedings of ICRA 2005, pp. 3193-3200.
11. R. A. Russell. *Odour Detection by Mobile Robots*, World Scientific Series in Robotics and Intelligent Systems, 1999.
12. R. A. Russell, D. Thiel, R. Deveza and A. Mackay-Sim. *A Robotic System to Locate Hazardous Chemical Leaks*, Proceedings of ICRA 1995, pp. 556-561.
13. G. S. Settles. *Sniffers: Fluid-Dynamic Sampling for Olfactory Trace Detection in Nature and Homeland Security—The 2004 Freeman Scholar Lecture*, Journal of Fluids Engineering, vol. 127, 2005, pp. 189-218.
14. M. Vergassola, E. Villermaux, and B. I. Shraiman. *'Infotaxis' as a strategy for searching without gradients*, Nature, vol. 445, January 2007, pp. 406-409.

Single Odor Source Declaration by Using Multiple Robots

Fei Li, Qing-Hao Meng, Member, IEEE, Jun-Wen Sun,
Shuang Bai and Ming Zeng

School of Electrical Engineering and Automation, Tianjin University
No. 92 Weijin Road, Nankai District, Tianjin 300072, P.R. China
qh_meng@tju.edu.cn, 86-022-27892367

Abstract. The single odor source declaration in indoor environments by using multi-robot system is addressed. A three-step odor source declaration method is put forward, which include robots convergence, odor concentration persistence judgment and odor mass throughput calculation. Initial experimental results in both artificial and natural indoor airflow environments by using three small mobile robots validate the feasibility of the proposed single odor source declaration method.

Keywords: Odor Source Declaration, Multiple Robots, odor mass throughput, indoor airflow
PACS: 07.07.Tw

INTRODUCTION

Odor source localization is widely used by many animals for searching for food, finding mates, exchanging information, and evading predators. Inspired by the odor source localization abilities of animals, in the early 1990s some researchers started to develop mobile robots with such abilities to replace trained animals [1]. It is expected that mobile robot based odor source localization will play more and more roles in such areas as judging toxic or harmful gas leakage location, checking for contraband (e.g., heroin), searching for survivors in collapsed buildings, humanitarian de-mining, and antiterrorist attacks.

Mobile robot based odor source localization is also called chemical plume tracing (CPT) [2]. Lilienthal and his colleagues [3] reviewed the research works in this field. Recently, Kowadlo and Russell [4] presented a survey of the existing methods. The realization of mobile robot based odor source localization can be divided into different phases. Hayes proposed a three-phase framework, which are plume finding, plume tracking and odor source declaration [5]. Li and his colleagues [6] presented a behavior-based CPT strategy, which was divided into four behavior types: finding a plume, tracing the plume, reacquiring the plume, and declaring the source location.

Compared with the single-robot search, multiple robots have at least two advantages: the expected search time can be decreased; and multi-robot systems provide a greater robustness against hardware failures. Multi-robot systems for odor source localization have not been well studied yet. Hayes [5] is probably the first researcher who used multiple robots to implement odor source localization with real robot hardware. Particle swarm optimization algorithm was tested using computer simulation by Jatmiko [7] and Marques [8], respectively. Li and Meng [9] proposed a probability particle swarm optimization (P-PSO) algorithm for multi-robot based odor source localization. Zarzhitsky and her colleagues [10] proposed an artificial physics-based multiple robots formation method to realize odor source localization via simulation, where the fluxotaxis tracing strategy was adopted to navigate the robots toward the emitter source. Byrne and his colleagues [11] described a system of miniature mobile robots and the algorithms used to demonstrate cooperative plume tracking and source localization.

To our knowledge, so far there have been few reports in literature addressing single odor source declaration with multiple real robots. This paper proposes a novel declaration algorithm for single odor source by using a multi-robot circular formation. The proposed algorithm consists of robots convergence, odor concentration persistence judgment and odor mass throughput calculation.

CP1137, *Olfaction and Electronic Nose: Proceedings of the 13th International Symposium,* edited by M. Pardo and G. Sberveglieri

MULTI-ROBOT BASED SINGLE ODOR SOURCE DECLARATION ALGORITHM

The first step is to see whether all the robots have converged in a relatively small area after tracing odor plume, see Fig. 1 (a). When the distance between each robot and the geometric center of all the robots is less than a set threshold r, it is assumed that the robots have converged in a small area. In such a case, the source declaration process will go to the second step. The geometric center, denoted by (x_{center} , y_{center}), is calculated as follows.

$$x_{center} = \frac{\sum_{k=1}^{K} x_k}{K}, y_{center} = \frac{\sum_{k=1}^{K} y_k}{K}, \quad (1)$$

where x_k and y_k are the horizontal and longitudinal coordinates of the kth robot, respectively, and K is the number of robots.

The second step consists of assessing the persistence of the odor concentration. This assessment assumes that an odor source can normally maintain a high concentration for a relatively long time. Normally a local optimum area cannot maintain such a long-lived high concentration, so the persistence judgment will exclude most local concentration optima. If the odor can maintain the high concentration for a period T_{MAX}, a possible source is assumed to be in the region, and the declaration process will go to the third step.

In the third step, the robots diverge towards a circle with radius of R (cf. Fig.1 (b)) and start moving around the suspected odor source (cf. Fig.1 (c)) to confirm the source.

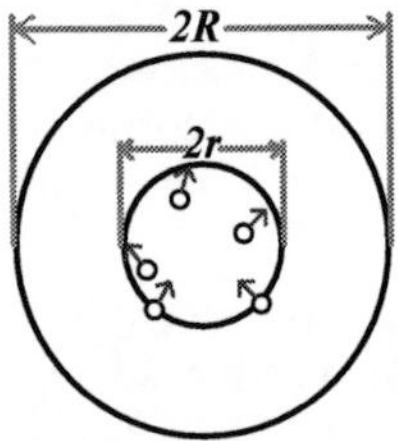

(a) robots converge

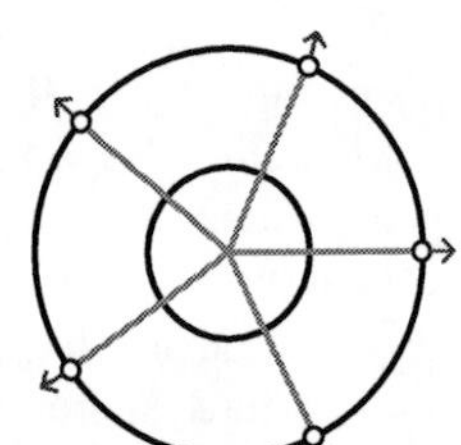

(b) robots move to a circle with radius of R

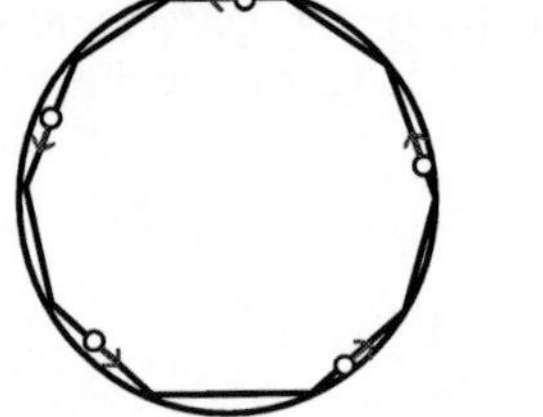

(c) robots start to move along the circumference

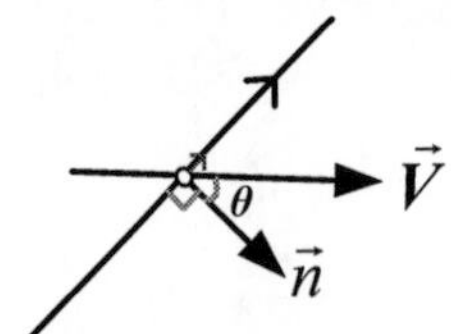

(d) mass throughput calculation of single robot

FIG. 1 Multi-robot based odor source declaration process

In theory, the following normal form of the Green theorem can be used to judge whether the suspected odor source surrounded by the robots is true or not [12].

$$\iint_A \nabla \cdot (c\vec{V}) dA = \int_{\partial A} (c\vec{V}) \cdot d\vec{L} = \int_{\partial A} (c\vec{V}) \cdot \vec{n} dL , \quad (2)$$

where c stands for the odor concentration; $\vec{V}$ is the fluid velocity; $c\vec{V}$ represents the odor mass flux, which measures the odor mass flowing through unit area per second; A is the control area; ∂A stands for the enclosed curve of the area; $\vec{n}$ is the outward-pointing normal of the infinitesimal curve dL.

The integral of the flux divergence enclosed by the robots can be rewritten as follows.

$$\begin{aligned} \int_{\partial A} (c\vec{V}) \cdot d\vec{L} &= \int_{\partial A} (c\vec{V}) \cdot \vec{n}(V_r dt) = \int_{\partial A} c(\vec{V} \cdot \vec{n}) V_r dt \\ &= \int_{\partial A} c(|V||n| \cos\theta) V_r dt \end{aligned} \quad ,(3\text{-a})$$

where V_r is the linear velocity of each robot. It can be seen from both (3) and Fig. 1 (d) that when the angle θ is bigger than 90°, the value $f(c,V,\theta)$ is smaller than zero.

Let $f(c,V,\theta) = c(|V||n|\cos\theta)$, (3-a) can be rewritten as follows,

$$\int_{\partial A} (c\vec{V}) \cdot d\vec{L} = \int_{\partial A} f(c,V,\theta) dL . \quad (3\text{-b})$$

Theoretically, equation (3) allows us to formally define the notion that a 2D control area containing the emitting source will have a positive mass flux divergence, while a control area containing a sink will have a negative mass flux divergence. However, the robots cannot sample all the locations along the circumference instantaneously in reality, so the integral of flux divergence surrounding non-odor source also might be positive. Considering the practical limitation that it takes time for the robots to sample all the locations along the circumference, instead of the flux divergence integral along the circumference, here we use the odor mass throughput per unit time at each location along the circumference, i.e.,

$$f(c,V,\theta)\Delta L , \quad (4)$$

where ΔL is the distance each robot moved per unit time along the circumference (2.5cm in our experiment). It can be verified that the results of (4) when the robots move around the odor source can not be negative, while the negative values will occur around non-source locations. By this criterion we can declare the real odor source.

EXPERIMENTS RESULT

The third step of the proposed odor source declaration algorithm was implemented in the Autonomous Robots System Lab of Tianjin University in January 2009. The lab has two windows and one door. The area of the lab is about 5m by 5m. Fig.2 (a) shows the experiment setting. A humidifier filled with 40% liquid ethanol was used as the odor source. The motion speed of each robot was 2.5cm/s. The circle radius *R* is 0.5m. The odor source was located at (0, 0), the initial locations of three robots were (0.5m, 0), (-0.25m, 0.43m), and (-0.25m, -0.43m), respectively. An electric fan located at (0, -1.8m) was used to produce artificial wind.

One of the three small mobile robots for the experiments is illustrated in Fig. 2(b). The robot is driven differentially. Each robot is equipped with an anemometer (Windsonic, Gill), an odor concentration sensor (TGS2620, Figaro), eight sonar sensors (L Series 40LPT16, Senscomp), eight infrared sensors (GP2D15, Sharp), and a wireless communication module (RPC module, Radiometrix). The robots were localized via an overhead CCD camera. The robots sent the wind and odor information to the workstation once per second, while the workstation sent motion command to each robot, both via wireless communication. The odor source declaration algorithm was run in the workstation.

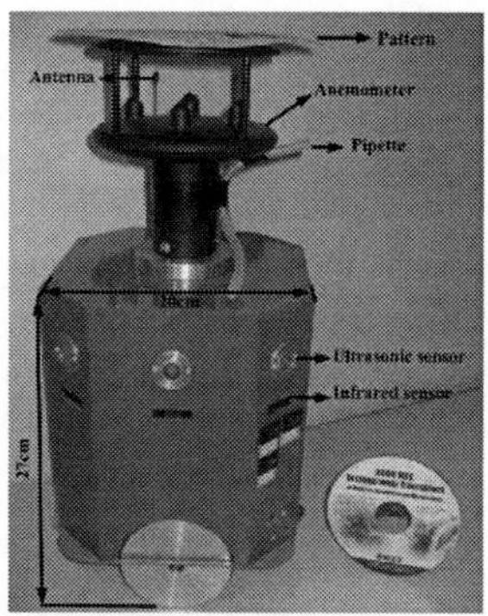

Fig. 2(a) Fig. 2(b)

FIG. 2(a) The experiment setup

FIG. 2(b) One of the mobile robots and onboard sensors

Three robots moved clockwise, and each robot moved about 120 degree along the circumference. Three different winds were used to verify the proposed algorithm, which are artificial changing-direction winds produced by the fan, and the natural wind mainly produced from the laminar airflow when the door and windows were open. The changing-direction wind was produced when the fan swung from one side to another with an angular spin speed of 10 degree/second.

Figure 3 to figure 6 illustrate the experimental results, where the sub-figures (a) and (b) are the concentration and wind direction data recorded by three robots, the sub-figures (c) are the calculated odor mass throughput using (4). The start angular position of the robots at the circumference can be found in Fig.2 (a), where the positive directions of x and y axes correspond to zero and ninety degrees, respectively. The position of the non-odor source is (0, 0.8). Three gas sensors were calibrated before the experiments. The measured airflow speed ranges are from 20cm/s to 220cm/s in the artificial wind environment and from 5cm/s to 60cm/s in the natural wind condition, respectively.

It can be found from the figures 3 to 6 that, in three different kinds of winds, the calculated odor mass throughputs surrounding the odor source are bigger than or equal zero; while the results can be negative when the robots move around the non-odor source.

CONCLUSION

Initial experiments in indoor artificial and natural airflow environments by using three small mobile robots moving with circular formation show that the odor mass throughputs calculation could be used to declare single odor source. The judgment criterion is that the odor mass throughputs surrounding the odor source can not be negative, while the results can be negative when the robots move around the non-odor source.

Next we will further study this identification method, and combine it with multi-robot based odor plume finding and tracing.

ACKNOWLEDGMENTS

The research work of this paper is sponsored by China '863' High-Tech Program (No. 2007AA04Z219), the National Natural Science Foundation of China (No. 60875053, 60802051) and the Program for New Century Excellent Talents in University (NCET).

REFERENCES

1. Ishida, H., Suetsugu, K., Nakamoto, T. et al, "Study of autonomous mobile sensing system for localization of odor source using gas sensors and anemometric sensors", *Sensors and Actuators A*, 1994, 45(2), pp. 153-157.
2. Farrell, J. A., Pang, S., Li, W., "Chemical Plume Tracing via an Autonomous Underwater Vehicle" in *IEEE Journal of Oceanic Engineering*, 30(2), 2005, pp. 428-442.
3. Lilienthal, A. J., Loutfi, A., Duckett, T., "Airborne Chemical Sensing with Mobile Robots", *Sensors*, 2006, 6(11), pp. 1616-1678.

4. Kowadlo, G., Russell, R. A., "Robot Odor Localization: A Taxonomy and Survey" in the *International Journal of Robotics Research*, 2008, pp. 869-894.
5. Hayes, A. T., Martinoli, A., Goodman, R. M., "Distributed Odor Source Localization" in *IEEE Sensors Journal*, 2002, 2(3), pp. 260-271.
6. Li, W., Farrell, J. A., Pang, S. et al, "Moth-Inspired Chemical Plume Tracing on an Autonomous Underwater Vehicle" in *IEEE Transactions on Robotics*, 2006, 22(2), pp. 292-307.
7. Jatmiko, W., Sekiyama, K., Fukuda, T., "A PSO Based Mobile Robot for Odor Source Localization in Dynamic Advection-Diffusion with Obstacles Environment: Theory, Simulation and Measurement" in *IEEE Computational Intelligence Magazine*, 2007, 2(2), pp. 37-51.
8. Marques, L., Nunes, U., de Almeida, A. T., "Particle swarm-based olfactory guided search", *in Autonomous Robots*, 2006, Vol. 20, pp. 277-287.
9. Li, F., Meng, Q. H., Bai, S. et al, "Probability-PSO Algorithm for Multi-Robot based Odor Source Localization in Ventilated Indoor Environments", *Lecture Notes in Computer Science, Springer,* 2008, pp. 1206-1215.
10. Zarzhitsky, D., Spears, D. F., "Swarm approach to chemical source localization" *in Proc of IEEE International Conference on Systems, Man and Cybernetics*, Hawaii, Oct. 10-12, 2005pp. 1435-1440.
11. Byrne, R. H., Adkins, D. R., Eskridge, S. E. et al, "Miniature Mobile Robots for Plume Tracking and Source Localization Research" *in Journal of Micromechatronics*, 2002, 1(3), pp. 253–261.
12. Hughes-Hallett, D., Gleason, A. M., *Calculus: Single and Multivariable*, Wiley, 1998.

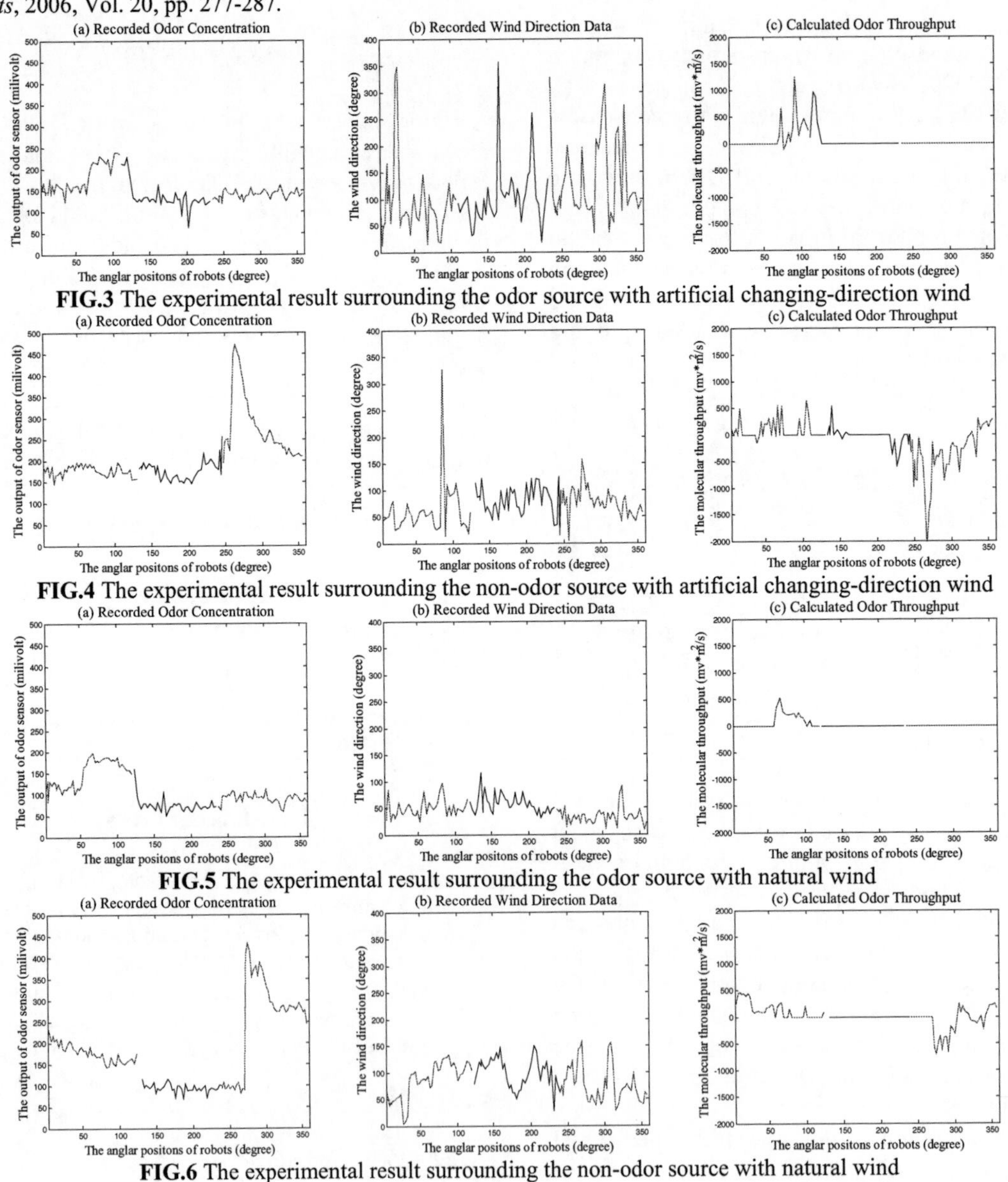

FIG.3 The experimental result surrounding the odor source with artificial changing-direction wind

FIG.4 The experimental result surrounding the non-odor source with artificial changing-direction wind

FIG.5 The experimental result surrounding the odor source with natural wind

FIG.6 The experimental result surrounding the non-odor source with natural wind

ACOUSTIC AND OPTICAL SENSING

High-Sensitive Chemical Sensor System Employing a Higher-mode Operative Micro Cantilever Sensor and an Adsorption Tube

Takashi Mihara*, Tsuyoshi Ikehara**, Jian Lu** , Ryutaro Maeda** ,
Tadashi Fukawa*** , Mutsumi Kimura*** , Ye Liu***, Toshihiro Hirai***

* *Future Creation Laboratory, Olympus Corporation, 2-3-1 Nishi-Shinjuku, Tokyo, 163-0914 Japan*
** *National Institute of Advanced Industrial Science and Technology, 1-2 Namiki, Tsukuba, 305-8564 Japan*
*** *Faculty of Textile Science & Technology, Shinshu University, Tokida 3-15-1, Ueda, Nagano, 386-8567, Japan*

Abstract. We report the basic functions and the sensitivity improvement of a chemical sensor system employing a poly-butadiene (PBD) coated micro cantilever sensor and a carbon-fiber-filled adsorption tube. The improvements of the sensitivity were carried out by two methods as 1) reduction of the volume in sensor camber by 1/30, 2) enlargement of the resonance frequency of the cantilever by 4.6 using a high speed analog-oscillation circuits and a low noise package. Using the 4th vibration mode (resonant frequency 715kHz) of a PBD coated cantilever (length 500µm, width 100µm , thickness 5µm) and a small sensor chamber with a volume of 3.5cc, the sensitivity was enhanced to be 2.4Hz/ppm for toluene in nitrogen carrier gas with 10 minute adsorption time, which was about 100 times larger than our previous sensor system.

Keywords: Chemical sensor, Micro cantilever, Adsorption tube, Poly-butadiene
PACS: 07.07.Df, 81.05.Lg, 82.80.-d, 85.85.+j

INTRODUCTION

The developments of chemical sensor systems to detect volatile organic compound (VOC) species have been received much attention for the environmental monitoring and an ultra fast medical diagnostics. To date, the widely used sensor to detect VOC has been metal oxide semiconductor (MOS) sensors; however, it was difficult to analyze the VOC elements due to the relatively poor selectivity. The new high sensitive sensors including a micro cantilever have been developed in these days for the integration with the electron devices [1, 2].

We have developed an integrated chemical analysis system with focusing on the micro resonant mass sensor for this purpose. We have reported the studies of mass sensitivity of micro cantilever with higher resonant modes [3, 4], sensing films using the co-polymer based elastic polymers [5], and first prototype sensor system [6]. In this study, we report the basic configuration, functions and the improvement of the detection sensitivity.

FIRST PRELIMINARY SYSTEM

Sensor System

A preliminary chemical sensor system comprises a miniature air pump, an adsorption tube, and a sample chamber with 100cc volume as shown in Fig.1. The miniature air pump pushes the mixed gas from inlet to the adsorption tube with heater. The outlet of the adsorption tube was connected to the sensor chamber in which three different kinds of sensors; a cantilever, Quarts Crystal Microbalance (QCM) and MOS sensors. The shift of the resonant frequencies on the cantilever and the QCM are acquired with a universal frequency counter (Agilent E53131A).

Cantilever Sensor

Silicon cantilevers were fabricated using micro fabrication techniques from a SOI wafer [3]. The length, width and thickness of the cantilever were 500µm, 100µm, and 5µm, respectively. A set of bridged piezoresistive stress gauges was formed at the root of the cantilever by the p-type layer in the n-type active layer employing thermal boron diffusion. The substrate under the cantilever was removed by deep RIE from the backside to eliminate the squeezed air damping effect. A gold pattern was formed on the upper surface of the cantilever to get an adequate adhesion characteristic with polymers.

Poly-butadiene (PBD) was selected as a molecular recognition layer among the elastomer-type polymers, which was sensitive to aromatic VOC gases including toluene and xylene; and possessed the fast response for an adsorption and desorption against the various VOCs [5]. The PBD thin film was deposited on the cantilever using the micro dispensing technique. The thickness of the film was about 1000nm. This micro dispensing technique is considered as one of the potential candidate capable to deposit the different materials on the surface of the cantilever array in future.

CP1137, *Olfaction and Electronic Nose: Proceedings of the 13th International Symposium,* edited by M. Pardo and G. Sberveglieri

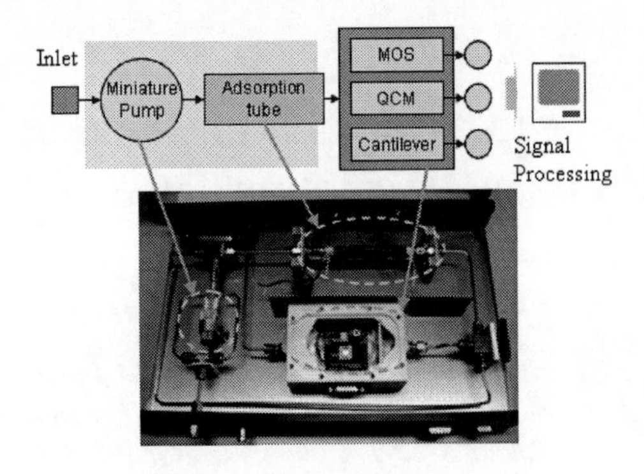

FIGURE 1. Preliminary chemical analysis system.

This PBD-coated cantilever was mounted on a ceramic dual-in-line package (DIP) in the first prototype system, and then a bulk PZT actuator plate was attached onto the bottom of the package as shown in Fig. 2. The sensing signal from the piezoresistive gauges was amplified and fed into analog oscillation circuit consisting of a front end differential amplifier (Amp), a second-order band pass filter (BPF), an automatic gain controller (AGC) and a phase shifter. The second vibration mode (frequency 153 kHz with a quality factor of 560) was used to observe the frequency shift.

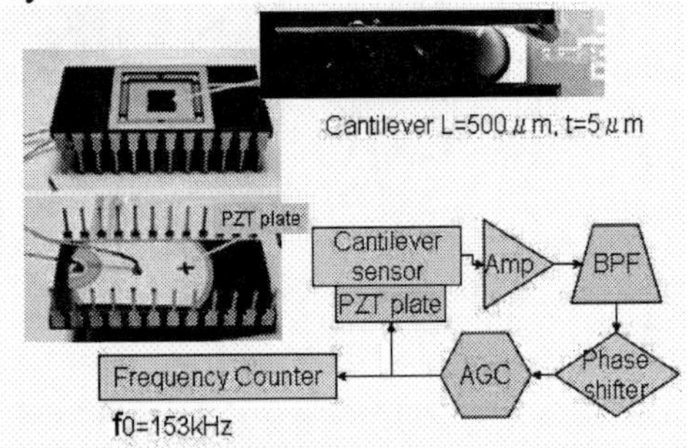

FIGURE 2. Cantilever mounted in a 28pin DIP and a block-diagram of the analog oscillation circuit.

Adsorption Tube

An adsorption tube was made of a 6.35mm diameter stainless steel tube, in which 0.32g of carbon-fiber was filled as shown in Fig. 3. A FeNi heater-wire was wound on the tube with a ceramic insulation sheet. This adsorption tube was capable to heat up to 500 degree in centigrade. The total surface area of the carbon fiber was estimated to be $800m^2$. The adsorption tube system was installed at the inlet of the sensor chamber. In order to separate the mixed VOCs , the temperature of this adsorption tube was gradually increased with 1.3 degree per second by the programmed DC power supply.

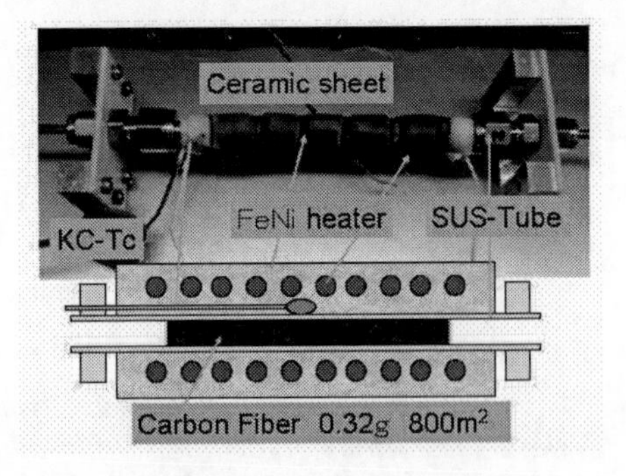

FIGURE 3. Adsorption tube with FeNi wire heater, which is filled by carbon fiber of 0.32g.

PERFORMANCE OF FIRST SYSTEM

Fig. 4 shows the frequency shifts of the cantilever when the adsorption tube is heated up after the adsorption of mixed gas of 1000ppm ethanol and toluene in nitrogen with the different adsorption times 10min, 30min and 60min, as well as the temperature of the adsorption tube. The frequency shift responses showed two peaks corresponding to first ethanol and second toluene peaks corresponding to the temperature of the adsorption tube. Two VOCs were separated thermally by the difference of boiling points and separation energies between VOCs and carbon.

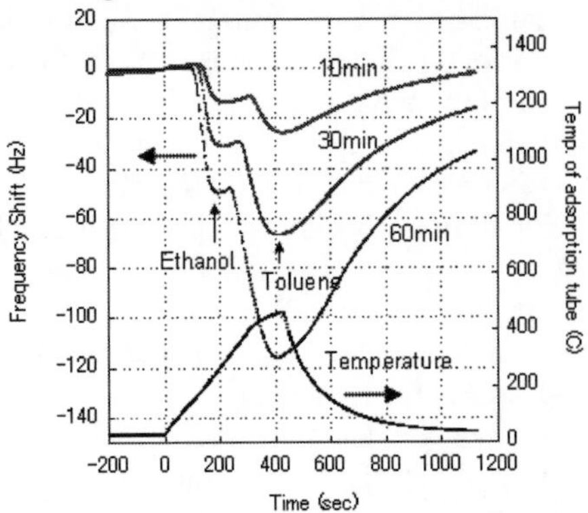

FIGURE 4. Responses of 153kHz cantilever sensor combined with adsorption tube with different adsorption times.

The frequency shift was approximately proportional to the time of adsorption. We confirmed the analysis and condensation functions by this carbon filled adsorption tube. The sensitivity of this system was estimated to be 0.023Hz/ppm with 10min adsorption time. We note that the sensitivity of the PBD coated cantilever without adsorption tube for diluted toluene was 0.04Hz/ppm with 153kHz.

IMPROVEMENT OF SENSITIVITY

Though we confirmed the basic functions on the analysis and detection of the several VOCs as described in previous section, the sensitivity of about 0.023Hz/ppm for toluene at 10min adsorption by this system, was not enough for actual applications. The improvement of the sensitivity has been carried out by two methods; 1) by reducing the sensor chamber and 2) by increasing the resonant frequency of the cantilever.

Reduction of Sensor Chamber Volume

The chemical sensor system shown in Fig. 1 had a relatively large sensor chamber of 100cc volume in order to contain three different sensors. This large sensor chamber made the response of the frequency change of cantilever slower when the concentrated VOC from the adsorption tube injected into a sensor chamber. The response time was measured to be about 60sec when the mass flow rate of the dry nitrogen carrier gas was 100sccm. We made a small sensor chamber of 3.5 cc volume, as shown in Fig. 5. Because the cantilever sensor chip was so small compare with QCM, it was easy to reduce the sensor chamber. Then the response time for this small sensor chamber was measured to be 2sec, which was 1/30 in comparison with that of the 100cc large sensor chamber.

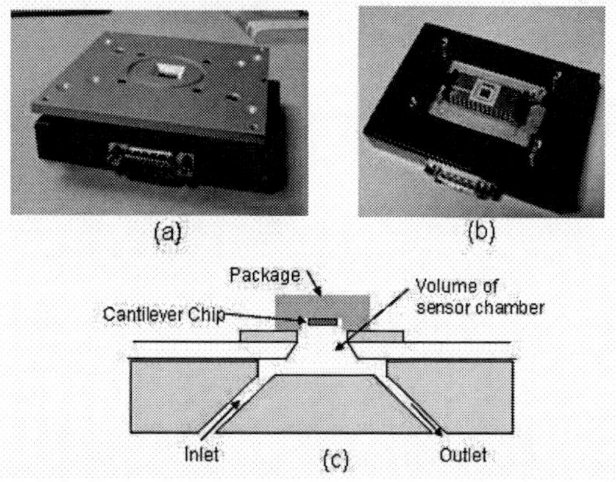

FIGURE 5. (a) sensor chamber, (b) packaged sensor and (c) cross section of small sensor chamber

Fig. 6 shows the frequency shifts of the cantilever combined with the heated adsorption tube for two different sample chambers, when a mixed gas of 1000ppm ethanol and toluene in nitrogen gas was adsorbed with 10min. A red line shows the response in a case of small 3.5cc sample chamber and blue line shows that of large 100cc one. The frequency shift responses showed two peaks according to the temperature of the adsorption tube. Red line with small chamber clearly showed sharper peaks with large peak values. Using this small sensor chamber, we found that both the sensitivity of total system and the analysis function were enhanced.

Consequently, the sensitivity of the 1000nm PBD coated cantilever (L=500μm, W=100μm, t=5μm, resonant frequency =153kHz) with 3.5cc small sensor chamber was estimated to be 0.4Hz/ppm for toluene after 10min adsorption, which was 17times larger than that with large sensor chamber.

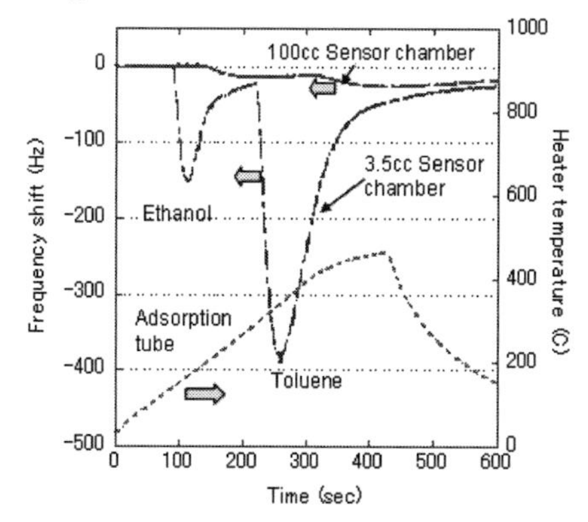

FIGURE 6. Responses of 153kHz cantilever sensor combined with adsorption tube in a case of small 3.5cc chamber in comparison with 100cc chamber.

Higher Resonant Frequency

The analog oscillation circuit of the cantilever in the first system as shown in Fig. 2 was made using operational amplifiers with relatively low cut-off frequency as high as 4MHz. However, to obtain an enough gain to oscillate the cantilever, the amplification bandwidth was lowered below much less than 1MHz. Then this circuit had a limitation of the operating resonant frequency. We designed and fabricated a high speed oscillation circuit with a high speed amplifier with cut-off frequency of 250MHz. This newly developed oscillation circuit was capable to operate the signal up to 1.5MHz.

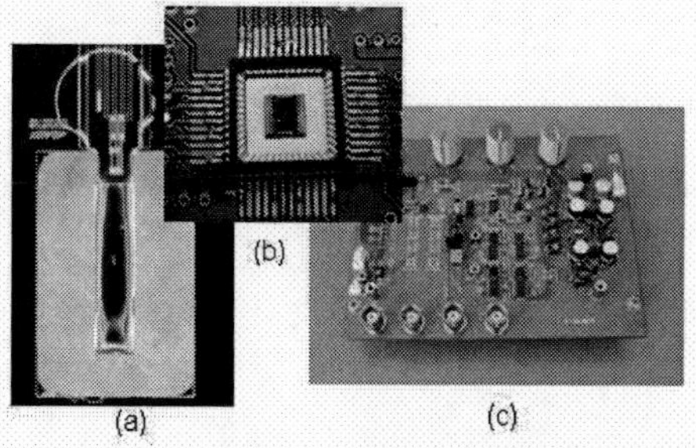

FIGURE 7. High sensitive sensor consists of (a) PBD coated cantilever, (b) cantilever-mounted QFP and (c) high speed analog oscillation circuit.

We also found that the DIP in Fig. 2 was not suitable for the high resonant mode operation of the cantilever for its large parasitic capacitance. Then we mounted the cantilever chip onto a quad flat package (QFP) with lower capacitance in place. Fig.7 shows a set of 715kHz operated sensor without adsorption tube, consisting of PBD (2μm) coated cantilever (L=500μm,

W=100μm), 3.5cc chamber, cantilever-mounted QFP and high speed oscillation circuit.

Fig. 8 shows the response of the frequency shift for PBD-coated cantilever with 4th resonant mode (715kHz) as a function of toluene concentration, in which the sensitivity of this cantilever operation was estimated to be 0.2Hz/ppm. The frequency changes were much larger than that of a QCM sensor coated with PBD. Fig. 9 shows the frequency shift of the same cantilever as a function of resonant frequency in different mode. This result clearly shows the sensitivity was proportional to the resonant frequency.

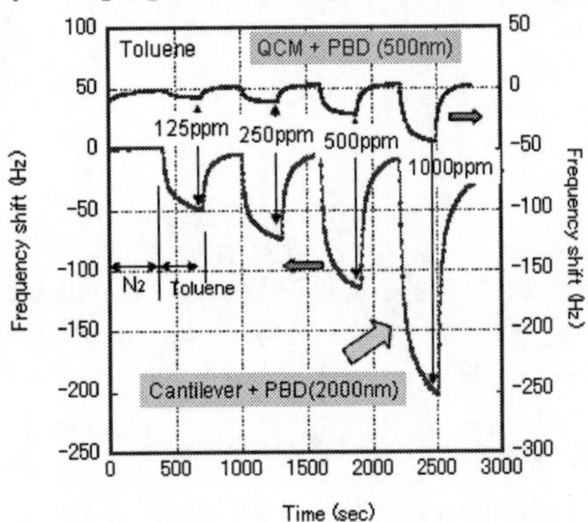

FIGURE 8. Frequency shift of the PBD coated 4th mode 715kHz cantilever and PBD coated QCM as a function of toluene concentration.

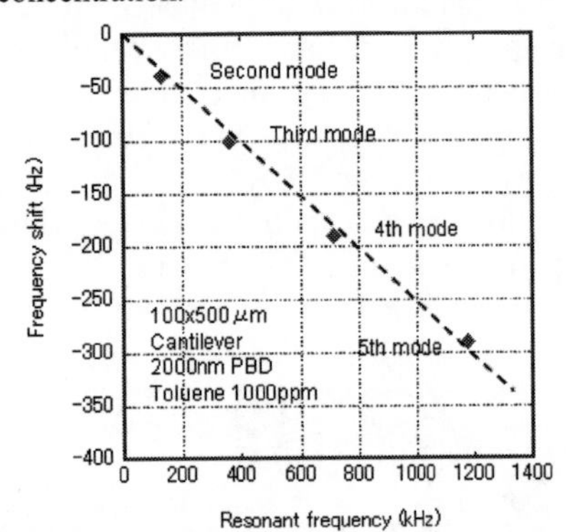

FIGURE 9. Frequency shift as a function of the resonant frequency for 1000ppm toluene diluted gas.

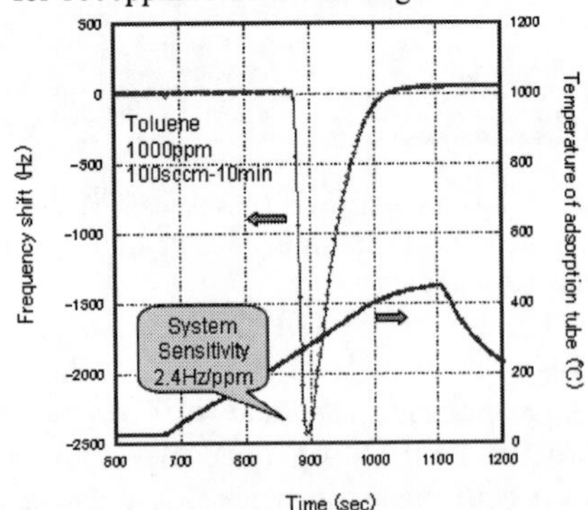

FIGURE 10. Detection peak from an adsorption tube for 10min with 1000ppm toluene in nitrogen.

Fig.10 shows the frequency-shift response of the same 715kHz cantilever by the separation from the heated adsorption tube with 10min-adsorption for 1000ppm toluene. This figure showed the corresponding sharp toluene peak. The sensitivity of this improved chemical sensor system for toluene with 10min adsorption time was about 2.4Hz/ppm which is about 100 times larger than one of first sensor system. This improvement was carried out by small sensor chamber and higher resonant frequency.

CONCLUSIONS

The basic functions and the sensitivity improvement of a chemical sensor system employing a poly-butadiene (PBD) coated micro cantilever sensor and a carbon-fiber-filled adsorption tube was presented. We confirmed the analysis function by an adsorption tube in this proposed system. Then the improvements of the sensitivity were carried out by two methods: 1) by reducing the volume of sensor camber by 1/30, 2) by using a higher resonance frequency of the cantilever at 4.6 times. Using the 4th vibration mode (frequency 715kHz) of a PBD coated cantilever (length 500μm, width 100μm and thickness 5μm), we confirmed that the sensitivity of this improved chemical sensor system for toluene with 10min adsorption time was about 2.4Hz/ppm which was about 100 times larger than that of first sensor system.

REFERENCES

1. I. Dufour, L. Fadel; "Resonant microcantilever type chemical sensors: analytical modeling in view of optimization" Sensor and Actuators B 91 (2003) pp353-361.
2. D. Lange, C. Hagleitner, A. Hierlemann, O. Brand, H. Baltes; "Complementary Metal Oxide Semiconductor Cantilever Arrays on a Single Chip: Mass-Sensitive Detection of Volatile Organic Compounds", Anal. Chem. 2002, 74, pp3084-3095.
3. T. Ikehara, J. Lu, M. Konno, R. Maeda, and T. Mihara: "High quality-factor silicon cantilever for high sensitive resonant mass sensor operated in air" J. Micromech. Microeng., 17. pp2491-2494 (2007)
4. J. Lu, T. Ikehara, Y. Zhang, R. Maeda, and T. Mihara, "Energy dissipation mechanisms in lead zirconate titanate thin film transduced micro cantilevers", Jpn. J. Appl. Phys. 45 (2006) pp.8795-8800
5. Y. Liu, T. Mihara, M, Kimura, M. Takasaki, T. Hirai: "Polymer film-coated quartz crystal microbalances sensor for volatile organic compounds sensing" Proc 24th Sensor Symposium pp309-312 (2007)
6. T. Mihara, T. Ikehara, J. Lu, R. Maeda, T. Fukawa, M, Kimura, Y. Liu, T. Hirai: "Integrated chemical sensor system employing micro cantilever and adsorption tube" Proc 12th International Meeting on Chemical Sensors, pp533-534(2008)

Optical sensor for the detection of explosives: example of a fluorescent material

Thomas Caron[o Π], Marianne Guillemot[o], Florian Veignal[o], Pierre Montméat[o], Eric Pasquinet[o], Philippe Prené[o], François Perraut[+] Jean-pierre Lère-Porte and Françoise Serein-Spirau[Π]

[o] CEA Le Ripault, BP 16,F-37260 MONTS, France
[+] CEA LETI, rue des Martyrs, F-38000 GRENOBLE, France
[Π] Ecole de Chimie, rue de l'Ecole Normale, F-34000 MONTPELLIER, France
Email: pierre.montmeat@cea.fr

Abstract. This paper deals with the detection of nitroaromatic explosives. A specific fluorescent material is devoted to the detection of ultra traces of explosives remaining on clothes. The sensor exhibits a large sensitivity for TNT or DNT vapors.

Keywords: gas sensors, explosive, fluorescence
PACS: 80

INTRODUCTION

During the last 10 years counterterrorism activities have been strengthened. Illicite substances detection has become a crucial point for civilians security. Various valid technologies for the analysis of explosives already exist: HPLC, CE, GC/MS, IR, XR imaging, NMR and IMS. However the cost, the portability and the complexity of many of these methods do not suit to an on-site monitoring device.

Interest is now focused on the development of portable, low cost and reliable detection devices. A suitable sensor would also be expected to analyse samples in a relatively short time.

EXPERIMENTAL AND METHODS

The selected material has a π-conjugaison which brings its fluorescence properties, and a cyclohexyl function which gives a spatial structure to the material [3]: diimine (figure 1).

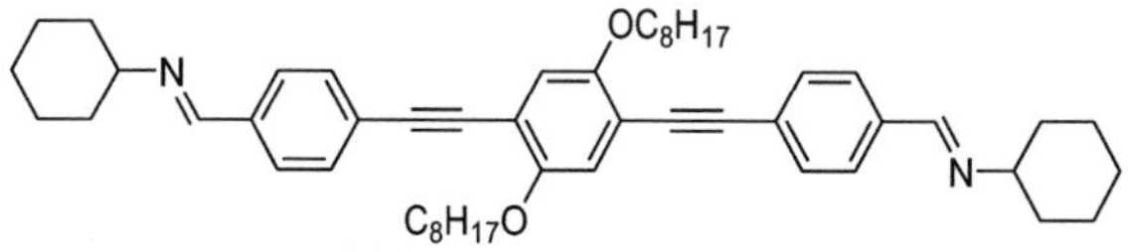

FIGURE 1. Diimine

The diimine behaves differently in presence or in absence of electron-poor compounds, such as nitroaromatics. When the NAC molecules adsorb on the diimine surface, it involves a decrease of the fluorescence intensity [5] (figure 2).

Diimine was evaluated for the detection of 2,4-dinitrotoluene (DNT) and trinitrotoluene (TNT). To estimate the performance of the sensor in real conditions, we have chosen to test it on explosives polluted clothes. Two ways of impregnation were investigated. Firstly, the NAC, in solid state, was deposited on a piece of cotton cloth and then removed by jiggling the cloth, so that only impregnated particules remained. Secondly, a clean piece of cloth was soaked in a solution of DNT or TNT in acetonitrile and then dried. Ultra traces of explosives were then generated,

CP1137, *Olfaction and Electronic Nose: Proceedings of the 13th International Symposium,* edited by M. Pardo and G. Sberveglieri

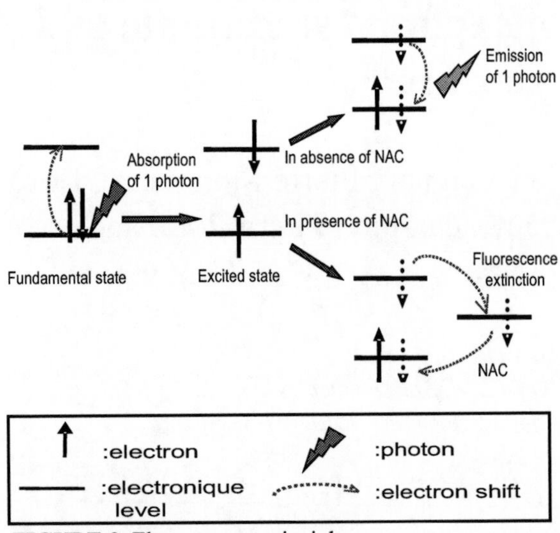

FIGURE 2. Fluorescence principle

The material was deposited on a glass substrate by spin-coating.

The detection device was composed of a gas analysis system based on a wave-guide.

Experiments were carried out as follows:

-30 min of stabilisation under ambiant atmosphere

-10 min of sampling at the surface of the polluted cloth

- 30 min under ambiant atmosphere again

The humidity rate was close to 50% at 20°C. The air flow was kept at 20L/h by the way of a peristaltic pump.

A cone-shaped apparatus was also added to the gas inlet to improve the sampling.

The sensor response was expressed as the percentage of the fluorescence inhibition after 10 min.

RESULTS

The sensor response to a DNT polluted cloth is plotted in figure 3. For both 10 min exposures, the signal is large: the fluorescence extinction is at least 13 %. The reversibility is rather fast since the recovery time is 20 min.

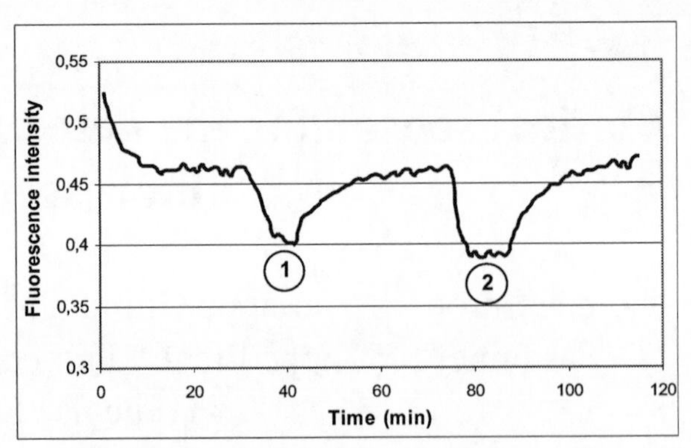

FIGURE 2. Detection of DNT on a piece of clothes. (1): without cone; (2): with cone

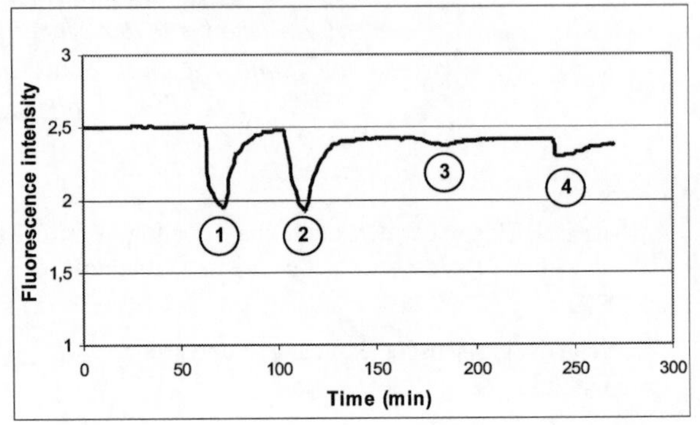

FIGURE 3. Detection of DNT and TNT by diimine. (1): solution of DNT without cone; (2): solution of DNT with cone; (3): solution of TNT without cone; (4): solution of TNT with cone

The impregnated piece of clothes leads to similar results. The signal is much more intense for DNT (22 %) than compared with TNT(5 %). The difference can be explained by the difference between the vapor pressures of both NACs. In the case of TNT, the reversibility is very poor, which may reveal a strong affinity between TNT and diimine. Weakly reversible and strong π interactions are involved in the process of detection.

All this demonstrates the outstanding sensitivity of the material to nitroaromatics explosives. The developed sensor allows to detect ultra traces of TNT or DNT vapors.

All results are reported in Table 1. As expected, the sensitivity of the sensor is improved by the way of the cone-shaped apparatus.

TABLE 1:Fluorescence inhibition (%)

Method of impregnation	without cone (%)	with cone (%)
Solid DNT	13	16
Solution of DNT	22	22
Solution of TNT	2	5

CONCLUSIONS

We have demonstrated the interest of a fluorescent material for the detection of ultra traces of explosives spread onto clothes without using a contact method.

The evaluation of the selectivity and the stability of the sensors is under investigation.

REFERENCES

1. 1. Sanchez, J. C., "Synthesis, Luminescence Properties, and Explosives Sensing with 1,1-Tetraphenylsilole-and 1,1-Silafluorene-vinylene Polymers", *Chem. Mater.*, Vol. 19 (2007) pp. 6459-6470
2. Bakaltcheva, I. B, "Multi-analyte explosive detection using a fiber optic biosensor", *Anal. Chim. Acta*, Vol. 399 (1999), pp. 13-20.
3. S. Clavaguera, "Conception, synthèse et caractérisation de matériaux fluorescents pour l'élaboration d'un capteur chimique d'explosifs", *Thesis, Montpellier* – France, October, 2007
4. US Environnemental Protection Agency, EPA/600/S-99/002, May 1999
5. J. Lakowicz, "Principles of fluorescence spectroscopy", 2nd Edition, Springer

A Portable Electronic Nose Based on Bio-Chemical Surface Acoustic Wave (SAW) Array with Multiplexed Oscillator and Readout Electronics

K.T. Tang[1], D.J. Yao[2], C.M. Yang[3], H.C. Hao[2], J.S. Chao[2], C.H. Li[1], P.S. Gu[3]

[1]*Dept. of Electrical Engineering, National Tsing Hua University, Hsinchu 30013, Taiwan*
[2] *Institute of NanoEngineering and MicroSystem, National Tsing Hua University, Hsinchu 30013, Taiwan*
[3]*Dept. of Chemistry, National Tsing Hua University, Hsinchu 30013, Taiwan*

Abstract. We have developed a new portable electronic nose based on a SAW sensor array and its readout electronics. The SAW array is based on 2×2 non-continuously working oscillators for sensors coated with different polymer/mesoporous carbon composite materials. Signals of the SAW array can be obtained by a readout PCB and a microprocessor. Experiments indicate good results for this portable system to perform gas detection and recogntion applications.

Keywords: electronic nose, surface acoustic wave, multiplexed oscillator, readout electronics, portable system
PACS: 43.38.Rh

INTRODUCTION

Portable electronic nose is desirable for many applications because of its size and convenience. Surface acoustic wave (SAW) sensors are generally known as high-resolution mass-sensitive transducers. They are composed of piezoelectric crystal plus at least one layer of chemically interactive material deposited on one of their surfaces in order to infer a given chemical sensitivity. The traditional way of a SAW-based electronic nose system is to read out the frequency by the use of an instrument such as spectrum analyzer or frequency counter. The large size and volume of these equipments strongly reduce the mobility of the system and the feasibility of portable applications. In this paper, we present our recent progress on a portable eNose in three directions: the SAW array, polymer/mesoporous carbon composite materials, and sensor signal readout electronics.

EXPERIMENTAL AND METHODS

Surface acoustic wave (SAW) sensor

In general, a bio-chemical sensor consists of two major components: a transducer and a sensitive coating. A transducer is required to transform the information to measurable signal.

K^2 gives the capacity of translation between electric and mechanical potential [1]. The value of K^2, which could be calculated by equation (1), depends on the properties of the piezoelectric substrate or on the experimental results due to the velocity shift under metallization. K is determined by

$$K^2 = \frac{e^2}{c\varepsilon}. \qquad (1)$$

where e, c, ε are the piezoelectric coefficient, elastic coefficient, and dielectric coefficient of the substrate. $LiNbO_3$ was chosen as the chip substrate for high acousto-electric effect, that is, sensitivity of the chip could be enhanced compared with other substrates.

Interdigital transducers (IDTs) were widely used for the electric signal excitation and detection of surface acoustic waves [2]. According to the design parameters from the simulation data of delta function model and cross field model, the SAW device was fabricated, using standard photolithographic technology, onto a $LiNbO_3$ piezoelectric substrate with IDTs of Cr/Au, 20nm/100nm, operating at 99.8MHz.(Figure 1). The distance L_{IDTs} between the centers of two IDTs was 3 mm, of which IDT aperture was 3 mm and the number of finger pairs was 30 with a period of 40μm, which is equal to SAW wavelength

CP1137, *Olfaction and Electronic Nose: Proceedings of the 13th International Symposium*, edited by M. Pardo and G. Sberveglieri

λ. After devices were fabricated, their electrical characteristics were measured by network analyzer. The center frequency of SAW sensor is 98.5MHz, insertion loss is around -8.53 dB, and sidelobe rejection is 26dB.

Polymer/Mesoporous carbon composite materials

We have synthesized mesoporous carbon material CMK-3 via a nanocasting route using mesoporous silica SBA-15 as a hard template. The furfuryl alcohol was impregnated into SBA-15 template and then carbonized under 900 ° C atmosphere. The silica template was removed by HF solution, and then CMK-3 was obtained. Besides, the hydrophobic surface of CMK-3 could be modified by hydrothermal method and the surface became more hydrophilic with more hydroxide groups. The structural properties of CMK-3 have been characterized by powder X-ray diffraction, transmission electronic microscopy, and nitrogen physisorption analysis. We have applied four different polymers (PNVP, P4VP, PS, and PVAc) with different polarity and functional groups. We have prepared polymer/CMK-3 solution for coating thin film on the SAW devices as sensing materials. The solvent has been choosed for the uniform dispersion of CMK-3 and polymer in the solution. The appropriate amount of the polymer/CMK-3 solution was spin-coated on the SAW devices.

Sensor Signal Readout Electronics

Figure 2 shows the block diagram of the readout electronics. Counter1 is the main element to calculate the SAW output frequency. Due to the stabilization time of the SAW sensor, the sampling starting time is set at 0.5 seconds after switching. Counter1 is selected to be 24bits with a total sampling time 0.12 seconds.

Counter2 controls the D flip-flop and temporarily stores the data in Counter1 for the 89C51 microprocessor to collect. The 89C51 collects data from the DFF, then calculates and stores the real SAW sensor frequency in its built-in memory. A reference sensor frequency is subtracted from these frequencies later to obtain the frequency shift of the sensor, then the result is displayed on the LCD module. The LCD display is designed so that its first row shows the real-time sensor number and frequency, and the second row shows the frequency difference between that sensor and the reference sensor.

System Test

The complete system is tested in a 1L 4-neck bottle chamber. The four windows are for wire connection, vacuum pump, air flow valve, and test gas injection. The test gas is accurately controlled by a micro-injector and estimated the concentration by PV=nRT. The experimental sensor coatings are P4VP and PNVP. The sensors are tested with methanol, ethanol, acetone, amine, and trimethylamine (TMA). Every data is repeated by at least 8 times for statistics. We use a higher concentration to monitor the maximum frequency shift of the membrance and to construct a unique corresponding radar plot based on the membrane to be the database of for future odor recognition. There are four steps in one cycle of gas testing: (1) The system operates stably under vacuum pumped for 10-15 minutes; (2) after the frequency stabilzes, turn off the vacuum pump. Inject the test solution and respond for 5 minutes; (3) turn on the vacuum pump and air flow volve to clean the chamber for 5 minutes; (4) turn off the air flow volve and return to step (1).

The total time of one cycle is about 20 minutes. When the solution is injected into the chamber, it is vaporized quickly beceause of it has been vacuum pumped. The response time should be less than 2 minutes.

RESULTS AND DISCUSSION

SAW array

By coating different polymers on the surface of different resonators, a SAW array is constructed. These SAW oscillators are controlled non-continuously by a multiplexing technique [3] as switching element, respectively.

When it is set to a negative voltage, the SAW oscillation does not work (switch off mode). When set to a certain positive voltage 5V, the oscillation is supported (switch on mode) as well as the phase position of the oscillator is set on a desired value. Before detection, the sensor array was exposed to the reaction chamber to arrive at stable state. After 3~4 min, the system was stable. We supposed one of 2x2 SAW had largest frequency to be reference, subtracted others from the reference, we observed the change with difference frequency between reference and others. The original and subtracted frequencies are the same trend, subtracted frequencies between start and stable state is about 1 kHz.

Polymer/Mesoporous carbon composite materials

In Figure 3(a), the PXRD pattern of SBA-15 shows the ordered p6mm symmetry porous structure, which is maintained in mesoporous carbon CMK-3. The periodicity of CMK-3 structure is smaller than SBA-15 due to the material shrinkage under 900 ° C carbonization process. The d-spacing of SBA-15 and CNK-3 is 11.1nm and 10.0nm respectively. This result is in accordance with the TEM image in Figure 3(b). The black circles in the figure are the intersection image of CMK-3 nanorods, and the gray-white region represents the pores between the carbon nanorods. We can see from the figure that the CMK-3 is the replica of SBA-15. The porous structure information of the CMK-3 can be characterized by nitrogen physisorption analysis. In Figure 4(a), $P/P0<0.05$ corresponds to the volume of nitrogen adsorbing in the micropores of CMK-3, and P/P0=0.05~0.25 shows the information of nitrogen single-layer adsorption formation. Thus we know that the CMK-3 is characterized by large micropore volume($0.44cm^3$) and large surface area ($1040m^2g^{-1}$). We can also know that CMK-3 have nanoscale porosity from the hysteresis loop between P/P0=0.4~0.6. The theoredical calculation shows that CMK-3 total pore volume is $1.16cm^3g^{-1}$. According to the results of PXRD, TEM, and nitrogen physisorption analysis, mesoporous carbon CMK-3 has ordered nanometer-sized mesopores, large pore volume, large microporosity and high surface area. When mixed with polymers, the resulting polymer/CMK-3 composites can also possess high surface area with polymers fully extended on the mesoporous carbon to facilitate the diffusion of the gaseous analyte molecules to interact with the polymers. Figure 4(b) shows the thickness of polymer/CMK-3 thin film is 0.5~10μm.

Powder X-ray diffraction, transmission electronic microscopy, and nitrogen physisorption analysis show that mesoporous carbon CMK-3 has ordered nanometer-sized mesopores with pore diameter of 3.5nm and has large surface area. For the preparation of polymer/CMK-3 solution, some factors should be considered for choosing solvent: the physical and chemical properties of polymer and solvent (molecular weight, hydrophilicity, hydrophobicity, polarity, etc) which affects the interaction of solvent-polymer-CMK-3, the rate of solvent evaporation. The solvent we chose facilitates the dispersion of CMK-3 in the sensing film. Besides, the weight percent of polymer and CMK-3 also affects the CMK-3 dispersion. Under all consideration, we can prepare thin polymer/CMK-3 sensing film with uniformly dispersed CMK-3 which extends polymer fully for faster sensing rate.

Sensor Signal Readout Electronics

The readout electronics in Figure 2 is implemented on a PCB together with 89C51 microprocessor board. The circuit is first tested with a fixed sinusoidal signal generated by function generator as to simulate the SAW signal. This signal serves as the input of Counter1. The frequency data can be obtained by reading the LCD display. These data is used to analyze the accuracy of the circuit.

We sweep the signal frequency from 98Mz to 99Mz with a step of 10kHz. The result is obtained from the LCD display. The error frequency (LCD displayed frequency – input frequency) is within ±20 Hz, which is equal to $\pm 2\times 10^{-5}\%$ based on 100MHz baseline frequency.

Next, the function generator is replaced by a real SAW sensor and an operational amplifier. The sensor is tested with ethanol for three consecutive times and the results are shown in Figure 5. In each test, the test chamber is vacuum pumped for 10 minutes, then 50 μL ethanol is injected into the chamber for the next 5 minutes, and for another 5 minutes the valve is open for air flow. The system shows good reproducibility for both the SAW sensor itself and the sensor readout electronics.

System test

The results of system test for two kinds of polymer-coated (P4VP and PNVP) SAW sensors responding to methanol, ethanol, acetone, amine, and TMA are summarized in Table 1. The average frequency shift, standard deviation, and percentage are shown in the table.

Radar plots of these two membranes and five gases are shown in Figure 6. The data in the figure is normalized. The shapes of the radar plots show the fingerprints of the membrane. These fingerprints can be used in the future for gas recognition.

CONCLUSIONS

We have demonstrated a portable electronic nose with SAW array coated with P4VP and PNVP. The sensor signals are obtained by non-continuous multiplexed oscillators, readout electronics, and microprocessor. Tests for each components and the whole system have shown good results for this portable system to perform gas detection and recognition applications.

ACKNOWLEDGMENTS

The authors would like to acknowledge the support of the National Science Council of Taiwan, under Contract NSC 97-2220-E-007-036. We also thank CSIST for the gas sensing instrument support of this research.

REFERENCES

1. Drafts B., "Acoustic wave technology sensors," *IEEE Transactions on Microwave Theory and Techniques*, Vol. 49, No. 4 (2001), pp. 795-802.
2. Eichinger L., "Accurate design of Low-noise high frequency SAW oscillators", in *Agilent EEs of EDA*. Agilent technologies. 2005.
3. Rapp M., "New miniaturized SAW-sensor array for organic gas detection driven by multiplexed oscillators", *Sensors and Actuators*, Vol. 65 (2000), pp169–172.
4. Enguang D., "Organic Vapor Based on SAW Resonatorand Organic Films," *IEEE Transaction on ultrasonics, ferroelectrics and frequency control*, Vol. 44, No. 2, 1997

TABLE 1. The summary of the system test results to different vapors.

Polymer	Δf \ Gas	Methanol	Ethanol	Acetone	Amine	TMA
P4VP	Average (kHz)	22.51	18.37	18.81	22.81	31.57
	Standard deviation (kHz)	2.7	1.33	1.85	2.3	2.8
	$\Delta f / f_0$ (%)	0.023	0.019	0.02	0.023	0.032
PNVP	Average (kHz)	76.85	73.87	52.45	188.9	130.6
	Standard deviation (kHz)	1.18	3.7	2.4	9.3	9.04
	Average (kHz)	0.079	0.075	0.054	0.193	0.134

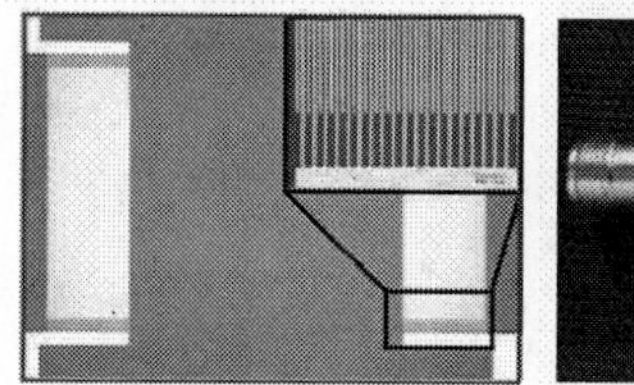

FIGURE 1. Photo of IDT after photolithography and SAW on the PCB.

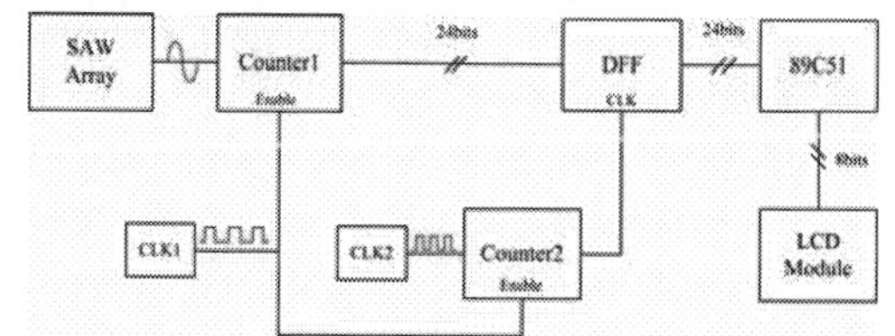

FIGURE 2. Block diagram of the readout electronics

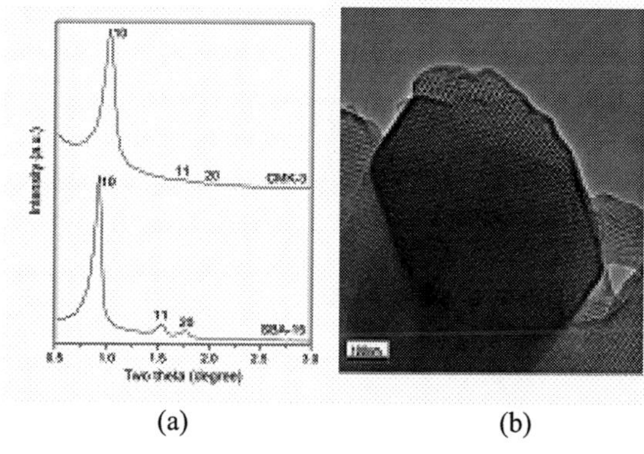

(a) (b)

FIGURE 3. (a) PXRD pattern of SBA-15 and CMK-3 (b) TEM image of CMK-3

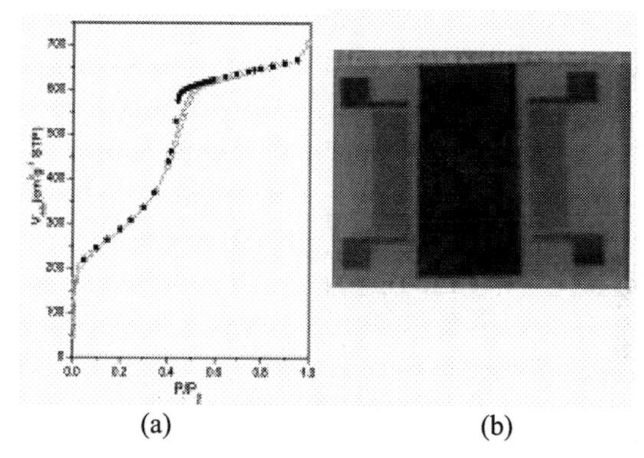

(a) (b)

FIGURE 4. (a) Nitrogen physisorption analysis of CMK-3 and (b) Optical microscopic photo of the SAW device

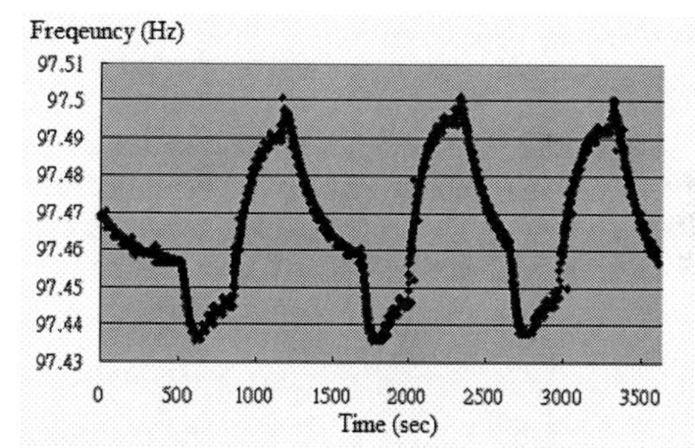

FIGURE 5. Ethanol test results for the SAW sensor with PCB sensor readout electronics and the 89C51 microprocessor

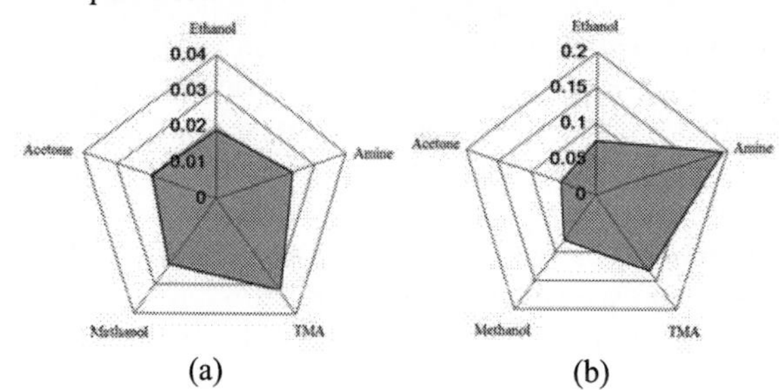

(a) (b)

FIGURE 6. Radar plots of (a) P4VP (b) PNVP

Porphyrin Electropolymers For Application In Hyphenated Chemical Sensors

L. Lvova, M. Mastroianni, E. Martinelli, C. Di Natale, A. D'Amico, D. Filippini, I. Lundström, R. Paolesse

Department of Chemical Science and Technologies, University "Tor Vergata", Rome, Italy
Faculty of Biology and Soil Science, St. Petersburg State University, St. Petersburg, Russia
Department of Electronic Engineering, University "Tor Vergata", Rome, Italy
Division of Applied Physics, IFM, Linköping University

Abstract. A series of pyrrole-substituted porphyrin monomers have been rationally prepared to tune the properties of the resulting polymeric film. Free-base porphyrins and their metallic complexes have been deposited onto Indium-Tin-Oxide (ITO) glass electrodes by electropolymerization technique. Electropolymers were characterized by UV-visible spectroscopy and Atomic Force Microscopy (AFM). Cyclic voltammetry has been utilised to study the electropolymerisation mechanism and to evaluate the polymer surface coverge parameters. The obtained porphyrin electropolymers have been exploited as sensing materials for hyphenated potentiometric and optical measurements with CSPT-potentiometric analytical system. Different food matrices, such as mineral waters, wines and vegetable oils, have been analysed by means of the resulting porphyrin based CSPT-potentiometric system.

Keywords: porphyrin electropolymers, chemical sensors.
PACS: 72.80.Le, 78.66.Qn

INTRODUCTION

Porphyrins are widely distributed in natural systems and they have been called "the pigments of life" due to the richness of their optical properties and to their essential biological functions. Inspired by these natural examples, porphyrins have been applied as sensing layers in a wide range of chemical sensors [1]. Moreover, the richness of properties makes porphyrin suitable for their exploitation in devices based on different transduction mechanisms.

For example porphyrins were shown to be the successful components of optical sensors in a sensing platform constituted by low-cost advanced optical equipments, called Computer Screen Photoassisted Technique (CSPT) [2-3]. On the other hand porphyrins have been widely used as ionophores in potentiometric ion selective electrodes [4] and electronic tongues based on such a devices have been developed [5].
Recently we have been interested in the development of sensing platforms where two different transduction mechanisms are hyphenated in the same porphyrin-based substrate [6]. This approach allows a significant increase in the chemical information that can be obtained from the device and it can boost the performances in terms of selectivity and sensitivity. Among the different deposition techniques available to obtain porphyrin thin films, electropolymerisation was applied for formation of porphyrin sensing materials. This technique is an elegant way to produce a stable conducting porphyrin polymer onto an electrode surface. Porphyrin electropolymers based on polypyrrole backbone are especially well studied. The direct electrochemical polymerization of pyrrole-substituted porphyrins leads to the well-structured multi-layer conducting film formation, which exhibits the electrochemical properties of the monomeric complex [7]. However it is necessary to rationally design the monomer structure and to characterize the resulting film to obtain sensors with optimized performance.
In the present work we summarise our recent achievements in rational design, preparation and characterization of new n-alkyl-(1-pyrrole) phenyl-substituted porphyrin monomers and their following application as sensing materials in hyphenated CSPT-potentiometric analytical system for analysis of foodstuffs.

CP1137, *Olfaction and Electronic Nose: Proceedings of the 13th International Symposium*, edited by M. Pardo and G. Sberveglieri

EXPERIMENTAL

The porphyrin monomers (Figure 1) were prepared according to the literature methods [8] and fully characterized by spectroscopic techniques. Cyclic voltammetry (CV) from 1-3 mM solutions of porphyrin monomers in acetonitrile (ACN) or dichloromethane (CH_2Cl_2), deoxygenated by bubbling with nitrogen and containing 0.1 M tetra-n-butyl-ammonium perchlorate (TBAP) supporting electrolyte, was carried out in a conventional three-electrode cell using a ITO working electrodes (Aldrich), with a nominal resistance of 30-60 Ω/sq, a platinum wire counter electrode, and a SCE (AMEL, Italy). Absorption spectra were measured on a Cary-50 Scan UV-visible spectrometer.

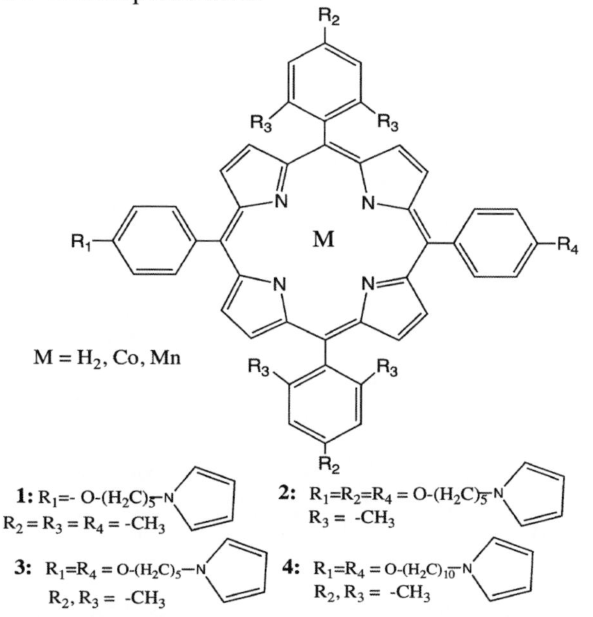

M = H_2, Co, Mn

1: R_1=- O-$(H_2C)_5$-N(pyrrole), $R_2 = R_3 = R_4$ = -CH_3

2: R_1=R_2=R_4 = O-$(H_2C)_5$-N(pyrrole), R_3 = -CH_3

3: R_1=R_4 = O-$(H_2C)_5$-N(pyrrole), R_2, R_3 = -CH_3

4: R_1=R_4 = O-$(H_2C)_{10}$-N(pyrrole), R_2, R_3 = -CH_3

FIGURE 1. Structures of studied porphyrin monomers.

Resulting electropolymers were soaked in 0.01M NaCl solution for 24 hours before first measurement. The sensor array was composed by 3-8 porphyrin electropolymers deposited on unique ITO modified glass slide. The separate electrical channels were realized by cutting the conductive ITO layer, and connected then to a high-impedance analog-to-digital potentiometer (Smartronix srl, Italy). The transparent flow-through cell (5x3x0.7 cm3, flow channel length of 3 cm and internal volume of 90 μL) permitted an easy integration of potentiometric sensor array within the cell, a simple replacement of individual sensors, as far as simultaneous registration of potentiometric and optical sensors response, Figure 2. The analyte was injected in 0.01M KCl carrier solution propelled with a peristaltic pump (Minipulse3, GILSON model M312) with 1 ml/min flow rate. The potentials of porphyrin electropolymer modified sensors were measured versus SCE placed in outlet chamber using a PC equipped high-impedance analog-to-digital converter (Smartronix, Italy). In order to register the optical response, the sensor array illuminated from the cell backside by 2.5" color TFT-LCD panel (Prime View International Co.). A webcam (Logitech Quickcam® for Notebook), with a detector operating at a resolution of 352*288 pixels was applied for optical signal detection. The MATLAB® software controlled the CSPT illuminating sequence, the video acquisition, and extracted the information (CSPT fingerprints). Fingerprints were then treated as spectral data together with potentiometric sensor responses and analyzed using multivariate techniques. Alcohol extracts of vegetable oils have been analyzed for classification purposes.

FIGURE 2. Picture of the designed flow-through cell containing sensor array and assembled with a backside-placed on a TFT-LCD screen and plug to potentiometer.

RESULTS AND DISCUSSION

Cyclic voltammetry technique has been applied for the formation of porphyrin electropolymers on ITO glass surface. As can be seen from Figure 3, the conductive film was grown at the ITO electrode surface during first 20 scans of **monomer 1** electropolymerisation. Two oxidation waves appear with anodic peak potentials of about 330 and 1100 mV respectively, clearly indicating the successful formation of a polymer film on the electrode surface. The oxidation wave at 330 mV (and quasi-reversible reduction at 250 mV) corresponds to the polypyrrole film formation, while wave at 1100 mV can be attributed to the doping/un-doping process of the polypyrrole backbone. Figure 3B shows the UV-visible absorption spectra of **monomer 1** in ACN and of ITO modified with electropolymerized **monomer 1**, taken in air. The spectral shape is similar and this result clearly indicates that porphyrin skeleton is accessible and not aggregated in the polymer, although a slight band broadening and red shift is present.

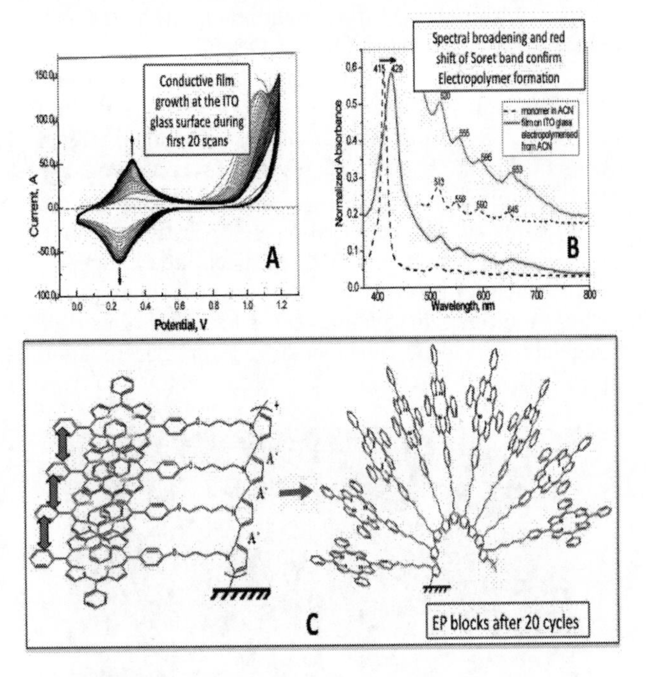

FIGURE 3. (A) Cyclic voltammogram showing repetitive anodic sweeps at ITO electrode immersed in a 3mM solution of monomer 1; (B) Absorption spectra of monomer 1 in ACN and spectra in air of electropolymer formed from monomer 1 solution in ACN on ITO glass. An expansion of the Q-bands region is included; (C) The schematic presentation of monomer 1 EP process.

The breakage of film formation after 20 cycles, as far as irregular morphology, detected by AFM (Figure 4) may be interpreted by the steric repulsion among porphyrin macrocycles, Figure 3C. As confirmed by a AFM-morphology study, a formation of a very thin and irregular electropolymer film composed of few molecular layers (about 10 nm thickness) was detected in case of monomer 2, Figure 4B. This is due to the possible concurrent participation of all the peripheral substituents in the electropolymerization process. The incorporation of more long and flexible spacer chain between the pyrrole unit and the porphyrin in monomers 3, 4 permitted to decrease the steric constraints and resulted in a formation of more ordinate conductive electropolymer films. A uniform, film was detected after free base monomer 4 electropolymerization (Figure 4C), having thickness of 300 nm. Cyclic voltammograms of electropolymerisation of Mn(III)- and Co(II)-complexes of monomer 4 are shown on Figure 5.

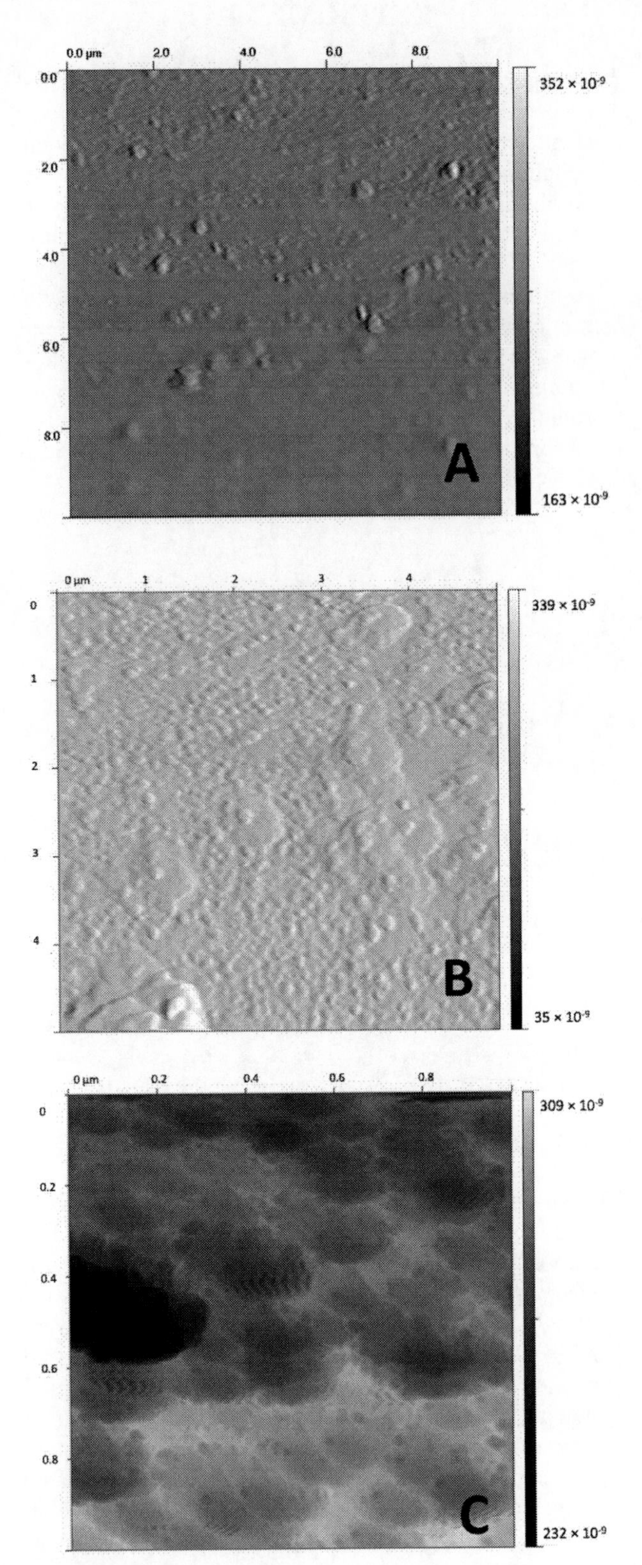

FIGURE 4. AFM morphology study of (A) monomer 1; (B) monomer 2; (C) free base monomer 4 electropolymers.

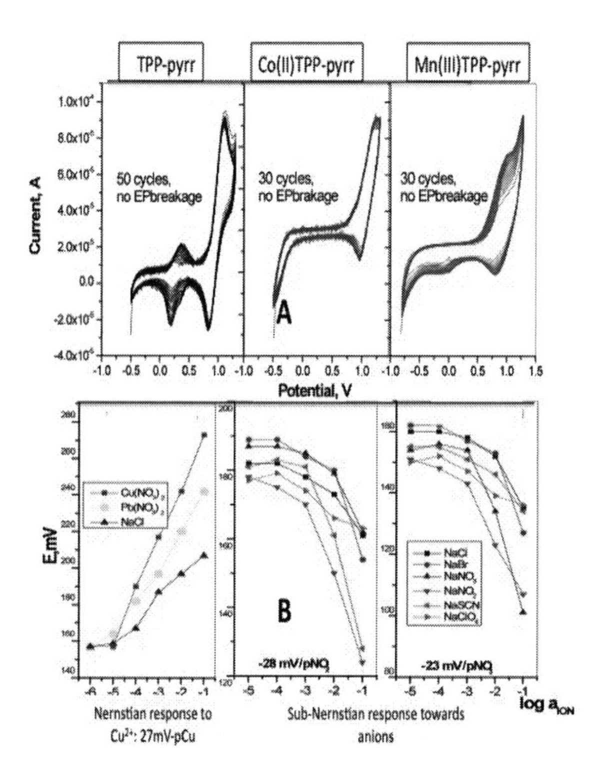

FIGURE 5. (A) Cyclic voltammograms showing electropolymerisation of monomer 4; (B) potentiometric responses of electropolymers formed with free base monomer 4 and its Co(II) and Mn(III) complexes.

As can be seen, the conductive films of free base monomer 4 was grown onto the ITO electrode surface during first 50 scans, while the presence of central Co and Mn ions complicated the electropolymerisation process, breaking the conductive film formation after 30 cycles.

The sensor array formed of pyrrole-substituted porphyrin electropolymers was applied for CSPT-potentiometric analysis of vegetable oils. Measurements were repeated 3 times in ethyl and methyl alcohols oil extracts in a random order. The optic and potentiometric responses of hyphenated porphyrin-based array towards vegetable oils extracts were then treated with PLS-DA first separately and then were fused in unique data set, Figure 6. 98% of oils were classified as belonging to determined classes, while only one of the commercial olive oils was falsely classified as seed oil.

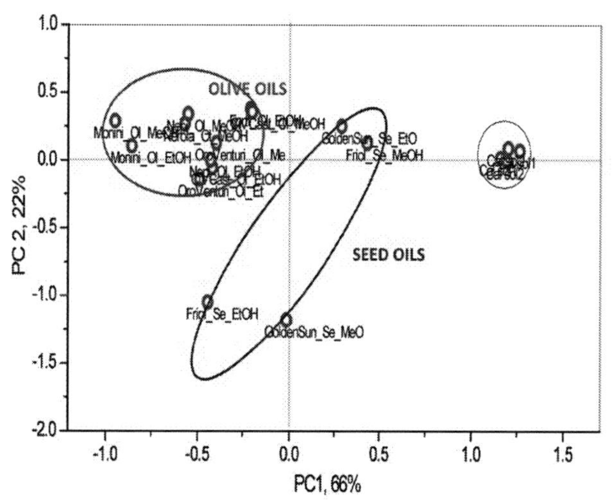

FIGURE 6. PLS-DA vegetable oil classification performed by CSPT-potentiometric analytical system.

CONCLUSION

The results obtained demonstrated the efficiency of porphyrin electropolymers for the development of hyphenated chemical sensors. Considering the simplicity of the integrated experimental apparatus, the hyphenated CSPT-potentiometric analytical system could represent a promising sensing platform with great potentials.

REFERENCES

1. R. Paolesse et al, "Porphyrin Based Chemical Sensors" in *Encyclopedia of Nanoscience and Nanotechnology*, eduted by H. Nalwa, American Science Publishers, 2004 pp. 21-43.
2. A. Alimelli, D. Filippini, R. Paolesse, S. Moretti, G. Ciolfi, A. D'Amico, I. Lundstrom, C. Di Natale, *Anal. Chim. Acta* **597**, 103–112 (2007).
3. C. Di Natale, E. Martinelli, R. Paolesse, A. D'Amico, D. Filippini, I. Lundstrom, *PLoS ONE* **3**, 3139-3144 (2008).
4. L. Gorski, E. Malinowska, P. Parzuchowski, W. Zhang, M.E. Meyerhoff, *Electroanalysis* **15**, 1229-1235 (2003).
5. R. Paolesse, L. Lvova, S. Nardis, C. Di Natale, A. D'Amico, F. Lo Castro, *Microchim Acta* **163**, 103–112 (2008).
6. S. Cosnier, C. Gondran, R. Wessel, F.P. Montforts, M. Wedel, *J. Electroanal. Chem* **488**, 83-91 (2000)
7. M. Mastroianni, L. Lvova, R. Paolesse, 5th ICPP Conference Proceedings, Moscow, Russia, 2008, p. 452.
8. K.M. Smith, in *Porphyrins and Metallo-porphyrins*, editd by K.M. Smith, Amsterdam: Elsevier, 1975

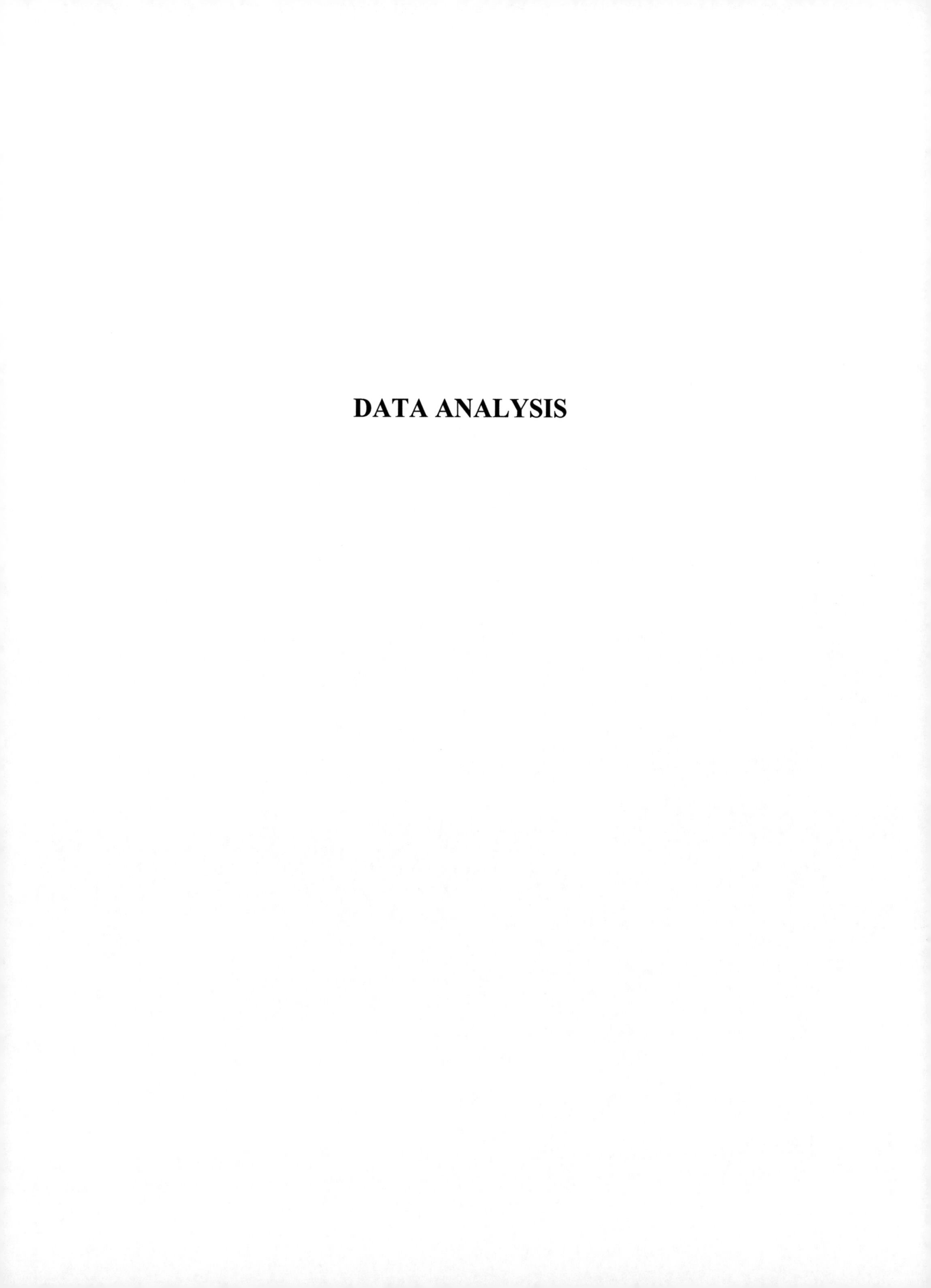

DATA ANALYSIS

Continuous Odour Measurement with Chemosensor Systems

Boeker, Peter*, Haas, T.*#, Diekmann, B.#, Schulze Lammer, P.*

University of Bonn
*Institute for agricultural engineering**
Institute of physics#
Nussallee 5, D-53115 Bonn, Germany
Email: boeker@uni-bonn.de, Tel.: +49-228-732387

Abstract. The continuous odour measurement is a challenging task for chemosensor systems. Firstly, a long term and stable measurement mode must be guaranteed in order to preserve the validity of the time consuming and expensive olfactometric calibration data. Secondly, a method is needed to deal with the incoming sensor data. The continuous online detection of signal patterns, the correlated gas emission and the assigned odour data is essential for the continuous odour measurement. Thirdly, a severe danger of over-fitting in the process of the odour calibration is present, because of the high measurement uncertainty of the olfactometry. In this contribution we present a technical solution for continuous measurements comprising of a hybrid QMB-sensor array and electrochemical cells. A set of software tools enables the efficient data processing and calibration and computes the calibration parameters. The internal software of the measurement systems microcontroller processes the calibration parameters online for the output of the desired odour information.

Keywords: Odour measurement, chemosensor, electronic nose, pattern recognition, olfactometry
PACS: 07.88 +y, 92.60 Sz

INTRODUCTION

Odour measurements have been undertaken by a lot of researchers in the past. The work was carried out in two main fields. The first field is the research on the aroma qualities of foodstuffs, e.g. fruits, vegetable oils or fishes [1-3]. The second field, to which this research paper belongs, is the odour measurement of emission sources, e.g. from waste processing, agriculture and waste water treatment [4-8]. The existing studies comprise the whole spectrum of chemical sensor technologies and data processing methods. One very important focus on the problem of odour measurement is still rarely addressed: the measurement uncertainty of the odour reference data [9].

The main reference method for odour is the olfactometry according to the European standard EN 13725. The measurement uncertainty of this method was evaluated with round robin tests and found to be in the range of 6 dB_{OD}. Therefore the uncertainty range around e.g. an odour concentration of 1.000 OU/m^3 reaches from 250 OU/m^3 to 4.000 OU/m^3. This high uncertainty requires an adapted methodology for the calibration of a chemosensor system with this data.

Figure 1 shows a scheme of cases distinguishing the different situations in the field of odour measurement. In most cases not the odourous gases, but correlating concomitant gases are measured. The condition of the correlation is the basis of the most odour measurement studies.

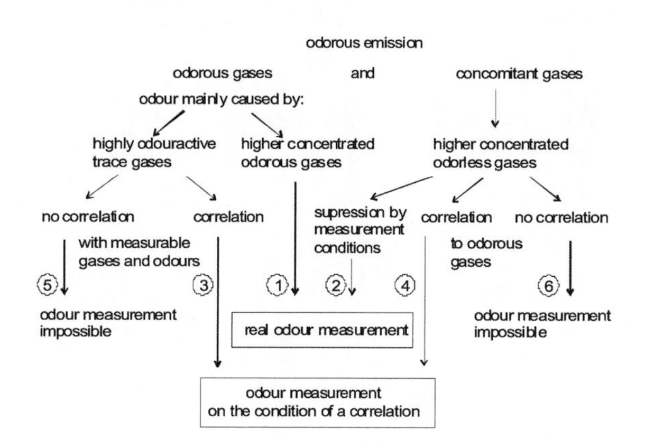

FIGURE 1. Case discrimination for the potentiality of odour measurements.

CP1137, *Olfaction and Electronic Nose: Proceedings of the 13th International Symposium,* edited by M. Pardo and G. Sberveglieri

EXPERIMENTAL SET-UP AND METHODS

The chemosensor measurement system used in the online odour measurements, the OdourVector, is build around a quartz crystal microbalance sensor array of 6 integrated sensors. The coatings on the sensors are specially processed stationary phases from gas chromatography. To enhance the sensitivity the system has an integrated pre-concentration unit. The unit is peltier-cooled and allows the stepwise thermal desorption of the trapped volatiles via precise temperature control. To broaden the range of detectable gases up to two electrochemical cells (e.g. for H_2S or NH_3) can be added. The measurement system has an internal microcontroller for data pre-processing and can be remote controlled via GSM-network. The collected measurement data is send daily or on request to the operators email account. Depending on selectable conditions, e.g. sensor signal threshold values, the measurement system can trigger the collection of odour samples in olfactometric bags.

The data processing of continuous long-term data is a challenge for itself. In a sound methodical odour measurement scheme it is of importance to identify corresponding sensor signal patterns. For each of these patterns a separate calibration with odour data is required.

Figure 2 demonstrates the method of data collection, pattern recognition, referencing of separate odour classes and the continuous processing of odour measurements.

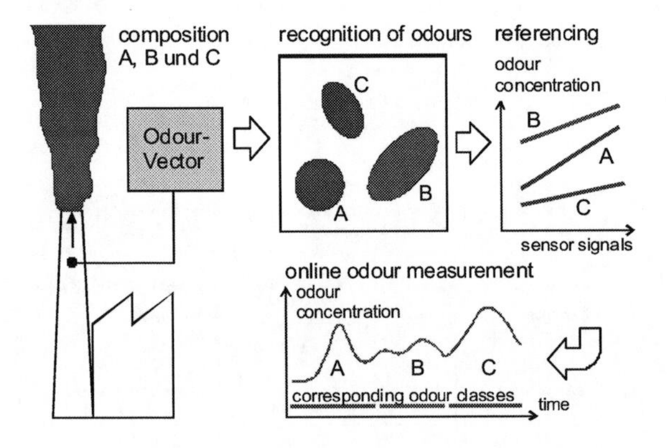

FIGURE 2. Scheme of the online odour measurement.

For the task of the identification of signal patterns in a long-term measurement we developed a software system, with which corresponding signals can be identified in a semi automatic manner.

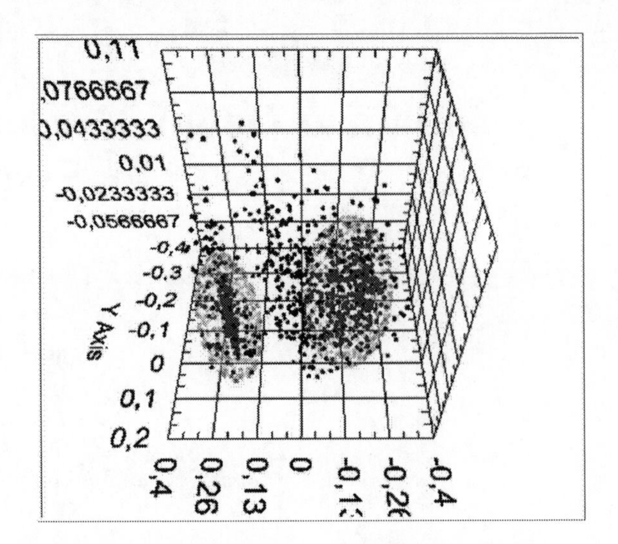

FIGURE 3. 3D-PCA plot with ellipsoids containing the selected data points.

In a first processing step the sensor signals of a preliminary time interval are fed into the software. Here a three-dimensional PCA is processed, using different, user selectable methods and sequences of normalisation and centering of the data.

With movable ellipsoids in the three-dimensional space the user then can isolate aggregations of data points (see Figure 3). These aggregations form the basis of typical chemical/odour classes as a basis for the following calibration with olfactometric reference data.

We found that this semi automatic method works better with real world data than automatic clustering methods. The human eye of an expert is a better classifier in the case of diffuse clusters with lot of transition data points in between.

RESULTS

Odour Measurements in a Waste Incinerator

Since 2004 a continuous odour measurement system is running as a permanent installation on one of the most modern waste incineration plants in Germany. The aim of the system is the supervision of charcoal odour filters. The operating company has set a limit of 300 OU/m^3 in the emission of the filters.

Figure 4 depicts the calibration phase of the continuous odour measurement system. The phase lasted three month to cover all the different phases of the exhaust emissions, including new filters and depleted filters.

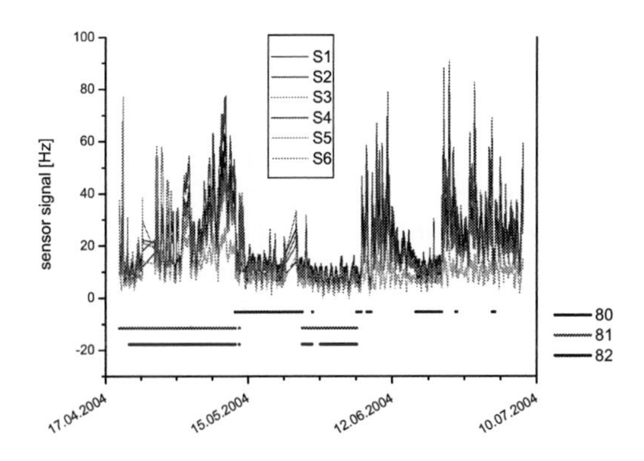

FIGURE 4. Calibration phase during 3 month in 2004. Three filters are indicated as being in action in certain time intervals. The sensor pattern change is visible.

During the first measurement phase olfactometric calibration was accomplished. In Figure 5 the single measurement values and a linear fit is shown. The measurement uncertainty is high for the reference values, as indicated with the error bars.

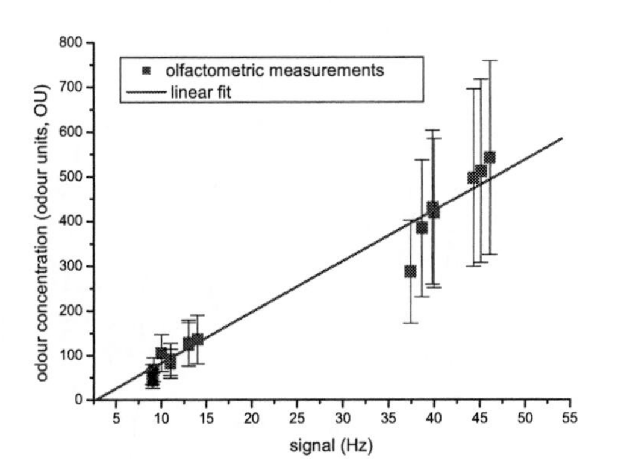

FIGURE 5. Calibration with olfactometric measurements according to the EN 13725..

In Figure 6 the course of the odour emissions, as calculated with the function above is displayed.

In the fourth year of operation, the odour measurement system has shown its ability to detect filter breakthrough and depletion. In Figure 7 the sensor signals during one and a half year is presented, with the depletion of the filters, as measured via the sensors signals.

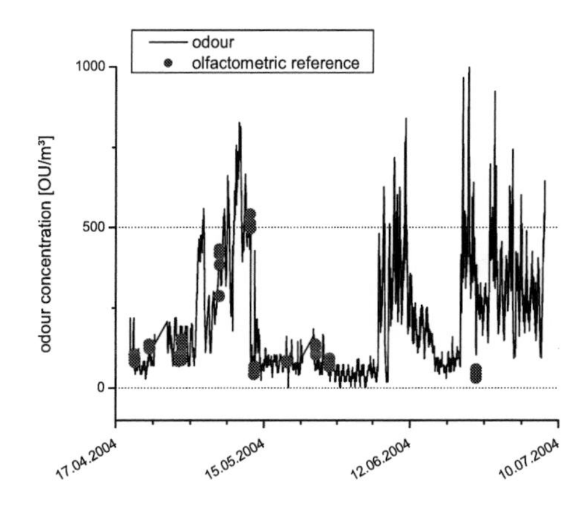

FIGURE 6. Odour emission prognosis and reference values.

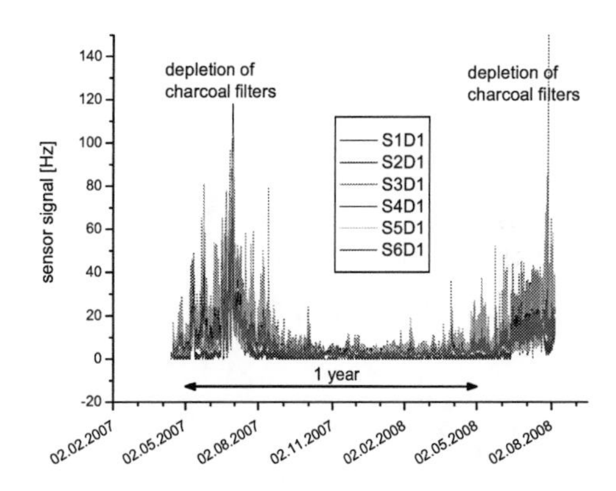

FIGURE 7. Course of sensor signals of the odour measurement system OdourVector during the monitoring of odour filters.

Odour Measurements in Sewage Canals

In 2005 another measurement campaign was undertaken in the sewage canals of a large town. The first results seemed to be very irregular with regard to the correlation between odour concentration and sensor data.

To exclude irregular measurements from the examination we choose the Mahalanobis distance as a measure. By self referencing between the complete class and every single measurement the outliers were identified. Without the outliers the linear fit of the remaining olfactometric references and the sensor signals proved to be satisfactory.

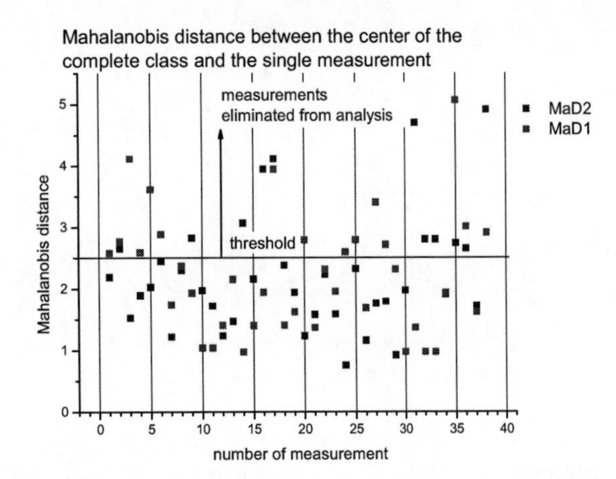

FIGURE 8. Mahalanobis distance of the reference measurements to the centre of the whole class.

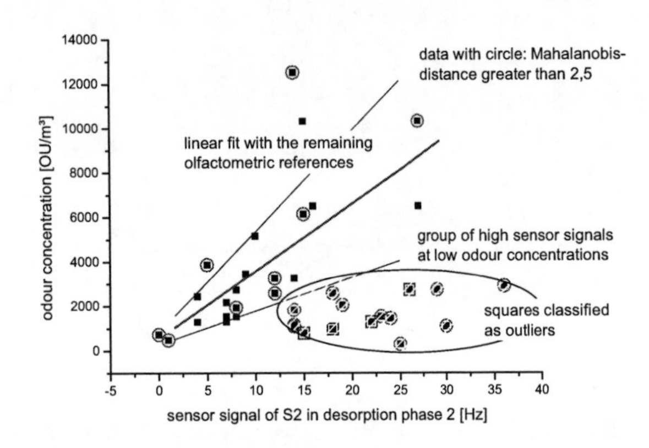

FIGURE 9. Linear fit of the remaining measurements after elimination of outliers via the Mahalanobis distance.

DISCUSSION

Although some progress has been made in the methodology of odour measurement, a lot of work is still to be done.

On the hardware side, the sensitivity and stability of the sensors needs improvement.

On the software or methodological side, it is not yet clear which mathematical method is best fitted to the particular problems of continuous odour measurement. The poor quality of odour reference measurements, as indicated by the high measurement uncertainty of the olfactometry (method EN 13725), prohibits the use of overly sophisticated fitting algorithms. We therefore suggest the use of simple linear fitting instead of higher order methods.

CONCLUSIONS

As there is no mandatory method for odour measurements with chemosensor systems it is desirable to establish some standards in this field. An expert workgroup organised by the German VDI/VDE is currently in the progress to formulate a guideline for chemosensor systems (so-called electronic noses).

ACKNOWLEDGMENTS

We thank the German Research Foundation (DFG) for the funding of the research.

REFERENCES

1. DiNatale C et al. (1996) Recognition of fish storage time by a metalloporphyrins-coated QMB sensor array. Meas. Sci. Technol. 7, No. 8, pp. 1103-1114
2. DiNatale C, Davide F, DAmico A, Nelli P, Groppelli S, Sberveglieri G (1996) An electronic nose for the recognition of the vineyard of a red wine. Sens. Actuator B-Chem. 33, No. 1-3, pp. 83-88
3. James D, Scott SM, O'Hare WT, Ali Z, Rowell FJ (2004) Classification of fresh edible oils using a coated piezoelectric sensor array-based electronic nose with soft computing approach for pattern recognition. Trans. Inst. Meas. Control 26, No. 1, pp. 3-18
4. Yuwono A, Boeker P, Schulze Lammers P (2003) Detection of odour emissions from a composting facility using a QCM sensor array. Anal. Bioanal. Chem. 375, No. 8, pp. 1045-1048
5. Nicolas J, Romain AC (2004) Establishing the limit of detection and the resolution limits of odorous sources in the environment for an array of metal oxide gas sensors. Sens. Actuator B-Chem. 99, No. 2-3, pp. 384-392
6. Stuetz RM, Fenner RA, Hall SJ, Stratful I, Loke D (2000) Monitoring of wastewater odours using an electronic nose. Water Sci. Technol. 41, No. 6, pp. 41-47
7. Persaud KC, Khaffaf SM, Hobbs PJ, Sneath RW (1996) Assessment of conducting polymer odour sensors for agricultural malodour measurements. Chem. Senses 21, No. 5, pp. 495-505
8. Hamacher T, Niess J, Lammers PS, Diekmann B, Boeker P (2003) Online measurement of odorous gases close to the odour threshold with a QMB sensor system with an integrated preconcentration unit. Sens. Actuator B-Chem 95, No. 1-3, pp. 39-45
9. Boeker P, Haas T (2007) The measurement uncertainty of olfactometry (Die Meßunsicherheit der Olfaktometrie). Gefahrst. Reinhalt. Luft 67, Nr. 7-8, S. 331-340

Improving Drift Correction by Double Projection Preprocessing in Gas Sensor Arrays

M. Padilla[1*], A. Perera[1], I. Montoliu[1], A. Chaudry[2], K. Persaud[3] and S. Marco[1]

[1] *Departament d'Electrònica. Universitat de Barcelona, Marti i Franquès 1, 08028 Barcelona, Spain/Institut de Bioenginyeria de Catalunya (IBEC), Baldiri i Reixac 13, 08028 Barcelona, Spain*
[2] *Protea Ltd, 11 Mallard Court, Mallard Way, Crewe Business Park, Crewe, Cheshire, CW1 6ZQ, UK*
[3] *School of Chemical Engineering and Analytical Science, The University of Manchester, PO Box 88, Sackville St, Manchester, M60 1QD, UK*

Abstract. It is well known that gas chemical sensors are strongly affected by drift. Drift consist on changes in sensors responses along the time, which make that initial statistical models for gas or odor recognition become useless after a period of time of about weeks. Gas sensor arrays based instruments periodically need calibrations that are expensive and laborious. Many different statistical methods have been proposed to extend time between recalibrations. In this work, a simple preprocessing technique based on a double projection is proposed as a prior step to a posterior drift correction algorithm (in this particular case, Direct Orthogonal Signal Correction). This method highly improves the time stability of data in relation with the one obtained by using only such drift correction method. The performance of this technique will be evaluated on a dataset composed by measurements of three analytes by a polymer sensor array along ten months.

Keywords: drift, Direct Orthogonal Signal Correction.
PACS: 82.47.Rs

INTRODUCTION

Instruments based on chemical gas sensors arrays plus data processing could be a cheaper and faster alternative for gas analysis than conventional analytic instruments like e.g. gas chromatographers. However their practical application is still limited. Among all the drawbacks that these instruments present, the main reason behind the limited application is, from the authors' point of view, the lack of time stability and the cost of recalibrations.

Gas sensor drift consists on a random temporal variation of the sensor response when it is exposed to the same gases under identical conditions. There are many different contributions to this lack of stability, like sensors aging, environmental, humidity variations, variations on flow rate, poisoning [1], etc. These factors can change both the baseline and the sensitivity of the sensors of an array in different ways, depending on the individual characteristics of every sensor. Therefore, on a long time operation the ability of the instrument to recognize analytes is degraded, since statistical models built in calibration phase become useless after a short period of time, in some cases weeks or few months. After that, the instrument must be completely re-calibrated, which is a time-consuming and expensive task, to ensure that the predictions remain valid. For this reason technological improvements or/and the use of some data processing for drift counteraction are required.

Several proposals are found in the literature in order to improve time stability by modifications in sensor technology and design [2] or by the use of different sensor operation modes [3]. Data processing methods for drift counteraction are based on very different approaches like Self Organizing Maps (SOM) [4], Wavelets decomposition [5], and linear methods like Component Correction (CC) [6], or Canonical Correlation Analysis (CCA) with Partial Least Squares (PLS) [7]. The application of such data processing methods extends the time of instrument usage between recalibrations, although these ones are still needed.

When drift is present, new measurements for the same composition appear shifted compared to their original location in feature space. Drift can also modify the structure of variance with respect to the one of the calibration set.

Non-adaptive signal processing techniques make a model of drift based on the calibration data set. Therefore, it is assumed that the effect of drift is time invariant. Consequently, the same correction applied on the calibration set is then applied on posterior data

CP1137, *Olfaction and Electronic Nose: Proceedings of the 13th International Symposium,* edited by M. Pardo and G. Sberveglieri

sets. However, it is known that drift is time variant. Therefore, continuous monitoring of the data is necessary and updating drift correction models with new data sets prevents the degradation of the correction. This is the base of adaptive algorithms for drift compensation. However, in some proposals, the quality of the adapted model usually degrades in time due to the effect sample misclassifications. Our proposal is based on an adaptation of the original calibration set to the new samples under analysis, instead of using recently classified samples as new calibration samples.

In this work, a very simple preprocessing technique, based on a double data projection (DP), is proposed to be used together with a linear method like Direct Orthogonal Signal Correction (DOSC) for drift correction. In this technique, the calibration model and the drift correction method have to be applied on the adapted versions of the data sets.

The performance of these drift correction techniques (DOSC and DP+DOSC) will be tested on a data set consisting on measurements of several analytes by a polymer sensor array during ten months.

METHODS AND EXPERIMENTAL

Double Projection

Double Projection is a preprocessing tecnique that adapts a training data set to the test data set that is to be classified. This method operates on data frame basis the size of which is given by the calibration set size. The test data frame is a time running window.

Let *XT* be the training data set matrix and *XV* the test matrix. *XT* and *XV* are *m* x *n*, being *m* the number of samples and n the dimension of feature space. This method consists on the projection of *XV* on *XT* to produce *XVp* (eq. 1). The projection matrix is made by using Moore-Penrose pseudoinverse (noted as X^{+}) (eq. 2). Next, data set *XT* is projected on the projected validation set *XVp,* resulting in *XTp*.

$$XVp=(XT\cdot XT^{+})\cdot XV \qquad (1)$$

$$XTp=(XVp\cdot XVp^{+)}\cdot XT \qquad (2)$$

Later, models for drift correction or/and classification are built from projected training set *XTp* and projected test set *XVp* is then processed.

Direct Orthogonal Signal Correction

The technique that will be used in this work for drift correction is Direct Orthogonal Signal Correction (DOSC) [7]. This method is a variant of the original Orthogonal Signal Correction (OSC) proposed firstly by Wold et al. [9] to be used use on NIR spectra correction. OSC and its variants are actually very commonly used in the field of Spectroscopy but they have found very little predicament for sensor arrays applications. The authors have been exploring the performance of OSC in this same dataset in a previous work [10].

All of OSC techniques are based on the removal of components of non-desired variance, for this reason we believe that all of them may be useful for drift compensation purposes. OSCs' methods main idea is to remove variance not correlated to variable/s in vector (or matrix) *Y*, which contains some extra data information. This is done by constraining the deflation of non-relevant information of *X,* so that only information orthogonal to *Y* should be removed. The inclusion of the condition of *orthogonality to Y* ensures that the signal correction removes as little information as possible.

The Dataset

The dataset on which the algorithm is validated has been measured at Osmetech plc. facilities. Three different gases at different concentrations levels were periodically measured by an array of 17 conducting polymer sensors during 10 months. The total number of samples in the dataset is 3362, which are assigned to 8 groups. The concentration and number of samples are shown in table 1.

TABLE 1. Measured compounds, concentrations and number of samples in the dataset.

Analyte	Samples	Concentration level
	447	0.01%
Ammonia	452	0.02%
	307	0.05%
	447	0.01%
Ammonia	452	0.02%
	307	0.05%
n-Buthanol	446	0.01%
	425	1.00%

For every sample, the full sensor sampling transient response is recorded. Every transient signal is measured at a sampling frequency of 1 Hz during 200s. At instant t=0s the analyte is introduced into the sensors chamber producing a change in every sensor signal, and at t=185s the analyte is removed and the sensors chamber is cleaned by introducing clean air into the chamber. All waveforms are baseline corrected so that starting baseline in all samples is common at time 0s. The complete transient waveform is introduced into the data matrix to consider also dynamic information, which provides additional discriminatory information [11]. To apply the proposed drift counteraction techniques, the data matrix is organized into a two-way matrix with dimensions: *number of samples x (sensors x transient time)* or 3362 samples x 3400 variables.

Methodology

To test the effectiveness of the drift counteraction techniques the complete dataset (3362 samples), has been divided in 9 sections (or subsets) of 369 samples each approximately, except first subset, which contains 410 samples. All samples are ordered in time. The first subset consists of samples measured during the first 16 days of experiments.

The calibration experiment has been repeated ten times. In each of them, 369 samples are randomly selected from the first subset, these ones forming the training or calibration set from which methods models will be built.

Time-stability evaluation is performed through validation with the remaining eight data subsets ordered along the time. Hence, usual statistical validation (test) techniques like k-fold, leave-one-out cross validation or random subsampling methods performed over the whole dataset are discarded, since they would use training samples from any time along 10 months, introducing information about the future to predict the past and thus providing over-optimistic and non real results.

For algorithms evaluation and comparison, a simple classifier has been used after data preprocessing with the selected methods. The classifier consists on a dimensionality reduction step using Principal Component Analysis, where dimensionality decreases from 3400 to 10 (capturing about 99% of total variability), and a k-NN classifier with k=3 nearest neighbors. The dataset has been classified in three classes, corresponding to the three chemical species. Time stability experiment has been repeated 10 times and the uncertainty in classification rate is given in standard deviation.

RESULTS

Figure 1 shows the evolution of the maximum value of the transient response of sensor 1, along time. The irregularity in the sensor response may be mostly due to sensors aging and also to the environmental changes that happen during the 10 months of measurements, since it is known that conducting polymer sensors are strongly affected by temperature and humidity changes. Therefore, in this figure the effect of drift and intra-class variability in the measurements is clearly seen, and also a clear correlation between the different traces is observed.

Figure 2 shows the PCA scores plot of the calibration set (solid symbols). In the figure, it can be seen that some classes apparently mix; there is scattering due to drift and also additional intra-class variability. The direction of the main dispersion component in all classes is quasi parallel, except ammonia 0.05%. Furthermore, in this data subset sensors are very correlated since only ten principal components captures more than 99% of the total variance. A posterior data subset of 369 samples, measured between days 118 and 150, has also been projected on the same PCA subspace (non-solid symbols). The distribution of the classes in this latter subset is similar to the one in the first subset; only ammonia shows an important deviation in its main dispersion directions. However, the whole subset is completely shifted from the location of the calibration set. Therefore, additional to variability and scattering due to short time noise/drift, there is a long-term drift effect that displaces and changes the data's structure of variance.

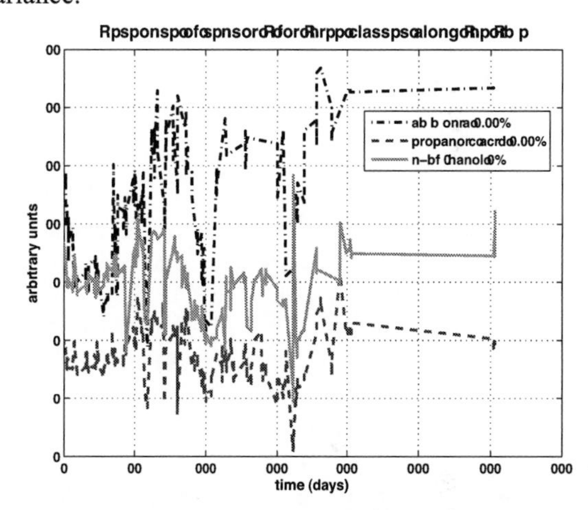

FIGURE 1. Responses of sensor 1 (maximum point of every transient signal) to three analytes along the time.

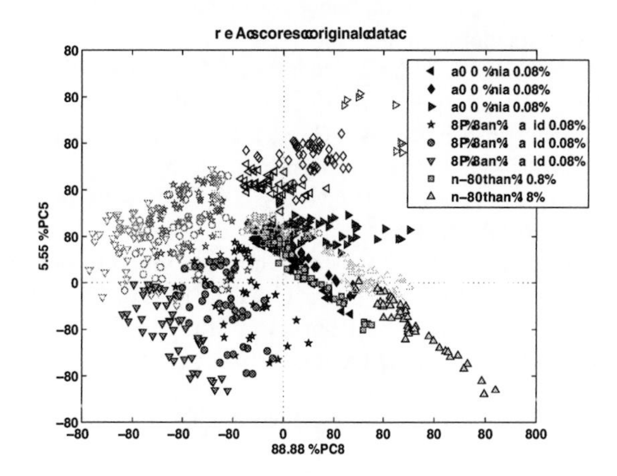

FIGURE 2. PCA scores of calibration data set (solid symbols) and a posterior data subset (samples from day 118 to 150) in non solid symbols.

After DP, DOSC is applied on the data, the PCA scores of resulting data are shown in figure 3. Here, the measurements are distributed in three lines according to the type of gas. Along every line, the corresponding analytes are grouped according to their concentration. The major sources of unwanted variance have dissapeared including drift, but some dispersion still remains.

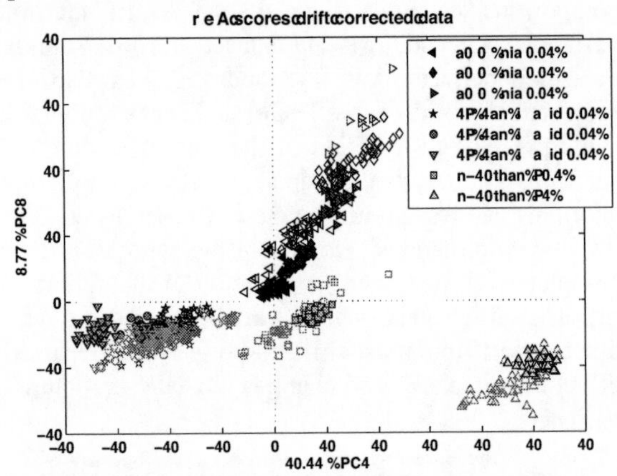

FIGURE 3. PCA scores of calibration data set (solid symbols) and a posterior drift corrected data subset (samples from day 118 to 150) in non-solid symbols.

To apply DOSC, it is necessary to define an information matrix Y. This Y matrix is a partition matrix. It consists on a binary matrix with dimensions *number of samples x number of groups* (369 samples x 8 groups) where a *'1'* placed at position *(i,j)* in matrix Y means that sample *i* belongs to class *j*.

It is also necessary to select a number of components to be extracted and a tolerance as a measurement of how strict the calculus of such components is. In general OSC methods tend to overfit, for this reason it is advised to choose few components to be extracted and not too high tolerance. We have chosen tolerance of 99% and number of extracted components 1.

Results of classification rates (CR) before and after the application of the drift correction methods are shown in figure 4. Here, it can be seen an improvement in the recognition of the type of gas when data is corrected by DOSC, but only until around day 70. After that day DOSC is totally unable to compensate drift. However, when applying DP before DOSC, results show better CR than for DOSC alone, not only until day 70 but even after. In fact CR is above 90% all along the time of measurements.

CONCLUSIONS

In this work a new technique (DP) is proposed to improve the performance of a drift correction method (DOSC). In the case shown in this work, the application of DP previously to DOSC not only improves the performance of DOSC, but it also compensates drift for longer periods of time. DP is thought for its use together with any other drift correction technique, especially those in which undesired components of variance are extracted, like Component Correction, OSC, etc.

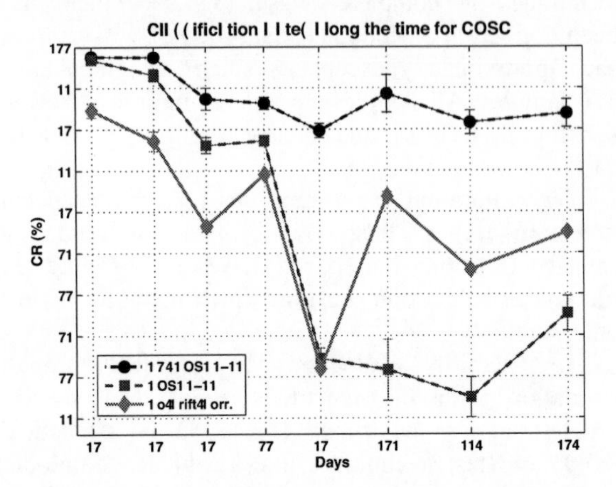

FIGURE 4. Classification rates along the time for drift correction and non-corrected data. Calibration set contains 369 samples.

ACKNOWLEDGMENTS

We want to acknowledge the partial funding of this work by NoE GOSPEL IST 507610 Fp6.

REFERENCES

1. P. Mielle, Sens and Act B, 34 (1996) 533
2. G. Muller, A. Friedberger, P. Kreisl, S. Ahlers, O. Schulz, T. Becker, Thin Solid Films, 436 (2003) 34
3. A. Gramm, A. Schutze, Sens and Act B, 95 (2003) 58
4. S. Marco, A. Ortega, A. Pardo, J. Samitier, IEEE Trans. on Inst and Meas., 47 (1998) 316
5. M. Zuppa, C. Distante, P. Siciliano, K.C. Persaud, Sens and Act B, 98 (2004) 305
6. T. Artursson, T. Eklov, I. Lundstrom, P. Martensson, M. Sjostrom, M. Holmberg, J. of Chem, 14 (2000) 711
7. R. Gutierrez-Osuna. Proceedings of the 7th ISOEN, Brighton, UK, July 20-24, 200.
8. J.A. Westerhuis, S. de Jong, A.K. Smilde, Chem. And Int. Lab. Sys. 56 (2001) 13
9. S. Wold, H. Antti, F. Lindgren, J. Ohman, Chem. And Int. Lab. Sys. 44 (1998) 175.
10. M. Padilla, A. Perera, I. Montoliu, A. Chaudry, K. Persaud and S. Marco. Proceedings of the 11th Conference on Chemometrics in Analytical Chemistry, Montpellier, France, July 1-4, 2008.
11. J. Samitier, J.M. Lopezvillegas, S. Marco, L. Camara, A. Pardo, O. Ruiz, J.R. Morante, Sensors and Actuators B-Chemical, 18 (1994) 308.

MS-Electronic Nose Performance Improvement Using GC Retention Times And 2-Way And 3-Way Data Processing Methods

Cosmin Burian[1], Jesus Brezmes [1,2], Maria Vinaixa [1,2], Eduard Llobet[1], Xavier Vilanova[1], Nicolau Cañellas[1] and Xavier Correig [1,2]

[1]*Department of Electronic Engineering, Universitat Rovira i Virgili, Avenida Paisos Catalans 26, 43007 Tarragona, Spain*
[2]*CIBER de Diabetes y Enfermedades Metabólicas Asociadas (CIBERDEM)*
cosmin.burian@urv.cat, jesus.brezmes@urv.cat

Abstract. We have designed a challenging experimental sample set in the form of 20 solutions with a high degree of similarity in order to study whether the addition of chromatographic separation information improves the performance of regular MS based electronic noses. In order to make an initial study of the approach, two different chromatographic methods were used. By processing the data of these experiments with 2 and 3-way algorithms, we have shown that the addition of chromatographic separation information improves the results compared to the 2-way analysis of mass spectra or total ion chromatogram treated separately. Our findings show that when the chromatographic peaks are resolved (longer measurement times), 2-way methods work better than 3-way methods, whereas in the case of a more challenging measurement (more coeluted chromatograms, much faster GC-MS measurements) 3-way methods work better.

Keywords: multi-way; PCA; PARAFAC; gas chromatography mass spectrometry; .
PACS: 02.50.Sk; 82.80.Bg; 82.80.Ms

INTRODUCTION

In this paper it is our intention, to evaluate whether a GC-MS configuration can improve the results of a standard MS-Enose using only a small fraction of the measurement time required in a typical GC-MS run (2-3 minutes versus 30 minutes or more).

To do so, m/z (mass spectra information) variation along a retention time axis is collected so that a final 3D matrix is created. In this matrix, each file is related to the measurement of a sample. Each sample measurement data is laid out in a plane with 2 axis: the time axis (columns) and the m/z axis ("tubes").

To process the 3D data matrix obtained, new 3D data analysis methods like PARAFAC [1,2], Tucker [3] , and N-PLS have been considered due to their proven advantages in different areas, such as spectroscopy [1], food chemistry [4] and environmental studies [5]. In these applications, these algorithms have been successfully employed to interpret multi-way data sets. It is our intention to compare, study and evaluate these algorithms in the experimental setup we have designed so that this work could be referenced as a case study for an electronic nose based on a GC-MS configuration.

Multiway data analysis, originating in psychometrics back in the sixties [6], is the extension of two-way data analysis to higher-order datasets, and is often used for extracting hidden structures and capturing underlying correlations between variables in a multiway array.The difference between two-way and multiway data analysis is the format of the data being analyzed, in which the former are higher-order generalizations of vectors and matrices.

EXPERIMENTAL

Twenty mixtures of nine isomers of dimethylphenol and ethylphenol were measured and analyzed by means of gas chromatography mass spectrometry (GC-MS). The nine isomers were chosen based on their theoretically similar mass spectra in order to have a challenging data set.

To design the experiment we looked at 2 key issues: the PCA of the 9 isomers mass spectra and their chromatographic retention times. To obtain the

CP1137, *Olfaction and Electronic Nose: Proceedings of the 13th International Symposium,* edited by M. Pardo and G. Sberveglieri

chromatographic retention time (figure 1), 9 solutions of 1% isomer in methanol were prepared and analyzed. In order to see which of the 9 isomers has most alike mass spectra we calculated a PCA of the theoretical mass spectra of each isomer (figure 2).

Based on this information we designed the experiment as shown in **Table 1.** Benzene acts as an internal standard, and having the highest concentration it is used in the normalization pretreatment of the data.

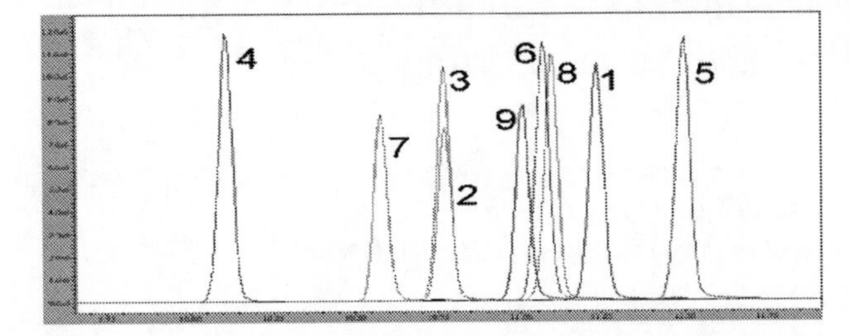

FIGURE 1 Chromatographic retention times for the 9 isomers as follows: 1) 2,3- diMe-phenol; 2) 2,4- diMe-phenol; 3) 2,5- diMe-phenol; 4) 2,6- diMe-phenol; 5) 3,4-diMe-phenol; 6) 3,5- diMe-phenol; 7) 2-Et-phenol; 8) 3- Et-phenol and 9) 4- Et-phenol; Isomers 2 and 3 and 6,8,9 are the most similar among them.

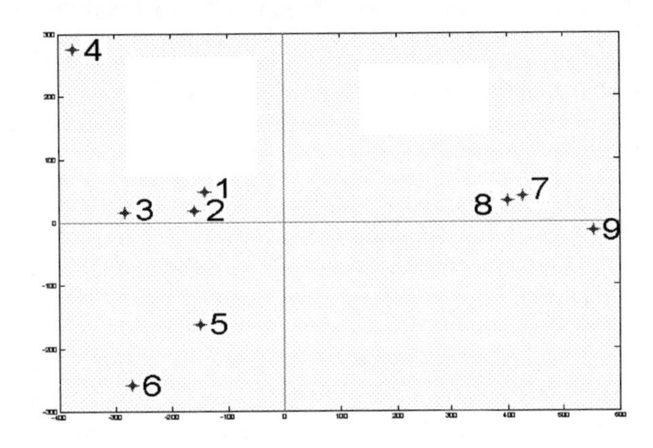

FIGURE 2 PCA of the mass spectra of the 9 isomers. Isomers 7, 8, 9 and 1, 2, 3 are the most similar among them

Two chromatographic methods were studied. Method one aimed towards well-resolved chromatograms, which are supposed to be easier to analyze. A temperature programmed separation was used, starting at 50°C, where the temperature was kept constant for one minute, until 180°C, where almost all the isomers were separated. Method two was designed to return coeluted peaks and therefore it was performed with an isothermal temperature of 190ºC. This method was supposed to give a more challenging dataset and to be executed in a much shorter time. The measurements were conducted through syringe injection of 1 μl per measurement, and ten repetitions for each method and solution were made.

RESULTS

The two resulting three-way (3D) data matrices (one for each method) were used for data analysis.

Data Processing

Peak alignment, mean centering and normalization were the pre-processing steps applied to the dataset matrices generated. For peak alignment, Recursive Alignment through Fast Fourier Transform (RAFFT) [7, 8] was employed, aligning the data in the initial three dimensional matrix. The alignment was done for each m/z chromatogram separately.

The normalization was made in two ways: between 0 and 1 and by internal standard, the benzene peak area. Finally the data was mean centered following the sample direction in all 5 sets.

These approaches allowed us to perform a complete study and comparison between each type of preprocessing used: Alignment or not aligned, 0 to 1 normalization, benzene normalization or not normalized and mean centered or not mean centered.

In order to analyze the 3D data matrix with two-way methods (such as PCA, PLS and PLS-DA), different ways of converting the dataset into a 2D matrix were studied. In 2D matrices sample measurement data is arranged in different files and each column represents different descriptors or variables.

First, to obtain the average mass spectra (AVMS) of each sample (MS matrix) from the 3D matrix we added the axis of the chromatographic separation time.

TABLE 1. Experiment design (the numbers represent percentage of substance against methanol)

component	Sol1	Sol2	Sol3	Sol4	Sol5	Sol6	Sol7	Sol8	Sol9	Sol10	Sol11	Sol12	Sol13	Sol14	Sol15	Sol16	Sol17	Sol18	Sol19	Sol20
2,3-Dimethylphenol	-	0.5	0.5	0.5	0.5	0.5	0.5	0.5	0.5	0.25	0.5	0.5	0.5	0.5	0.5	0.5	0.25	0.5	0.25	0.5
2,4-Dimethylphenol	0.5	-	0.5	0.5	0.5	0.5	0.5	0.5	0.5	0.5	0.25	0.5	0.5	0.5	0.5	0.5	0.5	0.25	0.5	0.25
2,5-Dimethylphenol	0.5	0.5	-	0.5	0.5	0.5	0.5	0.5	0.5	0.5	0.5	0.25	0.5	0.5	0.5	0.5	0.5	0.5	0.5	0.5
2,6-Dimethylphenol	0.5	0.5	0.5	-	0.5	0.5	0.5	0.5	0.5	0.5	0.5	0.5	0.5	0.5	0.5	0.5	0.5	0.5	0.5	0.5
3,4-Dimethylphenol	0.5	0.5	0.5	0.5	-	0.5	0.5	0.5	0.5	0.5	0.5	0.5	0.5	0.5	0.5	0.5	0.5	0.5	0.5	0.5
3,5-Dimethylphenol	0.5	0.5	0.5	0.5	0.5	-	0.5	0.5	0.5	0.5	0.5	0.5	0.25	0.5	0.5	0.5	0.5	0.5	0.5	0.5
2-Ethylphenol	0.5	0.5	0.5	0.5	0.5	0.5	-	0.5	0.5	0.5	0.5	0.5	0.5	0.25	0.5	0.5	0.25	0.25	0.5	0.5
3-Ethylphenol	0.5	0.5	0.5	0.5	0.5	0.5	0.5	-	0.5	0.5	0.5	0.5	0.5	0.5	0.25	0.5	0.5	0.5	0.25	0.25
4-Ethylphenol	0.5	0.5	0.5	0.5	0.5	0.5	0.5	0.5	-	0.5	0.5	0.5	0.5	0.5	0.5	0.25	0.5	0.5	0.5	0.5
B/Et-OH/acetone	2%	2%	2%	2%	2%	2%	2%	2%	2%	2%	2%	2%	2%	2%	2%	2%	2%	2%	2%	2%

On the other hand, summing all the m/z hits at a given scan we built the total ion chromatogram (TIC) and its corresponding matrix, the GC matrix.

By summation, the extra information brought by the third dimension is lost. In order to compensate for this and still allow the data to be treated with two-way methods two approaches were used: unfolding (UF matrix) and concatenation (MSGC matrix). Unfolding is done by taking the mass spectra of each chromatographic scan and pasting it where the previous scan ended. On the other hand concatenation keeps the extra information by concatenating the AVMS with the TIC in a single file of the 2D matrix.

2-D datasets were analyzed by Principal Component Analysis (PCA), Partial Least Squares Discriminant Analysis (PLS-DA), while the 3-D dataset was analyzed by Parallel Factor Analysis (PARAFAC), and multi-way PLS-DA.

PCA And PARAFAC Results

The PCA and PARAFAC results were evaluated using a clusterization merit figure (based on the relationship between intra-class and inter-class distances, higher meaning better clustering in the PCA or PARAFAC scores graph) and by success classification rates using a fuzzy ARTMAP neural network.

As expected, method one proved easier to classify than method three. Best classification results were obtained for method one (where the peaks are more resolved), with the MSGC concatenated matrix when the data is aligned, 0 to 1 normalized and mean centered. By contrast method 3, the most difficult one from the point of view of chromatographic separation (peaks more coeluted), gives better results using PARAFAC when the data is aligned, normalized between 0 and 1 and not mean centered.

Aligning the data for chromatographic shift and normalizing it between 0 and 1 improves classification accuracy almost in all of the cases. An interesting result is that in the case of a less challenging method (method 1, the most resolved GC) 2D algorithms works best, whereas in a more complicated case such as method 2 (coeluted peaks, shortest measurement time) they lose effectiveness in front of 3D methods.

For the Fuzzy Artmap classification of the PCA (2D) and PARAFAC (3D) coordinates the data was separated into training and testing sets, using a 50-50 training-testing ratio. After applying PCA and PARAFAC for the training set the scores were normalized between 0 and 1 and fed into a fuzzy ARTMAP neural network for training. The test set was then projected on to the already created PCA/PARAFAC model, and the scores were also normalized between 0 and 1 and used as the test set for the fuzzy ARTMAP neural network.

		Method 1		Method 2		
		unalign	align	unalign	align	
unprocessed	nunorm	1.50	1.01	0.71	1.29	3D
		1.03	1.03	0.94	0.94	MS
		1.81	2.13	1.06	1.65	GC
		1.39	1.24	0.96	1.18	MSGC
		1.85	2.18	1.01	1.64	UF
	0 to 1 norm	2.22	11.71	0.88	8.55	3D
		6.63	6.63	7.46	7.48	MS
		2.12	12.11	1.58	6.82	GC
		2.11	12.74	1.58	6.98	MSGC
		2.23	11.56	1.51	5.53	UF
	benzene norm	1.73	4.33	1.38	3.15	3D
		1.97	1.98	1.63	1.63	MS
		2.05	4.90	1.24	2.30	GC
		1.95	2.56	1.31	1.84	MSGC
		2.12	4.91	1.18	2.22	UF
mean centered	nunorm	1.90	3.78	0.26	3.07	3D
		1.03	1.03	0.94	0.94	MS
		1.87	2.18	1.07	1.66	GC
		1.41	1.26	0.96	1.18	MSGC
		1.96	2.29	1.02	1.64	UF
	0 to 1 norm	2.36	11.60	0.36	5.50	3D
		5.16	5.15	7.03	7.03	MS
		2.13	21.60	1.55	6.81	GC
		2.13	21.60	1.55	6.86	MSGC
		2.19	10.88	1.54	3.01	UF
	benzene norm	2.16	8.44	0.88	2.98	3D
		1.97	1.98	1.63	1.63	MS
		2.07	6.49	1.24	2.31	GC
		2.01	2.64	1.32	1.84	MSGC
		2.13	5.90	1.19	2.23	UF

FIGURE 3 Cluster analysis of 3D PARAFACT and 2D PCA of MS, GC, MSGC, and UF data. Intervariance-intravariance rapport. Higher values represent better clustering in the PCA-PARAFAC graph

The results were judged by means of success rate of the confusion matrix (figure 4). The results show that method 1 reaches better results when two-way analysis methods are employed while method 2 gives better results with three-way data analysis. For method 1, the best results were obtained by the PCA of the MSGC concatenated matrix when the data was aligned, mean centered and normalized between 0 and 1, scoring 100% success rate, followed by GC data with 96.36% success rate. In the case of the second method the best result was obtained using PARAFAC when the data was aligned, normalized from 0 to 1 and not mean centered, as well as using a PCA projection of the MSGC matrix when the data was just aligned.

PLS-DA And n-PLS-DA Results

The model and prediction performance were evaluated by means of the Root Mean Square Error of Cross-Validation (RMSECV), and the Root Mean Square Error of Prediction (RMSEP), respectively. RMSECV and RMSEP represent cross-validation error and prediction error respectively.

		Method 1 unalign	Method 1 align	Method 2 unalign	Method 2 align	
unprocessed	nunorm	69.09	80.00	30.91	74.55	3D
		80.00	72.73	83.64	81.82	MS
		34.55	96.36	16.36	54.55	GC
		60.00	92.73	18.18	90.91	MSGC
		65.45	72.73	21.82	38.18	UF
	0 to 1 norm	45.45	94.55	9.09	90.91	3D
		81.82	80.00	85.45	74.55	MS
		56.36	96.00	29.09	56.36	GC
		36.36	92.73	18.18	49.09	MSGC
		63.64	87.27	21.82	49.09	UF
	benzene norm	78.18	89.09	41.82	74.55	3D
		85.45	63.64	76.36	83.64	MS
		49.09	98.18	32.73	52.73	GC
		60.00	92.73	38.18	60.00	MSGC
		72.73	90.91	16.36	41.82	UF
mean centered	nunorm	76.36	94.55	63.64	72.73	3D
		29.09	27.27	41.82	30.91	MS
		61.82	63.64	40.00	38.18	GC
		49.09	67.27	16.36	70.91	MSGC
		50.91	69.09	36.36	32.73	UF
	0 to 1 norm	89.09	94.55	56.36	83.64	3D
		70.91	74.55	74.55	80.00	MS
		61.82	96.36	54.55	60.00	GC
		76.36	100.00	56.36	80.00	MSGC
		69.09	90.91	78.18	63.64	UF
	benzene norm	72.73	87.27	63.64	83.64	3D
		34.55	54.55	54.55	41.82	MS
		80.00	89.09	45.45	47.27	GC
		83.64	85.45	38.18	52.73	MSGC
		58.18	81.82	34.55	50.91	UF

FIGURE 4 Fuzzy Artmap classification success rate for PARAFAC (3D) and PCA (MS, GC, MSGC, UF)

The results present the sample prediction success rate for methods one and two for solutions 10 to 20, and are shown in figure 5.

We notice that method 2, in which peaks are more coeluted, gives worse results than method 1, where a temperature programmed separation was employed in order to achieve a better separation. We can clearly notice this in the chromatographic data sample success rate.

Even though the average mass spectra gives the best performance when classifying the samples, the combination of this information with the total ion chromatogram by means of unfolding or unifying the MS data with the TIC data almost always gives better results than MS or TIC data alone (specially in lower latent variable models). Between the unfolded data and the unified MSGC data the later one is appearing to give better results and presents the advantage of an easier model interpretation.

The multi-way PLS model seems to present consistent results on both methods, yielding a high prediction success rate in both datasets. The n-PLS algorithm presented the best results in the most difficult case, the second method, solutions 10 to 20, giving 100% success rate classification for 13 latent variables (figure 5).

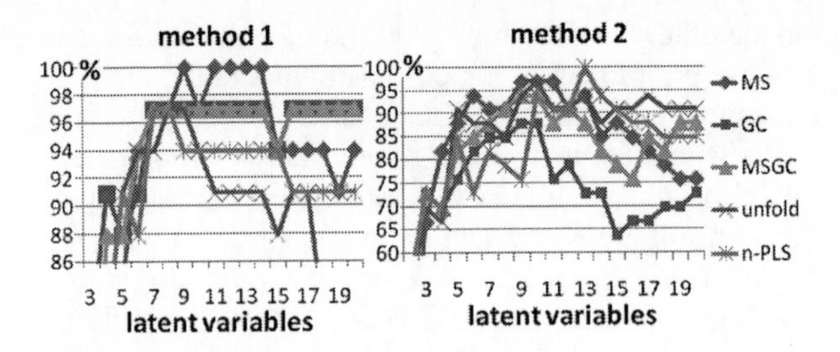

FIGURE 5 Prediction success rate for method 1 and 2

Conclusions

The experiment has shown that the addition of data from a chromatographic separation improves the results compared to using the Mass Spectra alone in most of the cases.

Treating the data with 2-way (PCA and PLS-DA) and 3-way (PARAFAC and n-PLS-DA) methods showed that working with a well resolved chromatogram 2-way methods work best, whereas for coeluted peaks (shorter GC runs) three-way methods are better suited.

From the point of view of data pretreatment, alignment is an important step, removing the possibility of sample classification based on signal drift or injection error.

Acknowledgments

This work has been funded by Ciberdem, The CIBER de Diabetes y Enfermedades Metabólicas Asociadas es an ISCIII initiative.

REFERENCES

1. R Bro, *Chemometrics and Intelligent Laboratory Systems* **38** 149-171 (1997)
2. N.Fabera, R.Brob, P.Hopke, *Chemometrics and Intelligent Laboratory Systems* **65** 119– 137 (2003)
3. P. Geladi, *Chemometrics And Intelligent Laboratory Systems* **7** 11-30 (1989)
4. V. Pravdova, C. Bouconb, *Analytica Chimica Acta* **462** 133–148 (2002)
5. P. Barbieri, G. Adami, *Chemometrics and Intelligent Laboratory Systems* **62** 89– 100 (2002)
6. Tucker, L. R. *The extension of factor analysis to three-dimensional matrices. In Contributions to Mathematical Psychology*. New York: Holt, Rinehart and Winston, 1964 , pp. 110-182
7. Wong, J.W.H, Cagney, G. and Cartwright, H.M. *Bioinformatics*, **21** 2088-2090 (2005)
8. Wong, J.W.H. and Cartwright, H.M. *Analytical Chemistry*, **77** 5655-5661 (2005)

Three-Dimensional Statistical Gas Distribution Mapping in an Uncontrolled Indoor Environment

Matteo Reggente and Achim J. Lilienthal

AASS Research Center - Learning Systems Lab
Örebro University - Sweden
E mail: matteo.reggente@oru.se, achim@lilienthals.de

Abstract. In this paper we present a statistical method to build three-dimensional gas distribution maps (3D-DM). The proposed mapping technique uses kernel extrapolation with a tri-variate Gaussian kernel that models the likelihood that a reading represents the concentration distribution at a distant location in the three dimensions. The method is evaluated using a mobile robot equipped with three "e-noses" mounted at different heights. Initial experiments in an uncontrolled indoor environment are presented and evaluated with respect to the ability of the 3D map, computed from the lower and upper nose, to predict the map from the middle nose.

Keywords: 3D-gas distribution, e-nose, gas sensing, mobile robots, kernel density estimation, model evaluation.
PACS: 01.30.Cc

1. INTRODUCTION

An increased quality of environmental monitoring is desired to protect the environment from toxic contaminants released into the air by vehicle emissions, power plants, refineries, to name but a few. Monitoring urban environments is typically done using immobile monitoring stations. Their total number and thus the number of sampling locations is limited by economical and practical constraints. Thus, the selection of monitoring/sampling locations becomes very critical, especially considering the time-varying, complicated local structure of the gas distribution. A further disadvantage of stationary air monitoring is that the monitoring stations are typically placed at expected "hot spots", close to busy roads, for example, and accordingly "background areas" are not monitored [1]. These issues can be addressed by mobile robots equipped with an "electronic nose", a combination we refer to as a mobile nose or "m-nose". An m-nose can act as a wireless node in a sensor network. With its self-localization capability and the ability to adaptively select sampling locations, m-noses offer a number of important advantages, among others: monitoring with higher resolution, the possibility of source tracking, integration into existing application, compensation for inactive sensors, and adaption to dynamic changes in the environment. Using mobile robots for air quality monitoring is addressed in the EU project DustBot, for example, in which robot prototypes are developed to clean pedestrian areas and concurrently monitor the pollution levels [2].

Gas distribution modelling is the task of deriving a truthful representation of the observed gas distribution from a set of spatially and temporally distributed measurements of relevant variables, foremost gas concentration (as used in this paper), but also pressure, and temperature, for example. Building gas distribution models is very challenging. One main reason is that in many realistic scenarios gas is dispersed chaotically by turbulent advection, resulting in a concentration field that consists of fluctuating, intermittent patches of high concentration [3]. In principle, CFD (Computational Fluid Dynamics) models can be applied, which try to solve the governing set of equations numerically. However, CFD models are computationally very expensive. They become intractable for sufficiently high resolution in typical real world settings and depend sensitively on accurate knowledge of the state of the environment, which is not available in practical situations. Here, we instead opt an alternative approach to gas distribution modelling and create a statistical model of the observed gas distribution, treating gas sensor measurements as random variables. Our approach creates a statistical model discretized to a grid map and it is "parameter-free" in the sense that it makes no assumptions about a particular functional form of the gas distribution. Previous approaches to statistical gas distribution mapping with mobile robots were largely restricted to mapping a 2D slice, parallel to the floor and level with the gas sensors on the robot (Sec. 2). The major contribution of this paper is the extension of Kernel extrapolation distribution mapping to three dimensions and its evaluation based on real world experiments in an uncontrolled indoor environment. This is an important step for gas distribution modelling since the gas distribution structure is essentially three dimensional. After a discussion of related work in the next section and the description of the hardware and set-up used for the monitoring trials (Sec. 3), we outline the 3D distribution mapping

CP1137, *Olfaction and Electronic Nose: Proceedings of the 13th International Symposium,* edited by M. Pardo and G. Sberveglieri

algorithm in (Sec. 4). Finally, we present first results and end with conclusions and suggestions for future work.

2. RELATED WORK

In urban environments, especially in areas with high population and traffic density, human exposure to hazardous substances is often significantly increased. High pollution levels exceeding air quality standards, have been observed in street canyons [4], for example. In a natural environment advective flow generally dominates gas dispersal compared to slow molecular diffusion. Since the airflow is almost always turbulent, the gas distribution becomes patchy and meandering [5]. As pointed out in the review of Vardoulakis et al. [4], just a few approaches to environmental monitoring with immobile sensing stations consider the complicated local structure of gas distribution. Acknowledging the need to refine the monitoring scale, Maruo et al. developed small inexpensive gas sensors for air pollution monitoring [6]. Addison et al. [7] propose a method for predicting the spatial pollutant distribution in a street canyon based on a stochastic Lagrangian particle model superimposed on a known velocity and turbulence field.

Gas distribution mapping with mobile gas sensors has been implemented and investigated by mobile robots equipped with an "e-nose" [8, 9, 10, 11, 12]. These approaches can be divided into two groups. Model-based approaches such as the one proposed by Ishida et al. [8] assume a particular model of the time-averaged gas distribution and estimate the corresponding parameters. Model- or parameter-free approaches deal with the fluctuating nature of the gas distribution either by recording individual concentration samples over a prolonged time (several minutes) [11, 12] or by statistically integrating subsequent measurements into a spatial grid [9, 10].

All the above mentioned approaches produce 2D gas distribution maps. In the field of mobile robot olfaction, the three-dimensionality of the environment is only taken into account in a few publications on gas source tracing [13, 14, 15]. Three-dimensional gas distribution mapping with a mobile robot has not been investigated before to the best of our knowledge.

3. EXPERIMENTAL SET-UP

An ATRV-JR robot equipped with a SICK LMS 200 laser range scanner (for localization) and three "electronic noses" was used for the monitoring experiments. The "electronic noses" comprise different Figaro 26xx gas sensors enclosed in an aluminum tube. These tubes are horizontally mounted at the front side of the robot at a height of 10 cm, 60 cm and 110 cm and actively ventilated through a fan that creates a constant airflow towards the gas sensors. This lowers the effect of external airflow or the movement of the robot on the sensor response and guarantees continuous exchange of gas in situations with very low external airflow. In this work, we address the problem of modeling the distribution from a single gas source. With respect to this task, the response of the different sensors in the electronic nose is highly redundant and thus it is suffcient to consider the response of a single sensor (TGS 2620) only.

The scenario selected for the gas distribution mapping experiments is to monitor an area of approx. $10 \times 3\text{m}^2$ in a long corridor with open ends and a high ceiling. This choice was motivated by the goal to monitor uncontrolled environments and even pedestrian areas. During our monitoring trials there was disturbance caused by people passing by and by the opening of doors and windows. The gas source was a small cup filled with ethanol or acetone. This source was placed roughly in the middle of the investigated corridor segment at a height of 1.6 m to prevent the robot from colliding with the source and ensure a substantially 3D gas distribution with the chosen analytes (which are heavier than air: ethanol $\approx$ 46 g/mol and acetone $\approx$ 58 g/mol). As a possible monitoring strategy, the robot followed either a random walk trajectory or a predefined sweeping path to cover the area of interest, using a fixed starting point.

In order to be able to relate the readings of the different electronic noses to each other, we perform a simple calibration by determining the baseline (response to clean air) and the maximum response in the actual experimental enviroment (but not in a controlled set-up) with the three e-noses positioned very close to each other. This is done by recording the respective minimum values R^n_{min} (baseline) and the maximum values R^n_{max} after a cup filled with the analyte was opened close to the noses. In the subsequent monitoring trial the raw readings R^n_i from nose n are scaled as

$$r_i = \frac{R^n_i - R^n_{min}}{R^n_{max} - R^n_{min}}. \tag{1}$$

Thus, we make the assumptions that each sensor was exposed to the same minimum and maximum concentration during the calibration process and that the sensors' response depends on the concentration in the same, monotonous way. Since the calibration is repeatedly carried out in the same environment and under the same conditions as the actual experimental runs, we avoid issues with long-term drift and mitigate drift issues due to different temperature and humidity in the trials.

4. 3D GAS DISTRIBUTION MODEL

In this section we introduce the basic ideas of the 3D Kernel GDM algorithm extending the 2D model of Lilienthal et al. [16], and describe briefly the Kernel DM+V algorithm [16] that models the distributions mean and the corresponding variance. The gas distribution mapping problem addressed here is to learn a predictive three dimensional model $p(r|\mathbf{x}, \mathbf{x_{1:n}}, \mathbf{r_{1:n}})$ for the gas concentration r at location $\mathbf{x}$, given the robot trajectory $\mathbf{x_{1:n}}$ and the corresponding concentration measurements $\mathbf{r_{1:n}}$. We consider the case of a single target gas, but in principle the proposed method can be extended to the case of multiple different odor sources as described in [17]. We also assume perfect knowledge about the position $\mathbf{x_i}$ of a sensor at the time of the measurement. To account for the uncertainty about the sensor position, the method in [18] could be used. To study how gas distribution in three dimensions we consider the concentration readings from multiple "e-noses" mounted at different heights. The central idea of kernel extrapolation methods is to understand gas distribution mapping as a density estimation problem that involves convolution with a kernel.The first step in the Kernel DM+V algorithm is the computation of the weights $\omega_i^{(k)}$, which represent the importance of each sensors measurement i at grid cell k:

$$\omega_i^{(k)}(\sigma_x, \sigma_y, \sigma_z) = \mathcal{N}(|x_i - x^{(k)}|, \sigma_x, \sigma_y, \sigma_z). \quad (2)$$

The weights are computed using a multivariate 3D-Gaussian kernel $\mathcal{N}$ evaluated at the distance between the location of the measurement x_i and the center $x^{(k)}$ of cell k. We use a diagonal covariance matrix Σ with elements σ_x, σ_y, σ_z, which defines the kernel extension along the three axis. Using Eq. 2, weights $\omega_i^{(k)}$, weighted sensor readings $\omega_i^{(k)} \cdot r_i$ and weighted variance contribution $\omega_i^{(k)} \cdot \tau_i$ are integrated and stored in temporary grid maps:

$$\Omega^{(k)} = \sum_{i=1}^{n} \omega_i^{(k)}, \quad R^{(k)} = \sum_{i=1}^{n} \omega_i^{(k)} \cdot r_i, \quad V^{(k)} = \sum_{i=1}^{n} \omega_i^{(k)} \cdot \tau_i, \quad (3)$$

$$\tau_i = (r_i - r^{k(i)})^2. \quad (4)$$

τ_i is the variance contribution of reading i and $r^{k(i)}$ is the model prediction from the cell $k(i)$ closest to the measurement point x_i. From the integrated weight map $\Omega^{(k)}$ we compute a confidence map $\alpha^{(k)}$, which indicates high confidence for cells if the estimate can be based on a large number of readings recorded close to the center of the respective grid cell:

$$\alpha^{(k)}(\sigma_x, \sigma_y, \sigma_z) = 1 - e^{-\frac{(\Omega)^{(k)}(\sigma_x, \sigma_y, \sigma_z)}{\sigma_\Omega^2}}. \quad (5)$$

σ_Ω is a scaling parameter that defines a soft margin which decides rather the estimate for a cell has high confidence or low confidence. By normalizing the map of weighted readings $R^{(k)}$ to $\Omega^{(k)}$ and linear blending with the best guess for the case of low confidence, we finally obtain the map estimates of the mean $r^{(k)}$ and the corresponding variance map $v^{(k)}$ as

$$r^{(k)}(\sigma_x, \sigma_y, \sigma_z) = \alpha^{(k)} \frac{R^{(k)}}{\Omega^{(k)}} + \{1 - \alpha^{(k)}\}\bar{r}, \quad (6)$$

$$v^{(k)}(\sigma_x, \sigma_y, \sigma_z) = \alpha^{(k)} \frac{R^{(k)}}{\Omega^{(k)}} + \{1 - \alpha^{(k)}\}v_{tot}. \quad (7)$$

The second terms in the equations are the best estimate for cells with a low confidence. $\bar{r}$ represents an estimate of the mean concentration for cells for which we do not have sufficient information from nearby readings, indicated by a low value of $\alpha(k)$. We set $\bar{r}$ to be the average over all sensor readings. The estimate v_{tot} of the distribution variance in regions far from measurement points is computed as the average over all variance contributions. The 3D Kernel-GDMV algorithm depends on seven parameters: the kernel widths $\sigma_x, \sigma_y, \sigma_z$, that govern the amount of extrapolation on individual readings according to three axis and the cell sizes c_x, c_y, c_z that determines the resolution at which different predictions can be made and σ_Ω.

5. RESULTS

Qualitative Comparison: In order to evaluate how well the model captures the true properties of the gas distribution we use two of the three noses (the lower and the upper one, see Sec. 3), to build a 3D model using the method described in the previous section. After that we slice the model and extract the layer corresponding to the height of the remaining middle nose ("3D@60cm"). From the readings of the middle nose we also build a 2D gas distribution map ("2D@60cm") and we compare it with the slice extracted from the 3D model. This evaluation method is visualized in Fig. 1. The first two maps at the top represent the mean distribution according to the models "2D@60cm" and "3D@60cm" obtained in a random walk experiment. These two maps display a structural similarity especially when comparing the high-concentration regions colour-coded in red. The two maps at the bottom of Fig. 1 represent the mean distribution obtained from the 2D models computed for the upper ("2D@110cm") and the lower nose ("2D@10cm").

Quantitative Comparison: As a measure of distribution similarity we use the Kullback-Leibler (KL) divergence or relative entropy [19] for probability functions:

$$KL(p|q) = -\int p(x) \ln \frac{q(x)}{p(x)} dx \quad (8)$$

where $p(x)$ is the "unknown" distribution (in our case "2D@60cm"), and $q(x)$ is the modelled distribution ("3D@60cm"). Since the gas distributions maps are not probability distribution we first normalize them so that the sum over all values equals to one. Then we compute the KL divergence for 14 layers of the 3D model for two different experiments, one with an ethanol source and one with an acetone source. As can be seen in Fig. 2, the minimum of the KL divergence was found exactly for the layer at the height of the middle nose in both experiments.

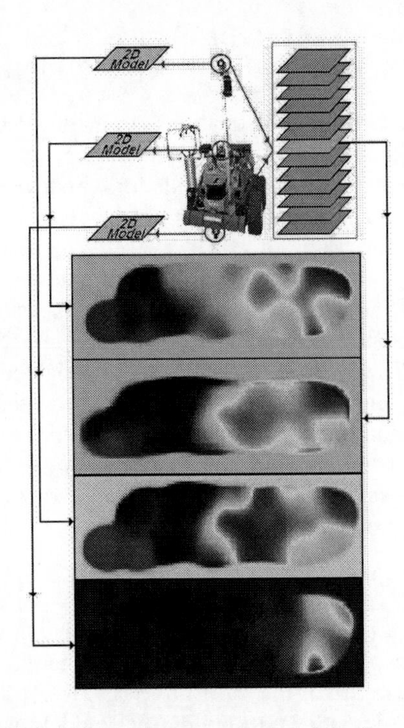

FIGURE 1. Top to bottom: picture of the "m-nose" prototype ("Rasmus") carrying the three electronic noses; mean of the 2D gas distribution map obtained from the middle nose ("2D@60cm"); from slicing the 3D model ("3D@60cm") obtained from the lower and the upper nose; 2D mean map from the upper nose ("2D@110cm"); from the lower nose ("2D@10cm").

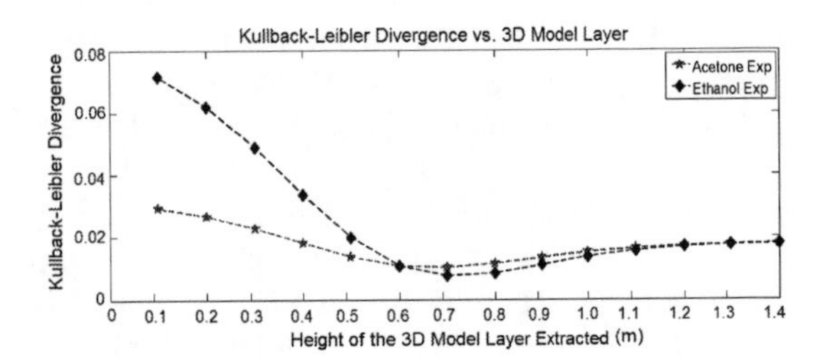

FIGURE 2. KL divergence for a random walk experiment with ethanol (blue line) and a sweeping experiment with acetone (red line) between 14 layers of the 3D model and the "2D@60cm".

6. DISCUSSION AND CONCLUSIONS

3D gas distribution modelling with a mobile robot in an uncontrolled environment is a challenging field of research. This is mainly due to the chaotic nature of the dispersed gas. Utilization of mobile robots to monitor pollution has a number of advantages reflected by an increasing interest in this field in the last ten years. In this paper we present a statistical method to build three-dimensional gas distribution maps (3D-DM). The mapping technique uses Kernel extrapolation mapping with a tri-variate Gaussian weighting function to model the decreasing likelihood that a reading represents the true concentration with respect to the distance in the three dimensions. The method is evaluated using a mobile robot equipped with three "e-noses" mounted at different heights. Initial experiments in an uncontrolled indoor enviroment are presented and evaluated with respect to the ability of the 3D map, computed from the lower and upper nose, to predict the map from the middle nose. This paper represents initial work and of course more trials are needed in different environments and with different analytes. Another interesting task is to integrate wind measurements, obtained by an anemometer to build an improved gas distribution model.

REFERENCES

1. K. Kemp, and F. Palmgren, *Annual Report, NERI, Roskilde* pp. 155–182 (1999).
2. DustBot - Networked and Cooperating Robots for Urban Hygiene, http://www.dustbot.org (2006–2009).
3. B. Shraiman, and E. Siggia, *Nature* pp. 639–646 (2000).
4. S. Vardoulakis et al., *Atmospheric Environment* **37**, 155–182 (2003).
5. P. Roberts, and D. Webster, "Turbulent Diffusion," in *Environmental Fluid Mechanics - Theories and Application*, 2002.
6. Y. Maruo et al., *Atmospheric Environment* **37**, 1065–1074 (2003).
7. P. Addison et al., *Environmental Monitoring and Assessment* pp. 333–342 (2000).
8. H. Ishida et al., *Sensors and Actuators B* **49** (1998).
9. A. Hayes et al., *IEEE Sensors Journal, Special Issue on Electronic Nose Technologies* **2**, 260–273 (2002).
10. A. Lilienthal, and T. Duckett, *Robotics and Autonomous Systems* **48**, 3–16 (2004).
11. A. H. Purnamadjaja, and R. A. Russell, "Congregation Behaviour in a Robot Swarm Using Pheromone Communication," in *Proc. ACRA*, 2005.
12. P. Pyk et al., *Auton Robot* **20**, 197–213 (2006).
13. H. Ishida et al., "Three-Dimensional Gas/Odor Plume Tracking with Blimp," in *Proc. ICEE*, 2004.
14. A. Rutkowski et al., "A Robotic Platform for Testing Moth-Inspired Plume Tracking Strategies," in *Proc.ICRA*, 2004.
15. H. Ishida.et al., "Three-Dimensional Gas-Plume Tracking Using Gas Sensors and Ultrasonic Anemometer," in *IEEE Sensors (2004)*, 2004, pp. 1175–1178.
16. A. J. Lilienthal et al., "A Statistical Approach to Gas Distribution Modelling with Mobile Robots - The Kernel DM+V Algorithm," in *submitted to IROS*, 2009.
17. A. Loutfi et al., *Robotica* p. online (2008).
18. A. J. Lilienthal et al., "A Rao-Blackwellisation Approach to GDM-SLAM, Integrating SLAM and Gas Distribution Mapping," in *(ECMR)*, 2007, pp. 126–131.
19. S. Kullback, and R. A. Leibler, *Annals of Mathematical Statistics* **2**, 79–86 (1951).

NANOSTRUCTURED SENSORS II

Smell Nanobiosensors: Hybrid systems based on the electrical response to odorant capture.Theory And Experiment

Eleonora Alfinito, Cecilia Pennetta and Lino Reggiani

Dipartimento di Ingegneria dell'Innovazione, Università del Salento, Via Monteroni, Lecce I-73100 (Italy)
CNISM- Consorzio Interuniversitario per le Scienze Fisiche della Materia

Abstract. Mammalian olfactory system is the bio-archetype of smell sensor devices. It is based on a very articulated mechanism which translate the odorant capture information performed by the olfactory receptors (ORs) into a code. Finally, the code is sent to the brain for aroma recognition. Our aim is to partially mimick this system to produce a biosensor on nanometric scale. The active part of the device is constituted of nanosomes containing specific ORs. Each nanosome is interfaced with nano-electrodes and the odorant capture is converted into an electric signal. Specifically, the electrical response is correlated with the conformational change that a single OR undergoes when it captures a specific odorant molecule. An array of nanodevices should be able to produce specific response profiles. In this paper we present a possible theoretical framework in which the experimental results should be embedded. It consists of the description of the protein in terms of an impedance network able to simulate the electrical characteristics associated with the protein topology.

Keywords: Nanobiosensors, olfactory receptors.
PACS: 87.15.hp, 87.15 Pc.

INTRODUCTION

The mammalian olfactory system is the most efficient biological sensor we know for odour recognition. In the nose, about 40 millions of neurons express one (or few) [1-2] different odorant receptor (OR) among more than 1000. Each OR is able to react to few odorants and the simultaneous activation of many neurons, at different level, produces a specific profile (coding) for each specific odour [3]. This is one of the most refined mechanism of pattern recognition and the perspective of using its facilities into a nanodevice, possibly easy to handle, is very attractive. This aim is addressed by the BOND (Bioelectronic Olfactory Neuron Device) project, recently proposed to the EC in the seventh framework programe. The project main goal is to construct an array of nanobiosensors whose active part consists of few specific ORs interfaced with nanoelectrodes. In such a way, various ORs, differently responding to the same odorant compound, should produce a specific odorant profile, like it happens *in vivo*.

Actually, *in vivo*, when the receptor captures the odorant molecule, it undergoes a conformational change and activates the G protein to which is connected inside the cell membrane. Eventually, cascade processes are produced which culminate in the signal transmission to the olfactory cortex. *In vitro*, only part of this chain will be reproduced [4-6]. In the present approach, the smell recognition is performed looking at the conformational change. As a matter of fact, we are interested to develop a device which monitors the odorant capture by means of the receptor impedance variation. The impedance variation is due to conformational change. In this work we report a set of experimental results which support the feasibility of the main concept. Experiments were carried out on two proteins pertaining to the GPCR (G protein coupled receptors) family, the light receptor bovine rhodopsin and the rat olfactory receptor I7. Then, we present and discuss the theoretical framework in which the results are embedded.

EXPERIMENTAL METHODS AND MATERIALS

Experimental - The debated question on the electrical conductivity of sensing proteins has been

CP1137, *Olfaction and Electronic Nose: Proceedings of the 13th International Symposium,* edited by M. Pardo and G. Sberveglieri

partially answered with some experiments, in particular on bovine rhodopsin [5]. Electrochemical impedance spectroscopy (EIS) measurements were performed on self–assembled multilayers of rhodopsin on gold substrate, at different stages of multilayer formation [5] (from (a) to (e), in Fig. 1). The results can be described in terms of Nyquist plots (A), as shown in Fig 1. Precisely, they can be interpreted by means of a simple electric analogue, the Randles cell (B). The same procedure was used for testing the immobilization of the rat olfactory receptor I7 [6]. The selective odorant detection of this receptor was tested onto three different aldehydes: two specific odorants, octanal (1) and heptanal (2) and one non-specific, helional (3). The analysis was performed by monitoring the variation of the polarization resistance, say R_P in the Randles cell, at different concentrations of the odorants (Fig. 2). In the circuit analog the polarization resistance is the passive element more sensible to the variation of odour concentration. The net result is a decreasing of R_P for increasing concentrations, with a maximal variation of 15% for octanal, going from a concentration of 10^{-12} M to that of 10^{-4} M.

The concomitance of these results enforces the proposal to monitor the ligand capture mechanism of a GPCR by means of the analysis of the variation of its electrical properties.

Methods - In order to explore the response of the single protein, we set up a theoretical framework in which the protein is mapped into a network of impedances. This kind of approach gives us the possibility to connect electrical and topological properties of the receptor. In order to draw up the network we take the tertiary structure of the protein from the protein data base (PDB) [7] or similar templates. Then, we associate a single node to each amino-acid. The position of the node is taken coincident with that of its C_α atom . Each couple of nodes is connected with a link when their distance is less than an assigned cut-off value, R_C. The meaning of R_C is that of an interaction range among amino-acids In principle, the value of R_C can substantially vary depending on the kind of interaction under consideration [8-10].

Finally, the network gives a sketch of the protein topology in a fixed conformation. This graph turns into an impedance network when an elemental impedance is attributed to each link. In the present case, following the experimental outcomes, the natural choice for the elemental impedance is to take the RC parallel circuit. As a matter of fact, the experimental curves (Fig. 1) move away from the ideal semicircle shape of the RC Nyquist plot, essentially for the presence of the Warburg (Z_W) and series impedances (R_S), which mainly pertain to the experimental environment. Therefore, we conclude that the electrical analogue of a single protein, in this context, mostly correspond to the R_P-CPE parallel circuit.

The elemental impedance between the i,j-th nodes is taken as [10]

$$Z_{i,j} = \frac{l_{i,j}}{A_{i,j}} \frac{1}{\rho^{-1} + i\varepsilon_{i,j}\varepsilon_0\omega}.$$

where $A_{i,j}=\pi\,(R_C^2 - l_{i,j}^2/4)$, is the cross-sectional area between two spheres of radius R_C centered on the i-th and j-th node, respectively; $l_{i,j}$ is the distance between these centers, ρ is the resistivity, taken to be the same for every amino-acid, with the indicative value of $\rho = 10^{10}$ Ωm; $i = \sqrt{-1}$ is the imaginary unit, ε_0 is the vacuum permittivity, ω is the circular frequency of the applied voltage. The relative dielectric constant of the couple of i-th and j-th amino-acids, $\varepsilon_{i,j}$, is expressed in terms of the intrinsic polarizability of each amino acid, as given in [10].

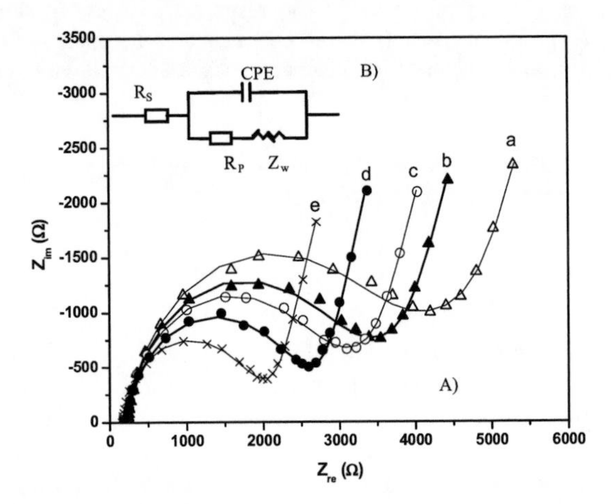

FIGURE 1. Electrochemical characteristics of rhodopsin SAM at different levels of assemblying (A) and the corresponding circuital analogue (Randles cell) (B). In figure 1A: (a) Step I: mixed SAMs modified gold electrode; (b) Step II: blockage with goa tIgG; (c) Step III:binding of neutravidin; (d) Step IV: immobilization of biotinylated antibody Biot-Rho-1D4; (e) after injection of 80 ng/ml rhodopsin membrane fraction [5].

By positioning the input and output electrical contacts on the first and last node, respectively, the network is solved within a linear Kirchhoff scheme and its global impedance spectrum is calculated in the standard frequency range 0.1 Hz ÷ 100 kHz. By construction, this network produces a parameter dependent Nyquist plot. However, since the purpose is to monitor the impedance variation of the protein due to the conformational change, the exact values of resistance and relative static permittivity are not of specific interest.

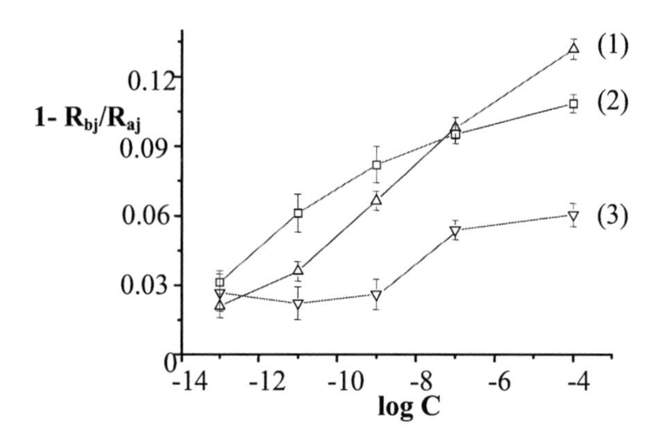

FIGURE 2. Dose response vs the log of concentration (expressed in Molarity) of rat OR-I7 to octanal (1), heptanal (2), helional (3). R_{bj} and R_{aj} indicate the polarization resistance before and after the odorant injection, respectively.

Materials - At present, the best known structure of the GPCR family is that of bovine rhodopsin, which is often used as a template for all the other members of the family. For this reason, first EIS measurements were carried out on this receptor. The knowledge acquired on this protein drove the experiments and the theoretical analysis on rat OR-I7. Here we present a comparative analysis of these two receptors. In order to set up our model, we need the tertiary structure of the protein in both its native and activated state. The tertiary structures of native and activated states for rhodopsin and OR I7 were obtained by means of the MODELLER software [11]. The template PDB entries were the same for both the proteins: 1JFP and 1LN6 for the native and activated state, respectively, complemented with 1F88, chain A [12]. Both 1JFP and 1LN6 were produced in the same experimental conditions. In Fig.3 the backbone of bovine rhodopsin and rat OR-I7, as obtained by this procedure, are reported.

RESULTS AND CONCLUSIONS

The network obtained from the protein mapping is highly irregular and its topology depends on the value of R_C. Best correspondence between network and protein is obtained with values of R_C in the range 5-15 Å [8-10].

Figure 4 reports the relative variation of the total number of links and of the static impedance associated with the native and the activated state as function of R_C, for rat OR-I7 (bottom) and rhodopsin (top). The best resolution, both for link number and impedance, is for R_C less than 15 Å.

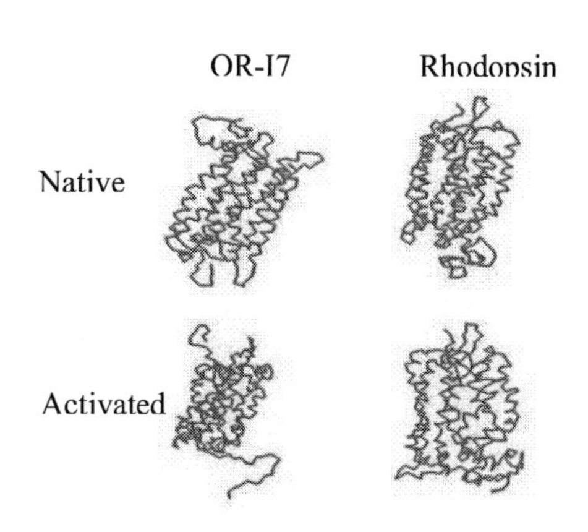

FIGURE 3. Sketch of the backbone of native and activated states of rhodopsin and rat OR-I7, as obtained with MODELLER (see text).

For values of R_C greater than 15 Å , the new links are mainly dangling bonds : They are more numerous in the more dilatated structures (the native for the OR-I7 and the activated for rhodopsin, see Fig. 3) and their increase does not correspond to a relevant variation of the impedance. Finally, for R_C comparable with the size of the protein, all the nodes are mutually connected and there is no difference, both of the link and static impedance values, between the native and activated state.

Figure 5 shows the Nyquist plot for the impedence of rhodopsin and OR-I7 in the native and activated state, for the specific value R_C=12 Å, which gives the best contrast, as shown in Fig. 4. The contrast of impedance is clearly better for rat OR-I7 than for rhodopsin and for both the proteins it is possible to conclude that it corresponds to a large conformational change.

In order to easily interprete the results, the real and imaginary part of the plots have been normalized to the maximal value of the static impedance of the configurations, the native for OR-I7 and the activated for rhodopsin.

The main conclusions we can extract from Fig. 5 are: i) there is a net change of impedance in going from the native to the activated state, for both the proteins; ii) OR-I7 and rhodopsin show opposite behaviour, when activated;

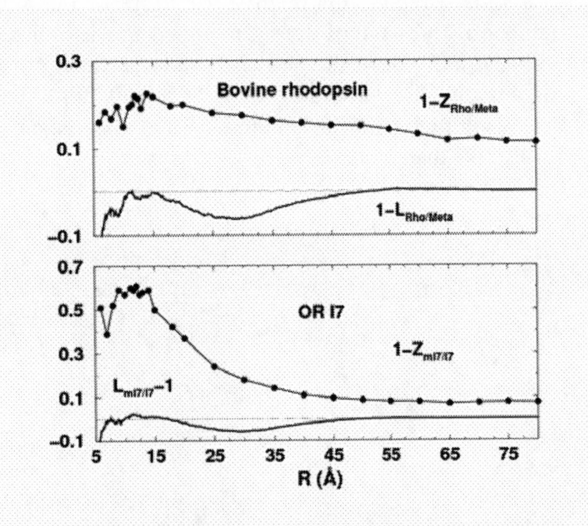

FIGURE 4. Links (continuous line and dots) and impedance (continuous line) relative variation for bovine rhodopsin (top) and rat OR-I7 (bottom) vs. the interaction radius.

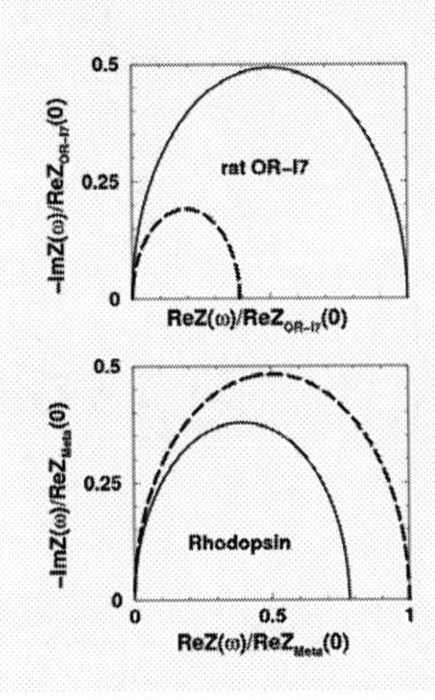

FIGURE 5. Theoretical Nyquist plot for two GPCR proteins: rat OR-I7 (top) and bovine rhodopsin (bottom). The interaction radius RC is 12 Å. The continuous and dashed lines refer to the native and the activated state, respectively.

iii) the zero-impedance (resistance) of OR-I7 in the activated state is 60% smaller than in the native state.

All these results pertain to an ideal structure perfectly responding, without thermal fluctuation and with a saturated concentration of odorant. Nonetheless, by comparing these outcomes with the experiments (Fig. 2), we find a good qualitative and quantitative agreement for the change of impedance following the sensing action of OR-I7. For what concerns rhodopsin, at present, the impedance spectrum is known only in dark (Fig. 1). In any case the essential features of the Nyquist plot are qualitatively well reproduced by the theoretical model.

We have investigated the conjecture that the conformational change due to the capture of a specific ligand can be monitored by means of a corresponding change of the protein electrical impedance. This conjecture has been tested on two different GPCR proteins, bovine rhodopsin and rat OR-I7, both with experimental and theoretical investigations. The amount of results confirms the conjecture and thus validates the proposal of constructing a smell nanobiosensor based on the sensing action of olfactory receptors.

ACKNOWLEDGMENTS

This work is carried out within the EC BOND (Bioelectronic Olfactory Neuron Device) Project, Proposal N_0 CP-FP 228685-2 BOND.

REFERENCES

1. R. Buck and L. Axel, *Cell* **65**, 175-181 (1991).
2. P. Mombaerts, *Curr. Opi. Neur.* **14**, 31-36 (2004).
3. P. Duchamp-Viret, M.A. Chaput and A. Duchamp, *Science* 284 , 2171-2174 (1999).
4. Q. Liu et al, *Biosens. Bioelectron.* **22**, 318-322 (2006).
5. Y. Hou et al, *Biosens. Bioelectron.* **21**, 1393-1402 (2006).
6. Y. Hou et al, *Biosens. Bioelectron.* **22**, 1550-1555 (2007).
7. H. M. Berman et al, *Nucleic Acids Research* **28**, 235-242 (2000).
8. M. M. Tirion, *Phys. Rev. Lett.* **77**, 1905-1908 (1999).
9. E. Alfinito, C. Pennetta and L. Reggiani, *Nanotechnology* **19**, 065202-1-12 (2008).
10. C. Pennetta et al. "Towards the Realization of Nanobiosensors Based on G Protein-Coupled Receptors", in Wiley-VCH Book series on Nanotechnologies for the Life Sciences, vol. 4, Nanodevices for the Life Sciences, edited by Challa S.S.R. Kumar, Wiley-VCH, Berlin, 2006, pp. 217-240.
11. A. Sali and T.L. Blundell, *J. Mol. Bio.* **234**, 779-815 (1993); A. Fiser, and A. Sali, *Bioinformatics* **19**, 2500-2501 (2003).
12. E. Alfinito, C. Pennetta and L. Reggiani, *J. Appl. Phys.*, in press.

CMOS Alcohol Sensor Employing ZnO Nanowire Sensing Films

S. Santra[1]*, S. Z. Ali[1], P. K. Guha[2], P. Hiralal[1], H. E. Unalan[1], S. H. Dalal[1], J. A. Covington[2], W.I. Milne[1], J. W. Gardner[2], F. Udrea[1]

[1]*Engineering Department, University of Cambridge, 9 J J Thomson Avenue, Cambridge CB3 0FA, UK*
[2]*School of Engineering, University of Warwick, Coventry CV4 7AL, UK*
**Corresponding author: email-Address: ss778@eng.cam.ac.uk, tel: + 44 1223 748311, fax: +44 1223 748348*

Abstract. This paper reports on the utilization of zinc oxide nanowires (ZnO NWs) on a silicon on insulator (SOI) CMOS micro-hotplate for use as an alcohol sensor. The device was designed in Cadence and fabricated in a 1.0 μm SOI CMOS process at XFAB (Germany). The basic resistive gas sensor comprises of a metal micro-heater (made of aluminum) embedded in an ultra-thin membrane. Gold plated aluminum electrodes, formed of the top metal, are used for contacting with the sensing material. This design allows high operating temperatures with low power consumption. The membrane was formed by using deep reactive ion etching. ZnO NWs were grown on SOI CMOS substrates by a simple and low-cost hydrothermal method. A few nanometer of ZnO seed layer was first sputtered on the chips, using a metal mask, and then the chips were dipped in a zinc nitrate hexahydrate and hexamethylenetramine solution at 90°C to grow ZnO NWs. The chemical sensitivity of the on-chip NWs were studied in the presence of ethanol (C_2H_5OH) vapour (with 10% relative humidity) at two different temperatures: 200 and 250°C (the corresponding power consumptions are only 18 and 22 mW). The concentrations of ethanol vapour were varied from 175 – 1484 ppm (pers per million) and the maximum response was observed 40% (change in resistance in %) at 786 ppm at 250°C. These preliminary measurements showed that the on-chip deposited ZnO NWs could be a promising material for a CMOS based ethanol sensor.

Keywords: Gas Sensor, SOI CMOS, Carbon nanotube, Zinc oxide nanowire
PACS: 07.07.Df

INTRODUCTION

Semiconductor gas sensors based on metal oxides are widely used to detect toxic, inflammable gases and vapours and have been well documented in the literature [1 – 4]. They are low cost and can detect a large range of gases with high sensitivity and reasonable response times. The working principle of these semiconductor gas sensors is mainly based on the change in conductivity of the material upon exposure to a target gas. The use of a micro-heater to raise the temperature of the sensing material can induce better sensitivity and quicker response times.

In recent years, significant interest has emerged in the synthesis of one-dimensional nanomaterials because of their potential application in various fields [5–7]. The large surface to volume ratio and hence high surface area of these nanoscale materials make them attractive for gas sensing. Among different metal oxides nanowire (e.g. SnO_2, TiO_2, WO_3 etc.), zinc oxide (ZnO) is one of the most promising, with its wide band gap (3.37 eV at room temperature) and large exciton binding energy. ZnO nanowires are n-type semiconductors with high thermal stability and, on the contrary to most oxide nanowires, can be grown at low temperatures via a hydrothermal method. ZnO thin or thick films, which are either deposited or grown on non-CMOS substrates, have already been extensively studied for gas sensing applications. In particular, they were used to detect ethanol (1 – 5 ppm (pers per million)) [8, 9], NO_2 (0.5-8.5 (ppm)) [10], humidity in the range 12 – 96.9% [11], H_2 (600 ppm) [10, 13, 14], NH_3 (30 ppm) [12, 13]. The integration of ZnO nanowires, with a fully compatible CMOS technology, would be highly desirable as it enables 'smart' smaller sensors to be fabricated at a lower cost than conventional gas sensor manufacture. The combination is also additionally useful as the sensor can be combined with on-chip circuitry for signal conditioning and to compensate for some of the short comings of the sensing material, i.e. drift, non-linearity, aging etc.

CP1137, *Olfaction and Electronic Nose: Proceedings of the 13th International Symposium,* edited by M. Pardo and G. Sberveglieri

Here, we present a micro gas sensor based on ZnO NWs deposited onto CMOS micro-hotplates and evaluated its sensitivity to ethanol vapour in air.

MICRO-HOTPLATE FABRICATION AND ZNO DEPOSITION

The micro-hotplates were designed in Cadence (5.0) and fabricated at a commercial (XFAB, Germany) foundry using a 1.0 μm, three metal, polysilicon SOI process. The interconnect metal (aluminum) of the SOI process was used to form a micro-heater. The micro-heater itself is used as a temperature sensor because its resistance increases as a function of temperature.

Interdigitated electrodes (aspect ratio 56) were formed using the top metal electrodes, and the passivation above them removed using the same process step that is used to expose the bond pads of the chip. The micro-heater was then thermally isolated from the substrate by a Deep Reactive Ion Etching (DRIE) back etch at a commercial MEMS foundry (Silex Microsystem, Sweden) to reduce the power consumption. This resulted in an aluminum micro-heater embedded within a thin membrane (ca. 5μm) of silicon dioxide, with a passivation of silicon nitride. The cross-sectional view and the optical microscope picture of the fabricated device are shown in Fig. 1 and Fig. 2 respectively. The heater is of 75 μm radius with the corresponding membrane size of 280 μm.

Aluminum interdigitated electrodes tend to form an oxide when exposed to air, and therefore do not provide good electrical contact with the sensing material. Therefore electroless plating was carried out to deposit nickel followed by gold. The process was carried out at the wafer level to remove oxides from the exposed Al layer and electrolessly plate Ni/Au on only the exposed aluminum pads (Pac Tech (Germany)).

The micro-heaters were characterized on several devices across the wafer and it was found that 300°C can be reached with around 27 mW of DC power (as shown in Fig. 3).

For sensing material growth, a thin ZnO seed layer was sputter deposited (~5 nm) on selective areas of our CMOS devices using a metal mask. Following seed layer deposition, these devices were dipped in an equimolar (25 mM) aqueous solution of zinc nitrate hexahydrate ($Zn(NO_3)_2.6H_2O$) and hexamethylenetramine (HMTA) and were kept at 90°C for two hours [15]. The devices were removed from the solution at the end of the growth, rinsed with DI water and dried under nitrogen flow. The chips were then annealed at 300°C for one hour. The advantage of this method is that ZnO NWs can be simultaneously grown on more than one micro-hotplate and hence can be extended as an inexpensive approach for wafer level fabrication.

The surface morphologies of the samples and size distribution of the nanowires were characterised using

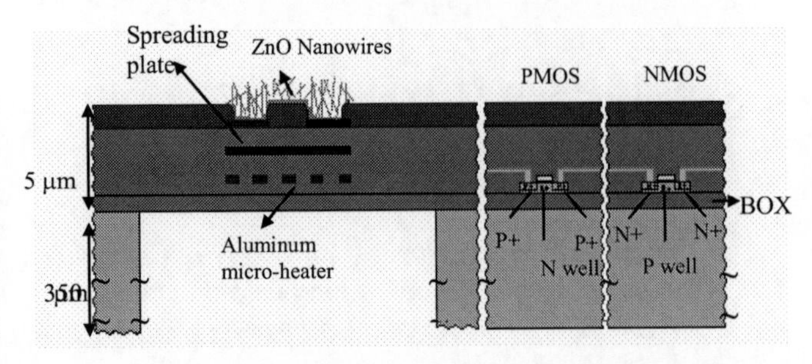

Fig. 1. Cross sectional view of the ultrathin (5 μm) SOI microhotplate and the CMOS electronic cells.

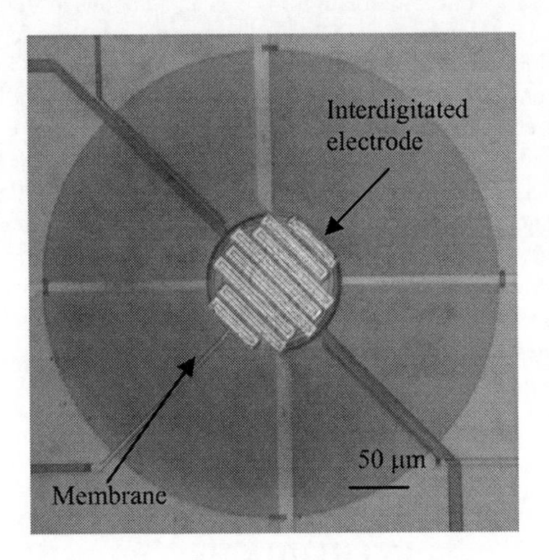

Fig. 2. An optical microscope picture of the fabricated micro-hotplate with interdigitated electrode.

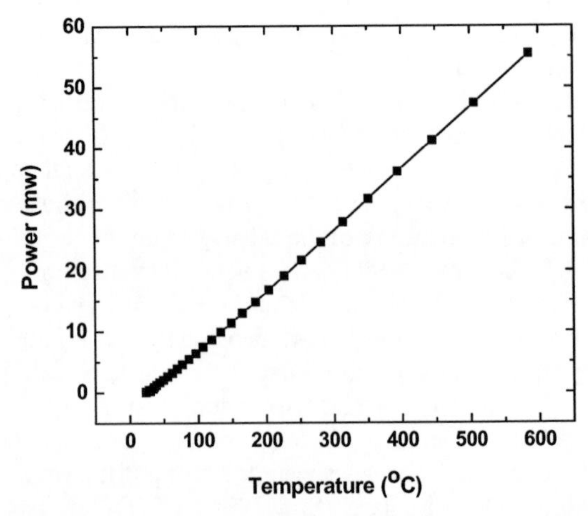

Fig. 3. Power vs temperature plot of the Al micro-heater.

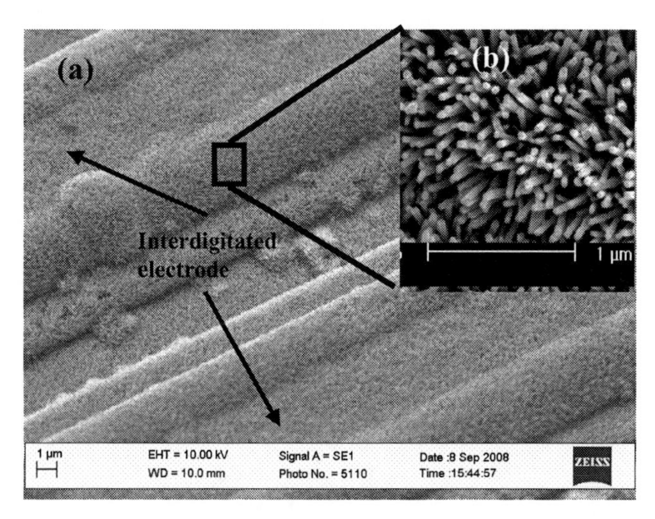

Fig. 4. (a) Top view of the SEM images of the ZnO NWs on interdigitated electrode, (b) zoom in view of NWs.

a field emission scanning electron microscope operated at 10 keV which is shown in Fig. 4. Nanowire length is ~400 nm and their diameter is ~60 – 80 nm. These nanowires are touching each other that provides the electrical paths between the pads of the electrode.

The current – voltage (I – V) characteristics of ZnO NWs was measured (shown in Fig. 5) in air. The linear I–V characteristic also confirms good ohmic contact between sensing material (ZnO NWs) and electrodes.

ETHANOL TEST RESULTS

Chemical testing was performed at the Sensor Research Laboratory (SRL), Warwick University. Here there are custom test facilties that can be used to expose sensors to specific vapour and water concentrations in clean air. Humidity was kept constant at 3000 ppm throughout.

The sensor was mounted on to a ceramic package (16 pin DIL, Spectrum Semiconductor, USA) and mounted in a stainless steel chamber kept at 30± 1°C within a Dri-bloc^{TM} heater. The chamber was connected to a National Instruments DAQ card within a PC so that the data acquisition was recorded automatically using Labview software.

The sensors were heated locally by the aluminium micro-heater, which is just underneith the sensing material area. Sensors were kept at two different temperatures (200°C and 250±3°C). The power consumptions to maintain these temperature is only 18 and 22±0.3 mW. The preliminary measurements were carried out at five different concentrations of ethanol vapour in air.

The ZnO NWs response at different ethanol concentrations (175 – 1484 ppm) were measured, as shown in Fig. 6. We found that the response of NWs to ethanol vapour is significant and takes place within a few minutes. Following the removal of ethanol vapor from the system, recovery of the ZnO NWs was observed to take few tens of minutes before their resistance gets back to the base line value.

The response of the sensor is defined in this work as [((Rb-Ra)/Ra)×100%] where Ra is the baseline resistance of ZnO NWs in presence of humid air and Rb is the resistance in presence of ethanol and humid air. Based on this definition the measured responses were calculated with different concentrations. This is shown in Fig 7.

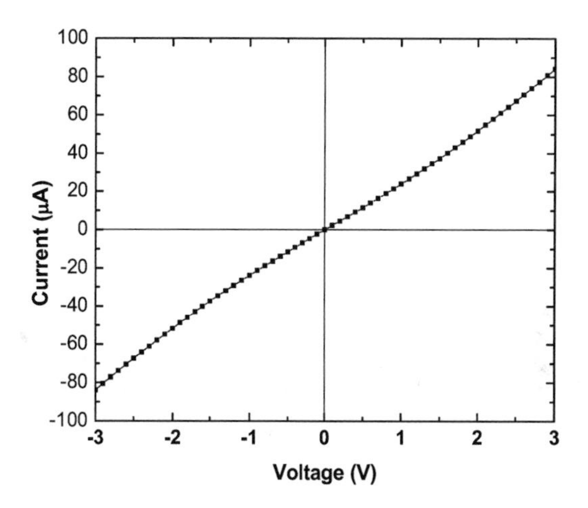

Fig. 5. I-V characteristics of ZnO NWs measured in air at room temperature.

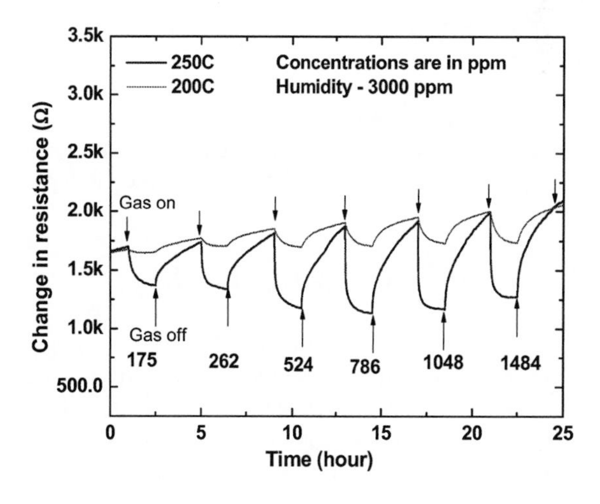

Fig.6. Change in resistance of the fabricated ZnO NWs ethanol gas sensor at different temperatures and different concentrations at a constant humidity of 3000 ppm.

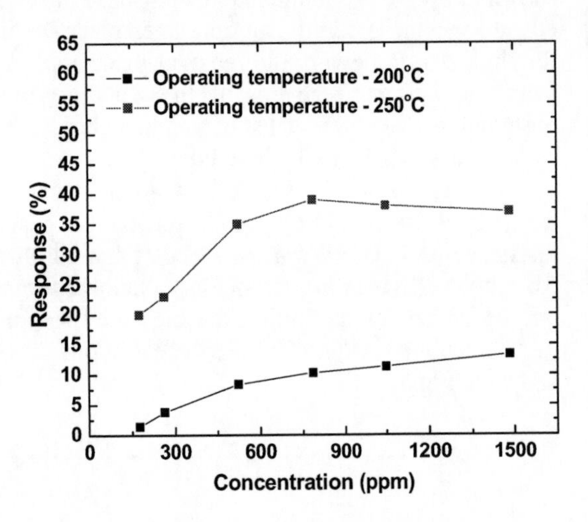

Fig.7. Response of the ZnO NWs measured as a function of ethanol concentrations.

DISCUSSIONS

It is well known that the sensing mechanism in case of most semiconducting oxide gas sensors is surface-controlled type [8, 9]. An oxygen molecule adsorbs on the surface of the ZnO NWs when it exposed to air. As a result an O–2 ion formed by capturing an electron from the conduction band. When these sensors are exposed to ethanol (reducing gas) at high temperature, the gas reacts with the surface oxygen species, which decreases the surface concentration of O–2 ion and increases the electron concentration. So the conductivity of the ZnO nanowires increases as can be seen in Fig 6.

In Fig. 7. we found that initially the fractional response increased with increasing ethanol concentration, but it flattens out at higher concentrations. More measurements are in progress to improve recovery time and explain the unexpected saturation effect.

CONCLUSIONS

This paper describes a method for the growth of ZnO NWs on SOI CMOS membranes. This method, as shown, is simple, economical, CMOS compatible and hence ideal for large scale wafer level production. The basic gas sensor device is a micro-hotplate structure containing an aluminium micro-heater embedded in a thin dielectric membrane, which allows us to achieve high temperatures with low power consumption. Following the growth of ZnO NWs, these sensors were exposed to different concentrations of ethanol vapor in air (RH 10%) and a maximum fractional response of 40% at 786 ppm was observed. These preliminary measurements show that on-chip ZnO NWs could potentially be used as a low-cost ethanol sensor.

ACKNOWLEDGMENTS

The work has been supported by Engineering and Physical Sciences Research Council (EPSRC) under the project no. EP/F004931/1. S. Santra acknowledges Mr. F. Courtney and Mr. F. C. Harun (SRL, Warwick University) for their help during gas test sytem operation.

REFERENCES

1. G. Korotcenkov, *Sens. Actuators B* **107**, 287 – 304 (2005).
2. N. Barsan, D. Koziej, U. Weimar, *Sens. Actuators B* **121**, 18-35 (2007).
3. K. Ihokura and J.Watson, The Stannic Oxide Gas Sensor: Principle and Application, Publisher city: CRC Press inc., (1994).
4. D. E. Williams, *Sens. Actuators B* **57**, 1-16 (1999).
5. X. Duan, Y. Huang, Y. Cui, J. Wang, C. M. Lieber, *Nature* **409**, 66-68 (2001).
6. X. Duan, Y. Huang, R. Agarwal, C. M. Lieber, *Nature* **421**, 241 - 244 (2003).
7. Y. Li, F. Qian, J. Xiang, C. M. Lieber, *Mater. Today* **9**, 18-27 (2006).
8. Q. Wan, Q. H. Li, Y. J. Chen, T. H. Wang X. L. He, J. P .Li C. L. Lin, *Applied Physics Letters* **84**, 3654 – 3656 (2004).
9. T.-J. Hsueh, S.-J. Chang, C.-L. Hsu, Y.-R. Lin, I.-C. Chen, *J. of the Electrochemical Society*, **155**, No. 9, K152 – K155 (2008).
10. A. Z. Sadek, S. Choopun, W. Wlodarski, S. J. Ippolito, K. Kalantar-zadeh, *IEEE Sensors Journal* **7**, No. 6, 919 – 924 (2007).
11. Y. Zhang, K. Yu, D. Jiang. Z. Zhu, H. Geng, L. Luo, *Applied Surface Science* **242**, 212 – 217 (2005).
12. G. S. Trivikrama Raol and D. Tarakarama Rao *Sens. Actuators B* **55** 166 – 169 (1999.
13. J X Wang, X W Sun, Y Yang, H Huang, Y C Lee, O K Tan and L Vayssieres *Nanotechnology* **17**, 4995-4998 (2006).
14. A.Z. Sadek, W. Wlodarski, Y.X. Li, W. Yu, X. Li, X. Yu, K. Kalantar-zadeh, *Thin Solid Films* **515**, 8705-8708 (2007).
15. H. E. Unalan, P. Hiralal, N. Rupesinghe, S. Dalal, W. I Milne, G. A J Amaratunga, *Nanotechnology* **19**, 255608 (5pp) (2008).

Cholesterol Biosensor Based On Nanoporous Zinc Oxide Modified Electrodes

Chia-Yu Lin[1] and Kuo-Chuan Ho[1,2★]

[1]*Department of Chemical Engineering, National Taiwan University, Taipei 10617, Taiwan*
[2]*Institute of Polymer Science and Engineering, National Taiwan University, Taipei 10617, Taiwan*
★*Corresponding author: Fax: +886-2-2362-3040; Tel: +886-2-2366-0739; E-mail: kcho@ntu.edu.tw*

Abstract. Nanostructured ZnO modified electrodes have been fabricated by co-electrodeposition with eosin Y onto F-doped SnO_2 (FTO) coated glass, and their application for cholesterol biosensor was studied. For cholesterol application, cholesterol oxidase (ChOx) was immobilized onto the ZnO/FTO electrode via physical adsorption. Cyclic voltammetric measurements showed that the ChOx/ZnO/FTO is sensitive to cholesterol and the optimal solution pH for the ChOx/ZnO/FTO electrode for detecting cholesterol was 6.7. Besides, the redox peak current densities are proportional to the square root of scan rate, or $\nu^{1/2}$, indicating the electron-transfer process is diffusion limited. The sensitivity, linear detection range, and the limit of detection for the ChOx/ZnO/FTO electrode are 3.01 mA•cm^2•M^{-1}, 1.1~4.83 mM, and 20 μM, respectively.

Keywords: Cholesterol, biosensors, Zinc oxide
PACS: 82.47.Rs

INTRODUCTION

Recently, the monitoring of cholesterol level in human body has attracted increasing attention since the elevated cholesterol level would cause hypertention, heart disease, arteriosclerosis, etc.[1]. To obtain better sensing performance of the enzyme-based cholesterol biosensor, various immobilization methods and matrics [2-4] for enzyme immobilization have been proposed. Among the propsed matrics, nanostructured ZnO film would be a good candiate for enzyme immobilization due to its high surface area, high isoelectric point (9.5), and biocompatibility [5]. Recently, the nanostructured ZnO films have been investigated for glucose [6], H_2O_2 [7], and urea [8] biosensor applications.

In this study, the nanoporous ZnO modified electrodes have been fabricated by co-electrodepositing ZnO and eosin Y onto FTO substrate, and subsequently desorbing eosin Y. The porosity created by eosin Y can provide the surface area for ChOx immobilization. ChOx/ZnO/FTO for cholesterol sensing application was also discussed.

EXPERIMENTAL AND METHODS

2-1 Preparation Of ChOx/ZnO-modified Electrode

ZnO/eosin Y hybrid films were prepared as described in the literature [10]. A F-doped SnO_2 (FTO) coated transparent conducting glass (Solaronix SA, 15Ω/□), with an electrode area of 1.5 cm^2, served as the working electrode. Before use, the FTO electrodes were washed by ultrasonically cleaning in the solution containing NH_4OH (28 %), H_2O_2 (35 %), and deionized water, with volume ratio of 1:1:5, for 30 min. After being cleaned, these electrodes were dried at room temperature.

The electrodeposition of ZnO/eosin Y hybrid films were carried out at a CHI 440 electrochemical workstation (CH Instruments, Inc., USA) with a conventional three-electrode system. The working electrode, reference electrode and counter electrode are FTO electrode, Ag/AgCl/KCl sat'd electrode and Pt foil, respectively. Electrodeposition of ZnO/eosin Y hybrid film was performed potentiostatically at -1.05 V (vs. Ag/AgCl/KCl sat'd) for 20 min in an O_2 saturated aqueous solution containing 100 μM eosin Y, 5 mM $ZnCl_2$, and 0.1 M KCl with stirring at 70 °C. After deposition, the electrodes were immersed into a dilute basic solution (pH 10.4) for 24 h to desorb eosin

CP1137, *Olfaction and Electronic Nose: Proceedings of the 13th International Symposium,* edited by M. Pardo and G. Sberveglieri

Y. After desorption process and being dried at room temperature, the desorbed ZnO modified electrode was incubated in a solution of cholesterol oxidase (EC 1, 1, 3, 6, ChOx) (1 mg/ml) prepared in 0.1 M phosphate buffer saline (pH 6.7) for 12 h to immobilize enzyme. Then the electrodes were washed with deionized water and dried at room temperature. The prepared ChOx/ZnO modified electrode was stored at 4 °C when not in use.

The phase of the nanostructured ZnO-modified electrode was determined by x-ray diffraction (XRD, X-Pert, the Netherlands).

2-2 Amperometric Detection Of Cholesterol

Scheme 1 shows the biochemical reaction, between ChOx and cholesterol, and the electron transfer mechanism involved in the detection process. For detection of cholesterol by using ChOx/ZnO-modified electrode, a suitable sensing potential in the limit current plateau region was determined between 1.5 and 0.8 V by the linear sweep voltammetry at a scan rate of 0.2 mV/s in the solution containing 5 mM $[Fe(CN)_6]^{3-/4-}$ and 50 mM PBS (pH 6.7, 0.9% NaCl) and 5 mM cholesterol. This suitable sensing potential was determined as 0.7 V. The current densities in the concentration range between 1.1 and 4.83 mM were collected and calibration curve for cholesterol was constructed. [1].

RESULTS AND DISCUSSION

3.1 Cyclic Voltammetry Analysis

Fig.1 shows x-ray diffraction pattern of the desorbed ZnO/FTO electrode.The prepared ZnO film shows a preferential orientation along the *c*-axis. Besides, as shown in Fig. 2, the surface morphology of the ZnO/FTO electrode was rough, and would provide high surface area for ChOx immobilization.

Figs. 3(A) and 3(B) show the cyclic voltammograms obtained for ChOx/FTO and ChOx/ZnO/FTO electrodes in PBS (50 mM, pH 6.7, 0.9% NaCl) containing 5 mM $[Fe(CN)_6]^{3-/4-}$ and various concentrations of cholesterol. Compared to FTO electrodes, ChOx/ZnO/FTO electrode can provide more surface area for ChOx immobilization, resulting in being more sensitive to cholesterol. Besides, the peak separation for $[Fe(CN)_6]^{3-/4-}$ mediator on the FTO/ZnO/ChOx electrode was higher than that of the bare FTO electrode due to the semiconducting nature of ZnO and its large thickness (about 0.9 μm, see Fig. 2(b)).

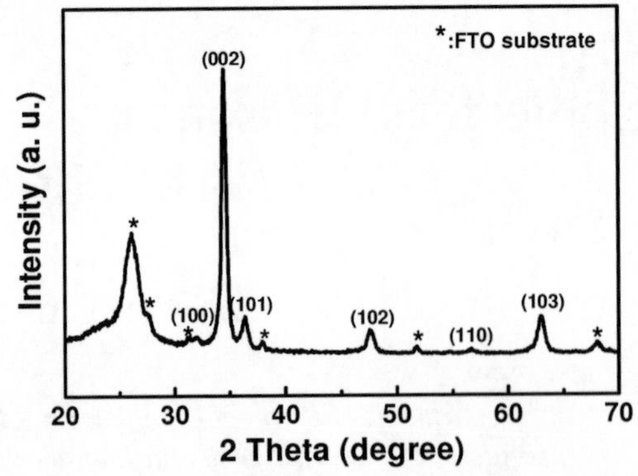

FIGURE 1. X-ray diffraction pattern of the desorbed ZnO/FTO electrode.

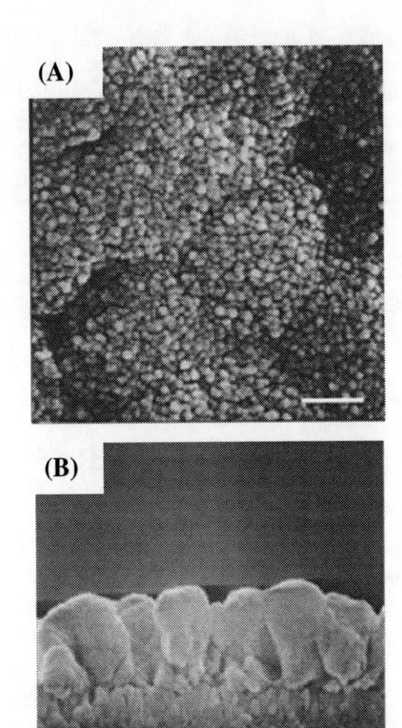

FIGURE 2. (A) The top-view and (B) the cross-view of SEM images of the ZnO/FTO electrode. Scale bar=0.5 μm.

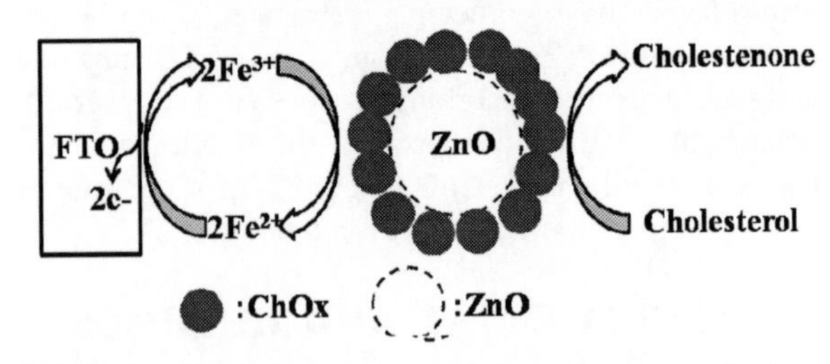

SCHEME 1. The biochemical reaction and the electron transfer at the ChOx/ZnO modified electrode.

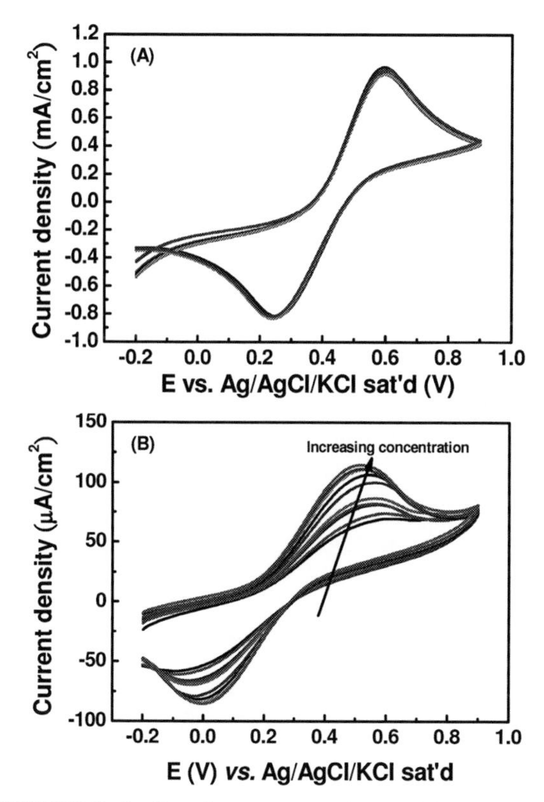

FIGURE 3. Cyclic voltammograms for (A) ChOx/FTO, (B) ChOx/ZnO/FTO electrodes in 50 mM PBS (pH 6.7, 0.9% NaCl) solution containing 5 mM $[Fe(CN)_6]^{3-/4-}$ and various concentrations of cholesterol ranging from 62.5 to 862 μM. Scan rate: 50 mv/s.

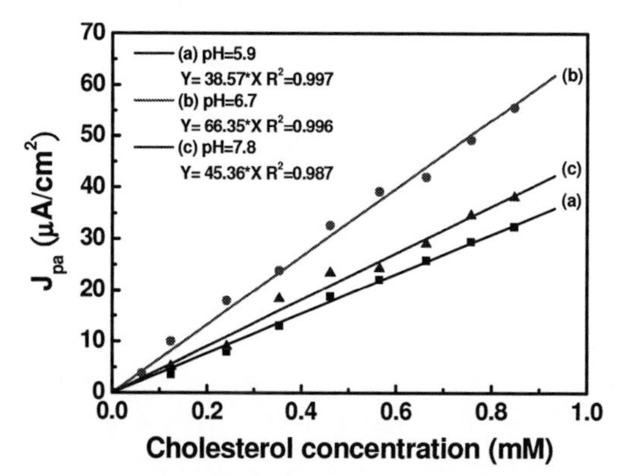

FIGURE 4. Peak current densities of the ChOx/ZnO/FTO elctrode in 50 mM PBS ((a) pH 5.9, (b) pH 6.7, (c) pH7.8, 0.9% NaCl) containing 5 mM $[Fe(CN)_6]^{3-/4-}$ and various concentrations of cholesterol ranging from 62.5 to 862 μM. Scan rate: 50 mv/s.

The enzyme activity is very sensitive to the solution pH, and therefore, the effect of solution pH on the current response of the ChOx/ZnO/FTO electrode to various concentrations of cholesterol in 50 mM PBS (0.9% NaCl) solution at pH 5.9, 6.7 and 7.8 was examined. As shown in Fig. 4, the ChOx/ZnO/FTO electrode showed higher sensitivity to cholesterol in a PBS solution with pH 6.7 than either 5.9 or 7.8. Therefore, the pH of PBS for later experiments was set at 6.7.

Fig. 5 depicts the cyclic voltammograms for the ChOx/ZnO/FTO electrode in a 50mM PBS solution (pH 6.7, 0.9% NaCl) containing 5 mM $[Fe(CN)_6]^{3-/4-}$ and 0.86 mM cholesterol at various potential scan rates. Since the peak current density increases linearly with the square root of the scan rate (ν), the electron transfer process was diffusion-controlled.

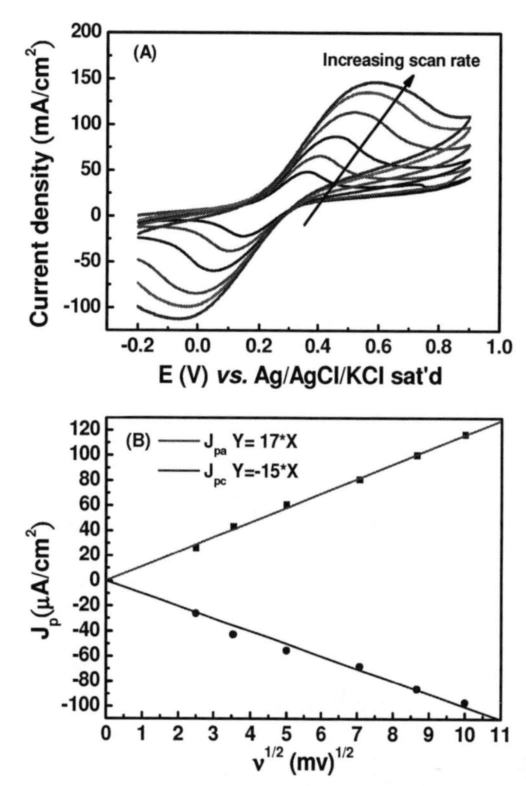

FIGURE 5. (A) Cyclic voltammograms for the ChOx/ZnO/FTO electrode in the 50 mM PBS (pH 6.7, 0.9% NaCl) solution containing 5 mM $[Fe(CN)_6]^{3-/4-}$ and 0.86 mM cholesterol at various potential scan rates of 6.25, 12.5, 25, 50, 75, and 100 mV/s. (B) J_{pa} & J_{pc} vs. scan rate$^{1/2}$ ($\nu^{1/2}$).

The current densities of the ChOx/ZnO/FTO electrode as a function of the cholesterol concentration with a sampling time of 200 s at each concentration level were measured and shown in Figure 6. It was found that the current density increases linearly with the increase in the cholesterol concentration. The sensitivity and the limit of detection, with signal-to-noise equal to 3, for the ChOx/ZnO/FTO electrode are 3.01 mA•cm^{-2}•M^{-1} and 20 μM, respectively.

Since the porosity of ZnO film can be controlled by adjusting the concentration of eosin Y, the finding of the optimal eosin Y concentration to improve the sensor sensitivity are under investigation. Besides, the interference and real sample tests are underway now.

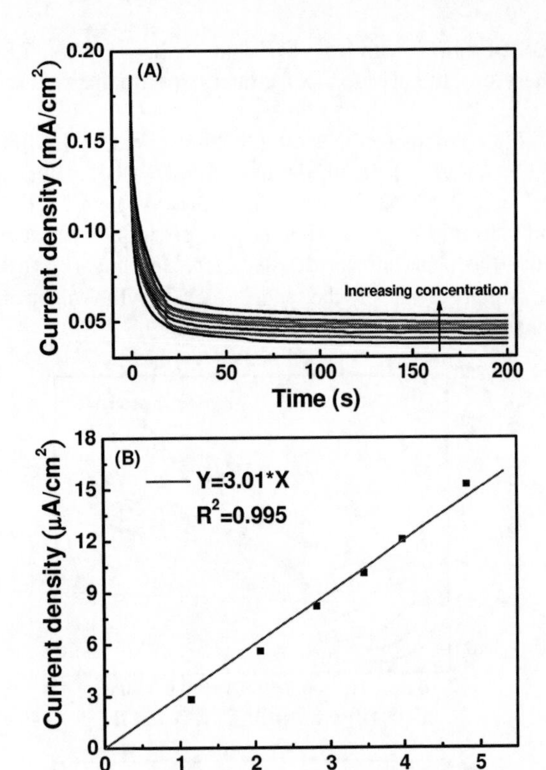

FIGURE 6. (A) The amperometric current response of ChOx/ZnO/FTO electrode to various cholesterol concentrations in 50 mM PBS solution (pH 6.7, 0.9% NaCl). The applied potential was set at 0.36 V. (B) The calibration curve between the sensing current densities vs. the cholesterol concentration.

CONCLUSION

The ChOx/ZnO/FTO electrodes have been successfully fabricated. The micro-porosity of the ZnO film, for attaching ChOx, was created by desorbing eosin Y. The sensitivity, linear detetion range, and the limit of detection (S/N=3) for the ChOx/ZnO/FTO electrode were 3.01 $mA \cdot cm^{-2} \cdot M^{-1}$, 1.1~4.83 mM, and 20 μM, respectively.

ACKNOWLEDGMENTS

This work was sponsored by the National Research Council of the Republic of China (Taiwan) under grant number NSC 96-2220-E-006-015 and NSC 97-2220-E-006-008.

REFERENCES

1. R. Forster, J. Cassidy, E. O'Donoghue, *Electroanalysis* **12**, 716-721 (2000).
2. G. K. Kouassi, J. Irudayaraj, G. McCarty, *J. Nanotechnology* **3**, 1-9 (2005).
3. J. Wang, *Anal. Chim. Acta* **399**, 21-27 (1999).
4. Z. Matharu, G. Sumana, S. K. Arya, S. P. Singh, V. Gupta, B. D. Malhotra, *Langmuir* **23**, 13188-13192 (2007).
5. J. X. Wang, X. W. Sun, A. Wei, Y. Lei, X. P. Cai, C. M. Li, Z. L. Dong, W. Huang, *Appl. Phys. Lett.* **88,** 233106 (2006).
6. X. Zhu, I. Yuri, X. Gan, I. Suzuki, G. Li, *Biosens. Bioelectron.* **22,** 1600-1604 (2007).
7. F. Zhang, X. Wang, S. Ai, Z. Sun, Q. Wan, Z. Zhub, Y. Xian, L. Jin, L. Yamamoto, *Anal. Chim. Acta* **519**, 155-160 (2004).
8. T. Yoshida, T. Pauporté, D. Lincot, T. Oekermann, H. Minoura, J. Electrochem. Soc. **150**, C608-C615 (2003).

Self-Heating in Individual Nanowires: a Major Breakthrough in Sensors Technology

J. D. Prades[a,ϒ], R. Jimenez-Diaz[a], F. Hernandez-Ramirez[b,ϒ], T. Fischer[d], A. Cirera[a], A. Romano-Rodriguez[a], S. Mathur[d], J. R. Morante[a,c]

[a]*EME/XaRMAE/IN²UB, Dept. Electrònica, Universitat de Barcelona, C/ Martí i Franquès 1, E-08028 Barcelona Spain*
[b]*Electronic Nanosystems S. L., Barcelona, Spain*
[c]*Institut de Recerca en Energia de Catalunya (IREC), C/ Josep Pla 2, B3, PB, E-08019 Barcelona, Spain*
[d]*University of Cologne, Greinstrasse 6, D-50939 Cologne, Germany*

Abstract. The major advantages of using self-heated individual nanowires as chemical gas sensors are presented and discussed. This novel strategy is based on the exploitation of dissipated power at the nanowire by Joule effect due to the bias current applied in conductometric measurements, which enables heating the tinny mass of these wires up to the optimum temperatures for gas sensing applications. Due to the nanoscale integration of the heater in the sensing material itself, the power required to operate these sensors is significantly reduced, if they are compared to the state-of-the-art technologies such as thin-film sensors with external microheaters. Furthermore, this strategy enables a reduction of the response time, improving the dynamic behavior of sensors obtained with current technologies. In summary, this approach represents a major breakthrough in sensor technology and it paves the way towards a new generation of fully integrated and autonomous electronic nano-noses.

Keywords: gas sensor, nanowire, self-heating, metal oxide
PACS: 85.85.+j, 81.07.-b, 81.16.Nd, 73.63.Bd, 73.63.Rt

INTRODUCTION

Two of the major challenges in the field of chemical gas sensors are (i) the development of new and more efficient strategies to heat them up to their optimum sensing temperature, and (ii) to increase the stability, and reversibility of their response. To solve the first problem, the common approach involves the scaling down of the devices and thus, the reduction of the power necessary to operate them. The second problem is usually circumvented by using sensing materials with well-defined and stabilized surfaces.

We have recently demonstrated that the use of individual nanowires could help to solve the both aforementioned problems. On the one hand, the current applied to individual nanowires in conductometric operation can be used to dissipate enough electrical power to self-heat their reduced mass [1] and thus, to reach the optimum working conditions for the detection of chemicals, avoiding the need of external heaters [2,3]. On the other hand, nanowires exhibit novel structural and electrical properties attributed to their reduced dimensions [4]. The negligible presence of grain boundaries and dislocations on their inside guarantees fast and reversible chemical transduction reactions at their surfaces upon exposure to gas molecules [5,6,7].

ϒ Current affiliation: Institut de Recerca en Energia de Catalunya (IREC)

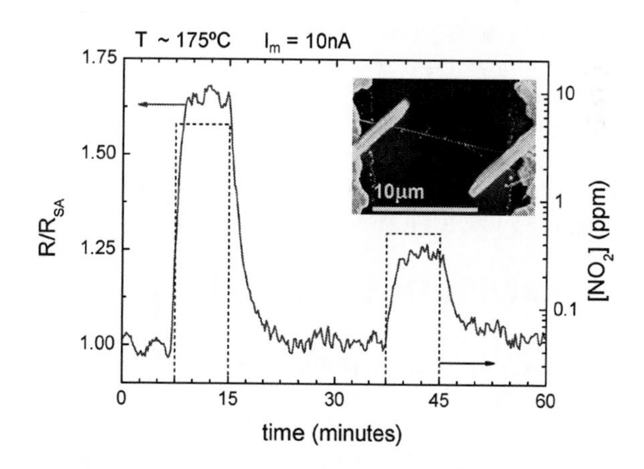

FIGURE 1. Response of one of the here-presented devices towards two pulses of NO_2 diluted in synthetic air. The self-heating current (I_m) required to reach the optimum sensing temperature (T) is indicated. (Inset) SEM image of the nanodevice (r_{NW} ~ 35±2 nm, typical interelectrode distances L_{NW} of few µm).

CP1137, *Olfaction and Electronic Nose: Proceedings of the 13th International Symposium,* edited by M. Pardo and G. Sberveglieri

It is noteworthy that the reduction of the sensing area of nanowire-based devices does not mean a decrease of their responses towards gases, since the modulation of the electrical resistance is directly proportional to the number of surface sites per unit area interacting with gas molecules, and regardless of the total effective surface. That is to say, the same relative gas response is monitored with a single nanowire prototype than the obtained with multiple nanowire-based devices contacted in parallel, provided that the nanowires' radii are the same in both cases [8].

This contribution will focus on the ultimate properties associated to the use of individual metal oxide nanowires as building components of advanced chemical sensors, and specially on the prospects opened by this novel operation methodology.

EXPERIMENTAL METHODS

Individual single-crystal SnO_2 nanowires, synthesized by catalyst supported chemical vapor deposition of a molecular precursor $[Sn(O^tBu)_4]$ [4], were transferred onto suspended silicon micromembranes equipped with Pt interdigitated microelectrodes. Individual nanowires (see inset in Figure 1) were electrically contacted to platinum microelectrodes by a lithography process using a FEI Dual-Beam Strata 235 FIB instrument [9]. Two and four-probe DC measurements were performed to take into account parasitic contacts effects [1] using an electronic circuit designed to guarantee and control low current levels (0.1 nA), avoiding any undesired fluctuation [10]. This enables the control of self-heating effects [1,2]. A microhetaer integrated in the micromembrane was used to calibrate the self-heating effect [3].

In the following R stands for the resistance value of the nanowire and R_{SA} indicates the reference resistance value in synthetic air (SA).

RESULTS AND DISCUSSION

Self-Heating Operated Gas Sensor

We demonstrated that the self-heating operated gas sensors are responsive to gases like NO_2, CO or to the oxygen content in air. In order to optimize the response to these different gas species, different current values were applied: values ranging from 100 pA to 300 nA allowed us setting the temperature from ambient to 300 ºC. Figure 1 and 2 shows the response of our devices to NO_2 and CO. In the case of NO_2 and CO it was possible to detect concentrations in the range of the hazardous limits [11]. The fact that the response of the wires towards different gases can be tuned by changing the bias current suggests that this technology should be good enough to develop electronic nano-noses based on differently heated nanowires.

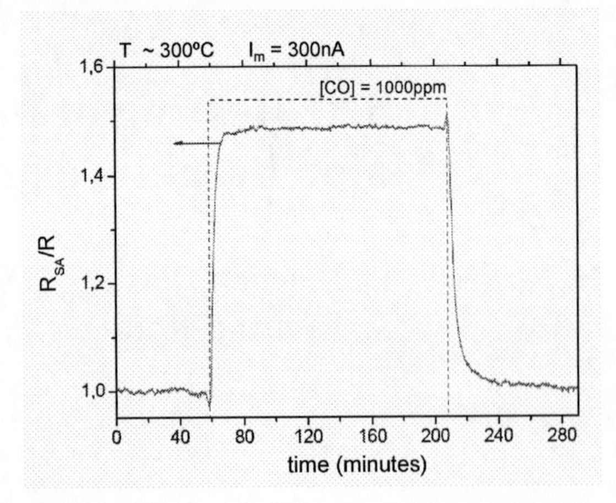

FIGURE 2. Response of one of the here-presented devices towards a long pulse of CO diluted in synthetic air. The self-heating current (I_m) required to reach the optimum sensing temperature (T) is indicated. A stable baseline is observed during the pulse.

An important consequence of the integration of the heater in the sensing nanomaterial is the dramatic reduction of the heating volume (from 10^6 μm^3 to 10^{-1} μm^3) and consequently, of the power requirements. According to our preliminary measurements, these nanowires could operate with less than 20 μW to both bias and heat them [2]. This estimation is more than two orders of magnitude lower than the power consumption of the microheaters.

Moreover, the response to gases can be used as an indirect method to estimate the effective temperature of the nanowire. This is an important progress in measuring technologies at the nanoscale and further details can be found elsewhere [3].

Sensor Stability, Reversibility and Reproducibility

The nanowires are single-crystalline materials with surfaces free of nooks and crannies which guarantee fast and reversible adsorption / desorption of gas molecules in comparison to the typical behavior experimentally observed in thin films sensors.

In Figure 2 we show an example of this assertion. The response of an individual nanowire at 300 ºC (applying I_m = 300 nA) towards a 1000 ppm of CO diluted in air was stable and fully reversible after exposure to the gas.

Ten devices were fabricated using SnO_2 nanowires with lengths (L_{NW}) between 5 and 15 μm, and radii (r_{NW}) between 35 and 45 nm, which showed reproducible electrical responses in experiments repeated along 4 weeks. During this time period, drifts in the baseline resistance (R_{SA}) of less than 10% were observed. No degradation of the contacts was observed, provided that working temperatures below 300ºC were used. This preliminary analysis of the reliability of these devices revealed attractive features for integrating these devices in future e-nose systems.

On the Response Dynamics

As far as the dynamic response is concerned, the use of individual nanowires represents an optimum scenario.

Firstly, the use of individual nanowires instead of bundles of nanowires or porous materials reduces the diffusion time of gas through the cavities of the material.

Second, the well defined, ordered and stable surfaces of these wires correspond to the most ideal scenario for the gas-surface interaction, where the dynamics are only limited by the reaction kinetics between surface atoms and gas molecules.

Third, from the thermal stabilization point of view, the mass of the nanowires is orders of magnitude lower than any conventional heater and therefore, there is no limit for this technology to achieve shorter stabilization times for faster pulsed operation.

CONCLUSIONS

We have demonstrated that the self-heating operated gas sensors based on individual nanowires present a number of advantages, such as the ultralow power consumption, the integration of the heater in the nanoscale, and the fast, reversible, and stable responses. The low energy requirements of these devices open the door to the development of fully autonomous sensor devices powered with energy scavenging systems. All these properties are also attractive for the future development of electronic nano-noses based on arrays of these devices implemented with self-assembly techniques, such as dielectroforesis, or other techniques like electrospinning or ink-jet printing.

ACKNOWLEDGMENTS

A Spanish patent application (P200900334) related to the above disclosed features has already been filled [12].

This work was partially supported by the Spanish Government [projects N – MOSEN (MAT2007-66741-C02-01), and MAGASENS], the UE [project NAWACS (NAN2006-28568-E), the Human Potential Program, Access to Research Infrastructures]. JDP and RJD are indebted to the MEC for the FPU grant. Thanks are due to the European Aeronautic Defense and Space Company (EADS N.V.) for supplying the suspended micromembranes.

REFERENCES

1. F. Hernandez-Ramirez, A. Tarancon, O. Casals, E. Pellicer, J. Rodríguez, A. Romano-Rodriguez, J. R. Morante, S. Barth and S. Mathur, *Phys. Rev. B* **76**, 085429 (2007).
2. J. D. Prades, R. Jimenez-Diaz, F. Hernandez-Ramirez, S. Barth, A. Cirera, A. Romano-Rodriguez, S. Mathur and J. R. Morante, *Appl. Phys. Lett.* **93**, 123110 (2008).
3. J. D. Prades, R. Jimenez-Diaz, F. Hernandez-Ramirez, S. Barth, P. Jun, A. Cirera, A. Romano-Rodriguez, S. Mathur, and J.R. Morante, *Int. J. Nanotechnol.*, submitted (2008).
4. (a) S. Mathur, S. Barth, H. Shen, J.-C. Pyun and U. Werner, *Small* **1**, 713-717 (2005). (b) S. Mathur and S. Barth, *Small* **3**, 2070-2075 (2007).
5. F. Hernández-Ramírez, A. Tarancón, O. Casals, J. Arbiol, A. Romano-Rodríguez, and J. R. Morante, *Sens. Actuators B: Chem.* **121**, 3-17 (2007).
6. F. Hernandez-Ramirez, S. Barth, A. Tarancon, O. Casals, E. Pellicer, J. Rodriguez, A. Romano-Rodriguez, J.R. Morante, and S. Mathur, *Nanotechnol.* **18**, 424016 (2007).
7. F. Hernandez-Ramirez, J. D. Prades, A. Tarancon, S. Barth, O. Casals, R. Jimenez-Diaz, E. Pellicer, J. Rodriguez, J. R. Morante, M. A. Juli, S. Mathur, and A. Romano-Rodriguez, *Adv. Funct. Matter.* **18**, 2990-2994 (2008).
8. T. Andreu, J. Arbiol, A. Cabot, A. Cirera, J.D. Prades, F. Hernandez-Ramirez, A. Romano-Rodriguez, J.R. Morante, "Nanosensors: Controlling Transduction Mechanisms an the Nanoscale Using Metal Oxides and Semiconductors", in *Sensors Based on Nanostructured Materials* edited by F. Arregui. Springer Science + Business Media, 2008, pp. 79-129.
9. F. Hernandez-Ramirez, A. Tarancon, O. Casals, J. Rodríguez, A. Romano-Rodriguez, J. R. Morante, S. Barth, S. Mathur, T. Y. Choi, D. Poulikakos, V. Callegari and P. M. Nellen, *Nanotechnol.* **17**, 5577-5583 (2006).
10. F. Hernandez-Ramirez, J. D. Prades, A. Tarancon, S. Barth, O. Casals, R. Jimenez-Diaz, E. Pellicer, J. Rodríguez, M. A. Juli, A. Romano-Rodriguez, J. R. Morante, S. Mathur, A. Helwig, J. Spannhake and G. Mueller, *Nanotechnol.* **18**, 495501 (2007).
11. World Health Organisation. Information available in http://www.who.int/peh/air/Airqualitygd.htm.
12. F. Hernandez-Ramirez, J. D. Prades, J. R. Morante, A. Cirera, A. Romano-Rodríguez, Spanish patent application No. P200900334 (2 February 2009).

INSTRUMENTATION: SAMPLING II

Development of a diagnostic aid for bacterial infection in wounds

A.M. Pisanelli,° K.C. Persaud,° Bailey A. °, M. Stuczen*, R.Duncan+ and K. Dunn +
°CEAS-Manchester (UK), *MMU (UK), +Wythenshawe Hospital, Burns Unit, Manchester (UK)

Abstract

Infection of wounds during hospitalisation often induces morbidity and sometimes mortality. The delay in patient recovery and subsequent increased length of hospital stay also has economic consequences. Standard techniques for microbiological detection are surface swabbing and wound biopsy culture. Surface swabbing is the most commonly used technique mainly because is quite inexpensive and is not invasive but can give only a representation of surface infection and analysis is also time consuming. Infected wounds are often characterised by an offensive odour that can be used as a diagnostic parameter. We report the results obtained by examining swabs and dressings taken from patients using a gas sensor array instrument developed as part of an EU funded project WOUNDMONITOR.

1. Introduction

Infected wounds often are characterised by an offensive odour [1]. The severity of this malodour is associated with the types of colonising micro-organisms, which usually includes both aerobic and anaerobic bacteria. The most common micro-organisms that induce wound infection such as Streptococcus *pyogenes,* Staphylococcus *aureus* and Pseudomonas aeruginosa, produce characteristic odours. There are a number of indicators of infection; these include the classic signs related to the inflammatory process and other subtle changes highlighted by Cutting and Harding [2]. The classic signs of infection include: localised erythema, pain, heat, cellulitis and abnormal smell. After a preliminary assessment, it is important to confirm this and identify the micro-organisms responsible of the infection and possible resistance to antibiotics. Methods used are:

a. Wound swabbing is the most common sampling method but is time consuming;

b. Quantative analysis (eg through wound biopsies). This can assist with the recognition of an increased bacterial burden but is invasive;

c. PCR is the most accurate but is quite expensive.

It is evident that an alternative faster, non time consuming method will be useful for screening for wound infections.

In this work we investigated the volatiles produced *in vitro* and *in vivo* by the most common bacteria responsible of wound infection in burns patients using a Hybrid Gas Sensor System containing both metal sensors and conducting polymers in order to assess if a this can be a valid aid for screening for wound infections.

2. Experimental and Methods

We carried out in *vitro* studies on pure bacteria cultures -

a) Pseudomonas *aeruginosa* (4 different strains)

b) Streptococcus *pyogenes* (2 different strains)

c) Staphylococcus *aureus* (7 different strains including two types of S. *aureus* MRSA) in order to identify bacteria markers.

GC-MS was carried out by using a Varian Saturn 2000 GC/MS.

Patient samples: swabs and dressings were analysed.

Microbial laboratory analysis was also carried out to identify bacterial infection from swab and dressing samples taken from patients.

Pre-concentration of the headspace volatiles from patient samples or bacterial cultures was performed with SPME fibers (Supelco 75μm Carboxen/ PDMS and 65μm PDMS/DVB) for 15′ at 37°C [3].

More than 150 swabs and dressings taken from patients at the Burns Centre-Wythenshawe Hospital, Manchester (UK) have been analyzed by using a Gas Sensor System containing commercial oxide sensors (Figaro, Applied Sensors) as well as thin film metal oxide sensors provided from CNR, Brescia (Italy) and Semiconductor Physics Institute Vilnius (Lituania). A hybrid sensor array system was also tested in addition containing conductive polymers from CEAS, Manchester (UK). The conductive polymers synthesized have a better selectivity to the volatiles that are markers of the bacteria activity.

Poly-3-hexanoylpyrrole (P4), poly-N-pentafluorobenzylpyrrole (N6) and poly-N-phenylpyrrole (N7) were polymerised with four different organic dopants; tosylate (TSA) for P4, tetraborofluorate (BF4), hexafluorophosphate (PF6), perchlorate (ClO4) and, for N6 and N7, triflate (OTF).

3. Results

3.1 Determination of volatile biomarkers

Gas chromatography-mass spectrometry (GC-MS) headspace analysis of the most common bacterial species found to be present in wounds has been used to determine the key volatile markers useful for sensor detection (Table 1).

It was possible to select sensors better able to discriminate between infected and uninfected wounds.

CP1137, *Olfaction and Electronic Nose: Proceedings of the 13th International Symposium*, edited by M. Pardo and G. Sberveglieri

Table 1 Identified Bacterial Markers

Bacteria	Volatile markers
P. *aeruginosa*	Ethanone 2-tetradecanol
S. *pyogenes*	2-nonanone 1-butanol
S. *aureus*	2-dodecanone acetic acid isovaleric acid

3.2 Analysis of patient samples

Microbial laboratory analysis was also carried out to identify bacterial infection from swab and dressing samples taken from patients. Table 2 shows preliminary results of microbiological analysis, which can be used to verify the performance of the developed instrumentation for identification of infected patient's samples.

Table 2 Microbial laboratory analysis results from samples taken from patients

Patient ID	Sample type	Pseud.	S. aureus	Strept.	Comment
011	Swab	-	-	-	
	Dressing	-	500	-	Infected
013	Swab	-	-	-	
	Dressing	-	-	-	
014	Swab	-	-	-	
	Dressing	-	1000	-	Infected
015	Swab	-	-	-	
	Dressing	-	-	-	
016	Swab	-	-	-	
	Dressing	-	-	-	
018	Swab	-	-	-	
	Dressing	-	-	-	
019	Swab	-	-	-	
	Dressing	-	-	-	
020	Swab	-	-	-	
	Dressing	-	-	-	
021	Swab	4160000	-	-	Infected
	Dressing	100000	-	-	Infected
022	Swab	120000	-	-	Infected
	Dressing	18000	550000	-	Infected

Figure 1 illustrates a Principal Component Analysis (PCA) obtained from the sensor array using SPME for preconcentration of the headspace from swab samples taken from patients. It is shown that patients 21-22, who were diagnosed infected from microbial laboratory analysis as shown in Table 2, are discriminated from the uninfected.

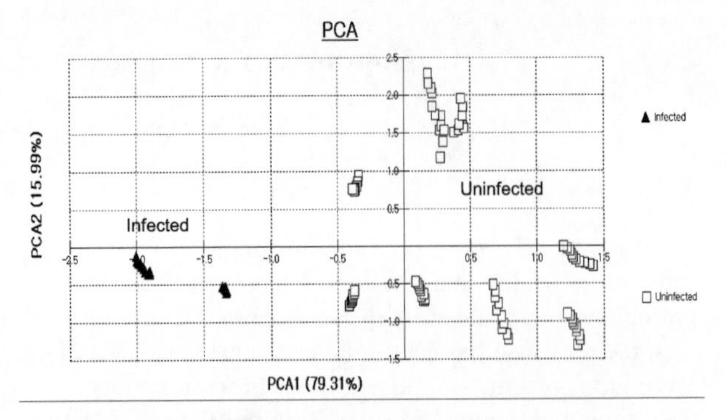

Figure 1 PCA results of data obtained from the Sensor Array/SPME system from swabs taken from patients

Dressing samples taken from patients also were analysed by using the same technique, and a PCA is displayed in figure 2. Of four patients whose dressings are infected from microbial analysis, three are discriminated while one patient is not classified due to the low number of microbial counts verified by microbial analysis.

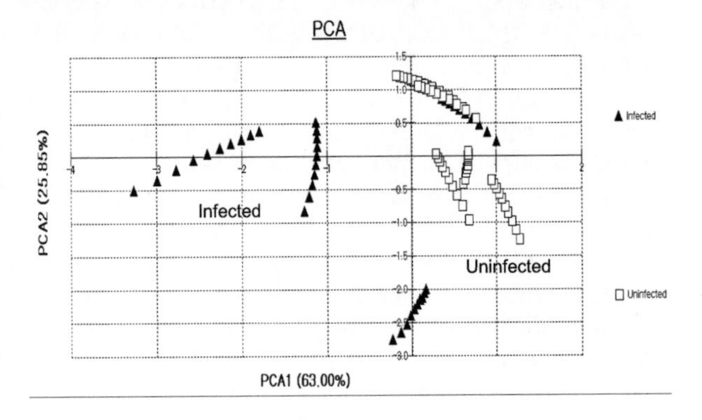

Figure 2 PCA result of data obtained from the system with dressing samples taken from patients

4. Discussion

The analysis of dressings appears to be more sensitive than swab samples. The PCA results of data obtained give indication that infected patients may be discriminated from uninfected patients, and the microbial laboratory analysis for samples taken from patients verifies the performance of the system. The system is currently being validated on a large patient sample.

5. Conclusions

An electronic odour recognition system has been developed for detection of bacteria types at the early stage of wound infection. GC-MS studies support the conclusion that typical infectious agents in clinical practice can be distinguished from each other by means of a limited set of key volatile products. To increase reliability of bacteria identification SPME pre-concentration was used for sampling of the headspace air and response to variable concentrations of volatiles emitted from the SPME fibre is processed for evaluation of output parameters of the sensor module. PCA analysis of the dynamic parameters of sensor responses to the headspace air of clinical samples collected by swabbing give indications that discrimination between classes of volatile products may be possible given suitable pattern recognition techniques. Microbial laboratory analysis using clinical samples taken from patients demonstrated acceptability of the system for discrimination between infected patients and uninfected patients. Preliminary results at this stage appear to be promising. However, more clinical validation for the instrument is needed in hospitals where patients with wounds are treated.

Acknowledgments

This project was funded by FP7-IST-2004-027859 WOUNDMONITOR and the work reported is a joint collaboration between the following partners:
Persaud K C, Pisanelli A M, Bailey A - University of Manchester, Manchester, UK
Falasconi M, Pardo M, Sberveglieri G - INFM, CNR, Brescia , Italy
Senulienė D, Šetkus A, Semiconductor Physics Institute, Vilnius, Lithuania
Rimdeika R - Kaunas Medical University Hospital, Kaunas, Lithuania
Dunn K - South Manchester University Hospital, Manchester, UK
Gobbi, M - Biodiversity Srl, Brescia, Italy
Schreiter, U , Preis, S - UST GmbH, Gera, Germany

References

1. Bowler, P. G. , Duerden, B. I. and Armstrong, D. G.. Wound Microbiology and Associated Approaches to Wound Management. Clinical Microbiology Reviews, (2001) Vol. 14, No. 2, pp. 244-269
2. Cutting, K. F. , White, R.J. Criteria for Identifying Wound Infection . British Journal of Community Nursing. 2004, 9 (3 Suppl), S6–S16.
3. Bailey, A.L., Pisanelli, A.M., Persaud, K.C. Development of conducting polymer sensor arrays for wound monitoring. Sensors and Actuators. B, Chemical (2008) 131, (1), pp. 5-9

Increasing Electronic Nose Recognition Ability by Laser Irradiation

Massacane Alejandra[1], Vorobioff Juan[1], Pierpauli Karina[1], Boggio Norberto[1, 2], Reich Silvia[3], Rinaldi Carlos[1, 2], Boselli Alfredo[1], Lamagna Alberto[1], Azcárate M. Laura [2,4], Codnia Jorge [4], Manzano Francisco[4]

1- Comisión Nacional de Energía Atómica (CNEA)- Gerencia de Desarrollo Tecnológico y Proyectos Especiales (Grupo MEMS) Av. General Paz 1499 – San Martín – Buenos Aires
2- Consejo Nacional de Investigaciones Científicas y Técnicas, Argentina
3- Escuela de Ciencia y Tecnología, Universidad de San Martín, Argentina
4- Centro de Investigaciones en Láseres y Aplicaciones CEILAP (CITEFA-CONICET), Argentina

Abstract: We present a method to increase the capability of an electronic nose to discriminate between a priori similar odours. We analyze the case of olive oil because it is well known that the characteristics of its aroma impair in many cases the discrimination between different kinds of olive oils especially when they are from similar geographic regions. In the present work we study how to improve the electronic nose performance for the above mentioned discrimination by the use of two IR laser wavelengths for vaporization.

Keywords: laser, e nose, olive oil,
PACS: 85.85+j

INTRODUCTION

Argentina has an increasing interest in assessing olive oil quality since it has become an important exportation product. Most indicators of this activity have improved in recent years, both for table olives and for olive oils, thus allowing Argentina to increase its market share and its positioning as a global producer. In 2006 Argentina was the 10th major producer of olive oil.

Odour is an important parameter determining the sensory quality of olive oils and it is therefore of interest to investigate if volatile compounds contributing to the characteristic odour can be measured as a quality indicator. In the last decades many efforts have been made to study the aromatic fraction of olive oils based mainly on chromatographic determinations [1]. However, these analytical techniques are time-consuming and require sophisticated equipment as well as skilled personnel.

The use of an electronic nose for quality evaluation as a means of olfactory sensing is becoming widespread due to its advantages of low cost, good reliability and high portability. Electronic noses based on different sensor technologies and using different recognition schemes have been employed for this task [2, 3].

On the other hand the high concentration of volatile compounds are not necessarily the principal contributors of odour. For example, Reiners and Grosch [4]reported a concentration of 6770 lg/g for trans-2-hexenal with an odour activity value of 16 whereas 1-penten-3-one with a much lower concentration of 26 lg/g had a higher odour activity value of 36. [5]

The aroma of virgin olive oil is characterized by various volatile compounds that include carbonyl compounds, alcohols, esters and hydrocarbons [6]. The C6 and C5 substances, especially C6 linear unsaturated and saturated aldehydes and alcohols, represent the most important fraction of the volatile compounds [5, 7]. Also, from a quantitative point of view, large amounts of these substances are generally found in high quality oils. These are the compounds which are responsible for the ''green flavor" of the virgin olive oil, while the esters are mainly related to the ''floral" sensory notes.

When samples of olive oil are analyzed with an electronic nose, the standard procedure is to put a fixed quantity in a vial and sense the headspace. The important drawback of this

CP1137, *Olfaction and Electronic Nose: Proceedings of the 13th International Symposium*, edited by M. Pardo and G. Sberveglieri

method is that the concentration of some compounds in the headspace may be quite different than their concentration in the liquid phase. It is for example the case for methanol and ethanol [1], whose concentrations in the vapor phase are significantly higher than their presence in the liquid and, furthermore, of no importance in order to define the olive oil characteristics. On the other hand, those substances responsibles of the organoleptic properties (such as hexanal, and trans-2-hexanal) which are abundant compounds [8, 9] in the oil due to their low volatility, are scarcely present in the headspace.

In the present work we study the relevance of improving the electronic nose performance for discriminating between different olive oils by the use of two IR laser wavelengths. Heating of the liquid sample by the use of an IR laser is a promising technique to modify the headspace either by volatilizing other organic compounds or by cracking them and thus improve the selectivity of the electronic nose overall response.

The role of the laser wavelength is particularly analyzed since the quality and the quantity of the chemical compounds incorporated to the headspace by laser heating depend on the irradiation techniques. We show that the increase in the electronic nose selectivity is dramatically changed by the use of laser irradiation and it is rather insensitive to the recognition pattern employed.

EXPERIMENTAL

We prepared 5 samples of two different olive oils bought at the local market. Henceforth denominated oil A and oil B. About 15 ml of oil was placed inside a 100 ml vial and the headspace was analyzed using a Cyrano 320^{TM} nose. The nose response is composed of the 32 sensors signals.

The three following analytical methods were implemented to measure the headspace of the olive oil samples.

Method I: The samples were analyzed without laser irradiation.

Method II: The oil samples, A and B, were irradiated with Nd:YAG laser (Continuum, Surelite I) pulses of 1064 nm at a repetition rate of 10 Hz. The pulse length was 5 ns, and the output energy was 80 mJ. The laser radiation was collimated, the area of the spot on the sample surface was 0.15 cm^2 and the laser fluence 2.14 J/cm^2.

Method III: The oil samples, A and B, were irradiated with a homemade pulsed TEA CO_2 laser [10] emitting in the P(20) line, (10.59 μm), of the 10.6 μm emission band. The pulse length was 100 ns and the output energy was 1.45 ± 0.04 J/pulse. The laser radiation was focused on the sample using a 50 cm focal length mirror. The area of the spot on the sample surface was 0.68 cm^2 and the laser fluence 2.14 J/cm^2.

It has to be stressed that the same protocols were applied to each method The protocols concerned not only the electronic nose parameters but also the ambient temperature and humidity since all measurements were performed in the same air conditioned laboratory.

The software provided by the Cyrano 320^{TM} e-nose allowed the processing of the raw data given by the sensors. Figure 1 shows typical data acquired by one of the 32 sensors. In order to compare the effects of the laser irradiation, the same olive oil sample, A, was measured under the three methods described above. As it can be seen in Figure 1, the signal-to-noise ratio (S/R) is considerably increased by the laser vaporization (blue and red lines) and the blue line corresponding to Method III shows the highest S/R ratio.

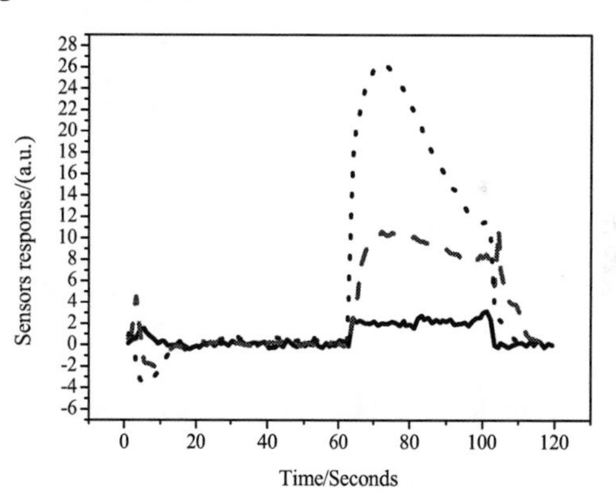

FIGURE 1: Time dependence sensor response. (—) Method I, (----) Method II, and (····) Method III.

RESULTS AND DISCUSSION

The software provided by the Cyrano 320 e-nose takes the ratio of the maximum value of each signal to the base-line value and provides a Principal Components Analysis (PCA). This type of analysis is not quite adequate when the different sensors may have distinct desorption times, as it was in our case and so, in order to take a better advantage of the information contained in each signal, we have taken the integral of each signal divided by the base line integral. This procedure takes into account the absorption and desorption rates.

Figures 2, 3 and 4 show the PCA results for Methods I, II and III, respectively. The reliability of the PCA is verified since, for all methods, the largest contribution to the total variance is given by the first two principal components. It can be straightforward deduced from Figures 2, that both olive oils are indistinguishable when using Method I. On the other hand, they are well discriminated when the sample is irradiated by any of the two IR láser and the headspace is subsequently measured (Method II and III). Figure 3 and 4 illustrate this fact.

In order to quantify the PCA discrimination we have used the silhouette value (from MatLab 7.0 Software). This parameter measures the existence of distinct clusters and it is defined for the score plane as:

$$S(i) = \frac{(\min(b(i,k)) - a(i))}{\max(a(i), \min(b(i,k))}$$

where *a(i)* is the average distance from the ith point to the other points in this cluster, and *b(i,k)* is the average distance from the ith point of cluster *i* to cluster *k*. The silhouette parameter ranges from -1 to +1 and the best discrimination corresponds to value 1.

Table I: silhouette values for PCA discrimination for the different methods.

Method I	Method II	Method III
0.486	0.869	0.954

According to these results the best discrimation of these olive oils is performed by Method III.

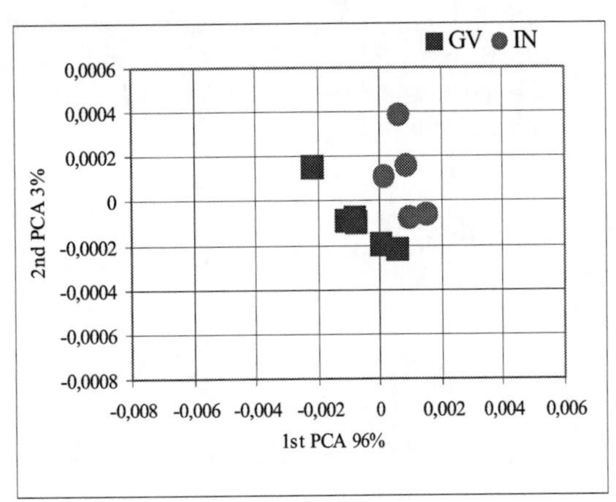

FIGURE 2: PCA for olive oils analyzed with Method I

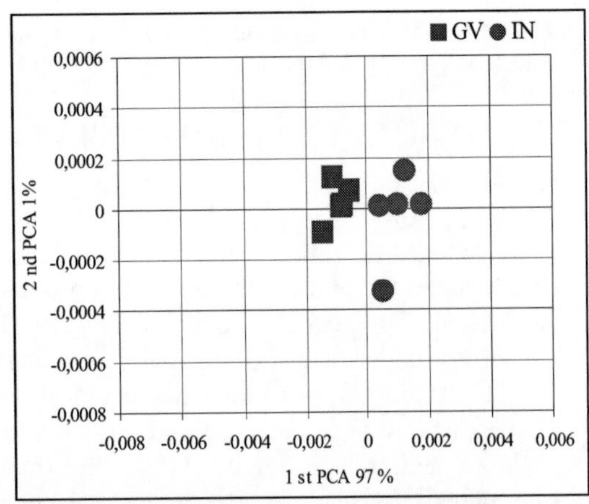

FIGURE 3: PCA for olive oils analyzed with Method II

CONCLUSIONS

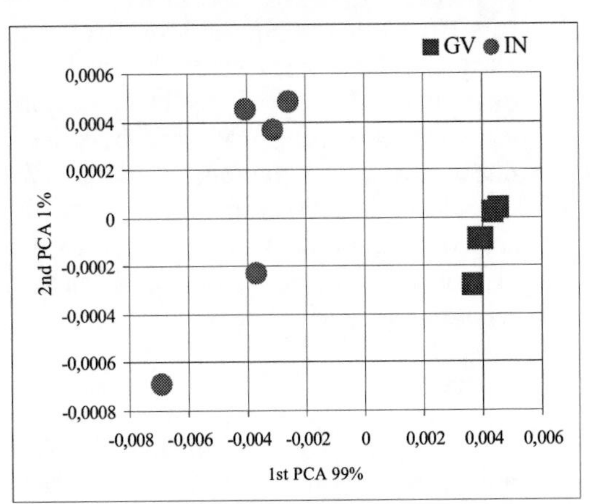

FIGURE 4: PCA for olive oils analyzed with Method III.

This work reports a method to improve the ability of an electronic e-nose to discriminate between *a priori* similar odours by laser irradiation of the sample. We have applied this technique to the case of virgin olive oils. We have also explored how different laser characteristics may improve this task.

We have concluded that the use of IR laser increases the sensitivity of the e-nose performance. Furthermore, the use of a CO_2 laser allows a better discrimination than the use of the Nd:YAG laser. The IR láser wavelength influence the discriminatoy capabilities of the method. It has also to be mentioned that, the signal to noise ratio (S/N) using the CO_2 laser is increased by an order of magnitude with respect to the S/N without laser vaporization effect.

A coming work is in progress where we anlalise the influence of the IR laser energy in the sensitivity of the method.

REFERENCES

[1]- Characterization of olive oil volatiles by multi-step direct thermal desorption–comprehensive gas chromatography–time-of-flight mass spectrometry using a programmed temperature vaporizing injector
Journal of Chromatography A, Volume 1186, Issues 1-2, 2008, Pages 228-235
Sjaak de Koning, Erwin Kaal, Hans-Gerd Janssen, Chris van Platerink, Udo A.Th. Brinkman

[2]- The potential of different techniques for volatile compounds analysis coupled with PCA for the detection of the adulteration of olive oil with hazelnut oil
Food Chemistry, Volume 110, Issue 3, 1, 2008, Pages 751-761
Sylwia Mildner-Szkudlarz, Henryk H. Jeleń

[3]- Evaluation of different storage conditions of extra virgin olive oils with an innovative recognition tool built by means of electronic nose and electronic tongue
Food Chemistry, Volume 101, Issue 2, 2007, Pages 485-491
M.S. Cosio, D. Ballabio, S. Benedetti, C. Gigliotti

[4]- Odorants of virgin olive oils with different flavor profiles. Journal of Agricultural and Food Chemistry, 1998, Volume 46, Issue 7, Pages 2754 -2763

[5]-Olive oil volatile compounds, flavour development and quality: A critical review.
Food Chemistry Volume 100, Issue 1, 2007, Pages 273-286
C.M. Kalua , a, , M.S. Allena, D.R. Bedgood, Jra, A.G. Bishopa, P.D. Prenzlera and K. Robardsa.

[6]- F. Lacoste, in: Proceedings of the User Forum of the NOSE network, Ispra, VA, 1999.

[7]-Monitoring of virgin olive oil volatile compauns evolution during olive malaxation by an array of metal oxide sensors
Food Chemistry, Volume 113, 2009, Pages 345 - 350
Esposto E., Montedoro J., Selvaggini R., Rico I., Taticchi A., Urbani S., Servili M.

[8]- Geographical origin and authentication of extra virgin olive oils by an electronic nose in combination with artificial neural networks
Analytica Chimica Acta, Volume 567, Issue 2, 2006, Pages 202-210
Maria S. Cosio, Davide Ballabio, Simona Benedetti, Carmelina Gigliotti

[9]- Comparison and integration of arrays of quartz resonators and metal-oxide semiconductor chemoresistors in the quality evaluation of olive oils
Sensors and Actuators B: Chemical, Volume 78, Issues 1-3, 2001, Pages 303-309
Corrado Di Natale, Antonella Macagnano, Sara Nardis, Roberto Paolesse, Christian Falconi, Emanuela Proietti, Pietro Siciliano, Roberto Rella, Antonella Taurino, Arnaldo D'Amico

[10]-Construcción de un láser de CO_2 TEA.
Anales AFA Volume 8, 1996, Pages 34-36
D. Petillo, J. Codnia, M. L. Azcárate.

Metal Oxide Gas Sensor Arrays: Geometrical Design and Selectivity

Frank Röck, Nicolae Barsan, Udo Weimar

Institute of Physical and Theoretical Chemistry, University of Tuebingen
Auf der Morgenstelle 15, 72076 Tuebingen, Germany, frank.roeck@ipc.uni-tuebingen.de

Abstract. Metal oxide gas sensors are commonly used in electronic noses. Their peculiarity, of altering the gas composition during the sensing process, is often not considered in planning the layout of the sensor chamber. However, for measurements with low flow rates (e.g. static headspace measurements) this effect can't be neglected. Results obtained with home made thick film sensors demonstrate the influence of consumption on the sensor signal. Depending on the sensor arrangement and the measurement conditions the selectivity of the whole system can be increased, respectively – for an inappropriate choice of the parameters – even decreased.

Keywords: Consumption, catalytic conversion, selectivity.
PACS: 07.07.Df

INTRODUCTION

Metal oxide gas sensors are based on the conductivity changes produced by the electrochemical oxidation or reduction of the analyte gases at the sensor surface.[1] This means that they are "destructive" gas sensors, which alter the gas composition during the sensing process.[2] The consumption, or, better, the relative gas concentration change in the sensor chamber, depends on a multiplicity of factors. On the one hand, the sensor itself with its size, temperature, kind of sensitive layer and electrodes, doping, etc. and, on the other hand, the ambient conditions like the flow rate, the volume of the sensor chamber, the humidity level, and, especially, the nature of the analyte gas. When using metal oxide sensors in sensor arrays, this fact should always be taken into account in the planning of the set up and the sensor chamber geometry; the first sensors in the flow path can be seen not only as sensing units but also as filters for the following ones. The effect is specific for each target gas and comparable with the use of a separate catalytic filter.[3, 4] In this work, the influence of the sensor position, respectively the sensor order is estimated. Two different target gases (methane and carbon monoxide) were chosen to demonstrate the boundaries of influence capabilities by the sensor arrangement.

EXPERIMENTAL AND METHODS

In a 4 channel gas mixing station defined compositions of cleaned air, humidity and combustible gases were obtained. The different channels were used for the following gases:

- purified air (produced by Schmidlin Ultra Zero Air Generator N-GT 1500; purity: hydrocarbons < 0,1 ppm; CO < 0,1 ppm; NO_x < 0,1 ppm; CO_2 < 5 ppm))
- humidified air (purified air was saturated with humidity by bubbling it through a water filled vessel at room temperature)
- CO from a gas cylinder (the concentration range was 50-1000 ppm)
- CH_4 from a gas cylinder (the concentration range was 50-1000 ppm)

After combining the different channels in a single tube, the test gas was purged under a defined flow rate through a standard Teflon measurement chamber (flow cross section 7 cm^2). In each experiment four identical sensors out of the same batch were heated up to 350 °C and placed in the test chamber. The samples used were undoped SnO_2 and 3% Pd doped SnO_2 sensors prepared by screen printing onto alumina substrates provided with platinum electrodes and platinum heater structures (details on the sensors are provided in ref. [2]). The composition of the off-gas was determined by an infrared gas analyzer (Innova Photoacoustic Gas Monitor 1312).

CP1137, *Olfaction and Electronic Nose: Proceedings of the 13th International Symposium,* edited by M. Pardo and G. Sberveglieri

In a second experiment it was avoided to purge the test gas through the sensor chamber in order to increase the consumption effect. Therefore, the passing sample gas is only transported by diffusion and convection into the sensor chamber. Attention was paid to increase the ratio between the amount of sensors and the sensor chamber volume.

A constant gas supply is ensured by coupling the chamber to the gas mixing system that delivers gas mixtures at a flow of 200 ml/min. It partly diffuses through the filter and is consumed by the sensors. In this setup the opening as well as the sensor chamber have small cross sections of 5 x 22 mm^2 only. We estimated that in this geometrical arrangement, each sensor consumption should be increased.

RESULTS AND DISCUSSION

Flow-Through Experiments

In a first approach the flow dependency of the consumption was investigated. In Figure 1 the results are shown for CO measurements with an initial gas concentration of 640 ppm and a relative humidity of 50%. For CO the uncoated substrate has the highest consumption followed by the Pd doped and the undoped sensors and one can also record a strong influence of the flow rate. It was found that the initial gas concentration has no influence on the percentage of the consumption and an increase in relative humidity only comes along with an insignificant decrease of the consumption.

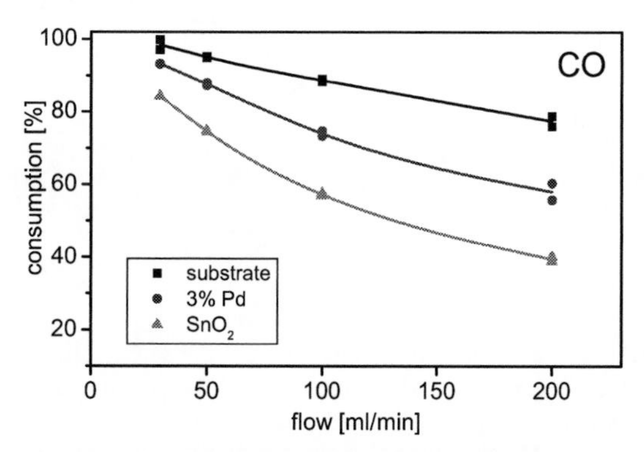

FIGURE 1. Consumption dependence on the flow rate. The measurements were performed with three different sets of four sensors (substrates without coating – 3% Pd doped SnO_2 layer – only SnO_2). The initial analyte gas concentration was 640 ppm and the relative humidity 50%. The figure shows the differences in consumption for CO.

Regarding CH_4, neither the uncoated substrate, nor the undoped SnO_2 sensor consumes any analyte gas (Figure 2). For the 3% Pd sensor the relative consumption is not as high as in the case of CO, but especially for low flow rates it is significant (more than 50% for a flow rate of 30 ml/min).

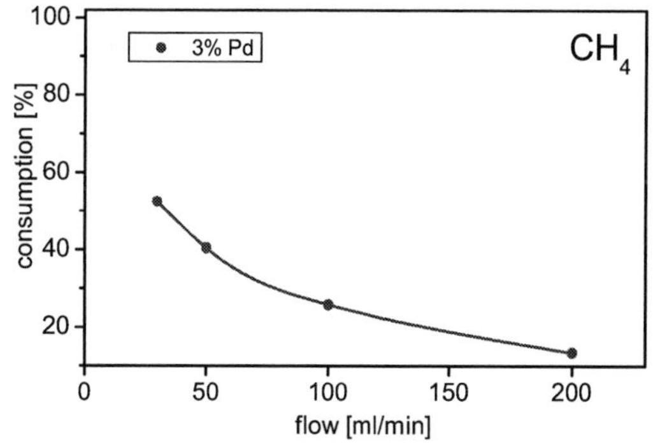

FIGURE 2. The measurements shown in Figure 1 were repeated with CH_4 (instead of CO). The initial analyte gas concentration was again 640 ppm and the relative humidity 50%. In this case only the Pd doped sensor has an effect on the gas composition.

Already by using the results of those experiments one can distinguish between CO and CH_4. In the case of undoped SnO_2 sensors only CO is consumed and, therefore, the last sensor in the flow direction should see a lower concentration compared to the first one. However, it is difficult to estimate the effect of turbulences within the chamber. For a laminar flow the maximal possible concentration gradient should be reached. But, in reality, convection causes always a gas mixing inside the sensor chamber and, therefore, the differences between the sensor signals are lower as expected (on the basis of previous measurements). Nevertheless, one can take advantage of the concentration gradient by changing the flow direction and compare the corresponding sensor signals. In Figure 3 an experiment with a flow trough volume of 50 ml/min is shown.

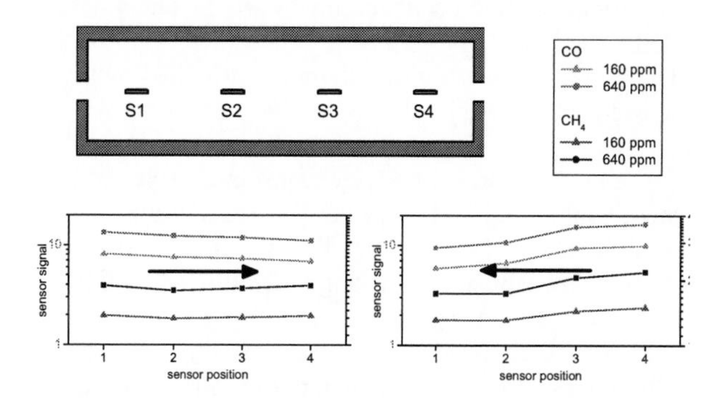

FIGURE 3. Sensor signal differences in dependency of the flow direction (flow rate of 50 ml/min) for undoped SnO_2 sensors. Each further sensor in the flow direction sees a lower CO concentration and consequently the sensor signal is lower as well.

Independent of the gas concentration, the same effect is observable. These can be made clearer by the help of a polar plot (Figure 4). The sensor signals of each sensor, each concentration (160 ppm and 640 ppm) and each gas were divided by the value recorded after changing the flow direction. It is obvious that the shape of the polar plot is depending on the test gas. The consumption effect causes a lower signal ratio for each subsequent sensor in the case of CO. For CH_4 this effect is inversed, which may be caused by changes in temperature along the sensor chamber. As expected, for a low flow rate the ability to distinguish between CO and CH_4 increases. Similar trends were also observed for 3% Pd doped sensors (not shown).

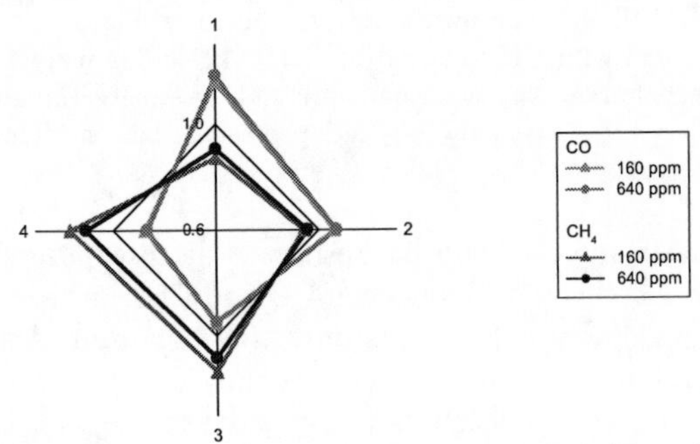

FIGURE 4. The polar plot was obtained by dividing the sensor signal of each individual sensor for both flow directions (compare Figure 3). The signal for CO decreases for each further sensor, whereas the effect is inversed for CH_4.

Diffusion Based Experiments

The initial experiments have clearly shown that with this concept it is possible to distinguish between different gases, id est gain selectivity. The best results were obtained by using low flow rates. Consequently, the effect should be increased by going one further step and use a diffusion controlled approach. Therefore, a new sensor chamber provided with 7 sensor sockets was used.

The new set-up allowed increasing the concentration gradient of the consumed gases along the chamber. As a result, identical sensors out of the same batch are sufficient to distinguish the test gases even without the need of changing the flow direction.

In a next step the sensor arrangement was investigated in regard of already existing selectivity differences of the sensors. Therefore, one doped (3% Pd) and one undoped SnO_2 sensor were placed in position 1, respectively, position 6. The results of the measurements are shown in Figure 5. Independently of the gas and its concentration, the undoped sensor (S6) is always showing higher sensor signals compared to the other one. Furthermore, the ratio between both sensor signals is very similar; for that reason it is not possible to discriminate between CO and CH_4.

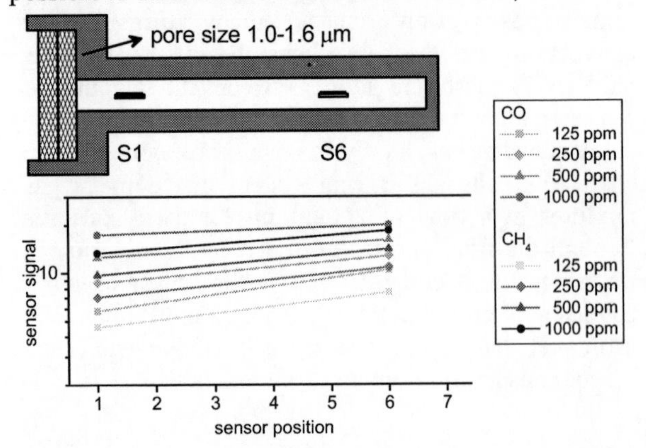

FIGURE 5. S1 is a 3% Pd doped sensor, whereas S6 is an undoped SnO_2 sensor. Differences in the consumption and in the selectivity of sensors with a different doping can be cancelled out by using them in the diffusion mode in a specific order. Therefore, it is not possible to distinguish between CO and CH_4.

If the sensor position is changed the difference in the sensor selectivity is increased. In Figure 6 the latter sensor in the diffusion direction (S6 – 3% Pd) shows higher sensor signals for CH_4 and lower sensor signals for CO compared to the first one.

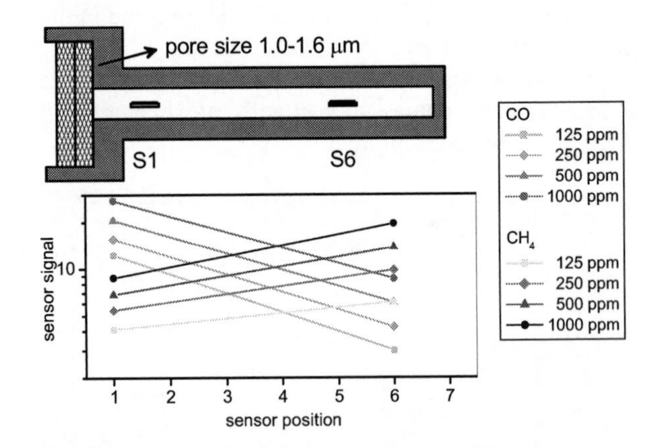

FIGURE 6. In comparison to Figure 5 the sensor order was turned around. Now the consumption effect amplifies the discrimination ability of the sensors. Between CO and CH_4 one can easily distinguish by comparing the sensor signals.

Only by comparing both sensor signals it is possible to determine if CO or CH_4 is present.

CONCLUSION

It was shown that the gas consumption of metal oxide sensors already alters the gas composition under standard measurement conditions. For common sensor arrays this is usually not intended and often ignored. However, this peculiarity can destroy or create selectivity.

Systematically exploited identical sensors out of the same batch are sufficient to distinguish between two sample gases. This can be reached by changing the flow direction during the measurement or by exclusive diffusion- and convection- controlled transport of gas molecules in the sensor chamber.

Furthermore, it was shown that the order of the sensor arrangement is responsible for an increase, respectively, a decrease of their particular selectivity differences.

ACKNOWLEDGMENTS

The authors acknowledge the financial support from the DFG project No. WE 3662/1-2, DUFT.

REFERENCES

1. N. Barsan and U. Weimar, "Understanding the fundamental principles of metal oxide based gas sensors; the example of CO sensing with SnO2 sensors in the presence of humidity," *Journal of Physics-Condensed Matter*, vol. 15, pp. R813-R839, May 28 2003.
2. W. Schmid, N. Barsan, and U. Weimar, "Sensing of hydrocarbons with tin oxide sensors: possible reaction path as revealed by consumption measurements," *Sensors and Actuators B-Chemical*, vol. 89, pp. 232-236, Apr 1 2003.
3. M. A. Portnoff, R. G. Grace, A. M. Guzman, P. D. Runco, and L. N. Yannopoulos, "Enhancement of MOS gas sensor selectivity by 'on-chip' catalytic filtering.," *Sensors and Actuators B*, vol. 5, pp. 231-235, 1991.
4. M. Fleischer, S. Kornely, T. Weh, J. Frank, and H. Meixner, "Selective gas detection with high-temperature operated metal oxides using catalytic filters," *Sensors and Actuators B-Chemical*, vol. 69, pp. 205-210, Sep 10 2000.

From Human to Artificial Mouth, From Basics to Results

Patrick Mielle [*1], Amparo Tarrega [1], Patrick Gorria [2], Jean Jacques Liodenot [2], Joël Liaboeuf [2], Jean-Luc Andrejewski [2], Christian Salles [1]

[1] *INRA, UMR FLAVIC, 17 rue Sully, 21065 Dijon, France,*
[2] *Plateform'3D, I.U.T Le Creusot, 12 rue de la fonderie, 71200 Le Creusot, France*
[*] *Patrick.Mielle@dijon.inra.fr, +33 3 80 69 30 86*

Abstract. Sensory perception of the flavor release during the eating of a food piece is highly dependent upon mouth parameters. Major limitations have been reported during in-vivo flavor release studies, such as marked intra- and inter-individual variability. To overcome these limitations, a chewing simulator has been developed to mimic the human mastication of food samples. The device faithfully reproduces most of the functions of the human mouth. The active cell comprises several mobile parts that can accurately reproduce shear and compression strengths and tongue functions in real-time, according to data previously collected in-vivo. The mechanical functionalities of the system were validated using peanuts, with a fair agreement with the human data. Flavor release can be monitored on-line using either API-MS or chemical sensors, or off-line using HPLC for non-volatile compounds. Couplings with API-MS detectors have shown differences in the kinetics of flavour release, as a function of the cheeses composition. Data were also collected for the analysis of taste compounds released during the human chewing but are not available yet for the Artificial Mouth.

Keywords: Mastication, simulation, aroma, release, food breakdown, mass spectrometry, electronic nose
PACS:

INTRODUCTION

Why should we simulate the human mastication by developing an Artificial Mouth, each of us is able to chew! This statement is of course true, but this is also the weak point: each human has a particular mouth, and a particular way of chewing, and thus receives his(her) proper sensory stimuli.

Our research aim to understand the sensory perception from our food intake. Before reaching our odour or taste receptors, the odorant or taste molecules should be released from the food during the chewing process.

Mastication is a complex process involving both the oro-facial muscles and the tongue. One of its main consequences is the food matrix deconstruction and impregnation with saliva, allowing flavor compounds to be released over time in the mouth.

During this process, flavor compounds are gradually released from the food into the mouth and thus attain the aroma and taste receptors, inducing our sensory perception.

Although temporal in-vivo studies enabled the direct release perception tests, numerous limitations were reported, such as major inter-individual differences, moderate intra-individual reproducibility and in addition the food sample must be acceptable to all the panelists.

The use of a chewing simulator to investigate in-vitro temporal analyses could overcome these limitations. Furthermore, time-intensity flavor release has been correlated to several masticatory parameters such as the chew rates, the number of chews, the work, the chewing time, the salivary flow and the masticatory performances [1], but it remained difficult to understand the individual effect of each of these parameters as in-vivo, many are correlated and vary simultaneously. So, the possibility of de-coupling these parameters using a chewing simulator is of a great interest to a better understanding of the flavor release phenomenon.

EXPERIMENTAL AND METHODS

The development of the Artificial Mouth itself has been described before [2, 3], so this paper will emphasis on the results and the comparison with the human chewing.

Chewing simulator tests: the principal functional parameters were adjusted separately: maximum bite force, maximum shearing force, biting speed, shearing speed, saliva flow and the number of cycles.

CP1137, *Olfaction and Electronic Nose: Proceedings of the 13th International Symposium,* edited by M. Pardo and G. Sberveglieri

The quantity of food was also varied. For the validation study, three peanuts were placed in the chewing simulator, and "chewed" samples were obtained after 4 and 8 mastication cycles. Different procedures were studied, varying in terms of the maximum bite force (24, 29 and 34 daN) or the shearing angle value (0, 1/8th, 1/4th tooth). Three replicates were performed for each procedure.

Human parameters measurement: all the parameters that are used in the Artificial Mouth were gathered from humans, using electromyography for the bite force, speed and acceleration. Bite sensors were used to calibrate the electrodes voltages for each panelist.

Human chewing tests: The subject group consisted of volunteers without any functional mastication problems. They were asked to chew three peanuts, and after 4 and 8 chewing cycles, to spit the sample into a coffee filter and rinse their mouth with water, which was also collected in the filter. The chewed sample dried for 24 h at room temperature. This procedure was carried out three times for each number of chews.

Particle size measurements: The degree of fragmentation of both in-vivo and in-vitro chewed samples was studied by measuring the weight of masticated peanuts that could pass through both 4 mm and 2 mm mesh sieves, and then the weight percentages were calculated of particles larger than 4 mm, particles between 2 and 4 mm and particles smaller than 2 mm [4].

Model cheeses: Different model cheeses were made, with different fat contents (20 and 40 %), dry matter (37 and 44 %) and salts (0.5 and 1.5 %) contents. Milk pH was 6.2 and 6.5.

Coupling with ion detectors: two types of atmospheric mass spectrometers (MS) were coupled:

- atmospheric pressure chemical ionisation (APCI-MS), in Dijon, France (figure 1) ;

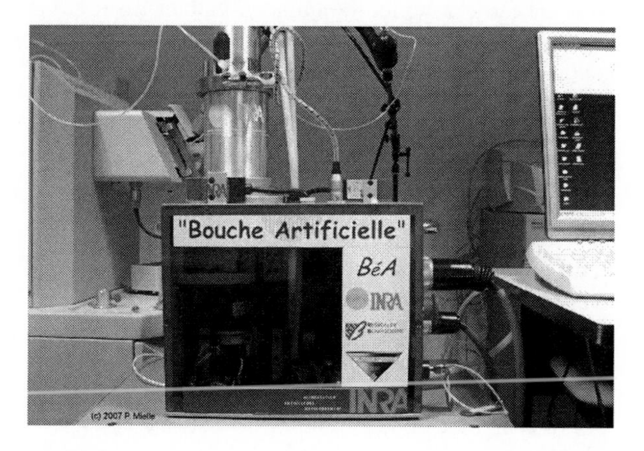

FIGURE 1. The chewing machine coupled to a APCI-MS.

- proton transfer reaction (PTR-MS), in Grignon, France (figure 2)

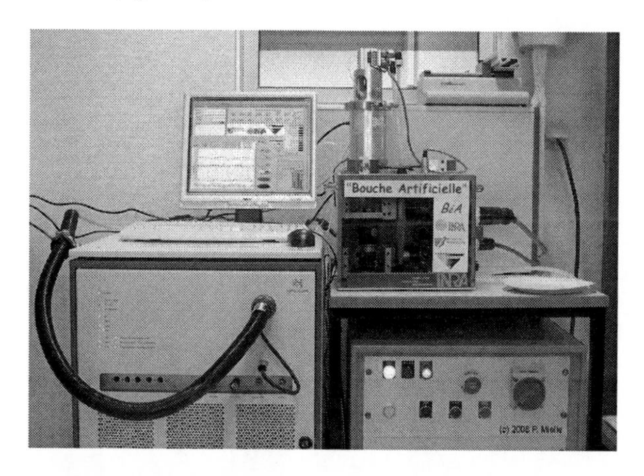

FIGURE 2. The chewing machine coupled to a PTR-MS.

The MS were used in the SIM collection mode, and a couple of ions were collected. For example, the ion of m/z 145 is a marker of the release of ethyl hexanoate, m/z 87 for diacetyl and m/z 115 for 2-heptanone.

The negative flow-rate (aspiration) was fixed respectively to 50 ml/min. for APCI-MS and 30 ml/min. for PRT-MS.

RESULTS

Chewing Efficiency

Panelists were asked to chew the samples for 4 and 8 masticatory cycles. As introduced before, the chewing efficiency was assessed by the particle size distribution in the bolus.

As expected, the scattering in the human mastication is quite high (figure 3).

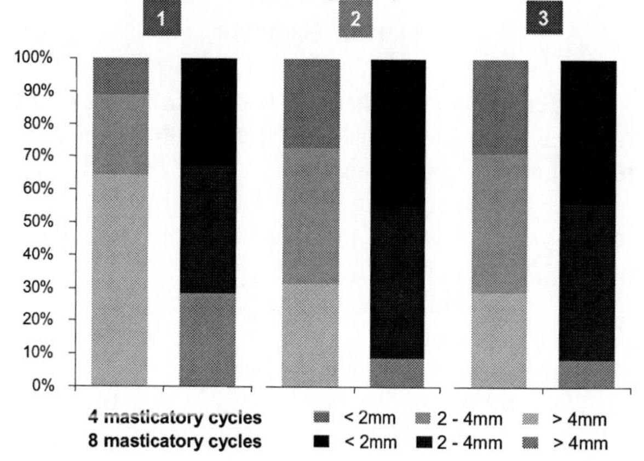

FIGURE 3. The Peanuts deconstruction by three panelists: particle size distribution for each panelist, for 4 (orange) and 8 (blue) masticatory cycles.

During each experiment, the mastication process in the artificial mouth was monitored by recording data: vertical position and force of both tongue and mandible, and angular position and shearing force of the mandible.

The parameters were set as follow: Mandible effort: 24, 29, 34 daN; Tongue effort: 30 daN; Shearing effort: 35 daN; Shear angle: 1/8th tooth.

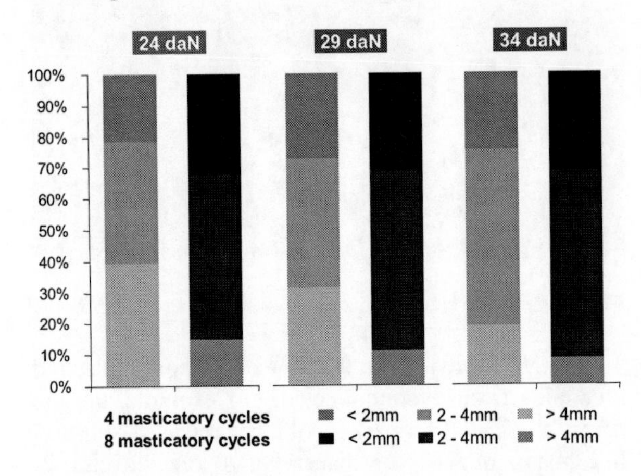

FIGURE 4. Peanuts deconstruction by the Artificial Mouth: particle size distribution versus mandible force, for 4 (orange) and 8 (blue) masticatory cycles.

It was shown that, if we want to simulate the chewing of panelist Nr 2, a force of 29 daN is required.

An increase in both the shearing angle and bite force resulted in an increase in the degree of peanut deconstruction. The effect of mandible force on the degree of breakdown depends on the shearing angle applied, and vice-versa. In general, the effect on particle size distribution of changing the mandible force was greater when no shearing was applied than when 1/8th and 1/4th tooth shearing were applied.

Aroma Release

Different model cheeses were chewed by the machine, with a artificial saliva flow-rate of 3 and 6 ml/min. The samples weight was 5 g ± 0.1 g.

The parameters were set as follow: Mandible and tongue effort: 30 daN; Shearing effort: 35 daN, Shear angle: 1/8th tooth.

The embedded sampling system has been previously described [5]. An example of the bolus after the experiment is given in figure 5.

We have shown on these samples that the time-intensity aroma release was actually different, in relation with the composition, i.e. different fat content 20 & 40 % (figure 6).

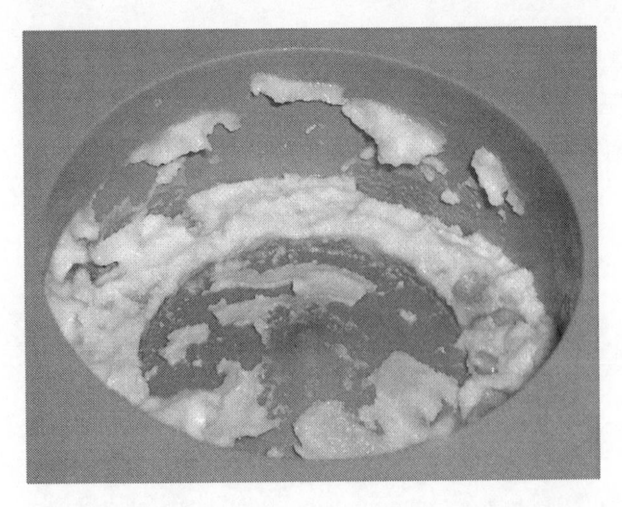

FIGURE 5. Picture of the cheese sample (5 g), after its chewing with artificial saliva at 3 ml/min..

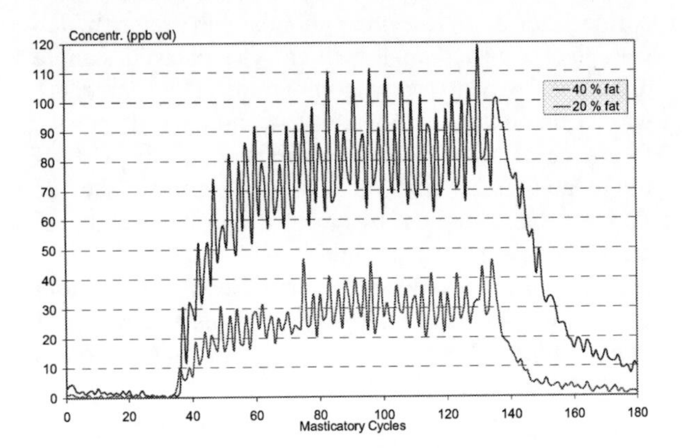

FIGURE 6. Release of ethyl hexanoate (m/z 145), as a function of the fat content, with a PTR-MS detector.

This behavior was not expected: in general, the aroma molecules, that are mostly hydrophobic, have a greater affinity with the fat, so the release should be lower when increasing the fat content. But indeed, with the fatter sample, the bolus is thinner, so it has a greater specific exchange surface with the headspace, leading to an increase of the release (hypothesis to be confirmed).

Taste Release

The taste compounds are extracted by the saliva during the mastication, and thus are permanently available and may be sampled at each cycle, either from the human or artificial mouth.

For the in-vivo analysis, a saliva sample was taken from the mouth with a cotton swab. After extraction and filtering, the concentration was assessed by HPLC.

An example of the kinetics of glucose release from French baguette, for four panelists, is given in figure 7.

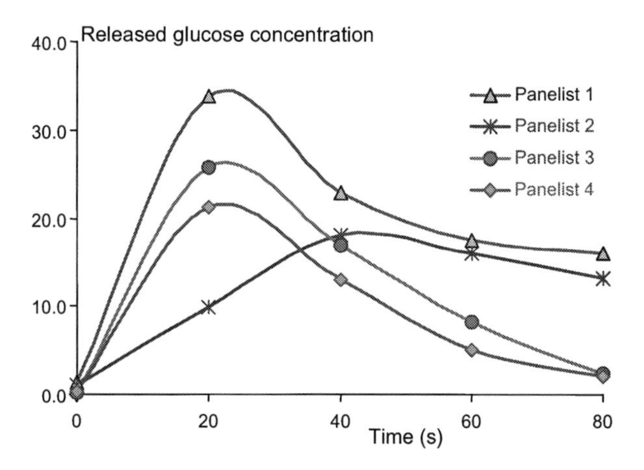

FIGURE 7. Glucose released from a bread, human mastication: discrete sampling with cotton swabs and HPLC analysis.

The automation of the liquid sampling is just being implemented in the Artificial Mouth. A first approach consists in a discrete sampling (after each cycle, the chewing is paused, then a small aliquot is withdrawn from the bolus using a 3 axes lab robot, figure 8). A special sampling head has been developed, which is actuated by a 2D air cylinder. The sample is then stored in a tray and analyzed off-line with a classical HPLC equipment. The absorption phase lasts less that one second.

FIGURE 8. The lab robot used to sample and store the liquid aliquots, together with the liquid sampling head (right).

The latest results will be shown in a couple of month.

We are also considering the use of liquid phase chemical microsensors (Electronic Tongue) [6], particularly for assessing the sodium release.

DISCUSSION

It was demonstrated that the forces and torques used during the human mastication are related at once to the appetite for the proposed dish, and the degree of the bolus deconstruction. This means that the masticatory parameters, together with the saliva flow-rate follows a complex time-intensity profile, that will be mimicked in the next generation of the Artificial Mouth. Artificial vision and sound capture will also be implemented.

The use of thickeners or texturing agents and the trend to sell "light" products have a extreme influence on the flavor release: a small variation in the formulation leads to a highly different sensory perception.

We have developed a functional system that is able to precisely reproduce the compressive and shear strengths of a human jaw causing the food breakdown with sufficient reliability when compared to in-vivo measurements. The continuous addition of saliva, the collection of both the non-volatile compounds released in saliva and the volatile released in the headspace during chewing, are also included.

ACKNOWLEDGMENTS

This work is supported by Conseil Régional de Bourgogne and was accredited by the Taste-Nutrition-Health competitive cluster Vitagora [7]. We particularly acknowledge our project partners, from Plateform'3D, Le Creusot, France for the modeling, 3D scanning and tooling of the prototype. Additional thanks to Raoul, Romain Kévin Loudenot and Sébastien Desroches.

REFERENCES

1. Pionnier, E., Chabanet, C., Mioche, L., Le Quéré, J. L., & Salles, C. Journal of Agricultural and Food Chemistry, 2004, 52, 557-564.
2. Salles C., Tarrega A., Mielle P., Maratray J., Gorria P., Liaboeuf J., Liodenot J-J., Journal of food Engineering, Vol. 82, Issue 2, September 2007, pp 189-198.
3. Salles C., Mielle P., Le Quéré J.-L., Renaud R., Maratray J., Gorria P., Liaboeuf J., Liodenot J-J., Proc. 11 th Weurman Symposium, Roskilde, Denmark.
4. van der Bilt, A. & Fontijn-Tekamp, F. A. Archives of Oral Biology, 2004, 49, 155-160.
5. Mielle P., Renaud R., Gorria P., Liodenot J-J., Liaboeuf J., Salles C., Guichard E. Proc. 11 th ISOENSymposium, Barcelona, Spain.
6. Auger J., Arnault I., Legin A., Rudnitskaya A., Sparfel G., Dore C. International Journal of Environmental and Analytical Chemistry, 2005, 85, (12-13), 971-980
7. http://www.vitagora.com/en/vitagora/news/a-ground-breaking-artificial-mouth--2342.aspx

SPECIAL SESSION:
BIOLOGICALLY INSPIRED COMPUTATION FOR CHEMICAL SENSING

Recent Developments in the Application of Biologically Inspired Computation to Chemical Sensing

S.Marco, A. Gutierrez-Gálvez

Department of Electronics, Universitat de Barcelona, Martí I Franquès 1, 08028-Barcelona, Spain
Artificial Olfaction Lab, Institut for BioEngineering of Catalonia, Baldiri i Rexach 13, 08028-Barcelona, Spain

Abstract. Biological olfaction outperforms chemical instrumentation in specificity, response time, detection limit, coding capacity, time stability, robustness, size, power consumption, and portability. This biological function provides outstanding performance due, to a large extent, to the unique architecture of the olfactory pathway, which combines a high degree of redundancy, an efficient combinatorial coding along with unmatched chemical information processing mechanisms. The last decade has witnessed important advances in the understanding of the computational primitives underlying the functioning of the olfactory system. In this work, the state of the art concerning biologically inspired computation for chemical sensing will be reviewed. Instead of reviewing the whole body of computational neuroscience of olfaction, we restrict this review to the application of models to the processing of real chemical sensor data.

Keywords: Computational Intelligence, Chemical Sensors
PACS: 07.05.Kf, 07.05.Mh, 07.07.Df

INTRODUCTION

Neuromorphic approaches for gas sensor arrays have been traditionally disregarded by the artificial olfaction community. However, recent neurophysiological findings[1] by Nobel Prize Buck and Axel in the olfactory system along with a growing body of knowledge from computational neuroscience[2] have made neuromorphic signal processing techniques a recent focus of attention.

Pioneering developments in this research area originate from the work of White and Kauer[345]. In their original work, an array of chemosensors provides time-varying inputs to a computer simulation of the olfactory bulb (OB) (1/1000 real size). The OB simulation produces spatio-temporal patterns of spiking that vary with odor type. These patterns are then recognised by a delay-line neural network.

Later on, Pearce et al.[6] addressed the issue of how efficient stimulus encoding may be carried out within the early stages of the olfactory system, in particular how a rate-coding scheme compares to the direct transmission of graded potentials in terms of the accuracy of the estimate that an ideal observer may make about the stimulus. They made use of a spiking neuronal model of the early stages of the olfactory system that is driven by fluorescent microbead chemosensors in order to compare these two coding schemes. Their results indicate how the charging time-constants present at the first stages of neuronal information processing within the olfactory bulb directly affects its ability to accurately reconstruct the stimulus.

In the last 5 years the most active research group in this area has been the PRISM laboratory at Texas A&M University leaded by R. Gutierrez-Osuna. In a series of three articles they have explored the role of the three first stages of the olfactory pathway to process chemical information. This processing is then mimicked and applied to gas sensor array data. The three stages are the following: 1)Convergence from Olfactory Receptor Neurons (ORN) to glomeruli, 2) periglomerular cells, and 3) granule cells.

Taking inspiration from the highly ordered convergence of Olfactory Receptor Neuron (ORN) axons to glomeruli, Perera et al.[7] have extracted a dimensionality-reduction technique based on the projection of sensor features according to their response across sensors. This technique is shown to outperform Principal Component Analysis (PCA) and Linear Discriminant Analysis (LDA) in classification rate in a gas sensor classification problem. It is also computationally efficient for high-dimensional

CP1137, *Olfaction and Electronic Nose: Proceedings of the 13th International Symposium*, edited by M. Pardo and G. Sberveglieri

problems, and suitable to tackle small sample set problems.

The second stage studied is the Periglomerular cell layer. These interneurons innervate the gomeruli connecting them laterally. The putative role of this layer of interneurons is hypothesized to be a kind of volume control for the incoming signals from the ORN layer. Raman and Gutierrez-Osuna[8] used a shunting lateral inhibitory network to compress concentration information. This network allows controlling the amount of stimulus concentration information that is retained by varying the width of the lateral connections. Going from a total removal of concentration information given by global connections and keeping only discriminatory information to the retention of most of the concentration information with local lateral connections. This network is used to control the concentration information of a MOS gas sensor array data.

The early olfactory connectivity has been also the focus of attention of Gill and Pearce[9]. They have used an array of optical micro-bead sensors to investigate the issues of development, organization and maintenance of connections in the early olfactory pathway. Two populations of micro-bead sensors: active (exposed to various odorants) and inactive (exposed only to air) were used to simulate the distribution of ORNs in the olfactory epithelium. Oja's Hebbian learning rule was used to develop activity-dependent weights between the sensor (receptor) layer and the GL layer; a Mexican hat function was used to model the lateral interactions between GLs mediated by PG cells. Similar to experimental findings on mice their results show segregation of the active and inactive ORN populations into separate GLs suggesting the influence of odorant-evoked activity in the organization and maintenance of OB connections. Further their results suggest that the lateral interaction between GLs through PG cells play an important role in realizing the topological organization of the ORN projections. However, this predicted role of PG cells has not been confirmed through experimental studies.

The third stage models the excitatory-inhibitory circuitry of the mitral and glomerular cells. This processing stage it is known to increase the separability of odor representations at the olfactory bulb. Inspired by this mechanism, Gutierrez-Galvez and Gutierrez-Osuna[10] proposed a new Hebbian/anti-Hebbian learning rule to increase the separability of sensor array patterns in a neurodynamics model of the olfactory system: the KIII. The KIII[11] is a neuro-dynamical model of the olfactory system capturing the behaviour of neural populations. It acts as an associative memory storing patterns at the level of mitral to mitral connections. In the proposed learning rule, a Hebbian term is used to build associations within odors and an anti-Hebbian term is used to reduce correlated activity across odors. The KIII model with the new learning rule is characterized on synthetic data and validated on experimental data from an array of temperature-modulated metal-oxide sensors.

Further work on the KIII model has been done by Fu et al.[12] used Freeman's KIII model to classify the response of eight MOS sensors to six typical volatile compounds in Chinese rice wines. Two rules are involved in the learning process: Hebbian learning, which increases the connection of coactive cells, and habituation that reduces the connection strength of those cells that are not active. The classification of patterns is done by considering the pattern obtained from the amplitude (standard deviation) of the Mitral cell oscillations. Then, the Euclidean distance between the output and the centroids of the data samples of each class is computed. The output is assigned to the class which centroid is closer. The classification performance of the KIII is used to show two interesting features of the system. First, the KIII is able to give a high classification rate despite introducing samples of different concentrations. Second, it is capable of obtain a high classification rate even with sensor drift. Three databases in the four months time difference were captured to test the KIII performance against sensor drift. These two features seem to be the effect of the associative memory that is able to recover patterns that are separated from the original pattern (due to concentration change or sensor drift) but still within the basin of attraction. Finally, the classification rate of the KIII is shown to outperform that of a Multi Layer Perceptron presented with the same classification task.

In two additional papers Gutierrez-Osuna's group has further investigated the role of the olfactory bulb to improve the contrast between odor representations and to remove the background in odor mixtures. Raman et al.[13] [used an spiking neuron model of the olfactory bulb with center on-off surround lateral interactions (granule cells) to enhance the initial contrast between odor representations and decouple odor identity from intensity. Gutierrez-Galvez and Gutierrez-Osuna[14] used the KIII model to combine the lateral inhibition with habituation occurring at the olfactory bulb. This model was able not only to increase the contrast between odor representations but also to remove the effect of odors previously presented to the system. In both articles the models were validated with temperature modulated gas sensor data.

In addition to the previous papers, Gutierrez-Osuna has studied also the advantages of information coding in the first stages of the olfactory pathway and applied them to gas sensor array processing. Raman et al.[15] built a model of the two first stages of olfactory coding: distributed coding with olfactory receptor neurons and chemotopic convergence onto glomerular units. The pattern recognition performance of the model is characterized using a database of odor patterns from an array of temperature modulated chemical sensors. The chemotopic code achieved by the proposed model is shown to improve the signal-to-noise ratio available at the sensor inputs while being consistent with results from neurobiology.

Raman et al.[16] have proposed a neuromophic approach to process optical microbead arrays signals. This processing approach seems particularly adequate for optical sensors due to the large number of sensors involved, as it is the case in the olfactory epithelium. Optical microbead arrays are composed of around 586 fibers coated with different functionalized beads that provides with a multidimensional response to different analytes. These signals are then processed in two stages. First, the signals are grouped depending on the response profiles of each fiber to different odours obtaining a spatial representation of the sensor responses and reducing the dimensionality of the pattern. This stage is inspired by the convergence of ORN axons based on the receptor they express from the olfactory epithelium to the glomeruli. The second stage is a center on-off surround lateral interactions of the clustered fibers activities that enhances the contrast of odour representations. This second stage mimics the lateral inhibition in the olfactory bulb at two different levels: glomerular through periglomerular cells, and mitral through granule cells. The pattern separation obtained by this method is compared with that obtained with PCA and LDA. The results show that their method outperforms all other methods. Most of the recent research of this group is in the area of neuromorphic processing is contained in two dissertations: Barani Raman[17] and Agustin Gutierrez-Galvez[18].

Regarding higher cortical areas, few works can be found. Pioggia et al.[19] used a cortical based neural network (CANN) to analyze data coming from a conducting polymer sensor based e-nose and a composite-array based e-tongue. The results are compared with a Mulit-layer Perceptron, a Kohonen Self-Organizing Map and a Fuzzy Kohonen self-organizing map. The comparison showed that the CANN model was able to strongly enhance the performance of both systems. For the CANN, the authors follow an implementation by Izhikevich. They use a CANN with 1000 neurons, consisting of 200 inhibitory neurons and 800 excitatory neurons. Cortical pyramidal neurons showing regular spiking behaviour were adopted for the excitatory subsection, following the neuron model of Izhikevich.[20] For the inhibitory neurons they selected a neuron model which exhibits a fast spiking property. Each neuron is connected to M different neurons in order to obtain a connection probability equal to 0.1. However, inhibitory neurons are connected only to excitatory neel urons. For learning, the following strategy was used: the synaptic weights of the connections arising from the inhibitory neurons remain unchanged during the learning process, while those regarding the connections from the excitatory neurons change according the STDP rule. Axonal delays are fixed in the range between 1 ms and 20 ms. Time resolution was set to 1 ms. The input data were sent to subset of the excitatory neurons. There appear a large number of neural groups that perform reproducible spike sequences with a precision of 1 ms. Finally a labelling procedure allows to associate a specific pattern to a neural group. Each stimulus that is used as an input pattern is able to select one group inside the network, showing that the network is able to classify. The problem faced was that of the analysis of olive oil samples from different Italy regions. A total of 540 experiments were performed. Validation was carried out by 5-fold cross-validation.

Finally, the work by Dominique Martinez and his co-workers[21] should also be highlighted. They propose also a Spiking Neural Network for Gas Discrimination in a Tin Oxide Sensor Array. The sensor array consist of tin oxide 16 micro hotplates. The array was exposed to Hydrogen, Ethanol, Carbon Monoxide and Methane. The architecture of the proposed network has different layers. First the current crossing every sensor is taken as the input for a Leaky Integrate and Fire Neuron. Then the sensor responses are transformed in a series of spikes. These spikes follow a synaptic level transformation by means of a double exponential function showing maxima for particular value of the spike timing. Finally the Post-synaptic potential of the output neurons is a weighted aggregation of the sypnaptically transformed spike timings. For the output neuron to fire, the post-sypnaptic potential has to cross a threshold. In order to train the weights for a correct classification the authors propose to use the TEMPOTRON algorithm by Gutïg and Sompolinksy[22] Nature Neuroscience, 2006). The authors find favorable results in classification when compared to Support Vector Mahines.

SUMMARY

The application of neuromorphic models to chemical sensor array data is a novel line of research that has been explored by few pioneering groups. We highlight the advances by White and Kauer, together with T. Pearce, as well as the work of R. Gutierrez-Osuna and his students at Texas A&M. These efforts are still largely unknown to the gas sensor array community and out of the mainstream of methods for data processing. Probably at the present stage of development, currently used techniques still largely outperform bioinspired methods. However, as far as we know, a careful comparison has not been performed yet. On the other hand, we feel that an important volume of effort is still needed to bring this topic to a mature status.

ACKNOWLEDGMENTS

This work has been funded by EC Fp7 NEUROCHEM grant agreement no. 216916 and Fp6 NoE GOSPEL FP6-IST 507610.

REFERENCES

[1] L. Buck and R. Axel, "A Novel Multigene Family May Encode Odorant Receptors: A Molecular Basis for Odor Recognition", Cell, 65, (1991) 175–187.

[2] Davis JL, Eichenbaum H (1991) Olfaction: A Model System for Computational Neuroscience. MIT Press, Cambridge.

[3] J. White, K.A. Hamilton, S.R. Neff, J.S. Kauer, "Emergent properties of odor information coding in a representational model of the salamander olfactory bulb". J Neurosci, 12: (1992) 1772-1780.

[4] J. White, J.S. Kauer, T.A. Dickinson, D.R. Walt, "Rapid analyte recognition in a device based on optical sensors and the olfactory system". Anal.Chem., 68 (1996) 2191-2202.

[5] J. White, T.A. Dickinson, D.R. Walt, J.S. Kauer, "An olfactory neuronal network for vapor recognition in an artificial nose", Biol. Cybern. 78, (1998) 245-251

[6] T.C. Pearce, P.F.M.J. Verschure, J. White, J.S. Kauer, "Stimulus encoding during the early stages of olfactory processing: study using an artificial olfactory system", Neurocomputing 38-40 (2001) 299-306.

[7] Perera, A., Yamanaka, T., Gutierrez-Galvez, A., Raman, B., Gutierrez-Osuna, R., "A dimensionality-reduction technique inspired by receptor convergence in the olfactory system", Sensors and Actuators B 116 (2006), 17.

[8] B. Raman, R. Gutierrez-Osuna, "Concentration normalization with a model of gain control in the olfactory bulb", Sensors and Actuators B 116(1-2), 36-42, 2006.

[9] Gill DS, Pearce TC "Wiring the Olfactory Bulb – Activity dependent models of Axonal Targeting in the Developing Olfactory Pathway". Reviews in Neuroscience 14: (2003) 63-72.

[10] A. Gutierrez-Galvez, R. Gutierrez-Osuna, "Increasing the separability of chemosensor array patterns with Hebbian/anti-Hebbian learning", Sensors and Actuators B 116, (2006) 29-35.

[11] Y. Yao, W.J. Freeman, "Model of biological pattern recognition with spatially chaotic dynamics" Neural Networks, 3, (1990) 153–170.

[12] J. Fu, G. Li, Y. Qin, W. J. Freeman, "A pattern recognition method for electronic noses based on an olfactory neural network", Sensors and Actuators B 125 (2007) 489-497.

[13] B. Raman, T. Yamanaka and R. Gutierrez-Osuna, "Contrast Enhancement of Gas Sensor Array Patterns with a Neurodynamics Model of the Olfactory Bulb", Sensors and Actuators B, 119, (2006) 547-555.

[14] A. Gutierrez-Gálvez, R. Gutierrez-Osuna, "Contrast enhancement and background suppression of chemosensor array patterns with the KIII model" Int. J. Intel. Syst., 21, (2006) 937–953.

[15] B. Raman, P. Sun, A. Gutierrez-Galvez and R. Gutierrez-Osuna , "Processing of Chemical Sensor Arrays with a Biologically-Inspired Model of Olfactory Coding",IEEE Transactions on Neural Networks, 17(4), 1015-1024, 2006.

[16] B. Raman, T. Kotseroglou, L. Clark, M. Lebl, R. Gutierrez-Osuna, "Neuromorphic Processing for Optical Microbead Arrays: Dimensionality Reduction and Contrast Enhancement", IEEE Sensors Journal 7 (2007) 506-514.

[17] B. Raman, PhD dissertation, "Sensor-based machine olfaction with neuromorphic models of the olfactory system" Texas A&M University.

[18] A. Gutierrez, PhD disseratation, "Coding and learning of chemosensory array patterns in a neurodynamics model of the olfactory system" Texas A&M University.

[19] G. Pioggia, M. Ferro, F. Di Francesco, A. Ahluwalia, D. De Rossi, "Assessment of bioinspired models for pattern recognition in biomimetic systems", Bioinspiration and Biomimetics 3 (2008) 11p.

[20] E.M. Izhikevich "Which model to use for cortical spiking neurons?" IEEE Trans. Neural Netw. 15 (2004) 1063–70.

[21] M. Ambard, B. Guo, D. Martinez, A. Bermak, "A Spiking Neural Network for Gas Discrimination Using a Tin Oxide Sensor Array," 4th IEEE Int. Symp.on Electronic Design, Test and Applications (2008) pp. 394-397.

[22] R. Gütig and H. Sompolinsky, "The tempotron: a neuron that learns spike timing-based decisions," Nature neuroscience, vol. 9, (2006) pp. 420–428.

Very Large Chemical Sensor Array for Mimicking Biological Olfaction

R. Beccherelli[1], E. Zampetti[1], S. Pantalei[1], M. Bernabei[2], K.C. Persaud[2]

[1]*Consiglio Nazionale delle Ricerche – Istituto per la Microelettronica e Microsistemi (CNR-IMM)*
[2]*University of Manchester, School of Chemical Engineering and Analytical Science*

Abstract. Olfactory receptor neurons (ORN) in the mammalian olfactory system, transduce molecular properties of the odorants into electrical signals and project these into the olfactory bulb (OB). In the biological system several millions of receptor neurons of a few hundred types create redundancy and the massive convergence of the ORNs to the OB, is thought to enhance the sensitivity and selectivity of the system. To explore this concept, the NEUROCHEM project will build a polymeric chemical sensor array consisting of 2^{16} (65536) sensors with tens of different types. To interface such a large sensor array, a topological array configuration with n rows and m columns, has been adopted, to reduce the total wiring connections to n+m. A method of addressing a single element in the array in isolation of the rest of the network has been developed. Over the array ten different conductive polymers with different sensing characteristics will be deposited by means of electrodeposition and inkjet printing. A smaller prototype of 64 elements has been investigated and the results are here reported and discussed.

Keywords: Electronic nose, sensor array, conductive polymers, ink-jet printing.
PACS: 07.07.Df.

INTRODUCTION

The mammalian olfactory system expresses a very large number of olfactory receptor neurons, greater than 10^6 of which there are greater than 100 different types. The NEUROCHEM (Biologically Inspired Computation for Chemical Sensing) project aims at emulating the high level of receptor redundancy in such systems. We report on the development of a fundamental part of the project which involves the development an artifact capable of providing a very large set of experimental responses to gases and odors. The sensor array technology consists of organic conducting polymers that allow a infinite number of sensor configurations to be made with broad but overalapping selectivity to different families of chemicals emulating the characteristics found in biological chemoreception.

EXPERIMENTAL AND METHODS

In the NEUROCHEM project, we are building an array featuring a 2^{16} (=65536) conductive polymer sensors array. A few tens of conductive polymers showing broad and overlapped specificity to different volatile organic compounds are repeatedly deposited over the array. Hence, a high degree of redundancy is implemented. However, such a large number N of sensing elements cannot be read by the usual $2N$ leads (or $N+1$ in case of a common electrode) as this number of external connections becomes unmanageable. It is much more convenient to arrange the elements in a matrix configuration having n rows and m columns, with $N=n{\bullet}m$. In this way each resistive element in the array shares its first terminal with all the elements belonging to the same row and the second terminal with the elements in the same column. The total wiring connections in this configuration are $n+m$, with a minimum of $2\sqrt{N}$ for a square array. The topology is illustrated in Fig. 1.

The resistance value R_{ij} of the generic element belonging to the ith row and j^{th} column can be read by use of analogue multiplexing techniques. However, parasitic paths are established in parallel to R_{ij} through multiple parallel combinations of the series of resistors on semi-selected row ($l{\neq}i,j$), semi-selected column ($i,k{\neq}j$) and non-selected rows and columns ($l{\neq}i,k{\neq}j$). These paths usually result in unacceptably large crosstalk in the reading of each resistor. Several techniques of interrogating resistive sensor arrays have been described in the literature [1,2]. These approaches differ in terms of performance, reading

CP1137, *Olfaction and Electronic Nose: Proceedings of the 13th International Symposium*, edited by M. Pardo and G. Sberveglieri

error, crosstalk, circuit complexity and applications [3,4].

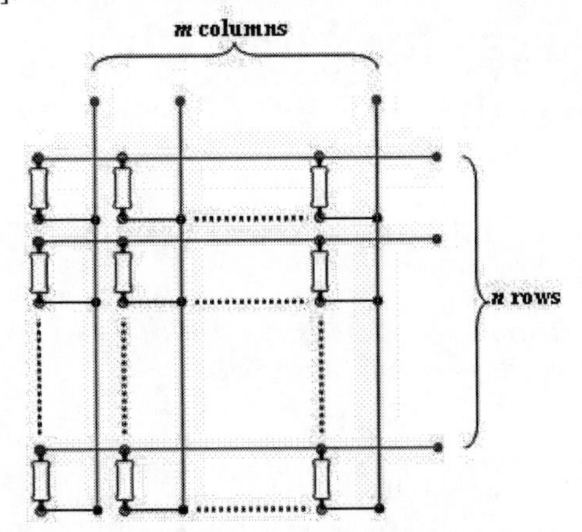

FIGURE 1: Topological arrangement of a sensor array.

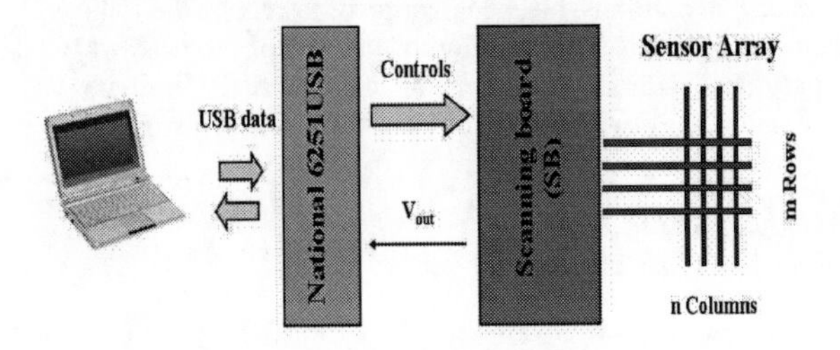

FIGURE 2: Electronic interface architecture.

Conductive Polymers (CP) have been chosen as sensing materials because CP gas sensors display great sensitivity to a wide range of volatile compounds, wide selectivity, have fast responses as well as low power consumption and they operate at room temperature [6].

RESULTS

Read-out electronics

A simplified version of the scheme of the scanning circuit we have implemented in NEUROCHEM is depicted in Fig. 3. In the picture there are three main blocks: two analogue multiplexers (*n* to 1 and *m* to 1 respectively) and *n* analogue switches.

The first analogue multiplexer (at the top) connects a load resistor R_L to the ith column, while the second multiplexer (in the bottom) connects the jth row to ground. The output voltage V_{out} is equal to $V_{cc}{\bullet}P$ where V_{cc} is the power source voltage and P is the resistor partition coefficient $(R_{ij}/(R_{ij}+R_L))$ when no parasitic low resistance parasitic paths exist (all the other resistive elements in the array are seen as open circuits).

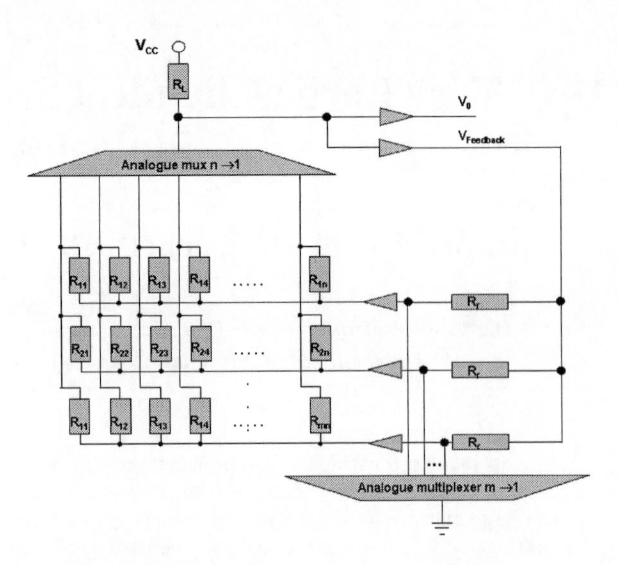

FIGURE 3: Scheme of electronic scanning circuit.

To reduce crosstalk occurring through unwanted parallel electrical paths, the non-selected rows are fixed to the output voltage value by feedback buffers (Fig. 3). The generic R_{ij} is then read by connecting through the analogue multiplexer to the ith row to ground and j^{th} column to V_{cc} through R_L. Using this read-out scheme the reading voltage depends only on V_{cc} and the values of R_L and R_{ij}.

To evaluate the electrical characteristics of the overall scanning circuit, a first fully functional prototype has been realized, capable of reading a *128x128* elements array (see Fig. 4). This preliminary system helps us to understand the problems related with managing a large number of sensors simultaneously.

An important feature of the acquisition system is the sampling rate. This is limited either by the sum of transient time constants or by the data transfer rate between the acquisition system and the PC. In our case the sample rate mainly depends on the circuit transient time constant, which is directly proportional to the equivalent parasitic capacitance C_{ij} and to $R_L//R_{ij}$. Accordingly, if the R_{ij} (sensor element resistance) increases the time constant increases too and then the final sampling rate decreases. In the case of the prototype scanning board, the equivalent capacitance C_{ij} is ≈2.8 nF. Considering a typical sensor resistance of 200 KΩ (depending on the conductive polymer in use) and R_L= 220 KΩ (commercial value) the time constant $\tau = C_{ij}{\cdot}(R_L// R_{ij})$ is about 293 μs. Hence, if we consider as stable a reading of the resistance after a time of 3τ, the sampling rate for one element is ≈1.14 kHz. The time needed to acquire *128x128* elements is in the order of 15 seconds. A full 256x256 system would have twice the capatitance and four times the elements. Hence a full acquisiton period requires 2•4•15 s = 120 s. This time is a limiting factor in sensor applications where the dynamics of the chemical interactions and of the sensor response are fundamental parameters. To overcome this problem the set of 65536 sensors has been subdivided in 16 array of *64x64* elements. At the same time the scanning system has been partinioned in 16 acquisition modules capable of reading the smaller arrays at the same time. This modular solution improves the parasitic capacitance of the single acquisition modules by a factor of 4 with respect to a full *256x256* system because of the reduction of the multiplexers needed, and of a further factor of 16 because of the parallelization, giving a total speed gain of 64. In this way the expected reading time for all the 65536 sensors is now about 1.8 seconds.

To obtain the best results, R_L value should be equal to R_{ij} (maximum transfer of active power). This operation is performed by using another analogue multiplexer (not shown in Fig. 3) that connects the optimum load resistor to V_{cc}. This optimal load resistor is chosen in a look up table and may be updated by software automatically. Morever, we implement a software programmable clock to to keep acquisition time to a minimum for each individual element. Hence, the circuit works always in the best conditions in respect to speed *and* sensitivity.

The *n x m* array signals are acquired by an electronic interface (EI), see Fig. 2, composed of a scanning board (SB) and an acquisition board (National 6251 USB) that convert analogue voltage to a numerical value.

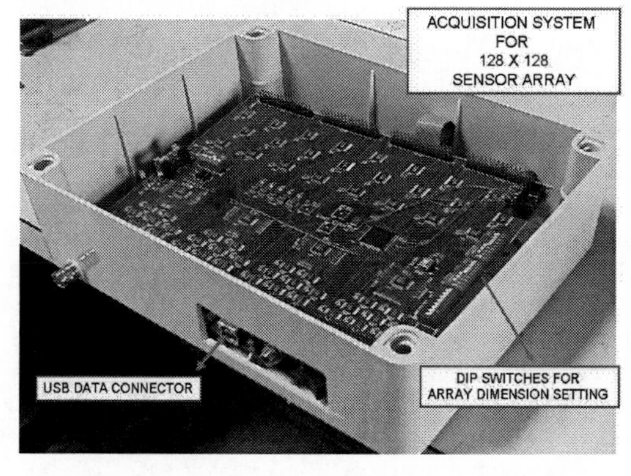

FIGURE 4: Preliminary prototype of scanning board for reading 128x128 elements.

Sensor array fabrication

In order to evaluate the feasibility of manufacture of the sensor array substrate using a clean room process and to investigate compatibility of the polymer deposition phase and the sensor assembly phase, a preliminary smaller *8x8* elements array version has been constructed. The array was fabricated using a 4" oxidized silicon wafer as a substrate. Two overlapping metal layers were deposited to build the rows and the columns of the sensor array, while a dielectric layer was deposited in between the two metal layers to ensure electrical insulation. The insulating layer is also needed to define the sensor region. This is defined by the intersection of the row and the column metal layers end with a simple interdigited structure capable of transducing the sensing polymer conductivity. This simple interdigited transducer is composed of 3 fingers having dimensions of 100 µm x 50 µm, and a gap of 50 µm. The sensing element pitch is set to 504µm. A stack of Cr-Au was used for the two metal layers (having total thickness of 150 nm each) and hard baked polyimide (SU8) 2.4 µm thick was used for the insulating layer. The array was processed by means of photolithographic techniques. A micrograph of the fabricated test array is shown in Fig. 5 and Fig. 6. In the final design the whole set of 65536 sensing elements has been split up into 16 separate modules made of 64x64 sensors each. The procedure of scanning the whole set involves simultaneously reading the smaller arrays modules while serially reading the elements of each array.

In order to keep the dimension of the single array small, the design rule of the process has been shrunk to 20 µm and the sensing elements pitch has been reduced to 250 µm. Moreover, to have an improved sampling rate, the interdigitated structure has been redesigned with 5 fingers, resulting in 30 squares in parallel *vs* the 6 squares in parallel in the 8x8 test version. The design allow to have the sensor array made of 4096 sensors in a square with side of 2.1 cm, comprising the pads necessary to bond the die to the scanning board through flip-chip technique or wire-bonding. The overall array outline and the single sensor layout are shown in Fig. 7 and Fig. 8.

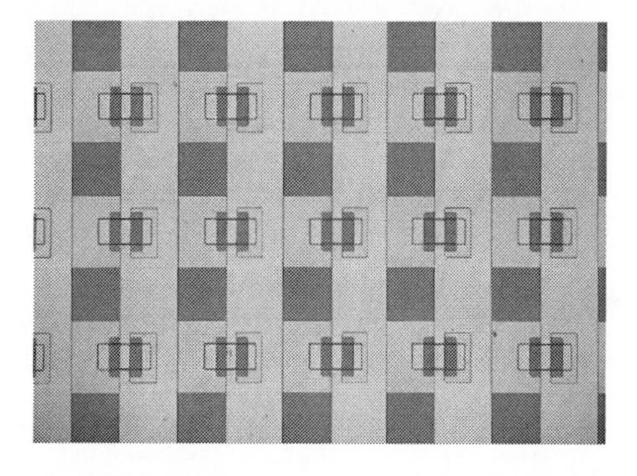

FIGURE 5: Microscopy photograph of a subset of a fabricated 8x8 test transducer. The figure shows only three rows and six columns.

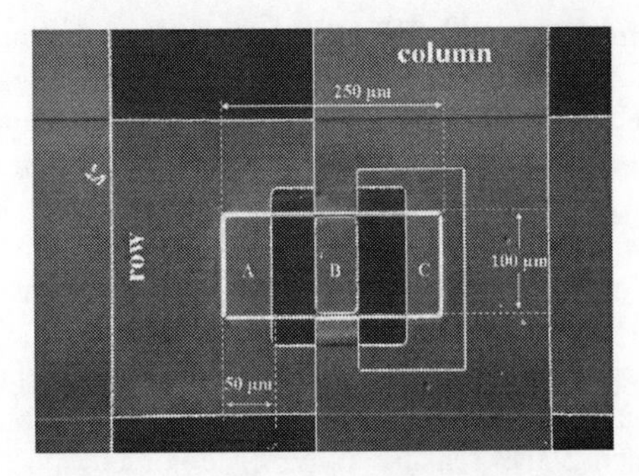

FIGURE 6: Microscopy photograph of a single transducer element, taken by means of differential interference contrast microscopy. A, B, C represent the fingers of the interdigitated transducer. The conductive polymer will be deposited to cover A, B, C and their gaps.

The polymers will be deposited on the interdigitated electrodes by means of electrochemical polymerization and ink-jet printing. The first technique is fast, clean and simple to implement, in fact it is possible to work at ambient pressure and temperature and in aqueous solutions. The polymer formed is already doped and the doping level and the film thickness can be controlled.

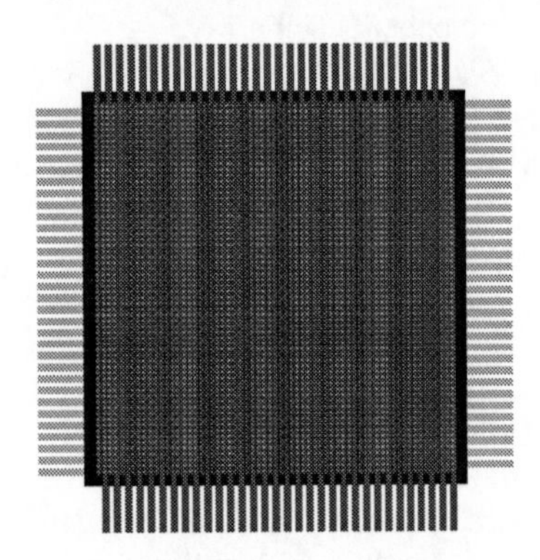

FIGURE 7: Single 64x64 sensor array. Occupied surface is 4.41 cm2.

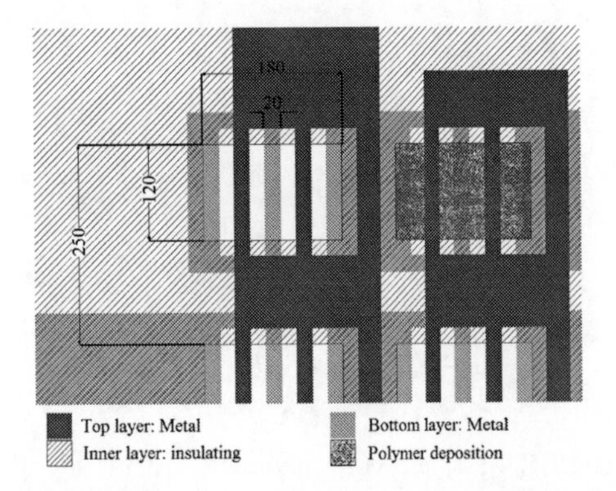

FIGURE 8: Detail of the single transducers. The figure shows the five fingers at the intersections between rows and columns. The interdigitated region will be deposited with conductive polymer by means of ink-jet printing.

Inkjet printing for depositing conductive polymers is very promising because it is easy to carry out, low cost, compatible with various substrates and fast. Different configurations can also be obtained with a very high grade of precision, without use of masks and with an efficient use of the material. Inkjet printing has been used to fabricate all organic transistors, polymer light emitting diodes, all polymer capacitors and other devices. Recent works on inkjet printed chemical sensor arrays based on conductive polymers show that in addition to the advantages noted above, the morphology of inkjet printed films may allow rapid diffusion of the vapor molecules into and out of the film, leading to response and recovery times faster than those obtained by the use of the other deposition technique[5]. Unfortunately, the inkjet printing depends on the polymers being soluble in compatible solvents of appropriate viscosity and this limits the number of polymers that can be used.

CONCLUSIONS

We have carried out the preliminary investigations to produce a model of a full scale sensor array. This will be used to optimize the manufacture, data acquisition and signal processing modules to be produced as part of the NEUROCHEM project.

ACKNOWLEDGMENTS

The research leading to these results has received funding from the European Community's Seventh Framework Programme (FP7/2007-2013) under grant agreement no. 216916.

REFERENCES

1. J.A. Purbrick, A force transducer employing conductive silicon rubber", Proc. 1st Int. Conf. on Robot Vision and Sensory Control 1981, pp. 73 – 80.
2. T. D'Alessio, "Measurement errors in the scanning of piezoresistive sensors arrays," Sensors and Actuators 72 1999 pp. 71–76.
3. Takao Someya et al., "Conformable, flexible, large-area networks of, pressure and thermal sensors with organic transistor active matrixes" PNAS 2005 vol. 102 , no. 35 pp 12321–12325.
4. R. Beccherelli, G. Di Giacomo "Low Cross-talk readout electronic system for arrays of resistive sensors Eurosensors XX M3B-P4, 17-20 September 2006, Göteborg, (SE).
5. Mabrook, M.F., Pearson, C. Petty, M.C., Inkjet-printed polypyrrole thin films for vapour sensing. Sensors and Actuators B: Chemical, 2006. 115(1): p. 547-551.
6. Persaud, K.C. Polymers for chemical sensing Materials Today, 2005, 8, (4) pp. 38-44.

A Bulb Model Implementing Fuzzy Coding of Odor Concentration

Malin Sandström, Thomas Proschinger and Anders Lansner

School of Computer Science and Communication, Royal Institute of Technology Stockholm, Sweden [msandstr, ala]@csc.kth.se

Abstract. It is commonly accepted that the olfactory bulb (OB) codes for odor quality by specific patterns of activated glomeruli. However, no such consensus has been reached for how the OB codes for odor concentration. We have constructed a model of the olfactory bulb which is able to generate a "fuzzy code" for odor concentration, while still coding for odor identity and showing synchronization of active mitral cells. The fuzzy code arises from competitive inhibition in the glomerular layer of the model. Fuzzy concentration coding could explain how the OB might encode odor concentration while still encoding odor quality according to the consensus view above.

Keywords: Computational model, Olfactory bulb, Odor Coding, Odor Intensity
PACS: 87.17.Aa, 87.18.Sn, 87.19.ls, 87.19.lt

INTRODUCTION

The vertebrate olfactory system is typically thought to be coding for odor quality by specific patterns of activated glomeruli [3,4,6-8]. The conventional view is to consider the glomeruli in the olfactory bulb merely as sites of specific ORN axon convergence that serve as signal-to-noise enhancing devices [1]. We propose an alternative view, the fuzzy concentration coding hypothesis, in which the dendritic processing that occurs in the glomerulus module works to create a fuzzy interval code over the stimulus concentration range, so that different mitral cells in the same glomerular column respond the strongest to different parts of the stimulus range. Such a code could be used to indicate the concentration of present odorants without sacrificing the ability to code for their identity.

MODEL AND METHODS

Model Components and Details

Our model (see Figure 1) includes a simplified ORN layer, a glomerular layer with each glomerulus modeled separately, and multicompartmental mitral (MT), granule (GC) and periglomerular (PG) cells. The model is structurally organized into glomerular columns similar to those found in recent staining experiments by Willhite et al [9].

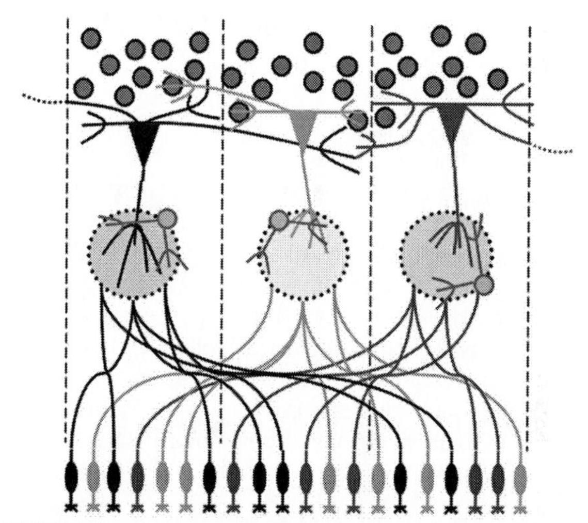

FIGURE 1. Schematic of the model, showing three glomerular columns with ORN groups, glomeruli, PG cells, mitral and granule cells.

Our glomerulus model consists of the glomerular compartments of n mitral cells, each receiving input from one ORN group (see below). Each mitral cell inhibits the other mitral cells in the same glomerulus, thorugh the PG cell population.

Our ORN group model is an aggregate of all receptor neurons that express the same olfactory receptor and have similar response properties with respect to their concentration-frequency characteristics. For ease of reference, the ORN groups are indexed in order of decreasing sensitivity with

CP1137, *Olfaction and Electronic Nose: Proceedings of the 13th International Symposium,* edited by M. Pardo and G. Sberveglieri

respect to odor intensity. Thus, the most sensitive ORN group is ORN0, and the least sensitive one is ORNn-1 (where n is also the number of mitral cells). The ORN groups belonging to different glomerular columns have different affinities K_d for the same odour stimulus.

Our bulb network model is written in parallel NEURON and consists of multicompartmental mitral and inhibitory cells (Hodgkin-Huxley type). The cells are divided into q glomerulus modules or columns of n mitral cells, p periglomerular cells and m granule cells. Throughout this paper, $q = 9$, $n = 25$ and $m = 25$. The periglomerular cell population is simplified to one cell per glomerulus module, representing the actions of the full population. The periglomerular cells in our model connect only within the glomerular column, while granule cells mediate inhibition between mitral cells in both the same and different columns.

The mitral and granule cell models were adapted from the bulb model of Davison et al [2], and from the parallelized version of the same model published to ModelDB by Migliore et al [5].

Assumptions Necessary for the Fuzzy Coding

The fuzzy coding hypothesis and model rests on two assumptions. Assumption 1: ORNs that express the same receptor have different response properties with respect to their concentration-frequency characteristics. This requirement is supported by the findings of Grosmaitre et al. [3], who reported significant differences in the responses from ORNs expressing the same odorant receptor. We will refer to ORNs that express the same receptor and have similar response properties as an ORN group. Assumption 2: The ORNs in an ORN group preferably innervate only a subset of MT cells connected to the same glomerulus module.

The result is a competition between the different mitral cells in a glomerulus, so that only a subset of cells will be active for a certain concentration (in this paper, they will be referred to as the "foreground" mitral cells). The competition between the mitral cells is mediated by PG cell inhibition within the glomerulus module. The identity of the active mitral cells gives an indication of the approximate odor stimulus strength (relative to the affinity K_d of the associated receptors for the odorant in question). Thus we get what we call a "fuzzy code" for concentration.

Connections

All mitral cells in the same glomerular column connect to the same periglomerular cell population unit, which in turn connects reciprocally to all mitral cells. Mitral and granule cell connections are also reciprocal; mitral and granule cells within the same column are connected with a high probability, and mitral and granule cells in different columns are connected with a low probability.

RESULTS

Model Behaviour

Because of the fuzzy concentration coding, only a few mitral cells are active in each glomerular column (see Figure 2).

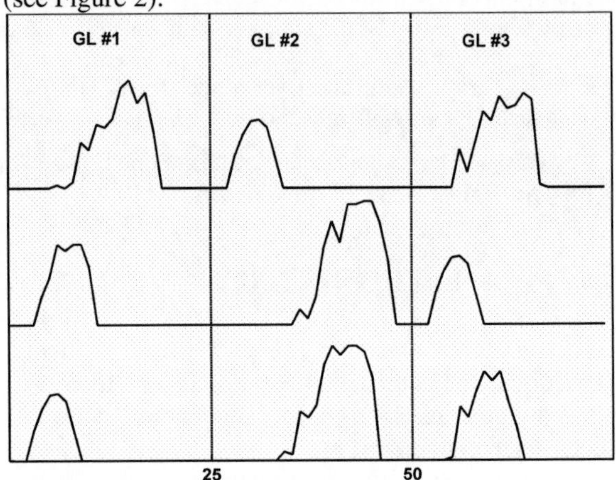

FIGURE 2. Schematic illustration of responses from mitral cells in three neighbouring glomerular columns (left-right, dashed lines indicate the borders between columns) in response to three different odour stimuli (top-bottom).

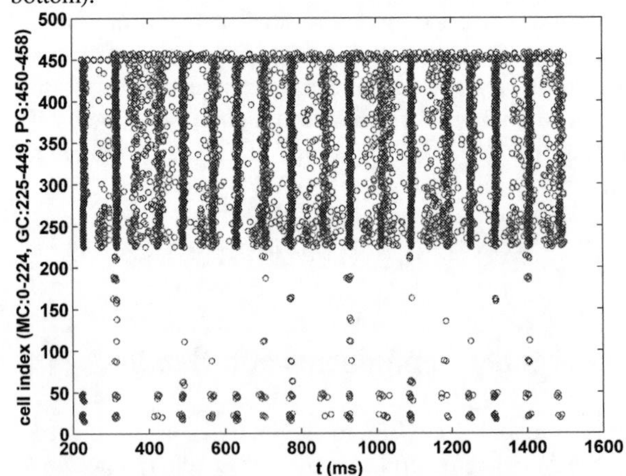

FIGURE 3. Spike raster from a typical simulation of nine glomerular columns, with 75% MC-GC connectivity within the same column and 25% connectivity between MC and GC in different columns. The two first glomeruli receive a medium-strong stimulus ($K_d/M = 5$), while the following seven receive a stimulus one hundred times weaker ($K_d/M = 500$).

Active mitral cells synchronize, but synchronization between columns is only partial (see Figure 3). The most strongly active glomerular column(s) suppresses the spiking activity of mitral cells in the other glomerular columns.

Influence From Connectivity

The degree of connectivity within and between columns affects the level of contrast between the glomeruli, also when the inhibitory weights have been adjusted so that the total mean inhibitory input to each mitral cell is the same (see Figure 4).

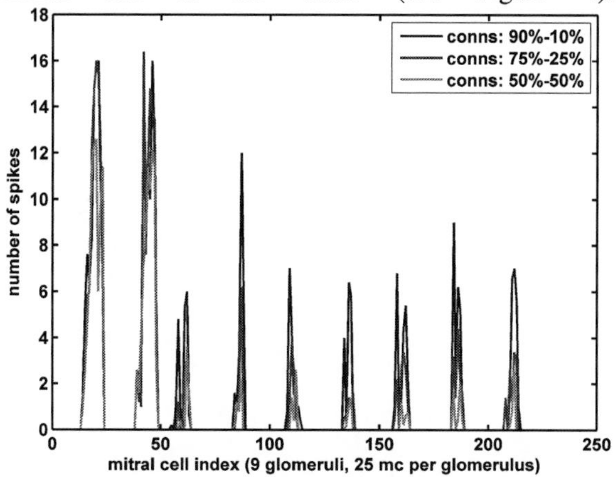

FIGURE 4. The mean number of spikes per mitral cell, from five runs equivalent in all but the (random) connections between mitral and granule cells, and three different connectivity scenarios. In blue: 90% connectivity within the column, 10% outside, in red: 75% connectivity within the column, 25% outside, in green: uniform 50% connectivity. For details on the stimulus used, see Figure 1.

Robustness of Responses/Phase Lags

In our model, interglomerular competition not only suppresses the relatively more weakly responding glomeruli. It also makes the responses from those glomeruli more unreliable, in the terms of similarity from trial to trial, judging from variability in responses and from mean phase lags of responses (see Figure 5).

Glomerular Column Interaction

The most strongly active glomerular column(s) suppresses the spiking activity of mitral cells in the other glomerular columns through their common granule cells (see Figure 6) – but also receives suppression from the less strongly active mitral cells in the other columns: both the number of responding mitral cells and the number of spikes per responding mitral cell is lowered. The result is a narrowing of the response profile of mitral cells in the strongest responding glomerular column, and a decreased spiking frequency for those cells that are still active. Most suppressed are mitral cells on the edges of the foreground population, i.e. the mitral cells that are least indicative of the input concentration. Very weakly responding glomeruli may be suppressed altogether, leading to a sparsening of the response.

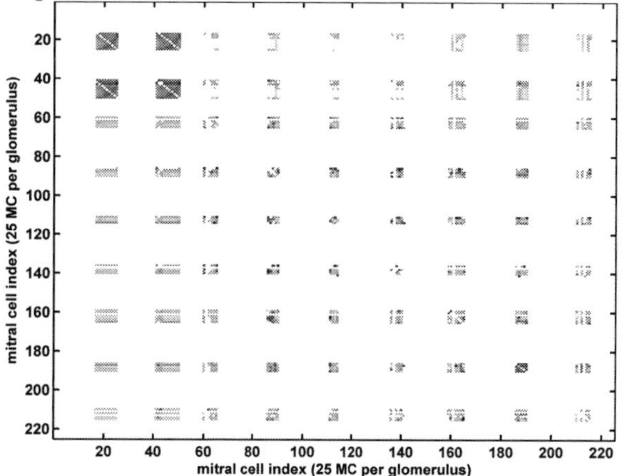

FIGURE 5. Sum over five 5.5s simulations (same type as shown in Figure 1-2) of mean phase lags between mitral cells. For clarity, we plot the logarithm of all values, scaled so that dark indicates short lags, and light/white indicates longer lags. Non-existing values (phase lags cannot be calculated for MC:s with too few or no spikes) have been set to twice the maximum existing value.

DISCUSSION & CONCLUSION

We conclude that our fuzzy coding hypothesis is possible to implement in a biologically realistic model, and that our model – from a qualitative perspective – behaves plausibly with respect to biology. A fuzzy concentration code might thus enable the OB to encode odor concentration while still encoding odor quality. However, more simulations and larger networks must be examined before we can draw conclusions on the qantitative behaviour of our model.

The fuzzy coding described in this paper originates in the glomerular layer, but its actions are reinforced and refined by the spiking cell network. Thus, the fuzzy code has a dynamic component affected by the lateral inhibition from the rest of the network; higher mean activity (and thus higher inhibitiory output) in other columns gives a sparser code in the receiving column, with fewer responding cells and fewer spikes per cell as a result.

In our upcoming planned work we will introduce a bulb connection matrix with synapses and weights determined by plasticity, in the form of a learning rule for odour processsing. We will also connect the bulb model to an olfactory cortex model, and further increase the size of the bulb model (by approximately a factor of ten) to run large-scale parallel simulations.

ACKNOWLEDGMENTS

The research leading to these results has received funding from the European Community's Seventh Framework Programme (FP7/2007-2013) under grant agreement no. 216916 (NEUROCHEM). We also acknowledge support from the EU project, the Swedish Foundation for Strategic Research, and The Swedish Governmental Agency for Innovation Systems (through the Stockholm Brain Institute; www.stockholmbrain.se).

REFERENCES

1. W.R. Chen and G.M. Shepherd, *J Neurocytol* **34,** 353-360 (2005).
2. A.P. Davison, *et al*, *J Neurophysiol* **90,** 1921-1935 (2003).
3. X. Grosmaitre, *et al*, *Proc Natl Acad Sci U S A* **103,** 1970-1975 (2006).
4. B. Malnic, *et al*, *Cell* **96,** 713-723 (1999).
5. M. Migliore, *et al*, *J Comp Neurosci* **21,** 110-119 (2006).
6. P. Mombaerts, *et al*, *Cell* **87,** 675-686 (1996).
7. B.D. Rubin and L.C. Katz, *Nat Neurosci* **4,** 355-356 (2001).
8. R. Vassar, *et al*, *Cell* **79,** 981-991 (1994).
9. D.C. Willhite, *et al*, *Proc Natl Acad Sci U S A* **103,** 12592-12597 (2006).

Learning from the Moth: A Comparative Study of Robot-Based Odor Source Localization Strategies

Sergi Bermúdez i Badia[1] *and Paul F.M.J. Verschure*[1,2]

[1]*SPECS, Universitat Pompeu Fabra, Tanger 135, 08018 Barcelona, Spain*
sergi.bermudez@upf.edu
[2]*Institució Catalana de Recerca i Estudis Avançats, Barcelona, Spain*

Abstract The odor search strategies of the moth have been researched since many decades. Many behavioral studies have described the behavior under well controlled conditions, making predictions on what the underlying mechanisms might be. However, it is almost impossible to asses these mechanisms directly since sensory and behavioral data on a freely behaving moth are very hard to obtain. Therefore, we propose a comparative study were the behavior of a robot is analyzed when controlled by a number of odor source localization models. Our results show that a system making use of stereo odor information outperforms some well-established chemical search models.

Keywords: Source localization, chemotaxis, autonomous robot, biologically based model

PACS: 40

INTRODUCTION

The problem of odor localization in insects has an enormous complexity. The best studied case of insect chemical localization is the moth. Female moths release sex attractant pheromones that are transmitted downwind. Male moths display a particular sensitivity to pheromone compounds released by the females and are capable to track pheromone plumes to its source [1]. This highly specific behavior is supported by specialized structures in the moth nervous system, the, so called, Macro Glomerular Complex of the primary olfactory center, the antennal lobe, which is dedicated to the detection of pheromone signals [2, 3].

So far, it has been very difficult to assess the details of moth chemotaxis directly because of the difficulty of visualizing a chemical plume without interfering with the flight behavior of the moth. Hence, it is still not quantitatively established whether the moth responds to a chemical gradient, filament contact or follows a more elaborated behavioral strategy. Experiments have been performed where a male moth was equipped with a third antenna with a wireless transmission system to approximate what it would sense [4, 5]. However, this is insufficient for a proper characterization of the relationship between chemical stimuli and behavior of the moth. It is known that there is a dependency of moth behavior on the structure of the pheromone plume [4-8]. In these experiments it was shown that a pulsed pheromone signal improves localization and induces faster flight.

In this study we propose to reverse engineer the problem by analyzing the behavior of a number of models under specific and controlled chemical localization conditions. We present a detailed evaluation of different models with detailed quantifications of performance.

METHODS

A custom circular robot with a diameter of 20 cm (FIGURE 1, left panel) was used in this study. It uses two active wheels placed on the axis of the robot, allowing in place rotation driven by geared motors (Parallax, Rocklin, USA). The wind direction was measured with a custom build sensor consisting of a wind vane that was fitted to a magnetic encoder (the angular position of the shaft was measured using a 2-axial magnetometer) that was placed near to the chemo sensor. The motor commands and sensory data are exchanged using a Bluetooth module (Bluetooth™ Class I, LinTech, Germany). The robot is controlled by the neural simulator program iqr [9] via a wireless link with a linux ground station. The olfactory board consists of a 6 grid array thin film metal oxide chemo-sensor (Alpha MOS SA, France) [10] that provides a broad spectrum of sensitivity to a wide range of volatile organic compounds while having relatively low power consumption (approx. 270 mW) and a high degree of miniaturization. The robot uses LI-PO rechargeable batteries (KOK 3270, Kokam, Korea) that provide up to 8 hours of running time [11, 12].

CP1137, *Olfaction and Electronic Nose: Proceedings of the 13th International Symposium*, edited by M. Pardo and G. Sberveglieri

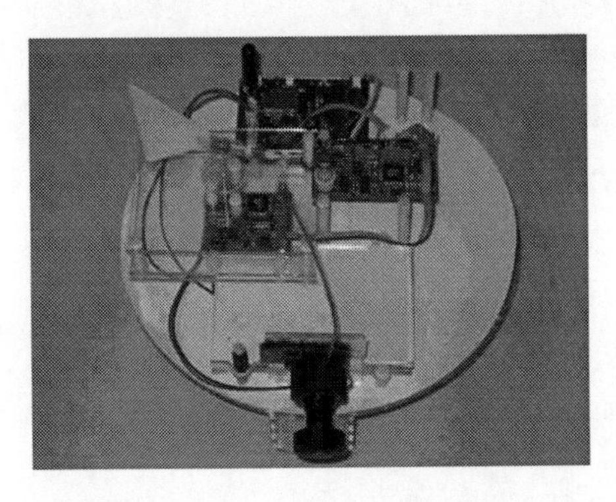

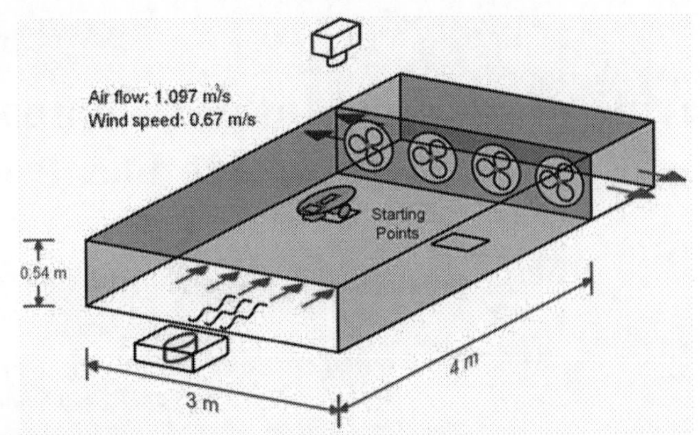

FIGURE 1. Experimental setup. Left panel: Mobile chemosensory vehicle. Visible are, from top to bottom, the control board with Bluetooth communication module, the chemo-sensor board and the wind direction sensor. Right panel: Structure of the wind tunnel. The wind tunnel is 4 m long, 3 m wide and 0.54 m high. 4 ventilators create negative pressure and suck the air into an exhaust tunnel. A custom made tracking system (see text for detailed description) tracks the robot within the wind tunnel. The odor source (blue) was placed in the middle of the entrance of the wind tunnel. The two squares indicate the two starting position for the experiments. Modified from Pyk et al. [10].

All of the mobile robot experiments were performed in a low-cost wind tunnel that was constructed from wood and transparent plastic sheets measuring 3 x 4 x 0.54 m (FIGURE 1, right panel). Four axial fans were installed at the wind tunnel exit in order to adjust for a uniform and symmetric velocity profile. A solution of fixed concentration of ethanol and distilled water (20% ethanol) was delivered using an ultrasonic release system (Mist of Dreams, XrLight, Zhongshong City, China) delivering about 0.8 ml/min of ethanol with an average air speed of 0.67 m/s.

The behavioral data was acquired in real-time with a custom-built general purpose video tracking system called AnTS. The AnTS tracking system receives its input from a CCD camera with a wide-angle lens fixed on the ceiling at about 3 m above the wind tunnel. To obtain an undistorted planar view of the arena, correction algorithms for perspective and wide-angle lens distortions were built into the AnTS tracking software. A 640 x 480 pixel image resolution was used to track the robot; this resulted in a spatial resolution of about 1 cm for the 3 x 4m wind tunnel at an update frequency of 15 Hz.

We used three different behavioral models: a Behavioral, a Braitenberg and aNeuronal based model.

- **Behavioral.** This model is solely based on two behavioral modes observed in the moth [13, 14]. Male moths tend to show a regular zigzagging behavior called casting when trying to intercept a filament of the pheromone plume. Once the plume is intercepted, moths make an upwind displacement in response to the pheromone contact, known as surge mode. Then, if the pheromone filament is lost, moths come back to the casting mode. Our particular implementation of this strategy is based on the one by Balkovsky et al. [15], where casting is characterized by an increase of the crosswind flight displacement over time if no chemical signal is detected.
- **Braitenberg Vehicle**. A Braitenberg like vehicle was considered as the simplest model that could make use of stereo information for odor localization [16]. In this case we implemented a classic Braitenberg vehicle, where the difference between the readings of two sensors with a spatial separation along a particular axis is transformed into motor commands that orient the robot towards the odor source.
- **Neuronal Model**. This model aims at exploiting known principles of the moth's behavior and its underlying neuronal substrate based on 3 components:

Odor modulated upwind progress. This is based on the neural substrate found in the Macro Glomerular Complex (MGC) of the Antennal Lobe (AL) of the moth, a glomerulus exclusively dedicated to the encoding of pheromone signals [2, 17]. The majority of the neurons in the MGC neurons (approx. 85%), most likely Projection Neurons (PN), are able to resolve odor pulses up to several Hz [17, 18]. Based on these neurons and behavioral studies [4-8], we designed a model that has a preference for pulsed signals as opposed to continuous stimulation. Therefore, the robot displays a short upwind surge when a contact with a pheromone patch occurs; and an inhibition of the upwind displacement when the moth is under continuous pheromone stimulation.

*Use of stereo information to modulate turning angle***.** Some Descending Neurons (DN) having dendritic arborizations in the Lateral Accessory Lobe (LAL) show a high/low firing rate state that switches depending on the difference of pheromone concentration in the insect antennae [19]. These neurons (flip/flop neurons) are synchronized with the change of orientation during the zigzag behavior in the

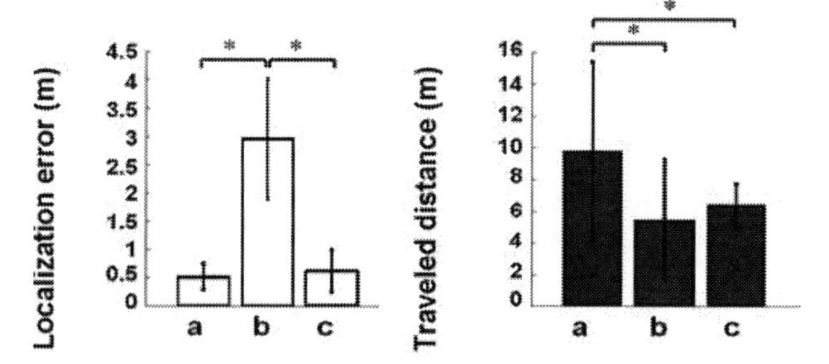

FIGURE 2. Comparison of the performance of the three proposed models (a – phenomenological, b – Braitenberg and c – neuronal). Left panel: Bar plots of the localization accuracy of the model. Right panel: Bar plots of the traveled distance (equivalent to energy consumption). Error bars indicate SD. See text for further explanation.

odor localization task [20]. Our model includes these flip/flop neurons to control the change in heading direction of the robot proportional to the difference of the sensed chemical concentration if above a certain threshold.

Self-steered counterturning. As suggested in previous studies [21], it seems that an internal oscillator could be used to control the timing of the zigzag turns observed in moths. Therefore, all the changes in heading direction generated by our model are performed synchronous with this internal oscillator.

RESULTS

In previous studies, the validity of the setup and sensor technology for chemical search was already demonstrated [10, 11]. The starting position of the robot in these experiments was at a distance of 3 m from the odor source and at 0.75 m from the walls, equally distributed at the left and right starting points of the wind tunnel (FIGURE 1, right panel).

The behavioral model (see methods) was tested for a total of 20 robot runs with a ratio of correct localization of 90%. Subsequently, the classic Braitenberg vehicle (see methods) was used for a total of 20 experiments. This model displayed a success rate of 10%. Although the robot was able to successfully detect the plume, it was incapable to follow it up to the source, and most of the runs do not pass the midline of the wind tunnel. Based on Kanzaki et al. [22], we added a looping behavior to the Braitenberg vehicle to ease the reacquisition of the odor plume when it is lost. Ten more robot experiments were performed with this modification. The addition of this reacquisition strategy increased the success rate from 10% (classic Braitenberg) to 40%. These experiments indicate that stereo odor information could be used in combination with higher level strategies to improve performance. Our neuronal model (see methods) shows a similar ratio of success to the behavioral one (85%).

If we consider the energy consumption (traveled distance) as a key factor for a successful behavioral strategy, we find that the least efficient behavior is displayed by the behavioral model, and that the Braitenberg vehicle is the most efficient. On this measure, the three models are statistically different ($p < 0.05$, 2-Sample t-test) (FIGURE 2, right panel). Since the robot moves at a constant speed, there is a correlation between time and distance traveled, consistent with the idea of a moth flying at a constant ground speed [1]. Nevertheless, the Braitenberg based model, which was the most optimal in distance terms, is very ineffective in the localization task and displays an error rate that is about five times larger than the one displayed by the other models (FIGURE 2, left panel). The lowest error is obtained by the phenomenological and neuronal models ($p < 0.05$, 2-Sample t-test), with no significant difference between them. From these results we conclude that our stereo-sensing neuronal model offers an optimal compromise between energy consumption, run time and localization error.

CONCLUSION AND DISCUSSION

We have presented a robot study of some well-established models of male moth chemical localization and introduced a novel neuronal model. We analyzed the performance of the models in robot experiments under controlled odor stimulation to study the effect and benefits of stereo odor information in the task.

The behavioral model displays a successful strategy with a low error rate but is not energy efficient (active search). This model has been shown in theoretical studies to be close to optimal when the search agent is inside the high probability plume area, but not when outside [15]. Some previous studies already investigated chemo-sensing Braitenberg vehicles with a passive delivery source, in a non-controlled air flow and show search times up to hours [23]. Our results suggest that the Braitenberg model is under-constraint, displaying the lowest success rate. However, when successful, this model presents the shortest trajectories in time and distance. Moreover, we demonstrated that a Braitenberg vehicle can be easily improved by combining it with a higher level strategy, i.e. plume reacquisition.

The neuronal model proposed in this study combines stereo sensing with behavioral and neural constraints derived from the moth. The localization error is indistinguishable from the phenomenological model, although it outperforms it in search time/traveled distance, i.e. efficiency. We will in future experiments evaluate its performance in outdoor robot chemical localization tasks.

AKNOWLEDGEMENTS

The research leading to these results has received funding from the European Community's Seventh Framework Programme (FP7/2007-2013) under the NEUROCHEM grant agreement no. 216916 and under the IST FET Programme (IST-2001.33066).

REFERENCES

1. Kennedy, J.S. and D. Marsh, *Pheromone-regulated anemotaxis in flying moths.* Science, 1974. **184**(140): p. 999-1001.
2. Christensen, J., et al., *Purification and characterization of the major nonstructural protein (NS-1) of Aleutian mink disease parvovirus.* J Virol, 1995. **69**(3): p. 1802-9.
3. Hansson, B.S., T.A. Christensen, and J.G. Hildebrand, *Functionally distinct subdivisions of the macroglomerular complex in the antennal lobe of the male sphinx moth Manduca sexta.* J Comp Neurol, 1991. **312**(2): p. 264-78.
4. Kuwana, Y., et al., *Synthesis of the pheromone-oriented behaviour of silkworm moths by a mobile robot with moth antennae as pheromone sensors.* Biosensors Bioelectronics, 1999. **14**: p. 195-202.
5. Rutkowski, A.J., et al. *A robotic platform for testing moth-inspired plume tracking strategies.* in *Proc. of the IEEE International Conference on Robotics and Automation (ICRA 2004).* 2004. New Orleans, USA.
6. Justus, K.A., Schofield, S.W., Murlis, J., and Cardé, R.T., *Flight behaviour of Cadra cautella males in rapidly pulsed pheromone plumes.* Physiol. Entomol, 2002. **27**: p. 58-66.
7. Mafra-Neto, A., and Cardé, R.T., *Effect of the fine-scale structure of pheromone plumes: pulse frequency modulates activation and upwind flight of almond moth males.* Physiol. Entomol, 1995. **20**: p. 229-242.
8. Willis, M.A. and T.C. Baker, *Effects of intermittent and continuous pheromone stimulation on the flight behaviour of the oriental fruit moth, Grapholita molesta.* Physiological Entomology, 1984. **9**: p. 341-358.
9. Bernardet, U., M.J. Blanchard, and P.F.M.J. Verschure, *IQR: A distributed system for real-time real-world neuronal simulation.* Neurocomputing, 2002. **44-46**: p. 1043-1048.
10. Pyk, P., et al., *An artificial moth: Chemical source localization using a robot based neuronal model of moth optomotor anemotactic search.* Autonomous Robots, 2006. **20**(3): p. 197-213.
11. Bermúdez i Badia, S., et al., *A Biologically Based chemo-sensing UAV for Humanitarian Demining.* International Journal of Advanced Robotic Systems, 2007. **4**(2): p. 187-198.
12. Bermúdez i Badia, S., P. Pyk, and P.F. Verschure, *A fly-locust based neuronal control system applied to an unmanned aerial vehicle: the invertebrate neuronal principles for course stabilization, altitude control and collision avoidance.* The International Journal of Robotics Research, 2007. **26**(7): p. 759.
13. Baker, T.C. and L.P.S. Kuenen, *Pheromone source location by flying moths: a supplementary non-anemotactic mechanism.* Science, 1982. **216**: p. 424-427.
14. Vickers, N.J. and T.C. Baker, *Reiterative responses to single strands of odor promote sustained upwind flight and odor source location by moths.* Proc. Natl. Acad. Sci. USA, 1994. **91**(13): p. 5756-60.
15. Balkovsky, E. and B.I. Shraiman, *Olfactory search at high Reynolds number.* Proc. Natl. Acad. Sci. USA, 2002. **99**(20): p. 12589-93.
16. Braitenberg, V., *Vehicles: Experiments in Synthetic Psychology.* 1984: MIT Press.
17. Christensen, T.A., et al., *Local interneurons and information processing in the olfactory glomeruli of the moth Manduca sexta.* J Comp Physiol [A], 1993. **173**(4): p. 385-99.
18. Lei, H. and B.S. Hansson, *Central processing of pulsed pheromone signals by antennal lobe neurons in the male moth Agrotis segetum.* J Neurophysiol, 1999. **81**(3): p. 1113-22.
19. Kanzaki, R., et al., *Physiology and morphology of projection neurons in the antennal lobe of the male moth Manduca sexta.* J Comp Physiol [A], 1989. **165**(4): p. 427-53.
20. Wada, S. and R. Kanzaki, *Neural control mechanisms of the pheromone-triggered programmed behavior in male silkmoths revealed by double-labeling of descending interneurons and a motor neuron.* J Comp Neurol, 2005. **484**(2): p. 168-82.
21. Kuenen, L.P.S. and T.C. Baker, *A non-anemotactic mechanism used in pheromone source location by flying moths.* Physiological Entomology, 1983. **8**: p. 277-289.
22. Kanzaki, R., S. Nagasawa, and I. Shimoyama, *Neural Basis of Odor-source Searching Behavior in Insect Brain Systems Evaluated with a Mobile Robot.* Chemical Senses, 2005. **30**(suppl_ 1).
23. Lilienthal, A.J. and T. Duckett, *Experimental Analysis of Gas-Sensitive Braitenberg Vehicles.* Advanced Robotics, 2004. **18**(8): p. 817-834.

BIOLOGICAL AND BIOMETRIC OLFACTION

Cluster Analysis of the Rat Olfactory Bulb Activity in Response to Different Odorants

M. Falasconi[a]*, A. Gutierrez[b,c], B. Auffarth[b,c], G. Sberveglieri[a], S. Marco[b,c]

[a] *Dept. of Chemistry and Physics, Univ. of Brescia & SENSOR Lab CNR-INFM, Via Valotti 9, 25133 Brescia, Italy;* [b] *Departament d'Electrònica, Universitat de Barcelona, Martí i Franquès, 1, 08028 Barcelona, Spain;* [c] *Artificial Olfaction Group, Inst. for Bioengineering of Catalonia (IBEC), Baldiri i Rexach 13, 08028-Barcelona, Spain*
** Corresponding author: Tel: +39 030 3715709; Fax: +39 030 2091271; email: matteo.falasconi@ing.unibs.it*

Abstract. With the goal of deepen in the understanding of coding of chemical information in the olfactory system, a large data set consisting of rat's olfactory bulb activity values in response to several different volatile compounds has been analyzed by fuzzy c-means clustering methods. Clustering should help to discover groups of glomeruli that are similary activated according to their response profiles across the odorants. To investigate the significance of the achieved fuzzy partitions we developed and applied a novel validity approach based on cluster stability. Our results show certain level of glomerular clustering in the olfactory bulb and indicate that exist a main chemo-topic subdivision of the glomerular layer in few macro-area which are rather specific to particular functional groups of the volatile molecules.

Keywords: olfactory bulb, 2-deoxyglucose mapping, olfactory coding, cluster analysis, cluster validity
PACS: 07.05.Kf, 07.05.Pj, 07.05.Rm, 29.85.Fj, 87.19.lt, 87.85.Ng, 89.75.Fb

INTRODUCTION

Odour stimulation of olfactory sensory cells in the mammalian nose involves odour molecules interacting with Olfactory-Receptor Neurons (ORN) located on the cilia of the Olfactory Epithelium (OE). It is known [1] that ORN expressing the same binding protein project onto the same roughly spherical synaptic bundles (glomeruli) of the Olfactory Bulb (OB). As a consequence, the activation of the glomerular layer is a bidimensional representation of the chemical features of the odorants. What it is not totally understood to this date is what chemical features are binded by the ORNs and in turn activate the different glomeruli. It is hypothesized that the ORN detect chemical features as functional groups or carbon chain length.

An extensive experimental study on the OB activity in response to different volatile organic compounds (VOCs) has been conducted in mice by Leon et al. [2] by mapping the uptake of ^{14}C radiolabeled 2-deoxyglucose (2DG). In about ten years of activity Leon and coworkers collected a database of OB responses to more than 300 different VOCs.

These studies show that odour stimuli in the rat produce spatial patterns of activity in the glomerular layer that are overlapping but that vary for different odours. The analysis of such activity patterns is a challenging problem. Odour patterns are complex, highly irregular, noisy, and may contain missing values due to the experimental procedure. For these reasons, automated exploratory data analysis techniques are required for objectively investigating the data structure.

In this work we used *fuzzy c-means* (FCM) cluster analysis to discover groups of pixels that have similar response patterns across the entire set of VOCs. In order to estimate the significance of the returned results we have developed an original validity criterion for fuzzy partitions [3] based on the evaluation of partition stability under bootstrap resampling of the data which allows to estimate the best partion of data.

The main objective is a deepen comprehension of the olfactory coding, i.e.: understanding whether and how glomeruli are clustered together according to their response profiles across the odorants and interpreting the retrieved valid clusters in terms of chemical properties of the VOCs.

EXPERIMENTAL AND METHODS

The available data (http://leonserver.bio.uci.edu/) consist of 470 activity maps of the OB glomeruli for

CP1137, *Olfaction and Electronic Nose: Proceedings of the 13th International Symposium*, edited by M. Pardo and G. Sberveglieri

different VOCs obtained by mapping the uptake of ^{14}C-2DG radiolabel over the entire OB layer. Such method provides a digital image (80x44 pixels) that represents a bidimensional map (ventral centered) of glomerular activity. Units in each matrix are standardized to z-scores relative to the mean and standard deviation of values across that matrix. Each map was vectorized and all background pixels (with zero variance) and missing values were removed leading to a final set of 1780 pixels.

The original database comprises measurements of the same chemical at different concentrations, thus the actual number of different VOCs is about 300. To eliminate the concentration effects and focus only on the chemical features of the molecule, the measurements of the same chemical at different concetrations were averaged together and replaced with a single map. Both data sets (D1=1780x470; D2=1780x308) were then analyzed.

Cluster analysis was performed by the FCM algorithm which outputs fuzzy partitions with a pre.specificed number of clusters. We used a novel paradigm for fuzzy partitions validity based on the concept of partition stability under data perturbation. Data can be perturbed in several ways, here we repatedly apply bootstrap sampling on pixels, so we named the method fBPSE (Bootstrap Partition Stability Estimation, where f stands for fuzzy) [3].

The fBPSE estimates and minimizes, over a series of partitons with increasing number of clusters K, the partition variability:

$$V(K) = \frac{1}{\binom{B}{2}} \sum_{i=1}^{B} \sum_{j>i}^{B-1} d\left(U_K(Y_i), U_K(Y_j)\right) \quad (1)$$

where B is the number of bootstrap samples (B=20). $V(K)$ is formally equivalent to the empirical bootstrap estimate of the variance, where $d(.,.)$ plays the role of a distance. Actually, $d(.,.)$ is similarity measure between two fuzzy partitions [4], being $U_K(Y_i)$ the membership matrix achieved by running the FCM algorithm on the i-th bootstrap sample Y_i. In our work we adopted the adjusted *fuzzy Rand index* [5].

RESULTS AND DISCUSSION

The mean and the variance of activity values for each pixel were calculated across the 308 different odorants of data set D2 (the results are very similar for D1).

The map of the mean activity (Figure 1) shows a region of high activity in the medial and lateral aspects of the OB while the ventral aspect seems to be almost inactive. Indeed, it seems that the lateral aspects of the OB are more active indepedently of the type of VOC under analysis. As pointed out by Leon et al. [2], the ventral aspect responds selectively to odorants with dense hydrocarbon features and with no oxygen-containing functional groups, however these are a minority in our database.

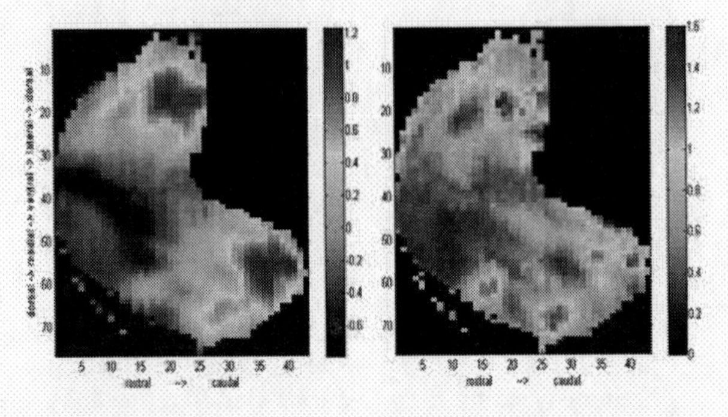

FIGURE 1 Distribution maps of the mean (left) and variance (right) activity values across the entire spectrum of 308 different chemicals of data set D2.

The average variance distribution of pixels (Figure 1) tells a different story. Some selected groups of pixels with higher variance – i.e. higher information content - emerge in the dorsa-medial and dorso-lateral regions of the OB whilst the ventral area has no variance. This observation partially confirms the previous one, but also strongly supports the hypothesis that there is a finer clustering structure inherent to the data that should be recovered by partitions with a higher K value.

We then performed FCM clustering and fBPSE validity from K=2 up to K=40. The difference of activity between the medio-lateral and ventral aspects is the main source of variance in the data set; this was confirmed by Principal Component Analysis (PCA) since the distribution map of PC1 is very similar to the left picture in Figure 1 (data not shown). Therefore, being the FCM method based on the intra-cluster variance minimization, the partition K=2 returns two clusters: the first cluster comprises the pixels in the medio-lateral aspects, the second one coincides with the ventral area.

Moreover all the achieved partitions show clusters of pixels that are paired in the medial and lateral aspects of the OB (see for example the clusters in Figure 3). This was an important consistency check for clustering because it is known that glomeruli are co-activated in the medio-lateral regions of the bulb.

By visually exploring the achieved partitions we observed a good qualitative agreement between certain clusters and the chemo-topic response model developed by Leon and coworkers. This was another argument in fovour of our clustering procedure.

Finally, we noted that some clusters, typically those matching modules with high specificity towards particular VOCs (e.g. carboxylic acids), were emerging in low K partitions and have a long life-time, i.e. they survive almost inaltered for several K values before splitting in smaller clusters. This indicates the presence of individually significant clusters, and then opens new perspectives for further validity studies of single clusters..

The cluster validity study aimed to determine the best number of clusters. However we have not found a clear evidence for this optimum value and our findings do not support the results previously obtained by Leon et al.

We found out that the best partition, i.e. most stable according to the fBPSE, was at K=3 (Figure 2) with a rather similar variability value for K=2.

Similar results were obtained on the two data sets D1 and D2, this indicates that the presence of measurements at different concentrations has negligible effects on the FCM outcomes and on the cluster stability.

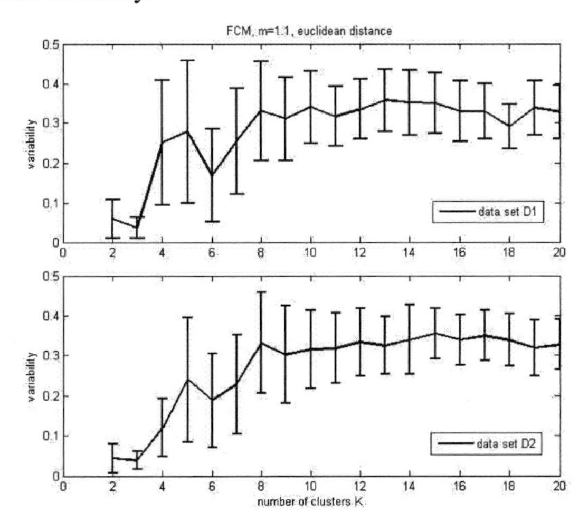

FIGURE 2 Cluster validity results by fBPSE on the data sets D1 and D2. The error bars report the standard deviation of V(K) calculated value over the 190 pairs of compared partitions.

Figure 3 shows the partition K=3 projected on the map. On the left figure we reported the crisp partition obtanied by converting the fuzzy outcome according to the maximum membership rule (pixel is assigned to the cluster having maximum membership). To give an insigth of the assignment of pixels belonging to the fuzzy cluster borders we showed on the right figurethe maximum membership value of each pixel (the brown regions represent the core pixels for which the membership is equal to one).

The high stability value seems associated with the big difference in the average activity between the lateral and ventral aspects of the OB. However, the partition K=3 shows a further split in the dorso-lateral part of the OB (clusters #1 and #2 in Figure 3) that may indicate and important difference between those areas.

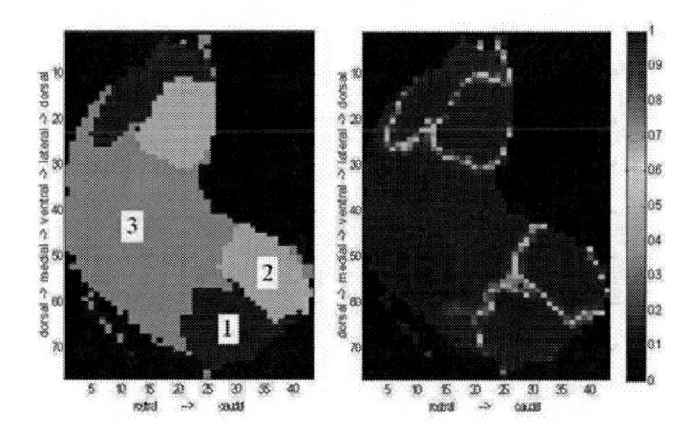

FIGURE 3 Best partition of data in K=3 clusters according to the fBPSE criterion: crisp partition corresponding to the fuzzy clustering (left), maximum value of the fuzzy membership of each pixel (right).

Our findings seem to indicate the existence of a main subdivision of the glomerular activity in three macro-areas.

Such topological representation of the glomerular layer in main zones of activity has been already proposed by Mori et al. [6]. According to Mori, glomeruli are parcelled into four zones in the OB and receive inputs from different regions of the olfactory epithelium. Zone I glomeruli are localized in the rostro-dorsal portion of the OB, while Zones II, III, and IV glomeruli are distributed progressively more ventrally and caudally. The hypothesis is that primary chemical features, such as functional groups, are responsible for the topological organization of glomeruli in these main zones. Secondary features of odorants like carbon chain length are represented by local arrangement of activated glomeruli within each zone.

The existence of such hierarchical structure may explain why we found out a best subdivision of data in three groups, which correspond approximately with the primary activity zones. A preliminar hypothesis is that Zone I corresponds to cluster #1, Zone II to cluster #2, while Zone III-IV are merged together in cluster #3. Why we did not find a subdivion in four clusters remains unexplained. A reasonable hypothesis is that this depends on the available data since many of the measured VOCs do not elicit glomeruli in the ventral region of the OB.

By investigating within each cluster the activation of groups of odorants sharing well defined chemical properties (Figure 4) we noted that: cluster #1 is mainly activated by ketones (including small aliphatic with less than 8 carbons) and aromatic compounds; in cluster #2 esters dominate; cluster #3 is mainly active for carboxylic acids and aldehydes (when looking at finer partitions, acids form a clear cluster in the dorsal aspect while aldehydes activate the medial region). This seems consistent with previously published works [2, 6].

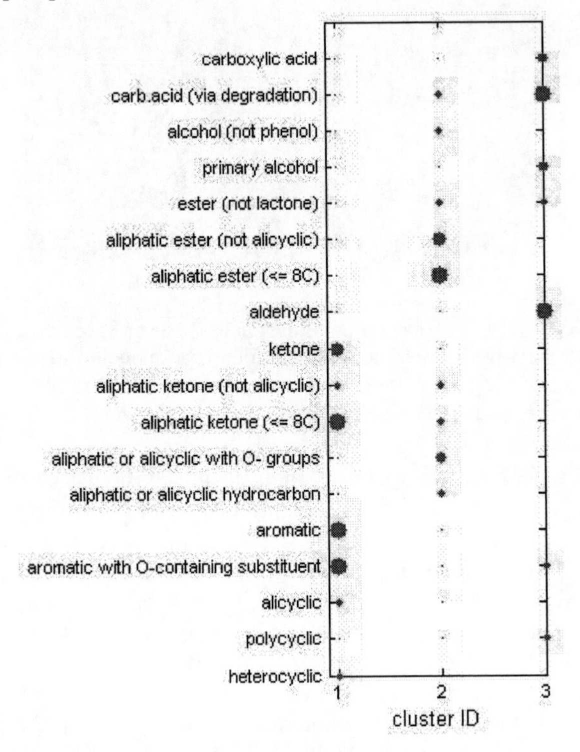

FIGURE 4 For each cluster are reported only the groups of odorants with the highest activation values (cluster ID refers to Figure 3). The bubble size is proportional to the average activity value across the group of odorants sharing the given chemical property.

The intrinsic limitations of our cluster validity method may provide another exaplaination why we did not recover the modular structure proposed by Leon et al. Both the clustering and the validity procedures are very challenging due to the high dimensionality of the data. For this reason we compared the results obtained in cluster analysis validity with all dimensions to those obtained reducing the dimensionallity (taking the first few PCs) of feature space. the outcome was very similar to that achieved with the complete feature set. We have also investigated whether the fBPSE method is biased towards low K values, at present we did not find any evidence for that by testing the method on synthethic data sets [3].

Finally, it can be argued that the current chemo-topic response model was build by Leon's group through a systematic investigation of the glomerular activity for selected VOCs. On the contrary, in this work we are clustering the glomeruli by taking all the chemicals at the same time. The model construction was more similar to a two-way clustering procedure, in which the pixels of the OB map are iteratively clustered by taking into account only subsets of chemical features. Therefore, further advances in this direction can be made by using two-way clustering methods.

CONCLUSIONS

In this work we applied fuzzy cluster analysis techniques to investigate the response patterns of rat's OB activity towards different volatile compounds aiming to understand the chemotopic organization of the OB. We also approached the challenging problem of validating the achieved partitions by using an original criterion based on fuzzy partition stability.

Our results suggest the existence of certain level of glomerular clustering in the olfactory bulb based on the activity behaviour across the measured VOCs. The results also indicate that exist a main chemo-topic subdivision of the glomerular layer in macro-area (zones) which are active for chemicals differing for the functional group. Further work is currently in progress to interpret the clustering results in terms of chemical features of the odour molecules.

ACKNOWLEDGMENTS

This work was supported by the European Network of Excellence GOSPEL "General Olfaction and Sensing Projects on a European Level" (FP6-IST-2002-507610).

REFERENCES

1. G. M. Shepherd, Smell images and the flavour system in the human brain, *Nature insight* **444** 316-321 (2006).
2. M. Leon, B. A. Johnson, Olfactory coding in the mammalian olfactory bulb, *Brain Research Reviews* **42** 23–32 (2003).
3. M. Falasconi et al., A stability based validity method for fuzzy clustering, *Pattern Recognition* (Submitted, Oct. 2008).
4. C. Borgelt, Resampling for fuzzy clustering, International Journal of Uncertainty, *Fuzziness and Knowledge-Based* **15** 595-614 (2007).
5. R.J.G.B. Campello, A fuzzy extension of the Rand index and other related indexes for clustering and classification assessment, *Pattern Recognition Lett.* **28** 833–841 (2007).
6. K. Mori, Y. K. Takahashi, K. M. Igarashi and M. Yamaguchi, *Physiol Rev* **86** 409-433 (2006).

Olfactory Mucosa Tissue Based Biosensor for Bioelectronic Nose

Qingjun Liu *, Weiwei Ye, Hui Yu, Ning Hu, Hua Cai and Ping Wang

Biosensor National Special Laboratory, Key Laboratory of Biomedical Engineering of Education Ministry, Department of Biomedical Engineering, Zhejiang University, Hangzhou, 310027, P. R. China
State Key Laboratory of Transducer Technology, Chinese Academy of Sciences, Shanghai, 200050, P. R. China
* E-mail: *qjliu@zju.edu.cn (Q. Liu) Tel.: +86 571 87952832; fax: +86 571 87951676.*

Abstract. Biological olfactory system can distinguish thousands of odors. In order to realize the biomimetic design of electronic nose on the principle of mammalian olfactory system, we have reported bioelectronic nose based on cultured olfactory cells. In this study, the electrical property of the tissue-semiconductor interface was analyzed by the volume conductor theory and the sheet conductor model. Olfactory mucosa tissue of rat was isolated and fixed on the surface of the light-addressable potentiometric sensor (LAPS), with the natural stations of the neuronal populations and functional receptor unit of the cilia well reserved. By the extracellular potentials of the olfactory receptor cells of the mucosa tissue monitored, both the simulation and the experimental results suggested that this tissue-semiconductor hybrid system was sensitive to odorants stimulation.

Keywords: Bioelectronic nose; Olfactory mucosa tissue; Cell-based biosensor; Light-addressable potentiometric sensor; Olfactory receptor cell
PACS: 07.07.Df

INTRODUCTION

Traditional electronic noses are techniques to mimics animals' smell to detect odors. The detection ability is mainly depends on absorbability or catalysis of their sensitive materials to special odors. At the end of last century, Professor Wolfgang Göpel and his colleagues first proposed utilizing olfactory neurons as sensitive materials to develop a bioelectronic nose [1, 2]. They suggested the biomolecular function units can be used to develop highly sensitive sensors (like the dog's nose to sense drugs or explosives) as one of independent trends for electronic nose.

Our group has reported an olfactory biosensor based on light-addressable potentiometric sensor (LAPS) to investigate the extracellular potentials of the primary cultured olfactory cells under stimulations of the odorants [3]. It proved that those receptor cells and olfactory bulb neurons cultivated on the surface of the sensor are sensitive to environmental changes.

In vertebrates, each of the about 10^7 neurons of the olfactory epithelium only expresses 1 of the 1000 genes of olfactory receptors. Each individual receptor can bind multiple odorants with distinct affinities and specificities but some olfactory receptors display relatively restricted odour specificity to a set of few chemically related molecules. Therefore, using the whole olfactory mucosa tissue, with natural station reserved for neuronal populations, as bio-detecting element of the olfactory biosensor is a potential trend for bionic bioelectronic nose.

In this study, we analyzed the basic recording theories of the olfactory tissue by LAPS, and then fixed rat mucosa on the surface of LAPS to monitor its extracellular potentials. Both the simulation and the experimental results suggest the tissue-semiconductor hybrid system is sensitive to odorants stimulation.

EXPERIMENTAL AND METHODS

Sensor System of the LAPS

LAPS is a surface potential detector. With light pointer illuminates on LAPS, the semiconductor absorbs energy and leads to energy band transition. If LAPS is biased in depletion, the width of the depletion layer is a function of the local value of the surface potential (Figure 1). In this study, the LAPS chip and detecting setup were similar to the system we have reported for the primary cultured olfactory cell [3].

CP1137, *Olfaction and Electronic Nose: Proceedings of the 13th International Symposium*, edited by M. Pardo and G. Sberveglieri

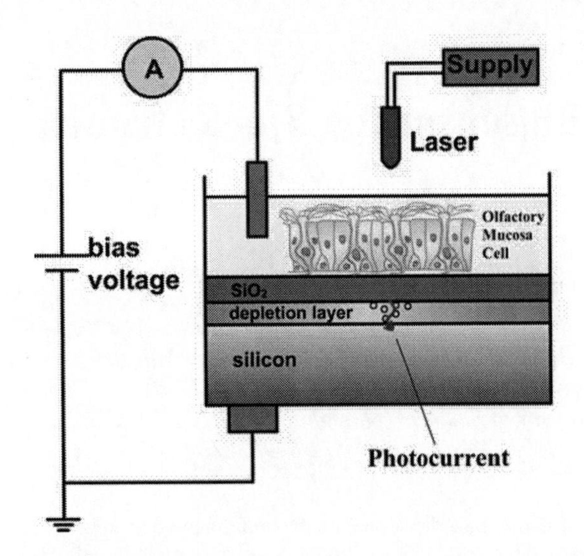

FIGURE 1. LAPS system of the olfactory mucosa tissue on the sensor surface.

The upper side of the LAPS chip was insulated with a layer of 30 nm SiO_2. A 1 μm layer of aluminum membrane was sputtered on the backside of the wafer to create an ohmic contact. During experiments, the modulated light (He-Ne laser, wavelength 543.5 nm, power 5 mW, Coherent Co.) was focused to less than 10 μm and illuminated on the desired cells. The fluctuation of the photocurrent was transmitted into peripheral potentiostat (EG&G Princeton Applied Research, M273A) through working electrode.

Model of the Tissue Potential

The mean electrical property of tissue on the semiconductor chips (Figure 2) is usually described by the volume conductor theory [4], as shown in Eq.1.

$$-\frac{1}{\rho}\left(\frac{\partial^2 V_{field}}{\partial x^2}+\frac{\partial^2 V_{field}}{\partial y^2}+\frac{\partial^2 V_{field}}{\partial z^2}\right)=j_{source}+j_{stim} \quad (1)$$

Where V_{field} raised from currents per unit volume j_{source} of cellular sources or j_{stim} of stimulation electrodes. ρ was the specific resistance. The curvature of the potential was proportional to current-source density with the coordinates x, y and z.

Neglecting the details of the current flow from tissue to bath and the stimulation, we described the sheet conductor model for LAPS recording. In the model, we described the shunting effect of the bath by an ohmic conductance per unit area g_{leak}, the tissue itself by a sheet resistance r_{sheet} and the substrate by a capacitance per unit area. Thus, we obtained Eq.2:

$$-\frac{1}{r_{sheet}}\left(\frac{\partial^2 V_{field}}{\partial x^2}+\frac{\partial^2 V_{field}}{\partial y^2}\right)+g_{leak}V_{field}+c_s\frac{\partial V_{field}}{\partial t}=j^{(2)}_{source} \quad (2)$$

Where V_{field} was the potential with an isotropic and homogeneous sheet resistance r_{sheet}.

We considered the activity of olfactory mucosa tissue. The x axis and the y axis of Eq.2 were in the plane of the tissue. The current-source density along the y direction was assumed homogeneous. Eq.3 obtained from Eq.2 without stimulation and with $c_s\frac{\partial V_{field}}{\partial t} << g_{leak}V_{field}$:

$$-\lambda^2_{sheet}\frac{d^2 V_{field}}{dx^2}+V_{field}=\frac{j^{(2)}_{source}}{g_{leak}} \quad (3)$$

Where λ^2_{sheet} equaled to $\frac{1}{g_{leak}r_{sheet}}$, written as Eq.4:

$$\lambda^2_{sheet}=\frac{1}{g_{leak}r_{sheet}} \quad (4)$$

For a constant $j^{(2)}_{source}$ in a range $-x_0 < x < x_0$ the field potential was given by Eq.5, where the amplitude $\frac{j^{(2)}_{source}}{g_{leak}}$ was in an infinite homogeneous sheet:

$$V_{field}(x)=\frac{j^{(2)}_{source}}{g_{leak}}\begin{Bmatrix} 1-\exp(-\frac{x_0}{\lambda_{sheet}})\cosh(\frac{|x|}{\lambda_{sheet}})|x|<x_0 \\ \sinh(\frac{x_0}{\lambda_{sheet}})\exp(-\frac{|x|}{\lambda_{sheet}})|x|>x_0 \end{Bmatrix} \quad (5)$$

Thus, this potential V_{field} can be recorded by LAPS.

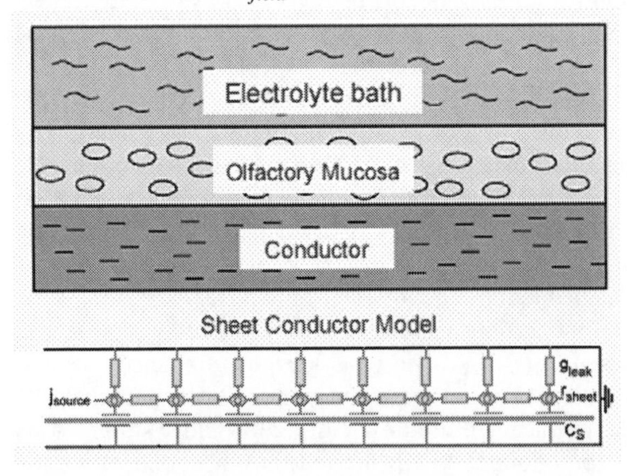

FIGURE 2. Geometry of tissue layer between electron conductor and electrolyte bath on LAPS, and the sheet conductor model for extracellular potentials recording.

Extracellular Potential Simulation

Extracellular potentials of the olfactory tissue on LAPS was simulated by the total transduction current was defined as:

$$I_{total} = -(I_{CNG} + I_{ClCa} + I_{NCX}) \quad (6)$$

In the formulas, lightface roman symbols represented constants; boldface symbols denoted dynamical state variables of the system; and italicized symbols denoted functions and auxiliary variables. We used the mechanistic mathematical model of the G-protein coupled signaling pathway responsible for generating current responses in olfactory receptor neurons to obtain I_{total} [5]. In order to construct the model of the tissue-LAPS, we substituted $j_{source}^{(2)}$ of Eq.6 with I_{total}, which varied with time.

$$V_{field}(x,t) = \frac{I_{total}(t)}{g_{leak}} \begin{cases} 1-\exp(-\frac{x_0}{\lambda_{sheet}})\cosh(\frac{|x|}{\lambda_{sheet}}) & |x|<x_0 \\ \sinh(\frac{x_0}{\lambda_{sheet}})\exp(-\frac{|x|}{\lambda_{sheet}}) & |x|>x_0 \end{cases} \quad (7)$$

Where $I_{total}(t)$ varied with time.

When recording, x was defined as the distance from the recording point. Thus, potential V_{field} varied with t and x. We used MATLAB to simulate the model and obtain the results.

Tissue Isolate and Fix onto Sensor Surface

The olfactory mucosa tissue was isolated from rat, and then rinsed and placed with cilia receptors side up on the surface of the sensor. After rinsing the mucosa with solution, the tissue was fixed by a plastic ring-shaped frame covered with a tightly stretched piece of mesh. At the same time, the expand cilia specialized for odor detection of the tissue was observed by the scanning electron microscope (SEM), where the olfactory receptor located in.

Odorants Stimulation

Odorants of acetic acid (CH_3COOH, a organic acid, with a distinctive pungent odor) was diluted to 1μM/ml, 25μM/ml, and 50μM/ml. After injected into the detection chamber, the detection chamber was washed with fresh medium without odorants. The minimum interval between the injections was 5 min to rule out the influence of remainder odorant and allow LAPS and tissue return to stady state.

RESULTS AND DISCUSSION

Olfactory Mucosa on LAPS

Olfactory receptor cells are bipolar nerve cells. From their apical pole the neurons extend dendrite to the epithelial surface, where they expand cilia, which are specialized for odor detection.

Figure. 3 shows olfactory mucosa isolated and fixed on the surface of LAPS for bioelectronic nose. Olfactory cilia formed a dense meshwork on the surface of the olfactory epithelium. Compared with isolated and primary cultured olfactory cells, the natural cilia of the mucosa tissue were very well preserved, based on basic structures of receptor cells' population without damaged.

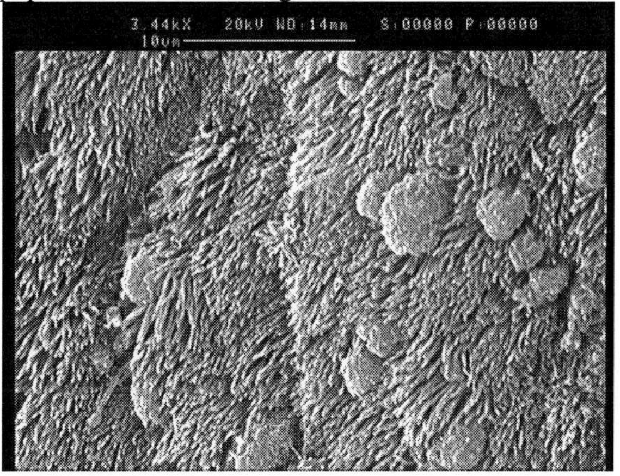

FIGURE 3. Olfactory mucosa tissue fixed on the surface of LAPS observed by SEM. Olfactory cilia form a dense meshwork on the surface of the olfactory epithelium.

Extracellular Simulation Result

We assumed the diameter of a single olfactory cell is 20 μm. So in Eq.7, the value of x_0 was 10 μm. The values of g_{leak} and λ_{sheet} were 13.3nS/μm^2 and 35 μm respectively, obtained from Fromherz [4]. Thus, Eq.7 can be written as:

$$V_{field}(x,t) = \frac{I_{total}(t)}{13.3} \begin{cases} 1-\exp(-\frac{10}{35})\cosh(\frac{|x|}{35}) & |x|<10 \\ \sinh(\frac{10}{35})\exp(-\frac{|x|}{35}) & |x|>10 \end{cases} \quad (8)$$

By MATLAB, we first got the values of $I_{total}(t)$ with the model established by [5]. Then, we used Eq.8 to get the simulation result shown in Figure 4.

We detected 100 s in time and 100 μm long in distance in the simulation, shown as distance axis and time axis, respectively. And the normal axis represented the voltage change. The recording point was assumed at 20 μm, in the center point of the recorded single cell. There was an obvious peak in the figure. The position of the peak was just the recording point, with the maximal amplitude. As the distance became father from the recording point, the amplitude became smaller. The phenomena may be explained by the distribution of potential. In the time axis, we can see the potential varies with time.

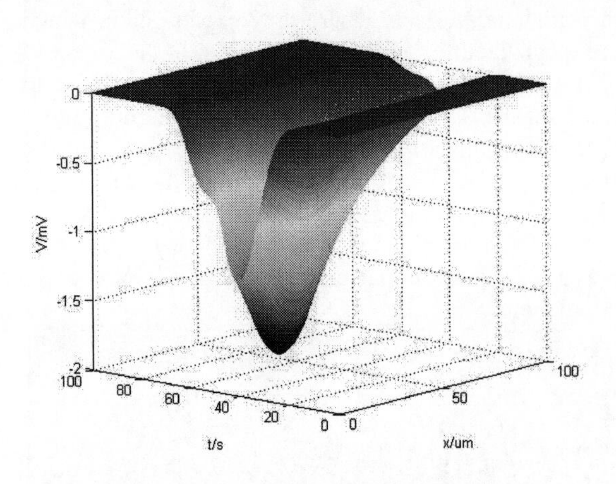

FIGURE 4. Extracellular potentials simulation result.

Odor Detection Result

Odorants of acetic acid was used as a stimulant for olfactory receptor cells to inspect the sensitivity of the sensor chip.

Figure 5 shows the experiment results recorded by our chip. Culture medium without odorants was pumped into micro-chamber, cells manifest very few potential changes, only sporadic spontaneous potential changes appeared, less then 10/min, and the amplitude is also very low. It only exhibited the presence of occasional spontaneous excitatory potentials. When adding acetic acid into micro-chamber, cells developed continuous tonic high frequency burst discharges; the amplitude is about 30-50 μV. The frequency and figure of the signal were similar to those extracellular potential recorded by the isolated and primary cultured olfactory cells [3]. Then, when culture medium without odorants was pumped into micro-chamber again, some potential changes were preserved. However, the amplitude was slowly decreased to the baseline lever of the pre-stimulation.

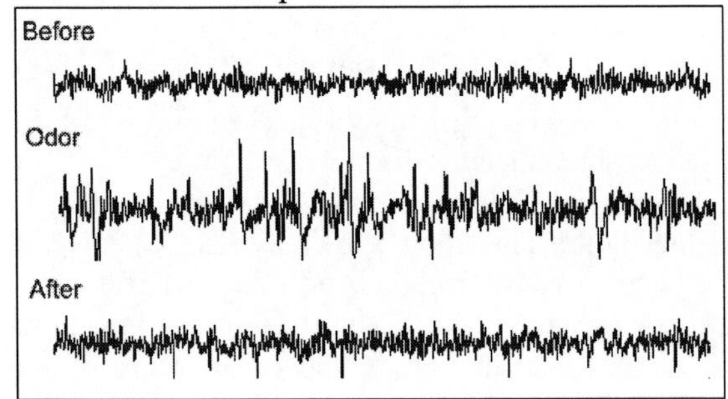

FIGURE 5. Responses of The extracellular potential changes of olfactory mucosa tissue to odor stimulation. The amplide of baseline before the stimulation is about 20 μV. The recording time of each signal is 60 sec.

We analyzed detail characters of the signals, most signals generated from stimulation contained a raise segment and a drop segment, which could drop even below the former potential and return even more slowly. However, the signals of spontaneous potentials, especially signals after the stimulation, often mainly contained only a negative segment below the baseline. The results were well according with the simulated signal. Both of them were spontaneous potential changes. In our next step, we will focuse on odorant stimulated potential changes.

In most vertebrates, transduction of an odour stimulus into a receptor current occurs on the cilia of the olfactory receptor cells [6]. As results described in this paper, it is possible to record neurons in an intact epithelium after the epithelium surgically removed. Therefore, with the epithelium was intact, and each olfactory receptor cell was in its native environment, studies of the bioelectronic nose will be very benefited form these liquid-phase stimuli recording to the olfactory tissue.

CONCLUSIONS

This article demonstrates an olfactory mucosa tissue based biosensor to investigate the response of the olfactory neurons under stimulations of the odorants. It has been proved by the stimulation of the acetic acid, that olfactory receptor cells in the mucosa tissue are sensitive to environmental changes. All these work suggests that the bionic designed tissue-semiconductor hybrid system can be used as a novel bioelectronic nose, with the natural stations of the neuronal populations and functional receptor unit of the cilia very well reserved.

ACKNOWLEDGMENTS

This work was supported by the National Natural Science Foundation of China (Grant No. 30700167, 60725102), the Project of State Key Laboratory of Transducer Technology of China (Grant No. SKT0702), and the Zhejiang Provincial Natural Seience Foundation of China (Grant No. Y2080673).

REFERENCES

1. C. Ziegler, W. Göpel, H. Hammerle, H. Hatt, G. Jung, L Laxhuber, H. L. Schmidt, S. Schutz, F. Vogtle and A. Zell, *Biosens. Bioelectron.* **13**, 539-571 (1998).
2. W. Göpel, *Sensors Actuators B* **65**, 70-72 (2000).
3. Q.J. Liu, H. Cai, Y. Xu, Y. Li, R. Li and P Wang, *Biosens. Bioelectron.* **22**, 318-322 (2006).
4. P. Fromherz, *Eur. Biophys. J.* **31**, 228- 231 (2002).
5. D. P. Dougherty, G. A. Wright and A. C. Yew, *Proc. Natl. Acad.Sci. USA,* **102**, 10415-10420 (2005).
6. S. J. Kleene, *Chem. Senses*, **33**, 839-859(2008).

Preliminary Modeling and Simulation Study on Olfactory Cell Sensation

Jun Zhou[1], Wei Yang[2], Peihua Chen[1], Qingjun Liu[1], Ping Wang[1]*

1 Biosensor National Special Laboratory, Department of Biomedical Engineering, Zhejiang University, Hangzhou 310027, P. R. China
2 School of Medicine, Institute for Neuroscience, Zhejiang University, 310058, P.R. China
**Correspondence addressed: Ping Wang (Email: cnpwang@zju.edu.cn)*

Abstract. This paper introduced olfactory sensory neuron's whole-cell model with a concrete voltage-gated ionic channels and simulation. Though there are many models in olfactory sensory neuron and olfactory bulb, it remains uncertain how they express the logic of olfactory information processing. In this article, the olfactory neural network model is also introduced. This model specifies the connections among neural ensembles of the olfactory system. The simulation results of the neural network model are consistent with the observed olfactory biological characteristics such as 1/f-type power spectrum and oscillations.

Keywords: Olfactory Sensory Neuron; Olfactory Neural Network; Olfactory Transduction
PACS: 87.19.It

INTRODUCTION

Animals discriminate and recognize numbers of chemical signals in their environment, which profoundly influence their behavior and provide them with essential information for survival. During this process, olfactory sensory neurons are the first energy transduction, which are responsible for both detecting odorant molecules and projecting neurons relaying information to the olfactory bulb. Because of this fundamental role in the olfactory process, an analysis of the physiology and membrane properties of these sensory cells is crucial to understand the function of the olfactory system. In order to elucidating the series of events leading from the detection of odorant molecules to the transmission of electrical impulses to the brain, the whole olfactory system model is introduced based on Freeman's KIII model.

EXPERIMENTAL AND METHODS

In microscopic research, much work has been undertaken in cells and ionic channels. Olfactory sensory neuron (OSN) dissociation and electrophysiological procedures have been described in detail[1]. Whole-cell patch-clamp method was provided to record voltage-gated ionic channel's current.

The single olfactory sensory cell modeling approach is based on a numerical reconstruction of the action potential by using Hodgkin-Huxley-type formalism. The change of the membrane potential (V) is given by

$$\frac{dV}{dt}=-\frac{1}{C_m}(I_i+I_{st}) \qquad (1)$$

Where C_m is the membrane capacitance, I_{st} is a stimulus current, and I_i is the sum of three ionic currents: I_{Na}, a fast inward sodium current; I_K, a slow outward potassium current; I_l, a time-dependent leak current. The ionic currents are determined by ionic gates, whose gating variables are obtained as a solution to a coupled system of three nonlinear ordinary differential equations. The ionic currents, in turn, change the membrane voltage V, which subsequently affects the ionic gates and currents. Rate constants of ionic gates were obtained by parameter estimation with an adaptive nonlinear least-squares algorithm. The numerical algorithm used to solve the differential equations is based on the Runge-Kutta method. All stimulation programs were implemented in software Matlab version 7.0.

The olfactory system consists of four main parts: the receptors (R), the olfactory bulb (OB), the anterior nucleus (AON) and the prepyriform cortex (PC), as

CP1137, *Olfaction and Electronic Nose: Proceedings of the 13th International Symposium,* edited by M. Pardo and G. Sberveglieri

shown in Fig.1, which specifies the connections among neural ensembles of the olfactory system and each node stands for a second-order ODE for the single ensemble dynamics. Different olfactory receptors (R) that are sensitive to different odorants can detect a large variety of odor molecules and send information through their axons to olfactory bulb (OB). The olfactory bulb is mainly composed of two kinds of cells, excitatory mitral cells (M) and inhibitory granule cells (G). The bulbar neurons send their axons by way of the lateral olfactory tract (LOT) to the AON and the PC. In the AON and PC, there are also excitatory (E) and inhibitory cells (I), and the negative feedback between these cells support oscillations in both AON and PC as in the OB but with different and incommensurate characteristic frequencies. The PC sends axons into the external capsule (EC) in the brain. In the other direction the AON and PC send axons back through the medial olfactory tract (MOT) to the OB.

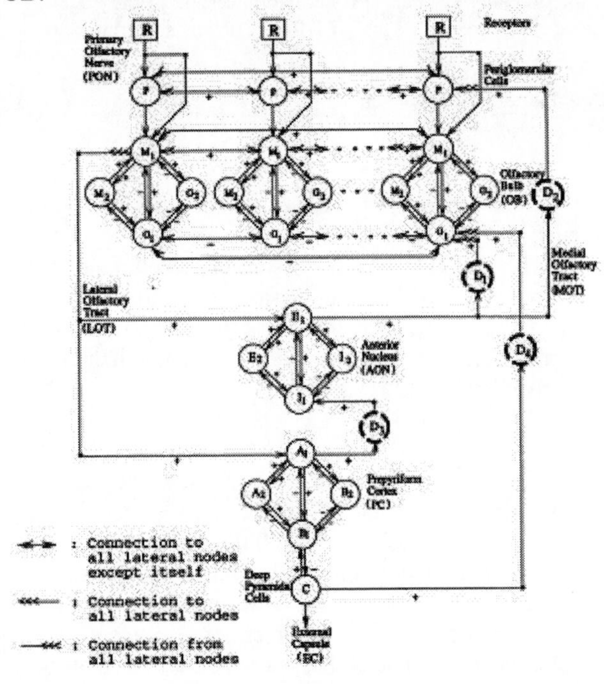

FIGURE 1. The topological diagram of the olfactory model

The dynamical behavior of each cell ensemble of the olfactory system can be governed by Eq. 2:

$$\frac{1}{a\cdot b}[\ddot{x}_i(t)+(a+b)\dot{x}_i(t)+a\cdot b\cdot x_i(t)]=\sum_{j\neq i}^{N}[W_{ij}\cdot y]+I_i(t) \qquad (2)$$

A general connecting equation[2], where a and b are derived from experimental rate constants. A single (double) dot over a variable means the first (second) derivative of the variable with respect to time. The variable x represents dendritic potential, as evaluated by extracellular measurements of EEGs. The variable y represents multineuronal pulse density as it is evaluated by multiunit recording. Both variables are continuous in the cortex. They are discretized by experimental measurement. Conversion from pulse density y to wave amplitude x is implicit in the synaptic weights. Conversion from wave amplitude x to pulse density y is implicit in the sigmoidal function, Q(x,q) (see Fig.2 with q= 5.0). This function was derived from the Hodgkin-Huxley system and holds for all parts of the central olfactory system [3]. The variable I stands for an external stimulus. Physiological experiments confirm that a linear second-order derivative is an appropriate choice. By the topological diagram of Fig.1, the n-channel olfactory model is implemented by Eq.2 .

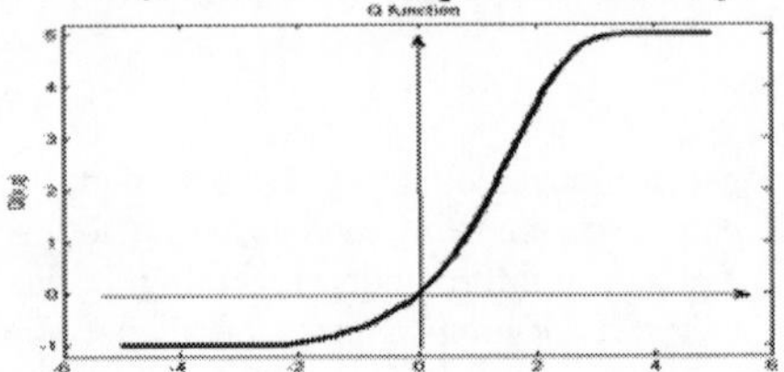

FIGURE 2. Asymmetric sigmoidal transformation function

RESULTS

Voltage-gated ionic channel model stimulation

The macroscopic currents related to numerous ion channels embedded in the cell membrane have been a topic already. These channels that are gated by voltage can be in either permissive or non-permissive state. Only when all gates of an individual channel are permissive, the channel is open and a current can pass. Rate constant α_i and β_i are voltage dependent and described using first-order kinetics. (see Eq.1) In our model, we limit the simulations to two voltage-gated channels including the fast inward sodium current and the outward potassium current. Given the maximum possible conductance $\overline{G_k}=12ns$, $\overline{G_{Na}}=19ns$. The resting membrane potential can be estimated by Nernst equation, which are $E_{Na}=60mV$, $E_K=-80mV$, $E_l=-10mV$.

$$\alpha_m=-0.1\times\frac{V+65}{\exp(-(V+65)/2)-1}$$

$$\beta_m=4\times\exp(-(V+60)/8)$$

$$\frac{dm}{dt}=\alpha_m(1-m)-\beta_m m$$

$$\alpha_h=0.07\times\exp(-(V+70)/10)$$

$$\beta_h=\frac{2}{\exp(-(V+37)/10)+1}$$

$$\frac{dh}{dt}=\alpha_h(1-h)-\beta_h h$$

$$\alpha_n = -0.01 \times \frac{V+35}{\exp(-(V+35)/24)-1}$$

$$\beta_n = 0.25 \times \exp(-(V+35)/40)$$

$$\frac{dn}{dt} = \alpha_n(1-n) - \beta_n n \tag{3}$$

Based on these findings, we define the currents as follows:

$$I_{Na} = \overline{G_{Na}} \cdot m^3 \cdot h \cdot (V - E_{Na})$$

$$I_K = \overline{G_K} \cdot n^4 \cdot (V - E_K)$$

$$I_l = \overline{G_l} \cdot (V - E_l) \tag{4}$$

The action potential of the single olfactory sensory neuron is as follows:

$$\frac{dV}{dt} = -\frac{1}{C_m}(I_{Na} + I_K + I_l + I_{st}) \tag{5}$$

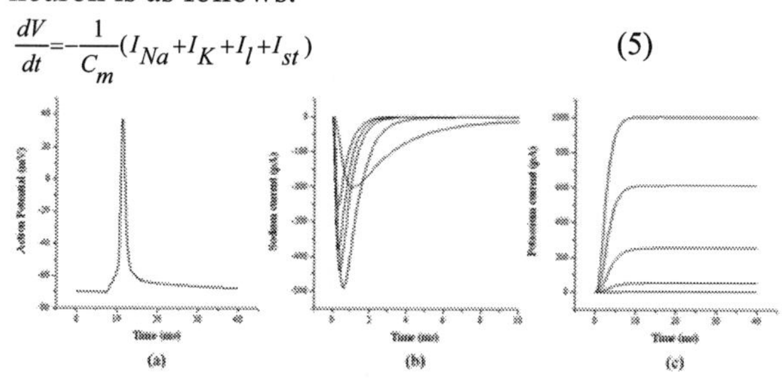

FIGURE 3. Simulation of action potential, sodium current and potassium current under different voltage clamps

Modeled olfactory sensory neuron started to fire action potentials from the resting membrane potential of -70 mV. As shown in Fig. 3a, action potential is elicited by a current pulse 5pA with 1ms.

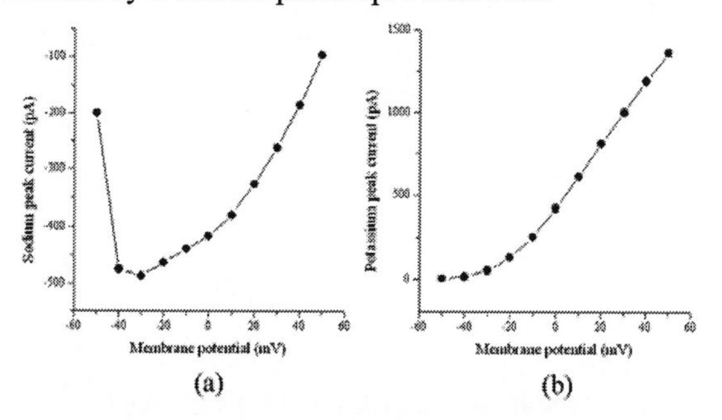

FIGURE 4. Simulation of sodium and potassium peak current under different voltage clamps

To examine the simulation of sodium current I_{Na}, voltage clamp experiments were implemented. The membrane potential was stepped from the holding potential of -80mV to the testing levels, varying from -50mV to +50mV in 20mV increments for 10 ms. The characteristic voltage-dependent activation of sodium channel was reproduced (Fig.3b) and the peaks of activated currents at each testing level were plotted (Fig.4a). Similar simulations were performed for I_K. Membrane potentials were stepped from -50mV to +50mV in 20mV increment for 40ms. Simulation results were shown in Fig.3c. The peak value of potassium current under each membrane potential were extracted and plotted in Fig.4b.

To understand the logic of olfactory information processing, one has to appreciate the coding rules generated at each level, from the odorant receptors up to the level of the olfactory cortex. Olfactory transduction starts with interaction between inhaled odorants molecules and olfactory sensory cells. Activation in the membrane of the sensory cells resulting in a transient influx of sodium current and efflux of long sustained potassium current in the generation of a graded receptor potential in the soma of the sensory cell[4]. Electrophysiological studies indicate that odorant sensitivity and the odorant-induced current are uniformly distributed along the cilia. From its basal pole, the sensory neuron sends a single axon through the basal lamina and cribiform plate to terminate in the olfactory bulb. The unmyelinated sensory neuron axons merge into densely packed fascicles to form the olfactory nerve, which transmits the electrical signals to the bulb. The olfactory bulb has also long been considered a model system including excitatory mitral cells and inhibitory granule cells. The cells are modeled as single-compartment Hodgkin-Huxley neurons as well. With a square-pulse input or a mimic “sniff” input, the simulation shows that mitral cells spike sparingly during nonstimulus periods and synchronize rapidly during stimulus periods and granule cells only spike during stimulus periods in accordance with LFP frequency recorded in zebrafish[5]. Fabio M. Simoes-de-Souza et al. also describe a biophysical model of the initial stages of vertebrate olfactory system which gives spatiotemporal patterns in the epithelium and bulb generated by the couplings due to the gap junctions and dendrodendritic synapses during odor stimulation[6] .

Olfactory Neural Network model stimulation

The synaptic connection weights cannot be directly measured from experiments, so there are many reports for describing weight optimization. The main purpose is to simulate the model more biologically in accordance with measured activity (endogenous EEG waves) and average evoked potentials (AEPs impulse response evoked by electrical stimulation). Here we use 8 channels KIII model and the parameters are taken from the paper [7]. Peripheral noise is added in the model to mimic the noise when olfactory receptor neuron stimulated by odors. Peripheral noise is distributed by Gaussian distribution. We also add central noise in AON and this noise is also distributed by Gaussian distribution.

Before stimulation from outside environment, the olfactory system is chaotic. The only responses from four nodes (P1, M11, G11, A1) are not in regular and distributed by noise type. After stimulation, large oscillations in all nodes are observed. From the phase portraits plotting, the system phase curves are in a complex boundary district. With the stimulation by a square wave with amplitude 1 from 768 second to 1024 second, the system starts from a fixed point zero and evolves to a limit cycle. When the input vanishes the system spirals back to zero state (shown in Fig.5). An important property of observed characteristics of EEGs is a 1/f-type power spectrum. Fig.6 shows the stimulation of corresponding log-log plot of power spectra of four nodes(P1, M12, G22, E1). The log power decreased nearly linearly with log frequency which is in accordance with biological observation.

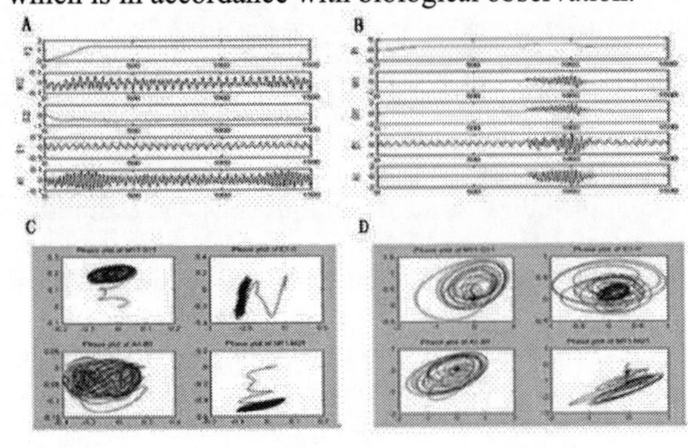

FIGURE 5. A). Nodes electrical signals output without stimulation B). Nodes electrical signals output with stimulation C). Phase portraits without stimulation D). Phase portraits with stimulation

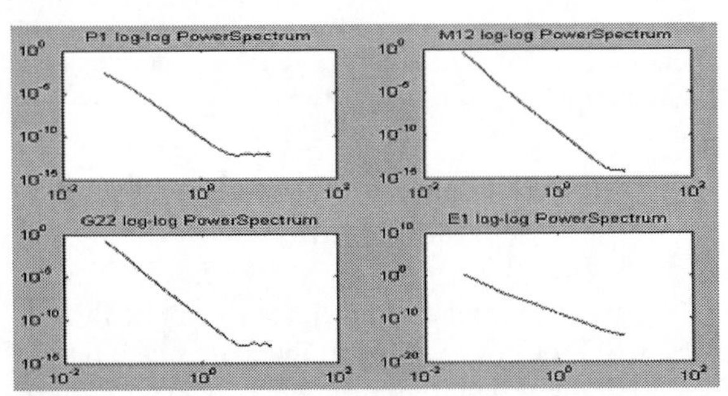

FIGURE 6. The corresponding log-log plot of power spectra of four nodes (P1, M12, G22, E1)

DISCUSSIONS AND CONCLUSIONS

A complete understanding of the olfactory transduction requires analysis from the cellular level to the systems. Here we considers olfactory sensory neuron model, olfactory bulb model and olfactory neural network model trying to provide a more clear route to the research of olfactory information processing. The olfactory neural network model specifies the connections among neural ensembles of the olfactory system. With model simulation, biological characteristics such as 1/f-type power spectrum, oscillations are remained. Though there are many models in olfactory sensory neuron and olfactory bulb, it still can not be incorporate in the whole olfactory neural model to mimic every level of the olfaction. Further work should be focused on how to modify the whole olfactory system model concerning every level model. Recently, a contextual model for axonal sorting into glomeruli[8] and olfactory system scheme model based on experimentally-observed, odour-specific spatiotemporal patterns of neural activity[9] are worth investigating in understanding of the mechanisms of early olfactory transduction.

ACKNOWLEDGMENTS

The authors thank the members of the laboratory for comments and helpful discussions. This work was supported by the National Natural Science Foundation of China (Grant Nos. 60725102, 30700167) and the Natural Science Foundation of Zhejiang Province of China (Grant No. R205502).

REFERENCES

1. A.M.Cunningham, P.B.Manis, Olfactory receptor neurons exist as distinct subclasses of immature and mature cells in primary culture. *Neuroscience,* 1999. 93(No. 4): pp. 1301-1312
2. Hung Jen Chang, Parameter optimization in an olfactory neural systems. *Neural Networks*, 1996. 9(No. 1): pp. 1-14.
3. Freeman, W.J., Mass action in the nervous system. 1975, New York: Academic Press.
4. Getchell TV, Responses of olfactory receptor cells to step pulses of odor at different concentrations in the salamander. *J Physiol*, 1978. 282: pp. 521-540.
5. Malin Sandtrom, Scaling effects in a model of the olfactory bulb. *Neurocomputing*, 2007. 70: pp. 1802-1807.
6. Fabio M., Biophysical model of vertebrate olfactory epithelium and bulb exhibiting gap junction dependent odor-evoked spatiotemporal patterns of activity. Biosystems, 2004. 73: pp. 25-43.
7. Huang Jen Chang, Biologically modeled noise stabilizing neurodynamics for pattern recognition. *J. Bifurcation and Chaos*, 1998. 8(2): pp. 321-345.
8. Felnsteln, P., A contestual model for axonal sorting into glomeruli in the mouse olfactory system. *Cell*, 2004. 117: pp. 817-831.
9. T.Schaefer, A., Spatiotemporal representations in the olfactory system. *Trends in Neurosciences* 2007. 30(3): pp. 92-100.

Applying Convolution-Based Processing Methods To A Dual-Channel, Large Array Artificial Olfactory Mucosa

J.E. Taylor, F.K. Che Harun, J.A. Covington, and J.W. Gardner

Sensors Research Laboratory, School of Engineering, University of Warwick, Coventry, CV4 7AL, UK

Abstract. Our understanding of the human olfactory system, particularly with respect to the phenomenon of nasal chromatography, has led us to develop a new generation of novel odour-sensitive instruments (or electronic noses). This novel instrument is in need of new approaches to data processing so that the information rich signals can be fully exploited; here, we apply a novel time-series based technique for processing such data. The dual-channel, large array artificial olfactory mucosa consists of 3 arrays of 300 sensors each. The sensors are divided into 24 groups, with each group made from a particular type of polymer. The first array is connected to the other two arrays by a pair of retentive columns. One channel is coated with Carbowax 20M, and the other with OV-1. This configuration partly mimics the nasal chromatography effect, and partly augments it by utilizing not only polar (mucus layer) but also non-polar (artificial) coatings. Such a device presents several challenges to multi-variate data processing: a large, redundant data-set, spatio-temporal output, and small sample space. By applying a novel convolution approach to this problem, it has been demonstrated that these problems can be overcome. The artificial mucosa signals have been classified using a probabilistic neural network and gave an accuracy of 85%. Even better results should be possible through the selection of other sensors with lower correlation.

Keywords: Convolution, Signal Processing, Electronic Nose, Chemical Sensors
PACS: 07.07.Df

INTRODUCTION

Over the past twenty years, significant advances have been made in the understanding of the mechanism by which odours are detected by the human olfactory system. This has led to the concurrent development of instruments designed to detect odours and commonly known as artificial or electronic noses (e-noses) [1].

However, sensor-based electronic noses today generally suffer from significant weaknesses that limit their ubiquitous application. Their sensing ability is heavily affected by a range of factors, including drift due to temperature, humidity variations and background electrical noise, sensor variations in production, aging and poisoning. These problems are increased by the frequent demand to detect very low concentrations (below PPM) of an odour in air [2, 3], making the design of an electronic nose difficult, even with expensive autosamplers and a carrier gas (e.g. zero-grade air, nitrogen and helium).

Faced with this challenge, novel instruments are being designed to tackle these issues and improve detection thresholds and classification success rates; examples include combining an electronic nose array with a commercial gas chromatography column or mass spectrometer unit.

More recently it has been suggested that chromatography plays a role in odour discrimination within the human olfactory system, known as 'nasal chromatography' [4]. This paper suggests that the aqueous layer covering the olfactory receptors in the

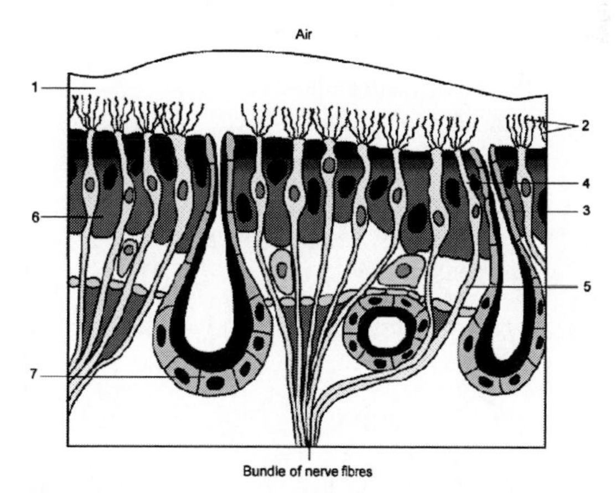

FIGURE 1. Schematic structure of the olfactory organ. (1) Mucous layer. Sensory cells consisting of olfactory cilia (2), cellular body (3), dendrite (4), and a neurite or axon (5). (6) Support cells. (7) Bowman's glands. [5]

CP1137, *Olfaction and Electronic Nose: Proceedings of the 13th International Symposium*, edited by M. Pardo and G. Sberveglieri

olfactory epithelium (Figure 1) acts as a retention layer to odours moving along and through it, functioning similar to a stationary phase coating channel in a gas chromatographic system. Hence different odours are partitioned and transported at different rates to olfactory receptor cells – leading to different temporal signatures. (Nb: this may even be further amplified by the use binding proteins to transport non-polar molecules through the mucous layer).

The signals produced by novel instruments mimicking this effect [6, 7] require a new approach to signal processing so that this information-rich environment can be fully exploited. One such method has been reported, where the spatio-temporal signals from matching spatially-separated sensors are combined and analysed using a convolution method [8]. This paper reports on the application of this novel processing approach to an artificial olfactory mucosa electronic nose.

EXPERIMENTS AND METHODS

The dual-channel artificial olfactory mucosa [7] has been realised by combining three large chemoresistive sensor arrays, employing composite polymer materials as the sensing layer (carbon black, Cabot corp. USA, mixed with 24 combinations of differing sensing polymers), and two plastic polymer-coated columns (Figure 3). Figure 2 shows the general arrangement of the system showing chemosensor arrays and two retentive columns.

Each large sensor array is laid out in a rectangular matrix configuration (12 columns × 25 rows) to reduce the pad count. This configuration allows us to get a total of 300 sensors per array, with a final chip size of 12 mm × 8.5 mm. This resistive based sensor is then coated with 24 different carbon black polymer composites to create a diversity of spatial responses (see Table 1). Multiple sensors are coated with each composite to provide redundancy that will be used to improve the accuracy of the data processing.

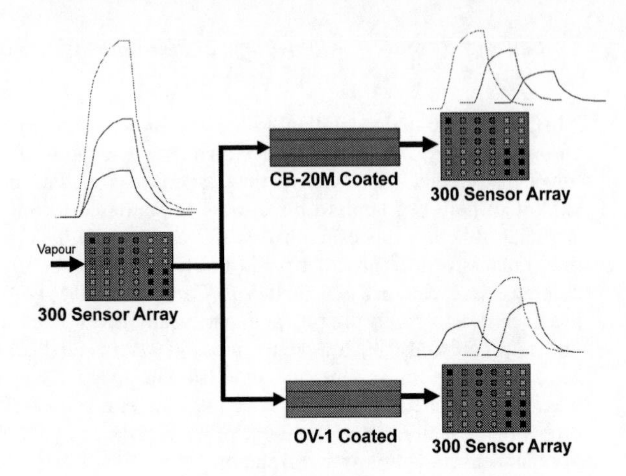

FIGURE 2. General layout of 900 sensor artificial olfactory system.

Each polymer composite solution was prepared by dissolving the polymer blend in 20 ml of solvent. Then the carbon black was added, and the solution shaken (Griffin & George flask shaker UK). Each group of 12

TABLE 1. Composition of sensing polymers.

No.	Polymer Composite	Polymer A (g)	Polymer B (g)	Carbon black (g)	Solvent (20 ml)
P1	Poly(stylene-co-butadiene) PSB	0.7	0	0.175	Toluene
P2	Poly(ethylene-co-vinyl acetate) PEVA	1.2	0	0.3	Toluene
P3	Poly(caprolactone) PCL	1.2	0	0.3	Toluene
P4	Poly(9-vinylcarbazole) PVC	0.7	0	0.175	Toluene
P5	Poly(ehtylene glycol) PEG	1.2	0	0.3	Toluene
P6	Poly(4-vinyl phenol) PVPH	1.2	0	0.3	Ethanol
P7	Poly(methyl methacrylate) PMM	1.2	0	0.3	Ethanol
P8	Poly(vinyl pyrrolidone) PVPD	0.7	0	0.175	Ethanol
P9	Poly(bisphenol A carbonate) PBA	0.7	0	0.175	Dichloromethane
P10	Poly(sulfane) PSF	0.7	0	0.175	Dichloromethane
P11	PSB 50% + PEVA 50%	0.35	0.6	0.2375	Toluene
P12	PSB 50% + PCL 50%	0.35	0.6	0.2375	Toluene
P13	PEVA 50% + PCL 50%	0.6	0.6	0.3	Toluene
P14	PEG 50% + PVPH 50%	0.6	0.6	0.3	Ethanol
P15	PSB 50% + PVC 50%	0.35	0.6	0.2375	Toluene
P16	PEVA 50% + PVC 50%	0.6	0.35	0.2375	Toluene
P17	PCL 50% + PVC 50%	0.6	0.35	0.2375	Toluene
P18	PMMA 50% + PSB 50%	0.6	0.35	0.2375	Toluene
P19	PMMA 50% + PEVA 50%	0.6	0.6	0.3	Toluene
P20	PMMA 50% + PCL 50%	0.6	0.6	0.3	Toluene
P21	PMMA 50% + PVC 50%	0.6	0.35	0.2375	Toluene
P22	PEG 50% + PVPD 50%	0.6	0.35	0.2375	Ethanol
P23	PVPH 50% + PVPD 50%	0.6	0.35	0.2375	Ethanol
P24	PBA 50% + PSF 50%	0.35	0.35	0.175	Dichloromethane

sensors were then deposited with an Iwata CP-30 airbrush by spraying through a mask with a set of (4 × 3) 200 μm square holes. The sensor resistance was controlled during deposition to have a value between 1 kΩ and 5 kΩ.

In order to mimic the 'nasal chromatograph' effect such as in the mammalian olfactory system [9], two retentive columns like nares but with different retentiveness were used (Figure 3). Each column was of similar size (0.38 mm × 0.25 mm × 2000 mm) and coated with a different coating. In this experiment, the first column is coated with a 5 μm thick layer of OV-1 (a non-polar stationary phase), while the second column is coated with a 5 μm thick layer of Carbowax 20M (a polar stationary phase). These two differently coated columns produce different retentive responses to the same analyte, producing a set of spatio-temporal signals.

In this experiment, we are discrimating between four essential oils: Lemon Grass, Cinnamon, Ylang Ylang and Lavender. Each vapour test had a duration of 450 s, consisting of 50 s of laboratory air followed by a 100 s pulse of essential oil vapour mixed with laboratory air, with a flow rate of 20 ml/min. The tests were performed at a temperature of 22°C, and each vapour was repeated 5 times.

FIGURE 3. Warwick stackable plastic retentive columns.

RESULTS

The data obtained from the experiment were analysed using the methods reported in [8]. This processing approach takes related signals, such as a signal from before the retentive channels (S_F) and another from after passing through a retentive channel (S_C), and combines them into a new characteristic signal using the convolution transform (Equation 1). This characteristic signal is then used for processing and analysis.

$$S_F(t) * S_C(t) = \int_{\tau=-\infty}^{\tau=\infty} S_F(\tau) S_C(t-\tau) d\tau \quad (1)$$

Samples of the relative responses from the sensor arrays are illustrated in Figure 4. With 3 sensor arrays, it is possible to pair up the arrays in three possible ways when performing the convolution: front array * OV-1 array (F*O), front array * carbowax array (F*C), or OV-1 array * carbowax array (O*C). Each of these pairings was considered.

Each sensor in the first array was paired with their counterpart in the second array. In cases where one signal was either saturated or too weak (less than a 1% response), the pair was censored and would not be used further in the processing process.

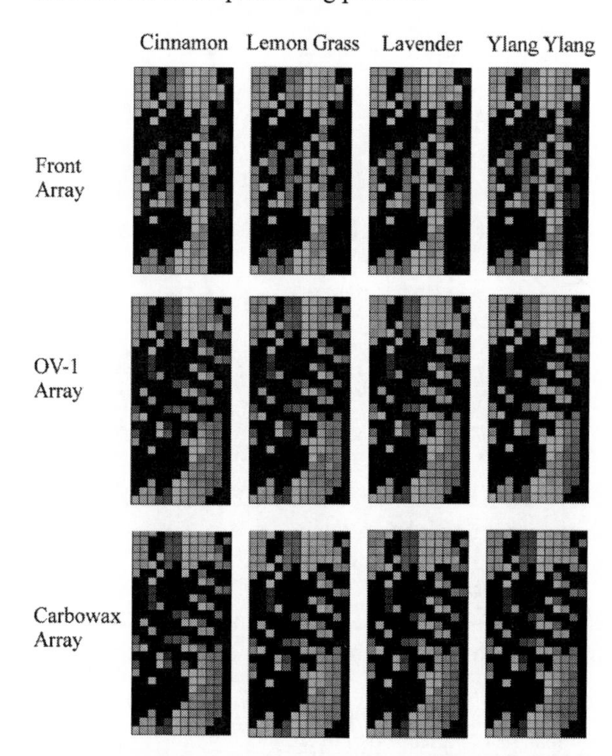

FIGURE 4. Illustration of relative response from each sensor array. Black cells show signal saturation, white cells show weak signals and greyscale shows relative peak magnitude of sensor response. It is evident that there is considerable correlation in the arrays.

The remaining pairs were then normalised and subjected to a convolution integral transform. This produced an array of characteristic signals that could be used for classification. The feature extracted from these signals was simply the area of the convolution integral. Figure 5 illustrates a sample of these signal arrays after normalisation.

These feature arrays were then classified using a probabilistic neural network (PNN). Due to the limited quantity of replicated data available, a bootstrap train and test method was used. A set of samples (1 of each test analyte) was omitted for testing, and the remainder used to train the PNN. The overall performance of the system was judged once all the 20 samples had been omitted once.

Due to the redundant (correlated) nature of much of the data, a subset of sensor results was used. A search-forward feature selection method was combined with a random sampling of sensor feature space.

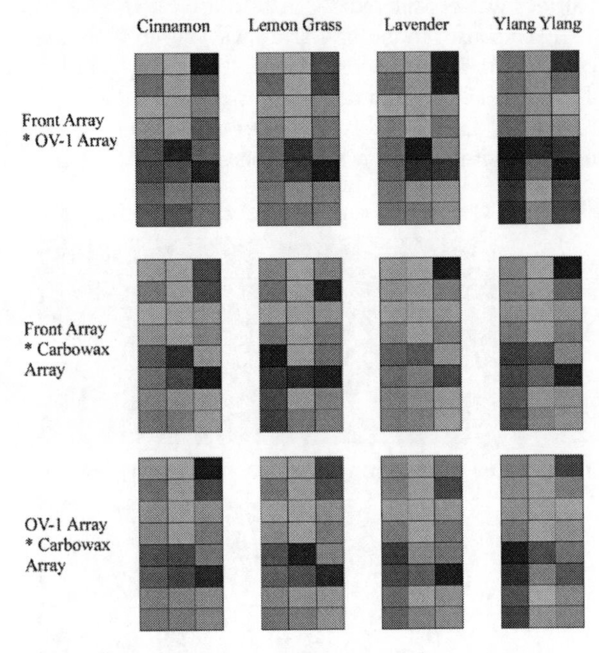

FIGURE 5. Illustration of normalized areas extracted from characteristic signals. Less correlation is observable than in the raw data.

Table 2 summarises the results obtained from optimization of the feature subset.

TABLE 2. Accuracy results from a PNN classified when utilising an optimal subset of sensors.

Array pair	Train and Test Accuracy (%)					Avg. (%)
F*O	50	75	75	75	100	75
F*C	100	100	100	75	50	85
O*C	75	75	75	75	25	65

CONCLUSIONS

A novel convolution integral transform based method, as proposed in [8], has been applied to a novel odour sensing instrument that combines 900 sensors with two mucous-like retentive channels. The unique design and size of this novel instrument has been exploited to classify different odours in air with an accuracy of about 90%.

The accuracy of the instrument can be substantially improved in a number of different ways: firstly, by the selection of other sensors that are less correlated to those employed here. Secondly, the choice of convolution feature (area) can also be extended to use height, shape etc. And thirdly, the process of selecting the sensor subset is basic and neither exhaustive nor even comprehensive - and more advanced search methods exist that utilise fewer restrictions in subset selections and sensor pairings. These methods are being explored and should provide further improvements to accuracy.

Interesting, it is not uncommon in biological olfaction for odour binding proteins (OBP) to capture and transport odorant molecules to the olfactory receptors (ORs). This mechanism may improve selectivity more than by simple diffusion alone and allow hydrophobic molecules to be sensed without the need for a non-polar mucous layer.

Finally, it should also be noted that the convolution approach has also be applied with some success to a single set of different sensors (cross-convolution), i.e. a conventional e-nose.

REFERENCES

1. J.W. Gardner and P.N. Bartlett, *Electronic Noses: Principles and Applications.* Oxford: Oxford University Press, 1999
2. M.A. Ryan *et al.*, "Monitoring Space Shuttle Air Quality Using the Jet Propulsion Laboratory Electronic Nose", *IEEE Sensors*, vol. 4, no. 3 (2004), pp. 337-347.
3. R.C. Young, W.J. Buttner, B.R. Linnell, R. Ramesham, "Electronic Nose For Space Program Applications", *Sensors and Actuators B*, vol. 93, (2003) pp. 7-16.
4. M.M. Mozell, M Jagodowicz, "Chromatographic Separation Of Odorants By The Nose: Retention Times Measured Across In Vivo Olfactory Mucosa", *Science*, vol. 181, (1973), pp. 1247-1249.
5. P. Vroon, *Smell: The Secret Seducer.* New York: Strauss and Giroux, 1997, pp. 28
6. S.L. Tan, "Smart Chemical Sensing: Towards a Nose-on-a-Chip," *Ph.D. thesis*, School of Engineering, University of Warwick, Coventry, UK, 2005
7. F.K. Che Harun, J.E. Taylor, J.A. Covington, J.W. Gardner, "Dual-channel Odour Separation Columns With Large Chemosensor Arrays For Advanced Odour Discrimination", *Proc. 12th International Meeting on Chemical Sensors,* 13-16 July 2008, Columbus, Ohio.
8. J.W. Gardner, J.E. Taylor, "Novel Convolution Based Signal Processing Techniques for a Simplified Artificial Olfactory Mucosa", *TRANSDUCERS 2007 Conference*, 10-14 June 2007, Lyon, France, pp. 2473-2476
9. J.A. Covington, J.W. Gardner, A. Hamilton, T. Pearce, S.L. Tan, "Towards A Truly Biomimetic Olfactory Microsystem: An Artificial Olfactory Mucosa". *IET Nanobiotechnology*, Vol. 1, (2007), pp. 15-21.

An Artificial Olfaction System Formed by a Massive Sensors Array Dispersed in a Diffusion Media and an Automatically Formed Glomeruli Layer

Corrado Di Natale[1], Eugenio Martinelli[1], Roberto Paolesse[2], Arnaldo D'Amico[1], Daniel Filippini[3], Ingemar Lundström[1]

1. Dept. Electronic Engineering, University of Rome Tor Vergata; Italy
2. Dept. Chemical Science and Technology, , University of Rome Tor Vergata; Italy
3. Division of Applied Physics, University of Linköping, Linköping; Sweden

Abstract. Optical imaging is a read-out technique for sensors that can easily provide advances in artificial olfaction implementing features such as the large number of receptors and the glomeruli layer. In this paper an artificial olfaction system based on the imaging of a continuous layer of chemical indicators is illustrated. The system results in an array of thousands of sensors, corresponding to the pixels of the image. The choice of Computer Screen Photoassisted Technology as a platform for optical interrogation of the sensing layer allows for the definition of a strategy for an automatic definition of the glomeruli layer based on the classification of the optical fingerprints of the image pixels. Chemical indicators are dissolved into a polymeric matrix mimicking the functions of the olfactory mucosa. The system is here illustrated with a simple experiment. Data are treated applying a lateral inhibition to the glomeruli layer resulting in a dynamic pattern resembling that observed in natural olfaction.

Keywords: Image sensor, metalloporphyrins, artificial olfaction.
PACS: 07.07.df

INTRODUCTION

One of the more evident differences between natural and artificial olfaction is represented by the number of involved sensors. Among the available sensor technology, optical imaging offers a simple approach for large sensor arrays development [1]. To this regard the research group of Walt and Kauer shown that a camera captured image of a bundle of chemically functionalize optical fibres tips can allow the contemporaneous measurement of a large number of sensors [2].

Recently, this approach has been fully exploited imaging a chemically sensitive surface formed by a continuous layer of chemical indicators [3]. In this situation, the image sensor provides a segmentation of the sensing layer into a number of elementary units corresponding to the pixel of the image. Eventually, since it is possible to evaluate the optical properties at the level of the single pixel, each pixel of the image may correspond to an individual sensor. To this regard, even low-resolution images may easily result in thousands of individual and independent sensing units. This approach offers also the possibility to further develop artificial olfactory structures introducing a layer of glomeruli. In the standard model of olfaction, glomeruli receive signals from olfactory receptor neurons (ORN) expressing the same kind of receptors, and then, they are likely characterized by the same chemical sensitivity [4]. Even in presence of an apparent random spatial distribution of ORNs a precise wiring between millions of ORNs and few hundreds of glomeruli is achieved [5].

In this paper we consider a collection of arbitrarily shaped regions of color indicators illuminated by a controlled source; the optical characteristics of each pixel of the image are measured by a camera providing the intensity of light in the three channels: red, green, and blue. The combination of illumination sequence and camera read-out results in a fingerprint encoding the optical properties of each single pixel. A classification of these fingerprints assigns pixels to classes, and each class contains pixels carrying the same color indicator. This behavior resembles the association between ORNs carrying the same chemical receptors into the same glomerulus. On the basis of this analogy it is straightforward to describe the layer of indicators as an artificial epithelium, Pixels of the

CP1137, *Olfaction and Electronic Nose: Proceedings of the 13th International Symposium*, edited by M. Pardo and G. Sberveglieri

image as artificial olfactory neurons, and the classes providing by the classifiers as an abstract representation of artificial glomeruli.

Another important characteristic of natural olfaction is related to the olfactory mucosa, the medium where the ORNs are embedded. Odorant molecules, diffusing through the mucosa, may be separated according to a gas-chromatographic principle [6]. To incorporate this property, the sensing molecules are dispersed in a polymer membrane that is sandwiched between two transparent sheets. One of the sheets is endowed with a hole allowing odorant molecules to freely diffuse through the polymer reaching the sensing molecules.

The properties of the system have here illustrated in a simple experiment aimed at recognizing pure compounds. A demonstration of the possibility to process glomeruli signals with bio-inspired signal processing is also described.

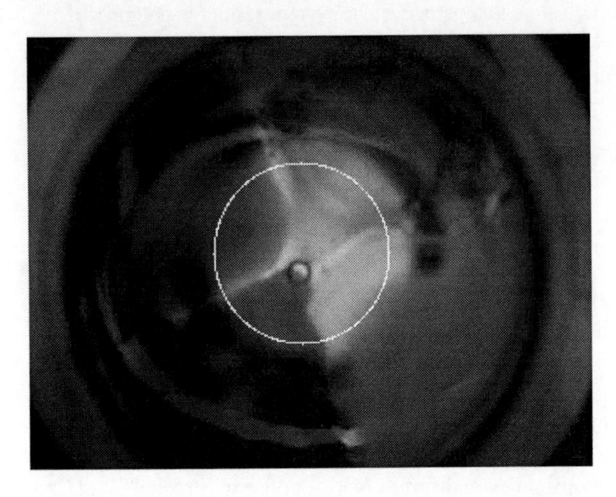

FIGURE 1. Image of the layer of porphyrins illuminated by white light, the white circle indicates the region of the image, containing 7845 pixels, considered in the analysis

EXPERIMENTAL

Color indicators used in this paper were the following tetrapyrrolic macrocycles (the acronyms used through the text are given in brackets): (5,10,15,20-tetraphenylporphyrin)zinc [ZnTPP], (5,10,15,20-tetraphenylporphyrin)manganese chloride [MnTPP], (5,10,15,20-tetraphenylporphyrin)cobalt [CoTPP], and (5,10,15,20-tetraphenylporphyrin)platinum [PtTPP].

The artificial epithelium was prepared placing drops of a tetrahydrofuran solution of a PVC membrane containing the tetrapyrrole indicator onto a plastic coverslip. The sensing film was overlaid with a PVC membrane and then covered with another coverslip. The upper coverslip was endowed with a hole to expose the polymeric film to gases and to allow odor molecules to freely diffuse into the film and interact with the indicator molecules. The appearance of the epithelium is shown in Fig. 1. The artificial epithelium was enclosed in a sample cell with transparent walls and gas inlet and outlet for odor delivery. The cell was placed in an optical path connecting a standard LCD computer monitor, used as a light source, and a webcam operating at a resolution of 320x240 pixels. This simple set-up, dubbed as Computer Screen Photoassised Technology (CSPT), allows optical fingerprints of color indicators and their interaction with guest molecules to be determined, where the fingerprints contain information about both absorption and fluorescence [7].

The optical properties of the artificial epithelium were fingerprinted with a three colors sequence, pure red, pure green and pure blue. For each color the camera recorded the intensity received in its three channels (red, green and blue). A total of 9 values thus define the optical fingerprint of each pixel.

The response to odors was measured in experiments where the artificial epithelium was exposed to vapors of ethanol, triethylamine, toluene, and butylamine at a concentration given by the saturated vapor pressure at room temperature diluted five times in a nitrogen flow (200 sccm total flow).

The response was evaluated illuminating the epithelium with the three pure colors. For each color the camera response in the three channels was acquired. The signals were simply summed to form the individual ORN signal. In order to evaluate only the changes induced by the exposure to odors in each experiment the signals recorded in the first measurement were subtracted from all of those following.

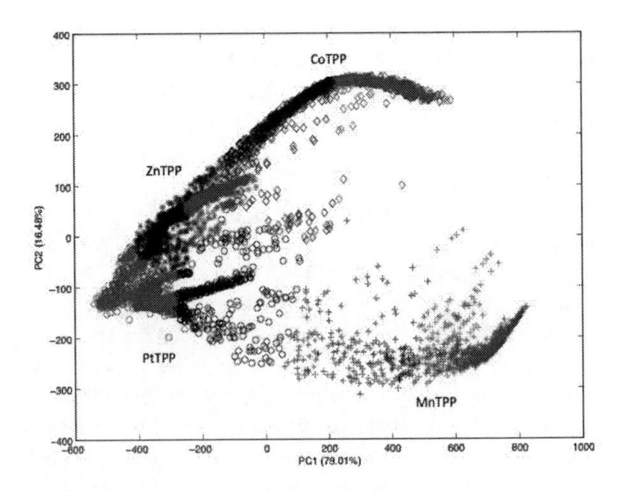

FIGURE 2. . Plot of the first two principal components of the fingeprints of Fig. 3. Different symbols indicate the four porphyrins (+ MnTPP, o PDTPP, * ZnTPP ¯ CoTPP).

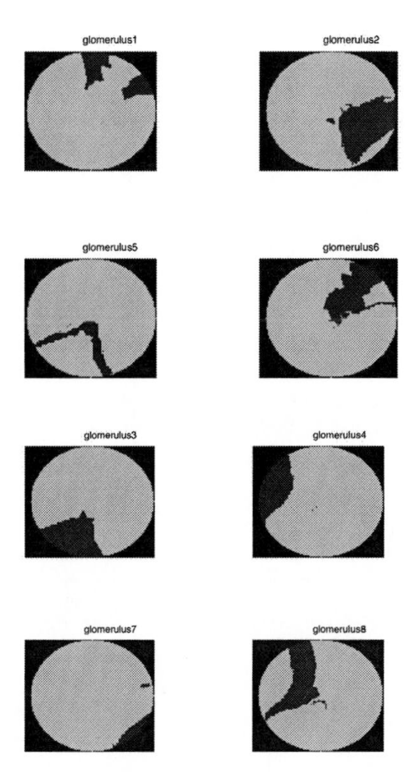

FIGURE 3. . Segmentation of the sensing layer into glomeruli convergence regions. The region of each porphyrin is divided in two glomerulus. ZnTPP: glomeruli 1 and 6; PtTPP: glomeruli 2 and7; MnTPP: glomeruli 3 and 5; CoTPP: glomeruli 4 and 8.

RESULTS

The test structure was interrogated by a transmission mode optical arrangement where a standard digital camera recorded the image of the artificial epithelium illuminated by a computer screen. The exposure of the indicators layer to the three color sequence creates a three colors optical fingerprint for each pixel of the image. Since the optical arrangement provides a uniform illumination only for the central portion of the whole image, the analysis was limited to the 7845 pixels occurring inside the white circle visible in Fig. 1. For each pixel, the fingerprint depends both on the nature of the porphyrinoid and on its amount dispersed in the polymer volume imaged by the pixel. Intermixing between the different indicator areas was avoided, but diffusion and drying effects created zones of different optical densities. The relationship between non-homogeneity and intrinsic pixels optical properties can be studied considering the Principal Component Analysis of the set of fingerprints. In figure 2 the plot of the first two principal components of the fingerprints of all the pixels is shown. It is worth to observe that in spite of the evident dispersion the character of each indicator is still preserved and then the classification of fingerprints into univocal classes is possible.

In this paper each pixel of the image is regarded as an artificial ORN and its optical fingerprint as the manifestation of the univocal nature of the kind of its receptors. The fingerprints are classified into a limited number of classes where each class contains artificial ORNs carrying the same type of receptor converge. The ensemble of classes is an abstract structure whose features are similar to that of the glomeruli layer found in natural olfaction. In this study a k-means classifier has been utilized. This is an unsupervised algorithm where the number of classes can be arbitrarily selected. In this particular case, 8 classes were chosen. With this number of classes artificial ORNs misclassifications are negligible.

Eventually, an artificial olfactory system composed by 4 different optically sensitive chemical reporters distributed in 7845 artificial ORNs, and converging to 8 glomeruli has been built.

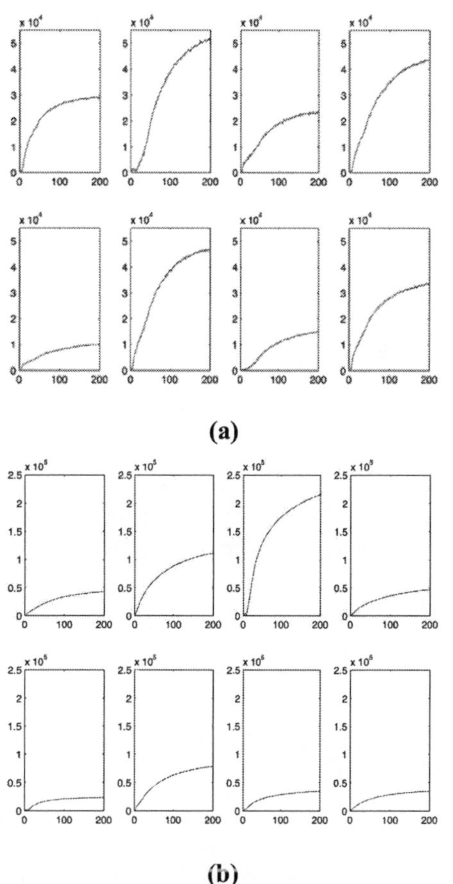

FIGURE 4. . Time evolution of glomeruli signals when the artificial epithelium is exposed to ethanol (a) and butylamine (b).

Figure 3 shows the membership of the artificial ORNs, distributed in the artificial epithelium, to the 8 different glomeruli. It has been necessary to use more than 4 glomeruli to fulfill the biological paradigm that a glomerulus receives signals from ORNs carrying the same kind of receptors.

Artificial glomeruli were modeled by software and any complex transfer function can be easily implemented both at inter- and intra- glomeruli level. Since a thorough description of such properties in the olfactory system is largely unveiled, the glomeruli are here described as nodes where the signals of all the artificial ORNs (pixels) convergent to the same glomerulus are instantaneously summed.

As an example of the behavior of artificial glomeruli signals in presence of a long time exposure to gas the responses to ethanol and butylamine are shown in Fig. 4. The time behavior illustrates the properties of the experimental arrangement that has been designed in order to introduce the diffusion of gas along the layer of chemically sensitive indicators.

In order to illustrate the potentialities of the introduced artificial system, glomeruli signals have been analyzed introducing a simple bio-inspired concept such as the lateral inhibition. Calculus have been performed considering the artificial glomeruli as nodes of an un-supervised spiking neural network model where lateral inhibition can be activated or not [8]. For each glomerulus, the sum of the signals of the artificial ORNs, shown in Fig. 4, has been transformed in a sequence of spikes according to a very simple integrate-and-fire methodology. The output of the eight glomeruli layer is represented by a dynamic principal component analysis where glomeruli layer outputs are plotted as trajectories in a principal components plane. Fig. 7 shows the trajectories when lateral inhibition is applied (Fig. 7a) and not (Fig. 7b). In both cases, the exposure to the four gases results in four distinct trajectories; Nonetheless, the application of lateral inhibition among glomeruli introduces an evident divergence of the trajectories providing a result that is rather similar to those observed in insects antenna lobe dynamics [9].

REFERENCES

1 D. Walt, Imaging optical sensor arrays, Curr. Opin. in Chemical Biology 6 (2002) 689-695
2. TA Dickinson, J White, JS Kauer, D Walt, A chemical detecting system based on a cross-reactive optical sensor array. Nature 382 (1996) 697-700
3. C. Di Natale, E. Martinelli, R. Paolesse, A. D'Amico, D. Filippini, I. Lundström, An experimental biomimetic platform for artificial olfaction, PLoS ONE 3 (2008) 3139
4. S. Korsching, Olfactory maps and odor images. Curr Opin Neurobiol 12 (2002) 387-392
5. P Mombaerts, Molecular biology of odorant receptors in vertebrates Annu Rev Neurosci 22 (1999) 487–509
6. PF Kent, MM Mozell, SL Youngentoub, P Yurco, Mucosal activity patterns as a basis for olfactory discrimination: comparing behavior and optical recordings. Brain Research 981 (2003) 1-11
7. D Filippini, A Alimelli, C Di Natale, R Paolesse, A D'Amico, I. Lundström, Chemical sensing with familiar devices. Angew. Chemie Int. Ed. 45 (2006) 3800–3803
8. W Gerstner, Spiking Neurons, in W. Maass and C. M. Bishop (eds.), Pulsed Neural Networks. MIT Press, 1999
9. R. Galan, S. Sachse, G. Galizia, A. Herz. Odor-driven attractor dynamics in the antennal lobe allow for simple and rapid olfactory pattern classification. Neural Computation 16 (2004) 999-1012

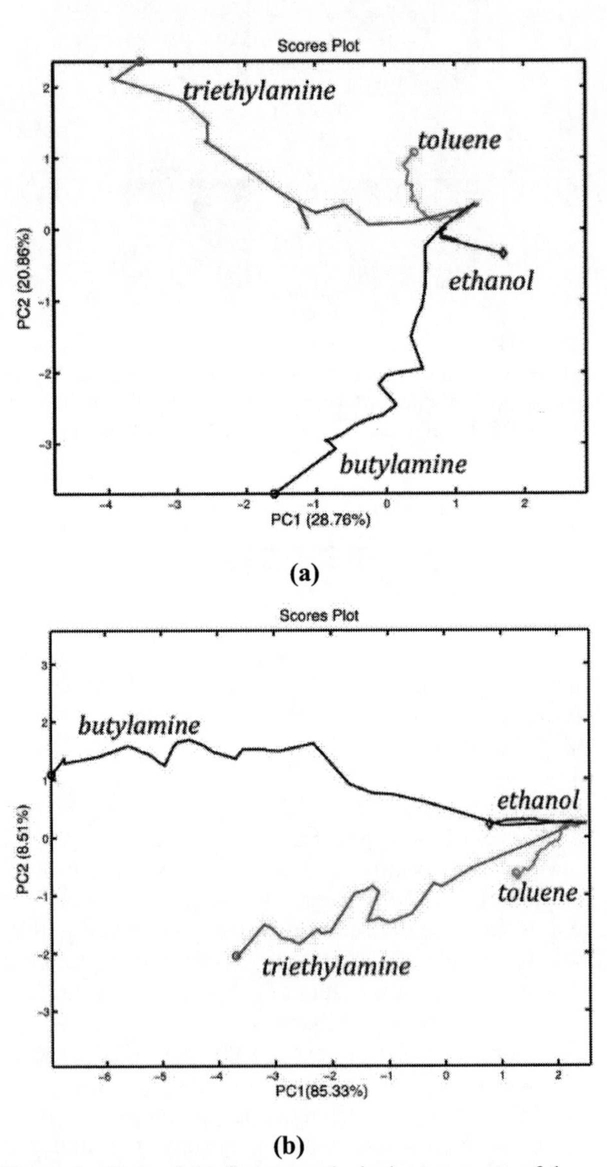

Figure 5. . Plots of the first two principal components of the glomeruli layer with (a) and without (b) the application of lateral inhibition.

NANOMEDICINE

Analysis Of Volatile Fingerprints: A Rapid Screening Method For Antifungal Agents For Efficacy Against Dermatophytes

Kamran Naraghi, Natasha Sahgal, Beverley Adriaans*, Hugh Barr*, Naresh Magan

Cranfield Health Applied Mycology Group, Vincent building, Cranfield University, Beds. MK43 0AL,
** Gloucestershire Hospitals NHS Trust, Great Western Road, Gloucester GL1 3NN, U.K.*

Abstract. The potential of using an electronic nose (E.nose) for rapid screening dermatophytes to antifungal agents was studied. *In vitro*, the 50 and 90% effective concentration (EC) values of five antifungal agents for *T.rubrum* and *T.mentagrophytes* were obtained by mycelial growth assays. Then, the qualitative volatile production patterns of the growth responses of these fungi to these values were incorporated into solid medium were analysed after 96-120 hrs incubation at 25°C using headspace analyses. Overall, results, using PCA and CA demonstrated that it is possible to differentiate between various treatments within 96-120 hrs. This study showed that potential exists for using qualitative volatile patterns as a rapid screening method for antifungal agents for microorganism. This approach could also facilitate the monitoring of antimicrobial drug activities and infection control programmes and perhaps drug resistance build up in microbial species.

Keywords: Volatile fingerprints, anti-fungals, dermatophytes, screening tools

INTRODUCTION

Dermatophytoses are common among skin diseases worldwide. Even in Europe, infections such as tinea capitis are an increasing problem (Hackett *et al.*, 2006). Although, resistance to antibiotics among dermatophyte species is very uncommon, determination of the *in vitro* susceptibility of dermatophytes, particularly for management of treatment failures, may prove helpful [1]. Many techniques, such as agar-based methods (dilution and diffusion) and broth dilution, have been used for antifungal susceptibility testing (AST) but there is no standard method for AST of dermatophytes [2]. However, dermatophytes grow slowly and thus agar-based methods are time-, labour-, and resource-intensive. There is a need for novel and quick alternative laboratory approaches.

All micro-organisms produce by-products as a result of their normal metabolism [3]. Some of these metabolic by-products, including alcohols, aliphatic acids, ketones and terpenes are volatile at low temperature and are known as volatile organic compounds (VOCs). Many VOCs have characteristic odours. Since the production patterns of VOCs are unique to certain micro-organisms (or disorders), they can potentially be used as biomarkers [4]. Quantitative analysis of VOCs has almost exclusively been based on gas chromatography-mass spectrometry which is relatively expensive, and requires skilled operators. However, rapid qualitative analysis of volatile fingerprints using sensor arrays including electronic nose devices for early detection/discrimination of infections has yielded promising results [5].

The aims of this work were to study the potential of using an E-nose as a rapid screening method for antifungal agents for controlling dermatophytes using the drugs and antioxidants by using volatile fingerprints. The objectives were thus to (a) identify the growth responses of two *Trichophyton* species, *T. rubrum* and *T. mentagrophytes*, to 50% and 90% effective concentrations values (EC_{50} and EC_{90} values) of five antifungal agents (itraconazole, griseofulvin, butylated hydroxyanisole, octyl gallate and n-propyl-p-hydroxybenzoate) and (b) use of an E.nose for discrimination between treatments for these fungal species.

CP1137, *Olfaction and Electronic Nose: Proceedings of the 13th International Symposium,* edited by M. Pardo and G. Sberveglieri

EXPERIMENTAL AND METHODS

Inocula and growth medium

Two *Trichophyton* species, *T. rubrum* (No. 115) and *T. mentagrophytes* (No. 224), were used in this study. These were human isolates and obtained from National Collection of Pathogenic Fungi (NCPF), Bristol, UK. Sabouraud dextrose agar (SDA) was prepared with 10 g l^{-1} mycological peptone (Oxoid, UK), 40 g l^{-1} glucose (Acros Chemicals, Belgium), and 15 g l^{-1} agar (No.3, Oxoid, UK). At the end, 0.05 g l^{-1} chloramphenicol (Sigma, UK) was added.

Antifungal agents

Two antifungal drugs, itraconazole (Janssen Pharmaceuticals, Belgium) and griseofulvin (Darou Pakhsh Co., Iran), and three antioxidants including butylated hydroxyanisole (BHA; $C_{11}H_{16}O_2$) (Sigma, US), octyl gallate (O-G; $C_{15}H_{22}O_5$) (Fluka Chemie GmbH, Germany) and n-propyl-p- hydroxybenzoate (P-P; $C_{10}H_{12}O_3$) (Sigma, US) were used in this study. All antioxidants are Generally Recognized As Safe (GRAS) compounds. For each drug, a stock solution with the concentration of 100 μg ml^{-1} was prepared by dissolving 2mg of anti-fungal agent in 20 ml of 99.5% dimethyl sulfoxide (DMSO) (Sigma, US). Stock antioxidant solutions (1 mM) in 20 ml of 99% ethanol was made. These were stored at 4°C.

E.nose system

An AppliedSensor 3320 E.nose (AppliedSensor Group, Sweden) was employed in this study. The core sensor technology is based on a hybrid array of 10 metal-oxide-silicon field-effect-transistor (MOSFET) sensors and 12 metal oxide sensors (MOS), and one humidity sensor.

Test procedure

Stock solutions were added to molten SDA (itraconazole, griseofulvin; 0.25, 0.50, 1.00 and 2.00 μg ml^{-1}). SDA was also amended with two antioxidants combinations (BHA + P-P, O-G + P-P) to obtain 20 and 40 mM. Triplicates of each treatment and each species were inoculated with a 0.5 mm diameter agar plug taken from growing cultures of *T.rubrum* and *T.mentagrophytes* on SDA. Plates were incubated at 25°C in dark and mycelial extension diameters measured daily for 14 days. However, no mycelial extension was detected on griseofulvin and antioxidants-amended plates after 14 days. Therefore, the test was repeated at 0.025-0.200 μg ml^{-1} griseofulvin. The radial extension rates were plotted against time.

The EC_{50} and EC_{90} values of all antifungal agents against *T.rubrum* and *T. mentagrophytes* were calculated by linear regression of the temporal extension rates and then plotting the relative growth rates (mm day^{-1}) to compare treatments.

For each treatment molten SDA media were amended with the calculated EC_{50} and EC_{90} values per species and poured in 9 cm plastic Petri plates. At least 30 replicate Petri plates of plain and amended SDA were inoculated with 0.25 ml of a 10^6 spore mL^{-1} suspension of each species, and 10 replicate blank SDA plates used as negative controls. They were incubated at 25°C for 96-120 hours. At each time point, four 2 cm diameter agar plugs from 5 replicate plates of each treatment were destructively sampled using a cork borer and placed in sample vials which were sealed with a septum and lid. After 1 hr equilibration, the headspaces were analysed by the 3320 E-nose. The 96-120 hours period represents the early stages of microscopic and visible growth, when identification is difficult.

Data analysis

Data were analysed using NSTSenstool (software package in the AppliedSensor 3320) to perform principal components analysis (PCA) on response parameters (mean-centred data) using the maximum peak response for the various sensors. These data were analysed by cluster analysis (CA) using Statistica 8.

RESULTS

Effective concentration values

Growth rates relative to the controls (data not shown) were used to determine the EC_{50} and EC_{90} values for antifungal agent concentrations. For all treatments the EC_{90} values were the highest concentrations used.

Early differentiation between inoculated antifungal treatments and controls

The main purpose of this research work was to rapidly (after 96-120 hours) study the growth responses of *T. rubrum* and *T. mentagrophytes* to EC_{50} and EC_{90} of the antifungal agents by analysing the volatile fingerprints. There was very good reproducibility of the response of 10 MOSFET sensors to replicates of *T. mentagrophytes* grown on SDA.

Figure 1 shows the PCA of the response of the hybrid sensor E-nose to treatments of inoculated amended SDA with EC values of itraconazole and antifungal-free SDA (as positive controls) together with blank SDA (as a negative control) and un-inoculated amended SDA at 2.00 μg ml^{-1} of itraconazole after 96 hours. There was clear separation between inoculated and un-inoculated itraconazole treatments, and negative controls implying that the *Trichophyton* species on amended SDA plates, as on the itraconazole treatments there was no growth after 96 hours. This accounted for 97.6% of the data described by the first two principal components (PC). After 120 hours growth there was also distinct clusters of the two dermatophytes growing on un-amended SDA clearly differentiated from other treatments (Figure 2).

Effect of anti-oxidant treatments

Analysis of the data related to the response of the E.nose sensors to different BHA + P-P treatments by PCA after 96 hours showed three clusters of *T. rubrum* and *T. mentagrophytes* growing on SDA, and *T. mentagrophytes* growing on amended SDA at the EC_{50} concentration of the antioxidants combination distinguished from other treatments which grouped together. This accounted for 94.9% of the data described by PCs 1, 2 and 9 (Figure 3). However, positive controls and replicates of *T. mentagrophytes* growing on SDA at EC_{50} of BHA + P-P were clearly differentiated from negative controls, and samples of un-inoculated SDA amended with antioxidants and inoculated SDA at EC_{90} of BHA + P-P which formed a mixed group. After 120 hours, examination of the PCA map showed four groups of dermatopyte species growing on SDA and that amended with the EC_{50} concentration of BHA+P-P per species, clearly separated from other treatments (Figure 4).

DISCUSSION

This is the first study to examine the potential of using an E.nose for qualitative screening of the responses of dermatophytes to antifungal agents. It was possible to discriminate between *T.rubrum* and *T. mentagrophytes* growing on un-amended solid media (SDA) from those inoculated on antifungal-modified media as early as 96 hrs after incubation by analysing their volatile production patterns.

In the presence of itraconazole, the species grown on control media and those amended with the EC_{50}/EC_{90} values were successfully differentiated after 96-120 hrs. This supports previous studies with four *Trichophyton* species which showed differentiation

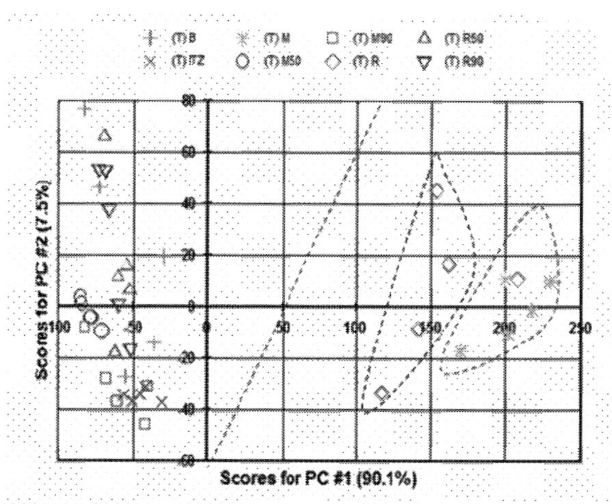

FIGURE 1. PCA score plot differentiating *Trichophyton* species grown on plain SDA and itraconazole treatments after 96 hours.(*Key*: B,blank SDA; ITZ, SDA amended at 2 μg ml^{-1} itraconazole; M, *T. mentagrophytes*; M50, *T.mentagrophytes* and EC_{50} itraconazole; M90, *T.mentagrophytes* and EC_{90} itraconazole; **R**, *T.rubrum*; R50, *T.rubrum* and EC_{50} itraconazole; R90, *T. rubrum* and EC_{90} itraconazole).

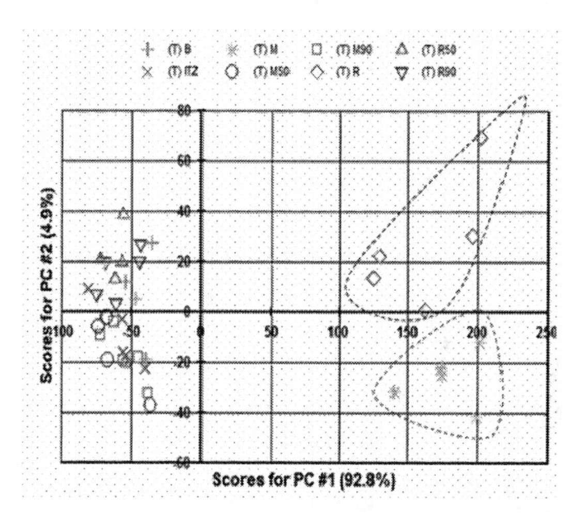

Figure 2. PCA score plot (120 hrs) separating positive controls and from itraconazole and blank SDA treatments. (*Key*: B, blank SDA; ITZ, SDA amended at 2 μg ml^{-1} itraconazole; M, *T.mentagrophytes*; M50, *T.mentagrophytes*/EC_{50}/itraconazole;M90,*T.mentagroy tes*/EC_{90}/itraconazole;**R**,*T.rubrum*;R50,*T.rubrum*/EC_{50}/ itraconazole;R90,*T.rubrum*/EC_{90}/itraconazole).

within 96 hours using qualitative volatile fingerprints [5]. The results also revealed a similarity in the volatile production patterns of controls and fungi inoculated on antifungal-modified media where growth was inhibited

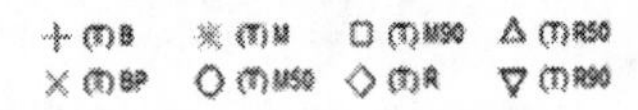

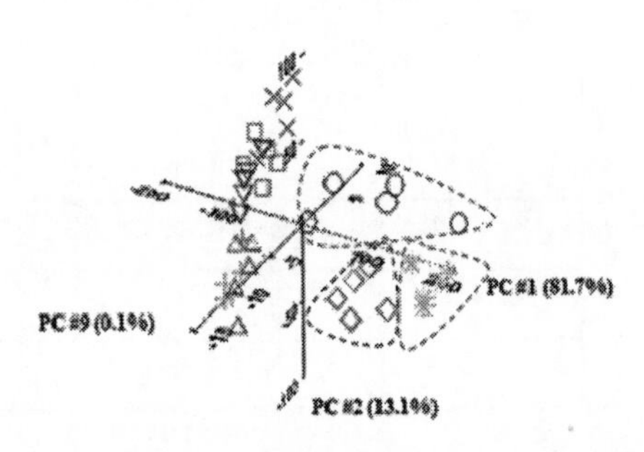

Figure 3. Non-linear PCA score plot (96 hrs) showing three clusters of *Trichophyton* species on amended SDA + antioxidant combination of BHA+P-P. (*Key*: B, blank SDA; BP, un-inoculated SDA at 0.250 mM BHA+PP;M,*T.mentagrophytes*;M50,*T.mentagrophytes* /EC_{50} BHA+P-P; M90, *T.mentagrophytes*/EC_{90} of BHA+P-P;**R**, *T.rubrum*;R50,*T.rubrum*/$EC_{50}$of BHA+P-P; R90, *T.rubrum*/EC_{90} of BHA+P-P).

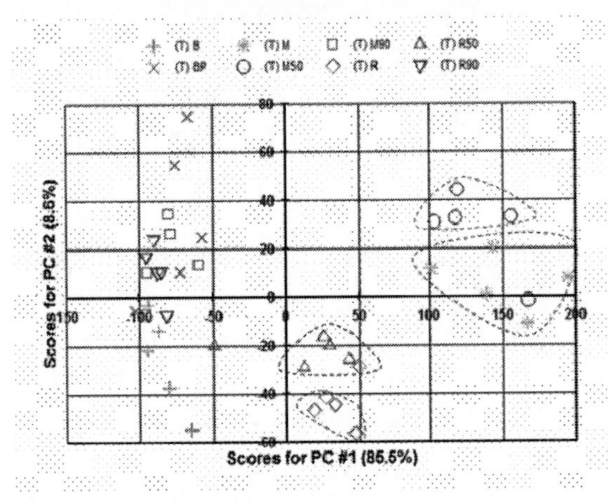

Figure 4. PCA score plot (120 hrs). Clusters of *T. rubrum* and *T. mentagrophytes* on various treatments. (*Key*: B, blank SDA; BP, un-inoculated SDA at 0.250 mM/BHA+PP;M,*T.mentagrophytes*;M50,*T.mentagrop hytes*/EC_{50} BHA+P-P; M90, *T.mentagrophytes*/EC_{90} BHA+P-P;**R**, *T.rubrum*; R50, *T.rubrum*/EC_{50} BHA+P-P;R90,*T.rubrum*/EC_{90}/BHA+P-P).

by the drugs within the same time. The results of the griseofulvin test suggest that culture age may affect the discrimination achieved. Differences were not clear after 96 hrs, but 120 hrs gave differentiation between EC_{90} treatments of griseofulvin and negative controls. .

Alternative antioxidants for dermatophytes control have not been studied. Working with a combination of BHA+P-P, the E.nose could successfully discriminate between 3-4 treatments of *Trichophyton* species growing on antifungal-free media and those modified with the EC_{50} values. The clear separation of negative controls and inoculated treatments at EC_{90} values from others within 96-120 hours indicated inhibition of growth similar to existing drugs (itraconazole, griseofulvin). The E.nose was able to differentiate between closely related treatments, notably those growing on controls and amended media at the EC_{50} value of the combination per species.

Previous studies showed that the nature of the culture media influences the production of volatiles by micro-organisms [4]. In contrast, Sahgal *et al.* [5] found that good discrimination by analysing the volatile fingerprints of four *Trichophyton* species growing on various solid media, regardless of the medium used, required at least 96 hours.

CONCLUSIONS

Most susceptibility testing of dermatophytes use two protocols (1, 2) which are time-consuming. In contrast, detection of VOCs by sensor arrays has many advantages. It is non-invasive, sensitive, and relatively inexpensive. This study showed the potential for rapid screening of novel antifungal compounds using volatile fingerprints and monitoring the build up of resistance to anti-microbial drugs. By examining the PCA maps it would be possible to identify when poor control is being achieved by having both positive and negative controls on a regular basis for comparison.

ACKNOWLEDGMENTS

We are grateful to Gloucestershire NHS Foundation Trust for financial support to Dr. Naraghi.

REFERENCES

1. Z.Cetinkaya, N.Kiraz, S.Karaca *et al.*, *European Journal of Dermatology* **15**, 258-261 (2005).
2. W.G.Merz, and R.J. Hay. Medical Mycology. In: *Topley & Wiilson's Microbiology and microbial infections*, 10 th ed. London: Hodder Arnold Ltd (2005).
3. J.M.Scotter, V.S.Langford, P.F.Wilson *et al.*,*Journal of Microbiological Methods* **63**, 127-134 (2005).
4. A.P.F.Turner and N. Magan, *Nature Reviews. Microbiology* **2**, 161-166 (2004).
5. N.Sahgal, B.Monk, M.Wasil *et al.*, *British Journal of Dermatology* **155**, 1209-1216 (2006).

Volatile Organic Compound (VOC) Analysis For Disease Detection: Proof Of Principle For Field Studies Detecting Paratuberculosis And Brucellosis

Henri Knobloch[1]§, Heike Köhler[2], Nicola Commander[3], Petra Reinhold[2], Claire Turner[1], Mark Chambers[3]

[1] *Cranfield University, Cranfield Health, UK*
[2] *Veterinary Laboratories Agency (VLA, Weybridge, UK)*
[3] *Institute of Molecular Pathogenesis in the 'Friedrich-Loeffler-Institut' (Jena, Germany)*
§Corresponding author
h.knobloch.s06@cranfield.ac.uk

Abstract. A proof of concept investigation was performed to demonstrate that two independent infectious diseases of cattle result in different patterns of volatile organic compounds (VOC) in the headspace of serum samples detectable using an electronic nose (e-nose). A total of 117 sera from cattle naturally infected with Mycobacterium avium subsp. paratuberculosis (paraTB, n=43) or Brucella sp. (n=26) and sera from corresponding control animals (n=48) were randomly and analysed blind to infection status using a ST214 e-nose (Scensive Ltd, Leeds, UK). Samples were collected under non-standardised conditions on different farms from the UK (brucellosis) and Germany (paraTB).

The e-nose could differentiate the sera from brucellosis infected, paraTB infected and healthy animals at the population level, but the technology used was not suitable for determination of the disease status of individual animals. Nevertheless, the data indicate that there are differences in the sensor responses depending on the disease status, and therefore, it shows the potential of VOC analysis from serum headspace samples for disease detection.

Keywords: electronic nose (e-nose), volatile organic compounds (VOC), paratuberculosis, brucella, field study
PACS: 07.07.Df, 68.03.-g, 87.19.xb

1. INTRODUCTION

Novel serum-based assays to monitor alternative non-immunological markers of infection are highly desirable, for instance in situations were false positive reactions are generated by cross reactive antibodies (eg: brucellosis) or were suitable immunodiagnostic markers are currently not available (eg: paratuberculosis) [1]. Analysis of the VOC profile in different biological media, eg. serum, urine or breath could be one approach to this problem. Discrimination between cattle experimentally infected with Mycobacterium bovis and non-infected control animals was possible using an electronic nose (e-nose) for analysis of VOC in the headspace above serum samples [2].

In the present study we investigated the possibility of detecting differences in serum headspace of non-standardised field samples from cattle infected with Brucella sp. or Mycobacterium avium subsp. paratuberculosis. Biological and methodological variation affecting the responses was assessed.

2. EXPERIMENTAL DESIGN AND METHODS

Four groups of samples from 2 different regions and with different disease status were obtained. Sera from 43 dairy cattle naturally infected with paraTB and from 24 control cattle from paraTB non-suspect herds were collected under field conditions on different farms in Germany. 26 sera from cattle naturally infected with brucellosis and 24 from non-infected control cattle were collected and provided by VLA, UK. Samples were kept at -80°C during storage and shipment.

For analysis, 1 mL of serum was thawed and dispensed into Nalophan bags. After 15 minutes of incubation at 25°C the headspace of the blinded samples was statically analysed in a random order using the conducting polymer-based ST214 e-nose (Scensive, Leeds, UK). The ST214 e-nose comprises 14 sensors including 1 reference sensor for generating a pattern corresponding to exposure of different VOCs obtained

CP1137, *Olfaction and Electronic Nose: Proceedings of the 13th International Symposium*, edited by M. Pardo and G. Sberveglieri

from serum headspaces. Replicates 3 to 5 were used for data analysis [3].

Univariate statistics such as multifactor ANOVA, linear regression and Spearman's rank correlation were applied to assess statistically significant differences associated with methodological and biological variation. Multiple range testing was used to investigate differences between diseased and healthy populations rather than a trend over all groups. Significant differences are indicated using different letters starting with "a" for the group with the lowest responses. The level of the least significant difference for all techniques applied was 5% (LSD; P≤0.05). SPSS (version 11.5) Statgraphics (version 4.0) and Matlab (version 2006b) including the PLS toolbox were used. Data are displayed as Box-and-Whisker plots and mean-centred heatmaps. Mean-centring is a way of normalising data. Here, the mean over all groups is calculated for a certain sensor and the mean or median of an individual group is then referred to the all-group-average. The sensor related means per group are then displayed as a colour from a spectrum; the red of which is above the sensor average while the blue end is below.

3. RESULTS

The samples were grouped into 4 subsets according to the disease status: 1= paraTB negative; 2= paraTB positive; 3= Brucella negative; 4= Brucella positive. According to multifactor ANOVA, almost all sensors significantly changed due to the infection status. Applying multiple range testing, significant differences between groups were found. For all sensors, group 2 (paraTB positives) showed the lowest responses ("a") while group 4 (Brucella positives) showed the highest responses ("c" or "d"). Groups 1 (paraTB negative) and 3 (Brucella negative) were found to be identical (e.g. sensors 2, 10, 11) or close to each other (almost all remaining sensors, see table 1)

TABLE 1: Results of multiple range testing for sensors. Group 2 consistently had lower responses than groups 1 and 3 while higher responses were seen for group 4.

Sensor	Group 1	Group 2	Group 3	Group 4
1	a	a	b	c
2	b	a	b	c
3	b	a	c	d
4	b	a	c	d
5	b	a	c	d
6	a	a	b	c
7	b	a	c	d
8	b	a	c	d
9	b	a	c	d
10	b	a	b	c
11	b	a	b	c
12	b	a	c	d
13	b	a	c	d

Multiple range test (LSD; p≤0.05); group 1= paraTB neg., group 2=paraTB pos, group 3= Brucella neg., group 4 = Brucella pos.

Data can also be displayed as Box-and-Whisker plots. By way of example, figure 1 shows responses of sensor 9 over all groups. Group 2 (paraTB positive) was below the control groups 1 and 3, while group 4 was above.

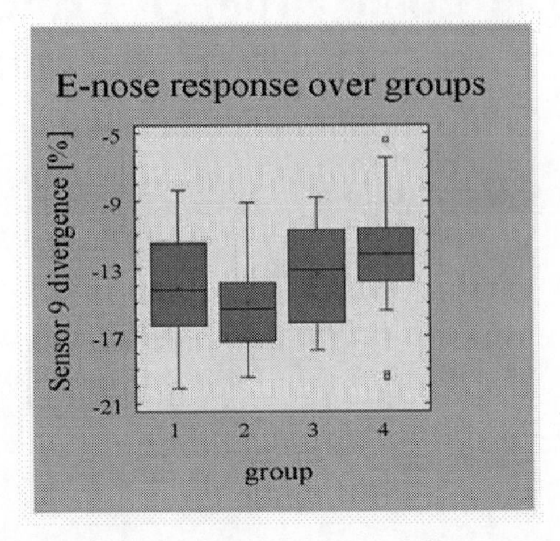

FIGURE 1: The response of sensor 9 is shown for all groups. Group 2 had lower divergences than control groups 1 and 3, while group 4 had higher.
Legend: group 1= paraTB neg., group 2=paraTB pos, group 3= Brucella neg., group 4 = Brucella pos.

Box-and-Whisker plots indicated a broad overlapping in sensor responses mainly due to methodological variation. This was also confirmed using multifactor ANOVA.

For summarising results, sensor responses were mean-centred and displayed as a heatmap (Figure 2). ParaTB positives showed "colder areas" especially for sensors 8, 9 and 12 while brucella positives had higher responses expressed as "warm areas" (see sensors 4 to 9, 12, 13).

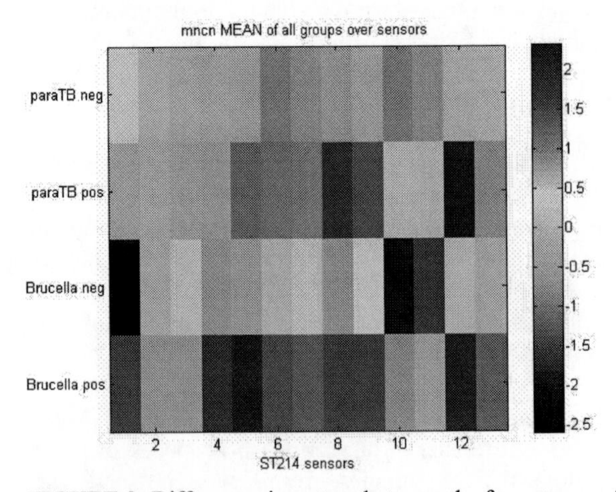

FIGURE 2: Differences in pattern between the four groups of samples. Lower responses were found for sensors 8, 9 and 12 (cold areas) for paraTB positives while for Brucella positives higher responses were found.

However, results were less clear compared to Box-and-Whisker plots or multiple range tests due to the fact that heatmapping does not discriminate the different types of biological and methodological variation present in the data.

4. DISCUSSION & CONCLUSIONS

The e-nose can differentiate sera from brucellosis and paratuberculosis infected animals and healthy animals at the population level.

Statistically significant population level differences were shown in the responses of various sensors between paraTB positives, paraTB negatives, Brucella negatives and Brucella positives. In this order, responses ranged from low to high. Samples from control animals lay between samples from the two infected groups indicating that differences in sensor responses due to the sample origin was lower than between the two infected groups developing into different directions.

Biological variation may be caused by inherent biological variability (age, sex, breed, and general physiological status) or environmental influences (diet, climate, season, husbandry practices, intercurrent infections). Eliminating these influences does not seem to be possible for a natural population. Therefore, they cannot readily be predicted, quantified or normalised to a sufficient extent to reduce their influence on the e-nose sensor response.

Furthermore, methodological variation during sample analysis (time and order of sample analysis, ambient temperature, sample storage conditions etc.) can also significantly affect e-nose sensor responses, and it is recognised that reproducibility of data between e-nose devices can be a significant problem.

Due to methodological variation the e-nose technology in its present form using this particular device would not be suitable for direct diagnosis of the disease status of an individual animal because of the broad overlap of sensor responses.

However, despite the limitations mentioned, the results of the study show the potential of VOC analysis for differentiation of infectious diseases in animals. Furthermore, disease specific volatiles need to be identified using methods such as GC-MS. E-nose methodology needs to be improved so that VOC analysis may ultimately be exploited as a novel diagnostic assay.

ACKNOWLEDGMENTS

This work was conducted as part of a research project (SE3221) funded by the UK Department for Environment, Food and Rural Affairs (Defra).

Serum samples were kindly provided by Heike Köhler and Nicola Commander (Defra project SE0310) and the staff from the Institute of Molecular Pathogenesis in the 'Friedrich-Loeffler-Institut' (Jena, Germany) and the Veterinary Laboratories Agency (VLA Weybridge, UK), respectively.

REFERENCES

1. H. Köhler, *Mag. Allatorv. Lapja*, **130**, Supplement I, 67-69 (2008).
2. R. Fend, R. Geddes, S. Lesellier, H.-M. Vordermeier, L.A.L Corner, E. Gormley, E. Costello, M.A. Chambers, *J Clin. Microbiology* **43** (4), 1745-1751 (2005)
3. H. Knobloch, C. Turner, A.D. Spooner, M.A. Chambers, Trace Gas and Headspace Analysis Using Electronic Nose Technology, *Sensors and Actuators B*, accepted 10/03/2009

Detection of Lung Cancer with Volatile Organic Biomarkers in Exhaled Breath and Lung Cancer Cells

Jin Yu [1], Di Wang [1], Le Wang [1], Ping Wang [1]*, Yanjie Hu [2], Kejing Ying [2]

1 Biosensor National Special Laboratory, Key Laboratory of Biomedical Engineering of Ministry of Education, Department of Biomedical Engineering, Zhejiang University, Hangzhou 310027, P. R. China
2 Zhejiang Sir Run Run Shaw Hospital, Zhejiang University, Hangzhou, 310027, P. R. China
**Correspondence should be addressed to Ping Wang (email: cnpwang@zju.edu.cn)*

Abstract. In patients with lung cancer, volatile organic compounds (VOCs) are excreted in exhaled breath. In this article, exhaled breath of 30 lung cancer paitients and 30 healthy people were collected, preconcentrated by solid-microextraction(SPME) and analyzed with gas chrom-atography and mass spectrometry (GC/MS). A predictive model composed of 5 VOCs out of 16 candidate VOCs detected in the lung cancer patients is constructed by discriminant analysis, with a sensitivity of 76.7% and specificity of 96.7%. We detected exhaled VOCs of 3 different lung cancer cell lines and human bronchial epithelial cell lines. 2-Tridicanone is considered the distinctive marker of lung cancer cells, which is found in lung cancer patients' exhaled breath as well. Compared to healthy people, patients with lung cancer had distinctive VOCs in their exhaled breath. The predictive model can work as diagnosis reference for lung cancer. VOCs found in lung cancer cell line help the cognition of the mechasim VOCs generating in lung cancer patients.

Keywords: Lung cancer, Volatile organic compounds (VOCs), Lung cancer cell line, Biomarkers
PACS: 87.19.xj

INTRODUCTION

Breath test is paid more attention to in recent decades. Volatile organic compounds act as important markers in breath test. Since Pauling discovered that human breath includes over 200 kinds of VOCs in 1971[1], scientists attemped to indicate diseases with certain VOCs. For example, acetone is regarded as the biomarker of diabetes in breath, proved physiologically and experimentally[2].

Lung cancer is one of the leading cause of cancer death , and only 5% patients survive with stage Ⅳ cancer[3]. The current diagnosis technique such as CT, fluorescence bronchoscopy is somehow uncomfortable and expensive. A noninvasive detection for lung cancer, especially in early stage is needed and breath test seems an optimised way. In 1999， M Phillips collected and analyzed the VOCs in human breath with gas chromatography/mass spectrometry (GC/MS) and found 22 VOCs as the biomarker of lung cancer, most of which are alkanes and alkanes derivatives[4]. In 2003, Phillips et al modulated his model to a combination of 9 VOCs[5]. Then they keep optimizing aim VOCs and statistic tools to discriminate pattients with primary lung cancer and healthy people[6][7]. Other works reported different VOCs as biomarkers of lung cancer in breath using different technique [8].

Until recently, all the observations could not be completely explained with clear pathphysiologic processes. One plausible hypothesis is relative to the oxidative stress and lipid peroxidation. Oxidative stress appears to increase in some cancers[9], and lipid peroxidation of fatty acids in cell menbranes generate s some alkanes and other organic though complex biochemical actions[10]. The study of VOCs in the microenvironment of cancer cells could help to study the mechanism how VOCs produced or increased in patients with lung cancer . In our previous work, 4 VOCs were found in the microenvironment exhaled by lung cancer cell excised from patients, supporting the basis of the breath detection[11].

In this article, a system of SPME combined with GC/MS is applied to determine the distinctive VOCs in exhaled breath of patients with lung cancer. We also detected the VOCs in the metabolic products of lung cancer cell lines as a supplement and validation of the breath test.

CP1137, *Olfaction and Electronic Nose: Proceedings of the 13th International Symposium*, edited by M. Pardo and G. Sberveglieri

EXPERIMENT AND METHODS

Subjects

30 patients with untreated primary lung cancer with an abnormal chest radiograph were selected from Sir Run Run Hospital. 30 volunteers were selected from the same hospital with normal chest radiograph and no cancer history.

Breath collection and assay

All of patients and controls were told to keep limosis for 12 hours before breath samples were collected. Every exhaled breath sample is collected in 5L Tedlar bags and to be detected within 12 hours. It is preconcentrated using SPME for 50min under 37 ° C circumstances then directly transferred to the GC/MS for analysis.

Cell line preparation

We use denocarcinoma cell line A549, squamouscarcinoma cell line SK-MEM-1, small cell lung cancer cell line NCIH 446 for the lung cancer cells, and the human bronchial epithelial cell line BEAS-2B for the control. All cells were cultivated at 37°C with 5% carbon dioxide for 7days. The culture medium was last changed 24 hours before detection. A static headspace extraction (SHS)-SPME method is used to adsorb the exhale VOCs from the cell culture medium for 100 minutes then transferred to the GCMS for analysis (Fig. 1).

FIGURE 1. VOCs preconcentration of cell line. Static head space –SPME (SHS-SPME) for 100 minutes, the extraction fiber is about 1cm ahead from the medium level.

Statistical analysis

The predictive model using discriminant analysis (SPSS 16.0) of the candidate VOCs classified the patients and controls. This model was cross-validated with a leave-one-out jackknife technique.

Results

VOCs from breath detection

Over several tens of VOCs were detected in the exhale breath both in patients and control breath. Individual difference is observable, so we have to draw conclusion from the statistic outcome. Table 1 showed 16 candidate VOCs detected with a frequency more than one forth of either group, most of which is alkanes and alkane derivatives.

TABLE 1. Candidate VOCs in breath. This table displays the breath VOCs with a frequency more than 25% in either patients or controls. The frequence is shown for each VOC.

Chemical structure	Patients	Controls
Decane	15	1
Undecane, 5-methyl-	12	0
Tetradecane	8	8
2-Tridecanone	10	5
Phenol, 2,4-bis(1,1-dimethylethyl)-	17	16
hexadecane	17	16
Tetradecane, 5-methyl-	13	2
Benzene, 1,1'-(1,3-propanediyl)bis-	9	1
2-Pentadecanone	9	4
Heptadecane, 7-methyl-	13	17
Hexadecanal	6	11
1-Hexadecanol	8	2
2-Nonadecanone	11	1
nonyl lactone	9	1
Eicosane	16	1
Heneicosane	17	0

The discriminant analysis (SPSS 16.0) is applied to construct a optimized model based on the data we have to classify patients with primary lung cancer apart from controls. The 5 VOC received of the optimized model is listed in Table 2.

TABLE 2. 5 VOCs identified as the optimized combined biomarkers of lung cancer by discriminant analysis.

Decane
Undecane,5-methyl-
2-nonadecanone
Eicosane
Heneicosane

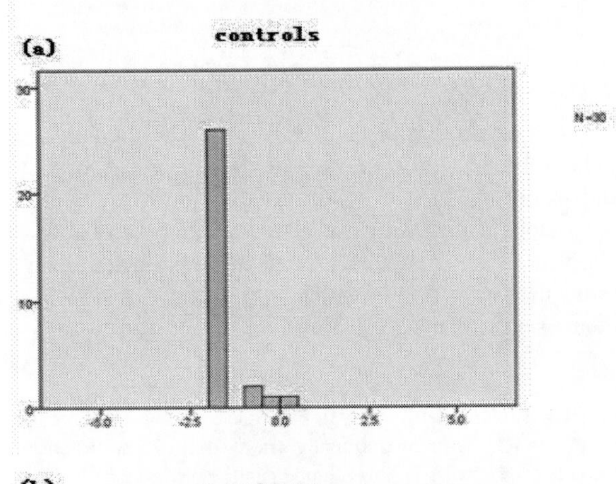

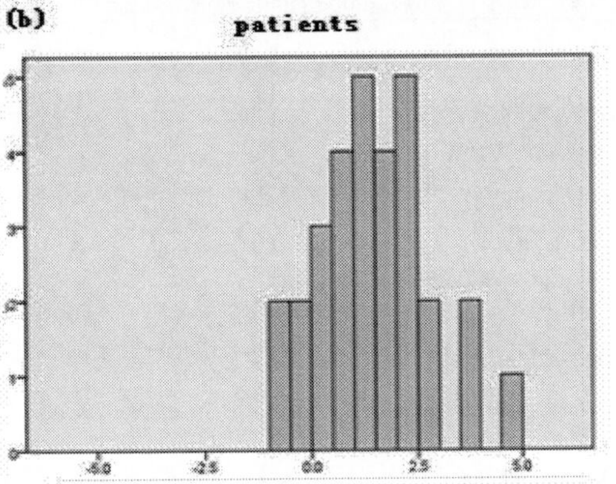

FIGURE 2. Classification of in patients and controls using discriminant function.

The x-axis is the discriminant function value of each case.The values of patients and controls are assumed to be above zero and below zero, respectively.

(a) Classified results for 30 controls, 29 out of 30 cases are below zero.

(b) Classified results for 30 patients, 26 out of 30 cases exceed zero.

The model for original cases has sensitivity of 86.7% and specificity of 96.7%, respectively. In cross-validation with a leave-one-out jackknife technique, each case is classified by the functions derived from all cases other than that case. The sensitivity is 76.7% and specificity is 96.7% in the cross-validation.

Cell line products detection

The chromatograms of the same cell line showed a good repeatbility. One distinctive VOC was revealed in the exhaled products of cancer cell as shown in fig3. 2-tridecanone was found in all 3 cancer cell line, although not with abundant content, compared to the blank in the BEAS-2B. 2-tridecanone can be recognized as the biomarkers of lung cancer cell line.

It is noticable that the marker we found in the lung cancer cell line microenvironment are included in the 16 VOCs we found in hunman breath.

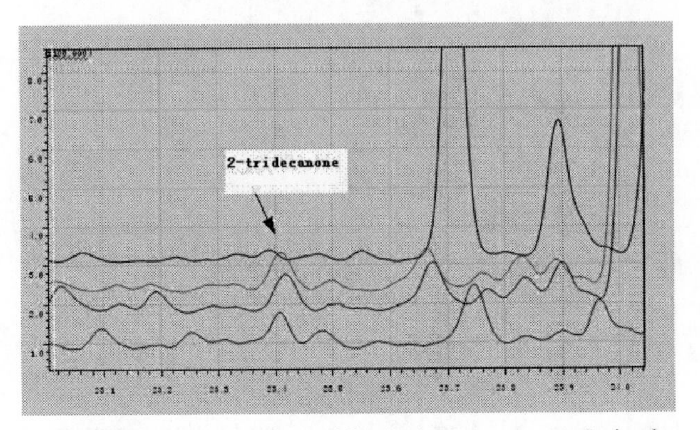

FIGURE 3. Part of combined chromatograms. It is the combined chromatogram of exhaled products of BEAS-2B, A549, SK-MEM-1, NCIH446 ordinaly from top to bottom. The peak pointed to is identifiied as 2-tridecanone. The same peak were detected in all the three different cancer cell line, while there are now peak of 2-tridecanone in the BEAS-2B,which is assumed as the control.

DISCUSSION

This study demonstrates findings that there are distinctive VOCs in patients breath with lung cancer, comparing to the normal people. Based on the candidate VOCs data we got in exhale breath detection, a model was constructed using discriminant anaysis to predict the lung cancer patients, with relatively high sensitivity and specificity. On the other hand, we analyzed the VOCs in the cell culture medium, and found 2-tridecanone is the distinctive marker of the lung cancer cell line.

Reported VOCs associated with lung cancer varied from study to study. The variations may due to the different method and technique for collection and assay. Anatomical dead space and discomfort may be the two key components in the breath collection process. A special apparatu for breath collection is recommended for less effect factor [5] [8].

Until recently, all the predictive VOCs models for lung cancer were educed from the empirical results. While the pathphysiologic processes remains unclear.

Oxidative stress is assumed to be connected with VOCs generation; lipid peroxidation of polyunsaturated fatty acids in membranes may be the source of VOCs excreted in breath [10]. Smoke, inflammation, chronic obstructive pulmonary diseases is reported to accrete the oxidative stress as well [10], but the predictive model with VOCs did not significantly affected by smoking[5][6]. Phillips proposed that hepatic cytochrome P450 is predominate factor for the activity of oxidative stress [6], while this hypothesis is not verified.

Our study of VOCs in the metabolic products may enlighten the cognition of the VOCs generating mechanism. We saw the correlation between VOCs in microenvironment in vitro and in human breath, though not completely coincide with each other. The VOCs found in cell microenvironment implied that cancer cells may generate the source VOCs under the unknown mechanism and excrete them in breath, making the cancer predictive model with VOCs in breath more persuasive.

In the development of a new predictive model of lung cancer, different multivariate analysis techniques were applied according to the characteristic of subjects and data. The small amount of subjects became the bottleneck of many studies trying to build a predictive model. Further work needs to be done with a larger and diverse population.

We expect to find a noninvasive way to detect and predict lung cancer. One attempt application is e-nose for lung cancer detection and diagnosis, with specific sensor or sensor array. Our work may provide the aim substances for the e-nose detection, which someday may be applied clinically.

CONCLUSIONS

In this article, we use the SPME-GCMS to analyze the VOCs in exhale breath of patients with lung cancer as well as the healthy people. The discriminant analysis is used to construct a model with 5 VOCs out of 16 candidate VOCs to predict lung cancer, with a relative high sensirivity and specificity. These 5 VOCs can work as the biomarkers of lung cancer in breath. We also find one distinctive VOC in the metabolic product of 3 kinds of lung cancer cell line, which is not detected in the human bronchial epithelial cell line. The VOCs found of cell line provide the corelation with VOCs detected in breath, which helps to explain the VOCs generation mechasim in lung cancer.

ACKNOWLEDGMENTS

This work was supported by the special foundation from Science and Technology Department of Zhejiang Province (Grant No. 2006C13021).

REFERENCES

1. L.Pauling and A.B. Robinson "Quantitive analysis of urine vapor and breath by gas-liquid parition chromatography", *Proc natl Acad Sci USA*, **68**, 2374-2376(1971).
2. M. Phillips and R.N. Cartaneo, "Increased breath biomarkers of oxidative stress in diabetes mellitus", *Clinica Chimica Acta* , **344,** 189-194(2004).
3. C. F. Mountain, "Revisions in the International System for Staging Lung cancer", *Chest,* **111,** 1710-1717(1997).
4. M. Phillips and J. Herrera, "Variation in volatile organic compounds in the breath of normal humans" ,*J Chromatogr B Biomed Sci Appl*, **353**, 75-88 (1999).
5. M. Phillips and R.N. Cartaneo, "Detection of lung cancer with volatile markers in the breath" , *Chest,* **123** ,2115-2123 (2003).
6. M. Phillips and N. Altorki, "Prediction of lung cancer using volatile biomarkers in breath", *Cancer Biomarker*, **3**, 95-109 (2007).
7. M. Phillips and N. Altorki, "Detection of lung cacer using weighted digital analysis of breath biomarkers", *Clinica Chimica Acta*, **393**,76-84 (2008).
8 D. Poli and P. Carbognani, "Exhaled volatile organic compounds in patients with non-small cell lung cancer: across sectional and nested short-term follow-up study", *Respiratory Research*, **6**, 71-81 (2005).
9. E. Hietane and H. Bartsch "Diet and oxidative stress in breast, colon and prostate cancer patients: a case-control study.", *Eur J Clin Nutr*, **48**,575-586 (1994).
10. W. Miekisch and J. K. Schubert, "Diagnostic potential of breath analysis-focus on volatile organic compounds", *Clinica Chimica Acta*, **347**, 25-39 (2004).
11. C. Xing and X. Fengjuan "A study of the volatile organic compounds exhaled by lung cancer cells in vitro for breath diagnosis" *Cancer*, **110-4** ,835-844 (2007).

Featuring Of Odor By Metal Oxide Sensor Response To Varying Gas Mixture

Arūnas Šetkus[1], Andrius Olekas[1], Daiva Senulienė[1], Matteo Falasconi[2], Matteo Pardo[2] and Giorgio Sberveglieri[2]

[1]*Sensors Laboratory, Semiconductor Physics Institute, A. Goštauto 11, Vilnius LT01108, Lithuania,*
[2]*SENSOR Laboratory, CNR-INFM, University of Brescia, Via Valotti 9, 25133 Brescia, Italy*

Abstract. In this report the responses of metal oxide sensors to varying amount of gas in air are used for multiparameter featuring of volatile compounds. The composition of the atmosphere is varied by time dependent release of gas from a SPME fiber into flow of synthetic air. Sensor outputs are described by sets of parameters related to the characteristics of gas injection and the sensor response kinetics. The sets of sensor parameters are processed by explorative data analysis (EDA) software comprising customizable feature, correlation and principal component analysis (PCA). This method is applied for distinguishing between four volatile compounds acetone, acetic acid, acetaldehyde and butyric acid, and is aimed at recognition of compounds emitted from infected wounds that is the subject of the WOUNDMONITOR project. The amounts of the target compounds can be evaluated by the EDA analysis if the magnitudes of signals are processed. Analysis of pure dynamic parameters results in separation between the types of the volatile compounds.

Keywords: Gas Sensors, Metal Oxide, Dynamic Response, Data Analysis, Medical Application.
PACS: 07.07.Df, 07.05.Kf, 68.47.Gh, 73.25.+i, 87.85.Ox, 87.19.xb

INTRODUCTION

Various attempts are made to mimic natural olfaction in development of electronic noses (e-nose). Though the sensors are the key elements in the gas recognition systems it is also important to improve the methods and modules used for gas sampling and delivery, signal preprocessing and data analysis. For example, attempts were made to improve e-noses by creating an artificial mucosa. We think that understanding of gas injection and delivery to the sensors can be useful for e-nose technology.

Odor recognition systems based on metal oxide (MOX) sensors are acceptable for detection of smell of organic materials. The problems appear in characterization of these materials due to changes in the smell source produced by living stuff such as bacteria. The featuring of mixture of volatile products is typically distorted by growth of bacteria. In our recent work [1], we used phenomenological model of the response mechanism to demonstrate possibility of splitting the qualitative and quantitative description of smell by parameters of transient response of MOX sensors. The present study deals with practical aspects of early identification of infection by smell within the frame of the FP6 STREP project WOUNDMONITOR.

EXPERIMENTS AND METHODS

An array was composed of commercial and home made metal oxide (MOX) sensors. The TGS type sensors, namely TGS2602, TGS2610 and TGS2620, were included in the array. The home made sensors were based on metal oxide thin films grown by two original methods: (a) the Rheotaxial (fused metal) Growth and Thermal Oxidation technique (RGTO) developed in Brescia University and (b) ultra-thin film growth by dc-magnetron sputtering in reactive gas atmosphere Ar_2:O_2=4:1 developed in the SPI. The SnO_2 films were modified with catalytic metals Au, Pt, Pd and Ru. The sensors were fixed within the tube-like test chamber of negligible dead volume. Diameter of the chamber and the tubing was equal to 6 mm. Details of sensor technology and the test system were described in our previous reports, e.g. [2].

Solid phase microextraction (SPME) fiber was used for injection of the target VOC. Limited amount of VOCs was emitted from the SPME into special section of the system and, consequently, carried to the chamber by constant flow (F) of synthetic air. Depending on the choice F was from 0 to about 200 sccm/min. It must be noted here that the construction

CP1137, *Olfaction and Electronic Nose: Proceedings of the 13th International Symposium,* edited by M. Pardo and G. Sberveglieri

is important for the e-nose but it does not change fundamental aspects of the results. Therefore these details are omitted in present report.

The odors were sampled by the SPME fiber from the headspace air with strictly controlled amount of pure VOC, namely acetone (CH_3COCH_3), acetic acid (CH_3COOH), butyric acid ($CH_3CH_2CH_2$-$COOH$), acetaldehyde (CH_3CHO), and mixed compounds. The list was based on the key VOCs of the most frequent wound infectious agents defined in the WOUNDMONITOR. The amount of individual VOC can be set between 0.1 ppm and 10 ppm.

The output parameters obtained from the sensor responses to VOCs were analyzed by the EDA software thoroughly described in previous report [3].

RESULTS

Typical responses of the sensors are illustrated in Fig. 1 in which the relative changes of the resistances are presented as functions of measurement time. The SPME fiber was pushed into the chamber at t=0.

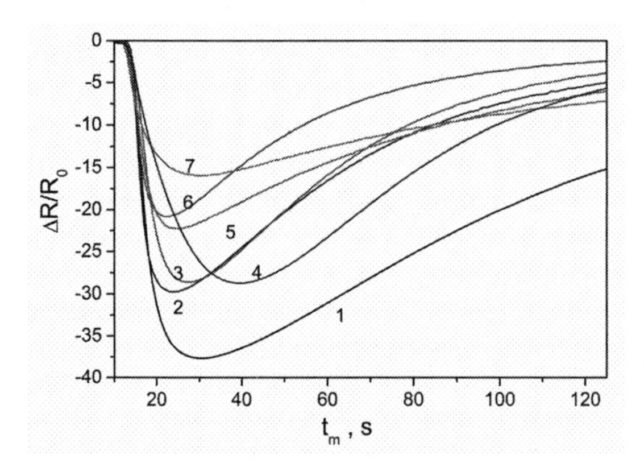

FIGURE 1. Responses of 7 MOX sensors to acetic acid introduced by SPME fiber (flow rate is 32 sccm/min).

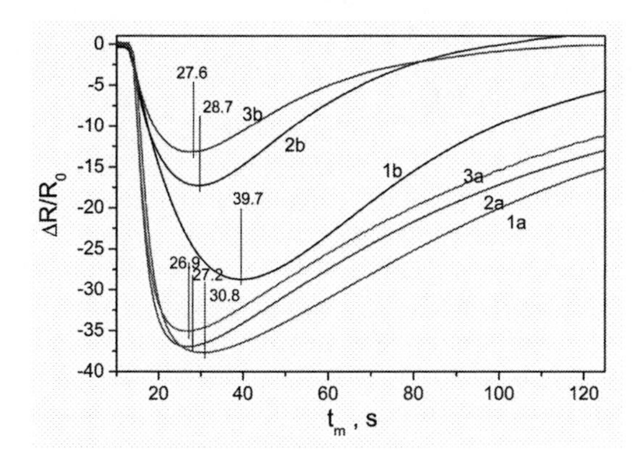

FIGURE 2. Responses of 2 MOX sensors ((a) TGS2602 and (b) SN30KA-LRu) to acetic acid introduced by the SPME fiber into the air flow (in sccm/min) 32 (1), 45 (2) and 56 (3). Labels at vertical lines are the time for the maximum.

The initial part of the transient signals corresponds to the time delay between introducing the SPME fiber into the chamber and the rise of the response.

The influence of F on the response is illustrated in Fig. 2 by typical results for comparatively faster sensor TGS2602 and slower one SN30KA-LRu. In Fig. 2, the response of the TGS2602 is less dependent on F than that of the SN30KA-LRu. An increase in F results in decrease of the response for comparatively slower sensors, i.e. the maximum signal and the maximum signal time significantly decrease with F. Since the faster sensor follows the VOC amount only the maximum signal time decreases with F for these sensors. The fall of the response signals in Fig. 2 also depends on F. The time of the return to the clean air state is frequently shorter for the slower sensors than for the faster ones as it can be seen in Fig. 2.

Parameters For The Featuring

The parameters for the data analysis in this work were defined to specific characteristics of gas kinetics in the test system. Some aspects of this approach are discussed in the next section.

It seems reasonable to suppose that different volatile compounds are released at individual rates from the SPME fiber. Therefore the shape of the response must be dependent of the origin of VOC. Typical responses of the sensor array to VOC are compared in Fig. 3.

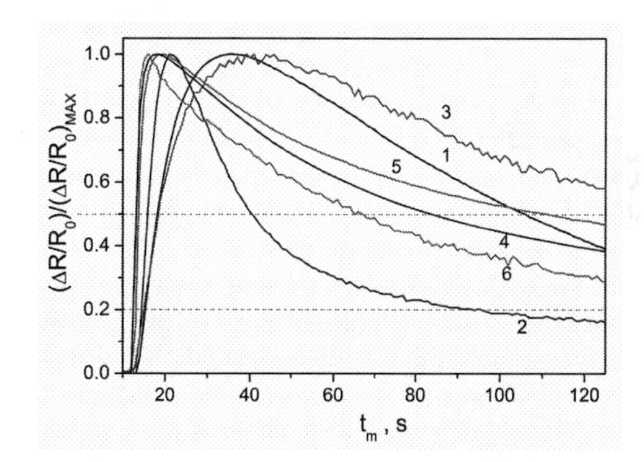

FIGURE 3. Responses of an array of 7 MOX sensors to the headspace air of *Staphylococcus aureus* in the chamber with flow of synthetic air F= 45 sccm/min. Missing labels means zero response.

The results in Fig. 3 represent arbitrary response signals divided by the magnitude of corresponding maximum signal. The transients of the response explicitly show differences in the kinetics of the response. It is clear that the time required to obtain the maximum signal is individual for different gases. It will be shown in the next section that this result can be predicted from simple consideration. Based on the same consideration, it also can be expected that the duration of the response of individual sensor have be dependent on the VOC type as in Fig. 3.

The slopes of the response transients in Fig. 3 are also individual for different VOC. The fact is supported by careful quantitative comparison of the derivatives of dependences in the increasing and decreasing parts of the signal. Nevertheless, it is clear that analysis of the decreasing part of the signal requires much less precision than the increasing part because the differences can be seen visually.

Based on the relationship between the gas characteristics and the sensor responses and the results similar to that in Fig. 3 we introduced three kinetic parameters in the present work. These parameters are the time at which the maximum signal is measure (t_{mx}), the length of the response defined by the time interval between the beginning of the signal rise and the almost clean air resistance at the end of the transient signal (t_l) and the slope at 66 % of the signal in the decreasing part of the signal (δ_{66}).

Assuming fast response of the sensors, the maximum response of sensor was accepted being parameter equivalent to the saturated resistance response of the corresponding sensor (Δ_R).

PCA Results

The sets of the sensor outputs were analyzed by the software called EDA that was originally developed in Brescia University [3]. Separation between the PCA clusters obtained by analysis of a few specific combinations of the parameters was compared in Figs. 4, 5 and 6.

The PCA plots obtained by processing of the data bases composed of only maximum responses Δ_R of the sensor array is shown in Fig. 4.

In Fig. 4, the results were obtained by analysis of the responses of the array to synthetic air with various amounts of separated volatile compounds. The amount of individual target VOC was intentionally set to be constant in single test and different in the series of the tests. Each point in the plot represents an individual test in the series and practically overlapping points represent the test with similar amount of corresponding VOC except the points on the right part of the plot. The spot with the overlapping points represent the lowest amounts of tested VOC.

An increase in the amount of any of the tested volatile compounds results in the shift of the points to the left on the score plots in Fig. 4. The shifts produced by the amount increase of each VOC can be associated with individual vector on the score plot. Therefore the clusters of the tested VOC look like several groups of points distributed along corresponding lines on the score plots. Each of the lines represents individual VOC in Fig. 4.

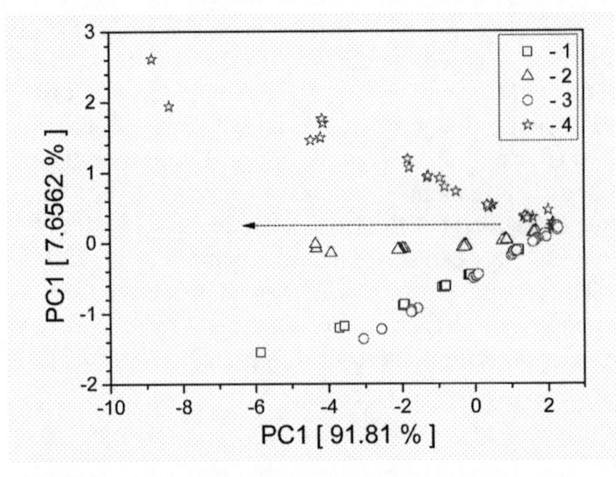

FIGURE 4. PCA plot obtained by featuring of VOCs by saturated responses of metal oxide sensors to various amounts of acetone (1), acetic acid (2), acetaldehyde (3) and butyric acid (4). The arrow shows direction in which the amount of VOC increases from 0.1 ppm to about 10 ppm.

The arrow in Fig. 4 shows the direction corresponding to an increase in the amount of VOC. The distance of any point from the starting points on the right along the corresponding imaginary lines is proportional to the amount of individual VOC.

It follows from Fig. 4, that the origin of the VOC can hardly be distinguished one from another if the amounts of any VOC falls within the interval of comparatively low quantities like < 3-4 ppm. The clusters of individual VOC overlap in the PCA score plots for these amounts if only the magnitudes of the response signals (i.e. Δ_R) are processed by the data analysis software.

The separation between the compounds can be significantly improved if the database of sensor outputs is composed of the parameters of dynamic response. The results obtained by the PCA analysis performed with only the dynamic response parameters, namely t_{mx}, t_l and δ_{66}, are illustrated in Fig. 5.

Compared to Fig. 4, the PCA score plots in Fig. 5 based on the dynamic parameters represent clear distinction between the data classes corresponding to individual type of VOC even at the lowest amounts. In Fig.5, some difference can be found even between the clusters representing acetone and acetaldehyde. The separation between the amounts of individual VOC is much lower than in Fig. 4. This result supports our recent suggestion about possibilities to separate qualitative and quantitative characterization of volatile compounds [1].

Processing of the combined database of all proposed parameters does not improve the situation significantly. The PCA score plot obtained by analysis of combined database including the parameters t_{mx}, t_l, δ_{66}, and Δ_R is shown in Fig. 6.

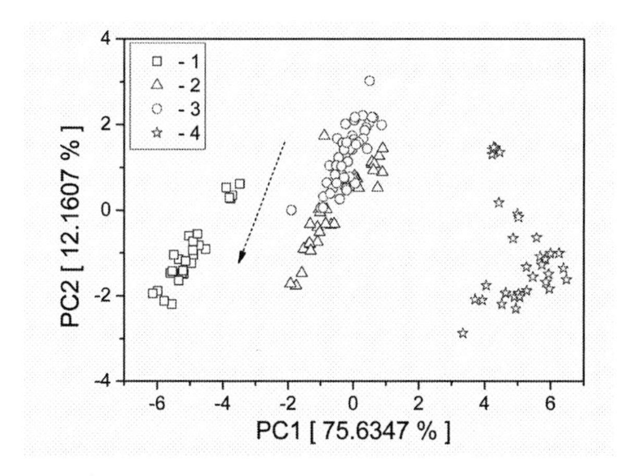

FIGURE 5. PCA plot based only on the kinetic parameters of the responses of metal oxides sensors to various amounts of the VOCs. Labels 1, 2, 3, 4 are defined in Fig. 4. The arrow is an eye guide for increase in the amount of VOC from about 0.1 ppm to about 10 ppm.

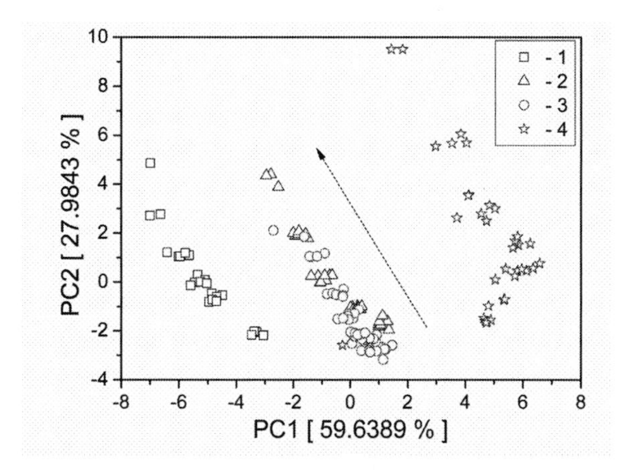

FIGURE 6. PCA plot based on combined data base including the kinetic parameters and the response magnitude obtained from array of metal oxides sensors exposed to various amounts of the VOC. Labels 1, 2, 3, 4 are defined in Fig. 4. The arrow is an eye guide for increase in the amount of VOC from about 0.1 ppm to about 10 ppm.

In the PCA plot in Fig. 6, separation of the amounts is clearly better than in Fig. 5 while separation between compounds is comparable. These results support our idea of two databases with kinetic and static parameters used for practically independent qualitative and quantitative analysis of VOCs.

DISCUSSIONS

Evaporation of sample VOCs from the SPME fiber creates varying concentration of gaseous components in any point in the chamber. A variation of partial pressure of a component in the chamber depends on individual evaporation energy (E_V), diffusion coefficient (D), molecular mass (M) and carrying flow rate (F). Since the gas channel is narrow it is reasonable to assume that the amount of VOC varies only along the axis of the chamber. Flow of carrying gas changes the distribution of the VOC amount along the chamber created by the diffusion. Assuming model of moving wall we can expect that the amount of VOC will be similar at any of the sensors at individual moments. A momentary amount of the target VOC at individual sensor can be approximated by the flux $P_0 \cdot S \cdot F$ where P_0 is partial pressure of a VOC at the surface of SPME fiber, S is the area of chamber cross-section. The kinetics of the response can be evaluated from the phenomenological model [4] which gives a change in the resistance $R(t)/R_0 \sim (a_0 + a_i \cdot \exp(-t/\tau_S))$. The response depends on the ratio of the time span t_F during which the flow passes a sensor to the time constant τ_S [4] which defines the sensor response time.

Assuming the length of sensor L, the time span can be evaluated by $t_F = L/F$. If $t_F/\tau_S >> 1$ ($\exp(-t_F/\tau_S) \to 0$), the response of the sensor will rise up to the maximum signal and it will correspond to the momentary amount of VOC. Under these conditions, an influence of F on the response is comparatively low. If F is high so that $t_F/\tau_S << 1$, the response of the sensor to the momentary amount of VOC is low (because $\exp(-t_F/\tau_S) \to 1$) and it is nearly independent of F. Only if $t_F/\tau_S \approx 1$, the response of sensor to VOC will be significantly changed by a change in F. These simple considerations explain both the results in Figs. 2 and 3 and the basis of the parameters for featuring of VOC. The approach of varying gas amount is not limited to the system construction and MOX sensors though knowledge of the details significantly simplifies the explanation.

ACKNOWLEDGMENTS

The work has been supported by the FP6 European project WOUNDMONITOR (contract no. IST-2004-27859).

REFERENCES

1. A. Šetkus, Ž. Kancleris, A. Olekas, R. Rimdeika, D. Senulienė and V. Strazdienė, *Sens. Actuators B: Chem.* **130**, 448-456 (2008).
2. A. Šetkus, C. Baratto, E. Comini, G. Faglia, A. Galdikas, Ž. Kancleris and G. Sberveglieri, *Sens. Actuators B: Chem.* **103**, 448-456 (2004).
3. M. Vezzoli, A. Ponzoni, M. Pardo, M. Falasconi, G. Faglia and G. Sberveglieri, *Sens. Actuators B: Chem.* **131**, 100-109 (2008).
4. A. Šetkus, Sens. Actuators B: Chem. 87, 346-357 (2002).

An Acetone Nanosensor For Non-invasive Diabetes Detection

L.Wang[1], X. Yun[2], M. Stanacevic[2], P.I. Gouma[1]

1Department of Materials Science and Engineering, SUNY, Stony Brook, NY 11794, USA
2Department of Electrical and Computer Engineering, SUNY, Stony Brook, NY 11794, USA

Abstract. Diabetes is a most common disease worldwide. Acetone in exhaled breath is a known biomarker of Type- 1 diabetes. An exhaled breath analyzer has been developed with the potential to diagnose diabetes as a non-invasive alternative of the currently used blood-based diagnostics. This device utilizes a chemiresistor based on ferroelectric tungsten oxide nanoparticles and detects acetone selectively in breath-simulated media. Real-time monitoring of the acetone concentration is feasible, potentially making this detector a revolutionary, non- invasive, diabetes diagnostic tool.

Keywords: Acetone Nanosensor
PACS: 81.05.Dz

INTRODUCTION

Diabetes is a most common disease. In 2000, according to the World Health Organization, at least 171 million people worldwide suffer from diabetes. Current diabetes diagnosis methods are invasive; while a few non-invasive Devices/methods are neither portable nor real-time monitors. On the other hand, among hundreds of gases and VOCs present in human breath, acetone has been identified to be a biomarker of diabetes, esp. type-1 diabetes [1]. Chemi-resistive gas sensors give a fast response to a presence of the targeted gaseous analyte (acetone in this case), they are inexpensive, and require simple electronics for the output display. Ferroelectric WO3 (ε-WO3) nanoparticles were chosen in this work to build nanosensors for diabetes detection because of their affinity to acetone.

EXPERIMENTAL AND METHODS

Ferroelectric WO3 (ε-WO3) nanoparticles were synthesized [2] using a method called flame spray pyrolysis (FSP). Using the facilities at the Particle Technology Laboratory of Prof. Pratsinis in ETH Zurich. In this method, a certain tungstic organic compound solution mixed with 10at% chromium dopants rapidly evaporates and decomposes by high-temperature flame and then oxidizes to WO3 ultra-fine particles. Cr dopants way, the ε -WO3 phase is able to stay above RT or even higher.

RESULTS

Like other stable phases, ε-WO3 can also be treated as the distortion of an ideal ReO3-like structure. However, different from other phases, such distortion does not occur only between adjacent [WO6] units, but also inside every unit. In the ε-phase the shifts in the negative z direction are larger than those in the positive z direction. Because of the inequality of shifts in the z direction, a net spontaneous polarization develops. This is the origin of ferroelectricity in the ε phase. So, ε-WO3 phase is a type of ferroelectric material that has a spontaneous electric dipole moment. On the other hand, acetone has a much larger dipole moment than any other gas commonly existing in human breath. As a consequence, the interaction between the ε-WO3 surface dipole and acetone molecules is expected to be much stronger than any other gas, potentially enabling selective acetone detection.

Sensor Sensitivity To Acetone

The nanosensor sensitivity obtained (see figure 1 below) suits well the detection requirements of acetone in human breath with concentrations of < 0.8 ppm for a healthy person and > 1.8 ppm for a diabetic patient. When consecutive cycles of acetone gas flows were introduced, the sensitivity did not change indicating good stability of the sensor. In addition, the sensor responds to acetone exposure very fast, in less than 10 seconds. Therefore, the ε-WO3 nanoparticle-based

CP1137, *Olfaction and Electronic Nose: Proceedings of the 13th International Symposium,* edited by M. Pardo and G. Sberveglieri

sensor is capable of real-time, fast-response, stable and highly sensitive detection of acetone gas.

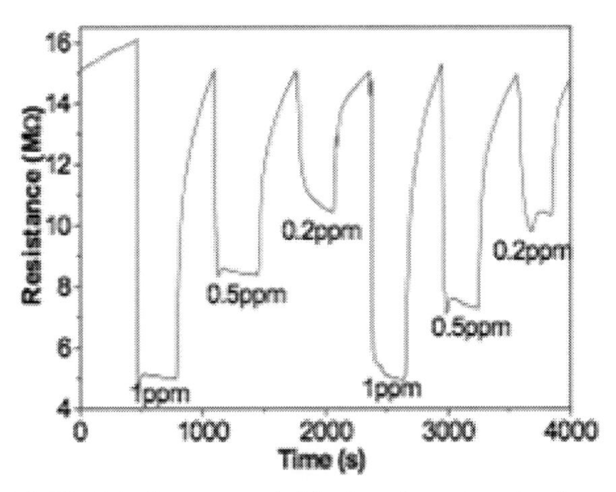

FIGURE 1. Sensor sensivitity to ppb acetone levels.

Sensor selectivity to acetone

The relative response of this sensor to a series of interfering gases, including CO, NH3, methanol, ethanol, CO2, NO and NO2, that are commonly present in human breath was also tested. The sensor has a very good selectivity to acetone at 400 °C.

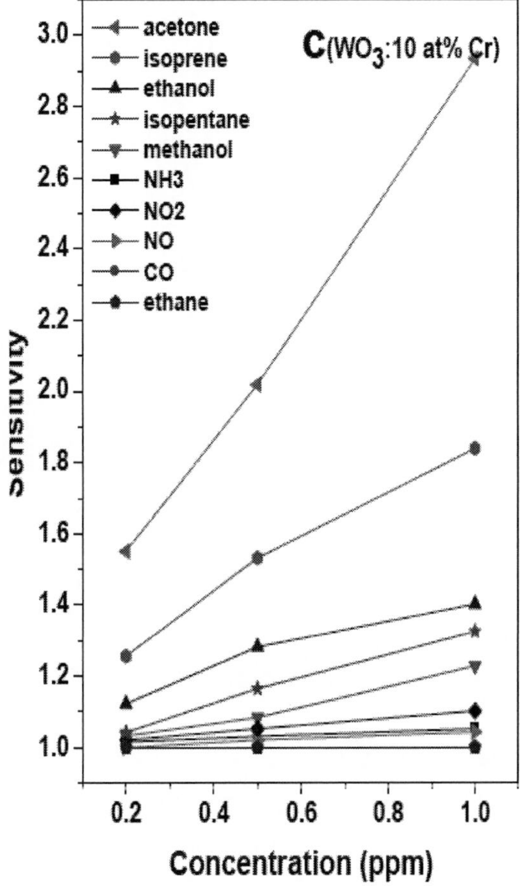

FIGURE 2. Sensor selectivity to acetone

To make the portable breath analyzer, the sensor was attached onto a heater and then embedded into a circuit board. Fig.3 shows the photograph of the prototype. The bottom-left part is the sensor. It will be isolated from the environment by a chamber and specially designed channel allows our human breath flow or controlled gas flow to go through the chamber and interact with the sensor (not shown here). Once the resistance of sensor lowers down below its threshold value, a green indicating LED will turn on. In this project, the baseline resistance of our sensor is 15MΩ, when it is exposed to 1.8ppm of acetone, the resistance decreases to 3.5 MΩ. Therefore, we set the resistance of comparative resistor as 3.5MΩ. After the measurement starts, a 6V power was supplied to the circuit and the sensor. The sensor was heat up to 400°C by the heater. After the sensor is stabilized ten minutes later, a gas flow of different concentrations (0.5ppm, 1ppm, 1.5ppm, 1.8ppm, 2ppm) of acetone in the air was introduced to the device. We have demonstrated that the green LED only lights up when the acetone concentration reaches 1.8ppm. The green LED remains off when we change acetone to other interfering gases listed in Fig. 2 up to 2ppm, which is much higher than actual concentrations of those gases in exhaled human breath.

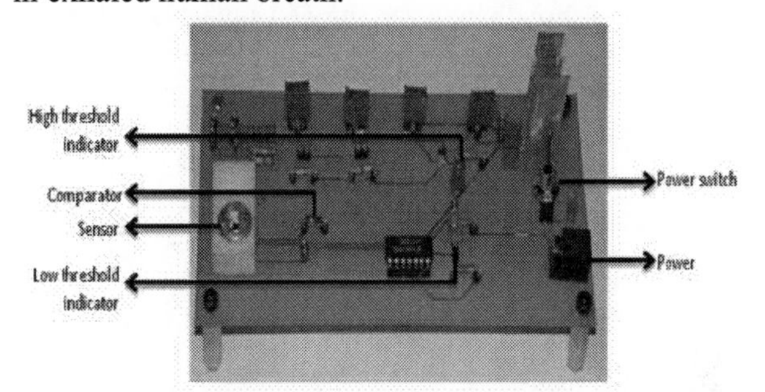

FIGURE 3. Photograph of designed portable prototype for diabetes diagnosis.

Although this device has not yet been utilized to analyze exhaled human breath, it is expected to serve as excellent acetone detector in human breath and finally a promising tool for non-invasive diabetes diagnostics

DISCUSSION

Acetone is a reducing gas. Its sensing mechanism involves with physisorption, chemisorption, and electron transfer processes. Since $W6^{+}$ and $Cr6^{+}$ ions are strong Lewis acids, they tend to easily adsorb acetone molecules which are Lewis bases [3]: In this and the following equations, g means gas molecules; s means surface state; a means adsorbate species.

$(CH_3)_2C{=}O(g) + W^{6+}(s) \rightarrow (CH_3)_2C{=}O \ldots W^{6+}(a)$

Surface acetone reversibly transfers to its isomer, enolate, which can react further with another acetone molecule to yield mesityl oxide. Chemisorption and accompanying electron transfer occur afterwards.

$CH_2{=}C(CH_3)OH \rightarrow W^{6+}(a) + W\text{-}O^-(s) \rightarrow$
$CH_2{=}C(CH_3)\text{-}OW(a) + OH\text{-}W(a) + e^-$

The above reaction processes still cannot explain the selectivity to acetone. Recently, attention has been paid on the surface chemistry of ferroelectric materials. Although conclusive theories have not been established, research based on LiNbO3 and some other materials has shown strong evidence that the dipole moment of a polar molecule may interact with the electric polarization of some ferroelectric domains on the surface [4]. This interaction would then increase the strength of molecular adsorption on the material surface. As a consequence, the interaction between the ε-WO3 surface dipole and acetone molecules could be much stronger than any other gas, leading to the observed selectivity to acetone detection.

CONCLUSIONS

A novel breathanalyzer for non-invasive detection of acetone- a biomarker for diabetes- has been developed. It utilizes a ferroelectric phase of WO3 which exhibits high sensitivity and selectivity to acetone vapor.

ACKNOWLEDGMENTS

NSF NIRT award.

REFERENCES

1. W.Q. Cao and Y.X. Duan, *Breath analysis: Potential for clinical diagnosis and exposure assessment.* Clin. Chem. 52(5): p. 800-811, (2006).
2. L. Wang, Ph.D. thesis, SUNY Stony Brook, 2008.
3. W.S. Sim and D.A. King, *Mechanism of acetone oxidation on Ag{111}-p(4x4)-O.* Journal of Physical Chemistry, 1996. 100(35): p. 14794-14802.
4. Y. Yun, et al., *Effect of ferroelectric poling on the adsorption of 2-propanol on LiNbO3(0001).* J. Phys. Chem. C, 2007. 111(37): p. 13951-13956.

INSTRUMENTATION: ELECTRONICS

Power Savvy Wireless E-Nose Network using In-Network Intelligence

Saverio De Vito*, Gianbattista Burrasca, Ettore Massera, Mara Miglietta, Girolamo Di Francia

ENEA, Physical Technologies and New Materials Dept., FIM/MATNANO Section
ENEA C.R. Portici, P.le E. Fermi, 1, 80055, Portici (NA), Italy
Contact author's e-mail: saverio.devito@portici.enea.it

Abstract. Fluid dynamics effects make single point of measure not a viable solution for toxic/dangerous gases detection both in complex indoor and outdoor environments. Efficient distributed chemical sensing needs reliable wireless battery operated devices capable to operate for long time performing detection and quantification of gases in complex mixtures. Single solid state sensor approach would suffer form interferents influences, enose platform would help to reduce this problem but power consumption is still an issue. Dramatic reduction in this parameter can be achieved by making an e-nose platform capable to perform local estimations avoiding unnecessary data transmission towards datasinks. Here we present the results of experimenting local sensor fusion algorithms implementation on the ENEA Tinynose platform, a wireless e-nose based on room temperature operating sensors and TinyOS over TelosB commercial platform. Results show how the w-nose was able to infer local qualitative and quantitative estimations for Acetic Acid and Ethanol test gases in an ad-hoc setup.

Keywords: Wireless Chemical Sensor Networks, Wireless Electronic Noses Networks..
PACS: 07.07.Df

INTRODUCTION

Irritant or toxic volatiles represents one of the major threats of indoor pollution in laboratories, industry, offices and other manned working environments. Furniture glues, drying paints, cosmetics and special (dry) cleaning products could represent a serious danger in houses. Small term exposure is known to significantly affect wellness and productivity causing eye-nose-throat irritation, headaches, loss of coordination and nausea. Worst, long term exposure even at small concentrations of particular volatile organics compounds (VOCs) is known to trigger much more serious illnesses such as damages to liver, kidney, central nervous system and even cancer [1]. Air quality monitoring scenarios such as indoor toxic or explosive gases monitoring, waste management plants odour monitoring, city air pollution monitoring, are thought to greatly benefit from the use of distributed intelligent sensing units. Fluid dynamic complexities can, in facts, hamper the outcome of single point of measure architecture in these scenarios. On the other hand, even distributed chemical sensing network when based on single solid state sensor will suffer from their lack of selectivity and will typically lead to qualtitative and quantitative errors (false alarm, false negatives). For effective chemical monitoring, we proposed in [2], the use of a group of self powered, networked sensing nodes each one equipped with a small array of chemical sensors, i.e. a wireless electronic noses network, cooperating together for extracting an olfactive "image" of the sensed environment. Multiple layer data fusion scheme should be implied in this scenarios to deal with spatio/temporal dependancies of the chemical signal [3]. Actually most e-nose architectures are not suitable for such usage, most of them are designed for fixed, single unit applications and lacks fundamental features andproperties for operating in distributed sensing scenarios. They have significant power consumption, have very limited or no local data processing capabilities; only a few meet small dimensions criterions and rely on wired connectivity. On the other hand, commercial wireless sensor networks motes have reached sound results in low power operations exploiting both microcontrollers and RF section features. During the last two years the availability of such platforms have suggested their possible use in networked, wireless chemical sensing scenarios.

CP1137, *Olfaction and Electronic Nose: Proceedings of the 13th International Symposium,* edited by M. Pardo and G. Sberveglieri

Shepherd et al., in 2006, were the first to show the development and operation of a network of wireless chemical sensing nodes equipped with a single LED/polymer based sensor [4]. During early 2007, our group proposed the first prototype wireless electronic nose architecture equipped with a small array of polymer sensors [2]. In the same year, Pan et al. proposed another wireless electronic nose architecture devised for outdoor use in livestock farms emissions monitoring scenario [5].
Recently we have presented the results of a first cooperative single gas plume detection performed by a wireless e-nose network in an ad-hoc measurement chamber [6]. Advanced sensors driving schemes as well as the use of Room Temperature (RT) operating sensors can help reducing power consumption in e-nose architectures. However, even in this cases, in a networked scenario, a significant amount of energy will be implied for unnecessary RF communications to the datasink where data fusion should be carried out. An example is the unnecessary multiple sensors data transmission to the datasink that occurs while no significant target gas concentration is actually detected. This can be avoided by impliyng single data fusion schmes on board giving the mote the capability to evaluate the significativity of the sampled data by locally inferring gas discrimination and concentration estimations. Of course, such a system could also compensate for low sensitivity and selectivity of RT operating sensors. Furthermore local situational awareness could also trigger the implementation of local reaction strategies such as increment of the sampling frequency in the near motes.
In this work we present the results obtained by equipping two Tinynose prototypes with an embedded sensor fusion component, actually a neural network. The two motes capabilities have been tested in a gas discrimination and quantification scenario by using Acetic Acid and Ethanol as example toxic/irritant gases. Results show that single motes has been able in most cases to correctly classificate the mock pollutant and estimate at least roughly the pollutant concentration. In this way the single mote has been able to decide whether to transmit or not the sampled data to the data sink for further cooperative processing, strongly affecting the Tx related power consumption. While recorded quantification errors prevents a precise local estimation even at relatively high pollutant concentration, it should be stressed that the use of dynamic features or tapped delay architecture will positively affects the overall performance at the cost of a small computational load increase [7].

EXPERIMENTAL, METHODS AND ARCHITECTURAL IMPROVEMENTS.

The ENEA Tinynose platform is basically built up by a small array of four room temperature operating polymeric sensors coupled to a commercial TelosB platform. Different non-conductive polymeric matrixes are filled by a dispersion of CB nanopowder (see Table 1). A custom electronic board provides signal conditioning on a resistance-to- voltage conversion basis. Custom sw components provide a full featured three layer architecture capable of hosting custom sensor fusion components supporting mesh shaped networking and featuring a GUI with recording and centalized remote controlling capabilities on single motes duty cycle. The basic sw/hw platform description can be found in [6].

TABLE 1: Chemical Sensors Array.

#	Polymer	Chemical Structure
1	Poly-(methylmethacrylate) (PMMA)	$CH_2C(CH_3)(CO_2CH_3)]n$
2	Poly-(2hydroxy-ethylmethacrylate) (PHEMA)	$H(NHCH_2CH_2)_nNH_2$
3	Poly-(styrene) (PS)	$(C_6H_{10}O_3)n$
4	Poly-(ethylenimine) linear(PEI)	Poly-(ethylenimine) linear(PEI) $[CH_2CH(C_6H_5)]n$

Two complete platforms have been developed and tested for cooperative acetic acid plume detection performances in an ad-hoc environmental monitoring simulation chamber simulating a forced ventilation scenario [6]. Here, their sensor arrays have been exposed for 20 minutes in a controlled chamber setup to four differentconcentrations of Acetic Acid (225, 450, 900, 1800 ppm) and Ethanol (200, 500, 1000, 2000 ppm) at 30% relative humidity. For each concentration, two complete exposure/ purge cycle have been performed. In this preliminary experiment, only steady state response was recorded. The $((R-R_{base})/R_{base})$ values were then used for the off-line training of two different three-layer back propagation neural networks (BPNNs) designed to operate with the two Tinynose prototypes sensor arrays. In our opinion, BPNNs represent an optimal choice, because of the trade-off between their capacity and their low footprint as regard as both memory and computating needs.
The 4-4-2 network architectures were designed and trained to detect and discriminate the presence of Acetic Acid or Ethanol (first output line) and estimate their concentrations level (second output line) using instantaneous sensor array responses. An optimal design would have required the use of three networks,

one for the discrimination task and the other two for concentration estimation purposes at the cost of a significant increase in the total footprint.

A neural network component, simulating network operative phase and equipped with the resulting weights and input normalizing vectors was then developed in NesC language relying on TelosB TIMSP430 floating point capabilities. The component have been embedded in the on-board measuring sw architecture and coupled to the normalized instantaneous sensors array readout. Actually the sampled voltage is converted back in resistance value on board for each sensor and then presented to the NN component. Discrimination output is made available to the user locally on each prototype by using two free TelosB LEDs.

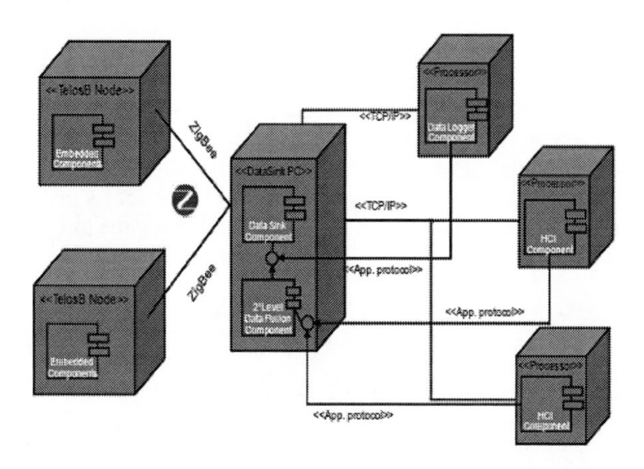

FIGURE 1. The 3 layer wireless networking architecture.

OVERALL PLATFORM RESULTS

The NN equipped single units have been tested in the same controlled chamber setup, sensors have been sampled each 15 sec and both sensors responses and local estimations have been transmitted and recorded at datasink for evaluation purposes. The Human to Computer Interface (HCI) provided a control for base resistance acquisition. Thresholds have been preliminary set on the network classification output, so that the unit can locally discriminate the presence of one of the selected analytes.

Here we show the results of estimations performed by single units during exposure phase to the two mock pollutants. Preliminary data exploration performed by PCA revealed that first principal component, was able to separate the response at the two pollutant at several concentrations level. Actually, the response magnitude to acetic acid for all sensors was significantly higher with respect to the ethanol related one, except when comparing lowest concentrarions of Acetic Acid with highest Ethanol concentrations responses.

In figures 2 and 3, the response of the embedded sensor fusion component is compared with the expected output. By using classification thresholds, Unit 1 proved capable to correctly discriminate, in most cases, the pollutant to which its array has been exposed during tests. Classification errors can be spotted when detecting Ethanol at lowest concentrations (see Fig. 2a). This result is related to the low sensors response and to amplification parameters of signal conditioning board.

As a result, the overall system seems not to be capable to timely discriminate the presence of ethanol under 200ppm, however the simultaneous presence of a nonzero concentration estimation can trigger classification rejection options and eventually the trasmission of data to the datasink.

As regards as concentration estimation, the unit seems to be able to discriminate at least three different concentration levels (low, medium, high) for both analytes. When focusing on dynamic performance (see Fig. 2a and 2b), in most cases we can appreciate a very fast response especially if considering that the expected filling time of measurement chamber is more than 1 minute (4 samples), however classification thresholds are met in most cases (16 over 18 test exposure cycles) on the first sample after pollutant onset with a mean T_{90} of about 55 seconds on the concentration estimation problem.

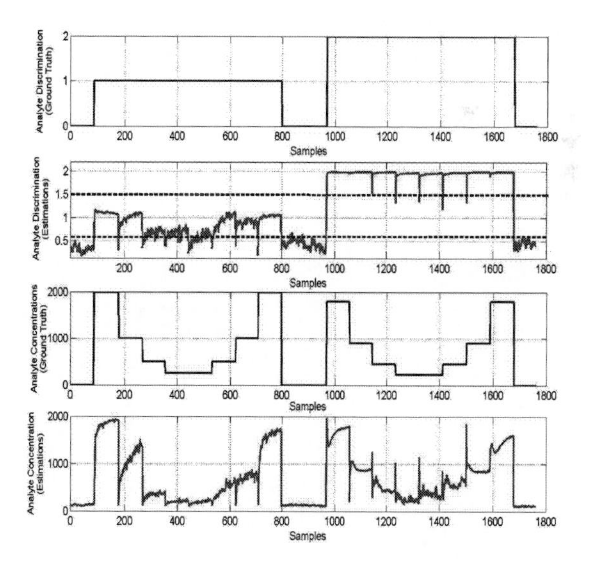

FIGURE 2a. Unit 1 is exposed to Ethanol (samples 90 to 800) and Acetic Acid (samples 967 to 1676). Comparison among expected output (blue) and Unit 1 NN estimations (red) in both classification (above) and quantification (under) problems.

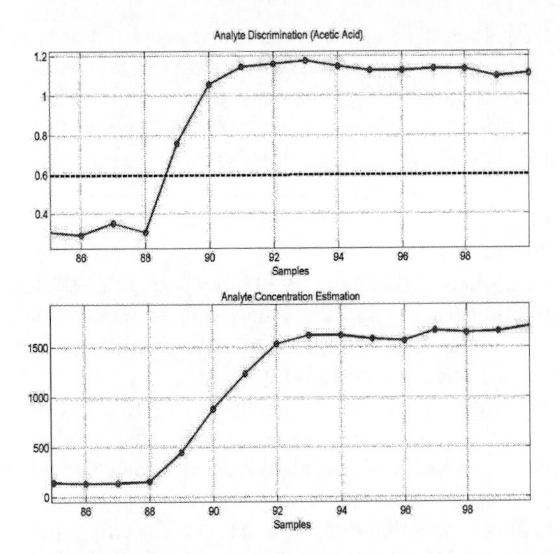

FIGURE 2b: Dynamic response of Unit 1 when exposed to Ethanol (2000 ppm). Lower threshold for Acetic Acid classification is reached at first sample after analyte onset.

Unit *2* shows slight performance degradation with respect to Unit *1;* this is particularly evident when looking *at* the very slow dynamic in the concentration estimation problem (Figures 3a and 4b). Furthermore, frequent spikes can be spotted in the overall system response due to electrical problems on two sensor. Problems in detection when dealing with lowest Ethanol concentrations (relatively to this experimental set up) are evident especially during the first samples after pollutant onset, so the critical Ethanol detection threshold of the current prototype seems confirmed. Anyway, it must be noted that classification estimations are almost perfect at steady state even for Unit 2.

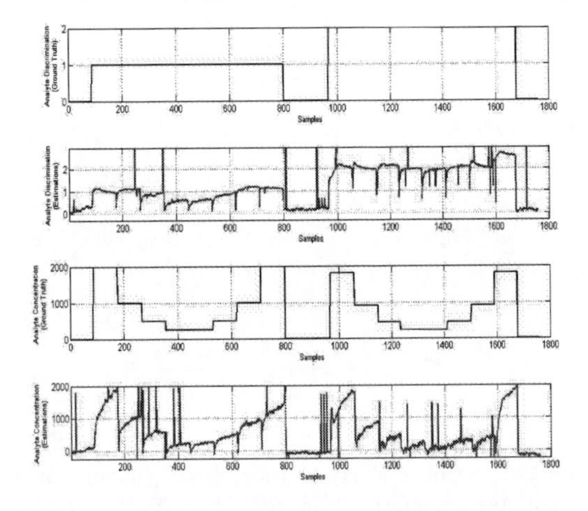

FIGURE 3a: Unit 1 is exposed to Ethanol (samples 90 to 800) and Acetic Acid (samples 967 to 1676). Comparison among expected output (blue) and Unit 2 NN estimations (red) in both classification (above) and quantification (under) problems.

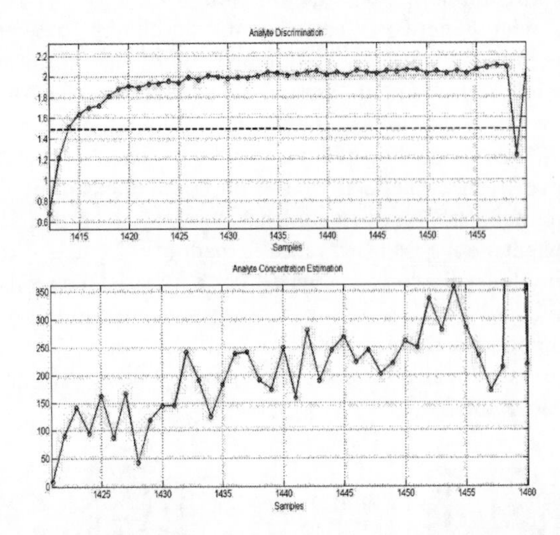

FIGURE 3b: Unit 2 NN responses to Acetic Acid (450 ppm. While discrimination dynamic performance are sufficient, concentration estimation reach target values very slowly. We expext this problem to be corrected by the use of a tapped delay architecture.

CONCLUSIONS

We have presented improvements to our prototype wireless e-nose network architecture, embedding in-network intelligence by using low footprint sensor fusion component. We have tested the single motes in a two pollutants qualitative and quantitative estimation problem targeting the elimination of unneeded data transmissions to the datasink for power consumption reduction purposes. Results are encouraging in order to proceed to test the developed feature for cooperative detection/quantification purposes in our ad-hoc wireless chemical sensor network test facility.

REFERENCES

1.IAQ – EPA Information Sheet,http://www.epa.gov/iaq/voc.html

2. S. De Vito et al., "Enabling distributed VOC sensing applications Toward Tinynose, a polymeric wireless enose", Proc. of XII Italian Conf. on Sensors and Microsystems, 2007, Eds. World Scientific.

3. Nakamoto et al., "Chemical Sensing in Spatio-Temporal Domain", Chem. Rev. 2008, 108, 680-704

4. R. Shepherd et al., "Monitoring chemical plumes in an environmental sensing chamber with a wireless chemical sensor network", Sens. And Act. B, 121, 2007, 142-149.

5. Leilei Pan et al., "An wireless electronic nose network for odours around livestock farms", Proc. of 14th International Conference on Mechatronics and Machine Vision in Practice, 2007, pp.211-216.

6. S. De Vito et al., "TinyNose: Developing a wireless enose platform for distributed air quality monitoring applications", Proc. of IEEE Sensors Conference, Lecce, Oct. 2008.

7. S. De Vito et al.,"Gas concentration estimation in ternary mixtures with room temperature operating sensors using tapped delay architectures", Sens. and Act. B Chem., 124, 2007, 309-316.

A distributed sensing platform for the detection of evaporated hazardous material released from moving sources

A J. Warmer[2], J. Mitrovics[1], P. Kaul[2], C. Becher[2]

[1]JLM Innovation, Beethovenweg 41, 72076 Tübingen, Germany, jan.mitrovics@jlm-innovation.de;
[2]FH-Bonn-Rhein-Sieg, von-Liebig-Str. 20, 53359 Rheinbach, Germany, peter.kaul@fh-brs.de

Abstract. Safety and security applications often need to gather data from distributed locations and a multitude of instruments and sensors.
We have developed a gas sensing platform that communicates via a wireless sensor network based on IEEE 802.15.4 and / or Ethernet. The data form this network is aggregated via a central server that feeds its information over TCP/IP into subsequent data fusion software.
The sensing platform has been equipped with metal oxide gas sensors in order to identify hazardous materials[1]. A number of these nodes have then been placed along a corridor that persons had to pass in order to enter a restricted area. The data from the chemical sensors were fused with tracking data from laser range scanner and video systems.These tests have shown that it was possible to allocate a chemical contamination of one person within a group of moving people and discriminate between various fire accelerating fuels and solvents. These functions were demonstrated outside the laboratory with a test corridor build in a tent during a military tech-demo in Eckernförde, Germany.

Keywords: Safety and security, data fusion, hazardous material, detection, metal oxide gas sensor, wireless sensor network, surveillance
PACS: 01.30.Cc

INTRODUCTION

Freedom of movement for people as well as freedom of coming together safely in open public events or utilities is vital for each citizen. The defence of this freedom against ubiquitous threats requires the development of intelligent security assistance systems that comprise state-of-the-art surveillance technology and work continuously in time. In our work we demonstrate core functions of an indoor security assistance system for real-time decision support that is based on a distributed sensor network.

Such a security assistance system based on multisensor data fusion has successfully been demonstrated in an EU project HAMLeT[2]. One aim of HAMLeT was the data fusion of person tracking data (originating from laser scanners) with additional attribute information in order to identify threats coming from hazardous materials (originating from chemical gas sensors).

The data fusion aspect combines a track of one or more persons with its or their sensor response.Two parameters can be calculated from the sensor response which help to identify the person carrying the hazardous material. One is the signal strengths the other is the time delay which are both dependent from the distance between the person and the sensor system. Analyte diffusion, convection and transportation by a ventilation air stream reduce the gas concentration and influence both parameters[3].

Fire on a ship is considered to be an uncontrollable threat. For this reason fuels and solvents which might be used e.g. in terroristic attacks should be identified before they are smuggled on board. Therefore a sufficient sensors system must be able to detect contaminations of these agents with low evaporation rates and low concentrations even if the suspicious person is moving within a group of people through the entrance gate.

A demonstrator was realised as a harbour entrance gate. In this particular project the early detection, localisation, and continuous tracking of individuals carrying hazardous material within a multiple person flow was the main goal.

CP1137, *Olfaction and Electronic Nose: Proceedings of the 13th International Symposium*, edited by M. Pardo and G. Sberveglieri

EXPERIMENTAL SETUP

Description Of Individual Nodes

Each node contains 4 metal oxide sensors (UST, Figaro and Applied Sensor) and a Sensirion temperature and humidity sensor.

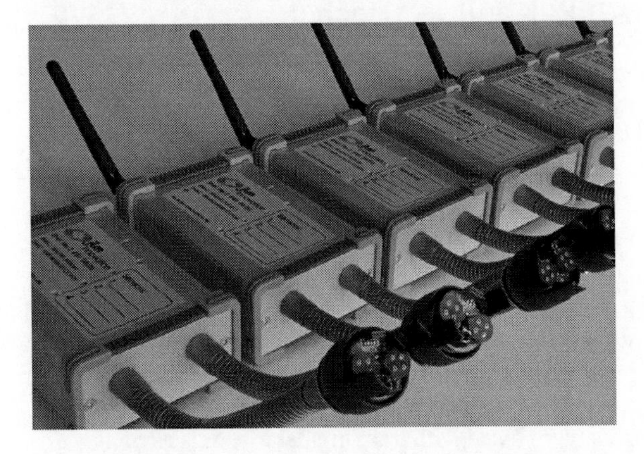

Figure 1: Sensor nodes with antenna and goose necks

A powerful ARM7 micro controller performs the measurements of the sensors and sends transmits the measurement results either via an IEEE 802.15.4 modem over the air or via an Ethernet interface over cable to the central server. The nodes may also be connected via a USB interface directly to a PC, e.g. to update the firmware. Various LED's show the currentstate of each node.

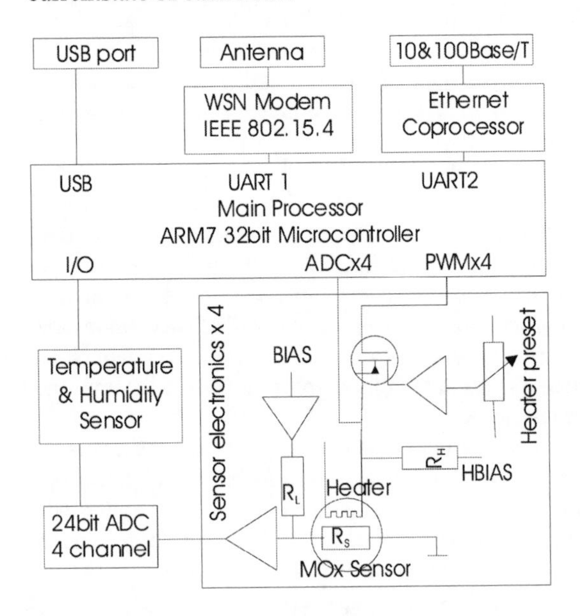

Figure 2: Block diagram of a sensor node

The voltage drop across the metal oxide sensors is measured with a 24bit ADC using a standard replaceable load resistor. The heater of each sensor can be controlled individually using a trimmer for preset and a PWM for software control. The heater resistance can be measured directly using an ADC input of the micro controller.

Description Of The Network

The network of sensor nodes is coordinated by a central server. The duty of the central server is to configure the individual nodes, synchronize the time on the nodes, collect the measurement data and forward it to subsequent data processing. The server itself makes its data available via a TCP/IP Telnet port. This makes it very easy to connect the complete network to data processing software like Matlab or LabView. Additionally MultiSens can be used to control the network and perform real time data analysis with PCA, PLS and other methods.[4]

We chose IEEE 802.15.4 as wireless networking standard, because of its ease of use and mesh capability. The network supports a large number of nodes and establishes routing and mesh architecture automatically. All sensor nodes are configured as routers and may forward messages from other nodes. The distance between two nodes may be several 100 m under good conditions.

A typical limitation of wireless mesh networks is low bandwidth and considerable delays as messages travel via multiple hops within the network. In a typical IEEE 802.15.4 network we found it difficult to transmit reliably more than approximately 40 messages per second. Instead of a tight communication between server and nodes, where the server would initiate all measurements and therefore control the timing, we let the nodes operate independent of the server. At the beginning the server configures the nodes and synchronizes the time bases of all nodes. The nodes then start sending measurement results at fixed time intervals on their own.

Since there may be considerable delays within the network, each measurement is amended by the time of its acquisition. The server may also resynchronize the time base of nodes without interrupting the operation.

Individual nodes may also be connected to the server via a 10BaseT Ethernet connection. The nodes expose a Telnet port. The central server establishes independent TCP/IP connections to the individual nodes.

MOx-Sensor Array

For the development of the sensor array in order to detect fuels and solvents as fire accelerating agents seven different MOx-sensors (UST: GGS 1530T, GGS 3430T, GGS 5430; Figaro: TGS 2602, TGS 2620; Applied Sensor: AS-MLK, AS-MLC) were available. Each was tested with six different analytes: ethanol, methanol, diesel, benzine, petroleum ether, and an acetone containing lack thinner. These agents were chosen because they are available in larger amounts on the free market. From the tested semiconducting sensors an array of four sensors was chosen, which gave the best results with regards to sensitivity and selectivity.
Synthetic air – controlled by a mass flow controller (MFC) is saturated with the analyte in a washing flask and then directed to the measuring chamber. On the measuring chamber a ventilator is attached which is sucking unfiltered air from the lab. The incoming turbulent airflow is measured with a handheld anemometer. The original analyte concentration is evenly diluted by a static mixer. Any lengthening of the measuring chamber or repositioning of the sensors did not significantly change the peak height of the sensors which shows the evidence for an even concentration in the whole volume of the measurement chamber. By controlling the flow through the mass flow controller and the speed of the ventilator the analyte concentration was adjusted to approximately 5 ppmV for pure substances. For mixtures like benzine and petroleum ether with known total vapour pressure the sum of the concentrations of all ingredients is also regulated to 5ppmV. Else the concentration was adjusted to 5 ppmV with regards to the most flighty component of the mixture.

Description Of The Test Corridor

The test aisle simulates an entrance area. It is build as a corridor of 10 m length containing a U-turn. Besides the necessary tracking system (three laser-range-scanner and two video systems) four chemical sensor nodes (L1-L4) are integrated. The sensor nodes had a power consumption of 3 W and were operated in wired as well as in unwired operation mode. They are hidden within in the ventilation system, which creates a laminar airflow across the corridor.
The chemical sensors alone lack the spatial and temporal resolution necessary to allocate a chemical signature to a moving person. The problem can be solved by data fusion of chemical sensor and tracking data and the use of probability based algorithms. This has already been demonstrated and was described in detail[5].

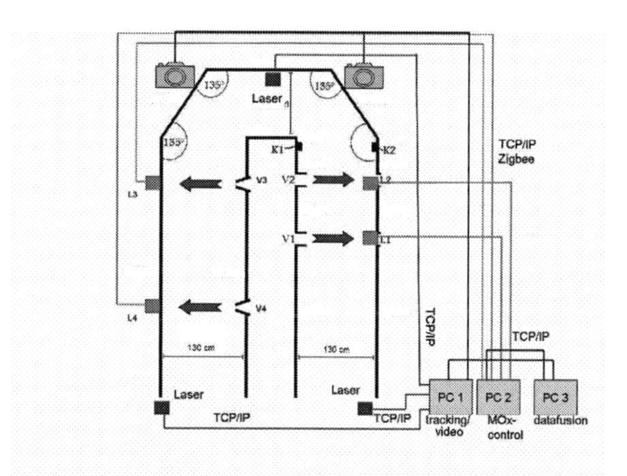

Figure 3: Schematic picture of the test corridor, L1 – L4: chemical sensors, blue arrows: moving air from the ventilation system, Laser: laser-range-scanner

The whole system was integrated into a tent, which was built up outdoor at the navy base in Eckernförde, just 100 m away from the seaside.
As moving analyte sources small, cloth covered vials were used. For ethanol a mass loss of 24 µg/sec at 20 °C was determined. If a person carrying this analyte source is moving through the corridor with normal walking speed the average time of exposure to the sensor array was one second or less.
During the demonstration up to six persons were walking through the corridor at the same time. Only one person was carrying an analyte source at the same time.

RESULTS AND DISCUSSIONS

The best results with regards to sensitivity and selectivity were achieved with an array consisting of four sensors: UST GGS5430T, Figaro TGS 2620, Figaro TGS 2602 and Applied Sensor AS-MLK.
Under lab conditions it was possible to identify the substance clearly at concentrations of 5 ppmV and less.

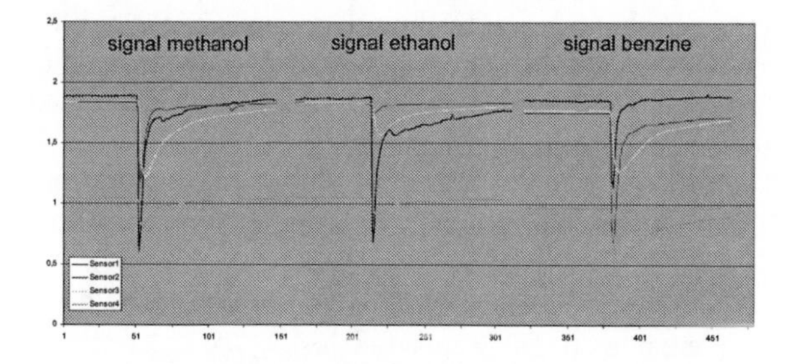

Figure 4: Sensor signal of the array of three single measurements with different analytes

For the three displayed substances clearly distinct patterns can be recognized. Even the methanol can be distinguished from its upper analogue ethanol. The sensor response is very fast (peak maximum within 2 seconds). After 30 seconds of recovery the sensor array is able to detect a new analyte cloud well enough for classification.
For the system demonstration with the test corridor a build in array was exposed to the before mentioned six fuels and solvents and additionally to acetone. From the results a PCA was created (figure 5 which shows the differences of the analytes clearly. One exception is acetone (white dots) which could not be distinguished from the lack thinner (green dots). The lack thinner mostly contains acetone. So the concentration of most volatile acetone in air will be the highest and therefore the amounts of the other components can be neglected. The System used this database for an online pattern recognition.

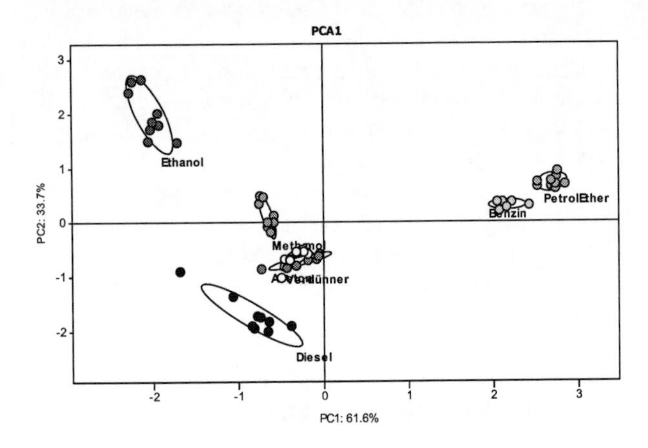

Figure 5: PCA of seven substances used for the demonstration in the test corridor at Eckernförde

A quick sensor response to an analyte source facilitates the allocation of the chemical signature to a moving person by the fusion algorithm. Therefore the delay was reduced by a strong, directed air flow (nozzle velocity 16 m/s) and a fast electronic sensor network. The quickest reading/data transfer rate is 100 ms. For the demonstrator a reading/data transfer rate of 500 ms was sufficient and thus reducing the requirements for data processing.
The following figure (figure 6) shows four pictures at different time stamps of a group of three persons moving through the corridor. The second person is carrying the analyte source. The thin black line behind the person icon represents the history of movement (tracking data). The system is identifying three different persons. Almost immediately after the second person is crossing the sensor node (S1), it is detecting a threat (indicated by colour change from green to red). Within the first time point the second person is identified as carrier (indicated by the red colour of the icon). At time point three the system has automatically identified the substance which is ethanol. In the middle of figure 9 the response of one sensor of each sensor node is displayed.

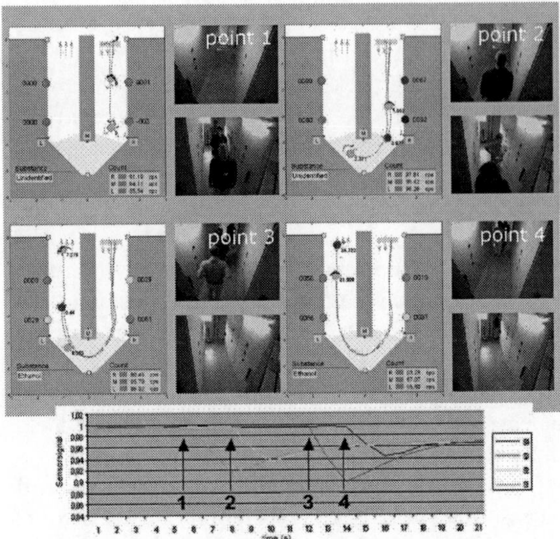

Figure 9: A group of three persons is moving through the demonstrator. The second person carries an ethanol source

Groups of varying size were moving through the demonstrator at different speeds, with or without varying analyte sources. In 95% of the 40 experiments when an analyte source was concealed carried through the corridor the analyte source was detected. False positives did not occur. Interferents like tobacco fumes, menthol chewing gums, coffee, wet clothes, or several perfumes only caused minor disturbances at the sensor system and also yielded no false alarms.
When an analyte was detected the allocation of the chemical signature to the moving person was correctly done in 95% of these cases. With a probability of 60% the system identified the correct chemical substances. In two of 40 cases the substance was identified wrongly. In the remaining runs the substance remained unidentified.
The problems with the identification of the substances can be attributed to the small data base for the classification system (n = 10). Another reason can be found in the big variations of the analyte concentrations created in the experimental procedures in the test corridor. These are caused by the various velocities of walking persons, the different distances of the source to the sensor and changing temperatures which influences the evaporation rate of the analyte source. These varying analyte concentrations cause problems for the data processing due to the non linearity of the sensor signals. At low concentrations

the initial slope of the sensor signals is strongly dependent on the different analytes whereas at high concentrations saturation effects occur.

CONCLUSIONS

The described concept of data fusion of tracking data together with additional sensor attributes – especially chemical sensor attributes – is a very powerful security assistance tool. To increase the correlation between a person track and its chemical signature a distributed network of chemical sensors is necessary. More chemical sensors yield more information with respect to time and space. The sensors should be placed in a way that the distances between the moving persons and the sensors vary. In this way, sensor signals can easier be correlated to different person tracks. Due to the need of a high number of chemical sensors they have to be small, cheap, robust, and they should have low maintenance costs.

ACKNOWLEDGMENTS

This work has been supported by HAMLeT²-team of the Research Establishment for Applied Sciences (FGAN) in Wachtberg, Germany. The system demonstration was funded and supported by the Federal Office of Defense Technology and Procurement (BWB).

The sensor signals must occur as fast as possible (one to two seconds, the shorter the best). With short time delays the fusion algorithms enable a better correlation to the track data. The used Network of semi conducting sensors for the detection of fuel and solvents – also in low concentrations- fulfils the demands perfectly. The realized ventilation system and the fast data transfer from the sensor nodes decreased the time delay and to increased the quality of the sensor signals.

Nevertheless, many design aspects still have to be improved. For example the problems with the analyte identification could be solved by using better statistical data processing methods in order to eliminate effects caused by non linearity of the sensors in the low and high concentration areas.

For future research aspects especially the detection of explosives or explosive related compounds (ERC) is of great interest. Their detection is a very challenging task, because most explosives have a very low vapour pressure and do not evaporate enough analyte molecules into air. For ERC detection the sensitivity of the chemical sensors must be extremely high for airborne analyte detection.

However, first experiments with advanced sensor technologies with higher sensitivities appear to be promising.

REFERENCES

1. U. Weimar., "Gas Sensing with Tin Oxide: Elementary Steps and Signal Transduction", *postdoctoral lecture qualification,* 2002, Universität Tübingen.
2. HAMLeT, "Hazardous Material Localisation & Person Tracking", *Proposal/Grant Agreement no. 204400, Preparatory Action on the enhancement of the European industrial potential in the field of Security research (PASR)*
3. A. Schmidt, "Bewertung von Halbleitergassensoren und QMB Sensoren zum Nachweis von beweglichen Analytquellen mit niedriger Abdampfrate", *bacherlor thesis,* 2007, Fachhochschule Bonn-Rhein-Sieg.
4. J. Mitrovics, "Auswerteverfahren für Gassensorarrays," *dissertation,* 2004, Universität Tübingen.
6. C. Becher, C. *et al*, Forschungsspitzen und Spitzenforschung edit: Zacharias, C. *et al*, Springer (Heidelberg, 2008), pp. 277-296.

A 10 ms-readout Interface For Experimental Resistive Sensor Characterization

A. Depari[1,3], A. Flammini[1,3], D. Marioli[1], E. Sisinni[1,3], E. Comini[2,3], A. Ponzoni[2,3]

[1] *Dept. of Electronics for Automation, University of Brescia, via Branze 38, 25123, Brescia, Italy*
[2] *Dept. of Chemistry and Physics, University of Brescia, Via Valotti 9, 25133, Brescia Italy*
[3] *CNR-INFM SENSOR Lab., via Valotti 9, 25133, Brescia Italy*

Abstract. Metal oxide gas sensors exhibit resistance values varying over a wide range, from tens of kilohms to tens of gigohms, depending on the chosen oxide and on the excitation parameters (voltage, temperature, gas exposure). Resistance-to-time converters (RTC) are widely used in electronic interfaces to such sensors thanks to the low-cost, low-noise and high-range characteristics. RTC main limit is in the variable and long measuring time, ranging from microseconds (tens of kilohms) to several seconds (tens of gigohms), impeding a fine analysis of fast transients. This work proposes a new approach based on combination of the RTC method with a new technique based on the least means square algorithm. The implemented prototype allows the sensor resistance estimation with 100 sample/s (T_{meas} = 10 ms) over the range 10 kΩ ÷ 10 GΩ with relative estimation error below 10% (below 1% in the range 47 kΩ ÷ 2 GΩ). Fast thermal transients of a SnO_2 nanowire sensor have been finely analyzed thanks to the new interface system.

Keywords: nanowire sensors, high range, low measuring time.
PACS: 07.07.Df

INTRODUCTION

Metal oxide based sensors show a resistive behavior exhibiting resistance values ranging from tens of kilohms up to tens of gigohms depending on the oxide preparation method and on the excitation conditions (e.g. voltage, temperature, gas exposure). Among metal oxides, a new class of nanostuctures, namely nanowires, has been recently prepared and proposed as building blocks for the development of nanoelectronic devices including gas sensors [1, 2]. The development of both innovative devices and techniques to analyze the sensor signal requires to finely examine the device response to different excitations such as bias-voltage or temperature pulses, gas exposure. Several solutions for the sensor readout interface have been proposed, based on traditional resistance estimation or on the resistance-to-time (RTC) conversion method [3-8]. The former solutions work using high-resolution A/D converter with a scaling factor changing system, in order to guarantee the best resolution for each resistance value in the considered range. However, such systems require difficult and expensive calibration procedures, especially when very high resistance values are considered. The RTC-based interfaces exploit the easiness of measuring times and intervals over a wide range of variation; therefore, no more scaling factors are needed to adapt the system to the resistance value. The main issue related to such solution is due to the non-constant and, especially with high resistance values, very long measuring time (up to tens of seconds with values on the order of tens of gigohms). Focusing on the need to have fast measurements, recent studies about carbon monoxide sensors have demonstrated the opportunity of a more detailed analysis of the fast transients, for example during the issue of heating pulses [9, 10]. For this reason, a new interface system for resistive sensors has been studied in order to obtain an isochronous and fast readout feature. Particularly, a low-cost electronic circuit is proposed to allow a regular sampling frequency on the order of 100 Hz (T_{meas} = 10 ms), still keeping the measuring range over six decades or more.

EXPERIMENTAL AND METHODS

The proposed interface system is based on the use of an integrator as in Fig. 1a, similarly to RTC schemes. Most of such schemes insert the sensor in an oscillating circuit, whose frequency varies according to the sensor resistance value. If a new sensor is considered, the best way to analyze its behavior, with respect to the selected compound, is to keep the sensor in a very stable

CP1137, *Olfaction and Electronic Nose: Proceedings of the 13th International Symposium*, edited by M. Pardo and G. Sberveglieri

environment (fixed temperature, humidity, excitation voltage…). For this reason, the proposed scheme adopts a constant excitation voltage V_{exc} [4]. The current I_s flowing through the sensor is converted in a voltage V_o varying in a linear way with a fixed slope α, depending on the sensor current itself (Fig. 1b).

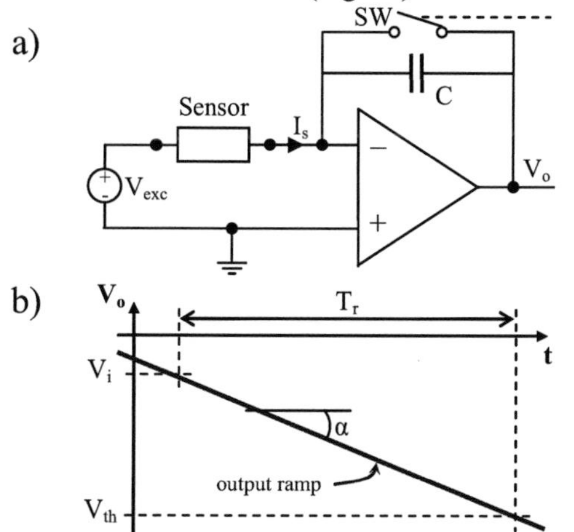

FIGURE 1. a) Scheme of the integrator circuit. b) Output signal V_o behavior.

The relation between the sensor resistive value R_{sens} and the slope α of the output voltage ramp V_o is shown in Eq. (1). The estimation of the ramp slope α can be performed in several ways. In the classical RTC systems, the time T_r required by the ramp V_o to reach a fixed and well-known voltage value V_{th} (threshold) is measured and the sensor resistive value R_{sens} can be estimated using Eq. (2), where V_i is the value of V_o voltage at the beginning of the measurement.

$$|\alpha| = {}^{V_{exc}}\!/_{R_{sens}C} \quad (1)$$

$$R_{sens} = \frac{V_{exc}}{|V_{th} - V_i|}\frac{T_r}{C} \quad (2)$$

The switch *SW* needs to be suitably driven in order to reset the output voltage to the initial value (in this case it is $V_i = 0$ V) and to allow continuous measurements of the time T_r. It should be noticed that to perfectly reset the integrator output V_o is not trivial, because very high insulation switches has an on-resistance on the order of kilohms. A circuit that solves this problem can be found in [4]. Otherwise, a two-thresholds (V_i, V_{th}) circuit can overcome uncertainty due to this aspect. A comparator can be used to detect when the output signal V_o reaches the threshold V_{th}. Using the comparator output signal, a digital electronic system can be used to easily estimate the T_r interval and to drive the switch *SW*. The choice of the V_{th} value is a tradeoff between the desired time resolution in the measurement and the time required to perform the estimation. In fact, the less the time T_r is (small V_{th}), the worse the resolution related to the time estimation is. On the contrary, if a high V_{th} value is chosen, the time T_r becomes bigger than the desired measuring time T_{meas}, when high sensor resistance values are considered. For example, if $C = 100$ pF, $V_{exc} = 1$ V, $T_{meas} = 10$ ms, and $V_{th} = 10$ V, the maximum R_{sens} value which can be estimated is 10 MΩ. If the threshold value is lowered to $V_{th} = 1$ V, then the measurement range is extended up to 100 MΩ. However, in the first case, the T_r value with $R_{sens} = 10$ kΩ is 10 μs, while in the second case it is only 1 μs, requiring a high-resolution timing measurement system (better than 10 ns). Nevertheless, even in case of using both thresholds according to the R_{sens} value, the problem in measuring resistances greater than 100 MΩ still exists. The proposed approach is to keep the RTC approach for small sensor resistance values adding new estimation methods if the threshold is not reached in the desired measurement time (high sensor resistance values). If the slope α of the ramp is too slow, the proposed solution is based on the estimation of α value by using an interpolation method starting from few samples of the ramp acquired in a limited time, less than the desired T_{meas}. In particular, the least mean square (LMS) interpolation method allows the determination of the slope of a line which minimizes the squared error with respect to the acquired experimental points, as shown in Fig. 2.

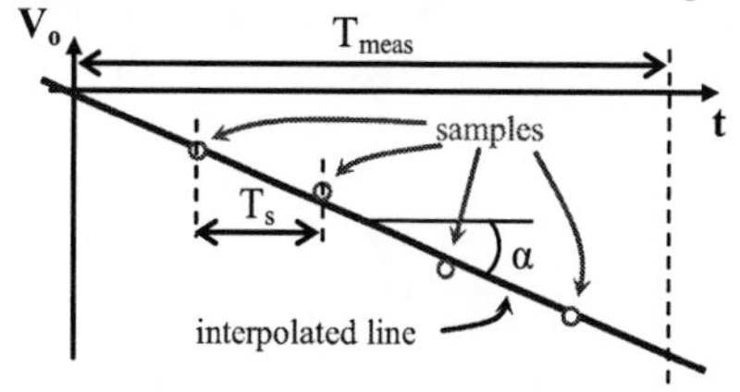

FIGURE 2. Least mean square interpolation algorithm applied to the integrator output ramp.

Depending on the measuring time T_{meas} and on the number N or the samples needed for the application of the LMS method, the sample frequency $F_s = 1/T_s$ can be determined considering that $N{\cdot}T_s < T_{meas}$. Theoretically, such method can be used for any resistive value, but actually there are limitations for its applicability both for high resistive values and for small ones. In fact, when high resistive values are considered, the slope of the ramp is very small and the variation of the V_o voltage within the measuring time T_{meas} can be on the order of the A/D converter resolution or of the noise present in the circuit. On the other hand, when small resistive values are considered, the ramp slope is very high and the limitation of the output range of the operational amplifier or the input range of the A/D converter can occur, as visible in Fig. 3. In such situation, the integrator output V_o reaches the negative saturation voltage V_{sat-} before the last sample is taken. For example, with the previous value for V_{exc}, C and T_{meas}, if we select $N = 5$, $Fs = 500$ Sa/s, and $V_{sat-} = -10$ V, the lower limit is about 10 MΩ.

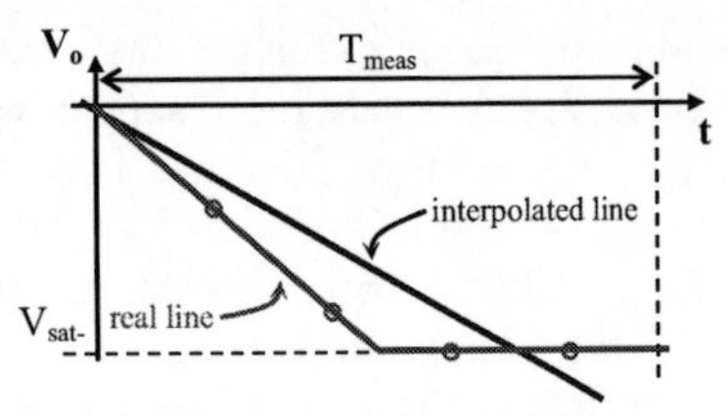

FIGURE 3. Least mean square interpolation fails when the output voltage reaches the saturation value.

Starting from the analysis of the limits of RTC and LMS methods, the authors have decided to combine such techniques realizing a measuring system with a wide range of applicability. A prototype to experimentally evaluate performances of the proposed method is shown in Fig. 4. It is based on an integrator whose output voltage V_o is used by two comparators tuned to different threshold values $V_{th\text{-}low}$ and $V_{th\text{-}high}$. In such a way, the RTC method can be applied with improved accuracy and/or range. In addition, V_o is also sampled by an ADC to allow the use of the LSM interpolation method when the RTC technique fails. The time estimation is performed by simple counters implemented in a programmable logic device (a Cyclone FPGA from Altera) which is also devoted to the control of the reset switch *SW*. In addition, the FPGA sends the measured data to a PC by means of an RS232 link and generates the correct trigger signal to control the A/D conversion within the measuring cycle. The A/D conversion is performed by a PCI acquisition board from National Instruments (NI-6110), with 12 bit of resolution.

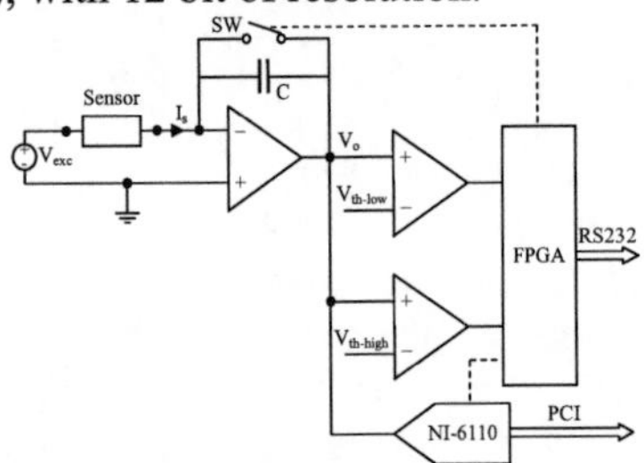

FIGURE 4. The proposed system, combining the RTC approach with the LMS interpolation method.

The choice of using an acquisition board is due to the need of better investigate the LMS approach, in order to suitably choose the sample rate F_s and sample number N for the specific application. In fact, using such acquisition board, it is possible to acquire samples with a high sample rate and then, offline and by means of a PC, try different combinations of N and real F_s (neglecting samples is equivalent to lower the sample rate). Once the suitable value for N and F_s has been chosen, a more compact and low-cost system including an A/D converter controlled by the FPGA could be designed. In the prototype, the R_{sens} estimation is performed by the PC, although the FPGA could execute calculations in real-time. In fact, if every cycle the FPGA acquires $N = 2^n$ ADC samples V_j, $j = 1..N$, the value of α can be estimated by means of a suitable N-tap FIR filter applied to the sequence V_j. Then, the value of R_{sens} can be easily computed starting from α value by inverting Eq. (1). If, in that cycle, both time T_i (related to the first threshold V_i interception) and time T_{th} (related to the second threshold V_{th} interception) are available, then R_{sens} can be computed applying Eq. (2), where $T_r = T_{th}\text{-}T_i$. It should be noticed that the R_{sens} estimation by means of Eq. (2) (RTC method), if available within T_{meas}, should be preferred. The estimation by means of α value computed according to the least mean square approach (LMS method) works properly only if the ramp does not saturate within T_{meas} and allows to complete the estimation before the ramp reaches the thresholds.

RESULTS

Experimental results have been conducted using the system in Fig. 4 using: $V_{exc} = 1$ V, $V_{th,low} = 1$ V, $V_{th,high} = 10$ V, $F_s = 10$ kSa/s, $N = 100$, $T_{meas} = 10$ ms, and power supply ±12 V. The voltage limitation for the LMS method ($V_{sat\text{-}}$ in Fig. 3) is not determined by the integrator output range, but by the input range of the NI-6110 acquisition board, which is ±10 V ($V_{sat\text{-}} = -10$ V). A set of commercial resistors in the range from 10 kΩ up to 100 GΩ has been used to characterize the measuring performances of the system. The resistors up to the nominal value of 10 MΩ have been measured using a multimeter (Fluke 8840A). Table 1 shows the estimation results obtained using three different approaches: the RTC technique, the LMS algorithm using all the 100 samples for every cycle (LMS_{100}), and the LMS algorithm using only 8 samples for every cycle (LMS_8). "Error" is the relative error computed as the difference between the estimated and the "true" value. The RTC approach estimation is computed considering the ramp time between the two thresholds, that is applying Eq. (2) with $V_i = V_{th\_low}$ and $V_{th} = V_{th\_high}$. In such situation, the upper operative limit of the technique is about 10 MΩ. The estimation error looks significant with resistance values on the order of 10 kΩ, due to the poor time resolution (50 ns) for the measure of the threshold interceptions. However, with higher resistance values, the RTC approach allows the resistance estimation with a relative error less than 0.5%. It should be noticed that, when available, the RTC method should be preferred thanks to the very low value of the relative standard deviation ("std"). The LMS technique lower limit is about 10 MΩ if the slope estimation is performed using 100 samples, whereas it is about 8 MΩ if only 8 samples are considered. In fact, in this case, samples are taken after 1 ms, 2 ms, … 8 ms from the beginning of the ramp; therefore, no matter if the saturation limit is reached after the last sample has been taken.

TABLE 1. Experimental results with commercial resistors.

R_{sens} "true" [MΩ]	RTC mean [MΩ]	RTC std [%]	RTC error [%]	LMS_{100} mean [MΩ]	LMS_{100} std [%]	LMS_{100} error [%]	LMS_8 mean [MΩ]	LMS_8 std [%]	LMS_8 error [%]
0.0102	0.0110	0.00	8.20						
0.0996	0.1005	0.02	0.92						
0.9998	1.0022	0.01	0.24						
9.9690	9.9931	0.01	0.24	10.075	0.01	1.06	10.072	0.02	1.04
100				100.85	0.06	0.85	100.89	0.24	0.89
1000				1009.3	0.51	0.93	1017.7	1.97	1.77
10000				9338.7	4.84	-6.61	10597	39.66	5.97
20000				16561	8.56	-17.19			
50000				34030	22.40	-31.94			

It should be noticed that the LMS_8 method leads to worse performances than the LMS_{100} one, both in terms of estimation error and measuring range. In fact, with very high resistance values (> 10 GΩ) 8 samples seem not to be enough to estimate with sufficient reliability the ramp slope. Moreover, even if 100 samples are used, the 12-bit resolution (5 mV) of the A/D converter leads to a significant error in the ramp slope estimation if high resistance values are considered (with a 10 GΩ resistance, the ramp decreases of only 10 mV in 10 ms). Furthermore, the system has been tested using a real sensor and examining its behavior when changing the heater power. The sensor used in this test is a tin oxide nanowire sensor (SnO_2). Fig. 5 shows the sensor response during such a test, where the heater voltage has been quickly changed from 1 V to 2 V and then again to 1 V, causing a change of the sensor working temperature. The R_{sens} values are obtained from the RTC estimation if it is available (resistance values up to 10 MΩ), otherwise using the LMS_8 approach. Details of the sensor response in Fig. 5 during the falling and rising transients are reported in Fig. 6 and Fig. 7 respectively.

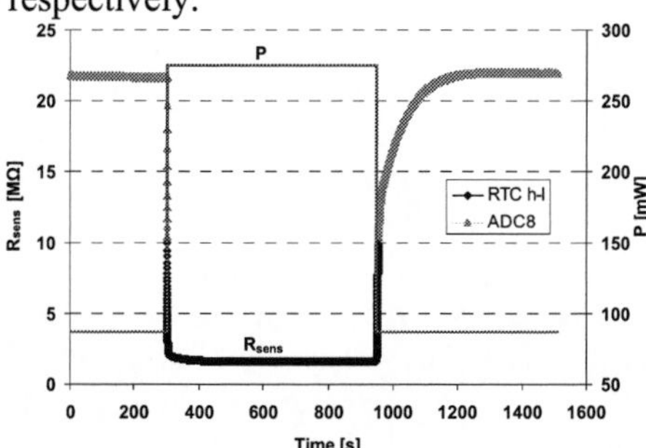

FIGURE 5. Sensor response to a fast variation of the power issued to the heater.

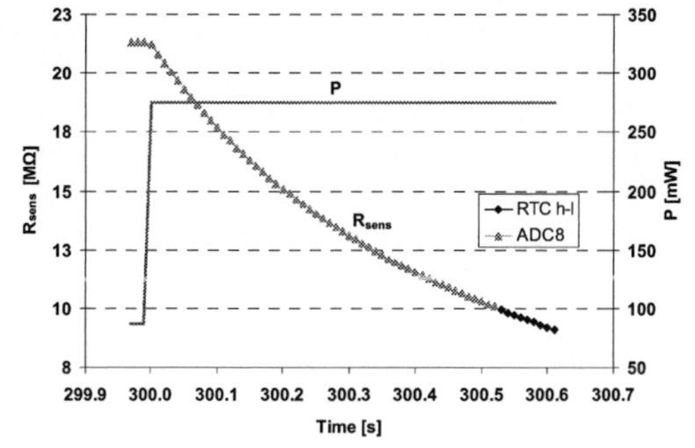

FIGURE 6. Particular of the falling transient of Fig. 5.

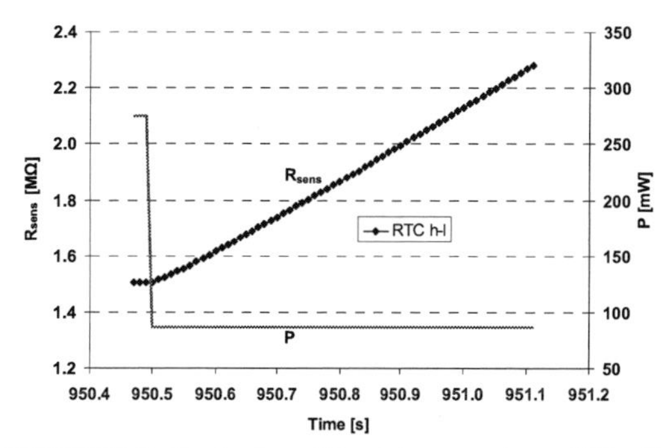

FIGURE 7. Particular of the rising transient of Fig. 5.

In Fig. 6 the point in which the method estimation change is visible (from LMS_8 to RTC) has been highlighted with a change in the line color. It should be noticed that the proposed method allows to track with regular sampling the sensor response in very fast transients, allowing a detailed analysis of the sensor behavior. In conclusion, the new proposed approach allows the sensor resistance estimation with a sampling rate of 100 Sa/s (T_{meas} = 10 ms) in the range from 10 kΩ to 10 GΩ with relative estimation error below 10% (below 1% in the range from 47 kΩ to 2 GΩ). Performances with high resistive values can be improved using an A/D converter with a resolution better than 12 bit.

REFERENCES

1. G.Sberveglieri et al., "Synthesis and characterization of semiconducting nanowires for gas sensing" Sens. and Act. B vol. 121 n. 1 (2007), 3-17.
2. Satyanarayana V.N.T. Kuchibhatla et al., "One dimensional nanostructured materials", Progr. Mater. Sci. n. 52 (2007), 699-913.
3. A. Flammini et al., "A low-cost interface to high value resistive sensors varying over a wide range", IEEE Trans. Instrum. and Meas. vol. 53 n. 4, (2004), 1052–1056.
4. A. Depari et al., "A low cost circuit for high value resistive sensors varying over a wide range", IOP Meas. Sci. and Tech. vol. 17 n. 2 (2006), 353-358.
5. A. De Marcellis et al., "Uncalibrated integrable wide-range single-supply portable interface for resistance and parasitic capacitance determination", Sens. and Act. B vol. 132 n. 2 (2008), 477-484.
6. C. Falconi et al., "Electronic interfaces", Sens. and Act. B vol. 121 n. 1 (2007), 295-329.
7. U.Frey et al., "Digital systems architecture to accommodate wide range resistance changes of metal oxide sensors", Proc. of IEEE Sensors 2008, 593-595.
8. A. Lombardi et al., "Integrated Read Out and Temperature Control Interface with Digital I/O for a Gas-Sensing System Based on a SnO2 Microhotplate Thin Film Gas Sensor", Proc. of IEEE Sensors 2008, 596-599.
9. A. Depari et al., "CO detection by MOX sensors exploiting their dynamic behaviour", Proc. of Eurosensors XXII (2008), 1070-1073.
10. A. Fort et al., "Behavior of MOX CO sensors during thermal transients", Proc. of IEEE Sensors 2008, 851-85.

Interface Circuit for Multiple-Harmonic Analysis on Quartz Resonator Sensors to Investigate on Liquid Solution Microdroplets

M. Ferrari, V. Ferrari, D. Marioli

Dipartimento di Elettronica per l'Automazione, Università di Brescia, Via Branze 38, 25123 Brescia, Italy and CNR-INFM SENSOR Lab, Via Valotti 9, 25133 Brescia, Italy
Phone: +39 030 3715899, Fax: +39 030 380014, e-mail: marco.ferrari@ing.unibs.it

Abstract. This work proposes an interface circuit which exploits a compact implementation of impedance measurement to innovatively analyze a quartz crystal resonator (QCR) sensor across a large number of harmonic overtones. The system measures the electrical admittance (real and imaginary parts) of the sensor, from which the series resonant frequency and the resonance damping are derived for each overtone. By probing the resonator at multiple harmonic modes, enhanced sensing capabilities can be conveniently achieved because a larger set of parameters can be measured with a single sensor. Experimental tests run with 5-MHz QCR sensors on which microdroplets of a sugar-water solution were deposited by a piezoelectric microdispenser show that the response patterns measured across different harmonics can be put in relation with the changes in the acoustic penetration depth into the loading medium.

Keywords: Multiple-Harmonic Analysis, Quartz Resonator Sensors, Liquid Solution Microdroplets.
PACS: 07.07.Df

INTRODUCTION

Multiple-harmonic analyses on quartz crystal resonator (QCR) sensors are recognized as important [1, 2]. By probing the resonator at multiple harmonic modes, enhanced sensing capabilities can be conveniently achieved because a larger set of parameters can be measured with a single sensor.

For multiple harmonic studies on QCR sensors, impedance measurements and transient analysis have been used [2, 3]. More recently, a dual-harmonic oscillator has been reported [4]. The parameters of interest to investigate properties of acoustic loads on the crystal surface, such as mass and viscoelasticity, are usually the series resonant frequency and the quality factor of the resonator [5, 6].

This work proposes an interface circuit which exploits a compact implementation of impedance measurement [7, 8] to innovatively analyze a QCR sensor across a large number of overtones.

THEORY AND MODELING

Under the one-dimensional assumptions that the oscillations propagate only in the direction of the crystal thickness, the loading is uniform over the crystal surface, and the lateral dimensions are large enough, a QCR sensor can be modeled as shown in Fig. 1, providing a first-order approximation around the fundamental, the third and the fifth harmonics [3]. The electrical admittance of the QCR can be expressed as follows:

$$Y_{QCR} = Y_{m1} + Y_{m3} + Y_{m5} + j\omega C_0^* \quad (1)$$

where:

$$Y_{m1} = (j\omega L_{T1} + R_{T1} + 1/j\omega C_{T1})^{-1} \quad (2)$$
$$Y_{m3} = (j\omega L_{T3} + R_{T3} + 1/j\omega C_{T3})^{-1} \quad (3)$$
$$Y_{m5} = (j\omega L_{T5} + R_{T5} + 1/j\omega C_{T5})^{-1} \quad (4)$$

represent the admittances of the motional arms for the fundamental, the third and the fifth harmonics respectively. The motional elements C_{T1}, L_{T1}, R_{T1}, C_{T3}, L_{T3}, R_{T3}, and C_{T5}, L_{T5}, R_{T5} represent the global behaviour of both the crystal and the load around the fundamental, the third and the fifth harmonic, respectively. L_T, C_T, R_T represent the equivalents of mass, compliance, and mechanical losses. The capacitance C_0^* represents the overall parallel capacitance including parasitic contributions.

The validity of the model of Fig. 1 stems from the fact that, depending on the resonant frequency considered, only one of the motional arms is actually active, while the other ones approximately behave as open circuits. This model can be conveniently expanded to a larger number of resonances by including additional motional arms connected in parallel. The acoustic, i.e. mechanical, load can be simply inductive in the case of pure mass accumulation, leading to the gravimetric regime, or complex when the mass loading is accompanied by significant damping, such as in liquids.

CP1137, *Olfaction and Electronic Nose: Proceedings of the 13th International Symposium*, edited by M. Pardo and G. Sberveglieri

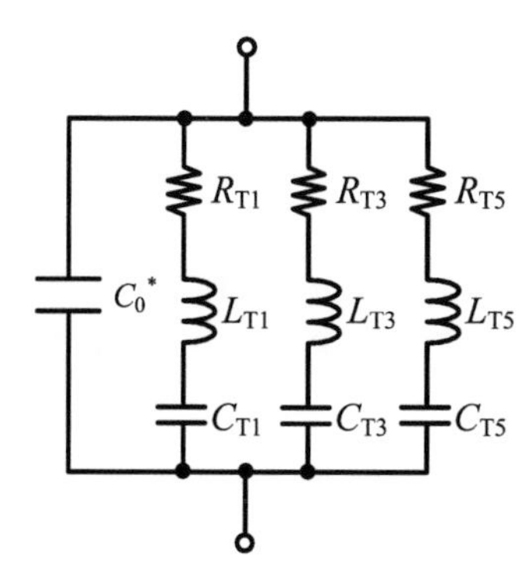

FIGURE 1. A first-order approximation to a general equivalent model of a quartz crystal resonator for the fundamental, the third and the fifth harmonics.

For the model shown in Fig. 1, the series resonant frequencies are given by:

$$f_{s1} = (4\pi^2 L_{T1} C_{T1})^{-1/2} \quad (5)$$
$$f_{s3} = (4\pi^2 L_{T3} C_{T3})^{-1/2} \quad (6)$$
$$f_{s5} = (4\pi^2 L_{T5} C_{T5})^{-1/2} \quad (7)$$

where L_T and C_T represent the total motional inductance and capacitance inclusive of the contributions coming from the load for each overtone.

The measurement of damping, related to the total motional resistance R_T or, equivalently, to the inverse of the quality factor Q, is also important to determine the losses, related to the viscoelastic properties of the load. When the load is not uniform on the crystal surface but is localized on a smaller area, the analytical treatment is less strightforward [9].

SYSTEM DESCRIPTION

The interface circuit exploits a compact implementation of impedance measurement to analyze a QCR sensor across a large number of overtones. The parameters of the sensor can be directly derived from the impedance spectra.

The block diagram of the interface circuit is shown in Fig. 2. A programmable direct-digital synthesis (DDS) chip is used to generate a sinusoidal excitation V_{exc} for the sensor and a pair of quadrature signals $V_{0°}$ and $V_{90°}$ at the same frequency. The block based on A_A and A_B works as a QCR-admittance to differential-voltage converter. The expression of the differential voltage (V_A-V_B) at the outputs of the amplifiers A_A and A_B is given by:

$$V_A - V_B = V_{exc} \cdot R \cdot Y_{QCR} \quad (8)$$

The differential-input multipliers M1 and M3 perform a synchronous demodulation of the differential voltage V_A-V_B with the signals $V_{0°}$ and $V_{90°}$.

Low pass filters (LPFs) provide the DC voltages V_{RE} and V_{IM} related to the real and imaginary parts of the QCR admittance as follows:

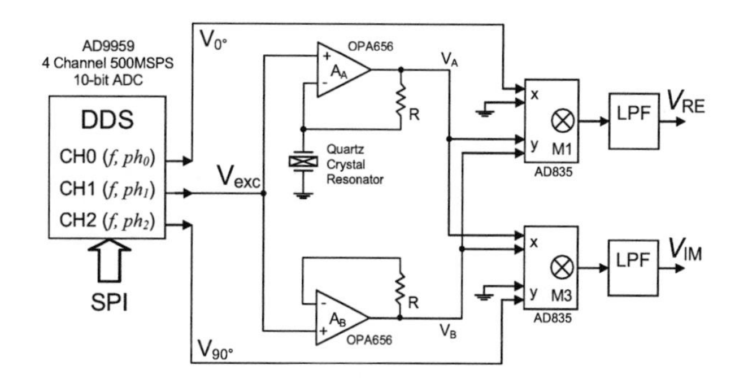

FIGURE 2. Block diagram of the interface circuit.

$$V_{RE} = \alpha \frac{V_{0° peak} V_{exc\,peak}}{2} \cdot R \cdot \mathrm{Re}(Y_{QCR}) \quad (9)$$

$$V_{IM} = \alpha \frac{V_{90° peak} V_{exc\,peak}}{2} \cdot R \cdot \mathrm{Im}(Y_{QCR}) \quad (10)$$

where: α is the scale factor given by the differential multipliers M1 and M3, and $V_{exc\,peak}$, $V_{0°peak}$, $V_{90°peak}$ are the peak amplitudes of the signals V_{exc}, $V_{0°}$ and $V_{90°}$ respectively.

A personal computer programs the DDS for frequency sweeps around the resonances (up to 65 MHz), and acquires the output signals V_{RE} and V_{IM}. The values of the series resonant frequencies are derived as the frequencies where the maxima of V_{RE} occur. At the series resonant frequencies, the admittances of the motional arms (2), (3) and (4) are real; it holds that $\mathrm{Re}(Y_{m1}) = 1/R_{T1}$ and $\mathrm{Re}(Y_{m3}) = 1/R_{T3}$ and $\mathrm{Re}(Y_{m5}) = 1/R_{T5}$. As a consequence, the values of the motional resistances can be calculated from the output signal V_{RE} measured at the resonances, according to (9).

EXPERIMENTAL RESULTS

Fig. 3 shows the interface circuit that was breadboarded in surface-mount technology by using the integrated circuits indicated in the block diagram of Fig. 2, and the following component value: $R = 1$ kΩ.

The AD9959 consists of four synchronized direct-digital synthesis cores that provide independent frequency, phase, and amplitude control on each channel, and four integrated 10-bit digital-to-analog converters (DACs). The device was configured for a single-tone mode with a 25-MHz external clock. Using the internal clock multiplier, the device provides up to $5 \cdot 10^8$ samples per second for each channel.

Each channel has a dedicated 32-bit frequency tuning word, leading to a tuning resolution of about 0.1 μHz, 14 bits of phase offset, and a 10-bit output scale multiplier.

The interface circuit was first tested with 5-MHz AT-cut crystals operating in air. The diameters of the crystal, top and bottom electrodes are 13.5, 8.2 and 6.6 mm respectively.

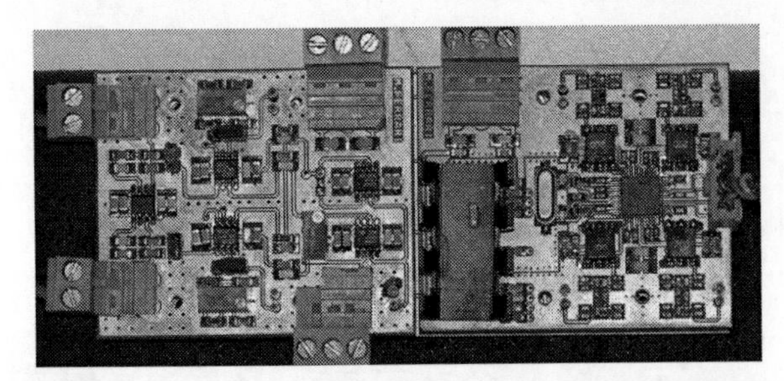

FIGURE 3. Interface circuit for multiple-harmonic measurements on quartz crystal resonator sensors.

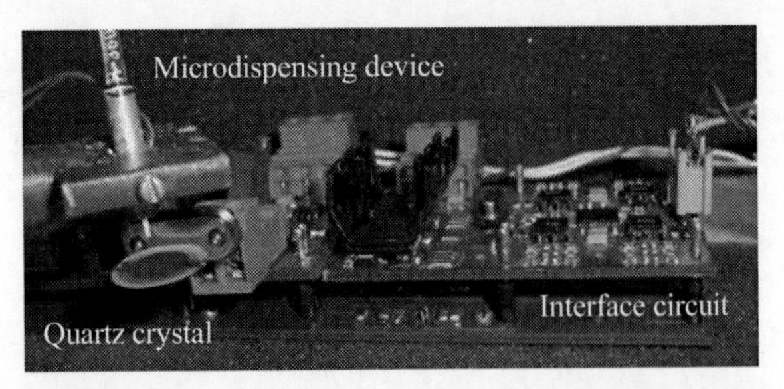

FIGURE 5. Experimental setup with the quartz crystal operating in air.

Fig. 4 shows the typical QCR-admittance circles measured around the fundamental, the third and the fifth harmonics with the crystal in air. The results are in good agreement with reference measurements taken on the crystals by a HP4194A impedance analyzer (IA).

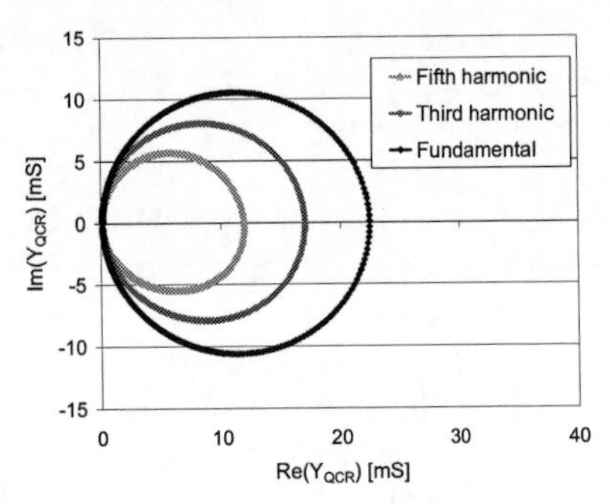

FIGURE 4. Measured admittance circles around the fundamental, the third and the fifth harmonics.

Tests were then done by depositing droplets of a solution of sugar in deionized water with a concentration of 0.4 wt.% on the center of the crystal surface. Depositions were done with a piezoelectric microdispensing device (Microfab MJ-AB-80) with a 80 μm diameter orifice. The experimental setup is shown in Fig. 5. The microdispensing device was driven with a squared pulse train waveform by means of a programmable function generator and a custom designed power amplifier.

Each driving pulse causes a single droplet ejection and has an amplitude of 50 V for 15 μs, with rise and fall times of 3 μs. The volume of a single droplet is estimated in the order of 150 pl. The adopted pulse train waveform is composed of 5 pulses, with 80-Hz driving frequency. In this condition, every deposition on the crystal surface is composed by 5 droplets of the sugar-water solution.

Fig. 6 shows the frequency and resistance shifts measured with the circuit relative to a sequence of six consecutive 5-droplets depositions. The time responses shown in Fig. 6 evidence both different steady-state results and different dynamics across the harmonics considered.

It can be observed that for each deposition there is an initial transient response, due to the water evaporation and film drying process, and then a steady-state is reached due to the residual film formed by the sugar on the crystal surface.

It can be noticed that increasing the number of depositions, and therefore the thickness of the film, the transient response increases and the film drying process becomes slower. Furthermore, increasing the order of the harmonic considered, and therefore decreasing the wavelength of the acoustic wave that probes the load, such phenomenon becomes more evident. This is probably due to the difference in the acoustic penetration depth in the load by different wavelengths, thereby probing the viscoelastic properties of the film to a different extent.

In Fig. 7 and Fig. 8 the steady-state frequency shifts and the resistance shifts as a function of the number of depositions are shown respectively. The steady-state resistance results, shown in Fig. 8, evidence that for the fundamental mode the response is negligible, while enhanced responses are obtained increasing the order of the harmonic considered. For the fundamental mode, due to the negligible dissipation, the load can be considered an acoustic thin and rigid film, confirmed by the linear behavior of the steady-state frequency shift shown in Fig. 7. Increasing the order of the harmonic considered, the load is probed by a shorter wavelength compared to the film thickness, which makes the effect of viscoelastic parameters significant.

CONCLUSIONS

An interface circuit which exploits a compact implementation of impedance measurement to innovatively analyze a QCR sensor across a large number of overtones has been developed. In particular, the system is able to operate over a range of the sensor resonant frequencies up to 65 MHz.

The proposed circuit conveniently achieves enhanced sensing capabilities through multi-frequency operation because a larger set of parameters can be measured with a single sensor, thereby increasing the information content and accuracy. The system measures the electrical admittance (real and imaginary parts) of the QCR sensor, from which the series resonance frequencies and the motional resistances across multiple overtones are derived.

Experimental tests run with 5-MHz AT-cut crystals on which microdroplets of a solution of sugar in deionized water were deposited by a piezoelectric microdispenser showed that the measured frequency and resistance shifts for the third and fifth harmonics versus the fundamental evidence both different steady-state results and different dynamics. These effects can be possibly related with the difference in the acoustic penetration depth in the load versus frequency.

The time responses and combined patterns of frequency and resistance steady-state shifts across different harmonics, at parity of deposited droplet volumes as determined by the microdispenser, are expected to provide a signature of the concentration of the solution. This could potentially offer a mean for solution analysis and solvent evaporation studies requiring very low sample amounts.

REFERENCES

1. M.Yoshimoto, S.Tokimura, K.Shigenobu, S.Kurosawa, M.Naito, *Analytica Chimica Acta*, 510 (1) (2004), pp. 15-19.
2. K.K.Kanazawa, *J. Elect. Chem.*, 524-525 (2002), pp. 103-109.
3. M.Edvardsson, M.Rodahl, B.Kasemo, F.Hook, *Anal. Chem.*, 77 (2005), pp. 4918-4926.
4. M.Ferrari, V.Ferrari, K.K.Kanazawa, *Sensors and Actuators A*, 145-146 (2008), pp. 131-138.
5. R.Lucklum, P.Hauptmann, *Electrochim. Acta*, 45 (22-23) (2000), pp. 3907-3916.
6. R.Lucklum, P.Hauptmann, *Sensors and Actuators B*, 70 (2000), pp. 30-36.
7. J.Schroder, R.Brongraber, F.Eichelbaum, P.Hauptmann, *Sensors and Actuators A*, 97-98 (2002), pp. 543-547.
8. R.Schnitzer, C.Reiter, K.C.Harms, E.Benes, M.Gröschl, *IEEE Sensors Journal*, 6 (5) (2006), pp. 366-381.
9. M.Rodahl, B.Kasemo, *Sensors and Actuators B*, 37 (1996), pp. 111-116.

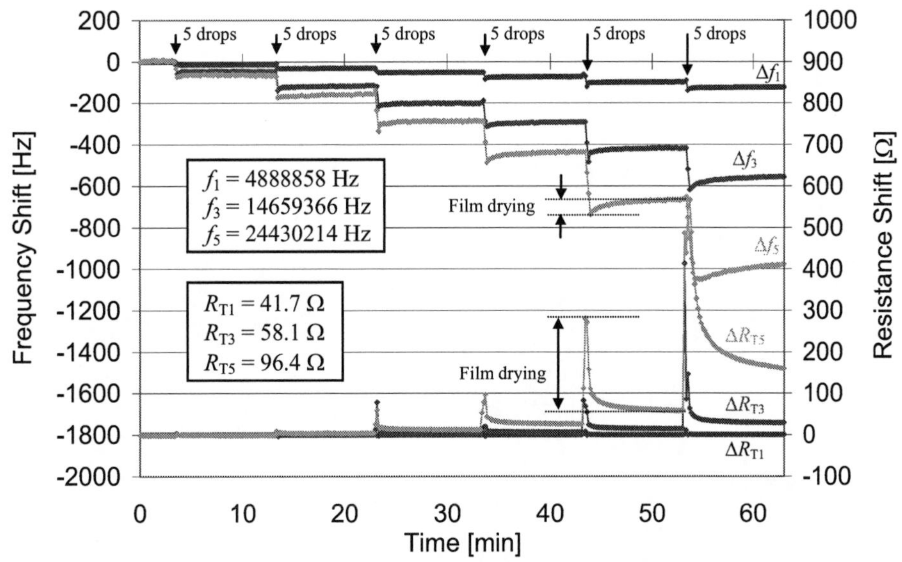

FIGURE 6. Frequency and resistance shifts measured with the circuit relative to a sequence of six consecutive 5-droplet depositions of solution of sugar in deionized water (0.4 wt.% concentration).

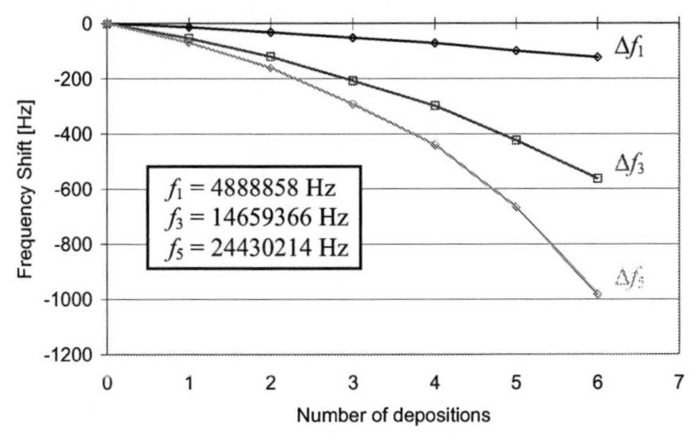

FIGURE 7. Steady-state frequency shifts as a function of the number of depositions of solution of sugar in deionized water (0.4 wt.% concentration).

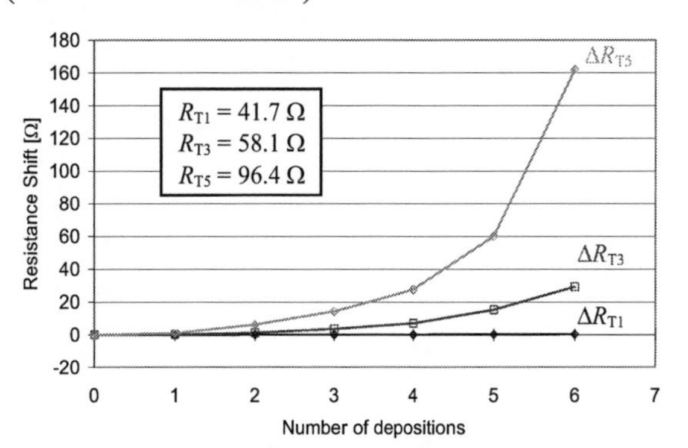

FIGURE 8. Steady-state resistance shifts as a function of the number of depositions of solution of sugar in deionized water (0.4 wt.% concentration).

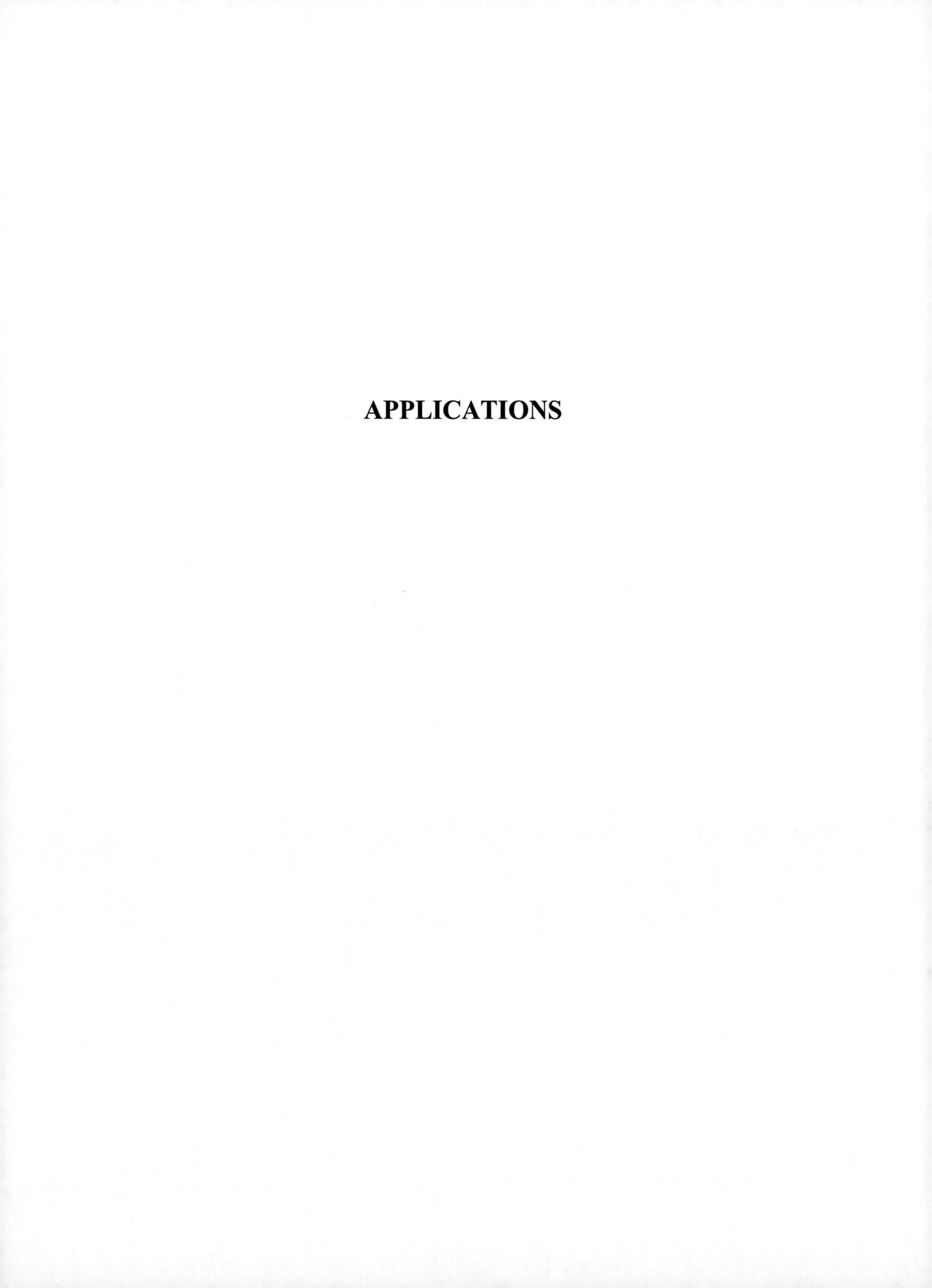

APPLICATIONS

Field Air Sampling and Simultaneous Chemical and Sensory Analysis of Livestock Odorants with Sorbent Tube GC-MS/Olfactometry

Shicheng Zhang[a, b], Lingshuang Cai [a], Jacek A. Koziel[a*], Steven Hoff[a], Charles Clanton[c], David Schmidt[c], Larry Jacobson[c], David Parker[d] and Albert Heber[e]

[a]*Department of Agricultural & Biosystems Engineering, Iowa State University, Ames, IA 50011, USA,*
[b]*Department of Environmental Science and Engineering, Fudan University, Shanghai 200433, China*
[c]*Department of Biosystems and Biobased Products, University of Minnesota, St. Paul, MN, USA*
[d]*College of Agriculture, West Texas A&M University, Canyon, TX, USA*
[e]*Department of Agricultural and Biological Engineering, Purdue University, West Lafayette, IN, USA*
zhangsc@fudan.edu.cn, lscai@iastate.edu, koziel@iastate.edu

Abstract. Characterization and quantification of livestock odorants is one of the most challenging analytical tasks because odor-causing gases are very reactive, polar and often present at very low concentrations in a complex matrix of less important or irrelevant gases. The objective of this research was to develop a novel analytical method for characterization of the livestock odorants including their odor character, odor intensity, and hedonic tone and to apply this method for quantitative analysis of the key odorants responsible for livestock odor. Sorbent tubes packed with Tenax TA were used for field sampling. The automated one-step thermal desorption module coupled with multidimensional gas chromatography-mass spectrometry/olfactometry system was used for simultaneous chemical and odor analysis. Fifteen odorous VOCs and semi-VOCs identified from different livestock species operations were quantified. Method detection limits ranges from 40 pg for skatole to 3590 pg for acetic acid. In addition, odor character, odor intensity and hedonic tone associated with each of the target odorants are also analyzed simultaneously. We found that the mass of each VOCs in the sample correlates well with the log stimulus intensity. All of the correlation coefficients (R^2) are greater than 0.74, and the top 10 correlation coefficients were greater than 0.90.

Keywords: GC-O, livestock, odor, field sampling.
PACS: 01.30.Cc

INTRODUCTION

Odor emissions from livestock facilities affect air quality in surrounding communities. Many volatile organic compounds (VOCs) have been identified, including acids, alcohols, aldehydes, amines, volatile fatty acids (VFAs), hydrocarbons, ketones, indoles, phenols, nitrogen-containing compounds, sulfur-containing compounds, and others [1, 2]. Compounds contributing to the livestock odor have been identified, such as VFAs, p-cresol, phenol, 4-ethylphenol, indole, skatole, and sulfur-containing compounds [3-7].

Livestock odor can be measured using dynamic forced-choice olfactometry, which relies on air sample collection in bags for subsequent evaluation with panelists. This method allows for quantification of the overall odor. However, it does not allow for identification of individual odorous compounds that might be significant to the overall odor controlling.

Gas chromatography (GC)-mass spectrometry (MS)-olfactometry offers the advantages of combining sensory assessment with the identification and quantification of compounds. Some researchers have reported using this method for identification of odorous compounds from swine facilities [3-7]. Rabaud et al [8] used thermal desorption-GC-Olfactometry/MS to identify and quantify odor compounds from a dairy. However, relatively few references exist on the relationship between livestock VOC concentrations and the odor character [9-11].

The focus of this research is to develop an odor characterization method for specific livestock odorants including their odor character, odor intensity, and hedonic tone and develop quantitative analysis method for the key odorous compounds responsible for

CP1137, *Olfaction and Electronic Nose: Proceedings of the 13th International Symposium*, edited by M. Pardo and G. Sberveglieri

livestock odor emissions using TD-MDGC-MS/O system.

EXPERIMENTAL AND METHODS

2.1 Thermal Desorption- Multidimensional GC–MS/Olfactometry (TD-MDGC–MS/O) system

Simultaneous chemical and sensory analyses of livestock odorants were completed using the thermal desorption- multidimensional GC–MS/Olfactometry (TD-MDGC–MS/O) system. The thermal desorption (TD) system is using a Model 3200 automated thermal desorption inlet for Agilent 6890 GC developed by Microanalytics based on a PAL® autosampler. The unique design of the Model 3200 system allows it to utilize a single-step desorption and sample introduction method that eliminates cryotrapping. This design allows the Model 3200 to desorb samples directly into the column interface, eliminating many of the problems associated with dual or two-step desorption such as those associated with the presence of trapped water in sorbent tubes.

Multidimensional GC–MS/O (from Microanalytics, Round Rock, TX, USA) was equipped with two columns in series connected by a Dean's switch. The non-polar pre-column was 12m, 0.53mm i.d.; film thickness, 1 μm with 5% phenyl methylpolysiloxane stationary phase (SGE BP5) and operated with constant pressure mode at 8.5 psi. The polar analytical column was a 25m×0.53 mm fused silica capillary column coated with poly (ethylene glycol) (WAX; SGE BP20) at a film thickness of 1 μm. The column pressure was constant at 5.8 psi. Both columns were connected in series.

System automation and data acquisition software were MultiTraxTM V. 6.00 and AromaTraxTM V. 7.02 (Microanalytics, Round Rock, TX, USA) and ChemStationTM (Agilent, Santa Clara, CA, USA). The general run parameters used were as follows: injector, 260 °C; FID, 280 °C, column, 40 °C initial, 3 min hold, 7 °C min^{-1}, 220 °C final, 10 min hold; carrier gas, GC-grade helium. The GC was operated in a constant pressure mode where the mid-point pressure, i.e., pressure between pre-column and column, was always at 5.8 psi and the heart-cut sweep pressure was 5.0 psi. The MS scan range was 33-280 amu. Spectra were collected at 6 scans s^{-1} using scan and selected ion monitoring (SIM) simultaneously. Electron multiplier voltage was set to 1000 V. MS tuning was performed using the default autotune setting using perfluorotributylamine (PFTBA) weekly.

Human panelists were used to sniff separated compounds simultaneously with chemical analyses. Odor caused by separated VOCs was evaluated with a 64-descriptor panel, intensity scale, and hedonic tone scale in Aromatrax software.

2.2 Sampling

Sampling sorbent tubes were constructed of 304 stainless steel and then double passivated with a proprietary process. They were packed with 65 mg Tenax TA. Silanized glass wool plugs and stainless steel screens were placed in the two ends of the tubes to hold the sorbent.

Before the first use, sorbent tubes were conditioned by thermal cleaning (260 °C for 5 hrs) under a flow rate of nitrogen of 100 mL min^{-1}. For subsequent uses, pre-conditioning at 260 °C for 30 min was applied.

Field air samples were taken using a SKC pump with the set flow rate of 70 mL min^{-1} for 1hr, were stored at 4 °C, and were analyzed within 7 days. The sampling flow rates were detected on-line using a Bios DryCal digital flow meter.

2.3 Standards and Calibration

Fifteen compounds were selected as the target compounds for this work. The selection was based on the previous studies relative to typical odorous volatile organic compounds emitted from livestock facilities (shown in Table 1)[1, 3-7]. Sulfur VOCs were not quantified due to the limitations of Tenax TA sorbent. Standard solutions were prepared by diluting stock standard solutions in methanol and were stored at 4 °C in dark. Stock standard solutions of VFAs and phenolics were prepared by adding certain weights of neat chemicals in a 40 mL pre-cleaned vial, and then filled the vial with a certain weight of methanol. Response factors for odorants were determined by direct injection of 1.0 μL of standard solution onto the GC column and measuring recovery of each odorant.

For sorbent tube analysis, 5 μL or 10 μL of the standard solution was spiked into a sorbent tube using an ATISTM Adsorbent Tube Injector System (Supelco). A nitrogen flow of 50 mL min^{-1} for 5 min with the block heater temperature of 75 °C was needed to transfer the target odorants onto the sorbent tubes.

RESULTS AND DISCUSSION

Using TD-MDGC-MS/O system, quantification of odorants concentration and odor intensity was performed simultaneously. Target compounds were separated in GC column and isolated compounds were split into mass detector and sniff port with the split ratio of 1:3. The concentration of the compounds was quantified with the mass detector, and the odor character, intensity, duration time, and hedonic tone was identified and quantified via the sniff port by the panelist (Table 2). Figure 1 shows the chromatogram and aromagram of a standard sample with 15 typical odorous VOCs. With the increase of the retention time, the start time of an odor event delayed much longer, up to 2.85 min. And the duration time also increased with the increase of retention time, which was called "lingering" of odor event. As a result, some odor events overlaid each other, especially for the compounds with retention times longer than 18 min. In order to quantify the odor event accurately, it is important to separate each odor event correctly.

TABLE 1. Summary of target odorous compounds quantified in this study, with the method linear range, and method detection limits (MDL).

No.	Compounds	MW	Retention time (min)	MS Ion(1)	Linear range (ng)	MDL (ng)
1	Acetic Acid	60.05	12.78	45, ***60***, 15	1.31-1135.54	3.59
2	Propanoic Acid	74.08	14.4	***74***, 28, 48	1.39-1202.28	0.57
3	Isobutanoic Acid	88.11	14.91	***43***, 27, 73	1.28-1108.45	0.30
4	Butanoic Acid	88.11	16.00	***60***, 27, 73	1.26-1089.11	0.38
5	Isopentanoic Acid	102.13	16.73	***60***, 43, 87	0.99-860.84	0.40
6	Pentanoic Acid	102.13	17.88	***60***, 73, 27	1.97-1710.08	0.82
7	Hexanoic Acid	116.16	19.68	***60***, 73, 27	2.08-1802.93	0.87
8	Guaiacol	124.14	20.06	***109***, 124, 81	0.74-637.90	0.03
9	Heptanoic Acid	130.19	21.38	***60***, 73, 41	2.28-1976.07	1.08
10	Phenol	94.11	22.13	***94***, 66, 39	1.35-1175.45	0.14
11	*p*-cresol	108.14	23.28	***107***, 77, 90	0.73-640.52	0.05
12	4-Ethylphenol	122.17	24.61	***107***, 122, 77	0.69-601.50	0.06
13	2-Aminoacetophenone	135.16	25.41	***120***, 135, 92	0.88-764.64	0.08
14	Indole	117.15	28.23	***117***, 90, 63	0.71-619.87	0.03
15	Skatole	131.18	28.88	***130***, 77, 103	0.71-615.09	0.04

Note: (1) The ions shown in ***bold*** italic type were used for quantification.

TABLE 2. Sensory analysis of typical standard solution

No.	Compound	Mass (ng)	Odor Character	Odor Intensity (%)	Hedonic Tone	RSD (%)
1	Acetic Acid	149.7	Acidic	30	-2	2.16
2	Propanoic Acid	161.1	Fatty acid, Body odor	30	-2	5.72
3	Isobutyric Acid	195.2	Body odor, Fatty acid	30	-2	3.03
4	Butyric Acid	137.7	Body odor, Fatty acid	50	-2	1.84
5	Isovaleric Acid	134.1	Body odor, Fatty acid	50	-2	4.25
6	Valeric Acid	197.4	Body odor,Acidic, Spicy	30	-2	9.55
7	Hexanoic	210.2	Acidic, Spicy	50	-3	9.79
8	Guaiacol	59.7	Burnt, Medicinal, Phenolic	30	-2	3.70
9	Heptanoic	237.4	Acidic, Spicy	50	-3	7.35
10	Phenol	133.8	Burnt, Phenolic	10	-1	0.57
11	p-cresol	68.9	Barnyard, Medicinal, Phenolic	30	-2	3.67
12	4-Ethylphenol	56.0	Burnt, Phenolic	30	-2	3.68
13	2-Aminoacetophenone	88.9	Taco Shell, Medicinal, Phenolic, Sweet	30	-1	2.05
14	Indole	42.6	Medicinal, Taco Shell, Barnyard, Sweet	30	-2	3.89
15	Skatole	34.1	Taco Shell, Medicinal, Sweet, Barnyard	30	-2	2.81

Method detect limit (MDL) was determined applying the U.S. EPA methodology[12]. The MDLs were defined as the minimum concentration of a substance that can be measured and reported with 99% confidence when the analyte concentration is greater than zero and is determined from analysis of a sample in a given matrix containing the analyte. The MDLs for our method were listed in Table 1, which were generally lower than those reported in other similar studies[13, 14].

Precision study was conducted by consecutive analysis of 3 tubes spiked with the same amount of a standard work solution. Values of repeatibility (% relative standard deviation values) are reported in Table 2. All of the odorants showed repeatibilities <20% that accomplished US EPA performance criteria[15]. To examine odorants breakthrough, two tubes were connected in series into the standards spiking system. Individual analysis of each tube showed that no significant breakthrough (measured as % odorant in the back tube) was observed for most of the standard odorants. Only some percentages of breakthrough were observed for low molacular compounds: acitic acid, proponoic acid, and isobutanoic acid. This is due to the weak adsorpobility of Tanx TA to low molacular compounds.

Based on above methods, sorbent tubes adsorbed of the standard solution with different concentration including 15 VOCs were analyzed using the TD-GC-MS/O system. We investigated the correlation of odor intensities to odorants mass in one tube. For the TD-GC-MS/O system used in this work, the make-up air flow rate is constant, so the correlation of odor intensities to odorants mass should be similar with that of odor intensities to odorants concentration. For many odorants used in the food and fragrance industry, there is a linear relationship between log olfactory intensity reported by the individual and the air concentration of the odorant present in air[16]. Zahn et al[9,10] also reported the total air concentration of VOCs emitted from swine manure correlate well with the log stimulus intensity. This relationship between perceived olfactory stimuli and intensity of sensation is referred to as the fundamental psychophysical law[17]. We found that the mass of each VOCs correlate well with the log stimulus intensity. All of the correlation coefficients (R^2) are greater than 0.74, and the top 10 correlation coefficients were greater than 0.90. Therefore, this confirmed with the fundamental psychophysical law.

CONCLUSIONS

The TD-MDGC-MS/O system could be used to estimate concentrations of VFAs and phenolic compounds associated with CAFOs odorous issue. Odor character, odor intensity, and odor hedonic tone can be assessed for separated target compounds simultaneously with chemical analyses. Concentrations of odorous compounds correlated well with the log stimulus intensity.

ACKNOWLEDGMENTS

The authors would like to thank the USDA for supporting this work via following grants: USDA-CSREES grant # 2005-35112-15336 “Odor emission and chemical analysis of odorous compounds from animal buildings” and USDA-CSREES grant # 2005-35112-15368 “Mass transfer modeling validation for gas and odor emissions from manure storages and lagoons”.

REFERENCES

1. Y. C. M. Lo, J. A. Koziel, L. S. Cai, S. J. Hoff, W. S. Jenks and H. W. Xin. *J. Environm. Qual.* **37**, 521-534 (2008).
2. S. S. Schiffman, J. L. Bennett and J. H. Raymer. *Agric. For. Meteorol.* **108**, 213-240 (2001).
3. J. A. Koziel, L. S. Cai, D. W. Wright and S. J. Hoff. *J. Chromatogr. Sci.* **44**, 451-457 (2006).
4. E. A. Bulliner, J. A. Koziel, L. S. Cai and D. Wright. *J. Air & Waste Manag. Assoc.* **56**, 1391-1403 (2006).
5. L. S. Cai, J. A. Koziel, Y. C. Lo and S. J. Hoff. *J Chromatogr. A* **1102**, 60-72 (2006).
6. K. M. Keener, J. Zhang, R. W. Bottcher and R. D. Munilla. *Trans. ASAE* **45**, 1579-1584 (2002).
7. L. L. Oehrl, K. M. Keener, R. W. Bottcher, R. D. Munilla and K. M. Connelly. *Applied Eng. Agric.* **17**, 659-661 (2001).
8. N. E. Rabaud, S. E. Ebeler, L. L. Ashbaugh And R. G. Flocchini. *J. Agric. Food Chem.* **50**, 5139-5145 (2002).
9. J. A. Zahn, A. A. DiSpirito, D. A. Laird, Y. S. Do, B. E. Brooks, E. E. Cooper and J. L. Hatfield. *J. Environ. Qual.* **30**, 624-634 (2001).
10. J. A. Zahn, J. L. Hatfield, D. A. Laird, T. T. Hart, Y. S. Do and A. A. DiSpirito. *J. Environ. Qual.* **30**, 635-647 (2001).
11. J. Greenman, M. El-Maaytah, J. Duffield, P. Spencer, M. Rosenberg, S. Saad, P. Lenton, G. Majerus and S. Nachnani. *J. Am. Dent. Assoc.* **136**, 749-757 (2005).
12. http://www.dnr.wisconsin.gov/org/es/science/lc/OUTREACH/-Publications/LOD%20Guidance%20Document.pdf, accessed on Dec. 13, 2008.
13. S. L. Trabue, K. D. Scoggin, H. Li, R. Burns and H. W. Xin. *Environ. Sci. Tech.* **42**, 3745-3750 (2008).
14. A. Ribes, G. Carrera, E. Gallego, X. Roca, M. J. Berenguer and X. Guardino. *J. Chromatogr. A.* **1140**, 44-55 (2007).
15. E.A. Woolfenden and W.A. McClenny. Compendium Method TO-17. Determination of volatile organic compounds in ambient air using active sampling onto sorbent tubes. EPA/625/R-96/010b. 2nd ed. USEPA, Research Triangle Park, NC. 1999.
16. A. Turk, and A. M. Hyman, “Odor measurement and control” in *Patty's industrial hygiene and toxicology*, edited by Clayton G.D. and F. L. Clayton, New York: John Wiley & Sons, 1991, pp. 842-881.
17. S. S. Stevens. *Psychol. Rev.* **64**, 153-181 (1957).

Industrial Applications of Electronic Nose Technology in the Textiles Industry

Tim Gibson[1]*, Rob Chandler[1], Viv Hallam[1], Claire Simpson[1] and Martin Bentham[2]

1) Scensive Technologies Ltd, Metic House, Ripley Drive, Normanton, West Yorkshire, WF6 1QT, UK
2) Courtaulds PLC, PO Box 54, Haydn Road, Nottingham NG5 1DH, UK

Abstract: Electronic nose technology has been available commercially for over 12 years but uptake in actual industrial applications has yet to be fully realised. We report 2 specific test protocols being used in the textiles industry that allow the direct measurement of anti-odour and anti-microbial capabilities of fabrics. Results will be shown for the standard anti-odour test which was specifically commissioned by Courtaulds PLC and which is being used by a number of manufacturers. The second test, which measures the anti-microbial and the anti-odour capabilities of fabrics simultaneously was developed in 2008. Results will be shown that clearly indicate both parameters are detected and proofs of anti-microbial capabilities will be given. These 2 tests will for the first time, enable the fulfillment of legislation that states for textile product claims, anti-odour and anti-microbial capabilities of fabrics must be scientifically substantiated.

Keywords: Electronic nose, anti-odour, anti-microbial, test protocol, industrial, application, product, claims.
PACS: 07.07.Df, 68.43._h, 72.80.Le, 89.20.Bb

INTRODUCTION

The textiles industry in Europe has dramatically changed with mass manufacture of clothing being shifted over to the far east in most cases. This has led to the high-value technical textile areas being the main focus and the incorporation of functionality in textiles and fabrics. Functionality in textiles includes such aspects as anti-odour finishes, anti-microbial capabilities and improved evaporation and water control capabilities. With these developments, methods of QC testing to check for functionality and allow the correct product claims have also advanced. Until now there has been no specifically designed test method to prove the actual anti-odour capabilities of textiles, except the use of human sniff panels. GC-MS can show anti-odour capabilities but they are prohibitively expensive for the industry to use as a routine QC method. Legislation is in place that forces the textile industry to substantiate any product claim made, this includes both anti-odour and anti-microbial capabilities. [1].

Electronic noses are designed to discriminate between complex odours made up of multiple volatile organic compounds (VOCs) and they have been shown to detect and differentialte between micro-organisms in a well substantiated manner [2, 3, 4]. The new Bloodhound® instruments are particularly suited for the detection of hydrophilic VOCs , which form the basis of 2 new tests for measurement of anti-odour and anti-microbial capabilities of textiles in a simple and cost effective manner.

EXPERIMENTAL AND METHODS

Two Bloodhound® ST214 e-noses were used in the work. These instruments are based on 14 organic semi-conducting sensors, one of which is internal, giving 13 outputs that record the responses to VOCs. The sensors are based on a variety of semi-conducting polypyrrole, polyaniline and polythiophene materials that are supported on interdigitated electrodes to give interactive chemo-resistors that respond to VOCs on contact. The VOC delivery systems are configured to give dynamic responses with a sensitivity down to ppb levels for VOCs.

Textile samples were supplied by Courtaulds PLC and consisted of a selection of cotton fabrics with anti-odour treatments included. Control cotton fabrics were also supplied for comparison. Anti-microbial treated fabrics were obtained as part of an industrial contract (confidentiality agreement in place).

Anti-odour and control samples were cut to A5 size swatches and incubated for a minimum of 16 hours at room temeprature (19-21°C) in synthetic sweat mixture (Scensive Technologies Ltd., ST1 Sweat Mix). Samples were removed from the ST1 atmosphere and aired for 1 minute (polyester) or 5 minutes (cotton) before sealing in a plastic sachet and inflating with 300ml of clean air. After 30 minutes at room temperature, the samples were sniffed 4 times using a Bloodhound® ST214 e-nose and

CP1137, *Olfaction and Electronic Nose: Proceedings of the 13th International Symposium,* edited by M. Pardo and G. Sberveglieri

the raw data results saved on the PC. Batching and exporting the saved data was done using the control software and the odour factors were calculated using Microsoft Excel. No sophisticated data processing is required for the anti-odour application, just a direct comparison of sensor responses.

Anti-microbial and control samples were cut to give $50cm^2$ size swatches and incubated at 120°C in foil covered glass containers to sterilise the samples. Samples were transferred acseptically into sterile plastic sachets and inoculated with either sterile LB broth or 10^5 cfu's per ml of either *Pseudomonas putida* or *Staphylococcus aureus* (ATCC 6538) in LB broth. Care was taken to ensure only the fabric samples absorbed the liquid (i.e. no broth as free liquid droplets in the sachet). Overnight incubation for 16 hours was carried out at 37°C, when 300ml clean air was introduced. After 30 minutes at room temperature each sample was sniffed 4 times and the data recorded. Data was exported as before, after which prinicipal compnent analysis (PCA) and linear discriminant function analysis (LDA) was carried out using XLStat 2006.

RESULTS

Anti-Odour Testing

The Bloodhound instruments used are sensitive down to ppb levels and can discriminate VOCs in the presence of water vapour. The sensor results for the bacterial VOC hepatanal in water is given in figure 1 down to 100ppb.

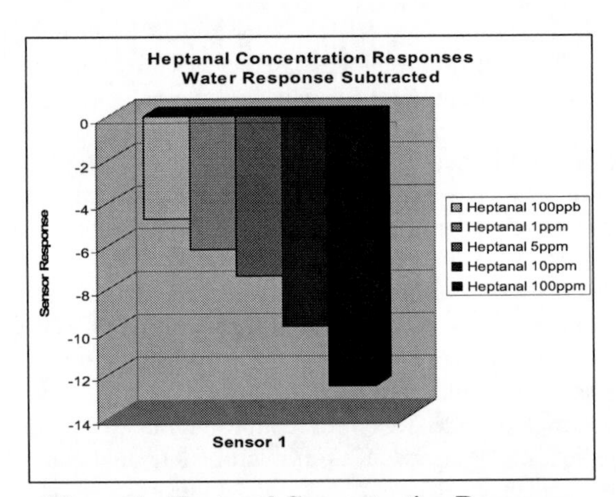

Figure 1. Heptanal Concentration Responses

This concentration effect is the basic principle used to discriminate the anti-odour capabilities of textiles. The sensor responses for a control cotton fabric and an anti-odour cotton fabric can be seen below from the screen of the Bloodhound software, figures 2 and 3.

The odour absorbing capability of the fabrics may be calculated from the divergence of the sensors from the baseline (peak height) and typical results for cotton fabrics are shown below in figure 4.

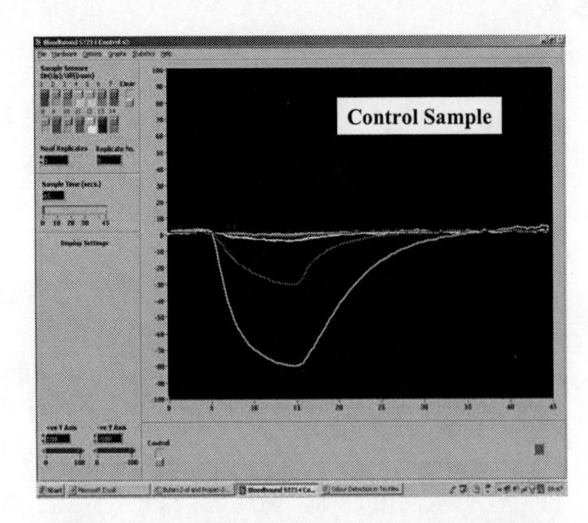

Figure 2. Responses for Control Cotton Fabric

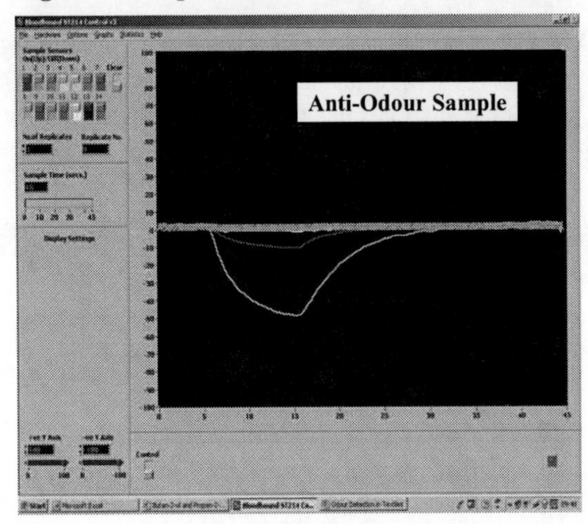

Figure 3. Responses for Anti-Odour Cotton Fabric

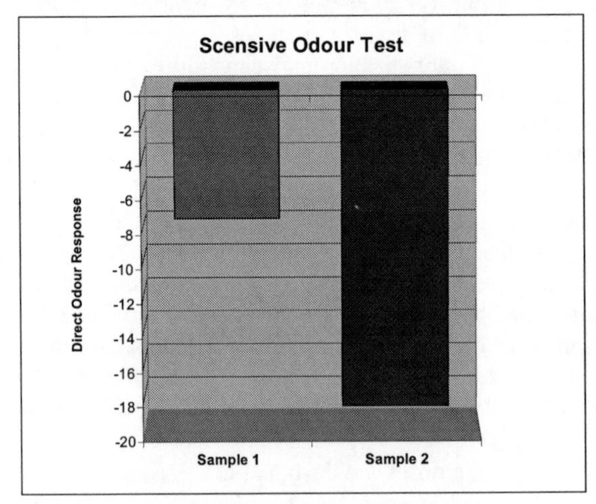

Figure 4. Anti-Odour Properties of Fabrics: Sample 1 Anti-Odour; Sample 2 Cotton Control

From the results in figure 4 the parameter designated the 'Odour Factor' may be calculated. This is a normalised average of the sum of the divergence of all responsive sensors, expressed as a fraction of the control fabric used. It will always be between 0 and 1, with the control fabric always being 1. So if a fabric has an excellent anti-odour capability the Odour Factor will be low. If it does not adsorb odours well then the Odour Factor will be high. Table 1 below shows the Odour Factors for the samples in figure 4.

Fabric	Odour Factor
Cotton Control	1
Anti-Odour Fabric	0.42

Table 1. Odour Factor for Anti-Odour Fabric

Anti-Microbial Testing.

Current anti-microbial testing protocols in the textile industry rely on plate counting of specific strains of micro-organisms after prolonged intimate contact of the micro-organism with the fabric unsder test in a nutrient medium. Any anti-microbial properties of the fabric are then directly measured, as the micro-organisms will not grow if an anti-microbial is present and active, whereas with no anti-microbial activity the micro-organisms proliferate. Extraction of living micro-organisms from the fabrics using a dilute non-ionic detergent mixture then allows serial dilution and plating out for counting the colonies formed after incubation. The test takes a minimum of 2 days and is quite labour intensive. One such test is the AATCC 100 test, which has become an industry standard.

Using the capability of the Bloodhound® instruments to detect bacterial odours, a test to detect the anti-microbial properties of fabrics has been developed. Validation of the test protocol by standard microbiological techniques has also be carried out to show that the odours measured were directly related to the bacterial growth and not to any other extraneous factor present. Results indicated that the odour of bacteria could indeed be used as a measure of their growth in broth cultures (LB broth, 37°C overnight and serially diluted), figure 5.

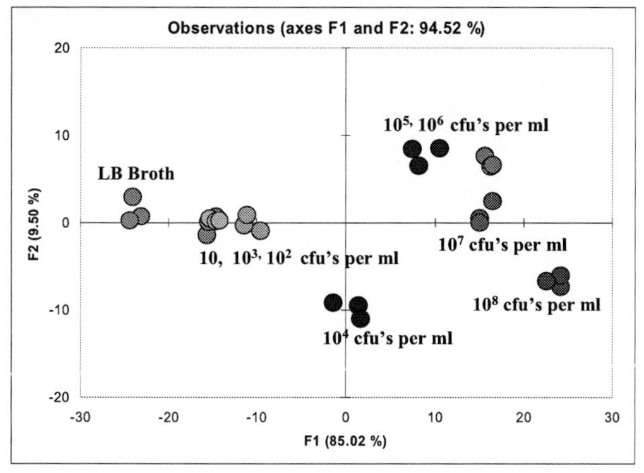

Figure 5. Linear Discriminant Analysis (LDA) of of *Staphylococcus aureus* concentrations in LB Broth

Two fabric swatches were then tested to see if the growth of bacteria on the fabric could be detected after an overnight (16 hour) incubation. Controls using sterile LB broth were also prepared for comparison, as no bacterial growth should occur in the control samples, figure 6.

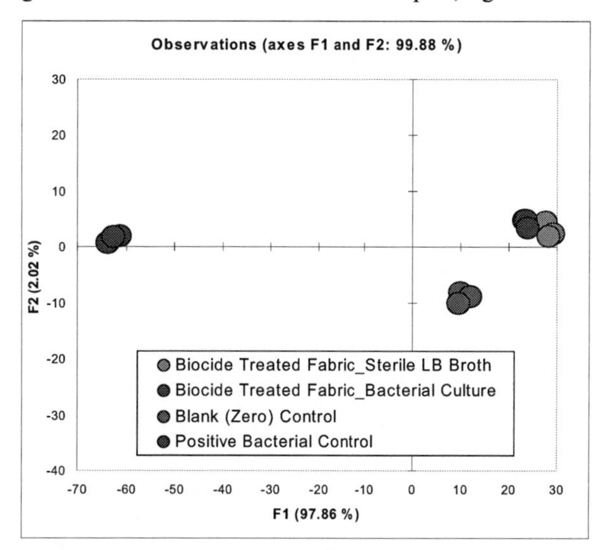

Figure 6. Anti-Bacterial Fabric and Control Fabric (Blank and Positive Controls) Incubated with LB Broth and *Pseudomonas putida* Suspension.

This result indicates clearly that the anti-microbial fabric is preventing growth of the micro-organisms based on their particular odour. The positive control, which allowed growth is well separated from the rest of the samples.

To provide solid proof of the methodology, a second series of tests where both plate counting of the bacteria as per the AATCC 100 test and the new e-nose sniff test were carried out in parallel on the same samples, figures 7 and 8 below.

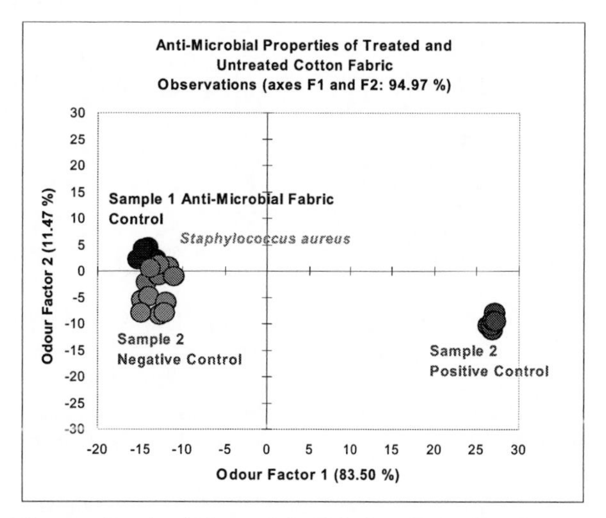

Figure 7. Anti-Bacterial Fabric (Sample 1) and Untreated Cotton Control (Sample 2) Incubated with LB Broth and *Staphyloccocus aureus* Suspension

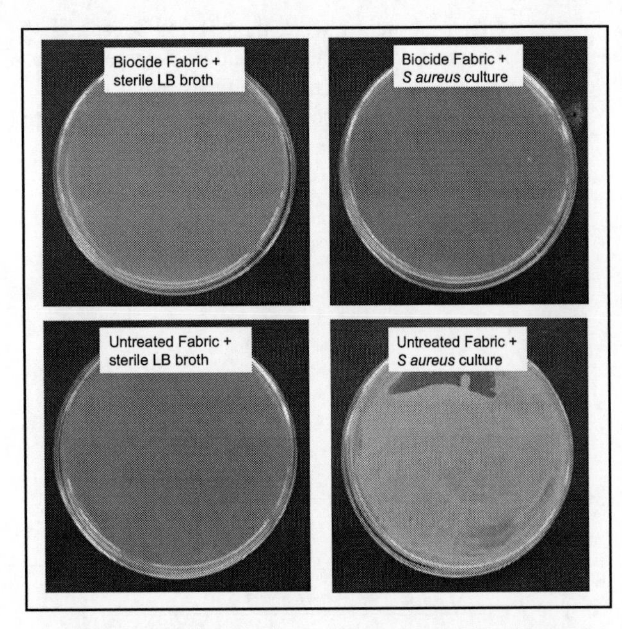

Figure 8. Plate Counts of Samples from Figure 7. Top Left = Sample 1 with LB Broth: Top Right = Sample 1 with *S aureus*: Bottom Left = Sample 2 with LB Broth and Bottom Right = Sample 2 with *S aureus*.

From these results it is very clear that the untreated cotton (positive control) is the only fabric that allows bacterial growth and is also the only sample that clusters a significant distance away from the rest of the other samples. This is positive proof that the bacterial growth is responsible for the distinctive odour that separates the clusters of results.

Using the Mahalanobis Distance as a measure of similarity it is possible to show an 'Odour Reduction Factor' as a measure of the effectiveness of the anti-microbial treatment or finish, table 2.

Sample	MD	ORF
Untreated Cotton Negative Control	0.0	100.0
Untreated Cotton Positive Control	1657.5	0.0
Anti-Microbial Cotton Fabric	116.9	92.9
Anti-Microbial Cotton Fabric	73.2	95.6

Table 2. The Mahalanobis Distance is a Measure of Similarity of Samples. Converting to a Percentage using the Negative and Positive Controls as 100% and 0% Respectively gives an 'Odour Reduction Factor' or ORF, which is the Effective Anti-Microbial Effect.

In the examples above the high MD of the untreated positive control sample where bacterial growth proliferated and the low MD's of the samples under test gave high ORF's upon calculation. A High ORF indicates anti-microbial activity is high and effective. Also the odour present is low and very similar to the negative control.

CONCLUSIONS

Two industrially relevant tests for technical textile functionality have been developed.

Anti-odour testing is now available to the textiles industry and is being used to check the anti-odour claims of manufacturers by various companies including Courtaulds PLC and Marks & Spencers PLC.

Anti-Microbial test capabilities have recently been developed and proven by bacterial plate counts. These are now being marketed as a rapid route for anti-microbial testing for the textiles industry. Discussions are taking place with a number of companies including world leading biocide producers.

Acknowledgments

Many thanks to Courtaulds PLC for supply of fabric samples and advice. Thanks also to Dr Paul Millner and the Biosensors and Biocatalysis Group at the University of Leeds, UK who allowed us to use the microbiological facilities to carry out the anti-microbial testing. Thanks also to Dr Malcolm Greenhalgh who assisted in the areas of biocidal materials and the legal position of product claims in the textiles industry.

References

1. Official Journal of the European Communities DIRECTIVE 98/8/EC OF THE EUROPEAN PARLIAMENT AND OF THE COUNCIL 16 February 1998 concerning the placing of biocidal products on the market

2. Gibson T D, Prosser O, Hulbert J N, Marshall R W, Corcoran P, Lowery P, Ruck-Keene E A and Heron S. Detection and Simultaneous Identification of Micro-organisms from Headspace Samples using an Electronic Nose. Sensors and Actuators **B44** (1997) 413-422.

3. Heron ST and Gibson TD. Rapid differentiation of microbial cultures in liquid media using an electronic nose. Eurosensors XII, Vols 1 and 2 (1998) Sensors Series 813 – 816.

4.Monitoring urinary tract infections and bacterial vaginosis. Krishna C. Persaud, Anna Maria Pisanelli, Phillip Evans and Paul J. Travers. Sensors and Actuators **B116** (2006) 116–120

Use of Sequential Injection Analysis to construct a Potentiometric Electronic Tongue: Application to the Multidetermination of Heavy Metals

Aitor Mimendia[a], Andrey Legin[b], Arben Merkoçi[a] and Manel del Valle[a,*]

[a] *Sensors and Biosensors Group, Chemistry Dept., Universitat Autònoma de Barcelona, Edifici Cn, 08193 Bellaterra, Barcelona, SPAIN*
[b] *Chemistry Dept, St. Petersburg University, Universitetskaya nab. 7/9, 199034 St. Petersburg, RUSSIA*
e-mail: manel.delvalle@uab.es

Abstract. An automated potentiometric electronic tongue (ET) was developed for the quantitative determination of heavy metal mixtures. The Sequential Injection Analysis (SIA) technique was used in order to automate the obtaining of input data, and the combined response was modeled by means of Artificial Neural Networks (ANN). The sensor array was formed by four sensors: two based on chalcogenide glasses Cd sensor and Cu sensor, and the rest on poly(vinyl chloride) membranes Pb sensor and Zn sensor. The Ion Selective Electrode (ISE) sensors were first characterized with respect to one and two analytes, by means of high-dimensionality calibrations, thanks to the use of the automated flow system; this characterization enabled an interference study of great practical utility. To take profit of the dynamic nature of the sensor's response, the kinetic profile of each sensor was compacted by Fast Fourier Transform (FFT) and the extracted coefficients were used as inputs for the ANN in the multidetermination applications. In order to identify the ANN which provided the best model of the electrode responses, some of the network parameters were optimized. Finally analyses were performed employing synthetic samples and water samples of the river Ebro; obtained results were compared with reference methods.

Keywords: Heavy metals, Artificial Neural Network, Potentiometric sensors, Sequential Injection Analyses (SIA), Automated Electronic Tongue.
PACS: 07.88.+y, 82.47.Rs, 92.40.Kc, 07.05.Mh, 43.60.Hj .

INTRODUCTION

Nowadays the uses and applications of heavy metals are quite extensive; many of the tools and technologies we use every day are based on them. At the same time, the risks that their use entails for the environment and health are well known. For this reason, there is a need to monitor their presence in the environment in order to prevent situations of risk and act if it happens.

We propose a technique for monitoring heavy metals, cost-effective, fast, capable of measuring several metals at the same time and easily adaptable to perform on-line analysis. Our research group has experience in developing Electronic Tongues (ETs) employing arrays of Ion Selective Electrodes (ISEs) in all-solid-state configuration and using their steady state potential [1]; in this work we employ a set of ISEs developed by St. Petersburg University, based on chalcogenide glasses and PVC matrices[2,3] for a quite novel application, the multidetermination of heavy metals, and incorporating transient recording as signal information.

Normally, our electronic tongues, devised for quantitative multideterminations, use Artificial Neural Networks (ANNs) as the data processing tool. For this purpose, a large amount of training samples are needed to generate the response model; therefore techniques for automated preparation and/or measurement of samples, like the flow techniques [1,4], have been proposed to be used with ETs. Besides saving time, a richer signal can be acquired when the sensors are used in flow conditions, i.e. the transient response to a pulse profile of sample arriving to the sensor array. This signal can be of high information content, of primary ions and also interfering ions, which can be better discriminated thanks to the kinetic resolution added. Given the neural network could not handle the complete kinetic profiles of the sensors used, we tried a reduced number of their frequency

CP1137, *Olfaction and Electronic Nose: Proceedings of the 13th International Symposium*, edited by M. Pardo and G. Sberveglieri

components, previously extracted using a Fast Fourier Transform (FFT). This method has already been used successfully in previous works [4].

EXPERIMENTAL AND METHODS

The automated ET used a potentiometric sensor array formed by four sensors, two of them are based on chalcogenide glasses (Cu-ISE and Cd-ISE) and the rest on PVC membranes (Pb-ISE and Zn-ISE) [3].

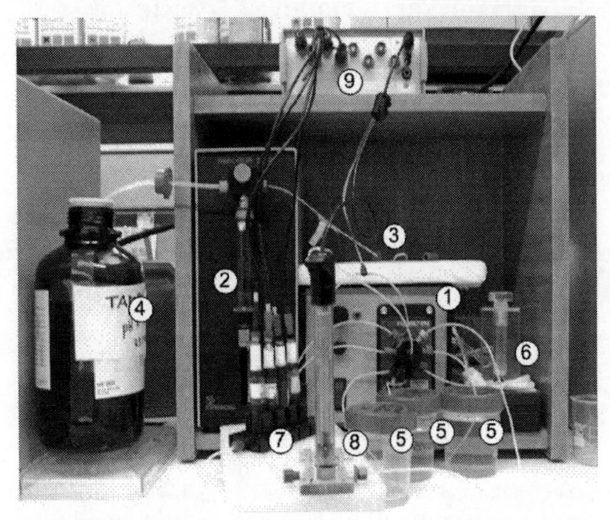

FIGURE 1: Sequential injection analysis (SIA) flow system used in this work. 1, 8-way valve; 2, Automatic Microburette; 3, Holding coil; 4, Buffer reservoir; 5, Stock Solutions of metals; 6, Mixing cell; 7, ISE Array; 8, Reference Electrode; 9, Multipotentiometer.

The developed Sequential Injection Analysis (SIA) system, shown in Figure 1, had two clearly different parts: the first part was the fluid system and consisted of an automatic microburette, a holding coil, a 8-way Hamilton MVP valve and a Perspex mixing cell with a magnetic stirrer. The second part was the measurement system that comprised the sensor array, a reference electrode and an 8-channel signal conditioning circuit connected to the National Instruments NI6221 Multifunction DAQ analog inputs. The whole system was controlled by a PC using a virtual instrument developed in Labview [5].

A set of 81 samples was prepared automatically by the SIA system, obtaining the corresponding transient pulse response for each ISE after a step-introduction of 1.6mL of sample. The studied ranges of concentrations were: for Cd^{2+}, 0-3.2ppm; for Cu^{2+}, 0-8.5ppb; and for Pb^{2+}, 0-3ppm, adapting them to the lower response range of the used sensors.

Given the Zn-ISE was not incorporated until the final stage of the study, the training process did not include this species. Despite that, the Zn-ISE was used as a generic source of information for the other metal ions.

The preprocessing of the data (MATLAB) was based on obtaining the Fourier coefficients of each transient and entering this as the input information in an ANN model. The whole configuration was then optimized, which entailed the setting of the number of Fourier coefficients to use, number of neurons in the hidden layer and the transfer functions in the hidden and output layers. Multiple ANN architectures and topologies were assayed employing Bayesian regularization algorithms.

Comparison graphs of predicted vs. expected concentrations for the three determined metals were built to check the prediction ability of the ANN. After this step the best configuration was chosen taking into account the slope, intercept and correlation coefficient (ideal values equal to 1, 0 and 1, respectively).

RESULTS

The Ion Selective Electrode (ISE) sensors were first characterized in their response; after this, the response model employing ANNs and kinetic resolution was built to perform multidetermination application.

Automated Sensor Flow Characterization

Sensors were first characterized in flow conditions with respect to one and two analytes, by means of high-dimensionality calibrations, thanks to the use of the automated flow system; this characterization enabled an interference study of great practical utility. In order to achieve 3D response surfaces of the electrodes and obtain the values of the potentiometric selectivity coefficients, the SIA system was used to prepare five measurement sequences, varying the concentration of both the primary ion and the interferent. In each sequence the concentration of the interferent ion was maintained constant but different to its magnitude in the other sequences.

Figure 2 illustrates the experimental points prepared automatically with the SIA system, as well as the 3D response surface which corresponds to the Nikolskii–Eisenman expression, fitted with non-linear regression methods [6]. It can be clearly seen how, with low concentrations of the primary ion, the curvature gets more pronounced as the concentration of the interfering ion increases. In the figure, sensors with good selectivity, as the Cu^{2+} sensor, showed a distinct feature, with almost linear plane shape; at the same time, sensors with a marked cross-response, the Pb^{2+} and Cd^{2+} sensors can be also identified, given their response is increased by both metals considered, primary ion and interferent.

A much more realistic meaning can then be given to the potentiometric selectivity coefficients obtained from modelling the surfaces compared with what would be obtained through classical methods (fixed interference method) Selectivity coefficients, $K_{X,Y}^{pot}$, are normally evaluated by means of the emf (electromotive force) response of the ISE in mixed solutions of the primary ion, X, and interfering ion, Y (fixed interference method), but in our case, both species, primary ion and interferent vary in concentration during the study.

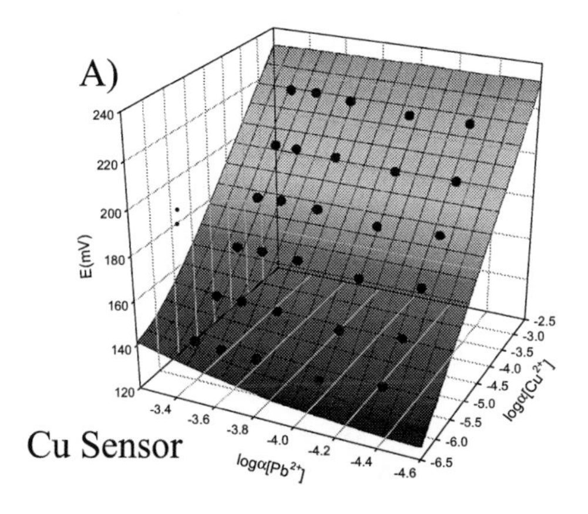

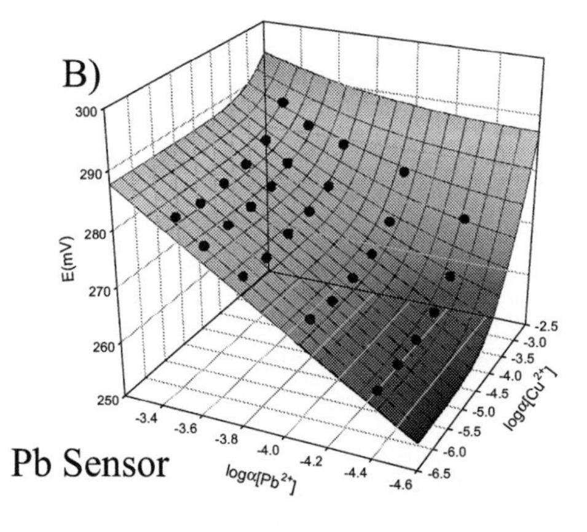

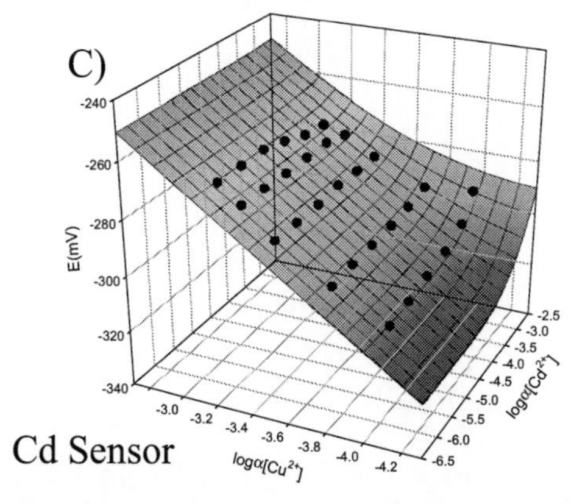

FIGURE 2: 3D plots corresponding to selectivity experiments of three of the sensors employed in the array, showing their cross response features.

TABLE 1. Potentiometric selectivity coefficient values obtained in the study.

Sensor and (Primary ion)	Interfering Ion	log $K^{pot}_{X,Y}$
Cd^{2+}	Cu^{2+}	1.4±0.1
Cd^{2+}	Pb^{2+}	1.2±0.1
Cu^{2+}	Cd^{2+}	-2.9±0.1
Cu^{2+}	Pb^{2+}	-2.8±0.1
Pb^{2+}	Cd^{2+}	-3.2±0.2
Pb^{2+}	Cu^{2+}	-2.8±0.4

Confidence intervals calculated at the 95% confidence level.

By examining the selectivity results, it can be said that with smaller values of $K^{pot}_{X,Y}$, potentiometric sensors demonstrate preference for the primary ion, X. Table 1 sumarizes the obtained potentiometric selectivity coefficients: while Cu^{2+} and Pb^{2+} sensors offer $K^{pot}_{X,Y}$ values lower than 1, Cd^{2+} sensor coefficients are greater. In conclusion the Cd^{2+} sensor is more selective to Pb^{2+} and Cu^{2+} than to its supposedly primary ion Cd^{2+}.

Application to the Multidetermination of Heavy Metals

The automated SIA system permitted the automated generation of the samples needed to build the response model; a total amount of 81 samples were used. The set of samples was divided into two subsets: a training subset, with 50 samples, this served to establish the response model; and a test subset, with 31 samples, which served to evaluate the model's predictive ability. Samples were selected randomly with the only precaution that samples with maximums and minimums had to be in the training subset, in order to avoid any extrapolation.

After an extensive study with varying configurations, a final model was optimized having 10 Fourier coefficients per channel, 8 neurons in the hidden layer, and *tansig* and *purelin* transfer functions in the hidden and output layers, respectively [8].

The results obtained for the external test subset, those samples not participating in training, are presented in Figure 3, where the calculated (obtained) concentration values are compared to the expected ones. The comparison slopes and intercepts, as summarized in Tables 2A-2C, are satisfactory, with slopes and intercepts nearly 1 and 0 respectively for the three studied cations and the different subsets of samples. The correlation coefficients between obtained and expected concentrations were also satisfactory, very close to 1.

Some dispersion was observed for Cd^{2+} determination, especially at the lower concentration levels. The reason may be attributable to the strong interfering effect that Pb^{2+} and Cu^{2+} ions produced to the Cd^{2+} sensor, as already determined in the sensor characterization step.

But thanks to the use of the kinetic dimension in the response, the determination of Cd^{2+} at the submicromolar level was possible, something not attainable with steady-state signals alone. In order to improve the results and reduce the dispersion observed, a more selective Cd^{2+} sensor would be needed. Another option could be to enlarge the sensor array, incorporating sensors from different nature or replicating the sensors in order to increase the available information.

To end the study, a quantitative determination of heavy metals in river samples was attempted, but the results are still deviated from ideal values probably due to matrix effects which are still being examined.

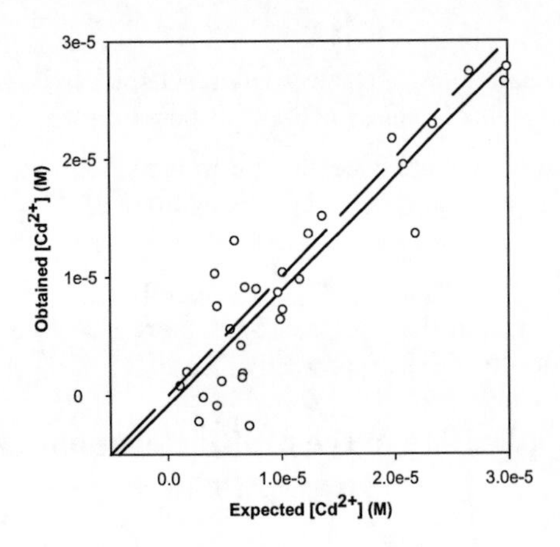

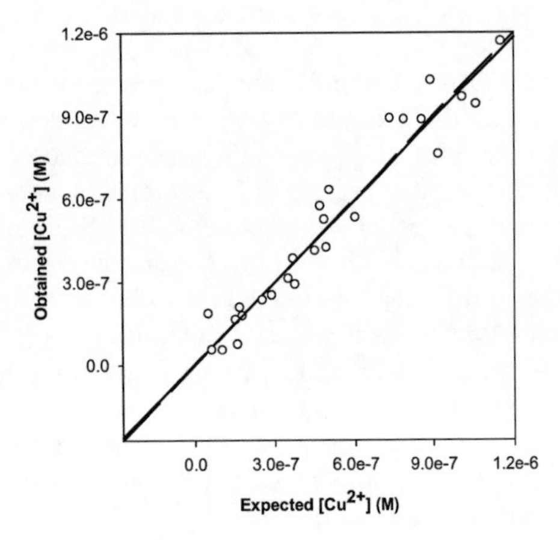

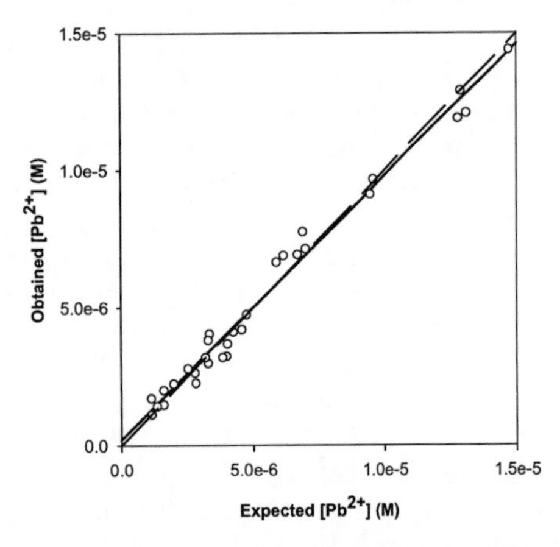

FIGURE 3: Comparison of obtained vs. expected results provided by the proposed ET for the test samples (those not participating in training) for Cd2+ , Cu2+, and for Pb2+ ions.

TABLE 2A. Fitted regression lines of the comparison between obtained vs. expected results provided by the proposed ET for the training subsets of samples and the three considered metal ions (intervals calculated at the 95% confidence level).

Training subset			
Ion	**Correlation**	**Slope**	**Intercept (μM)**
Cd^{2+}	0.999	0.99±0.01	0.10 ± 1.6
Cu^{2+}	0.999	1.0 ±0.01	0.002 ± 0.009
Pb^{2+}	0.999	1.0 ±0.01	0.016 ± 0.074

TABLE 2B. Fitted regression lines of the comparison between obtained vs. expected results provided by the proposed ET for the external test subset of samples and the three considered metal ions (intervals calculated at the 95% confidence level).

External test subset			
Ion	**Correlation**	**Slope**	**Intercept (μM)**
Cd^{2+}	0.91	1.02 ± 0.18	-0.8 ± 2.5
Cu^{2+}	0.972	0.959 ± 0.14	0.005 ± 0.008
Pb^{2+}	0.991	0.969 ± 0.05	0.2 ± 0.5

TABLE 2C. Fitted regression lines of the comparison between obtained vs. expected results provided by the proposed ET for a manually prepared set of synthetic samples and the three considered metal ions (intervals calculated at the 95% confidence level).

Manually prepared, synthetic samples			
Ion	**Correlation**	**Slope**	**Intercept (μM)**
Cd^{2+}	0.743	1.14±1.18	0.13 ± 8.5
Cu^{2+}	0.92	0.964 ±0.47	0.01 ± 0.22
Pb^{2+}	0.974	1.16 ±0.31	0.2 ± 1.1

CONCLUSIONS

The proposed ET achieves the goal of obtaining a tool to carry out quantitative determinations of heavy metals at low concentration levels. The kinetic resolution, as extracted using the FFT transform, permitted to obtain a response model, not achievable with steady-state information only. The ET approach for heavy metal detection is economic, versatile and can be easily adapted to perform on-line analysis. Further improvements of obtained results will involve the enlargement of the sensor array and the more in-deep characterization of matrix effects observed with the river waters tested.

ACKNOWLEDGEMENTS

This work was supported by European Community project FP6-IST No. 034472, "WARMER: Water risk management in Europe".

REFERENCES

1. M. Cortina et al., *Talanta* **66** 1197-1206 (2005).
2. A. Legin et al., *Sensors and Actuators B* **24-25** 309-311. (1995).
3. A. Rudnitskaya et al., *Microchimica Acta* **163** 71-80 (2008).
4. D. Calvo et al., *Analytica Chimica Acta*, **600** 97-104 (2007).
5. A. Durán et al., *Sensors* **6** 19-29 (2006).
6. Sigmaplot 2000, SPSS Inc., Chicago, IL, (2000).
7. G. Guilbault et al., *Pure Appl. Chem* **48** 127-132 (1976).
8. M.del Valle, "Potentiometric electronic tongues applied in ion multidetermination" in *Electrochemical Sensor Analysis*, edited by S.Alegret, A.Merkoçi, Elsevier, Amsterdam, 2007.

ANALYTICAL CHEMISTRY APPROACHES

Why do Ladybugs Smell Bad?
In-vivo Quantification of Odorous Insect Kairomones with SPME and Multidimensional GC-MS-Olfactometry

Lingshuang Cai,[1] Jacek A. Koziel,*[1] Matthew E. O'Neal[2]

[1]*Department of Agricultural & Biosystems Engineering, Iowa State University, Ames, IA 50011, USA; phone 515-294-4206; fax 515-294-4250; e-mail: koziel@iastate.edu*
[2]*Department of Entomology, Iowa State University, Ames, IA 50011, USA*

Abstract. Winemakers, small fruit growers, and homeowners are concerned with noxious compounds released by multicolored Asian ladybird beetles (*Harmonia axyridis*, Coleoptera: Coccinellidae). New method based on headspace solid phase microextraction (HS-SPME) coupled with multidimensional gas chromatography mass spectrometry – olfactometry (MDGC-MS-O) system was developed for extraction, isolation and simultaneous identification of compounds responsible for the characteristic odor of live *H. axyridis*. Four methoxypyrazines (MPs) were identified in headspace volatiles of live *H. axyridis* as those responsible for the characteristic odor: 2,5-dimethyl-3-methoxypyrazine (DMMP), 2-isopropyl-3-methoxypyrazine (IPMP), 2-sec-butyl-3-methoxypyrazine (SBMP), and 2-isobutyl-3-methoxypyrazine (IBMP). To the best of our knowledge this is the first report of *H. axyridis* releasing DMMP and the first report of this compound being a component of the *H. axyridis* characteristic odor. Quantification of three MPs (IPMP, SBMP and IBMP) emitted from live *H. axyridis* were performed using external calibration with HS-SPME and direct injections. A linear relationship (R^2>0.9958 for all 3 MPs) between MS response and concentration of standard was observed over a concentration range from 0.1 ng L^{-1} to 0.05 μg L^{-1} for HS-SPME-GC-MS. The method detection limits (MDL) based on multidimensional GC-MS approach for three MPs were estimated to be between 0.020 ng L^{-1} to 0.022 ng L^{-1}. This methodology is applicable for *in vivo* determination of odor-causing chemicals associated with emissions of volatiles from insects.

Keywords: *Harmonia axyridis*, SPME, Multidimensional GC–Olfactometry, Odor, Methoxypyrazines
PACS: 01.30.Cc

INTRODUCTION

The recent invasion and establishment of *Harmonia axyridis* (Coleoptera: Coccinellidae) in North America has resulted in a pest on several fronts. Extension entomologists have received numerous complaints from urban and rural homeowners complaining of larger numbers of adult *H. axyridis* gathering in windows and attics.[1] When disturbed the defensive response of adult *H. axyridis* includes reflexive bleeding and the release of noxious compounds. These compounds include but are not limited to MP.[2] MPs are very potent odorants and have a distinctive smell, similar to freshly cut green bell pepper or green peas. The human olfactory thresholds for MPs are extremely low, in the level of 2 ng L^{-1} in water.[3]

The larvae and adults are primarily predators and have been considered a significant source of biological control for another invasive pest, the soybean aphid, *Aphis glycines* (Hemiptera: Aphididae). The impact of this feeding by adult *H. axyridis* as a significant source of yield loss is not clear. A greater threat may be a loss in fruit quality, especially grapes, when harvested fruit is contaminated with adult *H. axiridis*. When processed into wine, MPs released from lady beetles have been identified as a fouling agent.[4] Allen et al. (1998) reported lower odor detection thresholds in white wine compared with red wine.[4] Pickering et al. found *H. axyridis* released MPs, particularly IPMP was the agent responsible for the wine taint.[5]

The concentration of MPs released by lady beetles (Coccinellids) is in the order of pg/beetle [6] and ng L^{-1} in wine.[5] Therefore it is necessary to develop highly sensitive extraction and analysis methods for qualitative and quantitative purpose at such low levels.

In this research, headspace (HS) SPME was used for extraction of volatiles released by live *H. axyridis*. This approach combines rapid sampling and sample preparation, olfactometry and multidimensional GC separation with conventional MS detector. The objective of this study was to (1) confirm if MPs are the sole source of noxious odors from *H. axyridis* using a novel approach - multidimensional GC coupled with olfactometry and to (2) determine the amounts of those characteristic odorants emitted from live *H. axyridis*.

CP1137, *Olfaction and Electronic Nose: Proceedings of the 13th International Symposium,* edited by M. Pardo and G. Sberveglieri

EXPERIMENTAL AND METHODS

Standards and Solutions

The three standards (IPMP, SBMP and IBMP) were used for quantification of the amount of MPs emitted from live beetles. An individual standard solution of 1 mg mL^{-1} of each MP was prepared in methanol. The external calibration standard solutions ranged from 0.1 ng L^{-1} to 0.05 µg L^{-1}.

Isolation of Characteristic Odorants with Multidimensional GC-MS-O

Multidimensional GC-MS-olfactometry (MDGC-O) system (Microanalytics, Round Rock, TX, USA) built on a 6890N GC / 5973 MS platform (Agilent Inc., Wilmington, DE, USA) were used for all analyses. The system was equipped with two columns in series connected by a Dean's switch. The non-polar pre-column was 12 m, 0.53 mm i.d.; film thickness, 1 µm with 5% phenyl methylpolysiloxane stationary phase (SGE BP5) and operated with constant pressure mode at 8.5 psi. The polar analytical column was a 30 m × 0.53 mm column coated with poly (ethylene glycol) (WAX; SGE BP20) at a film thickness of 1 µm. The column pressure was constant at 5.8 psi. Both columns were connected in series. System automation and data acquisition software were MultiTrax™ V. 6.00 and AromaTrax™ V. 6.63 (Microanalytics, Round Rock, TX, USA) and ChemStation™ (Agilent, Santa Clara, CA, USA). The general run parameters used were as follows: injector, 260 °C; FID, 280 °C, column, 40 °C initial, 3 min hold, 7 °C min^{-1}, 220 °C final, 10 min hold; carrier gas, GC-grade helium. Mass to charge ratio (m/z) range was set between 33 and 280. Spectra were collected at 6 scans sec^{-1} and electron multiplier voltage was set to 1400 V. The detection of trace three MPs was carried out using selected ion monitoring. m/z =137, 138 and 124 were used for quantification for IPMP, SBMP and IBMP, respectively. The MS detector was auto-tuned every day.

Sensory evaluations were made through the sniff port equipped with two capillary columns. The temperature for the sniff port capillaries was set to 220 °C to eliminate condensation. In addition, humidified air (Certified breathing air grade, 99.995% purity, Praxair, Inc., Danbury, CT, USA) was constantly delivered to the sniff port at 8.0 psi. This was done to maintain a constant humidity level for the panelists' mucous nasal membranes. The tip of the sniff port was equipped with a glass nose cone (SGE, Austin, TX, USA). Three trained panelists analyzed headspace volatiles from live *H. axyridis*. Panelist responses were compared based on odor character and odor intensity associated with separated compounds.

InVivo Headspace SPME of Compounds Released by Live H. Axyridis

H. axyridis were collected in Ames, Iowa in September, 2005, February and August 2006. Multiple sets of randomly-selected five live *H. axyridis* were then placed in screw-capped vials (40 mL, Supelco, Bellefonte, PA, USA) sealed with a PTFE-lined silicone septum and used for in vivo HS-SPME. Each vial with beetles was allowed to equilibrate for 24 h before HS-SPME at 30 °C. Headspace samples from life beetles only were considered for analyses, i.e., if the beetles died during sampling, the samples were discarded.

RESULTS AND DISCUSSION

Identification of Methoxypyrazines Released by Live H. Axyridis

According to previous studies, it is well known that pyrazines are secreted by lady beetles.[2-5] In this study, the four characteristic odors closely resembling the entire headspace of live beetles were identified as DMMP, IPMP, SBMP and IBMP. In order to identify the characteristic odors from live *H. axyridis*, three panelists analyzed headspace volatiles released by live *H. axyridis* through sniff port. The panelists were consistent identifying four 'characteristic' odors, and also describing them as 'moldy', earthy', 'green bell pepper', 'potato', 'peanut', 'nutty' that resulted from four MPs emitted from the headspace of live *H. axyridis*. The average odor intensity of four MPs for three panelists was 58% for DMMP, 71% for IPMP, 36% for SBMP and 59% for IBMP, respectively. The odor intensity of IPMP was the highest among other MPs. The reproducibility of the odor intensity of three panelists expressed as RSD were 19% for DMMP, 1% for IPMP, 15% for SBMP and 17% for IBMP, respectively.

SBMP was positively identified by matching the retention time of standard compound, matching mass spectrum of unknown compound with BenchTop/PBM and by matching the odor character, i.e. 'bell pepper', 'peanut', and 'potato'. One compound was consistently tagged by all panelists with the characteristic odor, i.e., 'roasted peanuts' and later tentatively identified as DMMP by the mass spectrum match greater than 90 % with BenchTop/PBM library. Seifert et al. reported 'roasted peanut' aroma and tentatively associated it with methyl MPs without specifically pointing to DMMP.[3] The release of IPMP, SBMP, and IBMP from dead beetles has been reported in previous studies.[2] However, we are not aware of any previous report of DMMP released by *H. axyridis*. Because pure DMMP is not commercially available, it could not be confirmed with a standard at this time. However, based on this preliminary chemical and sensory identification, it is important to consider DMMP as another important, fouling odor compounds that is emitted by live *H. axiridis*.

Previous studies suggested that IPMP is the most important component of *H. axyridis* 's aroma. Cudjoe et al. found IPMP was the most abundant MPs released by dead *Coccinella septempunctata*, *Harmonia axyridis* and *Hippodemia convergens* lady beetles (Coccinellidae).[2] Pickering et al. reported IPMP was detected at relatively high concentration and at levels above sensory threshold in grape juice used for wine fermentation and contaminated with live *H. axyridis*.[5] Pickering et al. also found that IPMP is responsible for the distinctive sensory characteristics of *H. axyridis* contaminated wines and found significant positive correlations between IPMP concentration and specific aroma attributes in wines.[5]

Multidimensional GC-MS-O

Odor and chemical separation of IPMP and other MPs from a complex matrix of insect volatiles can be challenging even with extended GC runs and other chromatographic tools. This makes it difficult to evaluate their odor impacts when analyzing the entire sample in a GC-MS-O mode. Thus, multidimensional GC-MS-O was used to (a) improve the isolation and separation of IPMP and other MPs from interferences, (b) to improve identification in the complex matrix, and (c) to separate and evaluate their odor impact. The dual-column GC system equipped a 'heart-cut' valve can divert (and isolate) a specific retention region with compounds and aroma of interest from the pre-column (non-polar) to the analytical column (polar) to enhance resolution and to minimize the interferences from coeluting compounds and aromas.

The instrument was first set to GC-FID-O mode with no heart-cut by utilizing the sniff port to identify specific GC pre-column retention times for which eluants exhibit characteristic odor. The description of odor released by *H. axyridis* that is often described as 'green bell pepper', 'roasted peanuts' or 'green peas'. Based on samples analyzed in GC-FID-O mode, the specific GC pre-column retention times associated with characteristic odors were then selected for activating the multidimensional GC-MS-O mode with the Dean's switch. At first, only three characteristic odors were identified by panelists in the GC-FID-O mode. Due to limited separation capacity of pre-column resulting in two of the MPs coeluting, i.e. SBMP and IBMP, the odor events were merged. When the pre-column heart-cut times were set from 9.00 to 13.00 min and a second replicate was analyzed, only heart-cuts (small segments) of chromatographic effluent were further separated on analytical column and analyzed simultaneously by the MS detector and a panelist at the sniff port. Resulting total ion chromatogram, FID chromatogram and aromagram of heart-cut effluent in MDGC-MS-O mode of volatiles released by *H. axyridis* is shown in Figure 1. As can be seen in Figure 1, the separation of IPMP and 2-ethyl-1-hexanol was much improved even though it was not a baseline separation.

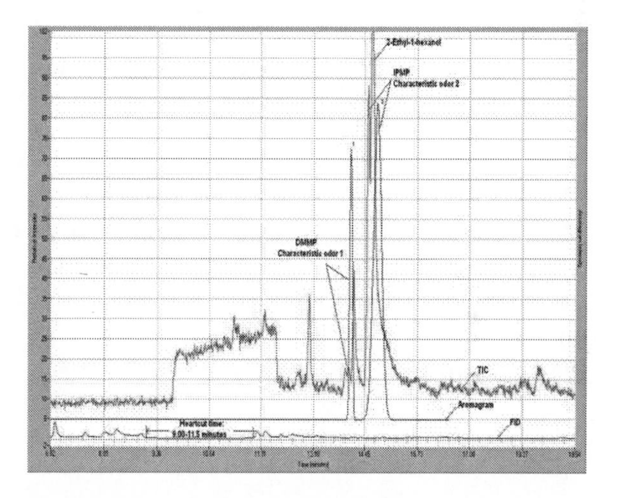

FIGHURE 1. Separations for MPs from the headspace of 5 live *H. axyridis* in MDGC-MS-O mode with heart-cut between pre-column and analytical column: comparison of the FID chromatogram, total ion chromatogram and aromagram isolating only characteristic odorants. Narrower heart-cut time range: 9.00-11.50 min was used to isolate aromas caused by IPMP and 2-ethyl-1-hexanol.

Estimation of IPMP, SBMP, and IBMP Releases Per Beetle Mass and Per Beetle

The MDGC-MS-O approach was used to quantify MPs released to headspace using SPME and *in vivo* sampling. The estimated amounts of three MPs emitted from live *H. axyridis* are presented in Table 1. The average amounts of three MPs per beetle mass (for n = 8 replicates for red beetles, with each replicate comprised of five beetles in a 40 mL vial) were 8.0569 ng g^{-1} for IPMP, 3.1680 ng g^{-1} for SBMP and 0.0811 ng g^{-1} for IBMP, respectively. The average amounts of three MPs per beetle mass (for n = 2 replicates for orange beetles, with each replicates comprised of five orange beetles in a 40 mL vial) were 0.4111 ng g^{-1} for IPMP, 0.6191 ng g^{-1} for SBMP and 0.0055 ng g^{-1} for IBMP, respectively. For pooled red and orange beetles, the average were 4.2340 ng g^{-1} for IPMP, 1.8965 ng g^{-1} for SBMP and 0.0091 ng g^{-1} for IBMP, respectively.

TABLE 1. Estimated amounts (ng g^{-1} and ng per beetle) of three methoxypyrazines emitted to vial headspace from live *H. axiridis* and detected by HS-SPME-MDGC-MS.

	Red beetles (n = 8)		**Orange beetles** (n = 2)		**All beetles** (n = 50)
	Mean (ng g^{-1})	**RSD** (%)	**Mean** (ng g^{-1})	**RSD** (%)	**Mean** (ng g^{-1})
IPMP	8.06	117	0.411	19.7	4.2340
SBMP	3.17	82.5	0.619	62.4	1.8965
IBMP	0.01	42.9	0.0065	50.7	0.0091
	Mean (ng)	**RSD** (%)	**Mean** (ng)	**RSD** (%)	**Mean** (ng)
IPMP	0.308	128.9	0.014	23.7	0.1612
SBMP	0.112	79.5	0.025	69.9	0.0688
IBMP	0.0006	51.6	0.0002	47.1	0.0004

CONCLUSIONS

In vivo HS-SPME combined with multidimensional GC-MS-O has a great potential for investigations of links between specific chemicals released by insects and the characteristic odors. In this research, 50/30 μm DVB/Carboxen/PDMS SPME fiber was used to extract headspace volatiles released by live *H. axyridis*. Thirty eight compounds were identified in headspace of live *H. axyridis* including four characteristic odorous compounds-DMMP, IPMP, SBMP and IBMP. We detected a previously unidentified MP (DMMP) that appears to be also a component of *H. axyridis*'s odor. We also provided the first evidence that IPMP is released to air and is also responsible for the characteristic odor of live *H. axyridis*. Quantification of three MPs, i.e., IPMP, SBMP and IBMP, emitted from live beetles was performed using external calibration curves by HS-SPME-MDGC-MS. Linear relationships (R^2 was > 0.9958 for all 3 MPs) was observed over a concentration range from 0.1 ng L^{-1} to 0.05 μg L^{-1}. The MDLs were estimated at 0.022 ng L^{-1}, 0.020 ng L^{-1}, 0.022 ng L^{-1} for IPMP, SBMP, and IBMP, respectively. These MDLs obtained with multidimensional GC-MS approach represent 52.2%, 52.4%, and 38.9% improvement compared to GC-MS approach. For the 0.1 ng L^{-1} concentration, the intra- and inter-day precision for the three MPs were less than 3.9 and 7.8 %. The HS-SPME-MDGC-MS method was applied to determine the amounts of three MPs emitted to headspace from live *H. axyridis* per beetle body mass and the average amounts of MPs were 4.2340 ng g^{-1} for IPMP, 1.8965 ng g^{-1} for SBMP and 0.0091 ng g^{-1} for IBMP, respectively.

ACKNOWLEDGMENTS

The authors would like thank Iowa State University for funding this study. Full description of this work is published as reference #7.

REFERENCES

1. M. Huelsman and J. Kovach, American Entomologist, *American Entomologist*, **50**, 163–164 (2004).
2. E. Cudjoe, T.B. Wiederkehr, and I.D. Brindle, *Analyst*, **130**, 152-155 (2005).
3. R.M. Seifert, R.G. Buttery, D.G. Guadagni, D.R. Black, and J.G. Harris, *J. Agric. Food Chem.*, **18**, 246-249 (1970).
4. M.S. Allen *et al*, *Chemistry of wine flavor*, New York: American Chemical Society, 1998, pp. 31–38.
5. G. Pickering, Y. Lin, A. Reynolds, G. Soleas, R. Riesen, and I. Brinndle, *Journal of Food Science*. **70**, 128-135 (2005).
6. J.R. Aldrich, W.S. Leal, R. Nishida, A.P. Khrimian, C.J. Lee, and Y. Sakuratani, *Entomologia Experimentalis et Applicata*, **84**, 127–135 (1997).
7. L. Cai, J.A. Koziel, and M.E. O'Neal, *J. Chromatogr. A*, **1147**, 66-78 (2007).

Carbon Nanotube Stationary Phase in a Microfabricated Column for High-Performance Gas Chromatography

Takashi Nakai*, Jun Okawa, Shuji Takada, Masaki Shuzo*, Junichiro Shiomi, Jean-Jacques Delaunay*, Shigeo Maruyama and Ichiro Yamada*

School of Engineering, The University of Tokyo, 7-3-1 Hongo, Bunkyo-ku, Tokyo 113-8656, Japan
**JST, CREST, 7-3-1 Hongo, Bunkyo-ku, Tokyo 113-8656, Japan*

Abstract. We report a microfabricated gas chromatography (GC) column that uses a thin layer of high-quality single-walled carbon nanotubes (SWNTs) as a stationary phase. This 1.0-m-long, 160-µm-wide, 250-µm-deep column has the highest separation efficiency reported to date for microfabricated columns having an SWNT stationary phase. Separation efficiency was evaluated with a Golay plot, in which the minimum of the height equivalent to a theoretical plate was 0.062 cm. The microfabricated column was able to separate n-alkanes having high boiling points under temperature-programmed conditions. Use of SWNTs as a stationary phase will be potentially useful for high-performance micro-GC.

Keywords: Carbon Nanotube, Stationary Phase, Gas Chromatography
PACS: 07.07.Df, 07.10.Cm, 82.80.Bg

INTRODUCTION

Conventional gas chromatography (GC) instruments using packed or capillary columns can separate most volatile compounds; however, these instruments are bulky, and their operations are power and time consuming. These drawbacks have hindered miniaturization of GC instruments for which a wide range of applications is foreseen [1]. Recent developments in the micro-fabrication of GC systems using micromachining technology have demonstrated the potential for reductions in size, power consumption, and analysis time compared to conventional GC systems [2–7].

A typical GC system consists of several modules, including a preconcentrator, a separation column, an oven heater, and a detector. The development and optimization of individual micromachined modules is the current trend in GC miniaturization [2]. Numerous studies have focused on miniaturizing the separation columns to develop micro-GC columns that can quickly separate samples with high efficiency [3–6]. One of the major challenges in developing micro-GC columns is the coating of the conventional polymer-based stationary phase [7,8]. Therefore, it is crucial to research and develop an efficient stationary phase for micro-GC columns.

Carbon nanotubes (CNTs) are attractive materials for the chromatographic stationary phase, and previous studies have demonstrated that CNTs have good separation ability due to chemical and thermal stability and a high surface-to-volume ratio [9–12]. These studies used a conventional steel, glass, or fused-silica column. A microfabricated column on a silicon on which CNTs were synthesized has also been reported [5,6]. These studies and reports have shown that CNTs are useful for quick temperature-programmed separation. Because CNTs grow selectively on a catalyst that can be patterned by lithography, they are better suited for integration into microfabricated columns compared to conventional polymers, which must be coated using conventional methods. Thus, CNTs have great potential for use as a stationary phase in micro-GC applications.

However, the previously reported use of CNTs as a GC stationary phase resulted in relatively low separation efficiency [5,9–12]. Moreover, gas separation columns with a CNT stationary phase suffer from poor separation of high-boiling-point compounds [5,11,12]. These problems are attributed to the thickness of the CNT layer.

Separation efficiency is determined by the degree of broadening of a solute band as it traverses the column. One of the sources of band broadening is solute mass transfer in the stationary phase. Thus, separation efficiency increases with a decrease in the thickness of the CNT stationary phase layer. A thin layer also promotes easy elution of highly retained

CP1137, *Olfaction and Electronic Nose: Proceedings of the 13th International Symposium*, edited by M. Pardo and G. Sberveglieri

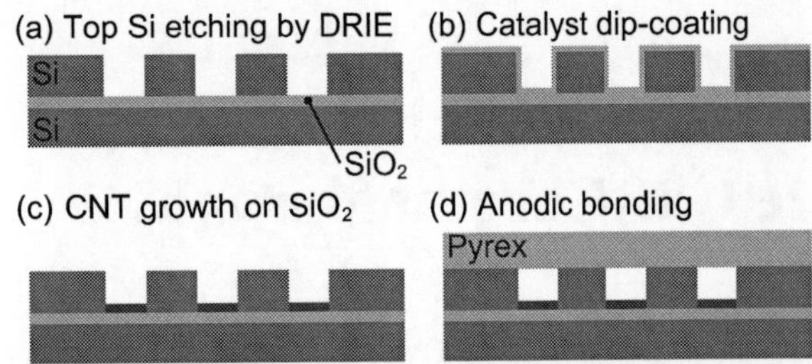

FIGURE 1. Fabrication sequence for micro-GC column with SWNT stationary phase.

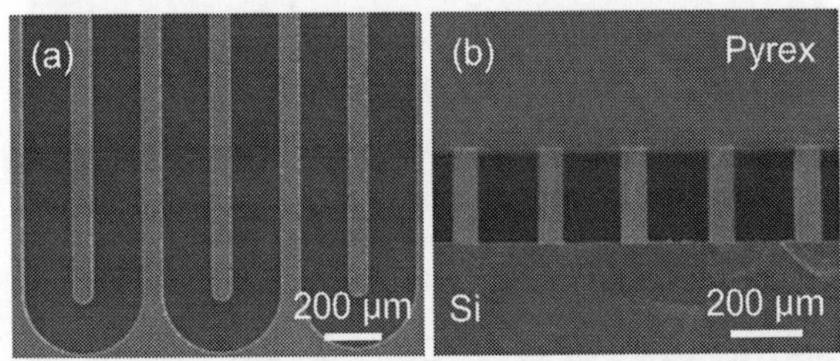

FIGURE 2. SEM images of (a) the microfabricated channel before SWNT growth and (b) the cross- section of the microfabricated column after anodic bonding.

solutes and high-boiling point compounds. Therefore, it is important to synthesize a thin layer of CNTs into microfabricated columns.

Our aim was to achieve high separation efficiency of CNT stationary phase micro-GC columns and to separate compounds with high boiling points. We used a single-walled carbon nanotube (SWNT) stationary phase synthesized into a microfabricated column. Separation efficiency was analyzed using Golay plot, and temperature-programmed chromatograms of n-alkanes having high boiling points are presented.

EXPERIMENTAL AND METHODS

Column Fabrication

Many microfabricated rectangular columns have been designed and fabricated by other research groups. We referred to previous work [3,4] and designed a standard open tubular GC column. The length, width, and depth were 1.0 m, 160 μm, and 250 μm. It has a serpentine structure while those in previous work [3,4] had a spiral one. According to a recent report [13], a serpentine GC column contributes less to band broadening than a spiral one and thus have better separation efficiency.

The low separation efficiency of previously reported CNT stationary phase columns can be attributed to the use of CNT films that are too thick, thereby increasing the resistance to mass transfer in the stationary phase, which causes band broadening [14]. A thin-film stationary phase improves separation efficiency at the cost of retention factors. Therefore, this approach is useful for improving the resolution for high-boiling-point compounds. Since CNTs have a high surface-to-volume ratio, a CNT stationary phase does not suffer from a decrease in retention factors due to a reduced film thickness. In the work reported here, CNTs were synthesized only on the bottom surface of the microfabricated column.

The steps in fabricating the open tubular GC column with a serpentine structure are illustrated in Figure 1. First, we prepared a diced SOI substrate (250/3/320 μm) and etched the top Si layer using deep reactive ion etching (DRIE) to fabricate a channel (Figure 1a). After the etching, the photoresist was removed from the substrate. Next, SWNTs were grown on the bottom surface of the channel by catalytic vapor deposition (CVD); the SWNTs grew selectively from the catalytic metal particles [15]. Since the catalyst loses its activity when it is directly deposited on a silicon substrate [16], we used an SOI substrate to obtain a well-defined oxide layer on the bottom surface of the channel.

We used molybdenum and cobalt as catalytic metals for the CVD process. They were dissolved (at the same weight concentration of 0.01 wt%) in ethanol to form a metal acetate solution. The solution was dip-coated on the SiO_2 surface at a constant pull-up speed of 4 cm/min at room temperature. The substrate was then dried in an oven at 400 °C for 5 minutes.

The CVD process was carried out in a quartz tube as follows (Figure 1c). The substrate was heated up to 800 °C in a furnace with a 300-sccm flow of argon gas containing 3% hydrogen (Ar/H_2 gas) so that the pressure inside the tube was kept at about 40 kPa. When the reaction temperature of 800 °C was reached, the gas flow was stopped, and ethanol vapor was introduced at a pressure of 1.3 kPa for 10 minutes.

Finally, a Pyrex cover was sealed on the SOI substrate by anodic bonding at 350 °C and 500 V (Figure 1d). For the tubing connection, the microfabricated column had a 250-μm-wide opening at each end, and a 30-cm-long, 150-μm-internal-diameter, 220-μm-outer-diameter deactivated fused silica capillary tube was inserted into each end, and the connections were sealed with an epoxy adhesive. Figure 2 shows SEM images of the microfabricated channel before CNT growth and of a cross-section after anodic bonding.

Experimental Setup

A commercial GC was used for testing the microfabricated column. The microfabricated column was used in place of the conventional one. Helium was used as the carrier gas. The samples were injected into the built-in split/splitless injection port, and the sample vapors were detected at the column outlet using a

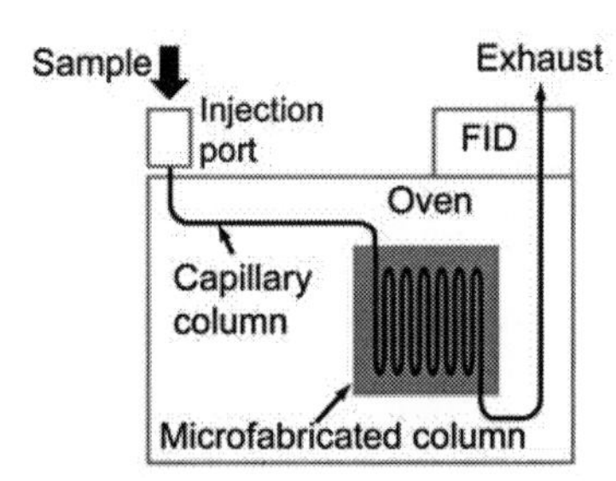

FIGURE 3. Schematic of the experimental setup used to record the chromatograms.

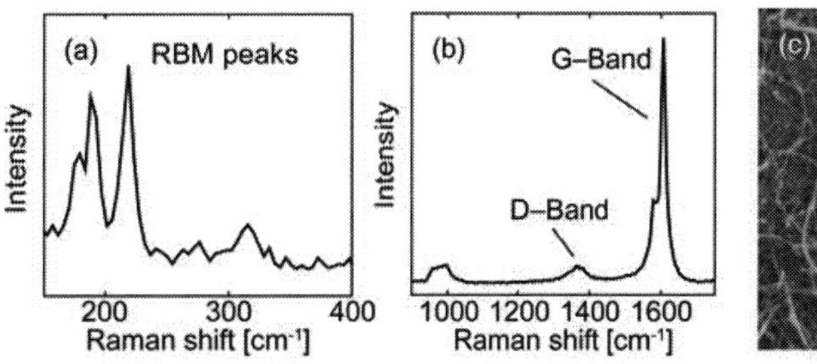

FIGURE 4. (a) Low-frequency and (b) high-frequency parts of resonance Raman spectrum for bottom surface of microfabricated channel after SWNT synthesis. (c) An SEM image of bottom surface.

built-in flame ionization detector (FID), as shown in Figure 3. The temperature of the built-in oven was controlled, and the column temperature was measured with a K-type thermocouple. All chromatograms were recorded using a data logger with a sampling frequency of 50 Hz and an A/D resolution of 16 bits.

The fabricated column was tested by separating n-alkane mixtures because CNTs are nonpolar phases [9,11]. For temperature-programmed separation, the injection volume was 0.1 μL for a mixture of n-alkanes with a split ratio of 140:1. No solvent was used. For separation efficiency evaluation, 0.05-μL samples of decane were injected with a split ratio of 140:1 using a column temperature of 40 °C. The inlet and detector temperatures were set at 200 °C. The average carrier gas velocities were calculated from holdup time measurements using methane injection.

RESULTS

Characteristics of SWNTs

The synthesized SWNTs were characterized by resonance Raman spectroscopy and scanning electron microscopy (SEM). The SWNTs on the channel bottom of the microfabricated column were analyzed by resonance Raman spectroscopy to estimate their diameters. Figure 4 shows a typical resonance Raman spectrum. The low-frequency part of the Raman spectrum is magnified in Figure 4a, revealing radial breathing mode (RBM) peaks, which is strong evidence of SWNT growth. The obtained RBM peaks indicate that the tube diameters were between 1 and 2 nm although only the SWNTs that were resonant with the excitation laser (488 nm) were detected. The split of the G-band peaks around 1590 cm^{-1} is also characteristic of SWNTs. This is highlighted in Figure 4b, where the high-frequency part of the spectrum is plotted. G-band peaks originate from in-plane vibrations of graphitic lattices, and D-band ones originate from defects in graphitic structures. The ratio of the G-band intensity to the D-band intensity was more than 20, indicating the growth of high-quality SWNTs.

Figure 4c shows an SEM image of the channel bottom surface where the resonance Raman spectrum was measured. The image suggests that the thickness of the SWNT film grown inside the microfabricated column was less than those in previous work [5]. In comparison to those of previous works, the SWNT film grown using our method was thin.

Column Efficiency

We evaluated the separation efficiency of our SWNT stationary phase microfabricated column by using a plot of the height equivalent to a theoretical plate (HETP), H, versus the average carrier gas velocity, u, (Golay plot), as shown in Figure 5. The HETP values were determined for decane at a column temperature of 40 °C. HETP was obtained by dividing the column length by the theoretical plate number. The theoretical plate number was calculated using the retention time for the eluted peak and its full-width at half maximum, which was estimated by Gaussian fitting. The retention factor was determined from the retention time and the holdup time, which was approximated by the travel time of methane. The retention factor was 0.6. The minimum HETP value, H_{min}, found at the optimal average carrier gas velocity, u_{opt}, gives the maximum number of theoretical plates N. The trend line in Figure 5 represents the regression curve calculated using [17]:

$$H = \frac{B}{u} + Cu + Du^2 \tag{1}$$

where B, C, and D represent the longitudinal diffusion, the resistance to mass transfer in the mobile and stationary phases, and the extra-column band broadening, respectively. At u_{opt}, H_{min} calculated using Equation 1 is 0.062 cm and N is 2500. The smaller the HETP value, the higher the separation efficiency. Note that the H_{min} value of 0.062 cm is smaller than that in a previous report [5] for micro-GC column with a glass cover and an etched silicon channel with an SWNT stationary phase. To the best of our knowledge, this separation efficiency is the highest yet reported for a micro-GC column with an SWNT stationary phase.

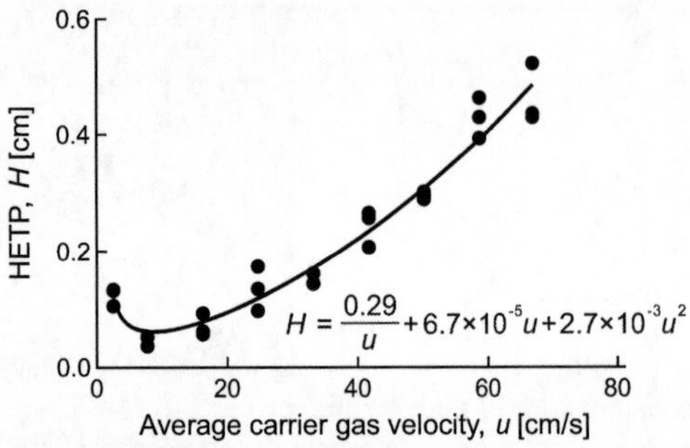

FIGURE 5. HETP versus average carrier gas velocity for micro-GC column with CNT stationary phase.

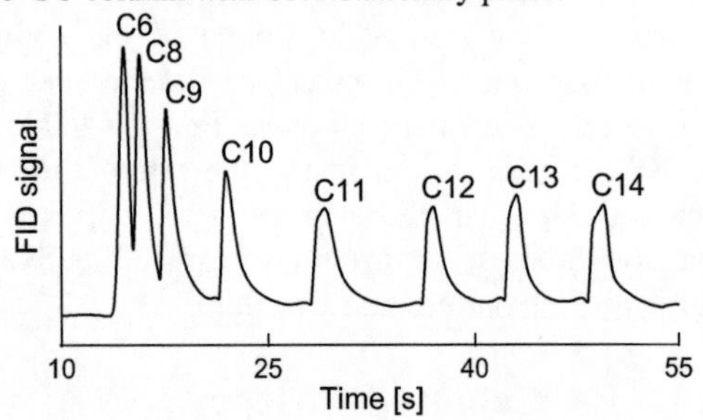

FIGURE 6. Temperature-programmed chromatogram of an n-alkane mixture. Nnumbers above peaks correspond to the alkane carbon numbers.

Separation Chromatogram

A chromatogram of the test mixture was recorded for the microfabricated column when the average carrier gas velocity was optimal under a temperature-programmed condition. The column temperature was increased from 30 to 105 ˚C at 1.4 ˚C/s. A chromatogram for the n-alkanes mixture is shown in Figure 6. All the n-alkanes were clearly separated. In previous work, the peaks for compounds having very high boiling points such as tetradecane exhibited significant tailing and band broadening [11]. Here, the peaks exhibited weaker tailing and less band broadening in spite of the much lower final column temperature (105 ˚C). These results demonstrate that use of an SWNT stationary phase enables the detection of compounds having high boiling points.

CONCLUSION

We successfully synthesized a thin layer of high-quality SWNTs on the channel bottom of the microfabricated column and evaluated the separation efficiency of the SWNT stationary phase microfabricated column. Using this column, we achieved the highest separation efficiency ($H_{\min}$ = 0.062 cm), reported to date, of the SWNT stationary phase microfabricated columns. We have demonstrated that the use of a thin layer of high-quality SWNT stationary phase results in high separation efficiency and separation of n-alkanes with high boiling points.

ACKNOWLEDGMENTS

This work was supported by JST, CREST and GCOE, MEXT. A part of this work was conducted in Center for Nano Lithography & Analysis, The University of Tokyo. We thank Prof. Yuji Suzuki for his valuable support in using the anodic bonder and Prof. Suguru Noda for his valuable support in using the Raman spectrometer.

REFERENCES

1. E. B. Overton *et al.*, *Field Analyt. Chem. Technol.* **1**, 87-92 (1996).
2. C. J. Lu *et al.*, *Lab Chip* **5**, 1123-1131 (2005).
3. M. Agah *et al.*, *J. Microelectromech. Syst.* **14**, 1039-1050 (2005).
4. G. Lambertus *et al.*, *Anal. Chem.* **76**, 2629-2637 (2004).
5. M. Stadermann *et al.*, *Anal. Chem.* **78**, 5639-5644 (2006).
6. V. R. Reid *et al., Talanta* **99**, 1420-1425 (2009).
7. M. A. Zareian-Jahromi *et al.*, *J. Microelectromech. Syst.* **18**, 28-37 (2009).
8. G. M. Gross *et al.*, *J. Chromatogr. A* **1029**, 185-192 (2004).
9. Q. L. Li and D. X. Yuan, *J. Chromatogr. A* **1003**, 203-209 (2003).
10. C. Saridara and S. Mitra, *Anal. Chem.* **77**, 7094-7097 (2005).
11. M. Karwa and S. Mitra, *Anal. Chem.* **78**, 2064-2070 (2006).
12. L. M. Yuan *et al.*, *Anal. Chem.*, **78**, 6384-6390 (2006).
13. A. D. Radadia *et al.*, *Proc. Transducers '07,* 2011-2014 (2007).
14. V. R. Reid and R. E. Synovec, *Talanta* **76**, 703-717 (2008).
15. S. Maruyama *et al.*, *Chem. Phys. Lett.* **360**, 229-234 (2002).
16. Y. Murakami *et al.*, *Chem. Phys. Lett.* **377**, 49-54 (2003).
17. G. Gaspa *et al.*, *Anal. Chem.* **50**, 1512-1518 (1978).

A novel Laser Ion Mobility Spectrometer

J. Göbel[1], M. Kessler[1], A. Langmeier[1]

[1] EADS Innovation Works, Dept. IW-SI, D-81663 München, Germany
Email: johann.goebel@eads.net

Abstract. IMS is a well know technology within the range of security based applications. Its main advantages lie in the simplicity of measurement, along with a fast and sensitive detection method. Contemporary technology often fails due to interference substances, in conjunction with saturation effects and a low dynamic detection range. High throughput facilities, such as airports, require the analysis of many samples at low detection limits within a very short timeframe. High detection reliability is a requirement for safe and secure operation. In our present work we developed a laser based ion-mobility-sensor which shows several advantages over known IMS sensor technology. The goal of our research was to increase the sensitivity compared to the range of ^{63}Ni based instruments. This was achieved with an optimised geometric drift tube design and a pulsed UV laser system at an efficient intensity. In this intensity range multi-photon ionisation is possible, which leads to higher selectivity in the ion-formation process itself. After high speed capturing of detection samples, a custom designed pattern recognition software toolbox provides reliable auto-detection capability with a learning algorithm and a graphical user interface.

Keywords: REMPI, IMS ion mobility spectrometry, machine olfaction, electric mobility explosive detections,
PACS: 07.05.Hd, 07.05.Rm, 42.65.-k, 42.62.Fi, 51.50.+v, 82.50.Hp, 82.50.Pt

INTRODUCTION

IMS technology is frequently used as a detector for toxic industrial chemicals (TICs), explosives, and narcotics. Its usage is wide spread from ports to airports and prisons or other national and international safety areas. IMS is a method that separates ionized analyte molecules on the basis of their mobilities in gas phase. The typical time of arrival spectra can be analyzed and matched with a pattern database to identify the analyte.

Experimental Set UP and Methods

Laser ionisation

In this work we present a laser ion mobility spectrometer prototype (LIMS) developed at EADS-IW. Here, in contrast to classical IMS systems, a laser, instead of a radioactive source e.g. ^{63}Ni or ^{241}Am [1] is utilized for the ionization of analyte, dopant and calibrant substances. We tested several fixed and variable wavelength pulsed laser systems as well as in-house developed specialized laser systems (active, passive Q-switched laser, OPOs). The laser system itself is one of the key elements in a LIMS and has to fulfill several requirements. First the system's laser wavelength has to be in the UV-range (200-280nm). Second the laser must be a pulsed system with at least 10 Hz repetition rate and intensities in the µJ range. Third the convergence and beam spot size should be small enough to guarantee good resolution in an IMS spectrum. Once the wavelength and photon energy of the laser radiation meet the requirements of an intermediate ionization step, a photon ionization step (REMPI: Resonance Enhanced Multi Photon Ionization) can occur. Due to this behavior, a high selectivity in the ionization step, with preference to the substances under investigation, can be achieved. Typical IMS with lower selectivity ranges could be expanded with laser ionization for improved selectivity ranges [2].

A typical REMPI spectrum resembles in general the UV absorption spectrum of a specific substance. See Figure 1 for a comparison of Benzene REMPI- and UV Absorption spectra. Using a defined wavelength for a certain substance or group of substances one can enhance the selectivity of the

CP1137, *Olfaction and Electronic Nose: Proceedings of the 13th International Symposium,* edited by M. Pardo and G. Sberveglieri

LIMS. The absorption spectra were recorded with a commercial UV-VIS spectrometer at 0.1nm.

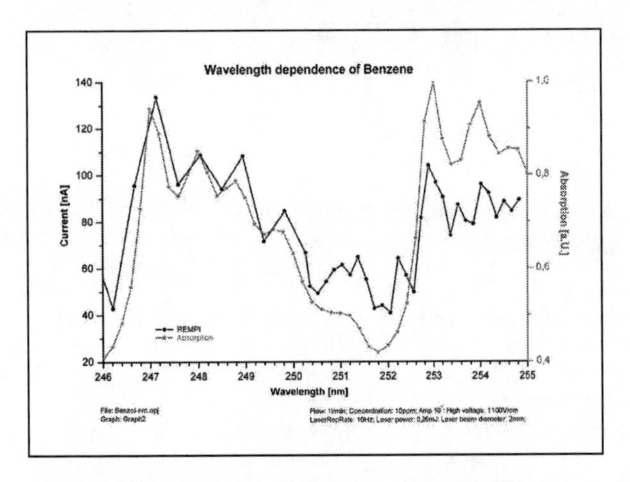

FIGURE 1 . Comparison of the wavelength dependence using REMPI and absorption spectra.

Drift tube

The planar drift tube design is driven by the ionization method via UV laser radiation. To achieve a maximum interaction between the analyte gas and the laser radiation, it is essential for the LIMS drift tube to maximize the laser path length in the ionization region. Therefore a planar, rather then a circular design is chosen (Figure 2).

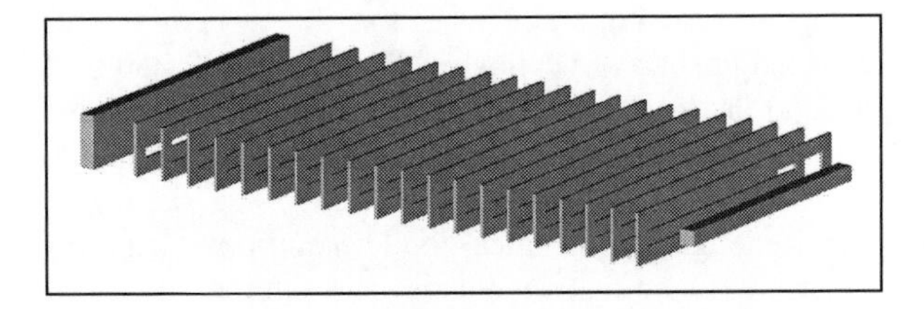

FIGURE 2. The LIMS planar drift tube in stacked design.

In contrast to the ^{63}Ni ionization, ions will only be created during the duration of the laser pulse. Additionally, employing a ^{63}Ni source entails the use of an ion gate to enforce a starting time for the ions. An electrical switching gate will always be slower and less sharp than a q-switched laser. An additional fuzziness occurs because of the mechanical dimensions of the gate itself. Disregarding the gate type, Bradburry-Nielson or Tyndall, the local starting position of the ions is wider. Therefore the ion cloud is enlarged, too.

In LIMS Technology the pulsed laser beam is utilized as a gate for the ionized analyte molecules, which start to travel toward the detector plate at the same instant from the geometrically well defined, equidistant area. It simultaneously triggers the measurement process, making the commonly used ion gates obsolete.

Any radioactive ion source has an energy distribution starting at low level eV up to several keV (e.g. ^{63}Ni up to 67keV with an average energy near 17keV). This variety of energy leads to a huge amount of different ions with a great chance of additional unwanted chemical reaction. Therefore the interpretation of the spectra will become more difficult at least in positive mode. IMS devices with radioactive ionization sources holds a very limited linear detection range, that means the detection device could be easily overloaded when the analyte concentration is to high or if the ionization limit of the ionization source is almost reached.

In laser ionization one can easily tune the flux of photons either by manipulate the Laser electrically at the pumping stage or optically in the laser path. Together with a proprietary detection and recognition software a very dynamic measurement system is possible and overcame one of the biggest problems in the IMS worlds.

E-Field

The electric field is formed by a stacked plates design. The alternating isolated and conducting layers plates create a linear electric field gradient. These plates in the planar drift tube geometry are about ten times wider than tall. The drift length at the moment is about 100mm. At the end of the drift a golden collector tube transforms the ion clouds into an electrical signal. Using this stacked and planarised design assembly opens the opportunity to vary the mechanically drift parameters effortlessly. Electrical field strength of 250 to 300 V/cm is used. An integrated voltage divider is connected to a high voltage power supply. Using a voltage divider in the MΩ range the needed power can be kept low. Direct in front of the collector tube an aperture grid is installed to avoid an induced signal from arriving ion clouds. An optimized finite element method (e.g. SimIon and Comsol) was utilized to simulate geometry and E-field of the design. To estimate the influence of the environment, especially of the conductive parts around the drift tube, the FEM model had to be expanded by this area.

Gas flow

Using a bidirectional gas flow yielded the best results, concerning sharp peaks, low impurities and short cleaning procedures. To keep the drift tube clean, a drift gas flow of about 100 to 400ml/min is introduced behind the collector. Before entering the drift cell the drift gas gets dehumidified and cleaned with a charcoal and silica based adsorbent material.

The gas flow carrying the analyte is generally lower (100 to 200ml/min) than the drift gas flow. The analyte gas inlet is direct in front of the ionisation chamber with its UV windows for the laser ionisation. The analyte molecules are homogeneously distributed all over the laser beam path inside the ionisation chamber with a laminar slow analyte gas flow. The exhaust for both, the analyte and the drift gas, is inside the ionisation chamber. Drift- and analyte gas flow are flow controlled via mass-flow-controllers (MFC) while entering the drift cell. The exhaust gas flow is monitored via a mass-flow-meter.

Material

IMS detection technologies are generally able detecting molecules in the lower ppb concentration range. Therefore contaminations caused by off gassing parts have to be prevented.

To avoid impurities the used material should not show gas emissions and shall not adsorb or react with the ambient vapours or molecules. During our research it turned out that the best materials are stainless steel and gold as conducting material. Both exhibit a high resistance to chemical attack and a long durability. Teflon™, Macor™ and Al_2O_3 ceramic are also good choices for the isolating material.. Macor™ is a mixture of mica (about 55%) and glass which can be used up to 800°C. Both, ceramics and the Teflon™ offer a good dielectric strength (Macor™ 60kV/mm, Al_2O_3 60kV/mm, Teflon™ 280V/mil = 11kV/mm).

Housing and system

Our LIMS drift tube is packed in a ruggedized, hermetically sealed housing with quartz windows for an external laser source. The UV laser beam passes through the housing and ionisation chamber. Using this assembly we can observe the laser beam and control its essential parameters during our measurements. We also note them with the captured IMS spectra. In future application there will be some parts of the laser system integrated inside the housing.

Software and GUI

Data is collected with a A/D-conversion board and processed on a Pentium Pro 2GHz laptop. The data acquisition system is based on an in-house developed LabView virtual instrument (VI) with graphical user interface (GUI). Aside from data acquisition, the LabView GUI allows control of flows, temperatures, voltage, and amplifier settings. Actual system parameters like drift- and analyte flow, their corresponding temperatures and humidities, voltage, system temperatures and pressure are monitored and recorded. In a subsequent post-processing step, single spectrum calibration and automatic explosive detection are performed.

Some of the results presented here are achieved using offline analysis with a Matlab based spectral analysis system. The GUI presented in Figure 3 is one part of our data analysis.

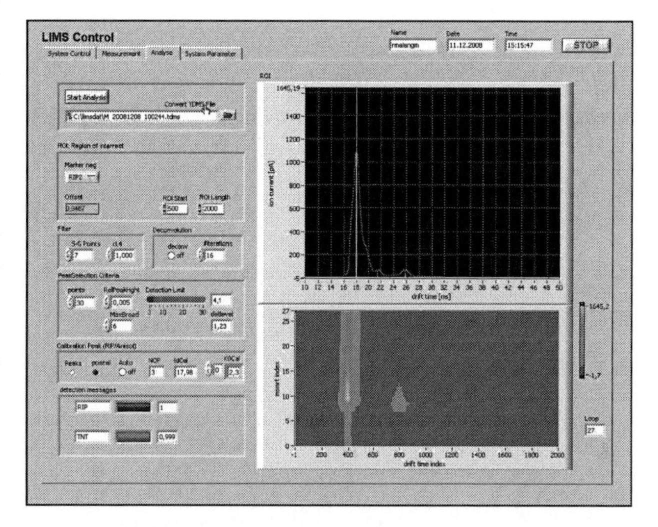

FIGURE 3. Data analysis panel of our Lab View GUI. On the left we put the analysis and detection control and on the right we can observe the spectra and the spectrogram.

Measurements and Data Processing

The spectra are sampled with a frequency of 50 kHz and 24 bits. The scan period was set at 100 ms matching the laser pulse rate. Peak intensities of analytes in the nanogram range are usually at 10^{-7} to 10^{-9} [A] and well above the noise level. Peak quality (resolution), after deconvolution, is around R=50, for TNT ($K_0 = 1.54$ cm^2/Vs) [4].

Usually raw single spectra are averaged over 10 scans before being recorded. Subsequent data processing includes background correction and trend reduction, noise filtering and deconvolution with a Richardson-Lucy algorithm. The pre-processed spectra are searched for peaks and calibrated. Afterwards peak classification is (currently) done with a modified nearest neighbour scheme.

RESULTS

The system can be self-consistently calibrated in positive and negative modes. In positive mode anisole ($K_0 = 2.011$ cm^2/Vs) and Pyridine ($K_0 = 2.011$ cm^2/Vs) are used as calibrant substance [3]. We obtain an estimated reduced ion mobility accuracy of ± 0.01 cm^2/Vs. We choose Naphthalene as a dopant which has a very high one-photon absorption cross-section at 266 nm of $2 \cdot 20^{-17} cm^2$ [5]. The solid naphthalene was dissolved in n-hexane for better dosaging. We desorbed naphtalene together with the analyte (explosive) samples from one common swab in amounts of 5 to 50 ng per sample. Naphthalene cations can be directly measured in positive mode. A sample positive mode LIMS spectrum is shown in Figure 4.

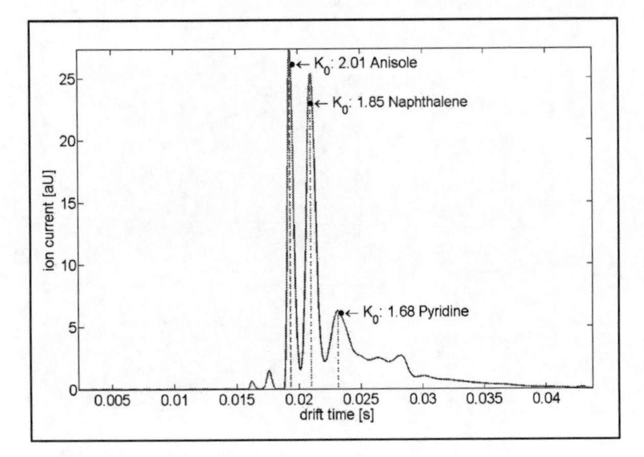

FIGURE 4. LIMS spectrum in the positive mode Anisole and Pyridine as calibration substances, and Naphthalene as a dopant.

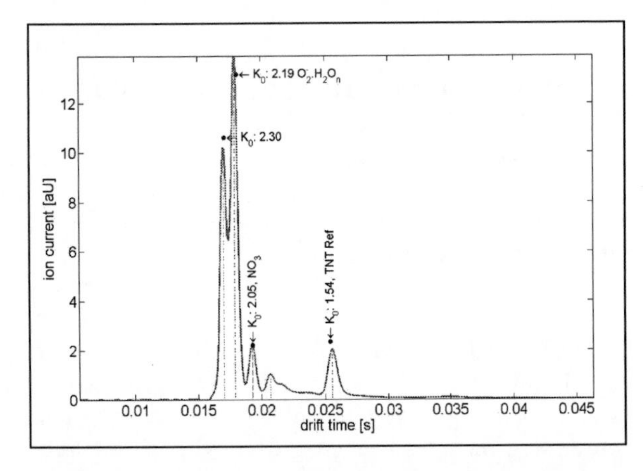

FIGURE 5. LIMS spectrum in the negative mode with TNT as reference substance at K_0 at 1,54 cm^2/Vs.

In the negative mode we currently utilize the reactant ion peaks (RIPs) for calibration. We observe 3 main RIPs at (K_0 = 2.30, 2.20 and 2.05 cm^2/Vs) which are preliminarily identified as (NO_n^-, CO_n^-), $(H_2O)_nO_n^-$ and NO_3^- peaks [6]. It is believed that the main RIP corresponds to the oxygen–water cluster $(H_2O)_3O_2^-$ [7], [6]. Figure 5 shows a negative mode LIMS spectrum with a TNT reference measurement. A future version of the system will feature a specific calibrant substance in negative mode as well.

The detection limit is determined by the last detectable 1 to 2 second integrated peak signal in a desorption history that is 3σ above noise level. It is defined as the amount of analyte desorbed during the measurement of that spectrum. To obtain this level it is necessary to initially desorb a larger amount of the analyte, and to track peak intensity as the concentration decreases. Typically a suitable amount for this procedure is 5 to 10 times the detection limit.

CONCLUSIONS

Laser based ion-mobility-spectrometry shows several advantages over radioactive based ion mobility sensors. Our studies show that using carefully adjusted designs in laser source and ion-sensor lead to higher selectivity and sensitivity of described technology. As in any other technologies LIMS has its several benefits and drawbacks. For field operation, one has to carefully decide on the budget, wavelength, selectivity, and sensitivity range requirements of the application. Continuing research in this technology will lead to sensors with real time capability, high sensitivity and better selectivity, which will suit the needs of future security, environmental, and medical applications perfectly.

ACKNOWLEDGMENTS

Part of the work at EADS Innovation Works was supported within the frameworks of the German nationally funded project MILAN, the European funded project SAFEE, and LIMS as an EADS internal project.

REFERENCES

1. R.H.St. Louis, H.H. Hill Jr., *Critical Reviews in Analytical Chemistry* Vol. **21**, No. **5**, 321 – 355 (1990)
2. D.M. Lubman, et al, *Analytical Chemistry*, Vol. **55**, No. **6**, 867 – 873 (1983)
3. G.A. Eiceman, Z. Karpas, *Ion Mobility Spectrometry 2 ed.*; Boca Raton: CRC Press, Taylor & Francis, 2005
4. R.G. Ewing, D.A. Atkinson G.A. Eiceman, G.J. Ewing, *Talanta* Vol. **54**, No **3**, 515-529 (2001)
5. R. R. Kunz, W. F. Dinatale and P. Becotte-Haigh, *International Journal of Mass Spectrometry*, Vol. **226**, No. **3**, 379-395 (2003)
6. C.J. Hayhurst, P. Watts and A. Wilders, *International Journal of Mass Spectrometry and Ion Processes*, Vol. **121**, No. **1-2**, 127-139 (1992)
7. C. Shumate, R.H. St. Louis and H.H. Hill, *Journal of Chromotography*, Vol. **373**, No. **2**, 141-173 (1986)

APPLICATIONS: FOOD

Measurement Of Beer Taste Attributes Using An Electronic Tongue

Evgeny Polshin[1, 2#], Alisa Rudnitskaya[2, 3], Dmitry Kirsanov[2], Jeroen Lammertyn[1], Bart Nicolaï[1], Daan Saison[4], Freddy R. Delvaux[4], Filip Delvaux[4], Andrey Legin[2]

[1]*BIOSYST/MeBioS, Katholieke Universiteit Leuven, W. De Croylaan 42, B-3001 Leuven, Belgium*

[2]*Laboratory of Chemical Sensors, Chemistry Department, St. Petersburg University, Universitetskaya nab. 7/9, St. Petersburg, 199034, Russia*

[3]*Chemistry Department, University of Aveiro, Campus Universitario de Santiago, 3810-193 Aveiro Portugal*

[4]*Centre for Malting and Brewing Sciences, Katholieke Universiteit Leuven, Kasteelpark Arenberg 22, 3001 Leuven, Belgium*

[#]*Corresponding author*

Abstract. The present work deals with the results of the application of an electronic tongue system as an analytical tool for rapid assessment of beer flavour. Fifty samples of Belgian and Dutch beers of different types, characterized with respect to sensory properties and bitterness, were analyzed using the electronic tongue (ET) based on potentiometric chemical sensors. The ET was capable of predicting 10 sensory attributes of beer with good precision including sweetness, sourness, intensity, body, etc., as well as the most important instrumental parameter – bitterness. These results show a good promise for further progressing of the ET as a new analytical technique for the fast assessment of taste attributes and bitterness, in particular, in the food and brewery industries.

Keywords: electronic tongue, potentiometric chemical sensors, beer, sensory attributes, STATIS, Bitterness.
PACS: 82.45.Rr, 82.47.Rs

INTRODUCTION

Brewing and aging of beer are complex processes during which several parameters have to be controlled to ensure reproducible taste and quality of the finished product. Those include physicochemical parameters that are measured instrumentally as well as taste and flavour attributes that are evaluated by the sensory panels. While methods for measuring some of the parameters such as density, colour or extracts are simple and rapid, determination of the others attribute like for instance bitterness is more difficult. Bitterness is a very important quality parameter in the production process of beer. This parameter is related to the amount of iso-α-acids which are responsible for bitter taste and bacteriostatic properties of beer and also play substantial role in enhancing the foam stability of beer as well as in the formation of off-flavors [1]. The recommended and widely used method of bitterness analysis in beer in terms of international bitter units (IBU) is carried out by a spectrophotometric measurement at 275 nm of an acidic solvent extract of beer [2]. This method is quite slow and requires extensive sample preparation including extraction with organic solvents. Sensory assessment of beer, as well as other foodstuffs gives by far the most realistic picture of the taste of a product as experienced by human. However, sensory analysis is also associated with several practical problems and has some drawbacks, namly high cost, irreproducibility and taste saturation of the panelist [3]. Thus, it is not surprising that significant efforts are being directed to the development of rapid methods of routine beer analysis including both chemical composition and taste attributes [4, 5].

The Electronic tongue (ET) multisensory systems are considered promising for taste assessment of the foodstuffs and beverages. The ET comprises an array of cross-sensitive (partially selective) chemical sensors of which the sensor responses are analyzed with pattern recognition techniques and multivariate calibration methods [6]. The objective of the present

CP1137, *Olfaction and Electronic Nose: Proceedings of the 13th International Symposium,* edited by M. Pardo and G. Sberveglieri

study is the evaluation of an electronic tongue based on potentiometric chemical sensors as a rapid analytical tool for measuring beer taste and for determination of bitterness.

EXPERIMENTAL

Samples

Fifty samples of Belgian and Dutch beers of different types along with reference data were provided by the Centre for Malting and Brewing Sciences, Katholieke Universiteit Leuven, Belgium. Those included dark, light and amber ales, lager beers, white (wheat) beers, lambic and trappist beers. Samples were stored in the dark bottles of 250 or 330 ml, in a dry dark cool room (1°C) before the experiments. Each bottle was opened prior to the measurements and used at the same day.

Sensory Evaluation And Bitterness Measurements

All beer samples were characterised by the trained sensory panel at the Centre for Malting and Brewing Sciences. The panel comprised of 7 members. Tasting sessions were carried out 3 to 4 times a week. In each session, 4 – 6 beers were randomly presented to the panelists. Tasting sessions were organised in a temperature controlled room (18ºC). Beer was cooled down to 7ºC and served in the dark glasses. Judges were asked to evaluate the beer foam, the aroma, the taste and finally the aftertaste. All attributes were ranked on the scale from 0 to 8 except global quality, for which a scale from 1 to 9 was used. A score of 0 meant that the particular flavour aspect was not present, whereas a score of 8 meant that this aspect was extremely strong. Judges used water to rinse the mouth and ate a piece of unflavoured bread between subsequent samples in order to eliminate fatigue and carry over effects. In total 42 attributes pertaining to the beer aroma, taste, mouth feel and appearance were evaluated. The bitterness was determined according to the method 9.8 of Analytica EBC [7] and results were expressed as European Bitter Units (EBU), which correspond to the concentration of iso-α-acids in beer (mg/l). Average values of two determinations were used for data analysis. The coefficients of variation of bitterness determination did not exceed 5%.

ET Measurements

The ET used in this work comprised of 29 potentiometric chemical sensors. Sensors with plasticized PVC membranes displaying sensitivity to organic anions and with chalcogenide glass membranes displaying red-ox response were used. Also a conventional glass pH electrode was employed. Sensor membranes were prepared according conventional procedure [8]. Measurements were carried out using a custom made high input impedance multichannel voltmeter connected to a PC. The sensor potentials were measured vs. a conventional Ag/AgCl reference electrode. Sample preparation consisted in beer filtration using a kieselguhr filter to remove CO_2 followed by dilution of the beer with distilled water with a ratio of 30 to 70 ml. Before each measuring session, sensors were conditioned during 3 minutes according to the described procedure. The beer sample was measured and the sensors washed with distilled water afterwards. Between measurements the sensors were washed with distilled water. Seven to nine replicates were run on each sample resulting into 440 measurements in total.

RESULTS AND DISCUSSION

Discrimination Of Beer Using The ET Data

A PCA was run on the ET data set. PCA score plot of the first and the second principal components for 14 different beers is depicted on Figure 1.

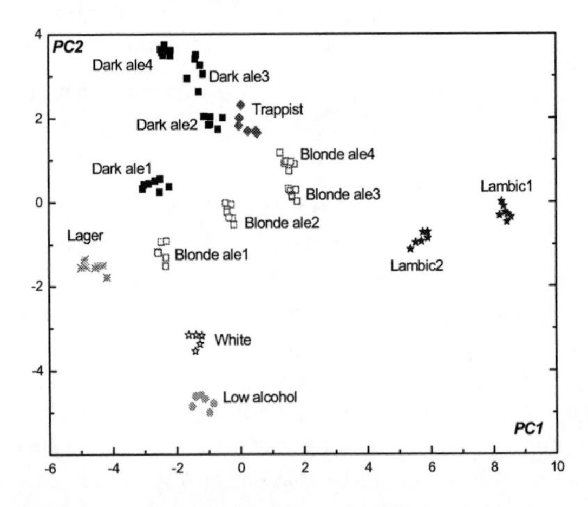

FIGURE 1. PCA score plot of several beer samples measured by the ET (X-expl. 55%, 25%).

The first two principal components account for 55 and 25 % of the total variance. As it can be seen from the plot all 14 beers of different types can be clearly discriminated by the ET. The alcohol content, the sugar content and the energetic value, is decreasing from strong dark and blonde beers on the left side of the plot to the lambic fruit beers on the right side of the plot. These beers are produced by spontaneous fermentation and with addition of fruit juice. They also have high content of polyphenols and low values of bitterness and pH, therefore they are quite different form the rest of the samples on the PCA score plot. The second principal component separates low alcohol, white, lager and blonde beers with low polyphenols content on the bottom side of the plot from the strong dark beers with high content of polyphenols (including dark trappist ale) on the top of the plot. Although, sample dark ale 4 has similar concentration of the polyphenols as wheat beer sample white.

Comparison Of Sensory And ET Data Sets

The analysis of the structure of the sensory panel data and the calculation of consensus average over the panellists was carried out using STATIS algorithm [9]. The correlation between sensory panel and ET data set was studied using Canonical Correlation Analysis (CCA). Four pairs of significant canonical variates were extracted. These four pairs were correlated with squared canonical correlation coefficients equal to 0.96, 0.91, 0.79 and 0.77 correspondingly. This indicates that a good description of sensory panel data by ET output is possible. Similarity maps for both sensory panel and ET data are shown in Figures 2 A and B respectively. The first canonical variate separates fruit lambic beers (samples 16 and 20) from the rest. The second canonical variate separates light beers with high scores of sulphury and DMS flavours and bitterness from dark ales with higher intensity, body, mouthfeel, sweetness and caramel.

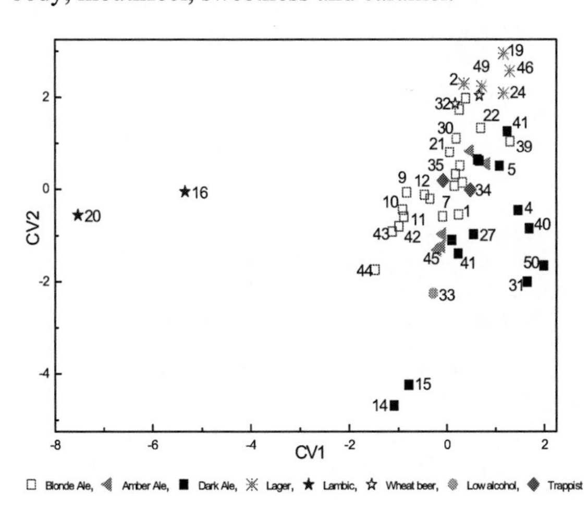

A

B

FIGURE 2. Similarity maps of the compromise average scores of the sensory panel and ET data calculated using CCA: (A) sensory panel data, (B) ET data.

Prediction Of The Sensory Attributes And Bitterness Using ET

Prediction of the sensory attributes of the foodstuffs from the instrumental response is a practically important task since it might potentially decrease the use of sensory panels and avoid several problems associated with them. Therefore, an attempt to perform such prediction of beer attributes from the ET measurements was performed. Calibration models were calculated using PLS regression using sensory data compromise values of the attributes as a reference. Models were validated based on a test set that consisted of one third of the ET data set. The replicate measurements of the same samples were included in either the calibration or the test set. Parameters of the predicted vs. measured curve for the test sample set are shown in Table 1. A good correlation was found between the ET output and 10 sensory attributes. Calibration models were also built by PLS regression using bitterness data as a reference, and validated based on a test set (Table 1). The bitterness values were ranged from 9.38 to 36.65 EBU for the whole set of fifty different samples. A correlation of 0.89 with relative error of prediction around 10% was found between the ET response and bitterness. It is important to note that the measurements of bitterness using ET is quite rapid and involves very simple sample preparation (beer degassing) that can be easily automated. The ET offers significant advantage over the conventional spectrometric method of bitterness determination that involves the use of the organic solvents and the extraction step and therefore is quite slow and cumbersome.

TABLE 1. Prediction of the sensory attributes and bitterness using the ET (test set validation).

Parameters	Correlation	Slope	Offset	RMSEP
Taste				
Intensity	0.85	0.70	1.70	0.28
Sour	0.74	0.70	-0.50	0.24
Sweet	0.96	0.80	0.08	0.29
Bitter	0.84	0.60	1.21	0.42
Mouthfeel				
Mouthfulness	0.77	0.53	2.10	0.39
Warming	0.82	0.66	0.50	0.64
Aftertaste				
Intensity	0.90	0.71	1.40	0.23
after bitter	0.79	0.58	0.41	0.47
sour	0.88	1.05	0.12	0.22
body	0.89	0.60	1.65	0.28
Bitterness (EBU)	0.89	0.83	3.3	2.5

CONCLUSIONS

The potential of the Electronic Tongue (ET) based on potentiometric chemical sensors to predict beer taste characteristics was evaluated. A PCA was used for data visualization and detection of groups in the data structure. The correlation between sensory panel and ET data set was studied using CCA. It was observed that the ET was capable to pick up the main differences between beer samples, according to the sensory data. Calibration models using ET data with respect to the sensory attributes and bitterness were made using Partial Least Square regression with test set validation. The electronic tongue was capable to predict 10 attributes of the beer taste out of 42 with a correlations of 0.74 – 0.97 and an RMSEP of 0.2 – 0.6. The electronic tongue is also able to predict bitterness of beer with comparatively high correlation, low offsets and RMSEP values. It was demonstrated that the ET is promising tool for the fast screening of beer taste quality and it is useful for determination of bitterness – one of the most important beer parameters.

ACKNOWLEDGMENTS

This research has been supported by the bilateral scientific cooperation BIL/05/47 between Russia and Belgium. Work of A. Rudnitskaya was supported by the postdoctoral fellowship SFRH / BPD / 26617 / 2006 by FCT, Portugal.

REFERENCES

1. R.J. Smith, D. Davidson, R.J.H. Wilson, J. Am. Soc. Brew. Chem. 56 (1998) 52.
2. Anonymous, Institute of Brewing Analysis Committee – Recommended Methods of Analysis. J. Inst. Brew., 1971, 77(2), 181–226.
3. A.Z. Berna, J. Lammertyn, S. Saevels, C. Di Natale, B.M. Nicolaï, Electronic nose systems to study shelf life and cultivar effect on tomato aroma profile, Sens. Actuators B 97 (2/3) (2004) 324–333.
4. Ferreira Guido L.; Curto A.; Boivin P.; Benismail N.; Goncalves C.; Araujo Barros A. Predicting the organoleptic stability of beer from chemical data using multivariate analysis. Eur Food Res Technol. 2007, 226, 57-62.
5. Wilson C.I.; Threapleton L. Application of artificial intelligence for predicting beer flavours from chemical analysis. In the Proc. of the 29th European Brewery Convention Congress, May 17-21, 2003, Dublin, Irland.
6. Legin A., Rudnitskaya A., Seleznev B., Vlasov Yu. (2002). Recognition of liquid and flesh food using an 'electronic tongue'. International Journal of Food Science and Technology, 37, 375–385.
7. Analytica-EBC, Verlag Hans Carl Getränke-Fachverlag: Nürnberg, 1998.
8. Legin A.; Rudnitskaya A.; Vlasov Yu. Electronic tongues: Sensors, Systems, Applications. In Integrated Analytical Systems, Comprehensive Analytical Chemistry XXXIX; Alegret S., Ed.; Publisher: Elsevier, Amsterdam, Netherlands, 2003, Vol. XXXIX, pp.437-486.
9. I. Stanimirova, B. Walczak, D.L. Massart, V. Simeonov, C.A. Saby, E. Di Crescenzo, STATIS, a 3-way method for data analysis. Application to environmental data, Chemometrics and Intelligent Laboratory Systems, 73 (2004) 219-233

Application Of A Potentiometric Electronic Tongue For The Determination Of Free SO_2 And Other Analytical Parameters In White Wines From New Zealand

Olga Mednova[1], Dmitry Kirsanov[1], Alisa Rudnitskaya[1], Paul Kilmartin[2], Andrey Legin[1]

[1] *St. Petersburg University, Laboratory of Chemical Sensors,St. Petersburg, Russia*
tel. +7 812 328 28 35, www.electronictongue.com
[2] *Chemistry Department, The University of Auckland, Private Bag 92019, Auckland, New Zealand*

Abstract. The present study deals with a potentiometric electronic tongue (ET) multisensor system applied for the simultaneous determination of several chemical parameters for white wines produced in New Zealand. Methods in use for wine quality control are often expensive and require considerable time and skilled operation. The ET approach usually offers a simple and fast measurement protocol and allows automation for on-line analysis under industrial conditions. The ET device developed in this research is capable of quantifying the free and total SO2 content, total acids and some polyphenolic compounds in white wines with acceptable analytical errors.

Keywords: Electronic Tongue, cross-sensitive chemical sensors, free SO_2 analysis, wine analysis .
PACS: 82.45.Rr, 82.47.Rs

INTRODUCTION

The composition of wine is influenced by many factors related to the specific production area: grape varieties, soil and climate, the yeasts employed and other wine making practices. A well-known problem in winemaking is oxidative spoilage of young white wines. Sulfur dioxide SO2 is the most common and important preservative (E220) in the food and beverage industry. In addition to antimicrobial activity, SO2 exhibits antioxidant properties and suppresses the activity of several oxidases and nonezymatic oxidative reactions.

In wine SO2 occurs in two forms: free and bound. Only free SO2 has antiseptic and reducing properties. However, at free SO2 levels between 15 and 40 mg/L most individuals begin to detect a distinctive burnt match odor. Moreover, consumption of high concentrations of sulfites may have adverse health effects, such as asthma, on humans. The FAO/WHO expert committee on Food Additives established an acceptable daily sulfite intake (ADI) of 0.7 mg/kg body weight, which may be exceeded by the consumption of half a bottle of wine. EC legislation now limits total SO2 concentrations to 160 mg/L for most red wines and 210 mg/L for most white wines. Thus, SO2 levels in wines should be accurately contolled to ensure sufficient SO2 is present for its beneficial role in winemaking, but no more than is necessary.

There is a number of standard methods for SO2 determination in wine such as the Ripper method, Paul method, aspiration method and so on. Although the accuracy of these methods is rather high they still suffer from a number of drawbacks: complexity, time consumption and lack of automatization possibilities.

An electronic tongue (ET) approach seems to be promising for SO2 determination in wines as it was earlier applied successfully for several tasks associated with wine quality [1-3]. The objective of the present research was a feasability study of the potentiometric ET system to determine free SO2 levels in different white wines produced in New Zealand. Another goal of the study was to try to find correlations with a set of other important parameters such as total acids content, polyphenolic compounds, etc.

EXPERIMENTAL AND METHODS

A potentiometric sensor array comprising 26 cross-sensitive electrodes with both polymeric and

CP1137, *Olfaction and Electronic Nose: Proceedings of the 13th International Symposium,* edited by M. Pardo and G. Sberveglieri

chalcogenide glass membranes was used. Sensor preparation details can be found elsewhere [4]. The experimental set-up is shown in Fig. 1.

36 different white wine samples of commercially produced Pinot Gris (4), Chardonnay (16) and Sauvignon Blanc (16) varieties were analyzed. To ensure the consistency of wine chemical composition the following procedure of sample preparation was used: once the bottle was opened, 20 ml of wine were taken for reference SO2 analysis, the rest of the bottle was poured in 5 plastic 50 ml containers for ET measurements (40 ml each), 1 plastic container for the FOSS WineScan analyzer and 2 small ampules for polyphenolic analysis with HPLC. All of these containers were frozen in liquid nitrogen and afterwards were stored at -80 deg C. Defreezing and thermostating of the sample at +22 deg C was performed right before the ET measurement for each sample. Reference SO2 analysis was performed with the standard aspiration method [5], which yields free, bound and total sulfur dioxide content. There were 5 replica ET measurements for each wine and the samples were diluted with Milli-Q water in the ratio 30 ml wine/70 ml water.

The data matrix from the sensor array was processed using The Unscrambler ® 9.7 (CAMO Software AS, Norway) using PCA and PLS1 algorithms.

FIGURE 1. Experimental set-up.

RESULTS

As a first step a calibration of the ET against reference data was obtained using the standard aspiration technique for SO2, performed using PLS1 regression for the whole data set. The observed correlation between ET response and reference free SO2 content was rather poor (R-square around 0.6). This is obviously because of the very narrow range of SO2 content which makes matrix effects in the sample dominating and thus hinders SO2 determination by the sensors of the array. So we tried to change an approach and randomly selected 13 different samples to cover the whole range 7-30 mg/L of SO2 to build a full cross-validation model. The parameters of the resulting model were pretty good and are presented in Table 1.

TABLE 1. Regression parameters for free SO2 calibration with 13 randomly chosen wines. PLS1, full cross validation.

	Slope	Offset	RMSE	R-square
Calib	0.926	1.232	1.653	0.926
Valid	0.902	1.635	2.022	0.893

This model was used to predict SO_2 content in the "unknown" samples, which were not employed in the calibration. The system performance was better for the samples with SO_2 > 15 mg/L (relative error of SO_2 prediction about 10%); for lower SO_2 content wines the relative errors were higher than 40% for some of the samples. This fact is probably due to the detection limit of the sensor array in the complex wine matrix. The detection limit for the HSO_3^- anion in pure water solutions for the sensor array under study was around 1 mg/L and in the real wine solutions the influence from other anions, e.g. tartaric acid, increase the detection limit significantly. Thus the ET system can be used to monitor free sulfur dioxide in a semiquantitative way, e.g. to see if the SO_2 content is higher than some prescribed value.

Also we have tried to estimate the correlation between the ET response and the total SO_2 content as obtained by the standard aspiration method. The corresponding model parameters are shown in Table 2.

TABLE 2. Regression parameters for total SO_2 calibration with 36 wines. PLS1, full cross validation.

	Slope	Offset	RMSE	R-square
Calib	0.868	1.232	1.652	0.926
Valid	0.851	1.634	2.022	0.893

In this case it was possible to build a PLS1 model with reasonable parameters even with the whole data set. A possible reason for this is that the set of complex equilibria is occuring in wine when binding SO_2. Even if the detection of free sulfur dioxide is hindered by different anions one can still estimate the total SO_2 content from the sensor response to the accompanying substances in equilibrium.

Parameters determined by the FOSS WineScan instrument such as total acids, volatile acids, pH, reducing sugars, were also studied for correlations with the ET response. Parameters of the PLS1 model for total acids are shown in the Table 3.

TABLE 3. Regression parameters for total acid calibration with 36 wine varieties. PLS1, independent test set validation (90 samples as a test set).

	Slope	Offset	RMSE	R-square
Calib	0.797	1.398	0.442	0.797
Valid	0.857	0.791	0.494	0.756

It is noteworthy that in this case an independent test set was used to validate the model, thus the corresonding parameters can be an estimation of real-life ET performance to measure total acids in unknown samples.

A very nice correlation was observed for pH (slope and R-square > 0.93 for calibration and independent test set validation) and this is fully expected as a pH glass sensor was included in the array.

There were no valuable correlations observed for the content of reducing sugars, which is not surprising because sensors within the array can only detect ionic spices and dissolved sugars will not be ionized under normal conditions.

TABLE 4. ET performance for the determination of polyphenolic compounds. PLS1, full cross validation.

	Slope	Offset	RMSE	R-square
Catechin – Sauvignon Blanc				
Calib	0.942	0.162	0.285	0.942
Valid	0.912	0.242	0.352	0.915
Catechin – Chardonnay				
Calib	0.949	0.169	0.444	0.949
Valid	0.906	0.319	0.591	0.913
Epicatechin – Sauvignon Blanc				
Calib	0.888	0.959	0.971	0.888
Valid	0.835	1.403	1.372	0.786
Epicatechin – Chardonnay				
Calib	0.355	5.521	2.909	0.355
Valid	0.2550	6.382	3.276	0.213
Caffeic acid – Sauvignon Blanc				
Calib	0.932	0.192	0.206	0.932
Valid	0.902	0.275	0.265	0.892
Caffeic acid – Chardonnay				
Calib	0.941	0.158	0.222	0.941
Valid	0.931	0.183	0.271	0.917
Coumaric acid – Sauvignon Blanc				
Calib	0.867	0.226	0.314	0.867
Valid	0.803	0.333	0.402	0.788
Coumaric acid – Chardonnay				
Calib	0.927	0.118	0.143	0.927
Valid	0.863	0.225	0.209	0.848

The HPLC data on the concentration of polyphenolic compounds present in wines were also used for ET calibration. When using the whole wine data set for calibration the corresponding R-square was lower than 0.7, probably due to a strong matrix effect. For further experiments we decided to build separate models for polyphenolic contents in two wine varieties: Chardonnay and Sauvignon Blanc. Four polyphenolic compounds: caffeic acid, coumaric acid, catechin and epicatechin were chosen for calibration.

Parameters for the corresponding models are presented in Table 4. The slopes and the R-square values for most of the models look reasonably good (>0.8), although the determination of epicatechin in Chardonnay wines is not possible by means of the suggested system. Also it should be kept in mind that full cross validation usually gives somewhat overoptimistic estimation of the prediction ability of the model. Further data processing is now in progress to receive more reliable results.

CONCLUSIONS

The developed ET multisensor system with potentiometric polymeric and chalcogenide glass sensors is capable of simultaneous determination of several quality parameters of white wines produced in New Zealand with acceptable precision. Further work is now in progress to improve both sensor performance and data processing.

ACKNOWLEDGMENTS

We thank the New Zealand Foundation for Research, Science and Technology (grant UOAX0404) for financial support for Olga Mednova's visit to New Zealand.

REFERENCES

1. Legin, A. Rudnitskaya, L. Lvova, Yu. Vlasov, C. Di Natale, A. D'Amico, *Anal.Chim.Acta*, 2003, 484, pp.33-44
2. Rudnitskaya, I. Delgadillo, A. Legin, S.M. Rocha, A.-M. Costa, T. Simões, *Chemometrics and Intelligent Laboratory Systems,* 2007, V.87, pp.50-56
3. M. L. Rodriguez-Mendez, V. Parra, C. Apetrei, S. Villaneuva, M. Gay, N. Prieto, J. Martinez, J.A. de Saja, *Microchim. Acta*, Vol.163, No. 1-2 (2008), pp. 23-31
4. Legin A., Rudnitskaya A., Vlasov Yu. "Electronic tongues: new analytical perspective for chemical sensors" in *Integrated Analytical Systems, Comprehensive Analytical Chemistry XXXIX*, edited by S. Alegret, Amsterdam, Elsevier, 2003, pp. 437-486
5. Iland, P. et al., "Techniques for Chemical Analysis and Quality Monitoring During Winemaking", Patrick Iland

Wine Promotions, (Campbeltown, South Australia, 2000).

Discrimination of Beef Samples by Electronic Nose and Pattern Recognition Techniques. Preliminary Results

P. Cornale, S. Barbera

Dipartimento di Scienze Zootecniche, Università di Torino, via L. da Vinci, 44 10095 Grugliasco (TO), Italy
paolo.cornale@unito.it – tel- +39 011 670 8713

Abstract. In this paper a study about the possibility of beef characterization with electronic nose is presented. Three beef classes were compared: Piemontese (PIE), Limousin (FRA) and meat from Argentine (ARG). 150 meat samples were put in glass vials and analysed with a commercial electronic nose instrument based on 10 metal oxide semiconductor sensors. Sensors response of beef classes seemed to be different. Different supervised and unsupervised pattern recognition procedures were applied to sensors signal: principal component analysis (PCA) and partial least square discriminant analysis (PLS-DA). Multivariate analysis pointed out promising classification and prediction results. Three clusters (according to the beef classes) can be clearly discriminated in PCA score plot. Statistical parameters from calibration, validation and prediction of PLS-DA model revealed themselves to be indices of a good model. These results demonstrate that electronic nose technology with multivariate analysis models is promising for the rapid determination of differences in meat aroma.

Keywords: Electronic nose, Meat, Aroma, PLS-DA.
PACS: 07.07.Df; 02.50.Sk

INTRODUCTION

In Europe, and particularly in Italy, widespread food qualification through an increasingly detailed characterization is gaining importance, especially for traditional foods. Researchers are constantly investigating new analytical, rapid, non-invasive, and cheap solutions to overcome this challenge. For instance, it is well known that aroma is a powerful driver of food choice.

Under these circumstances, the electronic nose technology is increasingly used both in scientific research and food industry.

An electronic nose is an array of chemical gas sensors with partial and different selectivity to an ensemble of chemicals. The overall response of these sensors allows, in combination with an appropriate pattern recognition method, the identification of odours and gases [1]. This technology is also used in the agricultural sector to evaluate odours of products and food like meat, grain, coffee, beer, cheese, sugar, mushrooms, etc. [2]. Moreover, gas sensors have been recently used to characterize products such as wine (distinguish different vineyards [3]), olive oil (classify varieties of olive and geographic origins [4]), and cheese [5].

In meat research field, these instruments are generally used to evaluate products' freshness and to monitor products' shelf life, by detecting chemicals produced by bacteria responsible of meat spoilage [6]. Previous publications show that over 1000 chemicals have been identified on volatile meat compounds [7]. Even if raw meat has only a blood-like taste, during cooking, wide mixtures of volatile compounds are developed mainly due to the Maillard reaction between amino acids and sugar, and the thermal degradation of lipids.

Aroma and meat flavour are subjected to variability due to both intrinsic and extrinsic factors [8]. Among them, breed and diet are of great importance as they induce different chemical compositions of meat, especially fat content and fatty acid composition. Machiels [9] observed significant influences of breed and diet on the development of odour compounds of cooked beef meat.

The aim of this study is to investigate the real capability of a MOS sensors based electronic nose in characterizing and discriminating different beef samples.

This paper is organized as follows: Section 2 describes meat sampling, electronic nose measurements and statistical analyses applied to sensors response. Section 3 explains and discusses

CP1137, *Olfaction and Electronic Nose: Proceedings of the 13th International Symposium*, edited by M. Pardo and G. Sberveglieri

obtained results. After looking at sensors signals, different multivariate models are compared. Finally, section 4 summarises the main contributions of this work and concludes the paper.

EXPERIMENTAL AND METHODS

Thirty packs of beef were purchased in three stores of the same big retailer in the N-W of Italy. Each pack contained a steak (M. longissimus dorsi) of adult male cattle, aged between 15 and 18 months. All packs were made of the same materials: a polyethylene terephthalate (PET) foam bowl containing the steak and packaged with plastic film. Ten steaks were Piemontese (PIE) meat, ten Limousin (FRA), and the remaining ten were imported beef from Argentina (ARG). Due to the well-know high variability of meat chemical composition, five meat samples were collected from each steak. Thus, a total of 150 samples were analysed in this experiment.

The used gas sensors array instrument was a portable electronic nose PEN2 (Airsense Analytics GmbH, Schwerin, Germany). The device consists of 10 MOS (metal oxide semiconductor) sensors positioned into a small chamber. Charcoal-filtered ambient air was used as reference gas ("zero gas" mode). The analysis was performed following the method proposed by Haugen [10]. For analysis, 2 g of raw meat samples were transferred to 40 ml glass vials with Teflon/silicon septa and screw cap. The samples were mantained in the fridge at 4°C during the preparation, and then put in an electric oven and incubated at 60°C for 20 min each. During the electronic nose measurements the headspace gas of samples was pumped into the sensors chamber for 20 s at a flow-rate of 150 ml/min. Reference air flushed the sensors for 120 s ("recovery time") after each sample. The 150 samples were divided in ten analytical sequences. To improve the robustness of the experiments, the samples were analysed in random order. During the analysis, the laboratory temperature was 22±1 °C. The electronic nose measurements lasted two days.

Data collected during the measurement are mainly represented by the sensors curve. It represents the conductivity change of the sensor against the time. The sensor signal, in fact, is expressed by the ratio G/G0 where G is the conductance of the sensor in presence of the sample, and G0 is the conductance of the sensor in reference air. The adopted sampling period was 1 s.

The feature extracted from the sensor signal (response) was determined by identifying the curve's peak (absolute response) and then by subtracting the baseline to that value.

Ten responses, one for each sensor, represent the variables (columns) in the data matrix considered for the statistical analysis, with rows corresponding to samples.

Different pattern recognition techniques [11] were applied on the collected data to explore relationships among classes of samples.

Principal component analysis (PCA) is probably the main unsupervised pattern recognition technique. PCA involves a mathematical procedure that transforms a number of possibly correlated variables into a smaller number of uncorrelated variables called principal components (PC). The first principal component accounts for as much of the variability in the data as possible, and each succeeding component accounts for as much of the remaining variability as possible, and so on. PCA allows exploring and visualizing hidden patterns in the data set.

PLS-DA (Partial Least Square Discriminant Analysis) is supervised classification method based on PLS regression. It differs from other multivariate classification systems since it assumes that a sample has to be a member of one of the classes included in the analysis. In PLS-DA, a dummy Y matrix is constructed with zeros and ones. The matrix contains a number of columns equal to the number of classes. For each observation (row), the matrix has value 1 in the column corresponding to the class it belongs to, and 0 in the remaining ones. Finally, a second matrix called X matrix contains the original data. In contrast to PCA that only uses the information of matrix X, PLS also takes into account the information of matrix Y.

All multivariate analyses were performed using the Unscrumbler 9.1 software (Camo ASA, Oslo, Norway).

RESULTS AND DISCUSSION

ARG, FRA and PIE meat samples show differences in the signals curves. From a first observation, the meat aroma response seems to be different. In particular, three sensors (W5S, W1S, and W2S) have greater variations than the remaining ones (with respect to baseline values). The W5S sensor reaches the highest peak in the PIE sample, followed by the ARG sample, and finally by the FRA one. Furthermore, ARG samples present highest value for the W1S and W2S sensors than FRE and PIE samples. However, only statistical analysis can detect the real importance of these sensors in beef discrimination process.

A first PCA procedure was carried out on all 150 samples to look at patterns in the data set. The results of the analysis show that the first two principal components explain about 100% of total samples variance. Three clusters (according to the beef classes) can be clearly discriminated from each other and no outliers stuck out of the groups. Even if FRA samples are more concentrative than PIE and ARG samples, there are not overlapped areas between clusters. Furthermore, the two factors "sequence" and "day of analysis" seem to have negligible effects on the sensors response since the PCA plot does not show any evident separation. The correlation loading shows that W5S sensor is the only and the most important variable defining the first PC. Mainly W1S sensor, but also W2S and W1W sensors, contributed positively to the definition of the second PC. Moreover, they are negatively correlated to W5C, W1C, and W3C sensors responses. The other sensors cannot be interpreted because they are very close to the centre of the plot. These preliminary results suggest that other statistical procedures can be applied to the data in order to build and to test classification and prediction models.

The 150 samples were divided in two sets. A first group, called "training set", had 75 samples analysed in the first day. It was used to build PCA and regression models. The second one, the "test set", corresponds to the other 75 samples, analysed in the second day, and it was used to test the models built on the training set by prediction procedures.

The PLS-DA is based on PLS regression. Since there were three classes in this experiment, a PLS2 method was used. Each class (beef) was represented by a dummy variable. By building a PLS2 model with all the dummy variables as Y matrix, the aim is to predict class membership from the X variables (sensors responses) describing the meat samples. In the development of the PLS-DA model, the variables were normalized divding them by the standard deviation value (1/SDev). Moreover, a full cross-validation procedure was used to evaluate the quality of calibration. The regression model performance was evaluated by the determination coefficient (R2), the root mean square error of calibration (RMSEC), and the cross validation (RMSECV). Finally, the test set was applied to evaluate the predictive ability of the model in classifying new meat samples.

The first four PCs were used in this model. They explained the 90.1% of the total calibration variance of the data. The score plot of the first two PCs is shown in figure 2. It clearly shows that the samples of the different classes are not overlapped and they can be separated. The correlation loading of PLS2 between X matrix (sensors response) and Y matrix (beef classes) shows that ARG is linked to the three variables (W1W, W1S, and W2S) and that they contributed negatively to the definition of the first PC. In contrast, PIE and FRA are positively linked to the first PC. However, while PIE is linked to the W5S variable in the positive part of the second PC, FRA was situated in the negatively part.

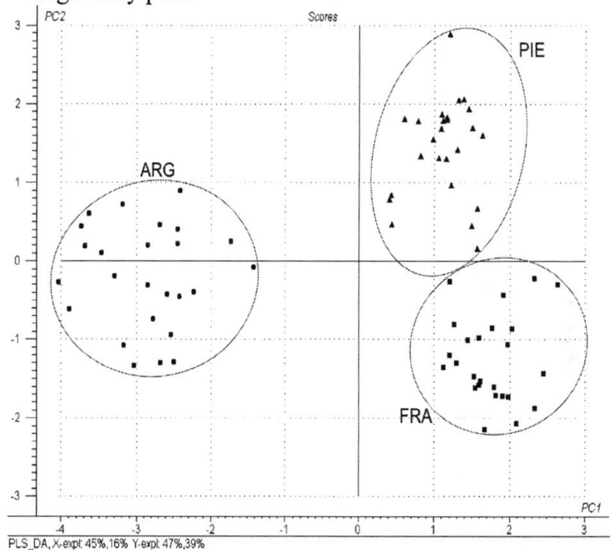

FIGURE 1. PLS-DA Score Plot of Training Samples.

Table 1 summarises the statistical parameters for the PLS2 regression model based on eight variables and four PCs. Both calibration and validation data revealed themselves to be indices of good model. R2 near 1 indicated strong correlation between sensors and categorised variables. Furthermore, low root mean square errors of calibration (RMSEC) and cross validation (RMSECV) were carried out. In addition, low RMSECV and SECV (standard error of cross

TABLE 1. Statistical Parameters from Calibration, Validation and Prediction of PLS-DA Model

	Calibration				Validation				Prediction			
	R^2	RMSEC	SEC	Bias	R^2	RMSECV	SECV	Bias	R^2	RMSEP	SEP	Bias
ARG	0.981	0.092	0.092	-4.54×10^{-7}	0.976	0.102	0.102	-4.59×10^{-5}	0.975	0.110	0.110	0.014
FRA	0.939	0.163	0.163	5.28×10^{-7}	0.922	0.183	0.184	3.75×10^{-4}	0.924	0.189	0.186	-0.042
PIE	0.920	0.184	0.186	-9.50×10^{-8}	0.899	0.201	0.203	-3.30×10^{-4}	0.897	0.214	0.213	0.027

validation) indicated quite good accuracy and precision of validation respectively.

In order to assess the usefulness of the final classification model, known data from the test set were predicted. The prediction procedure consists in feeding observed X values for new samples into the regression model to obtain predicted Y values. Statistical parameters of prediction (Table 1) showed values slightly lower than validation ones but still acceptable. The main results of the prediction includes the so-called "Predicted with Deviation" plot. The deviation value expresses a kind of 95% confidence interval around the predicted Y value.

Deviation and Y predicted values indicate the prediction performance. Samples with Ypred>0.5 and a deviation that does not cross the 0.5 line are predicted as members. On the other hand, samples with Ypred<0.5 and a deviation that does not cross the 0.5 line are predicted as non-members. Finally, samples with a deviation that crosses the 0.5 line cannot by safely classified (Camo, ASA, Norway). Figure 3 shows the "Predicted with Deviation" plot of ARG samples. It is possible to note that 100% of samples are correctly classified, even if one sample (arg38) has a Y predicted closer to 0.5 than to 1.

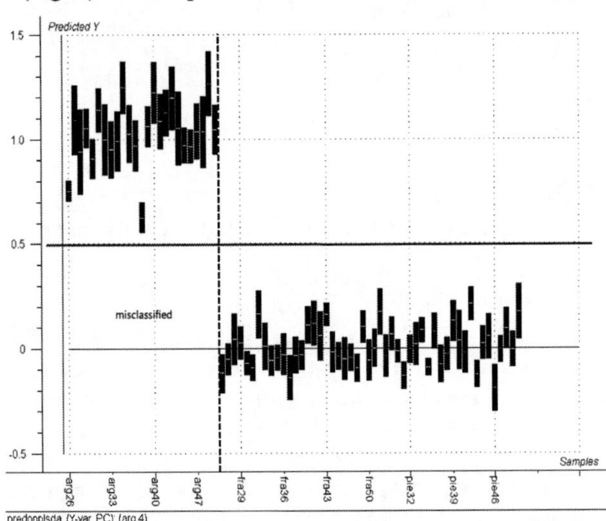

FIGURE 2. "Predicted with Deviation" Plot of ARG Samples.

The analysis of FRA and PIE plots showed some classification problems. Ypred of one sample (fra41) was close to 1 but its deviation value crossed the 0.5 line making uncertain the classification on the FRA class. The same occurred in the PIE class where three samples (pie27, pie45, and pie46) crossed the deviation threshold value of 0.5 and they were not safely classified. These data agreed with statistical parameters of prediction that pointed out best results for the ARG model than for FRA and PIE models.

CONCLUSIONS

An electronic nose based on ten metal oxide semiconductor sensors has been used on different beef meat samples. Thereafter, obtained sensors responses have been analysed with pattern recognition techniques (PCA and PLS-DA). Multivariate analyses showed that signals corresponding to different samples classes are significantly different.

These preliminary results demonstrate that electronic nose technology with multivariate analysis models is promising for the rapid determination of differences in meat aroma. PLS-DA model showed quite good classification and prediction results.

Electronic nose confirmed to be very useful in order to characterise beef samples. More studies are necessary to understand if electronic nose can also detect subtler difference in meat aroma.

ACKNOWLEDGMENTS

The authors wish to thank C.R.A.–Istituto Sperimentale per la Zootecnia of Torino for providing the gas sensors array instrument.

REFERENCES

1. J.W. Gardner and P.N. Bartlett, *Sens. Actuators, B, Chem.* **18**, 211-220 (1994).
2. E. Schaller, J.O. Bosset, F. Escher, *LWT* **31**, 305-316 (1998).
3. M. Penza and G. Cassano, *Food Chem.* **86**, 283-296 (2004).
4. M.S. Cosio, D. Ballabio, S. Benedetti, C. Gigliotti, *Anal. Chim. Acta* **567**, 202-210 (2006).
5. S. Benedetti and S. Mannino, *Compr. Anal. Chem.* **49**, 131-138 (2007).
6. S. Panigrahi, S. Balasubramanian, H. Gu, C. Logue, M. Marchello, *LWT* **39**, 131-145 (2006).
7. D.S. Mottram, *Food Chem.* **62**, 415-424 (1998).
8. R.A. Lawrie, *Lawrie Meat Science*, Cambridge: Woodhead, 2006, pp. 442.
9. D. Machiels, L. Istasse, S.M. van Ruth, *Food Chem.* **86**, 377-383 (2004).
10. J.E. Haugen, F. Lundby, J.P. Wold, A. Vederg, *Sens. Actuators, B, Chem.* **116**, 78-84 (2004).
11. T. Naes, T. Isaksson, T. Fearn, T. Davies, *A user-friendly guide to multivariate calibration and classification*, Chichester: NIR Pubblications, 2002, pp. 344.

Assessment of bitterness intensity and suppression effects using an Electronic Tongue.

A.Legin[1], A. Rudnitskaya[1,2], D. Kirsanov[1], Yu. Frolova[1], D Clapham[3], R Caricofe[4]

[1] *Laboratory of Chemical Sensors, Chemistry Department, St. Petersburg University, Russia*
[2] *Chemistry Department, University of Aveiro, Aveiro, Portugal*
[3] *GlaxoSmithKline Pharmaceuticals, Harlow, U.K.*
[4] *GlaxoSmithKline Pharmaceuticals, Research Triangle Park, U.S.A.*

Abstract. Quantification of bitterness intensity and effectivness of bitterness suppression of a novel active pharmacological ingredient (API) being developed by GSK was performed using an Electronic Tongue (ET) based on potentiometric chemical sensors. Calibration of the ET was performed with solutions of quinine hydrochloride in the concentration range 0.4 – 360 mgL^{-1}. An MLR calibration model was developed for predicting bitterness intensity expressed as "equivalent quinine concentration" of a series of solutions of quinine, bittrex and the API. Additionally the effectiveness of sucralose, mixture of aspartame and acesulfame K, and grape juice in masking the bitter taste of the API was assessed using two approaches. PCA models were produced and distances between compound containing solutions and corresponding placebos were calculated. The other approach consisted in calculating "equivalent quinine concentration" using a calibration model with respect to quinine concentration. According to both methods, the most effective taste masking was produced by grape juice, followed by the mixture of aspartame and acesulfame K.

Keywords: electronic tongue, potentiometric chemical sensors, instrumental bitterness assessment, analysis of pharmaceuticals
PACS: 01.30.Cc

INTRODUCTION

Most active pharmacological ingredients (API) have objectionable taste and in particular many are bitter. A bitter taste in the final formulation could cause poor compliance with the treatment regime by the consumer/patient. This is especially true with regard to paediatric formulations.

Masking the bitterness of active compounds to make pharmaceutical formulation more palatable has long been one of the important goals of drug development process. However, to be able to mask the taste of a bitter component, one must be able to evaluate its bitterness intensity first, which is usually done by a sensory panel. In addition the process of drug formulation could be facilitated by choice of the least organoleptically objectionable molecule or version of a molecule at the candidate selection stage. At this early stage sensory evaluation by a panel of the bitterness intensity of actives is a difficult exercise due to the lack of a full toxicology profile of some of these active molecules. Even once this profile is established there is a desire to reduce the amount of human tasting to the minimum required to achieve the desired objective.

An analytical instrument that is believed to be capable to replace sensory panels to some extent is an Electronic Tongue (ET). To date all ET's are based on arrays of chemical sensors combined with multivariate regression and pattern recognition tools [1]. ETs, based on potentiometric chemical sensor have been intensively studied as a potential tool for the assessment of bitterness intensity and bitterness masking effects. Applications of the ET to the detection and the suppression of the bitter taste of quinine hydrochloride by commercial bitterness-masking mixture of phospholipids and sucrose and aspartame were reported previously [2-3]. Discrimination of substances with different taste, ranking of several APIs according to the bitterness intensity and estimation of the bitterness masking by sweeteners, flavourings, soft drink and commercial bitterness-suppressing agents has also been reported [4-5].

The purpose of the present study was evaluation of the bitterness intensity of a new pharmacological compounds and determination of the effectiveness of various excipients in masking bitterness prior to FTIH (First time

CP1137, *Olfaction and Electronic Nose: Proceedings of the 13th International Symposium*, edited by M. Pardo and G. Sberveglieri

in human) studies using an Electronic Tongue based on potentiometric chemical sensors.

TABLE 1. List of the samples measured using the ET.

Bitter tasting substances		**Sweeteners**	
Substance	**Concentrations**	**Substance**	**Concentrations**
Quinine hydrochloride	0.4, 1.1, 3.6, 36, 360 mgL^{-1}	-	-
dihydrate	3.6 mgL^{-1}	Aspartame/Acesulfame K (ratio 2:1)	0.125, 0.25, 0.5%
Bittrex	1, 2, 4, 8 gL^{-1}	-	-
	0.087, 0.175, 0.35, 0.7 gL^{-1}	-	-
API	0.7 gL^{-1}	Aspartame/Acesulfame K (ratio 2:1)	0.125, 0.25, 0.5%
		Sucralose	20%, 40%
		Grape juice	50%

EXPERIMENTAL AND METHODS

Measurements with the ET were made in individual solutions of bitter tasting substances and their mixtures with potential masking agents. The bitter substances were quinine hydrochloride, bittrex and the API. The sweeteners and tasting masking agents were suclarose, mixture of asparatme with acesulfame K with ratio 2:1 and grape juice. A list of measured samples together with the concentration ranges is shown in the Table 1.

All the substances except grape juice were provided GlaxoSmithKline Pharmaceuticals, UK in the solid form as tablets or powders. Grape juice (100% juice reconstituted from the concentrate) was purchased in the local store. Concentrated aqueous stock solutions of the substances were prepared. Solutions with lower concentrations were prepared by dilution of the stock solutions. Doubly distilled water was used throughout the experiments for the solution preparation and washing the sensors. At least three replicates of each sample were run.

The ET used in this experiment comprised 11 potentiometric chemical sensors with plasticised PVC membranes. All sensors were made in the Laboratory of Chemical Sensors, of St. Petersburg University. Details of sensor preparation procedure and membrane compositions have been published elsewhere [1]. Sensor potential values were measured with 0.1 mV precision against the Ag/AgCl reference electrode using a custom made multi-channel digital high impedance voltmeter connected to a PC for data acquisition. A glass pH electrode was used to monitor and control the acidity of solutions.

Principal Component Analysis (PCA) was used for sample recognition. Calibration models with respect to the quinine concentrations were produced using Multilinear Regression (MLR). Logarithms of the quinine concentrations were used to calculate MLR calibration models. Validation was undertaken using leave-one-out cross-validation. All calculations were performed using The Unsrambler v.9.7 software by CAMO, Norway.

RESULTS AND DISCUSSION

Measurements with the ET were initially carried out in solutions of the bitter tasting substances i.e. quinine, bittrex and the API. PCA score plot of these samples are shown in the Fig. 1. Points forming clusters on the plot correspond to samples of the same concentration. The cluster numbers identify concentration (from the lowest to the highest).. The ET was capable of distinguishing all three substances as well as different concentration levels of all of them. PC1 corresponds to the concentration increase of all substances i.e. to the increase of the bitter taste intensity. PC2 mostly accounts for the difference in ET response to quinine and bittrex, and the novel compound.

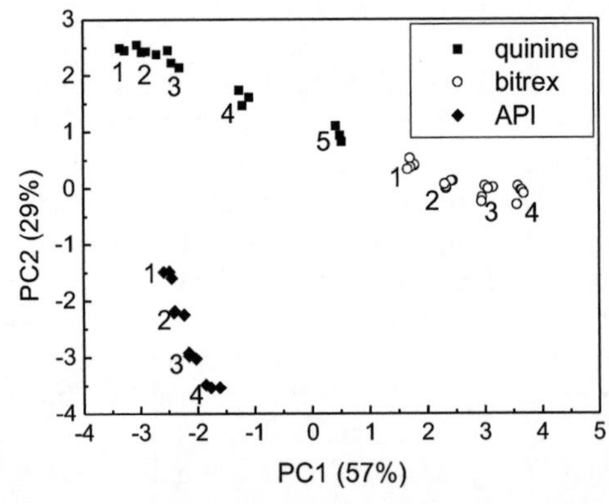

FIGURE 1. PCA score plots of the three bitter tasting substances. For the concentration ranges see Table 1.

An attempt to compare bitterness intensity of bittrex and the API with the quinine was performed. A calibration model with relation to the quinine concentration was calculated using MLR. Prediction accuracy for the validation samples was 5% for the measured concentration range. "Equivalent quinine" concentrations of bittrex and API solutions that were calculated using this calibration model are shown in Table 2. All bitrex solutions were found to be significantly more bitter than even the most concentrated quinine solution tested (0.36gL^{-1}.). Solutions of the API were predicted to have bitterness values at the low end of the concentration range measured. Human recognition threshold of quinine is considered to be about 2 mgL^{-1}. Therefore, all the API solutions would be expected to be perceived as bitter, especially at the higher concentration levels that correspond to the higher doses that may be administered in the clinical trial, but this bitterness should be moderate in intensity.

TABLE 2. Prediction of the bitterness intensity as an "equivalent quinine" concentration of the bittrex and API solutions

Substance	Real concentration, gL^{-1}	Predicted "equivalent quinine" concentration, gL^{-1}
Bittrex	1	2.3
	2	8.8
	4	23
	8	50
API	0.087	0.005
	0.175	0.008
	0.35	0.01
	0.7	0.03

Sweeteners and taste masking agents are commonly used to decrease the bitter taste of APIs and to make resulting preparation more palatable. Suppression of the bitter taste of the the API at the highest dose (0.7 gL^{-1}) by sweeteners was studied. Two approaches were employed to quantify this effect.

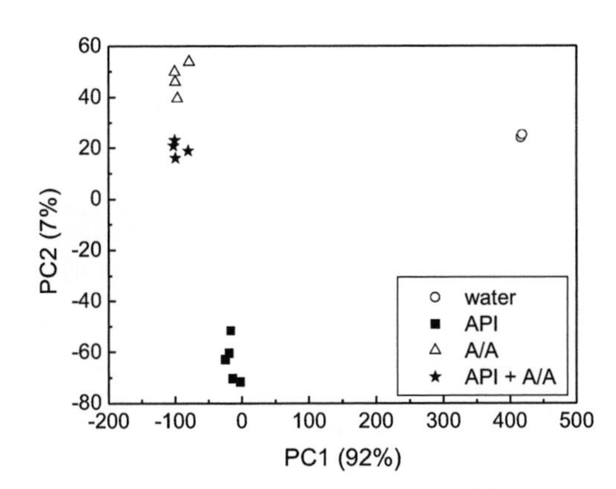

FIGURE 2. PCA score plots of water, individual API solutions individual aspartame and acesulfame K solutions (A/A) and mixed solutions of the API and of aspartame and acesulfame K (API + A/A).

The first approach consisted in expressing bitterness suppression as the Euclidean distances on the PCA score plots between API solutions and corresponding placebo, i.e. the distances are calculated between individual API solution and water, and between API solution with sweetener added and that of a solution containing the same concentration of sweetener alone. Concentrations used were 0.7 gL-1 for the API, 0.125% for the aspartame and acesulfame K mixture, 20% for sucralose and 50% for grape juice. Three separate PCA models were calculated for each type of sweetener. PCA score plot of the model for the API and mixture of aspartame with acesulfame K as a sweetener is shown in the Fig. 2. Mixed solutions of the API with aspartame and acesulfame K were situated on the score plot between corresponding individual solutions and quite close to the aspartame/acesulfame solution. This displacement of the sample containing sweetener in the direction of the pure sweetener may be considered as a measure of the bitterness suppression. Bitterness suppression effects expressed as distances for all three sweeteners are shown in the Table 3. It was found that all three sweeteners were capable of decreasing bitterness intensity of the compound. The biggest effect was observed for the grape juice, closely followed by the mixture of aspartame and acesulfame K. Concentration dependence of the bitterness suppression was not established using this method but this could be undertaken by running a series of such comparisons with different levels of masking agent.

TABLE 3. Euclidean distances on the PCA score plot between individual and mixed with masking agent API solutions and corresponding placebos. A/A stands for the mixture of aspartame with acesulfame K.

Solutions	Distance
API – water	436
API + A/A – A/A	28
API + grape juice – grape juice	24
API + sucralose - sucralose	108

Another approach to the assessment of the bitterness suppression effects consisted in predicting "equivalent quinine" concentrations of the mixed solutions. A Calibration model with respect to quinine concentration was calculated for this purpose as described above. However, in this case, measurements in mixed solutions of quinine at the concentration level 3.6 mgL^{-1} and mixture of aspartame and acesulfame K at the concentration levels 0.125, 0.25 and 0.5 weight % were included into the calibration data set. The resulting calibration model accounted for the ET response to the sweeteners and therefore was expected to be more adequate for predicting "equivalent quinine" concentrations in mixed solutions of the API and sweetener. Predicted "equivalent quinine" concentration values for mixtures of API with three sweeteners are shown in the Table 4. The same tendency was observed as with the distances (Table 3): the most effective bitterness masking agent was grape juice, then mixture of acesulfame and aspartame and the least effective was sucralose. Though, suppression effects of grape juice and aspartame/acesulfame K mixture appeared to be very similar according to the distance measure, grape juice was much more effective according to the "equivalent quinine concentration'. Taking into account recognition threshold of quinine, we can suggest that bitter taste of the API at the 0.7 gL-1 concentration should be completely masked by the addition of 0.5% solutions of aspartame/acesulfame K mixture or 50% of grape juice.

TABLE 4. "Equivalent quinine" concentrations for the mixed solutions containing 0.7 gL^{-1} of API and a sweetener. A/A stands for the mixture of aspartame with acesulfame K. Standard deviation of three measurements is shown in the brackets.

Sweetener concentration	"Predicted Equivalent quinine" concentration, mgL^{-1}
-	31 (12)
20% of sucralose	18 (6)
60% of sucralose	3.9 (0.9)
0,125% of A/A	3.3 (0.4)
0,25% of A/A	2.8 (0.3)
0,5% of A/A	2.0 (0.5)
50% of grape juice	0.0016 (0.0006)

CONCLUSIONS

An Electronic Tongue based on 11 potentiometric chemical sensors was applied to the quantification of bitterness intensity and suppression of the bitterness of a new active pharmacological compound. Measurements with ET were carried in the individual solutions of quinine hydrochloride, bittrex and the API at different concentration levels as well as in the mixtures of quinine hydrochloride or the API with sweeteners. Calibration with respect to quinine concentration was calculated by MLR using measurements in the individual solutions. Bitterness intensity expressed as "equivalent quinine concentration" was predicted for bittrex and API solution. Bittrex was found to be much more bitter than the API in all range of measured concentrations. Masking of bitterness achieved by three sweeteners (sucralose, a 2:1mixture of aspartame with acesulfame K and grape juice), was estimated using two methods. According to both methods, the most effective in taste masking was grape juice, followed by the mixture of aspartame and acesulfame K The least effective was sucralose. It is important to state that all the results concerning "taste assessments" are made solely on the basis of the sensor responses of the Electronic Tongue, without proper calibration of the ET against human assessments.

This work demonstrates applicability of the Electronic Tongue to the assessment of bitterness and the necessity to build the so-called "universal bitterness scale" for fast "unmanned" assessment of newly developed substances.

ACKNOWLEDGMENTS

Work of A. Rudnitskaya was supported by the postdoctoral fellowship SFRH / BPD / 26617 / 2006 by FCT, Portugal.

REFERENCES

1. Legin, A., Rudnitskaya, A., Vlasov, Yu. Integrated Analytical Systems (S. Alegret Ed.), Comprehensive Analytical Chemistry XXXIX, Elsevier (2003) pp. 437–486.
2. Tagaki, S., Toko, K., Wada, K., Ohki, T., "Quantification of suppression of bitterness using an electronic tongue", *J Pharm Sci*, Vol. 90 (2001), pp. 2042-2048.
3. Nakamura, T., Tanigake, A., Miyanaga, Yo., Ogawa, T., Akiyoshi, T., Matsuyama, K., Uchida, T., "The Effect of Various Substances on the Suppression of the Bitterness of Quinine–Human Gustatory Sensation, Binding, and Taste Sensor Studies", *Chem Pharm Bull,* Vol. 50, No. 12 (2002), pp. 1589-1593.
4. Zheng, J.Y., Keeney, M.P., "Taste masking analysis in pharmaceutical formulation development using an electronic tongue", *In. J Pharm*, Vol. 310 (2006), pp. 118–124.
5. Legin, A., Rudnitskaya, A., Clapham, D., Seleznev, B., Lord, K., Vlasov, Yu., "Electronic tongue for pharmaceutical analytics: quantification of tastes and masking effects", *Anal Bioanal Chem*, Vol. 380 (2004), pp. 36–45.

POSTER SESSION I

BIOSENSORS

Novel Nanocomposite-based Potassium Ion Biosensor

R. Xue and P.I. Gouma

Department of Materials Science and Engineering, State University of New York
Stony Brook, NY 11794-2275, USA

Abstract. Potassium ion (K+) is important in regulating normal cell function in the human body, specifically the heartbeat and the muscle function. Thus, it is important to be able to monitor potassium ion concentrations in human fluids. This paper describes a novel concept for a potassium ion biosensor that accurately, rapidly, and efficiently monitors the presence and records the concentration of potassium ions with high specificity, not only in serum and urine, but also in the sweat or even eye fluid. This specific biosensor design utilizes a nanomanufacturing technique, i.e. electrospinning, to produce advanced nano-bio-composites that specifically trace even minute quantities of potassium ions through the use of selective bio-receptors (ionophores) attached to high surface area nanofibers. Electroactive polymers are then employed as transducers to produce an electronic (rather than ionic) output that changes instantly with the change in K+ concentration. Such biosensors may be manufactured in a skin patch configuration.

Keywords: Potassium, biosensor
PACS: 81.05.Lg

INTRODUCTION

Helping to regulate the body's fluid levels and blood pressure are two of the mineral potassium's greatest functions. Typically, potassium levels in the blood are regulated by the kidneys and any excess potassium is eliminated in the urine [1].. There are many types of potassium sensors available. Among these, ion selective electrodes (ISE) are widely used as potassium sensors in biochemical and biophysical research to measure the potassium concentration in an aqueous solution [2]. In this paper, we present a novel approach where electrospinning is employed.

EXPERIMENTAL AND METHODS

Electrospinning [3], is a fiber forming process capable of producing non-woven mats of nanoscale dimensios, thus offering large surface area to volume ratio, and controlled pore size distribution. The electrospun fibers were collected on an aluminum foil to form fiber membranes. Scanning electron microscopy (SEM) was performed for all specimens using the FEG SEM LEO Gemini 1550.

RESULTS AND DISCUSSION

The "Band-aid"-type K+ monitor concept

We have invented [4] the concept of a "band-aid" type sweat test for measuring potassium ion levels fast, accurately, and potentially wirelessly. Measuring K+ concentration in sweat is expected to overcome the problems/artifacts often encountered during the blood drawing process [5] altering the level of (extracellular) potassium concentration measured in serum (see pseudohyperkalemia [5]). Based on the medical literature, there are known levels of normal potassium concentration in the sweat of humans (4.7-9.7mM/L for the male population; 7.6-15.6mM/L for the female population [6]). Using this knowledge, a convenient, non-invasive, maintenance-free, sensitive and inexpensive potassium ion monitor is described in figure 1.

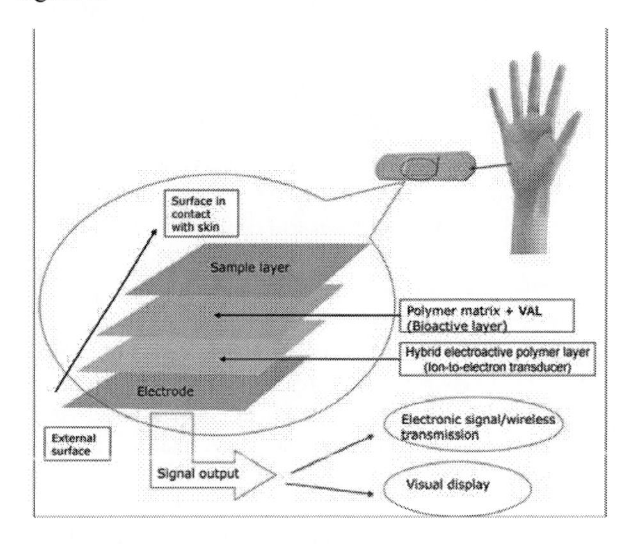

FIGURE 1. The "band-aid" K+ monitor concept

CP1137, *Olfaction and Electronic Nose: Proceedings of the 13th International Symposium,* edited by M. Pardo and G. Sberveglieri

Similarly to the drug sweat test [7], the patient will wear the patch for long periods of time so there is no need to activate the release of sweat, as is the case with the sweat test for cystic fibrosis which requires a large amount of sweat sample [8]. The first layer in contact with the skin consists of a non-woven carrier membrane on which the ionophore is positioned; this ionophore selectively binds with potassium ions and transfers them to the second membrane underneath, the ion-to-electron transducer based on electroactive polymer composites. Finally, a layer of electrodes are incorporated on the external surface of the flexible monitor (patch) to enable (wireless) transmission of the signal for (remote) data analysis, as well as potentially visual display of the electrolyte's concentration.

Sensor Components

Bio-active nanofibrous mats

Our recent work [9] has produced process maps for various precursor (polymer concentration) and process conditions (flow rate, voltage and working distance)

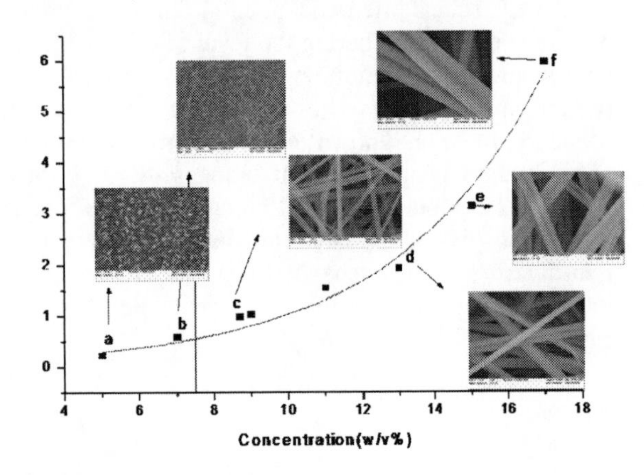

FIGURE 2: Bio-nanocomposites process maps.

Ion-to-Electron Transducer

By combining the transducer ability of electroactive polymers, such as polyaniline (PANI) [10], with the selective potassium ion detection offered by the sensing mat described above, chemical information (potassium ion concentration) is converted into electrical signal in the solid state. Thus our biosensor concept becomes a resistive solid-state chemodetector, giving a robust, accurate, and fast response to the changes in potassium ion concentration. It is also easily interfacing with electronic circuits and wireless platforms.

CONCLUSIONS

The novelty of using a nanomanufacturing technique, such as electrospinning, is to produce advanced nano-bio-composites that specifically trace even minute quantities of potassium ions through the use of selective bio-receptors and high surface area nanofibers. Electroactive polymers are then employed as transducers to produce an electronic (rather than ionic) output that changes instantly with the change in K+ concentration.

ACKNOWLEDGMENTS

This project is supported by a CEWIT/HSC nanomedicine initiative.

REFERENCES

[1]Cheng, J.F. et al " All-solid-state separated potassium electrode based on SnO2/ITO glass." Journal of the Electrochemical Society Vol 154, (2007) pp. 369-374.

[2] Zhu, Z. Q. et al "Fabrication and characterization of potassium ion-selective electrode based on porous silicon," Ieee Sensors Journal, Vol 7, No. 1-2, (2007) pp38-42

[3]Smita. Y. "Biodoped Ceramics for Resistive Biosensing," Master thesis, Stony Brook University, 2006.

[4]Gouma. P. "progress report on novel potassium ion biosensor project", Stony Brook University, October, (2008)

[5] Walker, H.K. et al, Clinical Methods: The History, Physical, and Laboratory Examinations, editors Stoneham (MA): Butterworth Publishers, (1990)

[6]http://www.liv.ac.uk/~agmclen/Medpracs/practical_2/practical_2.pdf

[7]http://www.drugpolicy.org/law/drugtesting/sweatpatch_/

[8] Vicky, A. et al. " Diagnostic and Therapeutic Methods: Sweat analysis proficiency testing for cystic fibrosis." Pediatric Pulmonology Vol 30, Issue 6, (2000) pp476-480

[9]Gouma, P. "progress report on novel potassium ion biosensor project" Stony Brook University, April, (2008)

[10] Lindfors, T.et al. "Stability of the inner polyaniline solid contact layer in all-solid-state K+-selective electrodes based on plasticized poly(vinyl chloride)" Analytical Chemistry Vol 76, (2004), pp4387-4394

A Research on Sour Sensation Mechanism of Fungiform Taste Receptor Cells Based on Microelectrode Array

Wei Zhang[12], Peihua Chen, Lidan Xiao, Qingjun Liu and Ping Wang[12]*

1 Biosensor National Special Laboratory, Key Laboratory of Biomedical Engineering of Ministry of Education,
2 State Key Laboratory of Transducer Technology, Chinese Academy of Sciences, Shanghai, 200050, China
Department of Biomedical Engineering, Zhejiang University, Hangzhou 310027, China
*Correspondence should be addressed to Ping Wang (email: cnpwang@zju.edu.cn)

Abstract. Taste receptor cells as the fundamental units of taste sensation are not only passive receivers to outside stimulus, but some primary process for the signals and information. In this paper, an innovation on acquisition of taste receptor cells was introduced and larger amount of cells could be obtained. A multichannel microelectrode array (MEA) system was applied in signal recording, which is used in non-invasive, multiple and simultaneous extracellular recording of taste receptor cells. The cells were treated with sour solutions of different pHs, and the relations between concentration of hydrogen and firing rate were observed. Firing rates on pH 7, pH 4 and pH 2 were approximately 1.38 $\pm$0.01 (MEAN±SE)/s, 1.61$\pm$0.07/s and 2.75$\pm$0.15/s.

Keywords: Taste receptor cells, fungiform papillae, Sour sensation, firing rate.
PACS: 87.19.It

INTRODUCTION

Taste is essential in daily life. Recent Works have been done to identify the molecular structure and electrophysiological function in taste mechanism and transduction. Different responses to sweet, salty, sour and bitter are well investigated[1-4]. However, many mechanisms are not completely clear yet. It has been proved that ion channels, ligand-gated channels, enzymes and GPCRs serve as receptors for sensory tastants, and trigger the downstream transduction events within taste cells. Voltage gated sodium and potassium channels are contributive to the generation of action potential of taste receptor cells. Depolarizations linked to an increase of the membrane conductance of taste cells have already been reported[5].

Some methods such as electrophysiological recording, immunohistochemical stain and RT-PCR are employed in studies. Patch clamp is considered as a standard intracellular recording method, in spite of its operation complexity and invasive to cells. Loose patch is a way closing to but outside the membrane, using the electrode to record the extracellular responses of single cell. Whereas, it is difficult for large amount acquisition of taste receptor cells and lacks an effective approach for extracellular electrophysiological recording.

In our previous work[6], it has been tried to culture taste receptor cells onto light addressable potentiometric sensors (LAPS) chip and the responses to a mixture solution of stimulus were recorded. Some specific components were extracted in frequency domain. In this paper, a method using microelectrode array for extracellular simultaneously recording of multiple taste receptor cells to sour solution is introduced, and the responses to different pHs in magnitude and firing rate are also presented.

EXPERIMENTAS AND METHODS

The MEA chip can be fabricated in CMOS-processing techniques. It needs sputtering with Cr thin film on Pyrex Wafer, then 300 nm Au was deposited and electrodes of different sizes (50 µm, 70 µm, 120 µm) were fabricated using wet etching technique. Passivated layer was deposited above the metal layer by PECVD. Au is a highly conducting material with durable biocompatibility. A 250 µL chamber was packaged on the chip for cell culture (as shown in Figure 1). A multichannel* MEA system was employed in the electrophysiological measurement. Its 16-channel data acquisition card and software were utilized in data recording, display and analysis.

CP1137, *Olfaction and Electronic Nose: Proceedings of the 13th International Symposium,* edited by M. Pardo and G. Sberveglieri

Taste receptor cells were isolated from adult Female Sprague Dawley rats. After tongue removed, approximately 1-1.5 mL elastase/collagenase enzyme mixture solution was injected uniformly under the tongue epithelium. After incubated for 40 min, the lingual epithelium was gently peeled off from the underlying tissue. And then the epithelium was pressed gently on a sieve with pore diameter of 75 μm. The taste bud could pass through the pore out of the fungiform papillas into tyrode solution. Single taste receptor cells dissociated and maintained for couple hours at 37 ℃ under humidified air with 5% CO_2 in fresh DMEM.

Cultured taste cells were fixed in 4% paraformaldehyde in 0.1 M PBS for 2 h at room temperature. After rinsed, the cells were treated with 0.3% Triton X-100 and normal goat serum (1:10 diluted in 0.01 M PBS) for 30 min to block nonspecific binding. Then it was incubated in polyclonal rabbit anti-gustducin overnight at 37℃. After rinsed with PBS, the cells were incubated in anti-rabbit-IgG (1:100 diluted) for 30 min and another 30 min in FITC at room temperature. Fluorescent images were captured with the Leica Confocal Scanning Microscope.

RESULTS

Taste Cell Immobilization on Chip

After a couple of hours, the cells could adhere to the surface of chip. In Figure 2(a), several bipolar cells could be seen located onto the electrodes. Taste cells are about 15-20 μm long with a width less than 10 μm. Sieving cells from the enzyme digested lingual epithelium is more effective in acquisition larger amount of cells than conventional suction using a fire-polishing pipette. There were 100-250 cells/cm^2 on chip.

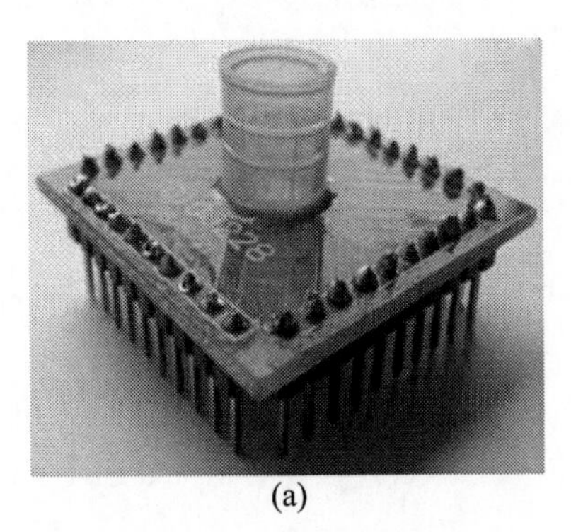

(a)

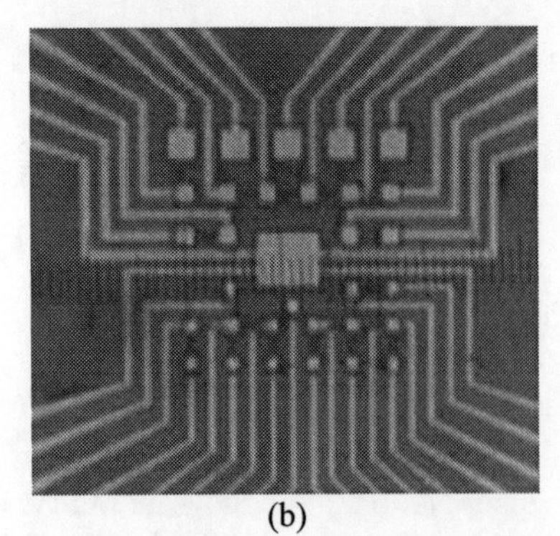

(b)

FIGURE 1. Schemes of the chip (a) and microelectrode array on chip (b).

Immunocytochemistry Results

In recent years, it is one of the most important discoveries on taste transduction mechanism study that the emergence of the Gustducin protein. It is a kind of G protein for gustataion transduction, which presented particularly in taste receptor cells. Ruiz has confirmed that there are Gustducin expression in dissociated TRCs , which argued it could reserved intact taste receptor and ion channels without innervation[7].

As shown in Figure 2(b), taste receptor cells can be seen clearly which spread through the taste bud. It is much brighter in the receptor region than other, which shows the evidence that cells cultured on chip are taste receptor cells. Negative control group was not shown here.

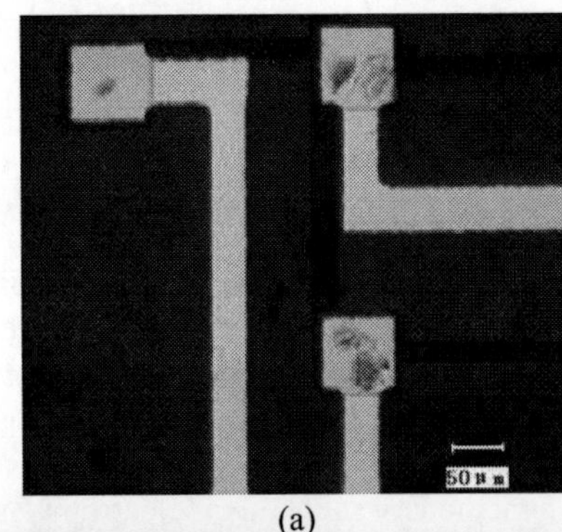

(a)

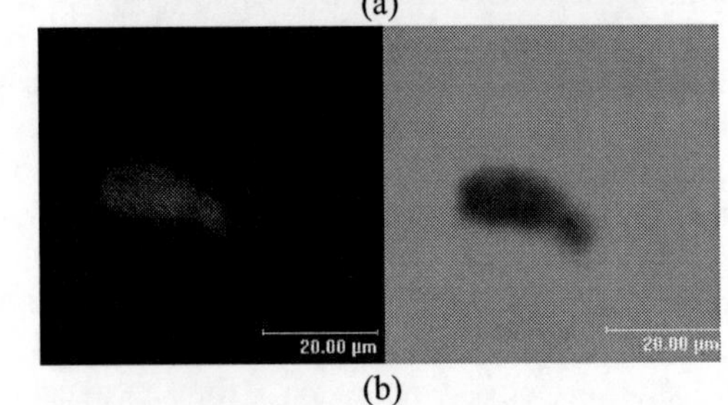

(b)

FIGURE 2: Taste receptor cells on the 50 μm electrode (a) and fluorescent image labeled in FITC (b).

Responses to Different Sour Stimulants

After incubation, taste receptor cells were treated separately in hydrochloric acid solution with pH 2 and pH 4. Tyrode solution with calcium was served as a treatment of pH 7. As shown in Figure 3, signals in control and the spikes responsed to pH 7, pH 4 and pH 2 were approximately 30-50 μV, 80-150 μV and 180-250 μV separately with a duration of 150-250 ms.

The firing rates changing were shown in Figure 4. Sour sensation responses behave corresponding to the change of pH. The frequency went higher when the concentration of hydrogen ion in solution increasing. Firing rates of pH 7, pH 4 and pH 2 were probably 1.38 ± 0.01 (MEAN±SE)/s, 1.61 ± 0.07/s and 2.75 ± 0.15/s.

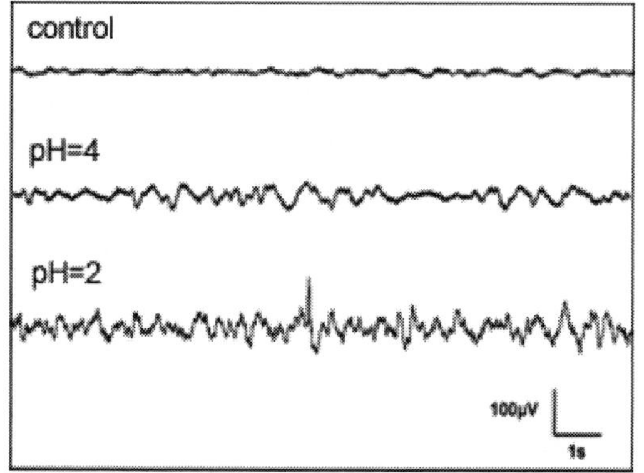

FIGURE 3: Responses of taste receptor cells to sour solutions of different pHs.

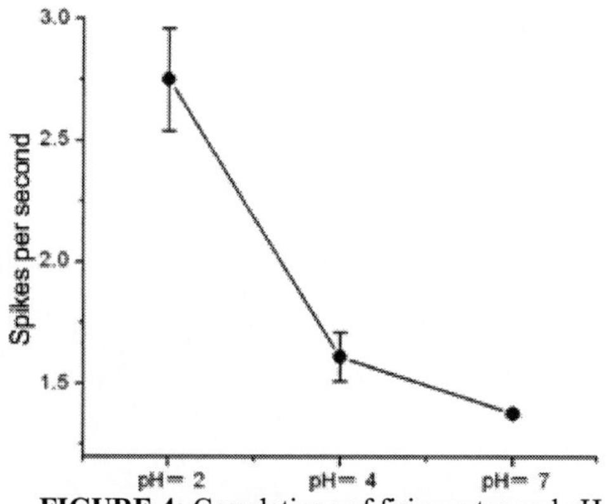

FIGURE 4: Correlations of firing rates and pH.

DISCUSSION

It suggestes that information derived from taste cells generating firings may provide the main components of taste information which is transmitted to innervation gustatory nerve fibers. The spikes firings recorded in MEA system is close to those recorded in mouse fungiform papillae via patch technology. Yoshida et al recorded the action potentials from the basolateral membrane of the taste cells isolated from mouse fungiform papillae[8]. The mean spontaneous firing rates were 2.59 ± 3.04/10s for taste cells and 2.78 ± 2.73/5s for fibers, and the best HCl stimulus onset were 9.02 ± 6.71/5s for taste cells and 82.4 ± 36.4/5s for fibers. That may be because single fiber innervate multiple taste receptor cells.

CONCLUSIONS

It presents a long-term, stable platform for multiple taste receptor cells extracellular recording, which provides a potential pathway to reveal the coding process of gustatory sensation and correlation of periphery and central gustatory system. Studies on extracellular responses to sour sensation have proved that firing rates of taste receptor cells depend on the hydrogen concentrations changing. Firings are seemed to be more frequent when cells were treated with higher concentrations solution with HCl.

Efforts have been done in the neuron pattern theory in central system recently. The correlative activities of taste sensitive neuron pairs were dependent on the relative response magnitudes in the nucleus of solitary tract, parabrachial nucleus and gustatory cortex. And the magnitude and firings might be involed in taste quality coding. Further works still need to be done to explain the taste transduction mechanism.

ACKNOWLEDGMENTS

This work was supported by the National Natural Science Foundation of China (Grant No. 30627002, No.60725102), and the National Natural Science Foundation of Zhejiang Province of China (Grant No. R205502).

REFERENCES

1. T.A. Cummings, J. Powell and S.C. Kinnamon, J Neurophysiol **70**, 2326-2336 (1993).
2. P. Avenet and B. Lindemann. Journal of Membrane Biology **124**, 33-41(1991).
3. T. A.Gilbertson, P. Avenet, S.C. Kinnamon and S.D. Roper, Journal of General Physiology **100**, 803-824 (1992).
4. H. Furue and K. Yoshii, Brain Res. **776**, 133-139 (1997).
5. M. Ozeki, Journal of General Physiology **58**, 688-699 (1971).
6. W. Zhang, Y Li, Q Liu, et al., Sensors and Actuators B **131**, 24-28 (2008).
7. C.J. Ruiz, Chemical Senses **26**, 861-873 (2001).
8. R. Yoshida, N. Shigemura, K. Sanematsu, et al., J Neurophysiol **96**, 3088-3095 (2006).

Detection Of Uric Acid Based On Multi-Walled Carbon Nanotubes Polymerized With A Layer Of Molecularly Imprinted PMAA

Po-Yen Chen[1], Chia-Yu Lin[1] and Kuo-Chuan Ho[1, 2, *]

[1]*Department of Chemical Engineering and* [2]*Institute of Polymer Science and Engineering, National Taiwan University, No. 1, Sec. 4, Roosevelt Road, Taipei, 10617 Taiwan*

Abstract. A molecularly imprinted poly-metharylic acid (PMAA), polymerizing on the surface of multi-walled carbon nanotube (MWCNT), was synthesized. The MWCNT was modified by a layer of carboxylic acid and reacted with EDC and NHS to activate the carboxylic acid, which was prepared for the purpose of bonding allyl amine and getting an unsaturated side chain (–C=C). The resultant structure is abbreviated as MWCNTs-CH=CH_2. It is well known that the vinyl group side chain provides good attachment between the MWCNTs and the molecularly imprinted polymer (MIP). The MIP based on PMAA was polymerized on the surface of MWCNTs-CH=CH_2 with the addition of uric acid (UA). The non-imprinted polymer (NIP) was polymerized without adding UA. The adsorbed amount of UA approached the equilibrium value upon 60 min adsorption. The adsorption isotherm was obtained by immersing 10 mg of MIP or NIP in 5 mL aqueous solution containing different concentrations of UA. The adsorbed amounts were measured via a UV-Vis spectrometer at a wavelength of 292 nm. From the adsorption isotherm, it is seen that the MIP particles possess a good imprinting efficiency of about 4.41.

Keywords: Multi-walled carbon nanotubes, molecularly imprinted polymer, uric acid.
PACS: 82.47.Rs

INTRODUCTION

Molecular imprinting technique is a very useful approach to fabricate a polymer matrix with molecular recognition sites, which are formed by the addition of template molecules during polymerization process. Molecularly imprinted polymers (MIPs) have been developed over a decade in many fields, such as chromatography [1], catalyst [2], drug delivery [3], artificial antibody [4], and sensing devices [5]. In the early years of MIP development, the applications of liquid chromatography were most concerned [6]. In this field, MIP particles were used as the stationary phase of liquid chromatography system, and the function was to improve the separation efficiency, especially for the enantiomers. The detection techniques for the adsorption or binding molecules in imprinted polymers could be achieved by electrochemical [7], piezo-electric, impedance and optical method. Among those detection methods, electrochemical method, especially the amperometric method, is the easiest and most economic way to fabricate a commercial sensor.

EXPERIMENTAL AND METHODS

Reagents And Apparatus

Trimethylolpropane trimethacrylate (TRIM), 2, 2'-azobis-isobutyronitrile (AIBN), were purchased from Aldrich (USA). Uric acid (UA), ascorbic acid (AA), N, N-dimethylform amide (DMF), acetonitrile, methacrylic acid, N-hydroxysuccinimide (NHS), allylamine, toluene, and phosphate buffered saline (PBS) table were obtained from Sigma (USA). MWCNTs were purchased from Nanotech Port Co. (Taiwan) and these MWCNTs were produced via the chemical vapor deposition (CVD, or sometimes known as catalytic pyrolysis) method. Deionized water (>18 MΩ) was produced by Purelab Maximum (ELGA). All chemicals were analytical reagent grade, and used as received. Cyclic voltammetry and amperometric measurements were carried out using potentiostat/galvanostat (CH Instrument, model CHI 440) and the compatible software. A Pt wire and Ag/AgCl/sat'd KCl were used as the counter and the reference electrode, respectively.

CP1137, *Olfaction and Electronic Nose: Proceedings of the 13th International Symposium,* edited by M. Pardo and G. Sberveglieri

Preparation Of MIP-MWCNT/PMAA Particles

Crude MWCNTs (500 mg) were added into a glass reactor containing 50 mL of HNO_3 and 150 mL of H_2SO_4 under ultrasonication for three days. After sonication, the mixture was filtered through a 0.2 μm aluminum oxide membrane and neutralized by passing through 500 mL of distilled water. The solid obtained from the top of membrane were further dried under 105 °C and resulted in carboxylic acid modified MWCNTs.

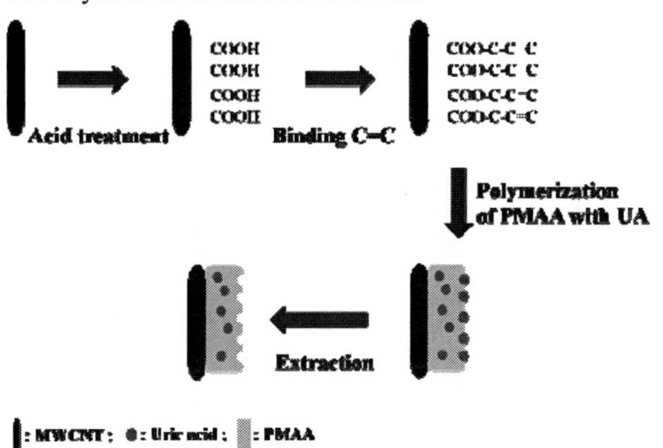

FIGURE 1. The preparation steps of MIP-MWCNT/PMAA particles.

The C=C modified MWCNT was synthesized via the reaction between carboxylic acid and amine functional group with EDC and NHS. Carboxylic acid modified MWCNTs (400 mg) were suspended in a 10 mL mixture of 0.03 M EDC, 0.06 M NHS and 0.03 M allylamine and the mixture was reacted in room temperature for 12 h. This experiment should be carried out under sonication to ensure that the MWCNTs were well suspended during the reaction. After the modification of C=C, the MIP-MWCNT/PMAA particles was prepared by precipitation polymerization method. The MWCNT-C=C particles of 100 mg were suspended in 30 mL of acetonitrile and 5 mL of toluene in a 250 mL round bottom flask and purged with nitrogen. The template (0.1 mmol) and MAA monomer (1 mmol) was dissolved by 10 mL of DMF and well mixed with a magnetic stir for 1 hr. Finally, the temperature was controlled at 70 °C for 24 hr during polymerization. After polymerization step, the resultant particles would contain PMAA and MIP-MWCNT/PMAA particles.

The MIP-MWCNT/PMAA modified glassy carbon electrode was used for the electrochemical studies and designated as MIP-GC/MWCNT/PMAA. The MIP-GC/MWCNT/PMAA electrode was made by dropping 2 μL of MIP-MWCNT/PMAA solution, which containing 2 mg particles suspended in 1 mL DMF. Then the electrode was put in a laminar flow hood to remove the solvent for further use.

RESULTS AND DISCUSSION

SEM Of The MIP-MWCNT/PMAA Particles

Figure 2 is the SEM images of MWCNT and MIP-MWCNT/PMAA particles. The average size of crude MWCNT was about 30 nm and the length was several micrometers. After polymerization, the average size was increased to 60 nm and this result revealed that the MIP layer was attached on the MWCNT successfully. In contrast, MWCNTs which have not been modified with C=C do not show the increasement on the size.

Adsorption Characterization Of MIP-MWCNTT/PMAA

For the MIP-MWCNT/PMAA, the adsorption of UA reached the equilibrium after 10 minutes and the adsorption isotherms of UA were shown in Figure 3. The adsorption capacity of MIP-MWCNT/PMAA was much higher than that of NIP-MWCNT/PMAA and the imprinting efficiency is about 4.41, as judged by the ratio of the maximum adsorption quantitiy of MIP to that of NIP. This indicates that the imprinted layer possesses recognition ability of UA and can be attributed to the complementary cavities created by UA templates. The two isotherms can be fitted by Langmuir and Freundlich models, as shown in Figure 4. The equilibrium data were analyzed using the Langmuir and Freundlich isotherms, given by Eqs. (1) and (2) respectively, where q_m is the maximum amount of adsorption (mg/g), K_L is the affinity constant (L/g) , K_f is a constant as a measure of adsorption capacity and n is a measure of adsorption intensity. It was observed that the equilibrium data were best fitted by Freunidlich model with good correlation coefficient values. This suggested that the adsorption between polymer and UA was a micropore (< 1nm) filling type. This is consistent with the adsorption type for the MIP-MWCNT/PMAA. Furthermore, from Table 1 below, the adsorption capacity and the intensity of MIP were larger than those of NIP.

Electrochemical Tests

The MIP-GC/MWCNT/PMAA electrode was swept from 0 to 0.6 V (vs. Ag/AgCl sat'd KCl) with a scan rate of 20 mV/s. 0.02 M PBS was used as a buffer solution and supporting electrolyte. Cyclic voltammograms (CVs), done under different adsorption times with 0.4 mM of UA + 0.02 M PBS, were shown in Figure 5. The oxidation peak potential of UA was observed around 0.4 V. According to the CVs, the adsorption amount of UA would reach saturation after 5 min of adsorption. This indicates that the current signal mainly comes from the adsorption quantity of UA.

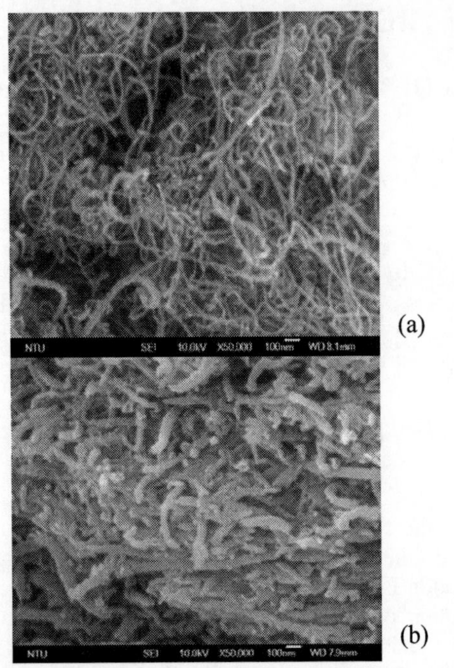

FIGURE 2. Scanning electron microscopy images of (a) MWCNT and (b) MIP-MWCNT.

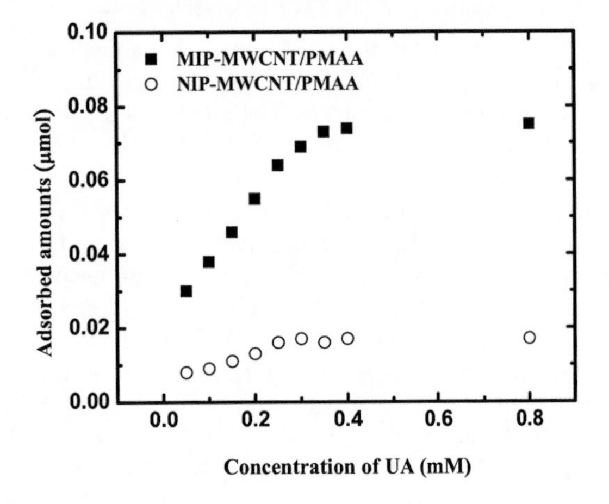

FIGURE 3. Adsorption isotherms of UA on MIP-MWCNT/PMAA and NIP-MWCNT/PMAA particles

$$\frac{C_e}{q_e} = \frac{1}{K_L q_m} + \frac{C_e}{q_m} \tag{1}$$

$$\ln q_e = \ln K_f + \frac{1}{n} \ln C_e \tag{2}$$

The MIP-GC/MWCNT electrode also shows high selectivity against the interference test of ascorbic acid (AA). The CVs of MIP-GC/MWCNT electrode tested in 0.04 mM AA and the background PBS solution are shown in Figure 6. The oxidation potentional of AA is about 0.2~0.3 V. After modification with MIP-MWCNT particles, the oxdation peak disappeared. This is ascribed to the cavities created by the template molecules, thus the existence of the affinity force between UA and functional monomer.

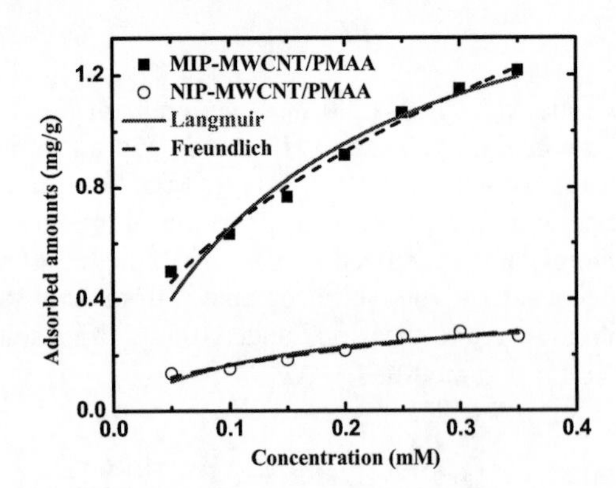

FIGURE 4. The adsorption isotherms fitted with Langmuir and Freundlich models.

Table 1. Isotherm parameters for UA adsorption

	Langmuir			Freundlich		
	q_m (mg/g)	K_L (L/g)	R^2	K_f	1/n	R^2
MIP	1.78	5.79	0.962	0.13	1.99	0.990
NIP	0.38	7.52	0.894	0.03	2.23	0.926

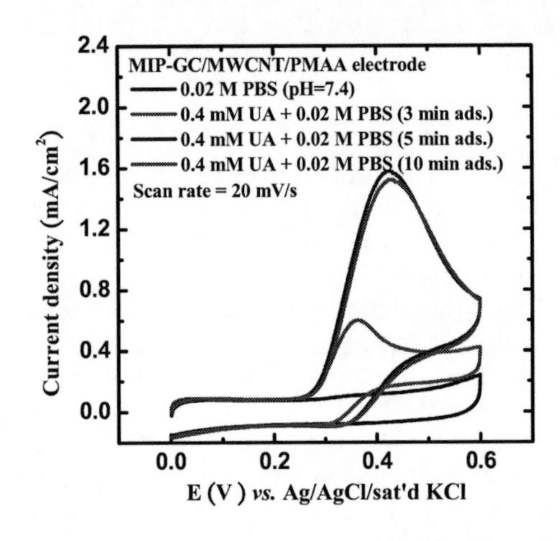

FIGURE 5. CVs of MIP-GC/MWCNT/PMAA electrode in 0.02 M PBS solution, 0.4 M UA + 0.02 M PBS and 0.4 M UA + 0.02 M PBS with different adsorption times of 3, 5 and 10 min.

Before performing amperometric detection, the potential for detection must be determined using a polarization curve. Figure 7 shows the linear sweep voltammetry (LSV) of the modified electrode at a sweeping rate of 0.1 mV/s. According to the result, a plateau was observed between 0.35 and 0.45 V, leading to the limiting current zone, which was resulted from the mass transfer limitation of UA. In a mass transfer controlled condition, the concentration of UA is near zero on the surface of the electrode due to the rapid reaction rate of UA. By applying the potential within the limiting current zone, the sensing current is proportional to the concentration of UA. Therefore, the sensing potential of this study was set at 0.4 V to obtain a steady-state current response of UA oxidation.

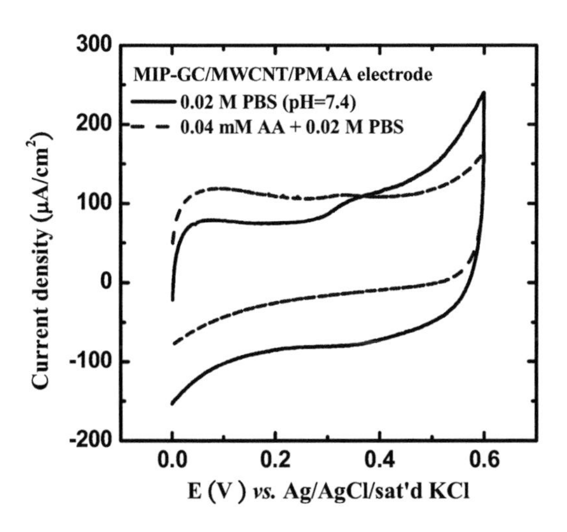

FIGURE 6. The interference test of MIP-GC/MWCNT/PMAA electrode under 0.04 mM AA.

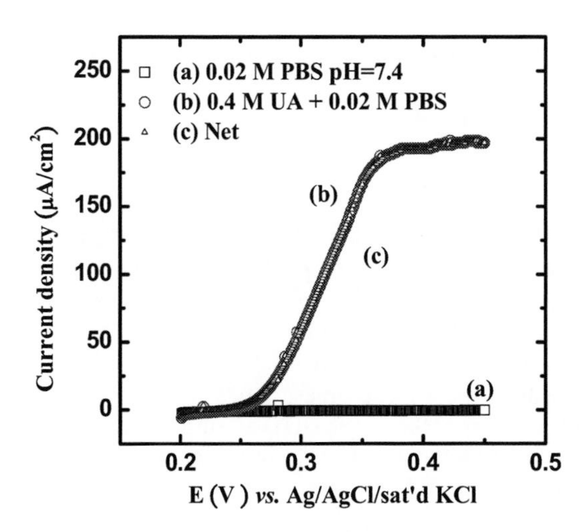

FIGURE 7. LSV for MIP-GC/MWCNT electrode including (a) background , (b) total , and (c) net current densities.

Amperometric experiments were performed to examine the current responses of the MIP modified electrodes. A steep increase of the UA concentration went from 80 to 800 μM and the potential was held at 0.4 V. An amperometric detection curve is shown in Figure 8, and a good linear relationship between the current density and the concentrations of UA were found between concentrations ranges of 80 to 500 μM of UA, with correlation coefficients of greater than 0.988. The current response would not increase when the concentration of UA is higher than 800 μM. This phenomenon is caused by the limited reactive surface area of the modified MIP electrode. According to figure 8, the sensitivities of the MIP and NIP modified electrodes are about 11.03 and 5.39 mA $M^{-1}cm^{-2}$ and the LOD is about 22 μM (S/N > 3).

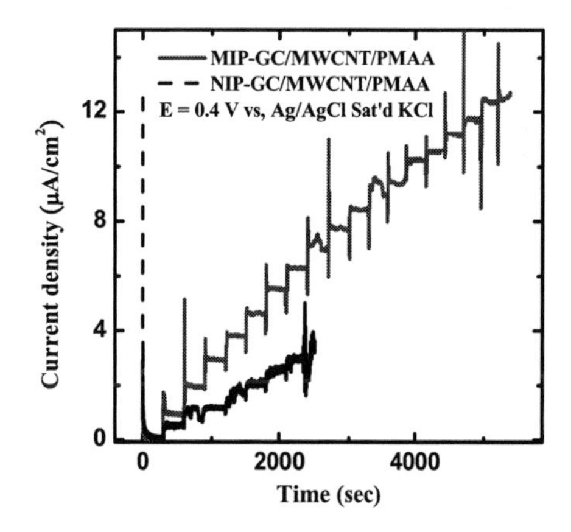

FIGURE 8. Amperometric detection of MIP-GC/MWCNT and NIP-GC/MWCNT electrodes at 0.4 V.

CONCLUSIONS

The MIP-MWCNT/PMAA particles were prepared and integrated into an electrochemical system for determining the concentration of uric acid. From the adsorption data, the MIP adsorbs more uric acid than NIP does and the imprinting efficiency is about 4.41. The MIP modified MWCNT can be deposited on the MWCNT electrode surface and used for the electrochemical detection for uric acid. The sensitivities of the MIP and NIP modified electrodes were about 11.03 and 5.39 mA $M^{-1}cm^{-2}$ and the difference is mainly come from the affinity cavities which were created by imprinted template. The MIP-GC/MWCNT/PMAA electrode also possesses high selectivity against ascorbic acid. The limit of detection of the MIP modified electrode can reach the level of micromolar and this performance is suitable for sensing UA in human's serum. The application of this study will benefit the development of electrochemical sensors, especially for developing a preconcentrator for UA analysis due to its high rebinding quantity and good selectivity with respect to UA.

REFERENCES

1. O. Norrlow, M. Mansson, and K. Mosbach, *J. Chromatogr.* **396**, 374-377 (1987).
2. K. Ohkubo, Y. Urata, S. Hirota, Y. Hinda, and T. Sagawa, *J. Mol. Catal. A-Chem.* **87**, L21-L24 (1994).
3. Rn. Karmalkar, MG. Kulkarni, and Ra. Mashelkar, J. Control. Release **43**, 235-243 (1997).
4. O. Ramstrom, L. Ye, and K. Mosbach, *Chem. Biol.* **3**, 471-477 (1996).
5. D. Kriz and K. Mosbach, *Anal. Chim. Acta* **300**, 71-75 (1995).
6. G. Wulff and S. Schauhoff, *J. Org. Chem.* **56**, 395-400 (1991).
7. M.C. Blanco-Lopez, M.J. Lobo-Castanon, A.J. Miranda-Ordieres, and P. Tunon-Blanco, *Trac-Trends Anal. Chem.* **23**, 36-48 (2004).

On The Catalytic Role Of MWCNTs For The Electro-reduction Of NO Gas In An Acid Solution

Chia-Yu Lin[1] and Kuo-Chuan Ho[1,2★]

[1]*Department of Chemical Engineering, National Taiwan University, Taipei 10617, Taiwan*
[2]*Institute of Polymer Science and Engineering, National Taiwan University, Taipei 10617, Taiwan*
★*Corresponding author: Fax: +886-2-2362-3040; Tel: +886-2-2366-0739; E-mail: kcho@ntu.edu.tw*

Abstract. The poly(3,4-ethylenedioxythiophene)/multi-wall carbon nanotubes (PEDOT/MWCNTs) modified screen-printed carbon electrodes (SPCEs) were fabricated and their catalytic properties towards nitric oxide (NO) gas, evolved via the disproportion reaction of nitrite in an acid solution, were studied. The thicker the PEDOT film was, the higher the peak current density (J_{pc}) was observed. However, further increase in the thickness of PEDOT (> 2 layers) would decrease the J_{pc}. Moreover, as the loading of the MWCNTs was increased, the peak potential for NO reduction shifted to more positive direction due to the promoted electron transfer by adding MWCNTs. Besides, due to the larger amount of the protonated nitrite ions in the acid solution, the sensitivity of the PEDOT/MWCNTs-modified SPCEs was enhanced at lower solution pH.

Keywords: Nitric oxide, nitrite, poly(3,4-ethylenedioxythiophene), multi-wall carbon nanotubes
PACS: 82.47.Rs

INTRODUCTION

The electrocatalytic oxidation of nitrite on the PEDOT/MWCNTs-modified SPCEs has been investigated previously [1]. To reduce possible interference from some easy-oxidizing biomolecules in bio-fluids, the electro-reductive biosensor is proposed.

EXPERIMENTAL AND METHODS

Preparation Of PEDOT- and PEDOT/MWCNTs-modified SPCEs

Electrochemical measurements were carried out at a CHI 440 electrochemical workstation (CH Instruments, Inc., USA) with a conventional three-electrode system. The electropolymerization of EDOT, on to three-electrode type SPCE, was carried out by cyclic voltammetric method in aqueous solution containing 0.01 M EDOT, 0.5 mM (2-hydroxypropyl)-β-cyclodextrin and 0.1 M $LiClO_4$ between 0 and 0.95 V (vs. Ag/AgCl) at a scan rate of 50 mV/s for 1 to 6 cycles. The working electrode, reference electrode and counter electrode of the SPCE are carbon electrode, Ag/AgCl electrode and Pt wire, respectively.

To prepare PEDOT/MWCNTs-modified SPCEs, 2 μl of the MWCNTs suspension was drop-casted on the surface of the electrode and the electrode was then dried at 60 °C. To optimize the amount of MWCNTs on the electrode, the above-mentioned procedure was repeated for several times. After deposition of the MWCNTs, the PEDOT film was then electrodeposited by the cyclic voltammetric method mentioned in last paragraph.

RESULTS AND DISCUSSION

In acid condition (pH<3), NO gas evolves via disproportion reaction from nitrite :

$$HNO_2 \leftrightarrow H^+ + NO_2^-,\ pK_a=3.3 \text{ at } 18\ ^\circ C \quad (1)$$

$$3HNO_2 \rightarrow HNO_3 + 2NO + H_2O \quad (2)$$

Therefore, nitrite ion can be detected indirectly by sensing the evolved NO gas.

Fig. 1 shows the cyclic voltammograms (CV) of bare SPCE, PEDOT- and PEDOT/MWCNTs-modified SPCEs in 0.5 M deaerated H_2SO_4 solution containing 1.3 mM nitrite. It was found that an increased J_{pc} was observed as the PEDOT or PEDOT-MWCNTs were deposited on the SPCEs. Besides, the PEDOT/MWCNTs-modifed SPCE showed higher J_{pc} than PEDOT-modified one, although peak potential shifted to more negative side.

CP1137, *Olfaction and Electronic Nose: Proceedings of the 13th International Symposium,* edited by M. Pardo and G. Sberveglieri

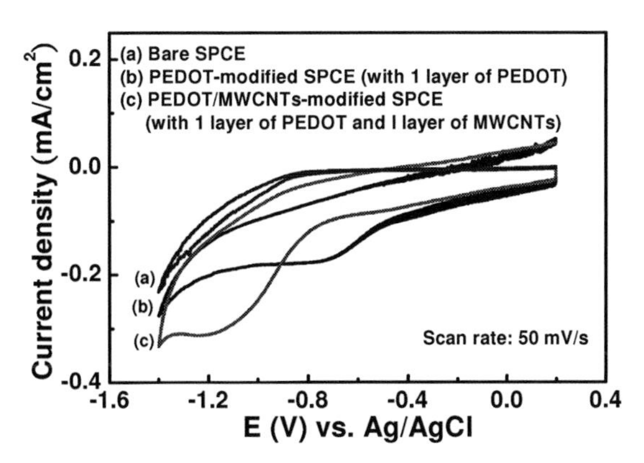

FIGURE 1. CVs of PEDOT and PEDOT/MWCNTs modified SPCE in 0.5 M H_2SO_4 containing 1.3 mM nitrite.

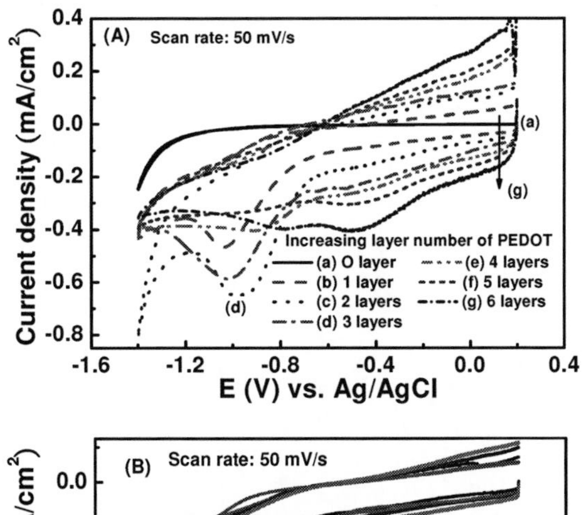

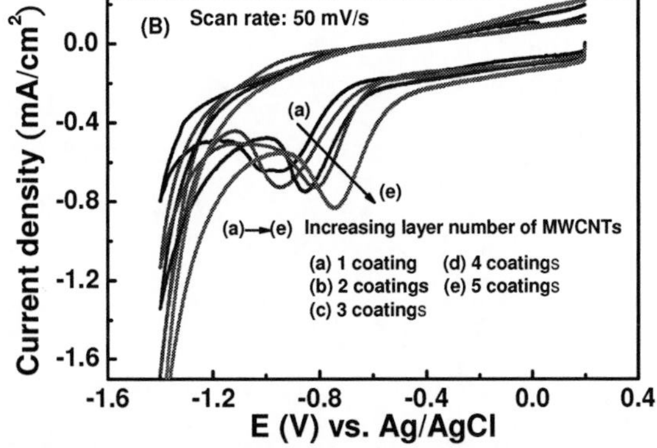

FIGURE 2. CVs of the SPCEs modified with (A) 1 layer of MWCNTs and various layers of PEDOT film, and (B) with 2 layers of PEDOT film and various coatings of MWCNTs in deaerated 0.5 M H_2SO_4 solution containing 1.3 mM nitrite.

Fig. 2(A) shows the effect of the number of PEDOT layer, controlled by adjusting the cycle number during electropolymerization, on the current resonse. It was found that as the layer number was increased from 0 to 2, the J_{pc} increased. However, as the layer number was further increased, the J_{pc} decreased, and the other peak at -0.49 V (vs. Ag/AgCl) appeared. Furthermore, the effect of the loading of MWCNTs was also examined. As shown in Fig. 2 (B), as the loading amount of MWCNTs, controlled by adjusting the drop-coating time, was increased, the peak potential shifted to more favorable side, which can be attributed to promoed direct electron transfer by adding MWCNTs. However, the PEDOT/MWCNTs film become unstable and fell off as the loading amount of MWCNTs was higher than 5 coatings. Therefore, the optimal layer number of PEDOT film and the loading of MWCNTs were chosen as 2 and 5.

The effect of solution pH on the current response of nitrite was also examined. The solution pH was adjusted by adding 0.5 M NaOH solution. As shown in Fig. 3, by increasing the solution pH, the J_{pc} decreased. At higher solution pH, the amount of the protonated nitrite ions decreases, and therefore the amount of the evolved NO gas decreased, resulting in the decrease in J_{pc}. Therefore, the PEDOT/MWCNTs-modified SPCE only shows response to NO gas.

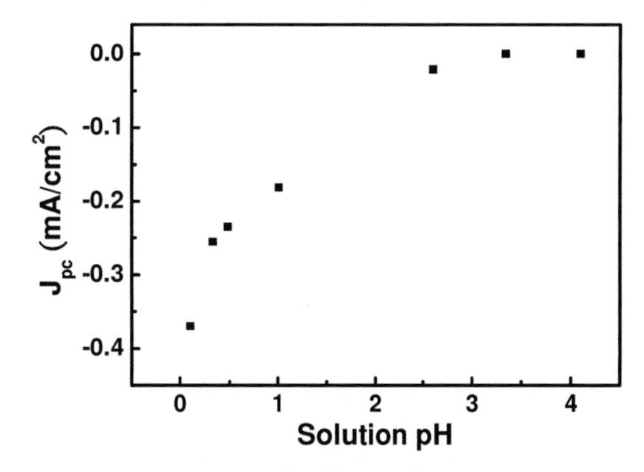

FIGURE 3. J_{pc} of the SPCE with 2 layers of PEDOT film and 5 coatings of MWCNTs in 0.5 M H_2SO_4 solutions containing 1.3 mM nitrite at different pHs.

CONCLUSION

The role of MWCNTs in thee PEDOT /MWCNTs-modified SPCEs was studied, and the use of this modified electrode for indirect amperometric detection of nitrite is underway.

ACKNOWLEDGMENTS

This work was sponsored by the National Research Council of the Republic of China (Taiwan) under grant number NSC 96-2220-E-006-015 and NSC 97-2220-E-006-008.

REFERENCES

1. C. Y. Lin, V. S. Vasantha, K. C. Ho, "Detection of nitrite based on a poly(3,4-ethylenedioxythiophene)-modified electrode," *Proc 12th International Meeting on Chemical Sensors*, Columbus, OH, July, 2008, pp. 806-807.

ELECTRONIC NOSES AND TONGUES INSTRUMENTATION: SOFTWARE / HARDWARE DESIGN

Multilayered Multisensor Based on Metal Oxide Nanostructures for Water Quality Monitoring

Dr. Serge Zhuiykov

Commonwealth Scientific Industrial Research Organization (CSIRO), Materials Science and Engineering Division, 37 Graham Rd., Highett, VIC. 3190, Australia
E-mail: serge.zhuiykov@csiro.au

Abstract. A new multisensor based on nanostructured semiconductor RuO_2 sensing electrode (SE) and Ag/AgCl, Cl^- reference electrode (RE) deposited on alumina substrate and coupled with specifically designed turbidity sensor has been evaluated for long-term stability during 12-month non-stop detection the main parameters of water quality: pH, dissolved oxygen (DO), temperature, conductivity and turbidity at a temperature range of 9-30°C. Sensors with RuO_2–SE have been showed stable Nernstian response to pH from 2.0 to 13.0 and were also capable to measure DO in the range of 0.6 - 8.0 ppm. The measurement results show good linearity and an excellent reproducibility was obtained during trial. Nernstian slope was 58mV/pH at a temperature of 23°C. Influence of the hydrogen ion (H^+) diffusion in nanostructured RuO_2 films on the output *emf* drift at pH measurements was also investigated. Turbidity and conductivity measurements revealed that multisensor is capable to measure both high and low ranges at different temperatures exhibiting high linearity of characteristics.

Keywords: Water Quality Monitoring, RuO_2, Multi-sensor, Nanostructures
PACS: 07.07.Df

INTRODUCTION

Recent development of integrated water quality monitoring sensors based on nanostructured RuO_2-SEs in CSIRO, CMSE Laboratories and ICT Centre, Australia has been resulted in multisensor assembly simultaneously measuring the main parameters of water quality: pH, temperature, DO, conductivity and turbidity [1,2]. Schematically this integrated water quality sensor is shown in Fig. 1. Screen-printed nanostructured RuO_2-SEs have been used in this assembly for pH, and DO measurement. SEs were deposited on the platinized alumina substrate of the sensor, which possesses an aperture for fixing specifically designed turbidity sensor. Choice of the platinized alumina was based on the fact the RuO_2 possesses very low affinity to ceramic substrate even if the alumina was rough [2]. Moreover, it was found that raw RuO_2 nano-powder is amorphous and required pre-treatment on order to stabilize its structure. The encouraging first results in 2007 stimulated the following more detailed investigations in CSIRO laboratory on RuO_2 nanostructures. CSIRO, CMSE Sensors and Sensor Networks team is working on the development of low-cost water monitoring sensors using nanostructured RuO_2-SE with improved characteristics. These *on-line* sensors will be used for different applications in wireless distributed sensor networks which are also under further development by CSIRO, ICT Centre. In order to investigate stability of characteristics of the multisensor based on RuO_2 nanostructures, long-term trial has been organized at the end of 2007. So far, the performance of multisensor has been monitoring 24 h per day, 7 days per week for more than 12 months. In fact, this trial still carries on in 2009. Thus, in this paper we report the properties of water quality sensors using nanostructured RuO_2-SEs after 12 months of service at temperatures below 30°C.

EXPERIMENTAL

All reagents for making nanostructured RuO_2–SEs are of high-purity analytical grade and were used as received. RuO_2–SEs were prepared from RuO_2 nano-particles (ABCR GmbH & Co.). Pt paste was supplied by Sigma-Aldrich Australia Pty. Ltd. Sensor alumina substrates were manufactured by Taylor Ceramic Engineering Pty. Ltd., Australia. The average particle sizes distribution of RuO_2 was analyzed on dynamic

CP1137, *Olfaction and Electronic Nose: Proceedings of the 13th International Symposium,* edited by M. Pardo and G. Sberveglieri

light scattering particle size analyzer LB-500X (HORIBA) and was found to be around ~360 nm [2].

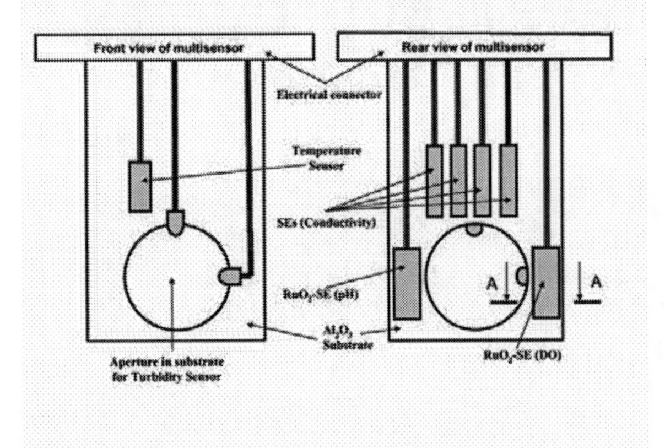

FIGURE 1. Schematic view of integrated water quality multisensor.

Fig. 2 illustrates cross-sectional view of the alumina substrate showing how RuO_2-SEs and protective glazing layer were deposited onto substrate. This figure also shows SEM images of RuO_2-SE, Pt film and protective glazing layer, respectively. Measurement of an average particle size distribution revealed that the majority of RuO_2 sub-micron particles were in the range of 250-550 nm.

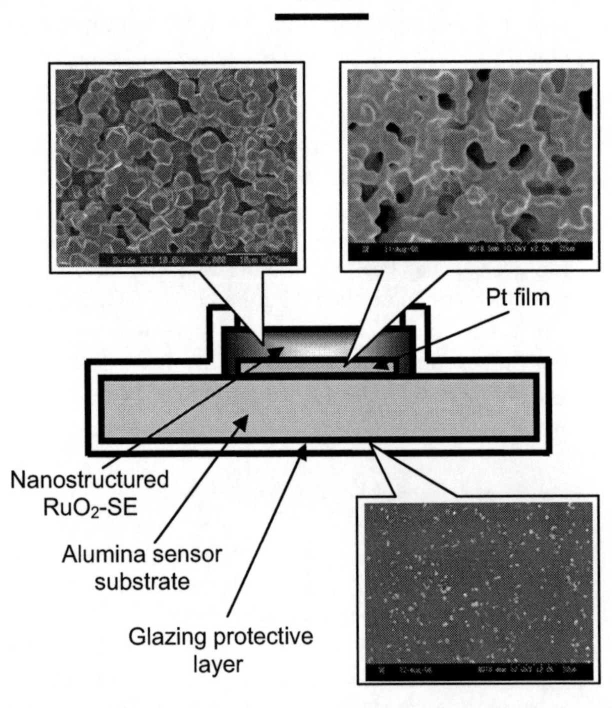

FIGURE 2. A cross-sectional view of sensor substrate.

Due to amorphous nature of raw RuO_2 nanoparticles, they were pretreated at 1000°C for 2 h in air for the structure stabilization. Pt current conductors of 5 μm in thickness were screen-printed onto alumina substrate and were sintered at 1000°C for 1 h in air prior to RuO_2 deposition. Then, the RuO_2 nanoparticles were applied onto platinized alumina substrate by using α–terpineol ($C_{10}H_{18}O$, 99.0%) and were sintered in two steps: heated to 400°C at rate 65°C/h and stabilized at 400°C for 2 h before heated to 965°C at rate 100°C/h in air ensuring the development of nanostructured RuO_2–SEs [1]. The size of RuO_2 nano-particles of the water sensor electrodes and the surface morphology of RuO_2–SEs was characterized by FE-SEM (JEOL JSM-6340F) and their composition was subsequently analyzed by a Wet-SEM (HITACHI, S-3000N) coupled with an energy-dispersive EDX, (HORIBA, EX-220SE).

RESULTS

Fig. 3 illustrates typical sensitivity curve, i.e., response potentials as a function of pH for multisensor based on nanostructured RuO_2–SE, pure Pt-SE and previously published results for thick-film micro-structured (10 μm in thickness) RuO_2-SE [3] and RuO_2-TiO_2 [4] at 23°C. Excellent reproducibility for the nanostructured RuO_2-SE was obtained. The Nernstian slope was 58 mV/pH at 23°C. The intercept at pH 0 gives the apparent standard electrode ($E^{0'}$) of about 640±10 mV vs Ag/AgCl-RE. Good agreement of Nernstian slopes and $E^{0'}$ values for the different RuO_2–SEs of the sensors were obtained. The slope was strongly followed the straight line suggesting that only one electron per oxygen molecule was involved. This could be substantiated to the existence of the following electrochemical reaction:

$$O_{2,aq} + e^- \leftrightarrow O_2^-{}_{,ads}.$$

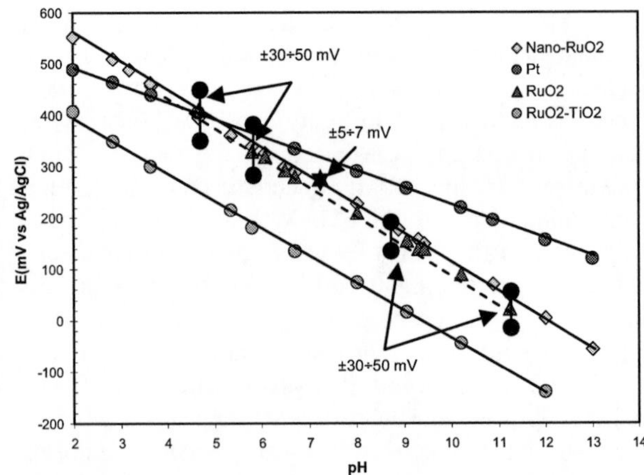

FIGURE 3. Measured *emf* as a function of pH for different materials of SE with standard deviations.

The standard deviation of the output *emf* was found to be around ±5-7 mV in whole pH measurement range. It is a big improvement compare to the previously published results [3], where the standard deviation was found to be ±30-50mV. This is apparently due to the nanostructure of the RuO_2-SE. Results of pH measurements involved pure Pt-SE revealed that the angle of the slope for the Pt-SE is less than the angle of the slope for the RuO_2-SE. Consequently, RuO_2-SE possesses better sensitivity within a whole pH measurement range. Reproducibility of the results between different days when the measurements were carried out was found to be ±2 mV. It was also reported that commercial Ru-containing electrode pastes used as SE in pH sensors are partially responded to the presence of halides, sulphate, bromide and carbonate anions in water [3]. This was mainly caused by the presence of lead (Pb) in available pastes.

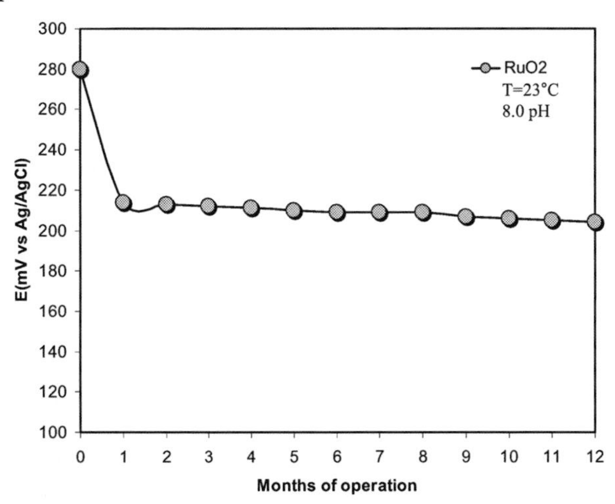

FIGURE 4. Long-term stability measurements of pH in water by multisensor based on nanostructured RuO_2-SE.

Results of the long-term stability test (Fig. 4) revealed that the lifetime of multisensor based on nanostructured RuO_2-SE is greater than 12 months and further testing of the long-term stability is currently in progress. This is also improvement compare to the published 6 months lifetime for thick-film SE based on Ru-containing pastes [3]. Substantial *emf* drift during the first month of the testing was due to the H^+ transport through SE, which is governed by H_2 trapping at the trap sites existing at the grain boundaries or mirco-pores of the nanostructured SE. So far two results were found to be important in these studies: (1) freshly prepared RuO_2-SEs are affected by the slow non-random *emf* drift owing to the hydrogen ions diffusion through SEs and neglecting the effects of the other ions; drift is significant for the first 6 days and is stabilized only after 15 days with negligible drift afterwards [5]; (2) active sites for the oxygen reduction are not limited to the triple boundaries, but extended to surfaces of RuO_2-SE [6]. After *emf* stabilization, the output signal kept stable during all following months of the testing showing the standard drift about ±1.5 mV/month.

Another interesting observation was made during SEM investigation of the surface of RuO-SEs after 12 months of operation in water at the temperature range of 9-23°C (Fig. 5). The *emf* of the sensor did not change much despite of the heavy colonies of the organic sediments deposited on RuO_2-SE during trial, sometime completely covering RuO_2 nano-particles or making net-like deposits.

FIGURE 5. SEM image of the surface of RuO_2-SE after 12 months of service in water quality monitoring multisensor.

The sensing performance of toward DO changes for the multisensor attached with nanostructured semiconductor RuO_2–SE was investigated at the standard buffer solutions, in which DO concentration was regulated by pumping N_2/O_2 gas mixture with fixed various O_2 concentrations through buffer solution. Although the changes of DO concentration in water achieved by this method are time consuming, the method kept pH of the solution intact. Multiensors with attached nanostructured RuO_2–SE were conditioned in buffer solution at least 2 h before potentiometric DO measurement. A linear response was obtained in *emf* variation for nanostructured RuO_2–SE vs Ag/AgCl, Cl^--RE as a function of dissolved oxygen from 0.6 to 8.0 ppm (log[O_2], -4.71 to -3.59) in water at pH 8.0. Results of these measurements at a temperature of 23°C and 8.0 pH of the solution are presented in Fig. 6. The straight line confirms fast electrode reaction involving one electron per oxygen molecule, as clearly shown in this figure. The output signal was stable with a Nernstian slope of -41 mV per decade. One of the typical reported interfering species is proton concentration in water, which has been shown above. Standard deviations of the output *emf* were found to be around ±5 mV in the whole DO measuring range. However, it was also discovered that although the gas flow controllers fixed pumping gas-mixture rate constant (100 cm^3/min) during all experiments, nevertheless ±1.5 mV deviation in DO measurement did occur. This suggested that apart from the fast electrode reaction (1), other slower electrode reactions are possibly also involved into DO measurement by nanostructured RuO_2-SE even at relatively neutral pH.

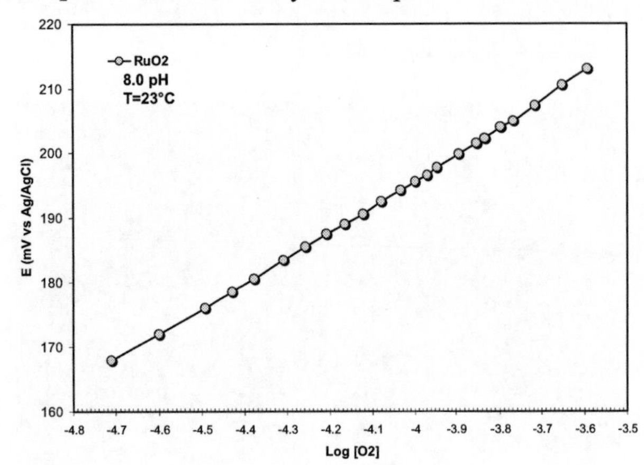

FIGURE 6. *Emf* variations vs measured DO in water for nanostructured RuO_2-SE at a temperature of 23°C.

CONCLUSIONS

Morphology of RuO_2–SE nano-composites and their use in the water quality multisensors have been investigated. Integration of these nano-composites as a SE for potentiometric measurement of pH and DO in the water quality monitoring sensors at different temperatures was demonstrated to be a promising concept. However, the surface chemistry has to be considered in order to obtain the reliable sensor data. pH measurements have shown that nanostructured RuO_2–SE reacts almost instantly (few seconds) to the pH changes at room temperatures and is capable to measure pH changes from 2.0 to 13.0. However, as water temperature cools down, the response/recovery rate of the sensor also slows down and is usually within 8-10 min at 9°C.

Moreover, nanostructured RuO_2-SEs have also exhibited a suitable response to dissolved oxygen in the 0.6 to 8.0 ppm range. The Nernstian slope of -41 mV per decade was obtained. During 12-month long-term stability trial water quality sensors attached with RuO_2-SE have demonstrated negligible sensitivity drift of measured parameters. Further investigation of long-term stability is now in progress. Overall the *emf* data for the sensor attached with RuO_2–SE have been demonstrated that the concept of potentiometric detection both DO and pH in water is virtually applicable to field measurements if the electrode pH-response drift is found and stabilized prior to the field trial. However, further investigation is necessary in order to obtain the reliable sensor data for using these sensors in wireless sensor networks.

ACKNOWLEDGMENTS

This work was partially supported by Research and Development Program of CSIRO Materials Science and Engineering Division and CSIRO ICT Centre, Australia – project "Integrated Water Quality Monitoring Sensors".

REFERENCES

1. S. Zhuiykov. Australian provisional patent application 2008902285 (9 May 2008).
2. S. Zhuiykov, *Electrochem. Comm.* **10**, 839-843 (2008).
3. R. Martinez-Manez et al, *Sens. Actuators B: Chem.* **120**, 589-595 (2005).
4. L.A. Pocrifka et al, *Sens. Actuators B: Chem.* **113**, 1012-1016 (2006).
5. S. Zhuiykov, *Sens. Actuators B: Chem.* **136**, 248-256 (2009).
5. S. Zhuiykov, *Ionics* (2009) – in press.

Uncalibrated Current-Mode Oscillator For Resistive Gas Sensor Integrable Applications

G. Ferri[*], V. Stornelli[*], A. De Marcellis[*], C. Di Carlo[*],
A. Flammini[+], A. Depari[+], D. Marioli[+]

[*]Dept. of Information and Electrical Engineering, University of L'Aquila
Monteluco di Roio, 67100, L'Aquila, Italy
[+]Dept. of Electronic for Automation, University of Brescia
25123 Brescia, Italy
e-mail: giuseppe.ferri@univaq.it

Abstract. We present a novel current-mode fully-integrable oscillating circuit for the interfacing of resistive gas sensors, using a reduced number of active blocks, in particular low voltage low power Second Generation Current Conveyors (*CCIIs*), to be used in portable applications. The proposed front-end, operating a Resistance-to-Time (*R-T*) conversion, can be utilized for AC-excited resistive sensors and does not need any initial calibration. Simulation and preliminary measurement results on a prototype board performed on the whole system have shown good linearity and agreement with theoretical expectations.

Keywords: Electronic Interfaces, Resistance-to-Time Conversion, Current Mode.
PACS: 85.40.-e.

INTRODUCTION

The development of miniaturized integrated electronic circuits and the microelectronic technologies progress have brought, in sensors environment, to the design of complete analog circuits implemented by both the sensor interface and the elaboration system. In this sense, the implementation of an appropriate front-end, which can be adapted to different kinds of gas sensors, through a suitable electronic topology, is one of the recent targets [1-4].

In resistive sensor analog interfaces, the use of oscillating circuits performing an *R-T* conversion is advisable when sensor baseline or its response to reagent substances changes of some order of magnitudes [5,6]. This avoids the use of front-ends with a wide output range only through the use of either scaling factors or high-resolution pico-ammeters [7-10]. In order to design suitable oscillators with low-voltage low-power performance for portable and compact sensor interfacing, the current-mode approach (and the use of *CCII*, the main analog current-mode device) is an open challenge [11,12]. In fact, electronic circuit trend is toward low-voltage low-power SoC and future compact electronic noses should follow same tendency [13,14].

In this paper, we propose a novel, compact, low-voltage, low-power, low cost interface based on a suitable *CCII*, designed at transistor level in a standard CMOS technology, employing an *R-T* conversion utilizing a reduced number of blocks. Simulation results as well as preliminary measurements, performed through a prototype laboratory board (using commercial component AD844 as *CCII* and sample resistances) have demonstrated the validity of the proposed interface and a good agreement with theoretical expectations.

CCII BASIC THEORY AND IMPLEMENTATION

The basic current-mode analog device is the *CCII*, whose block scheme is in Fig. 1a, which behaves as both current and voltage buffer. *CCII* operation is the following (Fig. 1b): if a voltage is applied to Y node, an equal voltage is obtained at X terminal and the current flowing into X node is equal (or opposite) to the current flowing into Z terminal [12]. Fig. 2 shows the internal equivalent circuit of a *CCII* where node parasitic elements have been highlighted.

CP1137, *Olfaction and Electronic Nose: Proceedings of the 13th International Symposium*, edited by M. Pardo and G. Sberveglieri

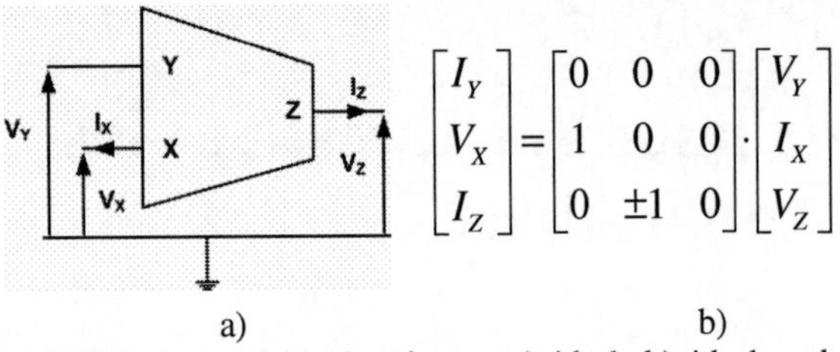

$$\begin{bmatrix} I_Y \\ V_X \\ I_Z \end{bmatrix} = \begin{bmatrix} 0 & 0 & 0 \\ 1 & 0 & 0 \\ 0 & \pm 1 & 0 \end{bmatrix} \cdot \begin{bmatrix} V_Y \\ I_X \\ V_Z \end{bmatrix}$$

a) b)

FIGURE 1. *CCII* block scheme: a) ideal, b) ideal nodes relationships.

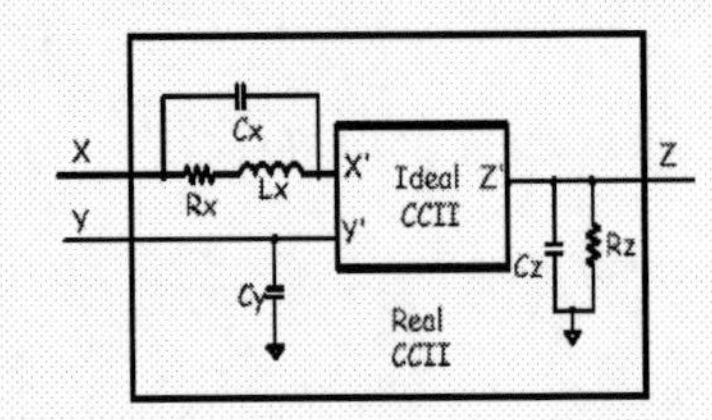

FIGURE 2. Internal equivalent circuit considering also its parasitic elements.

This is the paragraph spacing that occurs when you use the Enter key. In Fig. 3 the schematic of the designed *CCII* is shown [15]. The circuit has been implemented using a standard AMS CMOS 0.35μm process. It is formed by a differential input stage (M_1-M_7) with a class AB output stage (M_8-M_{11}; R_1,R_2; M_{16}-M_{17}) and a low voltage cascode current mirror (M_{12}-M_{15}; M_{18}-M_{21}). The class AB output stage allows to decrease X parasitic impedance, while cascode current mirror increases the Z one. The class AB output stage provides the current required by the X node load. This current is mirrored by using cascode current mirror that drives Z node load. The proposed *CCII* has a low voltage supply of ±0.75V and 118μW static power consumption. Bandwidth is about 10MHz. Parasitic components have been minimized. Voltage and current gains are very close to 1.

THE PROPOSED INTERFACE

Fig. 4 a) shows the basic block scheme of the proposed front-end. It is formed by three main blocks, highlighted in the same picture:

- a voltage integrator,
- a voltage buffer, in order to decouple input and output stages,
- a CCII-based hysteresis comparator (Schmitt trigger) [16].

The whole interface works as follows. The saturated output voltage of the Schmitt trigger ($V_{out}=\pm V_{SAT}$) represents both the periodic signal (from which it is possible to measure the period, proportional to sensor resistance) and the input signal for the voltage integrator, that gives the AC excitation voltage for the resistive sensor (due to the *CCII* voltage buffer operation). Output voltages $\pm V_{SAT}$, integrated by voltage integrator, generate a rising ramp when V_{out}=+V_{SAT} and a falling ramp if $V_{out}= -V_{SAT}$.
This triangular signal is compared with the voltage reference at Y node by the hysteresis comparator, so generating the square-wave voltage V_{out}, whose period T is proportional to R_{SENS}.
Fig. 4 b) shows the voltage signals at each node of the interface under the hypothesis of a constant R_{SENS} during the measuring operation.

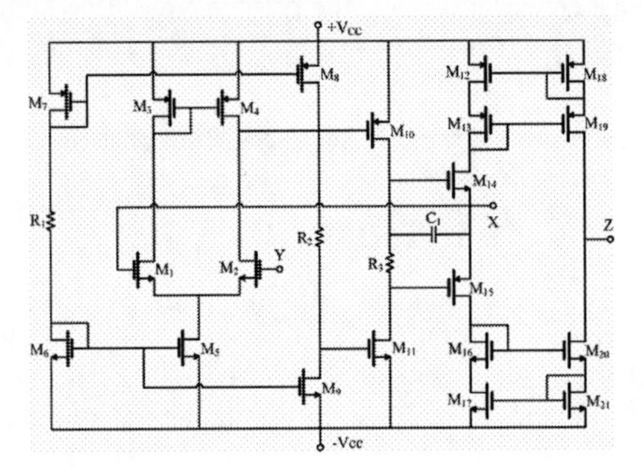

FIGURE 3. *CCII* schematic.

Through a straightforward analysis, considering ideal *CCII* behaviour, it is possible to determine the expression for the period T of generated output square wave signal, revealed at V_{out} node, as a function of the sensor resistance R_{SENS} as follows:

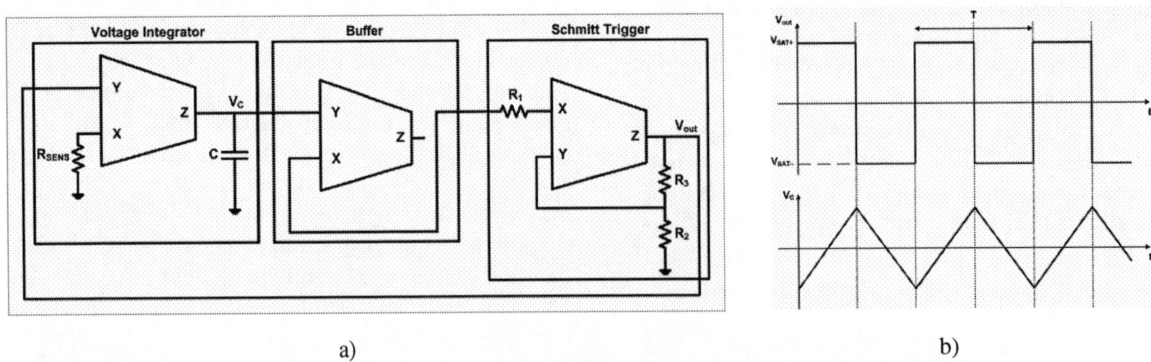

a) b)

FIGURE 4. a) Block scheme of the proposed interface; b) voltage behaviour at main interface internal nodes.

$$T = 4R_{SENS}C\left(\frac{R_2 - R_1}{R_2 + R_3}\right) \quad (1)$$

From eq. (1), circuit sensitivity can be opportunely set by choosing C, R_1, R_2 and R_3 values.

If we consider a real *CCII*, we must take into account its non idealities; in particular, in the following we will consider non unitary ratios between X and Y node voltages and between Z and X node currents:

$$\alpha = \frac{V_X}{V_Y} \qquad \beta = \frac{I_Z}{I_X} \quad (2)$$

and non-zero parasitic resistance at X and Z node, R_X and R_Z, respectively.

In this case, the relation between the generated output voltage period and the sensor resistance R_{SENS} is given by:

$$T = 4\frac{(R_{SENS} + R_X)C}{\alpha\beta} \cdot \frac{\alpha R_2 R_Z - (R_2 + R_3 + R_Z)(R_1 + R_X)}{R_Z(R_2 + R_3)} \quad (3)$$

In the aforementioned analysis, we have considered a purely resistive sensor. In some cases, it must be also taken into account a sensor parasitic capacitance, modelled in parallel to R_{SENS}. In this case, a straightforward computation gives the following expression for the output period:

$$T = 4R_{SENS}CG\left(1 - \frac{C_{SENS}}{2CG}\right) \quad (4)$$

being:

$$G = \frac{R_2 - R_1}{R_2 + R_3} \quad (5)$$

From eq. (4) it is evident that C_{SENS} contribution is negligible if the factor $2CG$ is designed to be much higher than the same capacitance value; otherwise its value, if constant, can be considered in calibration of C and G values. If either it is not possible to neglect the sensor parasitic capacitance or we want to estimate it, simple additional blocks as EX-OR gates must be added, according to the technique proposed in [10].

RESULTS

We have firstly performed simulations, in Cadence environment, using a CMOS 0.35µm standard technology for the internal implementation of the CCII. The following values for external passive components have been considered: R_1=5kΩ, R_2=50kΩ, R_3=10kΩ, C=100pF, so the circuit sensitivity for ideal *CCIIs* has been set to about 0.3ms/MΩ. Simulation results are in a good agreement with theoretical expectations, as depicted in Fig. 5 where the simulated period is compared with the theoretical one that takes into account *CCII* parasitics (eq. (3)). In this analysis the sensor capacitance has been ideally set to zero. From this figure it results that the relative error is below the 10% for R_{SENS} values ranging from 700kΩ and 70MΩ. Waiting for the layout design and the final chip fabrication, preliminary measurements on a prototype board (see Fig. 6), using the commercial component AD844 as *CCII*, have demonstrated the practical validity of the proposed interface and, once again, a good agreement with theoretical expectations. Data have been achieved using commercial resistors as resistive sensor, ranging from 18kΩ to 1.8MΩ. In this case, the interface sensitivity has been set to about 50µs/MΩ̃. Fig. 7 shows the measured periods compared with the theoretical ones, as a function of the sensor resistance, obtained with the following experimental values: R_1=470Ω, R_2=2.2kΩ, R_3=4.7kΩ, C=47pF, V_{EXC}=±15 V.

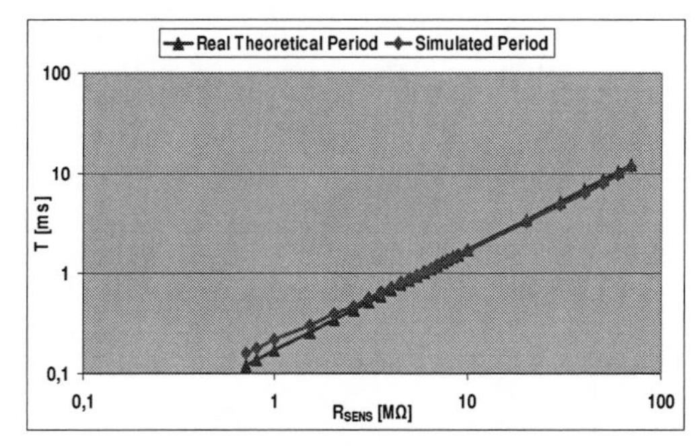

FIGURE 5. Simulated and theoretical output period T vs. R_{SENS}.

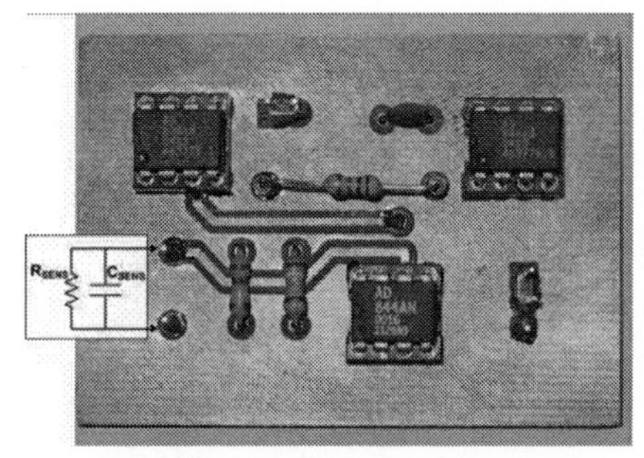

FIGURE 6. Prototype discrete element board.

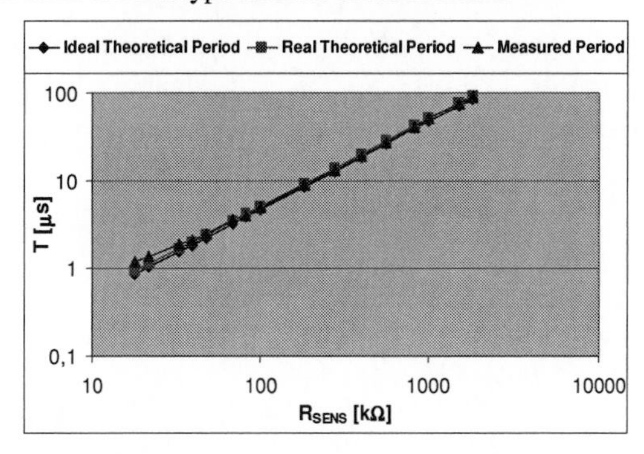

FIGURE 7. Measured and theoretical output period T vs. R_{SENS}.

Moreover, always with the same conditions except for the integrator capacitance value that has been here set at 100pF, the presence of a non-zero sensor

capacitance has been also investigated. In Table 1 the measured and simulated periods have been reported at different fixed sensor resistance, where sensor capacitance ranges from 1 to 10pF in three steps. These values have been chosen being them standard values for commercial sensors of the TGS Figaro family. As expected from eq. (4), the higher the parasitic sensor capacitance is, the worst the estimated periods are, with respect to theoretical ideal ones.
Actually we are planning to complete the interface analysis with some other experimental results, to reveal the presence of hydrogen into a closed chamber, that will be hopefully presented at the Conference.

TABLE 1. Measured periods and theoretical ones vs. R_{SENS} and C_{SENS}.

R_{SENS} [Ω]	C_{SENS} [pF]	Measured Period [μs]	Theoretical Period [μs]
	1	1.9	1.764
18 k	5.6	1.4	1.598
	10	1.2	1.440
	1	17.90	17.64
180 k	5.6	14.60	15.98
	10	12	14.40
	1	166	176.4
1.8 M	5.6	126	159.8
	10	104	144.0

CONCLUSIONS

A new current-mode oscillating circuit is proposed to estimate the resistance (and parallel parasitic capacitance) of chemical sensors. Simulation results and preliminary prototype measurements confirm the correct interface operation. The design of compact low-voltage low-power integrated *CCIIs*, at transistor level, in a standard CMOS technology, makes the proposed circuit particularly suitable for portable applications (e.g., consumer).

ACKNOWLEDGMENTS

The authors would like to thank Fabrizio Mancini and Ilaria Lucresi for their valuable help in implementing and testing the prototype board.

REFERENCES

1. C. Falconi, E. Martinelli, C. Di Natale, A. D'Amico, P. Malcovati, A. Baschirotto, V. Stornelli, G. Ferri, "Electronic interfaces", Sens. & Act. B, Vol. 121, 30 Jan. 2007, pp. 295-329.
2. L. Fasoli, F. Riedijk, J. Huijsing, "A general circuit for resistive bridge sensors with bitstream output", IEEE Trans. on Instr. and Meas. 46 (4), 1997, pp. 954-960.
3. A. Depari, M. Falasconi, A. Flammini, D. Marioli, S. Rosa, G. Sberveglieri, A. Taroni, "A New Hardware Approach to Realize Low-Cost Electronic Noses", Proc. of IEEE Sensors 2005, Irvine, USA, Oct. 31-Nov. 3, 2005, pp. 239-242.
4. A. Baschirotto, S. Capone, C. Di Natale, V. Ferragina, G. Ferri, L. Francioso, M. Grassi, N. Guerrini, P. Malcovati, E. Martinelli, P. Siciliano, "A portable integrated wide-range gas sensing system with smart A/D front-end", Sens. & Act. Vol. 130, Issue 1, March 2008, pp. 164-174.
5. S. Bicelli, A. Depari, G. Faglia, A. Flammini, A. Fort, M. Mugnaini, A. Ponzoni, V. Vignoli, "Model and experimental characterization of dynamic behaviour of low power Carbon Monoxide MOX with pulsed temperature profile", Proc. of I2MTC 2008, Canada, May 12-15, 2008, pp. 1413-1418.
6. A. Depari, G. Faglia, A. Flammini, A. Fort, M. Mugnaini, A. Ponzoni, E. Sisinni, S. Rocchi, V. Vignoli, "CO detection by MOX sensors exploiting their dynamic behaviour", Proc. of Eurosensors XXII, Dresden, Germany, Oct. 7-10, 2008, pp. 1070-1073.
7. A. Flammini, D. Marioli, A. Taroni, "A low-cost interface to high-value resistive sensors varying over a wide range", IEEE Trans. on Instr. and Meas., Vol. 53, No. 4, Aug. 2004.
8. K. Mochizuki and K. Watanabe, "A high-resolution, linear resistance-to-frequency converter", IEEE Trans. on Instr. And Meas., Vol. 45, June 1996, pp. 761-764.
9. A. De Marcellis, A. Depari, G. Ferri, A. Flammini, D. Marioli, V. Stornelli, A. Taroni, "Uncalibrated integrable wide-range single-supply portable interface for resistance and parasitic capacitance determination", Sens. & Act. B, Vol. 132, 16 June 2008, pp. 477-484.
10. A. De Marcellis, A. Depari, G. Ferri, A. Flammini, D. Marioli, V. Stornelli, A. Taroni, "A CMOS integrable oscillator-based front-end for high dynamic range resistive sensors", IEEE Trans. on Instr. and Meas., Vol. 57, No. 8, Aug. 2008, pp. 1596-1604.
11. G. Ferri, V. Stornelli, A. De Marcellis, A. Flammini, A. Depari, D. Marioli, "A novel low-voltage low-power second generation current conveyor-based front-end for high valued DC-excited resistive sensors", Proc. IEEE Sensors 2008, Lecce, October 2008.
12. G. Ferri, N. Guerrini, Low voltage low power CMOS current conveyors, Kluwer Academic Publisher, Boston, 2003, 226 pp., ISBN 1-4020-7486-7.
13. A.C.W. Wong, G. Kathiresan, C.K.T. Chan, O. Eljamaly, O. Omeni, D. McDonagh, A.J. Burdett,n C. Toumazou, "A 1V wireless transceiver for an ultra-low-power SoC for biotelemetry applications", IEEE Journal of Solid State Circuits, Vol. 43, n. 7, July 2008, pp. 1511-1521.
14. N. Van Helleputte, J.M. Tomasik, W. Galjan, A. Mora-Sanchez, D. Schroeder, W.H. Krautschneider, R. Puers, "A flexible system-on-chip (SoC) for biomedical signal acquisition and processing", Sens. and Act. A, Vol. 142, 2008, pp. 361-368.
15. G. Ferri, V. Stornelli, M. Fragnoli, "An integrated improved CCII topology for resistive sensor application", Analog Integrated Circuits and Signal Processing, Vol. 49 No. 3, Sept. 2006, pp. 247-250.
16. S. Del Re, A. De Marcellis, G. Ferri, V. Stornelli, "Low voltage integrated astable multivibrator based on a single CCII", Proc. PRIME 2007, Bordeaux, 2-5 July 2007, pp. 177-180.

Application Of Electronic Nose And Ion Mobility Spectrometer To Quality Control Of Spice Mixtures

U. Banach, C. Tiebe and Th. Hübert

BAM Federal Institute for Materials Research and Testing
Unter den Eichen 87, 12200 Berlin, Germany
ulrich.banach@bam.de

Abstract. The aim of the paper is to demonstrate the application of electronic nose (e-nose) and ion mobility spectrometry (IMS) to quality control and to find out product adulteration of spice mixtures. Therefore the gaseous head space phase of four different spice mixtures (spices for sausages and saveloy) was differed from original composition and product adulteration. In this set of experiments metal-oxide type e-nose (KAMINA-type) has been used, and characteristic patterns of data corresponding to various complex odors of the four different spice mixtures were generated. Simultaneously an ion mobility spectrometer was coupled also to an emission chamber for the detection of gaseous components of spice mixtures. The two main methods that have been used show a clear discrimination between the original spice mixtures and product adulteration could be distinguished from original spice mixtures.

Keywords: Electronic Nose, Ion Mobility Spectrometer, Spice Mixtures, Product Adulteration, Multivariate Data Analysis
PACS: 07.07.Df

INTRODUCTION

Electronic nose and an ion mobility spectrometer can rapidly distinct between slight variations in complex odors. This makes the techniques ideal for online process diagnostics and screening across a wide range of application areas [1]. Therefore it was tested, how complex odor detection and measurement systems can contribute to an enhancement in quality control and safety against product adulteration of spice mixtures.

EXPERIMENTAL AND METHODS

Spice Mixtures

Four spice mixtures were used to determine differences between original and product adulteration of spice mixtures. The spice mixtures were provided by Kahler-Spices Ltd., Berlin. The first spice was "Optima" for sausages. The product adulteration's consistence was a mix of 80 % original spice with 20 % garlic powder. A "Saveloy" spice mixture was the other. The product adulteration was mixed with 20 % curry spice.

Experimental Setup

An emission chamber (2.5 L) was used to simulate the gaseous head space atmosphere. Inside the chamber there were placed 20 g of each spice mixture to detect the gas composition and the aroma respectively. The spice mixtures were stored one hour at 25 °C in the emission chamber before the analyses were started. All tube connections of the chamber were closed after a spice mixture was placed inside. Later than the connections were opened for 30 minutes and the chamber was surged with 500 sccm/min purified air. The pump of the electronic nose initialized the flow. The gas flow was divided by a three way valve to connect the electronic nose and the ion mobility spectrometer.

Electronic Nose

An electronic nose was used to transform volatile components (odors) from spice mixtures by means of 38 sensitive elements (sensor array) into electronic signals and data patterns by a head space experiment. Conductivity at the sensor segments was measured every second to create signal patterns which allow

CP1137, *Olfaction and Electronic Nose: Proceedings of the 13th International Symposium,* edited by M. Pardo and G. Sberveglieri

sensitive recognition and quantification of the ambient gases (odor of the spice mixture). These data patterns are characteristic for specific odors, every has its own electronic "fingerprint". Data from electronic noses can be visualized directly, or attributed to various statistical methods (e.g. Linear Discriminant Analysis) in comprehensive patterns for visualization.

Ion Mobility Spectrometry (IMS)

An ion mobility spectrometer was used as a physical-chemical method to measure the velocity of gas-phase ions in an electric field at ambient pressure. Therefore vapour samples are ionised by β^--radiation and positive charged reaction ions of the type $(H_2O)_nH^+$ are gained. Those ions will react with volatile organic compounds from the sample to protonated monomers ($MH^+(H_2O)_n$) and proton bound dimers ($M_2H^+(H_2O)_n$) [2]. The results of these measurements are spectra which show a distribution of current against the drift time of ions in an electric field. The composition of the gas atmosphere of the emission chamber was measured three times.

RESULTS

The change in signal patterns during measurements reflects the dynamic of spice mixture sampling as well as possible effects on the spice composition. The Linear Discriminant Analysis (LDA), which is based on the data of spice mixtures by the electronic nose, shows reproducable and qualified results of three test runs. All four spice mixtures were clearly distinguished, so that even two product adulterations were detected (fig.1).

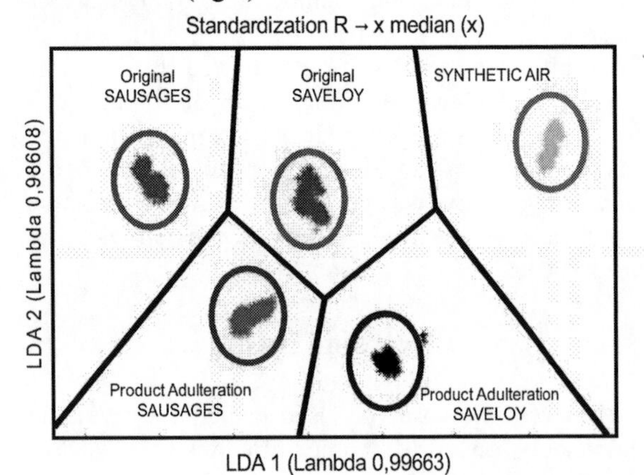

FIGURE 1. LDA-analysis of four spice mixtures and reference

It was also possible to differentiate signals of the original spice mixtures from the product adulteration after a comparison of the IMS spectra. The first comparison was realized due to the comparison of the spectra. A comparison of original and adulterated saveloy spice mixture composition has shown significant differences (fig. 2). Above all the intensity distribution and the peak shapes differs from each other. In the case of a further comparison of the sausage spice data has shown similar characteristics.

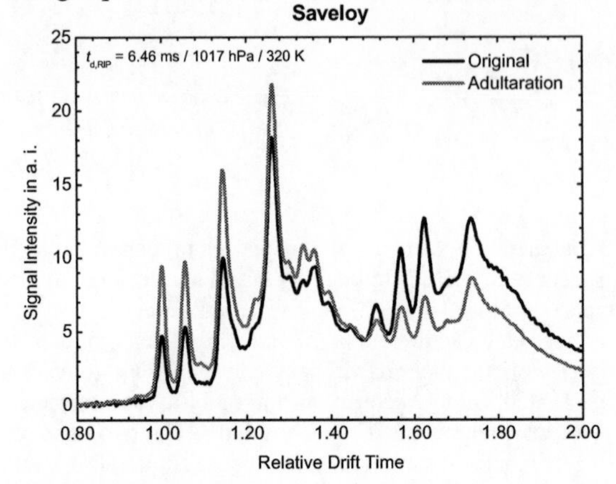

FIGURE 2. Two IMS spectra which show odor differences of saveloy spice mixtures.

A further step is the data analysis of IMS spectra done by hierarchical clustering methods. In this case of the analysis of sausage spice it was possible to differ the adulteration versus the original composition with a probability of 66 %. A distinction of the adulterated and original composition of the saveloy spice was possible with a probability of 75 %.

CONCLUSIONS

It could be demonstrated that the electronic nose and the ion mobility spectrometry provides good discrimination power for odor measurements of spice mixtures, which might find an technical application in an reproducable and qualitative test system to determine the original spice mixtures from product adulterations.

ACKNOWLEDGMENTS

We thank the Kahler-Spices Ltd., Berlin for providing the spice mixtures.

REFERENCES

1. F. Röck, N. Barsan and U. Weimar, *Chem. Rev.* **108**, 705-725 (2008).
2. Eiceman, G. A. et al Ion Mobility Spectrometry, Taylor & Francis Group (Boca Raton, 2005), pp. 3-7

A Laboratory Impedance Meter For Electrochemical Sensors

Ada Fort, Francesco Chiavaioli, Cristian Lotti, Marco Mugnaini, Santina Rocchi, Valerio Vignoli

Information Engineering Dept., University of Siena
Via Roma, 56 – 53100 Siena - Italy
Email: {ada,mugnaini,vignoli}@dii.unisi.it, web: leeme.dii.unisi.it

Abstract. In this contribution a vectorial impedance meter suitable for measurements on two-electrode and three electrode electrochemical sensors is described. The impedance meter is based on basic instruments that are commonly present in an electronics lab (a digital oscilloscope and a signal generator), on a personal computer (PC), and on an 'ad hoc' developed front-end circuit. It can perform impedance measurements up to 5 MHz with a 10% maximum magnitude error and a 10 degrees maximum phase error.

INTRODUCTION

Impedance measurement, or impedance spectroscopy, is one of the measurement techniques used both for studying the phenomena at the base of chemical sensor behavior, as well as for enhancing the sensor selectivity toward specific chemical species (see, e.g., [1]).

In particular, when electrochemical sensors are considered, this measurement technique is one of the most used, since it has demonstrated its capabilities both from a theoretical and a practical point of view (see again [1] and, e. g. [2] and the references therein). From a practical point of view, with this measurement technique the sensors are stimulated with low amplitude AC signals in a frequency range of a few MHz (that is with small perturbation signals, which do not disturb the electrode properties). The measured currents (for amperometric sensors) are typically in the nA-μA range.

With these frequency range and these characteristics of the signals to be measured it is possible to avoid the use of an expensive commercial vectorial impedance meter: in this contribution it is described an impedance meter based on basic instruments that are commonly present in an electronics lab (a digital oscilloscope and a signal generator), on a personal computer (PC) and on an 'ad hoc' developed front-end circuit. With this instrument it is possible to perform impedance measurements on two-electrode electrochemical sensors (that is to measure the current flowing between the counter electrode (CE) and the working electrode (WE) when a known variable voltage is applied between CE and WE), and on three-electrode electrochemical sensors (that is to measure the current flowing between CE and WE when a known variable voltage is applied between a reference electrode (RE) and WE). The measurement range is 100 Hz - 5 MHz with maximum magnitude error lower than 10%, and a maximum phase error lower than degrees.

EXPERIMENTAL METHODS

The Instrument

The developed measurement system is composed of a dedicated configurable front end electronics, of a digital oscilloscope (Tektronix Tek 1012B), of a signal generator (Agilent 33220A), and of a PC (fig. 1).

In fig. 1 it is also reported the sketch of a chemical sampling and control system. This latter must be integrated with the impedance meter to maintain the sensors, during measurements, in known and stable conditions in terms chemical atmosphere, umidity, and temperature. Details about the chemical sampling and control system used in this work can be found in [3].

The front end electronics (fig. 2) is based on several wide band ultra-low offset ultra-low bias current operational amplifiers (OA) of the same type (Analog Devices AD8610ARZ).

Among these OAs, one is used to drive the sensor CE with a signal $V_1(t)$ on the base of the signal $V_{in}(t)$ coming from the signal generator (that is a sinusoidal signal with selectable frequency, amplitude, and offset), whereas a

CP1137, *Olfaction and Electronic Nose: Proceedings of the 13th International Symposium*, edited by M. Pardo and G. Sberveglieri

second one is used to implement a gain-selectable I-V converter whose output $V_2(t)$ is a voltage signal proportional to the WE current.

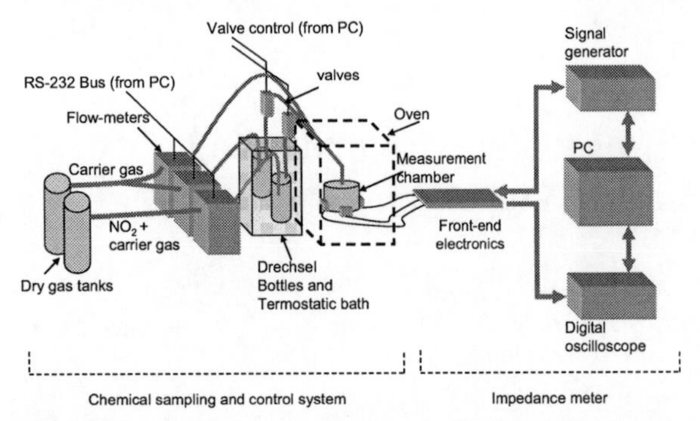

FIGURE 1. Block scheme of the impedance meter and of the associated chemical sampling and control system.

The wide band OA driving the sensor CE is arranged (by jumpers) in a buffer configuration when the impedance meter is used with two-electrode electrochemical sensors (fig. 2a), or as a feedback circuit that imposes a voltage $V_3(t) \approx V_{in}(t)$ between RE and WE when the instrument is used with three-electrode electrochemical sensors (fig. 2b).

With reference to fig. 2, for measurements on both two-electrode and three-electrode sensors the digital oscilloscope acquires the signals $V_1'(t)$ and $V_2(t)$ (the signals $V_1'(t)$ is obtained from $V_1(t)$ by a buffer circuits).

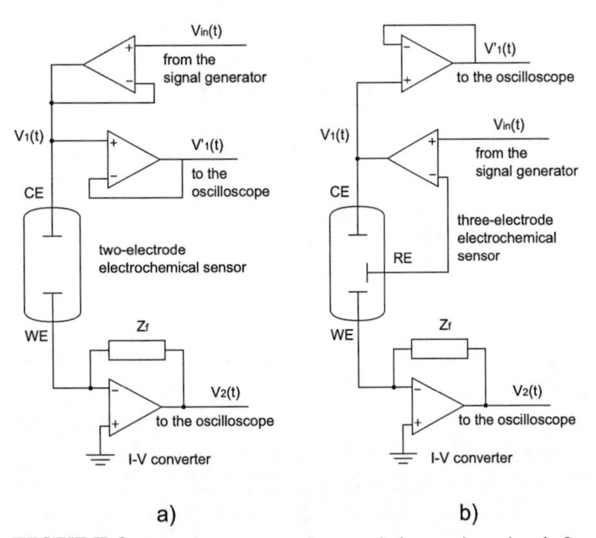

FIGURE 2. Impedance meter front end electronics: circuit for two-electrode (a) and three-electrode sensors (b).

The PC, that drives via a USB link the signal generator and the oscilloscope, is used to setup and control both the instrument calibration phase and the measurement phase (that is to set frequency calibration and measurement points, sinusoidal input amplitude and DC bias. It drives also throug a serial link the chemical sampling system. Finally, it processes the data acquired by the oscilloscope. This issues will be discussed in what follows.

IMPEDANCE MEASUREMENTS

For two-electrode sensors the quantity of interest Z_{2E} is the ratio of the voltage across CE and WE, and the WE current, whereas for three-electrode sensors the quantity of interest Z_{3E} is the ratio of the voltage across RE and WE, and the WE current.

Given the feedback impedance of the I-V converter Z_f, the unknown sensor impedances Z_{2E} and Z_{3E} are given by eq. (1) and (2), respectively:

$$Z_{2E}(\omega) = Z_f(\omega)\left[\frac{V_1'(\omega)}{V_2(\omega)}A(\omega) + B(\omega)\right] \tag{1}$$

$$Z_{3E}(\omega) = Z_f(\omega)\left[\frac{V_1'(\omega)}{V_2(\omega)}C(\omega) + \frac{V_{in}(\omega)}{V_2(\omega)}D(\omega) + E(\omega)\right] \tag{2}$$

In (1) and (2), $V_1'(\omega)$, $V_2(\omega)$, $V_{in}(\omega)$ are the phasors associated to the sinusoidal signals $V_1'(t)$, $V_2(t)$, $V_{in}(t)$, and $A(\omega)$, $B(\omega)$, $C(\omega)$, $D(\omega)$, $E(\omega)$ are correction coefficients accounting for gain, input impedance and output impedance of the OAs, that are treated as unknown frequency-dependent quantities. $A(\omega)$ and $B(\omega)$ are determined during the instrument calibration phase from two sets of impedance measurements, in the frequency range of interest, performed on two different calibration loads, whereas $C(\omega)$, $D(\omega)$ and $E(\omega)$ are determined from three sets of impedance measurements performed on three different calibration loads.

RESULTS

As anticipated in the introduction, with the proposed instrument it is possible, after calibration, to perform impedance measurements up to 5 MHz with a 10% maximum magnitude error and a 10 degrees maximum phase error. As an example, in fig. 3 the measured impedance magnitude (a) and phase (b) of a bipole consisting of a 100 kΩ resistor in parallel with a 100 pF capacitor are reported: the green curves are relative to data before instrument calibration, whereas the red curves to data after calibration. For comparison also the measurements obtained with a commercial instruments (Agilent 4294A), are reported (blue curves).

In fig. 4 the impedance data are reported that were obtained for a two-electrode amperometric sensor consisting of a Nafion proton exchange membrane (PEM), a Graphite WE, and a NiO CE. This sensor, that presents a satisfactory sensitivity to NO_2 (up to 90nA/ppm) and a very small cross sensitivity to CO and O_2 was described in detail in [3]. Each of the curves in fig. 4 was obtained by measuring the sensor impedance in

100 frequency measurement point linearly spaced in the range 100 Hz – 2 MHz, using a zero-offset 20 mV amplitude Vin(t) signal.

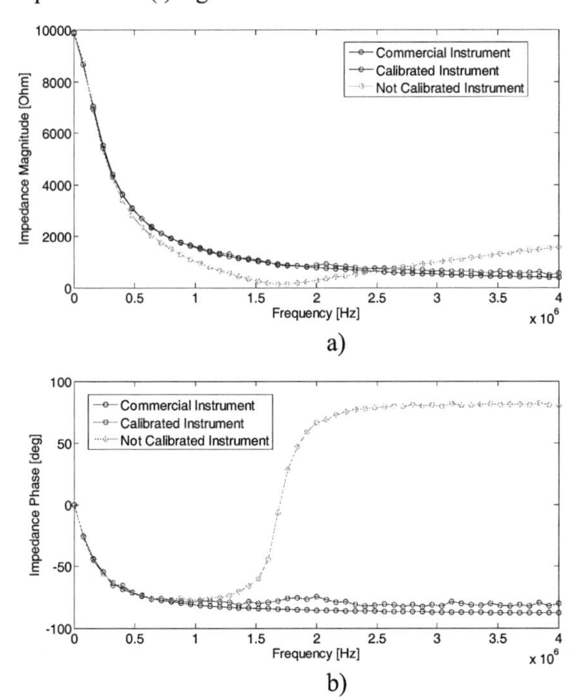

FIGURE 3. Impedance magnitude (a) and phase (b) of a 100 kΩ resistor in parallel with a 100 pF capacitor. In the picture not calibrated instrument, calibrated instrument and Agilent 4294A curves are shown.

In fig. 4 the data of two groups of 3 successive measurements are reported in red and in blue, respectively. The measurement in each group are repeated in a period of 30 min, in steady state chemical conditions. In detail, the red curves are relative to measurements with the sensor exposed to a 150 mL/min N_2 flow, with a 100% relative humidity (RH) at a controlled temperature of 30 °C.

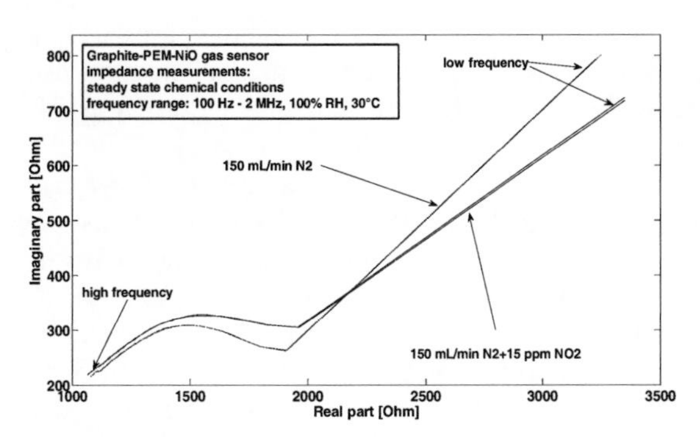

FIGURE 4. Two-electrode Graphite-PEM-*NiO* amperometric sensor impedance measurements executed in steady state chemical conditions in the frequency range 100 Hz-2 MHz. Red curves: sensor exposed to 150 mL/min N_2@ (100% RH, 30°C). Blue curves: sensor exposed to a mixture of 150 mL/min N_2 with 15 ppm NO_2@ (100% RH, 30°C).

On the other hand, for the blue curves the measurement conditions are the same in terms of temperature and RH, whereas the 150 mL/min total gas flow sent to the measurement chamber was a mixture of N_2 with 15 ppm NO_2.

The measurements reported in fig 4 can be explained in terms of the classic impedance model for this kind of sensors (fig. 5), where R_{PEM} is associated to the Nafion PEM, whereas C_{DL}, R_F and Z_{WB} are the double layer capacitor, the faradic resistor and the Warburg impedance associated to the interface WE-PEM [1].

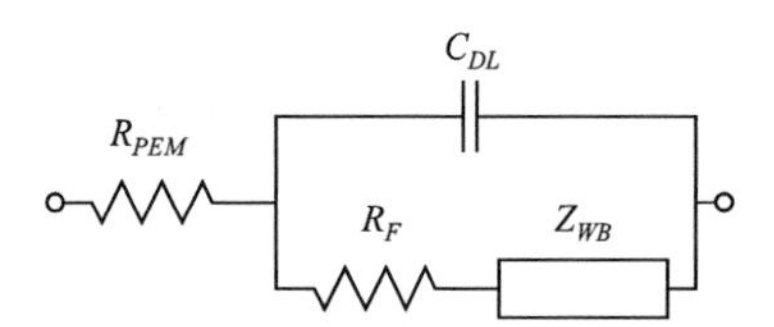

FIGURE 5. Equivalent impedance for the two-electrode Graphite-PEM-*NiO* amperometric sensor.

CONCLUSION

In this contribution a vectorial impedance meter based on basic instruments that are commonly present in an electronic lab was presented. The proposed apparatus is suitable for impedance measurements on two-electrode and three-electrode electrochemical gas sensors. The maximum errors in terms of impedance magnitude and phase (10% and 10 degrees, respectively), as well as the frequency measurement range (up to 5 MHz) makes it a valid alternative for this kind of application to expensive commercial vectorial impedance meters. The instrument cost can be further reduced by substituting both the signal generator and the digital oscilloscope with a DDS module and a DAQ board if adequate narrow band signal sampling techniques are used.

REFERENCES

1. Barsoukov E., Macdonald J. R. (Eds.), Impedance Spectroscopy: Theory, Experiment, and Applications, 2nd Edition, John Wiley & Sons, (New York, 2005).
2. Gabrielli C., Keddam M., "Contribution of electrochemical impedance spectroscopy to the investigation of the electrochemical kinetics", *Electrochimica Acta*, Vol. 41, No. 7-8 (1996), pp. 957-965.
3. Fort A., Lotti C., Mugnaini M., Palombari R., Rocchi S., Vignoli V., "A Two Electrode C - NiO Nafion Amperometric Sensor for NO_2 Detection", *Microelectronics Journal* (in press).

A New, Fast Readout, Interface For High-value Resistive Chemical Sensors

A. Depari[1,2], A. Flammini[1,2], D. Marioli[1], E. Sisinni[1,2], A. De Marcellis[3], G. Ferri[3], and V. Stornelli[3]

[1] *Dept. of Electronics for Automation, University of Brescia, via Branze 38, 25123, Brescia, Italy*
[2] *CNR-INFM SENSOR Lab., via Valotti 9, 25133, Brescia Italy*
[3] *Dept. of Electrical and Information Engineering, University of L'Aquila, Monteluco di Roio, L'Aquila, Italy*

Abstract. Metal oxide (MOX) gas sensors show a resistive value varying in a wide range, from hundreds of kilohms up to tens of gigohms. For this reason, oscillating circuits are often used to interface such sensors. Simple oscillating circuits allow a measuring time which is directly proportional to the resistive value; with large resistive values (gigohms), the output updating time is on the order of seconds. In addition, the sensor is excited with a square wave, but the variable frequency can affect the measurement accuracy. In this paper, a new oscillating circuit is proposed with a limited measuring time regardless the resistive value. This behavior is obtained by means of self-moving thresholds. The working principle is always based on the integration of a constant current flowing through the sensor and generating a ramp. Such ramp is compared with two threshold ramps with known slope and the maximum measuring time is limited by the slower threshold. The proposed circuit has been simulated and experimentally tested with commercial resistors (values between 1 MΩ and 100 GΩ).

Keywords: high resistive sensor, low measuring time, wide range.
PACS: 07.07.Df

INTRODUCTION

Metal Oxide gas sensors are often used thanks to their advantages: good sensitivity to relevant gases, low production costs, small size. They react to the presence of the gas by varying the conductivity, therefore they can be treated as resistive sensors. To interface such sensors, which can have resistance values varying from hundreds of kilohms to tens of gigohms, different solutions have been developed [1-6], also based on a resistance-to-time conversion (RTC). These oscillating circuits provide an output signal whose period directly depends on the sensor resistance value. In addition, due to the particular processes used to implement the sensor and to the presence of a heater element together with the sensitive film, a parasitic capacitance effect is created. Such capacitance, which is on the order of few picofarads, is usually modeled in parallel with the sensor resistive component; it can affect resistance measurement performance. Moreover, the capacitance estimation can be useful to extract further information from the sensor for diagnostic purposes or to better characterize new experimental sensors, based for example on nanowires structures [7]. The scheme in Fig. 1 has been recently proposed as a suitable interface able to estimate the sensor resistance value R_{sens} over a wide range together with the sensor parasitic capacitance C_{sens}. The working principle is based on the integration of the current flowing into the sensor by means of an operational amplifier (*Int*) with a very low input bias current. The triangular waveform at the integrator output is due to the use of a comparator ($Comp_{th}$), tuned on a suitable threshold V_{th}, which determines the correct switching of the sensor excitation voltage V_{exc} to realize the oscillator. The oscillation period T depends on the current flowing to the integrator and, as a consequence, on the sensor resistive value. Particularly, if the sensor resistance R_{sens} is on the order of some gigohms, the measuring time is more than one second. The effect of the parasitic capacitance can be taken into account by adding a second comparator ($Comp_0$) tuned on a different threshold. It should be noticed that, differently from $Comp_{th}$, $Comp_0$ is a simple threshold without hysteresis, therefore the noise has a big effect on this comparator. With the circuit depicted in Fig. 1, both sensor resistive and parasitic capacitance values can be estimated by measuring the period T and the

CP1137, *Olfaction and Electronic Nose: Proceedings of the 13th International Symposium,* edited by M. Pardo and G. Sberveglieri

duty-cycle T_{on}/T [3]. Such interface has been successfully implemented in a standard CMOS technology as an integrated circuit. It has shown good performances in terms of linearity, with a weighted least mean square (WLMS) error less than 5% in the range from 100 kΩ to 100 GΩ [8].

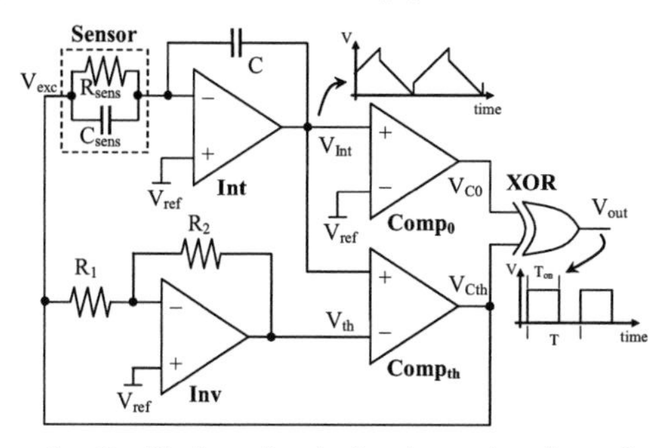

Figure 1. Oscillating circuit for the estimation of sensor resistive and capacitive components. The measuring time T increases as a direct function of sensor resistance value.

In conclusion, the main limits related to this circuit include unstable behavior of $Comp_0$ comparator due to noise and, especially, the variable and long measuring time. Therefore, this circuit seems unsuitable for applications in which the sensor response must be well characterized during fast transients, due to gas sensing or thermal profiles or the modification of the sensor excitation. In addition, new heating techniques, such as pulsed methods, can be investigated to obtain both a reduction of the power consumption and to extract more useful information from the sensor response [9, 10]. The aim of this work is to investigate a new circuit exploiting the RTC technique to estimate both sensor resistive and capacitive components, but keeping the measuring time in the order of tens of milliseconds.

THE NEW INTERFACE CIRCUIT

The proposed solution, shown in Fig. 2, is an RTC circuit directly derived on the one shown in Fig. 1. The main difference is in the comparator thresholds used to generate the output waveforms. In fact, in the new solution, such thresholds are not fixed voltages (see $Comp_0$ and $Comp_{th}$), but they are realized by means of two ramp voltages which move in the opposite direction with respect to the integrator ramp. In such a way, timings related to the oscillating circuit do not depend on the sensor resistive value only, but on the moving threshold slopes as well, as shown in the timing diagram in Fig. 3. By this solution, the two main limits of the previous circuit, that is the measuring time and the noise at the $Comp_0$ comparator level, are overcome. Referring to Fig. 2, three operational amplifiers have been added: Op_1 is a buffer to decouple the circuit from the capacitive effect of the cables connecting the sensor (useful also in the circuit in Fig. 1), Op_2 is the faster threshold ramp generator (the first one crossing the sensor ramp), and Op_3 is the slower threshold ramp generator (Op_2 and Op_3 are integrators, similar to *Int*). The two capacitors C_x and C_y are needed to generate the charge transfer effect on Op_2 and Op_3 during the slope commutation which determines the step on the threshold ramps (bootstrapping effect). The series of a diode D and a zener diode DZ are needed to make the threshold ramps start always from the same value $\pm V_t$ after the step due to the charge transfer (V_t is equal to the sum of the direct voltage of the diode D and the reverse voltage of the zener diode DZ).

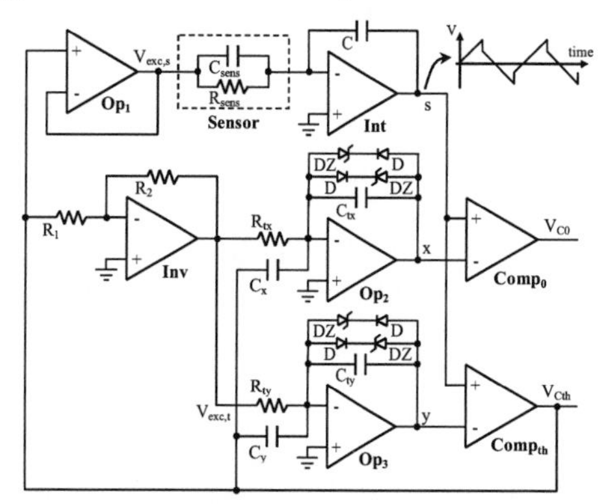

Figure 2. Scheme of the proposed circuit.

The sensor is supplied by a voltage $V_{exc,s}$ which derives from the $Comp_{th}$ output voltage V_{Cth} through the buffer Op_1. In the same way, the threshold ramp generators are supplied with a voltage $V_{exc,t}$ which derives from the $Comp_{th}$ output voltage V_{Cth} through the inverting amplifier *Inv*. If we suppose V_{Cth} to commutate between the two values $\pm V_{exc}$, then $V_{exc,s}$ and $V_{exc,t}$ voltages commutate, with opposite phase, between the same values $\pm V_{exc}$. Such hypothesis can be considered valid, for example, if rail to rail components are used to realize the comparators and the amplifiers. Both the moving thresholds (threshold ramps) have different slope and the maximum measurement time depends on the time T_y taken by the slower threshold ramp y, starting from V_t, to reach the sensor ramp s, as depicted in Fig. 3. It can be designed to be as little as necessary by means of a suitable choice of C_{ty} and R_{ty} values. However, the information obtained measuring the time interval T_y allows the estimation of the resistive component R_{sens}, but only under the hypothesis that the parasitic capacitance C_{sens} is neglected. If the parasitic capacitance cannot be neglected, the only measure of the time interval T_y leads to a significant error in the resistance estimation. The use of the faster threshold ramp x allows the estimation of the resistive component R_{sens} without being affected by the parasitic capacitance C_{sens}, which is furthermore estimated.

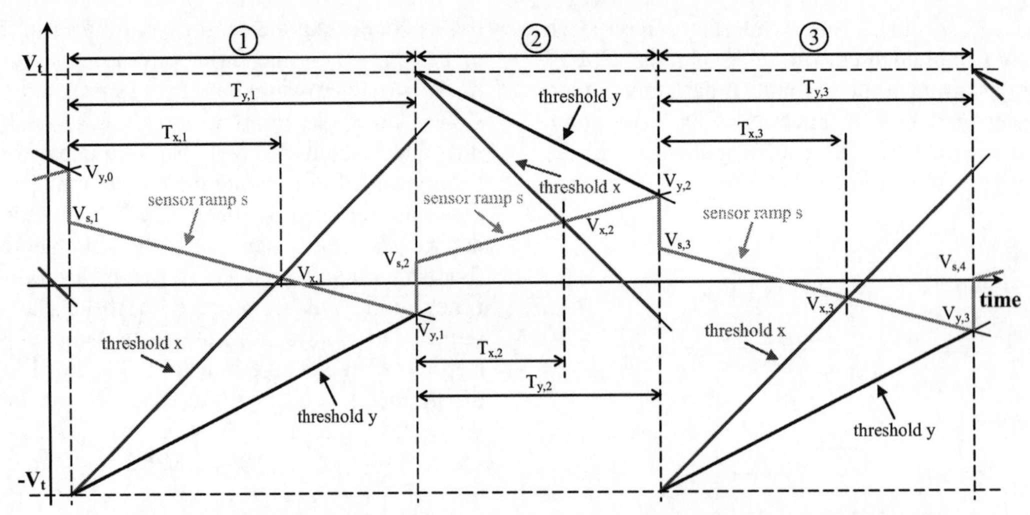

Figure 3. Time diagram of the proposed solution related to the circuit in Fig. 2.

It is the same concept which led to the use of the second comparator $Comp_0$ in the circuit in Fig. 1. Referring to the first cycle in Fig. 3 (decreasing sensor ramp s), the following four equations describing the circuit can be derived:

$$\frac{V_{exc}}{R_{sens,1}} T_{x,1} = -C(V_{x,1} - V_{s,1}) \quad (1)$$

$$\frac{V_{exc}}{R_{tx}} T_{x,1} = C_{tx}(V_{x,1} + V_t) \quad (2)$$

$$\frac{V_{exc}}{R_{sens,1}} T_{y,1} = -C_s(V_{y,1} - V_{s,1}) \quad (3)$$

$$\frac{V_{exc}}{R_{ty}} T_{y,1} = C_{ty}(V_{y,1} + V_t) \quad (4)$$

In such equations, if we suppose to measure $T_{y,1}$ and $T_{x,1}$, we have four unknown quantities: $R_{sens,1}$, $V_{s,1}$ which is the initial voltage value of the sensor ramp s (after the step due to the charge transfer effect on Op_1 caused by C_{sens}), $V_{x,1}$ which is the voltage where the sensor ramp s crosses the threshold ramp x, and $V_{y,1}$ which is the final voltage value of the threshold y and of the sensor ramp s (before the next cycle step due to C_{sens}). By inverting Eq. (1)-(4), we can find a solution for the four unknown quantities. The analysis of the other cycles in Fig. 3 leads to similar results; the estimation of sensor resistance during a generic cycle k is performed according to Eq. (5). Considering some circuital simplification, such as $C_{tx} = C_{ty} = C_t$, and $R_{tx} = \alpha \cdot R_{ty} = \alpha \cdot R_t$, with $0 < \alpha < 1$, Eq. (6) can be used to reckon $R_{sens,k}$ in a generic cycle k.

$$R_{sens,k} = \frac{C_{tx} R_{tx} C_{ty} R_{ty} (T_{y,k} - T_{x,k})}{C_s (C_{ty} R_{ty} T_{x,k} - C_{tx} R_{tx} T_{y,k})} \quad (5)$$

$$R_{sens,k} = \alpha \frac{C_t R_t}{C_s} \frac{T_{y,k} - T_{x,k}}{T_{x,k} - \alpha T_{y,k}} \quad (6)$$

The parasitic capacitance C_{sens} estimation takes advantage from the knowledge of the final value V_y of the sensor ramp s at the end of one cycle and the initial value V_s of the sensor ramp s at the beginning of the next cycle. In fact, the step that the signal s performs during the slope commutation is due to the effect of the parasitic capacitance, when the sensor supply voltage switches instantaneously (for example from V_{exc} to $-V_{exc}$). In general, considering the commutation between cycle k-1 and cycle k, Eq. (7) can be obtained. If all the simplifications described above are applied, Eq. (8) can be used to estimate the parasitic capacitance value C_{sens} in a generic cycle k.

$$C_{sens,k} = -C \frac{V_{s,k} - V_{y,k-1}}{V_{excs,k} - V_{excs,k-1}} \quad (7)$$

$$C_{sens,k} = \frac{C}{2C_t R_t} \left(2C_t R_t \frac{V_t}{V_{exc}} - \frac{1-\alpha}{\alpha} \frac{T_{x,k} T_{y,k}}{T_{y,k} - T_{x,k}} - T_{y,k-1} \right) \quad (8)$$

SIMULATIONS AND EXPERIMENTAL RESULTS

The new circuit described in the previous section has been simulated in order to evaluate its performances. In fact, the slope of ramp thresholds x and y should be chosen with great care. The slope of ramp y fixes the maximum measurement time, whereas the slope of ramp x should be chosen in order to measure with a sufficient resolution the time interval T_x, as well as the difference between T_y and T_x. Referring to Eq. (6), resolution in time measurement can affect both the numerator and the denominator of the fraction. With low values of R_{sens}, the term T_y-T_x is very small, whereas with high values of R_{sens}, the term T_x-αT_y can suffer from resolution problems. Starting from similar considerations, a small value of α favors the estimation of low values of R_{sens}, whereas a big value is more suitable if high R_{sens} values must be estimated. Fig. 4 shows the absolute value of the relative estimation error as a function of the value of R_{sens} and α if the timing resolution is 100 ns and the measurement time is about 40 ms. With these values, the estimation error is expected to be below 1% for 4 MΩ < R_{sens} < 100 GΩ. In our simulations and prototypes, α value has been set to about 0.2. Fig. 5 shows the effect of timing resolution on the estimation error (Matlab simulation). PSpice simulations have been conducted assuming the following values: α = 0.17, timing resolution of 100 ns, measuring time of 40 ms. We used a low input bias current operational amplifiers OPA350 and fast response comparators TLC3702, all from Texas Instruments, with a ±3.3 V power supply. Table I shows the obtained results. The estimation of R_{sens} have been performed using the weighted least mean square (WLMS) line, which is better than the usual least mean square line when a wide range of variation is considered. Except for the lowest resistance value, where the limit is the time resolution in the measurement (100 ns), as the high value of standard deviation clearly highlights, the relative linearity error is below 4% over more than five decades of resistance variation, from 1 MΩ to 100 GΩ.

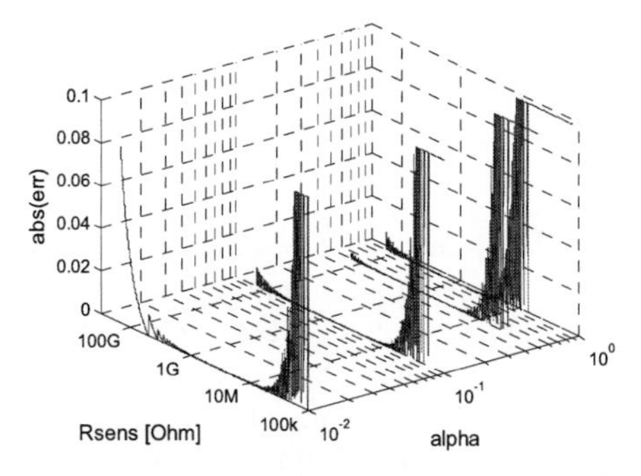

FIGURE 4. R_{sens} estimation error due to resolution in timing measurement (set to 100 ns) as a function of α.

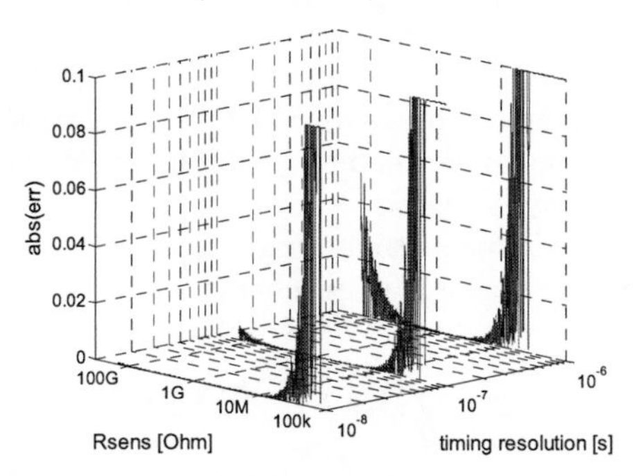

FIGURE 5. R_{sens} estimation error as a function of timing resolution (α set to 0.2).

The differences between Matlab and PSpice results shown in Table I depend on behavior of real components. Timing resolution is not the only source of non-ideality; with high value of R_{sens}, the current flowing in the integrator is very small, comparable with the bias current of the operational amplifier. In addition, PSpice takes into account the non idealities due to real component offset and propagation delay, while Matlab simply applies equations describing the ideal behavior. A discrete component prototype has been furthermore developed (same devices used in PSpice simulations). First experimental results are in agreement with PSpice simulation in the range from 1 MΩ to 50 GΩ.

TABLE 1. PSpice simulation results.

R_{sens} [MΩ]	<R_{sens}> WLMS [MΩ]	Std./<R_{sens}> [%]	err. WLMS [%]	err. Matlab [%]
1.00E-01	1.287E-01	87.2	28.7	100
1.00E+00	1.003E+00	5.3	0.3	9.7
1.00E+01	9.619E+00	2.1	-3.8	0.03
1.00E+02	1.021E+02	0.4	2.1	<0.01
1.00E+03	1.019E+03	1.7	1.9	<0.01
1.00E+04	1.038E+04	8.2	3.8	0.03
1.00E+05	9.671E+04	17.4	-3.3	0.5

REFERENCES

1. A. Flammini et al., "A low-cost interface to high value resistive sensors varying over a wide range," IEEE Trans. Instrum. and Meas., vol. 53, n. 4, pp. 1052–1056, Aug. 2004.
2. A. Depari et al., "A low cost circuit for high value resistive sensors varying over a wide range", IOP Meas. Sci. and Tech., vol. 17, n. 2, pp. 353-358, Feb. 2006.
3. A. De Marcellis et al., "Uncalibrated integrable wide-range single-supply portable interface for resistance and parasitic capacitance determination", Sens. and Act. B, vol. 132, n. 2, pp. 477-484, June 2008.
4. C. Falconi et al., "Electronic interfaces", Sens. and Act. B, vol. 121, n. 1, pp. 295-329, Jan. 2007.
5. U.Frey et al., "Digital systems architecture to accommodate wide range resistance changes of metal oxide sensors", Proc. of IEEE Sensors 2008, Oct. 26-29, 2008, Lecce, Italy, pp. 593-595.
6. A. Lombardi et al., "Integrated Read Out and Temperature Control Interface with Digital I/O for a Gas-Sensing System Based on a SnO_2 Microhotplate" Thin Film Gas Sensor Proc. of IEEE Sensors 2008, Oct. 26-29, 2008, Lecce, Italy, pp. 596-599.
7. G.Sberveglieri et al., "Synthesis and characterization of semiconducting nanowires for gas sensing" Sens. and Act. B, vol. 121, n. 1, pp. 3-17, Jan. 2007.
8. C. Di Carlo et al., "Integrated CMOS resistance to period converter with parasitic capacitance evaluation", Proc. of ISCAS2009, May 24-27, 2009, Taipei, Taiwan, in press.
9. S. Bicelli et al., "Model and experimental characterization of dynamic behaviour of low power Carbon Monoxide MOX with pulsed temperature profile", Proc. of I2MTC 2008, May 12-15, 2008, Victoria (BC), Canada, pp. 1413-1418.
10. A. Depari et al., "CO detection by MOX sensors exploiting their dynamic behaviour", Proc. of Eurosensors XXII, Oct. 7-10, 2008, Dresden, Germany, pp. 1070-1073.

A CMOS Integrable DDCCII-Based Readout System For Portable Potentiometric Sensors Array

G. Ferri[*], V. Stornelli[*], C. Di Carlo[*], A. De Marcellis[*], A. D'Amico[+], C. Di Natale[+] and E. Martinelli[+]

[*]*Dept. of Information and Electrical Engineering, University of L'Aquila Monteluco di Roio, 67100, L'Aquila, Italy*
[+]*Dept. of Electrical Engineering, University of Tor Vergata, Rome, Italy*
E-mail: giuseppe.ferri@univaq.it

Abstract. We present a very low voltage low power Differential Difference Amplifier based second generation current conveyor (*DDCCII*) for the readout of the voltage value coming from potentiometric sensors and to be used in a self-powered wireless sensor network for monitoring contaminants in hydric systems. Since the *DDCCII* is an extremely versatile block, it can be configured to give either an output voltage or a current directly proportional to the generated sensor voltage. The whole circuit has been designed in a CMOS 0.18μm standard technology with a total supply voltage of 1V and a standby current and power of about 10μA and 10μW, respectively.

Keywords: Current Conveyors, CMOS technology, potentiometric sensors.
PACS: 85.40.-e, 87.85.fk.

INTRODUCTION

Among the various classes of chemical sensors, ion-selective electrodes (ISE) are one of the most frequently used potentiometric sensors in laboratory analysis and industrial applications [1,2].

The principle of operation of the ion-selective electrodes is based on the measurement of their potential variation with respect to a reference electrode in "zero-current conditions". The potential of the ion-selective electrodes is a function of the activity of ionic species in a sample solution. The potential difference (voltage) between the working electrode and the reference electrode is the sensor signal and represents the information of the sample under measurement. Most of the time it is necessary to put these electrodes in remote environmental, difficult to reach with cabled power supplies, for a long period taking off the possibility of changing the system batteries. This means that the whole system should be able to restore his supply energy from the environment. The ability to scavenge energy from the environment and to use this energy for power remote distributed sensors, actuators, processors, memories and display elements is still an open task [3,4]. In order to realize a self-powered operation, a generic circuit must operate without batteries or any separate voltage source with low power consumption conditions and should be robust to power supply drift [5]. In this sense, analog circuits based on the current mode (CM) approach have been demonstrated (with respect to those based on op-amps) to be particularly suitable for low voltage and low power applications and also benefit of the following advantages: wide bandwidth, high slew rate and simpler circuitry [6,7]. In this work we propose a very low voltage low power Differential Difference Amplifier (*DDA*) based *CCII* (*DDCCII*) for the readout of the voltage value coming from a potentiometric sensor. This is the first step in the direction of developing a self-powered wireless sensor array network (showing very low power dissipation) for monitoring the presence of contaminants in waste treatment plants.

DDA-BASED DIFFERENTIAL CCII

The *CCII* is a single-ended device represented symbolically in Fig. 1. It can be implemented, as shown in Fig. 2, by a voltage buffer with a suitable current mirror. A *CCII* has three terminals, denoted with X, Y and Z. Ideally, if a voltage is applied to the

CP1137, *Olfaction and Electronic Nose: Proceedings of the 13th International Symposium*, edited by M. Pardo and G. Sberveglieri

Y node, the *CCII* will produce an equal voltage at the X node; furthermore, the current flowing into the X node is mirrored (equal or opposite) into the Z node. Equivalently, the *CCII* behaviour may be described by the following equations:

$$\begin{cases} I_Y = 0 \\ V_X = V_Y \\ I_Z = \pm I_X \end{cases} \tag{1}$$

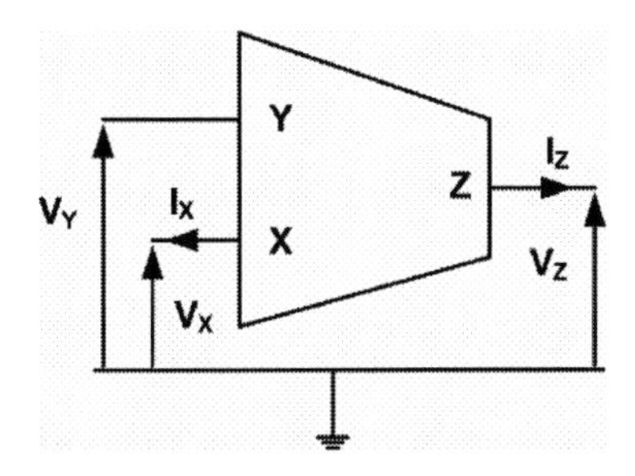

FIGURE 1. Circuit symbol block of a *CCII*.

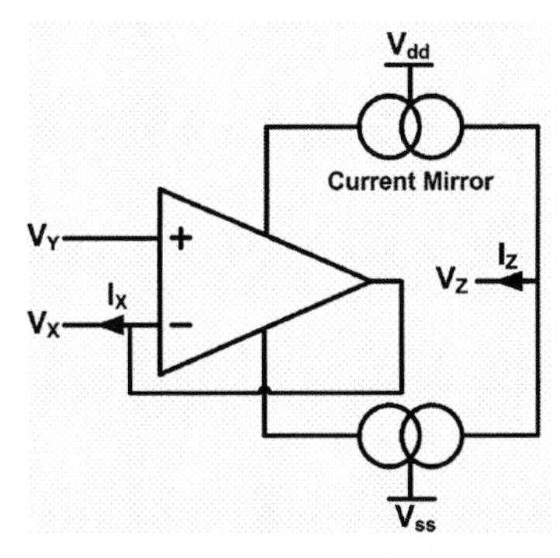

FIGURE 2. Single ended *CCII* internal basic implementation based on a voltage buffer and current mirror.

Concerning the Differential Difference Current Conveyor (*DDCC*) [8], it enjoys the advantages of the *CCII* and *DDA* such as larger signal bandwidth, greater linearity, wider dynamic range, simple circuitry, low power consumption, high input impedance and arithmetic operation capability. By considering the theory developed in [9], where it has been demonstrated that for lower open loop gain (typical case in low voltage and low power circuits) it is better to use a current mode approach instead of classical op-amp based circuits for implementing high accuracy amplifiers, in this work we have developed a fully differential *DDCCII* characterized by the following relationships:

$$\begin{cases} I_{Y_1} = I_{Y_2} = I_{Y_3} = 0 \\ V_X = V_{Y_1} - V_{Y_2} + V_{Y_3} \\ I_Z = \pm I_X \end{cases} \tag{2}$$

A block scheme and its internal representation of the *DDCCII* are shown in Fig. 3, while Fig. 4 shows its circuit schematic at transistor level and Fig. 5 the layout micrograph implemented with a CMOS 0.18 μm SMIC technology. In Table I the *DDCCII* component sizes are reported.

Y_1 and Y_2 terminals are high impedance terminals while X_1 terminal is a low impedance one due to the buffer output stages and the negative feedback. The differential input voltage applied between Y_1 and Y_2 terminals is conveyed to a single-ended differential voltage V_X ($V_X = V_{Y_{12}}$).

The input currents flowing into the X terminal are conveyed to the Z terminal, which is a high impedance node suitable for current outputs.

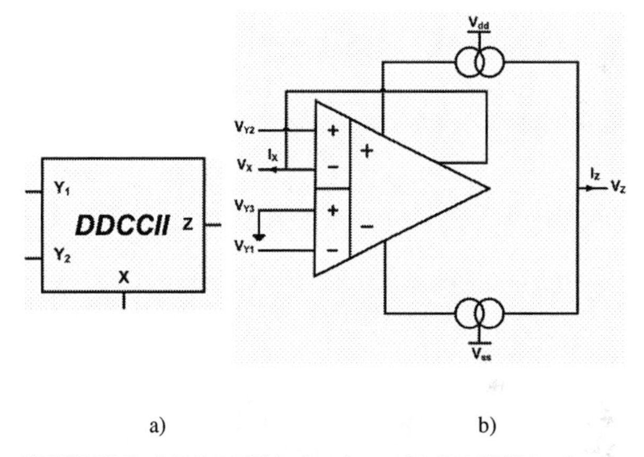

FIGURE 3. a) *DDCCII* block scheme; b) *DDCCII* basic internal implementation.

TABLE 1. Implemented DDCCII component sizes.

Component	Size
M1, M2, M3, M4	W=100μm, L=0.2μm
M8, M9	W=50μm, L=0.4μm
M10, M12, M15	W=50μm, L=0.2μm
M5, M6, M7, M11	W=2μm, L=0.2μm
M14, M16	W=0.5μm, L=0.2μm
M13	W=150μm, L=0.2μm
R_1	55kΩ
R_2	9kΩ
C	6pF

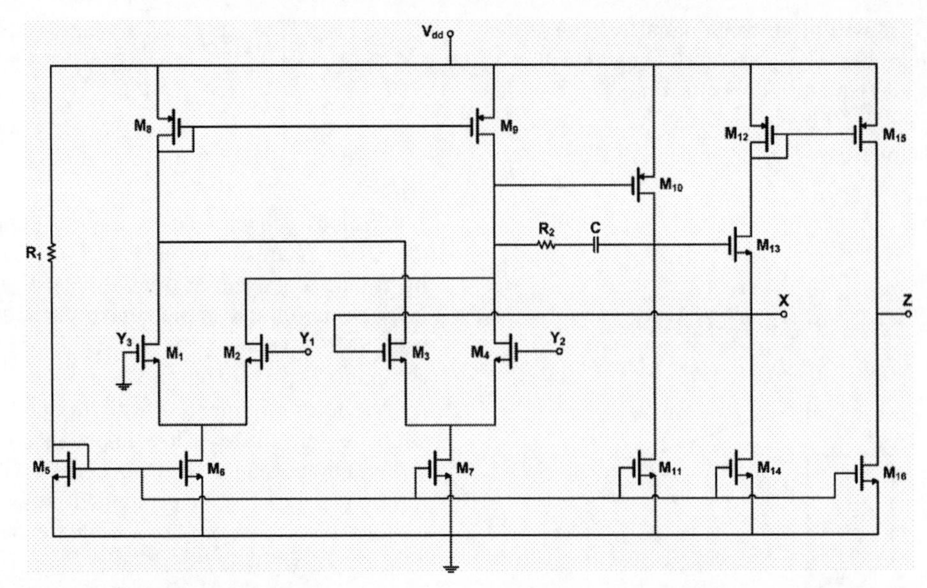

FIGURE 4. a) *DDCCII* block scheme; b) *DDCCII* basic internal implementation.

SIMULATION RESULTS AND APPLICATIONS

In table II the main characteristics of the low voltage low power *DDCCII* are shown. The *DDCCII* current and voltage DC transfer functions are also reported, in particular the voltage following between ports Y and X (see Fig. 6) and the current following between ports X and Z (Fig. 7). The asymmetrical output circuit topology causes a small difference between the two current curves at the low supply voltage that has been set to 1 V. The total standby current of the circuit is about 10 μA.

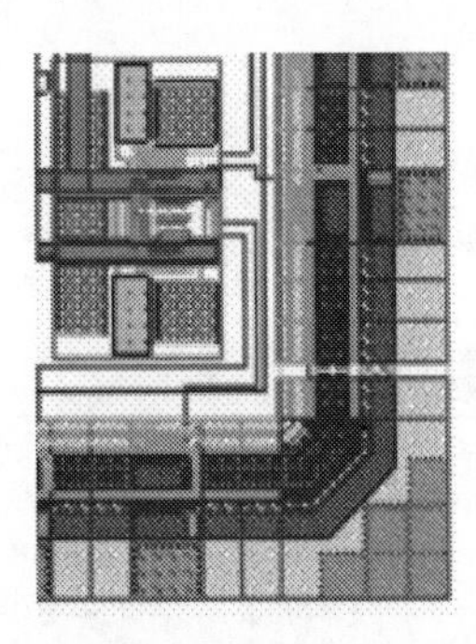

FIGURE 5. DDCCII implemented layout micrograph.

The designed *DDCCII* is planned to be used as the first element directly connected to the sensor electrodes as shown in Fig. 8. From this figure it results that the *DDCCII* can be used as a Variable Gain Amplifier (*VGA*) in order to set the system sensitivity. An important requirement for the *VGA* circuit is to provide high linearity for a wide range of signal swing. In this case the potentiometric sensors under studies show a voltage swing ranging from about 0 to 800mV.

This last characteristic will be a key feature for a further development of a sensorial system based on an array of potentiometric sensors where a microcontroller will communicate with a control unit using a wireless route (see Fig. 9).

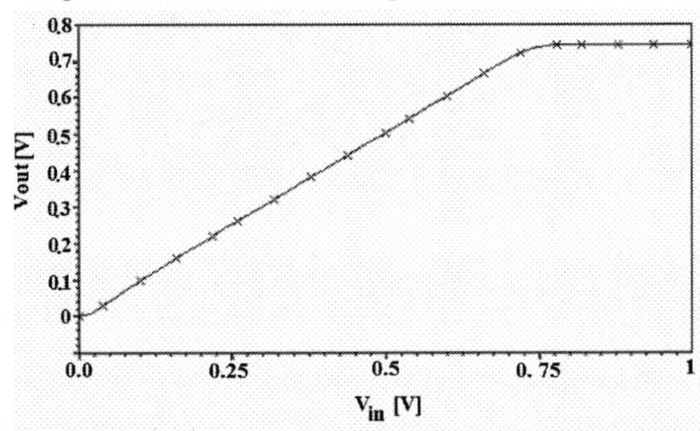

FIGURE 6. DC voltage transfer characteristic between ports Y_{12} and X.

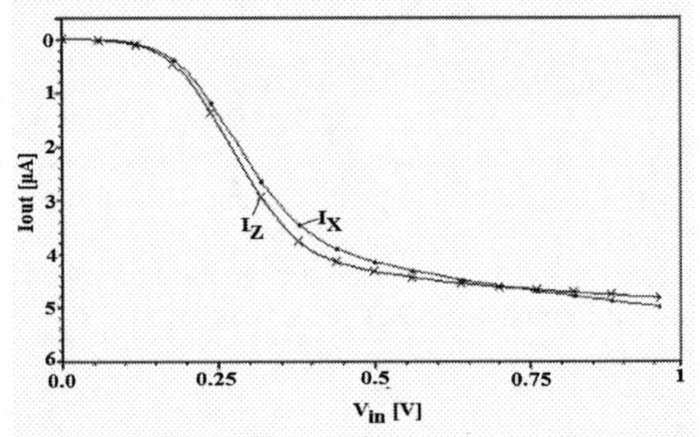

FIGURE 7. DC current transfer characteristic between ports X and Z.

TABLE 2. *DDCCII* main characteristics.

Data	Value
Voltage Supply	1V
Power Consumption	10μW
Voltage Gain (α)	0.95
Current Gain (β)	0.99
Parasitic R_X	0.1Ω
Parasitic L_X	0.01μH
Parasitic R_Z	210kΩ
Cut off frequency	3.5MHz
Layout area	$0.09mm^2$

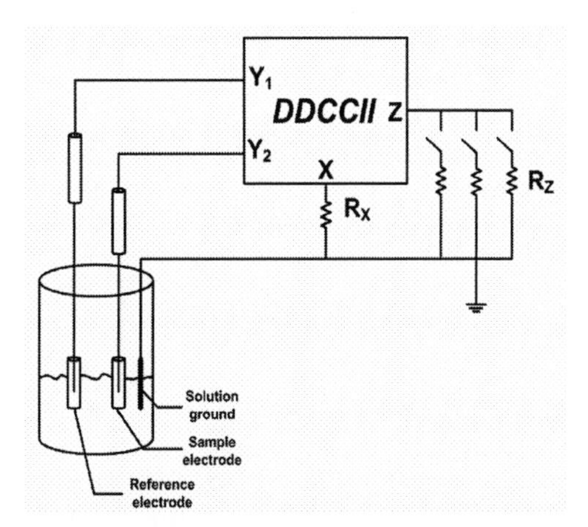

FIGURE 8. Ion-selective electrodes (ISE) interfacing circuit using the designed *DDCCII* in *VGA* configuration.

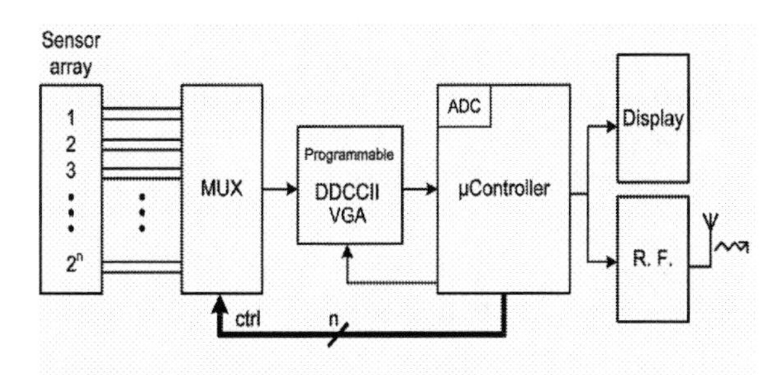

FIGURE 9. Sketch of the sensorial system where the *DDCCII* is planned to use as electronic interface.

CONCLUSIONS

We have presented a Differential Difference Amplifier based Current Conveyor for the readout of the voltage value resulting from potentiometric sensors. The use of the current conveyor allows implementing a first analog front-end with very low voltage low power characteristics, so suitable for self-powered wireless sensor network. Therefore, this interface represents the main part of a low cost artificial sensorial system that allows realizing a low price sensors array network, for monitoring contaminants in liquids, where the information of each single node is characterized by a response pattern coming from the potentiometric sensor array.

REFERENCES

1. M. Habara, H. Ikezaki, K. Toko, "Study of sweet taste evaluation using taste sensor with lipid/polymer membranes" Biosensors and Bioelectron, 2004, Vol. 19, No. 12, pp. 1559-1563.
2. F. Faridbod, M.R. Ganjali, R. Dinarvand, P. Norouzi, "The fabrication of potentiometric membrane sensors and their applications", Journal of Biotechnology, Vol. 6, December 2007, pp. 2960-2987.
3. H. Kulah, K. Najafi, "Energy scavenging from low-frequency vibrations by using frequency up-conversion for wireless sensor applications", IEEE Sensors Journal, Vol. 8, Issue 3, March 2008, pp. 261-268.
4. M.A. Spohn, S. Sausen, F. Salvadori, M. Campos, "Simulation of blind flooding over wireless sensor networks based on a realistic battery mode", 7th International Conference on Networking (ICN), 13-18 April 2008, pp. 545-550.
5. C. Falconi, E. Martinelli, C. Di Natale, A. D'Amico, P. Malcovati, A. Baschirotto, V. Stornelli, G. Ferri, "Electronic interfaces", Sensors and Actuators B, Vol. 121, January 2007, pp. 295-329.
6. C. Falconi, C. Di Natale, A. D'Amico, "Dynamic Op Amp Matching: a new approach to the design of accurate electronic interfaces for low voltage/low power integrated sensors systems", Proceedings of Eurosensors 2002, Prague, Czech Republic.
7. G. Ferri, N. Guerrini, Low voltage low power CMOS current conveyors, Kluwer Academic Publisher, Boston, 2003, 226 pp., ISBN 1-4020-7486-7.
8. S.C. Huang, M. Ismail, S.R. Zarabadi, "A wide range differential difference amplifier: a basic block for analog signal processing in MOS technology", IEEE Transactions on Circuits and Systems II, Vol. 40, No. 5, May 1993, pp. 289-301.
9. C. Falconi, G. Ferri, V. Stornelli, A. De Marcellis, A. D'Amico, D. Mazzieri, "Current mode, high accuracy, high precision cmos amplifiers", IEEE Transactions on Circuits and Systems II, Vol. 55, No. 5, May 2008, pp. 394-398.

A Study on Electronic Nose for Clinical Breath Diagnosis of Lung Cancer

Di Wang 1, Le Wang 1, Jin Yu 1, Ping Wang 1*, Yanjie Hu 2, Kejing Ying 2

1 Biosensor National Special Laboratory, Key Laboratory of Biomedical Engineering of Ministry of Education, Department of Biomedical Engineering, Zhejiang University, Hangzhou 310027, P. R. China
2 Zhejiang Sir Run Run Hospital, Zhejiang University, Hangzhou, 310027, P. R. China
**Correspondence should be addressed to Ping Wang (email: cnpwang@zju.edu.cn)*

Abstract. Lung cancer is one of the most deadly diseases and the leading cause of cancer death in both men and women. The high mortality in patients with lung cancer results, in part, from the lack of effective tools to diagnose the disease at an early stage before it has spread to regional nodes or has metastasized beyond the lung. The electronic nose combined with a diagnosis model which based on the biomarkers can provide a non-invasive and more convenient method. The article presented an improved e-Nose based on surface acoustic wave (SAW) gas sensors and an a diagnosis model to diagnose lung cancer.

Keywords: e-Nose, SAW, Lung Cancer.
PACS: 07.07.Df

INTRODUCTION

At present, the lung cancer has become the most common malignant tumor, the incidence rate rises unceasingly. In America, as know by the data, the livability is merely 14% in 5 years after being diagnosed with lung cancer, but if can be diagnosed and treated as soon as possible, the livability may rise to 48%. Therefore, it will be significant to develop an e-nose which can examine the breathing gas of the lung cancer patients.

In 1997, we published what to our knowledge is the first article concerning the diagnosis of diabetes using a gas sensors array [1]. In 1999, Phillips et al [2] first published their research results concerning the correlation between breath odors and lung cancer and indicated that 22 VOCs had been regarded as markers of lung cancer. These findings were all obtained using large-scale gas chromatograph in laboratory. In 2003, A gas sensor array composed of QMB used to detect the breathing gas of the lung cancer patients was bring forward by Corrado Di Natale [3]. But the QMB has a low sensitivity, and it is hard to find metal porphyrin thin films with different selectivity to compose the large-scale QMB sensor array. So, a new gas sensor array and detect-diagnose technology must be develop to increase the precision of lung cancer detection and realize the early diagnosis electronic nose.

A novel e-nose based on a kind of virtual array of surface acoustic wave (SAW) gas sensors is proposed in this article. The SAW sensor has a higher sensitivity compared to QMB [4], A technique of saw gas sensor combined with capillary which has a high separation efficiency was used to implement a large-scale virtual sensor array. The lung cancer patients were identified by a diagnosis model, an elementary result was obtained through the clinical experiment.

Validation of the biomarkers in breath diagnosis

30 patients with untreated primary lung cancer and 30 volunteers who have no cancer history were selected from Sir Run Run Hospital. The breath samples of the patients and controls were collected and detected by the GCMS. Several tens of VOCs were detected both in the patients and control breath. Table 1 showed 16 candidate VOCs, most of which is alkanes and their derivates.

The discriminate analysis(SPSS16.0) is applied to create a model using these VOCs to discriminate between patients with primary lung cancer and controls. The 5 VOCs in Table 2 was recruited in the best classification model.

The detailed process can be found in another paper, Jin Yu et al, Detection of Volatile Organic Biomarkers

CP1137, *Olfaction and Electronic Nose: Proceedings of the 13th International Symposium*, edited by M. Pardo and G. Sberveglieri

of Lung Cancer from Exhaled Breath and Lung Cancer Cells which has been submitted in this conference.

TABLE 1. Candidate VOCs of primary lung cancer in breath

Chemical structure	Patients	Controls
decane	15	1
Undecane, 5-methyl-	12	0
Tetradecane	8	8
2-Tridecanone	10	5
Phenol, 2,4-bis(1,1-dimethylethyl)-	17	16
hexadecane	17	16
Tetradecane, 5-methyl-	13	2
Benzene, 1,1'-(1,3-propanediyl)bis-	9	1
2-Pentadecanone	9	4
Heptadecane, 7-methyl-	13	17
Hexadecanal	6	11
1-Hexadecanol	8	2
2-Nonadecanone	11	1
nonyl lactone	9	1
Eicosane	16	1
Heneicosane	17	0

TABLE 2. 5 VOCs identified as the best combined biomarkers of lung cancer by discriminate analysis

1	decane
2	Undecane,5-methyl-
3	2-nonadecanone
4	eicosane
5	heneicosane

Design of the e-Nose system

The structure of the e-nose based on SAW gas sensors is shown in figure 1. The respiratory gas is enriched by a adsorption tube, desorption happens in the inlet of the capillary at a high temperature, then the VOCs is carried into the capillary to be separated by the carry gas. When the VOCs come out from the capillary, there will be a frequency change because the VOCs can attach to the surface of the SAW sensor independently by reason of condensation, then the PCA and image analysis are used for pattern recognition after the signal is obtained and processed.

The substrate of the SAW sensor(shown in figure 2) is made from 128° Y-X $LiNbO_3$. The center frequency of the sensor is 120MHz, the bandwidth is 0.9MHz, the quality factor is 133.5, and the insertion loss is 7.01dB. According to the formula of Mass deposit effect [5], SAW sensor's theory sensitivity n can be expressed by the ratio of the frequency shift $\triangle$f to the quality $\triangle$m of the VOCs absorbed on the surface of the sensor.

$$\Delta f = f_0^2 h\rho(k_1 + k_2 + k_3) = f_0^2 * \frac{\Delta m}{A} * (k_1 + k_2 + k_3)$$

$$\Rightarrow n = \frac{\Delta f}{\Delta m} = \frac{f_0^2 (k_1 + k_2 + k_3)}{A}$$

$$= \frac{(11.96 * 10^7 Hz)^2 * [(-17.30 + 0 - 37.75) * 10^{-9} m^2 \cdot s \cdot kg^{-1}]}{(0.5 * 10^{-3} m)^2}$$

$$= 3053 Hz / ng$$

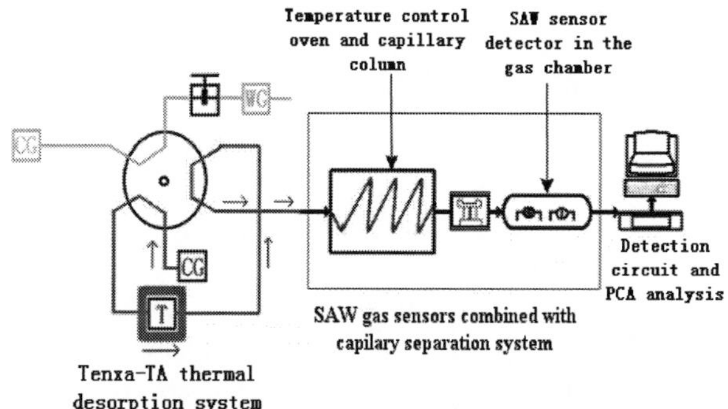

FIGURE 1. The structure of the e-Nose based on SAW gas sensors combined with capillary separation system

FIGURE 2: The SAW gas sensor

Eigenvalues such as retention time, peak-to-peak value and area of every peak detected can be easily obtained by given algorithms, the data of patients and healthy persons obtained by feature extraction is used to build database, then we can diagnosis the sample using the chosen model, and perfect the model.

The prototype machine is shown in figure 3. A supervisory computer is used to control the process of the detection and analysis the data , then give the diagnosis result. There is a air channel controller on the top of the machine to control the flow velocity of the carry gas.

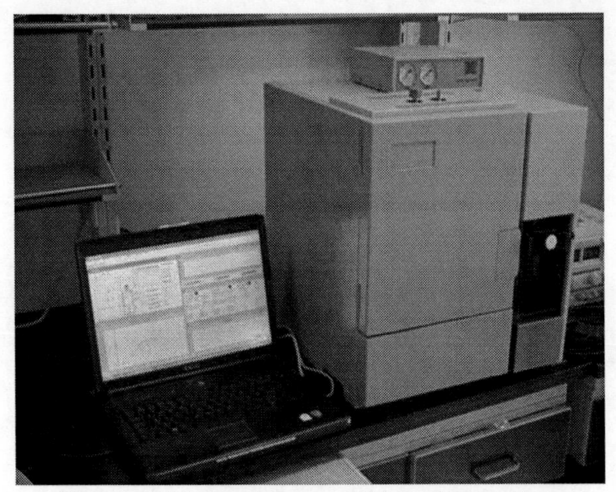

Figure 3: The prototype machine of the e-Nose

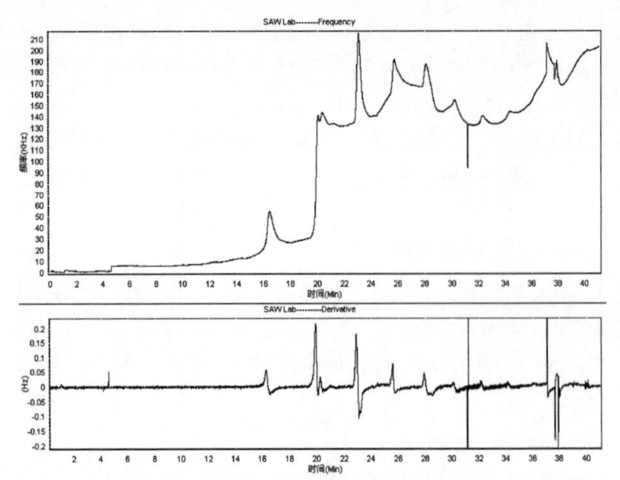

Figure 4: Spectrogram of the patient no.5

Result and Discussion

40 comparisons were carried on using the Tenxa-TA thermal desorption GC-FID system : When choosing the control group, tried to choose the healthy person who has the same gender, a proximate age, and the sample was collected at the same day in the same sampling environment. After comparing carefully, the characteristic peaks shown in table 3 were found in the exhaled gas of the patients relativing to the healthy persons.

The detect result of the e-nose system is shown in figure 4. The column oven of the e-nose was heat up to 250℃ at the speed of 5℃/min. The temperature of the capillary injector and outlet was set at 260℃. The working temperature of the detector was 10℃. The spectrograms of the patient no.3, no.4 and no.5 were measured. The spectrogram of the patient no.5 is shown in figure 4.

Retention time of 5 kind of gas in the Tenxa-TA thermal desorption GC-FID system and in the electronic nose system was compared. We found that the corresponding time was 17.7, 21.64, 22.45, 23.7 and 24.33 respectively after the retention time of the e-nose system plus 1.3, then they could correspond to the retention time 17.867, 21.809 and 24.331 given by the Tenxa-TA thermal desorption GC-FID system. According to this difference, and comparing to the data detected by the e-nose system, the retention time 9.52(turned to 10.82 after added with +1.3) of the patient no.4 corresponded to 10.870, the retention time 19.48(turned to 20.78 after added with +1.3) of the patient no.3 corresponded to 20.881,as shown in table 4.

Consequently, the e-nose system we designed can be used to detect the characteristic markers in the respiratory gas, and the result corresponds to the Tenxa-TA thermal desorption GC-FID system.

Table 3: Occurrence probability of the specific VOCs in the exhaled gas of the patient

Retention time, min	6.79	10.870	12.970	13.970	17.867	20.881	21.809
Appearance probability	7/40	15/40	10/40	4/40	21/40	21/40	24/40
Retention time, min	23.687	23.915	24.331	25.132	25.415	26.232	26.448
Appearance probability	6/40	10/40	20/40	5/40	3/40	5/40	6/40

Table 4: Comparision of the time of characteristic peaks between GC and e-nose

Time of characteristic peaks in GC, min	10.870	17.867	20.881	21.809		23.687	24.331
Time of characteristic peaks in e-nose, min	9.52	16.4	19.48	20.34	21.15	22.4	23.03
Plus 1.3 minutes	10.82	17.7	20.78	21.64	22.45	23.7	24.33

Table 5: The statistical result of the patients' respiratory gas detected by the e-nose									
Patient 3	16.85	18.84	19.02	19.3	19.48	19.85	20.35	20.85	
Patient 4	1.54	9.52	12.56	14.24	14.68	16.25	16.78	18.5	20.15
	20.26	20.45	21.22	22.2	22.45	23.03	23.28		
Patient 5	1.15	4.59	5.94	9.94	12.25	14.25	16.4	20.02	20.34
	21.15	22.4	23.02	24.17	25.68	28.04	32.27	34.33	
Healthy person 1	16.68	18.4	28.52	28.58	29.52	29.66	30.48	32	
Healthy person 2	16.75	18.45	19.9						
Healthy person 3	4.95	18.65							

The spectrograms of healthy persons and patients detected by the e-nose system were compared, the retention time of every peak is shown in table 5. The two peaks that appeared at 16.78min and 18.5min had the highest frequency of occurrence (colored with green), except the patient no.5, all the subjects have one of the two matters at least. Moreover, within the specific peaks of the lung cancer patients, the retention time of the peaks that appeared for two or more than two times is:16.4(2/3), 20.34(3/3), 21.15(2/3), 22.4(2/3) and 23.03(2/3) (colored with orange).

The results of respiratory detection were compared between patients with different types of cancer, and then we know that there are not only same components between them, but also different components, furthermore, the concentration is very different. The impact of smoking was considered into the test result, there were many more VOCs in the respiratory air of the healthy persons who smoked before being tested compared to the healthy persons who did not smoke, moreover, the quantity was much higher. So, smoke can greatly impact the whole experiment.

Conclusions

In this article, an innovative e-Nose system based on SAW gas sensors was introduced. Technique of GC combined with SAW sensor was adopted, two VOCs have a big relevance with the lung cancer was confirmed by experiment, and the result of the e-nose system corresponded to the Tenxa-TA thermal desorption GC-FID system, indicating that this electronic nose is able to examine the type and the content of the Characteristic gas. Besides, regarding the lung cancer breath gas examination, this electronic nose has merits such as a high sensitivity, a low cost as well as the simpleness of operating. Presently, the pathology research of breath diagnosis for lung cancer is waited to go further. On the other hand, the patients tested clinically were almost end-stage as a result of the absence of precise clinical early-stage lung cancer diagnosis method., The further work is gathering the breath sample of the high-risk group who have a long smoking history, and carrying on a long-term track and monitor, and finally develop an early diagnosis electronic nose.

Acknowledgments

This work was supported by the special foundation from Science and Technology Department of Zhejiang Province (Grant No. 2006C13021). We have to greatly acknowledge doctor Xing Chen for providing the experimental data.

References

1. Wang P, Tan Y, Li R, "A novel method for diagnosis diabetes using an electronic nose," Biosens Bioelectron, Vol. 12, (1997), pp. 1031–1036.
2. Phillips M, Gleeson K, Hughes JMB, et al, "Volatile organic compounds in breath as markers of lung cancer: across-sectional study," Lancet, Vol. 353,(1999), pp. 1930–1933.
3. Natale CD, Macagnano A, Martinelli E, "Lung cancer identification by the analysis of breath by means of an array of non-selective gas sensors," Biosensors and Bioelectronics, Vol. 18, (2003), pp. 1209-1218.
4. Wohltjen H, Dessy R, "Surface acoustic wave probe for chemical analysis," Anal Chem, Vol. 51, (1979), pp. 1458-1475.
5. Waldemar Soluch, "Design of SAW delay lines for senors," Sensors and Actuators A, Vol. 67, (1998), pp. 60-64.

Design and Implementation of a Low-Cost Non-Destructive System for Measurements of Water and Salt Levels in Food Products Using Impedance Spectroscopy

Rafael Masot[1], Miguel Alcañiz[1], Ana Fuentes[2], Franciny Campos[3], José M. Barat[2], Luis Gil[1], Roberto H. Labrador[1], Juan Soto[1], Ramón Martínez-Máñez[1]

[1]*Institute of Applied Molecular Chemistry (IQMA). Universidad Politécnica of Valencia (UPV)*

[2]*Department of Food Technology (DTA). Universidad Politécnica of Valencia (UPV) Camino de Vera, s/n 46022 Valencia, Spain*

[3]*Universidad Federal de Santa Catarina, Florianópolis, Brasil. e-mail: ramape@eln.upv.es (R.Masot)*

Abstract. The IQMA and the DTA have developed a low-cost system to determinate the contents of water and salt in food products as cured ham or pork loin using non-destructive methods. The system includes an electronic equipment that allows the implementation of impedance spectroscopy and an electrode. The electrode is a concentric needle which allows carrying out tests in a non-destructive way. Preliminary results indicate that there is a correlation between the water and salt contents and the module and phase of the impedance of the food sample in the range of 1Hz to 1MHz.

Keywords: Impedance spectroscopy, Electronic tongue, Food Control
PACS: 87.64.K-, 83.80.Ya, 82.45.Fk, 02.50.Sk

INTRODUCTION

Most of the methods applied in food industry to determinate the salt content use destructive techniques. New non-destructive measurement systems are being investigated [1].

Impedance spectroscopy allows the analysis of the properties of materials by applying alternate low level electrical signals of different frequencies and measuring the corresponding electric output signals. The relation between the voltage signal and the current signal is called complex impedance (module and phase) and it is frequency dependent.

The designed system allows the measurement of the impedance module and phase of a needle electrode inserted in the food sample at several frequencies in the range of 1Hz to 1MHz. Thus, many measurements can be obtained using a single electrode.

EXPERIMETAL AND METHODS

The sensor is a coaxial needle that incorporates two concentric electrodes (TECA X53159 of Oxford Instruments). The external part of the needle is made of stainless steel and acts as the reference electrode. The inner wire is made of stainless steel and it plays the role of working electrode. An epoxy resin is used to isolate both electrodes (Figure 1). An portable electronic equipment was designed to implement impedance spectroscopy with the before mentioned sensor.

Nine samples of minced pork loin were prepared with different salt levels. For each sample the impedance module and phase were measured at 20 frequencies in the range of 1Hz to 1MHz. Physicochemical analyses were done to asses the water and salt contents of the samples.

CP1137, *Olfaction and Electronic Nose: Proceedings of the 13th International Symposium*, edited by M. Pardo and G. Sberveglieri

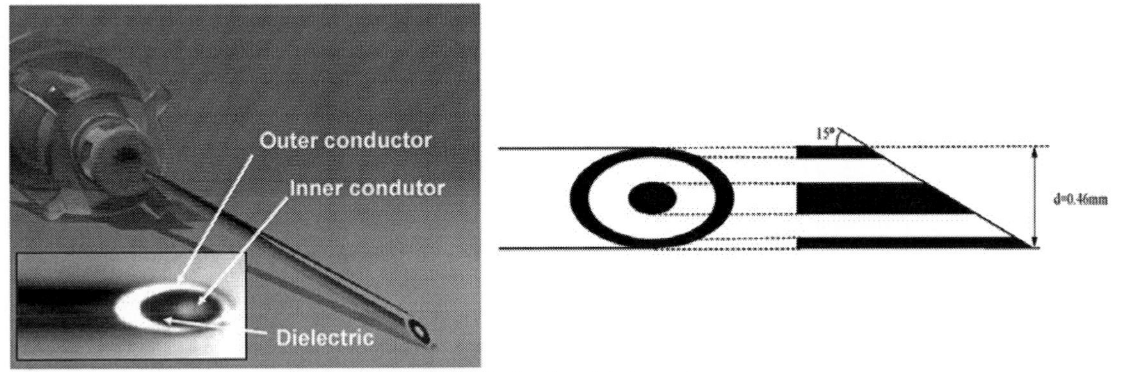

FIGURE 1. Coaxial electrode

A multiple regression was performed to evaluate which parameters (module or phase) and which frequencies were more relevant for the assessment of salt contents in the samples. The results of the multiple regression show that an R-squared of 97% could be achieved by selecting 7 frequencies for the module and 6 frequencies for the phase. The observed over predicted plot of the multiple regression is shown in Figure 2

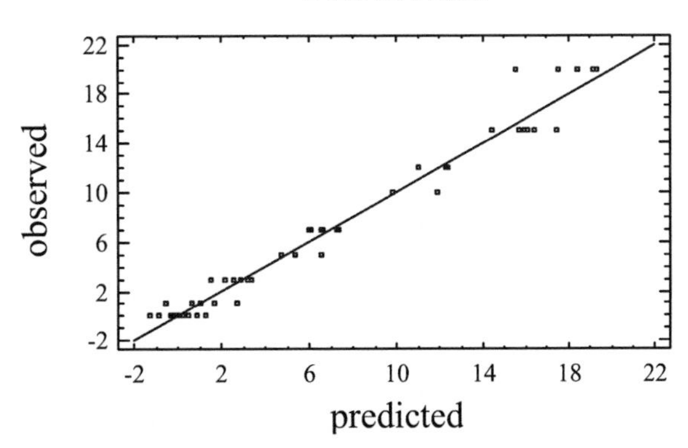

FIGURE 1. Observed over predicted plot of the multiple regression

CONCLUSIONS

A low-cost electronic equipment to implement impedance spectroscopy has been designed. The system has been used to asses the salt contents in meat. The electrode is a coaxial needle, which allows carrying out the tests in a non-destructive way. Results show that a very good salt level prediction can be achieved by applying impedance spectroscopy.

REFERENCES

1. García-Breijo, E. Barat, J. M.; Torres, O.; Grau, R.; Gil, L; Ibañez, J.; Alcañiz, M.; Masot, R. and Fraile, R. *Sensors and Actuators A: Physical,* 148, 63-67, (2008)

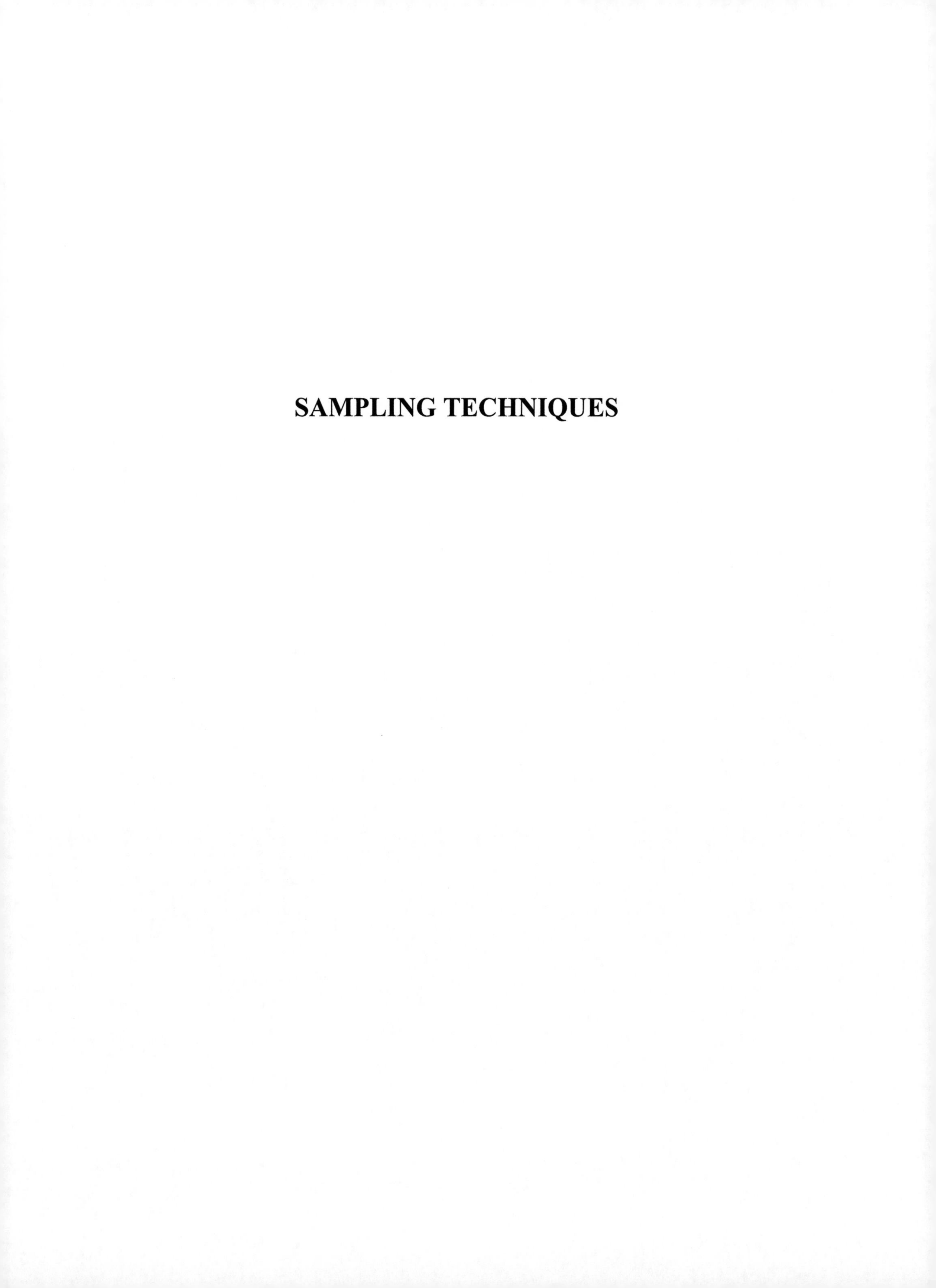

SAMPLING TECHNIQUES

Influence of the Adsorbent Material in the Performances of a Micro Gas Preconcentrator

E. H. M. Camara[(1, 2)], P. Breuil*[(1)], D. Briand[(2)], L. Guillot[(2)], C. Pijolat[(1)],
J. P. Viricelle, N. F. de Rooij[(2)]

(1) Ecole Nationale Supérieure des Mines, Centre SPIN, LPMG-UMR CNRS 5148, Saint-Etienne, FRANCE
(2) Institute of Microengineering, Ecole Polytechnique Fédérale de Lausanne, Neuchâtel, SWITZERLAND

Abstract. This paper presents the evaluation of different adsorbents for the improvement of the performances of a gas preconcentrator by targeting the adsorption of a large range of volatiles organics compounds (VOCs) The objectives of this work are to find the adequate adsorbent for a given gas target in specific experimental conditions and to select an efficient deposition process. Results related to the characterization of carbon nanopowders, carbon nanotubes (single walled (SWCNTs) and multi walled (MWCNTs)) and polymer (Tenax TA) for the development of a device for benzene preconcentration are reported. These results provide guidelines to define the right adsorbent for the preconcentration of benzene according to some specific criterions such as a large specific surface, a high adsorption capacity and low desorption temperature.

Keywords: Carbon Nanopowder, Carbon Nanotubes, Tenax TA, Benzene, Gas sensors, Preconcentrator.
PACS: 01.30.Cc, M 81.07.-b, 81.07.De, 82.35.-x, M 89.60.-k

1 INTRODUCTION

In the field of analytical techniques, the pre-processing of the gases is generally a very important step [1], especially to get a preconcentration effect in order to increase the sensitivity of the detection [2]. Indeed, in some applicative environments the concentration of gas is too small and therefore a preconcentration unit at the entrance of the analytical device is required. When a preconcenrator is used, the gas mixture to be analyzed flows through it and is accumulated during some time, then the mixture is desorbed by a temperature pulse and brought to the detector [3].

Requisite specifications, such as small system size, high performance, low power budget, and quick time response lead to the micro-fabrication of the key components of the analytical systems. Moreover, the miniaturization of the preconcentrators leads to superior heating rates, reduced power consumption and allows its integration in portable miniaturized analytical instruments, such as micro gas chromatographs and ion mobility spectrometers [4].

In a previous study [5, 6], a gas pre-concentrator based on micro-channel technology in porous silicon filled with carbon nanopowders dispersed in a fluidic solution have been developed. The present work aims to improve the performances of this device, with new approaches in the choice of materials, the realization of the devices and their operation. A series of alternative materials such as polymers, carbon nanotubes, whose surface areas, pores-size distributions, and structures can be varied to serve as conformal adsorbents, were evaluated for the improvement of the performances of the preconcentrator and to enlarge the applications by targeting the adsorption of a wider spectrum of VOCs [7].

The devices in this study have been developed targeting the preconcentration of benzene. This gas is considered as a representative for several types of applications relevant from atmospheric pollution control (BTEX compounds).

2 FABRICATION

2.1 Micro-channels realization

Micro-channels with three different designs have been realized with standard etching process as showed in figure 1.

CP1137, *Olfaction and Electronic Nose: Proceedings of the 13th International Symposium,* edited by M. Pardo and G. Sberveglieri

Figure 1: Schematic top view of the "Neutral", "Straight" and "Zigzag" designs, from left to right.

Following a photolithographic step with a thick photoresist, the inlet, outlet and reaction chamber of the micro-channel are DRIE etched in the silicon wafer to a depth of 325 µm (Fig. 2b). In order to improve the adhesion of the adsorbent that will be deposited afterwards in the micro-channels, 25 nm of thermal oxide were grown to make them become hydrophilic. Finally, the micro-channels were sealed with a Borofloat glass wafer by anodic bonding (Fig. 2c) and diced according to the dashed line showed in Fig. 2c revealing the inlet and outlet for the fluidic connectors. In same case, the glass cover was sealed after the adsorbent deposition.

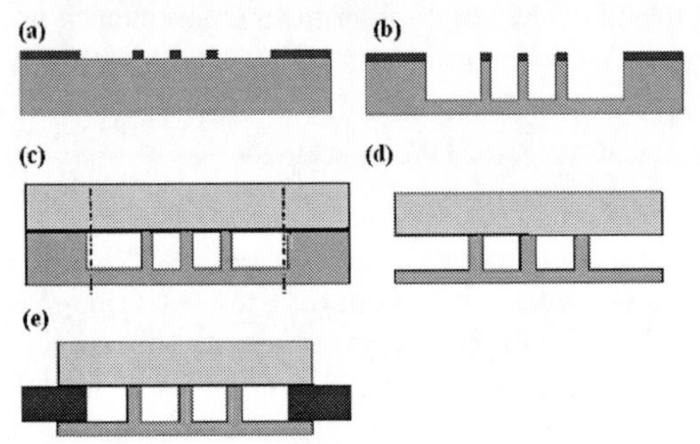

Figure 2: Fabrication process of the preconcentrator, which is illustrated here with the "Straight" type design.

2.2 Micro-preconcentrator implementation

In order to have an autonomous component, for either the thermal treatments of the adsorbent materials and the desorption of trapped gases, a platinum heater is integrated on the backside of the device by screen printing and the electrical connections are made with gold wires pasted on a platinum lacquer. This heating element is used to heat the device at temperatures up to 500 °C with a very homogeneous distribution of the temperature considering the good thermal conductivity of the silicon, as it was reported in our previous work [6].

The adsorbent insertion in micro-channels has been realized by two methods. The first one is a deposition using a fluidic solution [5], and the second one consists to deposit the adsorbent in a paste form before the sealing of the glass cover. It should also be noted that the choice of the deposition method is closely related to the adsorbent nature. For example, since the metallic capillary used as fluidic connection have an internal diameter of 220 µm, the micro-channels can not be filled using a fluidic solution with Tenax particles which have a size between 200 and 250µm.

Once the devices filled with the adequate adsorbent material, the fluidic connections are made with metallic capillaries (steel) having 220 µm of internal diameter and fixed with ceramic cement using a special procedure so as not to block the connectors during the sealing (Fig. 2e).

3. RESULTS AND DISCUSSION

The activation of carbon is an oxidizing reaction to release internal porosity created during the carbonization, to expand the pores and to create new micropores. This step must be controlled to avoid the total elimination of the carbon structure as the activation at high temperature (up to 900 °C) under oxygen can cause the total disappearance of carbon structure. That is why we have done the activation of carbon nanopowder at 500 °C during 3 hours in presence of oxygen.

3.1 Specific area measurement

The first tests were especially dedicated to the measurement of the specific area of the selected adsorbent materials namely carbon nanopowder, carbon nanotubes (SWCNTs, MWCNTs) and at last a polymer (Tenax TA) by nitrogen adsorption using the BET method. These results show that the specific area of the activated carbon nanopowder and the SWCNTs nanotubes are higher than the other three adsorbents (Table 1).

Table 1: Specific area and particle size of different adsorbents.

adsorbent	specific area (m^2/g)	particle size (µm)
Carbon nanopowder	99 +/-1	0.1
Activated carbon nanopowder	490 +/- 8	0.1
SWCNTs	399 +/- 5	2
MWCNTs	21 +/- 1	5
Tenax TA	35 +/- 2	250

However, a large specific surface does not mean necessarily a large adsorption capacity because the latter is closely related to the degree of affinity between the adsorbent and adsorbed gas. So the best solution is to evaluate the adsorption capacity of these adsorbents towards to benzene.

3.2 Benzene adsorption

The second series of experiments concern the evaluation of the adsorption capacity of the adsorbents by Temperature Programmed Desorption (TPD) technique coupled with a mass spectrometer.

The capacity of adsorption for benzene of the four adsorbents has been evaluated in similar conditions and in presence of humidity.

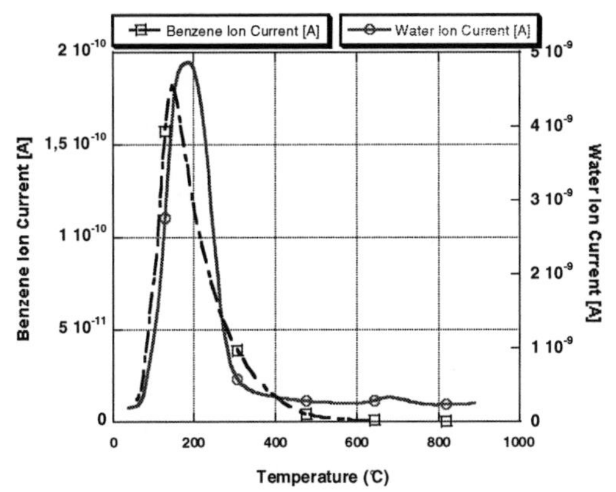

Figure 3: TPD spectrum of Carbon nanopowder after benzene adsorption.

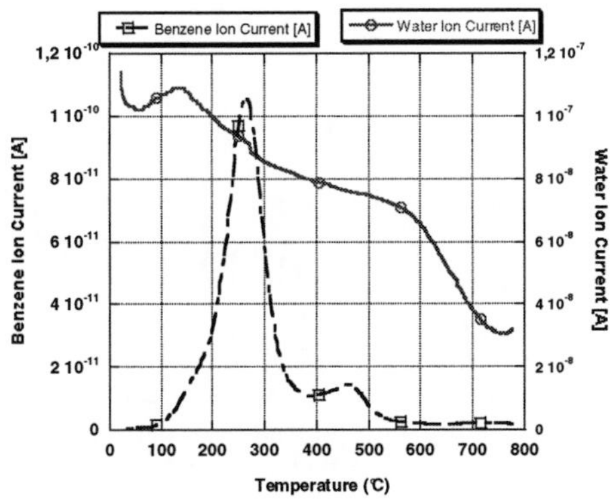

Figure 4: TPD spectrum of SWCNTs after benzene adsorption.

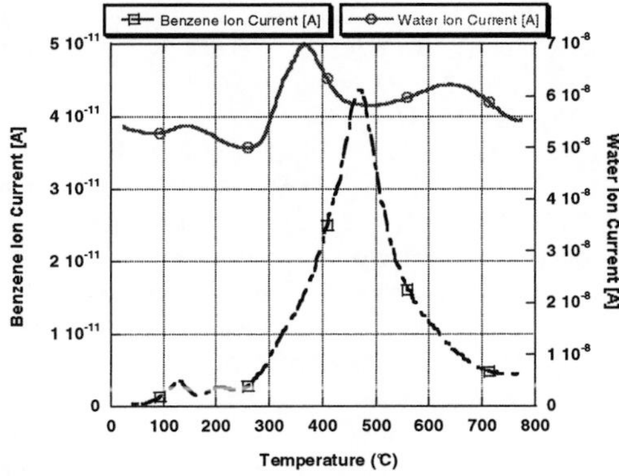

Figure 5: TPD spectrum of MWCNTs after benzene adsorption.

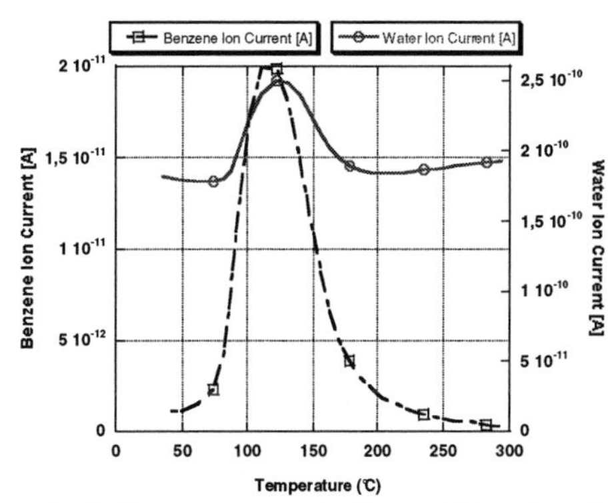

Figure 6: TPD spectrum of Tenax TA after benzene adsorption.

On the one hand, there is a significant desorption peak of water with the activated carbon nanopowder unlike the three other adsorbents which have a low affinity with water vapor. This result reflects a higher decrease of the adsorption capacity of the actived carbon nanopowder in a presence of water which is very harmful for the experimentation in ambient air.

On the other hand, the benzene desorption peak obtained with the activated carbon nanopowder and SWCNTs is about 5 times higher than the one obtained with MWCNTs and Tenax TA. This demonstrates a strong affinity of these two compounds with benzene (Figures 3, 4, 5 and 6).

However, it should be noted that the temperature of the maximum amplitude of the desorption peak varies with the adsorbent material. It is estimated at about 120, 180, 280 and 490°C respectively for Tenax TA, activated carbon nanopowder, SWCNTs and MWCNTs, respectively. The high desorption temperatures of the MWCNTs manifests an important activation energy of the desorption synonym with a strong link between the adsorbent and adsorbed gas which is bad for the reduction of the power budget and also for the reliability of the device.

3.3 Experiments with micro gas preconcentrator

The last experiments have been devoted to the demonstration of the preconcentration effect of these adsorbents when inserted in silicon micro-channels.

Our previous studies [5, 6] allowed us to determine the optimal conditions of operation of the micro gas preconcentrator such as an adsorption flow of 10 L/h, a desorption flow about 2 L/h, a heating rate of 160 °C/min, adsorption and desorption time of about 5 minutes. The maximum temperature at the desorption phase was defined according to the results obtained from the TPD experiments. The detector used for these tests is a Photon Ionization Detector (PID).

The preconcentration tests performed with the micro-channels filled with the activated carbon nanopowder confirm that the presence of water vapor reduce intensely the response of the PID. Indeed, the desorption peak obtained after the injection of 250 ppb of benzene in dry air during 5 minutes is twice higher than the one when in wet air at 50% of relative humidity (Figure 7). These experiments also showed that the temperature of the maximum amplitude of the desorption peak is substantially equal to that one obtained by TPD.

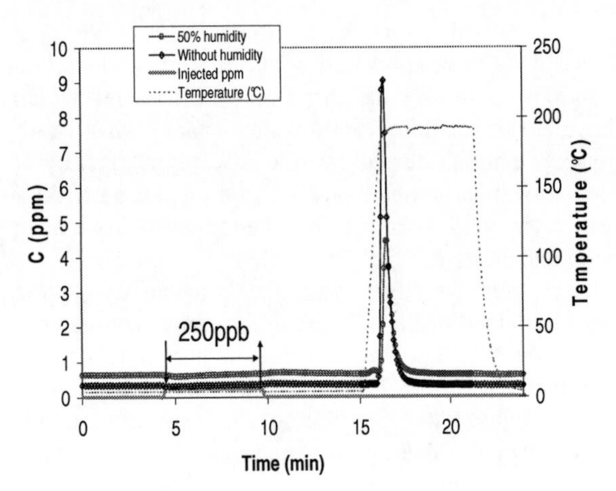

Figure 7: PID response of a silicon micro-channel filled with activated carbon nanopowder with and without humidity, when exposed to 5 min to 250 ppb of benzene and desorbed during 5 min at a temperature of 200°C.

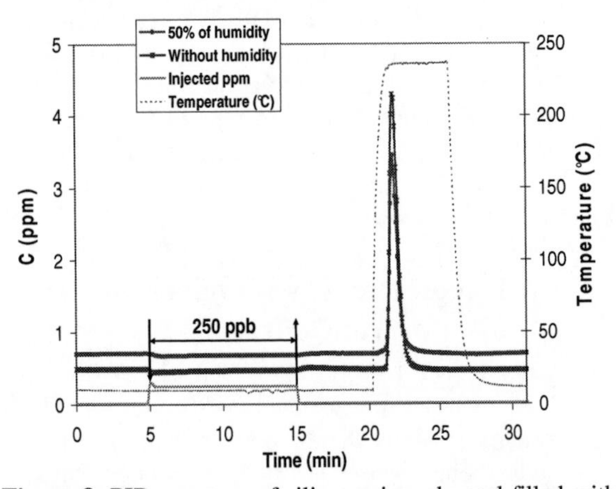

Figure 8: PID response of silicon micro-channel filled with SWCNTs with and without humidity when exposed to 5 min to 250 ppb of benzene and desorbed during 10 min at a temperature of 200°C.

The results obtained with micro-channels filled with SWCNTs have shown that the influence of water vapor in the performances of the micro-preconcentrator is lower than the one filled with activated carbon nanopowder despite a relative humidity of 50% (Figure 8). We have started preconcentration experiments with SWCNTs only recently and some optimization is underway.

4 CONCLUSIONS

Using TPD experiments and preconcentrator devices, different adsorbent materials have been evaluated. SWCNT has been chosen as the adequate adsorbent for preconcentration of benzene because of its high specific area and adsorption capacity and its low affinity with water vapor. Our new strategy is to focus on the functionalization of the SWCNTs in order to increase its affinity with benzene molecules.

Moreover, we are currently studying the preconcentration of other volatile compounds such as nitrobenzene. Because nitrobenzene is being strongly adsorbed on activated carbon nanopowder and nanotubes making its desorption very difficult even with a temperature of about 400 °C, alternative materials are required. To this end, the Tenax TA seems the ideal candidate for the preconcentration of nitrobenzene since the latter is completely desorbed below 250 °C.

REFERENCES

[1] Th. Becker, St. Mühlberger, Chr. Bosch, V. Braunmühl, G. Müller, Th. Ziemann, K.V. Hechtenberg, Air pollution monitoring using tin-oxide-based microreactor systems. *Sensors and Actuators B, 69 (2000) 108-119.*

[2] W. A. Groves, E. T. Zellers, G. C. Frye, Analyzing organic vapors in exhaled breath using a surface acoustic wave sensor array with preconcentration : Selection and characterization of preconcentrator adsorbent. *Analytica chimica Acta 371 (1998) 131-143.*

[3] C. Pijolat, D. Briand, Micro-preconcentrator for trace level detection of gases. *MST News N° 4/07 August 2007 pp 15-17.*

[4] P. R. Lewis, R. P. Manginell, D. R. Adkins, R. J. Kottenstette, D. R. Wheeler, Recent Advancements in the Gas-Phase MicroChemLab. *IEEE Sensors Journal, vol. 6(3) (2006), 784-795.*

[5] E.H.M. Camara, C. Pijolat, J. Courbat, P. Breuil, D. Briand, N.F. de Rooij, Microfluidic channels in porous silicon filled with a carbon adsorbent for gas preconcentration. *Transducers (2007) pp 249-252.*

[6] C. Pijolat, M. Camara, J. Courbat, J-P. Viricelle, D. Briand, N.F. de Rooij, Application of carbon nanopowders for gas micro-preconcentrator. *Sensors and Actuators B, 127 (2007) 179-185.*

[7] F. Zheng, D. L. Baldwin, L. S. Fifield, N. C. Anheier, Jr., C. L. Aardahl, and Jay W. Grate, Single-walled carbon nanotube paper as a sorbent for organic vapour preconcentration. *Anal. Chem. 2006, 78, 2442-2446.*

Methodological Variability Using Electronic Nose Technology For Headspace Analysis

Henri Knobloch [1§], Claire Turner [1], Andrew Spooner [1], Mark Chambers [2]

[1] *Cranfield University, Cranfield Health, Silsoe, UK*
[2] *Veterinary Laboratories Agency (VLA Weybridge)*
§Corresponding author
h.knobloch.s06@cranfield.ac.uk

Abstract. Since the idea of electronic noses was published, numerous electronic nose (e-nose) developments and applications have been used in analyzing solid, liquid and gaseous samples in the food and automotive industry or for medical purposes. However, little is known about methodological pitfalls that might be associated with e-nose technology. Some of the methodological variation caused by changes in ambient temperature, using different filters and changes in mass flow rates are described. Reasons for a lack of stability and reproducibility are given, explaining why methodological variation influences sensor responses and why e-nose technology may not always be sufficiently robust for headspace analysis. However, the potential of e-nose technology is also discussed.

Keywords: electronic nose (e-nose), temperature, mass flow, filters, sampling
PACS: 07.07.Df, 72.80.Le, 68.03.-g (fg), 68.47.Mn

1. INTRODUCTION

E-noses have been widely used for headspace and trace gas analysis from solid, liquid, and gaseous samples applying different sensing methods. In this paper, methodological variation due to changes in temperature, mass flow rates, filters and sampling methods will be addressed describing the reliability of the responses of conducting polymer (CP)- based e-noses.

2. METHODOLOGY

In this study, variability of responses of CP e-noses (ST214, Scensive Tech Ltd.) during the sampling process was assessed. The ST214 e-noses contain 14 conducting polymer sensors (including 1 internal reference).
Two sampling methods were tested using CP e-noses; firstly so called 'static sampling' using a flexible bag made from Nalophan (Kalle UK Ltd). Nalophan was used because it does not emit volatiles detectable by CP e-nose. Once the sample was placed in the bag, the other end of the bag was sealed. Hydrocarbon free air was added and after 10 minutes incubation, it was attached to the e-nose and the headspace was analysed. Secondly, in 'dynamic sampling', a container with a liquid sample is attached to the e-nose which then "inhales" the headspace. The pressure difference causes an inflow of ambient air into the liquid via an inlet which bubbles through the sample.

2.1 Effect Of Temperature

For assessing the effect of temperature, reverse osmosis water (ROW) and serum samples from clinically healthy cattle (breed "Holstein") were analysed at different temperatures. The bag volume was 0.7L and the sample volume was 0.5ml leading to a sample per bag volume ratio of 7x10-4. After 30 minutes of incubation, samples were analysed at different temperatures ranging from 21°C to 40°C.

CP1137, *Olfaction and Electronic Nose: Proceedings of the 13th International Symposium*, edited by M. Pardo and G. Sberveglieri

2.2 Mass Flow And Filters

Due to the lack of a flow controller in the CP e-noses, mass flow rates during static and dynamic sampling were analysed using a AWM3300V mass flow sensor (Honeywell Inc.) attached between the e-nose and the sample. Changes in signal pattern and intensity of the sensor responses after introduction in-line of a 0.45μm pore sized Sartorious Minisart or a 0.20μm Whatman Acrodisc LC13 were investigated. Small bags with a volume of 0.8L and static sampling were used to assess the mass flow stability across replicates at a constant temperature of 25°C. Selected ion flow tube mass spectrometry (SIFT-MS) was additionally used to assess the change in concentration of water, acetone, methanol and ammonia with and without filter use (0.20μm PVDF filter; Whatman, Acrodisc LC13), and whole mass spectra were taken to investigate qualitative differences of serum samples from clinically healthy cattle after passage through the filter. SIFT-MS is a real time mass spectrometry method that provides quantitative data on compounds present [1].

2.3 E-nose Characterisation

The temporal stability of two CP e-noses was assessed. Serum from clinically healthy cattle was analysed 4 times a day in 2h intervals using small bag static sampling (0.8L, sample volume 0.9mL). The temperature was kept constant at 25°C and bags were incubated for 15 minutes.
SPSS (version 11.5) was used to analyse data via linear regression and multifactor ANOVA. Least significant difference LSD was 5% (LSD, P≤0.05). The e-noses were compared with each other and across time. Differences arising from the two sampling methods (static and dynamic) and the replicate reproducibility were investigated. Principal Component Analysis (PCA) was used to compare both devices using MATLAB 2006b.

3. RESULTS

3.1 Effect Of Temperature

Temperature significantly influenced sensor responses and led to larger divergences during analysis of both media, ROW and serum. However, temperature increased the sensor response to ROW more than serum (data not shown).

3.2 Mass Flow And Filters

Comparing mass flow rates for both CP e-noses a difference of 20% was found. For the first device the mass flow rate was 200mL/min while for the second e-nose it was 240 ml/min.
On average, dynamic sampling decreased the original flow rates by 25% compared to static sampling which remained constant. Attaching a 0.45μm filter led to a drop of 33% in the original mass flow and together with dynamic sampling the mass flow dropped in total more than 50%. Using a 0.20μm filter led to a total decline of approximately 75% (see figure 1).

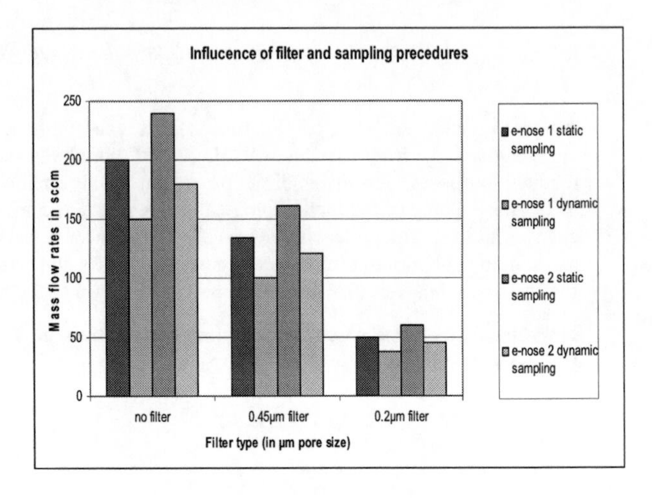

FIGURE 1: Both ST214 e-noses had significantly different mass flow rates. Different pore-sized filters significantly influenced the mass flow rates.

Due to bubbling effects during the sampling process dynamic sampling showed unstable flow rates while static sampling remained constant over replicates. However, a 4% reduction in flow rate was found for the first replicate and after analysing 72% of the headspace volume in the bag the flow rates decreased by 20% for both e-noses (mANOVA, P≤0.001, data not shown). Regarding the sensor pattern, qualitative changes due to filter application were investigated for sensors 1 and 7 to 11 using the 0.45μm filter and for all sensors using the 0.20μm filter. SIFT-MS analysis detected changes in the concentrations of cattle serum: water (-17%), ammonia (-19%), methanol (+126%) when analysed using the 0.20μm filter compared to no filter. Acetone concentrations remained unaffected. This and the SIFT-MS mass spectra (not shown) indicate and confirm quantitative and qualitative changes in the sample.

3.3 E-nose Characterisation

Using PCA, 99.97% of all variance in the signals was covered by components 1 and 2 indicating a clear discrimination between CP e-noses 1 and 2 (PCA plot not shown). Averaging replicates and plotting sensor responses over time, qualitative and quantitative differences in sensor responses of both ST214 e-noses were revealed (see figure 2A, 2B).

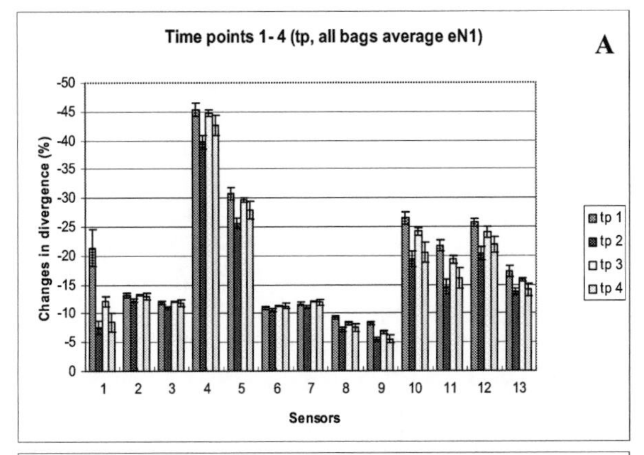

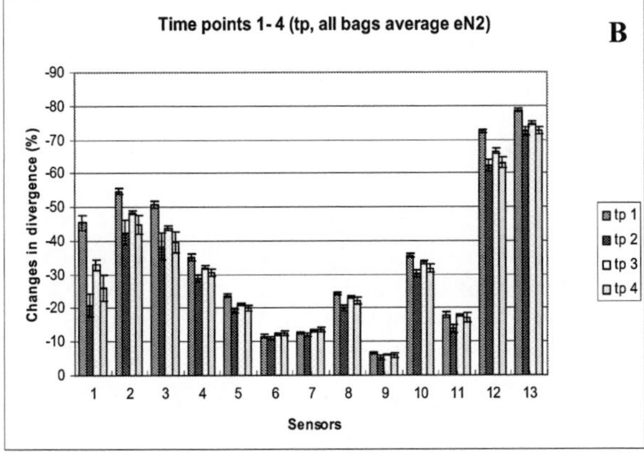

FIGURE 2: Replicate-averaged sensor responses of e-nose 1 (A) and e-nose 2 (B); y-axes have different scales 0 to -50 (A), 0 to -90 (B). Qualitative and quantitative changes in responses led to different patterns. Sine-shaped responses over all 4 time points (TP) were observed for both devices and over all sensors.

Considering the sensor responses over time, significant changes over 4 time points (TP) (similar to a sine wave) were observed. Changes in divergence for time point (TP) 1 were higher than TP 2. TP 3 had a bigger response again than TP 4 ($P \leq 0.05$). The changes were independent of device and sensor. Variation was lower using static sampling.

4. DISCUSSION

4.1 Effect Of Temperature

Variation in temperature led to significant changes in sensor response. Increasing temperature also changes relative humidity. Temperature and humidity are both causes of methodological variation and have been described by various authors. [2-5]. However, it is crucial to keep both factors controlled to ensure reproducibility during analysis. One approach for a temperature and humidity controlled flow cell was described by some authors [6,7] which may improve response reliability for e-noses without losing their portability.

4.2 Mass Flow And Filters

Using filters for CP e-nose protection and dynamic sampling as suggested by some authors [8-10] yields changes in signal intensity. SIFT-MS confirmed the results.

Pore size and mass flows are optimised to prevent water vapour destroying the sensing surface but at the same time they change the composition of the samples and therefore cannot be recommended. Filters lead to a drop in mass flow. For dynamic sampling, dilution of the samples was described by Misselbrook et al (1997) [11] which leads to a lack of reproducibility and sensitivity. In contrast, static sampling proved to be more stable during analysis and reduced methodological variation.

Difference in mass flow was observed between the same devices indicating a further lack of device reproducibility.

4.3 E-nose Characterisation

The two CP e-noses were qualitatively and quantitatively incomparable. Furthermore, temporal changes of sensor responses similar to a sine wave were observed. These changes are possibly due to semi-reversible changes on the sensor surface or insufficient desorption. The exact reason for this effect remains unclear and needs to be elucidated. 'Memory effects' affecting sensors have been reported before [3, 12].

5. CONCLUSION

Improvements in the devices, including the sensing materials (sensitivity to water vapour, memory effects), or temperature and flow control are urgently required for enhancing reproduciblility of CP e-nose analysis. A first step towards minimising methodological variation was achieved by using static sampling and analysis without applying filters. If the methodological sources of variance that plague CP e-noses are eliminated, an improved and optimised e-nose sensor with enhanced data analysis methods may yet yield a device that is capable of living up to its promise.

ACKNOWLEDGMENTS

This work was conducted as part of a research project (SE3221) funded by the UK Department for Enviornment, Food and Rural Affairs (Defra).

Serum samples were kindly provided by Petra Reinhold and the staff from the Institute of Molecular Pathogenesis in the 'Friedrich-Loeffler-Institut' (Jena, Germany) .

REFERENCES

1 Smith, D. and Španel., P. "Selected ion flow tube mass spectrometry (SIFT-MS) for on-line trace gas analysis", Mass Spectrometry Reviews 24 (2005): pp 661-700

2 Strike, D. J., Meijerink, M. G. H., Koudelka-Hep, M. "Electronic noses – A mini-review." Fresenius Journal of Analytical Chemistry vol. 364 (6)(1999): pp 499–505

3 Ampuero, S., Bosset, J.O.. "The electronic nose applied to dairy products: a review." Sensors and Actuators B Vol. 94(1) (2003): pp 1-12

4 Kashwan, K. R., Bhuyan, M., "Robust Electronic-Nose System with Temperature and Humidity Drift Compensation for Tea and Spice Flavour Discrimination", Asian Conference on Sensors and the International Conference on New Techniques in Pharmaceutical and Biomedical Research, 2005

5 Nake, A., Dubreuil, B., Raynaud, C.,Talou, T.. "Outdoor in situ monitoring of volatile emissions from treatment plants with two portable technologies of electronic." Sensors and Actuators B Vol. 106(1) (2005): pp 36-39

6 Falcitelli, M., Benassi, A., Di Francesco, F., Domenici, C., Marano, L., Pioggia, G. "Fluid dynamic for simulation of a measurement chamber for electronic noses." Sensors and Actuators B Vol. 85(1-2) (2002): pp 166-174

7 Di Francesco, F., Falcitelli, M., Marano, L., Pioggia, G. "A radially symmetric measurement chamber for electronic noses." Sensors and Actuators B Vol. 105(2) (2005): pp 295-303

8 Pavlou, A. K., Magan, N., McNulty, C., Jones, J.M., Sharp, D., Brown, J., Turner, A.P.F. "Use of an electronic nose system for diagnoses of urinary tract infections." Biosensors and Bioelectronics Vol. 17(10): (2002): pp 893-899

9 Fend, R., Geddes, R.,Lesellier, S., Vordermeier, H.-M., Corner, L. A. L. , Gormley, E., Costello, E., Hewinson, R. G., Marlin, D. J., Woodman, A.C., M.A. Chambers. "Use of an Electronic Nose To Diagnose Mycobacterium bovis Infection in Badgers and cattle." Journal of Clinical Microbiology Vol. 43(4) (2005): pp 1745-1751

10 Bastos, A. C., Magan, N. "Potential of an electronic nose for the early detection and differentiation between Streptomyces in potable water." Sensors and Actuators B Vol. 116(1-2) (2006): pp 151-155

11 Misselbrook, T. H. H., P . J ., Persaud, K . C.. "Use of an Electronic Nose to Measure Odour Concentration Following Application of Cattle Slurry to Grassland." Journal of Agricultural Engineering Research Vol. 66(3) (1997): pp 213-220

12 Kukla, A. L., Shirshov, Y..M., Piletsky, S.A.. "Ammonia sensors based on sensitive polyaniline films." Sensors and Actuators B Vol. 37(3) (1996): pp 135-140

Sampling Odor Substances by Mist-Cyclone System

Osamu Matsubara, Zhiheng Jiang, Shigeki Toyama

Tokyo University of Agriculture and Technology
Nakachou 2-24-16, Koganei, Tokyo

Abstract. Many techniques have been developed to measure odor substances. However most of those methods are based on using aquatic solutions(1),(2). Many odor substances specifically at low density situation, are difficult to dissolve into water. To absorb odor substances and obtain highest concentration solutions are key problems for olfactory systems. By blowing odor substances contained air mixture through mist of water and then separating the liquid from two-phases fluid with a cyclone unit a high concentration solution was obtained.

Keywords: cyclone, odor, aquatic solutions

PACS: 07.07.Df

INTRODUCTION

A great deal of concern has arisen recently regarding the detection on smell or odor substances detection. It is a key technique necessary for security, environment protect, medical processes and general industrial processes usage. More sensitive, more responsive and more reliable measure methods are demanded to those applications. The developments of odor detection always concern both investigations on sensing technologies and sampling procedures. In most cases, analysis of those substances depends upon whether or not a higher concentration solution can be obtained. In this research we used a mist-cyclone system and successfully got high concentration solutions of odor substances.

EXPERIMENTAL AND METHODS

Figure 1 is the image of mist-cyclone system used in this study. Odor substance (0.1ppm Eugenol vapor) contained air mixture was sucked into the duct, and mixed with mist there. The mist was generated in a sprayer unit by ultrasonic vibration equipment. Because of the huge contact surface odor substance dissolved into those drops rapidly and would achieve the saturate state. For every test cycle a fixed volume of gas mixture was supplied with a 45-liters sampling bag.

To separate the two-phase fluid a cyclone unit was used. From the bottom of cyclone unit a small amount of solution was obtained. The amount of liquid obtained depended upon suction speed utilized. The collected solution was analyzed through gas-chromatography method. For this analysis 2ml of solution was mixed with 0.5ml hexane and the upper layer was used. Thus the measured value of concentration would show a 4 times value of the solution.

To confirm the cyclone unit trapped all drops out of the gas-liquid mixture, several inner diameters with different inlet sizes of those units were tested.

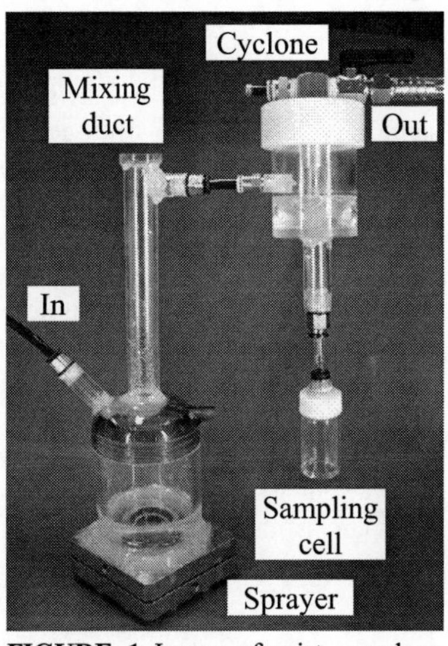

FIGURE 1 Image of mist - cyclone

RESULTS

Figure 2 and Figure 3 are chromatograms obtained individually by analyzing a typical solution of 100*μg/ml* eugenol/hexane, and the solution extracted from sampled liquid. On the left side appear several peaks which correspond to the component of hexanes. At the point of about 16 minutes identifies the peak of

CP1137, *Olfaction and Electronic Nose: Proceedings of the 13th International Symposium,* edited by M. Pardo and G. Sberveglieri

eugenol. By comparison of those areas, the concentration of eugenol aquatic solution would be about 3.1μM. This concentration value corresponded to a 3% recovery rate obtained from a volume of 45 liters 0.1ppm eugenol contained air mixture by using this mist-cyclone system.

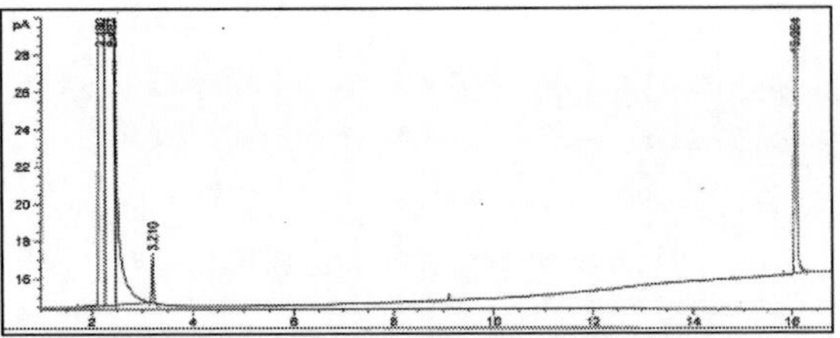

FIGURE 2: Chromatogram 100g/ml eugenol/hexane
Eugenol [pA*s]:99.7, RetTime: 16.064 min.

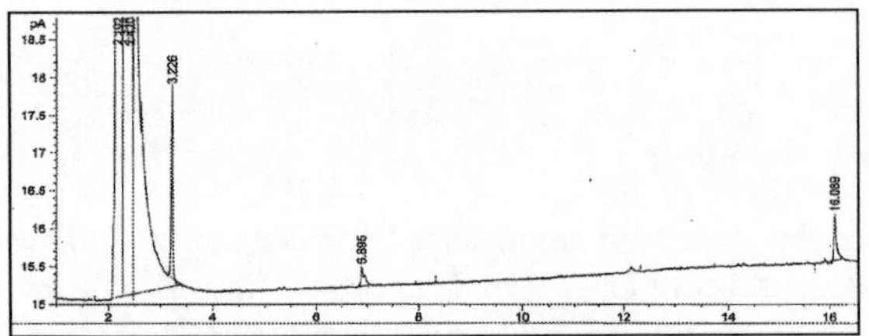

FIGURE 3 Chromatogram The solution extracted
Eugenol [pA*s]:2.046, RetTime: 16.089 min.

Figure 4 shows the relationship between the water sampling rate and the cyclone inlet velocity of the fluid by changing the cyclone internal diameter, the inlet internal diameter and the power of ultrasonic transducer.

Using the Low-power transducer which generats fewer amount of mist per minute, the water sampling rate kept almost a same value throughout all cyclone inlet velocities. As for using of the High-power transduser which generates larger amount of mist, the water sampling rate increased in accordance with the cyclone inlet velocity till it comes to 4000cm/s, and stays constant after then.

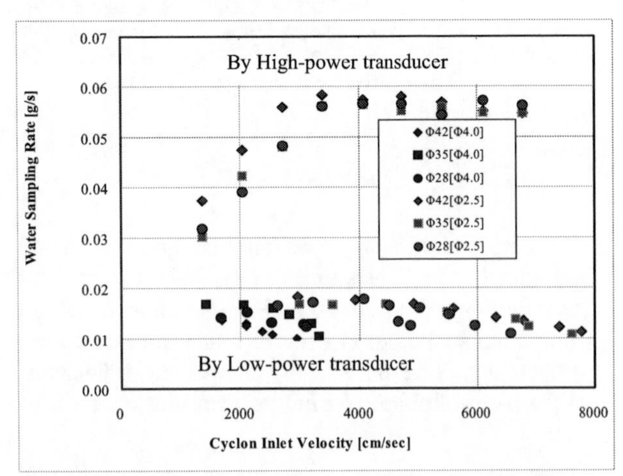

FIGURE 4 Cyclone Inlet Velocity vs. Water Sampling Rate

DISCUSSION

We have not yet found any presented data about the Henry constant of eugenol-water system. Therefore no evidence yet to say a 3.1μM would be the best result of dissolve 0.1ppm eugenol into water. But, it does be sufficient for us to detect the existence of those odor substances as long as utilizing some sensitive sensing procedures applicable now a day. And by using more sensitive method instead of using gas-chromatography would enable us to detect or distinguish odor substances by much less amount of solutions. For example, as some result reported(1), it would be possible to detect as less as 150 ppt of TNT-air mixture by using a special SPR equipment. It can be expected that combining the procedure used in this study with some sensing techniques the detection time could be reduced to several seconds and the required volume of gaseous mixture could be reduced to less than several milliliters for catching some particular chemicals. Also all of these apparatuses can be made into very small and compacted sizes to match those field and mobile usages.

The low value of recovery rate may indicate a low efficiency of dissolving gaseous mixture into solutions. On the other hand it also suggests that there is a big possibility of the limitation according to Henry's law. If this was the real reason about the value of recovery rate, increasing the mist generating rate would make great contribution to this value. This will be examed through our next search.

CONCLUSIONS

Because of the huge contact surface between gaseous mixture and water, a very efficient mass transfer process can be carried out to obtain aquatic solutions. Also it would be very easy to generate mist and achieve high stability and flexibility on its flow rate. Thus it would be developed to a low-cost, compact-sized, easy replacement unit, suitable to not only olfactory systems but also other applications.

REFERENCES

1. Praveen S, *et al*, "Novel DNP-KLH Protein Conjugate surface for Sensitive Detection of TNT on SPR Immunosensor," *Sensors and Materials*, Vol. 19, No. 5 (2007), pp. 261-273.
2. Yutaka M, *et al*, "Development of an oligo(ethylene glycol)-based SPR Immunosensor for TNT detection," *Biosensors and Bioelectronics*, Vol. 24, No. 2 (2008), pp. 191-197.

Field Air Sampling with SPME for Ranking and Prioritization of Downwind Livestock Odors with MDGC-MS-Olfactometry

Jacek A. Koziel*[1], Lingshuang Cai[1], Donald W. Wright[2], Steven J. Hoff[1]

[1]*Department of Agricultural & Biosystems Engineering, Iowa State University, Ames, IA 50011, USA; koziel@iastate.edu,515-294-4206.*
[2]*Microanalytics (a MOCON Company), Round Rock, TX, USA*

Abstract. Air sampling and characterization of odorous livestock gases is one of the most challenging analytical tasks. This is due to low concentrations, physicochemical properties, and problems with sample recoveries for typical odorants. Livestock operations emit a very complex mixture of volatile organic compounds and other gases. Many of these gases are odorous. Relatively little is known about the link between specific VOCs/gases and specifically, about the impact of specific odorants downwind from sources. In this research, solid phase microextraction (SPME) was used for field air sampling of odors downwind from swine and beef cattle operations. Sampling time ranged from 20 min to 1 hr. Samples were analyzed using a commercial GC-MS-Olfactometry system. Odor profiling efforts were directed at odorant prioritization with respect to distance from the source. The results indicated the odor downwind was increasingly defined by a smaller number of high priority odorants. These 'character defining' odorants appeared to be dominated by compounds of relatively low volatility, high molecular weight and high polarity. In particular, p-cresol alone appeared to carry much of the overall odor impact for swine and beef cattle operations. Of particular interest was the character-defining odor impact of p-cresol as far as 16 km downwind of the nearest beef cattle feedlot. The findings are very relevant to scientists and engineers working on improved air sampling and analysis protocols and on improved technologies for odor abatement. More research evaluating the use of p-cresol and a few other key odorants as a surrogate for the overall odor dispersion modeling is warranted.

Keywords: Odor, VOCs, livestock, air sampling, SPME.
PACS: 01.30.Cc.

INTRODUCTION

Livestock operations are sources of aerial emissions of gases, odor, and particulate matter. A large body of excellent analytical work has been reported during the past three decades relative to the volatile compounds emitted by confined animal feeding operations (CAFOs).[1,2] A variety of sampling and sample preparation techniques have been utilized in the extractions of scores, if not hundreds, of volatile compounds in these environments. These include acid traps, solvent extraction, sorbent tubes and thermal desorption, whole air sampling in canisters or sampling bags, and SPME.[1-3] A relatively small subset of previous studies involved actual field measurements downwind from these facilities.[2,3] Yet, the downwind impact of volatile compounds affects air quality and subsequently often results in nuisance complaints from an affected population. Included among these volatiles are a large number of compounds which are known to be potent individual odorants.[1] The challenge relative to the CAFO odor issue is to extract from this large field of 'potential' odorants, the compounds which carry primary responsibility for the downwind odor complaints relative to these operations.[2]

There is a popular 'school of thought' which states that there are no odorants emitted by CAFO environments which are sufficiently dominant to be utilized as quantitative odor markers. As a result, much of the odor assessment work to date has been restricted to qualitative assessment utilizing 'human' detectors in conjunction with techniques such as dynamic dilution olfactometry. Past and recent GC-Olfactometry (GC-O) work which has been carried out by these and other authors suggests that CAFO odor assessment should, in fact, be translatable to objective, instrument-based protocols such as those proposed by Pollien at al.[2-5] Wright et al. (2005) used the SPME and a GC-MS-O approaches for beef cattle and swine operations in

CP1137, *Olfaction and Electronic Nose: Proceedings of the 13th International Symposium,* edited by M. Pardo and G. Sberveglieri

Texas.[2] This work suggested that the key odorants that significantly contribute to the characteristic malodor of swine barn relative to distance separation from high density CAFOs are dominated by just a few compounds (i.e., 4-methyl phenol a.k.a. p-cresol, 4-ethyl phenol, isovaleric acid, 2'-aminoacetophenone, indole and skatole), which are characterized by relatively low volatility, high polarity and extreme odor potency.[2]

The identification of and quantification of the major key odorants downwind of CAFO's is needed to develop and evaluate effective technologies and approaches to control odor. Proper sampling and analysis protocols are needed to facilitate both of these tasks. There is absolute truth to the old adage that 'the analysis is only as good as the sample to which it is applied'. This consideration is especially pertinent to the question of environmental odor assessment in general and CAFO odor assessment in particular. For example, much of the odor monitoring work to date has been carried out utilizing sampling protocols which are based upon Tedlar™ (or alternate plastic) bags. Unfortunately, the propensity for plastic films to rapidly adsorb semi-volatile compounds from contained gas samples has been well documented.[4]

Other air sampling and sample preparation techniques have a potential for better sample recovery of odorous VOCs. Koziel et al. (2005) showed that the Carboxen/PDMS SPME coating and sorbent Tenax TA/thermal desorption are capable of recovering an average of 98.3% and 88.3%, respectively, of 11 odorous analytes from a standard gas mixture at 24 hrs sample preservation time at room temperature.[4] To date, relatively few published data exist on the quantitative use of SPME for characterization of ambient air.

In this research, we used SPME for field air sampling of odorants downwind from a swine CAFO in Iowa. In addition, we used SPME for far downwind odor impact of a beef cattle feedlot in Texas. The secondary objective was to compare these results with the odor prioritizations previously reported for beef cattle feedlots for shorter distances.[2]

EXPERIMENTAL AND METHODS

Multidimensional Gas Chromatography-Mass Spectrometry-Olfactometry

MDGC-MS-O is an integrated approach combining olfactometry and multidimensional GC separation techniques with conventional GC-MS instrumentation. A commercial, integrated AromaTrax™ system (from Microanalytics, Round Rock, TX) was used for the GC-olfactometry profiling work as presented below. The system integrates a conventional GC-MS (Agilent 6890N GC / 5973 MS with the addition of an olfactory port, MDGC control, flame ionization detector (FID) and olfactory data acquisition software. The SIM mode targeted H_2S, mercaptans, VFAs, phenolics, indolics, and phenones. Mass/molecular weight to charge ratio (m/z) range was set between 34 and 250 in the scan mode.

Air Sampling with SPME

SPME utilizing a Carboxen modified PDMS 75 μm and the PDMS 100 μm fibers was used for ambient air sampling in this odor profiling study. Before sampling, fibers were desorbed for 5 min at 260 °C, then wrapped in clean aluminum foil, enclosed in a clean jar, placed in a cooler with blue ice and carried to sampling site. SPME fibers were transported to the laboratory enfolded in clean, aluminum foil, placed inside a clean jar with a tight cover and then in a cooler with blue ice.

Swine Odor Sampling

SPME collections were carried out by exposing the fiber to ambient air at the source and several downwind locations relative to a commercial swine operation in central Iowa. All air samples were collected on the afternoon of November 9, 2004 at 1 m height and utilized variations in downwind distance for cross-comparison purposes (Figure 1). Samples were collected at the source (continuous barn exhaust fan) and at four locations downwind, i.e., approximately at 109, 159, 214, and 294 m, respectively, from the center of the emission site, at the tunnel end of the barns (Figure 1). Three rounds of samples consisting of 20-min sampling periods with one SPME fiber per location were collected consecutively. The first two rounds utilized the Carboxen/PDMS coating and the last one utilized the PDMS coating. In addition, one sample was collected with a PDMS coating at the pit fan. Wind was S-SW and steady during sampling. No other CAFOs were present upwind from this facility within at least 16 km. All SPME collections were carried out under ambient conditions.

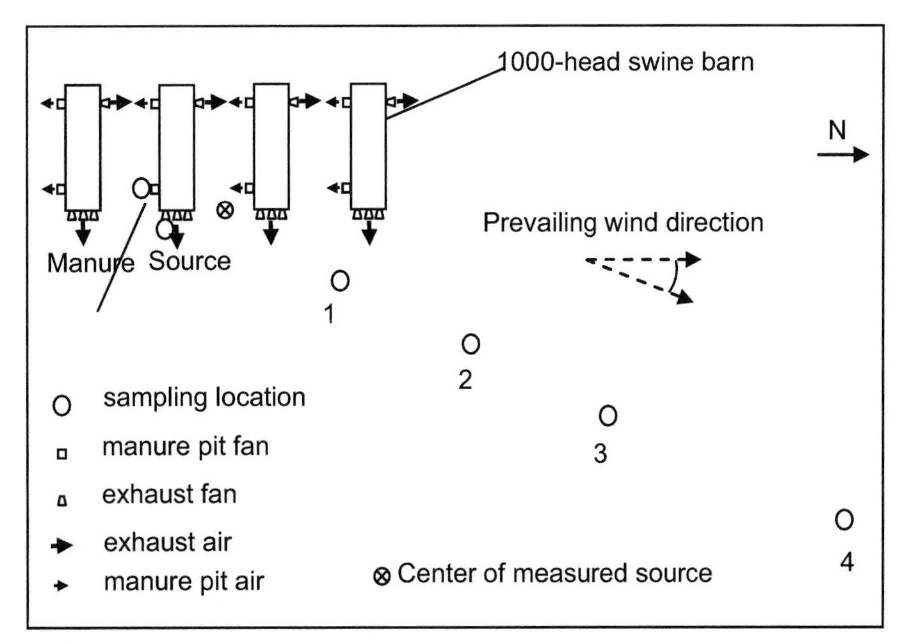

FIGURE 1. Schematic of field air sampling downwind from 4-barn swine finishing operation in Iowa with deep pit manure management system.

Beef Cattle Odor Sampling

Downwind sampling during the characteristic odor event was conducted on March 18, 2004 in Amarillo, Texas. The characteristic odor events occur a few times a year, typically within a few days following rain or snow-thawing. The subjective far-downwind perception of odor during these odor events is typically comparable to perception of odor at a large beef cattle feedlot, i.e., at the source. Two rain events occurred prior to this sampling event. On March 12 and 13, 1.5 and 0.5 mm of rain fell, respectively, followed by several days of cold weather. One day prior to this odor event, the ambient air temperature maximum increased by 5 °C from the day before to 25 °C, creating the appropriate conditions for the odor event to occur. For this event, 1-hr long sampling with Carboxen/PDMS 75 μm was completed between 8 and 9 P.M. at the Texas Agricultural Experiment Station grounds in Amarillo.

RESULTS

Swine Odor

Each air sample analysis resulted in simultaneous collection of a chromatogram and aromagram. The data shown emphasizes the relationship between the distance of the downwind separation from the source showing the two extreme locations, i.e., at the exhaust fan and 294 m downwind. As expected, locations at or near these source facilities appear to be characterized by greater odor complexity with a greater number and variety of individual odorants rising above their individual odor detection thresholds. Chromatograms and aromagrams for air samples collected in between, i.e., locations 1 to 3, were progressively less complex and consistent with the trend described above. The natural dilution effect associated with increasing distance from these sources had the effect of simplifying the resulting odor profiles, i.e., by reducing both the number of individual odorants detected and the relative intensities of those odorants that are detected. The total odor and the number of distinct odor/aroma events were generally decreasing with distance from the source, e.g., 32, 26, 18, 18, and 12 odors for series (II) at the source, location #1, #2, #3, and #4, respectively.

P-cresol (4-methyl phenol) with the characteristic "barnyard" odor character represented the dominant odorant relative to both near-source and at-distance downwind sampling locations. This was true for all 3 sample series and locations. This dominance was reflected in responses by the GC-O panelist to both perceived odorant intensity as well as perceived odor character. This prioritization of p-cresol relative to at-distance separation from the swine CAFO source is in agreement with earlier profiles developed for beef cattle CAFOs.[2] Relative to the near-site collection, only the dimethyl trisulfide (DMTS) homolog of the sulfide series caused a distinct individual odor response (i.e., 'onion' and 'fecal' character). There were no significant odor responses for H_2S or the lower MW organic sulfide compounds. The profile of odorants which were secondary to p-cresol in odor impact prioritization was found to be in good agreement with that previously shown for cattle CAFOs.[2] These included: isovaleric acid, 2'-aminoacetophenone ('taco shell, urinous'), 4-ethyl phenol, butyric acid and diacetyl.

Odor impact prioritization was estimated based upon the data presented above for near source and downwind from source (location #4). P-cresol and isovaleric acid were ranked as #1 and #2, respectively. They were followed by 2'-aminoacetophenone, and butyric acid, and guaiacol and DMTS for near and downwind locations, respectively. Somewhat surprisingly, in contrast to previous swine CAFO odor profile efforts, skatole and indole were not shown to be significant secondary odorants relative to this current series in downwind locations. It is assumed that this absence resulted from the extremely short SPME sampling times (20 min). Short exposure time bias relative to increasing molecular weight of volatiles is a well established characteristic of SPME sampling. These odor profile results were shown to be consistent with those previously reported by these authors for cattle CAFOs.[2] P-cresol was also #1 prioritization odor impact odorant for beef cattle feedlots.[2] These similarities serve as additional evidence supporting the suggestion that p-cresol is the odorant of greatest individual odor impact relative to either cattle or swine CAFOs.

Beef Cattle Odor

Samples were collected using Carboxen/PDMS 75 μm SPME and 1-hr sampling time. As many as 44 distinct odor events were recorded in one of the samples. Many of the important odorants were present, e.g., p-cresol, isovaleric acid, butyric acid, 4-ethyl phenol, and H2S. Acetic acid was one of the most abundant compounds detected. Sample #1 was significantly different than samples #2 and #3. The reason for this was likely differences in sample preservation during the transportation to the laboratory. These variations in replicates were likely the reason behind the apparent differences in odor analysis. Comparing of panelist responses to several characteristic odors and aromas collected in ambient air during an odor event in Amarillo. P-cresol was again the characteristic 'barnyard' odorant of the highest individual impact downwind, followed by butyric and isovaleric acids, and 4-ethyl phenol. It is remarkable to note that these samples were collected very far downwind from the nearest cattle feedyard (~16 km) and yet, the odor impact prioritization is very similar to those reported for much shorter distances (up to 2 km).[2] In addition, the ranking of odorants is consistent between two panelists analyzing three samples.

CONCLUSIONS

SPME was very useful in extracting livestock odorants from ambient air. It interfaced well with the GC-MS-Olfactometry system that, in turn, facilitated simultaneous chemical and sensory analyses. Based upon past and current GC-O based odor profile efforts, p-cresol appears to be the key 'character defining' odorant relative to downwind, distance separation from beef cattle and swine CAFOs. If these preliminary prioritizations can be proven consistent across a broader sampling of similar environments and analytical parameters, there will be increasing impetus for critical review of current sampling, analytical and odor abatement strategies. Particular attention appears to be warranted for p-cresol and other high priority semi-volatile odorants such as 4-ethyl phenol and 2'-aminoacetophenone due to their apparent odor impact prominence. SPME could be very useful as one possible alternative to current methods. Success in identifying this minimal critical odorant set from CAFOs simplifies the challenge of translating current, subjective, human 'detector'-based odor assessment protocols to objective, instrument-based alternatives.

ACKNOWLEDGMENTS

The authors would like to thank the Iowa swine producer for the access to the site. This work as published in reference #6.

REFERENCES

1. S.S. Schiffman, J.L. Bennett, and J.H. Raymer, *Agricult. Forest Meteorol.* **108**, 213 – 240 (2001).
2. D.W. Wright, D.K. Eaton, L.T. Nielsen, F.W. Kuhrt, J.A. Koziel, J.P. Spinhirne, and D.B. Parker, *J Agric Food Chem.* **53**, 8663-8672(2005),
3. G.L. Hutchinson, A.R. Mosier, and C.E. Andre, *J. Environ. Qual.* **11**, 288-293 (1982).
4. J.A. Koziel, J.P. Spinhirne, J.D. Lloyd, D.B. Parker, D.W. Wright, and F.W. Kuhrt. *J Air Waste Manage. Assoc,* **55**, 1147-1157 (2005).
5. P. Pollien, A. Ott, F. Montigon, M. Baumgartner, R. Munoz-Box, and A. Chaintreau, *J Agric. Food Chem.* **45**, 2630-2637 (1997).
6. J.A. Koziel, L. Cai, D. Wright, and S. Hoff, *J Chrom. Sci.*, **44**, 451-457 (2006).

Electronic Nose and Use of Bags to Collect Odorous Air Samples in Meat Quality Analysis

G. Sala[1], G. Masoero[1], L.M. Battaglini, P. Cornale, S. Barbera

Dipartimento di Scienze Zootecniche, Università di Torino
Via L. Da Vinci 44, 10095 Grugliasco (TO), Italy
[1] C.R.A. - P.C.M., via Pianezza 115, 10151 Torino, Italy
giacomosala@hotmail.com

Abstract. To test EN reliability and use of bags on meat, 17 bulls (one group of 9 and one of 8) fed similarly, except for a supplementary feedingstuff, were used. Samples were prepared according to the MCS protocol and repeated three times on different days for a total of 51 samples. Bags were used to collect raw and cooked meat air samples, and to test odour changes among samples analysed at different times. The first time analysis was performed immediately after collection then was repeated ,1 hour, 1 day and 1 week later. The Electronic Nose is very discriminant and clear differences were evident among raw, cooked and bags odorous profiles. The highest values were found in cooked samples and the broad range class (W5S) was the most representative. The EN also recognized the two tested feed treatments. In the cooked samples, all sensor responses decrease while time enhances, indicating a progressive chemical variation of the air composition in the bag, with a less correlation shown in the raw samples. When using bags, to avoid bias, is important to fix analysis in order to obtain useful results.

Keywords: Electronic Nose, Beef Meat, Bags, Aroma, Odour Profile
PACS: *87.85.jc*

1. Introduction

Meat flavour is an intrinsic aspect of quality parameters becoming an increasingly important consideration for the consumers. Electronic Nose (EN) devices are highly dedicated instruments which operate rapid and inexpensive measurements both in a qualitative mode comparing patterns, and in a quantitative mode. The development of chemical sensors, including hardware and software, makes it realistic to expect applications using this technique implemented on-line in the food and feed industry. Promising applications of EN on meat were exploited within spoilage, off-flavour, sensory analysis and fermentation processes experiments [1]. Several studies have shown the possibilities in the application of the EN to meat products. Gonzales-Martin *et al.* [2] and Otero *et al.* [3] tried to classify special Iberian cured ham into different commercial categories. Santos *et al.* [4] discriminated feeding regimen effects in pig meat and optimized some parameters of ripening time. They also concluded that different types of Iberian ham can be discriminated and identified successfully by the EN. Hansen *et al.* [5] used the EN to investigate properties of raw meat materials and reached a conclusion that the sensory quality of porcine meat loaf, based on measurements of volatiles in both the raw materials and the meat loaf, may be modelled in a predictive-causative multivariate analysis. Tikk *et al.* [6] used the EN to observe the warmed-over flavour in pork with positive results. This shows that the Electronic Nose could be useful in improving meat quality but it needs to be used in a specialized laboratory, it could be interesting and cheaper to perform EN analysis having collected the meat flavour in a bag then sending it to the laboratory. Gralapp *et al.* [7] used Tedlar bags to collect indoor odourous air samples from swine facilities. The time delay between capture and analysis of the meat aroma may not be so critical for meat as it was for tobacco smoke, as investigated by van Harreveld [8] who noted a great decay of signals in the bags after 12-30 hrs. The present work aims to test the use of EN reliability and bags in the investigation of aromatic meat traits. Bags were used to collect odorous air samples of raw and cooked meat and to test changes performing aroma analysis at different times after collection at: 1 hour, 1 day and 1 week.

2. Experiments and Methods

Seventeen beef *Longissimus dorsi* muscle portions, refrigerated at 4°C, were used as meat source. Nine Garonnaise bulls were fed without integration to control feed (Control) and eight were fed with a supplementary feedingstuff (Treatment) to increase food ingestion. From each portion a 1 cm thick circular sample (5.5 cm Ø) was obtained, according to the Meat Cooking Strinkage protocol proposed by Barbera *et al.* [9]. For each of 17 portions, analysis was repeated three times on different days for a total of 51 samples. An Electronic Nose PEN 2 (AIRSENSE Analeptics GmbH, Hagenower, Germany) with 10 metal oxide sensors (MOS), conveying at 400 mL/min air flow, was used to carry out measurements on bags and on raw and cooked samples, under glass bells (250 mL vol.). The 10 PEN2 sensors analysed 10 classes of chemicals compunds: 2 sensors analysed aromatic

CP1137, *Olfaction and Electronic Nose: Proceedings of the 13th International Symposium,* edited by M. Pardo and G. Sberveglieri

components (W1C and W3C) and one sensor for each of the following: broad range (W5S), hydrogen (W6S), aromatic-aliphatic (W5C), broad-methane (W1S), sulphur-organic (W1W), broad-alcohol (W2S), sulphur-chloride (W2W) and methane-aliphatic (W3S). Two glass bells were used: one for raw meat and one for cooked meat; each was equipped with an active charcoal filter and a Teflon tube connected to the sensor chamber of the EN. Four different air collection methods (Method), measuring at constant pressure, are applied to each sample: analysis of the head space of a glass bell before and immediately after cooking (Bell) and analysis of the meat aroma inside the bags after 1 hour, 1 day and 1 week (B1h, B1d, B1w).

Figure 1. Vacuum pump.

The trial start with a zero point trim to standardize the sensor conditions (oxidation-reduction). The air inside the empty glass bell is measured for 30s (control) then the raw sample is placed inside and measured for 60s. Then a Nalophan bag is connected and filled with "raw meat air" (volume ~ 3L). The bag is inflated under vacuum (Figure 1).

Straight after, the sample is cooked at 165°C for 10 min in an electric forced-air convection oven. The air is forced into the oven after passing through an active charcoal filter. The final internal sample temperature is about 70°C. The air inside another empty glass bell is measured for 30s (control) before the end of cooking. Straight after again, the cooked sample is put inside the bell and measured for 60s. Finally a new Nalophan bag is connected and filled with "cooked meat air". After 1 hour the bags are measured for 60s. The same bags are measured one day and one week later.

The Nalophan material was chosen because is airtight.

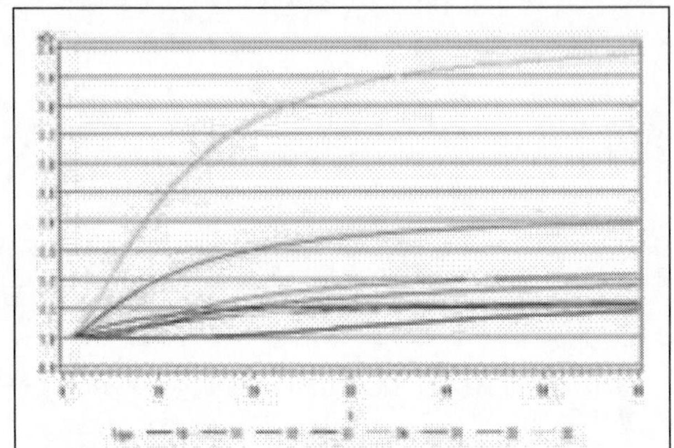

Figure 2. Trend for sensor W5S of different collection methods
10-Raw, 11-R.Bag 1h, 12-R.Bag 1d, 13-R.Bag 1w;
20-Cooked, 21-Bag 1h, 22-Bag 1d, 23-Bag 1w.

Room temperature and bags temperature were always controlled, before each measurement, and ranged between 21-23°C. Also air relative humidty was controlled, and ranged between 40% and 45%.

In the data matrix that EN supplies as output, each point of which is the conductance value per measurement second for each of the 10 sensors. PCA and GLM statistical analysis was performed on the interval between the 40th and 60th second of data by the SAS package (2001).

3. Results and Discussion

Figure 2 shows a typical trend detected by the EN. LSMeans resistivity on raw and cooked samples

Table 1. Collection method LSmeans (N=8099) - Test MC Bonferroni.

	Raw				Cooked				MSE
	Bell	**B1h**	**B1d**	**B1w**	**Bell**	**B1h**	**B1d**	**B1w**	
W1C	1.019^{A}	1.064^{B}	1.097^{C}	1.126^{D}	1.432^{A}	1.201^{B}	1.165^{C}	1.164^{C}	0.0097
W5S	1.862^{A}	1.625^{B}	1.780^{C}	1.762^{C}	2.665^{A}	2.071^{B}	1.958^{C}	1.817^{D}	0.0493
W3C	1.031^{A}	1.041^{A}	1.069^{B}	1.100^{C}	1.397^{A}	1.149^{B}	1.126^{C}	1.132^{C}	0.0061
W6S	1.002^{A}	1.011^{B}	1.006^{A}	0.971^{C}	1.154^{A}	1.069^{B}	1.009^{C}	0.968^{D}	0.0009
W5C	1.049^{A}	1.038^{B}	1.062^{C}	1.094^{D}	1.330^{A}	1.119^{B}	1.107^{C}	1.118^{B}	0.0034
W1S	1.131^{A}	1.155^{B}	1.204^{C}	1.187^{D}	1.871^{A}	1.430^{B}	1.277^{C}	1.210^{D}	0.0133
W1W	0.979^{A}	1.006^{B}	1.029^{C}	1.024^{C}	1.176^{A}	1.046^{B}	1.063^{C}	1.040^{B}	0.0031
W2S	1.073^{A}	1.110^{B}	1.166^{C}	1.109^{B}	1.952^{A}	1.383^{B}	1.206^{C}	1.116^{D}	0.0168
W2W	1.020^{A}	1.030^{B}	1.017^{A}	1.982^{C}	1.262^{A}	1.125^{B}	1.025^{C}	0.986^{D}	0.0015
W3S	1.006^{A}	1.020^{B}	1.011^{A}	0.971^{C}	1.309^{A}	1.123^{B}	1.011^{C}	0.970^{D}	0.0020

Air collection method means in the same row by Raw or Cooked with different letters are significantly different (A, B, C, D: $P<0.01$)

according to the different collection methods which are in table 1. The highest values were obtained when cooked samples are measured. The broad range class (W5S) is the most representative for every control method.. This meaning is due to a variation in the volatile composition during conservation, probably due to oxidative activity and adhesion of odorous molecules on the internal bag side. Among the "raw group", collection method values are lower and some sensors (W3C, W6S, W1W, W2S, W2W and W3S) fail to detect differences. This phenomenon is less evident in the "cooked group" where there is always a difference between the "Bell" collection method and the others: B1h, B1d and B1w. The related decay differences shown by all sensors after 1h denote an important effect of the Nalophan bag on the odorous volatiles of meat. The decrease proceeds, according to the time delay between collection and analysis, as found by van Harreveld [9] for tobacco smoke, with an higher decay of signals in bags after 12-30 hrs.

As well as the interaction between the different delays and the Nalophan bags, another important factor to explain variation in the measured compositon could be the temperature change of the air saved in the bags. During measurement the raw sample is under the bell, the sample temperature is increasing from 4°C to room temperature, which ranges from 21-23°C, in the same way while measuring the cooked sample, the sample temperature decreases starting from 70°C to the room temperature. Instead, when the Nalophan bag air sample is measured, its temperature is the same of the room temperature.

Figure 3 shows the PCA among different air collection methods where the different width of the black squares indicates their position in the third dimension.

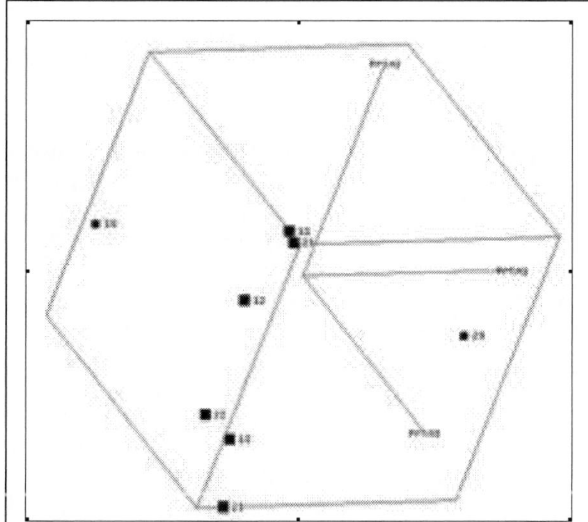

Figure 3. PCA by method of air collection:
10-Raw, 11-R.Bag 1h, 12-R.Bag 1d, 13-R.Bag 1w;
20-Cooked, 21-Bag 1h, 22-Bag 1d, 23-Bag 1w.

The first axis explains the 97% of the variance and the most discriminant odorous compounds classes are in decreasing order: the broad range W5S, the broad-alcohol W2S and the broad-methane W1S. It is not clear why the odorous profile at B1h is very similar in raw and cooked samples but obviously a comparison between cooked and raw should not be correct.

The EN could be employed in meat quality analysis as a qualitative method rather than a quantitative one, and it is important that the tested air collection methods are still discriminant.

The tested bulls were separated into two groups fed in a slightly different way and in figure 4 the PCA applied to method and feeding treatement is shown. The different width of the black squares indicates their position in the third dimension. The first axis explains the 97% of the variance and the most discriminant chemical classes are in decreasing order: the broad range W5S, the broad-alcohol W2S and the broad-methane W1S.

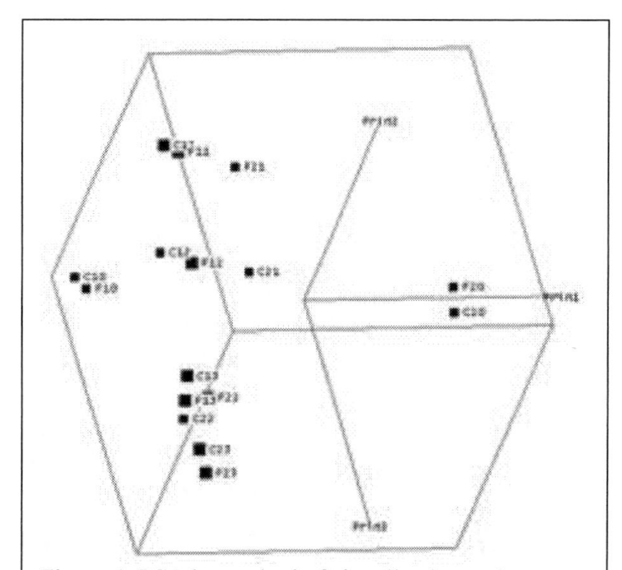

Figure 4. PCA by method of air collection and treatment:
Control, C10-Raw, C11-R.Bag 1h, C12-R.Bag 1d, C13-R.Bag 1w; C20-Cooked, C21-Bag 1h, C22-Bag 1d, C23-Bag 1w.
Treated, F10-Raw, F11-R.Bag 1h, F12-R.Bag 1d, F13-R.Bag 1w; F20-Cooked, F21-Bag 1h, F22-Bag 1d, F23-Bag 1w.

There is a very clear evidence that the methods are very similar to describe treatments and tough they are close in the graph, EN still discriminates the treatment (Table 2). These three sensors fail to discriminate raw samples under the bell but the other three sensors - aromatic W3C, hydrogen W6S and aromatic-aliphatic W5C – are able to a great significant discrimination.

It appears the experiment marks EN efficiency in the analysis of meat and measuring aroma with different delays detects both an obvious difference in air composition, due to the method, and also a difference in the treatments, if applied. The applied feed treatment was specific and with little feed differences, then a distinction in the air composition was not expected. To feed bulls with a small quantity of supplementary feedingstuff in order to increase ingestion, should not have caused any change on the aromatic composition of the meat.

This demonstrates the great efficiency of EN and its application to the operative field has to be well defined. In figure 4 the difference between treatments *intra* method are close to each other and are still significant. Using bags can introduce a bias due to a variable delay in measuring samples. Samples may be analysed immediately straight after they are collected. As shown in table 1, performing analysis just 1h after aroma sample release gets different odorous profiles. In fact, the cooked samples profiles have shown a significant negative correlation between the hours of delay and all the parameters measured by the sensors. Further examinations are important to investigate the reasons that the raw samples the two W5S and W2S sensors are not correlated. This negative correlation means that time causes a variation in the bag air composition and this has to be taken into account to avoid mistakes.

Table 2. Air collection Method by Treatment LSmeans (N=8099) for W5S, W1S nd W2S - Test MC Bonferroni.

		W5S		W1S		W2S	
		Cntr	Treat.	Cntr	Treat.	Cntr	Treat.
	Bell	1.865	1.859	1.128	1.135	1.075	1.072
Raw	B1h	1.664^A	1.583^B	1.170^A	1.139^B	1.126^A	1.093^B
	B1d	1.820^A	1.735^B	1.214^A	1.194^B	1.177^A	1.153^B
	B1w	1.773	1.750	1.199^A	1.173^B	1.126^A	1.090^B
Cooked	Bell	2.706^A	2.618^B	1.900^A	1.840^B	1.975^A	1.926^B
	B1h	2.141^A	1.904^B	1.461^A	1.397^B	1.403^A	1.361^B
	B1d	2.005^A	1.996^B	1.289^A	1.263^B	1.220^A	1.191^B
	B1w	1.832^a	1.801^b	1.221^A	1.198^B	1.131^A	1.098^B
MSE		0.0477		0.0130		0.0165	

Means by air collection method in the same row by sensor with different letters are significantly differents (a, b: P<0.05; A, B: P<0.01)

4. Conclusions

The correlation between meat flavour and productive traits or technological treatments could be assessed directly on the meat samples or on the air collected and measured after a delay, to reduce the analysis cost and to spread the use of the Electronic Nose.

The Electronic Nose confirm to be very useful in order to investigate global aroma characteristics and more efficient in cooked than in raw meat or in Nalophan bags. All the methods are effective also when applied to discriminate the use of very small quantities of supplementary feedingstuff. However, using bags can introduce bias due to the negative correlation among chemical categories measured by sensors and the delay in reading bag air after collection. The delay has to be chosen and strictly respected to obtain useful results.

Acknowledgements

The authors wish to thank the "Regione Piemonte - Direzione Sviluppo dell'Agricoltura - Programma Regionale di ricerca, sperimentazione e dimostrazione" for financial support and the Biotrade Company for the Taurus® supplementary feedingstuff.

References

1. Haugen, J. E., "Electronic nose in food analysis," *Adv. Exp. Med. Biol.* No. 488 (2001), pp. 43-57.
2. Gonzalez-Martin, I. *et al,* "Differentiation of products derived from Iberian breed swine by electronic olfactometry (Electronic Nose)," *Analytica Chimica Acta,* No. 424 (2000), pp. 279-287.
3. Otero, L. *et al*, "Detection of Iberian ham aroma by a semiconductor multisensorial system. *Meat Science*, No. 65 (2003), pp. 1175-1185.
4. Santos, J.P. *et al*, "Electronic nose for the identification of pig feeding and ripening time in Iberian hams," *Meat Science*, No. 66 (2004), pp 727-732.
5. Hansen, T. *et al*, "Sensory based quality control utilising an electronic nose and GC-MS analyses to predict end-product quality from raw materials," *Meat Science*, No. 69 (2005), pp. 621-634.
6. Tikk, K. *et al*, "Monitoring of warmed-over flavour in pork using the electronic nose - correlation to sensory attributes and secondary lipid oxidation products," *Meat Science*, No. 80 (2008), pp. 1254-1263.
7. Gralapp, A. K. *et al,* "Comparison of olfactometry, gas chromato-graphy, and electronic nose technology for measurement of indoor air from swine facilities," *Transactions of the ASAE*, Vol. 44, No. 5 (2001), pp. 1283-1290.
8. van Harreveld, A. P., "Odor concentration decay and stability in gas sampling bags," *J. Air Waste Manag. Assoc.*, No. 53 (2003), pp. 51-60.
9. Barbera, S. *et al,* "Meat Cooking Shrinkage: measurement of a new meat quality parameter," *Meat Science*, No. 73 (2006), pp. 467- 474.

On Site Generation Of Low Level Odorous Standards For Validation Of FTICR-MS Gas Detector In Ambient Air.

Hélène Mestdagh[1], Joël Lemaire[1], Michel Heninger[2], Julien Leprovost[2], Carine Cardella[3], Laurent Courthaudon[3], Nicolas Bouton[3]

1. Laboratoire de Chimie Physique, Bât. 350, Université Paris-Sud, 91405 Orsay, France
2. AlyXan, Bat 207B, Université Paris-Sud, 91405 Orsay, France, contact@alyxan.fr
3. Alytech, Centre Hoche, 3 rue Condorcet, 91260 Juvisy-sur-Orge, France, gasmix@alytech.fr

Abstract. Gas sensors and analyzers can be externally calibrated with standard gases. These gas cylinders are usually difficult to obtain when it comes to low concentration standards, and their lifetime may be questionable. Starting from high concentration and diluting on site to desired lower concentrations allows to set up multi-point calibrations of the analytical device, such as an electronic nose.
Volatile Organic Compounds (VOCs), including odorous chemicals, have been analyzed using Gas Chromatography (GC) often coupled with Mass Spectrometry (GC-MS), or specific olfactometric sensors. Proton Transfer Reaction (PTR) coupled with Fourier Transorm Ion Cyclotron Resonance (FTICR) MS is proposed to analyse low level of VOCs in air. FTICR MS is the most accurate and has the highest mass resolution of the MS techniques. B-Trap is a miniaturized FTICR instrument meant for real time VOCs analysis [1-4].

Keywords: Gas mixing, gas dilution, multi-point calibration, real time analysis, gas composition.
PACS: 37.10.Ty, 32.10.Bi, 07.75.+h, 81.70.Jb, 82.20, 06.20.fb

INTRODUCTION

GasMix generates on site multi-point gas calibration standards, according to ISO 6145-7.

The principle is based on the mixing and/or dilution of two to four gases by Mass Flow Controllers, which are controlled by software to deliver accurate flows of each gas. It is then possible to vary one gas concentration in a fixed matrix of other gases coming from different sources. Added standard method and internal calibration are also possible.

More than just a gas standard preparation device, GasMix operates on its own from a single injection to a fully automated and pre-programmed sequence, and thus can run unattended 24/7. It turns automaticcaly the gases on when purging to prepare a new standard concentration or mixture, and off during the time needed by the analyser to make the separation and detection, in case of long GC runs for example.

FTICR MS is a specific Mass Spectrometer widely spread in life sciences for the analysis of peptides and other heavy weight molecular compounds. The use of a permanent magnet allows to offer this technique for smaller molecules, such as VOCs. Among its benefits:

- Simultenaous separation and detection for broadband and real time analysis of complex mixtures of gases (typically 1s analysis cycles, or 1Hz).
- Accurate mass (+/- 0.01 uma) and high resolution (10,000 over the whole range) to assign accurately a molecular weigth for each peak, and thus easy identification of unknown compounds.
- Soft Chemical Ionization for no fragmentation and ease of interpretation (1 peak/compound), and for specific ionization of VOCs compared to O2 and N2 (increased sensitivity), by Proton Transfer Reaction (PTR) :
  $AH^+ + M \rightarrow MH^+ + A$ [7].
- Absolute quantification based on the reaction kinetics of the odorous molecules and ionic precursor.

B-Trap has been proven to be an additional technique in automative pipe exhaust, atmospheric and ambient air analysis [5], and polymer degradation studies [6].

CP1137, *Olfaction and Electronic Nose: Proceedings of the 13th International Symposium*, edited by M. Pardo and G. Sberveglieri

EXPERIMENTAL AND METHODS

B-Trap (AlyXan, Orsay, France) is equipped with a sniffer line to sample ambiant air containing the odorous VOCs. If the concentration of target analytes is lower than the detection limit of B-Trap (ppm), another sniffer sampling line is equipped with a membrane to preconcentrate the VOCs, and allow B-Trap to detect ppb levels. This technique is known as MIMS, Membrane Inlet Mass Spectrometry, for on-line and continuous flow preconcentration. The sniffer is placed above widely encountered products : whiteboard markers, a white correction fluid, sprays for plant and garden treatment, or for solid wood pieces of furniture.

GasMix (AlyTech, Juvisy-sur-Orge, France) allows to inject gas standards into B-Trap for accurate mass calibration and quantification validation.

RESULTS

GasMix is placed before the sniffer introduction line of B-Trap and the flow is forced in front of the vacuum chamber of the MS. The vent line allows to stabilize the inlet pressure at atmospheric conditions and vent the excess amount of gas coming from GasMix, the MS intake being 10^{-5} torr only during a fraction of a second.

A PDMS membrane (polydimethylsiloxane) can be installed before B-Trap which works as an in-line preconcentrator of VOCs compared to N2 and O2 [8]. Membrane Inlet Mass Spectrometry (MIMS) is a real time preconcentrator where analytes pass through the membrane continuously whereas the rest of the matrix flows along the membrane and is less injected into the instrument. The membrane is then also the physical barrier between the ambient air pressure and the MS vacuum chamber.

Mass calibration

After Fourier Transform, the instrument provides frequencies, which have to be transformed into masses. Before analysing unknown samples, a standard gas is injected with GasMix and obtained frequencies are adjusted to calibrate in mass. When recalculating, found masses are very close to exact masses:

TABLE 1. Mass Calibration.

Molecules	Exact masses (Da)	Measured masses (Da)	ΔM (mDa)
Water	19,0184	19,0171	-1,3
Acetonitrile	42,0344	42,0368	+2,4
Ethanol	47,0497	47,0525	+2,8
Acetone	59,0497	59,0502	+0,5
THF	73,0653	73,0635	-1,8
Benzene	79,0548	79,0529	-1,9
Dioxane	89,0602	89,0573	-2,9
Toluene	93,0704	93,0724	+2,0

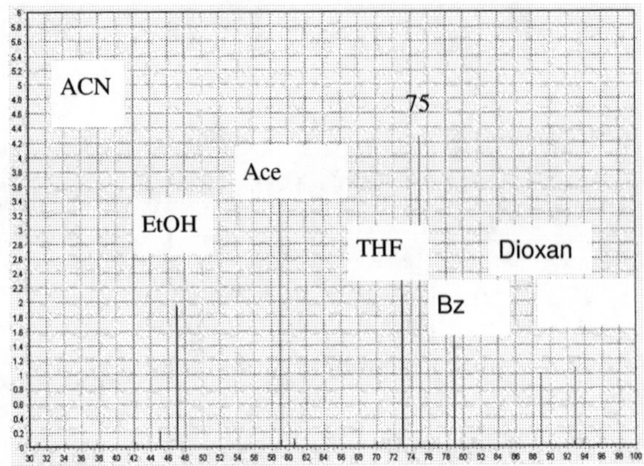

FIGURE 1. Obtained spectrum by PTR-FTICR of the calibrant.

If zooming in the m/z=75 region, the high resolving power of FTICR allows to separate two compounds of ΔM=0.0364 Da:

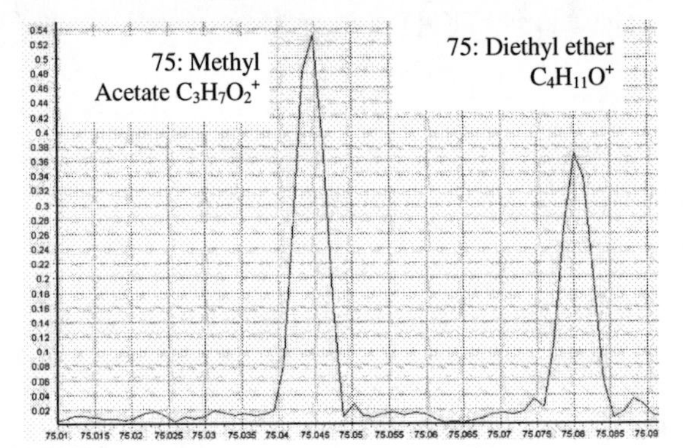

FIGURE 2. Zooming of figure 1 for m/z=75.

Headspace analysis

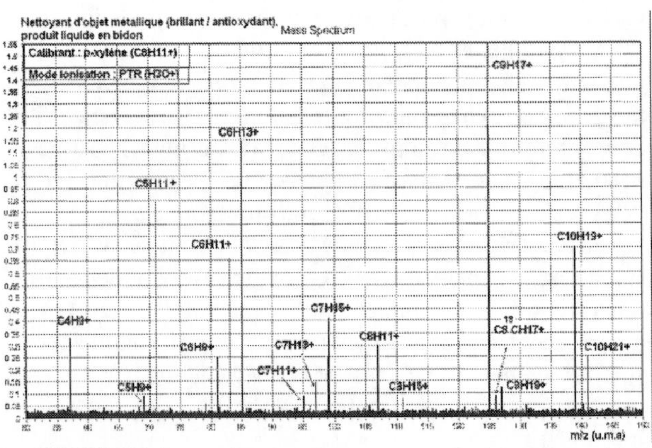

FIGURE 3: head space above a commercial anti-oxydant for metals.

Above an anti-oxydant for metals, a high number of components have been detected (see figure 3): butene, cyclopentene, pentene, benzene, cyclohexene and cyclohexadiene, hexene, toluene, 2-norbornene, heptyne, heptene, octyne, octene, trimethyl-benzene, nonyne, nonene, and even decyne and decene.

CONCLUSIONS

PTR-FTICR MS allows to detect, without prior separation by Gas Chromatography, VOCs present at trace levels in air. B-Trap is a miniaturized instrument that can be transported for on-site analysis. Applications for olfactometry have been developped with the introduction of a sniffer sampling line, equipped with an on-line membrane for preconcentration if necessary. MIMS is also possible in case of aqueous matrices, for solubilised organic compounds.

GasMix has been used for both mass calibration and linearity check of B-Trap. GasMix can be coupled with all gas analysers or detectors for validation purposes.

REFERENCES

1. Proton Transfer Reaction Mass Spectrometry, *Int. J. Mass Spectrom.*, **2004,** *239*, vol. 2-3.
2. Marshall, A. G., Milestones in Fourier transform ion cyclotron resonance mass spectrometry technique development, *Int. J. Mass Spectrom. Ion Processes,* **2000**, *200*, p. 331.
3. Mauclaire, G., Lemaire, J., Boissel, P., Bellec, G. & Heninger, M., MICRA : A compact permanent magnet Fourier Transform Ion Cyclotron Resonance mass spectrometer », *Eur. J. Mass Spectrom.,* **2004**, *10*, p. 155.
4. Heninger M., Clochard L., Mestdagh H., Mauclaire G., Boissel P., Lemaire J., FTICR MS transportable, *Spectra Analyse,* **2006**, *248*, p. 44.
5. De Gouw J., Warneke C., Measurements of volatile organic compounds in the earth's atmosphere using proton-transfer-reaction mass spectrometry, *Mass Spectrom. Rev.,* **2007**, *26*, p. 223
6. X. Colin, J. Verdu, Polymer degradation during processing, *C.R. Chimie* **2006**, *9*, p. 1380.
7. Dehon, C.; Gauzere, E.; Vaussier, J.; Heninger, M.; Tchapla, A.; Bleton, J.; Mestdagh, H. *Int. J., Mass Spectrom.* **2008**, *272*, 29-37.
8. Christian Janfelt, Helle Frandsen and Frants R. Lauritsen, Rapid Commun. Mass Spectrom. **2006**; 20: 1441–1446

SOLID STATE SENSOR TECHNOLOGY

A New Approach to Joining Dissimilar Ceramic Oxides for Chemical Sensors

Dr. Serge Zhuiykov

Commonwealth Scientific Industrial Research Organization (CSIRO), Materials Science and Engineering Division, 37 Graham Rd., Highett, VIC. 3190, Australia
E-mail: serge.zhuiykov@csiro.au

Abstract. Conventional joining of dissimilar oxides for sensing electrodes (SE) of chemical sensors has been pivotal to the development of various sensors and is vital to their further development. However, it is shown that the uncertainty (of a fundamental nature) in the properties of dissimilar oxides in SE causes the determination of their sensing characteristics to be ambiguous. Characteristics are different for such controlled parameters as pyrolysis temperature, crystal structure, particle's morphology and size, chemical and phase composition, the coefficient of thermal expansion (CTE), surface architecture, the bulk and surface stoichiometry and type and conductivity of additives. Here, we provide an alternative approach for joining dissimilar metal-oxides for chemical sensors SE. The approach relies on the development of at least one transient liquid oxide phase on the ceramic-SE interface. These results constitute key points relevant to selection oxides for joining, sintering temperatures and heating/cooling temperature rates.

Keywords: Ceramics, Oxides, Sensing Electrode, Nanostructures
PACS: 07.07.Df

INTRODUCTION

Here we show an alternative approach to conventional diffusion bonding method of joining dissimilar oxide materials for chemical sensors. This approach is based on selection of at least three (3) nanostructured oxides with formation an transient liquid oxide phase on interface. In contrast to the conventional joining dissimilar materials, new approach improves durability of ceramic-ceramic oxide joints, expands substantially the range of potential sensor applications (e.g. types of ceramic and eutectic sandwich microstructures) and, uses nanostructured metal oxides, which much more adaptable for sensors' miniaturization. Notwithstanding possible limitations of this approach in the specific sensor applications, it constitutes practical down-to-the-earth area of the chemical sensors development research. New approach has been utilized during development of the nanostructured $Bi_2Ru_2O_{7+x}$+RuO_2–SE and joining this SE to the ceramic substrate of the integrated water quality monitoring multi-sensor measuring pH, DO and conductivity at different temperatures. However, this approach is also applicable for joining various dissimilar oxide materials in the diverse chemical sensor structures.

EXPERIMENTAL AND METHODS

$Bi_2Ru_2O_{7+x}$+RuO_2–SE were prepared from SiO_2, Bi_2O_3 and RuO_2 nanostructures (ABCR GmbH & Co.). SiO_2, Bi_2O_3 and sensor alumina substrate were supplied by Sigma-Aldrich Australia Pty. Ltd. Measurement of an average particle size distribution revealed that the vast majority of RuO_2 nano-particles were in the range of 250-550 nm. SiO_2, RuO_2 and Bi_2O_3 nanoparticles were applied onto alumina substrate by using α–terpineol ($C_{10}H_{18}O$, 99.0%) and were sintered in two steps: heated to 400°C at rate 65°C/h and stabilized at 400°C for 2 h before heated to 965°C at rate 100°C/h in air ensuring the development of nanostructured composite $Bi_2Ru_2O_{7+x}$+RuO_2–SE enabling good contact with alumina substrate and with Pt current conductor. Chemical composition, crystal structure and obtained phase of RuO_2 were measured by power XRD (Rigaku, RINT 2100VLR/PC) using CuKα radiation in flat plane θ/2θ geometry between 10° and 80°. The size of RuO_2 nanoparticles of the water sensor electrodes and the surface morphology of RuO_2–SEs was characterized by FE-SEM (JEOL JSM-6340F) and

CP1137, *Olfaction and Electronic Nose: Proceedings of the 13th International Symposium*, edited by M. Pardo and G. Sberveglieri

their composition was subsequently analysed by a Wet-SEM (HITACHI, S-3000N) coupled with an energy-dispersive EDX, (HORIBA, EX-220SE).

RESULTS

Examination of the developed surfaces of $Bi_2Ru_2O_{7+x}$+RuO_2–SE indicated that joining of nanostructured matrix of SiO_2, Bi_2O_3 and RuO_2 resulted in the development of two additional to RuO_2 phases: $Bi_2Ru_2O_{7+x}$ and $Bi_2Si_2O_7$. Chemical composition, crystal structure and morphology of the developed $Bi_2Ru_2O_{7+x}$+RuO_2–SE as well as its sensing properties are published elsewhere. It is assumed that due to the lower melting point of $Bi_2Si_2O_7$ eutectic (~870°C) it was dissolved within $Bi_2Ru_2O_{7+x}$ structure (~965°C). These results suggested that $Bi_2Si_2O_7$ eutectic not only dissolved within $Bi_2Ru_2O_{7+x}$ structure but, as a phase with lower melting point, also filled pores on the alumina substrate enhancing bond developed between ceramic substrate and SE matrix.

Figure 1 illustrates main parts in the development of such joining. The initial matrix of nanostructured SiO_2 (~1723°C), RuO_2 (~1200°C) and Bi_2O_3 (~825°C) represents mixture of these oxides in the following molar ratio: 10/68/22. Information for establishment such molar ratio was obtained from the binary phase diagrams SiO_2-Bi_2O_3 and Bi_2O_3-RuO_2. Critical points for selection were to get eutectic phase, which should be heavier than RuO_2, must have melting point less than pure RuO_2 and, consequently, develops liquid phase first and acts as a flux for the desired oxide to be fixed on the top once solidification takes place.

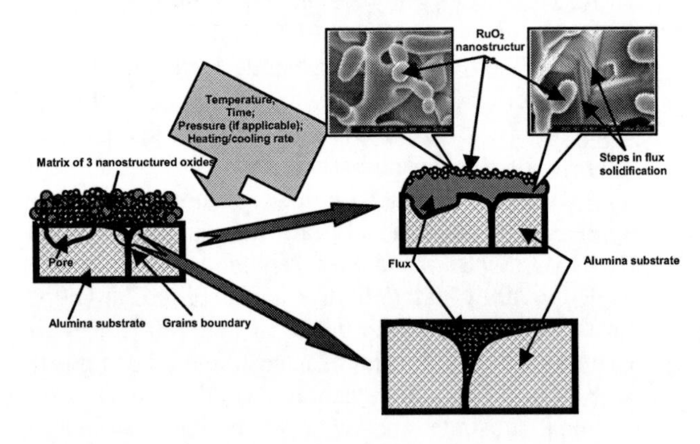

FIGURE 1.: Development joining between composite oxide SE and alumina substrate of the water quality monitoring chemical sensor.

Moreover, one of the oxides in the initial matrix SE must possesses melting point substantially lower than the melting point of other oxides, and the final composite material must be mixture of at least two phases. We also found that it is essential that the first metal in the developing new transient phase must be positioned to the left or at least be in the same column in the periodic table of elements compare to metals of other oxides in the composite flux structure. Temperature for joining should also be selected from the phase diagrams and should be above the solidus line for two-phase eutectic composite. This temperature provides sufficient heat for development RuO_2 nano-chains, as clearly shown in Figure 1. Once developed new transient phase melted and solidification takes place, it was evident that RuO_2 nano-chains, as more light structures, remained on the top of the flux. This figure also shows that as the temperature cools down, transient liquid phase (flux) crystallizes with appropriate steps during solidification representing gradual temperature changes. Absence of the microcracks in flux structure indicates that the cooling rate was selected appropriately for this particular oxides mixture.

Presented approach to joining of dissimilar oxide materials for chemical sensors based on diffusion bonding mixture of at least three (3) nanostructured oxides with formation an additional transient liquid phase on interface can be summarized as follows:

- CTE of the chosen oxides should be closely matched to CTE of the sensor substrate or materials of sandwich panel in multi-sensor assembly.
- Melting point of the new developing transient phase or solid solution should be chosen from the phase diagrams and should be above the solidus line.
- Selected melting point of the transient phase should be substantially lower than the melting point of each individual oxide in the oxides mixture.
- Metal cation in the chosen oxide, which is involved in the development of transitional liquid phase, should be positioned to the left in the periodic table of elements considering overall mixture of the metal cations in the composite oxide structure.
- One of the oxides in the nanostructured mixture should have melting point substantially lower than the melting point of other oxides.
- Size of the initial oxide nanostructures must be minimized.
- Weight of the developing transitional liquid phase must be heavier than the weight of depositing oxide SE.
- Molar ratio of the oxides mixture must ensure the development of one (two) transitional liquid phase plus depositing oxide phase.

The Influence Of ZnO Sensor Modification By Porphyrins On To The Character Of Sensor Response To Volatile Organic Compounds.

G.V. Belkova[1], S.A. Zav'yalov[2], N.N. Glagolev[1], A.B. Solov'eva[1]

[1]Semenov Institute of Chemical Physics, Russian Academy of Sciences, ul. Kosygina 4, Moscow, 117977 Russia
[2]Karpov Research Institute of Physical Chemistry, ul. Vorontsovo pole 10, Moscow, 105064 Russia

Abstract. The influence of modification of semi conductive ZnO and SnO_2 sensors by porphyrins and their metal complexes on the parameters of sensor response to volatile organic compounds, ethanol, acetone and benzene, was analyzed. The concentration of volatile organic compounds *Ci* varied in the range of 1-200 *ppm*. The sensor response was characterized by specific sensitivity $\gamma_i = 100\Delta R_i/(R_0 C_i)$, where $\Delta R = R_i - R_0$, R_i and R_0 are measurable and initial resistance of the sensitive sensor layer correspondingly. The modification of sensors by porphyrins caused changes in sensor response and first of all decrease of ZnO sensor temperature from 300 to 100°C at which the threshold of sensitivity is achieved (as used at this method C_i^{min}*=0.1 ppm*). It also caused a change in sing of the parameter γ, which was of importance for creating "electronic nose" sensor systems. At the same time the modification of ZnO sensor by metal complexes of porphyrin did not change sensor response

Keywords: gas sensor; thin films; porphyrin.
PACS: 81.05.Hd, 81.15.Cd, 81.15.Ef, 85.30.De, N 87.15.np.

INTRODUCTION

It was shown earlier [1], that the use of 5,10,15,20-tetraphenylporphyrin (H_2TPP) as the modifier of ZnO and SnO_2 sensors during creation of sensor systems for detection of volatile organic compounds (VOC) (ethanol, acetone and benzene) in the surrounding air caused changes in the character of sensor response and first of all decrease of ZnO sensor temperature from 300 to 100°C at which the threshold of sensitivity is achieved (as used at this method C_i^{min}*=0, 1 ppm*). It also caused a change in sing of the parameter γ, determined as $\gamma_i = 100\Delta R_i/(R_0 C_i)$, where $\Delta R = R_i - R_0$, R_i and R_0 are measurable and initial resistance of the sensitive sensor layer correspondingly.

In this work, the dependence of detected signal value on the porphyrin structure and the possibility of porphyrin metal complexes application for semi conductive gas sensors functional characteristics improving are discussed. Toward this end the modification of ZnO sensors by etioporphyrin – II (EP – II) and by Zn and Pd complexes of H_2TPP (Zn – TPP, Pd – TPP) and Cu complex of octaetilporphyrin (Cu – OEP) was carried out. Sensor characteristics of received systems during detection of ethanol, acetone and benzene vapor were analyzed. It turned out, that sensor response characteristics depend on the porphyrin structure. The modification of ZnO sensor by metal complexes of porphyrin turned out to be not effective in the view of sensor response.

EXPERIMENTAL AND METHODS

Thin sintered films [1, 2] were used as sensitive elements of semiconducting gas sensors based on ZnO, and polished S-5 quartz glass, as substrates. The substrates had a size of 1x5x5 mm^3. The surface of ZnO films was modified by metalfree etioporphyrin – II (EP – II) and by Zn and Pd complexes of H_2TPP (Zn – TPP, Pd – TPP) and Cu complex of octaetiporphyn (Cu – OEP) (Fig. 1). The deposition of porphyrins on to the sensitive layer was performed directly in an experimental unit, described in [1].

Experimental samples of ZnO sensors with different surface concentrations of porphyrins N_p in the range from 10^{-11} mol/cm^2 to 10^{-8} mol/cm^2) were prepared by varying the duration of deposition.

CP1137, *Olfaction and Electronic Nose: Proceedings of the 13th International Symposium*, edited by M. Pardo and G. Sberveglieri

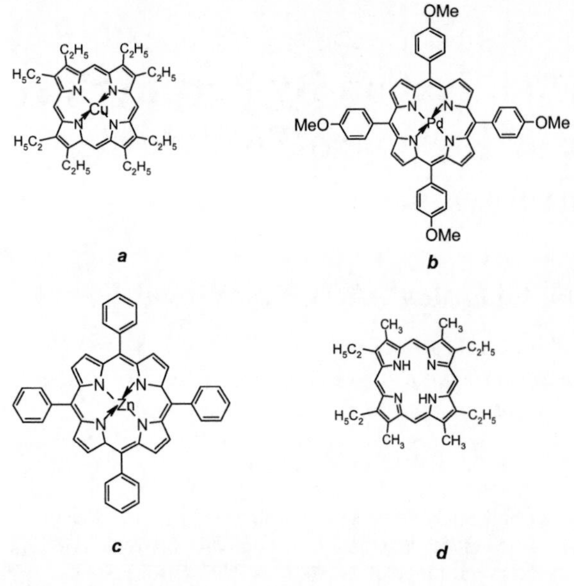

FIGURE 1. Porphyrins structure: a – Cu – OEP, b – Pd – TPP, c – Zn – TPP, d – EP – II.

The unit was equipped with a system for the introduction of purified VOC (ethanol, acetone, and benzene) vapors, whose concentration C_i could be changed from 1 to 200 *ppm*.

In all experiments, we analyzed changes in sensor resistance *R* as the composition of air in the unit and working temperature changed. The temperature was varied over the range 100–230^{0}C.

RESULTS

As it was shown earlier [1, 2], the initial specific sensitivities of the ZnO sensor with respect to the presence of VOC in air specified only at temperatures above 300^{0}C. All sensor effects were donor in character ($\gamma_i > 0$). And the specific sensitivity at low concentrations ($C_i \sim 1 - 10$ *ppm*) was an order of magnitude higher than at high VOC concentrations ($C_i \sim 100 - 200$ *ppm*) [1].

When the sensitive layer of the ZnO sensor was modified with EP – II molecules, as well as in the case of H_2TPP [1], sensor response to the introduction of VOC molecules into air could be fixed already at ~100^{0}C, and the character of the sensor response changed with change of deposited porphyrin concentration. The dependences of the specific sensitivity of the ZnO sensor at 200^{0}C at low ($C_i \sim 1 - 10$ *ppm*) and high VOC concentrations ($C_i \sim 100 - 200$ *ppm*) introduced into the system on the concentration of EP – II are shown in Fig. 2. The behaviour of dependences of values γ_i on the concentration of EP – II (N_p) in the range of low ($C_i \sim 1 - 10$ *ppm*) VOC concentrations is too much complex than during ZnO surface modification by H_2TPP. In the case of H_2TPP the donor character of sensor response was fixed during ethanol vapor adsorption and the acceptor character of sensor response – during acetone, and benzene vapors adsorption.

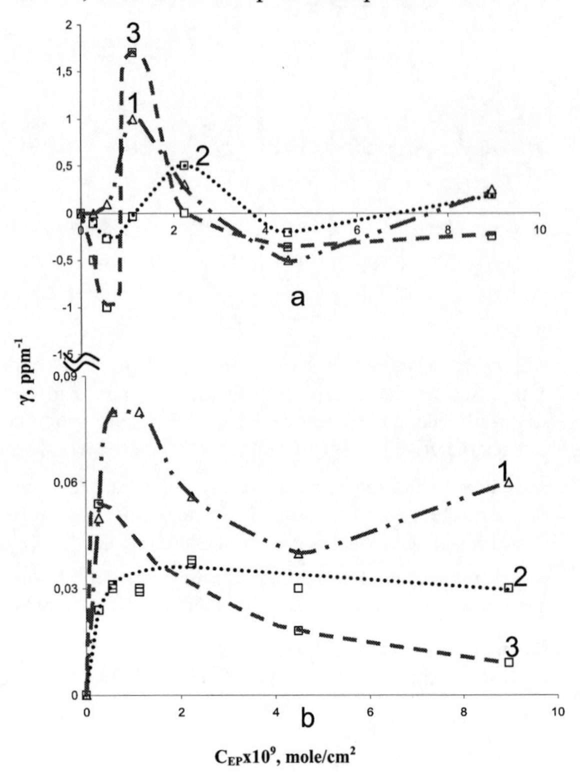

FIGURE 2. Dependences of the specific sensitivity of the ZnO based sensor with respect to (1) C_2H_5OH, (2) C_6H_6, and (3) CH_3COCH_3 on the concentration of EP - II at 200°C and VOC concentrations of (a) 1-10 and (b) 100-200 *ppm*.

As it follows from the data presented in Figs. 2 during ethanol vapor adsorption in the range of small concentration of EP – II ($N_P \sim 0.02 - 0.5 \cdot 10^{-9}$ *mole/cm^2*) the sensor response was not observed. The increase of *Np* concentration ($N_P \sim 1 \cdot 10^{-9}$ *mole/cm^2*) caused the donor character of sensor response. The further increase of *Np* concentration ($N_P \sim 2.5 \cdot 10^{-9}$ *mole/cm^2*) caused the change in the character of sensor response from donor to acceptor type and again to donor type ($N_P \sim 7.5 \cdot 10^{-9}$ *mole/cm^2*). The change in the character of sensor response from acceptor to donor type and again to acceptor type was observed also during benzene and acetone vapor adsorption (Fig. 2, curves 2 and 3 correspondently). At the same time absolute values of recorded responses were lower than corresponding values during the modification of ZnO-sensor by H_2TPP.

At high concentrations ($C_i \sim 100 - 200$ *ppm*) of all VOC response of ZnO-sensor, modified by EP – II (Fig. 2b) as well as during modification of ZnO-sensor by H_2TPP was donor type. At the same time values γ_i were by a factor of ten less than absolute values γ_i for the range of low ($C_i \sim 1 - 10$ *ppm*) concentrations of VOC vapors.

During the modification of ZnO-sensor surface by ZnTPP, as well as during modification of ZnO-sensor by H_2TPP, the character of sensor response to all VOC was donor type both in the range of low ($C_i \sim 1 - 10$ *ppm*) and high ($C_i \sim 100 - 200$ *ppm*) concentrations of VOC (Fig. 3).

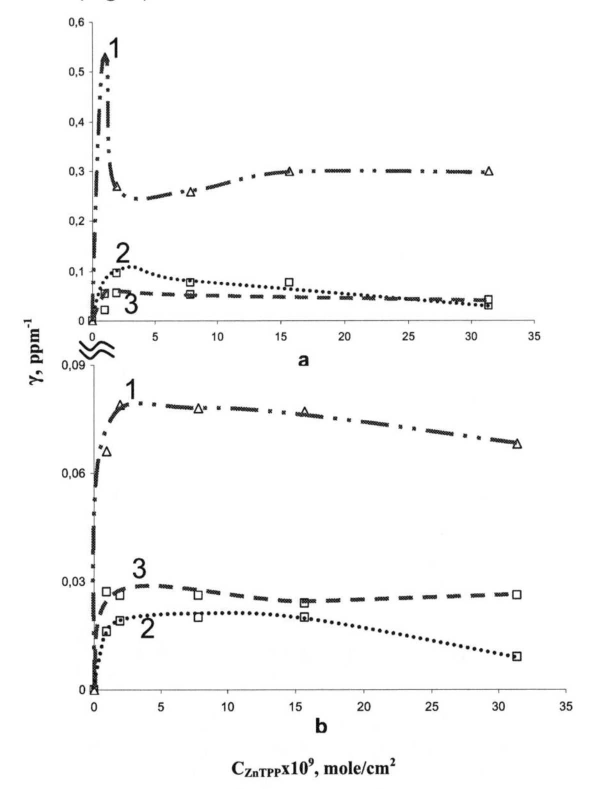

FIGURE 3. Dependences of the specific sensitivity of the ZnO based sensor with respect to (1) C_2H_5OH, (2) C_6H_6, and (3) CH_3COCH_3 on the concentration of ZnTPP at 200°C and VOC concentrations of (a) 1-10 and (b) 100-200 *ppm*.

It is necessary to point out that the maximum absolute values of ZnO-sensor sensitivity to all studied VOCs were achieved at EP – II and ZnTPP surface concentrations by a factor of ten more than in the case of H_2TPP. Using of CuOEP and PdTPP as the modifiers of ZnO-sensor did not exert influence on the parameters of sensor response.

DISCUSSION

All the totality of the obtained data related to the sensitivity of ZnO systems modified by porphyrins can be understood using stated earlier theory about the nature of sensor electron conductivity of polycrystalline ZnO films [2]. The nature of sensor electron conductivity is determined by the concentration of nonstoichiometric zinc atoms, which already at room temperatures are ionized to the Zn^+ state and play the role of sources of mobile charge carriers (electrons). It is also assumed that, in air close to surface Zn^+ ions at temperatures of 300°C and higher, dioxygen molecules chemisorbed from air transform into the charged form O^{2-} due to the capture of electrons from the bulk. The concentration of the charged form O^{2-} on the surface of ZnO increases as the temperature grows. It is possible to suggest that in the case of initial non modified ZnO films, when VOC molecules are introduced in air at 300^{0}C, these molecules displace oxygen chemisorbed in the form of O^{2-} and localized near Zn^+ surface centers. The negative charge of the surface then decreases, and the electrical conductivity of the sensitive ZnO layer increases ("donor response").

It was suggested in [1] that the sensitivity of the ZnO-sensor modified by H_2TPP, when VOC molecules are introduced in air, is determined by the presence of superstoichiometric Zn^+ ions on the surface of ZnO. They present in two active states $(Zn^+)_1$ and $(Zn^+)_2$, which manifest themselves in the formation of sensor effects.

We suggest that $(Zn^+)_1$ centers are geometrically and chemically more accessible to interactions with foreign molecules than $(Zn^+)_2$ centers. It was suggested that during the coordination of H_2TPP molecules on the ZnO surface near $(Zn^+)_1$ centers the formation of $(Zn^+)_1$– H_2TPP complexes occurs. Centers $(Zn^+)_1$ are localized along the axial axis of the porphyrin cycle of H_2TPP molecule due to the effective negative charge in the central part of the porphyrin cycle. The effective negative charge is conditioned with the presence of phenyl substituents positioned *meso* in H_2TPP [3]. In this case the charged O^{2-} can be localized as the superior ligand already at 100°C. We suppose that during the localization of H_2TPP molecules at $(Zn^+)_2$ centers the structure of $(Zn^+)_2$–H_2TPP complexes is different. In the vicinity of them oxygen is chemisorbed in the neutral form. Therefore the formation of "donor response" at the presence of ethanol vapor in [1] was connected with the displacement of oxygen chemisorbed in the form of O^{2-} near $(Zn^+)_1$–H_2TPP complexes by VOC molecules. During the chemisorption of benzene and acetone molecules acceptor sensor signals were observed. It is possible to suggest that during the chemisorption of acetone molecules in the vicinity of $(Zn^+)_2$–H_2TPP oxygen chemisorbed in the neutral form can be displaced. And the electron from the conducting band can be localized on the oxygen atom of carbonyl group. The acceptor effect caused by the chemisorption of benzene can be related with the formation of $\pi - \pi$ benzene – porphyrin complexes and with the formation of charged form of oxygen.

As was mentioned above during ethanol vapor adsorption in the range of small concentration of EP – II ($N_P \sim (0.02 - 0.5)\cdot 10^{-9}$ *mole/cm*^{2}) sensor response was not observed. At small concentrations ($C_i \sim 10 - 100$ *ppm*) of benzene and acetone vapors response of ZnO-sensor, modified by EP – II (Fig. 2a) as well as during modification of ZnO-sensor by H_2TPP was acceptor type.

The absence of sensor response to ethanol vapor during the modification of ZnO-sensor by EP – II could be explained in the following way. EP – II molecules have particularly plane structure. And the value of effective negative charge in the central part of the porphyrin cycle is less in comparison with H_2TPP [3]. It is possible to suggest that due to this factor during the localization of EP – II molecules on the surface of ZnO – sensor $(Zn^+)_1$–EP complexes described above are not formed. Apparently, during spattering of EP – II on to the ZnO - sensor $(Zn^+)_2$ centers are activated only and only $(Zn^+)_2$–EP complexes are formed. It causes the presence of sensor effects only to benzene and acetone vapors. As it follows from the data presented in Figs. 2 the further increase of *Np* concentration ($N_P \sim (1 - 2.5)\cdot 10^{-9}$ *mole/cm*^{2}) caused the donor response to ethanol vapor. The character of sensor response to benzene and acetone vapors changed from acceptor to donor type. Such behavior of sensor responses during increase of EP – II surface concentration could be understand assuming that, decrease of effective negative charge in the range of central part of the porphyrin cycle causes more effective EP – II molecules aggregation during their sputtering on to the surface of ZnO – sensor. The planar structure of EP – II molecules causes more density packing of EP – II molecules, than in the case of H_2TPP, during their aggregation in the process of sputtering. During such aggregation porphyrin molecules form “stacks”. The distance between them is 0.8 – 1 *nm* [4, 5]. It is possible to assume that the value of effective negative charge in the central part of the porphyrin cycle increase (the combination of effective negative charge of each molecule in aggregates) due to such porphyrin molecules packing. It cause the activation of $(Zn^+)_1$ centers with the formation of $(Zn^+)_1$ – aggregate EP complexes. During the chemisorption VOC molecules displace O^{2-}, which are localized near $(Zn^+)_1$ – EP complexes (donor character of sensor response).

At high VOC concentration during the modification of ZnO – sensor by EP – II molecules, as well as in the case of H_2TPP, donor character of sensor responses to all VOC were observed. It can be connected with VOC molecules condensation in the system of "canals" [6] – bound elements of unconfined space of EP – II layer and their transfer to the sensitive element surface is neutralized. The reduction of values of sensitivity γ_i during the VOC concentration increasing in the system can be connected with this fact (Fig. 2b). The change of sensor response character to all VOC vapors from acceptor to donor type during further increase of *Np* concentration (Fig. 2a) could be connected with structure reorganization of sputtered porphyrin layers. The formation of large EP – II aggregates occurs.

In the case of ZnO modification by ZnTPP weak donor effects were observed to all VOC. While during the modification by PdTPP and CuOEP the absence of sensor response was observed. This effect could be connected with electron structure of metalporfirins, which during the spattering on to the polycrystalline ZnO film can play a role of O^{2-} absorption centers on the film surface. Metal complexes of porphyrins (PdTPP and CuOEP) have four-axis planar structure, in which metal atoms Cu and Pd are situated centrosymmetrically and do not project from the plane of coordinate N atoms of porphyrin cycle. ZnTPP have pyramidal five – axis structure, in which four N atoms of porphyrin cycle form the base and Zn projects from the plane. In such system the coordination of donor ligand at the axial position is possible [7]. It is possible to suggest that ZnTPP, PdTPP and CuOEP differ in the capability of central ions to connect with O^{2-}. Only ZnTPP molecules capable to coordinate O^{2-} as the donor ligand. The formation of such complexes O^{2-} with PdTPP and CuOEP molecules is difficult.

ACKNOWLEDGMENTS

The work was supported by the Russian foundation for basic research grant № 08-02-00436.

REFERENCES

1. G.V. Belkova *at all*, *Russ. J. Phys. Chem. A.* **82**, 2323–2328 (2008).
2. V. Ya. Myasnikov *at all*, “Semiconductor Sensors in Physicochemical Investigations”, Moscow: Nauka, 1991[in Russian].
3. A.P. Hansen and H.M. Goff, *Inorg. Chem.* **23**, 4519-4525 (1984).
4. D.A. Dughty and C.W. Dwiggins, *J. Phys. Chem. (B)* **73**, 423-426 (1968).
5. M. Scarselli *at all, Surface Science* **601**, 5526-5532 (2007).
6. Yu. K. Tovbin, *J. Phys. Chem.* **82**, 1805-1820 (2008) [in Russian].
7. Gordon A. Melson, “Coordination chemistry of macrocyclic compounds”, New York: Plenum Press., 1979.

Reverse Biased Schottky Contact Hydrogen Sensors Based on Pt/nanostructured ZnO/SiC

Mahnaz Shafiei[1]*, Jerry Yu[1], Rashidah Arsat[1], Kourosh Kalantar-zadeh[1], Elisabetta Comini[2], Matteo Ferroni[2], Giorgio Sberveglieri[2], Wojtek Wlodarski[1]

[1]RMIT University, City Campus, GPO Box 2476V, Melbourne 3001, Victoria, AUSTRALIA
[2]SENSOR INFM-CNR University of Brescia, Via Valotti 9, 25133 Brescia, ITALY
**Email: mahnaz@ieee.org*

Abstract. Pt/nanostructured ZnO/SiC Schottky contact devices were fabricated and characterized for hydrogen gas sensing. These devices were investigated in reverse bias due to greater sensitivity, which attributes to the application of nanostructured ZnO. The current-voltage (*I-V*) characteristics of these devices were measured in different hydrogen concentrations. Effective change in the barrier height for 1% hydrogen was calculated as 27.06 meV at 620°C. The dynamic response of the sensors was also investigated and a voltage shift of 325 mV was recorded at 620°C during exposure to 1% hydrogen in synthetic air.

Keywords: Gas Sensors, Hydrogen, ZnO, Schottky Diode, Reverse Bias.
PACS: N 62.23.Kn

INTRODUCTION

Hydrogen gas sensing has attracted a great deal of attention due to the fact that it can be used as a clean and renewable source of energy in fuel cells. Hydrogen is explosive when mixed with air in the ratio of 4 to 75 vol.% [1] and is a major cause of corrosion, resulting in embrittlement.

Schottky contact based sensors with embedded metal oxide nanostructures has been shown to exhibit high sensitivity towards gases such as hydrogen. These types of sensors have been extensively used in chemical, food, and petroleum industries. Schottky contact based sensors with structures based on platinum or palladium and metal oxides such as SnO_2 [2, 3] and ZnO [4-7] thin films deposited on SiC operating at high temperatures [8] have been previously developed and investigated. Recently, nanostructured forms of ZnO have been used within Schottky contact devices to increase their efficiency due to large surface to volume ratio [9]. Almost all of these studies focused on the forward bias condition. However, nanostructures show interesting properties which allow the devices to be reverse bias. They show a lower breakdown voltage and generally have large lateral voltage shifts in their *I-V* curves as the free carrier concentration N_D increases when they exposed to reducing gases [10].

In this paper, we present gas sensing properties of nanostructured ZnO Schottky contact based sensors in the reverse bias condition towards hydrogen.

EXPERIMENTAL

N-type 6H-SiC wafers (Cree Co.) were used for the fabrication of the Pt/nanostructured ZnO/SiC devices. The thickness of wafers were approximately 250 μm. Wafers were etched in 10% HF for 15 s to remove the surface native oxide and then were diced into 3×3 mm^2 squares. The nanostructured ZnO were deposited onto the polished side of the SiC substrates by a vapour liquid solid (VLS) growth mechanism. A circular pad of Pt/Ti double metal layer with 1 mm diameter and 100 nm thickness each were deposited onto the unpolished side of the wafer to form the ohmic contact following the annealing process. A circular pad of Pt with diameter of 1 mm and thickness of 100 nm was deposited on the nanostructured ZnO by radio frequency sputtering to form the Schottky contact. After deposition, the devices were annealed in air at 450°C for 4 hrs and then at 600°C for another 2 hrs to form the ohmic contact. The schematic cross section of the developed devices is shown in Fig. 1.

CP1137, *Olfaction and Electronic Nose: Proceedings of the 13th International Symposium,* edited by M. Pardo and G. Sberveglieri

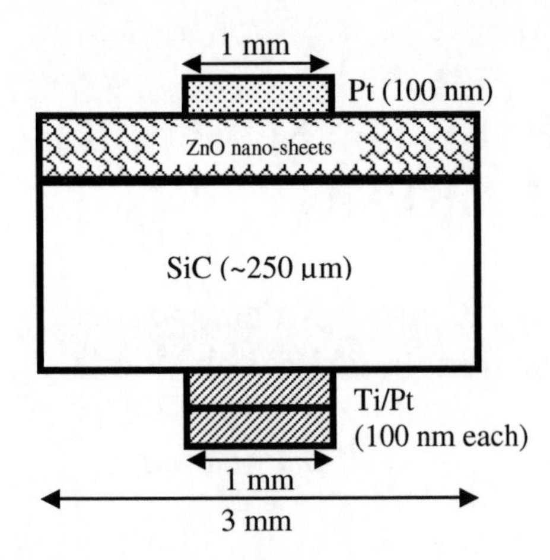

FIGURE 1. Schematic diagram of Pt/nanostructured ZnO/SiC sensors.

Gas sensitivity measurements of the sensors were performed in a test chamber made from Teflon, which was sealed in a quartz lid. The test chamber included an alumina micro-heater in close contact with the sensor, which controlled the operating temperature of the sensor. A gaseous mixture of analyte gas hydrogen and synthetic air (at zero humidity) were allowed to flow into the chamber by a mass flow controller. Electrical connections to the diode were achieved by physically contacting the sensor with a stainless steel base to the ohmic contact and needle contact to the Schottky contact. The *I-V* characteristics were measured at the Schottky contact using a sourcemeter (Keithley 2602). A programmable multimeter (Agilent 34410A) was used to record the dynamic response of the devices. Investigation into the influence of operating temperature between a range of 25°C to 620°C on the gas transducers were measured accordingly with respect to different concentrations of hydrogen gas.

RESULTS AND DISCUSSION

The morphology of ZnO nanostructures on the SiC substrate were characterised using scanning electron microscopy (SEM) as shown in Fig. 2. Randomly shaped and orientated 2D nano-sheets were found semi-uniformly distributed on the SiC surface. The thickness of the 2D nano-sheets was in the range of 40 to 80 nm.

Pt/nanostructured ZnO/SiC sensors were tested over a series of experiments to verify their stability. *I-V* characteristics of the sensors were measured with respect to increase in N_D by exposure to hydrogen gas at different temperatures. The Schottky contact device was biased with a constant reverse current of 1 µA. By exposing to gas, the reverse bias lateral voltage shift was observed to be significantly larger in the *I-V* curves than the lateral voltage shift in forward bias; which is a significant increase in sensitivity.

The catalytic dissociation of hydrogen molecules occurs at the Pt surface and H atoms diffuse through the thin Pt layer resulting in the adsorption of H atoms at the Pt/ZnO interface. These H atoms induce dipole-like charges at this interface which decreases the barrier height leading to the increase in bias current intensity and hence a greater lateral voltage shift in reverse *I-V* characteristics (Fig. 3). The changes in the slope of the linear portion of the *I-V* curve resulted from a decrease in the metal oxide (series) resistance [11, 12].

Fig. 4 shows the gas performance of the sensors as they were exposed towards 1% hydrogen in synthetic air over a range of operating temperatures between 25°C to 620°C. We investigated the gas response of these devices at 530°C and 620°C.

FIGURE 2. SEM image of nanosturctured ZnO as grown on the SiC substrates.

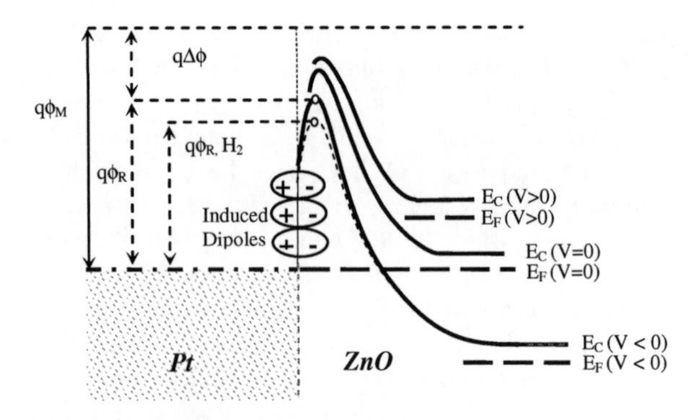

FIGURE 3. Energy-band diagram of the Pt/nanostructured ZnO Schottky contact under different bias conditions and upon exposure to hydrogen in reverse bias.

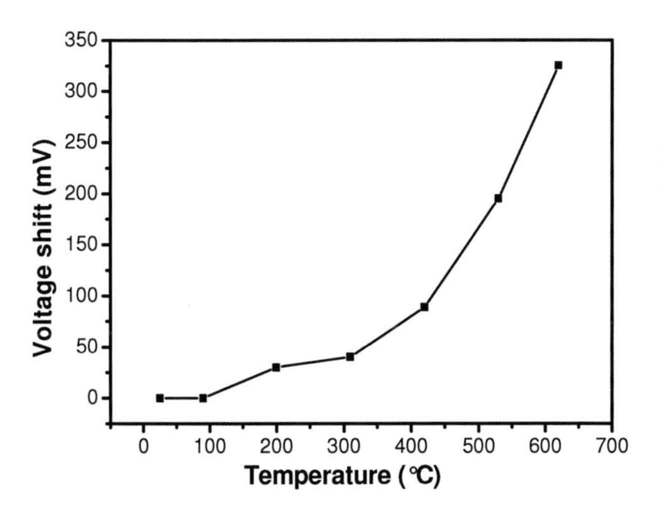

FIGURE 4. Voltage shift of Pt/nanostructured ZnO/SiC sensors measured over a range of different temperatures (1% hydrogen in synthetic air at constant reverse bias current of 1 μA).

Devices were tested at an operating temperature of 620°C due to its high stability performance and largest electrical response towards hydrogen. Fig. 5 shows reverse *I-V* characteristics of the sensors towards different concentrations of hydrogen at 620°C.

The following equations based on thermionic emission theory [13] are used for describing the reverse bias *I-V* characteristics of the Schottky diode:

$$|J_R| \approx A^{**} \cdot T^2 \cdot \exp\left[-\frac{q\left(\phi_B - \sqrt{q\xi_m / 4\pi\varepsilon_s}\right)}{kT}\right] \quad (1)$$

where, J_R is the reverse current density, A^{**} is the effective Richardson constant, T is the absolute temperature, q is the charge constant, ϕ_B is the barrier height, ξ_m is electric field applied across the Schottky contact, ε_s is the electric permittivity of metal oxide and k is the Boltzmann constant. We can simplify the barrier height in reverse bias as:

$$\phi_B = \left(\frac{kT}{q} \cdot \ln\left[\frac{A^{**} \cdot T^2}{J_R}\right]\right) + \sqrt{q\xi_m / 4\pi\varepsilon_s} \quad (2)$$

The change in the barrier height was calculated to be 27.06 meV at 620°C towards the exposure of 1% hydrogen.

The dynamic response was acquired by allowing different concentrations of hydrogen (0.06, 0.125, 0.25, 0.5 and 1 %) to be exposed to the sensors. Exposure time was for 5 min (Fig. 6). Recovery measurements were performed by purging the gas and re-exposing the sensors to synthetic air until the original baseline was restored. Voltage shifts of 132 mV and 325 mV were recorded for 0.06 and 1% hydrogen, respectively which was verified by *I-V* curves.

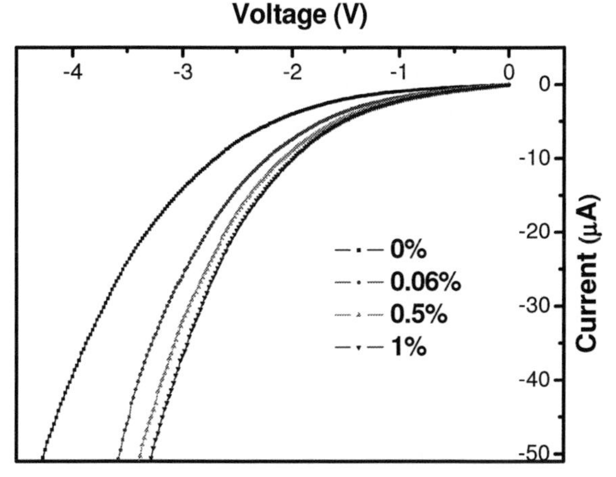

FIGURE 5. Reverse *I-V* characteristics of the Pt/nanostructured ZnO/SiC sensors towards 0, 0.06, 0.5 and 1% hydrogen gas at 620°C.

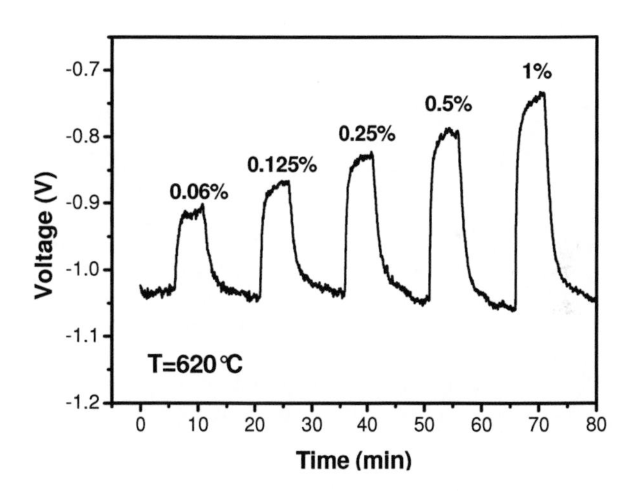

FIGURE 6. Dynamic response of the Pt/nanostructured ZnO/SiC sensors towards different hydrogen gas concentrations in synthetic air at 620°C and with a constant reverse bias current of 1 μA.

CONCLUSIONS

In this paper, Pt/nanostructured ZnO/SiC sensors were fabricated and characterised towards different concentrations of hydrogen at high temperatures. At 620°C, the effective change in the barrier height, in reverse bias, caused by exposure to 1% hydrogen in synthetic air was found to be 27.06 meV. The dynamic response of the reverse biased sensors at 620°C exhibited voltage shift of 325 mV when the sensors were exposed to 1% hydrogen.

REFERENCES

1. H. Eichert, and M. Fischer, *International Journal of Hydrogen Energy* **11**, 117-124 (1986).
2. GW. Hunter, PG. Neudeck, M. Gray, D. Androjna, L-Y. Chen, R.W. Hoffman, Jr., C.C. Liu and Q. H. Wu, *Mater Sci Forum* **338-342** (part 2), 1439-1442 (2000).
3. M. Shafiei, K. Kalantar-zadeh, W. Wlodarski, E. Comini, M. Ferroni, G. Sberveglieri, S. Kaciulis, and L. Pandolfi, *Int. Journal on Smart Sensing & Intelligent Systems* **1 No.3**, 771-783 (September 2008).
4. S. Kandasamy, W. Wlodarski, A. Holland, S. Nakagomi, and Y. Kokubun, *Appl. Phys. Letters* **90**, 064103-1-064103-3(2007).
5. S. Kim, B. S. Kang, F. Ren, K. Ip, Y. W. Heo, D. P. Norton, and S. J. Pearton, *Appl. Phys. Letters* **84**, 1698-1700 (2004).
6. S.J. Pearton, D. P. Norton, K. Ip, Y. W. Heo and T. Steiner, *Journal of Vacuum Science & Technology B* **22**, 932-948 (2004).
7. S.Pitcher, J. A. Thielea, H. Renb and J. F. Vetelino, *Sensors and Actuators B-Chemical* **93 (1-3)**, 454-462 (2003).
8. AL. Spetz, P. Tobias, A. Baranzahi, P. Martensson, and I. Lundstrom, *IEEE T Electron Dev.* **46**, 561-566 (1999).
9 K. Kalantar-zadeh and F. Benjamin, *Nanotechnology-enabled Sensors,* New York: Springer, 2008.
10.J. Yu, S. Ippolito, M. Shafiei, D. Dhawan, W. Wlodarski, and K. Kalantar-zadeh, *Appl. Phys. Letters* **94**, 013504 (2009).
11. S. Nakagomi, K. Okuda, and Y. Kokubun, *Sensors and Actuators B: Chemical* **96**, 364-371 (2003).
12. S. Kandasamy, A. Trinchi, W. Wlodarski, E. Comini, and G. Sberveglieri, *Sensors and Actuators B: Chemical* **111-112**, 111-116 (2005).
13. S. M. Sze and K. Ng Kwok, *Physics of Semiconductor Devices*, 3rd edition, New York: Wiley, 2007, pp. 170-172.

Gas Sensing Character of Polyaniline with Micro- / Nano-Fiber Network Structure

Chuanjun Liu, Kenshi Hayashi, Kiyoshi Toko

Department of Electronic, Graduate School of information Science and Engineering, Kyushu University 744, Motooka, Nishi-ku, Fukuoka 819-0395, Japan

Abstract. By directly chemical oxidative polymerization of aniline on Au electrode treated with Au nanoparticles (Au-NPs) deposition and 4-aminothiophenol (ATP) modification, polyaniline (PANI) fibers were obtained in-situ in micro-/nano- size with two dimensional network structures. It was found that the ATP modification played an important role for the formation of fibrous network and in the decrease in the contact resistance between the PANI film and the electrode surface. The PANI film showed a rapid response upon the exposure to various gases with low detection limit, which indicated that the developed PANI sensor could be applied in the high sensitive detection of hazardous gases.

Keywords: Polyaniline, Fiber, Gas sensor, High sensitivity
PACS: 61.46.km, 81.16.Be

INTRODUCTION

Although conducting polymers (typically polyaniline) have been widely applied in the field of electronic nose and gas sensor, problems such as low sensitivity, low detection limit and long response time should be improved if compared with other kinds of gas sensor (metal-oxide gas sensor). The preparation of conducting polymers with nanostructures provides a solution for this problem due to the unique electrical and sensing properties of nanostructures. For example, compared with the conventional prepared PANI, the sensors based on PANI nanofibers shows better sensitivity with faster response time because the large surface to volume ratio of nanostructures [1].

During the past ten years the nano-PANIs, such as nanofibers, nanowires, and nanotubes, have been prepared by a number of methods involving the chemical or electrochemical oxidative polymerization in the presence or absence of templates [2]. However, the application of nano-PANIs as chemical sensors prefers to a direct growth of nanostructures on the devices surface. The most typical PANI chemical sensor is generally prepared by casting PANI solutions or nanofiber suspensions onto the electrode surface [1,3,4]. The lack of alignment in the nanostructure is a problem that should be paid attention to [5]. Additionally, the influence of the contact resistance on the sensing character is aso a problem that shoud be considered but it is neglected in most studies [6].

In this work, we reported a novel formation process of PANI micro-/nano- fibers with two dimensional network structures by drop-casting a precursor solution of aniline and ammonium peroxydisulfate (APS) on electrodes deposited with Au-NPs and modified with ATP. The result indicated that the Au-NPs deposition and ATP-treatment played an important role for the formation of PANI fibrous network as well the lowering of the contact resistance between the PANI film and integrated electrodes. The sensing character of the prepared PANI sensor on various gases was investigated.

EXPERIMENTAL

The preparation of the PANI sensor is described briefly as follows. The Au electrode used was fabricated by photolithography with a gap width of 150 μm using 200 μm Au / 50 nm Ti deposited glass substrate (Corning 1737 glass). The electrodes were cleaned with TL1 (H_2O:NH_4OH:H_2O_2=5:1:1) and ozone process. Then a layer of gold was deposited on them (masked with a deposition area of ca. 0.28 cm^2) by sputtering equipment (VPS-050, ULVAC). The deposition current and time were 8 mA and 90 seconds, respectively. The Au-deposited electrodes were immerged in an ethanol solution of ATP (2 mg/ml) for 24 hours. On the other hand, solutions of aniline (3.2 mmol in 10 ml 1N HCl) and ammonium

CP1137, *Olfaction and Electronic Nose: Proceedings of the 13th International Symposium*, edited by M. Pardo and G. Sberveglieri

peroxydisulfate (APS) (0.8 mmol in 10 ml 1N HCl) used for rapid mixing polymerization were prepared separately [7]. The two solutions were mixed by a magnetic stirring for 30 s. Subsequently, 30 μL of the mixture was dropped on the ATP-treated area of the electrodes. The reaction was carried out in a refrigerator (4 ºC) for 2 hours. After washed with a large amount of deionized water, the substrates were dried either naturally or by N_2 flowing, which finally afforded a green, transparent, PANI thin film.

Laser Scanning Microscopy (LSM) (KEYENCE, VK-9700) was used to investigate the morphology of PANI fims because it can provide not ony the 3D coor viewing, but also the detaied information consisting of fim thickness, fiber diameter and height.

The gas sensing experiment setup is shown in Figure 1, in which a standard gas generator (PERMEATER PD-1B, GASTEC Co., Japan) was used. Permeation tube or diffusion tube was used as gas source to generate various gases with standard concentration.

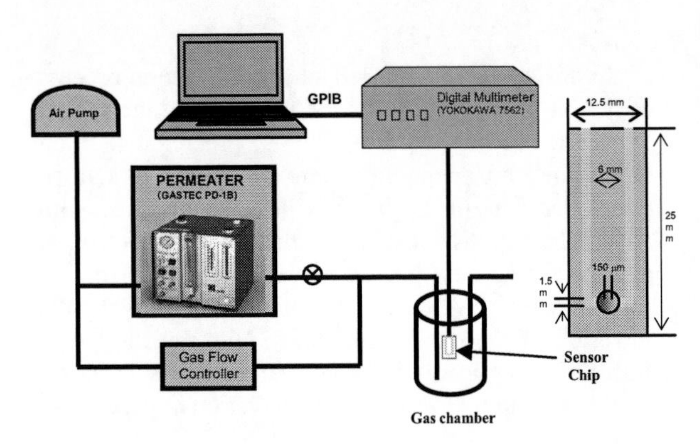

Figure 1: Gas sensing experimental

RESULTS AND DISCUSSION

Figure 2(a) shows a typical morphology of PANI prepared according to the described protocol. The PANI presented a fibrous network structure aligned with two dimensional ordering between the gaps of the Au electrode. The diameter of the fiber was about 300 ~ 500 nm and the length extended as long as 10 or 20 μm. The 3D color photos indicated a two layer structure: the fibrous network grew on a thin PANI film with a thickness of whole film lower than 1 μm and the fiber height was of proportion to its diameter. Fighure 2(b) shows the LSM photos of PANI film polyerizaed on Au-NPs deposited electrode but without ATP modification. A different morphology was clearly presented and no fiber structure was observed. This result demonstrated that the ATP modification greaty influenced the morohlogy of obtained PANI films.

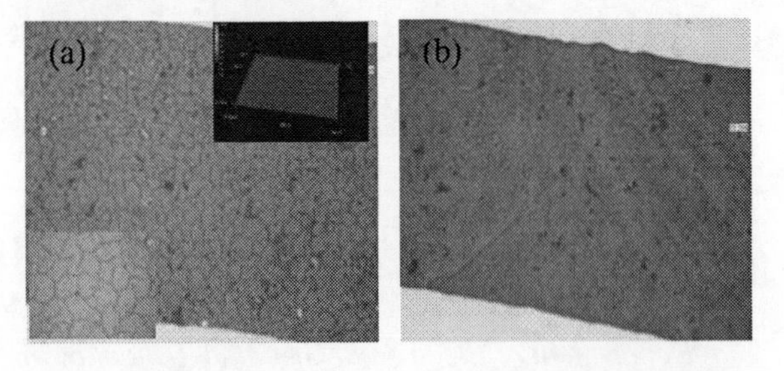

Figure 2: LSM photos of PANI film prepared on Au electrode with a gap width of 150 μm. (a) the electrode was treated both with Au-NP deposition and ATP modification (b) the electrode was treated with Au-NP deposition only. Scale bar: 13.7 μm.

The purpose of the Au deposition is to provide bonding sites of ATP on the electrode surfacee. Our previous studies [8,9] have demonstrated that the structure and conducting character of the Au-deposited layer were determined by the deposition current and time. A layer of Au nanoparticles (several tens of nanometers) can be formed by selecting appropriate deposition current and time. Especially, the deposition with 8 mA and 90 s can afford an almost insulating Au layer (the resistance lies in the order of giga ohm) with nanoparticle island structure separated by nanogaps. The insulating character of the Au-NPs deposited layer guaranteed that the sensing property of PANI film would not be affected by the Au deposition. At the same time the Au-NPs provided the bonding sites for ATP modification and the grafting of PANI nanofiber during the subsequent polymerization.

In order to confirm the grafting effect caused by the ATP modification, we compared the character of current vs. potential for PANI films prepared on Au-NPs deposited electrodes treated with and without ATP-modification. A linear dependence of the current on the applied potential confirmed an ohmic contact in

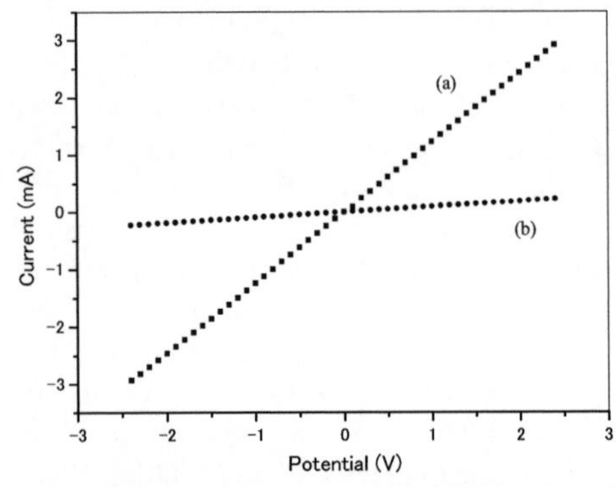

Figure 3: Current vs. potential for PANI films polymerized on Au-NPs deposited electrodes with (■) and without (●) ATP modification.

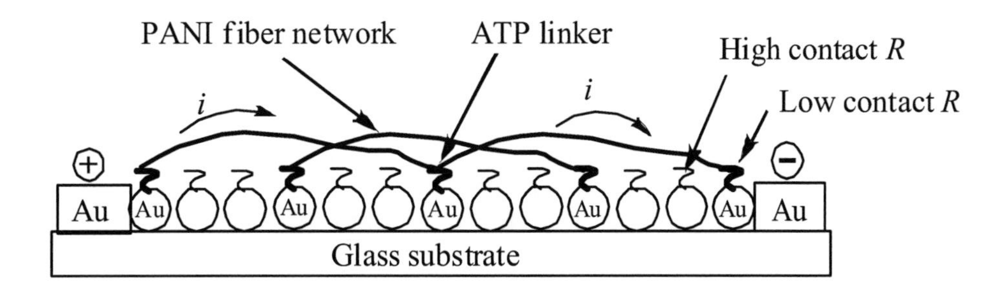

Figure 4: Assumed conducting mechanism in PANI film with 2D fibrous network. Due to the high conductive feature of nanofibers and the low contact resistance at the ATP-linked domains, the conducting path in the film was probably dominated by the 2D fibrous network.

both cases as shown in Figure 3. However, the film (a) showed a higher current response than the film (b). We suggested that two factors were probably responsible for this result: one was the high conductivity of PANI fibers with parallelly orientated network structure; another was the low contact resistance caused by the ATP modification on Au-NPs layer.

According to the above suggestion a conducting mechanism was assumed for the prepared PANI film as shown in Figure 4. In view of the high condutive charater of the nanofiber netwrok and the low contact resistence at the ATP-linked domains, it was possible that the electrical properties of the prepared film were mainly dominated by the PANI fibrous network, although the film showed a two-layer film structure.

Much work has been carried out on the gas sensing character of PANI films. Some limitations for the PANI-based gas snesors include relatively slow resonse times and small changes in conductivity upon exposure [10]. Figure 3 shows the response of the as-prepared PANI sensor on 5 ppm NH_3 gas. The sensor showed a very quick response (t < 20 s), which apparently preceded the responding rapidity of PANI nanofiber on NH_3 with much larger concentrations (100 s under 100 ppm) [1]. It was interest to find that the electrode (a) showed a higher sensitivity than the electrode (b), indicating that the ATP modification improved the sensitivity of the PANI sensor. Moreover, the resistance of sensor (a) showed a two-step increase upon the exposure to NH_3. Considering the structure character as shown in Figure 4, we suggested that the first rapid jump of resistance was corresponding to the response of PANI fibrous network on NH_3, and the second increase in resistance was corresponding to the response of inter layer PANI mat. Due to the introduction of ATP, the inter layer PANI fim probaby had a denser structure than the un-ATP treated one (sensor b), which made a hard diffusion of NH_3 into the film and thus the slow increase in resistance in the second step.

PANI generally shows intense and fast response on gases with strong chemical interaction such as protonation, depronation, and reduction. However, for the gases with weak physical interaction, such as swelling and conformation alignment, the response time and extent are not significant [1]. In the latter case, the response can only be obtained with long time as well as high gas concentration [11, 12]. Figure 6

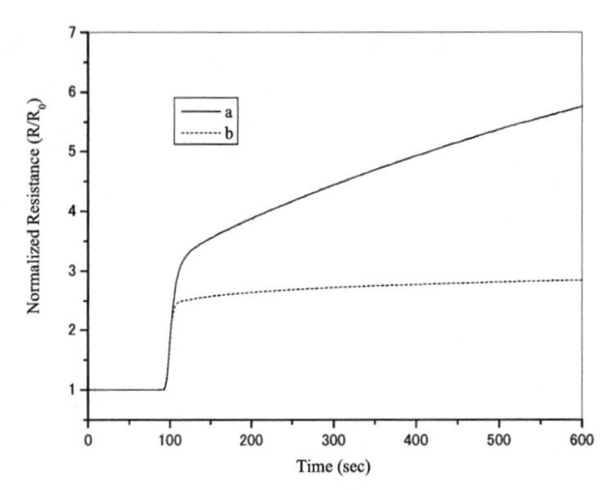

Figure 5: Response of PANI sensors to 5 ppm NH_3. The PANI films were polymerized on electrodes treated with both Au-NPs deposition and ATP modification (a) and Au-NPs deposition only (b).

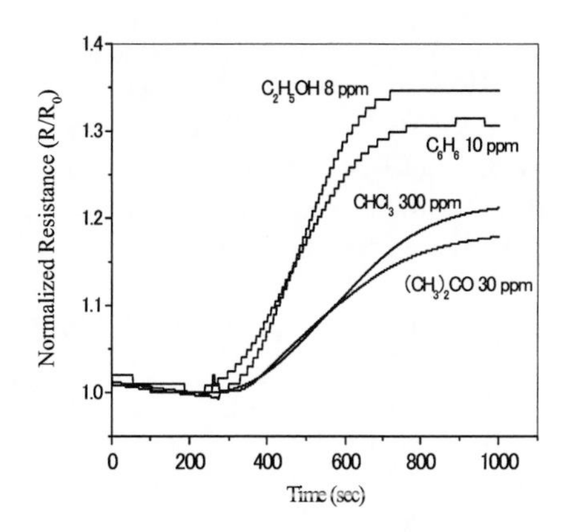

Figure 6: Response of the prepared PANI sensor electrode exposure to various gases.

shows the response of the prepared PANI sensor on gases including chloroform, ethanol, benzene and acetone. These gases were generated by the standard gas generator with relatively low concentrations. Nevertheless, enough response intensity could be obtained within several minutes. This result indicated that the prepared PANI sensor could be applied in high sensitive detection of hazardous vapors. Additionally, the two-step reponse as shown in the case of NH_3 was not observed for these gases, which probably indicated for gases with different characters, the response in PANI film with and without nanofiber structure was different too. The detailed investigation on mechanism is ongoing.

CONCLUSION

The ATP modification on Au-NPs deposited electrode not ony resulted in a novel fibrous network structure, but also lowered the contact resistance between the prepared PANI film and the electrode. The PANI sensor deveoped based on this technique showed rapid response character on various gases with low detection limit, which is promising to be applied in the high sensitive detection of hazardous vapors.

ACKNOWLEDGEMENTS

This work is supported by Knowledge Cluster Initiative (2nd Stage): R&D of bioelectronic technologies for safety and security and its application for sensing, from the Ministry of Education, Culture, Sport, Science and Technology of Japan.

REFERENCES

1. S. Virji, J. Huang, R.B. Kaner, B.H. Weiller, Nano. Lett. 4 (2004) 491-496
2. D. Zhang, Y. Wang, Mater. Sci. Engineer B 34 (2006) 9-19
3. A.L. Kukla, Yu.M. Shirshov, S.A. Piletsky, Sens. Actuators, B 37 (1996) 135-140
4. A.Z. Sadek, W. Wlodarski, K. Kalantar-Zadeh, C. Baker, R.B. Kaner, Sens. Actuators, B 139 (2007) 53-57
5. N-R. Chiou, C. Lu, J. Guan, L.J. Lee, A.J. Epstein, Nat. Nanotech. 2 (2007) 354-357
6. D.D. Sawall, R.M. Villahermosa, R.A. Lipeles, A.R. Hopkins, Chem. Mater. 16 (2004) 1606-1608
7. J. Huang, R.B. Kaner, Angew. Chem. Int. Ed. 43(2004) 5817-5821
8. K. Masunaga, M. Sato, K. Hayashi, K. Toko, Transducers '07 Digest of Technical Papers, pp. 1413
9. K. Masunaga, M. Sato, K. Hayashi, K. Toko, Proceedings of the 24th Sensor Symposium, pp.461
10. B. Timmer, W. Olthuis, A. Berg, Sens. Actuators, B 107 (2005) 666-677
11. A.A. Athawale, M.V. Kulkarni, Sens. Actuators B 67 (2000) 173-177
12. S. Sharm, C. Nirkhe, S. Pethkar, A.A. Athawale, Sens Actuators B 85 (2002) 131-136

Effect of Porosity on Response Behavior of Carbon Black-PMMA Conductive Composite Sensors toward Organic Vapors

Ehsan Danesh, Seyed Reza Ghaffarian*, Payam Molla-Abbasi

Department of Polymer Engineering, Amirkabir University of Technology, 15875-4413 Tehran, Iran
*Corresponding author; Email: sr_ghaffarian@aut.ac.ir, Tel: (+9821) 64542402, Fax: (+9821) 64543102

Abstract. We describe herein the fabrication of novel porous conductive composite vapor sensors characterized by different porosities and specific surface areas. These samples were obtained by dry-cast non-solvent induced phase separation (NIPS) method. We have studied porous composite structures by SEM, BET and water evaporation method. Testing to different concentrations of several organic vapors, the porous sensors showed improved sensitivities and response times, compared to their dense counterpart. Improved characteristics of the sensor response were related to better sorption properties of sensing film due to increased porosity and specific surface area obtained by this method of film fabrication.

Keywords: Conductive Composite, Vapor Sensor, Electronic Nose, Porosity, Phase Separation
PACS: 07.07.Df

INTRODUCTION

Recently, an increasing interest in the use of carbon-polymer conductive composites (CPCs), either as single gas sensors or as elements of electronic nose sensor arrays has emerged. These types of vapor sensors have been developed and studied extensively. Usually, CPCs are comprised of an insulator matrix (polymer) and a conductive filler (e.g., carbon-black). Swelling of polymer upon exposure to various gases and vapors causes the dispersed carbon particles to move farther apart from each other. Disruption of formerly formed conductive pathways in the film raises the resistance of the sensor, thereby providing an extraordinary simple mean for monitoring the presence of a vapor.

Compared to other vapor sensors, chemiresistors based on CPCs have shown some advantages including: great stability, acceptable lifetime, tunable selectivity, ability to work at room temperature, good reversibility and reproducibility, linearity of response curve, great processability, low power consumption and cost effectiveness, that make them attractive for use in commercial electronic noses. Nevertheless, some important performance parameters of these sensors are still not preferable. Much attention has been paid to modify polymeric sensors in order to endow them higher sensitivities and shorter response and recovery times. While modification of polymeric chemiresistors through optimizing filler type and concentration, and changing polymer identity by using specific chemical interactions are well developed [1-4], less attraction has been devoted to investigate effects of microstructure of composite film on sensing performance. The thought is that, one of the main issues related to microstructure of thin film composites is "porosity". It is well-known that in adsorptive sensors, characteristics of gas adsorption depend on the surface area of the sensing film.

Effect of porosity on sensor performance parameters has been reported broadly for various polymeric and non-polymeric based gas, vapor and humidity sensors [5-9]. To our knowledge, there is not any qualytative and quantitative study on the effect of porosity on the performance of carbon-polymer conductive composite vapor sensors. The goal of this work is to clarify the effect of thin film porosity on sensitivity, response and recovery time of CPC vapor sensors. Herein, we have exploited non-solvent induced phase separation (NIPS) method [10] to make porous conductive Carbon black-polymer composite vapor sensors. The response behavior of porous sensors has been compared with dense sensor films

CP1137, *Olfaction and Electronic Nose: Proceedings of the 13th International Symposium,* edited by M. Pardo and G. Sberveglieri

with comparable thickness, and the effect of porosity on sensing performances has been demonstrated.

EXPERIMENTAL

Materials

Polymer used in this study was poly(methyl methacrylate) (PMMA; M_W = 120,000; Sigma-Aldrich, Germany). The solvent and non-solvent were Ethyl acetate (EA; bp=350k; Merck, Germany) and 2-methyl-2,4-pentanediol (MPD; bp=470k; Merck, Germany). The conductive filler used in the composite was highly structured Printex XE2 (Degussa AG, Frankfurt, Germany) nano-sized carbon black. The average primary particle size of Printex XE2 is 30 nm and its specific surface area is more than 900 m^2/gr.

Composite Membrane Preparation

Here, we used a dry-cast phase separation method, to prepare microporous composite membranes. Homogenous dope solutions, consist of desired ratios of polymer (P), solvent (S) and non-solvent (N), were initially prepared. (e.g., P15S55 means 15 wt % of polymer, 55 wt% solvent and 30 wt % of non-solvent) Then, adequate amount of conductive filler was added to the ternary solution and the resulted suspension was mixed rigorously for a day by a stirrer at room temperature. For ensuring complete and homogenous dispersion of carbon particles in the mixture, the suspension was further sonicated for 15 min, by a Hielscher ultrasonic system (UP400S, Germany), and immediately was cast on a glass plate with a thickness of 250 micron with the aid of a film applicator (BYK Gardner, Japan), and was put on a balance place for drying. After complete evaporation of solvent, the film was immersed in 50 wt% aqueous methanol solution to extract MPD from the membrane and was dried gradually. Dense membrane was prepared by casting 200 μm of 20 wt% solution of PMMA in EA on a glass plate. In all cases the amount of carbon black in composite was 5 wt% of polymer.

Sensing Apparatus and Measurement

An automatic vapor generation system (AVGS), consist of an air compressor, a bubbler, MFCs, and a 3-way solenoid valve, controlled with a PC, has been prepared. Exact concentrations of a vapor were prepared by mixing adequate ratios of air saturated with the vapor and clean air. Then, the resulted vapor was introduced into a sensing chamber made of glass and Teflon. The sensing film was made by cutting dried membranes to 15 mm × 20 mm strips and depositing 2 fingers of silver ink separated by a gap of 5 mm, on the top surface of the conductive membrane strips. In order to determine the dynamic response of the sensor film to various concentrations of vapors, Cu pads were attached to sensors using silver paste and the resistance of the sensor was measured using a two-point configuration via a digital multimeter (PC5000, Sanwa Electronic Instruments Co., Japan). The sensors were placed into the chamber and background flow was passed over the sensors in order to establish a baseline. Then, sensors were exposed to controlled amount of vapor and changes in their DC resistance as a function of time were monitored. All the measurements were performed at 30 ± 2 °C.

Characterization of Conductive Composite Membrane

The membrane was freeze-dried for SEM observation. The dry membrane was immersed in liquid nitrogen, fractured, and coated with Au. The cross-section was viewed by an MV2300 Scanning electron microscope (Obduct CamScan Ltd., Cambridgeshire, UK). A single point BET (Brunauer–Emmett–Teller) method was used to determine the specific surface area of microporous composites (ChemBET3000; Quantachrome Corp., Odelzhausen, Germany). Also, the overall porosity was estimated by water evaporation method. The porous film was immersed in water for more than 2 weeks and weighed after saturation (m_{wet}), then it is removed from water, rinsed and dried in 80 °C oven under vacuum for a day and weighed again (m_{dry}). The overall porosity (P) will be:

$$P = \frac{d_p(m_{wet} - m_{dry})}{d_p(m_{wet} - m_{dry}) + d_w m_{dry}} \times 100\ \% \qquad (1)$$

Where d_w and d_p are water and polymer densities, respectively.

RESULTS AND DISCUSSION

Conductive Porous Membrane

In the dry-cast NIPS technique, an initially homogeneous ternary solution of polymer/volatile solvent/less volatile non-solvent, thermodynamically becomes unstable due to solvent evaporation and phase separates into polymer lean and polymer rich phases. The polymer-rich phase forms the matrix of the membrane, while the polymer-lean phase, rich in solvents and non-solvents, fills the pores. Figure 1

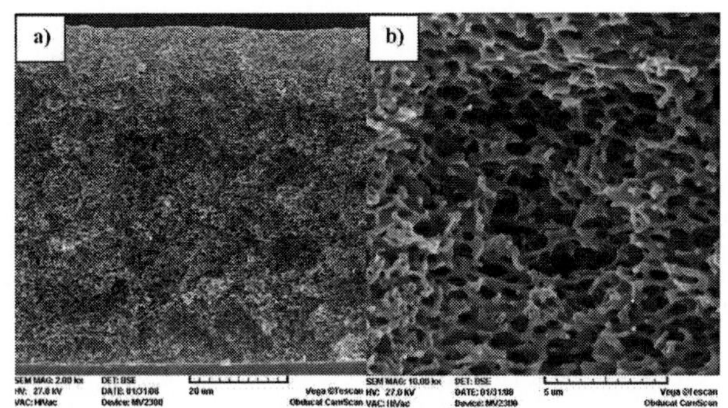

Figure 1 Cross-section of porous P15S50 composite membrane. a) Whole membrane cross-section. b) Structure in high magnification.

shows SEM image of the cross-section of P15S50 composite membrane. As can be seen from the images, phase separation of polymer resulted in a microporous membrane. Figure 1-a demonstrates a relatively isotropic pore structure throughout film thickness. Figure 1-b clearly shows interconnected highly porous structure of the system having 35 wt% of non-solvent with average pore diameter of 333 nm. Also, there wasn't any evidence of carbon agglomeration greater than 100 nm in the composite membrane. The thought is that carbon particles were forced to place in wall space between the pores; so the dispersion concentration of particles further improved. Resistivity data in Table 1, clearly demonstrates improved conductivity of porous membranes, compared to that of dense membrane ($\rho = 4.704$ kΩ-cm).

Table 1, also compares overall porosities and BET specific surface areas of the above mentioned three ternary systems. As non-solvent content in the solution increases, overall porosity increases. Furthermore, the specific surface area measured by BET method increases with increase in non-solvent content. The high surface area values were correlated to highly, on purpose-designed interconnected porous structure of the membranes. Porous structure and morphology of membranes can be tailored by controlling thermodynamic and kinetic of phase separation.

Response Behavior of Porous Sensors

In order to determine the effect of porosity on sensor performance parameters, we compared the sensor response of conductive dense composite, with those of conductive porous composites. Figure 2, illustrates normalized sensor response ($(R_{max}-R_b)/R_b$; where R_b and R_{max} are baseline resistance and maximum resistance signal of the sensor, respectively) of dense and porous membranes to 10 ppth of methanol vapor. First of all, it can be seen that introducing porosity, has greatly decreased the response times of the sensors. For all porous membranes, the responses reached the 90% of their maximum value below 200 sec; Although the thickness of all sensors were comparable, response times of the dense membrane were much higher (1300 sec to 10 ppth methanol vapor). We attribute this more rapid response to highly porous structure of the films. As the overall porosity and pore interconnection increases, the diffusion of vapors into polymer facilitates, thereby results in reduction of time needed for polymer to sorb an analyte, swell and reach equilibrium. We think that porosity decreases effective thickness and hence, greatly reduces the response time of the sensor. Because of similar mechanisms of sorption and desorption in polymers, reduction in recovery times of porous membranes can be explained based on porosity, too. While response times in the order of 200 seconds are still high for a real sensor, real-time sensors can be *easily* made by decreasing porous membranes' thicknesses down to few microns.

Secondly, it can be seen that the maximum response of the sensors also improved several times with introducing porosity to the membrane. So, it is evident that the sorption of vapors into polymeric sensors can be improved by increasing surface to volume ratio, which results in reduction of permeation length (Accordingly, reduction in effective thickness). This will greatly improve the sensitivity of the porous sensors compared with dense sensors. However, sensitivity of porous membranes did not increase monotonically with increase in non-solvent content and overall porosity. Figure 3 compares the maximum relative responses of our sensors toward headspace of four different vapors. Clearly, a maximum for P15S55 is observed. Explanation of this phenomenon with regard to porous microstructure and CPC's sensing mechanism has been presented in our upcoming papers.

Furthermore, figure 3 shows that porosity approach provides a method for making various polymeric vapor detectors from a limited number of initial chemical feedstocks (e. g., PMMA). These sensors can be used to fabricate the sensor array of an electronic nose (e-nose) system.

TABLE 1. Characteristic properties of three ternary samples

Ternary System	Non-Solvent Content (wt %)	B.E.T. Specific Surface Area (m^2/gr)	Overall Porosity (%)	Average Thickness (μm)	Average Pore Diameter (nm)	Resistivity [ρ] (kΩ-cm)
P15S60	25	17.2806	26.8	52	216	0.048
P15S55	30	22.8154	34.3	58	284	0.111
P15S50	35	29.7949	38.7	66	333	0.142

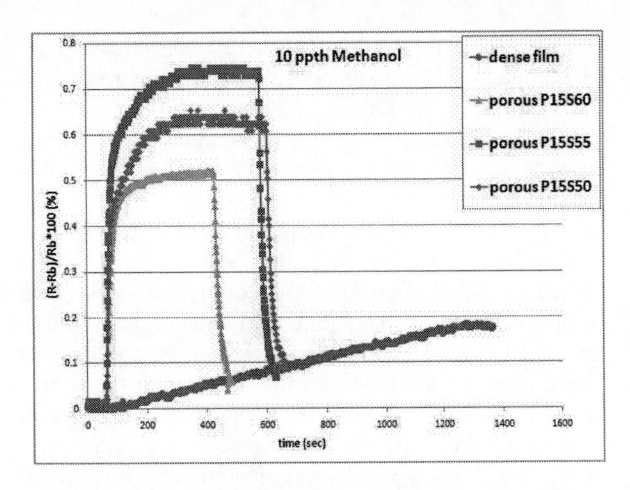

Figure 2 Comparison of response behavior of porous sensors and dense sensor toward 10 ppth Methanol.

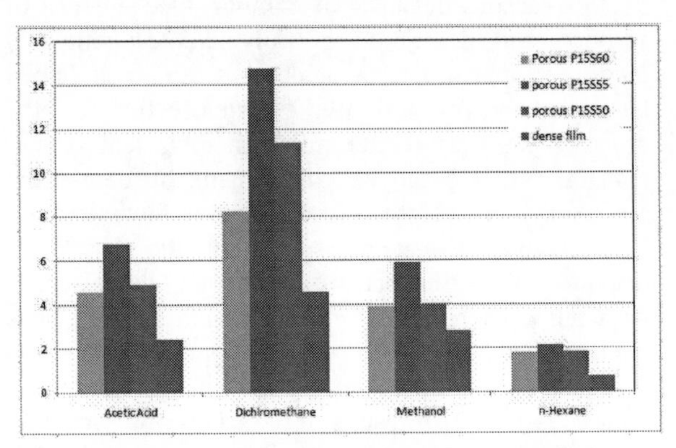

Figure 3 Comparison of the maximum relative responses of porous and dense sensors toward headspace of four different vapors: Acetic acid, Dichloromethane, Methanol and n-Hexane.

CONCLUSION

We have prepared porous conductive composite vapor sensors using non-solvent induced phase separation (NIPS) method. Difference in non-solvent content in the ternary solution affords sensing films with different porosities and specific surface areas.

We have pointed out the importance of porosity and specific surface area on the imperative sensing parameters, i.e. sensitivity and response time, by comparing sensor responses of porous films to their dense counterpart toward different concentrations of organic vapors. Near an order of magnitude improvement was obtained in response time of porous sensors in comparison to their dense counterparts. The maximum response of the sensors also improved several times with introducing porosity to the membrane compared to their dense corresponding samples. Although the vapor concentrations studied here are high, porous thin sensitive layers can be developed to detect vapor concentrations as low as several ppm in real-time.

Finally, it has been shown that porosity approach provides a method for making various polymeric vapor detectors from a limited number of initial chemical feedstocks to be used in sensing array of an electronic nose (e-nose) system.

REFERENCES

1. Lewis, N. S. et al, "Array-Based Vapor Sensing Using Chemically Sensitive Carbon Black-Polymer Resistors", *Chem. Mater.*, Vol. 8 (1996), pp. 2298-2312.
2. Tillman, E. S. et al, "Enhanced Sensitivity to and Classification of Volatile Carboxylic Acids Using Arrays of Linear poly (ethylenimine)-Carbon Black Composite Vapor Detectors", *Anal. Chem.*, Vol. 75 (2003), pp. 1748-1753.
3. Koscho, M. E. et al, "Properties of Vapor Detector Arrays Formed through Plasticization of Carbon Black-Organic Polymer Composites", *Anal. Chem.*, Vol. 74 (2002), pp.1307-1315.
4. Briglin, S. M., Lewis, N. S., "Characterization of the Temporal Response Profile of Carbon Black-Polymer Composite Detectors to Volatile Organic Vapors", *J. Phys. Chem. B*, Vol. 107 (2003), pp. 11031-11042.
5. McDonagh, C. et al, " Characterization of Porosity and Sensor Response Times of Sol–gel-derived Thin Films for Oxygen Sensor Applications", *J. Non-Cryst. Solids*, Vol. 306 (2002), pp. 138–148.
6. Garcı´a Salgado, G. et al, "Porous Silicon Organic Vapor Sensor", *Opt. Mater.*, Vol. 29 (2006), pp. 51–55.
7. Chen, X. et al, "Morphology and Gas-sensitive Properties of Polymer-based Composite Films", *Sens. and Actu. B*, Vol. 66 (2000), pp. 37–39.
8. Quercia, L. et al, "Influence of Filler Dispersion on Thin Film Composites Sensing Properties", *Sens. and Actu. B*, Vol. 109 (2005), pp. 153–158.
9. Bubb, D. M. et al, "Laser-based Processing of Polymer Nanocomposites for Chemical Sensing Applications", *J. App. Phys.*, Vol. 89 (2001), pp. 5739-5746.
10. Matsuyama, H. et al, "Light-Scattering Study on Porous Membrane Formation by Dry-Cast Process", *J. App. Poly. Sci.*, Vol. 86, (2002), pp. 3205–3209.

Modeling The Influence Of H_2O On Metal Oxide Sensor Responses To CO

A.Fort, M.Mugnaini, I.Pasquini, S.Rocchi, V.Vignoli

Department of Information Engineering, University of Siena
Via Roma 56, 53100, Siena,
Corresponding Author: ada@dii.unisi.it, +390577233608

Abstract. It is well know that the relative humidity largely affects the response of MOX gas sensors to the target gases. The influence of water vapor on MOX sensor operation has been deeply studied and many results can be found in the literature. Nevertheless the effect of water was not incorporated in the sensor models presented up to now. In this work the authors propose, on the basis of experimental evidence, a simplified model for SnO_2 sensors, able to account for the water contribution, when the target gas is CO. The authors start from a model already presented and tested for dry gases (CO and O_2), and add the water contribution, assuming that the direct reaction between CO and water can be neglected.

INTRODUCTION

The interaction of water with solid surfaces was deeply studied by many authors, e.g. [1] is a recent survey that diffusely treats also the adsorption of water on metal oxides. Goniakowski et al. in [2] described the adsorption of water on metal oxide surfaces and proposed adsorption and desorption mechanisms. Korotchenkov.et al. in [3] focused on the different adsorption mechanisms of water on semiconductor surfaces as a function of the involved measurement temperature. More specifically, some authors studied the influence of water on SnO_2 or other MOX conductivity sensor. In particular Barsan et al. in [4] and [5] studied the response of MOX sensors to CO in humid environment, proposing some chemical reaction paths. Others, e.g. Sohn et al. in [6] took into account the presence of water in the MOX sensor responses by means of simple correction functions describing the relationship between water concentration and conductance.

In any case, the water influence was not yet taken into account in a model able to predict the dynalmic behavior of the sensor when varying the operating conditions. In this work the authors tried to incorporate the influence of water in a previosly developed simplified model [7,8] based on occupied surface state desities. This model was obtained starting from the chemical-physical description of the sensor behavior, and adopting some simplifying assumptions. The result is a gray-box model, which gave satisfactory results for some commercial SnO_2 sensors, in a temperature range higher than 250 °C and for dry mixtures of N_2, O_2 and CO.

In this work the authors test the possibility of applying the same dynamic model to describe the sensor responses to humid CO and N_2 mixtures, accounting for water in a very simple way, that is assuming both that water adsorption is not the rate dominant kinetics in the considered experimental conditions, and neglecting the direct chemical reactions between adsorbed water and CO. Moreover non competitive CO and water adsorptions are considered. The results obtained in this work are very encouraging. The proposed simple model can be used to predict the main characteristics of the sensor responses when varying the relative humidity.

SENSOR RESPONSE MODELING

The dynamic response of some commercial MOX sensors (Figaro sensors) was studied by means of the model developed for high temperature operations (above 250 °C) and whose validity was tested also in pulsed heating power operations [7,8].

Assuming that the sensing film is a porous layer consisting of large grains connected by small contact areas, the model relates the film conductance to the absolute temperature, T, and to the density of the charged surface states, N_S, according to the following equation:

$$G(T, N_S) = G_0 \exp\left(-\frac{q^2 N_S^2}{2\varepsilon N_d kT}\right) + G_C \qquad (1)$$

In Eq. (1), G is the sensor conductance, N_d is the ionized bulk donor density (that is considered constant

CP1137, *Olfaction and Electronic Nose: Proceedings of the 13th International Symposium,* edited by M. Pardo and G. Sberveglieri

with temperature, (i.e. all donors always ionized), q is the charge of the electron, ε is the film dielectric constant, G_0 is a constant accounting for the sensor geometry and surface characteristics, and G_C is a constant term [9]. Equation (1) descends from the assumption that surface conduction depends on the Schottky barrier height, i.e. on the charge localized at the grain surface, and relates, as stated before, the conductance variation to the changes of temperature, and to the variation of N_S, which also depends on temperature.

The film behavior in presence of dry gas mixtures containing CO is described, assuming the following reactions at the surface[7]:

$$CO + S \underset{k_{-CO}}{\overset{k_{CO}}{\rightleftharpoons}} (CO - S) \quad (CO \quad adsorpion) \tag{2a}$$

$$(CO - S) \underset{k_{-CO^+}}{\overset{k_{CO^+}}{\rightleftharpoons}} (CO^+ - S) + e \quad (CO \quad ionization) \tag{2b}$$

where S indicates the surface adsorbing sites, $(X\text{-}S)$ indicates the adsorbed species X ($X = CO,\ CO^+$), whereas k_X and k_{-X} are the rate reaction constants.

The total charged surface state density N_S, in Eq. (1), is given by the charged species that are formed at the surface from adsorbed gases (in this case CO^+), but also by the density of intrinsic surface states occupied by an electron (defects trapping an electron from the conductance band), hereafter named N_{si}, whose density can vary depending on the temperature. In particular, the following Eq.(3) can be written:

$$N_S = N_{si} - [CO^+ - S] \tag{3}$$

where $[X\text{-}S]$ indicates the density of the adsorbed species X. N_S, obtained by Eq. (3), is used to evaluate G in Eq.(1).

It is considered that only the reaction involving charged species (2b) is relevant for the dynamic behavior. In fact, the adsorption reaction is assumed to be a fast phenomenon, so that, in practice, the adsorption reactions (2a) is always at the equilibrium. Under this hypothesis, and assuming first order kinetics, the dynamics of the surface states N_S as a function of temperature and chemical condition variations can be described by two first order differential equations as follow [7]:

$$\begin{aligned} \frac{dN_{si}}{dt} &= k_i n_s (N_i - N_{si}) - k_{-i} N_{si} \\ \frac{d[CO^+ - S]}{dt} &= k_{CO^+}[CO - S]_0 - k_{-CO^+}[CO^+ - S] n_s \end{aligned} \tag{4}$$

where N_i denotes the density of surface intrinsic states. As described in [7], the model parameters (G_o, G_c and the rate constants k_X) can be estimated from non-linear fitting of experimental data, obtained in different chemical conditions. The model parameters were estimated using the experimental data obtained by the measurement system described in [7], whose principal characteristics will be summarized in the following section.

A very simple way to account for the water vapor contribution is to consider that the kinetics of the chemical reactions involving water are not the rate determining ones. Moreover, being water vapor present in the atmosphere with a very large concentration (thousands of ppms), the usual assumption that the number of free electrons is related only to the semiconductor intrinsic properties (i.e. to the oxygen vacancies density), can not hold any more. Hence it is possible to incorporate the effect of humidity in the developed model by increasing the density of free electrons with increasing relative humidity, assuming a dissociative adsorption, as follows:

$$\begin{aligned} H_2O &\underset{k_{-H_2O}}{\overset{k_{H_2O}}{\rightleftharpoons}} OH^- + H^+ \\ OH^- + S' &\underset{k_{-OH-}}{\overset{k_{OH-}}{\rightleftharpoons}} (OH^- - S') \end{aligned} \tag{5}$$

where OH^- groups are acting as donors, and S' are the adsorption sites.

This is accomplished in the present work by varying the model parameters related to N_d (see the appendix of [7]).

EXPERIMENTAL

In this work, the presented experimental data were obtained by means of the measurement system described in [10]. This system is able to perforrm measurement with an array of 8 sensors and it is provided with a flexible sensor temperature control system, able to individually control the temperature of each sensor with a 5 °C temperature accuracy, for temperatures higher than 250 °C. The temperature profiles for the model calibration and validation were defined in a temperature range higher than 250 °C.

The sensors used in this study are five different commercial sensors: TGS2620, TGS2610 TGS2611, TGS2600 and TGS2442 sensors from Figaro Inc. [11]. For the measurements described in this work the charcoal filters packaged with TGS2442 sensors were removed. The measurement system comprises also a chemical sampling system to set both gas flow, gas humidity and gas temperature. The sensor responses used for parametric estimation were obtained by keeping constant in a measurement chamber both the gas flow (total flow 200 mL/min), and its composition, while forcing sensor thermal transients (temperature transient experiments). The experiments were conducted using three different temperature profiles with a temperature range from 250°C to 400°C. The selected profiles start with a 10 min period at a constant high temperature, this allows to bring the surface to a steady state condition, and hence to obtain repeatable measurement results. To evaluate the effectiveness of this method all the measurements were repeated three or four times. Satisfactory results were found, since a repeatibility (ratio of r.m.s. difference between different repetitions with the average response magnitude) in the 10% range was found.

RESULTS AND DISCUSSION

The sensors were at first tested in an inert gas to estimate the intrinsic states dynamics. In figure 1 the model outputs (normalized density of occupied superficial states *Ns'*, and conductance *G*) predicting the sensor behavior in N_2, are shown together with the experimental results for one sensor (TGS2620). Similar results in terms of model fitting error are found with all the tested sensors.

Once obtained the model for intrinsic semiconductor behavior , the sensor was tested in humid N_2, at different Relative Humidity (RH) values. This allows to estimate the increase of free electron density (N_d) due to adsorbed water. The results obtained by applying the simple modified model are shown in figure 2 (from RH=10% to RH=75%); here the results obtained with dry nitrogen are compared to the experimental and predicted sensor responses in humid nitrogen. From experimental data it was noted that already at 50% RH the sensor is nearly saturated, and responses vary only slightly by further increasing the RH level beyond this value. The results are satisfactory for all the tested sensors, since the model is able to capture the main characteristics of the sensor behavior, as it can be seen from the figure for sensor TGS 2620.

Successively, dry CO and N_2 mixtures were used to estimate the model parameters for CO adsorption and ionization. Some results obtained in this step are shown in figure 3. Comparing the responses to dry CO with those to humid and dry nitrogen, it can be seen that the dynamic behavior, especially at low temperatures, is very different. Finally CO and N_2 humid mixtures were used to validate the proposed simplified model. Results are reported in figure 4. It can be seen that the response to CO is considerably amplified by the presence of water and it increases with the RH level, even when the RH nearly saturates the surface (RH above 50%). For this reason a competition between water and CO adsorption seems not very likely.

From figure 4, it can be noted that there is a significant prediction error at low temperature (especially in the dynamics); this was expected, since the direct interaction between adsorbed water and CO is neglected in this model. Nevertheless, the general behavior of the response is well fitted, proving that the proposed modification of the sensor model can account for the most relevant aspects of the water effect, in the tested experimental conditions.

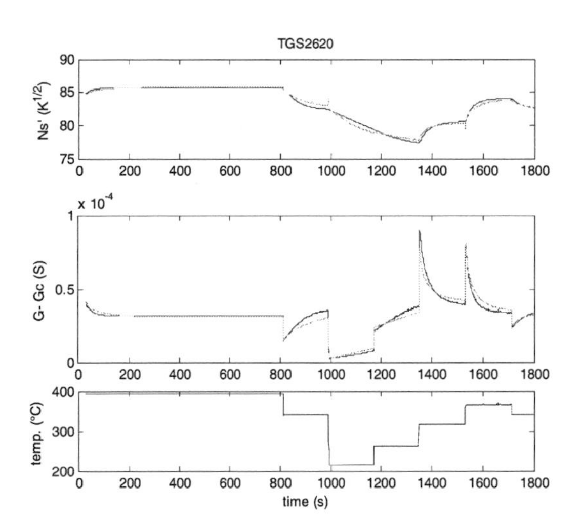

FIGURE 1. Response of a TGS2620 sensor in dry N_2, 200 ml/min flow. Light gray: experimental data, dark gray: model outputs. Upper plots: normalized surface state density. Center plots: Conductance. Lower plot: measured temperature.

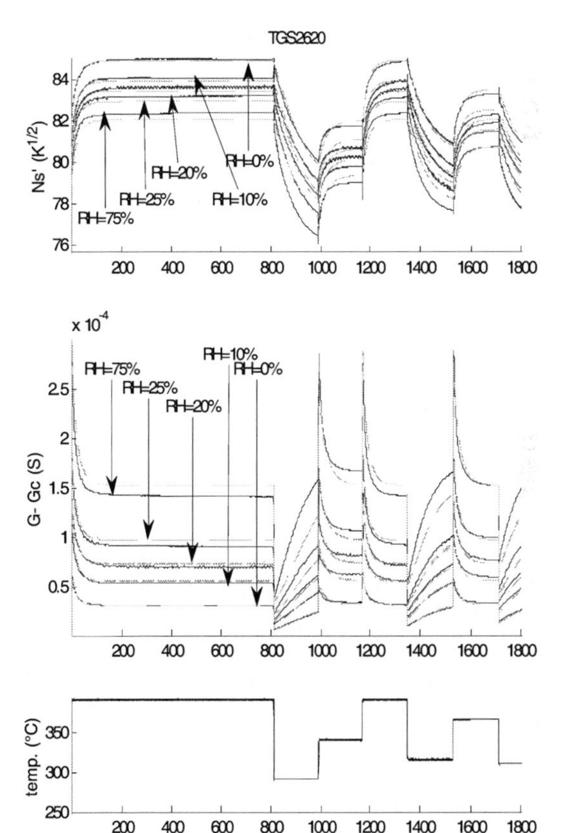

FIGURE 2. Response of a TGS2620 sensor in dry, and humid (RH from 25% to 75%) N_2, 200 ml/min flow, 30°C. Light gray: experimental data, dark gray: model outputs. Upper plots: normalized surface state density. Center plots: Conductance. Lower plot: measured temperature.

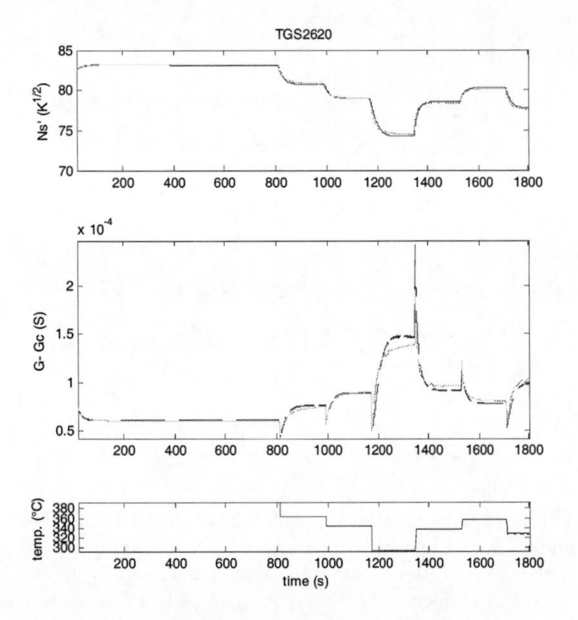

FIGURE 3. Response of a TGS2620 sensor in dry N_2 + 100 ppm CO, 200 ml/min flow. Light gray: experimental data, black: model outputs. Upper plots: normalized surface state density. Center plots: Conductance. Lower plot: measured tempertaure.

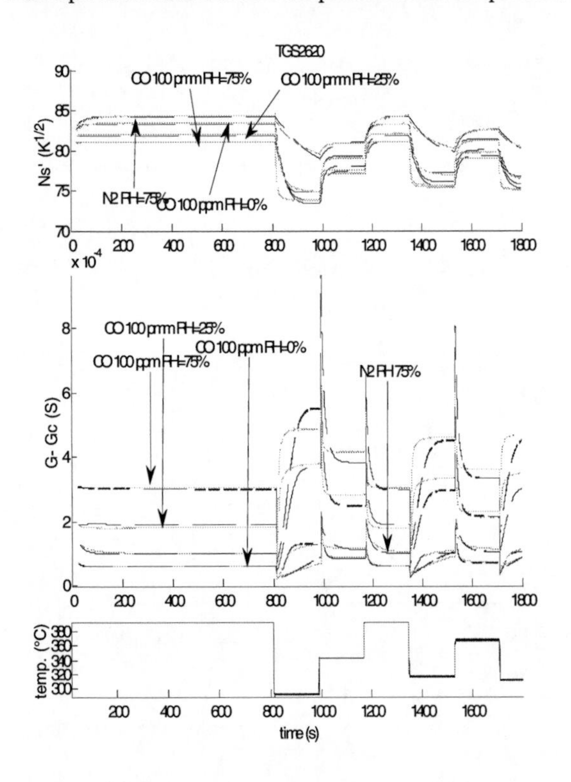

FIGURE 4. Response of a TGS2620 sensor in dry, and humid N_2 + 100 ppm CO (RH=25%, RH=75%), 200 ml/min flow, 30 °C. Light gray: experimental data, black: model outputs. Response of the same sensor in dry N_2, 200 ml/min flow 30 °C. Light gray: experimental data, dark gray: model outputs. Upper plots: normalized surface state density. Center plots: Conductance. Lower plot: measured temperature.

REFERENCES

1. Henderson M.A., "The interaction of water with solid surfaces:fundamentalk aspects revisited", *Surface Science Reports* 46 (2002).
2. Goniakowski J. *et al.*, "Mixed Dissociative and Molecular Adsorption of Water on the Rutile (110) Surface," *Surface Science, Vol.* 350 (1996), pp. 145-158.
3. Korotchenkov G. *et al., "*Electrical behavior of SnO_2 thin films in humid atmosphere," *Sensors and Actuators* B, Vol. 54, (1999), pp. 197–201.
4. Barsan N. *et al.*, "Conduction Model of Metal Oxide Gas Sensor ," J*ournal of Electroceramics,* Vol. 7, (2001), pp.143-167.
5. Barsan N. *et al.*, "CO sensing with SnO_2 thick film sensors: role of oxygen and water vapour," *Thin Solid Films,* Vol. 436, (2003), pp. 17–24.
6. Sohn J.H. *et al.* "Characterisation of humidity dependence of a metal oxide semiconductor sensor array using partial least squares," *Sensors and Actuators* B, Vol. 131, (2007), pp. 230–235.
7. Fort A. *et al*, "Simplified models for SnO_2 sensors during chemical and thermal transients in mixtures of inert, oxidizing and reducing gases," *Sensors and Actuators* B, Vol. 124, (2007), pp. 245–259.
8. Fort A. *et al.* "Surface State Model for Conductance Responses During Thermal-Modulation of SnO_2-Based Thick Film Sensors: Part II—Experimental Verification," *IEEE Trans. On Instrumentation and Measurement*, 55 (6), (2006), pp. 2107-2116.
9. Ding J. et al. "Surface state trapping models for SnO2-based microhotplate sensors," *Sensors and Actuators B: Chemical,* Vol. 77, (2001), pp. 597-613.
10. Burresi A. *et al.*, "Temperature profile investigation of SnO_2 sensors for CO detection enhancement," *IEEE Trans. on Instrumentation and Measurement,* 54, vol. 1, (2005), 79-86.
11. www.figarosensors.com

A High-Performance Nanocomposite Material Based on Functionalized Carbon Nanotube and Polymer for Gas Sensing Applications

L.C. Wang[1], K.T. Tang[2*], C.T. Kuo[3], S.R. Yang[4], Yuh Sung[4], H.L. Hsu[5], J. M. Jehng[5]

[1]Dept. of Materials Science and Engineering, National Chiao Tung University, Hsinchu 300, Taiwan
[2]Dept. of Electrical Engineering, National Tsing Hua University, Hsinchu 300, Taiwan
[3]Institutes of Materials and Systems Engineering, Ming Dao University, Chunghwa 510, Taiwan
[4]Physical Chemistry Section, Chung-Shan Institute of Science & Technology, Touyan 325, Taiwan
[5]Department of Chemical Engineering, National Chung Hsing University, Taichung 402 Taiwan

Abstract. The aim of this study is to develop a novel chemical gas sensing nanocomposite material. The traditional use for carbon nanotube in gas sensing polymer is to increase the composite's conductivity. However, we added functionalized carbon nanotube to fill the free volume of the sensing polymer films and enhance the gas absorption/desorption response time of these nanocomposites. These sensing materials were prepared by mixing functionalized multiwalled carbon nanotubes (MWNTs) and Poly (n, n dimethylamino propylsilsequioxane) SXNR polymer. These new materials were coated on the Surface Acoustic Wave (SAW) device, which is expected to increase its sensitivity in analyzing specific classes of vapors. The proposed materials showed an enhanced sensitivity upon exposure to ethanol and dimethyl methylphosphonate (Dmmp) vapors. Additionally, the performances of our nanocomposite film are much higher than those polymers without functionalized carbon nanotubes.

Keywords: Gas sensing, nanocomposite material, carbon nanotube
PACS: 81.07.De

INTRODUCTION

Carbon nanotubes (CNTs) are promising nanostructure materials in the past decade because of their unique optic, electric, magnetic, and mechanical properties. Many applications of CNTs have been explored, for instance, they have been used as scanning probe microscopy tips and electron field emission sources [1]. Several recent experiments on the preparation and mechanical characterization of CNT-polymer composites have also been conducted [2]. These measurements suggest modest enhancements in strength characteristics of CNT-embedded matrixes when comparing to bare polymer matrixes [3]. The advantages of CNTs are light weight, high aspect ratio, and large surface area, which are good for physical absorptions or chemical bonding. One of the widely investigated matrix materials is polymer in CNT-based nanocomposites. The atomic scale structures of CNT composites, especially the relative structures at the polymer matrix and CNT interface, are expected to explain many unique properties found in the nanocomposites.

Surface Acoustic Wave (SAW) sensors are very attractive due to their high sensitivity caused by confinement of acoustic energy into near-surface region of a piezoelectric substrate. Most of the SAW chemical sensors reported utilizes the ability of the surface wave to undergo perturbations in propagation velocity due to deposition of foreign mass by gas adsorption on the surface of the sensing device. On the other hand, the selectivity of traditional SAW gas sensor devices is very poor; SAW sensors measure the total mass of the adsorbed gas molecules irrespectively. Selectivity can be somewhat improved by using appropriate sensing films.

In this paper, we proposed a new composite material as SAW sensor sensing film. A thin film of gas-sensitive MWNT-COOH/SXNR polymer composite has been coated upon the surface of the SAW sensor. The meaningful vapor sensing results, at room temperature, of MWNT-COOH/SXNR polymer

CP1137, *Olfaction and Electronic Nose: Proceedings of the 13th International Symposium,* edited by M. Pardo and G. Sberveglieri

based nanocomposites will be extensively reported and discussed.

EXPERIMENTAL AND METHODS

Growth and Functionalization of CNTs

In this paper, we used the thermal chemical vapor deposition to prepare CNTs. The as-grown CNTs were soaked in the strong oxidants HNO_3 (300 ml) and slowly added H_2O_2 (200ml) to oxidize CNTs over 3 hours. The precipitate filtered was dried at 70℃ to form the hydrophilic CNTs (MWNT-COOH). The black solid that has COOH groups attached on MWNTs, could be obtained after completing the treatment procedures. The detail morphology and results of the as-grown and functionalized CNTs have been discussed in the earlier report [4].

Characterizations

Surface morphology of MWNTs was further carried out by field-emission scanning electron microscope (FESEM, JEOL JSM-4700) operating at 5–15 kV for a better resolution. In order to analyze and determine the functional groups of the modified MWNTs, Fourier transfer infrared (FT-IR) spectra were investigated with a PerkinElmer Paragon 500 spectrometer using a KBr pellet with scanning 50 times from 400^{-1} to 4000 cm^{-1} at room temperature.

Coating of CNTs-polymer composite film on the SAW sensor

The polymer SXNR (poly (n, n dimethylamino propylsisesquioxane)) is often used as chemical sensing polymer. We dissolved 0.015g of SXNR in 50 ml of ethanol solvent then added in 0.015g of functionalized MWNT-COOH and put the solution under ultrasonication for 3 hrs to assist the dispersion of MWNTs-COOH in the polymer solution. Before aggregation of CNTs occurred, the air spray coating method [5] was used to prepare the MWNTs-COOH/SXNR composite film on the surface of SAW sensor device. The spray coating equipment is shown in the fig 1. Nitrogen was used as the carrier gas to deliver the solution through the nebulizer, which formed aerosol and coated on the SAW device. After the solvent (ethanol) was evaporated, the residual on the device was the composite sensing film. After the coating process, spectrum analyzer was used to record the center frequency of SAW device, which then determined the thickness of the film. The frequency shift by spray coating in our experiment is about 200 KHz and the thickness of the film was calculated to be approximately 40 nm.

Gas sensing experiment setup

SAW one-port resonators operating at 156.6 MHz are used as passive acoustic elements. The resonator was mounted on a 6-pin round TO-8 package case, with electrical connections made by Au wire bonding. The resonators fabricated onto ST-cut quartz piezoelectric substrate have 0.3 μm thick Al patterned interdigital transducers (IDTs). The resonance SAW frequency shifts with the mass of adsorbed vapor molecules onto sensing layer of MWNTs-COOH/SXNR polymer. Figure 2 shows the experimental setup used to characterize the properties of the MWNTs-COOH/polymer composite film in SAW oscillating devices for sensing volatile organic compounds (VOCs). The standard gas generator (AID, model 360) heats organic liquid to produce vapor and reaches the stable temperature. Furthermore, the organic vapor concentrations were determined from the dimension of diffusion tubes. The dry air with a flow rate of 100 sccm controlled by MFC (mass flow controller) carried the organic vapor into the chamber, and reached the surface of the SAW gas sensor to result in a frequency shift. After completing the experiment, dry air was again used to purge the chamber. The SAW frequency shifts with the mass of adsorbed vapor molecules onto sensing layer of MWNT-COOH/SXNR composite. Data were collected every 1 s and stored in a PC by GPIB bus.

RESULTS AND DISCUSSION

Characteristics of the as-grown MWNT and functionalize MWNTs

Figure 3 is the FT/IR spectra of the as-grown MWNTs and functionalized MWNTs. Spectrum A and B show four peaks at 1400、1580、1723 and 3400cm^{-1} indicating the functionalized MWNTs. Due to the oxidation process; the as-grown CNTs have been functionalized with hydrophilic group. They were terminated with carboxylic acid groups (–COOH, IR frequency 1723cm^{-1}) or hydroxyl group (-OH, IR frequency 3400cm^{-1}).

Evaluation of the MWNT-COOH/SXNR Nanocomposite Film

The major problem of CNTs is their bad dispersion in polymers, and this is what was thought until now. The SEM image (Fig.4) disclosed that the surface of the as-grown MWNTs mixed with SXNR polymer film and the functionalize-MWNTs-SXNR composite film. The polymer films were microporous and rough. We also found that the as-grown MWNTs were difficult to disperse in the solvent. Figure 4a-b show the as-grown MWNTs would aggregate together in the polymer film even after stirred all day long. Figure 4c-d show the well dispersed MWNT-COOH/SXNR polymer composite solution and the uniform distribution of functionalized CNTs on the surface of polymer composite film even after the air–spray coating process.

Sensor Characteristics and Measurement Options

In this section, we demonstrate the efficiency of fluctuation-enhanced SAW sensing film. It has been reported that desorption of gas molecules from a sensing polymer membrane was a relatively slow process because of the van del weal force or strong hydrogen bond between the gas molecular and sensing polymer films [6]. This is also the main shortcoming of the traditional SAW gas sensor technology which the gas molecule exposed to the surface of the composite film, the binding time of the polymer and the reaction gas was excessively long.

In our study, we improved this phenomenon in our gas sensing results with the novel MWNT-COOH/SXNR composite films. As shown in Fig. 5(a) and (b), the time of ethanol and DMMP gas adsorption and desorption for SXNR films were extremely slower than that for MWNT-COOH/SXNR composite films. The process of adsorption and desorption of ethanol gas molecules on MWNT-COOH/SXNR composite films was found to be reversible and fast. It was also noted that MWNT-COOH/SXNR composite films reached thermodynamic equilibrium faster than SXNR films. The response time and recovery time of the MWNT-COOH/SXNR composite films to DMMP gas was also more rapidly than that of the SXNR films. In addition, the MWNT-COOH/SXNR composite films also revealed better ethanol gas sensitivity (stable frequency shift, Δf) about 5 KHz and the SXNR films were only 3 KHz. These sensors operate at room temperature, exhibit sensitivity to DMMP concentrations as low as 200 ppm, and have response and recovery times as low as 10 sec. This is mainly attributed to the occupation of some CNT stayed in the molecular holes of the polymer or the so-called free volume when the reaction gas is in contact with the polymer composite film. The sensing gas can't diffuse into the inside layer of polymer film and reduced the diffusion pathway. As the result, the equilibrium for adsorption and desorption would not be delay. Therefore, these new materials increase its sensitivity and the enhanced sensitivity upon exposure to ethanol and dimethyl methylphosphonate (Dmmp) vapors. Additionally, the performances of our nanocomposite film are much higher than those polymers without functionalized carbon nanotubes.

CONCLUSIONS

We have demonstrated the use of a novel gas-sensing material based on MWNT-COOH/SXNR nanocomposite film and investigated the influence of various concentrations of CNT within the film on the performance for the detection of ethanol and DMMP. With these results, we successfully reduced the response time of SAW sensor devices and increased the sensitivity as well as the expansion detection limit. The main improvement is to modify the nanomaterial by chemical reaction to increase the compatibility between the two components; therefore, the functionalized CNTs have the aerosol stability and would be disperse easily in the polymer material. These would enhance the performance and stability of the SAW sensor. The development of SAW sensors coated with MWNT-COOH/SXNR composite film for detection of DMMP and ethanol provides the best frequency shift or sensitivity.

ACKNOWLEDGMENTS

The authors would like to acknowledge the support of the National Science Council of Taiwan, under Contract No. NSC 96-2221-E-451-012 and NSC 97-2220-E-007-036. We also thank the Ministry of Economic Affairs (MOEA) and CSIST for the gas sensing instrument support of this research.

REFERENCES

1. de Heer, W. A., Chatelain, A., Ugarte, D.,"A Carbon Nanotube Field- Emission Electron Source ," *Science*, Vol. 270. No. 5239 (1995), pp. 1179 – 1180.
2. Vigolo B., Pénicaud A., Coulon C., Sauder C., Pailler R., Journet C., Bernier P., and Poulin P., "Macroscopic Fibers and Ribbons of Oriented Carbon Nanotubes," *Science*, Vol. 290 (2000), pp. 1331.
3. Andrews R., Jacques D., Rao A., Rantell T., and Derbyshire F., "Nanotube composite carbon fibers," *Applied Physics Letters*, Vol. 75, (1999), pp. 1329.

4. Hsu H.L., Jehng J.M., Sung Yuh, Wang L.C., Yang S.R., "The synthesis, characterization of oxidized multi-walled carbon nanotubes, and application to surface acoustic wave quartz crystal gas sensor," *Mater Chem Phys* Vol. 109, No. 1 (2008), pp. 148-155.
5. McGill R. A., Abraham M. H., Grate J. W., Am J., "Choosing polymer-coatings for chemical sensors," *Chemtech* Vol. 24, No. 9, (1994) pp. 27-37.
6. Sun L.X., Okada T., "Simultaneous determination of the concentration of methanol and relative humidity based on a single Nafion(Ag)-coated quartz crystal microbalance," *Analytica Chimica Acta*, Vol. 421, No. 1 (2000), pp. 83-92.

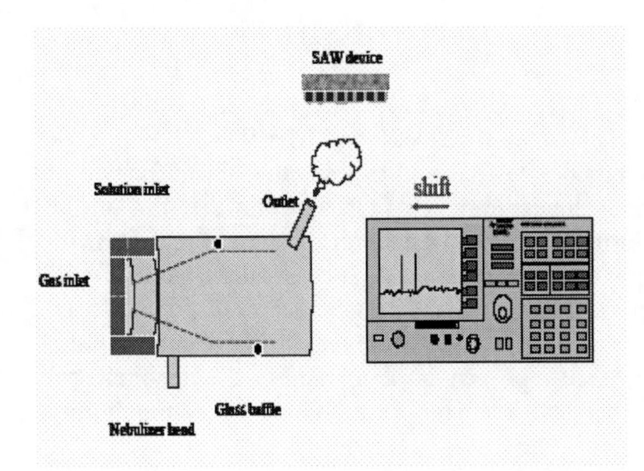

FIGURE 1. The setup diagram of the sensing film coating equipment

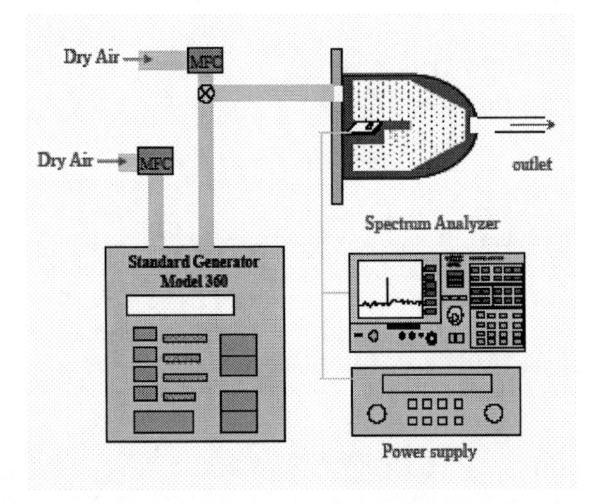

FIGURE 2. The setup diagram of the gas sensing system

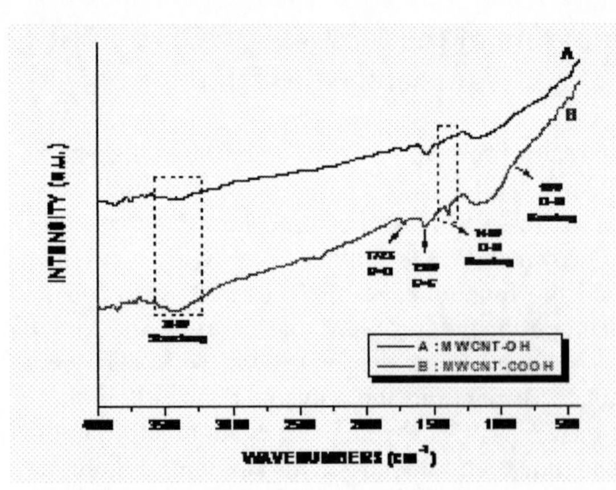

FIGURE 3. The FT/IR spectrum of the as-grown MWNTs and functionalizes MWNTs

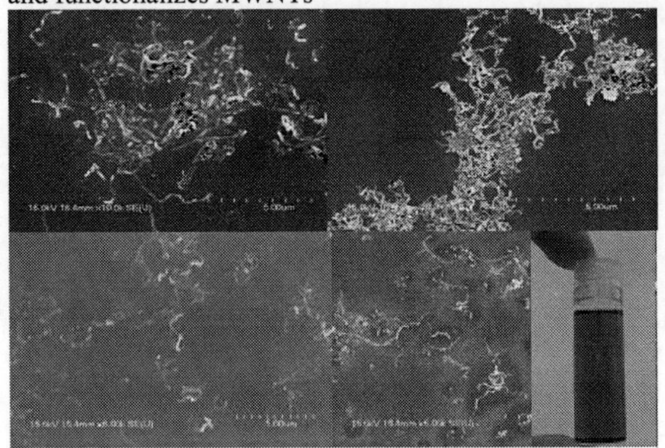

FIGURE 4. The SEM images of (a) (b) the surface of the SXNR polymer film and (c) (d) the functionalize-MWNTs-SXNR composite film

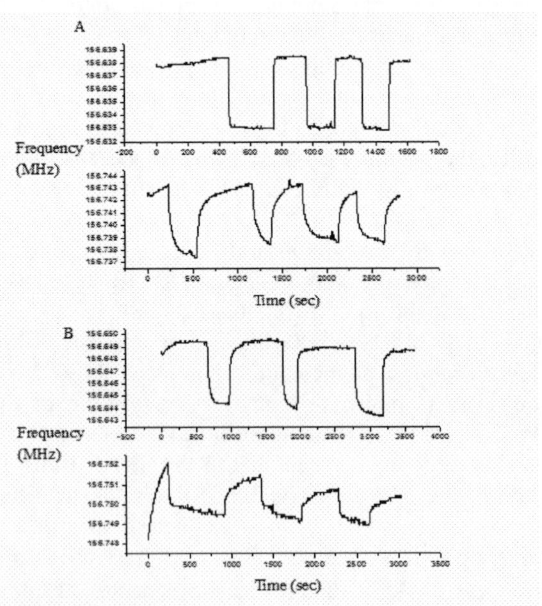

FIGURE 5. The frequency shift diagram of the functionalize-MWNTs-SXNR composite film and the SXNR polymer film to (a) ethanol gas (0.6%) (b) DMMP gas (200ppm)

Array of nanofibrous polyaniline-based sensors with different chemo-structural assembling

A. Macagnano[a], E. Zampetti[a], S. Pantalei[a], M. Italia[b], C. Spinella[b], A. Bearzotti[a]

a) IMM-CNR, Via Fosso del Cavaliere 100, 00133 Rome, ITALY
b) IMM-CNR, Via VIII Strada 5, 95121 Catania, ITALY
antonella.macagnano@artov.imm.cnr.it

Abstract. Among various attempts of mimicking olfactory system the present work focuses on the sensorial surface of mammalian olfactory cells. The aim of this research is to develop, in a single step, synthetic fibres mimicking the long, no motile cilia of the olfactory cells. Electrospun conductive nanofibrous layers of doped polyaniline, suitably blended to a group of polymers capable to carry the jet, have been easily assembled on chemoresistors (interdigital microelectrodes, IDEs). Such highly sensitive sensors have been able to detect gases in traces of relevant medical significance, mainly because of high density of sorption sites of the sensor. Changing the carriers, the range of sensitivity to gases could be tuned. The key role of the selected carrier polymers (polyethylenoxide of different molecular weights, PEO, polyvinilpyrrolidone, PVP, and polystyrene, PS) to the final sensor features has been highlighted by morphology analysis (texture analysis) and electrical measurements before and after gas interaction.

Keywords: nanofibres, sensor array, polyaniline blends, electrospinning
PACS: 07.07.Df, 81.16.-c

INTRODUCTION

The natural sensing strategy uses a massive increase of the interacting surface to increase the surrounding chemicals (guest) detection. Thus, structures as cellular cilia and microvilli are two clear examples of widening sorption area in biological approaches. In human smell sense, for example, the olfactory region of about 2.5 square centimeters area, involved to interact with odorant-gaseous molecules, consists of cilia projecting down out of the olfactory epithelium into a layer of mucus [1]. These cilia notably increase the surface area of the cell, and embedded within their plasma membrane are the sensory receptors and ion channels associated with chemical detection and cell excitation. The olfactory mechanism is in fact complex enough, and several factors are involved in detection the odorant molecules, such as the topographic layout of natural sensors through the sensory area, the diffusion time, the odorant binding proteins role [2]. However, in this work, an attempt to mimic a task of the design of this natural arrangement has been carried out in order to try to improve the features of an artificial sensor for gas sensing. In the last few years, increasing manipulation and production of nano-structures (nanowires, nanobeads, nanotubes) have been allowed by huge advances in nanotechnology. Electrospinning [3] is a simple, no expensive and handy process by which polymer nanofibres with diameters ranging from a few nanometres up to several micrometers (usually between 50 and 500 nm) can be produced and deposited at once using an electrostatically driven jet of polymer solution (or polymer melt). This is an interesting technology to produce materials with a huge increasing of sorption sites in only one step. Little modifications of the starting solution are sufficient to modify remarkably the fiber-tissue structure and then the sensing properties. In this work olfactory cilia-like structures have been fabricated adopting the electrospinning deposition and then investigated as gas sensors. Polyaniline (PANi) is one of the most appealing conductive polymers [4] because of the low cost of its synthesis, its conductive features when doped, and its thermal and environmental stability. Its sensing performances are well known in literature [5,6], overall to ammonia detection. Therefore, since high potentialities of sensing materials can be achieved employing nanostructured elements, different nanofibrous layers of PANi and polymer carriers (PVP, PS, PEO of different molecular weights) have been investigated as possible sensors of one gas having medical significance and hard to detect at comparable concentrations [7]. Many gases and

CP1137, *Olfaction and Electronic Nose: Proceedings of the 13th International Symposium*, edited by M. Pardo and G. Sberveglieri

volatile organic and inorganic chemical species, in fact, produced by metabolic processes within the body, are released in exhaled breath (often in only trace amounts). Medical conditions related to such metabolic exhaled breath constituents (i.e. NO, NO_2, NH_3, etc.) include tissue inflammation (asthma), immune responses (cancer cells or bacteria), metabolic problems (diabetes), digestive processes, liver, kidney and heart problems [8]. Suitable micro-chemoresistors have been designed and shaped (IDEs) to transduce gas sensing responses. Parameters like shape, size and length of the fibres, form and dimension of beads woven within the fibres, porosity, roughness and alignment in layer texture, mainly depend on several physical parameters that can be changed during the growing process (temperature and relative humidity, electrostatic forces and potentials, time of flight). Furthermore, physical properties of the polymer blend, such as viscosity, conductivity and density, imply significant changes in morphology. Thus, the selected carrier polymers are not inert scaffolds for PANi, but they are responsible of the features of the sensor as a whole (electrical parameters, selectivity and sensitivity ranges). In sensor design, in fact, the film texture is the key aspect, considering the role of it as a filter and as a gathering material at once. Here, the layer arrangements have been investigated by both scanning electron microscopy and a simple method of texture analysis [9].

EXPERIMENTAL AND METHODS

Chemical layers

Polyaniline emeraldine base (PANi-EB) (10,000), 10-camphorsulfonic acid (CSA), polyethileneoxide (PEO). (200,000 -900,000), polyvinilpyrrolidone (PVP) (40,000), polystyrene (PS) (192,000), chloroform ($CHCl_3$), ethanol (EtOH) were purchased from Sigma-Aldrich and used without further purification. Doped polyaniline (as literature) [10] was blended with PEO 200 (1.7% wt/wt) and PEO 900 (0.2% wt/wt) and PS (5% wt/wt) in chloroform, and PVP (4% wt/wt) in a 3:7 solution of EtOH and $CHCl_3$ in order to achieve suitable viscosities for electrospinning. Solutions were stirred at room temperature for 6 hours at least.

Electrospinning set-up

The electrospinning set-up [11] consisted of a glass syringe with a 1 cm long stainless steel needle (500 μm inner diameter) with a blunt tip connected to a high voltage generator and containing the polymer solutions.

The interdigital microelectrode transducer (IDE) was placed on the ground collector, consisting of a conductive rotating cylinder located at a distance of 8 cm from the needle's tip. The syringe was tilted at approximately 10° from horizontal to keep a hemispherical droplet at the tip. The generator consisted of a high voltage oscillator (100 V) driving high voltage (ranging from 1-50 kV) and a high power AC-DC (alternative current - direct current) converter. This feature allows producing polymer fibres in AC and DC mode. In the reported experiments 18 kV of electrostatic DC voltage was applied to the tip.

Electrical characterization

The IDE transducer was a photolithographically defined interdigital gold electrode implemented on oxidized silicon wafer. Electrical parameters were measured by Agilent 34401A Multimeter and Keitheley 6517A Electrometer. Dynamic measurements were carried out at room temperature using a 4-channel MKS 247 mass-flow controller, where NH_3, NO and NO_2 of known concentration (Rivoira, Italy) were diluted with N_2 (gas carrier).

Surface Analysis

Morphology of the nanofibres was analyzed by a FEG-Electron Microscope (SUPRA 35 High Resolution FESEM (Zeiss). An additional survey was got by a simple method of Texture Analysis (MATLAB 6.1) to define the quality of the whole layer (homogeneity).

RESULTS AND DISCUSSION

Morphology of the electrospinning nanofibrous layers of different blends of PANi were displayed by FEG-SEM images. Since layers resulted conductive enough, no metallic coating were used to carry out SEM survey. Changing carrier polymer, as expected, modified remarkably the tipology of layer scaffold. PANi/PEO 200 produced, in fact, fibers of different shape and size (diameter ranging between 100 nm up to 1 μm), with similar high porosity and roughness. Figure 1 depicts a rough fiber of PANi/PEO 200. On the contrary, PANi/PS (Fig.2) and PANi/PVP showed more homogeneous fibres with reduced thickness (down to 40 nm in PANi/PS), smoother and more aligned. The PANi/PEO 900 layer was characterized by great irregularity and porosity, having yet comparable fibre in shape and dimension (200-300 nm). However they appeared interwoven among them to form branched structures.

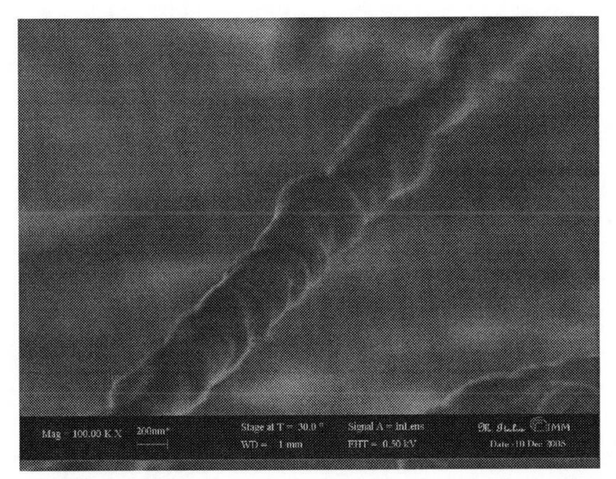

FIGURE 1. Picture of a PANi/PEO fiber, tilted at 60° and magnified 100000 X.

The whole layer coating the electrode has been investigated. At present, electrospinning deposition technique, infact, is the best solution to make continuous randomly or aligned nanofibers in one step, though the morphology of the produced nanofibres is the balance among a series of environmental (humidity, temperature) and physico-chemical parameters (viscosity, conductivity, density, polymer length and solubility).

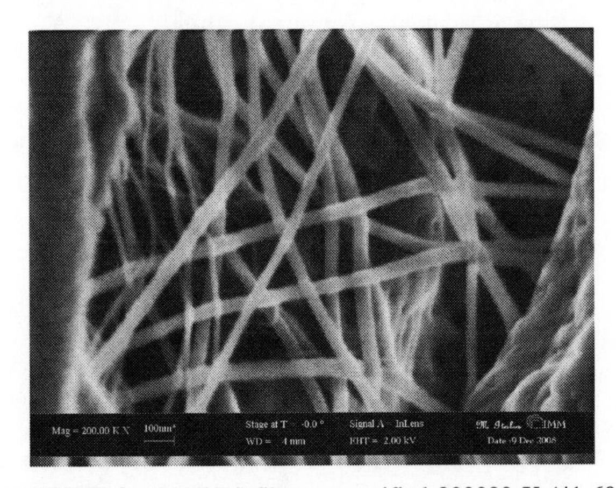

FIGURE 2. PANi/PS fibres, magnified 200000 X (41-60 nm thick)

The fiber yield (FY), defined as the ratio between the surface and number of non fibrous structures (e.g. beads, grains), strongly depends on all cited parameters (PP: Process Parameters). In order to study the importance of FY on sensor performances (e.g. sensibility, selectivity) the layer morphology has been investigated by a simple texture analysis which calculated the FY, dimension and density of the non fibrous structures in the resulting polymer textures (granularity). At fixed PP values, completely dissimilar fibre yields have been measured by the texture survey. The grain-beads distribution curves suggest the presence of regular grains evenly dispersed in the PANi/PS fibrous layer.

Different kinds of grain size distribution were reported in both PANi/PEO textures. Beads and grains seemed to be remarkably reduced in PANi/PVP. To test the goodness of the resulting chemoresistors for sensor applications, I-V curves were carried out at first. The initial conductance of the fibrous layers is depending on both the quantity and quality of the "texture" covering the electrodes and the adhesion to the metal electrode. Thus, electrical resistance of the sensor elements is related to individual fibre resistance (due to dimension and shape), fibre density (number of fibres per unit surface area) and the electrode coverage: it was ranging from 10^3 to 10^6 Ohm.

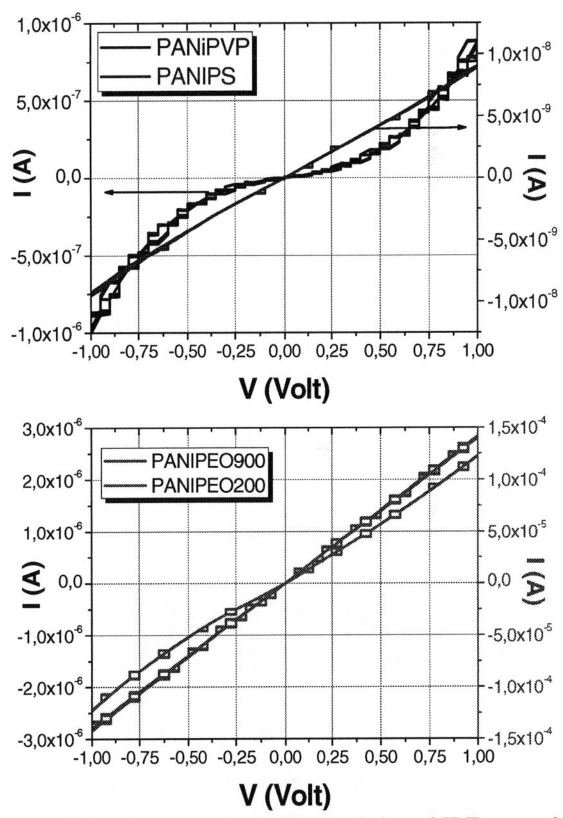

FIGURE 3. Current voltage characteristics of IDEs coated of nanofibrous layers in dry conditions (N_2 flow)

PANi/PEO (200-900) and PANi/PS showed linear behaviour along the selected voltage range (-1V to +1V).

This indicates a constant resistance value not depending on the current flowing into the device. PANi/PVP curve is slightly different, indicating a resistance value depending on the current flowing level. This effect could be due to the fibres modifications induced in PVP by Joule effect. Nevertheless to avoid instability during measurements, the operating voltage chosen for testing PANi/PVP was 0.2 V where I-V curve was more linear (Fig.3).

The change in any parts of the layer should cause a consequential change of the overall resistance of the device. The transient responses of PANi/PEO 200 to NH_3 show desirable performances as fast responses and recovery times to very low concentration (ppb). Same features are recorded for NO and NO_2, but with different sensitivities. Response curves of sensors depict the key effect of the blended polymer to PANi chemical interaction. Thus, if PEO (Langmuir-like curve shape) improves the sensitivity of the PANi to ammonia detection (resistance increasing), enhancing the sorption sites (capable to detect up to 15 ppb), PS and PVP get different ranges of sensitivity with different behaviour at very low concentration. These effects were supposed to be due to both different distributions of interacting PANi-sites throughin fibres and to a filtering effect of the polymer. However a divergent dynamic range of ammonia detection is observed in Fig.4.

When highly sensitive PANi/PEO were saturated of ammonia (about 6 ppm), PANi/PVP and PANi/PS had a linear behaviour letting suppose different sensing activities. Probably the lower performances of PANi/PEO 900 are due to an evident reduced coverage of the interdigital fingers by these kinds of fibres.

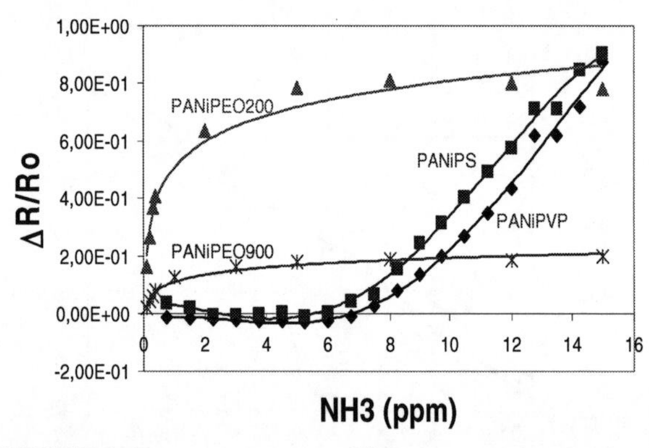

FIGURE 4. Response curves to NH3 concentration ranging from 15 ppb up to 15 ppm.

CONCLUSIONS

This work was a preliminary attempt to mimic structures like olfactory cilia possessing high surface to volume ratio. Since electrospinning deposition technique seemed to be a simple, no expensive and handy process by which polymer nanofibers could be produced, polyaniline fibres, blended with different polymers, were prepared and tested for gas sensing. The coverage texture resulted highly dependent on the kind of selected carrier polymer. This result was probably due to a series of parameters occurring during the fiber flight time, as well as polymer chemo-physical properties. However factors as concentration, solubility of PANi in several polymer solutions (such as the affinity to the carrier polymers), share in the fibrous layers construction. Thus morphologies remarkably diverse were obtained, ranging from grainy (PANi/PS) and roughness (PANi/PEO) fibres up to smoother and more homogeneous layers (PANi/PVP). Very good coverage of the electrodes, excepting PANi/PEO 900, as well as the adhesion, were yelded, producing a huge improvement to PANi sensing features. These chemosensors indicated the possibility to detect up to few ppb of NH_3 . However diverse range of sensitivities were got for tested different nanofibrous films, due to fundamental contribution of the carrier polymers to various tasks (shape texture, swelling effects, filtering and collecting). Therefore, it let suppose the chance to plan wider dynamic range of sensing, based on the same conductive polymer (i.e. PANi) having yet tunable array of sensitivity for real applications, getting more information about the whole system at once.

Major control of the parameters occurring during the deposition technique and the actual physico-chemical contribution of the polymer scaffolds to gas interaction are under deeper investigation.

REFERENCES

1. Nef, P., *News Physiol. Sci.,* **13**, (1998), pp. 1-5.
2. Lobel D., et al., *Chem. Senses* **21**, (2002), pp. 39-44.
3. Ramakrishna, S., et al., *An Introduction to Electrospinning and Nanofibres*, World Scientific Publishing Co. Pte. Ltd. (2005)
4. Persaud, K., *Materials Today*, **8** (2005), 38-44
5. Timmer, B., et al., *Sens. Actuators B: Chem.* **107** (2005), pp. 666-677.
6. Hosseini, S. H., et al., *Polym. Adv. Technol.*, **12**, (2001), pp. 482-493
7. Haynes, A. S., et al, *IEEE Sensors Journal*, **8**, (2008), pp. 701-705.
8. Turner, C., et al., *Rapid. Commun. Mass Spectrom.* **2**, (2008), pp. 526-532
9. Tuceryan, M., et al., "Texture Analysis" in: *The Handbook of Pattern Recognition and Computer Vision* by Chen et al., (1998), pp. 207-248, World Scientific Publishing Co.
10. Diaz-de-Leon, M., *Procee. The National Conference on Undergraduated Research (NCUR),* Lexington, Kentucky, March 15-17, 2001.
11. Macagnano, A., et al., *Adv. Sci. Tech.*, **58** (2008), pp.91-96

Comparison of the Gas Sensing Properties of Thin Film SnO_2 Produced by RGTO and Pore Wetting Technique

D.F.Rodríguez [a], B. Lerner [a], M.S.Perez [a], F.A.Ibañez [a], A.G.B.Leyva [b], J.A. Bonaparte [a], C.A.Rinaldi [a], A. Boselli [a], and A.Lamagna [a].

[a]*Comisión Nacional de Energía Atómica, Av. Gral. Paz 1499 (1650) Bs. As., Argentina*
[b]*Escuela de Ciencia y Tecnología-UNSAM. M. Irigoyen 3100 (1650) Bs.As., Argentina*
drodrig@cnea.gov.ar

Abstract. Two different techniques were used for manufacturing thin films of SnO_2: RGTO [1] and pore wetting [2-4]. From the first one, a microstructure formed by nanograins obtained and from the second, a microstructure consisting of nanotubes. These would permit understand the relationship between the microstructure and the electrical response of the thin film. The film of SnO_2 nanotubes shows good sensitivity to volatile gases and an unusual sensitivity even at room temperature.

Keywords: Sensor, SnO_2, Pore Wetting.
PACS: 85.35.-p

INTRODUCTION

Sensors based on SnO_2 are widely used to detect very low concentration of different gases. There are a lot of techniques to obtain SnO_2 films such as physical vapor deposition, magnetron sputtering [5], thermal evaporation [6], spray pyrolisis [7], laser pulses deposition [8] and chemical vapor deposition [9] .

A new generation of SnO_2 nanostructures have been produced recently, such as nanowires, nanobelts, nanorods, nanotubes and nanowhiskers [10, 11]. One of the main features of these nanostructures is the surface/volume relationship that makes them attractive to use as sensitive film gas sensors. The challenge now is to achieve a manufacturing process of nanostructures compatible with micromachining processes.

Throughout this process we used two different manufacturing techniques to develop thin films of SnO_2: RGTO and pore wetting. From the first one, microstructures formed by nanograins were obtained and from the second microstructures consisting of nanotubes were obtained. A comparison of the electrical response of the sensitive films obtained from different volatile gases was made.

EXPERIMENTAL AND METHODS

SnO_2 Thin films were obtained by Rheotaxial Growth and Thermal Oxidation (RGTO) [1] and pore wetting [2-4].

The RGTO technique is carried out in two steps. In first step Sn was deposited over silicon nitride for physical vapor deposition. In the second step, after the deposition of the Sn film, the samples underwent a thermal treatment at 873 K in wet air. The surface morphology of the films was examined with X-Ray Diffraction, Scanning Electron Microscope and Atomic Force Microscopy.

Figure 1 shows a typical morphology of the film where a microstructure formed by nanograins can be observed. The film shows a high specific surface area, which is a prerequisite for gas sensing applications.

CP1137, *Olfaction and Electronic Nose: Proceedings of the 13th International Symposium*, edited by M. Pardo and G. Sberveglieri

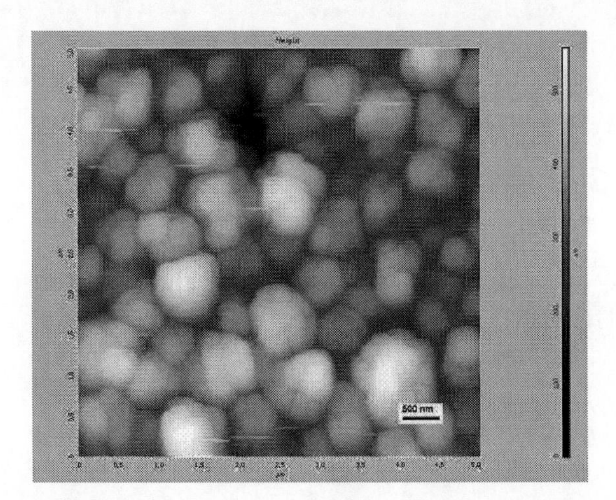

FIGURE 1. AFM micrograph obtained by RGTO techniques, spherical structure typical of SnO_2 is evident.

In the pore wetting technique we used $SnCl_4$ (analytical grade) and porous polycarbonate as template. The polycarbonate pores were filled using a syringe filtration device. The porous polycarbonate film was fixed to the silicon nitride surface and a thermal treatment at 873 K in air was performed. The resulting materials were nanotubes sintered onto the surface; no contamination was observed due to the total decomposition of the porous polycarbonate film and the electrical contacts were pressed during this treatment (see Figure 2).

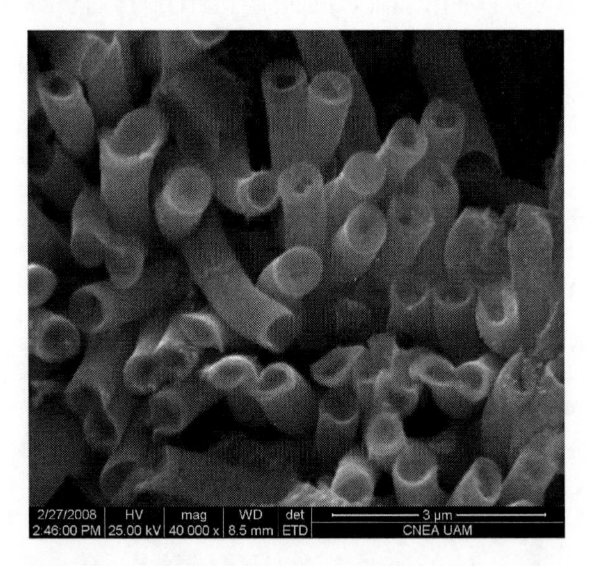

FIGURE 2. SEM micrograph obtained by pore wetting techniques.

For electrical measurement, sensors were located in a chamber (volume 3500 cc) where gases flow at a total and constant rate of 300 sccm. Changes in sensor electrical resistance were measured by voltmeter (Keithley 2000) and impedance analyzer (Solartron 1260A). Electrical measurements were performed at different temperatures, in presence of air and vapor isopropyl alcohol.

RESULTS

Electrical measurement of resistanceach

It is shown a typical electrical response to isopropyl alcohol vapor from the films of SnO_2 obtained by RGTO and pore wetting at temperatures higher than room temperature (Figure 3) and at room temperature (Fig. 4).

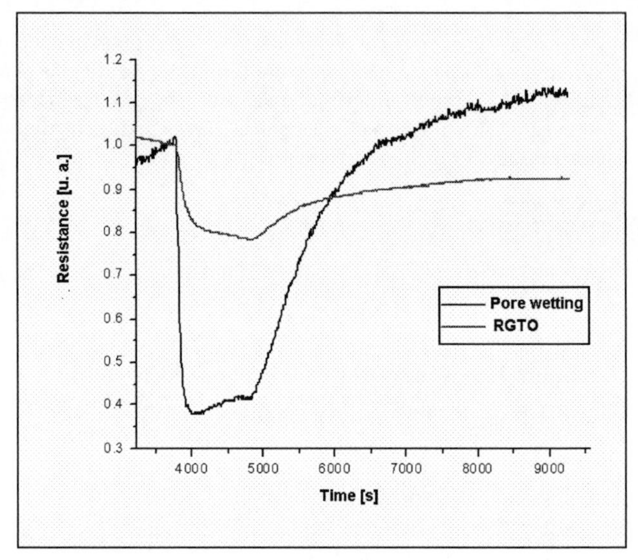

FIGURE 3. Comparison of responses from SnO_2 film obtained by RGTO and pore wetting at 482 K.

RGTO and pore wetting at temperature greater than the room temperature (Figure 3) and at room temperature (Fig. 4).

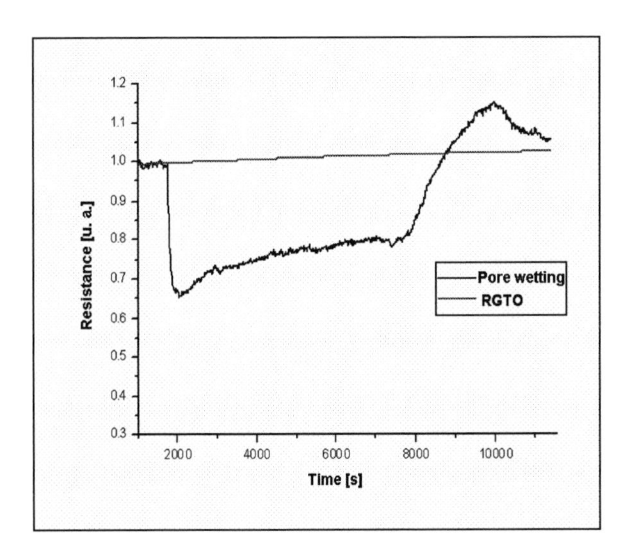

FIGURE 4. Comparison of responses from SnO2 film obtained by RGTO and pore wetting at room temperature.

Figure 3 shows that the sensitivity of the vapor isopropyl alcohol nanotubes film is greater than the RGTO film.

Figure 4 shows that the nanotubes film is sensitive to isopropylic gas even at room temperature.

Electrical measurement of resistanceach resistanceach

The frequency dependent properties of a material are generally described by complex impedance plots, where the impedance Z is given by:

$Z = Z' - i Z''$.

Z' and Z'' being the real and imaginary parts of the impedance, respectively.

Figures 5 and 6 show measurements of complex impedance in air and isopropyl alcohol vapor from both films at room temperature.

From the complex impedance spectra in Fig 5, we can say that a semi-circular arc corresponds to a distributed R-C element and the figures are practically the same for both air and isopropyl alcohol vapor. The equivalent circuit for complex impedance plots in Fig. 5 can be explained by a resistance (Rgb) and capacitance (Cgb) in parallel where Rgb is the grain boundary resistance and Cgb is the grain boundary capacitance. The optimum values for Rgb and Cgb are ~ 54 kΩ and 50 pF respectively.

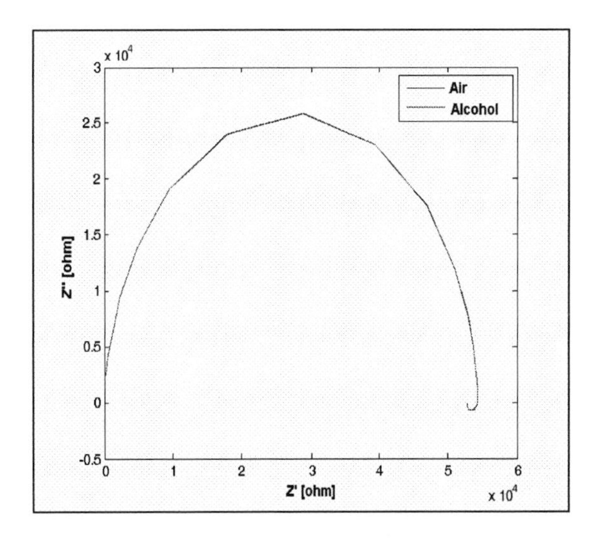

FIGURE 5. Complex impedance in air and isopropyl alcohol vapor from the RGTO film.

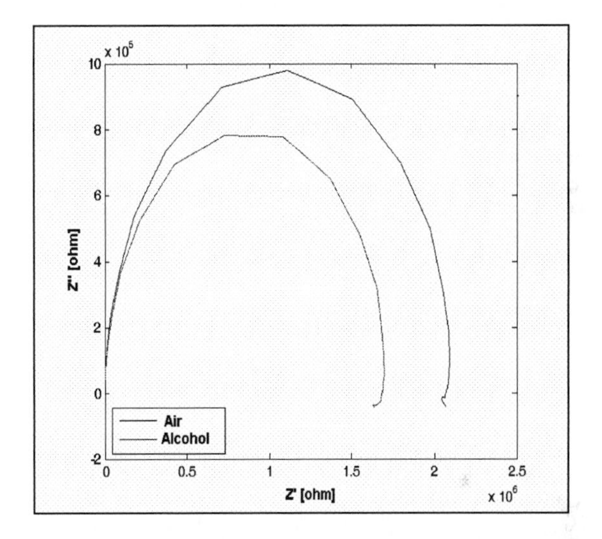

FIGURE 6. Complex impedance in air and isopropyl alcohol vapor from the RGTO film.

From the impedance spectra in Fig 6, we can also point out that a semi-circular arc corresponds to a distributed R-C element. However, the figures are different for air and isopropyl alcohol vapor. The equivalent circuit for complex impedance plots in Fig. 6 can also be explained by a resistance and capacitance in parallel. The optimum values for Rgb and Cgb in air are ~ 2000 kΩ and 8 pF respectively. In contrast, isopropyl alcohol vapor the optimum values for Rgb and Cgb are ~ 1700 kΩ and 10 pF respectively. In presence of isopropyl alcohol vapor, Rgb decreases and at the same time Cgb increases. The increase in the capacitance on exposure gas is attributed to a reduction in the width of the depletion region.

DISCUSSION

It is known that the microstructure of the sensing film depends on grain size among other variables. Currently, most of the semiconductor gas sensors based on SnO_2, TiO_2, Zn0, etc exhibit gas sensitivity only when these are heated to several hundred degrees. However, SnO_2 sensors fabricated by the hydrolysis of $SnCl_4$ exhibit sensitivity to the saturated vapor of alcohol at room temperature [12]. This can't only be explained by the size of the grains, since the SnO_2 sensors obtained from other techniques do not present sensitivity to alcohol vapor at room temperature. One of the possibilities is that a high surface/volume microstructure relationship was obtained by the hydrolysis of $SnCl_4$.

Two different techniques were used to prepare thin films of SnO_2: RGTO and pore wetting. From the first one a microstructure formed by nanograins was obtained and from the second, a microstructure consisting of nanotubes was obtained.

Only by using the nanotube sensors film sensitivity to isopropyl alcohol was reached at room temperature, highlighting the central role played by the microstructure of thin-film within its properties. Possibly, microstructures with a high surface/volume relationship would be more adequate for gas sensors, but it takes more work to know exactly what is the most adequate microstructure to enhance sensitivity.

CONCLUSIONS

Thin film of SnO_2 gas sensor with microstructures formed by nanotubes is sensitive to isopropyl alcohol vapor even at room temperature. This is an unusual behavior for a semiconductive sensor film.

It is necessary to highlight that the technique used in its fabrication, pore wetting, is compatible with the sensors microfabrication.

ACKNOWLEDGMENTS

Work partially supported by ANPCyT (PICT 2005 No. 35642).

REFERENCES

1. M G. Sberviglieri, *et al.*, *Semicon. Sci. Technol,* **5** 1231-1233 (1990).
2. P. Levy, A.G. Leyva, H. Troiani, R.D. Sanchez, *App. Phys. Lett*, **83** 5247-5249 (2003).
3. A.G. Leyva, *et. al.*, *J. of Solid State Chemistry* **11**, 3949-3953 (2004).
4. A.G. Leyva, P. Stoliar, M. Rosenbusch, P. Levy, J. Curiale, H. Troiani, R.D Sanchez, *Physica B* **354**, 158-160 (2004).
5. N. C. Oldham, C. J. Hill, C. M. Garland and T. C. McGill, *J. Vac. Sci. Technol. A.* **20**, 809-813 (2002).
6. N. Tigau, V. Ciupina and G. Prodan, *J. Cryst. Growth* **277**, 529-535 (2005).
7. G. Gordillo, L. C. Moreno, W. De La Cruz, P. Teheran, *Thin Solid Film,* **252**, 61-66 (1994).
8. T.K.H. Starke, S. V. Gary, *Sensors and actuators. B, Chemical* **88**, 227-233 (2003).
9. J. Sundqvist, M. Ottosson, A. Hårsta, *Chem. Vap. Deposition* **10**, 77-82 (2004).
10. Z.W. Pan, Z.R. Dai and Z.L. Wang, *Science,* **291** 1947-1949 (2001).
11. Z. R. Dai, J. L. Gole, J. D. Stout, Z. L. Wang, *The journal of physical chemistry. B.* **106** 1274-1279 (2002).
12. H. C. Wang, Y. Li, M. J. Yang, *Sensors and Actuators B: Chemical* **119** 380-383 (2006).

CMOS Integrated Carbon Nanotube Sensor

M.S. Perez [a1] , B. Lerner [a1], P.D. Pareja Obregon [b1] , P.M. Julian [b],
P.S. Mandolesi[b], F.A. Buffa [c], A. Boselli [a], A. Lamagna [a].

[a] *Grupo MEMS, Comision Nacional de Energia Atomica, Buenos Aires, Argentina*
[b] *Dpto. de Ing. Eléctrica y de Computadoras, Universidad Nacional del Sur, Bahía Blanca, Argentina*
[c] *INTEMA Facultad de Ingeniería, Universidad Nacional de Mar del Plata, Mar del Plata, Argentina*
blerner@cnea.gov.ar, Tel: 5411-6772-7931

Abstract. Recently carbon nanotubes (CNTs) have been gaining their importance as sensors for gases, temperature and chemicals. Advances in fabrication processes simplify the formation of CNT sensor on silicon substrate. We have integrated single wall carbon nanotubes (SWCNTs) with complementary metal oxide semiconductor process (CMOS) to produce a chip sensor system. The sensor prototype was designed and fabricated using a 0.30 um CMOS process. The main advantage is that the device has a voltage amplifier so the electrical measure can be taken and amplified inside the sensor. When the conductance of the SWCNTs varies in response to media changes, this is observed as a variation in the output tension accordingly.

Keywords: single wall carbon nanotube, sensor, chip.
PACS: 85.35.Kt

INTRODUCTION

Single wall carbon nanotubes (SWCNTs) have shown to be good sensing elements for pressure [1], gases and alcohol [2,3,4,5], and are potential candidates for sensors. SWCNTs have emerged as a viable electronic material for molecular electronic devices because of their unique physical and electrical properties.

System on a chip sensors are the integration of sensing elements, interfacing circuitry and measurement circuitry in a single integrated circuit. The major advantages of CS are reductions in cost and devices' dimensions, reduction in power consumption and response time. In order to demostrated that the integration process of the CMOS and the SWCNTs is possible, we used the sensor to investigate the effect of humidity on the electrical transport properties of SWCNTs.

EXPERIMENTAL AND METHODS

Carbon Nanotubes

The nanotubes used in this study have been obtained by the catalytic CoMoCAT [6], which employs a silica-supported Co-Mo powder to catalyze the selective growth of SWCNTs by disproportionation of CO. The nanotubes used in this study have an average diameter of 0.8 nm [7]. The SWCNTs grown by this method were purified by SWeNT (Southwest Nanotechnologies). The resulting nanotube material has an excellent quality, as verified by transmission electron microscopy (TEM) and scanning electron microscopy (SEM) (Fig.1). We used DEP process to do the carbon deposition 0.2 mg of SWCNTs were dispersed in ethanol and ultrasonicated for 20 minutes with a 100 W.

A droplet of this solution was put between the gap of the electrodes,an alternating current (AC) voltage of 5 V at a frequency of 1 MHz for 40 seconds was applied to the electrodes to generate the DEP force.

[1] M.S. Perez, B. Lerner and P.D. Pareja Oregon contributed equally to this work.

CP1137, *Olfaction and Electronic Nose: Proceedings of the 13th International Symposium,* edited by M. Pardo and G. Sberveglieri

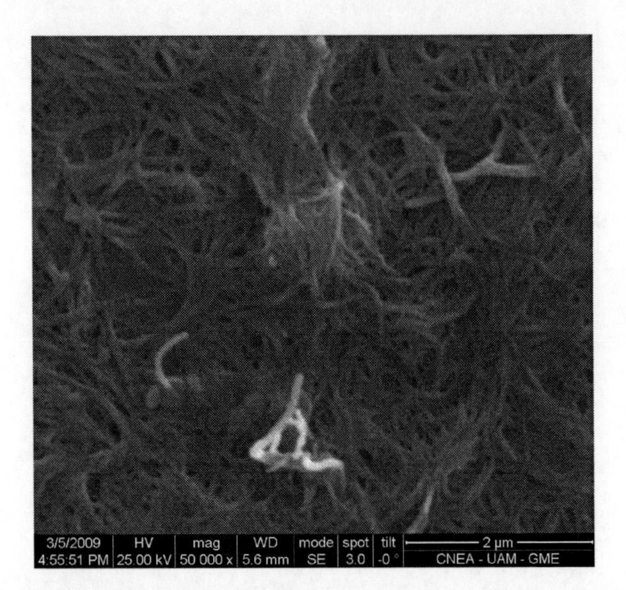

FIGURE 1. SEM micrography of SWCNTs.

Sensor

The top layer metal in the CMOS chip was designed to act as the electrodes for sensor fabrication.The Sensor integration was fabricated using a commercial CMOS process; this achieves the required resolution with no additional cost. The interface between the SWCNTs and the output signal was composed of a microelectronic circuit in charge of handling the signals inherent to the measurement, as well as generating the amplification needed to obtain an output signal with low noise level. The amplification was made through a transresistance amplifier circuit. The integrated circuit was fabricated by MOSIS in the standard process AMIS. The layout design was made using L-Edit

Humidity and Electrical Measures

Sensors were located in a chamber (volume 500 cc) where air flow at a constant rate of 300 sccm and humidity was controlled with a 0.1 % precision at 298 K. Current was supplied by Keithley 6221 current source and Voltage changes were measured with a Keithley 2000 multimeter.

RESULTS

Sensor Layout

The sensor is formed with an arrangement of free terminals connected directly with the external pads and one arrangement of terminals connected to an amplifier. Fig 2 show the dimensions of the terminals. The amplifier circuit comprises a current mirror, which reflects the input current of one of the pads on the nanotubes to generate an output voltage (Vout).

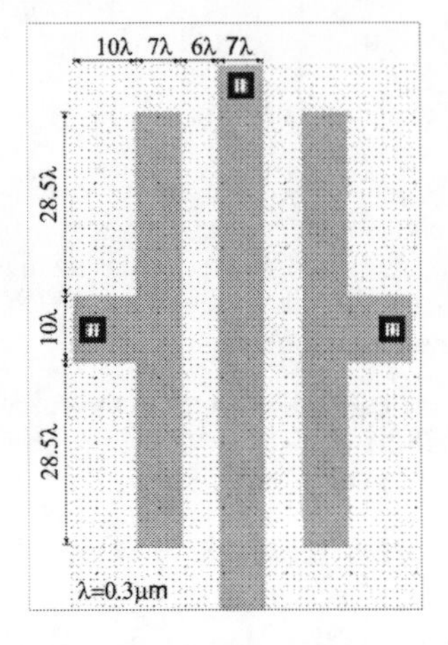

FIGURE 2. Dimensions of the terminal.

Electrical Measures

We showed that the conductance of SWCNTs in the sensor can be affected by environmental humidity.

We analyzed the effect of humidity over the current voltage characteristics of SWCNTs placed on the free terminals (Fig. 3).When the humidity increases, the conductance decreases up to a humidity value of 67%. With humidity values above 67% the conductance starts increasing again.

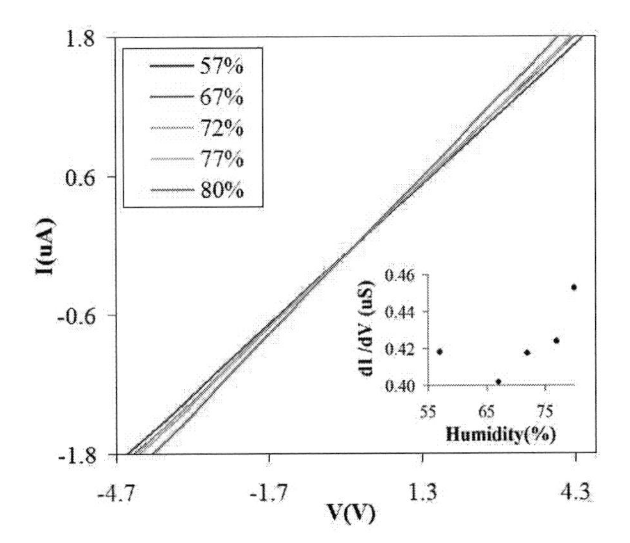

FIGURE 3. Effect of humidity over current voltaje characteristics of SWCNTs detected through the free terminals. Lower inset shows the conductance dI /dV as a function of humidity.

CONCLUSION

The feasibility of CMOS process with SWCNTs has been demonstrated in this work. We successfully developed and tested at different humidities a sensor prototype which includes an amplifier and free terminal electrodes with an excellent precision and at a very low cost.

ACKNOWLEDGMENTS

We acknowledge M. Di Federico, J. Bonaparte, D. Rodriguez and grupo MEMS staff for their assistance with device fabrication, measurements and software development.

REFERENCES

1. C.K.M.Fung, M.Q.H. Zhang, Z. Dong and W.J.Li. *IEEE Nanotechnology* **1**, 199-202 (2005).
2. M.L.Y. Sin, G.C.T. Chow, W.J. Li, P. Leong and K.W. Wong. *IEEE Transactions on Nanotechnology* **6**, 571-7 (2007).
3. D. R. Kauffman, A. Star. Angewandte Chemie International Edition **47**, 6550-6570 (2008).
4. M. L. Terranova, M. Lucci, S.Orlanducci, E. Tamburri, V. Sessa, A. Reale and A. Di Carlo. J. *J. Phys. Condens. Matter* **19**, 225004 (2007).
5. H. A. Pohl. "Dielectrophoresis: The Behavior of Neutral Matter in Nonuniform Electric Fields". Cambridge:Cambridge University Press,1987.
6. C. Fung, V. Wong, R. Chan and W. Li. IEEE Transactions on Nanotechnology **3**, 395-03 (2004).
7. D.E. Resasco, W. E. Alvarez,F. Pompeo,L. Balzano,J. E. Herrera,B. Kitiyanan, et al.J. Nanoparticle Res **4** 131-36 (2002).

Fabrication of NO_x gas sensors using In_2O_3-ZnO composite thin films

Chia-Yu Lin[1], Yueh-Yuan Fang[2], Chii-Wann Lin[2,3], Craig A. Jeffrey[4], James J. Tunney[4], Kuo-Chuan Ho[1,5★]

[1]*Department of Chemical Engineering, National Taiwan University, Taipei 10617, Taiwan*
[2]*Institute of Biomedical Engineering, National Taiwan University, Taipei 10617, Taiwan*
[3]*Institute of Electrical Engineering, National Taiwan University, Taipei 10617, Taiwan*
[4]*Institute for Chemical Process and Environmental Technology, National Research Council of Canada, Ottawa K1A 0R6, Canada*
[5]*Institute of Polymer Science and Engineering, National Taiwan University, Taipei 10617, Taiwan*
★*Corresponding author: Fax: +886-2-2362-3040; Tel: +886-2-2366-0739; E-mail: kcho@ntu.edu.tw*

Abstract. In_2O_3-ZnO composite thin films were fabricated and their NO_x sensing characteristics were investigated in this study. The content of ZnO in In_2O_3-ZnO thin film was controlled by adjusting the Zn^{2+}/In^{3+} molar ratio (r) during the film preparation. The results showed that the incorporation of ZnO into the In_2O_3 thin film can greatly improve the sensor reponse to NO_x at operation temperature below 200 °C. However, as the ZnO content was further increased, the grain of the composite film started to merge into big ones, and thus decreased the surface area, resulting in lower sensor responses. The detection limit (S/N=3) of In_2O_3-ZnO composite film (r=0.67) reached 12 ppb at 150 °C. Both pure In_2O_3 and In_2O_3-ZnO composit films (r=0.67) showed no response to CO gas.

Keywords: Nitric oxide, Indium oxide, Zinc oxide, Gas sensors
PACS: 07.07.Df

INTRODUCTION

Nitric oxide (NO) has received much attention due to its environmental and health-related importance. The emission of NO from vehicle exhaust produces smog which destroys the ozone layer [1]. Besides, NO is a component of acid rain upon conversion into nitric acid. According to the regulations set by the Occupational Safety and Health Administration (OSHA), the permissible exposure limit is set at 25 ppm (TWA). On the other hand, NO has been demonstrated as an essential messenger molecule in the body system [2]. For example, the exhaled NO concentration has been identified as a biomarker for airway inflammation such as asthma [3] and bronchiectasis [4]. It is, therefore, necessary to develop rugged and reliable NO gas sensors capable of making real-time measurements for public health and security applications.

For a long time, metal oxides have been known to be suitable materials for detection of pollutant gases in environmental conditions. Indium oxide (In_2O_3), an n-type semiconducting metal oxide, has been proposed as a gas sensing material for sensing ozone [5], H_2S [6], NH_3 [7], NO_2 [8], etc. To achieve higher sensitivity to the analytes, the normal operation temperature of the In_2O_3 based gas sensor is rather high (200~400 °C), thus consuming high power. Several methods have been adopted to improve the sensitivity and selectivity of the In_2O_3 based gas sensor, including the use of the nanostructured In_2O_3 [7, 8] and doping additives [9,10].

In this study, ZnO was used as the doping material to improve In_2O_3 based NO_x gas senor. Different amounts of ZnO were incorporated into In_2O_3 film via co-precipitating method. The effects of Zn^{2+}/In^{3+} (r) ratio during the preparation on the sensitivity of the resulted composite films toward NO_x were investigated. To take the instability of NO in air into acount, NO gas was noted as NO_x. The preliminary results showed that the incorporation of ZnO can improve the sensitivity of In_2O_3 film to NO_x gas at low operation temperature (<200 °C).

EXPERIMENTAL AND METHODS

2-1 Preparation Of In_2O_3-ZnO Composite Films

CP1137, *Olfaction and Electronic Nose: Proceedings of the 13th International Symposium*, edited by M. Pardo and G. Sberveglieri

Formation of $In(OH)_3$-$Zn(OH)_2$ was conducted using a chemical bath precipitation method in an aqueous solution of $InCl_3$ $Zn(NO_3)_2$ and hexamethylenetetramine (HMT) at 90 °C for 1 hr. To synthesize $In(OH)_3$-$Zn(OH)_2$ with different molar ratios, different amounts of 25 mM $Zn(NO_3)_2$ solution were added in a 25 mM aqueous solution of $InCl_3$, *i.e.* different Zn^{2+}/In^{3+} molar ratio (r). The molar ratio of Zn^{2+} and In^{3+} source to HMT was kept 1:1.

In_2O_3-ZnO composite films were fabricated by droping 60 μl of the above solutions on the gold interdigitated electrodes. Then the films were annealed at 500 °C for 30 min under pure O_2 atmosphere.

2-2 Gas sensing experiment

The electrodes were mounted onto a heater and inside a chamber equipped with gas-flow manifold and mass flow controller. The apparatus facilitated automated (PC-DOS) control of temperature and data acquisition of the film resistance using a 2-wire method with a Hewlett Packard 34401A digital voltmeter which permitted resistance (R) measurements in the range $0<R<120$ MΩ. Various concentrations of NO was tunned by mixing 484 ppm NO with a zero-grade air stream and were delivered to the chamber at a flow rate of 200 $cm^3 \cdot min^{-1}$.

The response, S, is defined by the following equation

$$S = (R_g - R_a)/R_a \tag{1}$$

where R_a and R_g are the resistances of the film in zero-grade air gas and in test gas, respectively.

The microstructure of In_2O_3-ZnO composites and the amount of ZnO incorporated into the composites were determined by scanning electron microscopy (JEOL 840 A, Japan) and energy disperse x-ray analysis, respectively. The thicknesses of the composite films were determined by cross-sectional SEM.

RESULTS AND DISCUSSION

3.1 Surface morphology of pure and ZnO-doped In_2O_3 films

The content of ZnO in In_2O_3-ZnO composite film was controlled by adjusting the Zn^{2+}/In^{3+} (r) ratio during the preparation, and the actual contents of ZnO in the composite film along with the film thickness are summarized in Table 1. It was found that the actual In/Zn molar ratio is much higher than that during the preparation, which is possibly resulted from much lower solubility constant of $In(OH)_3$ than that of $Zn(OH)_2$ (1×10^{-34} M^4 vs. 3×10^{-17} M^3, 25 °C).

The surface morphology of the ZnO-In_2O_3 composite film is quite different from that of the pure In_2O_3 film. As shown in Fig. 1 (a)-(e), the In_2O_3-ZnO composite films showed a bimodal grain size distribution with a coarse and a fine component. The amount of the coarse component and their size increased as the r was increased, while the grain size of the fine component remained almost the same.

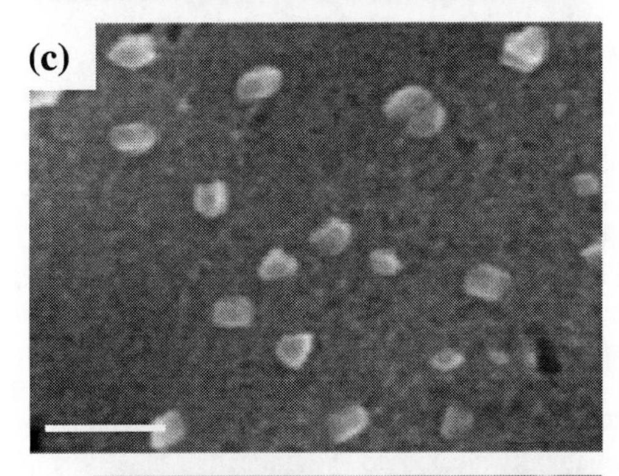

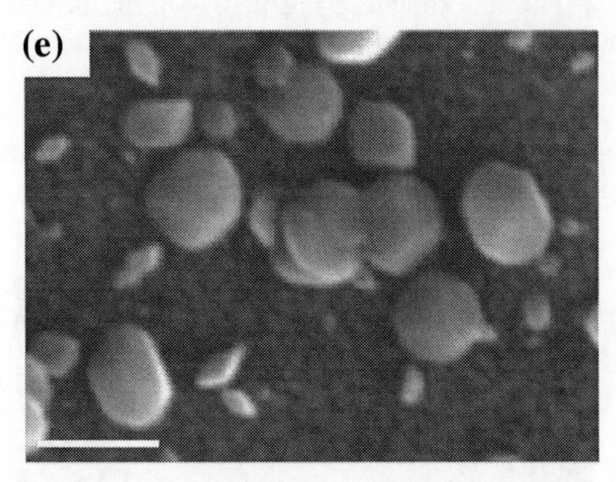

FIGURE 1. The SEM images for the In_2O_3-ZnO composite film with r values of (a) 0, (b) 0.33, (c) 0.5, (d) 0.67, and (e) 1, respectively. Scale bar: 500 nm.

TABLE 1. The film thickness and the actual In/Zn ratio of the In_2O_3-ZnO composite films with diferent rs.

	r values				
	0	0.33	0.5	0.67	1
Film thickness (μm)	3.83 ± 0.49	3.28 ± 0.62	4.00 ± 1.03	4.75 ± 0.41	4.00 ± 0.35
Actual Zn/In molar ratio	---	0.1	0.14	0.15	0.17

3.2 Gas sensing characteritics

Fig. 2 shows the gas sensing responses of the composite films to 10 ppm NO_x at different operation temperatures. It was found that the responses of the all composite film increased as the operation temperature was lowered. Besides, the Zn^{2+}/In^{3+} ratio (r) has great influence in the sensing responses of the composite film. At the operation temperatures lower than 200 °C, the sensor response increased as r was increased from 0 to 0.67. The increase in the sensor reponse could be attributed to the synergetic effect of In_2O_3 and ZnO or the evolution of the new phase ($Zn_xIn_yO_z$). However, as r was further increased to 1, the sensor response dramatically decreased dramatically. This decrease in the sensor response could be attributed to the morphology change. As shown in Fig. 3, the grains of the In_2O_3-ZnO composite film started to merge into big ones as the r was increased from 0.67 to 2. The increase in grain size would decrease the surface area of the sensing films, resulting in lower sensor response. Therefore, the optimal r used for later experiments was 0.67. It is also interesting to note that NO_x gas can act as reducing gas at higher temperature and as oxidizing gas at lower temperature, and the response transition temperatures are summarized in Table 2. It was found that the response transition temperature shifted as the r ratio was changed.

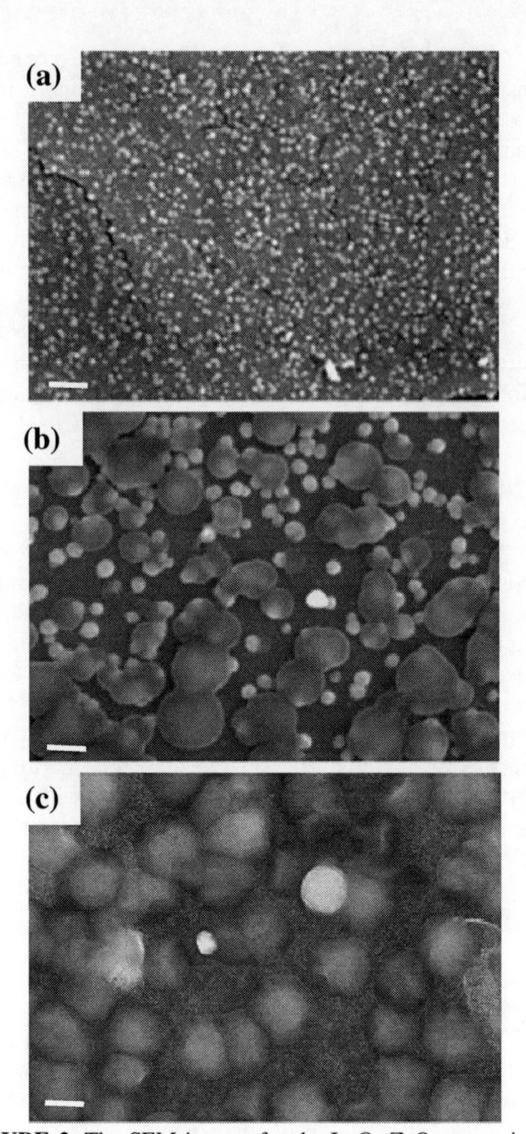

FIGURE 2. The SEM images for the In_2O_3-ZnO composite film with Zn^{2+}/In^{3+} ratios are (a) 0.67 (b) 1, and (c) 1.5, respectively. Scale bar: 1 μm.

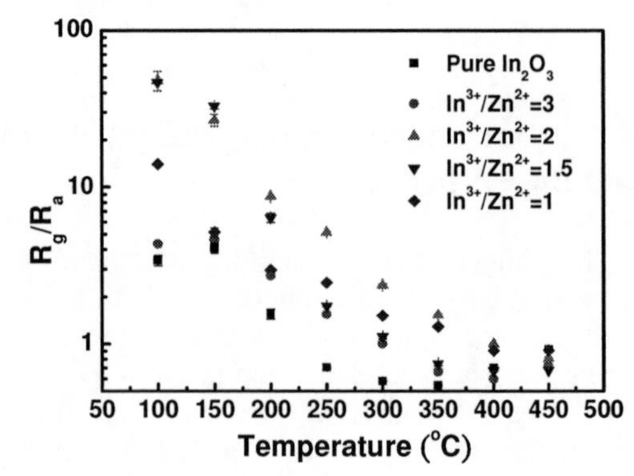

FIGURE 3. The changes in the resistance of the composite films as exposure to 10 ppm NO as a function of temperature and Zn^{2+}/In^{3+} ratio.

Figure 4 shows the dynamic response of the In_2O_3-ZnO film (r=0.67) to various concentration of NO_x gas at 150 °C. It was found that all the responses are linear with the increase in the concentration of NO_x, and the limit of detection, based on signal-to-noise ratio of 3, was found to be 12 ppb.

Table 2 Transition temperature range for the In_2O_3-ZnO composite films with different rs.

r values	Transition range (°C)
0.00	200 ~ 250
0.33	250 ~ 300
0.50	400 ~ 450
0.67	300 ~ 350
1.00	350 ~ 400

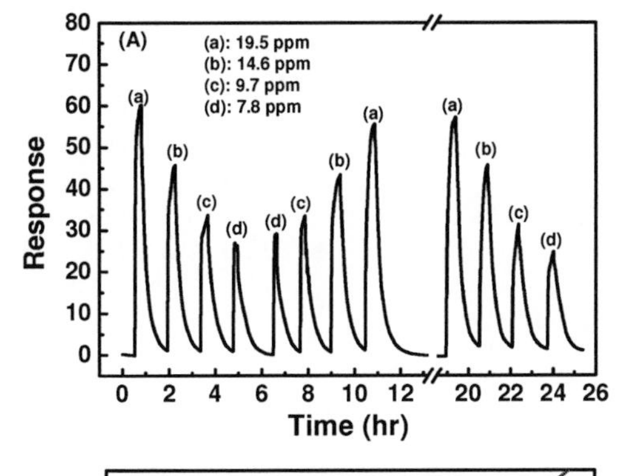

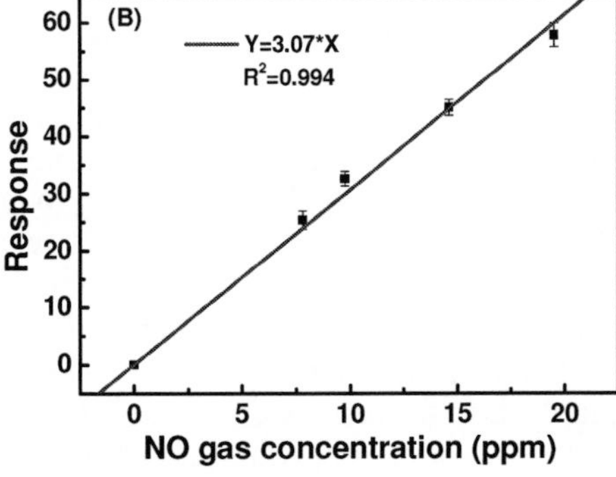

FIGURE 4. (A) The transient responses for In_2O_3-ZnO (r=0.67) composite film at 150 °C to various NO_x concentrations ranging from 7.5 to 20 ppm. (B) The resulted calibration curve.

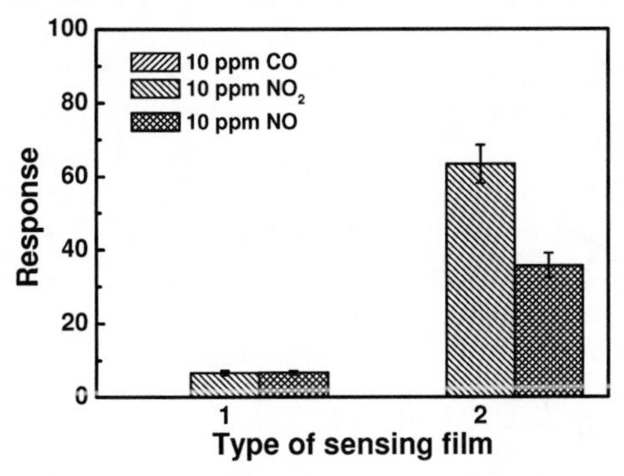

FIGURE 5. The responses of the pure In_2O_3 (#1) and In_2O_3-ZnO (r=0.67, #2) films against 10 ppm of NO_x, NO_2 and CO at 150 °C.

The interfering effect of CO and NO_2 were also examined. Figure 5 shows the responses of the pure In_2O_3 and the In_2O_3-ZnO (r=0.67) films against NO_x, CO, and NO_2 gases at 150 °C. It was found that both the pure In_2O_3 and In_2O_3-ZnO (r=0.67) films showed no repsonse to CO gas. However, NO_2 gas showed significant interfering effect for both pure In_2O_3 and In_2O_3-ZnO (r=0.67) films.

CONCLUSION

The In_2O_3-ZnO composite films were synthesized and their NO_x gas sensing characteristics were studied. The minute amount of ZnO addition can greatly improve the sensitivity of the composite film to NO_x gas at low temperature (< 200 °C). However, as the content of ZnO was increased (r > 0.67), the sensitivity for the resulted composite film decreased dramatically due to the morphological change. The results suggest that the gas sensors based on the In_2O_3-ZnO composite films have the potential to become candidate for environmental monitoring.

ACKNOWLEDGMENTS

This work was sponsored by the National Research Council of the Republic of China (Taiwan) under grant number NSC 96-2220-E-006-015 and NSC 97-2220-E-006-008.

REFERENCES

1. L. B. Kreuzer, C. K. N. Patel, *Science* **173**, 45-47 (1971).
2. B. I. Judutt, *The Role of Nitric oxide in Heart Failure*, Massachisetts: Kluwer Academic, 2004.
3. N. L. R. Han, J. S. Ye, A. C. H. Yu, A, F. S. Sheu, *Journal of Neurophysiology* **95**, 2167-2178 (2006).
4. M. J. Ratnawati, R. L. Henry, P. S. Thomas, *Pediatr. Pulmonol.* **41**, 929-936 (2006).
5. G. Sberveglieri, C. Baratto, E. Commi, G. Faglia, M. Ferroni, A. Ponzoni, A. Vomiero, *Sens. Actuators B: Chemical* **121**, 208-213 (2007).
6. Xu, J., Wang, X., Shen, J., *Sens. Actuators B: Chemical* **115**, 642-646 (2006).
7. N. Du, H. Zhang, B. Chen, X. Y. Ma, Z. Liu, J. Wu, D. Yang, *Adv. Mater.*19, 1641-1645 (2007).
8. C. S. Rout, K. Ganesh, A. Govindaraj, C. N. R.Rao, *Appl. Phys. A* **85**, 241-246 (2006).
9. L. Franciosco, A. Forleo, S. Capone, M. Epifani, A. M. Taurino, P. Siciliano, *Sens. Actuators B: Chemical* **114**, 646-655 (2006).
10. G. Sberveglieri, S. Groppelli, P. Nelli, A. Tintinelli,G. Giunta, *Sens. Actuators B: Chemical* **24-25**, 588-590 (1995).

The New Principle Of Sensor Differentiation By Electric Potential Bias In Metal Oxide Sensor Arrays

I. Kiselev[1], M. Sommer[1], V. V. Sysoev[2]

[1]*Forschungszentrum Karlsruhe, IMT, Hermann-von-Helmholtz-Platz 1, 76344 Eggenstein-Leopoldshafen, Germany.*
[2]*Saratov State Technical University, ul. Polytechnicheskaya 77, 410054 Saratov, Russia*

Abstract. We present a new method to differentiate the response of individual sensors in a metal oxide gas sensor array of E-nose type by application of varied electric potential biases and discuss how to employ the spatially distributed potential as a gas recognition pattern. The method is based on the impact, which the electric potential has upon the thermodynamic state of the metal oxide surface and on the gas adsorption processes. It is shown that the utilization of varied electric potentials yields a significant gas recognition power of E-nose, similar or even higher compared to the effect of conventional spatial temperature gradient differentiation. One advantage to use the potential distribution is the opportunity to have the whole continuous film as a sensor field, which would eliminate the influence of electrodes. The tests of a prototype of the multisensor array based on the considered method have shown its feasibility for getting high gas recognition and the capability to discriminate gas concentrations reliably.

Keywords: Metal oxide; sensor array; gas separation; device reproducibility; gas sensitivity.
PACS: 07.07.Df

INTRODUCTION

The concept of E-nose has been developed to overcome a lack of selectivity of single gas sensors and to attain a capability to classify complex gas mixtures such as aromas and odors [1]. The characteristics of the individual sensors composing the E-nose sensor array should be diverse as much as possible to ensure getting the sensor responses being non-correlated and to enable the instrument to reliably discriminate gases and gas mixtures, especially if the E-nose is exposed to gases of similar chemical nature. This is a particular concern for E-noses based on mono-type gas sensors as, for example, KAMINA (KArlsruhe MIcro NAse) which employs a metal oxide film segmented by electrodes to obtain a sensor array [2]. Therefore, specific techniques like the application of temperature gradient and gas-permeable coatings of varied thickness [2] are necessary to facilitate larger differences among the array sensor segments. The search for other reproducible and inexpensive methods able to differentiate properties of the sensor segments is of high interest for further E-nose development [1,3].

As the extensive dissemination of E-noses covers completely different fields of applications, a high reproducibility of the sensor arrays is required. This issue seems to be one of the key challenges, which is needed for a breakthrough of the technology [3]. Gas sensor microarrays based on a monolithic thin film segmented by electrodes seem to meet this requirement, because a number of chips is produced in a single fabrication process and the sensor segments are designed to be rather similar. However, even such multisensor microarrays still depend on non-controlled variations during the manufacturing process [3]. In particular, the factors, which are intentionally employed for sensor properties differentiation, as a varied temperature, are not sufficiently reproducible at the microscale of sensor segments. The sputtering of the current-injecting micro-electrodes results in a substantial and non-predictable doping of the sensing areas. It is obvious that enhancing control over the sensor segment differences would provide higher reproducibility of gas-discrimination models, since the gravity centers of representations of the sensor responses to different odors would be better separated in a pattern recognition coordinate space. Thus, there is a clear need to find additional reproducible factors to differentiate the properties of sensor segments, which have to have two-fold properties, 1) to allow a reliable enhanced gas discrimination and 2) to have a tolerance towards the electrodes without perfect ohmic

CP1137, *Olfaction and Electronic Nose: Proceedings of the 13th International Symposium*, edited by M. Pardo and G. Sberveglieri

contact to the film which therefore might not dope the oxide surface.

Here, we intend to employ a varied electrical potential bias as a new factor to differentiate sensor segment properties over the metal oxide array. Hence, the gas-characteristic patterns are based on the potential distribution. Certainly, several attempts were carried out in the past to affect the properties of metal oxide gas sensors by electric fields (see for instance [4]). But the most reports on the subject consider the electric field as the influence parameter, but not the potential, which actually differentiate the segments substantially.

It is still worth mentioning that adsorption of various species on a semiconductor surface could be expected to depend on the surface electrical potential relative to the surrounding. These processes are determined by a chemical potential of adsorbate particles at the surface [5] and therefore influenced by the operating temperature. That is why the temperature is frequently used for adjusting both the sensitivity and selectivity of metal oxide sensors. However, the chemical potential depends (linearly for charged particles) on the electrical potential as well. So, the latter one is a factor to control the adsorption processes, too.

The direct measurements show [6] that a positive potential applied to a metal oxide surface promotes a significant surface accumulation of electrical charge brought by oxygen ions. Further, it is shown that the potential distribution along the metal oxide surface is distorted after a voltage application (see for instance [4, 7]). Shape and amplitude of such a distortion depend on the presence of additives in the exposure air. Here we explain the potential distortions to be not caused by a migration of adsorbed species or bulk vacancies (that would not be even possible through the considered sensor microarrays), but we assume these distortions as a reflection of differences in the thermodynamic level of surface segments caused by differences in the electric potential. Indeed, film regions with different potentials get different states of adsorption balances, hence different resistances and charges, resulting in a non-linearity of the potential distribution over the whole area.

Here we discuss the efficiency of employing the electric potential as a factor to differentiate sensor segment responses to build stable gas-recognition models using Linear Discriminant Analysis (LDA).

EXPERIMENTAL AND METHODS

In the first part of this study the conventional chips of the KAMINA e-nose, which are described in details elsewhere [2], were employed. These contain sputtered metal oxide films segmented by parallel co-planar electrodes into 38 partial sensors. The size of the film is about 8x4 mm^2 with a thickness of about 0.1 µm; the size of electrodes is about 4x0.1 mm^2 x 1 µm. It is worth to note that all the electrodes go across the whole oxide sensing film. To attain a reasonable statistical validity, four chips, called hereafter as chips A-D, were taken into consideration. Chips A-C had a SnO_2:Pt film, while the chip D had a WO_3 one. All the chips were taken from different production batches. Not all the array segments were considered here because, 1) we intended to measure the segments having similar resistances to emphasize the role of the controllable differentiating factors; 2) previous measurements had shown that the potential distribution following a voltage application over the whole 38 array sensors stabilizes for a very long term that makes the observations to be hardly operable. Thus, 6 segments were considered in chips A-C and 4 in chip D.

The principle of differentiation of film segment properties with the help of electric potentials, and using potential distribution instead of conventional resistance profile as the gas-characteristic pattern, were realized as follows. Potentials 0 and +20 V to earth were applied to the electrodes limiting a film area while the potentials of electrodes within the film area were scanned. At the same time the temperature was kept constant over the film to be not served as a differentiation factor.

In principle, a measurement of electric potential needs no current and, hence, only virtual electrodes. So, as an approximation to a virtual-electrode instrument, a chip carrying the majority of electrodes merely touching the sensor field was fabricated (Fig. 1; chip E, hereinafter) to prove the functional capability of the potential distribution measurements.

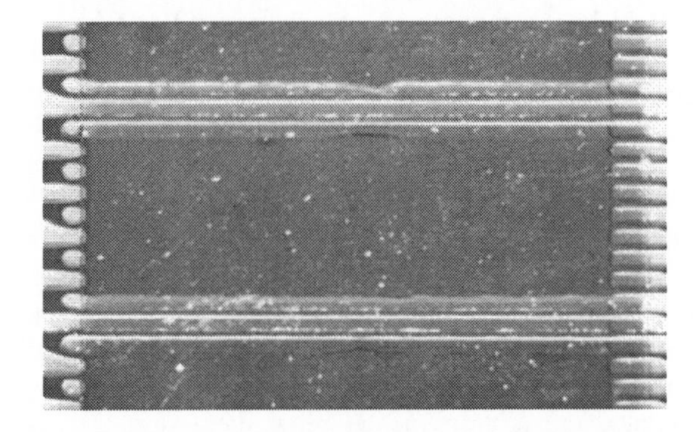

FIGURE 1. Micro-optical photo of the working area of the chip E. Long (4 mm) electrode strips serve for adjusting the electric potential drop over the film while short ones are used to measure the potential distribution.

Other parameters of the chip E including the size of the SnO_2 film were the same as of the chips A-C.

A KAMINA electronics unit [2] was used to measure segment resistances. The potentials were measured with a multimeter (Keithley 2001), supplied by a power source (Grundig PN-300). The chips were operated at maximal temperature of 300°C (A-C, E) and of 350°C (D). The operating temperature was applied to the whole sensor array under two options: 50 °C difference (so-called temperature gradient), or without temperature difference (<5 °C). The chips were exposed to synthetic air of 50% RH and its mixture with 2-propanol, 0.3-30 ppm range, or toluene, 5-40 ppm controlled concentrations.

As a benchmark to compare the gas-recognition power introduced by the "potential" method, conventional resistance pattern models were built for the chips B-D using the same sensor segments. In order to perform the examination close to a practical application and to estimate the gas recognition power on the background of disturbing variations, the LDA technique was chosen to discriminate the gas mixtures while the Leave-One-Out (LOO) method served to prove the recognition quality of the build LDA models. The resistance patterns were pre- normalized by median value of the sensor set.

RESULTS

The exemplary non-linear potential distribution over the active sensors of microarray at exposure to various atmospheres is shown in Fig. 2. As one can see, the potential distribution is substantially non-linear with a characteristic sagging over the intermediate electrodes even at the stationary state which was not observed in earlier reports [4,7] due to short measurement times. The shape of the potential distribution curve depends on the type of gas admixed to air. In general, the deviation of the curve from a straight line is suppressed with the growth of additive concentration. Yet, even a slight distortion of the potential distribution is still sufficient to be employed for the gas recognition as shown below.

The LOO method is the most often one to be utilized in practical applications for characterization of "quality" of LDA model recognition, therefore it was chosen here to compare the suitability of the methods. The LDA models built using the electric potential method were compared with ones based on conventional resistance signals of the microarray operated under the temperature gradient (Fig. 3). As one can see, the LDA model based on the electric potential recognizes the type of additives in general better than a "conventional" one. These results support the conclusion, that the method combining electric potential differentiation with the potential profile patterns could be employed for discrimination of gas mixtures at least equally to conventional one using the resistances and temperature gradient. The transition time of the potential profile response to a sharp-profiled change of the composition of incoming air is similar to that of the conventional method of measuring temperature-modulated resistance profiles, it is about 5 sec.

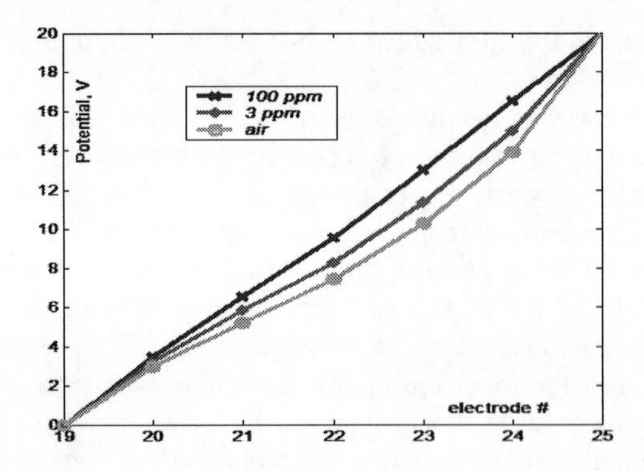

FIGURE 2. Typical stationary potential distribution over the sensor segments in sequence under the voltage application. The legend gives the concentrations of test gas, 2-propanol. Chip C.

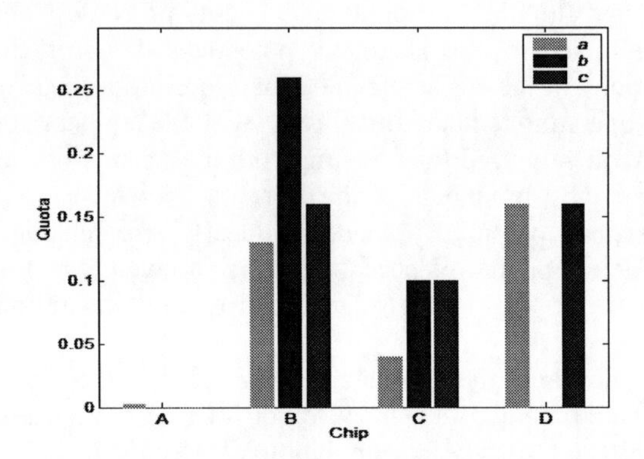

FIGURE 3. Misrecognition quota calculated by LOO for three LDA additive-discriminating models based on (a) the potential method, (b, c) the resistance profile modulated with the temperature gradient. The models in cases of (a) and (b) account for the same set of segments, while one in case (c) shows the recognition accounting the same number of segments but chosen over the whole array in order to achieve a maximal difference in their temperature. The training parameters were maintained same for the models. The model b was not built for chip D.

Important feature of using a potential profile is its stability towards changes in the air humidity. For example, changing humidity from 35 to 75% RH provides variation of the profile negligible compared with that caused by merely 1 ppm of 2-propanol.

The chip E was subjected to the same gas exposures as the other ones. The size of the film area used at this chip exceeded by approx. 20-25% the one of connected segments in chips A-C. The LDA analysis of gas recognition accounting for the potential profile over this film area shows the LOO recognition quota to be 100 %, higher than by any method employing the standard chips. Furthermore, the application of the temperature gradient to this chip resulted in further 35 % increase of the average characteristic Mahalonobis distances of classes in the LDA coordinate system. It shows that there is a good prospect for a combined use of the potential and the temperature gradient for segment differentiation. The potential profiles have also been found to be eminently sensitive to small concentrations of impurity gases (see the typical data drawn at the Fig. 4). Still, the influence of the measuring electrodes on the potential distribution is not excluded in this chip architecture. Therefore short stub electrodes located at the two edges of the film area exhibit the serrated potential profiles because of incomplete symmetry (Fig. 1).

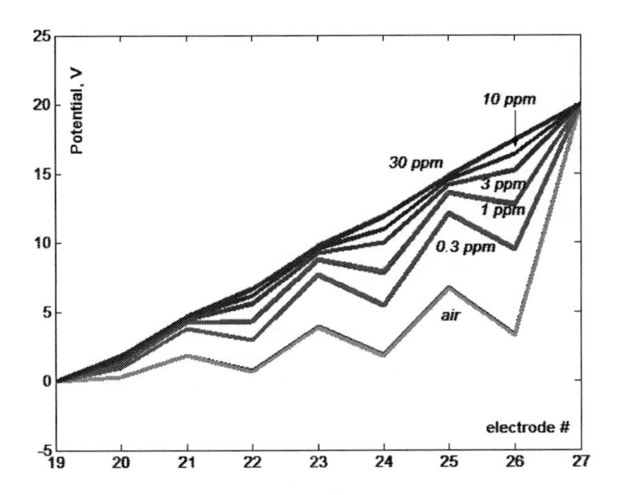

FIGURE 4. Potential profiles over the working area of the chip E observed as its response to the different concentrations of 2-propanol. The even and odd electrodes are located at the opposite edges of the metal oxide film.

DISCUSSION

The performed examination of the effect of the electric potential on the sensitivity of metal oxide films has an application related and a fundamental aspect: First, the possibility to develop new E-noses using potential measurements and secondly, the study of metal oxide surface adsorption state changes and its charging under application of a potential (see also [6]). Both issues are positively supported with the presented results. In particular, the data presented in Figs. 4 and 2 indicate a high sensitivity to low gas concentrations even at the sub-ppm level. The selectivity by higher gas concentration could be expected to be reduced. Still the minor differences characterizing the gas remained are sufficient for confident LDA gas-additive recognition. The further signal pattern normalization, like one derived previously in [8], should let the higher concentration to be recognizable as good as the lower concentrations.

The considered principles of employing the electric potential patterns and potential differentiation can be used also independently and under other arrangements. Requiring very low measuring currents, the potential measurements render many advantages against the conductivity method including short scanning times. Nonetheless, it is much more important in the aspect of reproducibility, that the electrodes can be virtualized, thus preventing the non-reproducible pollution of the sensor field with the electrode material during the fabrication process.

CONCLUSIONS

The combined methods of differentiating of sensing properties of segments at metal oxide sensor arrays through electric potential and reading out of the electric potential over the sensitive film as the characteristic pattern have been proved to enhance selectivity and sensitivity of gas recognition.

The enabled virtualization of measuring electrodes and the low requirements to the heating system eliminate the main factors of non-reproducibility attempting to produce E-noses reproducibly.

The received results confirm that the thermodynamic adsorption state of metal oxide surface depends largely on the applied electric potential.

REFERENCES

1. U. Weimar and W. Gopel, *Sens. Actuators B*, **52,** 143–161 (1998).
2. C. Arnold, M. Harms and J. Goschnick, *IEEE Sensors Journal*, **2,** 179-188 (2002).
3. G. Eranna. *et al*, *Crit. Rev. Solid State*, **29,** 111-188 (2004).
4. M. Liess, *Thin Solid Films*, **410,** 183-187 (2002).
5. T. Wolkenstein, *Electronic Processes on Semiconductor Surfaces During Chemisorption*, New York: Plenum, 1991, pp.125-158.
6. J. Goschnick, I. Kiselev and V. Simakov, "Time response behavior of segmented tin dioxide layers to stepwise changes of the electrical potential under air exposure", *MRS-2006 Fall Meeting*, http://www.mrs.org/s_mrs/doc.asp?CID=7201&DID=180202.
7. T. Sauerwald, D. Skiera and D. Kohl, *Thin Solid Films*, **490,** 86-93 (2005).
8. J. Goschnick *et al*, *Sens. Actuators B*, **116,** 85-89 (2006).

CO_2 Sensing Performances of Potentiometric Sensor Based on $Na_{1+x}Zr_2Si_2PO_{12}$ (0<x<3)

K. Obata* and S. Matsushima

Department of Materials Science and Chemical Engineering, Kitakyushu National College of Technology, 5-20-1 Shii, Kokuraminami-ku, Kitakyushu-shi, Fukuoka 802-0985, Japan
*E-mail: obata@kct.ac.jp (*corresponding author)*
(Tel.: + 81 – 93 – 964 – 7245, Fax: + 81 – 93 – 964 – 7308)

Abstract. CO_2 sensing characteristics were investigated under dry condition for potentiometric CO_2 sensors: air, Au | NASICON ($Na_{1+x}Zr_2Si_2PO_{12}$, x=0, 1, 2, 2.75) | Li_2CO_3-$BaCO_3$, Au, air + CO_2. The electromotive force (EMF) values of each NASICON sensor were linearly increased to the logarithm of CO_2 concentrations in the range of 250 to 2500 ppm at 450 °C. EMF values of each sensor continue to change toward negative potential during the operation exposing to 250 ppm CO_2 under dry condition. For the sensor using NASICON ($Na_{3.75}Zr_2Si_2PO_{12}$) with rich Na component, base-EMF (EMF value to fix a 250 ppm CO_2) could become fairly stable. In addition, from the use of a x-ray diffraction (XRD) measurements, the factor of EMF drifts caused the decrease of Na_2O activity in NASICON by decomposition of NASICON components.

Keywords: NASICON, CO_2 sensor, Li_2CO_3, Potentiometric sensor, EMF drift
PACS: 82.47.Rs

INTRODUCTION

There has been increasing demand for the measurement and/or control of CO_2 concentration for keeping indoor air quality clean. Compact CO_2 sensors such as potentiometric, capacitive, resistive and amperometric types have been proposed until now. Among these sensors, the potentiometric NASICON ($Na_3Zr_2Si_2PO_{12}$)-based sensor combined with an alkaline metal carbonate has become of great attractive lately because of an excellent selectivity to CO_2 gas, in addition to practical utility in gas sensitivity, response time and commercial price [1]. However, this type sensor includes a serious problem that should be improved to practical use [2, 3]. The problem is that the EMF of the sensor continues to decrease toward negative potential (EMF drift) during the operation under a constant condition. Recently, it was reported that base- EMF (EMF value to fix a 250 ppm CO_2) could become fairly stable by using NASICON ($Na_{3.75}Zr_2Si_{2.75}P_{0.5}O_{12}$) with rich Na component [4].

In the present study, we researched the effect of NASICON components on the long-term stability. It was fabricated four kinds of NASICON (x = 0, 1, 2, 2.75) - based sensors and compared those CO_2 sensing properties. Furthermore, the reactivity of NASICONs with Li_2CO_3 auxiliary phase was examined using an XRD measurement.

EXPERIMENTAL METHODS

Four kinds of NASICON (x = 0, 1, 2, 2.75) powders were prepared by a sol-gel technique using $Si(OC_2H_5)_4$, $Zr(OC_4H_9)_4$, $PO(OC_4H_9)_3$ and $NaOC_2H_5$. The precursor of each NASICON powder was compacted into a disk (9 mm in diameter and 1.2 mm thick) and then the disk was sintered at 1200 °C in air for 5 h. Li_2CO_3-$BaCO_3$ (1:2 in molar ratio) powders were prepared by calcining the mixture of Li_2CO_3 and $BaCO_3$ in air at 750 °C for 10 min.

NASICON sensors were fabricated by combining the NASICON disks with an auxiliary phase (Li_2CO_3-$BaCO_3$), as schematically drawn in Fig. 1. The reference electrode was prepared by using Au paste, followed by calcination at 800 °C in air for 2 h. On top of it, the sensing electrode was formed by applying Li_2CO_3-$BaCO_3$ binary carbonate. Then, the whole assembly was calcined at 750 °C in air for 5 min. The reference electrode was covered with a protective layer (an inorganic adhesive) to isolate from water vapor and CO_2 gas in atmospheric air. The NASICON sensors were fixed on the end of a quartz glass tube (9 mm in diameter) with an inorganic adhesive, as shown in Fig.1.

Gas sensing properties were measured in a conventional gas-flow apparatus equipped with a heating facility at 450 °C. The concentration of CO_2 was varied in the range of 250 to 2500 ppm CO_2. The sample gases were let to

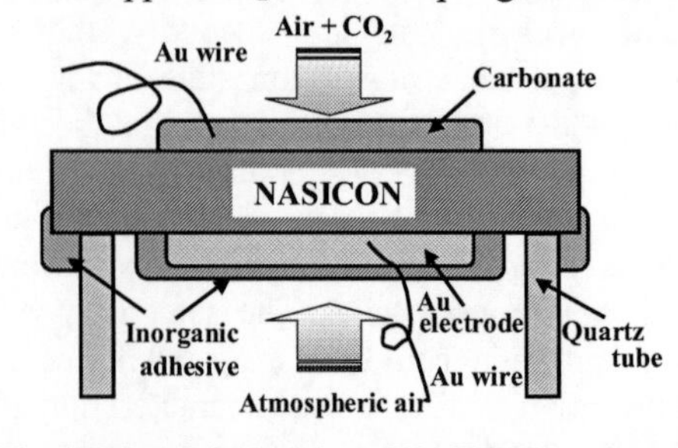

Fig. 1 Schematic drawing of NASICON sensor.

CP1137, *Olfaction and Electronic Nose: Proceedings of the 13th International Symposium,* edited by M. Pardo and G. Sberveglieri

flow over the sensing electrode at a rate of 0.1dm^3/min. The electromotive force (EMF) of the sensing sensor was measured with a digital electrometer.

Results and Discussion

Figure 2 shows the correlation between EMF values and CO_2 concentrations under dry condition at 450 °C for the sensors used NASICON (x = 0, 2, 2.75). NASICON (x = 1) sensor was not included in Fig. 2, because that the sensor was instable after aging for 250h, as described later in Fig. 3. When CO_2 concentration was increased, the EMF change (ΔEMF) of each sensor was estimated 64.7 mV, 69.7 mV and 66.3 mV, respectively. The EMFs were correlated linearly with the logarithm of CO_2 concentrations in the range of 250 to 2500 ppm. The theoretical ΔEMF_{CO2} of the potentiometoric NASICON (x = 2)-based sensor is expressed by using the next Nernstian equation [1]:

$$\Delta EMF_{CO2} = (RT/nF)\cdot \ln(P''_{CO2}/P'_{CO2}) \quad (1).$$

Where n is the number of electrons associated with the electrode reaction of CO_2, P the partial pressure, R the gas constant, T the absolute temperature and F the Faraday constant, respectively. Applying to Nernstian equation (1), the NASICON (x = 0, 2, 2.75) sensors were indicated n = 2.22, 2.06 and 2.16, respectively.

Figure 3 shows the relationship between aging time and EMF values of NASICON sensors at 450 °C in dry condition containing 250 ppm CO_2. When exposed to 250 ppm CO_2 for a long time, the EMF of each NASICON sensor decreased with aging time. EMF values of the sensors used NASICON (x = 0, 2, 2.75) sensors were stabled after aging above 50 h except for NASICION (x = 1) sensor. The electrochemical reaction of sensing and reference (Au) electrode can be explained by

$$2Li^+ + CO_2 + 1/2O_2 + 2e^- = Li_2CO_3 \quad (2)$$

$$2Na^+ + 1/2O_2 + 2e^- = Na_2O \text{ (in NASICON)}. \quad (3)$$

Therefore, the observed EMF can be given by the next Nernstian equation (4).

$$E = C + (RT/2F)\cdot \ln((a_{Li+})^2\cdot P_{CO2}\cdot a_{Na2O}/((a_{Na+})^2\cdot a_{Li2CO3})) \quad (4)$$

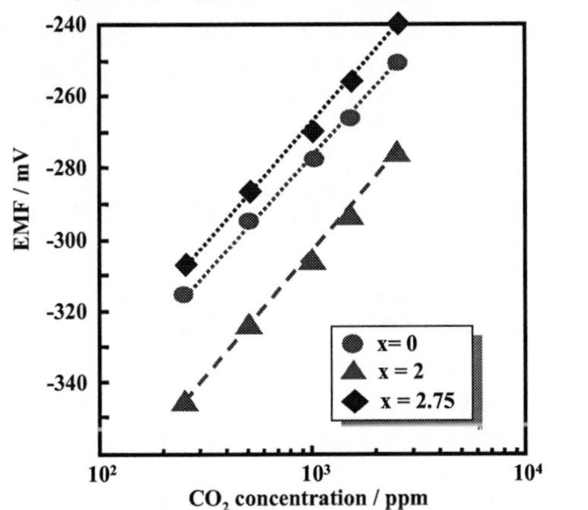

Fig.2 Correlation between EMF values and CO_2 concentrations under dry condition at 450 °C for NASICON (x = 0, 2, 2.75) sensors.

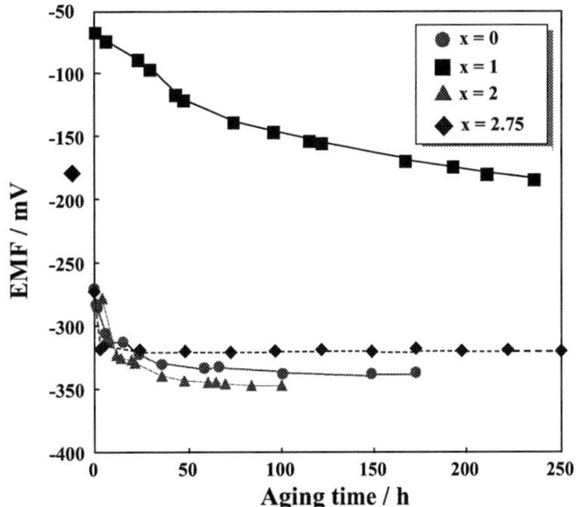

Fig.3 Relationship between aging time and EMF values of various NASICON sensors at 450 °C in dry condition containing 250 ppm CO_2.

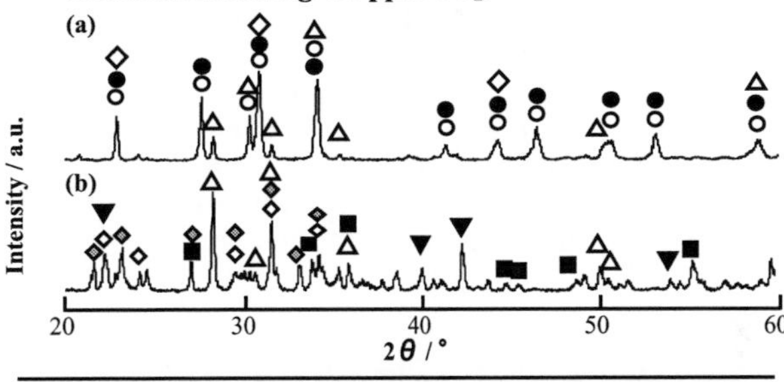

○: $Na_3Zr_2Si_2PO_{12}$ ◇: $Na_5Zr(PO_4)_3$ △: ZrO_2 ■: Na_2SiO_3
●: $Na_{3.4}Zr_2Si_{2.4}P_{0.6}O_{12}$ ◆: $Na_2Zr(PO_4)_2$ ▼: Li_2ZrO_3

Fig. 4 XRD patterns of (a) NASICON (x = 2.75) powder and (b) the mixture powder of NASICON (x = 2.75) and Li_2CO_3 after heated at 750 °C.

Here C is constant of standard electrode potential estimated from Gibbs energy and a is the activity of chemical species such as Li^+, Na^+, Li_2CO_3 and Na_2O (in NASICON), respectively. In consideration of Nernstian equation (4), it was suspected that EMF drift phenomena caused the decrease of a_{Li+} in Li_2CO_3 auxiliary or $a Na_2O$ in NASICON. In order to inspect this hypothesis, a mixture that consists of four NASICON powders and Li_2CO_3 powder was heated at 750 oC for 10 min in air. It was found that new crystal phases such as Li_2ZrO_3 and $Na_2Zr(PO_4)_2$ were formed instead of Li_2CO_3, $Na_3Zr_2Si_2PO_{12}$ and $Na_{3.4}Zr_2Si_{2.4}P_{0.6}O_{12}$ phases at the interface between NASICON (x=2.75) and Li_2CO_3 auxiliary by using XRD measurements, as shown in Fig. 4. From these results, the factor of EMF drifts might cause the decrease of Na_2O activity in NASICON by decomposition of NASICON components.

REFERENCES

[1] Yamazoe et al., *Solid State Ionics*, **86-88**, (1996) 987-993.
[2] H. Aono et al., *Sens. Actuators B*, **126** (2007) 406-414.
[3] P. Pasierb et al., *Sens. Actuators B*, **101**, (2004) 47-56.
[4] K. Obata et al., *Proc. the 12 th International Meeting on Chemical Sensors*, CBST25 (2008) p.86.

Tuning of the Sensitivity of Porous Silicon JFET Gas Sensors

G. Barillaro, L. M. Strambini, G. M. Lazzerini

Dipartimento di Ingegneria dell'Informazione: Elettronica, Informatica, Telecomunicazioni, University of Pisa, Via G. Caruso 16, 56122 Pisa - Italy
email g.barillaro@iet.unipi.it, tel +39 050 2217 601, fax +39 050 2217 522

Abstract. In this work, electrical tuning of the sensitivity of an integrated solid-state gas sensor is demonstrated. The sensor, namely PSJFET – Porous Silicon Junction Field Effect Transistor, consists of a *p*-channel JFET with an additional PS sensing gate on its top. The sensor current value is proportional to the NO_2 concentration in the environment, that is $I_{DS}= S\cdot[NO_2]$, at least in the range investigated (between 100 ppb and 500 ppb). Interestingly, and differently from most of gas sensors reported in the literature, the normalized sensor sensitivity $S=dI_{DS}/(I_{DS0}\cdot d[NO_2])$ can be effectively tuned by changing the voltage value of the electrical gate terminal of the JFET device. This feature allows the fabrication of gas sensors with superior performances: for example, it can be exploited to compensate for aging-induced degradation of the sensitivity during the sensor life-time. It is worthy of mentioning that, such an effect can be obtained without any increase of the sensor power dissipation, due to the high impedance of the gate terminal of the PSJFET.

Keywords: Porous Silicon, Integrated Gas Sensor, Sensitivity Tuning, CMOS-compatibility
PACS: 07.07.Df, 81.05.Rm, 82. 45.Vp, 85.30.Tv

INTRODUCTION

Nano-, meso- and micro-structured materials have been used for over two decades for gas sensor fabrication due to their large surface to volume ratio. This feature enables to translate surface effects, such as pollutant adsorption on the material surface, into bulk effects, such as the modification of the electrical properties of the whole material. Metal-oxides, polymers, porous silicon (PS), carbon nanotubes, and others, have been used so far [1].

Porous silicon is obtained from anodization of crystalline silicon in HF-aqueous electrolytes. The use of PS for gas sensor fabrication has been driven by the following main features: i) the possibility of changing the PS morphology, depending on the silicon substrate properties (i.e. doping type and concentration) and on the anodization parameters (i.e. current density and voltage, HF concentration, etc.), to obtain either micro-, meso- or nano-structured materials; ii) the high reactivity of the PS surface at room temperature; iii) the intrinsic compatibility of PS with CMOS industrial processes [2]. Therefore, PS enables the fabrication of small, low-cost and low-power-consumption CMOS-integrated gas sensors together with the necessary electronic circuits, as already demonstrated [3].

In ref. [3], integrated gas sensors were obtained by modification of a standard MOSFET (Metal Oxide Semiconductor FET) device upon integration of a meso-structured PS layer. The PS layer was exploited as a sensing material allowing to change the MOSFET electrical characteristics upon adsorption of pollutants into the PS-itself. The same approach was also used for the integration of other PS-based gas sensors by modifying different solid-state devices, for instance JFETs and diodes [4, 5]. Among the main advantages of such an approach there is the possibility of designing and fabricating gas sensors on the basis of well-known solid-state device, for example diodes or transistors, with increased electrical performances with respect to commonly reported conductometric sensors.

EXPERIMENTAL AND METHODS

The PSJFET is an integrated *p*-channel JFET with a meso-structured PS floating-gate on its top exploited as a sensing-gate. A sketch of the PSJFET structure (plan and cross-section views) is shown in Figure 1. By changing the polarization voltage of the electrical gate, that is the *n*-type substrate, it is possible to modulate the depletion region at the *p-n* junction and,

CP1137, *Olfaction and Electronic Nose: Proceedings of the 13th International Symposium,* edited by M. Pardo and G. Sberveglieri

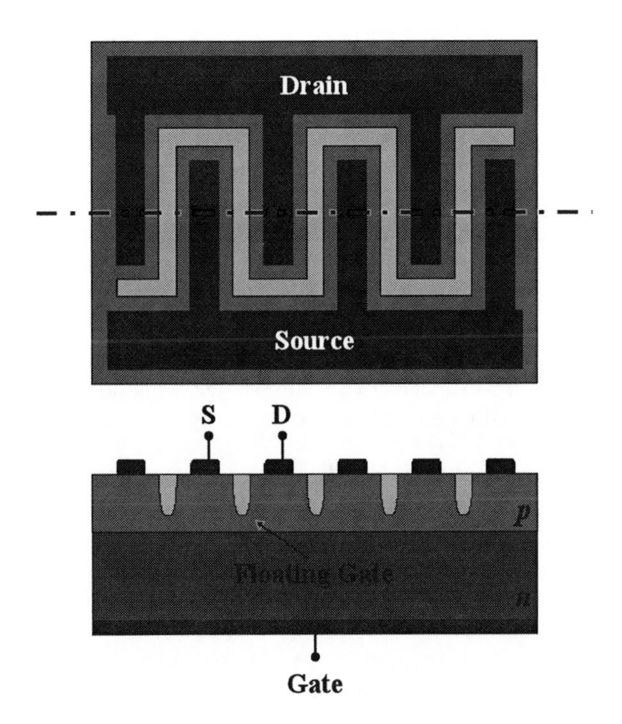

FIGURE 1. Schematic plan view (top) and cross-section (bottom) of a PSJFET sensor. Drain, source and gate electrical contacts of the JFET device are pointed out in the figure, along with the PS floating-gate between drain and source.

in turn, the p-channel section. On the other hand, adsorption of pollutants in the floating gate, that is the PS layer, modifies the PS unbalanced charge – as a function of the pollutant concentration – and, once more, the p-channel section, as a result of the space-charge width modulation at the p-Si/PS interface. In both cases, the device current I_{DS} changes according to the p-channel section modulation, for a given V_{DS} voltage.

The fabrication process of the PSJFET consists of the following main steps: 1) boron implantation and diffusion (in an n-type silicon substrate with doping of 10^{15} cm^{-3}) to form a p-type layer, 2.4 μm deep with a surface doping of 10^{17} cm^{-3}; 2) aluminium deposition on the front-side and patterning (1st mask) to define two interdigitated electrical contacts for the drain and source terminals; 3) aluminium evaporation on the back of the sample to define the electrical gate contact; 4) source and drain contacts protection by using a photoresist mask (2nd mask) hardbaked at 140 °C for 30 minutes; this step is crucial for making the resist layer able to withstand the electrochemical etching (anodization) in HF, which is necessary to the PS formation. The final step is the selective anodization of the p-type material, through the photoresist-free spaces, to produce the PS layer acting as sensing floating-gate, while leaving a crystalline p-channel underneath. The PSJFET fabrication process is straightforward and requires only two masks and a few technological steps.

Silicon chips with dimensions of 1 cm x 1 cm were processed. Several PSJFET sensors with different dimensions and characteristics (i.e. drain/source finger number, length and distance, PS layer length and width) were integrated on the same chip. Each chip was mounted on a TO12 metal-case and placed into a temperature-stabilized sealed chamber. The flow-through technique was used to test the PSJFET both in synthetic air and in NO_2, the latter at concentrations in the range between 100 ppb and 500 ppb using synthetic air as carrier gas. The electrical characterization was carried out by using a source-measure unit to record the I_{DS}-V_{DS} curves of the sensors. The polarization voltage V_{DS} was swept between 0 V and -5 V, with step of -0.05 V, while simultaneously monitoring the I_{DS} current flowing between drain and source. The effect of the electrical gate on the sensor current was evaluated by changing the V_{GS} voltage between 0 V and -1 V, with step of 0.1 V. All measurements were performed at room temperature.

RESULTS AND DISCUSSION

Figure 2 shows some typical experimental I_{DS}-V_{DS} curves of a PSJFET in synthetic air (blue traces) and at 300 ppb of NO_2 (red traces), for two different V_{GS} voltages (0 V and 0.5 V). According to the behaviour of a p-channel JFET, the current-voltage curves in synthetic air show a linear behaviour for low V_{DS} voltages (linear region) and a saturated behaviour for

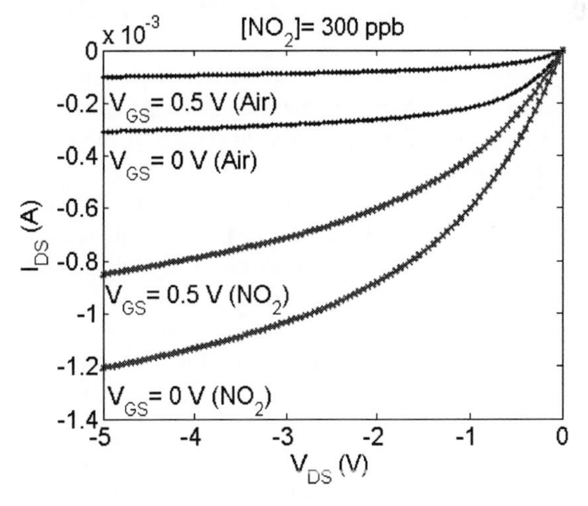

FIGURE 2. Experimental I_{DS}-V_{DS} curves of a PSJFET for two V_{GS} voltages in synthetic air and in the presence of 300 ppb of NO_2.

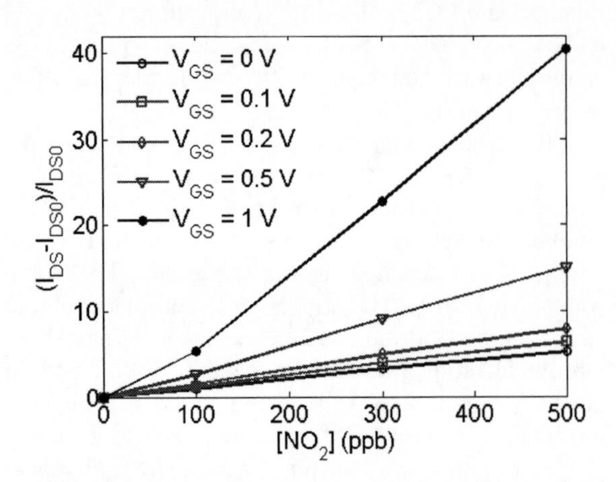

FIGURE 3. Relative current variation of a PSJFET for several NO_2 concentrations, as a function of the V_{GS} voltage value. The sensor current I_{DS} was measured with the PSJFET polarized in the saturation region, at V_{DS}=-3 V. The I_{DS0} value represents the sensor current in synthetic air.

high V_{DS} voltages (saturation region), for any gate voltage V_{GS}. Moreover, the current reduces as the absolute voltage value of the gate terminal increases. Upon injection of 300 ppb of NO_2 in the test chamber, the sensor still behaves as a *p*-channel JFET, with both a linear and a saturation region, even if the FET saturation is not as much evident as in air. However, it is clear that NO_2 injection gives rise to a significant increase of the PSJFET current, with respect to air in the same polarization (V_{DS} and V_{GS}) conditions. For instance, for V_{DS}= -5 V and V_{GS}= 0.5 V a variation of the I_{DS} current of about an order of magnitude occurs in consequence of the injection of 300 ppb of NO_2, with respect to air.

Among the different information that can be obtained from data of Figure 2, the possibility of changing the device current I_{DS} by tuning the voltage value V_{GS} of the gate terminal is definitely the most attracting one, as it can be used to compensate for fabrication errors. The more critical step of the whole PSJFET fabrication process is undoubtedly the PS layer formation. In fact, the electrical current flowing in the JFET *p*-channel mainly depends on the channel section itself. The thickness of the *p* crystalline layer underneath the PS, and thus the section of the *p*-channel, depends on the anodization time. Even if this time can be finely controlled, the resulting PS thickness can be still affected by a certain spreading. Consequently, two identical PSJFETs produced on the same chip can have a different conduction current, even in the same polarization condition. Therefore, the availability of an electrical gate terminal which allows to change the sensor current independently from the surrounding environment can be used to reduce process spreading effects, due to either the PS formation and/or other technological steps. This feature is not ordinary in gas sensors reported in the literature, where changes induced by pollutant adsorption in the electrical and/or optical properties of the material itself are usually investigated. Actually, in the PSJFET the PS layer is only exploited as a sensing material, in order to modify the electrical properties of a JFET device upon adsorption of molecules on the PS surface. Changes induced in the PS layer by the adsorption process itself, i.e. conductance variation, are not investigated. This allows the separation of adsorption-induced effects from electrically-induced effects.

Figure 3 shows the relative current variation $(I_{DS}-I_{DS0})/I_{DS0}$ of a PSJFET for several NO_2 concentrations, as a function of the gate voltage value V_{GS}. The sensor current I_{DS} was measured with the PJFET polarized at V_{DS}=-3V, that is in the saturation region. I_{DS0} represents the current value in synthetic air. The sensor current is reasonably proportional to the NO_2 concentration for all the tested gate voltages V_{GS}, at least in the concentration range investigated in this work. The linearity of the output characteristic (current *vs* concentration) is one of the key features of the proposed gas sensors, not always obtained in the literature. Moreover, Figure 3 shows that the PSJFET normalized sensitivity can be effectively tuned by changing the polarization voltage V_{GS} of the gate terminal. In fact, as the normalized sensitivity S is by definition $S=dI_{DS}/(I_{DS0}\cdot d[NO_2])$, its value is given by the slope of the curves in Figure 3, for a certain V_{GS} value. Therefore, the normalized sensitivity of the proposed sensor can be increased (reduced) by increasing (reducing) the gate voltage value. This feature can be exploited to compensate for aging-induced degradation of the sensitivity during the sensor life-time. It is worthy of mentioning that, such an effect can be obtained without any increase of the sensor power dissipation, due to the high impedance of the gate terminal of the PSJFET.

CONCLUSIONS

In this work, a PSJFET gas sensor that shows superior performances with respect to commonly fabricated sensors is reported. The PSJFET consists of an integrated *p*-channel JFET device modified with a PS layer on top of the conduction *p*-channel itself. The PS layer acts as a floating-gate and allows the modulation of the *p*-channel section, and in turn of the sensor current, upon NO_2 adsorption on the PS surface. On the other hand, the availability of an electrical gate terminal allows an effective tuning of both the sensor current and normalized sensitivity, thus enabling to compensate for fabrication-induced and/or aging-induced effects

REFERENCES

1. G. Jime´nez-Cadena, J. Riu*, F. Xavier Rius, *Analyst*, 132, pp. 1083–1099 (2007). |
2. T. L. Ritzdorf, G. J. Wilson, P. R. McHugh, D. J. Woodruff, K. M. Hanson, D. Fulton, *IBM J. Res. & Dev.*, 48 (1), pp. 65-77, (2005).
3. G. Barillaro, L.M. Strambini, *Sensors and Actuators B: Chemical*, 134 (2), pp. 585-590 (2008).
4. G. Barillaro, A. Diligenti, L.M. Strambini, E. Comini, G. Faglia, *Sensors and Actuators B: Chemical*, 134 (2), 25, pp.922-927 (2008).
5. G. Barillaro, A. Diligenti, G. Marola, L. M. Strambini, *Sensors and Actuators B*, 105 (2) 278-282 (2005).

Gas Sensing Performances of Copper Oxide Films and Quasi 1-D Nanoarchitectures

D. Barreca[1,*], E. Comini[2], A. Gasparotto[3], C. Maccato[3], G. Sberveglieri[2], E. Tondello[3]

[1] *ISTM-CNR and INSTM – Department of Chemistry – Padova University - Italy;* [2] *INFM-CNR – SENSOR Lab – Department of Chemistry and Physics - Brescia University, Italy ;* [3] *Department of Chemistry – Padova University and INSTM – Italy; *e-mail: davide.barreca@unipd.it*

Abstract. Supported copper oxide nanosystems were synthesized by Chemical Vapor Deposition (CVD) on Al_2O_3 substrates. A progressive evolution from polycrystalline Cu_2O nanodeposits to CuO samples with an entangled quasi 1-D morphology occurred upon increasing the growth temperature from 350 to 550°C. Gas sensing performances in the detection of Volatile Organic Compounds (VOCs; *e.g.* CH_3COCH_3, CH_3CH_2OH) revealed appreciable responses even at moderate temperatures, with characteristics directly dependent on the system composition and nano-organization.

Keywords: Copper Oxides; Films; Nanoarchitectures; Chemical Vapor Deposition; Gas Sensing.
PACS: 81.07.-b; 81.15.Gh; 07.07.Df

INTRODUCTION

Copper oxides (Cu_2O and CuO) are important multi-functional semiconductors for various applications [1-3]. Although many kinds of morphologies have been reported, such as wires, cubes, spheres, cages, whiskers [1,3], the use of nano-organized *p*-type Cu_xO (x=1,2) as gas sensors has not been thoroughly investigated. I the majority of works being focused on *n*-type materials [2]. In this context, nanosystems such as wires, rods and ribbons feature unique performances thanks to their large surface-to-volume ratio, along with the congruence of the carrier screening length with the lateral dimensions [2,3].

In this work, copper oxide nanosystems were synthesized by CVD, an amenable technique for nanomaterials production thanks to the possibility of controlling the Cu-O phase composition and morphology by simply varying the operating conditions [4]. For the first time, the gas sensing performances of CVD Cu_xO (x=1,2) deposits towards selected VOCs, interesting for environmental and food control tests, are presented and discussed.

EXPERIMENTAL AND METHODS

Copper oxide specimens were synthesized on polycrystalline Al_2O_3 by means of a cold-wall CVD reactor under O_2 atmospheres from a copper(II) β-diketonate (1,1,1,5,5,5-hexafluoro-2,4-pentanedionate, hfa) adduct with *N,N,N',N'*-tetramethylethylenediamine (TMEDA) [$Cu(hfa)_2$•TMEDA]. Further details on the system synthesis and characterization have been reported elsewhere [4,5]. The flow-through technique [5] was used to test the gas sensing properties, with an humidity level of 40% at p=1 atm, adopting the volt-amperometric method. A constant synthetic air flow (0.3 l/min) was used as carrier gas for the dispersion of the analytes in the desired concentration.

RESULTS AND DISCUSSION

The system structure was investigated by XRD. For T=350°C, the patterns were characterized by peaks at 2ϑ=36.3° and 42.2°, attributed to the (111) and (200) planes of cubic Cu_2O. Conversely, a T increase resulted in the predominance of signals at 2ϑ=38.8° and 48.8°, corresponding respectively to the (111) and ($\bar{2}02$) reflections of monoclinic CuO. The average nanocrystal size increased from 14 nm, at 350°C, to 26 nm, at 550°C.

Further insight into the system nano-organization were gained by FESEM analyses. At 350°C (Figure 1a), densely packed Cu_2O aggregates with a mean size of ≈50 nm, typical of an isotropic 3D growth mode [4],

CP1137, *Olfaction and Electronic Nose: Proceedings of the 13th International Symposium,* edited by M. Pardo and G. Sberveglieri

were observed. In a different way, at 550°C (Fig. 1b) quasi 1-D architectures, with lengths up to 1 μm and a mean width of ≈80 nm, were formed. These nanostructures appeared rather bent and strongly entangled, resulting in a high surface-area material. Their formation could be likely explained considering that, on increasing the growth T, the adsorbed species might progressively acquire a higher surface mobility, providing a more favorable path for the formation of the observed anisotropic structures [4].

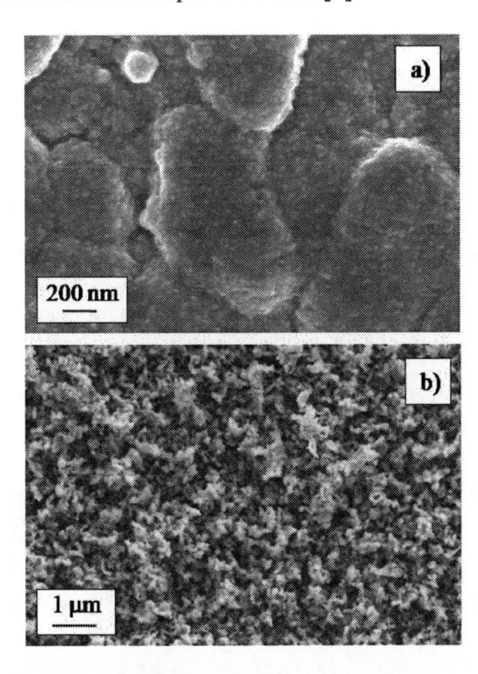

FIGURE 1. Plane-view FESEM micrographs for Cu_xO (x=1,2) samples deposited under O_2 at: a) 350°C; b) 550°C.

The obtained Cu_xO (x=1,2) nanosystems resulted sensitive to different kinds of VOCs. As an example, Figure 2 compares the dynamic response of a Cu_2O and a CuO sample in the detection of acetone. Upon exposure to the target gas, the current flowing through the samples displayed a decrease proportional to the analyte concentration, an opposite trend to that pertaining to *n*-type sensors [2,5]. In fact, in the present case, due to the adsorption and reaction between the reducing gas and the sensing oxide (electron donation), a decrease of the major *p*-type carriers, and hence of the current, occurred [3]. The sensor response, defined as the relative conductance variation $(G_{air}-G_{gas})/G_{gas}$, was ≈1 for a concentration as low as 10 ppm of acetone. The response time and recovery time were calculated as 150 and 500 s, for the CuO sample, and 200 and 500 s, for the Cu_2O one.

Notably, the observed current variations were systematically higher for the CuO sample **(b)** rather than for the Cu_2O granular film **(a)**, in contrast with previous studies [1]. In the present case, the situation was ascribed to the crucial role of the system morphology (Figure 1) [3], resulting in a higher surface-to-volume ratio and active sites density for the CuO specimen. Apart from enabling a more efficient target gas uptake, these features provide an increased depletion of charge carriers in the quasi-1D nanostuctures with respect to more compact systems [3,5]. These features determine the higher sensing response of the CuO specimen.

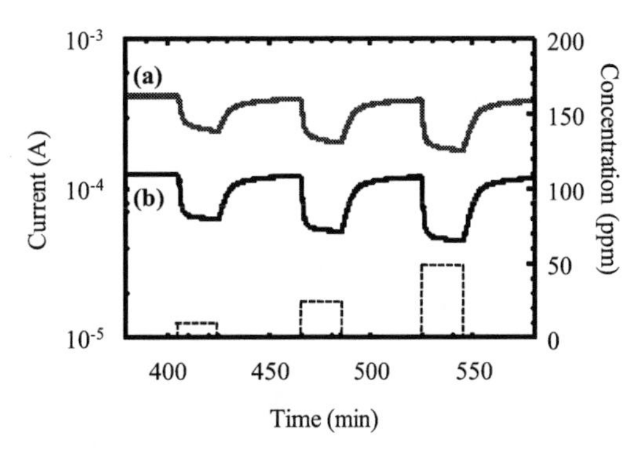

FIGURE 2. Dynamic response upon exposure to acetone concentration pulses of Cu_xO (x=1,2) samples deposited at: **(a)** 350°C; **(b)** 550°C (working temperature=300°C).

CONCLUSIONS

This work has been dedicated to the CVD synthesis and characterization of copper oxide nanosystems from $Cu(hfa)_2$•TMEDA on polycrystalline Al_2O_3, as well as to the investigation of their gas sensing properties. It is worth highlighting that no previous works on the topic have ever appeared in the literature up to date. The obtained data have showed the obtainment of Cu_2O or CuO as a function of the selected growth temperature, with a parallel morphological evolution from films to entangled quasi 1-D nanoarchitectures. Preliminary sensing measurements evidence appreciable responses even at moderate working temperatures.

REFERENCES

1. J. Zhang, J. Liu, Q. Peng, X. Wang and Y.Li, *Chem. Mater.* **18**, 867-871 (2006).
2. X. Wang, X. Q. Fu, X. Y. Xue, Y. G. Wang and T. H. Wang, *Nanotechnology* **18**, 145506/1-5 (2007).
3. X. Gou, G. Wang, J. Yang, J. Park and D. Wexler, *J. Mater. Chem.* **18**, 965-969 (2008).
4. D. Barreca, A. Gasparotto, C. Maccato, E. Tondello, O. I. Lebedev and G. Van Tendeloo, *Cryst. Growth Des.*, in press (2009).
5. D. Barreca, A. Gasparotto, C. Maccato, C. Maragno, E. Tondello, E. Comini and G. Sberveglieri, *Nanotechnology* **18**, 125502/1-6 (2007).

Photoexcited Individual Nanowires: Key Elements in Room Temperature Detection of Oxidizing Gases

J. D. Prades[a,ϒ], R. Jimenez-Diaz[a], M. Manzanares[a], F. Hernandez-Ramirez[b,ϒ], T. Andreu[a], A. Cirera[a], A. Romano-Rodriguez[a], J. R. Morante[a,c]

[a]*EME/XaRMAE/IN^{2}UB, Dept. Electrònica, Universitat de Barcelona, C/ Martí i Franquès 1, E-08028 Barcelona Spain*
[b]*Electronic Nanosystems S. L., Barcelona, Spain*
[c]*Institut de Recerca en Energia de Catalunya (IREC), C/ Josep Pla 2, B3, PB, E-08019 Barcelona, Spain*

Abstract. Illuminating metal oxide semiconductors with ultra-violet light is a feasible alternative to activate chemical reactions at their surface and thus, using them as gas sensors without the necessity of heating them. Here, the response at room temperature of individual single-crystalline SnO_2 nanowires towards NO_2 is studied in detail. The results reveal that similar responses to those obtained with thermally activated sensors can be achieved by choosing the optimal illumination conditions. This finding paves the way to the development of conductometric gas sensors operated at room temperature. The power consumption in these devices is in range with conventional micromachined sensors.

Keywords: gas sensor, room temperature, photoexcitation, ultraviolet light, nanowire
PACS: 85.85.+j, 81.07.-b, 81.16.Nd, 73.63.Bd, 73.63.Rt

INTRODUCTION

Gas detection, either for health or environmental applications, is a major area of interest in metal oxide nanowires, since their high surface-to-volume ratio makes their electrical properties extremely sensitive to surface-adsorbed molecules [1]. One of the main advantages of these nanosensors is the potential to detect extremely low concentrations of chemicals onto low power platforms [2] small enough to be used on microintegrated electronic nose systems [3].

Bulk metal oxide semiconductors require high temperatures to activate the chemical sensing reactions and accelerate their dynamics. Here, we demonstrate that light can be used to photoactivate SnO_2 nanowire-based gas sensors in order to detect sub-ppm concentrations of NO_2 at room temperature. These proof-of-concept devices (Figure 1) exhibit responses towards different concentrations of NO_2 fully equivalent to those obtained with an external microheater [4].

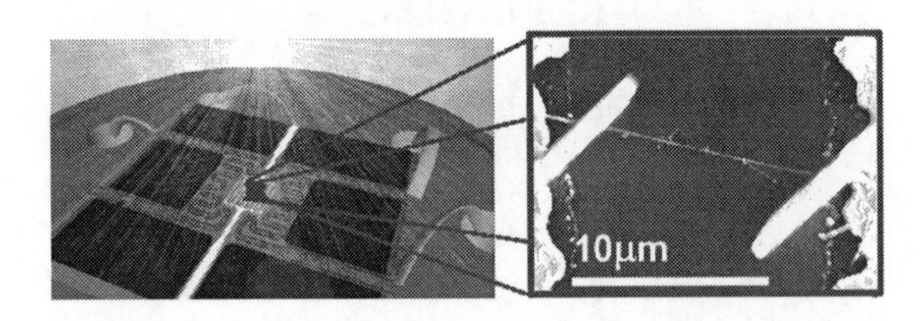

FIGURE 1. Schematic representation of the proof-of-concept device and SEM micrograph of a nanowire contacted in 2-probe configuration.

EXPERIMENTAL METHODS

Single-crystal SnO_2 nanowires, synthesized through catalyst supported chemical vapor deposition of a molecular precursor $[Sn(O^tBu)_4]$ [5], were transferred onto suspended silicon micromembranes with Pt interdigitated microelectrodes. Individual nanowires were electrically contacted to platinum microelectrodes in a lithography process using a FEI Dual-Beam Strata 235 FIB instrument [6]. A set of monochromatic LED sources (photon energies from $E_{ph} = 4.00 \pm 0.05$ eV to 0.75 ± 0.01 eV) were used to illuminate the devices. Two and four-probe DC measurements were performed to calibrate parasitic contacts effects [7] using an electronic circuit designed to guarantee and control low current levels (0.1 nA) and to prevent any undesired fluctuation [3] and self-heating effects [2].

ϒ Current affiliation: Institut de Recerca en Energia de Catalunya (IREC)

CP1137, *Olfaction and Electronic Nose: Proceedings of the 13th International Symposium,* edited by M. Pardo and G. Sberveglieri

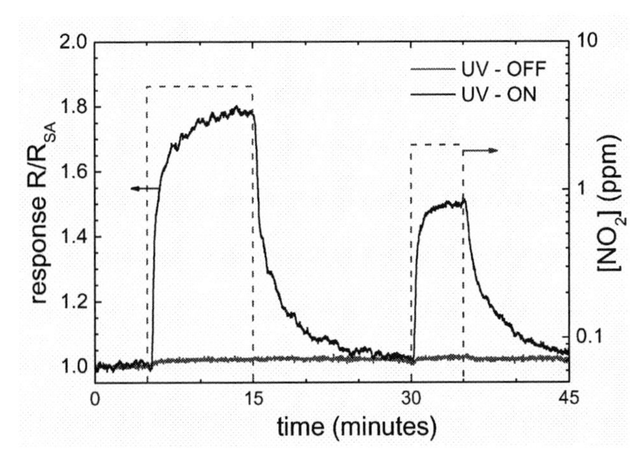

FIGURE 2. Response of a SnO_2 nanowire, operated in dark and UV illuminated ($E_{ph} = 3.67 \pm 0.05$ eV, $\Phi_{ph} = 30 \cdot 10^{22}$ ph m^{-2} s^{-1}), to NO_2 pulses of different concentration. Both measurements were obtained without external heating sources at room temperature (T = 25°C).

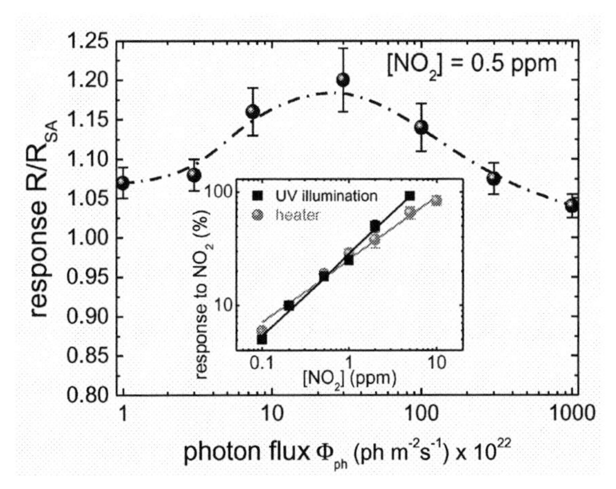

FIGURE 3. Sensor response versus the flux of photons (E_{ph} = 3.67 ± 0.05 eV). The maximum response clearly depends on the flux of photons. (Inset) Comparison of the response towards different concentrations of $[NO_2]$ of a nanowire based sensor operated with UV illuminated at room temperature (T = 25 °C) and operated with an integrated microheater (T = 175 °C).

RESULTS AND DISCUSSION

In dark conditions, individual SnO_2 nanowires exhibited low responses towards NO_2 at room temperature (T = 25 °C) without any noticeable recovery of the resistance baseline (Figure 2). On the contrary, the same devices displayed significant responses towards NO_2 pulses (concentrations from 100 ppb to 10 ppm) and fully recover the base line after few minutes under UV illumination (Figure 2). It is noteworthy that this detection range is appropriate for many safety applications [8].

According to first-principles calculations and Temperature Programmed Desorption experiments [9], NO_2 molecules adsorb onto surface oxygen vacancies (VO) with desorption energies above $E_{des} \geq 0.52$ eV, which corresponds to a thermal desorption process active above T ≥ 80°C. Thus, at room temperature (T = 25°C) desorptions are scarce and no recovery of the baseline after exposure to gas is expected.

Experiments show that the response to NO_2 strongly depends on the illumination conditions. First of all, photons with energies above the band gap of the nanowire lead to the highest responses. Second, the dependence of the response on the photon flux is significant and complex (Figure 3). Nevertheless, under the optimum conditions of energy and flux, the performance of these devices is similar to those obtained external microheaters (inset in Figure 3).

The response times of these devices are of only few minutes in the optimum illumination conditions. These values are fully in range with those obtained with heated devices. As far as the energetic efficiency is concerned, the power required to operate these UV LED-based sensors is comparable with the one needed to operate conventional microheater-based gas sensors (tens of miliwatt). The here-presented approach simplifies the layout of the underlying substrates since no integrated heater is required.

CONCLUSIONS

We demonstrate that the illumination of metal oxide nanowires can be used to photoexcite their electrical response towards gases such as NO_2 without the requirement of external heaters. These devices represent and important advance in room temperature gas sensors and are specially suited for explosive environments. Moreover, they are as energetically efficient as microheater-based devices, even at this early stage. All these properties are attractive for the development of future e-nose technologies.

ACKNOWLEDGMENTS

A Spanish patent application related to the above disclosed features has already been filled.

This work was partially supported by the Spanish Government [projects N – MOSEN (MAT2007-66741-C02-01), and MAGASENS], the UE [project NAWACS (NAN2006-28568-E), the Human Potential Program, Access to Research Infrastructures]. JDP and RJD are indebted to the MEC for the FPU grant. Thanks are due to the European Aeronautic Defense and Space Company (EADS N.V.) for supplying the suspended micromembranes.

REFERENCES

1. A. Diéguez, A. Vilà, A. Cabot, A. Romano-Rodríguez, J.R. Morante, J. Kappler, N. Bârsan, *Sens. Actutators B.* **68**, 94-99 (2000).
2. J. D. Prades, R. Jimenez-Diaz, F. Hernandez-Ramirez, S. Barth, A. Cirera, A. Romano-Rodriguez, S. Mathur and J. R. Morante, *Appl. Phys. Lett.* **93**, 123110 (2008).
3. F. Hernandez-Ramirez, J. D. Prades, A. Tarancon, S. Barth, O. Casals, R. Jimenez-Diaz, E. Pellicer, J. Rodríguez, M. A. Juli, A. Romano-Rodriguez, J. R. Morante, S. Mathur, A. Helwig, J. Spannhake and G. Mueller, *Nanotechnol.* **18**, 495501 (2007).
4. J. D. Prades, R. Jimenez-Diaz, F. Hernandez-Ramirez, S. Barth, J. Pan, A. Cirera, A. Romano-Rodriguez, S. Mathur, J. R. Morante, *Sens. Actuators B.*, submitted (2009).
5. (a) S. Mathur, S. Barth, H. Shen, J.-C. Pyun and U. Werner, *Small* **1**, 713-717 (2005). (b) S. Mathur and S. Barth, *Small* **3**, 2070-2075 (2007).
6. F. Hernandez-Ramirez, A. Tarancon, O. Casals, J. Rodríguez, A. Romano-Rodriguez, J. R. Morante, S. Barth, S. Mathur, T. Y. Choi, D. Poulikakos, V. Callegari and P. M. Nellen, *Nanotechnol.* **17**, 5577-5583 (2006).
7. F. Hernandez-Ramirez, A. Tarancon, O. Casals, E. Pellicer, J. Rodríguez, A. Romano-Rodriguez, J. R. Morante, S. Barth and S. Mathur, *Phys. Rev. B* **76**, 085429 (2007).
8. World Health Organisation. Information available in http://www.who.int/peh/air/Airqualitygd.htm.
9. (a) J. D. Prades, A. Cirera and J. R. Morante, *J. Electrochem. Soc.* **154**, H675-H680 (2007). (b) J. D. Prades, A. Cirera, J. R. Morante, J. M. Pruneda and P. Ordejón, *Sens. Actuators B* **126,** 62-67 (2007).

The Gas-Sensing Characteristics Of Percolating 2-D SnO_2 Nanowire Mats As A Platform For Electronic Nose Devices

V. V. Sysoev[1], I. Kiselev[2], T. Schneider[2], M. Bruns[2], M. Sommer[2], W. Habicht[2], V. Yu. Musatov[1], E. Strelcov[3], A. Kolmakov[3]

[1]*Saratov State Technical University, Polytechnicheskaya 77, Saratov 410054, Russia*
[2]*Forschungszentrum Karlsruhe, Hermann-von-Helmholtz-Platz 1, 76344 Eggenstein-Leopoldshafen, Germany*
[3]*Southern Illinois University at Carbondale, Carbondale, IL 62901-4401, USA*

Abstract. We describe gas-sensing characteristics of percolating SnO_2 nanowire (NW) mats employed in Electronic nose (E-nose) instrument. The current strategy is based on combining bottom-up technology of NWs growth and top-down fabrication of multisensor microarray according to KAMINA (KArlsruhe Micro NAse) E-nose architecture. Such issues of the NW-based multisensor systems are discussed as gas-sensing stability, gas sensitivity and gas classification using Linear Discriminant Analysis (LDA) pattern recognition technique.

Keywords: oxide, nanowire, gas sensor, multisensor system, electronic nose.
PACS: 07.07.Df; 68.47.Gh; 68.65._k; 73.63._b; 81.07._b; 85.85._j

INTRODUCTION

Until recently, the methods to fabricate sensor arrays for E-noses included mostly top-down processes as film sputtering, lithography and etching. These techniques allow a good microscopic control over the functional properties of the sensors which, however, goes down when nanometer-scaled structures are used as sensing elements. Nevertheless, the nanometer-scaled elements are the most prospective ones to improve the sensitivity of chemical sensors to gas concentrations in sub-*ppm* or *ppb* range that is a crucial issue for many possible applications of E-nose devices. On the other hand, the (bottom-up) grown nanostructures, as single-crystal nanowires, nanobelts, nanowalls, etc, can be large-scale fabricated via rather simple protocols from a liquid or vapor phases. The size domain of such nanostructures perfectly satisfies the requirements for high gas-sensing sensor arrays, and, therefore, they are considered as promising functional elements to be used in new generation of E-noses [1,2]. Here we combine earlier developed microelectronics technology of patterning films by coplanar electrodes to make a multisensor microarray KAMINA E-nose with SnO_2 NWs as sensing elements. The latter were grown by vapor-liquid-solid (VLS) process and placed as a mat over or beneath the electrodes in the microarray. The current state of studying the sensing performance of such NW-based microarrays is presented.

EXPERIMENTAL AND METHODS

The multisensor microarray KAMINA is described in details somethere else [3].

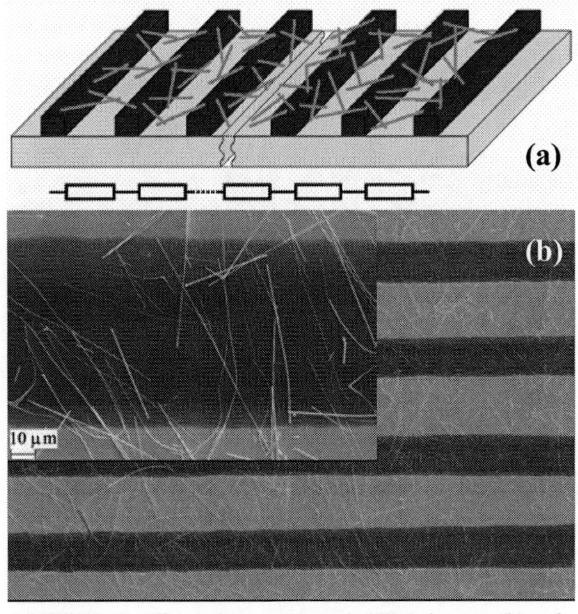

FIGURE 1. The scheme (a) and SEM image (b) of multisensor microarray employing the 2-D SnO_2 NW mat.

CP1137, *Olfaction and Electronic Nose: Proceedings of the 13th International Symposium,* edited by M. Pardo and G. Sberveglieri

The VLS protocol of SnO2 NWs growth is given in [4,5]. After the growth, the NWs are mechanically placed over Si/SiO2 substrates with prior or afterward patterning by multi-electrodes according to KAMINA chip architecture. The rear side of the substrate is equipped with four Pt film heaters to provide the operating temperature profile. The gas sensing characteristics presented are recorded using PC-controlled gas-mixing setup. For gas-recognition purposes, the transient values of conductances obtained under the change of gas mixture are removed from the consideration. For LDA processing, all the sensor signals are normalized as $R_i \to r_i = R_i / R_{med}$, where R_i is the sensor resistance of ith sensor segment; R_{med} the median resistance value over the whole microarray. The recognition power is determined by Mahalanobis distance between clusters related to gases in the LDA coordinate system.

RESULTS

It s shown that the sensor microarrays based on 2-D SnO_2 NW mats have a great gas sensitivity using pristine NWs without even any additional doping. The alcohol vapors are measurable at the sub-*ppm* level while CO at the *ppm* level (Fig. 2(a,b)). The NW mat-based sensors have a stable gas sensing performance and high recognition power (Fig. 2c). The latter one is caused by a large stochastic differentiation of segments due to interplay between potential barriers at the NW contacts and depletion of NW bulk. The combination of these two gas-sensing contributions makes unique percolation paths through the NW mats at each sensor segment and enhances the gas recognition power of the sensor array. We also employ a spatially varied operating temperature to additionally differentiate sensor segment gas responses. The further work is dealt with catalyst functionalization, size reduction and NW mat density control over the micro array.

ACKNOWLEDGMENTS

The authors thank G. Stengel, J. Benz, A. Serebrenicov, V. Hermann and U. Geckle for assistance in the fabrication and characterization of microarray chips. S.V.V. thanks the INTAS grant, YSF 06-1000014-5877, and Fulbright scholarship.

REFERENCES

1. E. Comini, C. Baratto, G. Faglia, M. Ferroni, A. Vomiero, G. Sberveglieri, *Progress in Materials Science* **54**, 1-67 (2009).
2. A. Kolmakov, *International Journal of Nanotechnology* **5**, 450-474 (2008).
3. J. Goschnick, *Microelectronic Engineering* **57-58**, 693-704 (2001).
4. V. V. Sysoev, B. K. Button, K. Wepsiec, S. Dmitriev, A. Kolmakov, *Nano Letters* **6**, 1584-1588 (2006).
5. V. V. Sysoev, J. Goschnick, T. Schneider, E. Strelcov, A. Kolmakov, *Nano Letters* **7**, 3182-3188 (2007).

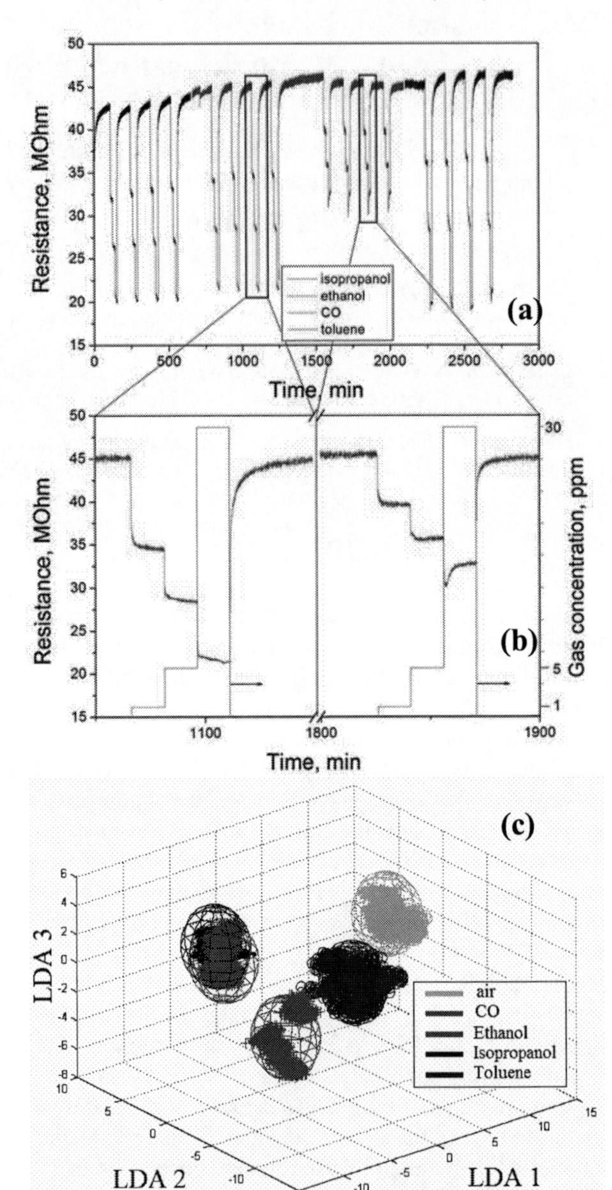

FIGURE 2. (a,b) - the change of median sensor segment resistance of NW–based multisensor system at exposure to four sample gas mixtures (CO, ethanol, isopropanol, toluene) with synthetic air; (c) – the gas classification by LDA of the same sample gas mixtures (4 exposures of each gas at 1, 5, 30 ppm concentration). The NW-based chip was operated under application of gradient of operating temperature over the substrate.

Development Of Single Wall Carbon Nanotube Based Sensor

B.Lerner [a], M.S.Perez [a], J.Bonaparte [a], D.J.Rodriguez [a], H.Pastoriza [b], A.Boselli [a] and A. Lamagna [a]

[a] Comision Nacional de Energia Atomica, Grupo MEMS.
Av. Gral. Paz 1499. (1650) San Martín, Bs. As. Argentina
[b] Laboratorio de Bajas Temperaturas, Centro Atómico Bariloche
Av. Bustillo 9500 , S. C. de Bariloche, Argentina
blerner@cnea.gov.ar. Tel: 5411-6772-7931

Abstract. Single wall carbon nanotubes (SWCNTs) have been found to be sensitive to gases, temperature and biological molecules, and have shown great promise as sensing elements. We developed SWCNTs sensors chemically functionalized with the COOH groups by oxidation and demostrated that the sensors are sensitive to changes in humidity.

Keywords: Single wall Carbon Nanotube Sensor.
PACS: 85.35.Kt

INTRODUCTION

SWCNTs have been shown to be good sensing elements for pressure, gases and alcohol, and are potential candidates for nanoscale biosensing systems because of their chemical and biochemical stability.

We used the SWCNTs sensor to investigate the effect of humidity on the electrical transport properties of SWCNTs.

EXPERIMENTAL AND METHODS

The nanotubes used in this study have been obtained by the catalytic CoMoCAT, which employs a silicasupported Co-Mo powder to catalyze the selective growth of SWCNTs by disproportionation of CO[1].

SWCNTs were sonicated in 3:1 concentrated sulfuric acid and nitric acid for 4 hs. By this method, the SWCNTs can be oxidized and COOH groups will be grafted along the sidewall and the tube ends of the SWCNTs. The SWCNTs were collected and washed thoroughly with DI-water until the pH value was 6–7. The SWCNTs were re-dispersed in buffer MES and drops deposited in a Silicon wafer. Then electrodes were fabricated on the Si-wafer by the lift-off process.

The Sensors were located in a chamber (volume 500 cc) where air flow at a constant rate of 300 sccm and humidity was controlled with a 0.1 % precision at 298K. Current was supplied by Keithley 6221 current source and Voltage changes were measured with Keithley 2000 multimeter.

RESULTS

Figure 1 illustrates the Raman spectra of the nanotubes before and after the oxidation using an excitation laser with wavelength 633 nm. A difference in the relative intensity of the D band (near 1300 cm-1) with respect to the main G band is evident. The increase in the D band has been previously reported and used as an indication of covalent side-wall functionalization, as it reflects the conversion of the hybridization of some C atoms on the nanotube wall from sp2 to sp3.

Figure 2 shows that SWCNTs-COOH sensors are sensitive to changes in humidities at negative voltages.

CP1137, *Olfaction and Electronic Nose: Proceedings of the 13th International Symposium,* edited by M. Pardo and G. Sberveglieri

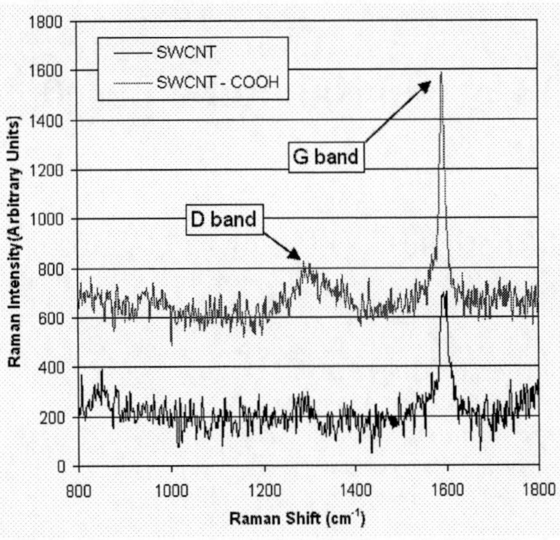

FIGURE 1. Raman spectra of SWCNTs and SWCNTs-COOH.

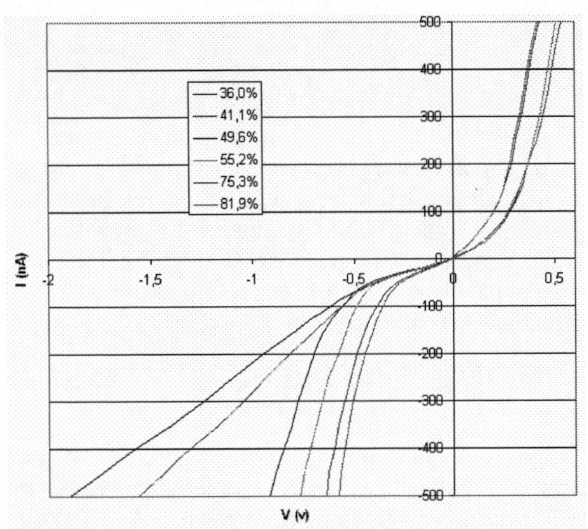

FIGURE 2. Effect of humidity over current voltaje characteristics of SWCNTs with COOH groups at 298K.

CONCLUSIONS

We successfully developed and tested SWCNTs-COOH sensors at different humidities.

ACKNOWLEDGMENTS

The authors thank Southwest Nanotechnologies (Norman, Oklahoma) for providing the purified CoMoCAT single-walled carbon nanotubes.

REFERENCES

1. F. Buffa and D.E. Resasco, Macromolecules 38, 8258-8263 (2005).
2. D. R. Kauffman, D.R. Star, Angewandte Chemie International Edition 47, 6550-6570 (2008).
3. D.E. Resasco, W.E. Alvarez, F. Pompeo, L. Balzano, J.E. Herrera, B. Kitiyanan and A. Borgna, J. Nanoparticle Res. 4, 131-136. (2002).
4. P.S. Na, H. Kim, H. So, K. Kong, H. Chang, B.H. Ryu, J. Lee, et al, Appl. Phys. Lett 87, 44-66 (2005).

Development Of Hot Surface Polysilicon-Based Chemical Sensor And Actuator With Integrated Catalytic Micropatterns For Gas Sensing Applications

E. Vereshchagina, J.G.E. Gardeniers

Mesa+ Institute for Nanotechnology, University of Twente
P.O. Box 217, 7500 AE Enschede, the Netherlands

Abstract. Over the last twenty years, we have followed a rapid expansion in the development of chemical sensors and microreactors for detection and analysis of volatile organic compounds. However, for many of the developed gas sensors poor sensitivity and selectivity, and high-power consumption remain among one of the main drawbacks. One promising approach to increase selectivity at lower power consumption is calorimetric sensing, performed in a pulsed regime and using specific catalytic materials. In this work, we study kinetics of various catalytic oxidation reactions using micromachined hot surface polysilicon-based sensor containing sensitive and selective catalysts. The sensor acts as both thermal actuator of chemical and biochemical reactions on hot-surfaces and detector of heats (enthalpies) associated with these reactions. Using novel deposition techniques we integrated selective catalysts in an array of hot plates such that they can be thermally actuated and sensed individually. This allows selective detection and analysis of dangerous gas compounds in a mixture, specifically hydrocarbons at concentrations down to low ppm level. In this contribution we compare various techniques for the local immobilization of catalytic material on hot spots of the sensor in terms of process compatibility, mechanical stress, stability and cost.

Keywords: gas sensor, actuator, polysilicon, catalytic oxidation, catalyst deposition
PACS: 82.47.Rs; 82.45.Rr; 82.45.Jn; 81.16.Dn

INTRODUCTION

In many industrial and domestic applications, it is important to measure the concentration of potentially dangerous gases such as hydrocarbons, sulfur compounds, carbon monoxide, carbon dioxide, amines and many others.

We propose to monitor the concentration of gases using a temperature actuation and sensing mechanisms in presence of a specific catalytic material. Temperature is an essential parameter for studying the kinetics and thermodynamics of highly exothermic or dangerous reactions on surfaces.

The physical principle of detection is based on a measure of heats (enthalpies) released during the oxidation reaction. The heat produced during this reaction is subsequently measured and is proportional to concentration of reacting species. We can shift the sensor-actuator performance towards the detection of a particular gas compound by selecting the optimal catalyst and temperature. In our approach, catalysts for various reactants are combined in an array (Figure 1). The main drawback of a multicomponent calorimetric analytical gas system is its cross-sensitivity to interfering gases [1]. This can be eliminated by sequential thermal actuation of integrated catalysts up to the ignition temperatures of the target gases. This provides enough selectivity, even between chemical compounds that have similar thermodynamic properties.

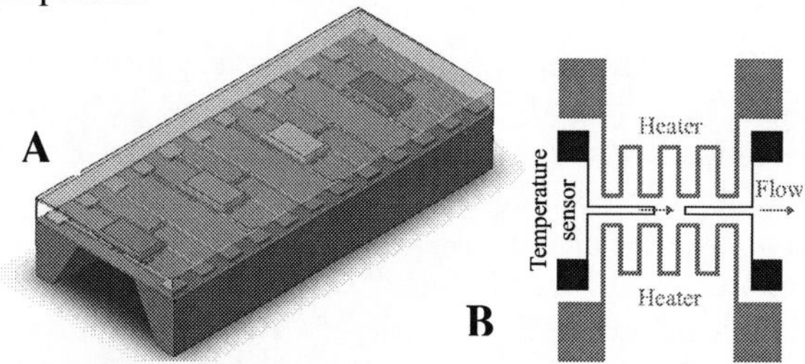

FIGURE 1. Different catalytic materials combined in an array on (optional) a gradient of reaction temperatures produced by heaters (A), top view of heaters /sensors (B).

Such a gas micro analytical system is a flexible platform for (bio-) catalyst screening, reaction kinetic studies or industrial optimization.

CP1137, *Olfaction and Electronic Nose: Proceedings of the 13th International Symposium,* edited by M. Pardo and G. Sberveglieri

EXPERIMENTAL AND METHODS

The prototype device consists of polySi++ heaters and temperature sensors integrated on a freely suspended silicon-rich silicon nitride membrane (Figure 2).

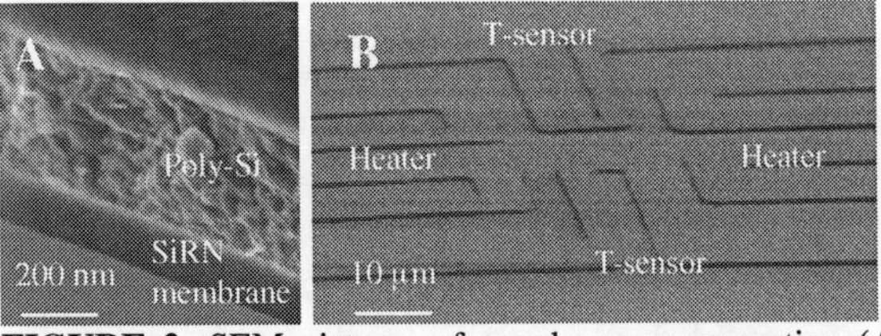

FIGURE 2. SEM pictures of membrane cross section (A) and fabricated heaters and temperature sensors (B).

Polysilicon heaters supply the required power homogeneously to the active sensor surface on which the catalytic material is deposited. Temperature sensors are located up and downstream of the gas flow. They measure reaction heats which are evolved in a catalytic oxidation and temperature gradient between gas inlet and outlet induced by other heat dissipation mechanisms (conduction of leads and chip holder, convection through the gas). Normally, the device is kept in a nonisothermal mode, so the increase in temperature on the surface due to oxidation is measured. The dependence between the reaction rate and corresponding increase in temperature is determined analytically before calibration experiments for different gas concentrations. The sensor signal is a result of temperature rise (fall) or difference in supplied power (in this case temperature of the system is not allowed to exceed a certain temperature limit).

RESULTS

The fabricated structures were electrically characterized and calibrated using the Cascade Microtech IV/CV Measurement Low Leakage Probe Station. The resulting sheet resistance for polysilicon thin films is in the range 80-140 Ω/□. The TCR is positive and ca. 0.5×10^{-3}/°C. We observed no significant hysteresis in measured IV characteristics.

We developed techniques for local deposition and patterning of gas sensitive materials on structured surfaces in submicron range. We fabricated patterns of highly porous γ-Al_2O_3 with finely dispersed Pt and Pd catalytic nanoparticles by combining sol-gel chemistry methods of catalyst preparation with soft lithographic [2,3] immobilization techniques (Figure 3). The prepared coatings show a strong adhesion to the support. Forming catalytic patterns on sensor membrane effectively distributes the stress and gives extra access for reactant gases to active centers, which contributes to a faster response and higher signal of sensor device.

Experimental results show a feasibility of using such system as a microcalorimeter for rapid identification of reaction heats and a gas sensor for detection presence or concentrations of dangerous gas compounds.

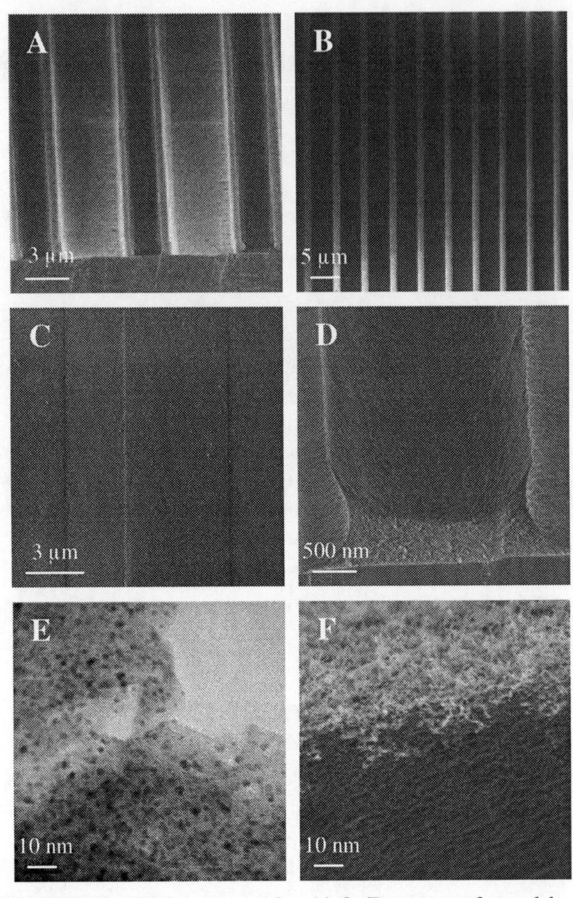

FIGURE 3. SEM images of γ-Al_2O_3/Pt pattern formed by various soft lithographic techniques on the membrane surface (A, B, C), sintering of patterned oxide (D), and TEM image of γ-Al_2O_3/Pt catalyst (E), CNF immobilized on patterned γ-Al_2O_3 support on a membrane(F).

ACKNOWLEDGMENTS

The project is financed by Dutch Technology Foundation (STW).

REFERENCES

1. Zanini, M. et al., *Sens. Actuators A,* 48(3) 187-192 (1995).
2. Rogers, J.A., *Mater. Today*, 48(2), 50-56 (2005).
3. E.Kim, Y.Xia, G.M.Whitesides, *J. Am. Chem. Soc.* 118, 5722-5731 (1996).

ANALYTICAL CHEMISTRY TOPICS

Analysis of Odor-Causing VOCs and Semi-VOCs Associated with Particulate Matter in Swine Barns Using SPME-GC-MS-Olfactometry

Lingshuang Cai, Jacek A. Koziel*, Yin-Cheung Lo, and Steven J. Hoff

Department of Agricultural & Biosystems Engineering, Iowa State University, Ames, IA 50011, USA; koziel@iastate.edu, 515-294-4206

Abstract. Swine operations can affect air quality by emissions of odor, volatile organic compounds (VOCs) and other gases, and particulate matter (PM). Particulate matter has been proposed to be an important pathway for carrying odor. However, little is known about the odor-VOCs-PM interactions. In this research, continuous PM sampling was conducted simultaneously with three collocated TEOM analyzers inside a 1000-head swine finish barn located in central Iowa. Each TEOM (tapered element oscillating microbalance) was fitted with total suspended particulate (TSP), PM-10, PM-2.5 and PM-1 preseparators. Used filters were stored in 40 mL vials and transported to the laboratory. VOCs adsorbed/absorbed to dust were allowed to equilibrate with vial headspace. Solid-phase microextraction (SPME) Carboxen/polydimethylsiloxane(PDMS) 85 μm fibers were used to extract VOCs. Simultaneous chemical and olfactometry analyses of VOCs and odor associated with swine PM were completed using a gas chromatography-mass-olfactometry (GC-MS-O) system. Fifty VOCs categorized into nine chemical function groups were identified and confirmed with standards. Five of them are classified as hazardous air pollutants. VOCs were characterized with a wide range of molecular weight, boiling points, vapor pressures, water solubilities, odor detection thresholds, and atmospheric reactivities. All characteristic swine VOCs and odorants were present in PM and their abundance was proportional to PM size. However, the majority of VOCs and characteristic swine odorants were preferentially bound to smaller-size PM. The findings indicate that a significant fraction of swine odor can be carried by PM. Research of the effects of PM control on swine odor mitigation is warranted.

Keywords: Odor, particulate matter, VOCs, swine, SPME.
PACS: 01.30.Cc.

INTRODUCTION

Livestock operations are sources of aerial emissions of gases, odor, and particulate matter In recent decades, intensive large-scale swine production has grown rapidly in the U.S. and other parts of the world. Most modern swine operations raise hogs in confinement buildings. The large number of animals raised in concentrated animal feeding operations (CAFOs) can affect air quality by emissions of odor, volatile organic compounds (VOCs) and other gases, and particulate matter (PM).

Previous studies focused mainly on total PM in swine housing. To date, still little is known about the odor-VOCs-PM interactions particularly for PM sizes of interest to regulatory agencies. In this study, headspace (HS) SPME combined with GC-MS-O system was used to identify VOCs and characterize the key odors adsorbed/absorbed on different size swine barn dust (PM-1, PM-2.5, PM-10 and the total suspended particulate (TSP)).

EXPERIMENTAL AND METHODS

Three collocated TEOM samplers were placed in one 2.4 m × 6 m pen reserved for this study. The building used for this study was designed to house pigs from 20 to 120 kg and during this study pigs averaged 60 kg. Pigs were present in all pens surrounding the pen containing the TEOM samplers.

Solid-Phase Microextraction

SPME extractions were performed with a manual fiber holder from Supelco. Screw-capped vials (40 mL) were used for storing used TEOM filters and for HS-SPME sampling. During SPME extraction the

CP1137, *Olfaction and Electronic Nose: Proceedings of the 13th International Symposium*, edited by M. Pardo and G. Sberveglieri

septum was pierced using the SPME needle and exposed the SPME fiber to the headspace for 3 hr. After extraction, the SPME fiber was removed from the vial and immediately inserted into the injection port of GC for analysis.

Gas Chromatograph –Mass Spectrometry-Olfactory System

Multidimensional GC-MS-O (from Microanalytics, Round Rock, TX, USA) was used for all analyses. The system integrates GC-O with conventional GC-MS (Agilent 6890N GC / 5973 MS from Agilent, Wilmington, DE, USA) as the base platform with the addition of an olfactory port and flame ionization detector (FID). Mass/molecular weight to charge ratio range was set between 33 and 280.

RESULTS

Identification of VOCs Associated with Swine Barn PM

A total of 50 different compounds were identified, of which 21 have never been reported to be present in swine barn dust in previous studies. Some odorous compounds that have not been reported include pentane, methyl mercaptan, trimethyl amine, 3-pentanamine, diacetyl, dimethyl sulfone, styrene, 2-pentyl furan, and 2'-aminoacetophenone. The fifty compounds identified cover a wide range of polarity and molecular weight (34.08-234.39) and belong to nine chemical classes: alkanes (4), alcohols (4), aldehydes (8), ketones (7), acids (8), amines and nitrogen heterocycles (8), sulfides and thiols (3), aromatics(7) and furans (1). The main chemical classes involved in odorous emissions from swine buildings previous identified were also identified in this study: volatile fatty acids, aromatics (4-methyl phenol and 4-ethyl phenol), nitrogen heterocycles (indole and skatole), thiols and mercaptans. Five of the compounds identified are classified as hazardous air pollutants (HAPs): styrene, N, N-dimethyl- formamide, acetamide, phenol and 4-methyl phenol.

Characterization of Odor and Comparison of Odor Intensities Between PM-1, PM-10 and TSP

Comparison of the mean total odor, total odor/M, total odor/TSA and total odor/M/TSA between PM-1, PM-10 and TSP is similar to those for VOCs. TSP carried much more total odor than PM-1 and PM-10. When total odor was normalized with the PM mass and the total surface area, the relative odor intensity of PM-1 was higher than that of PM-10 and TSP. This relationship was consistent with the VOC distributions discussed earlier in this paper.

Normalization of the odor intensity to PM mass and TSA resulted in distributions similar to those for VOCs and key odorants: PM-1 had much greater potential to be a carrier of odor than PM-10 and TSP, respectively, for all odorants except H2S. This could be due to the relatively low affinity of H2S to the Carboxen/PDMS fiber and low concentrations of H2S adsorbed by PM-1 below its published odor detection threshold of approximately 10 ppb.

CONCLUSION

The following conclusions were drawn from this study:

(1) HS-SPME coupled with GC-MS-Olfactometry is a novel and effective analytical tool for identifying VOCs and odor associated with swine barn PM.

(2) A total of 50 different compounds were identified using HS-SPME-GC-MS–O approach, 21 out of which have been reported to be present in swine barn PM for the first time. The 50 compounds covered a wide range of polarity and molecular weight and belong to nine chemical classes.

(3) Key malodorants associated with swine barn PM include methyl mercaptan, isovaleric acid, 4-methyl-phenol, indole and skatole. TSP adsorbed a much more absolute amount of those compounds and odors than PM-10 and PM-1, respectively. However, when absolute amounts of compounds and odors were normalized by the PM mass and the total surface area, the values (area count/M/TSA) of those compounds showed significant difference. PM-1 had a greater capacity for characteristic VOCs and odors relative to PM-10 and TSP.

ACKNOWLEDGMENTS

The authors would like to thank Iowa State University for funding this research, and the collaborating swine producer for hosting the collection of swine barn PM. This work was published as reference #1.

REFERENCES

1. L. Cai, J.A. Koziel, Y.C. Lo, and S.J. Hoff. *J. Chromatogr. A*, **1102**, 60-72 (2006).

138 Electrical Impedance Spectroscopy of a Pig Odorant Binding Protein immobilized onto gold interdigited microelectrodes: an ab-initio study

S. Capone°, L. Francioso°, P. Siciliano°, K. Persaud*, A.M. Pisanelli*

°Institute for Microelectronics and Microsystems (IMM-CNR), via Monteroni, university campus, 73100 Lecce (Italy)
**University of Manchester, School of Chemical Engineering and Analytical Science (SCEAS), Olfactory Research Group, PO Box 88, Sackville Street, Manchester, M60 1QD, (UK)*

In this work, an ab-initio study of the electrical response to odorants of a self-assembled monolayer of a pig OBP immobilized onto a miniaturized Si-substrate equipped with gold interdigitated electrodes (IDE), was started. Electrical Impedance Spectroscopy (EIS) was used as electrical characterization technique and a dedicated experimental set-up was arranged in order to carry out EIS measurements in controlled environment. The EIS data was fitted by using a fitting software based on Levenberg-Marquardt (LEVM) algorithm to determine the equivalent circuit of the system.

Keywords: Electrical Impedance Spectroscopy, Odorant Binding Protein

PACS: 87.15.Pc; 07.07.Df; 07.10.Cm

INTRODUCTION

Great expectations in gas sensing rise from a novel biomimetic approach that throws out the idea to use Odorant Binding Proteins (OBPs) as active elements in novel gas sensors[1]. The main feature of such proteins, is that the binding of the odorant molecule elicits a conformational change. Such conformational changes might cause a change in the electrical properties of the protein. In this work, we transfer a self-assembled monolayer of a porcine Odorant Binding protein (pOBP) onto a miniaturized Si-substrate equipped with gold interdigitated electrodes (IDE) and we characterized it by Electrical Impedance Spectroscopy (EIS) in air both in absence and in presence of odorant molecules (as ethanol).

2. EXPERIMENTAL AND METHODS

Porcine odorant binding protein (pOBP) is a monomer of 157 amino acid residues, purified in abundance from pig nasal mucosa[2]. Recombinant Pig Odorant Binding Protein (modified in the 2 position with a Cysteine Residue, so that the protein could be physically immobilised on a gold electrode surface, courtesy Paolo Pelosi, Un. of Pisa,) present in pig saliva was considered in this work. The Pig OBP was suspended in a working 10 mM phosphate buffer solution pH 7.4.

Miniaturized silicon substrates (1.5 mm x 1.5 mm sized) equipped with Au interdigitated electrodes (IDEs) were prepared using photolithography. A 1μl aliquot of Pig OBP was placed onto the Au IDEs. After immobilization the sample bonded onto TO-39 socket was put into a suitable brass test chamber for (EIS) measurements (fig.1).

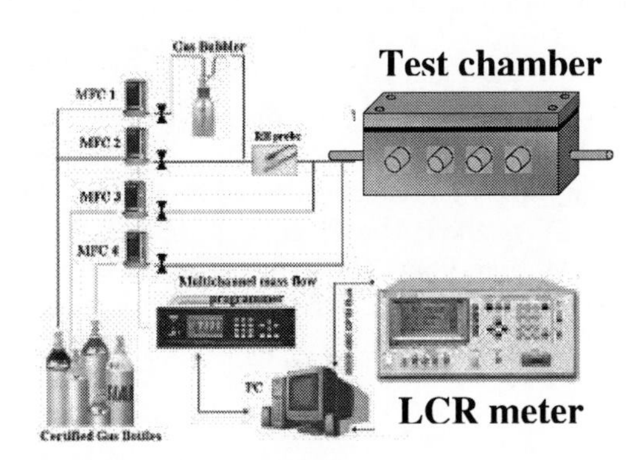

FIGURE 1. Electrical Impedance Spectroscopy set-up Parameters: 20 mHz ÷ 1 MHz, sinusoidal voltage, polarization potential 0 V, frequency modulation 20 mV.

CP1137, *Olfaction and Electronic Nose: Proceedings of the 13th International Symposium*, edited by M. Pardo and G. Sberveglieri

3. RESULTS

3.1 EIS analysis

It can be noticed that only a little variation in the impedance curves in air and in presence of the volatile is observed (fig.2). However, in this ab-initio study the Pig OBP response to such volatiles in air was not so significant and has to be confirmed by a more extensive experiments.

(A)

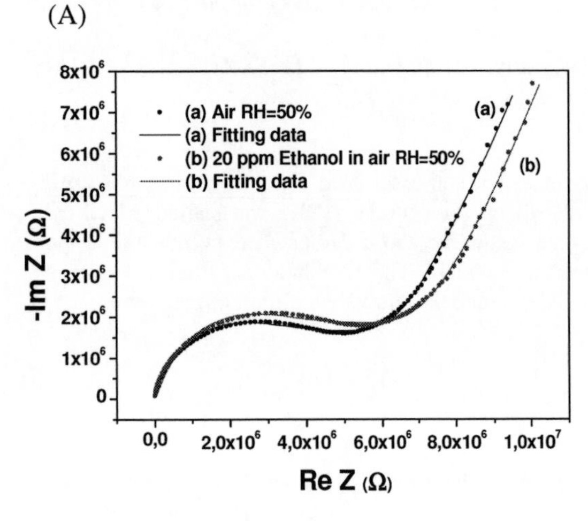

(B)

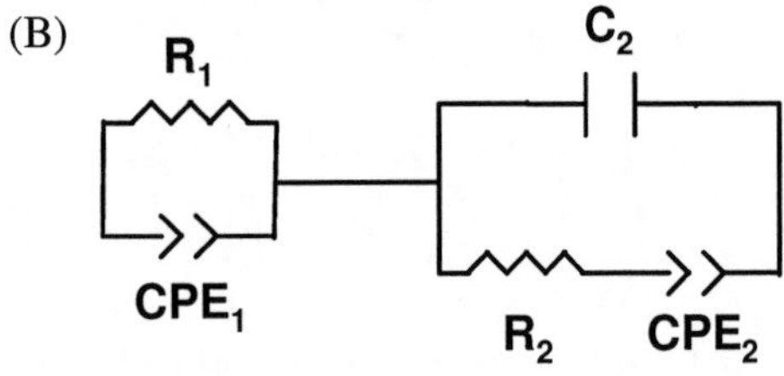

FIGURE 2. (A) Nyquist plots for a Pig OBP under exposure in humid air at RH=50% and under 20 ppm ethanol in humid air at RH=50%; (B) equivalent circuit.

The experimental impedance spectra were best fitted with the equivalent circuit shown in fig.5B by making use of a Complex NonLinear Least Squares (CNLS) method based on a Levenberg-Marquardt (LEVM) algorithm[3,4]. As one can see, a very good fit was obtained. In table 1 the fitting values calculated in LEVM for the elements of the equivalent circuit are listed.

TABLE 1. Fitting values calculated in LEVM for the elements of the equivalent circuit.

		Air RH=50%	20 ppm Ethanol in air RH=50%
R_1		4.32 MΩ	4.91 MΩ
CPE_1	Y_0	0.67 nF· (rad/s)$^{1-\alpha}$	0.71 nF· (rad/s)$^{1-\alpha}$
	α	0.6	0.6
C_2		5.41 pF	5.51 pF
R_2		2.09 MΩ	2.33 MΩ
CPE_2	Y_0	0.37 nF· (rad/s)$^{1-\alpha}$	0.35 nF· (rad/s)$^{1-\alpha}$
	α	0.7	0.7

4. CONCLUSIONS

An ab-initio study of electrical impedance spectroscopy (EIS) of a Pig Odorant Binding Protein exposed to air and target odorant species was started. An equivalent circuit model was developed and a very good fit for the impedance spectra were obtained. However, a lot of efforts have to be devoted both to the definition of a experimental measurement procedure and to understand the sensing mechanisms by which the odorant molecule interacts with the OBP layer by using the circuit model obtained by EIS characterization.

ACKNOWLEDGMENTS

We are grateful to GOSPEL (FP6-IST 507610) for financing this initial study, and to Prof. Paolo Pelosi, University of Pisa for the Pig Odorant Binding Proteins.

REFERENCES

1. M. P. Brown and K. Austin, *The New Physique*, Publisher City: Publisher Name, 2005, pp. 25-30.
2. F. Vincent, S. Spinelli, R. Ramoni, S. Grolli,, P. Pelosi, C. Cambillaud, M. Tegoni,, *J. Mol. Biol*, **300** 127–139 (2000)
3. D.W. Marquardt, *SIAM Journal on Applied Mathematics*, **11**, 431 (1963).
4. J.R. Macdonald, CNLS (Complex NonLinear Least Squares): Immittance, Inversion, and Simulation Fitting Programs – LEVM Manual. Version 8.08, Issue data: February, 2007D.W. Marquardt, SIAM Journal on Applied Mathematics, **11**, 431 (1963)

POSTER SESSION II

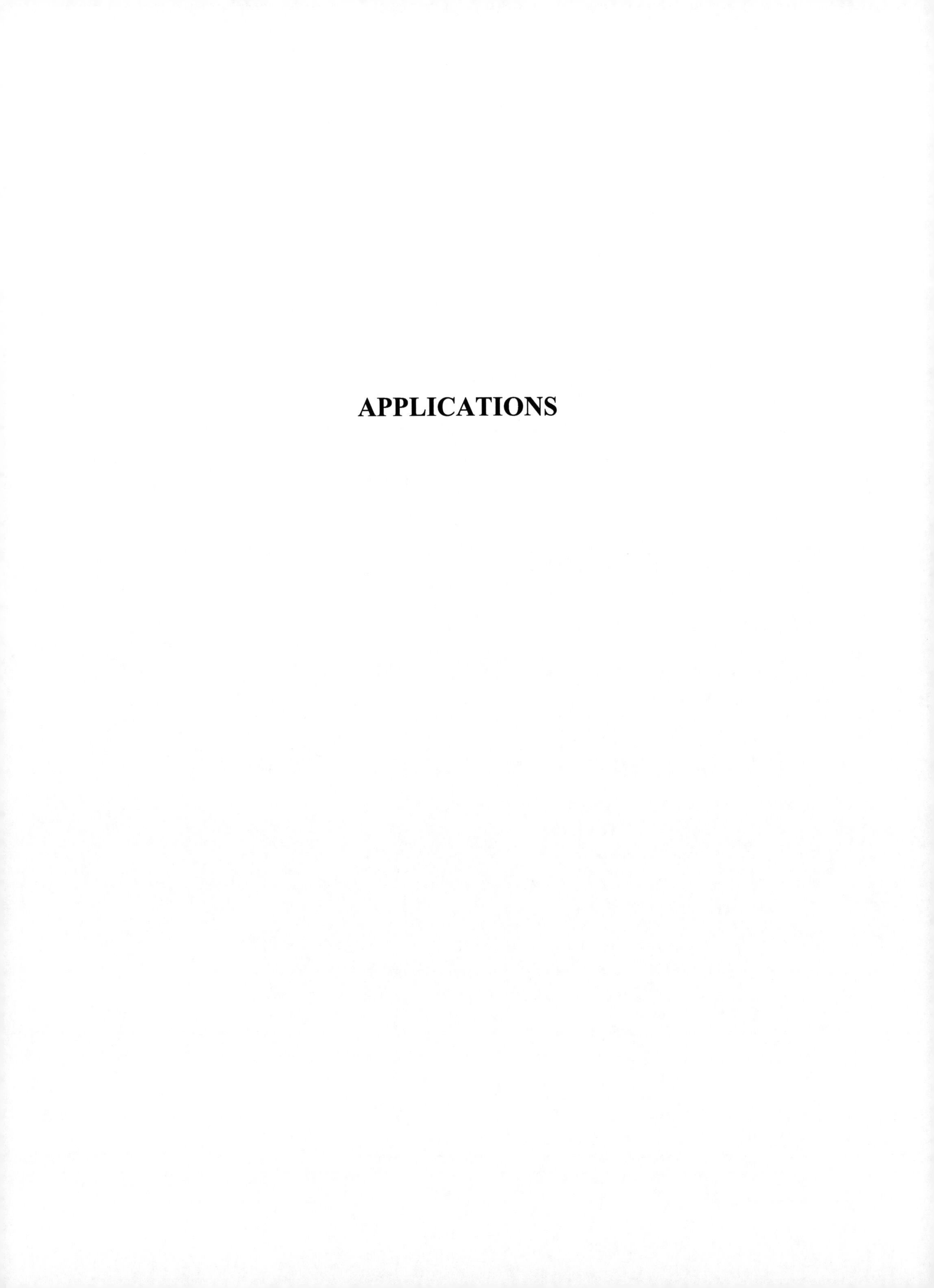

APPLICATIONS

Electronic Nose For Measuring Wine Evolution In Wine Cellars

J. Lozano[1], J.P. Santos[2], M.C. Horrillo[2], J.M. Cabellos[3], and T. Arroyo[3]

[1]*Grupo de clasificación de patrones y Análisis de Imágenes. Universidad de Extremadura. Av. Elvas s/n, 06071 Badajoz, SPAIN, e-mail: jesuslozano@unex.es*

[2] *Instituto de Física Aplicada. Consejo Superior de Investigaciones Científicas (CSIC). C/Serrano, 144, 28006 Madrid, SPAIN*

[3]*Dept. Agroalimentación. Instituto Madrileño de Investigación y Desarrollo Rural, Agrario y Alimentario (IMIDRA). Km 38.2 N-II, 28800 Alcalá de Henares, SPAIN*

Abstract. An electronic nose installed in a wine cellar for measuring the wine evolution is presented in this paper. The system extract the aroma directly from the tanks where wine is stored and carry the volatile compounds to the sensors cell. A tin oxide multisensor, prepared with RF sputtering onto an alumina substrate and doped with chromium and indium, is used. The whole system is fully automated and controlled by computer and can be supervised by internet. Linear techniques like principal component analysis (PCA) and nonlinear ones like probabilistic neural networks (PNN) are used for pattern recognition. Results show that system can detect the evolution of two different wines along 9 months stored in tanks. This system could be trained to detect off-odours of wine and warn the wine expert to correct it as soon as possible, improving the final quality of wine.

Keywords: gas sensors, principal component analysis, neural networks, wine evolution.
PACS: 07.07.Df, 07.05.Mh

INTRODUCTION

In recent years, efforts have been carried out to develop arrays of non-specific sensors coupled with pattern recognition methods (the so-called "electronic noses") for the identification and discrimination of aroma, recognition of adulteration and as an objective method to establish the wine quality [1-2].

The general scheme of an e-nose is formed by four main elements: an aroma extraction technique or air flow system which carry the volatile compounds from the samples to the next step; an array of chemical sensors which transform the aroma into electrical signals; an instrumentation and control system to measure the signal of the different sensors and the control and automation of the whole system. The pattern recognition system try to identify and classify the aroma of the measured samples into several classes previously learned.

EXPERIMENTAL

This paper presents the development of an electronic nose designed to use it in wine cellars for detection of wine evolution. Figure 1 shows the system installed in an experimental wine cellar.

FIGURE 1. View of the whole system installed in the wine cellar for the detection of wine evolution.

A novel sampling method based on static headspace with effluent transfer was developed to extract and carry the aroma directly from the wine tanks to the sensors cell. A Dreschel bottle with a blank solution of ethanol was used for calibration purposes.

The multisensor used included 16 sensor elements based on SnO_2, distributed in circular shape onto an alumina substrate. The temperature of operation was

CP1137, *Olfaction and Electronic Nose: Proceedings of the 13th International Symposium*, edited by M. Pardo and G. Sberveglieri

250ºC. The resistance of the sensors was measured with a digital multimeter (DMM) coupled to a 40-channel multiplexer connected to the personal computer through a GPIB interface. The system is fully automated and can be monitorized by internet. The process of data analysis starts after the sensor signals were acquired and stored into the computer and consists of Principal Components Analysis (PCA) and Probabilistic Neural Networks (PNN)

Samples of two varietal wines manufactured with grapes of the majority varieties in the Origin Denomination (O.D.) "Vinos de Madrid" were used for testing the discrimination capability of the system showed in this paper.

EXPERIMENTAL

The designed system were installed in an experimental wine cellar in Madrid (see fig. 1) for the on-line and in-situ monitoring of wine evolution in tanks. The tanks were filled with wine elaborated with Malvar and Grenache grapes of Madrid O.D.. The evolution of the wine is confirmed with chemical and sensory analysis.

The system were continuously working for 9 months after grape juice fermentation. Two measurements in each tank and another one for calibration were performed everyday. The data of the sensors were stored in disk and processed via internet. All data were normalized before analysis. For the data analysis, seven snapshots, separated at least one month among them, have been taken and used to represent the wine evolution.

Several wine samples were taken from the tanks to analyze with GC-MS and taste by a sensory panel. Analysis and tastes show that the wines have experimented a loss of quailty due to the oxidation and an increase of volatile acidity in the first two months after fermentation (stages 1 to 3), because SO_2 levels hasn't been corrected. In april, SO_2 and total acid have been corrected with the addition of SO_2 and tartaric acid respectively in order to prevent the decline in quality (between stages 3 and 4).

Principal Component Analysis was performed to check the discrimination capability of the system. In fig. 2 and 3 the PCA score plot for the measurements of each tank are shown. The clusters corresponding to the different stages of wine evolution are clearly separated. Several arrows have been added to the plots showing the evolution of the wine. Both, the spoiling and the correction of the wine also can be detected in between stages 2 and 4.

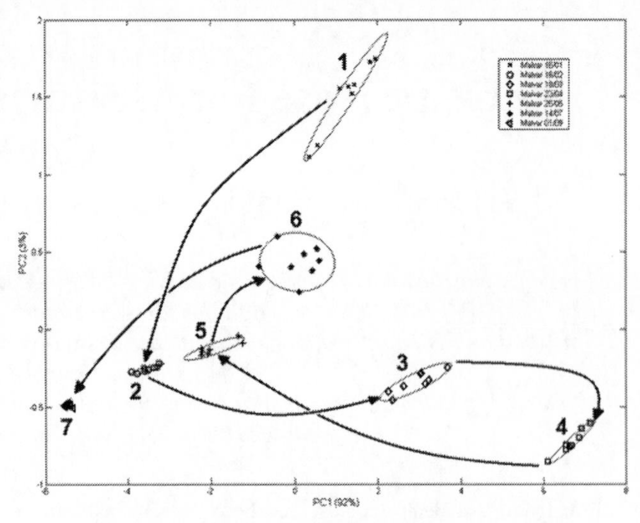

FIGURE 2. PCA score plot of measurements of Malvar wine (stored in tank A).

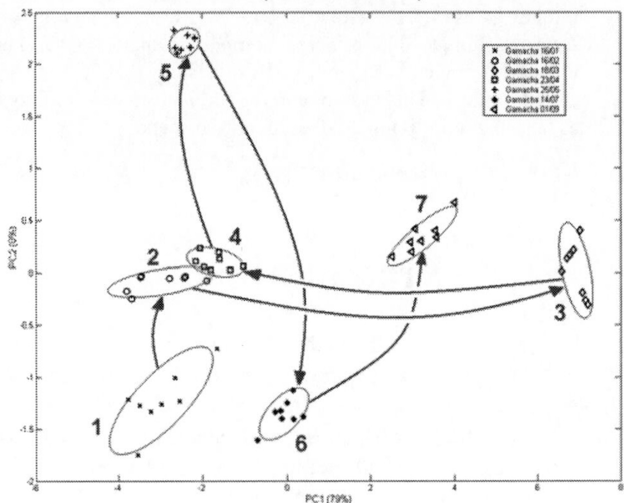

FIGURE 3. PCA score plot of measurements of Grenache wine (stored in tank B).

In the classification tasks, Probabilistic Neural Networks (PNN) were used. Seven classes were learned corresponding to the stages of the wine evolution during the 9 months of study. Leave One Out were used as validation method. A 100% of classification success (percentage of cases correctly classified in validation) were obtained using PNN.

REFERENCES

1. Gardner J. W. and Bartlett P. N., Electronic Noses: Principles and Applications, Oxford University Press, (Oxford, 1999).
2 Pearce T. C., Schiffman S. S., Nagle H. T., and Gardner J. W., Handbook of Machine Olfaction: Electronic Nose Technology, Wiley-VCH, (Weinheim, Germany, 2002).

Cavitands thin films as sensitive coating for explosives sensors

A. Bardet, F. Parret, M. Guillemot, S. Besnard, P. Montméat, C. Barthet and P. Prené

CEA Le Ripault
Laboratoire Synthèse et Formulation
BP 16
F-37260 Monts France

Abstract. This paper deals with the detection of explosive vapours with SAW gas sensors. The devices are coated with terbutylcalix(8)arene. The effect of the deposition technique on the sensor performances is discussed. The best perfrormances are obtained with dip coating.

Keywords: gas sensor, SAW, explosive.
PACS: 80

INTRODUCTION

With the increased use of explosives such as nitroaromatics in terrorist attacks the development of efficient, portable and low-cost explosive detection devices has become an urgent worldwide necessity.

In recent years, the chemical sensors present a growing interest because of their high sensitivity and selectivity for various explosive detections such as nitroaromatic compounds [1,2].

Surface acoustic wave devices (SAW) are interesting because of their high sensitivity to change of physical and chemical property at or near the transducer system surface. Chemical gas sensors using SAW devices are generally composed by two delay lines. The SAW delay lines are used as the sensing channel coated by a sensitive film and the reference one without any film on it [3].

In this paper, we focus on the detection of vapors of 2-4 dinitrotoluene (DNT) which is similar to trinitrotoluene (TNT). The sensors used are SAW devices coated with terbutylcalix(8)arene (Fig. 1). The effect of the deposition technique on the sensor performances will be discussed,

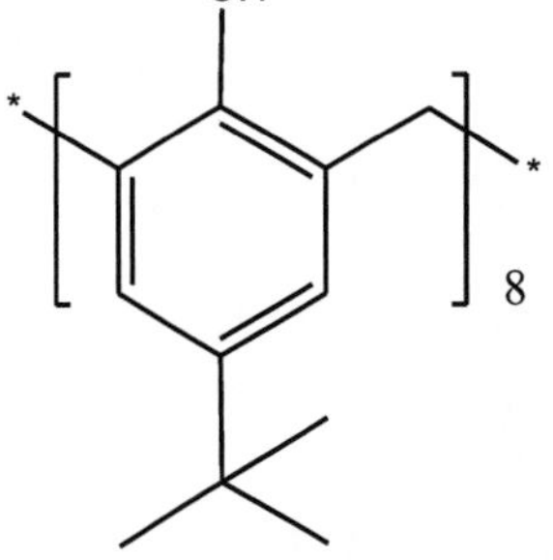

OH

8

FIGURE 1. View of terbutylcalix(8)arene

EXPERIMENTAL AND METHODS

The SAW device operates at 100 MHz.

Calixarene thin films are elaborated by spin coating, spray coating, dip coating and sublimation deposition methods. Each coating induces a 100 kHz frequency decrease.

Dry DNT vapors are generated with a specific testing bench. The generated concentration is closed to the vapor pressure of DNT (300 ppb [4]). The detection experiments consist in exposing the coated sensor to DNT for a 10 min duration at room

CP1137, *Olfaction and Electronic Nose: Proceedings of the 13th International Symposium,* edited by M. Pardo and G. Sberveglieri

temperature. Then the sensor response is expressed as the frequency shift observed for the exposure and the reversibility is the percentage of the recovered signal 1 hour after the exposure.

RESULTS

As shown in Fig. 2, when exposed to DNT, the frequency is slowly decreasing. It produces a 3000 Hz frequency shift for a 10 min exposure.

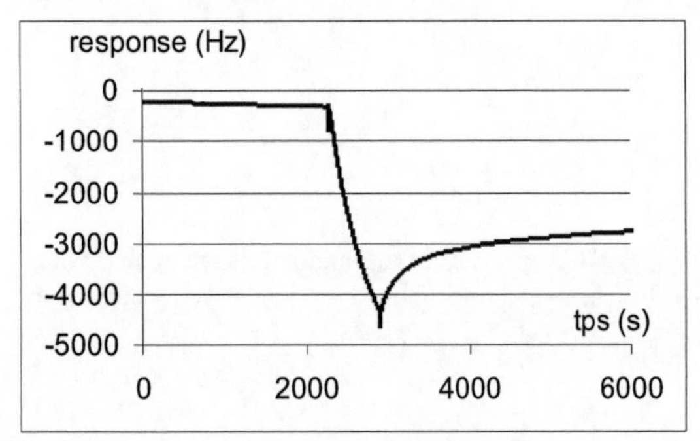

FIGURE 2. Response of the dip coated SAW to DNT

The same behavior is observed for the other coating. The only difference concerns the frequency shifts . The sensors responses are plotted in Fig. 3.

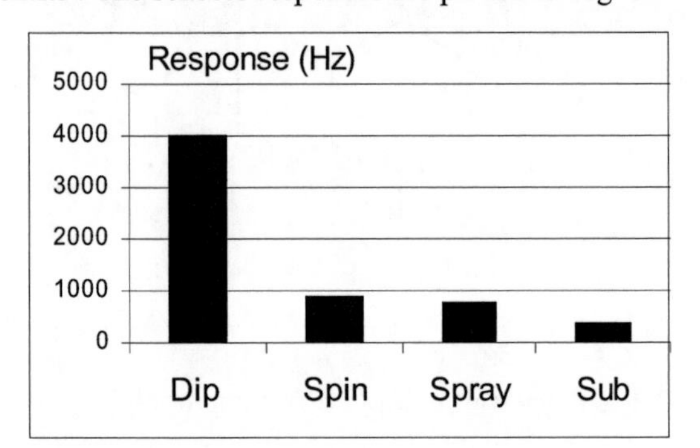

FIGURE 3. Response of the dip coated SAW to DNT

The best value is obtained for the dip coated film (4000 Hz) while the sublimation technique gives the smallest response (350 Hz). In any case, the sensor sensibility is significative. The difference of sensitivity between the coatings comes from the porosity of the film after the deposition process. The best porosity is obtained for the dip coated film while the sublimation leads to the most compact material.

CONCLUSIONS

In this article we have demonstrate the interest of calixarene thin films for SAW explosive sensors. At present the limit of detection, the selectivity and the stability of the sensor is under investigation.

REFERENCES

1. M. la Grone, *Proceeding of SPIE conference*, 3710, 1999, 409
2. F. Thery-Merland, *Proceeding of Eurosensors XVII*, 2003, 550
3. H. Wohltjen, *Analytical Chem.*, Vol 51, pp. 1458, 1979
4. US Environnemental Protection Agency, EPA/600/S-99/002, Mai 1999,

Early Discrimination Of Microorganisms Involved In Ventilator Associated Pneumonia Using Qualitative Volatile Fingerprints

Neus Planas, Catherine Kendall*, Hugh Barr*, Naresh Magan

Cranfield Health, Applied Mycology Group, Vincent building, Cranfield University, Bedfordshire MK43 0AL,
** Gloucestershire Hospitals NHS Foundation Trust, Great Western Road, Gloucester, GL1 3NN.*

Abstract. This study has examined the use of an electronic nose for the detection of volatile organic compounds produced by different microorganisms responsible for ventilator-associated pneumonia (VAP), an important disease among patients who require mechanical ventilation. Based on the analysis of the volatile organic compounds, electronic nose technology is being evaluated for the early detection and identification of many diseases. It has been shown that effective discrimination of two bacteria (*Enterobacter cloacae* and *Klebsiella pneumoniae*) and yeast (*Candida albicans*), could be obtained after 24 h and filamentous fungus (*Aspergillus fumigatus*) after 72h. Discrimination between blank samples and those with as initial concentration of 10^2 CFU ml^{-1} was shown with 24h incubation for bacteria and 48 h for fungi. Effective discrimination between all the species was achieved 72 h after incubation. Initial studies with mixtures of microorganisms involved in VAP suggest that complex interactions between species occur which influences the ability to differentiate dominant species using volatile production patterns. A nutrient agar base medium was found to be optimum for early discrimination between two microorganisms (*Klebsiella pneumoniae* and *Candida albicans*).

Keywords: Electronic nose, ventilator-associated pneumonia, volatile organic compound, volatile production patterns.

INTRODUCTION

Today, one of the main risks for patients in hospitals is that of associated diseases, i.e., nosocomial infections. One of these infections is ventilator associated-pneumonia (VAP) which is defined as a pneumonia occurring in patients after more than 48 hours of mechanical ventilation via endotracheal or tracheotomy tube. This may affect 8 to 28% of all the intubated patients [1, 2] and the mortality rate may arise from 24 to 50% where high-risk pathogens are involved.

Gram-negative enteric rods, *Staphylococcus aureus* and *Pseudomonas aeruginosa* are the predominant microorganisms responsible for this infection. However, many others microorganisms can cause VAP including *Streptococcus pneumoniae*, *Haemophilus influenzae* and *Acinetobacter baumanii* depending on the onset of the illness. There are other microorganisms considered opportunistic organisms such as *Candida albicans* and *Aspergillus fumigatus* which can play a role. *Staphylococcus* sp. and *P.aeruginosa*, and others, belong to the group of microorganisms which potentially can be multi-drug resistant. Some of the criteria for diagnosing VAP are not very specific and can lead to an inaccurate diagnosis of the disease. For these reasons, the mortality and morbidity of VAP can be very high (24-50%).

Many infections are known to produce characteristic odours due to the release of volatile organic compounds (VOCs) from the microorganisms involved. The development of sensor arrays which can reflect changes in qualitative volatile fingerprints has provided promising potential tools for the rapid and early detection of microbial infections at the point of care [3].

The aim of the present study was to examine the potential for using such qualitative volatile fingerprints as biomarkers of VAP infections enabling rapid, point at care diagnosis and to facilitate more effective treatment. Thus this work describes initial studies on

CP1137, *Olfaction and Electronic Nose: Proceedings of the 13th International Symposium,* edited by M. Pardo and G. Sberveglieri

the use of sensor arrays for the discrimination between five microorganisms and evaluation of the relative sensitivity for the detection of different concentrations of these microorganisms over 24-96 h periods. The media required for optimum detection and discrimination were also investigated.

EXPERIMENTAL AND METHODS

Species

In vitro studies were carried out with five microorganisms, two Gram negative bacteria (*Klebsiella pneumoniae* and *Enterobacter cloacae*), one Gram positive bacterium, (*Staphylococcus sp*), one yeast (*Candida albicans*) and one filamentous fungus (*Aspergillus fumigatus*). All of them, except *A.fumigatus,* were obtained from clinical isolates.

Inocula and optimal media

In order to calculate the required concentrations of cells (CFU ml^{-1}), suspensions were made using calibration curves previously obtained. The optical densities were measured with a spectrophotometer (Camspec MS350). Initial experiments examined different media in order to find which one could give better results in terms of discrimination between microorganisms. Three kinds of media (Oxoid, UK) were tested: Nutrient Broth (NB), Brain Heart Infusion Broth (BHIB) and Tryptone Soy Broth (TSB) at 1% concentration were examined. *K.pneumoniae* and *C.albicans* (10^4 CFU ml^{-1}) were inoculated in 100 mL of sterilised broth. All the inoculated samples and blanks were incubated at 37°C in shaken flasks for 3 days. Subsequent experiments were performed using 1% Nutrient Broth in order to minimise interfering volatiles from the liquid media and at the same time permit the growth of the species.

E nose system

Samples were tested using sensor arrays in the NST 3220 E nose from Applied Sensors (Sweden). This device consists of 12 MOS sensors, 10 MOSFET sensors and 1 humidity sensor. The headspace was analysed automatically from samples placed in the 12 position autosampler unit. Data were analysed with multivariate statistical methods such as principal components analysis (PCA) and cluster analysis (CA). In order to optimise the PCA results, replicates considered outliers were removed from the analysis if they were located far away from the mean group. The criteria for being considered an outlier was: samples with $\geq$ 25% response than the mean of the other replicates.

Sampling method

Five ml of each mixture of media and bacterial inocula were placed in sterilised 25ml vials and sealed using screw caps and septa after 24, 48, 72 and 96 h. The samples were left for one hour at 37°C in the incubator in order to generate volatiles from the samples in the headspace. Sterile media without inoculum was used as a blank sample.

RESULTS

Separation of standards

Standards such as diluted alcohols (isopropanol, acetone) and distilled water were measured used periodically to evaluate drift in the hybrid sensor array responses over time. Overall, this was found to be <15% over the experimental period.

Selection of optimal media for volatiles detection

NB was the only medium that achieved good discrimination between *K.pneumoniae* and *C.albicans* after 24 h incubation. TSB and BHIB only showed differentiation after 48 hrs incubation. The three different control media were classified in the same blank cluster (Figure 1).

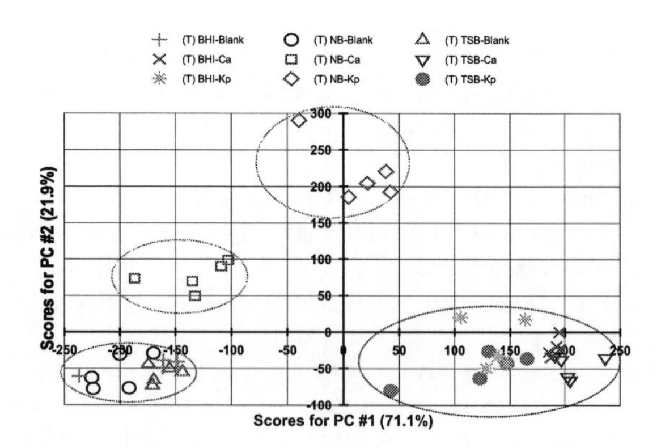

FIGURE 1. PCA plot showing the early discrimination between *C. albicans* (Ca) and *K. pneumoniae* (Kp) on 1% NB medium 24 h after incubation at 37°C. (Key: NB: Nutrient Broth; TSB: Tryptone Soy Broth; BHIB: Brain Heart Infusion Broth)

Discrimination between species with similar initial concentration

The volatile fingerprints produced after 24 to 96 h using the same initial concentration (10^4 CFU ml^{-1}) for all of five of the microorganisms was examined. After 24 h it was possible to visualise at least three clusters of microorganisms: one group with the two Gram negative bacteria, another with *C.albicans* (yeast) and a third group formed by blank samples, *A.fumigatus* and the *Staphylococcus* sp. After 48 hours, the PCA showed clearer differences between blank samples and inoculated samples and after 72 hours, at least six clusters (blank and five species) were observed (Figure 2). After ninety six hours the discrimination was less clear.

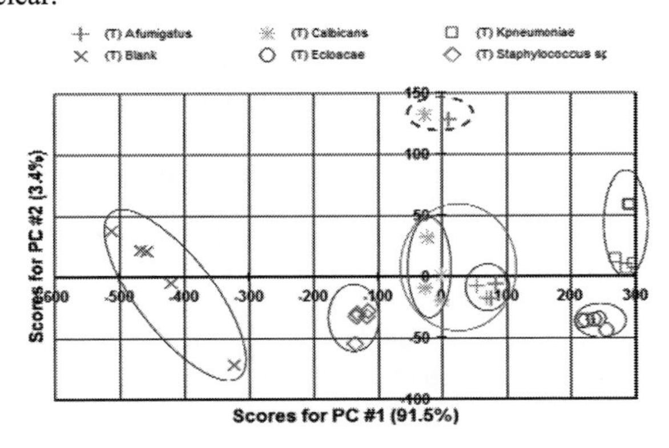

FIGURE 2. PCA score plot after 72 h at 37°C on 1% NB of five different microorganisms and blank samples. Dotted circle indicates two probable outliers

Relative sensitivity of detection

Studies were carried out to examine sensitivity of the sensor array to discriminate between different initial concentrations 10^2, 10^4 and 10^6 CFU ml^{-1} of the same microorganism. The threshold of detection of the E-nose system was seen to vary with the microorganism. There was not discrimination between blank samples and the lowest initial concentration (10^2 CFU ml^{-1}) after 24 h in the samples inoculated with fungal species. For the Gram negative species there was clear discrimination between samples after 24 h.

Figure 3 shows the dendrogram generated from cluster analysis of the data for *Enterobacter cloacae* after 24 incubation with three well defined clusters: one with blank samples, another with the lowest concentration (10^2 CFU ml^{-1}) and a third cluster with the higher concentrations (10^4 and 10^6 CFU ml^{-1}). After 48 hrs the discrimination between all the concentrations was less clear for Gram negative inoculations and in *C.albicans*. For *A.fumigatus* discrimination with the same initial concentration required 72 h of growth. With regards to the *Staphylococcus* strain there was not growth until 48h incubation. Due to that poor growth it was difficult to compare with the rest of bacteria and no conclusions can be made. A new type strain of *Staphylococcus* sp is now being used.

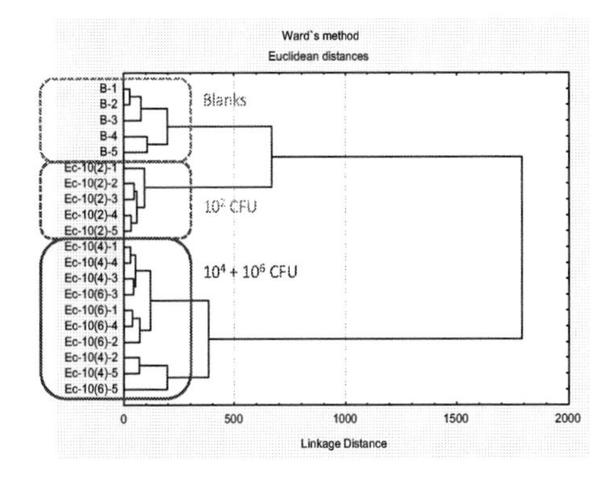

FIGURE 3. Dendrogram of three different concentrations of *E.cloacae* after 24 h incubation at 37°C on 1% NB. (Key: Ec: *Enterobacter cloacae*; 10(2), 10(4) and 10(6): 10^2, 10^4 and 10^6 CFU ml^{-1})

Mixture of microorganisms

Two experiments were performed by mixing different microorganisms: *C.albicans+E.cloacae* and *Staphylococcus* sp+*E.cloacae*; the latter experiment was performed twice. In all experiments blank samples were classified in a unique cluster as well as *C.albicans* and *Staphylococcus* sp samples alone. This did not happen with *E.cloacae* alone which was grouped with the rest of the samples in all cases and periods of incubation.

DISCUSSION

The choice of growth media is important. This was demonstrated by the experiments with *K.pneumoniae* and *C.albicans* on three different media: Nutrient Broth (NB), Tryptone Soy Broth (TSB) and Brain Heart Infusion Broth (BHIB) at 1% concentration. Effective discrimination between the species was successful after 24 h incubation on NB. Depending on the group of microorganisms under investigation it is essential to first optimise the choice of media due to the nutritional variation between species. Furthermore, it is not always true that traditional solid media for isolation of microbial species enhances volatile analysis. For example, Sahgal [4] found earlier discrimination (72 h incubation) between different dermatophytes using SBHIA (Sabouraud Brain Heart Infusion Agar) which is less commonly used in routine microbiology practice rather than SDA (Sabouraud Dextrose Agar).

Previous studies have suggested that the addition of specific enzymes to the liquid samples enhances volatile generation into the headspace and improves discrimination. This was successfully achieved with an enzymatic lipase-based cocktail for detection of *Mycobacterium tuberculosis* [5].

This study has shown that it is possible to discriminate between the main microorganisms which cause VAP in 24-48h with the same initial concentration. However, it was not until 72h when the maximum discrimination between blank samples and between all of the microorganisms was achieved. Due to the slow growth of *A.fumigatus* in comparison with the rest of the microorganisms, its cluster was grouped with the blanks after 24 h. Filamentous fungi grow slower than bacteria has been shown in previous studies with spoilage fungi [6]. With regards to bacteria and yeasts the discrimination between them was more rapidly achieved (24h after inoculation).

In terms of sensitivity the experiment performed by Magan *et al.* [7] studied the discrimination between three very close but high concentrations (10^6, $3.5x10^8$ and $8x10^8$ cells ml^{-1}) of *Pseudomonas aureofaciens* after 60 minutes incubation. The analysis showed that there was three-group formation: low concentration and milk control, medium and high concentration and the third the butanol control. These results are very closely paralleled in the current study although there were differences between the trials. In the present study it has been shown that the threshold of detection between blank and inoculated samples was achieved in different incubation times depending on the microorganism and initial concentration. As shown with *A.fumigatus* (10^2 CFU ml^{-1}) which required 72 h growth to discriminate inoculated from blank samples compared to only 24h growth with inoculation with a higher concentration (10^6 CFU ml^{-1}).

Few studies have been undertaken using mixtures of microorganism. However, one of these studies tested sputum samples from patients with a conducting polymer sensor array [5]. Samples consisted of three different bacteria (*Mycobacterium avium*, *M.tuberculosis* and *P.aeruginosa*) a mixture of these microorganisms and control sputum, all treated with lipase enzymes. They found clear discrimination between inoculated samples after 6 h at room temperature. The detection and discrimination of species in mixtures is a complex issue which is still being studied. We believe that with mixtures some may be more dominant than others (e.g. *E.cloacae*) which can be identified by their key volatile.

CONCLUSIONS

The best discrimination between *K.pneumoniae* and *C.albicans* was achieved after 24h incubation with 1% Nutrient Broth. However, the optimisation of volatile production needs further improvement, for example evaluation of the use of enzymes to enhance volatile generation.

It was possible to differentiate between some bacteria and yeast species after 24-48 h incubation at a threshold of $\geq 10^2$ CFU mL^{-1} as initial concentration. The discrimination between fungi required 72 h incubation. These findings show that potential exists for using this approach as a tool for the early detection of some bacterial infections in 24-48 hours with a relatively small amount of cells and using an easy and simple sampling method.

In mixtures of two different microorganisms one of them was usually seen to be predominant. This competitiveness of individuals in the mixture made interpretation of results more complex; however, the volatile pattern production could permit the identification of the predominant species and therefore guide selection of targeted therapy.

ACKNOWLEDGMENTS

We are grateful to Cranfield Health and Gloucestershire Hospitals NHS Foundation Trust for financial support.

REFERENCES

1. J. Chastre, J,Y, Fagon. *Am J Respir Crit Care Med.* **165**, 521-526 (2002).
2. M.H. Kollef. *Respiratory Care* **50**, 714-724.
3. A.P.F. Turner and N. Magan, *Nature Reviews: Microbiology* **2**, 161-166 (2004).
4. N. Sahgal. " Microbial and non-microbial volatile fingerprints: potential clinical applications of electronic nose for early diagnoses and detection of diseases", *Ph.D. Thesis,* Cranfield Health, Cranfield University 2008.
5. A.K. Pavlou, N. Magan, J. Jones, J. Brown, P. Klatser and A.P.F. Turner, *Biosensors and Bioelectronics* **20**, 538-544 (2004).
6. R. Needham. "Early detection and differentiation of microbial spoilage of bread using electronic nose technology", *Ph.D. Thesis,* Cranfield University 2004.
7. N. Magan, A. Pavlou and I. Chrysanthakis, *Sensors and Actuators B* **72**, 28-34 (2001).

An Experimental Methodology For The Analysis Of The Headspace Of In-Vitro Culture Cells

M. Santonico[1], G. Pennazza[2], A. Bartolazzi[3], E. Martinelli[1], R. Paolesse[4], C. Di Natale[1], A. D'Amico[1]

[1]Department of Electronic Engineering, University of Rome Tor Vergata; Via del Politecnico 1, 00133 Rome,Italy
[2] Faculty of Engineering, University Campus Bio-Medico, Rome;Via Alvaro del Portillo 21,00128 Rome,Italy
[3] Department of Pathology, St Andrea University Hospital, Rome; Via di Grottarossa 1035, 00189 Rome, Italy
[4] Department of Chemical Science and Technology, University of Rome Tor Vergata; Via della Ricerca Scientifica, 1, 00133 Roma, Italy

Abstract. Several examples of electronic nose applications to diagnose different forms of cancers have been presented in the recent years. Although clear relationship between sensors signals and the presence of the diseases have been provided, in-vivo experiments suffer of a number of non-controlled variables that makes uncertain the meaning of the experimental results. In this paper, an experiment aimed at measuring the volatile compounds from cultured tumoral cells lines is illustrated. Results suggest that each cell line, extracted from human tumors, is characterized by a proper volatile compounds pattern and that these patterns tend to clusterize according to the kind of tumor.

Keywords: electronic nose, cancer, cell culture
PACS: 07.07.D.f

INTRODUCTION

The application of electronic noses to the detection of pathologies is becoming an established field of research [1].

In particular, a number of researches focused on the diagnoses of cancer with the ambitious goal of providing a non invasive diagnostic technique that could improve early diagnosis capabilities. To this regard, examples of correlations between sensors arrays signals and the presence of pathologies have been provided for lung cancer in several experiments [2,3,4].

Recently we focused our attention to the detection of melanoma [5]. This is an important cancer form for which several anedoctical reports pointed out the detection capability of dogs [6].

It is worth mentioning that these *in vivo* studies were based on cross-sectional approaches and then they are rather sensitive to the experimental design and, as a consequence, hardly subject to generalization.

For instance, the interferences due to drugs uptake and life styles may influence the body odor and make the difference with regard to a control group. These problems, still unresolved, limit the reliability and reproducibility of clinical trials of electronic noses. In particular, it is still unresolved.

The following the ambiguity: if the changes of volatile compounds pattern are typical for each tumor form or it is rather a common phenomena occurring in all tumor forms.

In case of melanoma, the localization of the disease on a restricted part of the skin allows the application of differential measurement partially removing most of the problems previously outlined [5].

Nonetheless, a strong contribution to resolve these issues in order to obtain a relationship between the cancer tissue and the volatile compounds can be obtained by studying well-selected and sterile tumor cell lines able to form tumors *in vivo*.

Previous studies of cultured cells with electronic noses shown that tumor cells may be differentiated by their volatile products [7].

In this paper we were interested in studying the differentiation of lines of melanoma cells with respect to other tumor forms.

Here the preliminary *in-vitro* phase of the study is reported.

CP1137, *Olfaction and Electronic Nose: Proceedings of the 13th International Symposium*, edited by M. Pardo and G. Sberveglieri

EXPERIMENTAL AND METHODS

Five tumoral cells lines were studied. They were three melanoma cancers, a synovial sarcoma and a thyroid cancer. Cells were derived from the respective primary human tumors and cultured in standard conditions in Petri dishes with RPMI-1640 medium supplemented with 2 mM glutamine, 10% Fetal calf serum, penicillin and streptomycin (GIBCO BRL, Gaithersburg, MD) at 37°C and 5% CO_2 atmosphere. Experiments were carried out at the Department of Pathology of the St. Andrea Hospital in Rome.

Cells headspaces were measured with the electronic nose designed and produced at the University of Rome 'Tor Vergata'. It can accommodate up to eight quartz microbalances (QMB) sensors. Here a configuration with five sensors was used. QMB were AT-cut plates oscillating in the thickness shear mode at the resonance frequency of 20 MHz.

QMB sensors were functionalized by solid state layers of metalloporphyrins. All metalloporphyrins were metal complexes of 5,10,15,20-Tetrakis-(4-butyloxyphenyl)porphyrin; the metals, differentiating the sensors of the array, were cobalt, zinc, iron, tin, and chromium. These were the same porphyrins used in the *in-vivo* experiments [5].

Sensors were constantly exposed to a reference air and the exposure to the sample resulted in a negative shift of the resonant frequency. The differences between the frequencies measured in the steady conditions before and during the exposure are considered as the sensor feature and are utilized in all the following data treatment. Consequently, for each measured sample the electronic nose provides a pattern of five values.

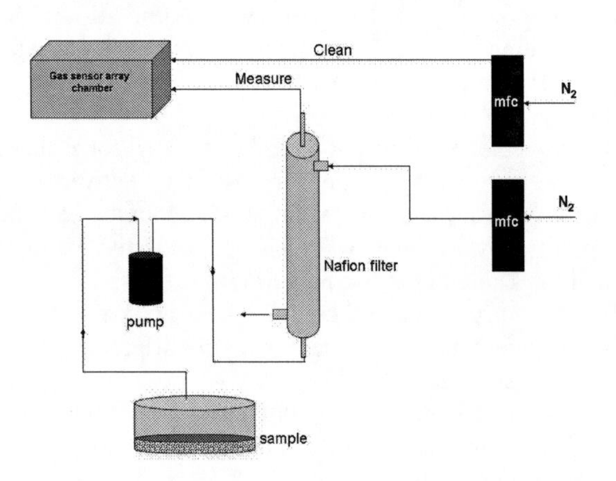

FIGURE 1. Sketch of the experimental set-up.

RESULT AND DISCUSSION

The main scope of this work was the optimization of a measurement strategy where most of the information contained in the volatile compounds pattern are retained. In Figure 1 the measurement set-up is shown.

One of the main problem of the measure was the changes of relative humidity due to variations of the water activity in the culture media. A Nafion filter was used (model MD-070 distributed by Perma Pure LLC) to reduce the influence of humidity keeping constant the relative humidity in the sample. Cells were kept in a Petri dish and a flow of nitrogen was used a carrier gas to clean the sensors and to dilute the cells headspace. Total nitrogen flow was kept constant at 200 sccm by a Mass Flow Controller. Dry nitrogen flow was also used to purge the Nafion filter.

To create a stable and reproducible headspace, a metallic cylinder (vol. 35 cm^3) fitting the diameter of the Petri dish was held for 10 minutes on the sample at room temperature. The headspace was then transferred by a pump through the Nafion filter and then to the electronic nose chamber.

Sensors steady-state signal shifts were considered as sensors features and used for further analysis. Data have been treated with Principal Component Analysis (PCA) calculated on the auto-scaled data set. For each tumoral cells lines three samples were prepared and measured immediately after the preparation. For comparison empty Petri dish and culture media were also measured before each measurement session. Three measurement sessions in three different days were held, each day a complete cycle of measurements was performed.

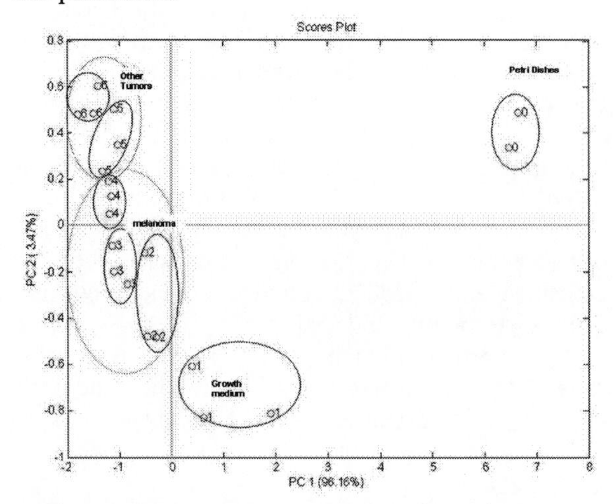

FIGURE 2. Plot of the first two principal components of the PCA model built on the whole measurement data set. The straightforward separation between empty Petri dish (along the first PC), and culture media versus cells (along the second PC) is visible.

Figure 2 shows the plot of the first two principal components. The plot is rather straightforward with a separation along the first Principal Component (96% of total variance explained). Along the second principal components (34% of explained variance) the culture medium is clearly separated from the cells cultures. Cells cultures data are also ordered evidencing a certain clustering according to kind of cellular lines.

In figure 2 same cells lines and controls are evidenced by a circle. The internal classes dispersion illustrates the non reproducibility of the measurement system. The magnitude of spread of classes is approximately of the same order of magnitude. Since in case of empty Petri dishes the non reproducibility is likely totally due to the electronic nose we can conclude that preparation and measurement methodology do not influence the performance of the experiment.

In order to better study the relationship between cells the PCA has been repeated omitting the controls data (empty Petri dish and culture media).

Figure 3 shows the plot of the first two principal components. The first principal component carries about 80% of the total variance. Along this axis the data are ordered according to the tumor kinds. Melanomas are segregated on the right hand side and there is also a clear tendence to separate the tumoral cell lines. Most of the spread inside each class occurs along the second principal component (13% of variance explained).

This result suggests that each cell lines is characterized by a peculiar volatile compounds profile and that these profiles do cluster together according to the kinds of cancers.

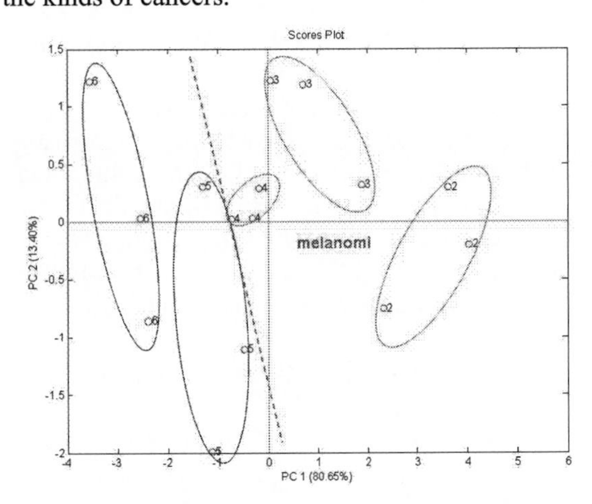

FIGURE 3. Plot of the first two principal components. Circles indicate data related to the same cellular line.

RESULT AND DISCUSSION

The processes occurring in an organism affected by cancer are complex, and traces of the presence of the disease can be also found in body compartment that are not directly interested by the tumor. For instance, in case of breast cancer, oxydative stress in cancer cells can release compounds that can be detected in the breath [8]. On the other hand, even in the cancer tissue accelerated metabolism and necrosis can give rise to aspecific volatile compounds pattern that do no allow the specific identification of the kind of cancer. For these reasons, it is important to identify, through their volatile compounds profile, the individual tumor cells. In this paper we shown that individual cell lines, extracted from humans, have a distinct volatile pattern and these patterns tend to cluster according to the tumor kind.

REFERENCES

1. A. P. F. Turner, N. Magan,*Nature review, Microbiology* **2**, 161-166 (2004).
2. C. Di Natale, A. Macagnano, E. Martinelli, R. Paolesse, G. D'Arcangelo, C. Roscioni, A. Finazzi-Agrò and A. D'Amico, *Bios. Bioelectr.* **185**, 1209-1218 (2003).
3. R. Machado, D. Laskowski, O. Deffenderfer, T. Burch, S. Zheng et al., *Am J Crit Care Med* **171**,1286-1291(2005).
4. P. Mazzone, J. Hammel, R. Dweik, J. Nia, C. Czich, D. Laskowski, T. Mekhail, *Thorax* **62**, 565-568(2007).
5. A. D'Amico, R. Bono, G. Pennazza, M. Santonico, G. Mantini et al., *Skin Research Techn.* **14** , 226-236 (2008).
6. D. Pickel, G. Mauncy, D. Walker, S. Hall, J. Walker, *Appl. Anim. Behav.* **89**, 107-116 (2004).
7. K. Gendron, N. Hockstein, E. Thaler, A. Vachani, W. Hanson, *Otolaryngology – Head and Neck Surgery.* **137**, 269-273 (2007).
8. M. Phillips, R. Cattaneo, B. Ditkoff, *Breast J.* **9**, 184-191(2003).

Detecting of Fruit Ripeness in the Orchard, Packing House and Retail Store of the Future

Benedetti S., Buratti S. and Mannino S.

Department of Food Science and Technology - University of Milan
Via Celoria, 2 – 20133 Milano Italy

Abstract. An electronic nose based on MOS type chemical sensors has been used to detect fruit ripeness. Different varieties of fruit classified in three different classes as unripe, ripe and overripe, were analysed. Starting from 10 sensors it was found that only one (W5S) is relevant to discriminate among fruit on the basis of their ripeness. The discrimination ability of the sensor array was studied separately for each fruit variety, as well as for the whole set. Multivariate statistical analysis was applied to classify fruits in an objective and simple way.

Keywords: Electronic nose, climacteric and non climacteric fruit, ripeness.
PACS: 07.07Df

INTRODUCTION

The ripeness control and monitoring is becoming a very important issue in fruit management since the sensory and storage properties are strictly related to the ripening stage. Many methods to monitor the ripeness of fruit have already been proposed and are principally based on rheological properties such as texture and firmness [1]. The main disadvantage of the majority of these techniques is that they require the destruction of the samples. This is why nowadays, prediction of ripeness is mainly based on practical experience or visual parameters as colour changes. In recent years, electronic nose technology opened the possibility to exploit information on volatile components of fruit providing real time information about quality and ripeness [2, 3]. The aim of this research was to evaluate the capability of a commercial electronic nose based on 10 MOS type chemical sensors to classify different varieties of fruit including climacteric (apple, peach apricot) and not climacteric fruit (cherry) on the basis of their ripening state. Classification was performed within each fruit variety but also to the whole set of fruit without any variety sorting.

EXPERIMENTAL AND METHODS

Two different cultivars of peach *(Prunus persica* L. [Batsch]) 'Springcrest' and 'Silver Rome', apple (*Malus domestica Borkh*) 'Stark delicious' and 'Fuji', sweet cherry (*Prunus alvium* L.) 'Mora di Cazzano' and 'Stella' and apricot (*Prunus armeniaca*) 'Petra' and 'Goldrich', grown in experimental and commercial orchards in Northen Italy, were selected for this work. Fruits were picked in order to obtain batches of different ripening stage. For each fruit variety every batch was composed of more than 10 fruit for each cultivar, selected for their homogeneous colour, uniform size and weight, absence of injury or spoilage. In order to classify fruits on the basis of their ripening state, classical analytical methods such as ethylene evaluation, soluble solid content, titrable acidity and firmness evaluation, were applied. After harvest fruit were allowed at 20 ± 1°C overnight and then used primarily for the electronic nose analysis and then for chemical analyses. A Portable Electronic Nose (PEN2 – Airsense, Schwerin, Germany) with 10 MOS type chemical sensors was used. Each fruit was placed in an airtight glass jar fitted with a pierceable Silicon/Teflon disk in the cap. After 2h equilibration, the measurement started. No sensor drift was experienced during the measurement period. Principal Component Analysis (PCA) and Linear Discriminant Analysis (LDA) were used for statistical elaboration.

RESULTS AND DISCUSSION

To monitor the capability of the electronic nose to detect differences in volatile profile of fruits on the basis of their ripening stage, for each variety examined three classes were defined, class 1 (unripe), class 2 (ripe) and class 3 (overripe) on the basis of traditional

CP1137, *Olfaction and Electronic Nose: Proceedings of the 13th International Symposium,* edited by M. Pardo and G. Sberveglieri

analytical methods such as colour evaluation, ethylene determination (only for climateric fruits), soluble solids content, titrable acidity and firmness evaluation. For each fruit variety the electronic nose responses related to the three mentioned classes were elaborate by PCA performed on covariance matrix and the results are reported in Figure 1.

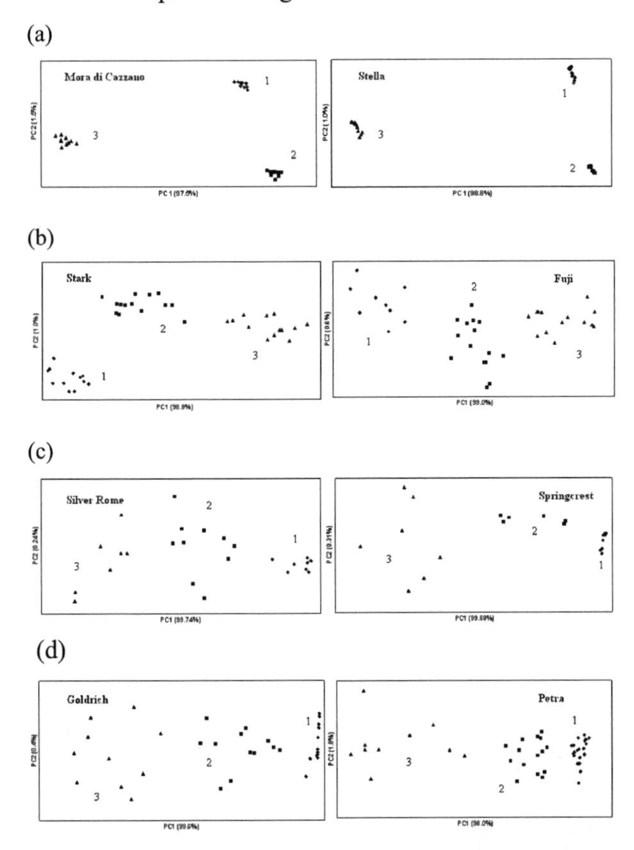

FIGURE 1: PCA score plot of the analysed samples (A=Sweet cherries; B=Apples; C=Peaches; D=Apricots): (◊) unripe, (■) ripe and (▲) overripe.

As it can be seen in the figure, for each single fruit variety almost all the variance is on the first principal component. Furthermore when data of each single variety are treated in separate, a clear distinction among unripe, ripe and overripe was found with exception of sweet cherries. For this variety the first principal component seems to discriminate mostly unripe and ripe fruit from overripe. Considering the loading plot (not shown) representative of all fruit varieties and of all cultivars, the W5S sensor was responsible for the discrimination of fruit on the first principal component showing that this sensor was particularly relevant in monitoring changes in volatile profile during ripeness. In order to characterize into the three classes the whole set of climacteric fruit without any variety sorting, the LDA was applied. Figure 2 shows how the first two functions discriminated among unripe (1), ripe (2) and overripe (3) fruits. The LDA results give 92.8% correct classification for the three classes and a cross validation error rate of 7.7% due to few unripe and ripe fruits that are not correctly classified.

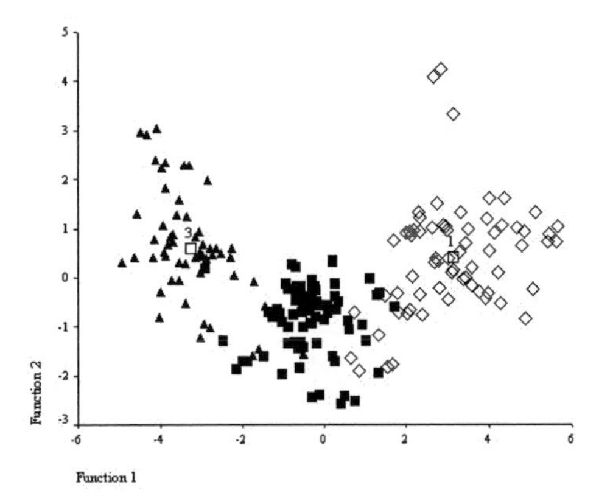

FIGURE 2: LDA Projections of the whole data set of climacteric fruits (apple, peach, apricot): (◊) unripe, (■) ripe and (▲) overripe.

LDA applied to non-climateric fruit (sweet cherry), figure not shown, give 95.0% correct classification for the three classes and a cross validation error rate of 8.3% due to unripe and ripe fruit that are not completely discriminated.

CONCLUSIONS

In this work it has been demonstrated that the electronic nose is able to assess the ripening stage of different variety of climacteric and non climacteric fruit suggesting that this instrument could be an analytical tool capable of give quick information about flavour evolution during fruit ripeness. In this perspective, even if the practical applicability of the electronic nose requires a more robust training and testing with more harvest season and with more fruit variety, results of our research suggest that in future, this instrument would be an useful tool to control the ripening stage of fruits directly in the orchard or in the packing house and retail store.

REFERENCES

1. J. Wang, B. Teng, and Y. Yu, *Eur.Food Res. Technol.* 218, 289-294 (2005).
2. J. Brezmes, E. Llobet, X. Vilanova, G. Saiz, X. Correig, *Sens. Actuators B* 69, 223-229 (2000).
3. G. Costa, M. Noferini, G. Fiori, A. Orlandi, *Acta Hort.* 603, 571-575 (2003).

The Electronic Nose: A Protocol To Evaluate Fresh Meat Flavor.

S. Isoppo, P. Cornale, S. Barbera

Dipartimento di Scienze Zootecniche, Università di Torino
Via Leonardo da Vinci 44, 10095 Grugliasco (TO), Italy
292676@unito.it - tel +39 011 670 8713

Abstract. An Electronic Nose, comprising 10 MOS, was used to carry out meat aroma measurements in order to define an analytical protocol. Every meat sample (*Longissimus Dorsi*) was tested before, during and after cooking in oven (at 165°C for 600 seconds). Analysis took place in these three steps because consumers perceive odor when they buy (raw aroma), cook (cooking aroma) and eat meat (cooked aroma). Therefore these tests permitted to obtain a protocol useful to measure aroma daily perceived by meat eater.

Keywords: Electronic nose, MOS, aroma, meat.
PACS: 06.60.-c; 06.60.Mr; 07.07.Df.

INTRODUCTION

The concept of meat quality varies dramatically according to the economic development of the human group. Nevertheless it seems reasonable to think that consumers have been always concerned with the safety, convenience and the sensory quality of meat. Sensory quality includes aroma. Aroma is, with color, the most important attributes consumers use to judge meat quality and acceptability [1]. Because of the relationship of odor to meat palatability it is important to gain a better understanding of this parameter [2].

Sensory analysis offer exhaustive information about meat odor, but there are expensive and subjective. Instead Electronic Nose is a device that can test odor rapidly and at low cost both in qualitative way, comparing pattern, and in quantitative mode by fitting specific aroma compounds. Several studies have already shown the possibilities in application of EN to classify meat products or to evaluate meat shelf-life. For example some Spanish AA [3] have devoted effort to classify the special Iberian cured ham into very different commercial categories. Others AA from Spain [4] have discriminated feeding regimen effects in pig meat, optimized some parameters of ripening time and they have also concluded that different types of Iberian ham can be discriminated and identified successfully by EN. The performance of an electronic nose with raw meat has been also tested. It has been able to predict the meat species (pork or beef) only analyzing aroma by EN [5]. Others AA used EN to identify spoiled beef and to evaluate poultry meat shelf life [6]. Instead there aren't many works about application of EN to investigate the properties of fresh meat aroma and in this case it is used VIALS method. Moreover this procedure takes place in standard status (60°C) but doesn't reply conditions in which consumers perceive meat aroma. Because meat eater perceives the odor of a steak in different conditions: when he buys (raw meat), cooks (meat cooking) and eats it (meat cooked).

In order to characterize meat aroma really perceived by consumer an analytical protocol it is necessary. The present work aims to define this procedure.

Thus EN multi- channel evaluation of meat samples will be carried out using a static and a dynamic modes. For the static mode the raw and the cooked samples are tested, while in the dynamic mode gas flow from oven is examined during cooking.

2. EXPERIMENTAL AND METHOD.

Several trials took place to define the best final protocol testing different operative conditions.

Longissimus dorsi, refrigerated at 4°C, was used as samples source. Every sample was prepared according to Meat Cooking Shrinkage (MCS) protocol [7] in order to obtain 1- cm thick circular steak (5.5 cm Ø).

Portable Electronic Nose (PEN2), (Airsense Analytics GmbH, Germany) was used. PEN2 comprises 10 Metal- Oxide Semiconductor (MOS) sensors. They are sensitive to several classes of chemicals: to aromatic (W1C and W3C), to chemicals broad range (W5S), to hydrogen (W6S), to aromatic-aliphatic (W5C), to methane (W1S), to sulphur-organic (W1W), to broad alcohols (W2S), to sulphur-

CP1137, *Olfaction and Electronic Nose: Proceedings of the 13th International Symposium,* edited by M. Pardo and G. Sberveglieri

chloride (W2W) and to methane-aliphatic (W3S) Moreover PEN2 has a sensors cleaning system.

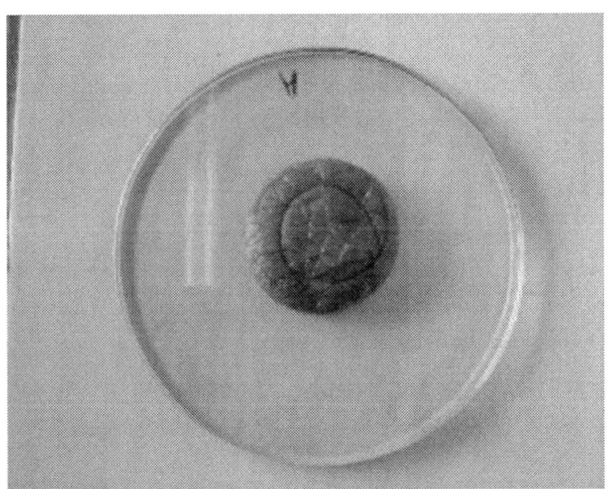

FIGURE 1. Meat Sample.

2.3 Meat Analysis.

Each sample analysis took place three times: before, during and after cooking. Raw steak, placed on a glass support, was put in a box (0.250 L) and then was analysed. Cooked meat was tested in the same way. Boxes (Fig. 2), one for raw sample (R box) and one for cooked sample (C box), were equipped with an active charcoal filter and tubes connected to EN. Cooking was performed for 600 seconds at 165°C, in electric forced air convection oven (F.lli Galli G. & P., Milano, Italy). The forced- air, before going into the oven, passed trough an active charcoal filter. The oven was equipped with a tube connected to PEN2.

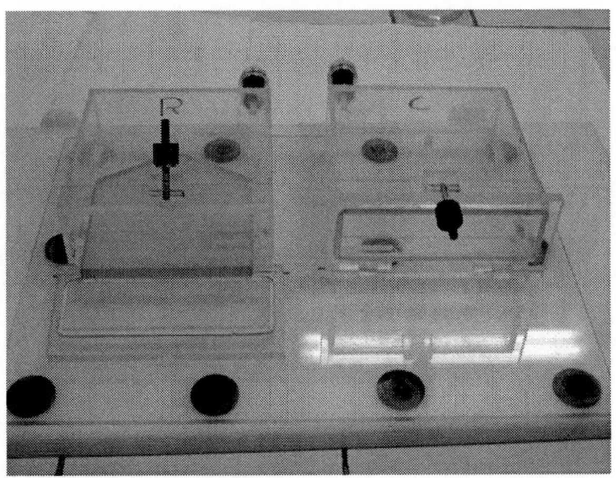

FIGURE 2.. Sample Boxes.

Every trial started performing a zero point trim to standardize the sensors conditions (oxidation-reduction). Then the first analysis step took place. This measurement included:

- 30 seconds white reference (air in the R-box);
- 30 seconds of sample aroma test.

Afterwards was performed a zero point trim. Then PEN2 was connected to a spill duct of oven, thus a 30 seconds measurement of the 165°C oven air was carried out. The air flow reached the EN at 30°C and 40-45% of RH.

Subsequently the sample was put in the oven and 600 seconds non stop aroma analysis took place. After cleaning phase this test was repeated. In this case a measurement, made up of 5 cycles, was adopted. Every cycle included:

- sample aroma test (20 s);
- PEN2 sensors cleaning (60 s).

After a zero point trim, PEN2 analysed the empty C- box air (for 30 seconds). At last the cooked sample, placed on a glass support, was put in C- box and tested for 30 seconds.

Whole trial was repeated twice whitout and with a filter to stop humidity (FP30/02 CA-S 0.2 μm, 7 bar, Schleichs & Schuell, Germany) between box/ oven and PEN2. At the beginning of every measurement phases filter was replaced.

3. RESULTS AND DISCUSSION.

The procedure included the measurement steps, previously indicated, to test aroma perceived by meat eater. Steaks were obtained according to the MCS method, because, in this way, we could do MCS and aroma analysis using the same sample. We introduced a filter because during measurement samples could emit steam, and MOS sensors could be damaged by excessive humidity.

We could observe filter stopped, unfortunately, some odour compounds (< 0.2 μm Ø). In fact, sensors signal was lower when it was used (Fig. 3). For this reason we decided to take away filter. Moreover humidity reached nose was moderate so filter wasn't strongly necessary.

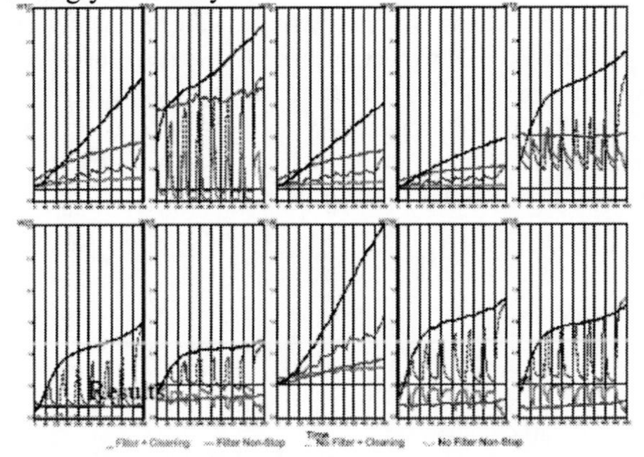

FIGURE 3. Sensor Trend During Cooking: Filter Effects.

During the cooking it was introduced cyclic measurement that comprised a cleaning phase, to avoid sensors saturation. This phenomenon could take place during non- stop measuring. However after cleaning there were still odor compounds on Metal- Oxide Semiconductor surface, and drift phenomenon was observed in W1C, W3C, W5C, W1W signal trend (Fig. 4). Then 90 seconds cleaning was introduced, in this way all residual compounds were removed.

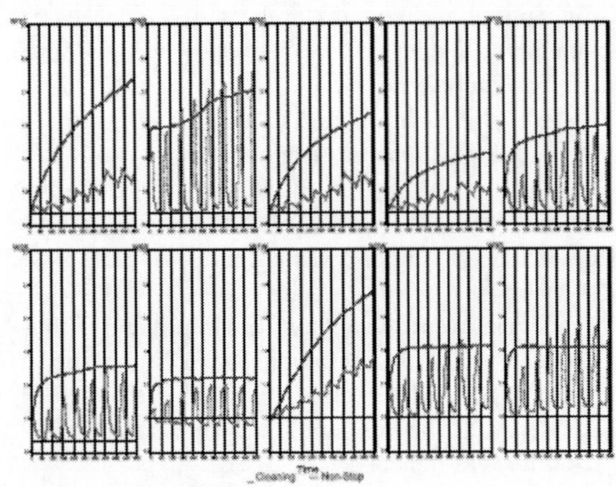

FIGURE 4. Sensors Signal Trend During Cooking: The Drift.

Consequently, according to these observations, we defined the best final protocol that comprised: 30 seconds white reference (air in the box); 30 seconds of sample aroma test. Afterwards was performed a zero point trim and 30 seconds white reference of the 165°C oven air is carried out. Subsequently the sample was put in the oven and 600 seconds measurement took place. The measurement comprised cycles made up of test (30s) and cleaning phase (90s). At last, after a zero point trim, there was measurement of empty box (30s) and then cooked sample test took place (30 s).

4. CONCLUSIONS

The electronic nose has been confirmed to be useful in meat analysis. However it has been rarely used to characterize odor like it is perceived by meat eater.
The protocol, we proposed, could permit to use EN to describe fresh meat aroma especially to measure it in the consumer conditions.

REFERENCES

1. S. Soncin, L.M. Chiesa, C. Cantonia and P.A. Biondi, *J. Food Comp. Anal.* **20**, 436- 439 (2007).
2. C. R. Calkins and J. M. Hodgen, *Meat Sci.* **77**, 63- 80 (2007).
3. I. Gonzalez-Martin, J. L. Perez-Pavon, C. Gonzalez- Perez, J. Hernandez-Mendez and N. Alvarez-Garcia, *Anal.Chim. Acta* **424**, 279- 287 (2000).
4. J. P. Santos, M. Garcia, M. Aleixandre, M. C. Horrillo, J. Gutierrez, I. Sayago, M. J. Fernandez and L Ares, *Meat Sci.* **66**, 727- 732 (2004).
5. F. Winquit, E. G. Horsten, H. Sundren and I. Lundastrom, *Meat Sci.* **4**, 1493- 5000 (1993).
6. T. Rajamaki, H. L Alatomi, T. Ritvanen, E. Skytta, M. Smolander and R. Ahvenainen, *Food Control* **17**, 5- 13 (2006).
7. S. Barbera and S. Tassone, *Meat Sci.* **73**, 467- 474 (2006).

Threshold detection of aromatic compounds in wine with an electronic nose and a human sensory panel

José Pedro Santos[1], Jesús Lozano[2], Manuel Aleixandre[1], Teresa Arroyo[3], Juan Mariano Cabellos[3], Mar Gil[3], Maria del Carmen Horrillo[1]

[1]*Instituto de Física Aplicada (CSIC), Madrid, Spain*
[2]*Grupo de clasificación de patrones y Análisis de Imágenes. Universidad de Extremadura. Badajoz, Spain*
[3] *Dept. Agroalimentación. Instituto Madrileño de Investigación y Desarrollo Rural, Agrario y Alimentario, Madrid, Spain*
josepe@ifa.cetef.csic.es

Abstract. An electronic nose (e-nose) based on thin film semiconductor sensors has been developed in order to compare the performance with a trained human sensory panel. The panel had 25 members and was trained to detect concentration thresholds of some compounds of interest present in wine. Typical red wine compounds such as whiskylactone and white wine compounds such as 3-methyl butanol were measured at different concentrations starting from the detection threshold found in literature (in the micrograms to milligrams per liter range). Pattern recognition methods (principal component analisys and neural networks) were used to process the data. The results showed that the performance of the e-nose for threshold detection was much better than the human panel. The compounds were detected by the e-nose at concentrations up to ten times lower than the panel. Moreover the e-nose was able to identify correctly each concentration level therefore quantitative applications are devised for this system.

Keywords: Threshold detection, human panel, electronic nose
PACS: 07.07.Df, 07.05.Mh

INTRODUCTION

Electronic noses have aroused in last years as a powerful tool in many applications. Their performances are usually compared with respect to analytical methods as gas chromatography-mass spectrometry or human sensory panels. Comparison with human panels have been performed mostly in food quality applications [1,2,3] with a qualitative approach. There are few examples of quantitative applications [4] and are outside this field.

In the present work a sensory panel and an electronic nose have been trained in parallel with several aromatic compounds present in wine in order to compare their performances in threshold detection and concentration quantification. This threshold is the minimum concentration of an aroma in water perceived for at least 50 % of the members of a sensory panel. Other term related to the olfactory threshold detection is the difference threshold detection, the minimum amount of an aroma that has to be added to a product already containing this aroma in order to produce an appreciable sensory change. This concept is especially useful in dealing with compounds in wine.

EXPERIMENTAL AND METHODS

Two typical compounds of white and red wines have been measured at increasing levels from the olfactory detection threshold. The compounds chosen were 3-methyl butanol (a fermentation aroma compound) for white wines and whiskylactone (aromatic compounds associated with aging in wood) for red wines. The amounts of compound added to

CP1137, *Olfaction and Electronic Nose: Proceedings of the 13th International Symposium*, edited by M. Pardo and G. Sberveglieri

each wine ranged from the olfactory detection threshold to eight times the threshold found in literature [5, 6]. Those thresholds, their odor description and their concentrations in the test wines are shown in table I. All compounds were of analytical quality and were provided by Sigma-Aldrich and Merck. It can be noted the great disparity in concentrations and odor thresholds.

TABLE 1. Odor thresholds, odor descriptions and concentrations in the test wines of the studied compounds

Compound	Odor threshold (μg/l)	Odor description	Concentration (μg/l)
3-Methyl butanol	29930	Oil	185000
Whiskylactone	67	Wood, sweet fruit	8.5

A group of 25 persons with previous experience in wine analysis were trained in recognizing 45 aromatic compounds from wine. Sensory evaluations were realized under Spanish UNE norms related to methodology, sensory analysis vocabulary, tasting room and selection and formation of tasters between others. All compounds were presented to the tasters at their threshold concentration in water and in the test wines.

An electronic nose based on a tin oxide array has been employed. The array was composed by 16 thin film tin oxide sensors with thickness between 200 and 800 nm doped with small amounts of chromium and indium to increase their selectivity. Sensors were deposited by reactive sputtering onto an alumina substrate. Details of the preparation can be found elsewhere [7]. The array was place in a 20 cm^3 stainless steel cell kept at 250 °C.

Two sampling methods were used: headspace and purge and trap. In the first case 10 ml of sample was kept in a Dreschell bottle at 30 °C for 20 minutes. The headspace was carried by an inert gas for 10 minutes to the sensor cell. The carrier used gas was nitrogen 99.998% purity at a constant flow of 200 ml/min. With the purge and trap method 1 ml of sample were placed in the vessel of a Tekmark 3000 purge and trap concentrator. The sample was purged with pure helium into a Tenax trap for 10 minutes and then desorbed at 250 °C with a nitrogen flux.

The resistance of the sensors were measured with a Keithley 2001 digital multimeter (DMM) connected to a Keithley 7001 multiplexer controlled by a personal computer. The sensor responses were calculated as the relation between the equilibrium resistance value in pure nitrogen, and the equilibrium resistance value in the presence of the sample.

The sensors were calibrated with ethanol in a weekly basis to compensate for drift. The experimental setup is shown in figure 1.

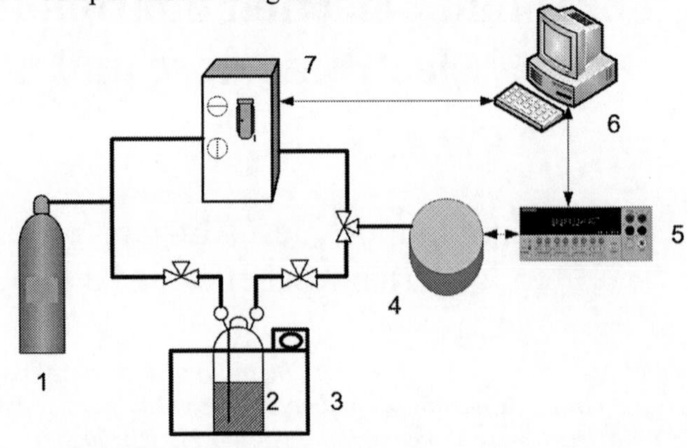

FIGURE 1. Measurement setup: 1 nitrogen bottle, 2 Dreschell bottle with sample, 3 thermostatic bath, 4 sensors cell, 5 multiplexer and DMM, 6 personal computer, 7 purge and trap.

Principal component analysis (PCA) was performed to the data to reduce the dimensionality. A Probabilistic neural network (PNN) was used for classification purposes. Cross validation was applied to check the performance of the network [8].

RESULTS

Sensory Panel

The analysis of the sensory panel is summarized in table II. The table gives the percentage of panel members that detects each compound at each concentration. No compound was detected at the olfactory detection threshold. The compounds were detected above eight times the threshold.

TABLE 2. Percentage of panel members that recognized the compounds at each concentration level. T = odor threshold concentration (see table 1).

Compound	T	2xT	4xT	8xT
3-Methyl butanol	11	15	30	32
Whiskylactone	17	25	42	46

Electronic nose

Data were preprocessed before PCA analysis. The preprocessing involved autoscaling and centering:

$$r_i = \left(r_i - \bar{r}_i\right)/\sigma_i \qquad (1)$$

where is the mean and σ_i is the standard deviation of sensor i response over the input data. The distribution of values for each sensor across the entire database is set to have zero mean and unit standard deviation.

The PCA plot for 3-methyl butanol in white wine with headspace extraction is shown in figure 2. The variance explained by each principal component is in brackets.

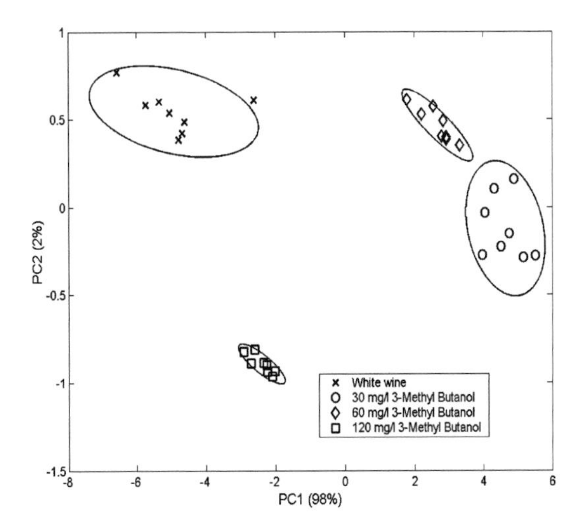

FIGURE 2. PCA plot for several concentrations of 3-methyl butanol in white wine with headspace extraction.

The datasets appears clearly separated. Figure 3 shows the PCA plot for the same compound measured with purge and trap extraction method.

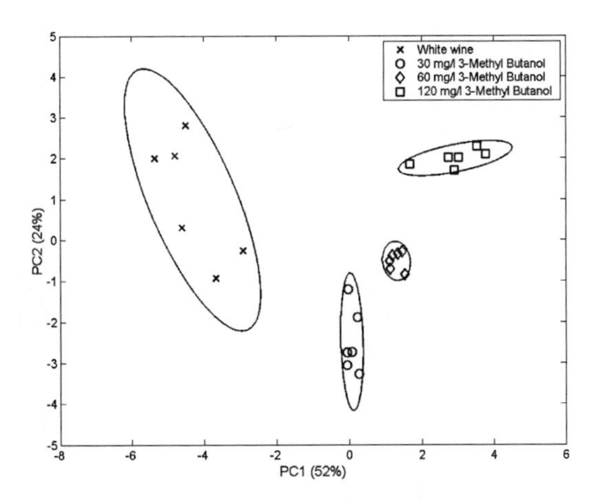

FIGURE 3. PCA plot for several concentrations of 3-methyl butanol in white wine with purge and trap extraction.

As in the other case datasets are clearly separated.

Figures 4 and 5 show the PCA plots for whiskylactone with headspace and purge and trap extraction methods respectively. As in the other compound datasets are separated except an slight overlap between the intermediate concentration datasets for purge and trap extraction method.

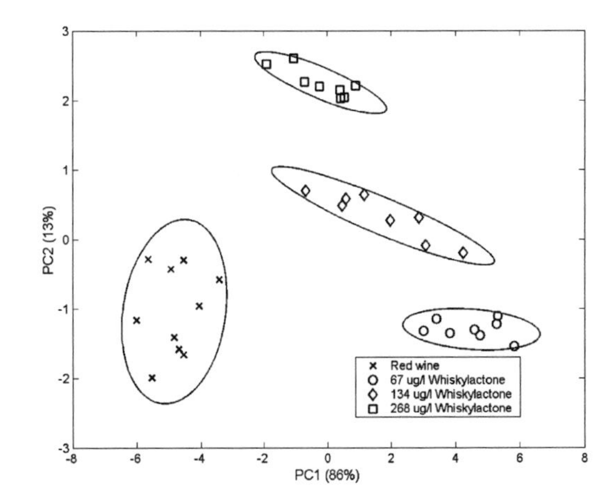

FIGURE 4. PCA plot for several concentrations of whiskylactone in red wine with headspace extraction.

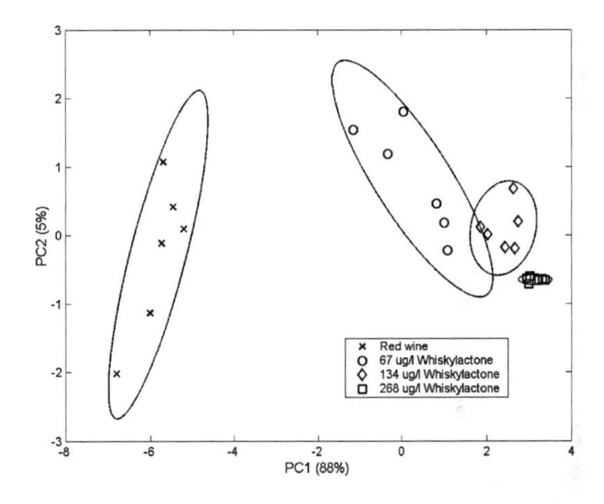

FIGURE 5. PCA plot for several concentrations of whiskylactone in red wine with purge and trap extraction.

Data from the first three principal components were used to train and validate the PNN. The classification success rate of the PNN was 100% both qualitatively (differentiation of blank wine from wine with added compounds) and quantitatively (all measurements were correctly assigned to their concentration class).

DISCUSSION

In all cases the electronic nose exhibited a better performance than the human sensory panel in the discrimination of aromatic compounds. The first threshold data are correctly separated from wine data for all compounds. Higher concentration datasets are separated as well. In the case of the sensory panel the compounds are detected only at concentrations above eight times the threshold. In the case of 3-methyl butanol, this is probably due to the high amount of the compound already present in wine relative to its odour threshold. The electronic nose, however, performed very well with a 100 % quantitative success rate with both extraction techniques. In the case of the whiskylactone, not detected by the panel even at eight times the threshold, the electronic nose performed well both qualitative and quantitative despite the low concentrations involved (the first odour threshold is only 67 μg/l). It appears evident that the performance of the electronic nose is not directly related neither to the odour threshold concentration of a compound nor to the ratio of its real concentration in wine to its threshold concentration.

CONCLUSIONS

The use of electronic noses in the determination of compound thresholds is advantageous over the human sensory panel. They can give a fast and accurate response, both qualitatively and quantitatively even in a complex matrix as wine.

ACKNOWLEDGMENTS

This work was supported by the Spanish Science and Technology Ministry under the project TIC2002-04588-C02-01.

REFERENCES

1. Di Natale C., Paolesse R., Burgio M., Martinelli E., Pennazza G.and D'Amico A., "Application of metalloporphyrins-based gas and liquid sensor arrays to the analysis of red wine", Analytica Chimica Acta, 513 (2004), pp. 49-56..
2. Panagou, E.Z. , Sahgal, N. , Magan, N. , Nychas, G.-J.E., "Table olives volatile fingerprints: Potential of an electronic nose for quality discrimination", Sensors and Actuators B: 134 (2008), pp. 902-907..
3. Luzuriaga, D.A. , Korel, F. , Balaban, M.Ö., "Odor evaluation of shrimp treated with different chemicals using an electronic nose and a sensory panel", Journal of Aquatic Food Product Technology, 16 (2007), pp.57-75.
4. Poprawski, J.a , Boilot, P.b , Tetelin, F., "Counterfeiting and quantification using an electronic nose in the perfumed cleaner industry", Sensors and Actuators B: 116 (2006), pp. 156-160.
5. Aznar M., López R., Cacho J., Ferreira V., "Prediction of aged red wine aroma properties from aroma chemical composition. Partial least squares regression models", *J. Agric. Food Chem.*, 51 (2003), pp. 2700-2707.
6. Guth H., Chemistry of Wine Flavor, Waterhouse, A. L., Ebeler, S. E., Eds. American Chemical Society: (Washington, DC 1998), pp. 39-52.
7. M.C. Horrillo et al., "Design and development of an electronic nose system to control the processing of dry cured Iberian hams monitored via Internet" in Electronic noses and olfaction 2000. Institute of Physics publishing. (Bristol and Philadelphia, 2000) pp.75-80
8. Derks E. P. P. A., Beckers M. L. M., Melssen W. J., Buydens L. M. C., "Parallel processing of chemical information in a local area network—II. A parallel cross-validation procedure for artificial neural networks", *Computers Chem.* 20 (1995), pp. 439-448.

Serum Headspace Analysis With An Electronic Nose And Comparison With Clinical Signs Following Experimental Infection Of Cattle With *Mannheimia Haemolytica*

Henri Knobloch[1§], Claire Turner[1], Mark Chambers[2], Petra Reinhold[3]

[1] *Cranfield University, Cranfield Health, UK*
[2] *Veterinary Laboratories Agency (VLA, Weybridge, UK)*
[3] *Institute of Molecular Pathogenesis in the 'Friedrich-Loeffler-Institut' (Jena, Germany)*
§Corresponding author
h.knobloch.s06@cranfield.ac.uk

Abstract. Electronic noses (e-noses) have been widely used for medical applications or in the food industry. However, little is known about their utility for early disease detection in animals. In this study, 20 calves were experimentally infected with Mannheimia haemolytica A1. Blood serum was collected from 7 days before infection to 5 days after infection and headspace of sera was analysed using the ST214 (Scensive Tech. Ltd., Leeds, UK) e-nose. Differences between pre- and post infection status were investigated and a temporal profile of sensor responses was compared with body temperature over the course of infection. A similar profile for sensor responses and body temperature indicated the e-nose was detecting a genuine physiological response following infection.

Keywords: electronic nose (e-nose), acute phase proteins (APP), animal model, gram-negative bacterial infection, Mannheimia haemolytica A1, host response
PACS: 07.07.Df, 68.03.-g, 87.85.Ox, 87.85.fk, 87.19.xb

1. INTRODUCTION

The early detection of biological markers ('biomarkers') for disease detection and identification as emitted by a particular microorganism or as a result of the host response is a valuable yet challenging area of investigation. Some biomarkers have been associated with infectious or metabolic processes such as ammonia for kidney and liver diseases or acetone for changes in blood glucose and diabetes [1]. For diagnosis, different biological samples (i.e. blood, urine, exhaled breath, faeces) can be used to generate volatile organic compounds (VOC) for trace gas analysis. One approach of assessing changes in VOC composition is using electronic nose (e-nose) technology. Some studies in the area of medical diagnostics have already been conducted including rapid detection of tuberculosis, Helicobacter pylori and urinary tract infections [2-5]. In veterinary medicine, some studies have demonstrated the potential of e-nose technology and VOC analysis for disease detection in general, by discriminating serum samples obtained from badgers and cattle infected with Mycobacterium bovis from those obtained from non-infected controls [6].

E-noses consist of an array of non-specific gas sensors generating a pattern of responses following exposure to the VOCs present in the headspace of samples. Usually, complex data analysis is necessary to match changes in VOC mixture with disease.

Mannheimia haemolytica is a gram-negative bacterium causing respiratory infections in cattle that may also lead to systemic disorders or even septicaemia. As a consequence, the regulation of electrolyte composition and acid-base balance might be altered, and the body odour may be significantly influenced by the changed pattern of VOCs present in the body.

Therefore, the potential of rapid VOC analysis using e-nose technology for detecting this gram-negative bacterial infection in vivo was assessed.

CP1137, *Olfaction and Electronic Nose: Proceedings of the 13th International Symposium*, edited by M. Pardo and G. Sberveglieri

Blood samples were collected before and after an induced infection with Mannheimia haemolytica and were analysed for both (i) changes in sensor responses and (ii) the rectal temperature as a surrogat marker for non-specific host response and inflammation. Univariate data analysis was applied to investigate changes in the sensor response due to the infection status, and to eliminate confounding methodological variation overlapping with biological variation and masking changes in sensor responses due to the infection status.

2. ANIMALS, MATERIALS AND METHODS

2.1 Animals

Twenty conventionally cross-bred calves aged 2 to 3 months were housed in 4 groups (each group consisted of 5 calves) under controlled conditions according to the guidelines for animal welfare in the European Union.

Daily clinical observation confirmed their healthy status before being included in the experiment. All phases of the study were performed in a specialised veterinary institute (Federal Research Institute for Animal Health, Germany) under supervision of a veterinarian, and had ethical approval.

2.2 Study Design

This study was designed as an intra-individually controlled study in order to decreases the number of animals used for experimental purposes for ethical reasons. Thus, each animal was carefully and consecutively characterised before challenge (non-infected status or baseline data) as well as after challenge (infected status) by identical examinations and techniques. Daily clinical examination was performed in each animal throughout the full study (including monitoring of respiratory rate, nasal secretions, ocular secretions, rectal temperature, appetite, and body weight).

2.3 Experimental Challenge

All calves were inoculated with Mannheimia haemolytica biotype A serotype 1 (M. haem. A1) intratracheally as described by Schimmel (1987) [8]. Ten ml of the culture containing 1.5 - 2.0 x 109 cfu per ml were administered per calf and day on two consecutive days (the interval between the two inoculations was 30 hours). Surviving calves (n=14) were euthanized for necropsy at 5 d after first inoculation.

2.4 Collection Of Blood And Serum Preparation

In each animal, two blood samples were collected before experimental infection (7 days and one hour pre infection), and 8 blood samples were obtained post infection: 3h, 6h, 12h, 24h, 48h, 3d, 4d, and 5d after the first bacterial inoculation. Jugular venous blood was collected using plastic syringes for serum production (S-Monovette®-Serum, Sarstedt AG & Co, Nuembrecht, Germany). Serum was harvested by centrifugation (20 minutes at 1500g) and stored at -80°C until analysed.

2.5 E-nose Headspace Analyses

For e-nose analysis, the conducting polymer based ST214, Scensive Tech. Ltd., Leeds, UK was used. This contains an array of 14 sensors including one reference sensor. After thawing, the serum samples were dispensed into Nalophan bags which were sealed and incubated at 25°C for 15 minutes. The headspace generated was analysed by attaching the bag to the e-nose and responses of replicates 3 to 5 were used for data analysis since they provide a maximum of stability during static sampling process as described [9].

2.6 Statistical Analyses

Linear regression (Pearson) and multifactor analysis of variance (mANOVA) was used to identify linear correlations between variables (sensor responses and methodological or biological variables) and to separate biological from methodological variation. For e-nose data analyses, divergence, i.e. maximum amplitude of signal in sample analysis, was used. Multiple range testing was used to investigate differences between groups (e.g. over time). The level of significance for all statistical methods applied was $P \leq 0.05$.

3. RESULTS

3.1 Clinical Course Of The Study

After inoculation of M. haem. A1, four calves died 24-48 hours, and two 48-72 hours, after infection. The remaining calves (n=14) showed clinical symptoms such as loss of appetite or changes in behavior (moribund and unresponsive). The respiratory rate and rectal temperature started increasing 6 hours after infection. The maximum temperature was measured 48 hours after infection.

Both rectal temperature (Figure 1) and breathing frequency (data not shown) remained significantly increased until the end of the study compared to pre-infection data. Surviving calves (n=14) were euthanized for necropsy (i.e. 5 d after first inoculation).

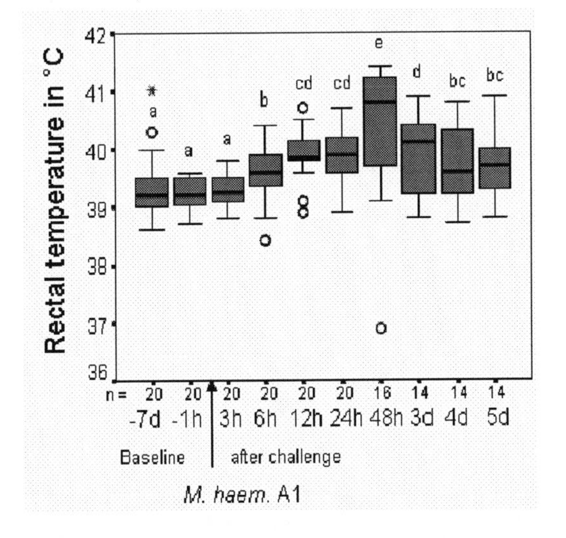

FIGURE 1: Rectal temperature significantly increased after M. haem. A1 infection and reached a maximum 48h after challenge.

Legend: Different letters indicate significant differences between time points (mANOVA).

3.2 E-nose Analyses: Changes Over Time

The sensor responses were analysed for significant changes over time. Sensors 3, 6 and 7 showed overall differences between pre- and post challange samples. Divergences of sensors 2 to 5, 10 and 11 decreased after challenge showing a temporal profile. Multiple range testing confirmed the findings and found significantly decreased divergences at days 2 and 3 post-infection. Figure 2 illustrates a typical temporal profile of a sensor response during course of the experiment.

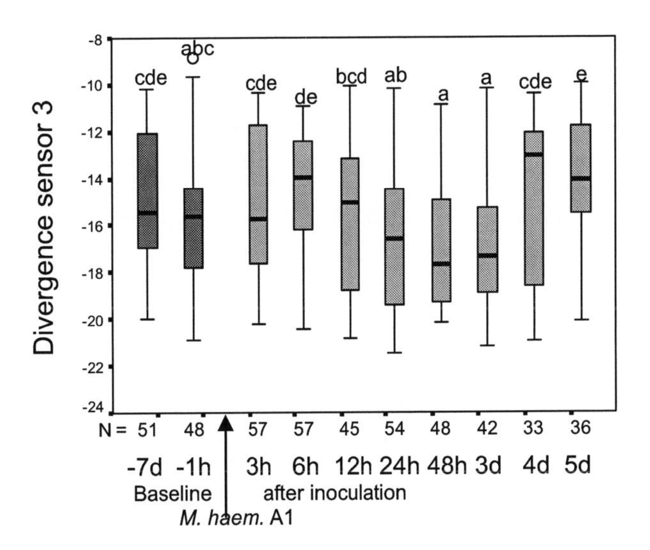

FIGURE 2: Sensor responses changed significantly over time. Minimum Divergences were obtained at days 2 and 3 after challenge.

Most sensors were significantly affected by methodological influences such as the time and day of analyses and temperature of the room in which the e-nose was housed (data not shown). The subject (animal) did not influence the profile (data not shown). Comparing surviving individuals with the ones that died, no differences between both groups was found.

3.3 Correlation Between E-nose Results And Rectal Temperature

Comparing e-nose responses and rectal temperature (the latter as an indicator for the severity of the disease), significant correlations were found between 2 sensor responses (sensors 8 and 9) only and rectal temperature. However, e-nose responses followed the the trend of rectal temperature over time for almost all sensors (see profiles above). An increase in rectal temperature was associated with a decrease in sensor response.

4. DISCUSSION

The main objectives of this study were to assess the possibility of using e-nose technology for detecting changes in VOC composition and to underline the findings by comparing them to another independently obtained surrogate marker of disease (rectal temperature), using experimentally induced M. haem. A1 infection as the model.

A temporal profile was observed for various e-nose sensor responses and the rectal temperature. This indicates significant changes (i) in the headspace composition of the blood sampled at different time points during the study and (ii) between pre- and post-infection time points. It generally indicated that VOC analysis has real potential for disease diagnostics. However, the e-nose technique is not able to elucidate the nature of the molecules leading to the change in sensor responses and furthermore this particular device did not allow discrimination between individual animals since methodological variation led to unpredictable changes in sensor responses [7]. The changes in sensor responses in this trial may be due to the presence of compounds liberated by the bacteria themselves, and/or the host inflammatory response to infection, as indicated by increased rectal temperature. Temperature and sensor responses showed the same temporal profile with a temporal shift of approximately 24h, and 2 sensors were found to correlate directly.

5. CONCLUSIONS

E-nose technology was found to have potential for analysing serum headspace VOCs and non-specifically identifying infections, e.g. M. haem. A1. Temporal changes in e-nose sensor response associated with infection with M. haem. A1, and during the course of the infection, could be found at the group level and were found to correspond to changes in body temperature, demonstrating the e-nose was tracking real physiological responses to infection. Further methodological improvements may enable discrimination between individuals. Further studies are required to identify VOCs that could be used as specific biomarkers for certain infections or diseases. The developments in analytical methodology including e-nose technology hold promise for an objective test for rapid diagnosis of infections of livestock.

ACKNOWLEDGMENTS

This project was conducted as part of a wider research project funded by the UK Department for Enviornment, Food and Rural Affairs (Defra). The authors gratefully thank the Friedrich-Loeffler-Institut (Jena; Germany) for supporting this project.

REFERENCES

1 C. Turner, B. Parekh, C. Walton, P. Španěl, D. Smith, M. Evans, *Rapid Comm. in Mass Spectrometry* **22**(4), 526-532 (2008)
2 R. Fend, A.H. Kolk, C. Bessant, P. Buijtels, P.R. Klatser, A.C. Woodman, *J Clin. Microbiology* **44**(6), 2039-2045 (2006)
3 A.K. Pavlou, N. Magan, D. Sharp, J. Brown, H. Barr, A.P.F. Turner, *Biosensors and Bioelectronics* **15**(7-8), 333-342 (2000)
4 A.K. Pavlou, N. Magan, C. McNulty, J.M. Jones, D. Sharp, J. Brown, A.P.F. Turner, *Biosensors and Bioelectronics* **17**(10), 893-899 (2002)
5 R. Fend, R. Geddes, S. Lesellier, H.-M. Vordermeier, L.A.L Corner, E. Gormley, E. Costello, M.A. Chambers, *J Clin. Microbiology* **43**(4), 1745-1751 (2005)
6 A. Voss, V. Baier, R. Reisch, K. Von Roda, P. Elsner, H. Ahlers, G. Stein, *Ann. Biomedical Eng.* **33**(5), 656-660 (2005);
7 A.D. Spooner, C. Bessant, C. Turner, H. Knobloch, M.A. Chambers, Diagnosis of Mycobacterium bovis in wild badgers using a combination of SIFT-MS and multivariate data analysis, RSC's *The Analyst (submitted)*
8 D. Schimmel, Archiv für Experimentelle Veterinärmedizin **41**, 463-472 (1987)
9 Knobloch, H., Turner, C., Spooner, A.D., Chambers, M.A; Trace Gas and Headspace Analysis Using Electronic Nose Technology, Sensors and Actuators B; accepted 10/03/2009

Long Term Stability Of Metal Oxide-Based Gas Sensors For E-nose Environmental Applications: an overview

Anne-Claude ROMAIN, Jacques NICOLAS

University of Liège – Department "Environmental Sciences and Management", Avenue de Longwy, 185, 6700 ARLON - Belgium
acromain@ulg.ac.be

Abstract. The e-nose technology has enormous potentialities for in site monitoring of malodors. However a number of limitations are associated with the properties of chemical sensors, the performances of the signal processing and the realistic operation conditions of environmental field. From the experience of the research group in the field, the metal oxide based gas sensors (Figaro type) are until now the best chemical sensors for long term application, more than one year of continuous working in the field. To be usable for malodors measurement in the field, the e-nose has to deal with the lack of long term stability of these sensors. The drift and the sensors replacement have to be considered. In order to appraise the time evolution of the sensors and the effect on the results of an electronic nose, experimentation has been performed during three years on two identical sensor arrays. The two arrays contain the same six Figaro sensors and are in the same sensor chamber of the e-nose system. Both arrays have worked continuously during three years without break.

Keywords: e-nose, tin oxide sensors, drift, environmental application.
PACS: 07.88.+y

INTRODUCTION

The Artificial olfaction system is a very promising tool to monitor the malodor in the field. Usual measurement techniques of odor use human olfaction or conventional analytical techniques. The first category represents the real odor perception but is not applicable to measure continuously bad odors in the field. The second class of techniques gives the mixture composition but not the global information representative of the odor perception. The e-nose has the potentialities to combine "the odor perception" and the "monitoring in the field". The instrument, based on non-specific gas chemical sensors array combined with a chemometric processing tool provides a suitable technique for in site monitoring of malodors. The research group in Arlon has more than ten years experience in the measure of environmental malodors in the field. Published studies report attractive results [1,2]. This technique has probably the best potentialities to answer to the expectations of the various actors of the environmental problems in relation with the odors annoyance [3]. However a number of limitations are associated with the properties of chemical sensors the performances of the signal processing and the realistic operation conditions of environmental field. From the experience of the research group in the field, the metal oxide based gas sensors are the best chemical sensors for long term application, more than one year of continuous working. However, as a result of harsh environmental conditions, of hardware limitations and of olfactory pollution specificities, odour real-time monitoring with an electronic nose is always a real challenge. The instrument has to cope with several specific drawbacks. In particular, it has to automatically compensate the time drift [4] and the influence of ambient parameters such as temperature or humidity [5]. This paper is focused on the time drift and the long term stability of the metal oxide gas sensors (Figaro sensors). Sensor drift is a first serious impairment of chemical sensors. The sensors alter over time and so have poor repeatability since they produce different responses for the same odour. That is particularly troublesome for electronic noses [6]. The sensor signals can drift during the learning phase [7]. Another frequent problem encountered in the field and particularly in highly polluted atmosphere is a sensor failure or an irreversible sensor poisoning. Clearly, life expectancy of sensors is reduced for real-life operation with respect to clean lab operation. Sensor replacement is generally required to address such issue, but, after

CP1137, *Olfaction and Electronic Nose: Proceedings of the 13th International Symposium,* edited by M. Pardo and G. Sberveglieri

replacement, odours should still be recognised without having to recalibrate the whole system. But commercial sensors are rarely reproducible.

1. EXPERIMENTAL AND METHODS

In order to appraise the time evolution of the sensors and the effect on the results of an electronic nose, experimentation was performed during three years on two identical sensor arrays. The two arrays contain the same six Figaro sensors (see Table 1) and are in the same sensor chamber of the e-nose system. Both arrays worked continuously during three years without break.

TABLE 1. the six selected Figaro sensors.

Sensor	Application (from the manufacturer)
TGS 822	Organic solvents (ethanol, benzene, acetone, . .)
TGS 880	Volatiles vapour from food (alcohols)x
TGS 842	Natural gas, methanex
TGS 2610	Propane, butane
TGS 2620	Hydrogen, alcohols, organic solvents
TGS 2180	water

To try to compensate the sensor drift, three types of solutions were tested for our applications. The usual way of minimising drift effect is to consider as useful response the difference between the base line, obtained by presenting the sensor array to pure reference air, and the signal obtained after stabilisation in the polluted atmosphere. However, such solution requires operating by cycling between reference air and tainted air, which is not convenient for on-site applications. That requires carrying in the field heavy gas cylinders. Alternatively, generating the reference air by a simple filtering of ambient atmosphere gives rise to only partial drift compensation and to a lack of purity of the reference gas, which increase the data dispersion. Posterior global drift counteraction algorithms could be applied either for each individual sensor or by correcting the whole pattern, using multivariate methods. First the main direction of the drift is determined in the first component space of the multivariate method, such as Principal Component Analysis (PCA), or by selecting time as dependant variable of a Partial Least Square regression (PLS). The drift component can then be removed from the sample gas data, correcting thus the final score plot of the multivariate method [8].n order to appraise the time evolution of the sensors and the effect on the results of an electronic nose, experimentation was performed during three years on two identical sensor arrays. The two arrays contain the same six Figaro sensors (see Table 1) and are in the same sensor chamber of the e-nose system. Both arrays worked continuously during three years without break.

2. RESULTS

2.1. Drift

Figure 1 shows drift compensation of a commercial tin oxide sensor (TGS2620, FigaroTM) by multiplicative factor estimated from calibration measurements.

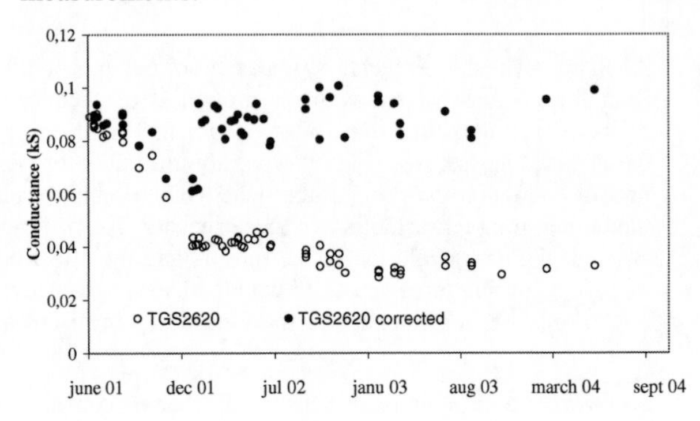

FIGURE 1. Drift correction of the sensor TGS 2620 by multiplicative factor estimated by calibration gas measurements.

The multiplicative factor seems to be able to compensate the drift of this sensor.

In order to test the performance of the method with this calibration gas on malodour measurements, linear discriminant analysis (LDA or fisher discriminant analysis) is applied for two different environmental odours: compost odour and printing house odour. Measurements are obtained with an array of 5 tin oxide sensors.

Linear discriminant analysis (LDA) was used with 5 features (5 sensor conductance values) and 63 observations collected within a 22- month's period. Two classes were considered (compost class and printing house class) The Fisher-Snedecor F-ratio of intergroup/intragroup variances was chosen as classification performance criterion. Higher the F value is, better is the separation of the classes.

TABLE 2. Evaluation of the classification without correction or with correction models (by the F criterion, F-ratio of intergroup/intragroup variances)..

Method	F
No correction	33
Correction by sensor (individual multiplicative factor)	56
Correction of the sensor array "PLS"	26
Correction of the sensor array "PCA"	18

2.1. Sensor replacement

Figure 2 shows a PCA score-plot in the plane of the two first components. It concerns 260 observations, 3 classes (ethanol, background air and compost odour), 5 features and a 2-year period. After the replacement of the sensors in the array with the same trade mark references, the previous calibrated model is no longer applicable for the same odorous emissions: all the observation points are shifted to another part of the diagram.

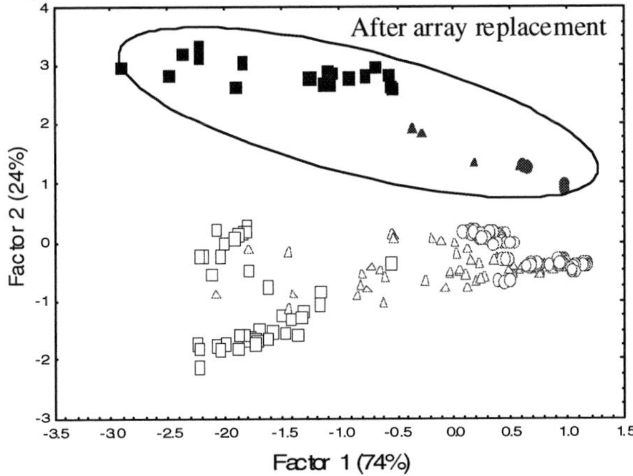

FIGURE 2. Illustration, by a PCA score plot of the shift of the measurements after the replacement of the TGS sensors

Again, correction routines including algorithms for handling shift related to sensor replacement can be successfully applied. For the above example, illustrated in figure 2, the classification performances were severely reduced after array replacement. The percentages of correct classification were 40%, 100% and 33% respectively for ethanol vapour, background air and compost emission.

After individual sensor correction (with the same procedure individual multiplicative sensor than for drift correction), each classification rate reaches 100% (table 3).

TABLE 3. Percentage of correct classification after replacement of sensors with and without correction

Source	% of correct classification	
	Without correction	**With correction**
Ethanol	40	100
Air	100	100
Compost	33	100

3. CONCLUSIONS

Univariate sensor correction gave the best results for complex data like malodours measurements.

With real-life measurements, it is indeed very difficult to identify a single direction in a multivariate space that is only correlated to sensor drift. So, for each sensor, an individual multiplicative factor was calculated by estimating the drift slope for a standard gas.

To meet the requirements of this environmental use of artificial olfaction system, the signal processing method must be simple, but not simplistic, and capable of generalisation. It must be tolerant to hardware weaknesses and adapted for application in the real life.

REFERENCES

1. Bourgeois, W., Romain, A.-C., Nicolas, J., Stuetz, R. M., "The use of sensor arrays for environmental monitoring: interests and limitations," J. Environ. Monit., Vol. 5, No. 6 (2003), pp. 852-860.
2. Romain, A.-C., Nicolas, J., Wiertz, V., Maternova, J., Andre, P., "Use of a simple tin oxide sensor array to identify five malodours collected in the field," Sensors and Actuators B: Chemical, Vol. 62, No. 1 (2000), pp. 73-79.
3. Romain, A.-C., Delva, J., Nicolas, J., "Complementary approaches to measure environmental odours emitted by landfill areas," Sensors and Actuators B: Chemical, Vol. 131, No. 1 (2008), pp. 18-23.
4. Romain, A.-C., Andre, P., Nicolas, J., "Three years experiment with the same tin oxide sensor arrays for the identification of malodorous sources in the environment," Sensors and Actuators B: Chemical, Vol. 84, No. 2-3 (2002), pp. 271-277.
5. Romain, A.-C., Nicolas, J., Andre, P., "In situ measurement of olfactive pollution with inorganic semiconductors : Limitations due to humidity and temperature influence," Seminars in Food Analysis, Vol. 2, (1997), pp. 283-296.
6. Romain, A.-C., Nicolas, J., "Three years experiment with the same tin oxide sensor arrays for the identification of malodorous sources in the environment," Sensors and Actuators B: Chemical, Vol. 84 (2002), pp. 271-277.
7. Holmberg, M., Davide, F. A. M., Di Natale, C., D'amico, A., Winquist, F., Lundstrom, I., "Drift counteraction in odour recognition applications: Lifelong calibration method" Sensors and Actuators B: Chemical, Vol. 42, No. 3 (1997), pp. 185-194.
8. Artursson, T., Eklov, T., Lundström, I., Martersson, P., Sjöström, M., Holmberg, M., "Drift correction for gas sensors using multivariate methods", Journal of chemometrics, Vol. 14, (2000), pp. 711-723.

Perfume Fragrance Discrimination Using Resistance And Capacitance Responses Of Polymer Sensors

John Paul Hempel Lima[1, 3*], Thomas Vandendriessche[3], Fernando J. Fonseca[1]
Jeroen Lammertyn[3], Bart M. Nicolai[3] and Adnei Melges de Andrade[1, 2]

[1] EPUSP-PSI, Universidade de São Paulo, Av. Prof. Luciano Gualberto, trav. 3 n. 158, 05508-900, Sao Paulo, Brazil
[2] IEE, Universidade de São Paulo, Av. Prof. Luciano Gualberto, n.1289, 05508-900, Sao Paulo, Brazil
[3] Department of Biosystems -MeBioS, Catholic University of Leuven, W. De Croylaan 42, B-3001 Leuven, Belgium
**johnpaul@lme.usp.br*

Abstract. This work shows a comparison between electrical resistance and capacitance responses of ethanol and five different fragrances using an electronic nose based on conducting polymers. Gas chromatography - mass spectrometry (GC-MS) measurements were performed to evaluate the main differences between the analytes. It is shown that although the fragrances are quite similar in their compositions the sensors are able to discriminate them through PCA (Principal Component Analysis) and ANNs (Artificial Neural Network) analysis.

Keywords: conducting polymers, electronic noses, ANN, GC-MS, perfume
PACS: 73.61.Ph

INTRODUCTION

Fragrance and perfumery are multi-billion industries producing different aromas and scents by mixing synthetic or natural compounds. Perfume oils are complex substances containing up to 2000 different compounds and it is very important to objectively characterize and recognize these different oils aiming at detection of counterfeit products and to perform quality control in perfume production.

Electronic noses (E-nose) have been extensively used to detect different aromas, odors and scents, in food, cosmetics and medical diagnostics. Particularly, electronic noses based on conducting polymers have been extensively used since these materials can operate at room temperature, exhibit a good sensitivity and are easily processed [1]. Several scientific publications report resistance measurements while others use capacitance as the electrical response. The majority of the research in this field uses principal component analysis (PCA) and/or artificial neural networks (ANN) to discriminate between different samples [2].

The objective of this research was to discriminate fragrances based on the electrical resistance and capacitance responses of a conducting polymer based electronic system.

EXPERIMENTAL AND METHODS

Conducting Polymers and Sensors

32 pairs of ITO interdigitated electrodes onto glass slides were used to monitor the electrical characteristics of the polymeric materials.

The sensors were produced by spin coating and layer-by-layer techniques using different conducting polymers, as described elsewhere [3, 4, 5]. The polymers used in this work were polyaniline, poly(o-methoxyaniline), poly(3-hexylthiphene), polypyrrole and PEDOT:PSS.

Nine sensors (E-nose) were kept inside an in-house made cylindrical analysis chamber and AC measurements were performed using an impedance analyzer Solartron SI-1260 at 1 kHz with 2 V_{pp}.

Five different aftermarket fragrances resembling the commercial perfumes were obtained from a local shop; "CK One™", "Tazo™", "Lagoa™", "Eternity LE™" and "Kris™". Ethanol (99.5%), methanol (99.5%) and water were also used as analytes during these experiments.

Air was used to purge the chamber and as carrier gas bubbled into the analytes to transport them to the analysis chamber. Three cycles were repeated at each sequence to observe drift and degradation phenomena

CP1137, *Olfaction and Electronic Nose: Proceedings of the 13th International Symposium*, edited by M. Pardo and G. Sberveglieri

of the sensors. The flow rate was maintained constant at 1,600 mL/min during the injection of the air/analyte mixture.

The absolute values of capacitance and resistance were recorded and normalized using equation 1.

$$X_{norm} = \left(\frac{X_j - X_{ini}}{X_{ini}} \right) \quad (1)$$

where X_j is the *jth* measurement and X_{ini} its initial value.

As indicated by Kermani et al. [6] it is possible to minimize the effect of compound concentrations by dividing the response of each sensor at a specific time by the average response of all the sensors at the same instant. For some analysis we also normalized the dataset using equation 2.

$$X_{norm,j} = \left(\frac{X_j}{\sum_N Xi} N \right) \quad (2)$$

where X_j is the value of the *jth* sensor and N is the number of sensors; all values at a specific instant t.

SPME-GC-MS analysis

For each fragrance an aliquot of 10 µL was placed into a 20 ml glass vial which was sealed using crimp-top caps with TFE/silicone septa seals. Volatile identification was conducted on an Agilent 6890 gas chromatograph (GC) (Agilent Tech., USA) equipped with an autosampling device (MPS2, Gerstel Multipurpose sampler, Ger.) and an Agilent 5973 network mass selective detector (MS) (Agilent Tech., USA). Prior to solid phase microextraction (SPME), the fragrance samples were incubated for 30 min at 30°C. Headspace volatiles were extracted by exposing a divinylbenzene-carboxen-polydimethylsiloxane SPME fiber (DVB-CAR-PDMS, 50/30µm film thickness; Sulpeco Inc., Bellefonte, PA, USA) to the vial headspace for 5 min at 30°C. Subsequently, the volatiles were desorbed for 3 min into the injection port of the GC maintained at 250°C. Chromatography was performed on an Optima-5-MS capillary column (5% diphenyl/ 95% dimethyl polysiloxane; 30m x 0.25mm x 0.25µm, Macherey-Nagel, Ger.). The volatiles were injected in split mode (split ratio 50:1) and helium was used as a carrier gas (1.2 mL/min). The GC temperature programme started at 35°C (3 min), was then first ramped to 175°C at a rate of 2°C/min followed by a second ramp to 230°C at 10°C/min. Finally the temperature was kept for 5 min at 230°C. The total runtime was 83.50 min. Mass spectra in the 10 to 400 m/z range were recorded at a scanning speed of 3.66 scan/sec and ionization energy of 70eV. The chromatography and spectral data were evaluated using ChemStation Software (Agilent Tech., USA). The experimental spectra were compared with those of the National Institute for Standards and Technology (NIST98, Version 2.0, USA) data bank or with authenticated standards. For each fragrance five replicates were analyzed.

Data Analysis

Principal component analysis (PCA) and artificial neural networks (ANN) were executed using MATLAB 6.1 (The Mathworks, USA) software.

A multilayer perceptron architecture was selected because of popularity for pattern recognition, cluster analysis and discrimination [2]. MLP was used in a feed-forward format with the backpropagation algorithm and an adaptative learning rate. The dataset was split in two subsets of 70% for training and 30% for testing. The selection of data for both sets was performed randomly and each training session was repeated 100 times. It was selected to have nine neurons at the input layer, N in the hidden layer and eight in the output layer. A multidimensional vector was used as the network output with the number of the dimensions equal to the number of the classes to be classified, where each vector dimension was assigned to a class in a canonical form. (1 when the input belonged to a class and 0 if not). The number of neurons in the hidden layer varied from 5 to 40 and also the activation function of the hidden neurons between *logsig* and *tansig*. The maximum number of epochs was limited to 5000. The performance of the neural networks was measured by the number of correctly classified samples divided by the number of the samples in the test set.

RESULTS AND DISCUSSION

Electrical Resistance and Capacitance Measurements

Fig 1 shows a typical response of the E-nose made with nine different polymeric sensors under a sequence of three cycles. One can see that each sensor produces a different response in both intensity and absolute value.

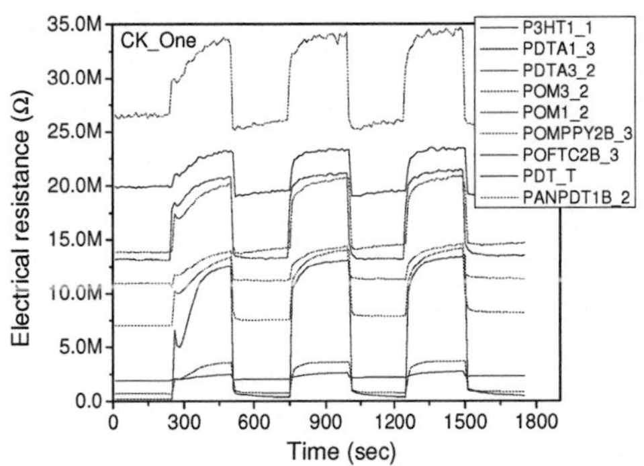

FIGURE 1. Typical electrical resistance responses of different sensors for air and air with an analyte repeated three times.

Although capacitance measurements are less susceptible to noise, the observation of its response revealed a lower intensity variation when compared to the resistance (Table 1).

TABLE 1. Average of normalized (N) variations of all sensors for different analytes using equation 1; sensitivity loss (S).

Analyte	N. Resist. (%)	S (%/day)	N. Cap. (%)	S (%/day)
Tazo	270	13.9	-3.2	16.6
CK One	555	19.7	-8.0	13.7
Ethanol	821	15.5	-9.3	16.5
Eternity	308	11.8	-4.4	14.5
Kris	239	24.1	-4.4	37.9
Lagoa	116	16.6	-1.0	14.3

The sensors also showed residual drift varying from 2 to 15% between measurements and also loss of sensitivity between different days of measurements (Table 1). This loss is mainly related to the sensor saturation and to the decrease of doping level in the polymeric matrix [1].

Data Analysis

PCA was carried out on 6 samples including ethanol and five different fragrances. All data used were obtained at the same day to avoid a "virtual discrimination" due to degradation effects of the sensors.

The PCA plot of normalized capacitance values (fig. 3) showed a better clustering (i.e., small intra-fragrance variability compared to the inter-fragrance variability) when compared with PCA plot of resistance values (fig. 2). It is also interesting to notice that the fragrances "Tazo" and "CK One" showed the largest distances among all clusters for both resistance and capacitance PCA plots.

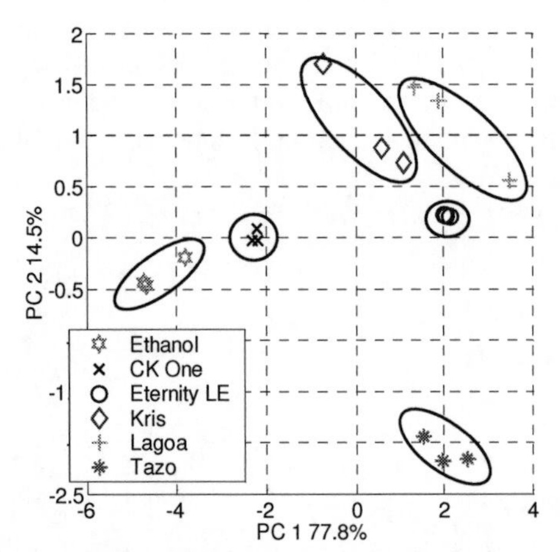

FIGURE 2. PCA plot of resistance measurements of fragrances and ethanol.

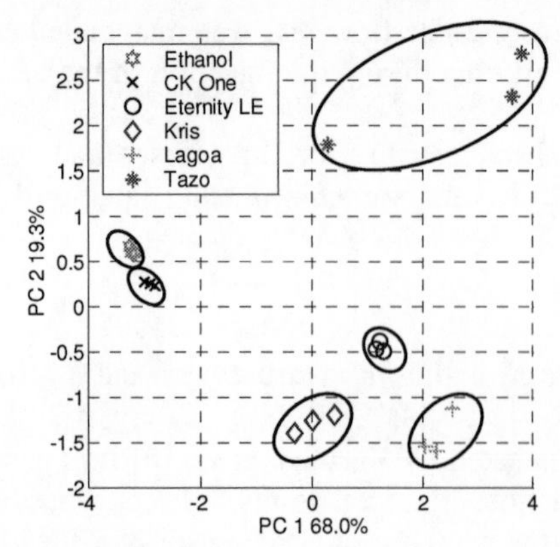

FIGURE 3. PCA plot of capacitance measurements of fragrances and ethanol.

By applying the normalization technique (eq. 2) described by Kermani, the discrimination capability (fig. 4) decreased when compared to the conventional technique. Apparently the concentration of individual compounds contributes to a better discrimination both in capacitance and in resistance.

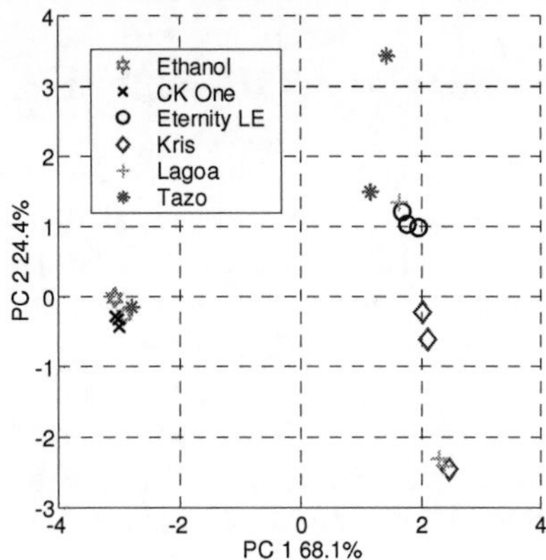

FIGURE 4. PCA plot of normalized capacitance measurements of fragrances and ethanol.

Another way of discriminating different analytes is through the use of artificial neural networks. Among the main parameters of a multilayer perceptron architecture the number of neurons in the hidden layer, the number of layers and the activation function of each neuron are the most important ones.

There is a small influence of the number of neurons and the activation function on the classification score. For the data obtained from the five fragrances, methanol, ethanol and water, the best architecture found was *tansig* as activation function, 9-30-8 where 9 are the inputs, 8 the outputs and 30 the number of neurons in the hidden layer.

It is important to evaluate the reproducibility of this result because in a commercial application, the dataset is randomly chosen as well as the initialization of the ANN parameter values. Our best result found was 91.4% of corrected classifications but the average of 100 randomly training procedures showed an average of 67% of corrected classifications in the test set.

The ANN analysis using capacitance values did not show a high classification score nor did the ANN analysis using the normalized resistance and capacitance obtained from equation 2. The performance of the ANNs for these cases was 15% and 40% worse than for resistance values, respectively. The normalization technique plays an important role on the performance of the MLPs and for our case the influence of individual concentrations increased the discrimination power of the ANN architectures.

SPME-GC-MS Measurements

For all five fragrances a total of 54 compounds could be identified. The 25 peaks with the highest relative peak area are shown in Fig. 5. It is interesting to notice that the five fragrances could be discriminated from each other based on the presence of compounds specific for certain fragrances and also based on concentration differences (figs. 5 and 6).

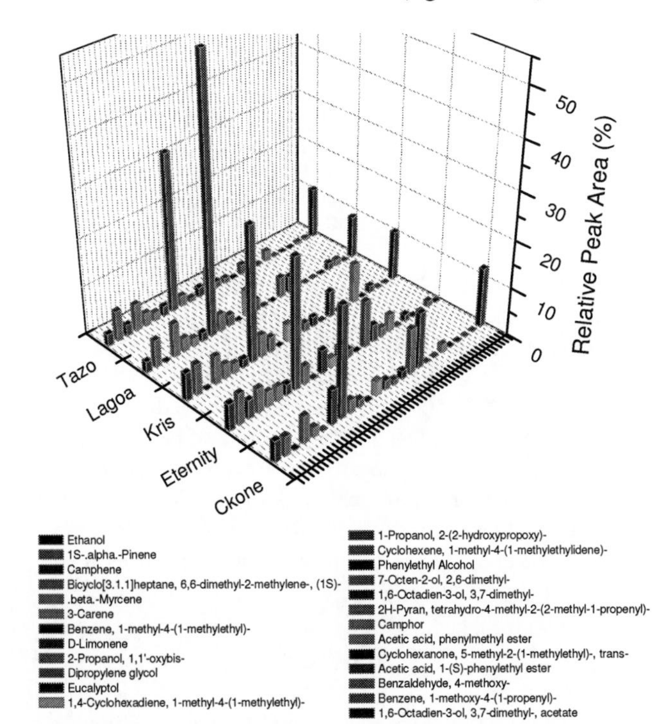

FIGURE 5. 25 peaks with the highest relative peak area of the five different fragrances.

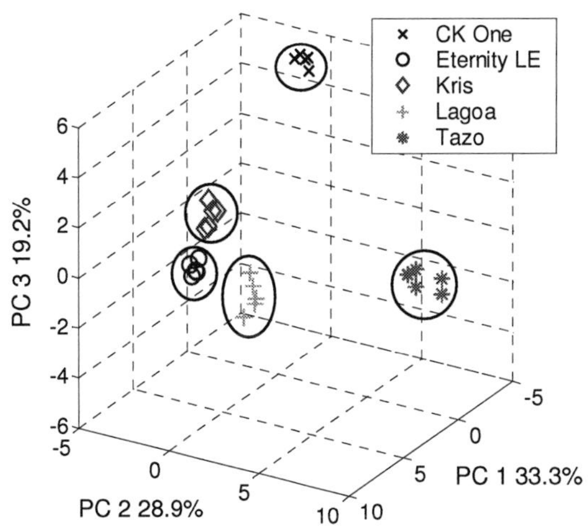

FIGURE 6. PCA plot of SPME-GC-MS measurements of five fragrances.

CONCLUSIONS

This paper showed a comparison between electrical resistance and capacitance measurements of conducting polymers as an electronic nose for different fragrances. The resistance relative variation was found to be one hundred times higher than the capacitance relative variation. Although this important difference, it was possible to use both types of electrical information for discriminating the different fragrances. The normalization technique that minimizes gas concentration effects reduced the PCA and ANN discrimination capabilities. Furthermore, the same fragrances were also analyzed with SPME-GC-MS resulting in identification of 54 different compounds. The PCA analyses of both techniques showed that the sensors discriminate similarly to the GC-MS technique.

ACKNOWLEDGMENTS

The authors wish to thank Elfie De Kempeneer for GC-MS support, FAPESP proc. 03/07454-5 and CNPQ/PNM for financial support.

REFERENCES

1. Bai, H and Shi, G, *Sensors*, **7**, 267-307 (2007).
2. Jurs, P. C. et al, *Chemical Reviews*, **100**, No. 7, 2649-2678 (2000).
3. Braga, G. S. et al, Proc 6th Brazilian MRS Meet, Natal, Brazil, Oct. 2007, F528.
4. Lima, J. P. H. and Andrade, A. M., ", Proc 7th Brazilian MRS Meet, Guarujá, Brazil, Oct. 2008, H550.
5. Lima, J. P. H. et al, Proc 12th BWSP, S Jose dos Campos, Brazil, Apr. 2005, th117.
6. Kermani, B. G. et al, *IEEE Trans-Bio-Eng*, Vol. 46, No. 4, 429-439 (1999)..

Exhaust Gas Sensor Based On Tin Dioxide For Automotive Application

Arthur VALLERON [a,b], Christophe PIJOLAT [a], Jean-Paul VIRICELLE [a], Philippe BREUIL [a], Jean-Claude MARCHAND [a] and Sébastien OTT [b]

[a] *Ecole Nationale Supérieure des Mines, Centre SPIN, 158 cours Fauriel, 42000 Saint-Etienne*
[b] *Renault SAS, Engineering Materials Department*

The aim of this paper is to investigate the potentialities of gas sensor based on semi-conductor for exhaust gas automotive application. The sensing element is a tin dioxide layer with gold electrodes. This gas sensor is able to detect both reducing and oxidizing gases in an exhaust pipe with varying selectivity depending on the temperature in the range 250°C-600°C.At low temperature 350-400°C, the sensor detects nitrogen dioxide while it is more sensitive to carbon monoxide at temperatures exceeding 500°C.

Keywords: Gas sensor, Tin Dioxide, Exhaust, Automotive.
PACS: 07.07.Df

INTRODUCTION

The new legislation obliges car manufacturers to reduce and control emissions. Hence, there are actually many researches on exhaust gas sensors. Most of them are based on devices issued from oxygen lambda sensors, mixed potential type sensors [1] using a solid electrolyte (often yttria stabilized zirconia, YSZ) as sensing element. On the contrary, there are few studies on this topic, using resistive gas sensors based on semi-conductor. This kind of gas sensor, in particular, tin dioxide sensors, is well known and is used in many areas such as, for example, process control, environmental control and domestic security [2]. So, in this study, we developped a robust gas sensors based on tin dioxide, able to work in exhaust harsh conditions. Its performances are studied on a synthetic gas bench.

EXPERIMENTAL AND METHODS

Sensors are made by screen-printing on an alumina substrate, one side of the substrate being used for the sensing layer and the other for the heating resistance. The thickness of the tin dioxide layer is about 40µm.

The sensors are tested in a bench which is able to submit sensors to high gas temperature (250°C-450°C) and high gas speed (v=5,6m/s), with a gas flow of 80L/min. Two kinds of tests are performed.

Static Tests

In"static tests", the sensor temperature is increased step by step, and for each temperature, sensors are subjected to a carrier gas flow (12% O_2), then the target gas : CO in the range [50-1000ppm] and NO_2 in the range [50-350ppm]. The sensor's conductances in carrier gas flow (G0) and in presence of pollutants (G) are measured. The sensitivity is represented by the relative response of the sensors : (G-G0)/G0 for reducing gases, and (R-R0/R0) for oxidizing gases, with R0 and R the sensing element resistances under carrier and target gas.

Dynamic Tests

These tests consist to study response time of sensors for responses to CO and NO_2. Sensors are successively subjected to a "base" gas and then mixed to the target gas at a certain frequency. For CO experiments, the "base" gas is constituted of O_2, H_2O, CO_2, NO and C_3H_8. 1000ppm of CO target gas is added periodically. For NO_2 experiments, "base" gas contains O_2, H_2O and CO_2, and NO_2 is injected as a target gas.

CP1137, *Olfaction and Electronic Nose: Proceedings of the 13th International Symposium*, edited by M. Pardo and G. Sberveglieri

RESULTS AND DISCUSSION

Static Tests

The responses of the sensors to CO versus temperature are shown in Figure 1. The gas temperature was set at 250°C. The sensitivity to a fixed concentration continuously increases with varying rate versus the temperature.

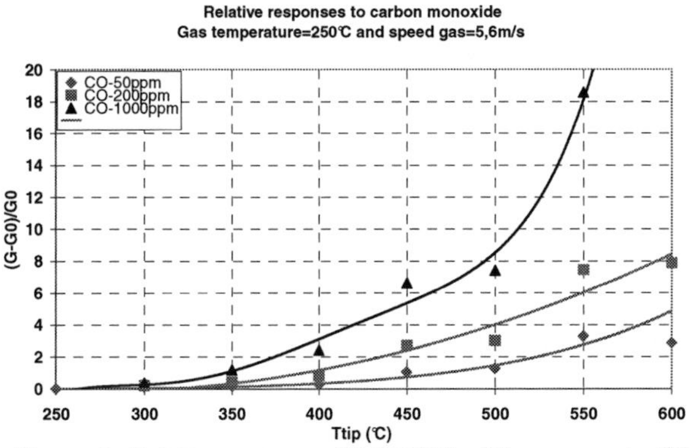

Relative responses to carbon monoxide
Gas temperature=250°C and speed gas=5,6m/s

Figure 1. Relative response (G-G0)/G0 of the sensors to CO 50ppm, 200ppm and 1000ppm as a function of sensors temperature (Ttip).

The responses of the sensors to NO_2 are shown in Figure 2. Contrary to CO, the response curves versus temperature present a maximum of sensitivity. The responses are high in the range 300°C-500°C with a maximum at 350°C.

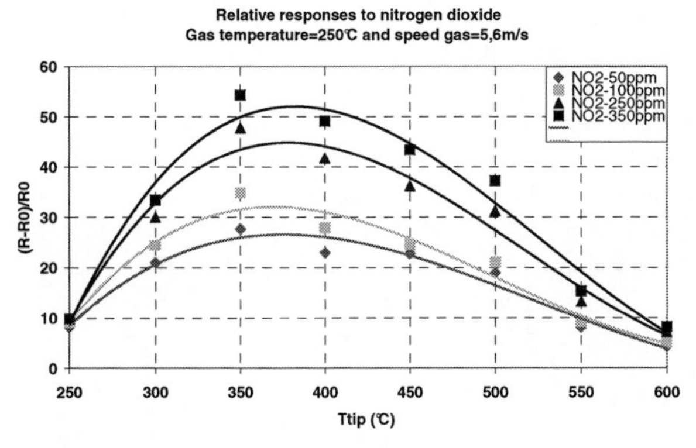

Figure 2. Relative response (R-R0)/R0 of the sensors to NO_2 50ppm, 100ppm, 250ppm and 350ppm as a function of sensors temperature (Ttip).

As expected, we observe a reducing behaviour for CO and an oxidant one for NO_2.

The difference of behaviour is very interesting because it gives the opportunity to detect CO at high temperatures (550-600°C), while NO_2 can be detected at 350°C.

Dynamic Tests

CO responses in dynamic tests at 600°C are shown in Figure 3. Both response time and recovery time are approximatively one second. It is worthy to note, that even with the presence of NO and C_3H_8 in the "base" gas, we have a significant response to 1000ppm CO.

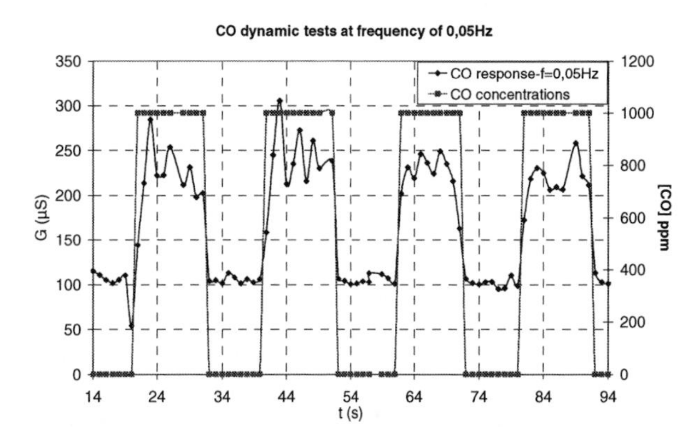

Figure 3. CO response in dynamic tests (f=0,05Hz).

For NO_2, (not shown here) response time is around 3 seconds but the recovery time is slower (>15s at 300°C).

For automotive application, detection of CO at 600°C is possible, unlike NO_2 detection at 300°C, for which the recovery time may be too long. A solution could consist in a temperature pulse to recover the base signal.

CONCLUSIONS

Thus, SnO_2 sensors are able to detect CO at a temperature above 500°C on a first hand, and to detect nitrogen dioxide at a temperature below 400°C on a second hand. These performances can be used to make a device comprising two sensors based on tin dioxide and which is able to detect the two gases simultaneously [3]. Another solution is to use only one sensor but with temperature regulation for cycling monitoring of CO and NO_2 depending on the temperature.

REFERENCES

1. S.Zhuiykov and N.Miura, Sensors and Actuators B,vol. 121 (2007), pp. 639-651
2. M.Kamionka, P.Breuil and C.Pijolat, Materials Science and Engineering C26 (2006), pp. 290-296
3. D.Guy-Bouyssou, C.Pijolat, J.P. Viricelle, P.Breuil, Capteur pour une détection multigaz, INPI n°FR0757033, le 10/08/2007

Electronic Nose To Detect Patients with COPD From Exhaled Breath

Adriana Velásquez, Cristhian M Durán, Oscar Gualdron, Juan C Rodríguez, Leonardo Manjarres

University of Pamplona, Pamplona (N.S), Colombia
Km. 1 Vía Bucaramanga (El Buque)
{geniaz14, ingleomanca}@gmail.com, {cmduran, ju4n, oscar.gualdron} @unipamplona.edu.co, (057+7) 5685303

Abstract. To date, there is no effective tool analysis and detection of COPD syndrome, (Chronic Obstructive Pulmonary Disease) which is linked to smoking and, less frequently to toxic substances such as, the wood smoke or other particles produced by noxious gases. According to the World Health Organization (WHO) estimates of this disease show it affects more than 52 million people and kills more than 2.7 million human beings each year. In order to solve the problem, a low-cost Electronic Nose (EN) was developed at the University of Pamplona (N.S) Colombia, for this specific purpose and was applied to a sample group of patients with COPD as well as to others who were healthy. From the exhalation breath samples of these patients, the results were as expected; an appropriate classification of the patients with the disease, as well as from the healthy group was obtained.

Keywords: Electronic Nose, COPD, PCA, Pulmonary Disease, Gas sensors
PACS: 07.05.Hd, 07.05.Kf, 07.07.Df, 07.05.Rm

INTRODUCTION

Chronic Obstructive Pulmonary Disease (COPD) is one of the leading causes of morbidity and mortality in both industrialized and developing countries. Patients with COPD are usually who smokers or those who have stopped smoking and are over the age of 45, with a history of shortness of breath with physical exertion and a chronic dry cough indicating oppression in the lungs.

The current mechanism of spirometry, enables COPD to be diagnosed. The Spirometry measures the patient's lung capacity, but it is not specific; that is, it shows a variable that is measured and may be the same or very similar to that of a patient with asthma or pneumonia. Therefore, the diagnosis is in the hands of the medical professional in order to decide whether or not the patient is suffering from the disease in question or other diseases. Given these conditions, it is normal to find misdiagnosed cases with treatments that do not respond appropriately. Taking into account the aforementioned aspect and also the large number of population affected, it became necessary to develop specific equipment for the detection of this very specific disease. Also, it establishes a global standard for the characteristics and possible indications of the presence of this illness.

Nowadays, many investigations have been reported by various research institutions that list and focus on respiratory diseases using electronic noses, but very few have been specifically about COPD.

In this article we show the results obtained with the first prototype of the Electronic System non-invasive, multisensorial apparatus called "DEPOC" (Registration pending). The system was developed at the University of Pamplona for the classification of people with healthy lungs and those suffering from COPD. A sample group of patients was divided into categories. The "DEPOC" is capable of showing those with certain levels of the illness as well as those who do not have it. The two-dimensional PCA plot showed that Smell-sensory Attribute prints of healthy patients could be distinguished from those of subjects with COPD.

EXPERIMENTAL AND METHODS

Figure 1 illustrates the Electronic Nose that contains a hermetic chamber where the sample of exhaled breath passes through six commercial gas sensors (i.e. Figaro and FIS, detailed in Table 1). The mechanism is composed of a disposable nozzle fitted into the main chamber, and can be easily replaced.

CP1137, *Olfaction and Electronic Nose: Proceedings of the 13th International Symposium*, edited by M. Pardo and G. Sberveglieri

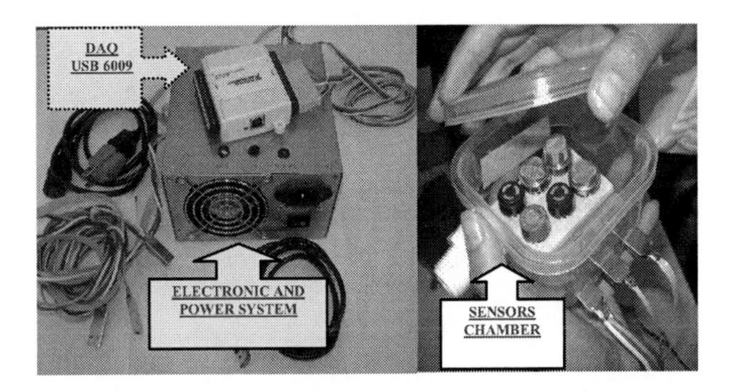

FIGURE 1. EN developed at the University of Pamplona.

TABLE 1: Gas Sensors.

Amount	Sensor	Application
1	SP-MWO 729	General Intention, Kitchen Control
1	TGS 826	Ammonia
1	SP-15A 921	Propane, butane
1	TGS 821	Hidrogen
1	SP-53 729	Hydrocarbons
1	TGS 822	Organic dissolvents agents

The patient must wait at least 20 minutes after having eaten any food or drink to take the test (See figure 2). Then, the system will indicate when to exhale. Once a short intake of breath has occurred, the exhalation should be long and complete (approximately 2 seconds). The exhaled breath is introduced to the chamber, and the information is acquired with the board "USB-DAQ-6009" from National Instruments.

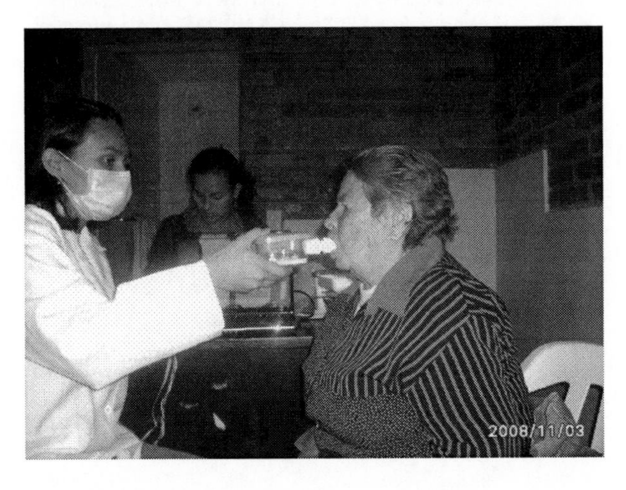

FIGURE 2. Photographs taken at the Macarena Geriatric Home.

After 3 minutes the system indicates that it has completed the data acquisition. This is repeated at least 5 times to get a good sample of the person's deep lung exhalation and to verify the repetitiveness of the action.

Once the samples are stored, they are processed through a simple graphic interface for the user. Samples from patients were acquired at different places in Colombian, such as; The Erasmus Meoz University Hospital (Cucuta, Northern Santander), Geriatric homes: The Macarena (Chía-Bogota), The Nazareth Home for Geriatrics (Zipaquirá-Cundinamarca) and San Francisco de Asis, another Geriatric home (Usaquén-Bogota).

TABLE 2: Healthy volunteers and patients with COPD (Erasmus Meoz University Hospital, Geriatric Homes: Narazeth, the Macarena and San Fransisco De Asis).

Healthy volunteers	Age	Samples	Patients with COPD	Age	Sam-ples
(a)	23	8	(A)	58	5
(b)	30	5	(B)	62	5
(c)	31	5	(C)	69	5
(d)	52	8	(D)	72	5
			(E)	74	5
			(F)	76	5
			(G)	77	5
			(H)	85	5
Total Number		26			40

All samples were carried out under the supervision of the doctors and nurses on duty. An approximate number of at least 5 exhalations were taken, per patient at an intervals of 4 minutes. Eight patients are mentioned in Table 2, who were previously diagnosed with COPD. For the comparison with healthy volunteers, 4 non-smokers were used as a pattern for the healthy sample. The patients were selected according to medical history from different institutions. For example, patient (G) from Table 2 has developed COPD, and patient (D) on the same table has been recently diagnosed with COPD and his health is relatively better.

RESULTS

Figure 3 shows the results when applying the PCA technique to the data set acquired, which clearly identifies the two categories of samples, Healthy (Right side) vs. COPD (left side).

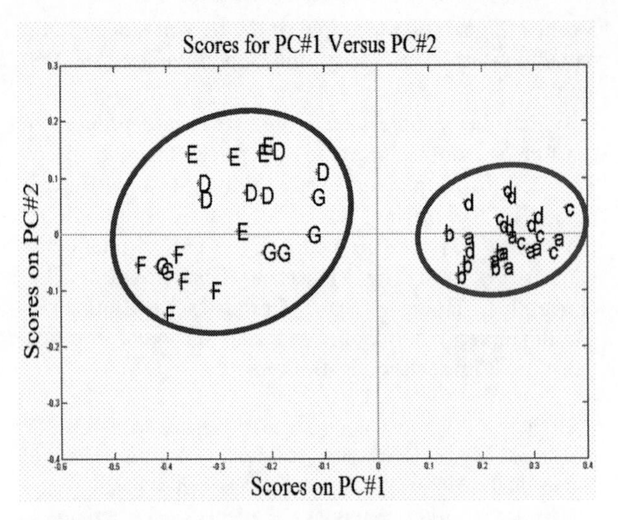

FIGURE 3. PCA plots a total variance of 83.68%, where the samples with COPD are enclosed in red circle.

Figure 4 illustrates how the system not only distinguishes between samples with COPD and healthy ones but also classifies the samples with the disease, into the varying stages of COPD according to medical tests (i.e. the patient's Medical History). Therefore, the EN is able to identify the level of development of the disease (as is shown in patient (G) with COPD).

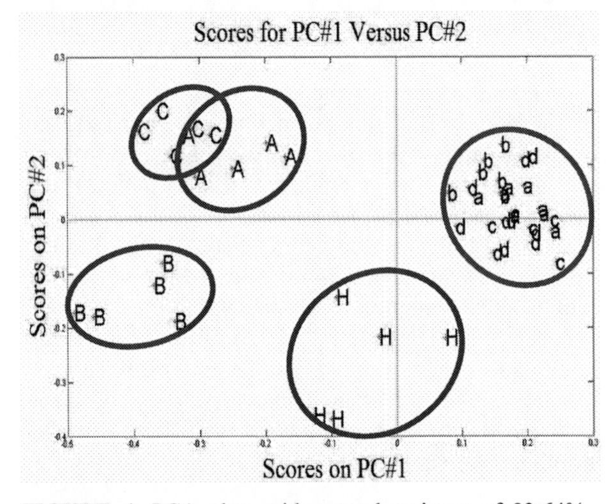

FIGURE 4. PCA plots with a total variance of 82.64%, where the samples with COPD are enclosed in red dots.

DISCUSSION

To identify and categorize the stages of COPD correctly, it is necessary to abide by a standard and currently there is no world standard. Therefore, this analysis was conducted based on the experience of the physicians of the region.

CONCLUSIONS

This contribution, presents a novel Electronic Nose to identify the samples of exhaled breath of a person with COPD. Also, this EN can classify the stages of the disease. Exhaled breath is the most important point of study for subsequent analysis in this broad field of development for investigative equipment in the detection of lung disease.

Although, the results have not yet validated with classic techniques such as GC-MS, we propose to carry out these test with future samples of patients.

ACKNOWLEDGMENTS

We would like to thank the Erasmo Meoz University Hospital and lung specialist, Dr. Fabio BERBESI. Also, the Geriatric homes: Macarena, Nazareth, San Fransisco of Asis, and their directors, Brother Carmelo FST, Dr. Maria Ismelda Moreno and Brother Fausto FST, respectively.

REFERENCES

1. Boschetto, Piera, "Chronic Obstructive Pulmonary Disease (COPD) and occupational exposures", *Journal of Occupational Medicine and Toxicology*, 2006, pp.1-6.
2. Dragonieri, S, *et al*, "An electronic nose in the discrimination of patients with asthma and controls," *Journal of Allergy and Clinical Immunology,* Vol.120, No. 4, (2007), pp.856-862.
3. Blatt, R, *et al*, "Lung Cancer Identification by an Electronic Nose based on an Array of MOS Sensors", *Neural Networks, 2007. IJCNN 2007. International Joint Conference on,* Orlando, FL, Aug 12-17, 2007, pp. 1423-1428.
4. Chen, X, *et al*, "A study of an electronic nose for detection of lung cancer based on surface acoustic wave sensors and image recognition method". *Measurement Science Technology*, Vol: 16, (2007), pp.1535–1546.
5. Natale D.C, *et al,* "Lung cancer identification by the analysis of breath by means of an array of non-selective gas sensors". *Biosens Bioelectron,* Vol: 18, (2003), pp. 1209–1218.

Electronic Nose for Quality Control of Colombian Coffee Through the Detection of Defects in "Cup Tests"

Juan C. Rodríguez, Cristhian M. Duran, Adriana X Reyes

University of Pamplona, Pamplona (N.S), Colombia
Km. 1 Vía Bucaramanga (El Buque)
{ju4n, cmduran, areyes}@unipamplona.edu.co, (057+7) 5685303

Abstract. This article presents a preliminary study on the analysis of samples of Colombian coffee for the detection and classification of defects (i.e. using "Cup Tests"), which was conducted in Almacafé quality control laboratory Almacafé in Cúcuta, Colombia). The results obtained show the application of an Electronic Nose (EN), called "A-NOSE", used in the coffee sector for the cupping tests. The results show that e-nose technology can be a useful tool as quality control of coffee grain.

Keywords: Colombian Coffe, Sensors chemical, Electronic Nose, Data acquisition, Data processing
PACS: 07.05.Hd, 07.05.Kf, 07.07.Df, 07.05.Rm

INTRODUCTION

Colombia is one of the largest producers of coffee in the world. It is the major producer of the Arabic variety which is considered the highest-quality grain. Colombian Coffee has had a prominent position in history for many reasons; it has earned the label of one of the best in the world. Its quality is reflected in many aspects, and is characterized by amazing smoothness, aroma and flavor.

There is an important literature on the use of Electronic Noses to discriminate different samples of coffee, but none has reported a study on the quality control of Colombian Coffee. In this paper, on the contrary, we will give a description of an EN applied to detect defects in Cup Test of Colombian coffee.

EXPERIMENTAL AND METHODS

In this experiment we used the equipment "A-NOSE", a Electronic Nose developed at "University of Pamplona (Colombia)". This equipment has a matrix of 8 semiconducting metal-oxide gas sensors of Figaro and FIS type. Table 1 shows the gas sensors used.

Initially some measurements were performed with green coffee beans, but the sensors did not react in this case. Then coffee roast and ground were tested and hot water was added, which provided a good response from the sensors. The first test measurements were used to determine the conditions for the measurement process.

TABLE 1: Sensor array description

Amount	Sensor	Application
1	SP-12A	Flammable Gases
1	SP-31	Organic Solvents
2	TGS-813	Combustible gas
1	TGS-842	Methane, natural gas
1	SP-AQ3	Air Quality Control
1	TGS-813	Combustible Gases
1	ST-31	Organic Solvents
1	TGS-800	Air Quality, Smoke, Benzene

Later a set of measurements were carried out with some types of coffee: "Excelso UGQ2, "Excelso Europa", "Pasillas", and coffee with defects (black, white, vinegar). Once the measures were done, the technique of Principal Component Analysis (PCA) was applied and a feed-forward backpropagation neural network, as a data processing technique.

Parameters Used In the Measurement Processes

The coffee beans analyzed, were roasted and ground, where 2 gr of ground coffee were used for each measure, 8 gr of water were added to facilitate

CP1137, *Olfaction and Electronic Nose: Proceedings of the 13th International Symposium,* edited by M. Pardo and G. Sberveglieri

the extraction of volatiles with a total weight of 10 gr for the ground coffee and water.

The water temperture was 60 °C and room ambient temperature of 24 °C.

Times Used In the Measurements

Concentration: 5 minutes. Volatile collecting in the concentration chamber.

Measuring: 5 minutes. The volatiles are dragged from the concentration chamber until the sensor chamber acquires the smell-print olfactory.

Repose Time: 5 minutes. Recovery the gas sensors. The time to each measure with coffee samples was of 15 min.

RESULTS

The first analysis was made with 2 types of good coffee, which were "Excelso UGQ" that complies with standard 1260; it's an export coffee where sample up to 72 defective grains can be foud in a sample 500 grs. The other kind of good coffee was manually selected to provide "Healthy Coffee" beans of Pasillas.

These two types of coffee were compared with 2 samples of pasillas, which were tested and detected in cup test. According to the taster's opinion, one of these coffee samples had "Restful Coffee" and the other "Fermented Coffee". Of these two defects the most common is restful coffee, caused by prolonged storage or poor storage conditions. The fermentation is presented by some of the following factors: Lack of water during the development of the plants, delays between the collection, processed and others.

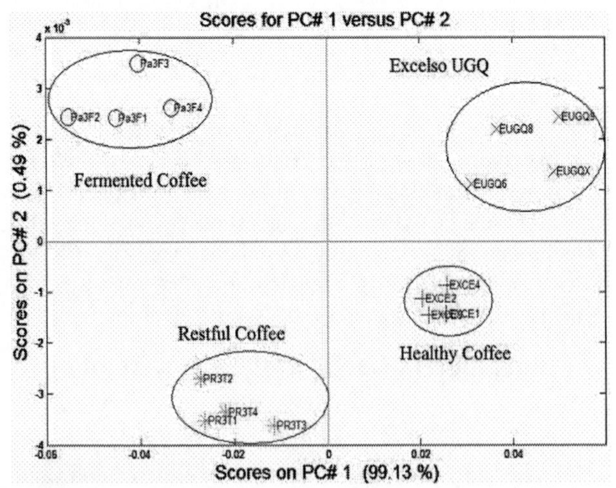

FIGURE 1. PCA graph with different types of cluster (coffee with defects and without defects)

Figure 1 illustrates the results of PCA which were applied to the measurements mentioned above. The samples of good coffee, which were "healthy coffee" and "Excelso UGQ", are located on the right side of the graph. The coffee samples with defects in the cup, such as restful coffee and fermented coffee, are located on the left side. Each group is composed of 4 samples, for a total of 16 measures (for not being harvest season, there was difficulty in obtaining a larger set of samples).

Additional to the PCA a feed-forward backpropagation neural network was applied and the validation was obtained with the "leave-one-out" method, where a classification rate of 100 % was achieved.

CONCLUSIONS

In the preliminary study carried out in the coffee sector, in Almacafe, Cúcuta-Colombia, it was found that an Electronic Nose can be used for quality control in the coffee sector, since it can classify different samples of Excelso coffee, coffee with defects in the cup and others.

ACKNOWLEDGMENTS

In this article we would like to give our most sincere thanks to Dr. Fabio Botero, Director of Almacafe (Cúcuta-Colombia), for his kindness and colaboration in carrying out these tests.

REFERENCES

1. Gardner, J.W, "Application of the electronic nose to the discrimination of Coffee," *Sensors and Actuators B*, Vol. 6, No. 1-3, (1992), pp.71-75.
2. Pardo, M, "Electronic Nose for Coffee Quality Control," *IEEE Instrumentation and Measurement, Technology Conference,* Budapest, Hungary, May 21-23, 2001, pp. 123-127.
3. Falasconia, M, "The novel EOS[835] electronic nose and data analysis for evaluating coffee ripening," *Sensors and Actuators B*, Vol. 110, No. 1 (2005), pp. 73–80.
4. Pardo, M, "Coffee Analysis with an Electronic Nose," *IEEE transactions on instrumentation and measurement,* vol. 51, No. 6, December 2002, pp. 1334-1339.
5. Tan, T, T, "Quality Control of Coffee Using the FOX4000 Electronic Nose," *ISIE'97*, Guimarges, Portugal, 1997, pp.140-145.
6. Web of Almacafe (Colombia). http://www.cafedecolombia.com/quienessomos/almacafe/almacafe.html.

Potential application of the electronic nose for shelf-life determination of raw milk and red meat

Aziz Amari [1], Noureddine El Barbri [1], Nezha El Bari [3], Eduard Llobet [2], Xavier Correig [2] and Benachir Bouchikhi [1]

[1] Sensor Electronic & Instrumentation Group, Faculty of Sciences, Physics Department, Moulay Ismaïl University, B.P. 11201, Zitoune, Meknes, Morocco
[2] MINOS, Universitat Rovira i Virgili, Avda. Països Catalans, 26, 43007 Tarragona, Spain
[3] *Biotechnology Agroalimentary and Biomedical Analysis Group, Faculty of Sciences, Biology Department, Moulay Ismaïl University, B.P. 11201, Zitoune, Meknes, Morocco*

Abstract. The present study describes the performance of an electronic nose in food odor analysis. This methodology was successfully applied to odor characterization of milk stored at 4 °C during 4 days and of beef and sheep meat stored at 4 °C for up to 15 days. The electronic nose sensor system coupled to PCA as a pattern recognition technique, is able to reveal characteristic changes in raw milk and red meat quality related to storage time. Additionally, a bacteriological method was selected as the reference method to consistently train the electronic nose system for both beef and sheep meat analysis.

Keywords: electronic nose, bacteriological analysis, shelf life estimation, milk, read meat
PACS: 07.05.Hd; 07.05.Kf; 07.05.Mh

INTRODUCTION

The control of milk quality and freshness is of increasing interest for both the consumer and the agro-food industries. Classical analytical techniques such as gas chromatography-spectrometry (GC-MS) and gas chromatography-olfactometry are however time-consuming, expensive, and laborious which can hardly be done on-site or on-line [1].

On the other hand, the spoilage of meat is a sensorial quality and may consist in the occurrence of off-odors and off-flavors or discoloration [2]. In red meats, the off-odors resulting from bacterial activity determine their shelf-life witch may be defined as the time elapsing between production and spoilage. It is known that a post-mortem endogenous enzymatic activity within muscle tissue can contribute to significant changes during meat storage [3]. This is why many efforts are underway or have been conducted to develop different types of sensors for meat quality or safety applications [4].

It has been shown that an electronic nose system can be used as a powerful method to control the quality of foods in some specific applications [1, 5].

The purpose of this paper is to evaluate the electronic nose performance as an instrument for the quality/ safety control of raw milk and red meat stored at 4 °C.

EXPERIMENTAL AND METHODS

Sample preparation and sampling

The raw milks used in the first experiment were obtained from a dairy exploitation site in Meknees city (Morocco). Fresh milk samples, immediately after reception, were placed in plastic bottles (100 mL/bottle) and introduced in a refrigerator kept at a constant temperature of 4 °C to be analyzed at 1, 2, 3, and 4 days of storage. The sampling system consisted of a dynamic headspace sampling. For the measurement procedure, a volume of 100 mL of milk in a 200 mL glass vial kept at the temperature of 35 °C, was stripped by means of the 1000 sccm nitrogen flow for 10 min. In this way the volatile compounds were directly transferred by the carrier gas into the sensor chamber. The vials had two small holes in their cover, to allow the headspace to be analyzed with the electronic nose equipment. Each time that a new set of milk was analyzed, new vials were employed.

CP1137, *Olfaction and Electronic Nose: Proceedings of the 13th International Symposium,* edited by M. Pardo and G. Sberveglieri

In the second experiment, two different types of meat species representative of Moroccan production and purchased from a local market were analyzed. The samples from different animal species (beef and sheep) were cut into pieces of the same weight (10 g ± 1 g) immediately after reception, placed in plastic bags (bags for freezing food) and introduced in a refrigerator kept at a constant temperature of 4° C ± 1° C. For each measurement, a meat sample was taken from the refrigerator and put inside a 500 ml glass bottle. The sampling bottles were sealed with septum and held at room temperature (22 °C ± 2 °C) for 50 min in order to reach a stable composition of the headspace. Measurements were performed each day for up to 2 weeks. Every day two replicate samples were withdrawn from the refrigerator to undergo microbiological analysis and six replicate samples were employed for electronic nose analysis.

Red meat microbiological population enumeration

A 25 g sample of each product of meat was taken aseptically and placed in a sterile stomacher bag containing 225 mL of 0.1 % (wt/vol) peptone water (PW, Oxoid Ltd., Hampshire. England). The sample and the PW were stomached for 2 min. Decimal dilutions were prepared using the same diluent. These dilutions were subsequently plated on the surface of a Plate Count Agar (PCA, Oxoid Ltd.). The plates were incubated at 30 °C for 2 days. The total viable counts (TVC) were obtained by enumerating the colonies present, and calculated as log10 colony forming units (cfu)/g of the sample.

Electronic nose system

An electronic nose system was employed to obtain the smell patterns from the headspace of meat samples. This electronic nose system contains an array of six tin oxide based Taguchi gas sensors obtained from Figaro Engineering, a temperature sensor (National Semiconductors LM35DZ), and humidity sensor (PHILIPS H1). The sensor array used comprised six Tagushi gas sensors [1,6]. The sensor array was mounted in a flow type gas chamber equipped with two ports for the gases (inlet and outlet) and heater wires. The electrical conductance of the sensors varied in the presence of reducing/oxidizing gases.

The variation of the sensors' conductivity is acquired and then digitized using a data acquisition board (PCL 812PG, Advantech). A program in LabVIEW was developed to control the data acquisition process. Data processing and pattern recognition are decisive factors in order to obtain a versatile instrument able to reliably recognize a wide variety of odors. MATLAB 6.5 software is used for pre-processing and data analysis.

Feature extraction and pre-processing

The features used for data analysis are extracted from the temporal responses of the sensor array (i.e. from the temporal evolution of G). For every sensor within the array and measurement performed, four representative features from the response signal are extracted [7].

- G_0 : the initial conductance.
- G_s : the steady-state conductance.
- dG/dt : the dynamic slope of the conductance corresponding to 30 % and 70 % of steady-state conductance. This corresponds to a phase where a fast increase of sensor conductance is observed.
- A : the area below the conductance curve in a time interval defined between 30 % and 80 % of steady-state conductance. This area is estimated by the trapeze method.

Data analysis

The Principal Components analysis (PCA) method is a powerful linear unsupervised pattern recognition method, which is usually successfully used in gas sensor applications [8]. Generally, the PCA method reduces the dimensionality of a multivariate problem. In fact, this method consists in extracting features by projecting the high-dimensional data set in a dimensionally reduced space constituted by the uncorrelated and orthogonal eigenvectors of the covariance matrix computed from the sensor responses, called principal components. The magnitude of the single eigenvector or percentage of ''information'' is expressed by its eigenvalue, which gives a measure of the variance related to that principal component.

RESULTS AND DISCUSSION

Milk Analysis

Fig. 1 shows the evolution of the steady-state conductance as characteristic parameter. We note that the 822 and 825 TGS sensors showed the highest responses of the sensors of all the samples. However, the response of the sensors changes with storage days. It is shown that this characteristic parameter increase with storage days.

This experiment was devoted to study a dynamical evolution of the clusters related to different degree of rancidity during a fixed time. In order to see whether the portable electronic nose was able to distinguish between different spoilage states, a PCA analysis was applied to the database. The analysis was done using all the responses that were obtained in the different measurements for the raw milk. With PCA, the three first principal components allow us to well represent 75.23 % of the information in the database. Fig. 2 reports in a three-dimensional plot (PC1-PC2-PC3), the data resulting from measurements collected during a period of 4 days for raw milk. From the picture it is clearly evident as the 3D PCA plot allows distinguishing the first, the third and the fourth storage days in which the milk has different rancidity values. It is remarkable how PCA analysis allows us to identify a preferential direction according to differences in aging, giving an indication on the dynamic evolution of the system.

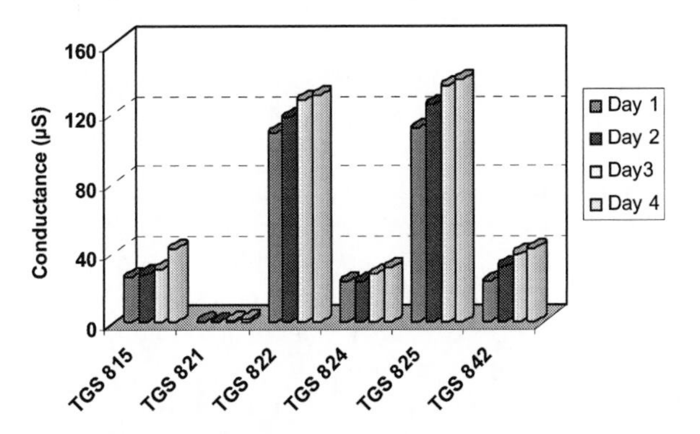

FIGURE 1. Time variation of the steady-state conductance.

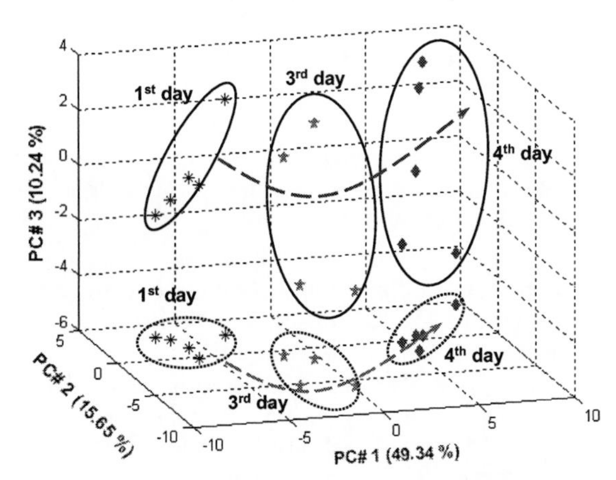

FIGURE 2. PCA plot of raw milk in different days.

The fact that samples appear ordered along the first principal component according to the number of storage days is a good result. The first principal component explains the main variance in the data (i.e., in the response of the sensors), and Fig. 3 shows that this variance is well correlated with the number of storage days. In other words, the sensor array employed seems appropriate to envisage an application where the main goal would be to predict the number of days of cold storage undergone by raw milk. These scores show a monotonic increase during the period of storage. A polynomial fit of the scores, whose equation is shown in fig. 4, suggests that the value of the scores on the first PC could be used to predict the period of storage and, therefore, the decay of raw milk.

Red Meat Analysis

The results of the bacterial analysis performed on the red meats, which were used in the development of the classification models with the electronic nose, are presented in Fig. 3. This figure shows the evolution of log10 cfu/g developed in the two types of meat as a function of the days of cold storage. It was decided that a quality criterion corresponding to 6 log 10 cfu/g for TVC should be applied to discriminate between unspoiled and spoiled samples in this study, because this is the general microbiological safety guideline applied for food quality. A similar behavior was observed for the two types of analyzed meat. The TVC shows a slight variation in the first five days followed by a very fast increase between days five and ten, and finally in the last days of conservation the curves show a tendency to stabilize.

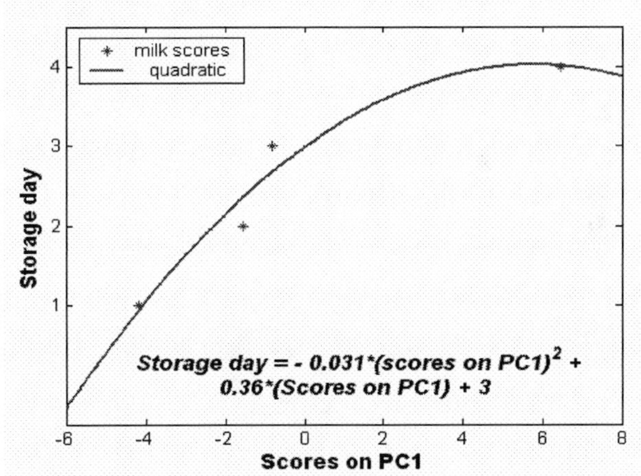

FIGURE 3. Scores evolution on PC1 with the period of storage and polynomial fitting.

The deterioration of meat quality starts from day four or five, when meat is stored at 4 °C. The meat, which contains bacteria at a level of less than 5 log10 cfu/g, is said to be acceptable for consumption. However, the meat is spoiled and not suitable for consumption when viable counts exceed 6 log10 cfu/g. At this stage, a putrid smell was released from the meat. It was found that the TVC increased at a slightly faster pace in sheep than in beef meat. This can be explained by the fact that sheep meat contains more proteins and lipids than beef meat. It was found also that the threshold of consumption acceptability (i.e., microbial counts < 6 log10 cfu/g) is reached at day seven for beef meat and at day five for sheep meat.

In parallel to the bacteriological analysis, an analysis employing the electronic nose system was made. In the first step a PCA was used as an unsupervised classification method to visualize the resemblance and the difference among the different measurements in the datasets. In the second step, the method was employed to compare the deterioration speed between beef and sheep meats.

Fig. 4 shows the score plot of the data in the PC1-PC2 plane for beef and sheep meats. For beef meat (Fig. 4-a) it can be noticed that the measurements cluster together in three different groups. However, the bacteriological analysis shows that beef meat stored 7 and 8 days should be considered as unspoiled and spoiled, respectively. So, it is difficult to reach a correct determination of shelf life for beef meat by using a PCA method. For sheep meat (Fig. 4-b) it can be noticed that the dataset measurements are grouped in two groups. In this case the PCA method shows a good separation between unspoiled and spoiled sheep meats. So, the method could be used for the rapid shelf-life determination of sheep meat.

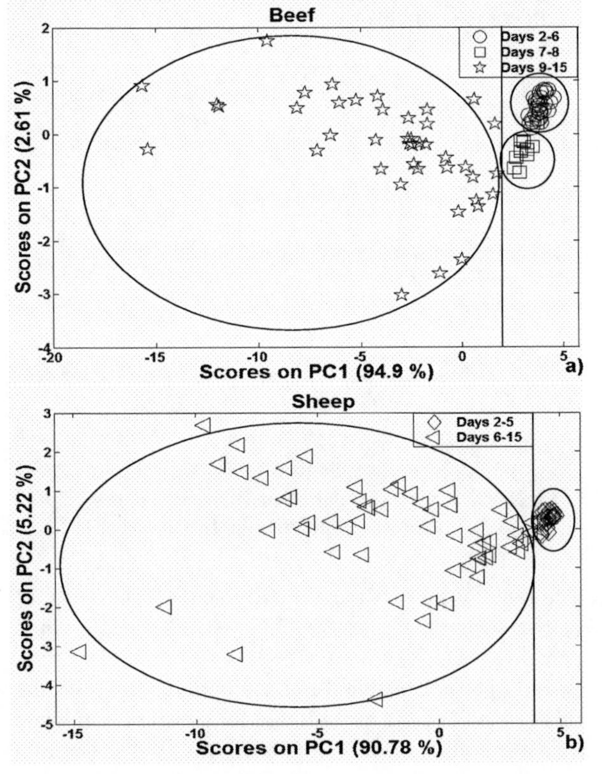

FIGURE 4. Scores plot of a PCA on beef meat (a) and sheep meat (b).

CONCLUSIONS

We have shown the feasibility of our prototype electronic nose based on an array of semiconductor gas sensors to detect off-odors from both raw milk and red meat. The sensor array coupled to appropriate feature extraction and pattern recognition methods was found applicable to obtain selective discrimination of our samples of different aging levels.

The proposed approach is not limited to the examples given but could be used to determine the shelf life of other agro-food products.

ACKNOWLEDGMENTS

Part of this work has been funded by the AECID under project no. A/9167/07.

REFERENCES

1. Brudzewski, K. *et al*, "Classification of milk by means of an electronic nose and SVM neural network," *Sensors and Actuators B*, Vol. 98, 2004,
2. Borch, E. *et al*, "Bacterial spoilage of meat and cured meat products," *Int. J. Food of Microbiol.*, Vol. *33*, No. 1 (1996), pp. 103–120.
3. Schreurs, F. J. G. "Post-mortem changes in chicken muscle," *Worlds Poultry Science Journal*, Vol. *56*, No. 4 (2000), pp. 319–346.
4. Yano, Y. *et al*, "Application of a microbial sensor to the quality control of meat freshness," *Talanta*, Vol. 54, No. 2 (2001), pp. 255–262.
5. Gardner, J. W. and Bartlett, P. N. "A Brief History of Electronic Noses," *Sensors Actuators B*, Vol. 18, (1994), pp. 211–220.
6. O'Connell, M. *et al*, "A practical approach for fish freshness determinations using a portable electronic nose," *Sensors Actuators B*, Vol. 80, No. 2 (2001), pp. 149–154.
7. Amari, A. *et al*, "Monitoring the Freshness of Moroccan Sardines with a Neural-Network Based Electronic Nose," *Sensors*, Vol. 6, No. 10 (2006), pp. 1209-1223.
8. Gardner, J. W. "Detection of vapours and odours from a multisensor array using pattern recognition Part 1. Principal component and cluster analysis," *Sensors Actuators B*, Vol. 4, (1991), pp. 109-115.

Comparison Between Sensing Systems for Ammonium Detection And Measurement In Soil

F. De Cesare [a*], E. Zampetti[b], S. Pantalei[b], A. Macagnano[b]

[a] *Dept. of Agrobiology and Agrochemistry, University of Tuscia, Via S. Camillo De Lellis - 01100 Viterbo, Italy*
[b] *Institute of Microelectronics and Microsystems, CNR, Via Fosso del Cavaliere, 100 - 00133 Rome, Italy*
**Email: decesare@unitus.it; phone: +39-0761357338; fax +39-0761357242*

Abstract. Usually, ammonium in soil is carried out through steam distillation of ammonia obtained after alkalinization of soil extracts and further back titration of the collected solutions. Alternatively, ion selective electrodes (ISE) specific for ammonium ions can be used, in order to measure their concentration in aqueous soil extracts. The aim of this study is to assess the possibility to use, alternatively to the previous techniques, two kinds of chemical sensors able to measure NH_3, such as an interdigital microelectrode (IDE) coated of conductive polymer and a sensors array, usually named electronic nose (EN), based on quartz crystal microbalances (QCMs) covered with functionalized polymers. These sensors were chosen on the base of their ability to detect NH_3 in sample headspace (specifically or aspecifically, respectively). Therefore, NH_4^+ in solution was converted to NH_3 by alkalinizing soil extracts. Sensors were calibrated at first against known concentrations of NH_4^+. Results were compared with those obtained with an ISE for NH_4^+.

Keywords: Electronic nose; interdigital sensors; ion selective electrodes; soil; ammonium; ammonia.
PACS: 82.47.Rs; 07.07.Df; 07.88._y; *91.62.La.

INTRODUCTION

Nitrogen (N) is an essential nutrient for all organisms. In soil, N can be found in several forms, both organic and inorganic. The main inorganic forms of N in soil are NH_4^+, NO_3^- and NO_2^-. The importance of these forms, specifically NH_4^+ and NO_3^-, lays in their being the main and more direct forms absorbed by organisms. Nitrate and nitrite are highly soluble in soil and then more bioavailable but also more leachable, since as anion they are repulsed from negatively charged soil colloid surface. On the contrary, NH_4^+ can be highly adsorbed onto soil particles, owing to its positive charge, and it can also be entrapped into soil minerals. To measure the bioavailable NH_4^+ in soil, it must be previously desorbed from soil particles, by extraction with neutral salts solutions (KCl or K_2SO_4) [1], able to exchange the adsorbed NH_4^+ with K^+. Extracts can then be measured in their NH_4^+ content using different procedures. These comprise colorimetric methods, titration methods and the use of ISEs [1]. Currently, colorimetric methods are used after NH_4^+ extraction and its convertion to NH_3 in strong alkaline solutions (NaOH) and reaction with different reagents to form a chromophore (indopheol blue method [1], Nessler method [1], salicylate method [2]). Differently, NH_4^+ can be converted to NH_3 in strong or mild alkaline solutions (NaOH or K_2CO_3, respetively), which after diffusion into the atmosphere is collected in a boric acid solution and then titrated with H_2SO_4 or determined colorimetrically (steam distillation method [1, 2] and microdiffusion method [1], respectively). Alternatively to these methods, ISEs can be used to measure NH_4^+ in soil extracts. Relative to the previous methods, the use of ISEs is easier to use and not time consuming. Two kinds of ISEs are commercially available, which can measures either NH_4^+ or NH_3. In this latter case, a previous convertion of NH_4^+ to NH_3 by alkali addition is required. However, in all the methods where strong alkali (e.g. NaOH) is added directly to soil extracts, it is probable that alkali-labile organic N compounds are hydrolysed and that they release additional NH_4^+, then overestimating the measures [1]. For this reason an ISE for NH_4^+ was preferred as reference method.

Interdigital sensors are well suited to quantitative and qualitative analyses of gases and vapours. The surfaces of the sensors are usually coated with substances commonly adopted in gas chromatography. The signals for gas analysis are the changes in electrical conductance and capacity. IDEs are capable, in fact, to measure both the dielectric properties (capacitive transducers, IDCs) and the conductance (resistive transducers, IDEs) of the material placed both among their fingers and above. IDEsare used in telecommunications, biotechnology, chemical sensing, dielectric imaging, acoustic sensors, and microelectromechanical systems (MEMS) applications [3]. In the present study an IDE transducer coated with a polymer selective for ammonia was used.

CP1137, *Olfaction and Electronic Nose: Proceedings of the 13th International Symposium*, edited by M. Pardo and G. Sberveglieri

Electronic noses (ENs) consist of a sensors array able to perceive volatile (VOCs) or semivolatile organic compounds (SVOCs) present in sample headspace, i.e. the "odour", and to transduce the consequent chemical or physical changes of the sensing films into electric signals. ENs have been involved in several applications: biomedicine [4], agroindustry and food quality [5], environmental and industrial analyses [6, 7]. Sensors in the arrays may be specific for single analytes, or for a chemical class of compounds or completely aspecific. The EN purposely set up in this study comprised both specific and aspecific polymers for NH_4^+ detection.

The aim of this study was to compare the capacity of these three sensors, based on different principles potentiometry, chemoresistivity and gravimetry, to perceive and measure NH_4^+ in soil.

EXPERIMENTAL AND METHODS

Samples Preparation

Soil used for detection of ammonium was a clay loam forest soil. Before extraction soil was treated with NH_4Cl at increasing concentrations (250 mM, 500 mM and 2 M) for 1 h under shaking, in order to saturate with NH_4^+ all soil sites available for cation binding, and then to simulate soil with different NH_4^+ concentrations. Then, soil was filtered on paper discs deprived of NH_4^+; the recovered soil was then extracted with 2 M KCl (1:10 w/v) under shaking for 1 h, then filtered, and the solutions were tested for NH_4^+. Controls without previous NH_4^+ saturation (native soil) were extracted with 2 M KCl only. All the measurements were carried out in a thermostated incubator (Stewart S160D) set at 25±0.5°C.

Ion Selective Electrode

The measurements of NH_4^+ concentrations in both standards and soil extracts were carried out by an ISE for NH_4^+ (Crison) in beakers under stirring, and after adjusting ionic strength with 1 M MgSO4 (1:10 v/v) to uniform ionic strength in all samples. The potentials detected by ISE after contact with NH_4^+ in soil extracts were referred to a reference curve, where NH_4^+ concentrations (from 1 µM to 1 M) in ultrapure water and their relative potentials were plotted in a logarithmic scale, in order to calculate the NH_4^+ concentrations in the extracts. The measurements of NH_4^+ were performed directly on soil extracts without any further treatment, if not specified. When the potential of soil extracts exceeded the maximal value in the calibration curve, soil extracts were diluted prior to the adjustment of the ionic strength in order to have potentials within the range of the reference curve.

Interdigital Microelectrodes

The IDE transducer had interdigital chromium electrode designed on oxidized silicon wafer (5640 µm fingers length, 150 µm fingers width and 150 µm fingers spacing between them, respectively). IDE was coated of a nanostructured fibrous layer of doped polyaniline (polyaniline emeraldine base - PANi-EB), 10 kD molecular weight (Mw) and polyethylene oxide (PEO) 200 kD Mw, deposited by electrospun technique. After alkalinization of soil extracts with MgO for 10 min under stirring of the various solutions placed in sealed flasks, dynamic measurements were carried out using a 4 channels MKS 247 mass flow controller, where N_2 (carrier) was mixed with samples headspaces (5% v/v) prior to measurements. The concentrations of NH_4^+ in the solutions were calculated by the reference of the electric resistance variations (Ohm), measured by IDE exposed to the atmosphere (NH_3) of the same solutions, to a calibration curve obtained by plotting NH_4^+ concentrations in ultrapure water (ranging from 50 nM to 5 µM) versus the electric resistance changes induced in IDE by exposition to their relative headspaces. Dilutions were carried out in samples exceeding the reference curve.

Electronic Nose

The EN used in the present study was comprised of an array of 7 sensors consisting of QCMs coated with diverse organic films. Specifically, 5 polymers (polymethylphenylsiloxane, polyethylenimine, polyacrilonitrile, 6-polyamide, PANi-EB) with different physical and chemical features (polarity, basicity, hydrophobicity) and 2 different metallo-porphyrins (Zn, Mn), deposited as thin films onto QCMs oscillating at 20 MHz, located in a 10 ml measuring chamber. The NH_4^+ present in both soil extracts and standards placed into sealed flasks was measured after its conversion to NH_3 similarly to what described for IDE sensors. EN measurements were carried out by diverting a 200 sccm N_2 flux into the flasks through a 4 channels Mass Flow Controller MKS 247. Frequency variations of QCMs measured at equilibrium between gases or volatile analytes adsorbed onto coating films and those present in the chamber's atmosphere were processed through Principal Component Analysis (PCA). At first, a calibration curve was obtained by processing through PCA the variations in frequency (kHz) measured by EN when exposed to the headspaces of standard solutions in ultrapure water containing known concentration of NH_4^+ (ranging from 100 μM to 5 mM) in order to obtain PC1 values, which were then plotted against NH_4^+ concentration. Then, the variations in frequency (kHz) measured by EN exposed to headspaces of extracts with unknown NH_4^+ contents were processed by using the same algorithm used to carry out the PCA of measures relative to standard solutions. Then, the reckoned PC1 values relative to soil extracts were used in the equation of the calibration curve to determine NH_4^+ content of the extracts. Soil extracts were suitably diluted when relative PC1 values exceeded the reference curve.

RESULTS

The reference curve of NH_4^+ concentrations vs. their relative potential measurements obtained by ISE showed a linear response on logarithmic scale (Fig.1).

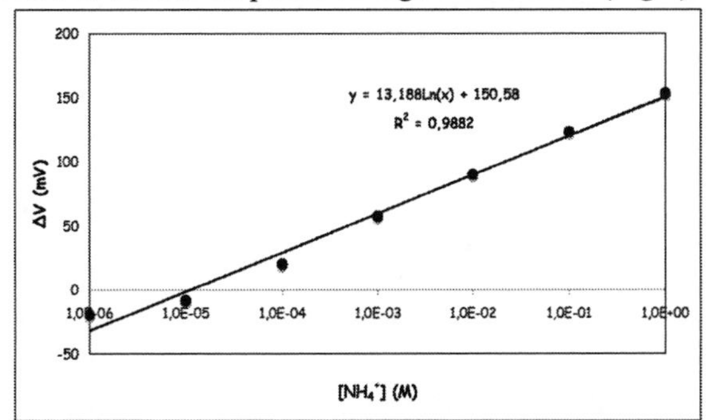

FIGURE 1. Calibration curve obtained by plotting different concentrations of NH_4^+ against their relative potential differences measured by ISE

In the reference curve obtained by IDE a progressive linear increase in electric resistance of IDE was observed at increasing NH_4^+ concentrations on a logarithmic scale (Figure 2). In measurements performed with EN, the relationship between NH_4^+ concentrations and their relative frequency shifts was

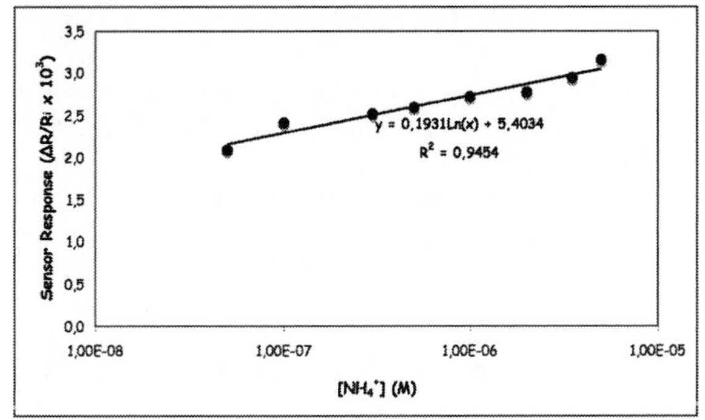

FIGURE 2. Calibration curve obtained by plotting different concentrations of NH_4^+ against their relative potential differences measured by IDE

obtained by processing at first the frequency shifts by PCA (Figure 3) and then fitting PC1 values (carrying 90.2% of the data variance) vs. increasing concentrations of NH_4^+ standard solutions alkalinized with MgO. Since PC1 is the axis of PCA plots containing the greatest variance and it also separates various concentrations of NH_4^+ in different clusters, its values were used to obtain the reference curve.

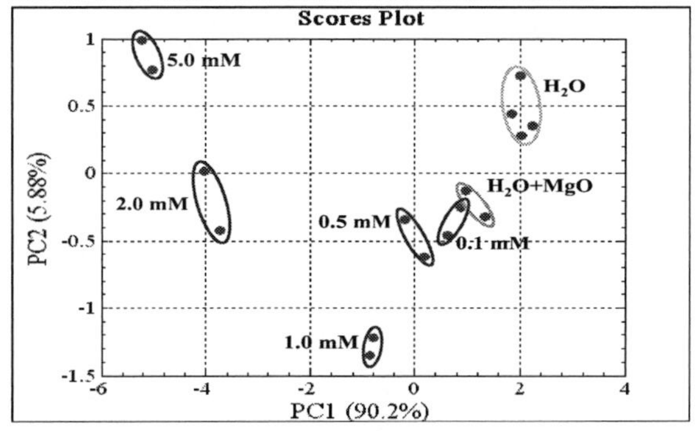

FIGURE 3. Scores Plot of the first two principal components as resulting from the calibration data set

The resulting relationship had a polynomial negative double exponential trend (Figure 4):

$$y = A_1 e^{-\frac{x}{C_1}} + A_2 e^{-\frac{x}{C_2}} + y_0 ;\ r^2 = 0.99032 \quad (1)$$

In the reference curves obtained by IDE and EN, lower NH_4^+ concentrations ranges were used relative to those used in the calibration by ISE. This fact was due to the different (higher) sensitivity of the former sensors relative to the latter one.

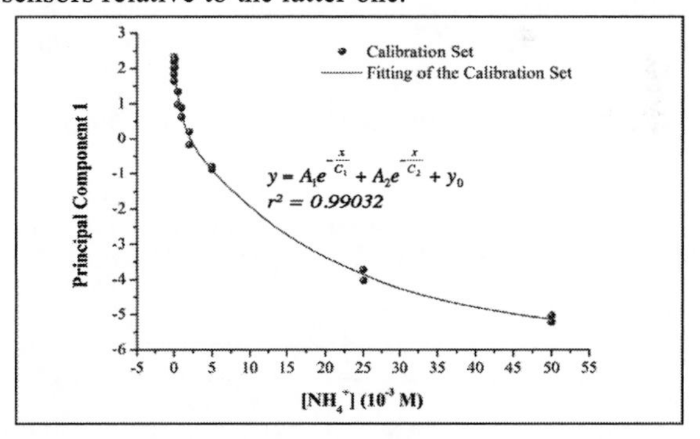

FIGURE 4. Plot of the PC1 versus the ammonium concentration, as resulting from the PCA analysis (•). Data have been fitted to build a calibration curve (grey line).

Ammonium content measured by ISE in extracts from variously treated soils exceeded the calibration curve. Then extracts were diluited to include their NH_4^+ contents in the reference curve. Since both IDE and EN sensors showed a higher sensitivity to NH_3 derived from the conversion of NH_4^+ in alkaline solution than that showed by ISE versus NH_4^+, soil extracts were diluted much greatly than in ISE measurements.

Therefore, although all sensing systems tested in this study responded with increasing electric signals to increasing concentrations of NH_4^+, the range of NH_4^+ concentrations practicable with different sensors was 1.5 order of magnitude lower with EN and even 4.5-5 orders lower with IDE than with ISE. This means that different sensors can be used for different purposes or in different experimental or environmental conditions. Therefore, in the presence of high NH_4^+ concentrations, sensors with lower sensitivity (ISE) are recommended, while in opposite conditions a higher sensitivity is required (EN < IDE). Neverthless, if diluiting is a straightforward practice to obtain samples which can be measured also with high sensitive sensors, concentrating the samples, to achieve values of the measured parameter within the reference curve, is a procedure which can have some controindications such as analytes precipitation. Therefore, highly sensitive sensors are recommended; in this case, EN and IDE are preferable to ISE.

Table 1 reports NH_4^+ content in soil extracts measured with different sensors. Ammonium concentration measured by ISE in soil extracts, previously alkalinized and measured with EN, showed that some NH_4^+ was still present in solution after alkalinization, as it was presumed by the mild alkaline pH (9.5) obtained after MgO addition. Then the conversion of NH_4^+ to NH_3 was not exhaustive. The addition instead of NaOH would result in about pH 12 solution, which would induce 10^6 times higher NH_3 concentration, i.e. almost all NH_4^+ would be converted to NH_3, that is a desirable condition. Nevertheless, the use of NaOH, instead of MgO, is not recommended in soil extracts for its effect on hydrolysis of N-containing organic compounds for soil extracts.

TABLE 1. Ammonium concentrations (µg g^{-1} of dry soil) in the various soil samples measured with different sensors and diverse methods. Data are reported ± their relative coefficient of variation.

Samples	ISE		IDE	EN
	/	MgO_{EN}	MgO	MgO
Soil	268±3	9.6±0.5	102±13	142±3
Soil +250 mM NH_4^+	622±1	59±11	135±10	1026±11
Soil +500 mM NH_4^+	547±1	28±16.1	118±8	2232±18
Soil +2 M NH_4^+	557±1	0.8±0.3	141±15	9476±22

On the base of data reported in Table 1, EN seemed to be the most reliable sensor. It showed an increase in NH_4^+ (NH_3) concentration detectable in soil which was significantly correlated with the amount of NH_4^+ added to soil ($r^2 = 0.999$, $P < 0.001$). The results obtained with ISE, i.e. the higher NH_4^+ content in extracts of soil with a lower NH_4^+ saturation (250 mM) than in those with a greater saturation, can be probably due to the interference of K^+ present at high concentration in soil extracts (2 M). This interference is higher when NH_4^+ is lower (greater aspecific signal) and decreases at increasing NH_4^+ concentration (greater specific signal). Regardless the absolute value in these treated soils, this sensor measured 621.8 µg of NH_4^+ g^{-1} of dry soil, which was only 2.3 times that measured in native soils without NH_4^+ addition, compared to 7.2 times increment measured by EN. Furthermore, the amount of NH_4^+ measured in untreated soil was not only greater than that detected by EN, but also than values commonly reported in most of the published studies, where they were instead comparable to measurements by EN. Differently, although IDE showed the highest sensitivity to NH_4^+ (NH_3), by perceiving up to 50 nM NH_4^+ in soil extracts, it also showed fast saturation at very low NH_4^+ concentrations. This fact determined that although soil extracts were diluted 5000 times they were still over the calibration curve, then generating similar values in NH_4^+ content.

CONCLUSIONS

Electronic nose, although it was neither specific nor the most sensistive sensing detector, it seemed to be the most reliable perceptive system. This feature may depend on the presence of an array of sensors with different sensitivities. Therefore, as well as the analytes of interest will interact differently with various sensors, also the interferents will bind differently to them. The processing by PCA of these presumably (statistically) different fingerprintings of signals, especially if compared with those of standards in calibration curves, will allow to discard the contribution due to background noise or aspecific interferences, thus obtaining information more strictly dependent on the analyte measured. Therefore, EN can be reliably used for measuring NH_4^+ content in soil.

REFERENCES

1. M. R. Carter and E. G. Gregorich, *Soil Sampling and Methods of Analysis,* Boca Raton: CRC Press, 2007.
2. K. Alef and P. Nannipieri, *Methods in Applied Soil Microbiology and Biochemistry*, London: Academic Press, 1995.
3. A. V. Mamishev *et al.*, *Proc. IEEE* **92**, 808-845 (2004).
4. A. D'Amico *et al.*, *Sensors Actuat. B-Chem*, 130, 458-465 (2008).
5. A. Macagnano *et al., Sensors Actuat. B-Chem*, **111-112**, 293-298 (2005).
6. Q. Ameer *et al., Sensors Actuat. B-Chem*, **106**, 541-552 (2005).
7. J. N. Barisci *et al., Sensors Actuat. B-Chem*, **84**, 252-257 (2002).

Variation Of Odour Profile Detected In The Floral Stages of *Prunus Persica* (L) Batsch Using An Electronic Nose

Messina Valeria[1]; Radice Silvia[2]; Baby Rosa[1] and Walsöe de Reca Noemí[1]

[1]*CINSO-CITEFA-CONICET. Juan Bautista de La Salle 4397 (B1603 ALO), Buenos Aires-Argentina*
2CEFYBO-CONICET-UBA. Paraguay 2155 Piso 16 (C1121ABG) Buenos Aires, Argentina

Bees use signals from plants to identify worthwhile visits. They learn quickly to differenciate mainly their floral odor than their colour. In some species the flowers remain open, intact and turgid until they are pollinated (anthesis) after which they are no longer attractive to pollinators (post-anthesis). Pollinators use fragrance for distance orientation, approach, landing, feeding and associative learning. The aim of this work was to study the variation of odor profile between anthesis and post-anthesis produced in flowers of different cultivars of *Prunus Persica* (L.) batsch, using an electronic nose since odor is a communication between flowering plants and bees. Visual results on field showed that peach flowers are generally more visited in the anthesis stage. Among all the analysed cultivars, Forastero cultivar was the only one visited in this floral stage. Statistical analysis of the electronic nose data showed that doped semiconductuvtive SnO_2 sensors could differenciate between stages (anthesis and post-anthesis) only in case of Forastero cultivar.

Keywords: Electronic nose, Floral stages, Odour.
PACS: 00 and 80

INTRODUCTION

Fruit set requires pollination and fertilization in non parthenocarpic species. Cross-pollination is the most frequent in the Prunus genus. Pollination in peach and nectarine flowers *Prunus Persica* (L.) Batsch is relatively simple because only one ovule per ovary needs to be fertilized. Cultivars are self-fertile in a great proportion, so plants can develop seeds and fruits when pollen is transferred from anthers of a flower to the stigma of the same flower or on a different flower of the same plant. Self-sterile varieties require pollenizers and insect pollinators. Bees are mainly effective insect pollinators because they eat pollen and nectar almost exclusively, visit many flowers of the same species during a single trip and have hairy bodies that easily pick up pollen grains. Bees use signals from plants to identify worthwhile visits. In some species the flowers remain open, intact and turgid until they are pollinated after which, they are no longer attractive to pollinators. In this case the negative involved facts are cessation of nectar, scent production, change in colour, wilting, permanent flower closure and petal drop [1]. Odor is an ancient medium of chemical communication between flowering plants and animal pollinators [2]. Pollinators use odor for distance orientation, approach, landing, feeding and associative learning [1, 3]. Floral scents are mixtures of small, volatile organic compounds varying in molecular weight, vapour pressure, polarity and oxidation state. Several authors [4, 5] studied the therpenoid emission pattern of several fruit species. This investigation revealed that fruit trees are monotherpene emitters and that the amount and composition of flower emission varies mainly between pome and stone fruits. The main portion of emitted monotherpenes was a mixture of a-pinene and camphene which on average counted for over 60 % of the emitted carbon from cherry flowers. The b-Myrcene only represented 15 % of the emitted volatiles while limonene, b-pinene, linalool and camphor all together constitute about 17 % of the total hydrocarbon rates. The aim of this work was to study the variation of odor profile between anthesis and post-anthesis produced in flowers of different cultivars of *Prunus Persica* (L.) batsch using an electronic nose, since odor is a communication between flowering plants and bees.

CP1137, *Olfaction and Electronic Nose: Proceedings of the 13th International Symposium*, edited by M. Pardo and G. Sberveglieri

EXPERIMENTAL AND METHODS

Prunus Varieties

Flowers of Prunus persica (L) batsch with Barcelo, Dixiland, Summerprince and Forastero cultivars, were collected at the Buenos Aires Province on August 2007 in the anthesis and post-anthesis stages. All these cultivars are frequently gathered in the fruit orchards of this region.

Electronic Nose

An electronic nose (EN) MOSES II (Modular Sensor System) containing two modules of gas sensors, one of them composed of eight quartz microbalance sensors (QMB) and the other by eight pure and doped semiconductive SnO2 sensors (MOx) were used. Doping with different elements increases selectivity for different gases. The adopted configuration results very flexible for general purposes and convenient for a wide range of applications [6].

Samples

Samples of flowers in the anthesis and post-anthesis stages were placed in five 10 mL glass vials equipped with a screw cap and silicon septum. Samples were stabilized at 40°C for 15 min (incubation time) in an 86.50 Dani headspace sampler and introduced into the MOSES II. Synthetic air was employed as carrier gas with a flow of 30 mL/min.

Statistical Analysis

Data of sensors response were analyzed as the mean values of the maximum minus the minimum response of the eight sensors QMB and the eight MOx sensors for the anthesis (A) and post-anthesis (P) flowers of each cultivar. Post Hoc using Tukey Test ($P \leq 0.05$) was employed [7]. Statistical software SPSS v.12. was applied for the analysis.

RESULTS

For a better understanding of the statistical analysis, results were described in four plots for each sensor.

Quartz Microbalance Sensors (QMB)

Statistical analysis showed that results obtained with QMB sensors (S) of the electronic nose have shown a significant difference only for the Forastero cultivar ($P < 0.001$). Figure 1A-B shows the mean values of the maximum minus the minimum responses of the eight QMB sensors from flowers of Barcelo, Dixiland, Summerprince and Forastero cultivars in the anthesis (A) and post-anthesis (PA) stages. Sensor number 3 (S3) is the only one differentiating stages of Forastero cultivar among the other cultivars.

On the otherhand Post Hoc analysis showed that only Forastero cultivar results were different from the others.

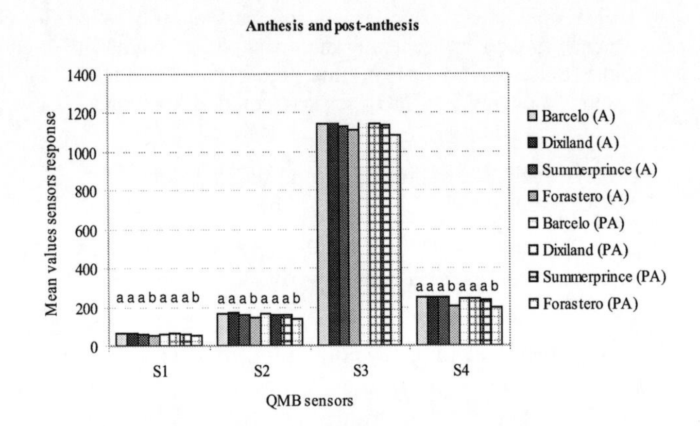

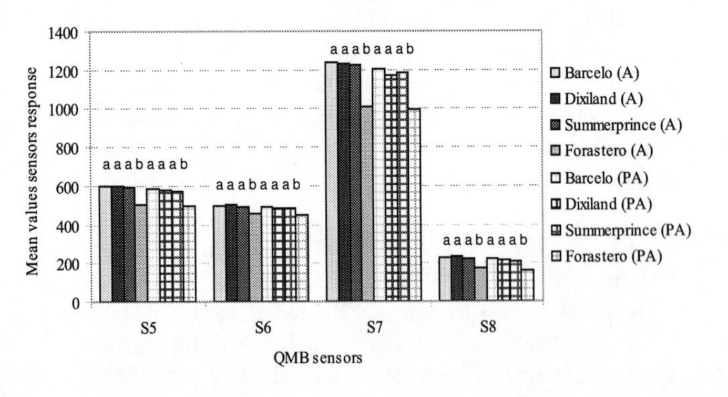

*Mean values with different letters indicate significant differences ($P < 0.05$) related to cultivar.

FIGURE 1A-B. Mean values of the maximum minus the minimum responses of the eight Qmb sensors from flowers of Barcelo, Dixiland, Summerprince and Forastero cultivars in the anthesis (A) and post-anthesis (PA) stages.

Doped Semiconductive SnO_2 Sensors (MOx)

Results obtained with all the sensors showed significant differences among cultivars ($P < 0.0001$), floral stage ($P < 0.001$) and considering the interaction of both variables ($P < 0.01$). On the other hand, results showed that Barcelo, Dixiland and Summerprince cultivars were similar in the anthesis and post-anthesis stages. Forastero cultivar showed differences from values of all the other cultivars and beetwen the own flower of the two tested floral stages (Figure. 2A-B).

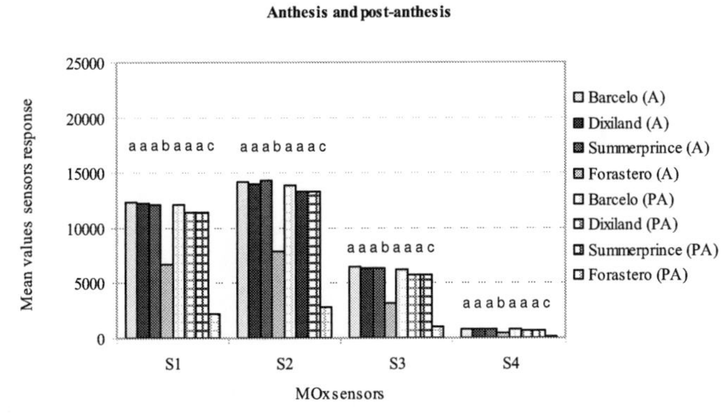

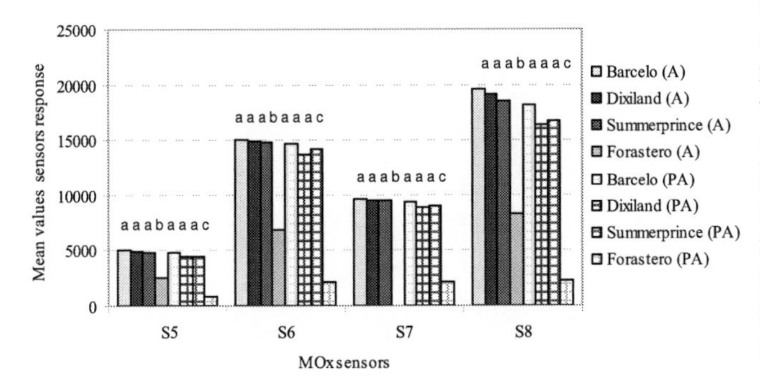

*Mean values with different letters indicate significant differences ($P < 0.05$) related to interaction between cultivar and floral stage.

FIGURE 2A-B. Mean values of the maximum minus the minimum response of the eight MOx sensors from flowers of Barcelo, Dixiland, Summerprince and Forastero cultivars in the anthesis (A) and post-anthesis (PA) stages.

DISCUSSION

Scientific literature does not report till now, studies on the variation of floral odor related to the bees attraction in the Prunus species. In a previous paper, Radice et al. (2008) described that visual exploratory analysis in fields with flowers of Prunus armeniaca (L.) Giada and the Prunus salicina Lindl showing that the Prunus armeniaca (L.) Giada was frequently visited by bees, not occuring the same for the other mentioned cultivar. On the otherhand the analysis of volatile compounds in both cultivars showed that therpenes were present in the Prunus armeniaca (L) and benzaldehyde compounds were present in the Prunus salicina Lindl.

Several authors [8, 9, 10] have reported that benzaldehyde is considered as a repellent for bees and therpenoid emission pattern are involved in the anthesis composition acting as an attraction for bees.

Analysis of the Prunus persica (L) batch cultivars with the electronic nose did not show differences in the odor profile between the anthesis and post-anthesis floral stages for Barcelo, Dixiland and Summerprince cultivars. These cultivars are self-fertile so, bees presence is not essential. Only Forastero cultivar exhibited significant differences between the floral stages (using MOx sensors). Visual field analysis showed that bees only visited flowers in the anthesis stage in case of Forastero cultivar. This cultivar is male-sterile [11]. This behaviour was related to those which have been observed in the Prunus armeniaca and the Prunus. Salicina. So, it is probable that the Forastero cultivar has to change the odor of flowers to attract the visit of bees and consequently to improve the pollination work. On the other hand, it is known that variation in the chemical composition of odor and in the emission rate may temporally vary according to circadian rhythms and post-pollination changes.

CONCLUSIONS

Results of this work enables to find the differences of flowers odor emanation in the anthesis and post-anthesis stages for different cultivars through the evaluation of electronic nose data. This equipment demonstrated to be a useful tool to separate the floral stages. In the following, it will be considered to identify the volatile compounds present in the different cultivars in both stages which are responsible of the attraction or rejection of bees.

REFERENCES

1. Van Doorn, W.,Effects of pollination on floral attraction and longevity, *Journal of Experimental Botany,* Vol.48 (1997), pp 1615–1622.
2. Pellmyr, O., Insect reproduction and floral fragrances: keys to the evolution of the angiosperms?, *Taxon,* Vol. 35(1986), pp 76-85.
3. Williams, N., Floral fragrances as cues in animal behavior. In Handbook of Experimental Pollination Biology, Scientific and Academic Editions, (New York, 1983), pp 20-36.
4. Baraldi, R., *et al.,*Volatile organic compound emissions from flowers of the most occurring and economically important species of fruit trees, *Phys. Chem. Earth, Part B,*Vol. 24 (1999), pp 729–732.
5. Radice, S. Thesis 2005. Biología floral y reproductiva del cultivar Forastero (*Prunus persica* [L] Batsch.) *Rosaceae, Prunoideae*, en estaciones crecidos sobre pies francos o clonales macro y micropropagados. Facultad de Ciencias Exactas y Naturales. UBA.
6. Walsöe de Reca, *et al,* Electronic nose: an useful tool for monitoring the environmental contamination. *Sensor & Actuators B*, Vol. 69, No.3 (2000), pp 214-218.
7. Weber, D, *et al.*, A first course in the design of experiments. Editorial CRC, (New York, 2000), pp 320.
8. Dobson, H, Floral volatiles in insect biology. In Insect-Plant Interactions, Bernays, ed (CRC Press, Boca Raton, FL, (1994), pp. 47- 81.
9. Vogel, S., Duftdrüsen im Dienste der Bestäubung. Über Bau und Funktion der Osmophoren, Abh. Math Naturw. Kl., (Akad. Wissensch. Mainz, (1963),pp 1-165.
10. Dudareva, N.,*et al*, Floral scent production in *Clarkia breweri.* III Enzymatic synthesis and emission of benzoid esters, *Plant Physioll*, Vol. 998, No 116 (1998), pp 599-604.
11. Radice S, Giordani E, Nencetti V, Bellini E. 2008. Phenological expression in *Prunus salicina* Lindl. genotypes and its relation with insect attraction and pollination. IX International Symposium on Plum and Prune genetics, breeding and pomology. March 16-19, 2008. Palermo, Italia.

Close-To-Practice Assessment Of Meat Freshness With Metal Oxide Sensor Microarray Electronic Nose

V. Yu. Musatov[1], V. V. Sysoev[1], M. Sommer[2], I. Kiselev[2]

[1]Saratov State Technical University, ul. Polytechnicheskaya 77, 410054 Saratov, Russia
[2]Forschungszentrum Karlsruhe, IMT, Hermann-von-Helmholtz-Platz 1, 76344 Eggenstein-Leopoldshafen, Germany

Abstract. In this report we estimate the ability of KAMINA e-nose, based on a metal oxide sensor (MOS) microarray and Linear Discriminant Analysis (LDA) pattern recognition, to evaluate meat freshness. The received results show that, 1) one or two exposures of standard meat samples to the e-nose are enough for the instrument to recognize the fresh meat prepared by the same supplier with 100 % probability; 2) the meat samples of two kinds, stored at 4 °C and 25 °C, are mutually recognized at early stages of decay with the help of the LDA model built independently under the e-nose training to each kind of meat; 3) the 3-4 training cycles of exposure to meat from different suppliers are necessary for the e-nose to build a reliable LDA model accounting for the supplier factor. This study approves that the MOS e-nose is ready to be currently utilised in food industry for evaluation of product freshness. The e-nose performance is characterized by low training cost, a confident recognition power of various product decay conditions and easy adjustment to changing conditions.

Keywords: food freshness; metal oxide sensor array; industrial application; LDA.
PACS: 07.07.Df; 87.56.Fc

INTRODUCTION

Food product quality control is one of the most actual and promising areas for e-nose application [1]. However, the real extent of employment of artificial olfaction devices in this area is disproportionately low due to a chemical composition complexity of food aromas. This complexity results in a set of specific questions from the food industry, which requires well-established and concrete answers to let a wide usage of e-noses. These questions could be summarized as follows:

- For how long should a user train the e-nose to obtain a stable recognition?
- How steady is the product of training (hereafter, the "model") to variations of input parameters, for instance, different storage conditions, different vendors of product, etc.?
- How long does a trained model preserve it's withstand ability against unavoidable drift of sensors and surrounding conditions?
- Which expenses would it take to maintain a model able to account for possible condition variations at the long term?

In the present report we address these issues in relation to application of e-nose for evaluation of meat freshness. In earlier performed works in this area (for recent examples, see [2-4]) it was shown, that the e-nose units are generally able to recognize the state of meat during a storage. However, the reported data are based on certain sets of samples and conditions, which do not provide enough knowledge in relation to above questions. Here, we have implemented the KAMINA e-nose (Research Center Karlsruhe, Germany), to the task and evaluated its performance according to practical issues required by meat industry users.

EXPERIMENTAL AND METHODS

The commercially available KAMINA e-nose is described in details somewhere else [5]. The sensing unit of the instrument is the 38-element sensor array chip carrying a thin metal oxide film segmented by co-planar electrodes. The measurements were conducted with three e-noses operated in parallel to achieve a reasonable level of statistical reliability of results. The exposures were performed in thoroughly filtered air generated with Zero-Air-Generator TG 12-Up (Sylatech GmbH, Germany). The purity of this air is only slightly inferior to that of conventional synthetic

CP1137, *Olfaction and Electronic Nose: Proceedings of the 13th International Symposium,* edited by M. Pardo and G. Sberveglieri

air. The dry air was delivered into an air-proof glass bell of ca. 200 ml volume containing a meat sample at the flow rate of 1.7 l/min. Following a pass through the bell, the smell-contaminated air was evenly distributed to lines feeding the three e-noses.

Retail available minced pork of different suppliers was used in the measurements. The meat was a subject to exposure in equal samples of approx. 1 g which appeared to provide a sufficient decrease of 2-20% of sensor resistance. The sensor exposure to air containing a meat smell was maintained to continue about 5 min which ensured a stable level of resistance. The sensor array resistance patterns recorded at such stable conditions have been employed for gas-recognition data analysis. The initial meat was divided into two parts to be stored at two temperatures, 4°C and 25°C. These meat pieces are called hereafter as samples of I and II types, respectively. The measurements with samples I were carried out at 1, 2 ("Good" meat, hereafter), 3, 4 ("Middle" meat), and 7, 8 ("Bad" meat) days of storage. The measurements with samples II were carried out at 1, 2, 3, 4 days of storage (the meat stored up to 3, 4 days was classified as the "Bad" one). At first experiment, the e-noses were exposed to meat of one supplier three times per day with 3 hour intervals between the exposures. At the second experiment, the e-noses were exposed to meat of four suppliers once a day.

In order to minimize the odor concentration effect, sensor signals of the microarray were divided by array median value prior feeding the pattern recognition algorithm. LDA discrimination and Leave-One-Out (LOO) [6] methods were used to separate the classes corresponding to odor of the meat at different storage age and to test the recognition power. The latter one tends usually to rise with further addition of class samples because the condition variations enclosed in classes are better comprised. The behavior of recognition power *versus* the number of data included into training serves here as a characteristics of the LDA model completeness or stability.

RESULTS

Storage stage recognition for meat from one supplier and one storage temperature

Figure 1 represents the LDA model of data recorded with one of the e-noses. One can see, that the meat smell at different days of meat storage is well separated. There are 6 models of meat freshness recognition which correspond to three e-noses and two storage temperatures. The criterion for the power of class separation in the LDA model is the mean distance between the class centers and LDA-space origin. The mean characteristic distances of all the 6 models are shown in the Table 1 in Mahalonobis units.

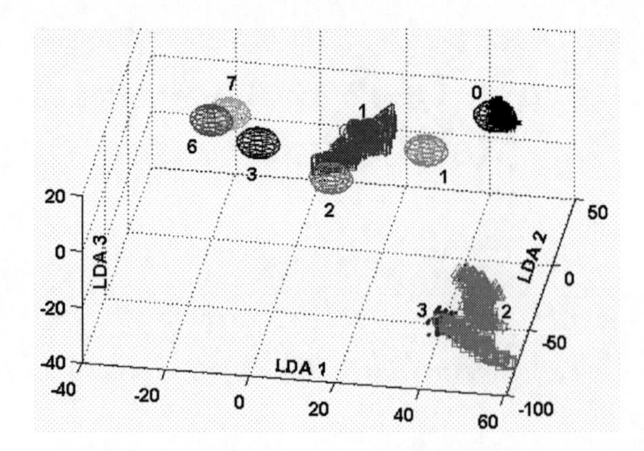

FIGURE 1. LDA-classes (ellipsoids) of the per-days model corresponding to the chip A and meat probes I. Point clouds represent the e-nose data related to measurement of probes II. Numbers placed around the ellipsoids and points correspond to storage days (0 => 1st day, etc).

TABLE 1. Characteristic distances of models.

	Storage temperature of model samples		
Chip	4°C	25°C	4°&25°C
A	15.0	31.6	9.8
B	27.5	61.1	17.4
C	20.5	64.8	13.2

A characteristic distance of about 2.5 would correspond to a LDA model with two ellipsoids enclosing 99% of class data which merely touch each other. Thus, as one can see, all the 6 models have perfectly separated classes, i.e. each of three employed microarray chips is capable to confidently differentiate the smell emitted by meat at various storage stages.

The Table 2 demonstrates the enhancement of stage recognition in the LDA models during LOO test under addition of the data into a model training pool.

TABLE 2. Recognition enhancement with the growth of N - number of series in a training pool.

N	1	2	3
Chip	Storage temperature: 4°C		
A	0.97	1.00	1.00
B	0.93	0.89	1.00
C	0.91	1.00	1.00
Chip	Storage temperature: 25°C		
A	1.00	1.00	1.00
B	1.00	1.00	1.00
C	0.83	0.90	1.00

The data given in the table show averaged recognition quota, defined as a ratio of correctly recognized test data to all test data, over all possible permutations of training and test series (exposures). It could be seen, that test samples are highly recognizable by the LDA model based even on the first training series only. The recognition quota is almost perfect using 2 series and comes to 100 % following the 3 training series. The important applied conclusion here is, that the KAMINA e-nose user needs to apply rather low expenses for training to obtain a perfect discrimination of meat freshness if the samples are from the same supplier.

Cross-recognition of the meat samples stored at different temperature

Still the question arises, can the LDA model built on data related to storage of meat of kind I be used to evaluate a freshness of meat of kind II correctly? As one can see in the Figure 1, the sensor microarray signals related to meat of both kinds are mutually recognized at earlier stages of meat storage but not at the late ones. One reason might be, that the different storage temperature facilitates a growth of different kinds of bacteria which is in accordance to human assessment. The samples II at the 1st day perfectly, and at the 2nd day to some degree, could be classified by LDA into the area corresponding to samples I measured after storage for 1 and 2-4 days respectively. However, the samples II at late stages of storage exhibit the signal to be located far away in the LDA space from the model path characterizing the samples I. In agreement with the above assessment, Table 1 indicates a worse smell discrimination of the combined LDA model built on data received for both samples I and II. In addition, the stabilization of recognition power using the combined LDA model is not as good as one obtained using the separate cases.

Here it is worth to introduce a generally useful tool for estimation of the training process speed. This is just a curve of LOO recognition quota enhancement as a function of data number included into building the LDA model. The example of such a curve calculated for chip A is drawn in the Figure 2. This function allows us to estimate the stability and recognition power of the LDA model. As far as the LDA model combined for both meat samples I, II attempts to cluster different smells into the same classes at the late storage states, the recognition quota curve needs a plenty of exposures to be stabilized and does not tend to reach a level appreciably higher than 0.9. In contrast, the LDA model built without data of the sensor microarray exposure to the samples II at late stages quite quickly reaches the absolute recognition level similar to behavior of the LDA model built on data related to samples I. That indicates a consistence of meat smell at earlier stages of storage at different

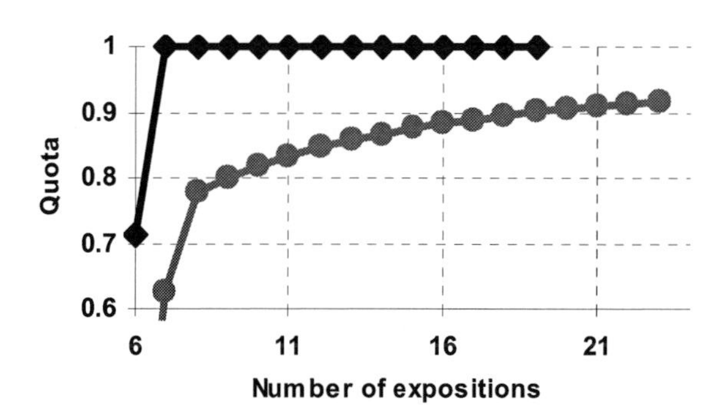

FIGURE 2. LOO-recognition quota as function of data included into the training pool. Red line corresponds to the model combining data obtained for samples I, II; the blue line - to the same training data except for ones obtained at the late stages of samples II storage. Chip A.

temperatures (which still include the initial staleness). However at late stages of storage under 4°C and 25°C, the smell differences become to be more pronounced.

Recognition of meat samples from various meat suppliers

The next question to be addressed is: how do differences in production and/or pre-treating of meat influence the recognition of its spoilage? To follow this issue, we have tested a meat of the same kind, but prepared by four other suppliers, under the same methodology as above using the same 3 sensor microarrays. To distinguish this meat from the previous one, the samples are called hereafter as 2-I, 2-II types (where 2 means a second set). The Figure 3 shows clusters of microarray data in response to samples 2-I relatively to the LDA model built using the data of exposure to samples I. As one can see, the data corresponding to meat from different suppliers at the second set turned out to be shifted relative to the LDA classes of the first series and to each other.

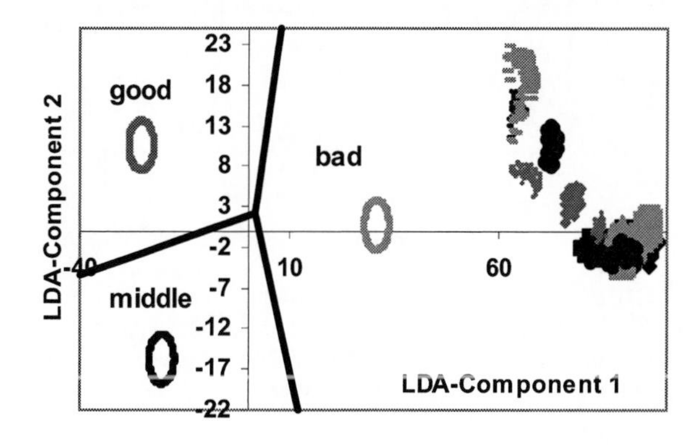

FIGURE 3. Mapping of the response data to samples 2-II in the space of LDA model built on that of samples I. Clusters of a same color differ in the meat suppliers.

To obtain a model for supplier-independent stage recognition, the data related to meat from different suppliers were sequentially included into the LDA training pool improving the LOO recognition quota as drawn in the Figure 4. In fact, the user-accepted level of 3% probability of error is reached following training of the LDA model by employing the samples from two suppliers in case of meat 2-I or that from three suppliers in case of meat 2-II. The 100 % recognition is obtained when the model accounts for data to the meat samples from all the four suppliers.

DISSCUSION

Usually, the more classes needed to be recognized makes it more difficult to reach the stable LDA model through a training process. Here the LDA models, which relate to microarray response to samples I, II per each day, are still better separated and getting to be stable faster than the LDA models distinguishing Good, Middle and Bad meat classes. Example of such a model built for the chip A is given in the Figure 1. Because the meat is spoiled continuously, the more distinct classes allow the better recognition. Using the first series only, the recognition the LDA model quota amounts to 0.96 and goes up to 1.00, when adding further series in the pool. The change of meat state during the storage follows a continuous and stable "path" under imaging in the LDA coordinate system (see the Figure 1). Therefore, there is an opportunity to classify the freshness of meat by matching the sensor microarray data to a signal position at this mentioned "path" rather than to several stages of storage. This approach is under development now.

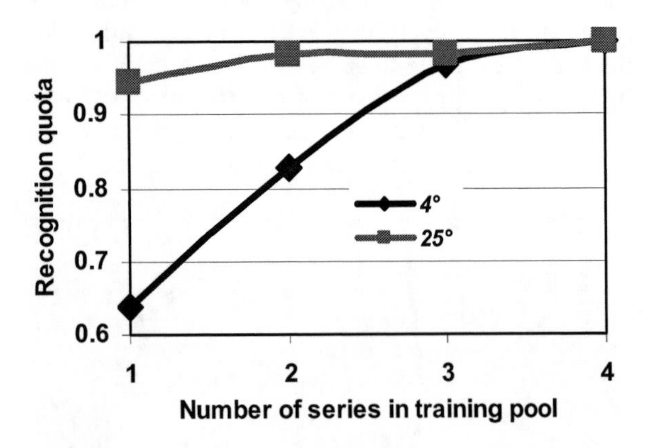

FIGURE 4. This LOO-recognition enhancement with inclusion of data related to meat samples from a number (1-4) of suppliers into a training pool. The curves are averaged over the three KAMINA microarray chips.

It is worth mentioning, that using pure air is essential in the food assessment measurements. From our former experience, the e-nose training at ambient room air needs much more significant number of measurements to achieve some level of stable LDA model, because the metal oxide sensors are very sensitive to air contaminations at the ppm or even ppb level concentrations [7]. Therefore, the use of synthetic (or carefully cleaned) air seems to be very important to successfully apply the e-nose for most of applications in food industry.

CONCLUSIONS

It is shown, that the KAMINA e-nose meets requirements of food industry and can be employed for evaluation of meat freshness under practical applications. If we consider a meat from a same supplier, the sensitivity and recognition power of MOS microarray chips is sufficient to train the corresponding pattern recognition model (LDA) using just one exposure per each stage of meat storage. However, the differences of smell from a meat prepared by different suppliers are still relatively high. Therefore, in order to achieve a reasonable level of meat freshness recognition for unknown suppliers, an analyst has to take at least four samplings of the variety of suppliers for the e-nose training. But the question for how long the trained LDA model could function properly is still remained to following studies.

ACKNOWLEDGMENTS

The authors thank G. Stengel, J. Benz, A. Serebrenicov for technical assistance in the fabrication of multisensor microarray chip and KAMINA unit; T.G. Kuznetsova and E.B. Selivanova for initialization of the study. V.Yu.M. thanks the DAAD and Russ. Ministry for Higher Education for "Michail Lomonosov" scholarship. V.V.S. thanks the INTAS postdoc grant, YSF 06-1000014-5877.

REFERENCES

1. E. Kress-Rogers and C. J. B. Brimelow, *Instrumentation and sensors for the food industry*, Boca Raton: CRC Press 2001.
2. N. El Barbri *et al*, *Sensors* **8**, 142-156 (2008).
3. K. Tikk *et al*, *Meat Science* **80**, 1254-1263 (2008).
4. I.M. Chernuha *et al*, *Myasnya Industrya*, **3** 9-12 (2008) (in Russian).
5. J. Goschnick, *Microelectron. Eng.* **57** 693-704 (2001).
6. R. Henrion, G. Henrion, *Multivariate Datenanalyse*, Berlin: Springer-Verlag 1995, pp. 67-102.
7. J. Goschnick *et all*, *Sens. Actuators B* **106** 182–186 (2005).

Predictive Detection of Tuberculosis using Electronic Nose Technology

Tim Gibson *[1], Arend Kolk[2], Klaus Reither[3], Sjoukje Kuipers[2], Viv Hallam[1], Rob Chandler[1], Ritaban Dutta[1], Leonard Maboko[3], Jutta Jung[3] and Paul Klatser[2].

1) Scensive Technologies Ltd., Metic House, Ripley Drive, Normanton, WF6 1QT, UK
2) KIT Biomedical Research, KIT (Koninklijk Instituut voor de Tropen / Royal Tropical Institute), Meibergdreef 39 1105 AZ AMSTERDAM, The Netherlands
3) Mbeya Medical Research Programme, P.O Box 2410, Mbeya, Tanzania

Abstract: The adaptation and use of a Bloodhound® ST214 electronic nose to rapidly detect TB in sputum samples has been discussed in the past, with some promising results being obtained in 2007. Some of the specific VOC's associated with *Mycobacteria tuberculosis* organisms are now being discovered and a paper was published in 2008, but the method of predicting the presence of TB in sputum samples using the VOC biomarkers has yet to be fully optimised. Nevertheless, with emphasis on the sampling techniques and with new data processing techniques to obtain consistent results progress is being made Sensitivity and specificity levels for field detection of TB have been set by WHO at a minimum level of 85% and 95% respectively, and the e-nose technique is working towards these figures. In a series of experiments carried out in Mbeya, Tanzania, Africa, data from a full 5 days of sampling was combined giving a total of 248 sputum samples analysed. From the data obtained we can report results that show specificities and sensitivities in the 70-80% region when actually predicting the presence of TB in unknown sputum samples. The results are a further step forward in the rapid detection of TB in the clinics in developing countries and show continued promise for future development of an optimised instrument for TB prediction.

Keywords. Electronic nose, tuberculosis, volatile, biomarkers, prediction, diagnosis.
PACS. 07.07.Df, 72.80.Le, 87.19.xb,

INTRODUCTION

Rapid detection of TB is usually done using ZN staining, whch is the WHO recommended method for *Mycobacteria tuberculosis.* However the efficacy of the technique in the field is quite low, with positive detection of all types of TB infection being under 50%. The following is a quote from the Foundation for Innovative New Diagnostics (FIND) web site about ZN microscopy:

" *the sensitivity of this technology is low: it can only detect roughly half of all active cases of tuberculosis when properly used, and in areas struck by the HIV epidemic, where co-infections are frequent, as well as in children, the sensitivity is even lower. Moreover, though routinely described as a simple technology, microscopy is actually complex, and highly dependent on the training and diligence of the microscopist, requiring multiple examinations. Furthermore, under programmatic conditions it may take days rather than hours to complete, with the consequence that many patients drop out during the diagnostic process".* So it can be seen that ZN microscopy is not the ideal diagnostic tool for TB and new, simple, low cost diagnostic methods are required.

Early work on VOC detection for TB diagnosis was carried out by Cranfield University in collaboration with KIT Biomedical Research at the Royal Tropical Institute in Amsterdam, using a Bloodhound® BH114 instrument to evaluate if electronic nose technology could be used to detect TB in clinical samples. This early study proved quite successful with sensitivity and specificity being recorded at 89% and 91% respectively [1, 2]. Further trials at KIT, in direct collaboration with Scensive, used a newer instrument, a Bloodhound® ST214 and these were reported in Shanghai at the World Congress on Biosensors in May 2008, where newer sampling methods had reduced the limit of detection down to 10^3 cfu's per ml.

The continued study of TB detection by VOC measurement has now allowed a method of prediction of the presence of TB in sputum to be initially evaluated. This is the first step towards an actual diagnostic test for TB in unknown samples using the VOCs biomarkers produced for detection.

EXPERIMENTAL AND METHODS

Individual sputum samples were prepared by adding 0.5 ml sputum to 0.5 ml NaCl solution in a standard sputum cup and mixing for several minutes to break up the sample. Samples were then incubated at 37°C for 30 minutes and then sniffed 3 times by the ST214, with an injection time of 7 seconds and the data recorded on the

CP1137, *Olfaction and Electronic Nose: Proceedings of the 13th International Symposium,* edited by M. Pardo and G. Sberveglieri

PC in two different formats for analysis. These were processed using linear discriminant function analysis (LDA) with XLStat, a statistical add-in package linked to Microsoft Excel.

The more sophisticated data processing involved extraction of features from the full data sets collected as text files. A new software suite produced 'in house' enabled four feature parameters to be extracted and used, with specific slopes of absorption and desorption at selected time slots being added to the peak heights and integrated areas.

Predictive results were produced by combining all the data sets collected over the 5 days and randomly extracting a total of 49 samples to be used as a prediction set. This raw data did not 'see' any processing whatever. The remaining 199 samples made up of 154 TB negative and 45 TB positive samples was used as the training set. All TB negative samples were both microscopy negative and culture negative and all TB positive samples were microscopy positive and culture positive to ensure the most definitive sample sets. The prediction set was inserted into the process as a set of random samples to be classified by the software, with the results being expressed as a grid of correctly / incorrectly classified TB –ve and TB +ve results. Specificity is the number of TB-ve correctly identified by the system and sensitivity is the number of TB+ve correctly identified by the system.

RESULTS

Data processing of the samples using LDA gives a scatter plot of the total sample set which shows some significant overlap between the 2 classes. This is not suprising as similarity between sputum samples is seen. The infiltration of the TB negative samples into the TB positive is the most obvious effect here, figure 1.

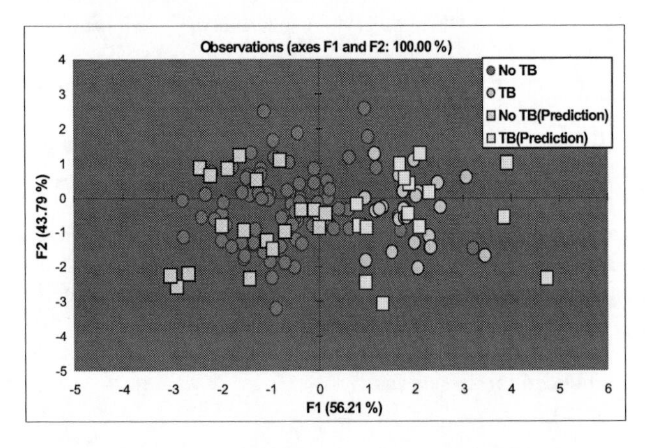

Figure 1. Scatter Plot between TB Negative (Green) and TB Positive (Yellow) Samples including a Prediction Set (Light Green and CreamYellow).

The results for the prediction of samples are shown in table 1.. Here 10 randomly selected sets of raw data from the TB negative and TB positive pools were extracted before any data processing, i.e. the prediction sets acted as sample 'unknowns'.

	Specificity	Sensitivity	Total
1	71.05%	72.73%	71.43%
2	76.32%	81.82%	77.55%
3	84.21%	81.82%	83.67%
4	73.68%	90.91%	77.55%
5	76.32%	81.82%	77.55%
6	73.68%	72.73%	73.47%
7	65.79%	81.82%	69.39%
8	71.05%	81.82%	73.47%
9	76.32%	72.73%	75.51%
10	81.58%	81.82%	81.63%
Overall	75.00%	80.00%	77.50%

Table 1. 10 Randomly Selected Prediction Sets Classified by the Training Set using LDA on 4 Extracted Features from the Raw Data.

The results are very promising bearing in mind that the Bloodhound® ST214 instrument used was not optimised for TB VOCs, since these were not known at the time of sampling. Also no sample pre-concentration was used and the VOCs were just taken directly from the headspace of the 1:1 diluted sputum samples.

CONCLUSIONS

The predictive detection of TB using electronic nose technology is feasible and results to date indicate between 70-80% specificity and sensitivity of identification of live, active TB infections using sputum as a sample matrix. The use of VOCs as diagnostic biomarkers is beginning to become accepted in the medical profession and it is very clear now that TB has a distinct odour [3].

Further work is necessary to develop more sensitive and specific sensors that will allow optimisation of the sensor heads for known TB volatiles. Pre-concentration of VOCs before detection is likely to greatly assist in the correct diagnosis of TB by VOC detection.

ACKNOWLEDGMENTS

Many thanks to FIND for financial support and the permission to use the copyrighted material from their website.

REFERENCES

1. Detection of *Mycobacterium tuberculosis* (TB) in vitro and in situ using an electronic nose in combination with a neural network system Alexandros K. Pavlou *et al* Biosensors and Bioelectronics **20** (2004) 538–544

2. Prospects for Clinical Application of Electronic-Nose Technology to Early Detection of *Mycobacterium tuberculosis* in Culture and Sputum Reinhard Fend *et al* J Clinical Microbiol **44** (2006) 2039 – 2045

3. The Scent of *Mycobacterium tuberculosis.* Mona Syhre and Stephen T Chambers. Tuberculosis **88** (2008) 317-323.

Classification of Odours for Mobile Robots Using an Ensemble of Linear Classifiers

Marco Trincavelli, Silvia Coradeschi and Amy Loutfi

AASS Research Center, Örebro University
Fakultetsgatan 1, 70182 Örebro
E mail: name.surname@aass.oru.se

Abstract. This paper investigates the classification of odours using an electronic nose mounted on a mobile robot. The samples are collected as the robot explores the environment. Under such conditions, the sensor response differs from typical three phase sampling processes. In this paper, we focus particularly on the classification problem and how it is influenced by the movement of the robot. To cope with these influences, an algorithm consisting of an ensemble of classifiers is presented. Experimental results show that this algorithm increases classification performance compared to other traditional classification methods.

Keywords: Odour Classification, Mobile Robotics
PACS: 01.30.Cc

INTRODUCTION

Mobile robots equipped with an array of gas sensors can be a valuable tool in scenarios like inspection of hazardous areas or environmental monitoring [1], particularly in cases where toxic contaminants are involved. Mobile robots can play an important role in assessing the presence of dangerous substances, identifying their character, providing a map of their distribution and if possible quantifying their concentration [2]. One essential feature of such a robot is to be able to discriminate substances in a fast and reliable way. Up to date the classification of odours with an array of gas sensors has been done in systems that perform a controlled sampling process (three-phase sampling) and then process the data offline [3]. For mobile robotics application the bulky and expensive equipment needed for controlling the sampling process is often unsuitable. Therefore it is more convenient for the array of sensors to be exposed directly to the external environment and sample it continuously. Under these conditions, however, the signal from the array does not have the three characteristic phases of the controlled sampling process but shows a series of sensor responses which vary depending on the interaction between the nose and the odour plume. An example signal collected during an experiment is shown in Figure 1. Only few works have dealt with odour classification problem in such conditions [4, 5].

Since the gas diffusion in uncontrolled environments is dominated by turbulence [6], the odour plume has a very articulated shape with many patches and meanders. Therefore the interaction of the nose with the odour plume is complicated and hard to model. Despite the difficulties in providing an exact model of this interaction we can investigate if some controllable factors, like the movement of the robot or the location of the experiment, introduce some regularity in the data set. The main contribution of this work is to discover such regularities and exploit them to enhance the classification performance. The approach proposed is the creation of an ensemble of specialized classifiers. The performance of the proposed method is then compared to two standard classification algorithm, namely the Support Vector Machine (SVM) and the Radial Basis Function (RBF) [7].

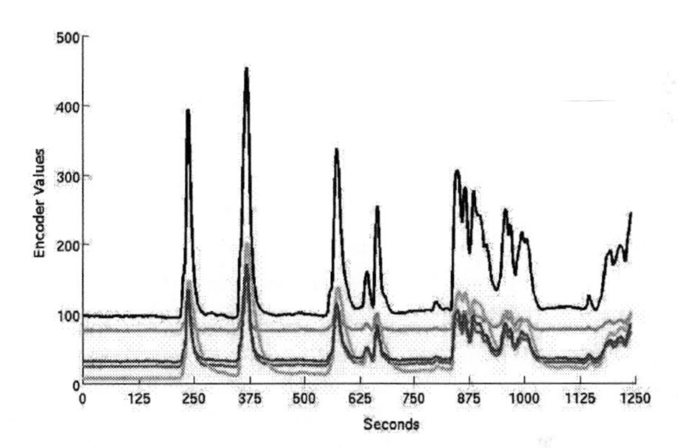

Figure 1. Signal collected during an experimental run.

EXPERIMENTAL AND METHODS

The robot used in the experiments is an ATRV-JR all terrain robot equipped with the Player Robot Device Interface [8]. Player provides both the interface to the

CP1137, *Olfaction and Electronic Nose: Proceedings of the 13th International Symposium,* edited by M. Pardo and G. Sberveglieri

sensors and the actuators, and high level algorithms to address robotic tasks such as localization (*amcl* driver) and navigation (*vfh* and *wavefront* drivers). The robot is equipped with an electronic nose, an actively ventilated aluminum tube containing an array of five metal oxide gas sensors, mounted in front of the robot at a height of 0.1 m on the ground. The sensors present in the array are listed in Table 1 together with their target gases.

Table 1. Gas sensors used in the electronic nose.

Model	Gases Detected	Quantity
Figaro TGS 2600	Hydrogen, Carbon Monoxide	2
Figaro TGS 2602	Ammonia, VOC, Hydrogen Sulfide	1
Figaro TGS 2611	Methane	1
Figaro TGS 2620	Organic Solvents	1

The experiments have been performed in three different locations using four different moving strategies. In all the experiments the robot was moving with a speed of 0.05 m/s. The odour source was a cup full of the analyte placed on the ground. The first location that has been considered is a large close room in which the robot followed a sweeping trajectory with two orthogonal orientations that we name N-S and E-W. The second set of experiments has been carried out in a small classroom whose door has been left open. In this environment the robot performed two different kinds of spiral path: a spiral without any stops from the beginning to the end of the experiment and a spiral with stops when an odour is detected, at which point the robots stands static until enough information is obtained to perform a classification. The last experimental location was a courtyard with an uneven grass surface. In this case the robot performed a spiral movement stopping when a gas is detected similar to the one performed in the classroom. Table 2 summarizes the five different experimental configurations. The experiments have been repeated multiple times with three different substances, namely ethanol, acetone and isopropyl, that are the target substances for our classification problem. During one experimental run multiple responses were collected, for a total of 592 responses evenly distributed among the three classes.

The classfication algorithm is articulated into five phases, namely baseline subtraction, signal segmentation, feature extraction, dimensionality reduction and classification. The baseline is the value that a gas sensor gives as output when it is exposed to clean air. This value depends on temperature, humidity and long/short term drift [9]. The baseline is subtracted from the output value of the sensors in order to limit the effect of these factors using differential baseline subtraction. The baseline value is measured for 60 s at the beginning of every experiment, when the odour source is closed and the robot is standing still. After performing this first transformation the signal is smoothed using an average filter of dimension 5 in order to suppress the noise due to sampling and quantization. The smoothed signal is then segmented into three different phases, namely baseline, rise and decay according to the value of the first derivative. The segmentation procedure can be easily explained using a finite state machine as shown in Figure 2.

Table 2. Summary of the experimental conditions in which the data have been collected.

Data Set	Location	Moving Strategy	Number of Runs
1	Large Room	Sweep N-S	15
2	Large Room	Sweep E-W	15
3	Classroom	Spiral	18
4	Classroom	Spiral with Stops	72
5	Garden	Spiral with Stops	16

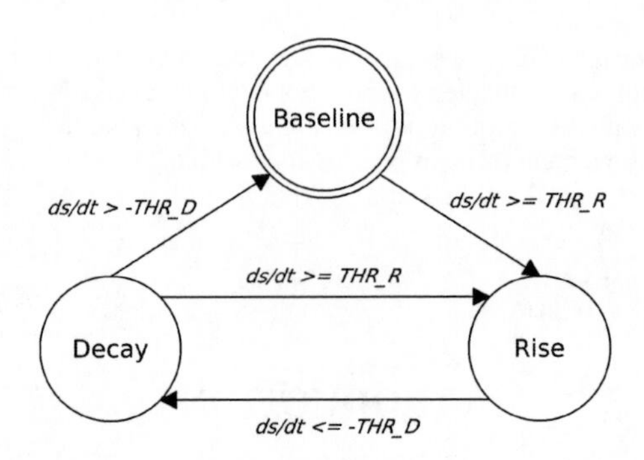

Figure 2. Finite State Machine that illustrates the segmentation algorithm.

In this figure the first derivative is denoted as ds/dt and the threshold for the rise and decay are THR_R and THR_D respectively. Two different thresholds are needed since the rise and decay phase are best described using a first-order model and the time constant for the rising phase is smaller [10]. A complete response to an odour patch is considered to be the ensemble of a consecutive rise and decay phase. The isolated response is then passed to the feature extraction module that fits a 2nd degree polynomial to the points in the response. The choice of the polynomial degree that is more suitable to the sensor response that consists of a consecutive rise and decay. Moreover a parabula provides a sufficient degree of fitting without overfitting the signal. The feature vector is built by concatenating the 3 coefficients obtained by fitting each of the five sensors, obtaining a 15 dimensions vector. The dimensionality of the feature vector is then reduced applying the Linear Discriminant Analysis

(LDA) technique, that projects the vectors into a lower dimensional space in which the distance between samples of different classes is maximized and the distance between samples of the same class is minimized [7]. In contrast to the Principal Component Analysis (PCA) which searches for the directions in which the variance of the data is maximized neglecting the fact that samples belong to different classes, the LDA searches for the directions on which to project the data taking into consideration the membership of the samples to the different classes. The first two LDA Components are kept and passed to the subsequent classification algorithm. The classification algorithm we propose is an ensemble of Linear Probabilistic Discriminative Models [7]. This model provides as output the posterior probabilities that the sample belongs to each of the classes. In particular, we want to investigate the possibility of breaking down the odor classification problem into a set of simpler classification tasks/subproblems. These tasks can be solved efficiently by a set of linear classifiers. The global solution is then obtained recombining the solutions of the simpler tasks by marginalizing with respect to the subproblem according to the following equation:

$$P(C_k|\mathbf{x}) = \sum_{m=1}^{M} P(C_k|\mathbf{x},m)P(m|\mathbf{x}) \tag{1}$$

Where $P(C_k|\mathbf{x})$ is the probability that sample $\mathbf{x}$ belongs to class C_k, $P(C_k|\mathbf{x},m)$ is the probability that sample $\mathbf{x}$ belongs to class C_k given that sample $\mathbf{x}$ belongs to subproblem m and $P(m|\mathbf{x})$ is the probability that sample $\mathbf{x}$ belong to subproblem m. This method requires the training of $M+1$ classifiers: one for each of the subproblems and one for assigning the membership of the sample to the subproblem.

RESULTS

Figure 3 shows a plot of the two LDA components when the LDA is calculated with respect to the analyte. We can notice how the acetone samples are well clustered, while the ethanol and isopropyl samples are partially overlapping. This makes the classification problem nontrivial.

If instead of considering the LDA projection with respect to the analyte, we consider the LDA projection with respect to the experimental conditions in which the responses has been collected. This result displayed in Figure 4. Here we can observe clearly three data clusters that correspond to the different moving strategy of the robot, namely sweeping, spiraling and spiraling with stops. In Figure 5 the two LDA components with respect to the substance to classify for each single subgroup are displayed.

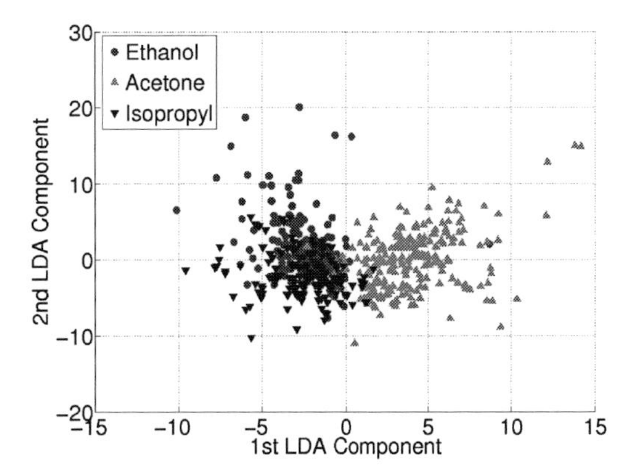

Figure 3. LDA Plot of the full classification problem.

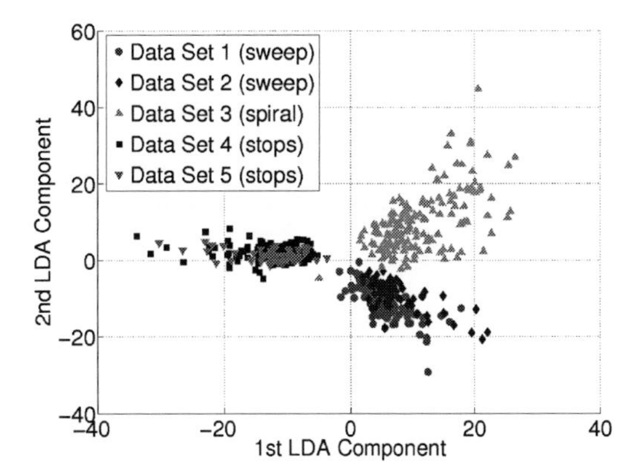

Figure 4. LDA plot of the data set classification problem.

The classification problem is much simpler in each of the groups than in the whole training set. Therefore an ensemble of *four* Linear Probabilistic Discriminative Models is used: one that calculates the probability $P(m|\mathbf{x})$ that a sample has been collected using moving strategy m, where m can take three values that correspond to sweeping, spiraling or spiraling with stops movement. The other three classifiers calculate the probability $P(C_k|m,\mathbf{x})$ that sample $\mathbf{x}$ belongs to class k, with k that can be ethanol, acetone or isopropyl, given that $\mathbf{x}$ has been collected performing movement m. The four decisions are then ensembled according to Equation (1) in order to obtain $P(C_k|\mathbf{x})$. The sample is then assigned to the class with the highest $P(C_k|\mathbf{x})$.

The obtained classifier has been tested with a 20 fold cross validation. As a term of comparison a SVM and a RBF have been trained and evaluated performing also a 20 fold crossvalidation on the whole data set. The performance of the classifiers together with the average training time is reported in Table 3. Table 4 reports the classification rates of the classifiers that form the ensemble. The proposed ensemble of classifiers clearly outperforms both the SVM and the RBF that operate on

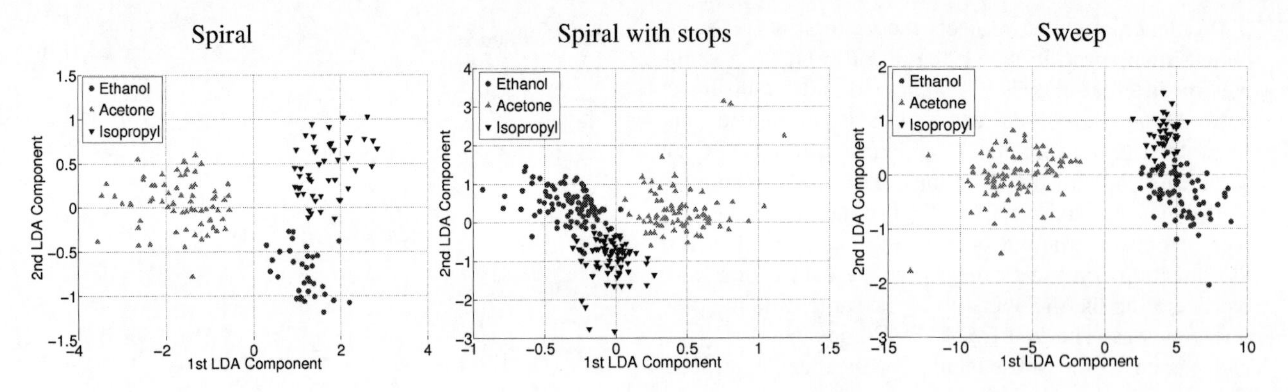

Figure 5. LDA plot of the classification problem when a single subgroup is considered.

the whole data set. Moreover the ensemble is also much quicker to train than the SVM.

Table 3. Average and confidence interval (95% confidence level) of the classification rate for the ensemble of classifiers and for the reference classifiers (SVM and RBF). In the third column the average training time for the three classifiers is reported.

Classifier	Accuracy	Avg. Train Time
Ensemble	89.36 ± 2.58	0.02 *s*
SVM	75.41 ± 4.59	14.52 *s*
RBF	68.64 ± 3.55	0.08 *s*

Table 4. Average and confidence interval (95% confidence level) of the classification rate for the classifiers that are composed into the ensemble.

Classifier	Accuracy
Movements	98.67 ± 0.93
Sweep group	88.47 ± 3.94
Spiral group	93.46 ± 4.93
Spiral with stops group	93.15 ± 2.91

CONCLUSIONS

Odour classification is a fundamental ability for a robot that has to monitor the pollution or explore an area and discover hazardous gases. In order to achieve reliable classification it is important to understand the factors that influence the shape and the properties of the sensors response. In this paper we have shown that the movement strategy of the robot clearly affects the properties of the signal and by taking into account this movement in the classification performance can be enhanced. A next step is to analyze exactly which properties of the signal response are affected by the different movement strategies. Also future work will continue the analysis of the mobile robotic platform and the environment in which the data is collected in order to find other patterns that can further improve the classification performance.

REFERENCES

1. M. Trincavelli, M. Reggente, S. Coradeschi, A. Loutfi, H. Ishida, and A. J. Lilienthal, "Towards Environmental Monitoring with Mobile Robots," in *Proceedings of the IEEE International Conference on Intelligent Robots and Systems (IROS 2008)*, 2008, pp. 2210–2215.
2. A. J. Lilienthal, A. Loutfi, and T. Duckett, *Sensors* **6**, 1616–1678 (2006).
3. R. Gutierrez-Osuna, A. Gutierrez-Galvez, and N. U. Powar, *Sensors and Actuators B: Chemical* **93**, 57–66 (2003).
4. M. Trincavelli, S. Coradeschi, and A. Loutfi, "Classification of Odours with Mobile Robots Based on Transient Response," in *Proceedings of the IEEE International Conference on Intelligent Robots and Systems (IROS 2008)*, 2008, pp. 4110–4115.
5. D. Martinez, O. Rochel, and E. Hughes, *Autonomous Robots* **20**, 185–195(11) (June 2006).
6. P. J. W. Roberts, and D. R. Webster, "Turbulent Diffusion," in *Environmental Fluid Mechanics - Theories and Application*, edited by H. Shen, A. Cheng, K.-H. Wang, M. Teng, and C. Liu, ASCE Press, Reston, Virginia, 2002.
7. C. M. Bishop, *Pattern Recognition and Machine Learning (Information Science and Statistics)*, Springer, 2006, ISBN 0387310738.
8. B. Gerkey, R. T. Vaughan, and A. Howard, "The Player/Stage Project: Tools for Multi-Robot and Distributed Sensor Systems," in *Proceedings of the IEEE International Conference on Advanced Robotics (ICAR)*, 2003, pp. 317–323.
9. Figaro Gas Sensor: Technical Reference, 1992. http://www.figaro.com.
10. A. J. Lilienthal, and T. Duckett, "A Stereo Electronic Nose for a Mobile Inspection Robot," in *Proceedings of the IEEE International Workshop on Robotic Sensing (ROSE)*, Örebro, Sweden, 2003.

Use Of Electronic Noses For Continuous Monitoring And Recognition Of Environmental Odours In Presence Of Multiple Sources

Laura Capelli, Selena Sironi, Sandra De Luca, Renato Del Rosso, Paolo Céntola

Politecnico di Milano, Olfactometric Laboratory, Department of Chemistry, Materials and Chemical Engineering "Giulio Natta", Piazza Leonardo da Vinci 32, 20133 Milano

Abstract. This paper describes an experimental monitoring conducted in order to study the olfactory impact on a small town located in the North of Italy, where the coexistence of more different factories is the cause of an olfactory nuisance problem. Three electronic noses were first trained to recognize the characteristic odours of the plants considered as the major sources of odour emissions in the studied area, then installed on the territory with the aim of continuously analyzing the ambient air in order to detect the presence of odours and to identify their origin. The results of the study demonstrate how specific electronic noses developed for environmental applications, once suitably trained, can successfully be applied for the continuous monitoring of odours and their identification in presence of multiple sources. In this specific case, the continuous odour monitoring at the first receptor, i.e. the nearest dwelling to the industrial zone, resulted in the detection of odours for 11.6% of the total monitoring period. Moreover, the qualitative classification of the electronic nose installed at this receptor enabled to identify the major source of odour nuisance, which turned out to be the aluminium foundry.

Keywords: Odour impact determination, environmental application, odour source, training
PACS: 07.07.Df

INTRODUCTION

One of the main issues of our research activity is the development of electronic noses for the continuous monitoring of environmental odours at specific receptors, i.e. directly where the presence of odours is lamented. For this purpose, the instrument should be able to continuously analyze the ambient air at specific receptors and, in real time, to detect the presence of odours and to recognize them, by attributing the analyzed air to a specific olfactory class. The work required for the development of a similar system is composed of two interconnected fundamental activities, i.e. the instrument design and the definition of its utilization procedures. The experimental methods adopted therefore are laboratory tests with air samples of known odour quality and odour concentration and field tests, represented by continuous monitoring campaigns of limited duration.

This paper describes one of these experimental monitoring campaigns. In this specific case the continuous monitoring was conducted in order to study the olfactory impact on a small town located in the North of Italy, where the coexistence of more different factories is the cause of an olfactory nuisance problem.

For this purpose, three electronic noses were used. The instruments were first trained to recognize the characteristic odours of the plants considered as the major sources of odour emissions in the studied area, then installed on the territory with the aim of continuously analyzing the ambient air in order to detect the presence of odours and to identify their origin. The monitoring positions were chosen with the purpose of obtaining a significant portrayal of the olfactory impact on the monitored area, considering its geography, the actual meteorological conditions, the reciprocal location of the five plants and the presence of receptors [1].

EXPERIMENTAL METHODS

Electronic Nose Description

The instruments used for this study have been developed in collaboration with Sacmi s.c.a.r.l. [2]. The system includes a pneumatic assembly for

CP1137, *Olfaction and Electronic Nose: Proceedings of the 13th International Symposium*, edited by M. Pardo and G. Sberveglieri

dynamic sampling (pump, electro-valve, electronic flow meter), a thermally controlled sensor chamber with 35 cm^3 of internal volume and an electronic board for controlling the sensor operational conditions. Each instrument has been equipped with an array of six different thin film MOS (Metal Oxide Semiconductor) sensors, which makes the system sensitive to a large spectrum of volatile compounds, and a humidity sensor.

The instrument remote control and the data acquisition can be performed by an external personal computer through the standard communication port RS232 or a USB port. Two specific software have been developed for the electronic nose data processing: the Nose Pattern Editor (NPE), which is used for data pre-processing and for multivariate statistical analysis, and the Nose Pattern Classifier for pattern recognition and data classification.

Training

The training of the electronic nose is a very important and delicate phase, necessary in order to create a complete database that the instrument uses as a reference for the subsequent pattern recognition. For the training, it is important to consider all the odour typologies to which the instruments may be subject during the monitoring.

In this case, five plants were identified as the possible contributors to odour nuisance in the studied area:

- a cast iron foundry (P1);
- a plant for the production of paints (P2);
- a waste treatment (mechanical-biological treatment and incineration) plant (P3);
- a plant for the production of polyester resins (P4);
- an aluminium foundry (P5).

After an in-depth study of the plants production cycles, the major odour sources to be sampled for the electronic nose training were identified. The training involved the execution of two campaigns for the collection of gas samples and their subsequent analysis by dynamic olfactometry [3] and by electronic nose. Moreover, some ambient air samples were collected while odours from the monitored plants were not perceivable, in order to create a reference olfactory class "neutral air".

Monitoring

The monitoring started Thursday, 12th June 2008, and ended Friday, 4th July 2008.

During this period, a first electronic nose (EN1) was installed in an office at the eastern boundary of P4. A second instrument (EN2) was positioned at a receptor, i.e. the nearest dwelling to the industrial zone. Finally, a third instrument (EN3) was installed at a second receptor, a farm located in the eastern part of the industrial zone, in order to monitor the ambient air in this area (Figure 1).

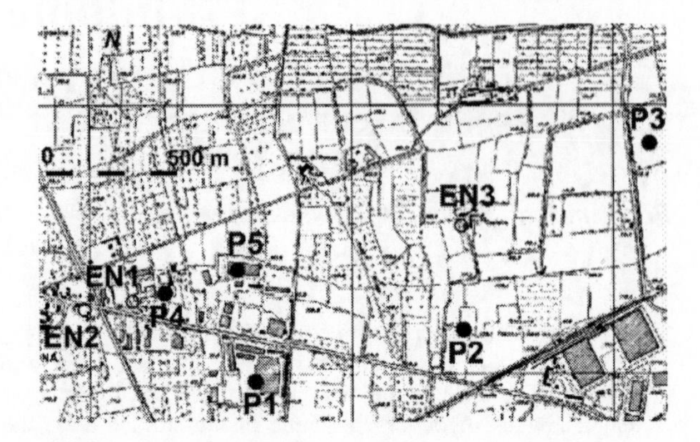

FIGURE 1. Localization of the electronic noses and of the plants.

Auxiliary Data For Odour Impact Determination

For a more accurate evaluation of the instrumental responses, some auxiliary data were considered

First, the hourly wind speed and wind direction data registered by a nearby meteorological station relevant to the monitoring period were analyzed.

Moreover, the inhabitants of the dwelling where the electronic nose EN2 was installed were asked to fill in a specific form indicating all the episodes when they perceived odours in their opinion attributable to the industrial zone at issue.

RESULTS

Graphical Representation Of Results

The monitoring results are represented by large tables that report the olfactory class attributed to the analyzed air for each measurement carried out during the monitoring period.

A synthetic representation of the qualitative air classification results can be given in graphs that report in abscissa the measurement date and hour, and in ordinate the olfactory class attributed to the analyzed air, i.e. the odour origin recognized by the electronic nose.

As an example, Figure 2 represents the results of the qualitative classification of the air analyzed by

EN2 at the first receptor, i.e. the first dwelling in direction west with respect to the industrial zone. In the upper line, Figure 2 also illustrates the odour episodes reported by its inhabitants.

The large number of measurements during which the electronic nose EN2 recognized the presence of odours different from "neutral air" while the house owners didn't might be due by the fact that the electronic nose, which analyzes the air continuously, has a higher and more regular sampling frequency than the frequency of the house occupants observations. Moreover, the instrumental detection limit may be lower than the house occupants odour detection threshold.

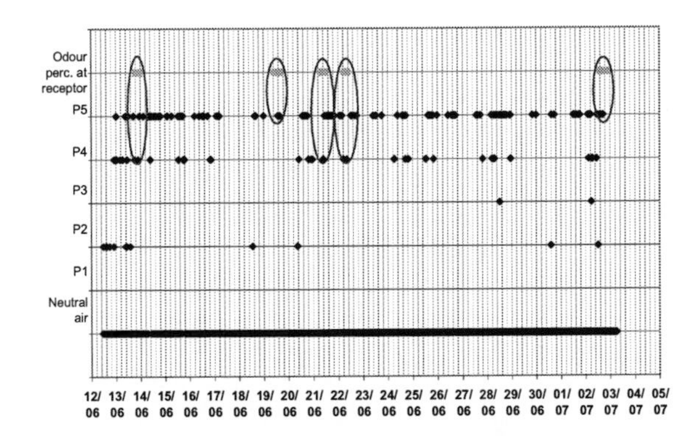

FIGURE 2. Quality of the ambient air analyzed by EN2 at the first receptor and odour episodes signalled by inhabitants.

Odour Impact Determination

Based on the instrumental responses it is possible to determine the odour impact caused by the considered facilities on the three monitoring positions.

More in detail, it is possible to evaluate the relative frequency of the odour detections, i.e. the number of measures attributed by the electronic noses to the considered olfactory classes with respect to the total number of measures. These results are illustrated in Figure 3 and reported more in detail in Table 1.

EN1 detected the presence of odours for 10% of the monitoring period. In this case, odours from facility P4 were detected most frequently (8.7%), probably due to the proximity of EN1 to this facility (EN1 was practically installed at the fence line of P4).

EN2 detected the presence of odours from the monitored plants for 11.6% of the time. Even though the nearest facility to this first receptor is P4 (see Figure 1), in this case, the majority of the odour detections (8.1%) were attributed to the facility P5, i.e. the aluminium foundry. This indicates that the odour emissions relevant to P5 are far more significant that those due to P4, proving therefore the aluminium foundry to be the major contributor to the presence of an olfactory nuisance on the monitored zone.

Finally, the percentage of odour detections registered by EN3 at the second receptor turned out to be extremely low (1.6%) and is therefore negligible.

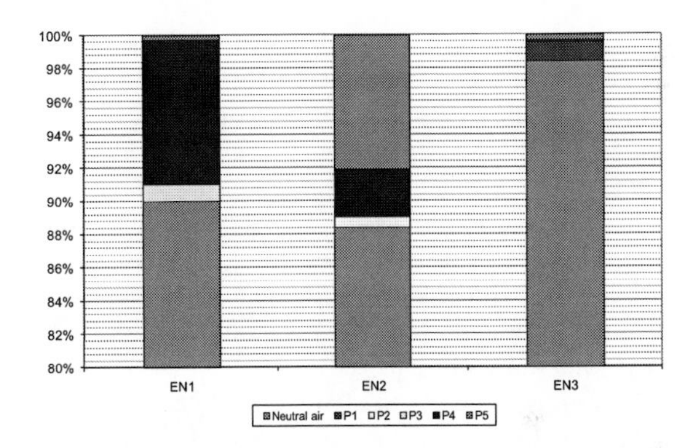

FIGURE 3. Odour relative recognition frequencies in the three monitoring positions.

TABLE 1. Odour relative recognition frequencies in the three monitoring positions

	Neutral air	P1	P2	P3	P4	P5
EN1	90.0%	0.0%	0.0%	1.0%	8.7%	0.3%
EN2	88.4%	0.0%	0.6%	0.1%	2.8%	8.1%
EN3	98.4%	1.2%	0.0%	0.1%	0.0%	0.4%

Evaluation Of Results

For a more accurate evaluation of the olfactory impact relevant to the monitored industrial zone, different data have to be compared. For this purpose, Table 2 shows the olfactory classes (i.e. the odour provenance) detected by EN2, installed at the first receptor, and the data of wind speed and wind direction in correspondence of the odour episodes signalled by the people living at the receptor.

It is possible to observe a good correspondence between these data.

Given the lack of specific regulation in this sector, in order to evaluate the acceptability of the odour impact on the monitored area it is possible to refer to the German guideline "GIRL – Geruchsimmisions-Richtlinie" on odour immissions, even if this guideline is generally applied to another odour investigation technique called "field inspection" [4]. This guideline fixes the limit of yearly "odour hours" that can be perceived at receptors to 10% for residential and mixed areas, while this limit is increased to 15% for industrial or agricultural areas [5].

Considering the above mentioned limits, the percentage of odour detections registered by EN1 and EN3 can be evaluated as acceptable, while the continuous odour monitoring at the first receptor by EN2 highlighted the presence of a critical situation in this specific monitoring position.

TABLE 2. Comparison between odour perceptions at receptor, olfactory classes recognized by EN2 and meteorological data (wind speed and direction)

No. of odour perception	Date	Hour	Intensity	Olf. class recognized by EN2	Wind provenance direction	Wind speed (m/s)
1	13/06/2008	20.00 - 23.00	4	P5 / P4	93 - 28	2.2 - 4.1
2	19/06/2008	11.00 - 12.00	3	P5	119	1.8
3	21/06/2008	9.00 - 10.30	2	P5 / P4	72	1.6 - 1.7
4	22/06/2008	7.00 - 9.00	3	P4	60 - 96	1.1 - 2.1
5	02/07/2008	21.00 - 24.00	4	Neutral air	181 - 300	0.5 - 0.6
6	03/07/2008	11.00 - 14.00	4	P4 / P5	150 - 185	1.2 - 1.7

Conclusions

The results of the study demonstrate how specific electronic noses developed for environmental applications, once suitably trained, can successfully be applied for the continuous monitoring of odours and their identification in presence of multiple sources.

In this specific case, the continuous odour monitoring at the first receptor, i.e. the nearest dwelling to the industrial zone, resulted in the detection of odours from the industrial zone for 11.6% of the total monitoring period.

Moreover, the qualitative classification of the electronic nose installed at this receptor (EN2) enabled to identify the major source of odour nuisance in this area, which turned out to be the aluminium foundry (P4).

REFERENCES

1. S. Sironi, L. Capelli, P. Céntola, R. Del Rosso, *Sens. Actuators B Chem.* **124**, 336-346 (2007).
2. M. Falasconi, M. Pardo, G. Sberveglieri, I. Riccò, A. Bresciani, *Sens. Actuators B Chem.* **110**, 73-80 (2005).
3. EN 13725:2003, *Air quality - Determination of odour concentration by dynamic olfactometry*, Comité Européen de Normalisation, Brussels, Belgium.
4. L. Capelli, S. Sironi, R. Del Rosso, P. Céntola, M. Il Grande, *Atm. Env.* **42**, 7050-7058 (2008).
5. LAI (Länderausschuss für Immissionsschutz), 1998. *Geruchsimmissions - Richtlinie (GIRL - Odour Emissions Guidelines)*, Ministry of Environment, Environmental Planning, and Agriculture of the State Nordrhein-Westfalen, Germany.

Evaluation of Fish Spoilage by Means of a Single Metal Oxide Sensor under Temperature Modulation

A. Perera[a], A. Pardo[a], D. Barrettino[c], A. Hierlermann[c], S. Marco[b]

[a]*Sists. d'Instrumentació i Comunicacions SIC, Universitat de Barcelona, Martí i Franquès 1, 08028-BCN, Spain*
[b]*Artificial Olfaction group, Inst. for Bioengineering of Catalonia (IBEC), Baldiri i Rexach 13, 08028-BCN, Spain*
[c]*ETH Zurich, Physical Electronics Laboratory, Wolfgang-Pauli-Strasse 16, 8093 Zurich, Switzerland*

Abstract. In this paper the feasibility of using metal oxide gas sensor technology for evaluating spoilage process for sea bream (Sparus Aurata) is explored. It is shown that a single sensor under temperature modulation is able to find a correlation with the fish spoilage process

Keywords: Gas Sensors, Electronic Nose, spoilage process, temperature modulation
PACS: 07.07.Df

INTRODUCTION

The most standard and accepted way of determining fish quality is by means of sensory analysis. Like any human based assessment, this implies a number of issues ranging from the impossibility of automated measurements and lack of objectiveness to poor reproducibility among others. It is therefore desirable to develop alternative freshness evaluation methods providing better features in addition than objective judgments.

Gas sensors based technology offers an interesting option for fish freshness evaluation with a low cost and relatively fast evaluation when compared to other alternatives. Additionally, an interesting option for expanding the amount of information from metal oxide gas sensor is based in the modification of the working temperature of the sensing material [1].

CMOSSens sensor is a prototype of a metal oxide gas sensor designed by the ETH (Zurich, Switzerland) and developed under the European project by the same name IST-1999-10579. It integrates CMOS circuitry with a metal oxide sensor, capable of applying fast temperature modulation profiles to the sensing material. The application of this sensor to food quality evaluation is an aim of the mentioned project.

In this paper, we measure the fish spoilage process by means of a single CMOSSens metal oxide gas sensor, submitted to a given temperature modulation profile. Before the analysis with CMOSSens sensor, the use of an array of commercial gas sensor for fish spoilage measurement is presented. From both analyses, it will be shown that a single sensor working at different temperatures and a commercial gas sensor array are able to respond to the tissue spoilage, with good performance.

A number of features are characteristic of the fish spoilage. Its general aspect presents dull pigmentation with absence of shines or reflections; the body presents a soft consistency and is perceived as flaccid, a slight pressure by the finger leaves a mark, etc [2].

However, one of the main characteristic features of the spoilage process is the odour emitted by the fish. Fresh samples present a fresh, light, pleasant water weeds characteristic odour, while spoiled fish have unpleasant, acrid, acid, putrid, ammonia-like intense odour.

As the off-odours emitted by the fish are related to the fish spoilage process, off-odour analysis is considered as potential method for fish freshness assessment. The characteristic odour from the process will be related to the biochemical changes occurring in the fish.

The analysis of the fish spoilage process by head-space gas chromatographic technique shows that the decomposition of the fish generates sulphur and nitrogen compounds. They are mostly responsible of the bad fish smell perceived by the human nose and their analysis permits an estimation of freshness [3].

Apart from gas-chromatography (GC), there exist a variety of methods for measure fish spoilage. Sensory evaluation is the preferred and most reliable method. However, GC methods and sensory method are both expensive and slow. Other faster methods as physical-

CP1137, *Olfaction and Electronic Nose: Proceedings of the 13th International Symposium,* edited by M. Pardo and G. Sberveglieri

electrical measurements or automatic texture measurement machines show high variability and not enough performance.

Gas sensor array approach to this problem aims the characterization of chemical mixture patterns instead of the detection of specific compounds. Several authors have published, in recent years, papers devoted to fish freshness evaluation [4,5,6]. Among the literature reviewed, only one [7] used a metal oxide sensor array for the freshness evaluation of capelin.

EXPERIMENTAL

We have focused our attention on the freshness assessment of cultured sea bream (Sparus Aurata). The increase on the demand of this product and the presence of important farming industries in Europe motivates the research on new methods of freshness assessment for this particular fish specie.

For this study, sea bream samples were cultured in a fish farm located in the Mediterranean (CRIPE S.A., Ametlla de Mar, Spain). Fish samples were slaughtered by hypothermia and dispatched in a polyurethane box with ice to Universitat de Barcelona. Immediately after reception, fish were cleaned, filleted, stored in six 500ml Pyrex bottles and connected to the measurement system. Bottles with the samples were immediately stored in a fridge and the experiment data acquisition proceeded without delay.

For the results presented, two different experimental setups have been used. The use of one given experimental setup depends on whether the measurement was done with the CMOSSens sensor or with the isothermal sensor array

For the isothermal sensors test a PC-controlled fluidic system has been designed together with an automated and temporized acquisition system.

Sensor array is built with a set of six different commercial MOX gas sensors. Three sensors from FIS (Japan) (SP-31 (general purpose sensor), SP-32 (alcohol) and SP-MW1 (humidity)), and three from Figaro (Japan) (TGS-882 (organic solvents), TGS-880 (alcohol and other volatiles) and TGS-823 (Alcohol, toluene, xylene, and other solvent vapours)). All sensors were powered at 5V and conditioned by means of a half-bridge. Also a temperature sensor AD590JH was used to monitor the temperature inside the chamber.

The sensor chamber consists of a custom-designed Teflon cover. This cover encloses the sensors just over the printed circuit board. The volume of the chamber was of 210ml.

Signal acquisition and valve control was implemented by a micro-controller based acquisition system. This micro-controller (ADuC812) was programmed for performing periodic readouts of the sensor array (fs =1s) and saving the results to personal computer via a RS-232 interface. A personal computer controlled the acquisition system by means of a software in Labview (National Instruments, Austin, USA).

The structure of the system allowed vessel sampling under a controlled period of time. The protocol of the measurement was programmed to repeat groups of cycles of six hours. On each group all the bottles were sequentially addressed.

The sampling cycle was 30 minutes long: 20 minutes for sampling clean air and 10 minutes for sampling air from the bottle. This cycle is repeated for all the bottles until the whole group is sampled. The flow through the sensor chamber during the complete experiment has been 0.5 l/m. The complete experiment took 144 hours.

The experimental setup for isothermal sensor array was modified in order to allow arbitrary temperature modulation profiles to be applied to the CMOSSens sensor. This metal oxide sensor offers limited power consumption, on-board signal acquisition, a digital control for the temperature of the heater and on-board tests [8]. The sensor achieves a maximum temperature of 400ºC at a 5.5V supply voltage. An ipNose instrument [9] was adapted to interface CMOSSens protocol and setup to take control of the scheduling of the experiment. The system's software controlled the application of the temperature profiles and the acquisition of data from the CMOSSens sensor.

The internal volume of the sensor chamber is of 0.35ml approximately and the flow through the sensor chamber during the complete experiment was reduced to 0.3l/m.

In this case the CMOSSens device was located outside the refrigerator and connected to the bottles by Teflon tubing. There have been six bottles/containers and an additional channel sampling air from the fridge. The different channels were selected using the electro-valve manifold already present in ipNose.

The measurement cycles were substantially shorter as the sensor chamber was smaller. Each cycle consisted of 4 minutes air intake from the bottles. The purge of the system was performed with 28 min of clean air sampling. Cycling all bottles lasted for 224 minutes (3h 44m). The ordering of the intakes from bottles/containers was randomized.

A temperature modulation waveform was used for sensor excitation in all CMOSSens experimental stages. The temperature waveforms present a periodicity of 16s. Because the sampling frequency is 2 Hz, this feature vector contains 32 points.

RESULTS

In this section we present an analysis of the spoilage process.

For isothermal experiment, each channel (bottle) was sampled 28 times at intervals of 6 hours. Feature matrix was computed by using differential measurement of each sensor i, $x_i = \rho_i^{max} - \rho_i^{ini}$, where $\rho_i^{max,ini}$ represents the maximum and initial conductance of sensor i respectively.

The normalization applied to the feature matrix will play an important role in the way we extract information from the sensor array. We have applied a normalization based on scaling all measurements from the same sensor to have unit variance and zero mean.

The corresponding principal component analysis (PCA) computed by using the resulting features from normalization is shown in the figure 1. In this case features are normalized to have zero mean and unit variance. This PCA is computed with the bottle number two. It can be observed that certain S shape path is traced by the features as the sample evolves. The model stability and the repeatability of the spoilage process can be explored by projecting in the same PCA subspace the features obtained from channels one and three.

If we project the other two bottles using this model we obtain very similar results. This suggests that the position along the path in the PCA plot gives a measure of the spoilage degree.

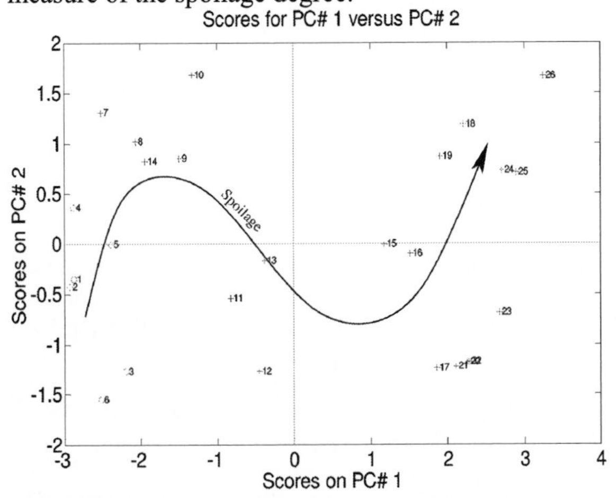

FIGURE 1. Pattern-way autoscaling. Absolute Features (x6h)

In the CMOSSens experiment, we aim the use of a single sensor under temperature modulation instead of a sensor array.

With an identical handling, five fillets of Sea Bream were held in five glass bottles and stored in the fridge during 17 days at a temperature of 5±2°C.

IpNose was programmed to execute a total number of 1524 continuous acquisitions among the length of the experiment. Each acquisition applied a 4 minutes sampling of each bottle followed by a purge cycle of 16 minutes. All six channels (bottle) were analyzed 254 times. The head generation time was of two hours. For each acquisition, the last waveform corresponding to the last cycle of temperature modulation within the last 4 sampling minutes was taken as a feature vector.

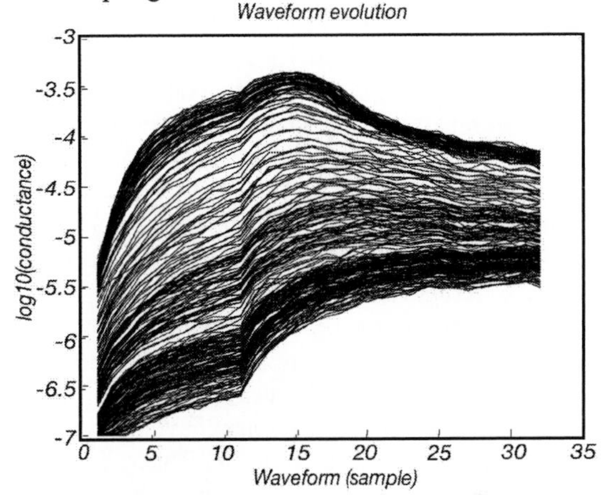

FIGURE 2. Raw evolution of signals

It can be seen that there is a continuous deformation of the waveform pattern suggesting that there is a maximum on the sensibility to spoilage volatiles. The change in the shape of the waveform has a strong chemical meaning as the metal oxide sensing layer presents different sensitivities to different compounds at different temperatures. The maximum response of the sensor is smoothly displaced from the end of the waveform (sample 32, maximum temperature) to a point located in the temperature ramp section.

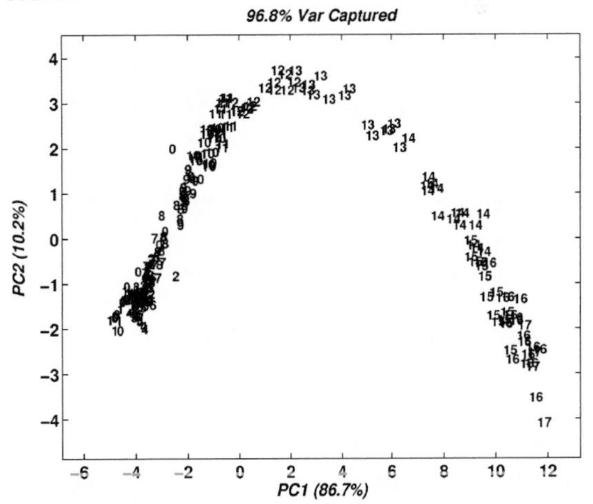

FIGURE 3. PCA scores of the evolution of filleted fish (normalised). Number indicates days since capture.

The evolution in the PCA plot projection of dataset mean centered with unit variance is shown in figure 3

It can be observed that while at the beginning of the process the signals are basically dominated by perturbations (probably temperature and humidity changes within the fridge), from day five the signals depart clearly from the initial point, indicating the spoilage of the fish. While the main indication of the change is associated to a change in the intensity of the odour (1st principal component), there is also a change in the nature of the odour as it is revealed in the evolution of the 2nd principal component. With slight differences regarding the shape and speed, several bottles show the same behaviour. Our interpretation is that a certain position in the PCA plane can be attributed to a certain state of degradation during the spoilage. However, this state is achieved at different times for different fillets.

From the present results it is possible to try to predict the time since capture for a particular fish. A PLS model has been built by using data from the 1rst channel. Half dataset (112 samples) have been used in building the model and the other half have been used for validation. A number of 4 latent variables were found optimum for describing the evolution. The correspondence between the real time and the predicted time in validation is shown in figure 4. The RMS error in prediction accounts for 1.5 days.

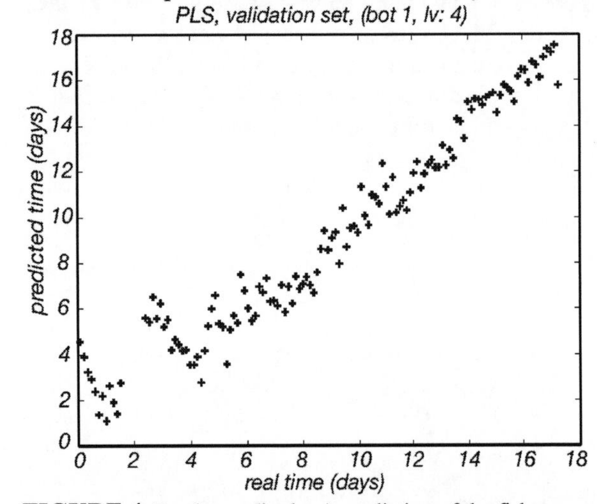

FIGURE 4. Freshness (in days) prediction of the fish.

CONCLUSIONS

In this work we have demonstrated that with a single sensor and the appropriate signal processing it is possible to find a correlation with the spoilage process of the fish samples. We have validated this approach with a commercial sensor array in isothermal mode and with a single sensor modulated in temperature.

The single sensor modulated in temperature was able to predict the storage day with an error of 1.5 days. The shape of the waveform due to the temperature modulation showed similar correlation than results found with the commercial sensor array, yet the head regeneration time was half the regeneration time with the commercial sensor array.

Further work should consider other species and check the validity of the models with samples coming not from fish farms but from open sea.

ACKNOWLEDGMENTS

The authors wish to acknowledge Sr. Guillermo Bores (CRIPE S.A.) for their kind collaboration providing the fresh samples. This work has been performed under funding by the IST-1999-10579, TIC1998-0978-C03-03 and DPI-2001-3213-C02-0.

REFERENCES

1. P. K. Clifford, et al, "Characteristics of semiconductor gas sensors II. transient response to temperature change", Sensors and Actuators 3 (1983) 255–281,
2. D. D. Johnson, et al, "Effects of percentage Brahman and Angus breeding, age-season of feeding and slaughter end point on meat palatability and muscle characteristics", J. Anim Sci. 68 (7) (1990) 1980–1986.
3. P. M. Schweizer-Berberich, et al, "Characterisation of food freshness with sensor arrays", Sensors and Actuators B, Chem. 18 (1994) 282–290.
4. C. DiNatale et al, "Comparison and integration of different electronic noses for freshness evaluation in cod-fish fillets", Sensors and Actuators B, Chemical 77 (2001) 572–578.
5. D. A. Luzuriaga, et al, "Electronic nose odor evaluation of salmon fillets at different temperatures", International Symposium on Olfaction and Electronic Nose, 1999, pp. 177–184.
6. D. Newman, et al, "Odor and microbiological evaluation of raw tuna: Correlation of sensory and electronic nose data", International Symposium on Olfaction and Electronic Nose, 1999, pp. 170–176.
7. R. Ólafsson, et al, " Monitoring of fish freshness using tin oxide sensors",in: J. W. Gardner, P. N. Bartlett (Eds.), Sensors and Sensory Systems for an Electronic Nose, Kluwer Academic Publ., 1992, pp. 257–272.
8. 10. M. Graf, et al, "Monolithic metal-oxide microsensor system in industrial cmos technology micro electro mechanical systems", MEMS-03 Kyoto. IEEE The Sixteenth Annual Int. Conference on, 2003, pp. 303 –306.
9. A. Perera, et al, "A portable electronic nose based on embedded PC technology and GNU/Linux: hardware, software and applications", IEEE Sensors Journal 2 (3) (2002) 235–246.

SAW sensors for chemical detection of ultra trace explosive compounds

A. Bardet, M. Guillemot, A. Wuillaume, C. Barthet, P. Montméat, P. Prené

A. Bardet, M. Guillemot, A. Wuillaume, C. Barthet, P. Montméat, P. Prené
CEA Le Ripault
Laboratoire Synthèse et Formulation
BP 16
F-37260 Monts France
Email : pierre.montmeat@cea.fr

Abstract. This paper deals with the use of SAW gas sensors for the detection of explosives vapors. The developped sensor exhibits a large sensitivity towards DNT, TNT and EGDN.

Keywords: gas sensors, SAW, explosives.
PACS: 80

INTRODUCTION

Chemical sensors for the explosives detection have attracted increasing attention over the past 10 years. There is a demand for means of quick and reliable detection of these substances particularly to anticipate terrorist attacks and for landmine detection. Nitroaromatics and nitric esters are among ones of the most common explosives.

Surface acoustic wave devices (SAW) are interesting because of their high sensitivity to change physical and chemical properties at or near the transducer system surface. Chemical gas sensors using SAW devices are generally composed by two delay lines. The SAW delay lines are used as the sensing channel coated by a sensitive film and the reference one without any film on it [1].

In this paper, we focus on the detection of vapors of trinitrotoluene (TNT), 2,4 dinitrtoluene (DNT) and ethylene glycol dinitrate (EGDN). We use SAW devices coated with a porous polymer dedicated to the detection of explosive as sensors [2].

EXPERIMENTAL AND METHODS

The SAW device operates at 100 MHz.

Sensitive thin films are elaborated by spray coating. The coating induces a 100 kHz frequency decrease.

TNT, DNT or EGDN vapors are generated at room temperature with a specific testing bench. Bulbing is used to generate explosive vapors. 1 g of each explosive is used, with a flow of 20 L/h. by the way of mass flow controllers and valves, the delivrated vapor can be diluted from 2 to 10 times. The resulting concentration is checked by first bulbing the explosive vapor in a reactor full up with CH_3CN and then analyzing the explosive solution by spectroscopy.

The evaluation of the sensors performances is performed under dry air.

The detection experiments consist in exposing the sensor to explosive at a given concentration for DNT for 10 min duration at room temperature.

CP1137, *Olfaction and Electronic Nose: Proceedings of the 13th International Symposium*, edited by M. Pardo and G. Sberveglieri

RESULTS

The generated concentrations are presented in Table 1.

TABLE 1. Delivrated vapors

Explosive	Higher concentration
TNT	5 ppb
DNT	140 ppb
EGDN	20 ppm

The values are in agreement with the typical values of the vapor pressure of the common explosives compounds [3].

The response of a SAW sensor when exposed to various concentrations of DNT is presented in Fig. 1.

The signal is large for each concentration. As expected the material has a large affinity for explosive vapors. The evolution of the response as a function of the concentration is plotted in Fig 2. The limit of the sensor response can be extrapolated: a response of 10 Hz leads to a concentration of 1 ppb DNT.

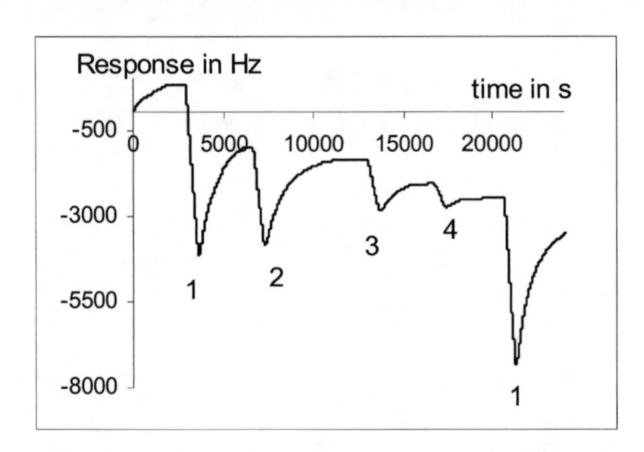

FIGURE 1. Response of the SAW to DNT (1): 140 ppb, (2): 70 ppb, (3): 28 ppb and (4): 14 ppb

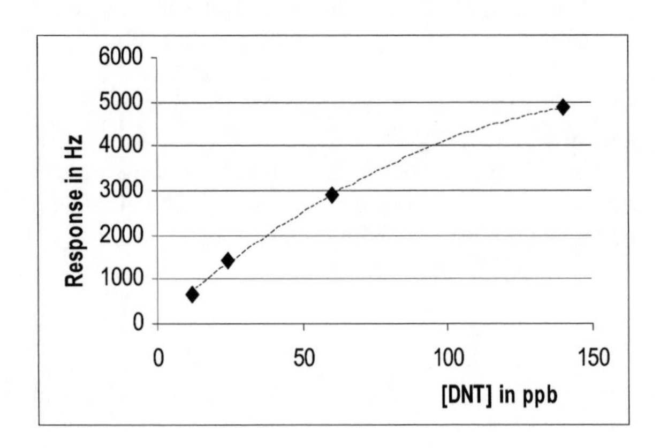

FIGURE 2. Effect of the concentration of DNT on the SAW response

The effects of each explosive are plotted in Fig. 3. Here again, each vapor leads to a significant signal. The largest variation is observed for DNT (5000 Hz) whereas the lowest one is for TNT (300 Hz). EGDN leads to an intermediate value: 1700 Hz. It is also interesting to notify that the reversibility and the response time depend on the target: for both nitroaromatics those characteristics are poor, they are excellent for EGDN.

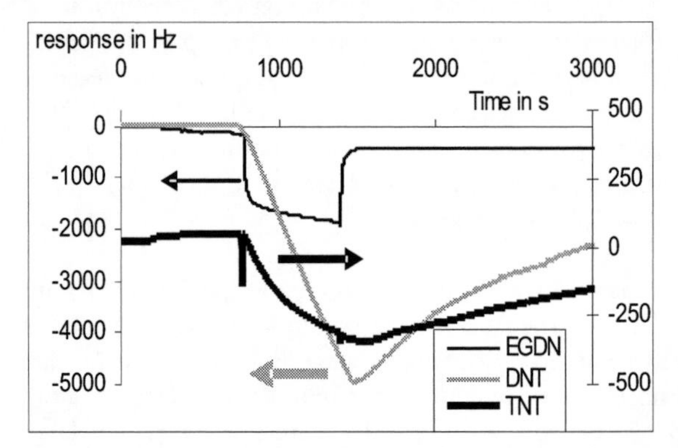

FIGURE 3, Responses of the SAW to EGDN (20 ppb), DNT (140 ppb) and TNT (5 ppb)

CONCLUSION

In this article we have demonstrated the interest of SAW sensors for the detection of ultra traces of explosive vapors. The sensor exhibits a large sensitivity towards DNT, TNT and EGDN.

To compare the effect of the vapors in terms of sensitivity needs much more experiments : we have to plot the evolution of the response as a function of the concentration for TNT and EGDN.

We also investigate the selectivity, the response time and the stability of the sensor.

REFERENCES

1. H. Wohltjen, *Analytical Chem.*, Vol 51, pp. 1458, 19
2. 2. J. S. Yang, *J. Am. Chem. Soc.*, 120, 1998
3. 3. J. Nakamura, *8th ISADE*, Ottawa, Canada, June 6-10, 2004

Mineral Water Taste Attributes Evaluated By Sensory Panel And Electronic Tongue

Zoltan Kovacs[a], Laszlo Sipos[b], David B. Kantor[a], Zoltan Kokai[b], Andras Fekete[a]

Corvinus University of Budapest, Faculty of Food Sciences, [a]Department of Physics and Control, [b]Sensory Laboratory H-1118, Somloi ut 14-16., Budapest, Hungary

Abstract. The objective of the current research was to determine relationships between sensory evaluation and measurement results obtained by electronic tongue for mineral waters. Furthermore, the purpose was to predict the sensory characteristics of the mineral waters measured by the electronic tongue and to determine taste differences that cannot be detected by the sensory evaluation. Two mineral waters were definitely different from the others according to the sensory attributes based on profile analysis. With the electronic tongue measurements the PCA and CDA analysis were found to be able to discriminate mineral waters having chemical composition similar to each other. Very good correlation was found between the sensory attributes and the electronic tongue measurements. However, the results of the measurements performed with the electronic tongue showed a more accurate discrimination of the different mineral waters than the sensory evaluation.

Keywords: Electronic tongue, Sensory panel, Profile analysis, Mineral water, Taste attributes.

PACS: 82.47.Rs

INTRODUCTION

In several regions of Europe the quality of the tap water considerably changed because of the different pollutants. The quality is usually suitable for human consumption but the sensory properties are not always acceptable. Thus, the role of mineral water products has been changed. The consumption is increasing and the sensory properties have a more important role in selecting the appropriate mineral water.

The sensory evaluation of tap waters and mineral waters is an emerging research field and only a limited number of publications deal with this topic. The first results concerning tap water samples were published in the USA in the middle of the 1980's [1, 2]. Flavor profile analyses have been used to assist in detecting, controlling, and understanding off-flavors in drinking water. A flavor profile panel has been selected and trained for the method. The classical analytical methods provide reliable results concerning chemical ingredients. However, some chemicals cannot be detected by the conventional analytical methods [3].

The development of artificial sensor techniques resulted in a rapid test method of the sensory attributes. Electronic tongue system was developed and it seemed to be suitable for the estimation of the qualitative and quantitative properties of mineral waters [4, 5]. Nowadays, scientific papers on liquid food measurements and the evaluation of sensory properties are of great importance [6, 7, 8].

The objective of the current research was to determine relationships between the sensory evaluation and measurement results obtained by electronic tongue for mineral waters. Furthermore, the purpose was to predict the sensory characteristics of the mineral waters measured by the electronic tongue and to determine taste differences that cannot be detected by the sensory evaluation.

EXPERIMENTAL AND METHODS

Materials

Seven different still mineral waters (without any additive) were tested. A French premium brand which contains low amount of minerals (named: *A*) and six Hungarian mineral waters from different regions of Hungary (named: *B*, *C*, *D*, *E*, *F* and *G*). Two of them have high amount of minerals (named: *B* and *G*). The *B* sample contains carbon dioxide from natural source. Two water samples contain very low amount of sodium (named: *A* and *E*). Samples were measured directly from the bottle without any pre-treatment.

CP1137, *Olfaction and Electronic Nose: Proceedings of the 13th International Symposium*, edited by M. Pardo and G. Sberveglieri

TABLE 1. Short name of the mineral waters and the concentration (in mg/L) of certain anions, cations and Total Dissolved Solids (TDS) in the tested waters.

Name	TDS	$[CO_3H^-]$	$[SO_4^{2-}]$	$[Cl^-]$	$[Na^+]$	$[Ca^{2+}]$	$[Mg^{2+}]$
A	489.0	360.0	12.6	6.8	6.5	80.0	26.0
B	1803.0	1098.0	40.0	69.0	196.0	193.0	41.8
C	1042.0	540.0	126.0	81.0	72.0	134.0	40.0
D	627.0	327.0	108.0	24.0	18.0	82.0	41.0
E	610.0	451.0	< 10	7.0	6.9	83.0	39.9
F	520.0	400.0	-	3.0	21.0	63.0	26.0
G	1600.0	1110.0	38.0	17.0	37.0	280.0	57.0

Instrumentation

An αAstree II (Alpha-MOS, Toulouse, France) potentiometric electronic tongue connected with LS 16 autosampler unit was used to measure the samples. The electronic tongue had seven ISFET (ion sensitive field effect transistor) based potentiometric chemical sensors for food application (ZZ3401, BA3401, BB3401, CA3401, GA3401, HA3401 and JB3401). Due to the polymer membrane coating sensors were sensitive on the organic acids, salts, mono- and disaccharides being in the mineral waters. Naturally, the sensors have cross-sensitivity for tasting chemicals which are typically found in food stuffs and beverages. Sensor potential was measured versus a standard Ag/AgCl 3M KCl reference electrode (Metrohm AG). Measurement time was 180 s and the sensor cleaning time was 30 s. Each sample was measured nine times. Sensors were cleaned with distilled water between the measurements till stable potential was obtained. Measurements were made under constant room temperature (23–24 °C). Sample volume was 100 mL. The measurement sequence of the different samples was randomly determined.

An intelligent pH meter was used for pH measurement and specific conductivity was measure by Seven Multi™ with Conductivity Expansion Unit for conductivity measurement.

Sensory analysis

Profile analysis is one of the most comprehensive and reliable procedures for taste evaluation [9, 10, 11]. Its main advantage is that it provides with comparison between products by completely describing the sensory attributes of the tested samples. This method is also recommended by the American Water Works Association [12]. The sensory tests were conducted in the Sensory Laboratory of Corvinus University of Budapest, Faculty of Food Science. The test room was well ventilated, protected from direct sunlight, provided with artificial light, so the circumstances for the evaluation were constant [13]. The sensory test was implemented by the means of computers, organized in a LAN structure. The data were collected and evaluated with the ProfiSens software developed by Technical University of Budapest and the Sensory Laboratory of Corvinus University of Budapest [14].

The profile analysis was performed in two steps with multiple overlapping of the different mineral waters because of the relatively large sample number. Samples A, B, E, F and G were analyzed in the first profile analysis test and samples A, D, E, F and G in the second test. The applied sensory attributes determined by consensus groups are shown in Table 2.

TABLE 2. Sensory attributes determined by the consensus group.

Attributes	Minimum	Maximum
Carbonation	weak	strong
Bubble quantity	few	many
Metallic taste	weak	intense
Sweet taste	weak	intense
Salty taste	weak	intense
Bitter taste	weak	intense
Aftertaste	weak	intense
Mineralization	low	high
Softness in Mouthfeel	not soft	soft

Besides the profile analysis the triangle test was applied in research [15]. This latter one is one of the most sensitive tests. In this case the differences between two samples were determined. The question was whether a definite difference exists between the samples at all. One can detect very small differences with the triangle test. Therefore, the difference was analyzed between the mineral waters by the above mentioned method, as well. At a definite time two different mineral water samples were tasted and tested.

RESULTS AND DISCUSSION

The evaluation of the profile analysis was performed by canonical discriminant analysis. The result of the first test is shown in the Figure 1. Two mineral waters were definitely different from the others according to the sensory attributes based profile analysis. Discriminant function 1 shows the differences between the sample *B* and the other samples. This was caused by the natural bubble content of the mineral water. The high standard deviation of this group shows that the bubble content decreased during the tasting. This results show that the sample *G* is different from the other samples according to all of the sensory attributes.

During the second profile analysis test only the *G* sample was found to be significantly different from the other mineral water samples.

The results of the triangle tests demonstrated also definite differences in the case of samples *B* and *G*. However, sample *C* was also distinguished from the others in some tests.

The canonical discriminant analysis and principle component analysis were used for the evaluation of the results obtained by the electronic tongue. The typical parameters of water quality such as conductivity and pH were also measured with the tested waters. Conductivity, pH, PC1, PC2, Can1 and Can2 values for each water type are shown in Table 3.There was not found any clear correlation between parameters shown in Table 3. However, the PCA and CDA analysis are able to discriminate mineral waters having chemical composition similar to each other.

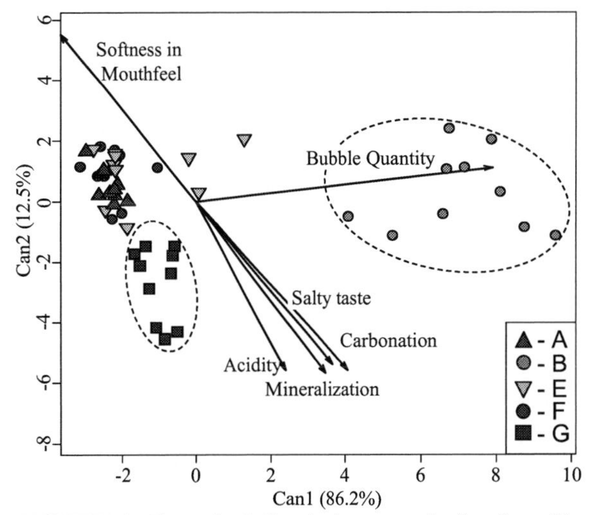

FIGURE 1. Canonical discriminant analysis of profile analysis by sensory panel

TABLE 3. Values of conductivity (cond., in μScm^{-1}), pH the principle component scores (PC1, PC2) and the canonical discriminant function scores (Can1, Can2) of results obtained by the electronic tongue.

Name	Cond.	pH	PC1	PC2	Can1	Can2
A	1082	7.43	-396.17	-26.76	25.95	0.95
B	3630	6.64	122.08	197.22	-30.65	2.72
C	2320	6.82	-53.54	94.82	0.72	-6.49
D	1350	6.98	-165.34	25.78	17.13	-10.36
E	1306	7.57	-363.02	-41.28	21.39	4.10
F	985	7.71	-428.97	-53.00	22.06	8.90
G	3070	6.39	248.58	197.29	-45.81	1.18

The canonical discriminant analysis was used for the evaluation of the results obtained by the electronic tongue (Figure 2). Discriminant function 1 shows the differences between the sample group B and G relative to the other samples. The latter ones are connected to CA, BA, GA, BB and ZZ sensors. The difference among the other samples was not distinguished by sensory evaluation. However this difference is distinguished by the electronic tongue and it is based on the discriminant function 2. This is mainly because of the results obtained by the JB sensor.

It is interesting to note that according to the profile analysis the sample B was definitely distinguished from the other samples. However, with the electronic tongue measurements the sample G was considerably different from the other samples. This is the effect of the carbon dioxide with the profile analysis. Since, the carbon dioxide was eliminated from the G sample for the measurements by the electronic tongue.

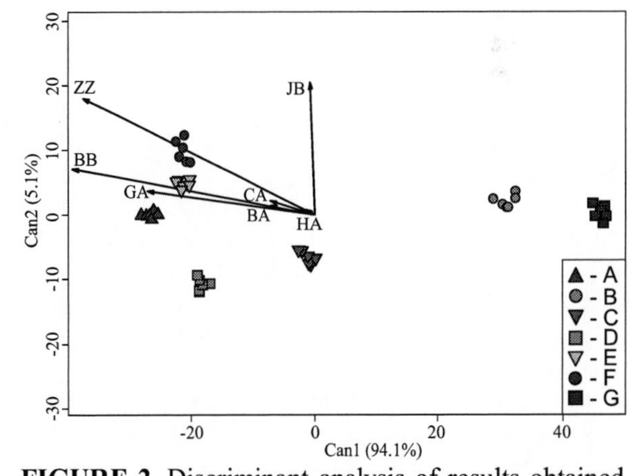

FIGURE 2. Discriminant analysis of results obtained by the electronic tongue.

On the basis of Figure 2 a seven dimensional space can be predicted with respect to the distances between group centers to show the threshold for the sensory evaluation. Therefore, the electronic tongue is able to distinguish taste differences being under the human sensory threshold.

Partial least square regression was used to find relationship and to predict the sensory attributes from the results obtained by the electronic tongue. The determination coefficients of the models with cross validation are shown in Table 4.

TABLE 4. Correlation between sensory evaluation and electronic tongue measurements (PLS-R).

Profile analysis No. 1.	r^2	**Profile analysis No. 2.**	r^2
Bubble quantity	0.604	Bubble quantity	0.670
Salty taste	0.974	Salty taste	0.995
Carbonation	0.965	Carbonation	0.993
Mineralization	0.958	Metallic taste	0.983
Acidity	0.903	Sweet taste	0.681
Softness in		Bitter taste	0.951
Mouthfeel	0.988	Aftertaste	0.977

Very good correlation was found between the sensory attributes and the electronic tongue measurement. It is interesting to note that relatively low determination coefficients (0.604 and 0.670) were found for the bubble quantity of both profile analysis. These results confirm that the disturbing character of carbon dioxide was eliminated with electronic tongue measurements.

CONCLUSIONS

Two mineral waters were definitely different from the others according to the sensory attributes based profile analysis.

With the electronic tongue measurements there was not found any clear correlation between typical parameters of water quality and principle components or canonical discriminant functions. However, the PCA and CDA analysis are able to discriminate mineral waters having chemical composition similar to each other.

On the basis of our experiences seven dimensional spaces can be predicted with respect to the distances between group centers to show the threshold for the sensory evaluation. Therefore, the electronic tongue is able to distinguish taste differences under the human sensory threshold.

Consequently very good correlation was found between the sensory attributes and the electronic tongue measurement. However, the results of the measurements performed with the electronic tongue showed a more accurate discrimination of the different mineral waters than the sensory evaluation.

ACKNOWLEDGMENTS

This work was supported by OTKA Foundation Grant Number M045745.

REFERENCES

1. S. W. Krasner, M.J. McGuire and V.B. Ferguson, *J. Am. Water Works Assn.* **77**, 34-40 (1985)
2. I.H. Suffet, B.M. Brady, J.H.M. Bartels, G. Burlingame, J. Mallevialle, and T. Yohe, *Water Sci. Technol.* **20**, 1-9 (1988)
3. R. Devesa, C. Fabrellas, R. Cardeñoso, L. Matia, F. Ventura and N. Salvatella, *Water Sci. Technol.* **49**, 145-151 (2004)
4. Legin, A. Rudnitskaya, Y. Vlasov, C DiNatale, E. Mazzone and A. D'Amico, *Electroanal.* **11**, 814-820 (1999)
5. R. Martínez-Máñez, J. Soto, E. Garcia-Breijo, L. Gilb, J. Ibáñez and E. Llobet, *Sens. Actuators, B* **104**, 302-307 (2005)
6. R. N. Bleibaum, H. Stone, T. Tan, S. Labreche, E. Saint-Martin and S. Isz, *Food Qual. Preference* **13**, 409-422 (2002)
7. D. B. Kantor, G. Hitka, A. Fekete and Cs. Balla, *Sens. Actuators, B* **131**, 43-47 (2008)
8. Dalmadi, D. B. Kántor, K. Wolz, K. Polyák-Fehér, K. Pásztor-Huszár, J. Farkas and A. Fekete, Prog. *Agric. Eng. Sci.* **3**, 47–66 (2007)
9. ISO 11035:1994 Sensory analysis - Identification and selection of descriptors for establishing a sensory profile by a multidimensional approach.
10. M. C. Meilgaard, G. V. Civille and B. T. Carr, *Sensory Evaluation Techniques*, Boca Raton: CRC Press, 1999, pp. 161-170.
11. L. Sipos, M. Ladányi and Z. Kókai, *Acta Alimentaria*, in press (2009)
12. American Water Works Association, *Manual: Flavor Profile Analysis: Screening and Training of Panelists*, Denver: AWWA, 1993
13. ISO 8589:2007 Sensory analysis - General guidance for the design of test rooms.
14. K. Kollár-Hunek, J. Heszberger, Z. Kókai, M. Láng-Lázi and E. Papp, *J. Chemom.* **22**, 218-226 (2008)
15. ISO 4120:2004 Sensory analysis - Methodology - Triangle test.

An Automated Electronic Tongue for In-Situ Quick Monitoring of Trace Heavy Metals in Water Environment

Wei Cai, Yi Li, Xiaoming Gao, Hongsun Guo, Huixin Zhao and Ping Wang*

Biosensor National Special Laboratory, Key Laboratory of Biomedical Engineering of National Education Ministry, Department of Biomedical Engineering Zhejiang University, Hangzhou 310027, China
** Tel.: +86-571-87952832. E-mail address: cnpwang@zju.edu.cn (Ping Wang).*

Abstract. An automated electronic tongue instrumentation has been developed for in-situ concentration determination of trace heavy metals in water environment. The electronic tongue contains two main parts. The sensor part consists of a silicon-based Hg-coated Au microelectrodes array (MEA) for the detection of Zn(II), Cd(II), Pb(II) and Cu(II) and a multiple light-addressable potentiometric sensor (MLAPS) for the detection of Fe(III) and Cr(VI). The control part employs pumps, valves and tubes to enable the pick-up and pretreatment of aqueous sample. The electronic tongue realized detection of the six metals mentioned above at part-per-billion (ppb) level without manual operation. This instrumentation will have wide application in quick monitoring and prediction the heavy metal pollution in lakes and oceans.

Keywords: Electronic Tongue, Trace Heavy Metals, MEA, MLAPS.
PACS: 82.47.Rs.

INTRODUCTION

Heavy metals, such as Zn(II), Cd(II), Pb(II), Cu(II), Fe(III) and Cr(VI) are contaminants that do a lot of harm to human beings even if they are present in extremely minute quantities. All forms of ecological systems are affected to varying extents by heavy metals[1]. Traditionally, water samples are collected and physically transported to laboratories for analysis, which leads to high cost and low efficiency. And the contamination of samples during the stages of storage and transmission and the long time delays associated with these procedures may result in unacceptable data. Thereby on-line monitoring is in great requirement [2]. Most of modern trace heavy metal analysis methods, such as atomic absorption spectrometry (AAS) and inductively coupled plasma atomic emission spectrometry (ICP-AES), are too expensive, power-inefficient and bulky for field work [3]. While electrochemical stripping voltammetry (SV) exhibits the merits of its fast multi-element capabilities, simple operation, high sensitivity, relative small volume and low cost of equipment, as well as its suitability for on-line, in-situ and automatic application in heavy metal detection [4].

Microelectrode array (MEA) has been widely applied to electrochemical heavy metals detection. MEA offers many advantageous features (e.g. high current density, high signal-to-noise ratio and independence from hydrodynamics) over conventional macroelectrodes [5]. It allows low cost in batch production, has good uniformity, demands only a small amount of sample solution, enables relatively rapid and sensitive analysis. For SV purposes, mercury is still the common sensing material that offers the widest electroactivity domain because of its large overpotential toward hydrogen. Thus, mercury thin films are electroplated on a conducting base, which consists of an array of microdisks, to form the Hg microelectrode array. Gold, platinum and carbon are among the materials that have been extensively used as the conducting base for MEA [6].

Light-addressable potentiometric sensor (LAPS) is a kind of novel sensor [7]. Vlasov and Legin et al. have developed a series of ion-selective chalcogenide glass sensors with outstanding chemical stability, high available sensitivity, durability, and potential reproducibility which are suitable for the detection of heavy metals [8]. The chalcogenide glass sensitive membranes are resistant to attack by aggressive and corrosive media and possess chemically stable surface characteristics. So these sensors can be applied for environment monitoring of wastewater, industrial control processes, and natural water, etc.

Combining MEA for the detection of heavy metals such as Zn(II), Cd(II), Pb(II) and Cu(II) with MLAPS for Fe(III) and Cr(VI), the electronic tongue instrumentation realized stand-alone real-time in-situ and fully-automatic measurements of trace heavy metals in water environment. The work represented good results of quantitative analysis of Zn(II), Cd(II), Pb(II), Cu(II), Fe(III) and Cr(VI) with sensitivity at ppb level.

CP1137, *Olfaction and Electronic Nose: Proceedings of the 13th International Symposium*, edited by M. Pardo and G. Sberveglieri

EXPERIMENTAL AND METHODS

Design of the Control Part

Design of the automated electronic tongue including the control part which contains several pumps, valves, tubes and etc. is shown in Fig. 1. There are two measurement cells (MEA cell and MLAPS cell) in the system. All the actions of the electronic tongue are controlled by a program in the computer. In the pretreatment, the aqueous sample is picked up by the pump through the tube and filtered by a 0.45μm percolation film so that some big molecules are removed. Then the treated sample runs through the valve into the measurement cell in which buffer solution is added. When pH of the sample is adjusted to the intended value, the first measurement begins. A second measurement with standard sample injected into the cell is needed in order to obtain the calibration curves of the heavy metals. After the two measurements, the sample is drained to the waste vessel and de-ionized water is pumped into the cell to do the cleaning so that the next loop is ready to begin. Both cells follow the course mentioned above.

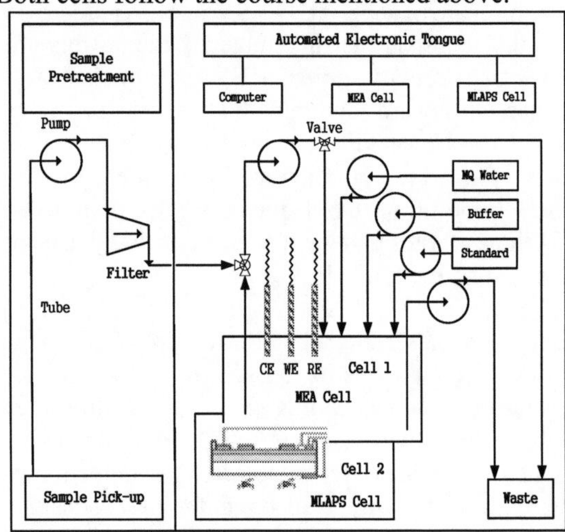

FIGURE 1. Scheme of the automated electronic tongue.

Design of the Sensor Part

MEA Cell

A micro-fabrication process was developed to produce the MEA. The fabrication procedure, as illustrated by a cross section of a completed device in Fig. 2, was begun by thermally oxidizing a 100 mm silicon wafer to a thickness of 1000 nm. Then a 100 nm Cr layer and a 250 nm Au layer were deposited on insulating layer of silicon dioxide by magnetron sputtering. After that, the conductor layer was protected and insulated by a 1000 nm thick layer of polyimide. A layer of photoresist was patterned acting as a mask, the unwanted polyimide was removed during the subsequent wet etching and the active Au surfaces were exposed. Once the etching of the polyimide had been completed, the photoresist mask was stripped away from the surface in the appropriate solvent. Thus AuMEA region was defined. Finally, the wafer was diced in 6 mm side chips, each chip containing an array of 30×30 Au microdisks of 10 μm diameter separated by 150 μm from each other. Chips were individually mounted, bonded and encapsulated on printed circuit board using epoxy resin. A Pt foil as the counter electrode (CE) and an Ag/AgCl foil as the reference electrode (RE) were attached on the other side of printed circuit board and also encapsulated using epoxy resin. After mercury deposition was carried out on the Au, the MEA was ready to detect heavy metals such as Zn(II), Cd(II), Pb(II) and Cu(II).

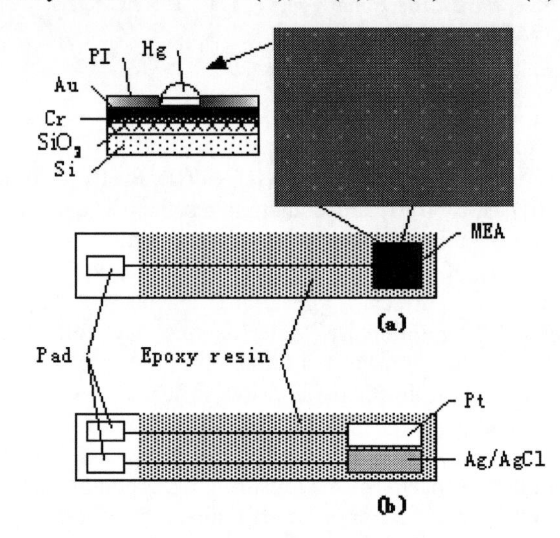

FIGURE 2. Structure of the sensor: (a) on one side is silicon-based Hg-coated Au microelectrodes array; (b) on the other side is Pt and Ag/AgCl electrodes.

MLAPS Cell

The LAPS chip was n-type silicon with a specific resistance of 15 Ω·cm. A thin gold film was prepared on the back of the chip as an ohmic contact. The front of the chip was covered with 100 nm SiO_2 layer on the silicon followed by 50 nm layer of Si_3N_4. The ion selective materials sensitive to Fe(III) and Cr(VI) were purchased from Analytic System Inc. of St. Petersburg University in Russian (Andrey Legin). These sensitive materials were set as targets in the preparation of thin film by pulsed laser deposition (PLD) technique. Two chalcogenide films respectively sensitive to Fe(III) and Cr(VI) were prepared in the same LAPS. Because of the poor adhesion between the Si_3N_4 and the gold, the intermediate layer of Cr with 4 nm thicknesses was used to improve it. The scheme of the MLAPS is shown in Fig. 3. Both chalcogenide glass thin-film sizes are 3 mm × 4 mm in area and 300 nm in thickness. Besides two thin-film fields, other parts on the MLAPS were covered with epoxide resin. The dimensions of the MLAPS chip are 20 mm× 10 mm. For LAPS, the potential response is equivalent to changing the direct current bias voltage in outside circuit. So different ion concentration can cause the drift of I-V curve and accordingly we can obtain the ion concentration in solution by calculating the drift voltage. In this way, the concentrations of Fe(III) and Cr(VI) were measured by using the MLAPS.

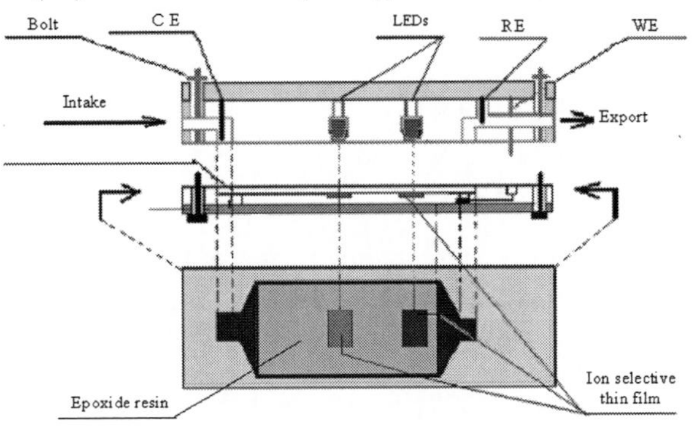

FIGURE 3. Scheme of the MLAPS.

Design of the experiment

All reagents were prepared from analytical reagent grade chemicals. De-ionized water (≥18 MΩ·cm) from a Millipore system was used for dilution and cleaning. The standard sample solutions were supplied by the Second Institute of Oceanography, China. The mixed muriatic solution of Zn(II), Cd(II), Pb(II) and Cu(II) was measured with electrochemical method of differential pulse anodic stripping voltammetry (DPASV) in the MEA cell. The solutions of Fe(III) and Cr(VI) were measured with current-voltage (I-V) scanning method in the MLAPS cell.

RESULTS

Detection of Zn(II), Cd(II), Pb(II), Cu(II) with MEA

Rather good sensitivity was achieved with a 150s deposition time in DPASV. Quantitative analysis was realized by means of standard addition method. The standard sample added was consisted of Zn(II), Cd(II), Pb(II) and Cu(II) whose concentrations were 80 μg/L, 3 μg/L, 3 μg/L and 10 μg/L. After four additions, voltammograms for Zn(II), Cd(II), Pb(II) and Cu(II) were shown in Fig. 4. The voltammograms contain zinc peaks at around -1.05 V, cadmium peaks at around -0.62 V, lead peaks at around -0.45 V and copper peaks at around -0.05 V. Measurements showed good results with their linear ranges separately in 10-600 μg/L, 1-100 μg/L, 1-200 μg/L and 2-300 μg/L. And their detection limits respectively were 3.4 μg/L, 0.2 μg/L, 0.5 μg/L and 1.8 μg/L.

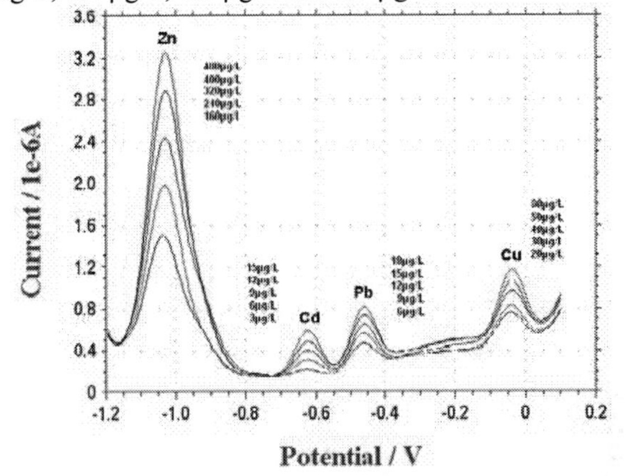

FIGURE 4. Voltammograms of Zn(II), Cd(II), Pb(II) and Cu(II).

Detection of Fe(III) and Cr(VI) with MLAPS

The I-V scanning curves of the MLAPS were measured in different aqueous solutions with different contents. The Fe-LAPS was measured in concentrations of Fe(III) ranging from 10^{-3} M to 10^{-5} M while the Cr-LAPS in concentrations of $K_2Cr_2O_7$ ranging from 10^{-3} M to 10^{-5} M. The horizontal voltage biases corresponding to the inflection point of both I-V curves were plotted versus the logarithm of ion concentration in Fig. 5 and Fig. 6, which were also the calibration curves of the MLAPS.

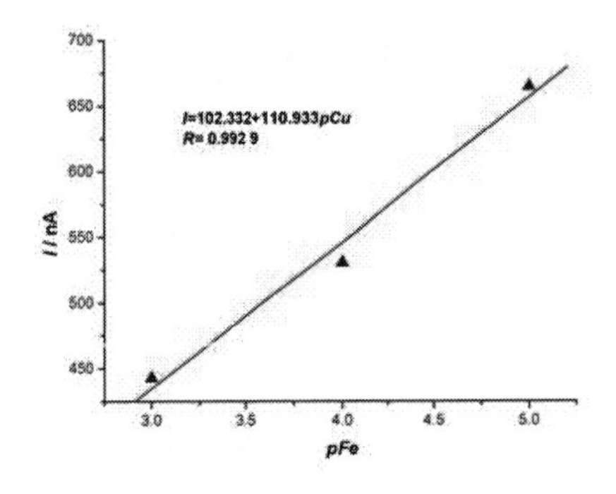

FIGURE 5. Calibration curves of the Fe-LAPS.

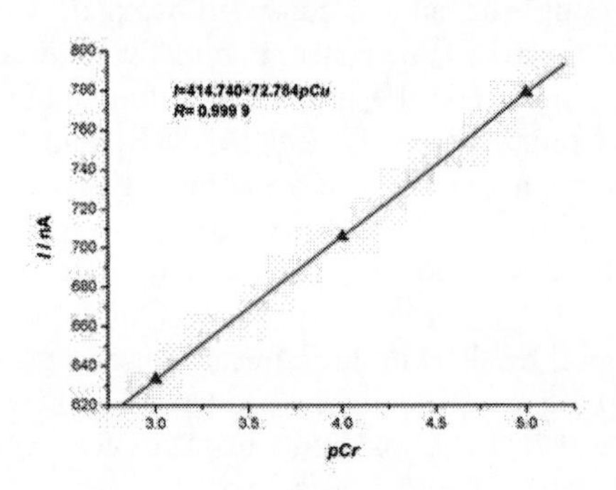

FIGURE 6. Calibration curves of the Cr-LAPS.

Prototype of the Instrumentation

The prototype of our automated electronic tongue instrumentation was given in Fig. 7. It contains two MEA cells and two MLAPS cells completed the measurements of different trace heavy metals respectively. The pumps and valves are installed in the nether box. The second cells of MEA and MLAPS are retained for the further development to measure more species of trace heavy metals.

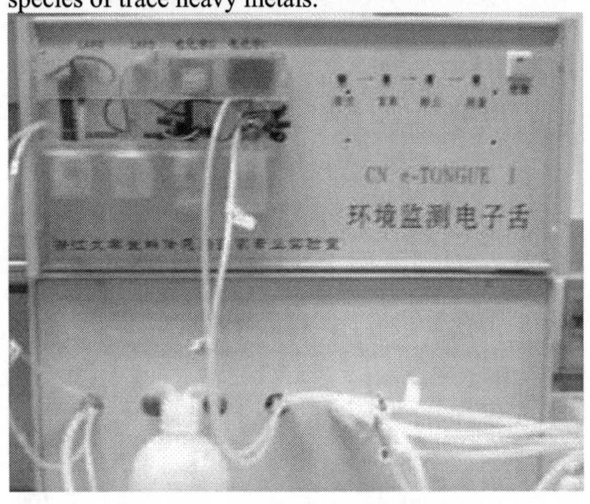

FIGURE 7. Prototype of the automated electronic tongue instrumentation.

DISCUSSION

We didn't measure all the six heavy metals in one cell, because there are still some difficulties to use MEA and MLAPS in the same solution. If they are integrated in one cell to detect, the accuracy of detection would decrease. The targets for PLD purchased from Analytic System Inc. of St. Petersburg University in Russian didn't cover all the six metals. We look forward to research on it and develop more thin films sensitive to heavy metals on the surface of MLAPS and realize the simultaneous detection of more species of heavy metals on-line.

CONCLUSIONS

We realized real-time and in-situ measurements of Zn(II), Cd(II), Pb(II), Cu(II), Fe(III) and Cr(VI) with technologies of MEA and MLAPS. The electronic tongue consisted of the control and sensor part has the virtue of portability, low cost and capability for fast, automatic and continuous determination of trace heavy metals. One measurement could be completed within 15 min with only 10 ml sample. The detection of Zn(II), Cd(II), Pb(II) and Cu(II) was completed with DPASV in the MEA cell while Fe(III) and Cr(VI) with I-V scanning in the MLAPS cell. In the future, we will try to overcome problems to improve the sensitivity, take other species of trace heavy metals such as Hg(II) into account and miniaturize the electrodes and the whole instrumentation.

ACKNOWLEDGMENTS

This work was Supported by grants from Ministry of Science and Technology of China (973 Program) (Grant No. 2009CB320303). The authors gratefully acknowledge the graduated Dr. Shaofang Zhou, Dr. Hong Men and Dr. Weijun Hu for their works.

REFERENCES

1. E.P. Achterberg, "Stripping voltammetry for the determination of trace metal speciation and in-situ measurements of trace metal distributions in marine waters", Analytica Chimica Acta, 1999(400), pp.381–397
2. H. Suzuki, "Microfabrication of chemical sensors and biosensors for environmental monitoring", *Materials Science and Engineering C*, 2000(12), pp. 55-61
3. D. Desmond, "An environmental monitoring system for trace metals using stripping voltammetry", *Sensors and Actuators B*, 1998 (48), pp. 409-414
4. K.Z. Brainina, "Stripping voltammetry in environmental and food analysis", *Fresenius J Anal Chem*, 2000 (368), pp. 307-325
5. D. Desmond, "An ASIC-based system for stripping voltammetric determination of trace metals", *Sensors and Actuators B*, 1996 (34), pp. 466-470
6. R.J. Reay, "Microfabricated electrochemical analysis system for heavy metal detection", *Sensors and Actuators B*, 1996 (34), pp. 450-455
7. B. Stein, "Spatially resolved monitoring of cellular metabolic activity with a semiconductor-based biosensor", *Biosensors and Bioelectronics*, 2003 (18), pp. 31-41
8. A.M. Rudnitskaya, "Detection of ultra-low activities of heavy metal ions by an array of potentiometric chemical sensors", *Microchim Acta*, 2008 (163), pp. 71-80

"Low-cost Electronic nose evaluated on Thai-herb of Northern-Thailand samples using multivariate analysis methods"

Paisarn Daungjak na ayudhaya[1,2], Arrak Klinbumrung[1], Krisanadej Jaroensutasinee[2], Sirapat Pratontep[3] and Teerakiat Kerdcharoen[4]

[1]*Major of Physics, School of Science and Technology, Naresuan University Phayao Campus 19 Moo 2 Maeka, Muang, Phayao, 56000, Thailand.*
[2]*Computational Science Program, School of Science, Walailak University, 222 Thaiburi, Thasala, Nakhonsithammarat, 80160, Thailand.*
[3]*National Nanotechnology Center, National Science and Technology Development Agency, Thailand Science Park, Pathumthani 12120, Thailand*
[4]*Center of Intelligent Materials and Systems (CiMS), Department of Physics, Faculty of Science, Mahidol University, Rama 6 Road, Bangkok 10400, Thailand*
Author's contact: paisarnd@nu.ac.th

Abstract. In case of species of natural and aromatic plant originated from the northern Thailand, sensory characteristics, especially odours, have unique identifiers of herbs. The instruments sensory analysis have performed by several of differential of sensing, so call 'electronic nose', to be a significantly and rapidly for chemometrics. The signal responses of the low cost electronic nose were evaluated by principal component analysis (PCA). The aims of this paper evaluated various of Thai-herbs grown in Northern of Thailand as data preprocessing tools of the Low-cost electronic nose (enNU-PYO1). The essential oil groups of Thai herbs such as Garlic, Lemongrass, Shallot (potato onion), Onion, *Zanthoxylum limonella* (Dennst.) Alston (Thai name is Makaen), and Kaffir lime leaf were compared volatilized from selected fresh herbs. Principal component analysis of the original sensor responses did clearly distinguish either all samples. In all cases more than 97% for cross-validated group were classified correctly. The results demonstrated that it was possible to develop in a model to construct a low-cost electronic nose to provide measurement of odoriferous herbs.

Keywords: Low cost electronic nose, principal component analysis, Thai-herbs.

INTRODUCTION

The protocol of electronic nose [1-4] comprised of gas sensor arrays and a pattern recognition tool is a flabbergasting instrument with neuroscience information. Since a metal oxide semiconductor for electronic nose can be conducted using low-cost, small size, and portable devices with a fast response, and a short time for investigated compared to traditional method, i.e. gas chromatography methods. This methods usefully to a wide various application in filed as food-beverage quantity control, and environmental monitoring. An electronic nose system is a sensor-based technology, which considers the headspace of volatiles and creates unique odour profile. These patterns of profiles are signature of the particular set of aromatic compounds. Electronic nose technologies have been investigated for discrimination of different quality of food-fruit products, were performed apples, oranges and other applications [4-17]. The aromatic of Thai herbs, a potential indicator of the physiological condition, can be used to develop consistence and reproducible to evaluate herbs quality from harvest to consumer. Several scientists have most used chemical laboratory (GC-MS, gas chromatography-mass spectrometry instrument) to identify volatile organic compounds.

The objectives of this study were: (1) investigate the effectiveness of electronic nose to classify the change of Thai herbs. (2) To study principal component analysis (PCA) techniques to obtain the enNU-PYO1 instrument could be able to recognized of species of Thai herbs.

CP1137, *Olfaction and Electronic Nose: Proceedings of the 13th International Symposium*, edited by M. Pardo and G. Sberveglieri

EXPERIMENTAL AND METHODS

2.1 Sample preparation

The material "Thai" herbs were obtained from natural product from the northern of Thailand. Makaen and all of Thai herbs were located only in the area of Phayao-Nan province. The fresh of herbs were inspected to enclose that they were uniform and non-damaged. All of samples were evaluated at the picked day and were measured by e-nose during the period of October-December, 2008. Samples kept 200 gram, were put in a vial containing 50 cm^3. The number of days and samples were limited by practical circumstances through data collection during 4 P.M.-9 P.M., five day per week. The value of room temperature is 22±3 °c and the percentage of relative humidity is 38±5.

2.2 Hardware and software of Low-cost electronic nose

This instrument was conducted using the low-cost electronic nose (enNU-PYO1) equipped with sensor arrays, glass chamber and data acquisition card. Commercial sensor arrays were performed with a gas sensors array consisting of AF63, TGS813, TGS821, TGS825, TGS826, TGS2600, TGS2601, TGS2602, TGS2610, TGS2620 and SHT15 temperature-humidity sensors. The 10 MOS sensors (manufactured by Figaro™ Company) and SHT 15 from Sensirion® Company were placed in a circular shape inside a Teflon cover of 10 cm in diameter. The volume of chamber has 1000 ml. In Figure 1, show schematic diagram of low-cost electronic nose (enNU-PYO1). We used a static headspace sampling technique. The headspace sampling method is an easy-rapid method to use. Although many samples can be measured within short time, the supplied vapor concentration is not know and varies during the vapor supply.

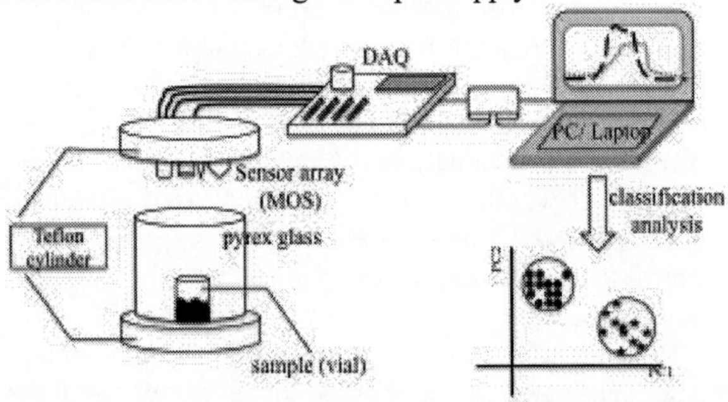

Figure 1. Schematic diagram of enNU-PYO1 to evaluated and classified for Thai herb agriculture products. [23]

The vapor concentration at the outlet of vial gradually changes until it reaches the liquid equilibrium. The vapor-concentration profile sometimes influences the waveform of the sensor response, which is a convolution of flavor profile and the sensor impulse response. The fundamental static system measures the steady-state response of each sensor. In course of measurement sensors were exposed to the flow of prepared odor mixture from vial. The exposure was 10 minute long. Changes of sensor resistance, during exposure of the array to the samples, were considered as the response of the sensors. During all experiments were conducted at room temperature. A USB 6008 data acquisition card (DAQ) was used for online data gathering and controlled using Labview© (National instruments Inc). During the measurement phase, DAQ recording the intensity of smell in term of conductivity changes that the sensors experienced. When the measurement was completed, the acquired profiles were properly collected for use later.

2.3 Data analysis

In this system, sensor array consisting several types of its were found to be sensitivity and selectivity for volatile organic compounds (VOCs) to Thai herbs. From each sensor responses are related to amount of chemical composition that interaction depends on its sensing material. Variations of Thai herbs odours have been evaluated with each of selective sensor. The intelligent quotient of our two kinds of commercials SnO_2 gas sensor arrays to be used as e-nose for discrimination between of ambient environment. In the analysis food have been evaluated, measuring the sensors responses after exposure to species and various kind of Thai herbs. We used the following pre-processing that normalization of individual sensor over the array with fractional relation as:

$$S_{ij} = \frac{R_{ij} - R_o}{min(R_o)} \quad (1)$$

where R_{ij} and R_o are the values of resistance at the end of sample vapor responses exposures and in the ambient reference.Considering the value of fractional relative change in resistance, $(R_{ij}-R_0)/R_0$, was chosen as an appropriate feature to be extracted from the sensor response to optimizing the data input matrix for the subsequent pattern recognition (PARC) algorithm. PCA is a linear feature extraction method that reduced of the i-dimensional data set in a dimension smaller than i [19-22]. In spite of the limits due to its linear characteristic, PCA is a very useful classification technique widely used in the gas-sensing area observed.

RESULTS AND DISCUSSIONS

Based on the results, the application with enNU-PYO1 electronic nose and statistical analysis method,

PCA, could be appropriate tool either for identification of cultivar to accelerate selection process or to distinguish crude herbs of different spices.

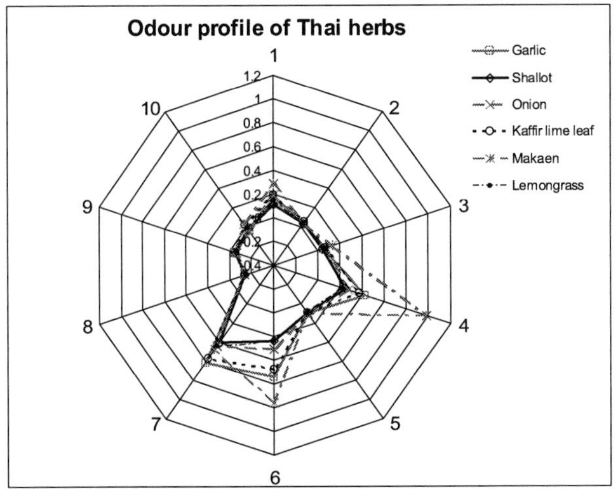

Figure 2. Signal responses of 12 MOS sensors to 6 samples odours of Thai herbs.

In the MOS sensor, we analyzed the response of the array to the aromatic from Thai herbs. The typical response sensors were be evaluated by 10 sensors shown in Fig 2. In the preprocessing data, Fractional measurments of signal responses were compensated with R_o. PCA analysis was performed on the normalized matrix of sensor response to varience of samples through the covariance matrix. This process reduces the nonlinear concentration dependence and improves performance of the classification technique. For the normalized of varies of odours response; this procedure improves the cluster become clearly analysis since in the related PCA plot the three cluster become clearly revealed shown in Fig 3. Pattern classification with principal component space was represented. We can also note that over more 97% of the total variance within data.

There is no doubt that sensory evaluation test have considerable advantages over the commonly used gas sensor arrays, in terms of adaptability and sensitivity. With regard to adaptability, when an enNu-PYO1 is constructed, the number and the type of gas sensors that have to be corporate must be specified. The result of application works were similar to vistualisation of human sensory system. First, we constructed an array of different chemically type of commercials gas sensors and temperature-humidity sensor. The sensors array have shown good sensitivity to some of volatile compound from Thai-herbs, which interest in quality control and to discriminates for complex flavour coming from different beverage-food product, reconstruction procedure, quality control to postharvest packaging for argiculture product.

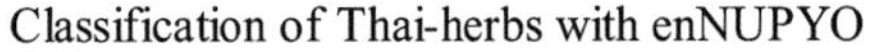

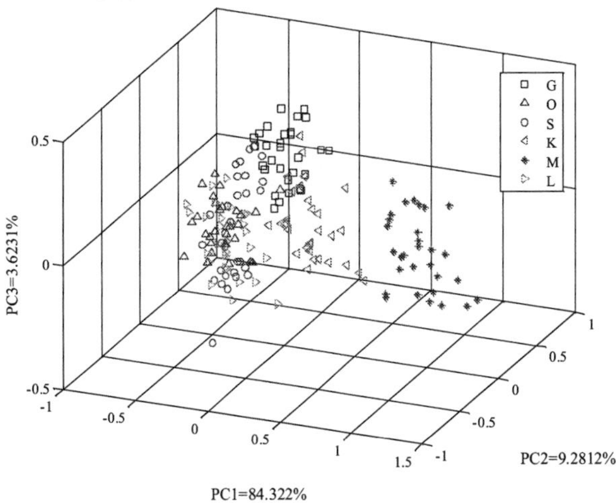

Figure 3. Showed 3D of principal component analysis. Classification of Thai-herbs, the meaning of G-Garlic, S-Shallot, O-Onion, K-Kaffir lime leaf, M-Makaen and L-Lemongrass were evaluated with enNU-PYO1.

CONCLUSIONS

These results confirm that the use of Low-cost electronic nose with limited amount of sensors is a powerful tool for rapidly classifying volatile compounds to Thai herbs. The active research that is presently being study of pattern classification is aimed in particular in the specification and sensitivity of the sensors, studying in application work, classification to stench from toxic , ripeness and fermentation of Thai argiculture product. The enNU-PYO1 prototype were investigates and evaluated agriculture farming applications, were presently is aimed in particular at increasing the specificity and sensitivity of sensor for assessing volatile compound in term of ingredient and aromatic of food-beverage, revelatory. The working system of the array relative to application in order to provide the bioinformatics science, application of pattern classification to be used.

ACKNOWLEDGMENTS

Daungjak P. acknowledges the scholarship support by Thailand Graduate Institute of Science and Technology (TGIST), NSTDA and Center of Intelligent Materials and Systems (CiMS), Mahidol University which is also gratefully acknowledge. In particular, the author would like to thank Dr. Teerakiat and Dr. Siripat for fruitful discussions and School of Science, Naresuan University Phayao campus for financial support to ISOEN2009 international conference.

REFERENCES

1. U.Weimar, K.D. Schierbaum, W. Göpel and R. Kowalkowski, Pattern recognition methods for gas mixture analysis : application to sensor arrays based upon SnO_2, *Sensors and Actuators B: Chemical,* Vol 1, (1990) pp. 93-96.
2. T. Maekawa, J. Tamaki, N. Miura, N. Yamazoe, development of SnO_2-based ethanol gas sensor, *Sensors and Actuators B: Chemical,* Vol 9 (1992) pp. 63-69.
3. C. Di Natale, R. Paolesse, A. D'Amico, Food and beverage quality assurance, in ; T. Pearce, S. Schiffman, H. Nagle, J. Gardner (Eds.), Handbook of Machine Olfaction, Wiley-VCH, 2003, pp. 505-524.
4. Szczurek, A. and Maciejewska, M. Relationship between odour intensity assessed by human assessor and TGS sensor array response. *Sensors and Actuators B: Chemical*, Vol 106, No 1, 2005, pp. 13-19.
5. Maekawa, T., K. Suzuki, et al.. "Odor identification using a SnO2-based sensor array." *Sensors and Actuators B: Chemical*, Vol 80, No 1, 2001, pp. 51-58.
6. F. Sinesio, C. Di Natale, G. Quaglia, F. Bucarelli, E. Moneta, A. Macagnano, R. Paolesse, A. D'Amico, Use of electronic nose and trained sensory panel in the evaluation of tomato quality, *J.Sci.Food Agric*, Vol 80, 2000, pp. 63-71
7. Gomez, A. H., Hu, G., Wang, J. and Pereira, A. G. Evaluation of tomato maturity by electronic nose. *Computers and Electronics in Agriculture,* Vol 54, No 1, 2003, pp. 44-52.
8. Gonzalez Martin, Y., Luis Perez Pavon, J., Moreno Cordero, B. and Garcia Pinto, C. Classification of vegetable oils by linear discriminant analysis of Electronic Nose data. *Analytica Chimica Acta,* Vol 384, No 1, 2006, pp. 83-94.
9. Dutta, R., Kashwan, K. R., Bhuyan, M., Hines, E. L. and Gardner, J. W. Electronic nose based tea quality standardization. *Neural Networks,* Vol 16, No 5-6, 2003, pp. 847-853.
10. Dutta, R., Hines, E. L., Gardner, J. W., Udrea, D. D. and Boilot, P. Non-destructive egg freshness determination: an electronic nose based approach. *Measurement Science and Technology,* Vol 14, No 2, 2003, pp. 190-198.
11. Gomez, A. H., Wang, J., Hu, G. and Pereira, A. G. Electronic nose technique potential monitoring mandarin maturity. *Sensors and Actuators B: Chemical*, Vol 113, No 1, 2006, pp. 347-353.
12. Natale, C. D., Olafsdottir, G., Einarsson, S., Martinelli, E., Paolesse, R. and D'Amico, A. Comparison and integration of different electronic noses for freshness evaluation of cod-fish fillets. *Sensors and Actuators B: Chemical*, Vol 77, No1-2, 2006, pp. 572-578.
13. Pardo, M., Niederjaufner, G., Benussi, G., Comini, E., Faglia, G., Sberveglieri, G., Holmberg, M. and Lundstrom, I. Data preprocessing enhances the classification of different brands of Espresso coffee with an electronic nose. *Sensors and Actuators B: Chemical*, Vol 69, No 3, 2000, pp. 397-403.
14. Sobanski, T., Szczurek, A., Nitsch, K., Licznerski, B. W. and Radwan, W. Electronic nose applied to automotive fuel qualification. *Sensors and Actuators B: Chemical*, Vol 116, No 1-2, 2006, pp. 207-212.
15. Strathmann, S., Pastorelli, S. and Simoneau, C. Investigation of the interaction of active packaging material with food aroma compounds. *Sensors and Actuators B: Chemical,* Vol 106, No 1, 2000, pp. 83-87.
16. Pathange, L. P., P. Mallikarjunan, Marini, Richard P.O'Keefe, S., and Vaughan, D., "Non-destructive evaluation of apple maturity using an electronic nose system." *Journal of Food Engineering*, Vol 77, No 4, 2006, pp. 1018-1023.
17. Brezmes, J., E. Llobet, Vilanova, X., Saiz, G. and Correig, X., "Fruit ripeness monitoring using an Electronic Nose." *Sensors and Actuators B: Chemical*, Vol 69, No 3, 2000, pp. 223-229.
18. Sarrazin, C., J.-L. Le Quéréa, Gretschb, C. and Liardonb, R. "Representativeness of coffee aroma extracts: a comparison of different extraction methods." *Food Chemistry*, Vol 70, No 1, 2000, pp: 99-106.
19. Yu, H., J. Wang, Yao, C., Zhang, H., and Yu, Y. "Quality grade identification of green tea using E-nose by CA and ANN." LWT - *Food Science and Technology*, Vol 41, No 7, 2008, pp: 1268-1273.
20. Haddad, R., Carmel, L. and Harel, D. A feature extraction algorithm for multi-peak signals in electronic noses. *Sensors and Actuators B: Chemical*, Vol 120, No 2, 2007, pp. 467-472.
21. Jurs, P. C., Bakken, G. A. and McClelland, H. E. Computational Methods for the Analysis of Chemical Sensor Array Data from Volatile Analytes. *Chemical Reviews*, Vol 100, No 7, 2000, pp.2649-2678.
22. Meilgarrd, M., Civille, G.V, and Carr, B.T. (1987). Sensory evaluation techniques. Florida, USA: CRC Press.
23. Daungjak P, "Pattern Classification of coffee odor with electronic nose and Human sensory test", Ph.D. Thesis (manuscript), Walailak University, 2009.

Use of an Electronic Tongue System to Detect Methylisoborneol in Distilled Water

Guilherme S. Braga, Leonardo G. Paterno, Fernando J. Fonseca

Escola Politécnica da Universidade de São Paulo (EPUSP)
Av. Prof. Luciano Gualberto, travessa 3, 158, São Paulo, SP, Brasil
gbraga@lme.usp.br

Abstract. An electronic tongue (ET) system consisting of conducting polymer sensors was employed to detect 2-methylisoborneol (MIB) in distilled water. MIB is a tainting compound and known to cause undesirable tastes and odours in water and aquaculture farming. Samples of distilled water with different concentrations of MIB were analysed in order to evaluate the capabilities of the ET system. Principal component analysis (PCA) showed that for higher concentrations of MIB (50 and 100ng.L^{-1}) the ET could separate the tainted samples into distinct clusters. Clusters of untainted and tainted samples with lower concentrations (bellow 10ng.L^{-1}) overlapped and resulted in a single cluster. Nevertheless, close grouping between repeated tests indicated that the ET system response is reproducible.

Keywords: electronic tongue system, chemical sensors, nanostructured films, nanotechnology, water, methylisoborneol.
PACS: 01.30.Cc, 07.07.Df

INTRODUCTION

Eutrophic conditions and cyanobacterial blooms are known to affect the water quality mainly because the respective cyanobacterial metabolities cause undesirable tastes and odours to marine and fresh water, surface water supplies, extensive, and intensive aquaculture systems [1, 2]. Among the off-flavour compounds, geosmin (GEO) and 2-methylisoborneol (MIB) [1, 3] are usually found in water and are associated to an earthy-musty taste and odour. Even water contaminated by a low level of these compounds is unpleasant to humans [1]. This is a major concern for the water treatment utilities [3] which must control their presence in water at low levels.

Several analytical methods are employed for monitoring these off-flavours compounds [1], including gas chromatography-mass spectroscopy (GC/MS). GC/MS is highly selective and sensitive [1] but is expensive and time-consuming, and must be operated by skilled operators. Furthermore, it can not be used in real-time monitoring.

As an alternative methodology for liquid analysis, the use of chemical sensor arrays known as electronic tongue (ET) has gained considerable importance. The array is formed by chemical sensors of different sensitivity and selectivity which together display a unique "fingerprint" for the sample they analyse. Different sensoactive materials, such as conducting polymers are usually employed in combination with suitable electrodes. The electrical response of these materials is highly dependent on the environmental conditions. Changes on their electrical conductivity can be associated to the presence of specific chemicals or changes on the physico-chemical properties of the medium.

In the present contribution, we employ a conducting polymer-based ET system [4, 5, 6] to detect MIB in distilled water at various concentrations. This study aimed to evaluate the performance of the ET and its potential application in water quality monitoring. [7].

EXPERIMENTAL AND METHODS

Electronic Tongue

The electronic tongue system was composed by an array of 5 non-specific chemical sensors whose electrical responses depend on the chemical composition of the medium [8, 9]. With the aid of pattern recognition tools, such as principal component analysis (PCA), the responses of the sensor array can

CP1137, *Olfaction and Electronic Nose: Proceedings of the 13th International Symposium*, edited by M. Pardo and G. Sberveglieri

be precisely correlated and used to identify and quantify the presence of different chemical substances [10]. The chemical sensors are formed by interdigitated microelectrodes (IMEs, 50 digits pairs with a width of 10 μm) [8, 10] produced by conventional photolithography coated with nanostructured polymeric films deposited via layer-by-layer assembly [11]. Details on film preparation can be found elsewhere [10]. Each sensor is formed by a different combination of conducting polymers (polyaniline and its derivatives, polypyrrole) and common polyelectrolytes (sulfonated polystyrene, sulfonated lignin, and polyacrylic acid).

Data Acquisition and Analysis

Distilled water was used to prepare MIB samples at various concentrations (10, 50 and 100ng.L^{-1}) and were stored at 6°C before the test period (tainted samples). A control sample (pure distilled water) was also used during the tests and stored at 6°C (untainted sample).

The electrical response of each sensor in different water samples was measured using an impedance analyzer (Solartron SI 1260). The system was connected to a PC through a GPIB interface and the measurements and data acquisition were performed by a platform created in LabView software.

Measurements were taken after signal stabilization, usually 1 min after immersion of the sensors into the water sample. Between each evaluation, the sensors were rinsed by soaking in ultrapure water under stirring for 5 minutes. Untainted and tainted samples were analyzed in a random sequence. All measurements were carried out at 25°C. The data were analyzed using a pattern recognition tool, principal component analysis (PCA). PCA score plots were made by using a pre-designed algorithm and the MatLab (version 6.1) software.

RESULTS AND DISCUSSION

A previous study was conducted to evaluate the dependence of sensor's response on the frequency of the applied field (range 100Hz to 10kHz). As a typical result, the impedance spectra for a polyaniline-based sensor submitted to different concentrations of MIB are showed in Figure 1A and 1B. It is noted that the electrical resistance of the sensor decreases as the concentration of MIB increases. The electrical responses of the sensor to the tainted and untainted samples are clearly different (no overlap of curves in a wide range of frequency). On the other hand, capacitance curves are overlapped and there is no distinction between the tainted and untainted samples. The same behavior was observed with the other sensors.

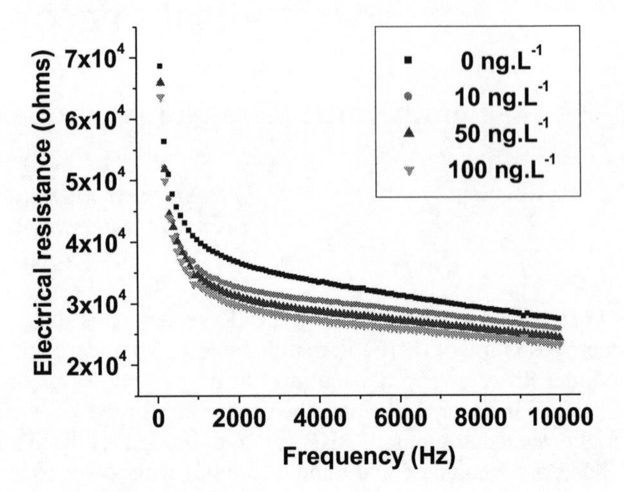

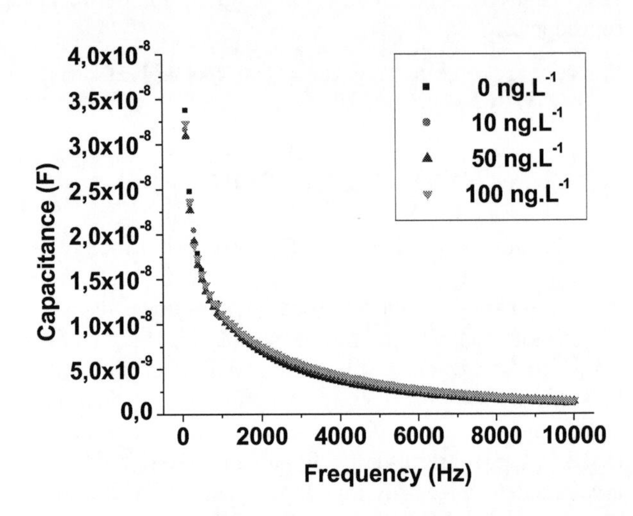

FIGURE 1. Impedance spectra, electrical resistance (A) and capacitance (B) for a polyaniline based sensor submitted to samples at various MIB concentrations as quoted (at 25°C).

An inspection on the impedance spectra for each sensor,showed that the discrimination between untainted and tainted samples is the highest at 1 kHz. In Figure 2 is shown the electrical fingerprints provided by the sensor array when it is submitted to samples of different MIB concentrations. For all sensors one could observe that the electrical resistance decreases as the MIB concentration increases. However, the signal intensity depends on the type of sensor.

A PCA plot was obtained from statistical treatment of data collected with the sensor array submitted to different MIB samples, Figure 3. The first principal component PC1 detains almost 90% of the total data variance, which means that MIB tainted and untainted samples may be distinguished almost exclusively by inspecting the data projected in the horizontal axis. Additionally, it is possible to note a trend in the location of points in the plot: as MIB concentration increased, the points shifted to the left along the axis. This feature indicates a straight relationship between the measured electrical resistance and MIB concentration, which confirms the ability of the ET in quantifying this analyte. Additionally, the PCA plot reveals that the detection limit for MIB using this sensor array is about 10 ng.L^{-1}. Untainted and 10ng.L^{-1} tainted samples are grouped in the same cluster. Nevertheless, close grouping between repeated tests indicated that the ET system response is reproducible. Thus, it may be concluded that the ET system is capable to differentiate between untainted and MIB tainted distilled water samples, specially in concentrations higher than 10ng.L^{-1}.

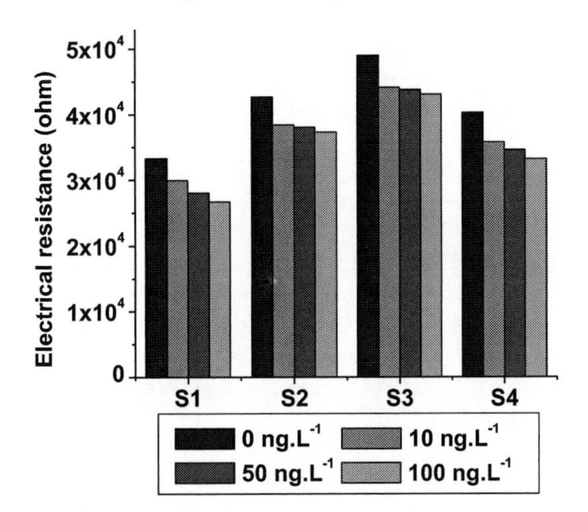

FIGURE 2. Electrical fingerprints (electrical resistance) for MIB samples at various concentration, as quoted, obtained with the sensor array operated at 1kHz.

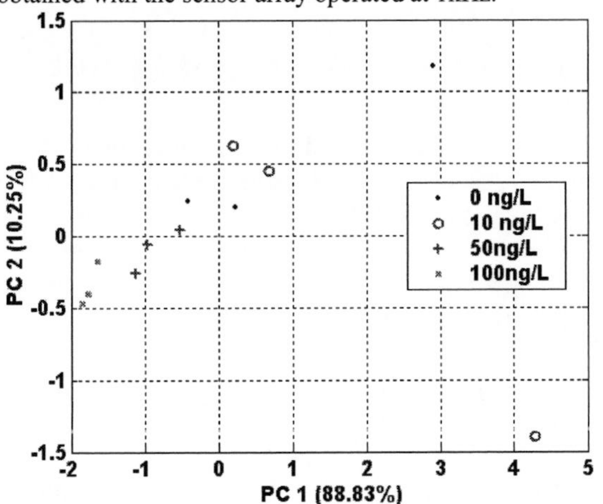

FIGURE 3. PCA plot for MIB tainted and untainted samples obtained with the sensor array operated at 1 kHz.

CONCLUSIONS

An array of chemical sensors, electronic tongue, was employed to detect and discriminate MIB samples in distilled water at various concentrations. Measurements of the electrical resistance of the sensors made from different conducting polymers could be used in combination with principal component analysis to detect MIB at concentrations as low as 10ng.L^{-1} in distilled water in a repetitive manner. Our results corroborate that non-specific sensor array is a suitable technique for detecting MIB in distilled water.

Further studies employing different polymeric systems and other off-flavour compounds are currently under investigation in our laboratory.

ACKNOWLEDGMENTS

The authors wish to thank FAPESP and CNPq for the financial support.

REFERENCES

1. Zhang, L., Hu, R., Yang, Z., *Water Research*, Vol. 40 (2006), pp. 699-709.
2. Smith, J.L., Boyer, G.L., Zimba, P.V., *Aquaculture*, Vol. 280 (2008), pp. 5-20.
3. Alves, R.; "Techniques of evaluation of taste and odor in drinking water: analytical method, sensory analysis and consumers perception", Master dissertation, EPUSP, São Paulo (2005).
4. Toko, K., Hayashi, K., Yamanaka, M., Yamafuji, K., *Tech. Digest 9th Sensor Symp.* 1990, pp. 193-196.
5. Winquist, F. , Wide, P., Lundstrom, I., *Anal. Chem. Acta*, Vol. 357, No. 1-2 (1997), pp. 21-31.
6. DiNatale, C., et al., *Sens. Actuators*, Vol. 34 (1996) pp. 539-542.
7. Iliev, B., Lindquist, M., Robertsson, L., Wide, P., *Fuzzy Sets and Systems*, Vol. 157 (2006), pp. 1155-1168.
8. Riul Jr., A., et al., *Langmuir*, Vol. 18 (2002), pp 239-245.
9. Medeiros, E.S., Paterno, L.G., Mattoso, L.H.C., in: C.A. Grimes, E.C. Dickey, M.V. Pishko (Eds.), Encyclopedia of Sensors, American Scientific Publishers (Pennsylvania, 2005) Vol. 10, pp. 1-36.
10. Wiziack, N.K.L., Paterno, L.G., Fonseca, F.J., Mattoso, L.H.C., *Sensors and Actuators B*, Vol. 122, No. 2 (2007), pp. 484-492.
11. Lvov, Y.M., Decher, G., *Crystallography Reports*, Vol. 39, No. 4 (1994), pp. 628-647.

Use of an Electronic Tongue System and Fuzzy Logic to Analyze Water Samples

Guilherme S. Braga, Leonardo G. Paterno, Fernando J. Fonseca

Escola Politécnica da Universidade de São Paulo (EPUSP)
Av. Prof. Luciano Gualberto, travessa 3, 158, São Paulo, SP, Brasil
gbraga@lme.usp.br

Abstract. An electronic tongue (ET) system incorporating 8 chemical sensors was used in combination with two pattern recognition tools, namely principal component analysis (PCA) and Fuzzy logic for discriminating/classification of water samples from different sources (tap, distilled and three brands of mineral water). The Fuzzy program exhibited a higher accuracy than the PCA and allowed the ET to classify correctly 4 in 5 types of water. Exception was made for one brand of mineral water which was sometimes misclassified as tap water. On the other hand, the PCA grouped water samples in three clusters, one with the distilled water; a second with tap water and one brand of mineral water, and the third with the other two other brands of mineral water. Samples in the second and third clusters could not be distinguished. Nevertheless, close grouping between repeated tests indicated that the ET system response is reproducible. The potential use of the Fuzzy logic as the data processing tool in combination with an electronic tongue system is discussed.

Keywords: electronic tongue system, nanostructured films, chemical sensors, mineral water, Fuzzy logic.
PACS: 01.30.Cc, 07.07.Df

INTRODUCTION

Multi-sensor systems also known as electronic tongue (ET) [1, 2, 3] have been intensively studied over the years, especially regarding foodstuff analysis [4]. As an alternative to more established technologies such as GC/MS chromatography, ET provides a faster and cheaper method to assess taste characteristics of food samples.

The electronic tongue system used in this work is an array of non-selective chemical sensors whose electrical capacitance may vary according to the chemical composition of the medium they are immersed in [5]. The chemical sensors are made up of nanostructured conducting polymer films deposited onto interdigitated microelectrodes (IME) [5, 6]. The electrical and optical properties of conducting polymers are very sensitive to small changes in their surrounding environment, such as pH, ionic strength, and so on [7]. With the aid of pattern recognition tools, such as principal component analysis (PCA), these changes can be precisely measured and used to identify and quantify the presence of different chemical substances [6]

Fuzzy logic [8] is another recognition tool that can be used with ET systems. It is considered a kind of artificial intelligence (term applied to technologies that mimics knowledge similar to humans) [9] known for its capacity in converting numerical inputs into linguistic variables. However, few reports concerning the use of ET systems and Fuzzy logic are found in the literature [4, 10-12].

In the present work, PCA and Fuzzy logic are used in combination with an electronic tongue system to analyse water samples from different sources (tap, distilled and three brands of mineral water). The performance and accuracy of those pattern recognition tools are evaluated.

EXPERIMENTAL AND METHODS

The ET was comprised by an array of 8 non-specific sensors. The sensors were made of interdigitated microelectrodes (IMEs, 50 digits pairs with a width of 10 μm) produced by conventional photolithography coated with nanostructured polymeric films deposited via layer-by-layer assembly [13]. Details on film preparation can be found elsewhere [6]. Each sensor was based on a different combination of conducting polymers (polyaniline and its derivatives, polypyrrole, poly (3,4-ethylenedioxythiophene)) and common

CP1137, *Olfaction and Electronic Nose: Proceedings of the 13th International Symposium,* edited by M. Pardo and G. Sberveglieri

polyelectrolytes (sulfonated polystyrene, sulfonated lignin) and nickel(II) phthalocyanine.

The electrical response (capacitance) of each sensor in different water samples was measured at 1 kHz, using an impedance analyzer (Solartron SI 1260). The system was connected to a PC through a GPIB interface and the measurements and data acquisition were performed by a platform created in LabView software.

Measurements were taken after signal stabilization, usually 1 min after immersion of the sensors into the water sample. Between each evaluation, sensors were rinsed by soaking in ultrapure water under stirring for 5 minutes. Tap, distilled and 3 brands of mineral water were analyzed in a random sequence. All measurements were carried out at 25°C. A total of 12 samples per each type of water were evaluated.

These data were analyzed using two pattern recognition tools: PCA and Fuzzy logic program. PCA score plots were made by using a pre-designed algorithm and the MatLab (version 6.1) software. To create the Fuzzy program, the collected data were divided in two groups: (1) the developing group, with the first 3 measurements and (2) the testing group with the other 9 measurements. The Fuzzy program was developed using MatLab (version 6.1) software, on Sugeno method. It has 8 inputs (the ET has 8 sensors) and 5 outputs (distilled water, tap water and 3 brands of mineral water). The input domains (minimum and maximum values) were defined and based on the electrical capacitance of each sensor (developing group). After the program development the accuracy was evaluated using the data from the testing group.

TABLE 1. Fuzzy logic analysis of tap, distilled and 3 brands of mineral water samples. Developing group, measurements were used to create the program and testing group, measurements carried out to test the program accuracy.

Developing group

Water sample \ Results	DI	Tap	Mineral water A	Mineral water B	Mineral water C	
DI	**20%**	0	0	0	0	
Tap	0	**20%**	0	0	0	
Mineral water A	0	6.13%	**13.87%**	0	0	
Mineral water B	0	0	0	**20%**	0	
Mineral water C	0	0	0	0	**20%**	
Accuracy						***93.87%***

Testing group

Water sample \ Results	DI	Tap	Mineral water A	Mineral water B	Mineral water C	
DI	**20%**	0	0	0	0	
Tap	0	**16.67%**	3.33%	0	0	
Mineral water A	0	3.33%	**16.67%**	0	0	
Mineral water B	0	0	0	**20%**	0	
Mineral water C	0	0	0	0	**20%**	
Accuracy						***93.37%***

RESULTS AND DISCUSSION

PCA plot was obtained from statistical treatment of data collected with an array of 8 sensors coated with different polymeric systems. The result is presented in Figure 1, showing the response for different water samples. More than 86% of the total data variance is retained with the first and second components (PC1 and PC2), which means that water samples could be distinguished by inspecting the positioning of each cluster. However, the discrimination of water samples could be performed exclusively along the horizontal axis. Additionally, water samples were grouped in three clusters. One with the distilled water samples (1), another with samples of tap water and one brand of mineral water (2) and the third with the other two brands of mineral water (3). PCA analysis did not allow the ET to distinguish between tap water samples and one brand of mineral water (A). The same could be observed in cluster 3, where the two brands of mineral water were classified as being equal. Additionally, close grouping between repeated tests indicated that the ET system response is reproducible.

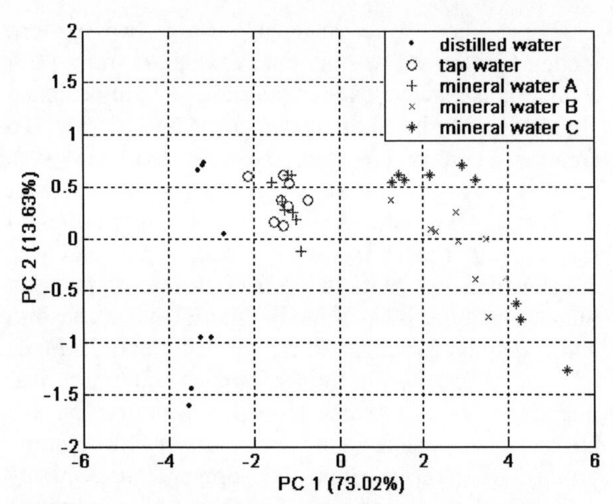

FIGURE 1. PCA plot obtained with the ET operated at 1 kHz for distilled, tap and mineral water samples.

On the other hand, the Fuzzy logic analysis resulted in a better discrimination of the water samples. Table 1 presents the hit ratio for each water sample obtained with the Fuzzy logic. Considering that 5 types of water samples were analyzed and that the maximum accuracy is 100 %, each water sample contributes with 20 % for the total score. Looking at the data, the ET could correctly classify 4 in 5 water types, which is a better performance than that provided by the PCA. Using the data from the testing group the program reached an accuracy of 93.37%. However, samples of mineral water A were wrongly classified as tap water, due to their similar electrical responses.

CONCLUSIONS

In the present work we have evaluated the use of PCA and Fuzzy logic as pattern recognition tools in combination with an electronic tongue system to analyze water samples.

By using PCA, the ET system showed a reproducible response but was not able to distinguish tap water samples from one type of mineral water. On the other hand, the ET presented a better performance with the Fuzzy logic. The Fuzzy program created had a higher accuracy (93.37%) and could correctly classify 4 in 5 types of water. However, some samples of tap water and one brand of mineral water (brand A) were misclassified.

In conclusion, when the data was analyzed with the Fuzzy logic program, the ET system could differentiate between water samples in a better way than that obtained with PCA

ACKNOWLEDGMENTS

The authors wish to thank FAPESP and CNPq for the financial support.

REFERENCES

1. Toko, K., Hayashi, K., Yamanaka, M., Yamafuji, K., *Tech. Digest 9th Sensor Symp*. 1990, pp. 193-196.
2. Winquist, F. , Wide, P., Lundstrom, I., *Anal. Chem. Acta*, Vol. 357, No. 1-2 (1997), pp. 21-31.
3. DiNatale, C., et al., *Sens. Actuators*, Vol. 34 (1996) pp. 539-542.
4. Iliev, B., Lindquist, M., Robertsson, L., Wide, P., *Fuzzy Sets and Systems*, Vol. 157 (2006), pp. 1155-1168.
5. Riul Jr., A., et al., *Langmuir*, Vol. 18 (2002), pp 239-245.
6. Wiziack, N.K.L., Paterno, L.G., Fonseca, F.J., Mattoso, L.H.C., *Sensors and Actuators B*, Vol. 122, No. 2 (2007), pp. 484-492.
7. Medeiros, E.S., Paterno, L.G., Mattoso, L.H.C., in: C.A. Grimes, E.C. Dickey, M.V. Pishko (Eds.), Encyclopedia of Sensors, American Scientific Publishers (Pennsylvania, 2005) Vol. 10, pp. 1-36.
8. Zadeh, L.A. , *Information and Control*, Vol. 8 (1965), pp. 338-353.
9. YEN, J., LANGARI, R., ZADEH, L.A.; Industrial Applications of Fuzzy logic and Intelligent Systems, IEE Press, New York (1995).
10. Kim, JD., et al., *Talanta*, Vol. 70, No. 3 (2006), pp. 546-555.
11. Martinez-Manez, R., et al., *Sensors and actuators B*, Vol. 104, No. 2 (2005), pp. 302-307.
12. Rong, L. , Ping, W., Wenlei, H., *Sensors and actuators B* , Vol. 66, No. 1-3 (2000), pp. 246-250.
13. Lvov, Y.M., Decher, G, *Crystallography Reports*, Vol. 39, No. 4 (1994), pp. 628-647.

Screening Cereals Quality by Electronic Nose: the Example of Mycotoxins Naturally Contaminated Maize and Durum Wheat

Anna Campagnoli, Vittorio Dell'Orto, Giovanni Savoini, Federica Cheli

Department of Veterinary Sciences and Technologies for Food Safety - University of Milan
Via Trentacoste, 2 - 20134 Milan, Italy
anna.campagnoli@unimi.it

Abstract. Mycotoxins represent an heterogeneous group of toxic compounds from fungi metabolism. Due to the frequent occurrence of mycotoxins in cereals commodities the develop of cost/effective screening methods represent an important topic to ensure food and feed safety. In the presented study a commercial electronic nose constituted by ten MOS (Metal Oxide Sensors) was applied to verify the possibility of discriminating between mycotoxins contaminated and non-contaminated cereals. The described analytical approach was able to discriminate contaminated and non-contaminated samples both in the case of aflatoxins infected maize and deoxynivalenol infected durum wheat samples. In the case of maize data two sensors from the array revealed a partial relation with the level of aflatoxins. These results could be promising for a further improvement of electronic nose application in order to develop a semi-quantitative screening approach to mycotoxins contamination.

Keywords: Electronic Nose, Mycotoxins, Maize, Durum Wheat
PACS: 07.07.Df

INTRODUCTION

Aflatoxins and deoxynilvalenol are toxic compounds produced by the secondary metabolism of toxigenic fungi, mainly *Aspergillus* and *Fusarium* genus. The occurrence of these toxins is frequent in food commodities, including cereals for human and animal consumption. Inadequate storage conditions of cereals could result in improved growth of toxigenic fungi and their metabolites synthesis including mycotoxins, dangerous for human and animal health. Because of the highly heterogeneous distribution of mycotoxins in contaminated alimentary matrices, the availability of cost-effective analytical methods, which enable high sample throughput from the same lot, is highly needed. From this point of view, head space sensor array (electronic nose) appears to be a promising technology. The underlying hypothesis for developing electronic nose-based sensors for cereal safety/quality evaluation is that the growth and metabolism of mycotoxins producing fungi is associated with the production of off-flavours and changes in volatile compounds composition potentially detectable by the instrument. The electronic nose uses in cereals quality estimation were reported by several authors. Specific applications on volatiles detection/estimation as indicators of fungal occurrence and their metabolites presence; and as fungi taxonomic markers has been decrypted by Abramson [1], Balasubramanian [2], Cheli [3], Jelen [4], Magan & Evans [5], Falasconi [6], Olsson [7], Paolesse [8]. The authors demonstrated that the mycological evaluation of cereals and a partial quantification of level of mould contamination were possible. The use of electronic nose in this field has to be supported by appropriate statistical tools. Chemometric approaches as the pattern recognition techniques, are the most frequently adopted. There is evidence from some authors as Olsson [9] and Benedetti [10] that combining instrumental approach with statistical analysis it could be possible to use volatile compounds to predict the presence and level of mycotoxins contamination.

This paper reports results of studies to investigate head space sensor array capacity in classification of aflatoxins and deoxynivalenol natural contaminated and non-contaminated maize and wheat samples, respectively.

EXPERIMENTAL AND METHODS

For the study two types of samples were used: thirty maize (*Zea mais*) natural contaminated samples containing aflatoxins and thirty wheat natural contaminated (*Triticum turgidum*) samples containing deoxynivalenol. Maize samples were analyzed by a commercial direct competitive enzyme-linked immunosorbent assay (ELISA) (AgraQuant ELISA kit, Romer Labs, Inc., Union, MO, USA) for the determination of total aflatoxins (B1, B2, G1 and G2)

CP1137, *Olfaction and Electronic Nose: Proceedings of the 13th International Symposium*, edited by M. Pardo and G. Sberveglieri

content. Wheat samples were analyzed for the content in deoxynivalenol (DON) by High Performance Liquid Chromatography/Mass Spectrometry (HPLC/MS). For the electronic nose analysis a PEN2 (Airsense Analytics GmbH, Schwerin, Germany) commercial available instrument equipped with an enrichment and desorption unit EDU2 (Airsense Analytics GmbH, Schwerin, Germany) was adopted. The head space sensor array of PEN2 was composed of ten Metal Oxide Semiconductors (MOS) sensors. In operating conditions (high temperature) interactions between volatiles and each sensor surface occur and induce changes in conductance of sensors. The change of conductance are recorded as function of time and graphically reported by the dedicated software. MOS sensors details of PEN2 are reported in TABLE 1.

For the experiment three grams of each sample were placed in a 12 ml airtight glass vials. The headspace inside the vials was equilibrated for 24h at room temperature (24±2°C). Afterwards, head space of each sample was exposed to a thermal desorption period and finally analyzed by the electronic nose. Statistical analysis based on application of Principal Component Analysis (PCA) and Linear Discriminant Analysis (LDA) was performed by SAS software 8.2 (SAS Institute. Cary, NC, USA).

TABLE 1. Sensors details and applications of electronic nose PEN2.

No. in array	*Sensor name*	*Description*	*Reference*
1	W1C-aromatic	Aromatic compound	Toluene, 10 ppm
2	W5S-broadrange	Broad range sensitivity react on nitrogen oxides and ozone very sensitive with negative signal	NO2, 10 ppm
3	W3C-aromatic	Ammonia, used as sensor for aromatic compounds	Benzene, 10ppm
4	W6S-hydrogen	Mainly hydrogen, selectively (breath gases)	H2, 100 ppb
5	W5C-arom-aliph	Alkanes, aromatic compounds, less polar compounds	Propane, 1ppm
6	W1S-broad-methane	Sensitive to methane (environment) ca. 10 ppm. Broad range, similar to No. 8	CH4, 100 ppm
7	-sulphur-organic	Reacts on sulphur compounds H2S 0.1 ppm. Otherwise sensitive to many terpenes and sulphur organic compounds, which are important for smell, limonene, pyrazine	H2S, 1 ppm
8	W2S-broad-alcohol	Detects alcohols, partially aromatic compounds, broad range	CO, 100 ppm
9	W2W-sulphur-chlor	Aromatics compounds, sulphur organic compounds	H2S, 1ppm
10	W3S-methane-aliph	Reacts on high concentrations >100 ppm, sometime very selective (methane)	CH4, 10 ppm

RESULTS

Analyses of total aflatoxins content estimation in maize samples performed by ELISA showed in 24 maize samples an aflatoxins concentration in a range of 6ppb-100ppb, while 6 samples resulted under the detection limit of the assay (3 ppb).

DON content in wheat samples performed by HPLC/MS showed a concentration range of 400ppb-2400ppb in 22 samples, while 8 samples resulted non-contaminated (detection limit: 50ppb).

The response curves from the ten sensors of electronic nose from both, maize and wheat samples, stabilized after from 80 to 90 sec and therefore the signal of the sensors at 90 sec. was recorded and used in the statistical analysis. PCA and LDA were used in order to investigate the potential of discriminating between contaminated and non-contaminated samples.

Electronic Nose Analysis of Maize Samples

In the case of maize samples PCA showed that the first two components were able to explain 98.04% of total data variance (first principal component explained 76.57% of variance; while second principal component 21.47%). Sensors W1W (Sulphur-organic) and W5S (Broadrange) appeared to be the most important to distinguish between aflatoxins containing maize and free maize samples and in some way associated to aflatoxins content (R^2=0.74 and R^2=0.63 for W5S and W1W respectively). The corresponding plot of the two principal components is given in FIGURE 1.

FIGURE 1. PCA score plot from maize samples showing a discrimination between aflatoxins contaminated and non-contaminated samples.

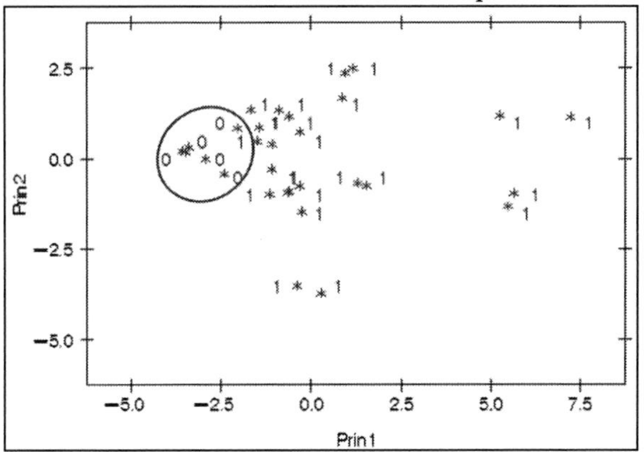

To classify the maize samples into contaminated and non-contaminated groups, the LDA classification methods, was applied with leave-one-out cross-validation. LDA performed on the same data set showed similar results of PCA. LDA gave a recognition percentage of 100% of response in prediction (TABLE 2).

TABLE 2. LDA classification of maize samples when leave-one-out cross-validation was applied.

		n. and % predicted group	
	Group size	Contaminated	Non-contaminated
Contaminated	24	24 (100%)	0 (0%)
Non-Contaminated	6	0 (0%)	6 (100%)

Electronic Nose Analysis of Wheat Samples

The volatiles profile of wheat samples investigated by PCA showed differentiation between contaminated and non-contaminated samples by DON. First and second principal components accounted for 59.77% (35.73% and 24.04% first and second principal components respectively) of total variance in the data (FIGURE 2). Sensors W6S (Hydrogen) and W3C (Aromatic) showed the major eigenvectors value in first principal component but relation between sensors value and contamination level was not evident.

LDA applied as in the case of aflatoxins contaminated maize samples, gave a recognition percentage of 100% of correct response for fitting and for leave-one-out cross-validation, with values of 0% for both the fitting error rate and the cross-validation error rate (TABLE 3).

FIGURE 2. PCA score plot from wheat samples showing a discrimination between deoxynivalenol contaminated and non-contaminated samples.

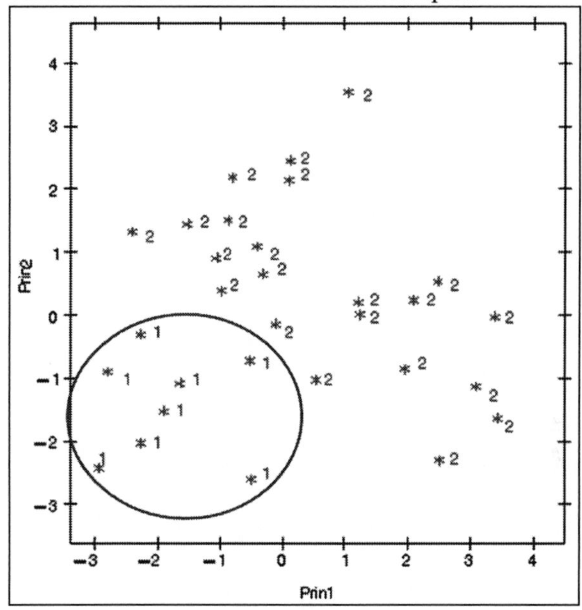

TABLE 3. LDA classification of wheat samples when leave-one-out cross-validation was applied

		n. and % predicted group	
	Group size	Contaminated	Non-contaminated
Contaminated	22	22 (100%)	0 (0%)
Non-Contaminated	8	0 (0%)	8 (100%)

CONCLUSIONS

The results of the study demonstrated that the electronic nose use associated to multivariate statistical analysis can classify between cereals samples contaminated and non-contaminated by mycotoxins. The descripted approach revealed its applicability both in the case of maize and durum wheat. Therefore electronic nose may be applied for screening of commodities contaminated with fungal toxins, in order to select samples which must undergo further accurate quantitative analysis.

In the case of maize data two sensors from the array revealed a partial relation with level of aflatoxins. These results could be promising for a further improuvement of electronic nose application in order to develop a semi-quantitative approach to mycotoxins contamination.

Some preliminary results of this study were presented at the "Feed Safety International Conference". November 27-28, 2007 Namur-Belgium.

REFERENCES

1. D. Abramson, *et al.*, *J Stored Prod Res,* **41**, 67-76 (2005).
2. S. Balasubramanian, *et al.*, *LWT,* **40**, 1815-1825 (2007).
3. F. Cheli, *et al*, *Ital. J. Food Sci*, **20,** 447-462 (2008).
4. H. H. Jelen, *et al*, *J Agr Food Chem*, **51**, 7079-7085 (2003).
5. N. Magan, and P. Evans, *J Stored Prod Res*, **36**, 319-340 (2000).
6. M. Falasconi, *et al*, *Sensor Actuat B-Chem*, **108**, 250-257 (2005).
7. J. Olsson, *et al*, *Int J Food Microbiol*, **59**, 167-178 (2000).
8. R. Paolesse, *et al*, *Sensor Actuat B-Chem*, **119**, 425-430 (2006).
9. J. Olsson, *et al*, *Int J Food Microbiol*, **72**, 203-214 (2002).
10. S. Benedetti, S, *et al*, J Food Protect, **68**, 1089-1092 (2005).

Gas sensor characterization at low concentrations of natural oils

H. Sambemana, M. Siadat, M. Lumbreras

LASC/ University of Metz
ISEA- 7, rue Marconi, 57070 Metz
siadat@univ-metz.fr; lumbre@univ-metz.fr

Abstract. Inhalation of essential oils can be used in aromatherapy due to their activating or relaxing effects. The study of these effects requires behavioral measurements on living subjects, by varying the nature and also the quantity of the volatile substances to be present in the atmosphere. So, to permit the evaluation of therapeutic effects of a variety of natural oils, we propose to develop an automatic diffusion/detection system capable to create an ambient air with low stabilized concentration of chosen oil. In this work, we discuss the performance of an array of eight gas sensors to discriminate low and constant concentrations of a chosen natural oil.

Keywords: Odorant atmosphere, gas sensors, sensibility threshold, recovery time.
PACS: 00.07.07.Df

INTRODUCTION

Nervous disorders, such as states of heightened excitability, anxiety or stress, as well as reduction of the capacities caused by aging, are severe problems with important consequences. Recently a considerable number of investigations have been carried out to evaluate activating or sedative effects of a variety of essential oils and odorants after inhalation. These studies are composed of neurophysiological as well as behavioral measurements in both animal and human subjects [1, 2, 3]. However, these inhalation effects must be linked to the concentration of the absorbed volatile substances. But in our knowledge, there is no such quantitative studies reported in the literature. The main difficulty in this investigation is due to the impossibility to measure the inhaled quantity. To pass around this problem, it is necessary to carry out the behavioral experimentations in a chamber where the physical and chemical parameters (natural oil concentration, temperature, humidity…) of the ambient atmosphere are closely controlled.

This paper presents the preliminary results of a system which is designed to create an atmosphere containing selected natural oil at a constant concentration. The main function of this system is to quantify the volatile compounds of an essential oil (E.O.) diffused at low concentrations; it will be further used in a feedback loop to control the E.O. diffusion.

Essential oil of pine at several concentrations is used in this work to test the performance of the experimental system, particularly concerning the quantification threshold and the stability of the gas sensors used.

EXPERIMENTAL AND METHODS

Our goal concerns the development of a gas diffusion/detection system performing two functions: generation of chosen concentrations of an essential oil in a closed atmosphere, and the discrimination of these concentrations. Therefore, our experimental set-up is composed of an electronic nose [4] for volatile compounds quantification and an essential oil diffusing unit.

The volatile compound quantification part is essentially based on commercial metal oxide gas sensors (Figaro and FIS). Eight sensors are selected for their sensitivities to aromatic and alcohol substances (TGS 816, 880, 882, 2600, 2602, 2620 and SP-AQ1, SP-MWO) and also for their robustness. To provide same experimental conditions for all the sensors, they are uniformly placed in a circular sensor cell. In presence of gas, the resistance variation of the gas sensors is calculated from the collected voltage variation [5].

The diffusion system is based on an automatic double synthetic air line (1 and 2). The synthetic air

CP1137, *Olfaction and Electronic Nose: Proceedings of the 13th International Symposium,* edited by M. Pardo and G. Sberveglieri

line (1) is sent bubbling in a bottle containing the essential oil at a liquid state. The air leaving the bottle is charged with the volatile molecules of the E.O. and keeping the same flow rate as the inlet flow. This gas line is joined to the synthetic air line (2) to form a total flow rate fixed at 100ml/min. With this diffusion method we can generate different concentrations by varying the flow rate of the gas line (1). Consequently, all the concentrations used in this study are represented in percentage corresponding to the flow rate of the gas line (1) over the total flow rate. We would underline that with this diffusing principle, the E.O. molecules are not deteriorated compared with a diffusing system using a heating principle. In order to select the more appropriate gas sensors, the charged air is sent directly to the sensors cell and not into the experimentation chamber where will be placed living (animal or human) subjects.

Pine natural oil was used in the 1% to 15% range (1%, 3%, 5%, 7%, 10%, 13% and 15%). A human panel has tested these concentrations. Notes from 0 to 10 were respectively given for no-smell until very strong smell (Table 1). The olfactory results indicate that the 1% concentration is not detected, actually literature reports that biological and physiological effects can still be occurred [6]. The highest concentration (15%) obtains the higher score of the olfactory test; so in real conditions biologists would certainly work under this concentration value.

TABLE 1: Olfactory score (from 0 to 10) obtained from a panel of five persons.

	[C] Pine	1%	3%	5%	7%	10%	13%	15%
Olfaction perception scale	Person 1	0	2	4	7	7	7	7
	Person 2	0	2	3	5	6	8	7
	Person 3	0	2	3	4	7	7	8
	Person 4	0	3	4	5	7	6	7
	Person 5	0	3	4	5	7	7	8

RESULTS

The first step of our work consists in establishing the procedure for the gas exposure and the sensor regeneration phases. Several protocols have been studied to choose the recovery procedure which permits the best regeneration of the sensors, whatever the value of the gas concentrations. In a second step, after selecting the experimental protocol, repetitive measurements have been realized to control the reproducibility and then to sudy the sensitivity of the gas sensors.

Protocol set-up

Metal oxide sensors are known to be subject to a more or less important drift after the gas exposure. So for continuous use, the regeneration phase is very important in order to avoid a baseline shift. Our first procedure was composed of two equivalent phases: 10 minutes of gas exposure, 10 minutes of regeneration. Sensors were heated at 5V, the recommended heating voltage value. Results showed important drifts of the sensor baselines essentially when high concentrations were used, attributed to a high adsorption at the sensitive layer of the sensors. To ameliorate the sensor regeneration, we tried to activate the regeneration phase by heating the gas sensors at 6 Volts during the five first minutes followed by ten minutes at 5 Volts. The elevation of the heating voltage during the regeneration phase, has not given better results in term of the sensor recovery (figure 1). Therefore, we decided to work at constant heating voltage (5V, less harmful for the sensors), keeping the gas exposure time at 10min but increasing the recovery time to 20min. This procedure offers a good regeneration of the sensors for all the chosen E.O. concentrations; it was used to follow this work.

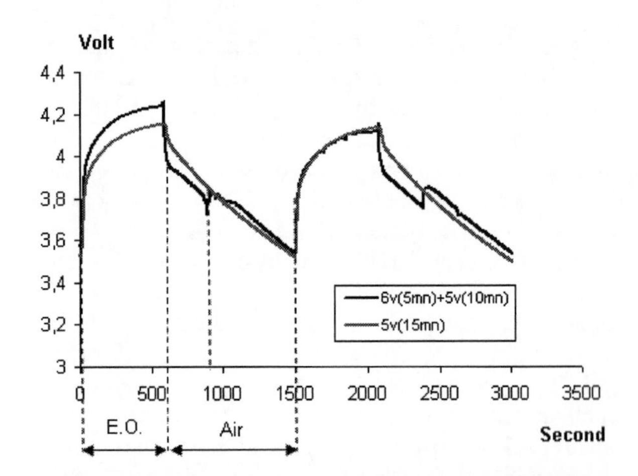

FIGURE 1: Example of one sensor response (SP-MW0) under 20% of pine E.O. using two different regeneration protocols.

Sensor sensitivity and reproducibility

After the measurement protocol establishment, we have studied the behaviour and the reproducibility of the sensors along with the essential oil concentrations. Figure 2 shows the time response of one sensor (SP-MWO) to all the selected concentrations. We can observe the good sensitivity of this sensor to each concentration, especially at the lowest ones. The other sensors showed the same behavior with more or less sensitivities or rapidity.

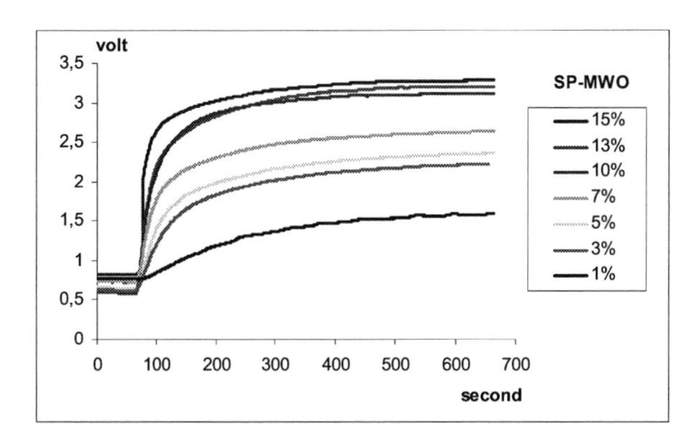

FIGURE 2: The SP-MW0 sensor responses to all the selected concentrations of pine oil.

The figure 3 represents the eight sensor responses to a fixed concentration (5%) of pine oil. The SP-MW0 sensor gives a very rapid and significant response. The TGS 2600, 2620 and SP-AQ1 sensors have also a notable response but they are less rapid. Concerning the TGS 880, 882 and 816 sensors, the responses are less important than the sensors described below. The TGS 2602 sensor presents a permanent saturation, inducing a very weak response. These results confirm the possibility of the concentration identification with the chosen sensor array.

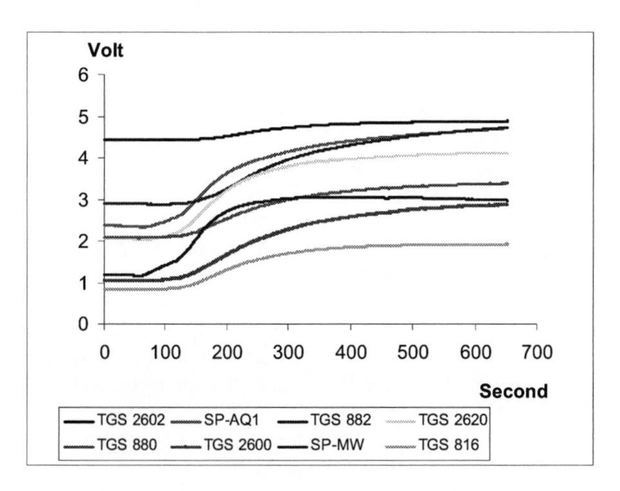

FIGURE 3: Sensor responses to the 5% concentration of pine oil.

After the sensor response consideration, we have done repetitive measurements to test the reproducibility of all the sensors. For each concentration, we have alternated gas exposition and regeneration phases. We can observe this reproducibility for 2% and 7% of E. O. concentration in Figure 4. Recorded signals show a good reproducibility for all the sensors: whatever the concentration value, the baselines of the sensors are always recovered. Only the TGS 2602 does not give good results and presents permanent saturation behaviour in spite of several adaptations of its measurement circuit. This sensor must not be taken into account.

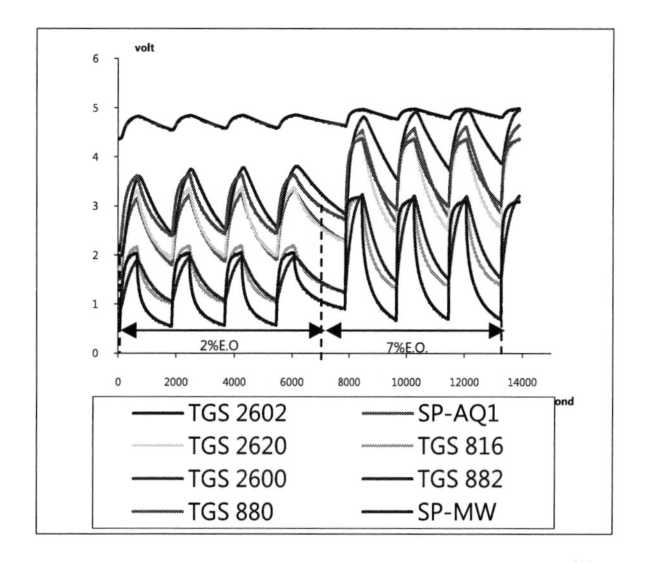

FIGURE 4: Reproducibility and sensitivity of the array sensors for two concentrations.

Comparing the amplitude of the signals, we note an increase from 2% to 7% concentration as it was shown before in figure 2.

To show the possibility of the discrimination of all the studied E.O. concentrations, sensor responses are presented in terms of the relative resistance variation R_0/R_S, (R_S: the stabilized resistance during gas exposure; R_0: the initial one, just before gas introduction). The variation of this parameter is shown along with the essential oil concentration in Figure 5 for the eight sensors. All the measurements were made more than ten times and by using several samples.

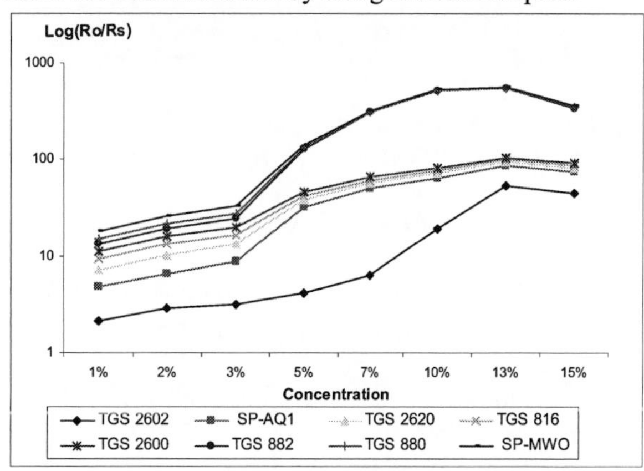

FIGURE 4: Sensitivity of the sensors in terms of the relative resistance in function of the E.O. concentration.

This figure presents the mean value of the relative resistance obtained from repetitive measurements made for each concentration. We note an increase of the relative sensor response when increasing the essential oil concentration. All the curves present two parts corresponding to different behaviours. The first part occurs at low concentrations (1 to 3%); in this range Log (R_S/R_0) varies linearly along with the concentration. For higher concentrations the variation of the sensitivity is more important: the relative resistance of some sensors increases by more than thirty orders of magnitude when increasing the E.O. concentration in the range of 1 to 13%. For the 15% concentration a saturation behaviour is observed for all the sensors: the sensitivity decreases.

The maximum error is about 0.5% due to the good stability of the sensors.

In order to verify the ability of our sensor array to follow a sudden change in the oil concentration, we have exposed the sensors to different and successive concentrations without regeneration phases. In the figure 5, the FIS SP-AQ1 and SP-MW0 responses are presented, they show a notable variation which can be easily evaluated.

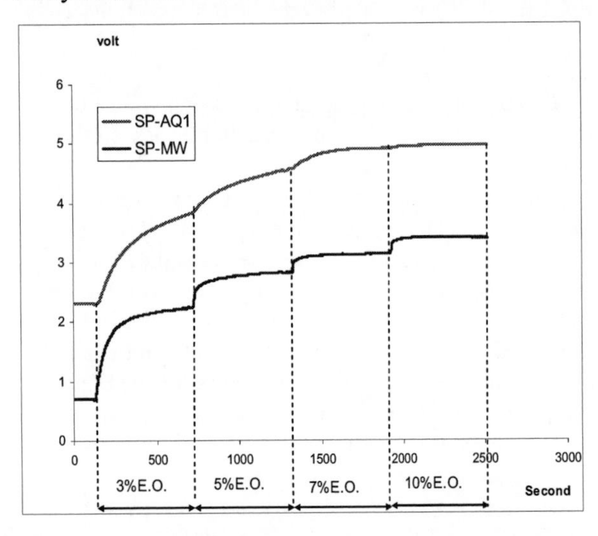

FIGURE 5: Successive expositions of the sensors to different E.O. concentrations.

CONCLUSIONS

This array study shows the capacity of the chosen sensors to detect the variation of natural oil quantities. We can easily differentiate several levels of a selected diffusion range, with a good reproducibility. For all the sensors, the responses increase notably with the concentration of the essential oil.

We are currently studying the detection of other essential oils in order to select the best sensors reacting to each chosen oils.

In a second step, we have planned to control the uniformity of the gas diffusion in the experimental chamber, in order to start experiments with alive subjects.

Future studies should also take up suggestions [7] to investigate effects of odorant mixtures, as well as those of single odor molecules, and also the effects of natural odor products compared to synthetic products.

ACKNOWLEDGMENTS

The authors would like to thank Carlos Etienne Donfack for his helpful assistance during the experimental work.

REFERENCES

1. J. L. Styles, "The use of aromatherapy in hospitalized children with HIV Disease" in *Complementary Therapies in Nursing &Midwifery,* 3, 1997, pp. 16-20.
2. L. Jackie and al , "Odor Provocation Test for Laryngeal Hypersensitivity" in *Journal of Voice,* 22, 2006, No. 3, pp. 333-338.
3. J. Lehrner, G. Marwinski, S. Lehr, P. Johren, L. Deecke, "Ambient odor of orange in a dental office reduces anxiety and improves mood in female patients" in *Physiological & Behavior,* 71, 2000, pp. 83-86.
4. P. Strobel A. Lfakir, M. Siadat, M. Lumbreras, "A portative gas recognition system based on metal oxide gas sensor array" in *IEEE Sensors journal*, 1, 2004, pp. 123-126, ISBN 0-7803-8692-2.
5. C. Delpha, M. Siadat, M; Lumbreras, "Discrimination and identification of a refrigerant gas in a humidity controlled atmosphere containing or not carbon dioxide: application to the electronic nose" in *Sensors and Actuators B*, 98, 2004, pp. 46-48.
6. D. Ryan, P. Prenzler, A. Saliba, G. Scollary, "The significance of low impact odorants in global odour perception" in *Trends in Food Science and technology*, 19 (2008), pp. 283-389.
7. J.S. Jellinek, "Aroma-chology: a status review" in *Perfume Flavor,* 19, 1994, pp. 25-49.

Technical Standard For Multigassensors

Th. Hübert and U. Banach

BAM Federal Institute for Materials Research and Testing
Unter den Eichen 87, 12200 Berlin, Germany
thomas.huebert@bam.de

Abstract. The guideline draft VDI/VDE 3518 describes the state-of-the-art of multigassensor technology and can support the use of multigassensors by supplying technical specifications and assistance for practical applications. It is divided in 5 parts concerning terms, structure and classification, testing of multigassensors, odor perception with electronic noses, specific applications and hints for appropriate applications. The first part briefly is presented contains and explanations and definitions of important terms, a description of construction and working procedure of multigassensors, a classification according to application categories and functionalities, minimum requirements on multigassensors, characteristic parameters and a scheme for assessment in quality grades.

Keywords: Multigassensor, Standardisation, Codes and Standards
PACS: 07.07.Df

INTRODUCTION

Codes, standards and guidelines give common accepted rules in industry and a benchmark for correct technical procedures. They can support quality, comparability and safety use of this type of sensors. Technical standards are necessary for conformity assessment and certification in order to determine that agreed relevant requirements are fulfilled. Thus they will increase confidence of customers for this new technology.

In the framework of VDI - the Association of German Engineers, Society for Measurement and Automatic Control, a group of experts has been preparing a guideline on multigassensors [1].

Standards for detectors for flammable and toxic gases are already well established and may be used for multigassensors too [2, 3]. But they do not consider specific properties and procedures of multigassensors and give no rules for many applications.

BASIC CONCEPT OF VDI/VDE 3518

Multigassensors as commercial available products were considered in context to the various intended applications and to the required performance. Therefore a diversification and classification according to application categories and functionalities is suggested as a basis for statement of applied testing methods and assessment in quality grades.

CONTENT OF VDI/VDE 3518-1

The first part of the guideline is subdivided in sections, which are briefly described in the following.

Terms and Definitions

Multigassensor is a measuring system in order to differentiate, identify or quantify one or more gaseous components in a gas mixture or distinguish between gas mixtures.

Electronic nose is such a multigassensor, which gives information about the odor of a gas mixture by using odor reference data obtained from a human olfactometric panel.

Further used terms to elucidate properties, application and classification of multigassensors are defined also.

Structure and Procedure

Multigassensors contain in many cases units for sampling and gas conditioning, sensor elements or a sensor array of different types and measuring principles, units for signal possessing, data storage and transfer. A characteristic of multigassensors technology is the application of various mathematical

CP1137, *Olfaction and Electronic Nose: Proceedings of the 13th International Symposium,* edited by M. Pardo and G. Sberveglieri

methods for signal analysis and pattern recognition to allow visualization, classification and quantification of measuring results.

Multigassensors were supplied for specific applications. Therefore a conditioning or training phase of hard- and software has to be performed in order to get sufficient selectivity or sensitivity.

Classification

The classification of multigassensors follows in two ranges. One of them, are the application categories, like comfort increasing quality of live, diagnosis of status, process run and malfunctions, monitoring of technical processes and safety (security) for persons and technical facilities. The other range, are the functionalities of multigassensors such as the ability to differentiate, identify and quantify gases or gas mixtures, which implies different performance of multigassensors.

Minimum Requirements

Multigassensors shall be specified for the gasses and mixtures to detect, measuring ranges, uncertainties and the environmental conditions for use. They shall be constructed in such a way that permits a reliable and safe use. This includes an unambiguous readout of results and malfunction information, as well as conformity with essential health and safety requirements (CE mark).

Performance Parameters

The performance of multigassensors will be estimated according to parameters like trueness, reproducibility, measuring range, scope of application, detection limit, resolution, selectivity, sensitivity, robustness, responds and recovery time and live time of sensors and device. Additionally, the operation requirements on a user, needed training and measuring time were considered.

Assessment

For an assessment of multigassensors it is necessary to classify a multigassensors according to the agreed application category and the functionality first. In a result of following testing of sensor properties, then an assessment to quality grades can be carried out and confirmed that the multigassensor fulfils all demands, or fulfils with minor or only with remarkable limitations.

OUTLOOK

This draft will become after further discussion and qualification a national German standard. Just as could it contribute to standardization efforts in European scale.

ACKNOWLEDGMENTS

The authors are grateful to the members of VDI/VDE/GMA committee 2.62 “Multigassensors” who contributed all to this guideline draft and J. Berthold additionally for organizational support.

REFERENCES

1. VDI 1000:2006, Establishing guidelines, Principles and procedures.
2. EN 60079-29-1:2008, Explosive atmospheres - Gas detectors - Performance requirements of detectors for flammable gases.
3. EN 45544-1:1999, Workplace Atmospheres - Electrical Apparatus used for the Direct Detection and Direct Concentration Measurement of Toxic Gases and Vapours - Part 1: General Requirements and Test Methods.

Correlation Of An E-Nose System For Odor Assessment Of Shoe/Sock Systems With A Human Sensory Panel

Stephan Horras[1], Alessandra Gaiotto[2], Maria Mayer[3], Peter Reimann[1], Andreas Schütze[1]

[1]*Laboratory for Measurement Technology, Saarland University, Saarbrücken, Germany;*
[2]*Prüf- und Forschungsinstitut Pirmasens, Pirmasens;* [3]*Forschungsinstitut Hohenstein, Boennigheim;*
Email: s.horras@LMT.uni-saarland.de

Abstract. Evaluation of strength and quality of smell is today still primarily done with human sensory panels. For a range of applications, technical systems for an objective smell assessment would provide a great benefit in R&D and also day-to-day application. The project presented here specifically addresses the problem of assessing the strength and unpleasantness of smell caused by sweat in shoes and socks by an E-nose system. The ultimate goal is to provide a tool for developing improved shoe/sock systems with optimized materials.

The main approach to achieve this goal is to find a correlation between the assessment of a human sensory panel and the complex sensor response patterns of an E-Nose system to appraise the smell of sweat in shoes and socks. Therefore a range of test persons wear shoes and socks under defined ambient conditions in a controlled test environment as well as during everyday use. Afterwards the smell of the shoes and socks is both measured with the E-Nose system and assessed by a human sensory panel. We report here the results of the first larger test series and the identified correlation between the E-Nose system and the human assessment of the smell of sweat.

Keywords: data acquisition, neural networks, sensors (gas)
PACS: 07.05.Hd, 07.05.Mh, 07.07.Df,

INTRODUCTION

Smell is an important information source for many living creatures warning of dangers or aiding in finding food or sexual partners [1]. For humans smell also has the very important aspect of providing a sense of well-being in a pleasant environment or exactly the opposite if unpleasant or obnoxious smells are present. Especially for the interaction between humans, body smells influence sympathy or antipathy unconsciously much more then we generally take into account. Body odors are therefore a very important aspect in our everyday lives.

Suppression or mitigation of unpleasant odors is therefore a key issue for shoe and clothes manufacturers in Europe, especially to achieve a competitive advantage over low cost suppliers from the Far East. The evaluation of different materials and especially their many combinations with regard to the strength and character of smells after sweating is currently a very time consuming process: only complex studies with human sensory panels allow a more or less objective assessment of influence of different materials and of different construction characteristics of the shoe/socks system. In this context shoes present an exceptionable problem as they usually are not or cannot be washed so that microbes

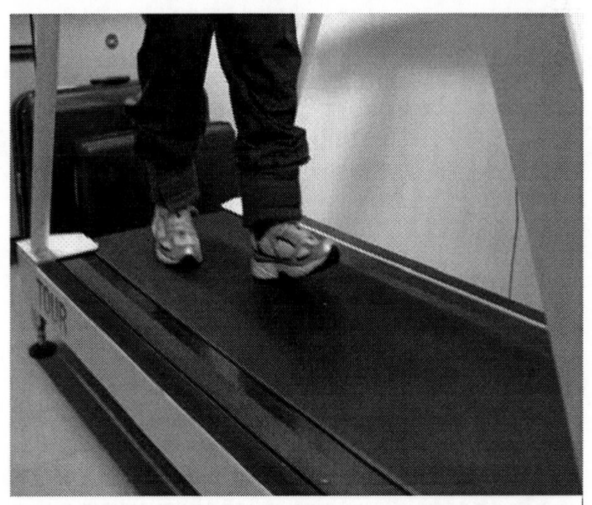

FIGURE 1. A test person at PFI on a treadmill placed in a climate controlled chamber to produce a reproducible amount of sweating and, thus, a reproducible odor for the assessment.

CP1137, *Olfaction and Electronic Nose: Proceedings of the 13th International Symposium,* edited by M. Pardo and G. Sberveglieri

and other substances can accumulate and lead to quite severe smells. Thus the aim of this project is the development and evaluation of a mobile test system for an objective evaluation of the strength and unpleasantness of odors caused by sweat in shoes and socks. The system is later to be used for research and development for the shoe and clothes industry to develop better materials and material combinations, in quality control during production and for settling costumer complaints [2].

EXPERIMENTAL AND METHODS

The two project partners Prüf- und Forschungsinstitut Pirmasens (PFI) and Hohenstein Institutes (HI) were responsible for a large range of tests with commercially manufactured shoes and socks and also for the assessment of the smell with human sensory panels. In a first series of tests 144 experiments were made with three different shoes and four different socks both under controlled and repeatable laboratory conditions as well as during everyday use. These tests form the basis for the correlation between the human nose and our E-nose system. Shoes and socks are worn by test persons and afterwards assessed by the human sensory panel and measured with the E-nose system.

Test persons at PFI were required to run or walk for one hour on a treadmill (Fig. 1) in a climate controlled chamber to sweat the shoes/socks under defined and reproducible ambient conditions. Test persons at HI on the other hand were wearing the shoes/socks for four hours under everyday conditions. The shoes and socks as well as the tested combinations of shoes and socks were the same for both partners.

The sensory panel assesses the smell of both shoe and socks individually in the categories overall olfactory sensation (pleasant to unpleasant on a scale from -2 to 2) and intensity of the smell of sweat (on a scale from 1 to 5). Furthermore, a detailed assessment of the categories sweet, sour and overpowering odor is done on a scale from 1 to 4 each. The sensory panels always consisted of 6 persons, 3 men and 3 women. The assessment of the panels resulted in average grades which were used as reference or nominal values for the correlation with the E-Nose system. Most experiments resulted in a standard deviation of the grade for smell of sweat, which was our main parameter, of .8 showing that the assessment by the sensory panel is not a very exact process.

The portable E-nose system mainly combines a test chamber, a sensor and electronics module and a PC for data evaluation and interpretation, shown in Fig. 2 [3]. A shoe or sock (or other object) to be tested is placed in the glass test chamber, which can easily be cleaned

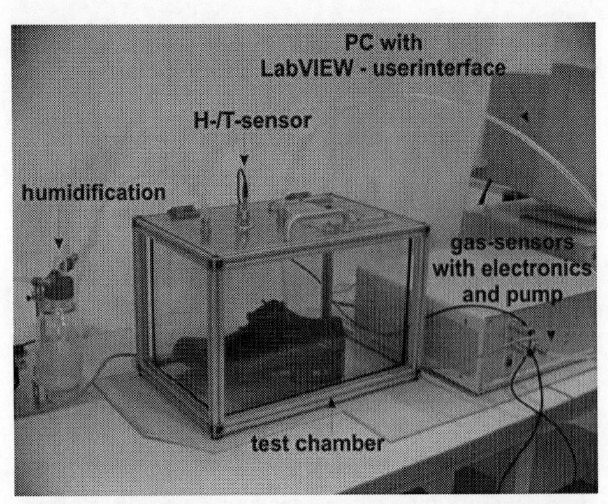

FIGURE 2. E-nose test setup in the laboratory at HI. A worn shoe is tested after wearing for four hours under everyday conditions. The different parts of the system are indicated.

after the experiments to suppress residual effects. To suppress unwanted influence by changes in relative humidity and interfering gases, the chamber is continuously flushed with pressurized clean air, which is humidified to almost saturation. In addition, the test chamber can be placed in a temperature chamber to further suppress the influence of changes in the ambient. The E-nose system itself is based on four semiconductor gas sensors, which are temperature cycled to achieve high selectivity and good stability for field tests without permanent recalibration of the sensors [3, 4].

The sensor electronics for temperature cycling and data acquisition is based on our PuMaH-module, a stand-alone microcontroller board for each individual sensor controlling the temperature cycling and signal read-out [5] with a common backplane for power supplies and data interfaces. It is connected via USB interface to a standard laptop computer for signal evaluation and interpretation.

The measurements with the E-nose system currently takes about 20 min, in which approx. 30 temperature cycles with a duration of 40 sec each are measured. To set the optimal duration for the E-nose measurements in terms of both quality and speed of the classification, it is important to determine when the difference between the patterns of two successive temperature cycles is small enough, i.e. the interpretation of the extracted features leads to the same classification result. Fig. 3 shows several sensor response patterns of a single measurement run of a worn shoe. While the upper plot contains the absolute response patterns, the lower plot shows the same patterns, but normalized to the mean value of the corresponding temperature cycle. While the absolute value patterns still have significant changes after 20 temperature cycles, i.e. after 15 min, the normalized

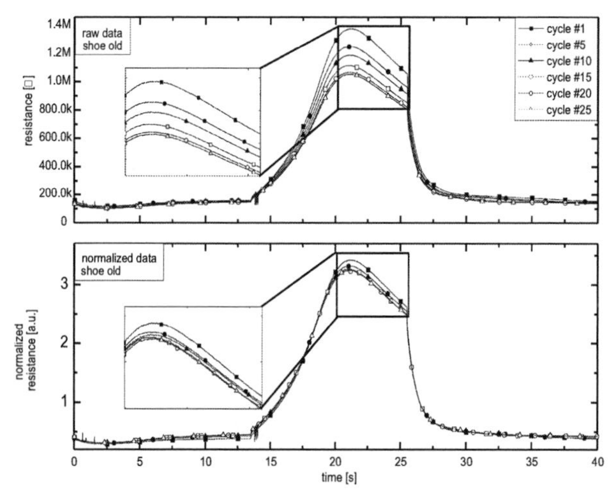

FIGURE 3, Response patterns of one semiconductor gas sensor (UST 1330, [6]) when exposed to the odor from a worn shoe. The upper plot shows the raw data, the lower plot the normalized data (all data divided by the average resistance over the complete temperature cycle). Normalization leads to very similar patterns for the different cycles, in principle allowing fast interpretation of the pattern long before the sensor reaches equilibrium.

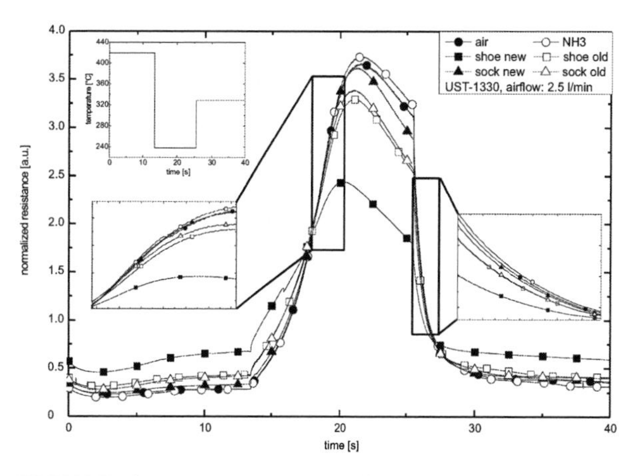

FIGURE 4, Response patterns of one semiconductor gas sensor when exposed to air, NH_3 (for calibration) and the odors from new and worn shoes and socks, respectively. Inserts show details of the sensor response. Note that the patterns for worn shoes and socks are quite similar to each other, while the patterns for new shoes and socks are distinctively different and also different from each other.

patterns reach equilibrium after only about 10 cycles, i.e. half the time. In addition, normalization of the sensor response partially compensates sensor drift [4]. The feature extraction for further evaluation is performed with the normalized sensor response patterns.

In future, the measurement time should be shortened to a few minutes for routine operation. This advantage of the temperature cycling approach was already shown in other projects, for example for the classification of vapors with micro gas sensors [7].

Both the E-nose measurements and assessment by the human sensory panel are repeated one day and six days after wearing to discover how the smell of the sweated shoes and socks changes over time. Shoes and socks are stored in plastic bags during that time to suppress unwanted influences from the outside during that duration.

RESULTS AND DISCUSSION

Preliminary tests with a standard E-nose system indicated that assessment of the smell of sweat seemed possible [2]. First experiments with our setup had shown that a separation of air, new and worn shoes, respectively, is possible [3]. Furthermore, Fig. 4 shows that the patterns of a worn shoe and a worn sock are similar compared to each other and distinctively different from the patterns obtained from a new shoe and sock, respectively. This indicates that the system is suitable for the detection of smell of sweat. The patterns show significant differences in the zoomed areas like different slopes, mean values or absolute maxima. These features are used to classify different shoes and socks after use dependent on the smell assessment by the human sensory panel. We have tested the classification with typical methods like linear discriminant analysis (LDA) or artificial neural networks (ANN) [8].

We tested the overall correlation between the human sensory panel and the prediction of a simple artificial neural network (ANN). Only eight relevant features, which were selected based on experience and manual assessment of the patterns, were extracted from the response patterns for the evaluation to prevent overtraining the network. The ANN was a simple three layer feedforward type with 8 hidden neurons and was trained with a backpropagation algorithm using a commercial software tool [9]. Fig. 5 shows the overall correlation for all temperature cycles for a single sensor over the 48 different experiments with shoes and socks during everyday use. It is evident, that even for this simple way of data evaluation and interpretation a good correlation is achieved.

In order to check the possibility of using less data for training and a purely linear signal evaluation approach, we selected only the data obtained from shoes, which were assessed by the human panels with values of 1, 2 and 3 ±1/6, respectively. The data were gathered in three groups for the average smell values of 1, 2 and 3. The LDA projection yielded three almost linearly separated groups along on discriminant function. Thus, the LDA could also be used for smell prediction using a linear relation between the DF value and the smell level. The resulting projection is shown in Fig. 6 for training and test data of all shoes.

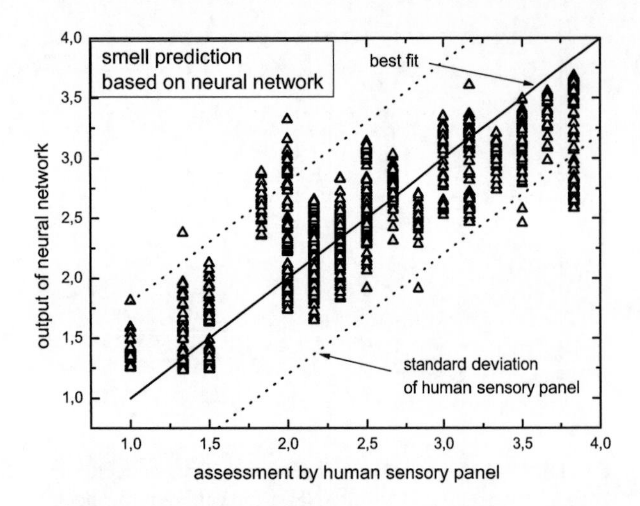

FIGURE 5, Correlation between prediction of the strength of smell of sweat with the E-nose system and the assessment of a human sensory panel. Data are based on 48 experiments with three different shoes and four different socks. The prediction is based on an artificial neural network (ANN). A three layer feedforward network was used with 8 input, 8 hidden and a single output neuron, the network was trained using a backpropagation algorithm [9].
The prediction is performed individually for each temperature cycle (approx. 28 cycles per test). A good general correlation is evident, which could be further improved using data from additional sensors and/or more features from the sensor response pattern.

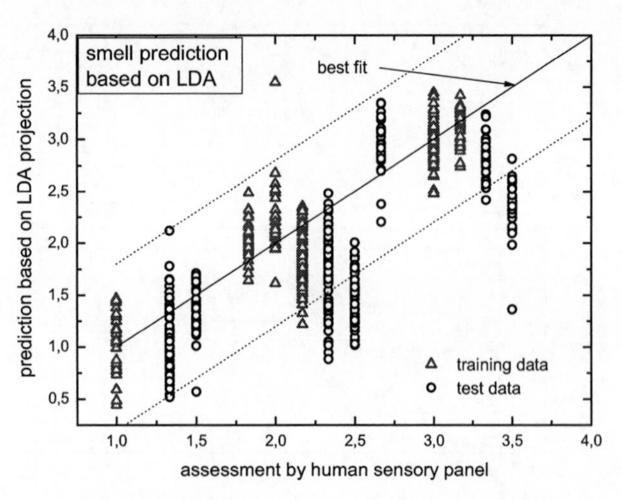

FIGURE 6, Correlation between prediction of the strength of smell of sweat with the E-nose system and the assessment of a human sensory panel. Data are based on 24 experiments with three different shoes. The prediction is based on Linear Discriminant Analysis (LDA). The LDA projection was calculated for three groups containing all samples with a smell assessment of 1, 2 or 3 (±1/6) by the human sensory panel, leading to a nearly linear separation of the groups along one discriminant function (DF). The smell prediction is then calculated fro the DF value using a linear function.
The prediction is again performed individually for each temperature cycle (approx. 28 cycles per test).

CONCLUSION

The hardware of a portable E-Nose system with semiconductor gas sensors for objective evaluation of smell of sweat of shoes and socks was developed. For this purpose shoes and socks were worn by test persons and afterwards measured with the E-Nose system and assessed by the human sensory panel. The assessment of the sensory panel was correlated with the sensor signals of the system using both ANN and LDA projection. In further tests with different sensors and more elaborate data evaluation using more features and more sophisticated evaluation techniques, we expect to further improve the correlation. In additional test series, the system will be checked with completely different shoes and socks in order to prove that the system is able to provide objective assessment of the smell of sweat for this specific application.

ACKNOWLEDGMENTS

Financial support for this research project (201ZN) of the research community "Prüf- und Forschungsinstitut Pirmasens e.V." in the program "Industrielle Gemeinschaftsforschung (IGF)", IGF-version ZUTECH, by the Federal Ministry of Economics and Technology via the AiF is gratefully acknowledged.

REFERENCES

1. Pearce, T.C., Schiffman, S.S., Nagle, H.T., Gardner, J.W. (eds.), Handbook of Machine Olfaction – Electronic Nose Technology, WILEY-VCH (2003).
2. Bartels, V.T., Umbach, K.H., Schmidt-Fries, U., "Untersuchung der Schweißgeruchsbildung in Textilien mit Hilfe einer elektronischen Nase," Melliand Textilberichte 84(10), (2003), pp. 872-874.
3. Horras, S., Fricke, T., Conrad, T., Engel, M., Schütze, A., „Objective evaluation of smell of sweat for optimal development of shoe/socks systems based on an electronic nose", ISOEN2007, St. Petersburg, Russia
4. Schütze, A., Gramm, A., Rühl, T., IEEE Sensors Journal, Vol. 4, No. 6 (2004), pp. 857-863.
5. Conrad, T., Hiry, P., Schütze, A., "PuMaH - a temperature control and resistance read-out system for microstructured gas sensors based on PWM signals", *Proc. IEEE Sensors Conf.*, Irvine, CA, Nov. 2005.
6. UST Umweltsensortechnik GmbH, Geraberg, Germany, http://www.umweltsensortechnik.de
7. Kammerer, T., Ankara, Z., Schütze, A., "GaSTON – a versatile platform for intelligent gas detection systems and its application for fast discrimination of fuel vapors," *Proc. Eurosensors XVII*, Guimarães, Portugal, Sept. 2003.
8. Gutierrez-Osuna, R., IEEE Sensors J Vol. 2, No. 3 (2002), pp. 189-202.
9. NeuroShell 2.0, Ward Systems Group Inc., USA, http://www.neuroshell.com

167 Detection of NH_3, DMA and TMA by a metal oxide gas sensor array for fish freshness evaluation

S. Capone°, L. Francioso°, P. Siciliano°, I. Elmi*, S. Zampolli*, G. C. Cardinali*

°*Institute for Microelectronics and Microsystems (IMM-CNR), via Monteroni, university campus, 73100 Lecce (Italy)*
**Institute for Microelectronics and Microsystems (IMM-CNR), via P. Gobetti 101, 40129 Bologna Lecce (Italy)*

A gas sensor array based on SnO_2 and In_2O_3 thin films prepared by sol-gel method and deposited on Si-micromachined substrates was configured. The array was calibrated to the detection of low concentration of some gas species (NH_3, DMA, TMA) considered markers of fish deterioration. The work aims to select a suitable multisensor system fulfilling the task of gas detector in a gas-chromatographic miniaturized system developed in "GoodFood" FP6 Integrated Project (FP6-IST 508774-IP).

Keywords: gas sensor array, fish freshness

PACS: 07.07.Df; 07.10.Cm

INTRODUCTION

The monitoring of fish freshness by reliable and fast analysis techniques perfomed in novel portable devices is an urgent requirement of fish industry. At present sensory evaluation of fish is the most used method for freshness evaluation. Other microbiological, chemical, biochemical, as well as other instrumental methods, provide information about fish freshness[1]. The use of a system of different gas sensors in a so-called electronic nose approach, has been also reported in literature[2].

2. EXPERIMENTAL AND METHODS

Two Si-micromachined chips, each containing four microsensors based on SnO_2 and In_2O_3 thin films respectively were used for the gas-sensing tests. The gas sensing layers were prepared by sol-gel technique and deposited by spin coating on the Si-micromachined substrates. The micromachined chip sized 5 mm x 5 mm was realized by advanced KOH anisotropic etching[3]; each microhotplate has a side of 1000 μm and a thickness of 0.70 μm. The four sensors of each chip, equipped with Pt heaters and gap electrodes, were characterized at four different temperatures (250°C, 300°C, 350°C, 400°C) in order to modulate the gas sensing properties in the sensor array. Specific measurements to the fish spoliage marker species (NH_3, DMA, TMA) in low concentration in humid air (R.H.=30%) were carried out. In particular, three different concentrations of NH_3 (4, 2, 1 ppm), DMA (1, 0.5, 0.25 ppm), and TMA (1, 0.5, 0.25 ppm) were considered. All the sensors were biased at 1 V and the electrical current monitored under exposures to gas and reference humid air.

3. RESULTS

The gas-sensing measurements confirmed a good response and sensitivity of SnO_2-based sensors towards first to TMA and secondly to DMA, especially at high temperature (~400°C). Regarding ammonia (NH_3), a competition between among a increase and a decrease of the sensor conductance was observed in the range of low gas concentrations. This effect is clearly dependent on working temperature; the sensor response decreases at increasing working temperature tending to an evident inverse response at high working temperatures. In_2O_3 showed also good gas-sensing properties towards TMA and DMA again an interesting negative response towards NH_3 at high temperature (~400°C). A comparison between SnO_2 and In_2O_3-based sensors as results of the laboratory tests is reported in fig.1.

The gas-sensing results demonstrate that the SnO_2 and In_2O_3-based sensors working at different temperatures could be good candidates for the

CP1137, *Olfaction and Electronic Nose: Proceedings of the 13th International Symposium,* edited by M. Pardo and G. Sberveglieri

detection of NH_3, DMA, TMA in a multisensor system dedicated to fish freshness analysis.

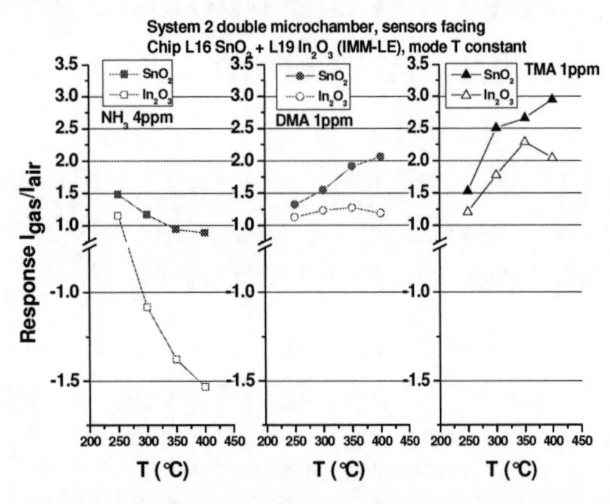

FIGURE 1. Comparison between the SnO_2 and In_2O_3-based sensors: responses to 4 ppm NH_3, 1 ppm DMA and 1 ppm TMA in humid air (RH=30%).

However, in order to fulfill this task, above all the multisensor system has to distinguish among the different chemical species to be detected. The experimented unexpected inverse responses of the SnO_2 and In_2O_3-based sensors towards ammonia is a not well understood phenomena and needs more investigations. In order to evaluate the discrimination ability of an array based on SnO_2 and In_2O_3 sensors, all the data were processed by Principal Component Analysis (PCA). This allowed also to select the best sensor array configuration, identifying those sensors that result redundant. The PCA score plot showed a good separation between ammonia and ammines group (DMA, TMA). For each gas, data clusters related to the different gas concentrations spread along preferential directions of gas concentration increment (fig.2).

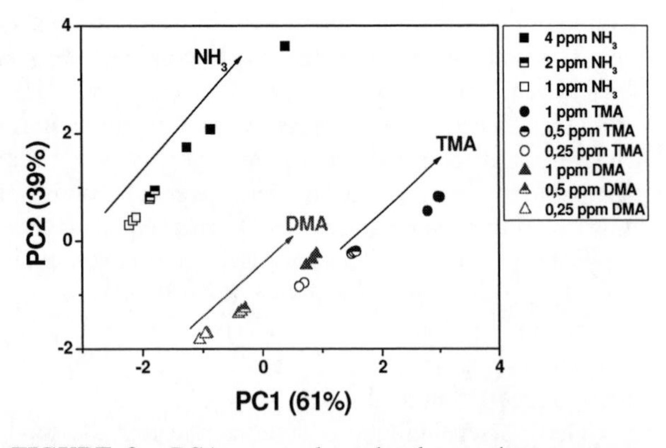

FIGURE 2. PCA score plot related to a 4-sensors array consisting of a couple of SnO_2 sensors working at 250°C and 400°C and on a couple of In_2O_3 sensors working at 250°C and 400°C

By considering that an electronic nose analyses a complex gas mixture as a whole without separating its components, the integration of a gas chromatographic column providing a separation among the different volatile species according to a suitable temperature gradient is an attracting approach that can potentially overcome the problem of the poor selectivity of semiconductor gas sensors[4]. This approach was considered in the framework of "GoodFood" FP6 Integrated Project (FP6-IST 508774-IP), where a multisensor miniaturized gas-chromatographic system prototype for the evaluation of fish freshness was developed[5].

4. CONCLUSIONS

Two different Si-micromachined chips containing four gas sensors based on SnO_2 and In_2O_3 thin films deposited by sol-gel technique were considered. A series of gas-sensing tests to different gas concentrations of NH_3, DMA, TMA were carried out in controlled environment at different sensor working temperatures (250°C, 300°C, 350°C, 400°C). A suitable sensor array was arranged. The results of Principal Component Analysis, applied to responses data, showed a potential ability of the array to distinguish among the considered gases (NH_3, DMA, TMA). The natural extension of this work is the use of the as-configured sensor array as multisensor detector in a superior microsystem prototype integrating a packed gas-chromatographich column. This microsystem was developed in the context of "GoodFood" FP6 Integrated Project, to which such work refers.

ACKNOWLEDGMENTS

We are grateful to GoodFood (Food Safety and Quality Monitoring with Microsystems) project (FP6-IST 508774-IP) for financing this work

REFERENCES

1. K. Mitsubayashi, Y. Kubotera, K. Yano, Y. Hashimoto, T. Kon, S. Nakakura, Y. Nishi, H. Endod, *Sensors and Actuators B,* **103** 463–467 (2004)
2. M. O'Connell, G. Valdora, G. Peltzer, R. Martın Negri, Sensors and Actuators B **80**, 149–154 (2001)
3. I. Elmi, S. Zampolli, G.C. Cardinali, *Sensors and Actuators ,B* **131** 548-555 (2008)
4. S. Zampolli, I. Elmi, J. Stürmann, S. Nicoletti, L. Dori, G.C. Cardinali, *Sensors and Actuators* B, **105** 400-406 (2005).
5. I. Elmi, S. Zampolli, L. Masini, G. C. Cardinali, M. Severi, L. Francioso, P. Siciliano, *Proc. of IEEE Sensors Conference*, Lecce, October 27th-29th, 2008

Detection of *Helicobacter pylori* infection by examination of human breath odor using electronic nose Bloodhound-214ST

E.P. Shnayder, M.P. Moshkin

Institute of Cytology and Genetics, SB RAS, 630090, Novosibirsk, Lavrentieva ave 10, Russia

D.V. Petrovskii

Institute of Animal Systematics and Ecology, SB RAS, 630091, Novosibirsk, Frunze str. 11, Russia

A.I. Shevela, A.N. Babko, V.G. Kulikov

Center of New Medical Technologies, ICB&BM, SB RAS, 630090, Novosibirsk, Pirogova str. 25/4, Russia

Abstract. Our aim was to examine the possibility of use e-nose Bloodhound-214ST to determine presence or absence of *H.pylori* infection using exhalation samples of patients. Breath samples were collected twice: at baseline and after oral administration of 500 mg of urea. *H.pylori* status of patients was confirmed by antral biopsy. Using two approaches for the data analysis we showed the possibility to distinguish *H.pylori* free and infected patients.

Keywords: helicobacter pylori, breathe odor, electronic nose Bloodhound-ST214
PACS: 01.30.CC

INTRODUCTION

By working out of a method of scent diagnostics it is possible to use 2 approaches: to develop specific detectors for each disease or to adapted of the universal detector (e-nose as an example) for determination of the different diseases. There are several demonstrations of using BloodHound e-noses (114 or 214) for scent recognition of the bacterial contamination of the different biological fluids [1,2,3]. Here we examine the possibility to use e-nose as detector, for identification of the stomach *H.pylori* infection.

EXPERIMENTAL AND METHODS

26 patients of both sexes with previous suspicion on *H.pylori* infection were studied just before esophagogastroduodenoscopy. Sampling of breath air hove been done in the morning on the starved patients. *H.pylori* status of patients was confirmed by antral biopsy. Since *H.pylori* has high urease activity we analyzed expired air collected to tedlar bags (Supelco®) before and 5 min after oral administration of 500 mg of urea in 25 ml water. All samples immediately analysis by an e-nose Bloodhound-214ST (Scensive Technologies Ltd.). We were used two approaches for the data analysis. First, Discriminate Analysis was performed for the Bloodhound data previously treated by soft of the device. Since urea load lead to production of the *H.pylori* derived ammonia in breath air [4], in second approach we used the difference between maximum deviation of the detector that was most sensitive to ammonia and detector that was least sensitive to ammonia.

CP1137, *Olfaction and Electronic Nose: Proceedings of the 13th International Symposium*, edited by M. Pardo and G. Sberveglieri

RESULTS

Using the fist approach we have been able to graphically discriminate analyzed breath samples on the 4 groups:

- exhalations of *H.pylori* free patients before urea;
- exhalations of *H.pylori* free patients after urea;
- exhalations of *H.pylori* infected patients before urea;
- exhalations of *H.pylori* infected patients after urea.

Well discriminated reaction to urea both at patients with and without *H.pylori* occurred. The discrimination infected and healthy patients was better for breath samples collected before urea administration.

Using the second approach we have found that breath samples of *H.pylori* free and infected patients showed good distinguishing after urea administration. The difference between most and least sensitive to ammonia detectors was positive and very variable in *H.pylori* free patients (2,20±2,26, mean±SD). And it was negative and low variable in infected patients (-0,87±0.29, $p<0,001$ in comparison with *H.pylori* free group). Only 2 patients from *H.pylori* free group were classified incorrectly. But the reliability of an antral biopsy is no more than 80%. So, we can not exclude about 20% of infected patients in the *H.pylori* free group. This mixing, perhaps, explains almost 10 higher variability of the differences between most and least sensitive to ammonia detectors in no infected group.

CONCLUSIONS

Our previous data show the possibility to distinguish *H.pylori* free and infected patients based on Bloodhound-214ST analysis of the exhaled odor. But increasing of studied groups is necessity for finally conclusion about applied perspectives this methodology.

REFERENCES

1. Gibson, T.D., et al., "Detection and simultaneous identification of microorganisms from headspace samples using an electronic nose," *Sensors snd Actuators*, Vol. 44 (1997), pp. 413-422.
2. Pavlou, A., Turner, A.P.F., Magan, N., "Recognition of anaerobic bacterial isolates *in vitro* using electronic nose technology," *Lett App .Microbiol.*, Vol. 35, No. 5 (2002), pp. 366-369.
3. Fend, R., et al., "Use of an Elecrtonic Nose To Diagnose *Mycobacterium bovis* Infection in Badgers and Cattle," *J Clin Microbiol.*, Vol. 43, No. 4 (2005), pp. 1745-1751.
4. Kearney, D.J., Hubbard, T., Putnam, D., "Breath Ammonia Measurement in *Helicobacter pylori* Infection, "*Digestive Diseases and Sciences*, Vol. 47, No. 11 (2002), pp. 2523-2530.

Classification of fuels using multilayer perceptron neural networks

Sérgio T. R. Ozaki, Nadja K. L. Wiziack, Leonardo G. Paterno, Fernando J. Fonseca

Department of Electronic Systems Engineering, Polytechnic School, University of São Paulo
Avenida Professor Luciano Gualberto, travessa 3, 158, 05508-900, São Paulo – SP, Brasil
sergio.ozaki@poli.usp.br

Abstract. Electrical impedance data obtained with an array of conducting polymer chemical sensors was used by a neural network (ANN) to classify fuel adulteration. Real samples were classified with accuracy greater than 90% in two groups: approved and adulterated.

Keywords: Neural network, fuel adulteration, interdigitated electrodes, poly(3-methylthiophene), poly(3-hexylthiophene), MLP.
PACS: 07.07.Df Sensors (chemical, optical, electrical, movement, gas, etc.)

INTRODUCTION

Fuel adulteration is a major concern in Brazil. The local governmental agency detects from 2 to 6% of problematic samples yearly, which is a lot considering Brazil's market size. Besides, as gasoline may have different properties depending on its source and the myriad of adulteration possibilities is even more vast, array of sensors based on "global selectivity concept" [1,2] seems to be more suitable methodology to detect problems in fuel.

The global selectivity concept encompasses the cross-sensitivity of non-specific chemical sensors and the use of multivariated data analysis methods as a way to provide "fingerprints" for samples of different chemical composition. The chemical sensors can employ different types of sensoactive materials, whose electrical responses are dependent on the physico-chemical characteristics of the media they get in contact with. Conducting polymers are *per excellence* suitable sensoactive materials, since their electrical conductivity is highly influenced by the environmental conditions and they can be easily processed in the thin film form by different techniques. Moreover, a vast number of conducting polymers with different structures and chemosensitivity are available.

In the present contribution, we employ an artificial neural network (ANNs) with multi-layer peceptron (MLP) [2,3] to classify data of fuel samples. ANNs are distributed computing systems composed by units connected by weighted links, simulating the structure and functioning of the brain. Data from real samples obtained with an array of conducting polymer sensors were analyzed. Impedance data of each sensor were used to discriminate and classify adulteration of gasoline and ethanol fuels using principal component analysis (PCA) and ANN.

EXPERIMENTAL AND METHODS

Fuel samples collected from gas stations and previously analyzed by a certified laboratory were discriminated and classified with an array of conducting polymers sensors [4,5]. The sensors were made of poly(3-methylthiophene) (PMTh) and poly(3-hexylthiophene) (PHTh) films deposited by chronopotenciometry and chronoamperometry [5]. The array was composed by four sensors with polythiophene films and one plain sensor. Impedance data were collected with the sensor array immersed in 25 mL fuel samples at 25°C by using a Solartron 1260A impedance analyzer.

Data were initially processed by PCA in order to discriminate groups of samples (40 fuel samples). Afterwards, a MLP ANN was developed (mais detalhes) and employed to classify samples in two groups: approved or adulterated.

CP1137, *Olfaction and Electronic Nose: Proceedings of the 13th International Symposium,* edited by M. Pardo and G. Sberveglieri

RESULTS

Impedance spectroscopy data of different mixtures of gasoline and ethanol and ethanol and distilled water were taken. The sensors respond to differences in the fuel composition. Figure 1 shows the response of a polythiophene sensor, which could be modeled by an equivalent circuit (inset) as proposed by Taylor [6]. All circuit elements are affected (Figure 2) by changes on the fuel composition. Thus, data of different properties such as dielectrics, double-layer, and interface characteristics could be used to analyze the fuel samples.

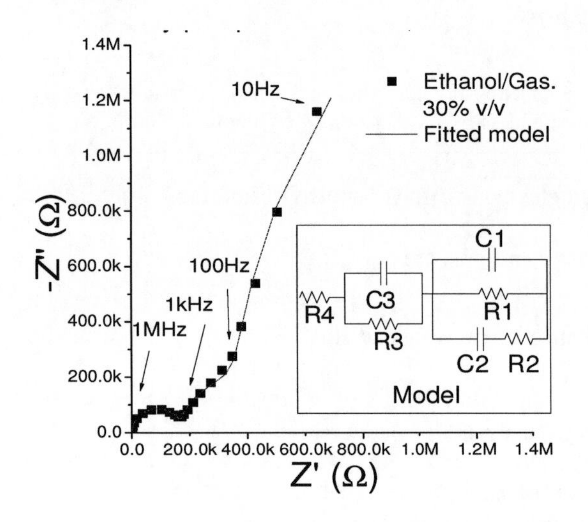

FIGURE 1. Impedance spectrum (Nyquist plot) of a polythiophene sensor submitted to a fuel sample, as quoted. The inset shows a model equivalent circuit for the sensor.

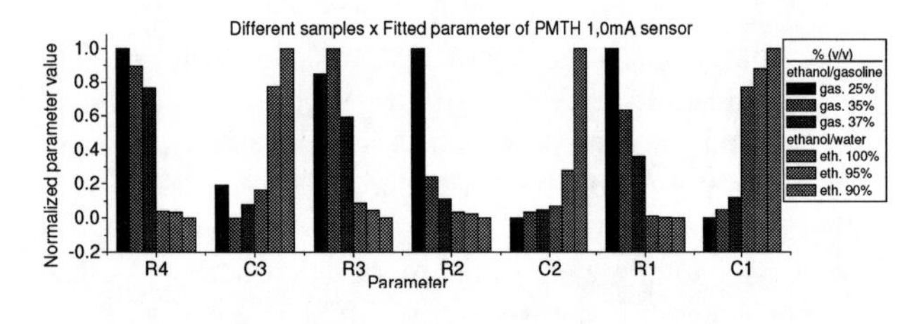

FIGURE 2. Circuit elements of the polythiophene sensor as a function of the fuel sample composition.

The PCA analysis (Figure 3) shows that approved fuels tend to group together while adulterated ones are sparser. The MLP ANN analysis was done with several configurations (learning rate, units in hidden layer, etc) and using the 5 fold cross-validation technique. Performances better than 90% were achieved.

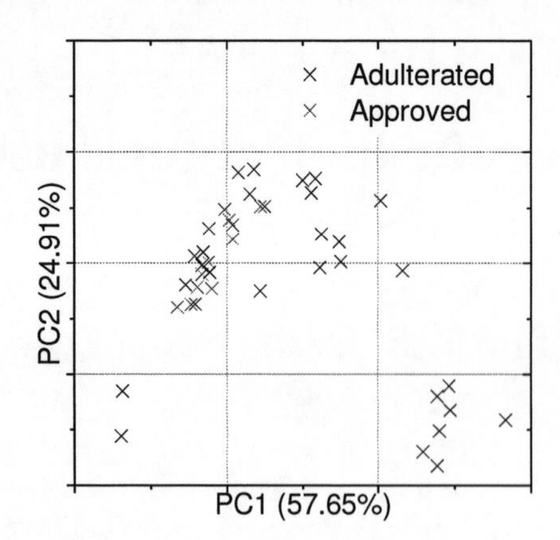

FIGURE 3. PCA plot of 40 gasoline samples analyzed with the conducting polymer sensor array.

CONCLUSIONS

The use of an ANN allowed to discriminate and classify real fuel samples according to their chemical composition, in two groups: approved and adulterated. Moreover, better performances can be achieved improving the ANN algorithm which is under current investigation.

ACKNOWLEDGMENTS

We thank FAPESP and CNPq for their financial support. We also thank Alessandra A. Correa Forner and Embrapa for the preparation of the polythiophene.

REFERENCES

1. Toko, K, "A taste sensor," Meas. Sci. & Tech., vol. 9, no. 12 (1998), pp. 1919-1936.
2. A. Riul, R.R. Malmegrim, F.J. Fonseca, L.H.C. Mattoso, An artificial taste sensor based on conducting polymers, Biosens. Bioelectron. 18 (2003) 1365–1369.
3. S.Haykin, Neural Networks: A Comprehensive Foundation, Prentice-Hall, Englewood-Cliffs, NJ, 1999.
4. Wiziack N. K. L. et al; "Effect of film thickness and electrode geometry on the performance of taste sensors made of nanostructured conducting polymer films" Sensors and actuators B, v. 122, p 489-492, 2007.
5. Wiziack N. K. L.; Integration and development of liquid sensors (ET) and gas sensors (EN) based in conducting polymers and different microelectrodes for fuel analyzes: Phd thesis in progress at University of Sao Paulo.
5. Correa, AA et al, "Weak ferromagnetism in poly (3-methythiophene) (PMTh) Synt. Met, vol. 121, n. 1-3, 1836-1837, 2001.
6. Taylor, D.M.; Macdonald A.G.; "AC admittance of the metal-insulator-electrolyte interface" J. phys. d: Appl. Phys, v.20, p.1277, 1987.

Olfactory, chemical and e-nose measurements to characterize odors emission of construct materials for the implementation of the European construction products directive (CPD) on a Belgian level.

Anne-Claude ROMAIN[1], Christophe DEGRAVE[1], Jacques NICOLAS[1], Marc LOR[2], Kevin VAUSe[2], Karla DINNE[2], Frederick MAES[3] and Eddy GOELEN[3]

[1].*University of Liège – Department "Environmental Sciences and Management", Avenue de Longwy, 185, 6700 ARLON – Belgium,* [2] *Belgian Building Research Institute (BBRI), Belgium*, [3] *Flemish Institute for Technological Research (VITO), Belgium*

Contact: acromain@ulg.ac.be

Abstract. Standardization work on test methods for dangerous substances in the field of construction products is currently ongoing at European level in CEN/TC 351 and EOTA PT9. A Belgian research project, with three partners, is going on to optimize current evaluation methods for VOCs (including SVOCs and VVOCs). This project examines also methods for the determination of particle emission from building materials, methods for evaluating the microbial resistance and also methods for odors determination. Different products like flooring materials are placed in three emissions chambers of different sizes (each partner has its own emission chamber) to evaluate their VOC's emission. Chemical analyses are performed by all the partners following the ISO-16000 standard series (by TD-GC-MS). This paper presents first results of ongoing odor measurements with sensory methods and e-nose. Keywords: e-nose, tin oxide sensors, drift, environmental application.

PACS: 07.07.df

INTRODUCTION

T research group is equipped with a real size chamber (50 m³). In this Belgian research project, in addition to the VOC's measurements, we are responsible for the odor measurement part.

For fifteen years, our research group "Environmental monitoring" is active in the metrology of malodors in the environment and especially in the development of e-noses for odor environmental real life applications. Olfactometry measurements (for instance with dynamic olfactometer according to the European Standard EN 13725 to measure odor concentration) and chemical analysis are performed in complement of the e-nose technology.

The indoor application has several characteristics similar to the outdoor approach: temperature and humidity variations, complex gaseous mixture, odor character, need of continuous monitoring (the CPD requires a monitoring of the material emissions on 28 days in the emission chamber). But there are two major differences for the odor evaluation: in the indoor community, current methods evaluate the odor intensity of the material more than the odor concentration and the total VOC's concentrations are usually lower than for malodors emitted by industrial, livestock or waste treatment plants.

1. EXPERIMENTAL AND METHODS

This paper presents the different sensorial approaches to characterize odor from construction products. Dynamic olfactometry measurements (to evaluate odor concentration in ouE/m³) and intensity measurements with the comparison to a scale of 5 n-butanol concentrations (dynamic – by a lab-made dynamic system- and static – with vessels following NF X 43-103 Standard) are explained and first results are shown. Chemical analyses of several materials emissions are also exposed.

An e-nose has been developed to monitor the emissions on 28 days. Firstly, experiments are investigated in the lab to test the ability of the metal oxide gas sensors array to discriminate different

CP1137, *Olfaction and Electronic Nose: Proceedings of the 13th International Symposium*, edited by M. Pardo and G. Sberveglieri

flooring materials, placed in 500 cm^3 glass vessels, with temperature and humidity conditions similar to those of the chamber. For these preliminary tests, the e-nose measurements are realized by cycles of reference air and headspace material samples. After this preliminary step, the e-nose has worked continuously (without reference air cycle) during 28 days with PTFE tubing placed inside the emission chamber and connected directly to the inlet of the e-nose.

2. RESULTS

Sensory results are shown in regard to chemical analyses. First continuous e-nose signals are exposed and PCA analysis is used to explore these first data sets.

ACKNOWLEDGMENTS

This study is supported by the Belgian Science Policy (Belspo); we thank also the Odometric Company for his support.

REFERENCES

1. Romain, A.-C., Delva, J., Nicolas, J., Complementary approaches to measure environmental odours emitted by landfill areas, Sensors and Actuators B: Chemical, 131, 1 (2008) 18-23.
2. Lor M. "How to evaluate the impact of building materials on health & indoor environment?" *10th Workshop Odour and emissions from plastics materials*, Brussels, October 2008.

Electronic Nose Aided Verification of an Odour Dispersion Model for Composting Plants' Applications

Riccardo Artoni[1,2,*], Luca Palmeri[2], Alberto Pittarello[3] and Maurizio Benzo[3]

1. Granular Research Group, DIPIC, University of Padova, via Marzolo 9, 35131 Padova (PD) Italy.
2. Environmental Systems Analysis Lab (LASA), University of Padova, via Marzolo 9, 35131 Padova (PD) Italy.
3. OSMOTECH S.r.l. via Taramelli 12, 27100 Pavia.
**corresponding author: riccardo.artoni@unipd.it*

Abstract. The dispersion of odour from a composting plant was calculated with the CALPUFF modeling system, where site specific meteorology and geophysical informations were taken into account. The odour emissions, both from forced and free-convection sources, were measured by means of dynamic olfactometry and implemented in the model. The results obtained from the model were verified with a MOS sensor based Electronic Nose equipped with a preconcentrator, placed in two target sites 50m and 250m far from the plant. Odour episodes, detected by electronic nose, were compared with model's forecast; a procedure for tuning the model parameters is needed, in order to reproduce Electronic Nose measurements.

Keywords: odour dispersion, field measurements, model verification
PACS: 92.60.Sz, 07.88.+y

INTRODUCTION

The study of the nuisance induced by dispersion of odours in the atmosphere is an issue of growing importance for regulatory purposes. Some models have been developed for the specific scope of determining the dispersion of odours[1]. Among the regulatory models developed for the dispersion of ordinary pollutants, the CALPUFF modeling system [2] seems to be a useful tool for many reasons, e.g. non-stationary transport modeling, complex topography and meteorology issues, etc. This study is a comparison between the prediction of the dispersion model as regards nonstationary odour dispersion and field measurements made with an Electronic Nose, in the case of a composting plant.

EXPERIMENTAL AND METHODS

The CALPUFF modeling system was implemented with site-specific meteorology from various sources (ground observations, radiosonde data). Emission sources, both forced (biofilter) and free convection (cumuli of material) ones were introduced in the model by means of emission estimates derived from dynamic olfactometric measurements(EN13725). The period under study was choosed to be a month for the purpose of comparing the results with field measurements.

For the scope of this work, it is necessary to monitor odour far from the plant: considering the presence of area sources, it was preferred to use a MOS Sensor based Electronic Nose, equipped with a preconcentrator to augment its sensitivity to low concentrations. It was choosen to place the E-Nose in two positions (50 m and 250 m far from the plant), and imposed measurement cycles of 12 minutes.

RESULTS

A complete comparison with model results and E-Nose measurements is a very complex task. Full analysis involved transformation of measurements into odour units and analysis through PCA allowing to recognize contribution from different sources.

As regards the first E-Nose placed within 50m of the plant (in the north direction), the results are somewhat discouraging, indicating that at such a distance the model fails to represent the odour dynamics. This may be due to mistreatement of the specific meteorology, but also related to the actual situation in the plant, at that hour. As regards the second E-Nose, placed 250m far from the plant in the SW direction, results behave better, and some odour

CP1137, *Olfaction and Electronic Nose: Proceedings of the 13th International Symposium*, edited by M. Pardo and G. Sberveglieri

episodes are well predicted by the model, as depicted for example in Fig. 1.

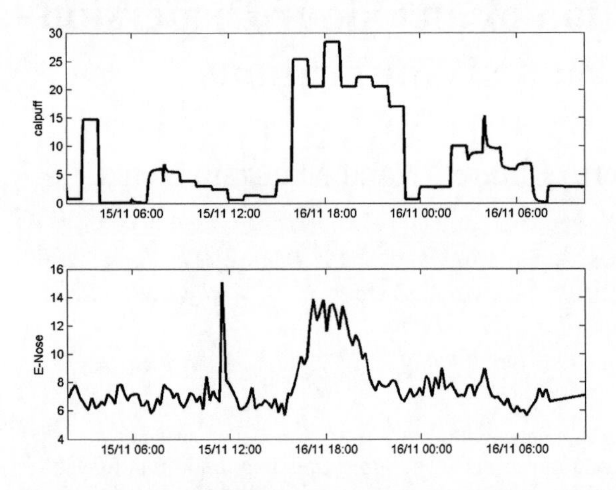

FIGURE 1. Example of odour episode predicted by the model (concentrations in Oue/m^3).

CONCLUSIONS

In conclusion, this work compares the predictions of a dispersion model with field measurements made with an E-Nose for a composting plant. Results for a distance of 250m far from the plant show that the model is able to represent odour episodes with some approximation. Further study is needed to understand failure at low distances, parameter tuning, and to produce a quantitative analysis of the results.

REFERENCES

1. Schauberger, G., Piringer, M., Petz, E.,Atm. Env. (2000), 34 (28), 4839-4851.
2. Scire J.S., Strimaitis, D.G., Yamartino, R.J., 2000. A User's Guide for the CALPUFF Dispersion Model. Earth Tech, Inc. Concord (MA-USA) 521pp.

Design and Implementation of an Electronic Nose System for the Determination of Fish Freshness

Rafael Masot[1], Miguel Alcañiz[1], Edgar Pérez-Esteve[2], José M. Barat[2], Luis Gil[1], Ramón Martínez-Máñez[1], Eduardo Garcia-Breijo[1]

*Institute of Applied Molecular Chemistry (IQMA). Universidad Politécnica of Valencia (UPV)
** Department of Food Technology (DTA). Universidad Politécnica of Valencia (UPV)
Camino de Vera, s/n 46022 Valencia, Spain. e-mail: ramape@eln.upv.es (R. Masot)

Abstract. An electronic nose has been designed to evaluate fish freshness. The e-nose consists of 15 commercial sensors, a home-made data acquisition system and a PC application that communicates with the data acquisition system. The equipment has been conceived so that the sensors can be configured in non-standard operation modes in order to improve the sensors discrimination capabilities..

Keywords: Electronic nose, Fish freshness, Gas Sensor
PACS: 87.19.lt, 02.50.Sk, 83.80.Ya

INTRODUCTION

The electronic noses are used to detect a wide variety of odours from different systems and processes as fish freshness [1]. These electronic noses include normally several standard or home-made sensors, and it is the combination of the signals of these sensors what allows the detection of a certain odour.

The designed e-nose includes 15 standard sensors. Each one of the sensors detects a specific gas. The system has been designed in a way that allows the adjustment of the measurement parameters (voltages and timings of heater and sensor resistors); so that measurements can be done out of the standard test conditions. The idea is to establish if some odours produce a characteristic response in sensors when they are working in non-standard test conditions.

EXPERIMENTAL AND METHODS

The 15 sensors were selected among commercial sensors for different gases (hydrogen, carbon monoxide, butane, methane, etc.). The measurement cycle for each sensor includes several steps: warming the heater, powering the sensor and measuring the sensor voltage. This measurement cycle is repeated constantly during the test. In order to give flexibility to the equipment, the data acquisition system has been designed based on a master-slave architecture (Figure 1); where each slave processor controls the measurement cycle of one sensor and the master processor collects the data from the 15 slaves and sends them to the PC.

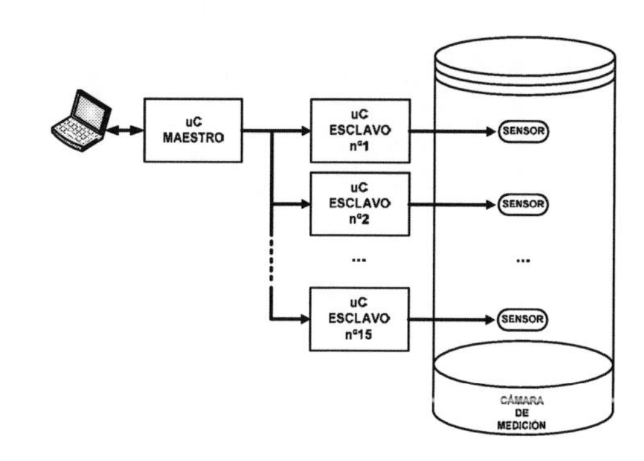

FIGURE 1. Electronic equipment.

CP1137, *Olfaction and Electronic Nose: Proceedings of the 13th International Symposium*, edited by M. Pardo and G. Sberveglieri

A gas management system (Figure 2) has been designed including two chambers: a concentration chamber (where the samples are introduced) and measurement chamber (where the sensors are placed). The system includes a pump that helps the gas circulation within the system. Two manual valves allow different configurations of the system (closed loop, cleaning cycle or external gas measurement).

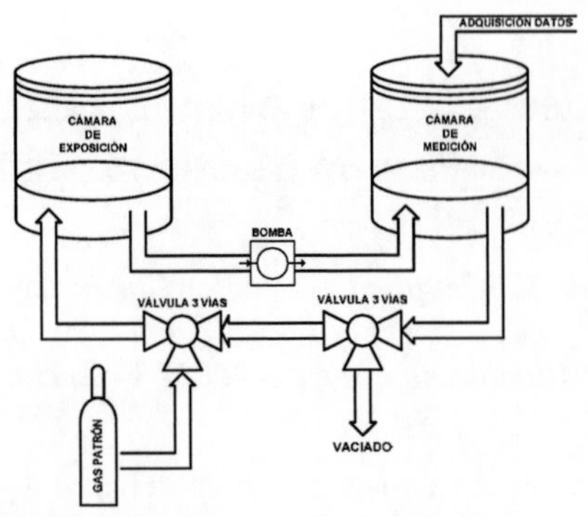

FIGURE 2. Gas equipment

The software allows the configuration of the test set-up for each sensor: voltage applied to the heater, voltage applied to the sensor, heater warming and cooling times, sensor powering times and measurement time.

RESULTS

The results of the tests corresponding to a fish sample at different freshness levels are displayed in Figure 3. Some of the sensors produce a characteristic response and this response depends on the freshness level of the fish sample. Using multivariate techniques discrimination between the different freshness levels can be achieved.

FIGURE 3. Sensors response for different freshness level

CONCLUSIONS

An electronic nose has been designed and standard tests have been carried out to determinate fish freshness. Changes in the sensors set-up parameters should be implemented to evaluate the improvements of the discrimination levels of the tests.

REFERENCES

1. El Barbri, N.; Llobet, E.; El Bari, N.; Correig, X. and Bouchikhi, B. *Materials Science & Engineering C*, 28. 666-670 (2008).

Application of Electronic Tongue in Identification of Soybeans

Camila Gregorut [a], Josemeyre B Silva [b], Nadja K L Wiziack [a], Leonardo G Paterno [a], Mercedes C C Panizzi [b], Fernando J Fonseca [a]

[a] *Departamento de Engenharia de Sistemas Eletrônicos, EPUSP, Universidade de São Paulo, Av. Prof. Luciano Gualberto, 158 trav. 3 – 05508-970 – São Paulo, SP, Brasil*
[b] *Embrapa Soja, Rod. Carlos João Strass, Distrito de Warta, Caixa Postal 231 – 86001-970 – Londrina, PR, Brasil.*
Email: cagregorut@usp.br

Abstract. Soybean is a product of great importance in the global economy and recognized by its great nourishing value with high protein content. In this work, a conducting polymer-based electronic tongue (ET) is employed to identify and discriminate five different soybeans cultivars with genetically distinct characteristics. Combination of electrical measurements and data analysis (PCA and PLS), permitted the ET system to discriminate the five different types of soybeans in accordance with a previous analysis performed by a human sensory panel.

Keywords: Electronic Tongue, Soybeans.
PACS: 82.35.Cd

INTRODUCTION

Soybean is a very versatile grain that originates various products and byproducts. The possibility to produce different types of soybeans is of interest to the food industry since they provide distinct taste features and protein contents [1].

Taste evaluation of soybean cultivars is usually done by humans (sensory analysis) whereas the chemical composition is assessed by analytical techniques, such as chromatography and mass spectrometry. In spite of their effectiveness, both methods require trained operators and are time-consuming. Alternatively, taste assessment can be carried out with chemical sensors arrays, known as electronic nose and electronic tongue (ET). These systems present as main advantages the capability of assessing taste characteristics of foodstuffs in short-time and are easy to be operated, even by unskilled personnel. They are also less expensive. For example, the ET has been widely used in the analysis of beverages, including coffee, tea, and mineral water [2]. In a previous investigation we have observed a straight correlation between the response of an ET and the sensory analysis for different sweetener formulations [3].

In the present contribution we have evaluated the ability of a conducting polymer-based ET in distinguishing five different types of soybeans cultivars with genetically distinct characteristics. The soybean cultivars under analysis were genetically designed for production of grains to be consumed as beverage. The study has two distinct goals: 1) production of grains to improve the taste attributes (cultivars 2, 3 and 4), and 2) grains to increase the protein content (cultivars 1 and 5).

EXPERIMENTAL AND METHODS

The soybeans samples from five different cultivars were prepared as raw extracts by cooking an weighted amount of beans in an autoclave and grinding it in a mortar. The extracts were then diluted with distilled water, in a proportion 1:10, v/v and evaluated with the ET. For each cultivar, at least five samples were prepared and measured at random.

The ET system was based on an array of eight individual chemical sensors made of conducting polymer films deposited onto interdigitated gold electrodes. The films were prepared via the layer-by-layer technique using an automated system described previously [4] and employing commercially available materials, including polyaniline and derivatives, polythiophene, polypyrrole, Ni(II) phtalocyanine, sulfonated, polystyrene, poly(allylamine hydrochloride) and sulfonated lignin [3]. The variety

CP1137, *Olfaction and Electronic Nose: Proceedings of the 13th International Symposium,* edited by M. Pardo and G. Sberveglieri

of polymers and other materials ensures the cross-sensitivity to the sensor array.

The electrical capacitance of each individual sensor when immersed into the different soybeans samples was measured with an impedance meter from HP, model 4263A LCR Meter. Data resulted from ET analysis were processed by principal component analysis (PCA) and partial-least square regression (PLS).

RESULTS

The electrical response of each individual sensor results from the interaction between the polymeric film and the chemicals present in the sample. The extent of such an interaction will depend on the polymer's chemical structure and the frequency of the electrical potential. At 1 kHz, sensors signal is due only to changes on the polymer's conductance [5]. Thus, higher cross-sensitivity is attained when sensors made of different polymers are operated at 1 kHz.

The sensory analysis performed by humans was able to distinguish the five soybeans samples according to taste attributes. The human evaluation has also detected significant differences in cultivars 2, 3, and 5. In a similar manner, the ET was also capable to distinguish the soybeans samples, as shown by the PCA plot in Figure 1. More than 90% of the data variance is provided by PC1 and PC2. Three groups of soybeans cultivars are distinguished along PC1 (cultivars 1, 2, and 4), whereas cultivars 2, 3, and 5 are distinguished along PC2 direction. Samples form cultivar 1, which presents the highest protein content are located in a more isolated region in the plot. It is also noted a clustering of samples from a same cultivar measured in different days. These data indicate that the sensors responses were systematically reproduced in different days of experiments. This performance ensures that the analysis was reliable and that the ET could be used repeated times with constant performance.

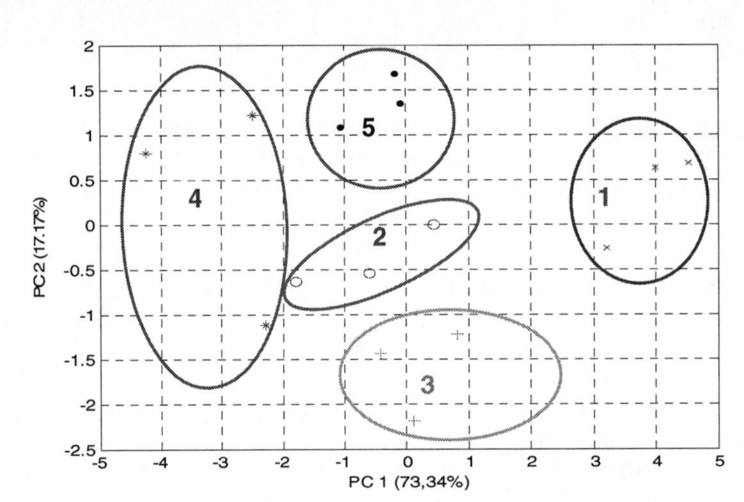

FIGURE 1. PCA plot for the 5 soybean cultivars obtained from data collected with the ET at 1 kHz.

CONCLUSIONS

The taste characteristics of five different types of soybeans cultivars were evaluated by a human sensory panel. With similar performance, an electronic tongue system based on conducting polymer sensors was able to detect differences in the soybeans samples and, consequently, could distinguish them in five different groups. The response of the ET was reproducible in different days of experiments enabling its use for repeated times with constant performance. Although partial, these results are encouraging and indicate a possible aplication for the ET based on conducting polymers.

ACKNOWLEDGENTS

The authors thank CNPq, CAPES and FAPESP for the financial support.

REFERENCES

1. J.B. Silva et al, *Pesquisa Agropecuária Brasileira* **42**, 1779-1784 (2007).
2. A. Riul Jr. et al, *Synthetic Metals* **132**, 109-116 (2003).
3. D. Dyminski et al, *Sensor Letters* **4**, 1-4 (2006).
4. G. S. Braga et al, *Materials Science and Engineering C* **28**, 555-562 (2008).
5. D. M. Taylor and A. G. MacDonald, *J. Phys. D: Appl. Phys.* **20**, 1277 (1987).

Electronic Nose: A Promising Tool For Early Detection Of *Alicyclobacillus spp* In Soft Drinks

I. Concina[a], M. Bornšek[b], S. Baccelliere[c], M. Falasconi[a], G. Sberveglieri[a]

[a]*SENSOR Laboratory, CNR-INFM, Via Valotti, 9 – 25133 Brescia, Italy*
[b]*Pivovarna Union d.d., Pivovarniška ulica 2 - 1000 Ljubljana, Slovenia*
[c]*ARPAV, Dipartimento Regionale Laboratori, S. L. di Padova, via Ospedale, 22 - Padova, Italy*

Abstract. In the present work we investigate the potential use of the Electronic Nose EOS835 (SACMI scarl, Italy) to early detect *Alicyclobacillus spp* in two flavoured soft drinks. These bacteria have been acknowledged by producer companies as a major quality control target microorganisms because of their ability to survive commercial pasteurization processes and produce taint compounds in final product. Electronic Nose was able to distinguish between uncontaminated and contaminated products before the taint metabolites were identifiable by an untrained panel. Classification tests showed an excellent rate of correct classification for both drinks (from 86% uo to 100%). High performance liquid chromatography analyses showed no presence of the main metabolite at a level of 200 ppb, thus confirming the skill of the Electronic Nose technology in performing an actual early diagnosis of contamination.

Keywords: Electronic Nose, food quality control, *Alicyclobacillus spp.*
PACS: 82.47.Rs

INTRODUCTION

In 1982, spoilage of aseptically filled apple juice from Germany was attributed to a new type of thermophilic acidophilic bacteria, later classified to a new genus, named *Alicyclobacillus spp.* (ACB), now acknowledged as a major quality control target microorganisms by beverage industry. These bacteria are able to survive commercial pasteurization processes and produce off-flavors in final product [1]. Functional waters were until recently due to acidic pH considered to be only susceptible to spoilage by yeasts, mycelial fungi and lactic acid bacteria, whose contamination is prevented by pasteurization and filling under aseptic conditions.

Since the contamination by ACB can origin from several potential sources, the identification of the critical point in production line is almost impossible. At present beside microbiological, analyses which last 3-5 days, the diagnosis of ACB's presence is performed by means of cromatographic techniques or biological assays based on determination of chemical markers (especially guaiacol, recognized as the predominan metabolite) produced by ACB. However, these techniques are valuable only when these secondary metabolites have been already produced, i.e. after ACB has irreparably spoiled the drinks.

In the present work we investigate the potential use of the Electronic Nose EOS835 (SACMI scarl, Italy) [3], already successfully used to reveal food microbial contamination [4], to early detect *Alicyclobacillus spp* in two commercial soft drinks.

EXPERIMENTAL AND METHODS

Two commercial flavoured beverages, naturally contaminated by ACB, were submitted as received to EN analysis. The beverages were low energy non-carbonated drinks, both of them containing 3% of fruit juice (peach/aloe vera and pear/lemon balm extract). These beverages are typically shelf-stable, high-acid, non-carbonated products packaged in PET containers

The EN EOS835, equipped by an array of six thin film semiconductor metal oxide sensors, was coupled to a static headspace sampling system. Uncontaminated and contaminated samples were analysed randomly; before any analysis, samples were sniffed by a non trained human panel and no anomalous odours were detected.

High Perfomance Liquid Chromatography (HPLC) analyses were carried out in a VARIAN serie 90 chromatograph coupled with a UV/DAD detector (220 nm), with a detection limit of 200 ppb..

CP1137, *Olfaction and Electronic Nose: Proceedings of the 13th International Symposium,* edited by M. Pardo and G. Sberveglieri

Explorative data analysis was performed by Principal Component Analysis (PCA).

Supervised classification was carried out by various classifiers: k-Nearest Neighbours (kNN) with k=1, Linear Discriminant Analysis (LDA), and Support Vector Machines (SVM) with a linear kernel. The classification error was estimated by using 5-fold cross-validation.

RESULTS AND DISCUSSION

Figure 1 shows the PCA biplot for beverage No. 1 (feature extracted: R/R0): uncontaminated and contaminated samples are well distinguished. The loading plot shows two classes of sensors that separate the objects on the PC2. Similar results were obtained for beverage No.2.

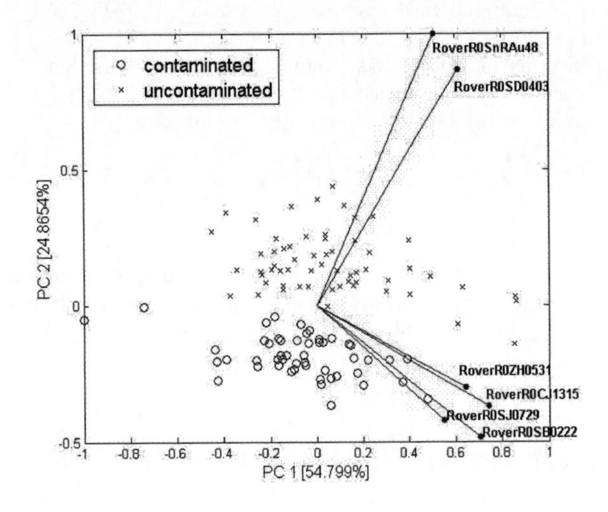

FIGURE 1. PCA score plot of data, showing the discrimination between uncontaminated and contaminated samples.

kNN classifier gave a 100% of correct classification for beverage No. 1, whereas results for the same drink obtained by DFA and SVM (Figure 2) showed 98% and 100%, respectively, of correct classification (by considering 6 sensors and PC 1-PC2). Data of beverage No. 2 were classified by taking into account 6 sensors and PC 1, 2 and 3. kNN correctly recognized the 87% of samples, whereas a percentage of 86% was found by applying both DFA and SVM.

HPLC analyses (data not shown) were carried out in order to check the presence of guaiacol in soft drinks. No peaks were detected attributable to this metabolite at the actual detection limit of 200 ppb.

By growing, the contaminant can alter the relative percentage composition of the original volatile profile, making the EN able to recognize the presence of ACB even when guaiacol is not yet producing in relevant amount, so that traditional analytical techniques, though intrinsecally more accurate than EN, could not identify the ACB presence. Moreover, though the guaiacol olfactory threshold is generally recognized as low (from 2 ppb to 30 ppb), an untrained human panel was not able to identify the spoiled products.

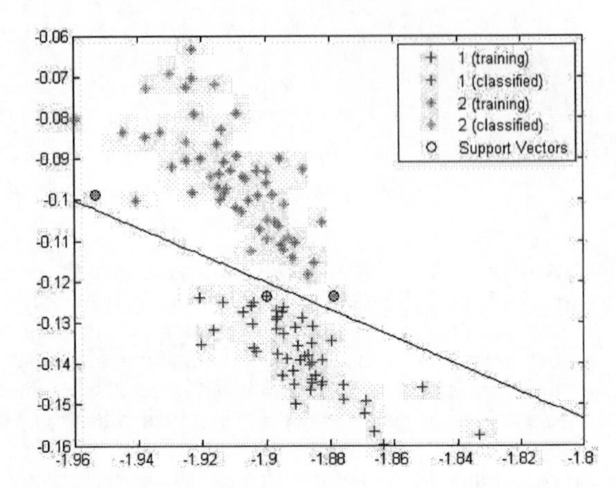

FIGURE 2. Classification of beverage No. 1 by SVM.

It is noteworthy that no sample pre-treatments are needed, thus allowing a simple and fast analysis. Moreover, tests were performed on naturally contaminated matrices, ready to be sealed on the market, thus demonstrating the effectiveness of EN technique.

CONCLUSIONS

EN EOS835 performed a trustworthy diagnosis of the presence of ACB, which was detected before starting a detectable guaiacol production. The technique is faster, simpler and cheaper than traditional analytical approaches.

Though EN technology still needs to be improved, in order to overcome some recognized limits (as poor selectivity and long-term stability of sensors), Electronic Noses are reliable candidates as microbial monitoring tool for industrial applications.

REFERENCES

1. M. Walker and C. A. Phillips, *Int. J. Food Sci. and Tech.* **43**, 250-260 (2008).
2. M. Peris and L. Escuder-Gilabert, *Anal. Chim. Acta* **638**, 1-15 (2009)
3. M. Falasconi, M. Pardo, G. Sberveglieri, I. Riccò, A. Bresciani, *Sensor and Actuators B* **110**, 73-80 (2005).
4. I. Concina et al., *Food Control* (2008), in press doi:10.1016/j.foodcont.2008.11.006.

DATA ANALYSIS

Could We Apply a NeuroProcessor For Analyzing a Gas Response Of Multisensor Arrays?

V. V. Sysoev[1], V. Yu. Musatov[1], A. A. Maschenko[1], A. S. Varegnikov[1], A. A. Chrizostomov[1], I. Kiselev[2], T. Schneider[2], M. Bruns[2], M. Sommer[2]

[1]*Saratov State Technical University, Polytechnicheskaya 77, Saratov 410054, Russia. E-mail: vsysoev@sstu.ru*
[2]*Forschungszentrum Karlsruhe, Hermann-von-Helmholtz-Platz 1, 76344 Eggenstein-Leopoldshafen, Germany*

Abstract. We describe an effort of implementation of hardware neuroprocessor to carry out pattern recognition of signals generated by a multisensor microarray of Electronic Nose type. The multisensor microarray is designed with the SnO_2 thin film segmented by co-planar electrodes according to KAMINA (KArlsruhe Micro NAse) E-nose architecture. The response of this microarray to reducing gases mixtured with a synthetic air is processed by principal component analysis technique realized in PC (Matlab software) and the neural microprocessor NeuroMatrix NM6403. It is shown that the neuroprocessor is able to successfully carry out a gas-recognition algorithm at a real-time scale.

Keywords: Electronic nose, neural processor, multisensor array, pattern recognition
PACS: 07.07.Df; 07.05.Mh

INTRODUCTION

The humans percept a smell information coming from the surrounding world with the help of neural mechanisms, which analyze multi-dimensional images generated by olfaction receptors [1]. This biological approach inspired [2] the development of a machine olfaction device, which is frequently called in literature as Electronic Nose or E-nose [3]. This instrument includes (*i*) the array of gas sensors, which generate signals in the presence of target gas or gas mixture, and (*ii*) pattern recognition technique(s), which sort out and discriminate the multi-dimensional sensor signals into classes corresponding to prevoiusly measured reference gases. The processing algorithms are based on extracting and analyzing statistical features as well as on employing artificial neural networks (ANNs) (for recent reviews, see [4,5]). The ANN algorithms benefit of neurological related performance as learning from experience, making generalizations from similar situations and judging situations where poor results were achieved in the past [6]. Especially, the problem of shifting references, as always present in realistic environments, can be diminished by ANNs, which are able to adjust themselve automatically during an operation. So, using ANNs do not require *a priori* knowledge about statistics underlying the sensor array signals and could be considered as the most natural way to design the E-nose instrument fully mimicking the biological counterpart.

In most cases, the ANNs are emulated by software developed for conventional personal computers (see, for example, [7]). But such implementations are often insufficient to meet the real-time requirements of many industries. Therefore, hardware platforms are recently developed to perform basic operations with ANN algorithms which employ a parallel processing and are capable of increasing the speed compared to conventional digital processors. Here, we consider using a hardware ANN processor, "dual-core" NeuroMatrix® NM6403 [8], for carrying out a gas recognition task based on signals generated by a single-chip multisensor array of KAMINA type [9].

EXPERIMENTAL AND METHODS

The Multisensor Microarray And Data Acquisition

The multisensor microarray employed in this study was fabricated according to earlier reported protocols [9-11]. In brief, the SnO_2:Pt film was deposited by r.f. magnetron sputtering on the front side of $SiO_2/Si/SiO_2$ substrate, over of approx. 4x8 mm^2 square, followed by the sputtering of Pt co-planar electrode strips, 1 μm width, and thermoresistors. The electrode architecture

CP1137, *Olfaction and Electronic Nose: Proceedings of the 13th International Symposium,* edited by M. Pardo and G. Sberveglieri

allows a segmentation of the metal oxide thin film to have the sensor array consisting of up to 38 chemiresistors (or sensor segments). The rear side of the substrate is equipped with four meander heaters. In order to modify the film sensor properties over the microarray, a gas-permeable SiO_2 membrane with a gradually changed thickness (in this particular case, of approx. 2-10 nm) was deposited on the top of the film using ion beam assisted deposition technique [12,13].

Following the fabrication, the chip was wired into PGA-120 holder and housed in the KAMINA unit [9] for measurements. In order to further differentiate the sensing properties of the SnO_2:Pt film segments, the substrate was inhomogeneously heated up to maintain a temperature gradient of approx. 7 °C/mm over the chip substrate or the operating temperature difference of ca. 290-360 °C. The sensor segment conductance between each pair of electrodes was read out with 1 Hz frequency per a whole microarray. The sensor signals are initially processed by the KAMINA unit electronics to be transferred through RS232 interface to PC for storage and visualization.

The gas response measurements were performed in a PC-controlled gas-mixing setup. In this exemplary case, the sample gases were vapors of isopropanol, acetone and CO mixed with a humidified, 50 rel. %, synthetic air (21 % of O_2, 79 % of N_2) to yield the 1 ppm of vapor concentration. The time of exposure to the test gas mixtures was ca. 10 min to ensure the conductance to come to stationary values. Between the exposures to the test gas mixtures, the chip was flushed with synthetic air for 15 min.

For gas-recognition purposes, the transient values of sensor segment conductances received under the change of atmosphere were removed from the consideration and only the stationary conductance values were employed. To minimize the effect of possible long-term drift of conductance and the dependence on the gas concentration, all the signals were normalized prior feeding to the ANN as $R_i \rightarrow r_i = R_i / R_{med}$, where R_i is the sensor resistance of ith sensor segment; R_{med} the median resistance value over the whole microarray.

The NeuroMatrix Neural Network Microprocessor

The NeuroMatrix neuroprocessor is described in details elsewhere [8]. In brief, it consists of 32-bit RISC core (a scalar processor) and 64-bit vector co-processor to support operations with vectors. The scalar processor has 8 address registers and 8 general registers and supports the following functions: matrix-by-matrix, matrix-by-vector multiplication, vector addition, saturation function for vector data and other vector operations support. The processor performs the operations over data vectors for one clock cycle. Each data vector is a 64-bit word of packed integer data, $\vec{D} = \{D_1 ... D_k\}$, where k is the number of elements. The bit length of each data in a packed word can be of any value in the range from 1 to 64 bits.

Results And Discussion

To classify the multisensor microarray signals into the groups corresponding to the test gas mixtures with the help of NM6403 neural microprocessor, we have realized Principal Component Analysis (PCA) algorithm based on correlation or Hebb's ANN with self-organization [14]. The number of neurons in the ANN corresponds to that of principal components employed for analysis (Fig. 1).

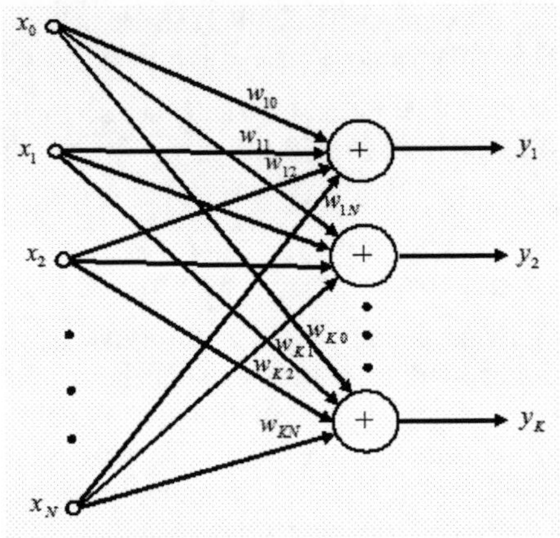

FIGURE 1. The ANN realized to perform the PCA pattern recognition method.

The ANN generates the principal components according to the Senger's rule: $y_i(k) = \sum_{i=0}^{4} w_{ij}(k) x_j(k)$, where $x_j(k)$, y_i, w_{ij} are inputs, outputs and weights, respectively. We have utilised four principal components covering a dispersion of 99.99 %.

The same ANN was realized in the NM6403 neuroprocessor and emulated in the conventional PC-based Matlab software. It is worth to mention that all the data in the ANN realized with neuroprocessor were 32-bit words packed into 64-bit words by double. The neurons had a linear activation function.

The received signals (480 points, consisting of 120 ones for each gas atmosphere including the synthetic air), were divided into two parts. The first one (320 signals) was employed for training of the neuroprocessor ANN algorithm, while the second one was a sampling for testing the algorithm's recognition ability (160 signals). The teaching procedure was limited by 150 epoches with an approximation constant of $1\cdot10^{-3}$. The recognition capacity was estimated as a percent ratio of correctly determined signals relative to all test signals under processing.

Because the neuroprocessor operates only with integer data, all the real weights and entry signals were converted to this data type by multiplication with factor K. The reduction of data from the real type to the integer one brings the error $\delta = \left| X_{real} - X_{integer} \right| / X_{real}$, where X_{real} and $X_{integer}$ are weight values calculated and converted to the integer type, respectively. It was found that the factor K=1000 is enough to keep the error, δ, below 0.3 %.

The Fig. 2 shows the results of PCA gas recognition of the test gas mixtures carried out with the help of the neuroprocessor. As one can see, the data related to different gases are clearly grouped by the PCA algorithm separately from each other allowing one to easy discriminate the gases.

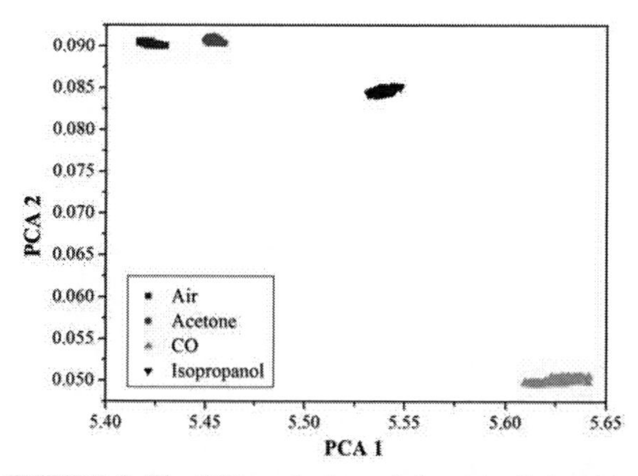

FIGURE 2. The PCA analysis carried out to distinguish the gas mixtures with the neuroprocessor. The reducing gases (CO, isopropanol, and acetone) are present in the mixture with synthetic air at 1 ppm concentration.

These results are compared with the ones received under modeling the same ANN in Matlab software at the conventional PC. This comparison is shown in the Fig. 3 as a related error $\Delta = \left| X_M - X_N \right| / X_M$ *versus* the multisensor array signal data. Here, X_M and X_N are the values of the first principal component calculated in Matlab and with the neuroprocessor, respectively. As one can see, the realization of the ANN algorithm with the neuroprocessor gives the error, Δ, not exceeding 1 % which still could be improved by application of factor higher than K=1000.

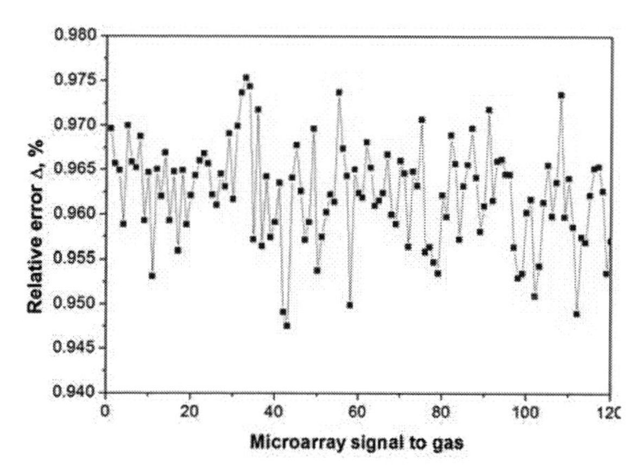

FIGURE 3. The relative error of PCA recognition (of CO exposure, as example), $\Delta = \left| X_M - X_N \right| / X_M$, carried out by ANN modeled with the neuroprocessor (XN) relative to that modeled with the Matlab software in a conventional PC (X_M).

CONCLUSIONS

It is shown that the neuroprocessor is feasible to be employed for gas recognition analysis of signals generated by the multisensor system of Electronic Nose type. These results could find an application in further development of E-noses combining chips carrying out multisensor arrays and hardware pattern recognition techniques.

ACKNOWLEDGMENTS

The authors thank G. Stengel, J. Benz and A. Serebrenicov for assistance in the fabrication of multisensor microarray chip and KAMINA unit. S.V.V. thanks the INTAS postdoc grant, YSF 06-1000014-5877. M.V.Yu. thanks DAAD and Russ. Ministry for Higher Education for the scholarship in frames of "Michail Lomonosov" program.

REFERENCES

1. G. M. Shepherd, *Nature* **444**, 316-321 (2006).
2. K. Persaud, G. Dodd, *Nature* **299**, 352-354 (1982).
3. T. C. Pearce, S. S. Schiffman, H. T. Nagle, J. W. Gardner (eds.), *Handbook of machine olfaction: electronic nose technology*, Weinheim: Wiley-VCH, 2003.
4. S. M. Scott, D. James, Z. Ali, *Microchim. Acta* **156**, 183-207 (2007).
5. P. C. Jurs, G. A. Bakken, H. E. McClelland, *Chem. Rev.* **100**, 2649-2678 (2000).
6. M.R.G. Meireles, P.E.M. Almeida, M. G. Simoes, *IEEE Trans. Industr. Electron.* **50**, 585-601 (2003).

7. S. Osowski, K. Brudzewski, T. Markiewicz, J. Ulaczyk, *J. Sens. Studies* **23**, 533–557 (2008).
8. P. A. Chevtchenko, D. V. Fomine, V. M. Tchernikov, P. E. Vixne, *Proceedings of SPIE* **3728**, 242-252 (1999).
9. J. Goschnick, *Microelectr. Engineering* **57-58**, 693-704 (2001).
10. V. V. Sysoev, I. Kiselev, M. Frietsch, J. Goschnick, *Sensors* **4**, 37-46 (2004).
11. V. V. Sysoev, N. I. Kucherenko, V. V. Kissin, *Tech. Phys. Letters* **30**, 759-761 (2004).
12. J. Goschnick, M. Frietsch,; T. Schneider, *Surf. & Coat. Technology* **108-109**, 292-296 (1998).
13. M. Bruns, M. Frietsch, E. Nold, V. Trouillet, H. Baumann, R. White, A. Wright, *J. Vacuum Sci. & Technology A* **21**, 1109-1114 (2003).
14. D. O. Hebb, *The organization of behaviour: a neuropsychological theory*, Mahwah: Lawrence Erlbaum, 2002.

Two Analyte Calibration From The Transient Response Of Potentiometric Sensors Employed With The SIA Technique

Raul Cartas[a], Aitor Mimendia[a] Andrey Legin[b], and Manel del Valle[a,*]

[a] *Sensors and Biosensors Group, Chemistry Dept., Universitat Autònoma de Barcelona, Edifici Cn, 08193 Bellaterra, Barcelona, SPAIN*
[b] *Chemistry Dept, St. Petersburg University, Universitetskaya nab. 7/9, 199034 St. Petersburg, RUSSIA*
e-mail: manel.delvalle@uab.es

Abstract. Calibration models for multi-analyte electronic tongues have been commonly built using a set of sensors, at least one per analyte under study. Complex signals recorded with these systems are formed by the sensors' responses to the analytes of interest plus interferents, from which a multivariate response model is then developed. This work describes a data treatment method for the simultaneous quantification of two species in solution employing the signal from a single sensor. The approach used here takes advantage of the complex information recorded with one electrode's transient after insertion of sample for building the calibration models for both analytes. The departure information from the electrode was firstly processed by discrete wavelet for transforming the signals to extract useful information and reduce its length, and then by artificial neural networks for fitting a model. Two different potentiometric sensors were used as study case for simultaneously corroborating the effectiveness of the approach.

Keywords: neural network, wavelet transform, muli-analyte calibration, potentiometry.
PACS: 82.47.Rs, 43.60.Hj, 07.05.Mh

INTRODUCTION

Multivariate signals are definitely a consolidated trend in analytical chemistry despite their complexity, which makes mandatory the application of appropriate chemometric tools for their handling. In the field of electrochemical sensors for liquids, an already consolidated approach for dealing with these kinds of signals is the electronic tongue. An electronic tongue is a system that has been widely defined throughout the literature [1]; it uses a non-specific potentiometric or voltammetric sensor array that responds non-selectively to a series of chemical species along with some of the existent signal processing techniques.

Non-selective sensors placed in a multi-component solution produce complex signals which contain information about the different components diluted in the solution. This is an advantage rather than a disadvantage since the richer content of information recorded provides additional chemical information which in turn is used to differentiate the multiple analytes present.

Processing of complex signals has been addressed using tools such as multiple linear regression, principal component regression, partial least squares, non-linear partial least squares [3] and artificial neural networks (ANNs) [3, 4]. From this set of tools, ANNs have outstood due to its ability to model linear and non-linear responses obtained from the sensors. Despite this good characteristic, ANNs becomes unpractical when the amount of departure information for modeling is too large. To surpass this problem, data pre-processing using methods such as principal component analysis (PCA) [3], Fourier transform [5] and Discrete Wavelet Transform (DWT) [6, 7] have been tried in order to compress the signals before modeling, while preserving relevant information.

In this work, we apply the principles of a potentiometric electronic tongue to solve a binary mixture of heavy metals; the present approach takes advantage of the cross sensitivity showed by non-specific electrodes to simultaneously resolve a binary analyte mixture using the information recorded from one sensor only. The transient signals after step inputs of sample were pre-processed using DWT for reducing their length and extract their significant features; next, the compressed information was input to an ANN to construct the calibration model [5, 6]. The efficiency of the approach is showed by the modeling of both analytes in two different study cases attempted for this purpose, one using a Pb^{2+} sensor and another using a Cd^{2+} sensor.

CP1137, *Olfaction and Electronic Nose: Proceedings of the 13th International Symposium,* edited by M. Pardo and G. Sberveglieri

EXPERIMENTAL AND METHODS

Reagents, Apparatus And Procedure

Discrimination of two heavy metals was the goal of this application; 45 different binary mixtures of cadmium and lead in ranges [0, 2.44×10^{-4}] M for Cd^{2+} and [0, 4.97×10^{-5}] M for Pb^{2+} were programmed for being automatically prepared and injected into a cell by a sequential injection system (SIA) available at our laboratories [5, 8, 9]. The cell incorporated two ion selective electrodes (ISEs) with cross response from heavy metals, one based on a potentiometric Pb^{2+} PVC membrane with trioctylphosphine oxide as ionophore and the other based on a chalcogenide glass selective to Cd^{2+} [2] plus an Ag-AgCl reference electrode. Transient recordings for each electrode were simultaneously obtained for a recording time of 60 s per sample in steps time of 0.1 s. All reagents used were analytical grade or similar.

Information Processing And Modeling

Both matrices of potentials were processed to reduce the length of their data points. Compression was accomplished by discrete wavelet transforming the recordings and keeping only the approximation coefficients while discarding the detail ones. The full set of discrete wavelets available in MATLAB's Wavelet Toolbox was tested with decomposition levels ranging from 1 up to 9 in order to find the one that worked the best in terms of retaining most of the original information with the fewer components for correctly modeling the next step.

The calibration model was based on an artificial neural network (ANN) built with MATLAB's Neural Network Toolbox and trained using the approximation coefficients previously obtained. Structures with one and two layers, different number of neurons and different transfer functions in their hidden layers were tested; the goal was to train a model able to generalize to unknown data.

RESULTS

The departure universe of data for building the calibration model consists of two input matrices (one per ISE) formed by the transient recordings (600 points length) obtained with the 45 samples, plus a target matrix formed by the corresponding binary mixtures of heavy metals for each sample. The length of each recording makes mandatory to compress the data before building the calibration model with the ANN. The sets of recordings from each electrode are shown in Figure 1.

The mother wavelet and decomposition level used in our application were chosen after processing the recordings and evaluating the ratio of compression, the energy retained by the approximation coefficients after each decomposition, and the degree of similarity between the original recording and the one reconstructed from the approximation coefficients.

Compression ratios ranged from 1.9:1 up to 300:1 for the all the possible combinations of discrete wavelets available in MATLAB and decomposition levels from 1 up to 9. The energy retained by the approximation coefficients after each decomposition was over 99% of the total energy from the previous level for all cases, which means that almost no high-frequency components are contained in the raw recordings. Degree of similarity was evaluated by running a test described in [7], this test is intended to compare two discrete signals of the same length and output a value between 0 and 1 depending on the degree of similarity, the result is 0 when two signals

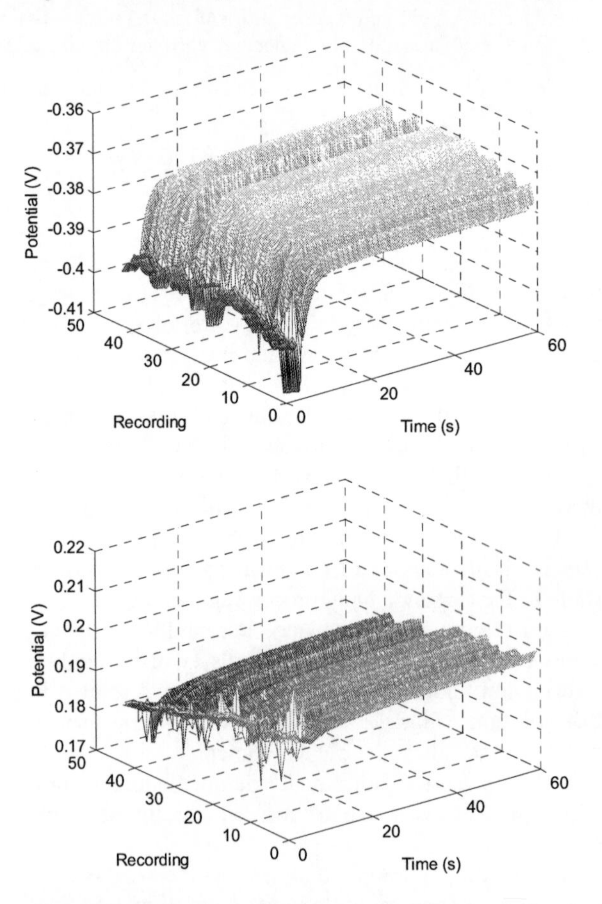

FIGURE 1. Transients recorded from the electrodes after the step introduction of the standards. Upper plot corresponds to Cd-electrode response and lower plot to Pb-electrode response.

have nothing in common and increases its value as similarity does. Results from this test ranged from 0.7466 for signals reconstructed with 6 coefficients up to 0.9994 for signals reconstructed with 314 coefficients. Those wavelets and decompositions levels whose reconstructed signals yielded values for similarity below 0.9 were discarded, as a result, only levels from 1 up to 6 were considered for evaluation.

All trained networks were feedforward type with one or two hidden layers and varying number of neurons. Structures with one output for modeling each metal and two outputs for the simultaneous resolution of Cd^{2+} and Pb^{2+} mixtures were built. Training was done using backpropagation with Bayesian regularization algorithm. Data was randomly split into two groups for training and testing purposes, 75% of the total information was used for training and the remaining was used for testing the network's generalization capability. For convenience, input and output training sets were first normalized in the interval [-1, 1] to facilitate the convergence of the learning algorithm. Normalization of testing sets was done according to extreme values obtained from training sets. The sum of squared errors (SSE) was used for tracking error during training and convergence goal was set to 0.09. Generalization capability was evaluated using the SSE and the degree of correlation obtained by linear regression analysis between the obtained and the expected values.

For one and two layers structures no good results were obtained for testing although training was fulfilled. For the one output case, the neural network with one layer structure didn't show good generalization capability for any number of neurons ranging from 6 up to 30. Error goal was met in 50% of training cases and correlation coefficient for testing was below 0.6 and even showing certain negative values.

The network structure that worked the best in our application had a single output, two hidden layers suiting non-linear transfer functions, 10 neurons in the first hidden layer, 4 neurons in the second one and linear transfer function in its output layer. The simultaneous determination of the analytes was accomplished with two parallel networks, one per metal. Figures 2 and 3 show the comparative graphs between the predicted and expected concentration values for Cd^{2+} and Pb^{2+}.

Regarding the approximation coefficients resulting after wavelet transformation and the decomposition levels left for evaluation, level 6 was also discarded since none of the set of coefficients obtained from this level for any of the discrete wavelets helped the network to converge. Decomposition with levels 1 to 4 were also discarded since training using these coefficients lasted too long, although the reconstruction was better displaying degrees of similarity closer to 1. As a result, only coefficients from level 5 were tested with the chosen network structure.

The wavelet that worked the best for our purpose is the biorthogonal wavelet of decomposition order 3 and reconstruction order 1, the number of approximation coefficients retained and used as input to the ANN for modeling is 23, and the mean value for the degree of similarity is 0.98. To corroborate the effectiveness of the network and wavelet processing chosen, 25 additional trainings with a 9-fold cross validation were performed for each electrode's recordings. Tests were run with random split of input and output data into five parts, four fifths were taken for training and one fifth for testing. As previously, training data was normalized before testing data. Error goal was met in all cases and minimum correlation coefficient was above 0.995 for training and 0.816 for testing; average values for SSE and correlation coefficient obtained with electrode 1 are shown in Table 1, Table 2 presents the results obtained with electrode 2. Uncertainties indicated correspond to 95% confidence interval of the 25 replicate training cases.

CONCLUSIONS

The present work demonstrates a procedure for simultaneous quantitative determination of cadmium

TABLE 1. Mean values for SSE and correlation coefficient (R) of the obtained vs. expected comparison lines of the cross validation process using the Cd^{2+} electrode. Uncertainties intervals calculated at 95% of confidence level.

Analyte	Training		Testing	
	SSE ($mol^2 \cdot l^{-2}$)	**R**	**SSE ($mol^2 \cdot l^{-2}$)**	**R**
Cd^{2+}	0.098 ± 0.0003	0.996 ± 0.000997	0.544 ± 0.062	0.917 ± 0.019
Pb^{2+}	0.0977 ± 0.0005	0.995 ± 0.0001	0.715 ± 0.044	0.844 ± 0.029

TABLE 2. Mean values for SSE and correlation coefficient (R) of the obtained vs. expected comparison lines of the cross validation process using the Pb^{2+} electrode. Uncertainties intervals calculated at 95% of confidence level.

Analyte	Training		Testing	
	SSE ($mol^2 \cdot l^{-2}$)	**R**	**SSE ($mol^2 \cdot l^{-2}$)**	**R**
Cd^{2+}	0.0978 ± 0.0005	0.996 ± 3e-4	0.725 ± 0.049	0.926 ± 0.004
Pb^{2+}	0.0976 ± 0.0004	0.995 ± 1e-4	0.259 ± 0.029	0.934 ± 0.031

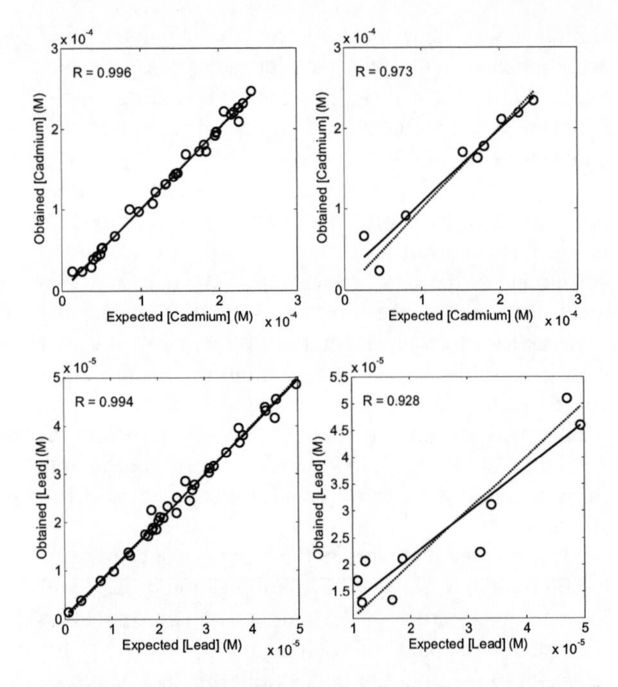

FIGURE 2. Comparison of obtained vs. expected concentration for cadmium and lead using the recordings from the Cd^{2+}-electrode. Dashed line corresponds to ideality and solid line corresponds to the fitting. Plots at left are the results for training, plots at right are the results for testing.

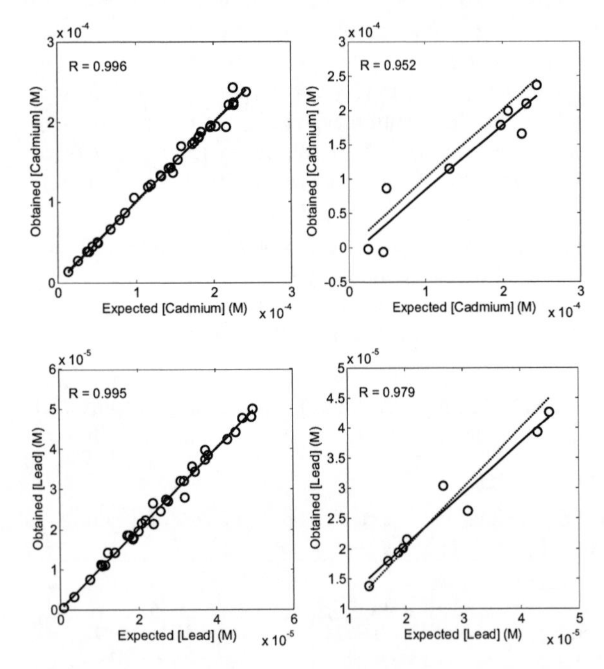

FIGURE 3. Comparison of obtained vs. expected concentration for cadmium and lead using the recordings from the Pb^{2+}-electrode. Dashed line corresponds to ideality and solid line corresponds to the fitting. Plots at left are the results for training, plots at right are the results for testing.

and lead using the transient recording from one single electrode. Our approach takes advantage of the interference produced on a sensor for building the calibration models; so, any of the Cd^{2+} or Pb^{2+} electrode can be used to determine the concentration of the heavy metals diluted in the same solution.

The calibration model is built by first extracting significant features employing the DWT, then evaluating the information obtained after wavelet transformation for getting a good compression/reconstruction relationship, and finally building the appropriate ANNs as calibration tools. The DWT processing of the recordings has permitted the reduction of the acquired data from each electrode in a factor ca. 26. Evaluation of the compressed signals using the degree of similarity was useful although it was not sufficient for estimating the amount of representation of the original data; the final estimation was made based on the capability of the network to fit a model for each analyte using the information from only one electrode.

From the results obtained with the trained networks, we can conclude that it is possible to model more than one analyte using the kinetic information recorded from a non-specific sensor after immersion in a multi-analyte solution, since this information is formed by the crossed response of the sensor to the analyte plus the interferents. This application intends to show the feasibility of building potentiometric electronic tongues for more than one analyte using responses from fewer sensors but exploiting their dynamic resolution.

ACKNOWLEDGMENTS

This work was supported by Spanish Ministry of Education, project TEC2007-68012-c03-02/MIC and by CONACyT (México) through PhD scholarship for one of the authors (R. C.).

REFERENCES

1. Y. Vlasov et al, *Pure Appl. Chem.* **77**, 1965-1983 (2005).
2. A.V. Legin et al, *Russ. J. Appl. Chem.* **78**, 89-95 (2005).
3. E. Richards et al, *Electroanal.* **14**, 1533-1542 (2002).
4. F. Despagne et al, *Analyst* **123**, 157R-178R (1998).
5. D. Calvo et al, *Anal. Chim. Acta* **600**, 97-104 (2007).
6. L. Moreno-Barón et al, *Anal. Lett.* **38**, 2189-2206 (2005).
7. L.Moreno-Barón et al, *Sens. Actuators, B* **113**, 487-499 (2006).
8. M.Cortina et al, *Anal. Bioanal. Chem.* **385**, 1186-1194 (2006).
9. D. Calvo et al, *Sens. Actuators, B* **131**, 77-84 (2008).

Multidetection Of Anabolic Androgenic Steroids Using Immunoarrays and Pattern Recognition Techniques

D. Calvo[1,2], J.P. Salvador[3,4], N. Tort[3,4], F. Centi[1,2], M.P. Marco[3,4], S. Marco[1,2,4]

[1] *Intelligent Signal Processing Group, Department of Electronics, University of Barcelona, Martí i Franques 1, 08028 Barcelona, Spain*
[2] *Artificial Olfaction Group, Institute for Bioengineering of Catalonia, Baldiri Reixac 13, 08028 Barcelona, Spain*
[3] *Applied Molecular Receptors Group, IQAC-CSIC, Jordi Girona 18-26, 08034 Barcelona, Spain.*
[4] *CIBER of Bioengineering,Biomaterials and Nanomedicine*

Abstract. A first step towards the multidetection of anabolic androgenic steroids by Enzyme-linked immunosorbent assays (ELISA) has been performed in this study. This proposal combines an array of classical ELISA assays with different selectivities and multivariate data analysis techniques. Data has been analyzed by principal component analysis in conjunction with a k-nearest line classifier has been used. This proposal allows to detect simultaneously four different compounds in the range of concentration from $10^{-1.5}$ to 10^{3} mM with a total rate of 90.6% of correct detection.

Keywords: Immunoarray, anabolic androgenic steroid, multidetection, pattern recognition, k-nearlest line
PACS: 02.50.Sk, 07.05.Kf, 07.07.Df, 87, 87.14.Lk

INTRODUCTION

Anabolic androgenic steroids (AAS) are compounds that mimic the endogen compound, testosterone. The use of AAS for growth promotion in farm animals was prohibited in 1981 by the European Union (Directive 81/602/EEC). Regarding to sports, the World Antidoping Agency (WADA) publishes a list of prohibited drugs [1], where AAS are included.

Several methodologies have been developed for the determination of AAS to control the use of these substances [2, 3]. Due to the difficulties related with this type of analysis such as throughput, sample treatment and cost per analysis, alternatives based on immunochemical methods are sometimes used. ELISAs (Enzyme-linked immunosorbent assays) are based on the specific recognition by antibodies against small molecules, such as AAS. ELISA can provide an accurate measurement, rapid results and it is capable to measure a large number of samples [4, 5].

The main problem of ELISAs for the determination of similar compounds such as AAS, is the cross-reactivity of different AAS in different assays. Statistical analysis can help to know which is the contribution of each analyte in each assay, in order to discriminate the target analyte. One possibility is to apply multivariate data analysis techniques. These techniques can be used for solving problems in the selectivity (cross-recognition) as well as to perform multidetection using high dimensionality data (for instance an immunoarray) This strategy is well-known in artificial olfaction[6]. This strategy can improve the possibilities of the single ELISA assay. In fact, it has been already reported for analysing mixtures of two pesticides using immunoarrays [7, 8].

The aim of this work is to combine an immunoarray and multivariate data analysis techniques to perform AAS multidetection (Figure 1). In order to carry out this objective, the Principal Component Analysis (PCA) in conjunction with k-nearest line (k-NL) classifier was used. K-nearest line classifier is a modified version of k-NN for elongated clusters.

EXPERIMENTAL AND METHODS

Materials

(a) ELISA: Polystyrene microtiter plates were purchased from Nunc. Washing steps were carried out using a SLY96 PW microplate washer (SLT Labinstruments GmbH). Absorbances were read using a SpectramaxPlus (Molecular Devices) at a single wavelength mode of 450 nm. The competitive curves were analyzed with a four-parameter equation using the software SoftmaxPro v2.7 and GraphPad Prism v 4

CP1137, *Olfaction and Electronic Nose: Proceedings of the 13th International Symposium,* edited by M. Pardo and G. Sberveglieri

(GraphPad Software Inc). Data presented corresponds to the average of at least two well replicates. Immunochemicals were obtained from Sigma Chemical Co.. Gestrinone was purchased from Sequoia Research Products, Ltd. Other chemical reagents were purchased from Aldrich Chemical Co.

(b) Buffers: PBS is 10mM phosphate buffer 0.8% saline solution, and unless otherwise indicated the pH is 7.5. Coating buffer is 50 mM carbonate-bicarbonate buffer pH 9.6. PBST is PBS with 0.05% Tween 20. 2X PBST is PBST double-concentrated. Citrate buffer is a 40mM solution of sodium citrate pH 5.5. The substrate solution contains 0.01% TMB (tetramethylbenzidine) and 0.004% H_2O_2 in citrate buffer.

(c) Analytes: Stanozolol (St), Boldenone (B), a-Boldenone (a-B), Methylboldenone (MB), Androstandiendione (ADD), 19-nortestosterone (NT), Methyltestosterone (MT), Progesterone (P), Testosterone (T), Androstandione (A1), Pregnenolone (Preg), Estrone (E1), Cholesterol (Ch), Dihydrotestosterone (DHT), Dexamethasone (D21P), Norstanozolol (Norst), 16b-hydroxystanozolol (16OH), Trembolone (Tr), 3'hydroxystanozolol (3OH), Ethynylestradiol (EES), Gestrinone (G), Tetrahydrogestrinone (THG), Estradiol (E2) and Norethandrolone (NEth).

ELISA Assay

(a) General Procedure: Microtiter plates were coated with the coating antigens BSA conjugates in coating buffer; 100 μL/well in all cases overnight at 4°C. The following day, the plates were washed four times with PBST, and the standard curves prepared (0.064nM-1000nM different AAS) were added to the microtiter plates (50μL/well) followed by the corresponding dilution of antiserum, previously optimized in lab (in PBST; 50μL/well) and incubated for 30 min at room temperature (RT) [9]. The plates were washed again as before, and a solution of anti-IgG-HRP (1/6000 in PBST) was added (100μL/well) and incubated for 30 min more at RT. After a new washing step, the substrate solution was added (100μL/well) and the enzymatic reaction was stopped after 30 min at RT with 4N H_2SO_4 (50μL/well). The curves were fitted to an equation according to:

$$Y = [(A - B)/1 - (x/C)D] + B \qquad (1)$$

where A is the maximal absorbance, B is the minimum absorbance, C is the concentration producing 50% of the maximum absorbance (or IC_{50}), and D is the slope at the inflection point of the sigmoid curve.

(b) Crossreactivity studies: Stock solutions of different AAS were prepared (10mM in dimethyl sulfoxide) and stored at 4ºC. Standard curves were prepared in the same buffer for each ELISA format in the same range that the curve was made and each IC_{50} was determined in the competitive experiment described above. The cross-reactivity (CR) values were calculated according to the following equation:

$$CR = \frac{IC_{50(\text{Reference Assay})}}{IC_{50(\text{AAS Tested})}} \cdot 100 \qquad (2)$$

Statistical Methods

In chemical sensing, it is well known that concentration changes produce cluster scattering. While some normalization methods can correct this effect, this is more pronounced in non-linear sensors. In this work, a modified version of k-NN is presented, specially designed to operate with cluster elongation due to concentration scatter. In this section, a description of the procedure is given

(a) Preparation of data and line formation: The data obtained from the cross-selectivity study was used for performing the multianalyte aproach. We selected some analytes that had no missing data for the 8 immunosensors. That is necessary in order to perform the data analysis. Thus, we had 64 assays (4 analytes x 8 immunosensors x 2 replicates).

First, we performed the mean of each pair of replicates. In order to build the k-nearest line classifier additional synthetic samples were added by linear interpolation within the available training data, in such a way that a sequence of points predicts the expected evolution due to concentration changes. When dense enough this sequence of points is a linear piece-wise line in feature space that spans the expected feature vector positions for a range of concentrations. The concentrations interpolated were from $10^{-1.5}$ to 10^{3}mM with increments of 0.1 (in logarithm units of mM). Thus, we obtained 46 samples per each ELISA assay.

To improve the linearity, the Logit transformation was applied.

$$x = \log(y/(1 - y)) \qquad (3)$$

After that, the data was autoscaled and normalized.

(b) Principal components analysis was used for visualization and for dimensionality reduction.

(c) The novel k-nearest line is a supervised method for classifing samples of unknown concentration. There is only one line per class/analyte. Its main principle is to assign the class of one sample by the majority vote of its k nearer lines. The nearest line is identified by the label of the nearest neighbour. The second nearest line is identified by the nearest neighbour label, excluded all data points belonging to the first line. This process goes on for additional nearest lines identification.

In order to evaluate the predictive ability of the classifier we used a leave-one-out validation. This methodology uses all samples, excluding one, as a training set and that sample is used as a test set. After that, the model is constructed again using all samples without the next test sample. This process is repeated N times, where N is the number of test samples. In this work we use as test samples the original samples. The leave-one-out was applied from the generation of new samples to the final classification in order to obtain a good predictive model. The results were presented in the confusion matrix that contains information about the obtained and the expected detection.

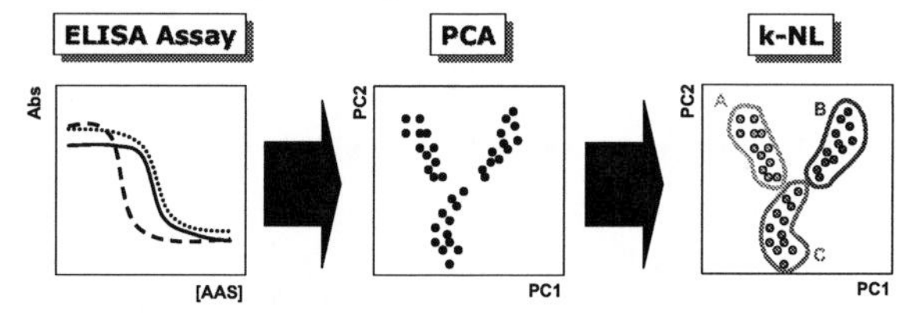

FIGURE 1. Scheme of the process for classifying AAS.

Data organization, mathematical transformations and graphics were performed using Matlab® R2007a from The MathworksTM. PCA was carried out using the special functions provided by PLS_Toolbox 3.0 from Eigenvector Research Inc. for Matlab use.

RESULTS

ELISA Assay

Different ELISA assays have been developed for the determination of AAS (table 1). The concentration of immunoreagent was previously optimized. For the determination of B, the assay As138/hA_B-BSA achieved a detectability value of 23.77nM. For the determination of MB exists different assays with different behaviour: As140/hB_MB-BSA, As142/hB_MB-BSA, As143/hA_T-BSA and As143/hB_B-BSA, with a detectability of 4.33, 0.79, 10.51 and 3.98nM respectively. In the case of the analysis of St, we have As147/hA_St-BSA and As147/hB_St-BSA, and IC_{50} of 2.60 and 0.38nM, respectivaly. Finally, for the determination of THG, the assay employed is As170/hG-BSA with an IC_{50} of 1.39 nM.

Regarding to cross-reactivity studies, each assay has a different profile of recognition of the different analytes of interest. The immunosensor As138/hA_B-BSA recognizes B and ADD equally. The assays developed for MB shows similar recognition for MT and for MB. Within these assays plus the As143/hA_T-BSA, the recognition of St is also important. In the case of the St ELISAs, the change of the coating antigen causes different profile. Cross-reactivity values of 100, 21, 45, 18 and 51% were obtained for St, MB, NorSt, 16OH and 3OH respectively. On the other hand, values of 100, 21, 2.5, 1.5 and 2% for the same analytes were obtained using the As147/hB_St-BSA. Finally, for THG detection, a specific assay can recognize partially G and NEth with a 20 and 62%, respectively.

Multianalyte ELISA Assay

We considered all assays as a unique experiment, but in fact, each assay was made independent. Finally, four analytes (DHT, MB, P and St) and the data from the 8 sensors were selected for carring out the aproach.

After data pre-processing, PCA captured 80.04% of the total variance in the first principal component, the second one captured 11.95% and the third captured 5.27%. We used these three PCs for performing the k-nearest line classifier, because these PCs had an accumulated variance higher than 95%.The Figure 2 shows the scores plot. There are two main tendencies, one with values PC1<1.5 where the four analytes are overlapped and the second with higher values of PC1, where we can distinguish visually the four analytes.

One can see in the scores plot that the analytes were distributed as a lines. We selected the k value as k=1 in order to obtain a robust model. Using that value, we considered the classifier as a nearest line aproach. The distance used for evaluating the nearest neighbours was the Euclidean distance in the scores space. To facilitate the interpretation of the results, we calculated the rate of the correct classification for each concentration and the confusion matrix.

TABLE 1. ELISA assays employed.

Assay	Target	As dilution	BSA-conj conc. (μg/mL)	IC_{50} (nM)
As138/hA_B-BSA	B	1/64000	0.0625	23.77
As140/hB_MB-BSA	MB	1/16000	0.0625	4.33
As142/hB_MB-BSA	MB	1/32000	0.125	0.79
As143/hA_T-BSA	MB	1/16000	0.125	10.51
As143/hB_B-BSA	MB	1/16000	0.5	3.98
As147/hA_St-BSA	St	1/2000	0.625	2.60
As147/hB_St-BSA	St	1/32000	0.039	0.38
As170/hG-BSA	THG	1/16000	0.125	1.39

TABLE 2. Confusion matrix. Between brackets there is the number of samples.

Total 90.6% (64)	Obtained DHT	Obtained MB	Obtained P	Obtained St
Expected DHT	75% (12)	12.5% (2)	12.5% (2)	0.0% (0)
Expected MB	0.0% (0)	100% (16)	0.0% (0)	0.0% (0)
Expected P	0.0% (0)	0.0% (0)	100% (16)	0.0% (0)
Expected St	12.5% (2)	0.0% (0)	0.0% (0)	87.5% (14)

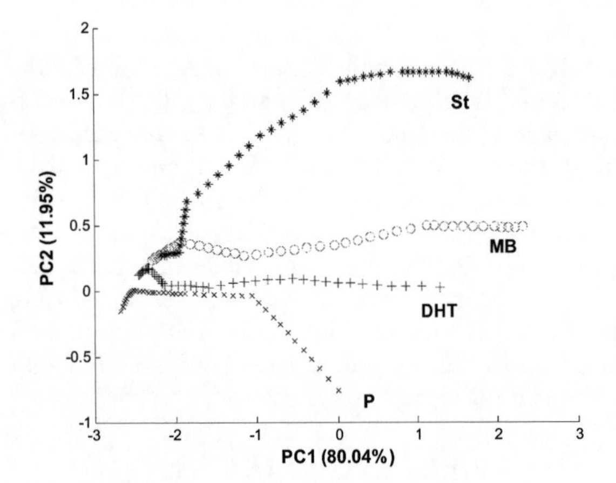

FIGURE 2. Scores plot. PC1 vs PC2.

As shown in the table 2, the total rate using leave-one-out validation was 90.6%. The samples that had MB and P were completely correct classified. The St analyte had a rate of 87.5%, it had been confused with DHT (at low concentrations). The worst classified analyte was the DHT, that had a 75% of correct classification. DHT was confused with MB (at high concentrations) and P.

The combined rate of correct classification for each concentration shows that if the concentrations were from $10^{1.9}$ to $10^{-0.2}$mM the analytes were completely recognised. There were wrong detections only at lowest and highest concentrations.

CONCLUSIONS

ELISA technique have been used in this work for multidetection of anabolic androgenic steroids. Eight cross-selectivity immunosensors were prepared and characterized. These assays have been used for knowing the different profile of recognition of the target anaytes. After data preprocessing using statistical tools, the PCA has shown the data structure. The structure visualization gave us an idea about if multidetection of AAS was possible and how we could perform it.

A k-Nearest Line procedure was implemented. Line formation was accomplished by producing a dense line representation by adding synthetic samples between consecutive training samples (once ordered in analyte concentration).

Finally, the multianalyte ELISA aproach was performed using 1-N line classifier and using the leave-one-out validation. Our proposal carried out multidetection of four different compounds in the range from 10^{3} to $10^{-1.5}$mM. We can conclude that muldetection of anabolic androgenic steroids using ELISA data is possible, especially at concentrations, from $10^{1.9}$ to $10^{-0.2}$mM.

ACKNOWLEDGMENTS

This work has been supported by the Ministry of Science and Education (Contract No NAN2004-09415-C05-02 and DEP2007-73224-C03-01). Financial support for this work was also provided by the Networking Biomedical Research Center on Bioengineering, Biomaterials and Nanomedicine (CIBER-BBN) through project MICROPLEX. The AMR group is a consolidated Grup de Recerca de la Generalitat de Catalunya and has support from the Departament d'Universitats, Recerca i Societat de la Informació la Generalitat de Catalunya (expedient 2005SGR 00207).

REFERENCES

1. WADA, The 2008 Prohibited list. International Standard, World Anti-Doping Agency 2008 http://www.wada-ama.org/en/prohibitedlist.ch2
2. W.Schanzer, P.Delahaut, H.Geyer, M.Machnik, S.Horning, *J. Chrom. B* **687** 93-108 (1996)
3. C.Van Poucke, C.Van Peteghem, *J.Chrom. B* **772** 211-217 (2002)
4. J.P.Salvador, J.Adrian, R.Galve, D.G.Pinacho, M.Kreuzer, F.Sánchez-Baeza, M.P.Marco, M.P.a.D.Barceló, "Chapter 2.8. Application of bioassays/biosensors for the analysis of pharmaceuticals in environmental samples" in *Comprehensive Analytical Chemistry*, Elsevier, 2007, pp. 279-334
5. J.P.Salvador, F.Sanchez-Baeza, M.P. Marco, *Anal. Chem.* **79** 3734-3740 (2007)
6. S. M.Scott, D.James, Z.Ali, *Microchim. Acta* **156** 183-207 (2007)
7. G.Jones, M.Wortberg, B.D.Hammock, D.M.Rocke, *Anal. Chim. Acta* **336** 175-183 (1996)
8. S.Reder, F.Dieterle, H.Jansen, S.Alcock, G.Gauglitz, *Biosens. Bioelectr.* **19** 447-455 (2003)
9. J.P.Salvador, F.Sánchez-Baeza, M.P.Marco, *Anal. Biochem.* **376** 221-228 (2008)

Blind Source Separation For Ion Mobility Spectra

Marco S[1,2], Pomareda V[1,2], Pardo A[1], Kessler M[3], Goebel J[3], Mueller G[3]

[1]*Department of Electronics, University of Barcelona, smarco@el.ub.es, apardo@el.ub.es, vpomareda@el.ub.es*
C/Martí i Franqués, nº1, planta 2, 08028, Barcelona, Spain.
[2]*Artificial Olfaction Lab, Institute of BioEngineering of Catalonia.*
Institut de Bioenginyeria de Catalunya (IBEC), C/ Baldiri Reixac 13, 08028 Barcelona, Spain.
[3]*Department LG-SI 2, EADS Innovation Works, EADS Deutschland GmbH.*
EADS Innovation Works , Dept. IW-SI - Sensors, Electronics & Systems Integration , 81663 München/Germany.

Abstract. Miniaturization is a powerful trend for smart chemical instrumentation in a diversity of applications.. It is know that miniaturization in IMS leads to a degradation of the system characteristics. For the present work, we are interested in signal processing solutions to mitigate limitations introduced by limited drift tube length that basically involve a loss of chemical selectivity. While blind source separation techniques (BSS) are popular in other domains, their application for smart chemical instrumentation is limited. However, in some conditions, basically linearity, BSS may fully recover the concentration time evolution and the pure spectra with few underlying hypothesis. This is extremely helpful in conditions where non-expected chemical interferents may appear, or unwanted perturbations may pollute the spectra. SIMPLISMA has been advocated by Harrington et al. in several papers. However, more modern methods of BSS for bilinear decomposition with the restriction of positiveness have appeared in the last decade. In order to explore and compare the performances of those methods a series of experiments were performed.

Keywords: Ion Mobility Spectrometry (IMS), Blind Source Separation (BSS), Multivariate Analysis, SIMPLISMA, MCR, Non-Negative Matrix Factorization (NMF).
PACS: 02.50.Sk, 07.05.Kf, 07.07.Df

INTRODUCTION

Ion Mobility Spectrometry (IMS) is a leading technology for the detection of chemical warfare agents, explosives and narcotics [1]. IMS technology has only a moderate chemical selectivity, and it gets worse when the system dimensions are scaled down. Moreover, in real world scenarios, IMS spectra may become messy featuring overlapping peaks and bad signal-to-noise ratio. In those conditions, blind source separation techniques are an option for spectra deconvolution. They may fully recover the concentration profiles and the pure spectra of the constituents. In this work, a series of laboratory experiments have been conducted to explore the performance of different blind source separations. In particular, SIMPLISMA [2], Multi Curve Resolution (MCR-ALS) [3] and Non-Negative Matrix Factorization (NMF) [4]. The results show that MCR-ALS and NMF are able to recover pure spectra and concentration profiles even for high degrees of peak overlapping. While SIMPLISMA has been applied in several occasions in the past, no comparative analysis with those other techniques has been reported.

CP1137, *Olfaction and Electronic Nose: Proceedings of the 13th International Symposium*, edited by M. Pardo and G. Sberveglieri

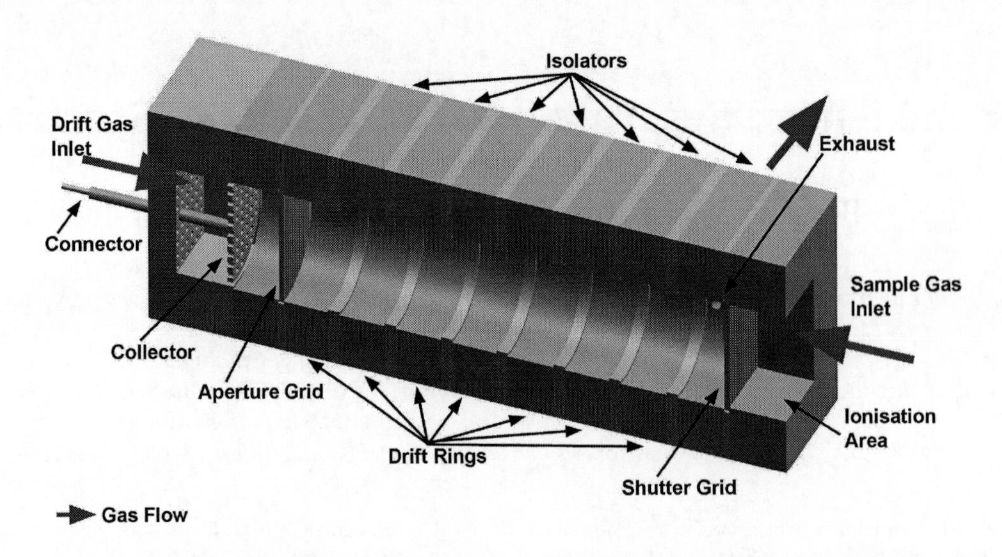

FIGURE 1. Scheme of an Ion Mobility Spectrometer.

RESULTS

Spectra for Anisole and Chlorbenzene were sequentially recorded in a Laser IMS in different instrument set-ups so to obtain spectra with different degrees of overlap. Principal Component Analysis (PCA), SIMPLISMA, MCR-ALS, NMF were applied to the recorded signals stored in a datasets containing both chemicals. Inner product has been considered as figure of merit to compare the recovered spectra and the concentration profiles with the pure ones. Figure 2 shows spectra in two extreme cases: a) Overlapping, b) Non-overlapping. Figure 3 shows the comparison of the various methods in terms of the angle. Best results are obtained with MCR-ALS and NMF. PCA fails unless the spectra are originally orthogonal.

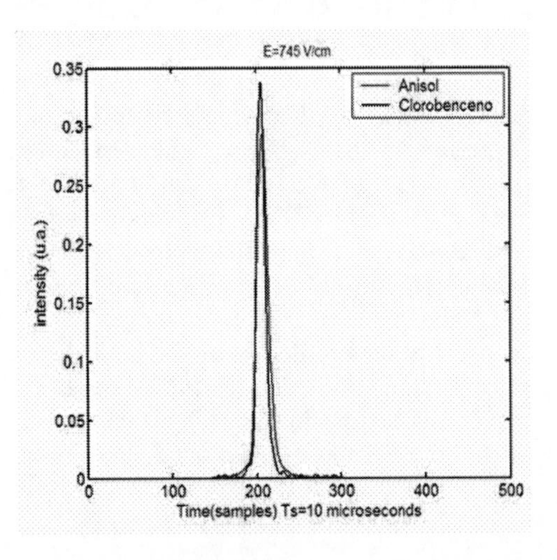

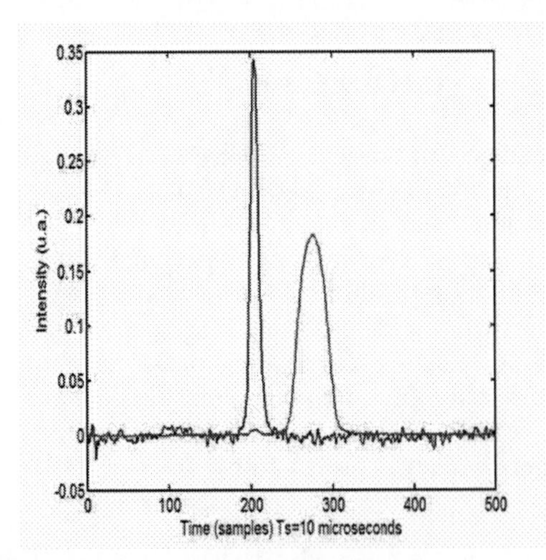

FIGURE 2. Spectra of Anisole and Chlorbenzene for different conditions.

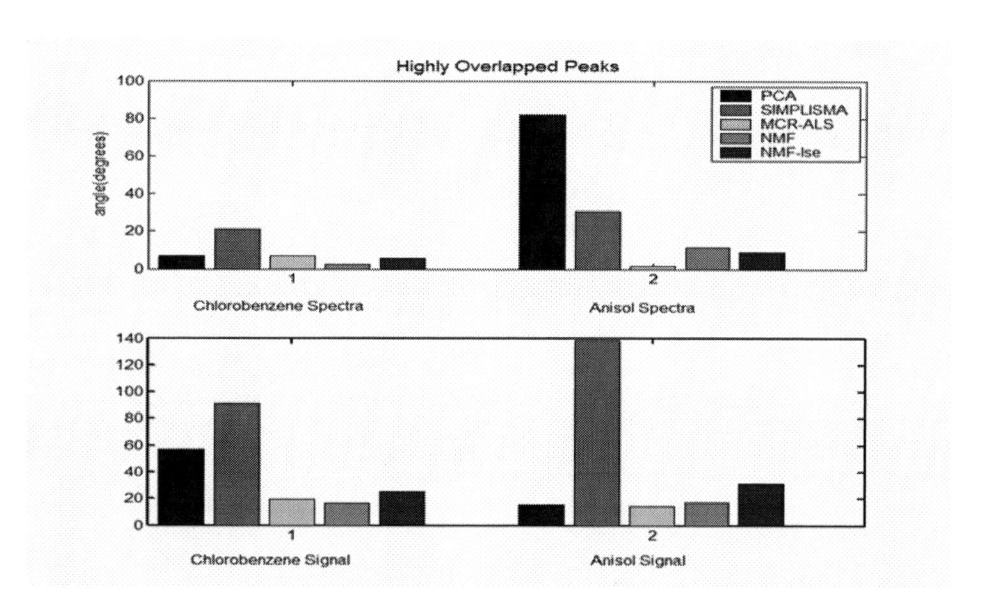

FIGURE 3. Angle (degree) between the real and recovered mobility spectra (top) and the real and recovered concentration signal (bottom).

ACKNOWLEDGMENTS

This work has been partially funded by the Sixth Framework Programme Network of Excellence GOSPEL FP6-IST 507610.

REFERENCES

1. G.A. Eiceman and Z. Karpas, *Ion Mobility Spectrometry: 2nd edition*, Boca Raton: Taylor and Francis, 2005.
2. P. de B. Harrington, E.S. Reese, P.J. Rauch, L. Hu, D.M. Davis, *Applied Spectroscopy* **51**, 808-816 (1997).
3. A. de Juan and R. Tauler, *Analytica Chimica Acta* **500**, 195-210 (2003).
4. D.D. Lee and H.S. Seung, *Nature* **401**, 788-791 (1999).

Sensor Drift Compensation Algorithm based on PDF Distance Minimization

Namyong Kim[1] , Hyung-Gi Byun[1] , Krishna C. Persaud[2] and Jeung-Soo Huh[3]

[1]*School of Electronics, Information & Communication Eng. Kangwon National University at Samcheok, Korea*
[2]*School of Chemical Engineering & Analytical Science, University of Manchester, Manchester, U.K.*
[3]*Dept. of Materials Science & Metallurgy, Kyungpook National University, Daegu, Korea*

Abstract. In this paper, a new unsupervised classification algorithm is introduced for the compensation of sensor drift effects of the odor sensing system using a conducting polymer sensor array. The proposed method continues updating adaptive Radial Basis Function Network (RBFN) weights in the testing phase based on minimizing Euclidian Distance between two Probability Density Functions (PDFs) of a set of training phase output data and another set of testing phase output data. The output in the testing phase using the fixed weights of the RBFN are significantly dispersed and shifted from each target value due mostly to sensor drift effect. In the experimental results, the output data by the proposed methods are observed to be concentrated closer again to their own target values significantly. This indicates that the proposed method can be effectively applied to improved odor sensing system equipped with the capability of sensor drift effect compensation

Keywords: Sensor Drift Compensation, RBFN, PDF, Odor Sensing System.
PACS: S 07.05.kf

INTRODUCTION

Using an odor sensing system, it is desirable to discriminate between chemicals and compare one sample with another. The ability to classify pattern characteristics from relatively small pieces of information has led to growing interest in methods of sensor recognition. The Radial Basis Function Network and other neural networks have recentely been applied to odor classification problems [1,2].

To use the RBFN as a classifer, the parameters such as centers, widths and weights have to be optimized in the training phase before the classification. In the testing phase, the unlabeled input data are usually in a drifted state largely due to the effect of sensor aging, or sensor contamination [1,3].

With the trained and then fixed weights of the RBFN, the output of the system in the testing phase is shifted and dispersed to a great extent, so the fixed weights are no longer effective in odor classification. Our idea is to adjust the RBFN weights even in the testing phase in order for testing phase output distribution to follow and match the distribution of the previous output data obtained from the training phase.

METHODS

Adjustment of the RBFN Parameters in the Training Phase

For the tuning of RBFN parameters the stochastic gradient (SG) method has been applied to RBFN [4] and a more enhanced RBFN-SG algorithm with optimal convergence coefficients has been developed for odor sensing system [5].

The RBFN-SG algorithm adapts all the free parameters of the network using gradient descent of the instantaneous output error power. An input vector having L elements is defined as

$$X_k = [x_k, x_{k-1}, \ldots, x_{k-L+1}]^T . \quad (1)$$

Let the error be denoted by $e_k = d_k - y_k$, where d_k is the desired target value and y_k is the RBF output, all at the training time k. The output of RBFN-SG with B Gaussian basis functions is

$$y_k = \sum_{j=1}^{B} v_k^j w_k^j \quad ,$$

CP1137, *Olfaction and Electronic Nose: Proceedings of the 13th International Symposium,* edited by M. Pardo and G. Sberveglieri

$$v_k^j = \exp(\frac{-\left\|X_k - c_k^j\right\|^2}{[\sigma_k^j]^2}) \quad (2)$$

w_k^j is defined as the j -th element of the RBFN weight. The width, center and weight of hidden unit j are adapted according to the normalized RBFN-SG algorithm [5].

ED Minimization Algorithm for Weight Adjustment in the Test Phase

Recentely, Erdogmus introduced an information theoretic framework based on Kullback-Leibler (KL) divergence minimization for training adaptive systems in supervised learning settings using both labeled and unlabeled data [6,7]. As another measure of divergence, the Euclidean distance (ED) between two PDFs contains only quadratic terms to utilize the tools of information potential [8]. Based on ED of two PDFs, the interactions between the testing phase output data and the previously acquired training phase output data are possible to compensate sensor drifting effects on the system output in the testing phase without the aid of labeled target information.

Given a set of the previously obtained training phase output data $Y_N = \{y_1, y_2, \ldots, y_N\}$ and a set of the testing phase output data $Z_M = \{z_1, z_2, \ldots, z_M\}$, the Euclidian distance between the pdf f_Y of the trained output data during the training phase and the pdf f_Z of the current system output data during the testing phase can be minimized with respect to the RBFN system weight W as

$$\underset{W}{Min}\left(ED[f_Y, f_Z]\right) = \underset{W}{Min}\left(\int f_Y^2(\xi)d\xi + \int f_Z^2(\xi)d\xi - 2\int f_Y(\xi)f_Z(\xi)d\xi\right). \quad (3)$$

Estimating the data PDF nonparametrically is carried out by Parzen window method using a Gaussian kernel [8] as follows:

$$f_Y(\xi) \cong \frac{1}{N}\sum_{i=1}^{N} G_\sigma(\xi - y_i). \quad (4)$$

where $G_\sigma(\cdot)$ is typically a zero-mean Guassian kernel with standard deviation σ (kernel size).

Then the integrals of the multiplication of two PDFs in (3) become

$$\int f_Y^2(\xi)d\xi = \frac{1}{M^2}\sum_{i=1}^{M}\sum_{j=1}^{M} G_{\sigma\sqrt{2}}(y_j - y_i), \quad (5)$$

$$\int f_Z^2(\xi)d\xi = \frac{1}{N^2}\sum_{i=1}^{N}\sum_{j=1}^{N} G_{\sigma\sqrt{2}}(z_j - z_i), \quad (6)$$

$$\int f_Y(\xi)f_Z(\xi)d\xi = \frac{1}{MN}\sum_{i=1}^{M}\sum_{j=1}^{N} G_{\sigma\sqrt{2}}(y_j - z_i). \quad (7)$$

Equation (6) and (7) are defined as Information Potential (IP) $IP(z,z)$ and $IP(y,z)$, respectively. Then the minimization of the cost function with respect to RBFN weight W is as follows:

$$W_{new} = W_{old} - \mu\frac{\partial P}{\partial W}, \quad (8)$$

where $P = IP(z,z) - 2 \cdot IP(y,z)$.

The gradient is evaluated from

$$\frac{\partial P}{\partial W_k} = \frac{1}{2M^2\sigma^2}\sum_{i=k-M+1}^{k}\sum_{j=k-M+1}^{k}(z_j - z_i) \cdot G_{\sigma\sqrt{2}}(z_j - z_i)\cdot(V_i - V_j) - \frac{1}{MN\sigma^2}\sum_{i=k-M+1}^{k}\sum_{j=1}^{N}(y_j - z_i)\cdot G_{\sigma\sqrt{2}}(y_j - z_i)\cdot V_i. \quad (9)$$

By inserting (9) into (8), we have

$$W_{k+1} = W_k - \mu_{test}\cdot\left[\frac{1}{2M^2\sigma^2}\sum_{i=k-M+1}^{k}\sum_{j=k-M+1}^{k}(z_j - z_i) \cdot G_{\sigma\sqrt{2}}(z_j - z_i)\cdot(V_i - V_j) - \frac{1}{MN\sigma^2}\sum_{i=k-M+1}^{k}\sum_{j=1}^{N}(y_j - z_i)\cdot G_{\sigma\sqrt{2}}(y_j - z_i)\cdot V_i\right]. \quad (10)$$

Equation (10) is the proposed algorithm readjusting weights of the RBFN odor classifier in the testing phase after supervised learning is completed.

RESULTS AND DISCUSSION

Data was collected an odor sensing system which had an array of conducting polymer sensors mounted on a ceramic substrate together with associate electronics developed by Krishna Persaud, University of Manchester.

Measurements of chemicals were repeated to collect patterns from solvent vapors over periods of 4 weeks. Figure 1 shows that there was some variability in the responses of the sensor array during the 4 weeks measurement, which may be due to causes including drifts.

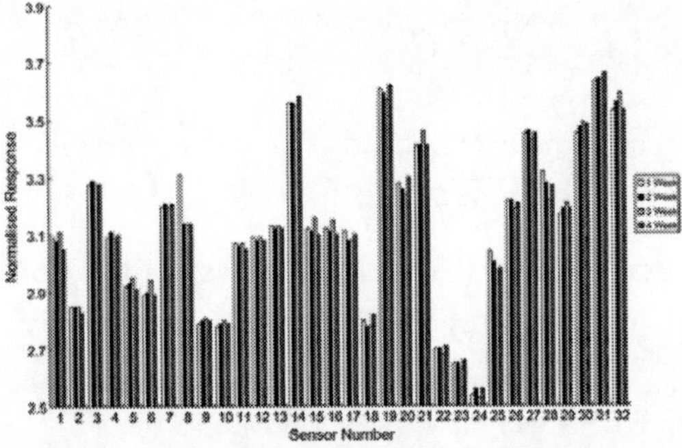

Figure 1. Normalized response patterns from 32 sensors in an array

For the RBFN classifer, patterns from solvent vapors with associated concentrations (1 and 10% acetonitrile (ac1, ac10), 1% acetone(ae), 1% butanone(bu), 10% methanol(me), 1 and 10% propanol(pr1, pr10), and water(wa)) are used. For network training, eight centers for each each class (B=8) were chosen from total of 528 patterns in the weeks 1 and 2 data sets. Initial centers and widths are calculated by fuzzy c-means algorithm. The convergence coefficient for width adjustment μ_s is set to 0.001. The constant a_c for center is 1400 and a_w is 400 for weight update. After having trained the RBFN using weeks 1 and 2 data sets obtained from the solvent vapors, the RBFN was applied to the unlabeled test data of 412 patterns from weeks 3 and 4 to evaluate sensor drift effects and the performance of drift compensation by the proposed ED minimization method.

In Figure 2, the trained and tested output results are illustrated. As expected, the trained output data are very well aligned at each target value, but the drifted output in the testing phase is significantly dispersed and shifted from each target value due mostly to sensor drift effect.

For the proposed algorithm in the testing phase, the number of block data N and M = 66, kernel size σ =0.5, convergence parameter μ_{test} = 0.00001 are used. The test phase results, depicted in Figure 3, show that the proposed method has compensated the sensor drift effect significantly.

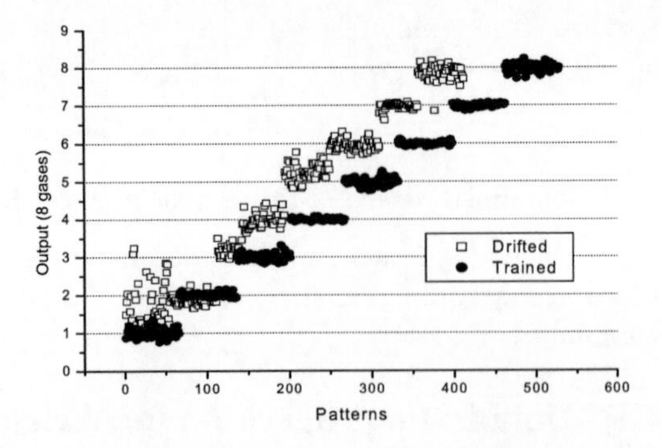

Figure 2. Trained and tested (drifted) output results for 8 solvent vapors with associated concentrations (1: ac1, 2: ac10, 3: ae, 4: bu, 5: me, 6: pr1, 7: pr10, 8: wa). X-axis is the number of input patterns. 528 and 412 input patterns are used in the training and testing phase, respectively.

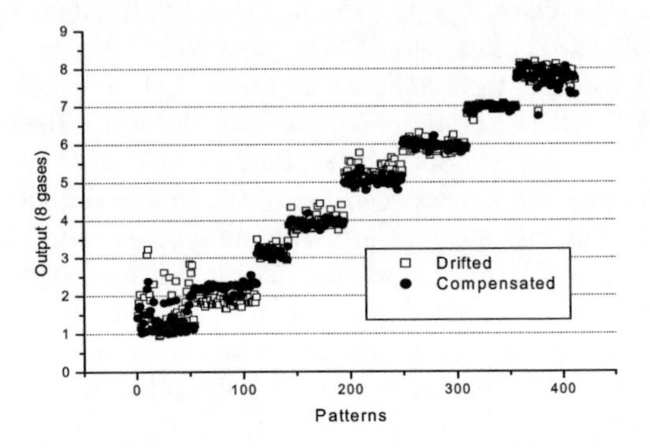

Figure 3. Compensated and tested (drifted) output results for the 8 solvent vapors

For close examination probability distributions of each output data set are illustrated in Figures 4 and 6. The probability distribution of the trained output data shows in Figure 4 how much correctly the RBFN parameters are adjusted in the training phase. The output data are closely concentrated on each target value by means of the normalized RBF-SG algorithm.

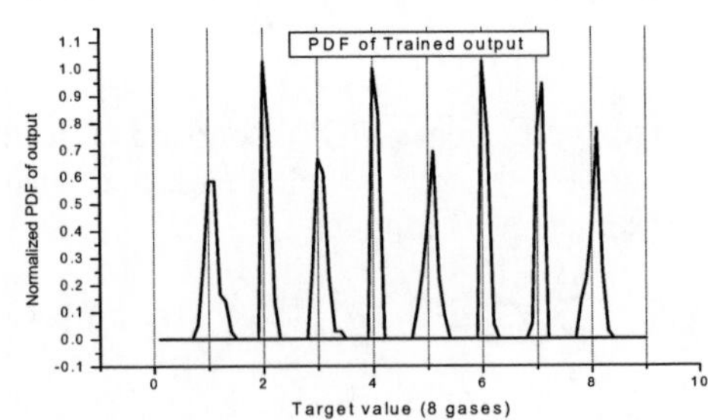

Figure 4. Probability density for trained output data

In the testing phase after a period of time of sensor aging and contamination, the RBFN parameters are no longer effective in classification for unknown test gases as depicted as a form of PDF in Fig. 5.

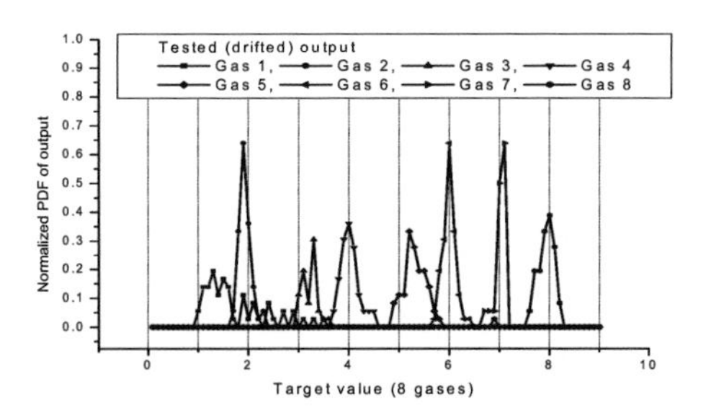

Figure 5. Probability density for tested output data

The output PDF for gas 1 is severely spread and trespasses even on the region for gas 4. And the output data for gas 3 and gas 5 are shifted to the right and overlapped with gas 4 and gas 6 output data, respectively. These defects make the classification of odors wrong or impossible. In Figure 6 the output result from the proposed compensation algorithm as a form of PDF is shown.

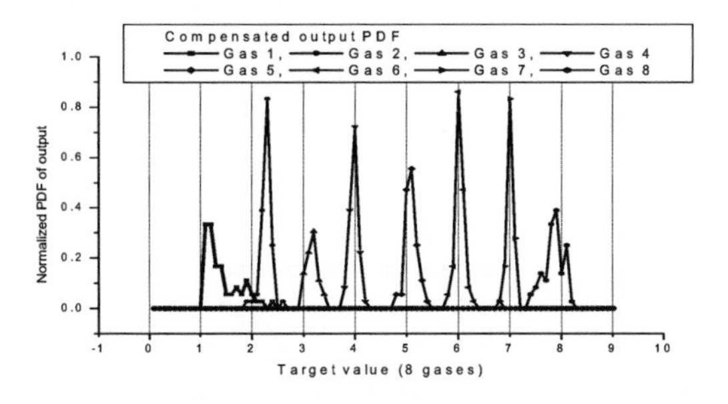

Figure 6. Probability density for compensated output data.

The proposed method readjusts the RBFN weights in order for the distribution in Figure 5 to match the PDF in Figure 4. All output data are observed to be concentrated closer to their own target values compared to the output PDF of the testing phase. Though gas1 output data are not gathered around target value 1 completely, they are located much closer to target value 1 than in the case of the testing phase. Also less output data trespass into the region of gas 2, and no data points of gas 1 appear in the region of gas 3. Data for gas 3 and gas 4, as well as for gas 5 and gas 6, are completely separated having very narrow and sharp PDF forms.

CONCLUSION

In this paper, a new method of compensation for sensor drift effect of odor sensing systems using a conducting polymer sensor array is proposed. The adaptive RBFN is used as a classifier in which the parameters such as centers, widths and weights are optimized by using the normalized RBFN-SG algorithm in the training phase before the testing phase. The trained output data are very well aligned at each target value, but the drifted output in the testing phase using the fixed weights of the RBFN are significantly dispersed and shifted from each target value due mostly to sensor drift effect. By using the proposed method that readjusts the RBFN weights based on Euclidian PDF distance minimization criterion in order for the distribution of the current output data to match the PDF of the well-trained and stored output data, the testing phase output data are observed to be concentrated closer again to their own target values significantly. This implies that the proposed method can be effectively implemented for the improved odor sensing system equipped with the capability of sensor drift effect compensation.

ACKNOWLEDGMENTS

This work is partly supported by PMI2 and Basic Research Program of KOSEF[R01-2007-000-20122-0].

REFERENCES

1. Gardner, J. W. *et al*, Electronic Noses Principles and Applications, Oxford University Press (1999)
2. Byun, H. *et al*, "Application of adaptive RBF networks to odor classification using conducting polymer sensor array", *Proc. ISOEN 2000*, Brighton, 2000, pp. 121-126.
3. Zuppa, M *et al*, "Drift counteraction with multiple self-organising maps for an electronic nose", Sensors and Actuators B 98 (2004), pp. 305-317.
4. Cha, I *et al*, "Interference cancellation using radial basis function networks", Signal Processing 47 (1995), pp. 247-268.
5. Kim, N *et al*, "Normalization approach to the stochastic gradient radial basis function network algorithm for odor sensing system", Sensors and Actuators B 124 (2007), pp. 407-412.
6. Kullback, S. Information Theory and Statistics, Dover Pubications (New York, 1968)
7. Erdogmus, D *et al*, "Supervised training of adaptive systems with partially labeled data", *Proc. Of International Conference on ASSP,* Apr. 2005, pp. 321-324.
8. Principe, J.C. *et al*, "Information Theoretic Learning", In: S. Haykin, Unsupervised Adaptive Filtering, 2000, pp. 265-319.

The Improved Data Processing Method For Electronic Tongue Based On Multi-Ion LAPS

Yi Li, Wei Cai, Xiaoming Gao, Hongsun Guo, Huixin Zhao and Ping Wang*

Biosensor National Special Laboratory, Key Laboratory of Biomedical Engineering of Education Ministry, Department of Biomedical Engineering, Zhejiang University, Hangzhou, 310027, P.R. China
**E-mail address: cnpwang@zju.edu.cn (Ping Wang)*

Abstract. We present an improved data processing method for multi-ion detection and feature recognition of electronic tongue for *in situ* application. Light Addressable Potentiometric Sensor (LAPS) with its peripheral units can be easily integrated into portable applications and sensor networks. Sensitive components, such as Pb-Ag-As-S and Cu-Ag-As-Se chalcogenide glasses are deposited onto the surface of LAPS. I-V curves have been studied, and features of different ions were extracted and successfully classified. Methods of Principle Component Analysis (PCA) and Relevant Vector Machine (RVM) were effectively applied to provide a new approach for electronic tongue to recognize the varieties of data sequences, which benefit electronic tongue a lot for double checking the validity of data sequences.

Keywords: Chalcogenide Glass LAPS, Electronic Tongue, Relevant Vector Machine, Feature Recognition
PACS: 82.80.Fk

INTRODUCTION

It is well known that Atomic Adsorption spectrum (AAS), Induced Coupled Plasma Mass spectrum (ICP-MS), Ion Chromatography and Stripping Voltammetry (SV) are inconvenient in situ conditions outside analytical laboratories. Similar with the principle of Ion Selective Electrode (ISE), LAPS is more suitable for field trace detection of heavy metal ions in contaminated water with great sensitivity.

In many cases, single sensor can hardly reach higher detection limit nor solve the interference problem of multi-ions. Electronic tongue consisting of kinds of chalcogenide glass ISEs can exhibit high sensitivity and selectivity to many heavy metal ions[1, 2]. The pulsed laser deposition (PLD) technique has been proved to serve as an advanced technique in depositing thin films on LAPS[3].

We design a micro-structural thin film sensor array based on different chalcogenide glass (ChG) materials. Four different types of thin films were firstly deposited on the surface of LAPS by means of PLD technique. The ChG thin film sensor array can determine simultaneously H^+, Cu^{2+}, Pb^{2+} ions in solutions using electronic tongue principle.

Electronic tongues represent an array for operation in liquid phase to analyze the environment. They are not only a sensor but also more a measurement strategy. In one array, there are sorts of partially selective individual element. Classifiers and regression methods are applied to the complex signals of biomimetic electronic tongues. Parameter classifiers such as Artificial Neutral Network (ANN) and K-cluster have enhanced the intelligence of biomimetic system to generalize the learning results to unknown situations and made the sensor smarter. Partial Least Squares (PLS) approach is applied to analyze properties of cross-sensitivity between sensors[4,5], which improves the detection limit. We present a method of classifying acquired data itself from the derative features of I-V curve using RVM, which is noise-tolerant and fits for distributed sensor networks.

Relevance Vector Machine, as an improved tool of Support Vector Machine (SVM) avoiding the limitations, was added as a module of electronic tongue system for recognition and classification[6]. The motivation inside this approach is that we can infer a classification model which is accurate and only uses a small number of relevant basis functions for prediction[7]. The functions were automatically selected from a potentially large initial set. Sampling dataset is to be trained in Bayesian model, which leads to much sparser matrix resulting on test data and maintaining comparable generalization error.

CP1137, *Olfaction and Electronic Nose: Proceedings of the 13th International Symposium*, edited by M. Pardo and G. Sberveglieri

EXPERIMENTAL AND METHODS

Sensor Design

Light Addressable Potentiometric Sensor is chosen as the core electrochemical sensory chip in a node for its high Signal-to-Noise-Ratio and energy efficiency. Sensor arrays are able to sense the concentration of heavy metal ions after the chalcogenide glass deposits on the sensor surface. Sensor arrays as electronic tongue, can improve not only the sensing ability for singular target but also the cross-sensitivity of complex targets.

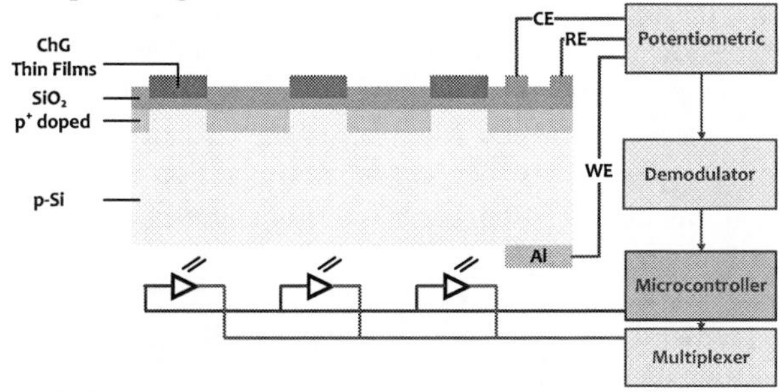

FIGURE 1. Micro-structural electronic tongue based on LAPS and its measurement system

In Fig.1, different chalcogenide membranes are embedded into the oxide layer to avoid misplacement with heavily doped layer. Aluminum alloy is used as the ohmic contact at the bottom of the silicon, while counter electrode (CE) and reference electrode (RE) are isolated and electrically contacted from the top of the chip. The frequencies of LEDs are modulated by microcontroller through multiplexer. AC signal of the measure system is demodulated and transferred to the microcontroller. Photocurrent gathered from the non-sensitive region will be damped since the depletion region could hardly form at the interface between heavily doped silicon and thick oxide layer. The compositions of lead and copper chalcogenide glasses are respectively Pb-Ag-As-S and Cu-Ag-As-Se, which are purchased from the Chemistry Department of St. Petersburg State University. Laser pulsed deposition (PLD) is carried out for the complex stoichiometry of the chalcogenide glass materials.

Experiment Design

All solutions were prepared at the background of 0.01M sodium chloride because the purpose of the study was to determine heavy metals in the freshwater or rivers. The pH values of all solutions were adjusted to 5 by addition of HNO_3. The mixed nitrate solution of Pb^{2+} and Cu^{2+} with the concentration of 10^{-4}, 10^{-5}, 10^{-6}, 10^{-7} mol/L, respectively, was demarcated by the Second Institute of Oceanography, China. The bias voltage was applied between the aluminum pad and the platinum counter electrode. The photocurrent signal containing several frequency components that correspond to selected sites of LAPS was transmitted into the demodulator through working electrode (WE).

RESULTS

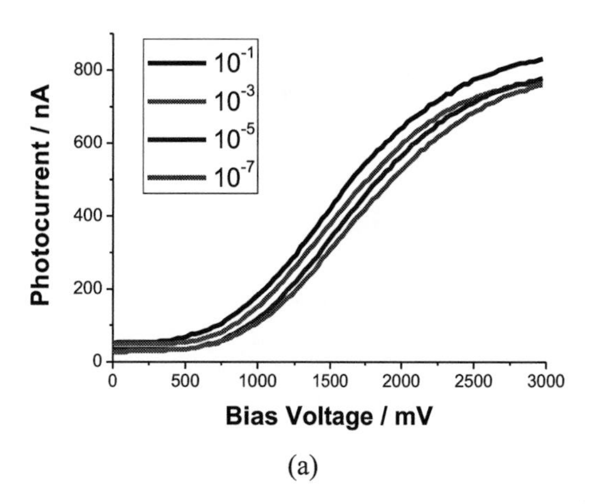

(a)

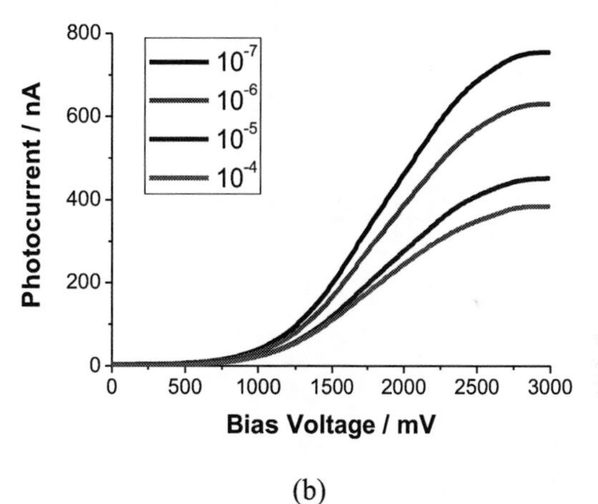

(b)

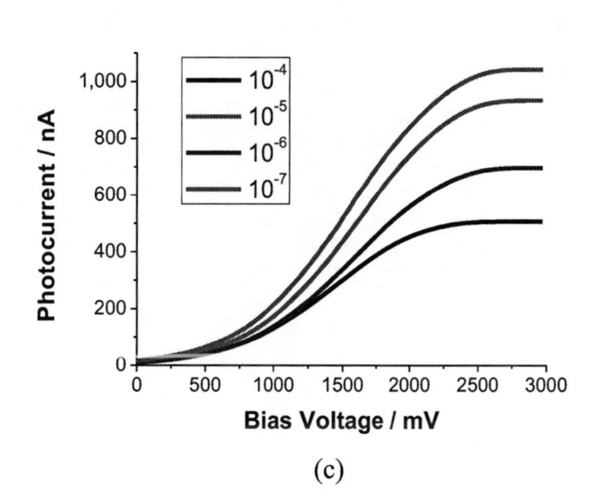

(c)

FIGURE 2. The response of H^+, Cu^{2+} and Pb^{2+} sensitive LAPS (a) Curve of concentration gradient of H^+ solutions from 10^{-7} mol/L to 10^{-1} mol/L. The electrolytes have been tuned nearly the same. (b) and (c) Curves of concentration gradient of Cu^{2+} and Pb^{2+} mixed solutions from 10^{-7} mol/L to 10^{-4} mol/L.

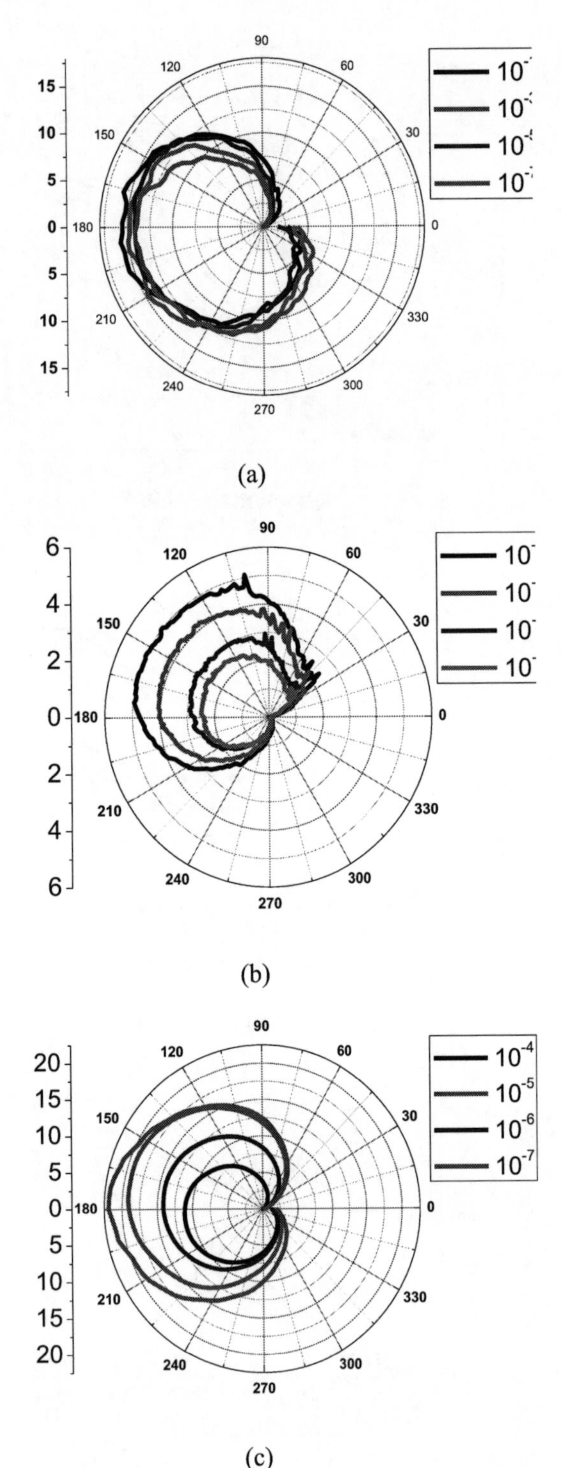

FIGURE 3. Plot the first derivative of H^+, Cu^{2+} and Pb^{2+} ion I-V curves in polar coordinate before normalization, in which curves show a little phase shift in (a) and good in morphology in all three patterns.

The total pH, pCu^{2+} and pPb^{2+} response results of LAPS are shown in Fig.2. The raw data of copper and lead ion detection shows great differences of gradient concentrations. However, the impedance of solutions also contributes to the current of our system because the actual potential on the depletion region is decreasing compared with the bias. The carriers generating in the depletion region descend with descending the photocurrent.

The interference effect (log K) between Pb^{2+} and Cu^{2+} ions has also been studied. Cu^{2+} ions interfered by lead-LAPS have a K value of -4.5, meanwhile Pb^{2+} ions interfered by copper-LAPS have a K value of -5.6. It is shown that sensitivity of thin films is good enough to measure the corresponding ions in the same solution.

Phase congruency filter based on median filter is used to preprocess the first-ordered derative data and then transformed from 0-3V Cartesian coordinate to 0-360° polar coordinate in Fig.3. It is interesting that the peaks exhibit diverse directions. The patterns plotted on polar coordinate are for convenient observation. Features are extracted from the derivated sequence, which are experimental demonstrations to identify special ions with certain patterns.

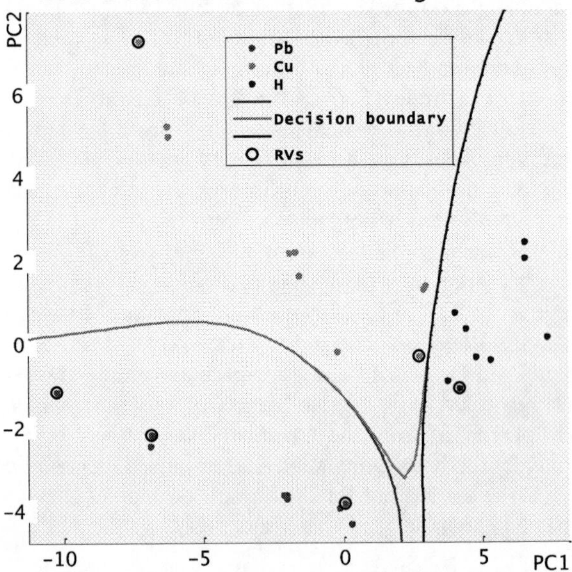

FIGURE 4. PCA plot of two main components of I-V curves from the LAPS electronic tongue: PC1 (69.7%) and PC2 (23.6%) after descending dimensions of 100 point sequence.

The response time of ChG-LAPS varies with membrane compositions due to the diffusion coefficients of the potential-determining ions from the solution to the active sites on the surface in modified surface layer[8]. While the parameters of wafer process changes, the state of charge space and the grain interface vary. Stable and transient potential at the membrane interface is established diversely which determines the feature of I-V curve. In our LAPS design, the reason that I-V curves have distinguish pattern might ascribe to the PLD process of chalcogenide glass thin film, since other steps of the process make homogeneous effects to the whole wafer and the properties of variant target materials differ.

RVM algorithm is applied to classify these three kinds of points with noise tolerance. Points inside black circles are selected as Relevance Vectors (RVs), which are used to determine the decision boundaries. Training data is randomly selected from experimental data set. Linear spline function with no bias[9] is the most effective one in our experiment, while kernel length is optimized to 5.68. The result shown in Fig.4 gives rise to classifying data with loose limitations. Apart from morphological characters of curves, we can easily make the decision of the classification without more data.

DISCUSSION

The species of ions can be identified through RVM after the dimension of the data has been descended through PCA. For an electronic tongue, overfitting becomes one of the disadvantages in ANN and SVM methods. Although support vectors are less than training instances, SVM is fully dependent on the instances of deterministic sampling data. The time consuming is increased as the amount of sampling data increased. Another problem that SVM confronted with was that the likelihood function was not provided in this conventional method. Sorts of noises were inevitably involved in sampling data. In our data set, the experiment is easily influenced by kinds of matters such as temperature turbulence, electrolyte intersection, or fluid disturbance. With sparse Bayesian learning, RVM is a more stable method for generalization usage.

However, RVM needs relatively longer training times in our practice. Because it yields sparse matrix, the computation time is typically much less.

ACKNOWLEDGMENTS

This work is supported by grants from Ministry of Science and Technology of China (973 Program) (No. 2009CB320303). The authors gratefully acknowledge the technical support on Pulsed Laser Deposition from Dr. Renguo Song and master student Cuiling Xue, Zhejiang University of Technology.

REFERENCES

1. Y. Vlasov and E. Bychkov, "Ion-selective chalcogenide glass electrodes," *Ion-Sel. Electrode Rev*, *9* (1987), 5–93.
2. Y. Vlasov, A. Legin, A. Rudnitskaya, A. D'Amico and C. Di Natale, "Electronic tongue - new analytical tool for liquid analysis on the basis of non-specific sensors and methods of pattern recognition," *Sensors & Actuators: B. Chemical*, *65* (2000), 235-36.
3. Y. Mourzina, M. Schöning, J. Schubert, W. Zander, A. Legin, Y. Vlasov, P. Kordos and H. Lüth, "A new thin-film Pb microsensor based on chalcogenide glasses," *Sensors & Actuators: B. Chemical*, *71* (2000), 13-18.
4. A.V. Legin, V.A. Babain, D.O. Kirsanov and O.V. Mednova, "Cross-sensitive rare earth metal ion sensors based on extraction systems," *Sensors & Actuators: B. Chemical*, *131* (2008), 29-36.
5. Y. Vlasov, A. Legin and A. Rudnitskaya, "Cross-sensitivity evaluation of chemical sensors for electronic tongue: determination of heavy metal ions," *Sensors & Actuators: B. Chemical*, *44* (1997), 532-37.
6. P. Ciosek and W. Wróblewski, "Sensor arrays for liquid sensing–electronic tongue systems," *The Analyst*, *132* (2007), 963-78.
7. C. Bishop and M. Tipping, Variational relevance vector machines (San Francisco: Morgan Kaufmann Publishers, 2000), 46–53.
8. V. Vassilev and S. Boycheva, "Chemical sensors with chalcogenide glassy membranes," *Talanta*, *67* (2005), 20-27.
9. S.R. Gunn, "Support Vector Machines for Classification and Regression," *ISIS Technical Report*, *14* (1998).

Active Chemical Sensing With Partially Observable Markov Decision Processes

Rakesh Gosangi and Ricardo Gutierrez-Osuna*

Department of Computer Science, Texas A & M University
{rakesh, rgutier}@cs.tamu.edu

Abstract. We present an active-perception strategy to optimize the temperature program of metal-oxide sensors in real time, as the sensor reacts with its environment. We model the problem as a partially observable Markov decision process (POMDP), where actions correspond to measurements at particular temperatures, and the agent is to find a temperature sequence that minimizes the Bayes risk. We validate the method on a binary classification problem with a simulated sensor. Our results show that the method provides a balance between classification rate and sensing costs.

Keywords: Active sensing, Chemical sensors, and Partially Observable Markov Decision Processes.
PACS: 07.07.Df

I. INTRODUCTION

Previous research has shown that modulating the working temperature of metal-oxide sensors can give rise to gas-specific temporal signatures that provide a wealth of discriminatory and quantitative information [1]. A number of empirical studies with various temperature waveforms (e.g. rectangular, sine, saw tooth, and triangular) and stimulus frequencies have been published [2-4], but only a handful of authors have approached the problem in a systematic fashion. Kunt et al. [5] developed a computational method to optimize the temperature profile in binary discrimination problems. The authors used a wavelet network to obtain a dynamic model of the sensor from experimental data, followed by an optimization procedure that found the temperature profile that maximized the distance between the two gas signatures. More recently, Vergara et al [6] proposed a system-identification method for optimizing temperature profiles. In their method, a pseudo-random binary sequence was used to drive the sensor heater while the sensors were exposed to various chemicals. The authors then estimated the frequency response of the sensor to each individual chemical, and selected a subset of the most informative frequencies. Both approaches, however, required that the temperature program be optimized off-line. Here we propose an active-sensing approach that can optimize the temperature profile on the fly, that is, as the sensor collects data from its environment. The method can also determine when sensing should be terminated in order to make a final classification; this is achieved by comparing the cost of measuring the sensor response at additional temperatures against the expected reduction in Bayes risk from those additional measurements. These capabilities are important not only to improve detection performance, but also to meet the increasing power constraints of real-time embedded applications as well as extend sensor lifetimes.

We model the problem as a decision-theoretic process, where the goal is to determine the next temperature pulse to be applied to the sensor based on information extracted from the sensor response to previous temperature pulses. Our method operates in two stages. First, we model the dynamic response of the chemical sensor to a sequence of temperature pulses as an Input-Output Hidden Markov Model (IOHMM) [7]. Then, we formulate the process of finding the ideal sequence of temperature pulses as a POMDP [8]. By assigning a cost to each temperature pulse and a cost for misclassifications, the POMDP is able to balance the total number of temperature pulses against the uncertainty of the classification decisions.

The paper is organized as follows. In section II we formulate the problem and show how IOHMMs can be used to model the dynamic response of a sensor. Section III describes the optimization of temperatures as an active sensing problem with POMDPs. Section IV provides experimental results on a dataset from a simulated metal-oxide sensor. The article concludes with a brief discussion and directions for future work.

CP1137, *Olfaction and Electronic Nose: Proceedings of the 13th International Symposium*, edited by M. Pardo and G. Sberveglieri

II. PROBLEM STATEMENT

Consider the problem of classifying an unknown gas sample into one of M known categories $\{\omega^{(1)}, \omega^{(2)}, \ldots, \omega^{(M)}\}$ using a metal-oxide sensor with D different operating temperatures $\{\rho_1, \rho_2, \ldots, \rho_D\}$. To solve this sensing problem, one typically measures the sensor's response at each of the D temperatures, and then analyzes the complete feature vector $\boldsymbol{x} = [x_1, x_2, \ldots, x_D]^T$ with a pattern-recognition algorithm [9]. Though straightforward, this "passive" sensing approach is unlikely to be cost-effective because only a fraction of the measurements are generally necessary to classify the chemical sample. Instead, in active classification we seek to determine an optimal sequence of actions $\boldsymbol{a} = [a_1, a_2, \ldots, a_T]$, where each action corresponds to setting the sensor to one of the D possible temperatures (or terminating the process by assigning the sample to one of the M chemical classes). More importantly, we seek to select this sequence of actions dynamically, based on accumulating evidence. Our proposed solution to this problem is based on Ji and Carin [10].

A. Modeling the Sensor

Given a chemical from class $\omega^{(c)}$, we model the steady-state response of the sensor at temperature ρ_i with a Gaussian mixture:

$$p(x_i|\omega_c) = \sum_{m_i=1}^{M_i} \alpha_{i,m_i}^{(c)} N\left(x_i \middle| \mu_{i,m_i}^{(c)}, \Sigma_{i,m_i}^{(c)}\right) \tag{1}$$

where M_i is the number of mixture components, and $\alpha_{i,m_i}^{(c)}, \mu_{i,m_i}^{(c)}, \Sigma_{i,m_i}^{(c)}$ are the mixing coefficient, mean vector and covariance matrix of each mixture component for class $\omega^{(c)}$, respectively. Given a sequence of actions $[a_1, a_2, \ldots, a_T]$, we assume that the sensor transitions through a series of states $\boldsymbol{s} = [s_1, s_2, \ldots, s_T]$ to produce a corresponding observation sequence $\boldsymbol{o} = [o_1, o_2, \ldots, o_T]$. Each state s_i represents a mixture component in eq. (1) and is therefore hidden. Following Ji and Carin [10], we model the sensor dynamics with an IOHMM, a generalization of the traditional hidden Markov model (HMM) [11]. An IOHMM conditions the next state in a sequence not only on the previous state (as in a first-order HMM) but also on the current input to the sensor. In our case, this additional input consists of sensing actions (i.e. temperature steps).

Formally, an IOHMM can be defined as a 6-tuple $\{S, A, O, \pi, \tau, \phi\}$ where S is a finite set of states, each state corresponding to a mixture component in eq. (1), A is a finite set of discrete actions, each action corresponding to selecting one of D sensor temperatures, O is a set of observations, each corresponding to the sensor's response at a given temperature, $\pi(s)$ is the initial state distribution, $\tau(s'|s,a)$ is the state transition function, which describes the probability of transitioning from state s to state s' given action a, and $\phi(o|s)$ is the observation function, which describes the probability of making observation o at state s. We train a separate IOHMM for each individual chemical class, i.e. by driving the chemical detector with a random sequence of actions in the presence of the chemical, and recording the corresponding responses; for details see [7].

III. ACTIVE CHEMICAL SENSING AS A POMDP

We define a POMDP as a 7-tuple $\{S, A, O, b_0, T, \Omega, C\}$, where S, A, and O are the finite set of states, actions and observations from the IOHMMs respectively, $b_0(s)$ is an initial belief across states, $T(s'|s,a)$ is the probability of transitioning from state s to state s' given action a, $\Omega(o|s)$ is the probability of making observation o at state s, and $C(s,a)$ is the cost of executing action a at state s. These POMDP parameters can be obtained directly from the IOHMM as follows:

- Initial belief: $b_0(s) = p(\omega^{(c)})\pi^{(c)}(s);\ s \in S^{(c)}$
- State transition: $T(s'|s,a) = \tau^{(u)}(s'|s,a)$; $s, s' \in S^{(u)}$; zero otherwise[1].
- Observation model: $\Omega(o|s) = \phi^{(c)}(o|s);\ s \in S^{(c)}$

The POMDP stores information about the state of the system in a belief state $b_T(s)$, a probability distribution (across states from all the IOHMMs) *given* the initial belief $b_0(s)$ and the history of actions $[a_1 \ldots a_T]$ and observations $[o_1 \ldots o_T]$:

$$b_T(s) = p(s|o_1 \ldots o_T, a_1 \ldots a_T, b_0) = p(s|o_T, a_T, b_{T-1}) \tag{2}$$

The second equality above reflects the fact that $b_T(s)$ is a *sufficient statistic* for the history of the system, which allows us to update $b_T(s)$ incrementally from its previous estimate $b_{T-1}(s)$ by incorporating the latest action a_T and observation o_T:

$$b_T(s') = \frac{p(o_T|s', a_T)\sum_s p(s'|a_T, s) b_{T-1}(s)}{p(o_T|a_T, b_{T-1})} = \frac{\Omega(o|s')\sum_s T(s'|s,a) b_{T-1}(s)}{\mu} \tag{3}$$

where the denominator $\mu = p(o_T|a_T, b_{T-1})$ can be treated as a normalization term to ensure that $b_T(s')$ sums up to 1, and all terms in the numerator are known from the POMDP/IOHMM model.

[1] This ensures that transitions from the IOHMM of one class onto another class are not allowed, since we assume that the chemical stimulus does not change over time.

Using this POMDP formulation, the problem becomes one of finding a policy that maps belief states into actions so as to minimize the expected cost of sensing. We consider two types of actions:

- Sensing actions ($a = \rho_i$), which correspond to setting the sensor to temperature ρ_i. Sensing actions have a cost of $c(s, a = \rho_i) = c_i$, which reflects the fact that certain temperatures may be more expensive (e.g. draw more power).
- Classification actions ($a = \hat{\rho}_c$), which assign the sample to a particular class. Classification actions are terminal; their cost is $(s, a = \hat{\rho}_u) = c_{uv}$ $(\forall s \in S^{(v)})$, which represents a misclassification penalty whenever $u \neq v$.

A. Finding the Sensing Policy

Unfortunately, the problem of finding an exact solution for a POMDP policy is P-SPACE complete and therefore intractable for most problems. Moreover, a standard POMDP solution allows repeated actions (measuring the response of the sensor at the same temperature multiple times), which is undesirable in our case. For these reasons, we employ a myopic policy [10] that only takes sensing action if the cost of sensing (c_i) is lower than the expected future reduction in Bayes risk. Given belief state $b_T(s)$, the expected risk of a classification action is:

$$R_C\big(b_T(s)\big) = \min_u \sum_v c_{uv} \sum_{s \in S^{(v)}} b_T(s) \qquad (4)$$

where u corresponds to the class with minimum Bayes risk $\left(\sum_v c_{uv} \sum_{s \in S^{(v)}} b_T(s)\right)$. In turn, the expected risk of a sensing action is:

$$R_S(b_T(s), a) = \sum_{\forall o} \min_u \left(\sum_v c_{uv} \sum_{s' \in S^{(v)}} \sum_s p(o|s', a) p(s'|s, a)\, b_T(s)\right) \qquad (5)$$

which averages the minimum Bayes risk over all observations that may result from the action. Hence, the utility of sensing action a can be computed as:

$$U(b_T(s), a) = \left[R_C\big(b_T(s)\big) - R_S(b_T(s), a)\right] - c_a \qquad (6)$$

If $U(b_T(s), a) < 0$ for all sensing actions, then the sensing costs exceeds the expected reduction in risk $[R_C(\cdot) - R_S(\cdot)]$, and a classification action is taken. Otherwise, the action with maximum utility is taken.

IV. EXPERIMENTAL RESULTS

We validated the method on a synthetic dataset of metal-oxide sensor responses. Following [12], we modeled the temperature-conductance response using a Gaussian function. We also modeled the sensor dynamics with a first-order linear filter, resulting in:

$$G(T(t)) = \alpha G(T(t-1)) + (1-\alpha)\left(k_1 e^{-\left(\frac{T(t)-T_0}{\sigma}\right)^2} + k_2 T(t)\right) \qquad (7)$$

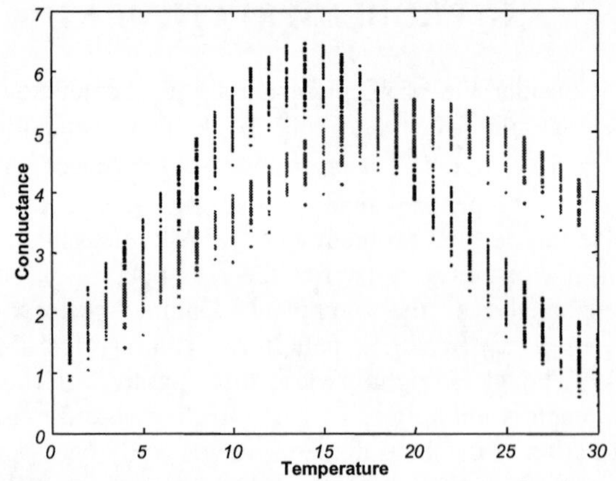

FIGURE 1. Conductance versus temperature for the two chemical classes.

where $T(t)$ is the sensor temperature at time t, and $G(T(t))$ is the conductance of the sensor at temperature $T(t)$, T_0 is the temperature at which the sensor conductivity is maximum, k_1, k_2 and σ are parameters that capture the steady-state properties of the sensor, and α captures history effects.

We evaluated the method on a problem with two chemicals and a sensor with 30 different temperatures. Sensor parameters were as follows: $\alpha = 0.2$, $k_1 = 6.0$ and $k_2 = 0.2$ for both classes; $\sigma = 10$ and $T_0 = 14$ for $\omega^{(1)}$; $\sigma = 15$ and $T_0 = 20$ for $\omega^{(2)}$. The temperature-dependent response of the sensor to the two chemicals is shown in FIGURE 1. These results were obtained by running the sensor with a random temperature sequence and recording the corresponding responses; thus, the spread at each temperature illustrates the effect of the sensor dynamics.

Training data for each analyte consisted of 40 random temperature sequences, 60 temperature pulses per sequence. Two IOHMMs (one per class) were trained; the number of Gaussian components in eq. (1) was set to $M_i = 4$. FIGURE 2 shows IOHMM predictions against the sensor response in eq. (7). These results show that the IOHMM can capture the temperature dependence and dynamics of the sensor.

The model was tested on 80 samples, 40 from each class. Each sample was generated by randomly selecting an initial temperature $T(0)$ unknown to the POMDP, and initializing the sensor response to $G(T(0)) = \left(k_1 e^{-((T(0)-T_0)/\sigma)^2} + k_2 T(0)\right)$. Classification costs c_{uv} were assumed uniform ($c_{uv} = 1$ if $u \neq v$; 0 otherwise). FIGURE 3 shows classification rate and average length of the temperature sequence as a function of feature acquisition costs c_i. For $c_i = 0.025$, the system achieves 100% classification rate with an average sequence length of 2.9 temperatures. For $c_i = 0.5$ the system performs at chance level (50%), and essentially produces a classification after measuring the response at a single temperature –this happens because sensing costs become too high compared to misclassification costs. Between these two extremes, the POMDP provides a balance between sequence length and classification rate: as feature acquisition costs increase relative to misclassification costs, the POMDP selects increasingly shorter sequences at the expense of classification rates.

V. DISCUSSION AND CONCLUSION

We have presented an active sensing approach for metal-oxide sensors that is capable of selecting operating temperatures in real-time. The problem is formulated as one of sequential decision making under uncertainty, and is solved by means of a POMDP. We have validated the method on a binary classification problem using synthetic data from a computational model of metal-oxide sensors that captures their temperature-selectivity dependence and history effects. Our results show that the POMDP is able to balance sensing costs and classification accuracy: higher classification rates can be achieved by increasing the length of the temperature sequence. The method also appears to be robust to the particular choice of sensing and classification costs, since classification rates degrade smoothly as a function of these parameters. Future work will validate the method using experimental data. The results presented here assumed uniform sensing costs, but the method can also be used to penalize high temperatures, and as a result reduce power consumption and increase sensor lifetime.

REFERENCES

1. A. P. Lee and B. J. Reedy, "Temperature modulation in semiconductor gas sensing," *Sensor Actuator B,* vol. 60, pp. 35-42, 1999.
2. R. Gutierrez-Osuna, S. Korah, and A. Perera, "Multi-Frequency Temperature Modulation For Metal-Oxide Gas Sensors," in *Proc. ISOEN.* Washington, DC, 2001.
3. A. Ortega et al., "An intelligent detector based on temperature modulation of a gas sensor with a digital signal processor," *Sensor Actuator B,* vol. 78, pp. 32-39, 2001.
4. X. Huang et al., "Gas sensing behavior of a single tin dioxide sensor under dynamic temperature modulation," *Sensor Actuator B,* vol. 99, pp. 444-450, 2004.
5. T. A. Kunt et al., "Optimization of temperature programmed sensing for gas identification using micro-hotplate sensors," *Sensor Actuator B,* vol. 53, pp. 24-43, 1998.
6. A. Vergara et al., "Optimized temperature modulation of micro-hotplate gas sensors through pseudorandom binary sequences," IEEE *Sensors J,* vol. 5, pp. 1369-1378, 2005.
7. S. Fu et al., "Audio/visual mapping with cross-modal hidden Markov models," *IEEE Trans Multimedia,* vol. 7, pp. 243-252, 2005.
8. L. P. Kaelbling, M. L. Littman, and A. R. Cassandra, "Planning and acting in partially observable stochastic domains," *Artif Int,* vol. 101, pp. 99-134, 1998.
9. R. Gutierrez-Osuna, "Pattern Analysis for Machine Olfaction: A Review," *IEEE Sensors J,* vol. 2, pp. 189-202, 2002 2002.
10. S. Ji and L. Carin, "Cost-sensitive feature acquisition and classification," *Pattern Recogn,* vol. 40, pp. 1474-1485, 2007.
11. L. Rabiner and B. Juang, "An introduction to hidden Markov models," *ASSP Magazine,* vol. 3, pp. 4-16, 1986.
12. S. Wlodek, K. Colbow, and F. Consadori, "Signal-shape analysis of a thermally cycled tin-oxide gas sensor," *Sensor Actuator B,* vol. 3, pp. 63-68, 1991.

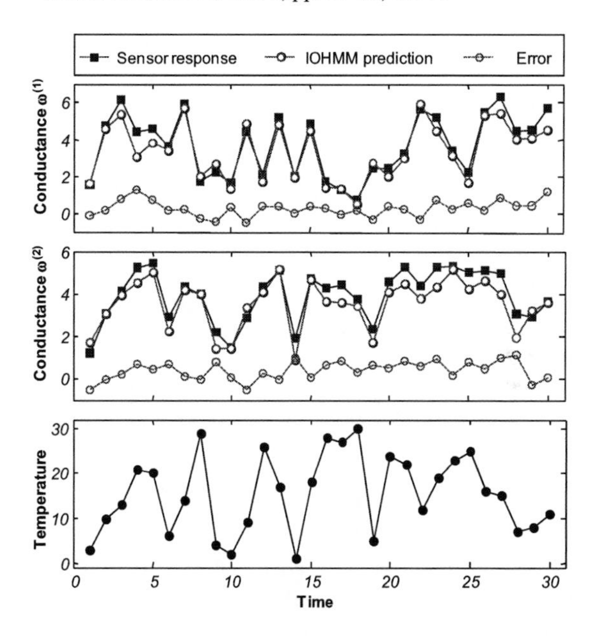

FIGURE 2. Simulated response of the sensor model in eq. (7) for two classes, IOHMM predictions and residuals. The same random temperature sequence was used in all cases for comparison purposes.

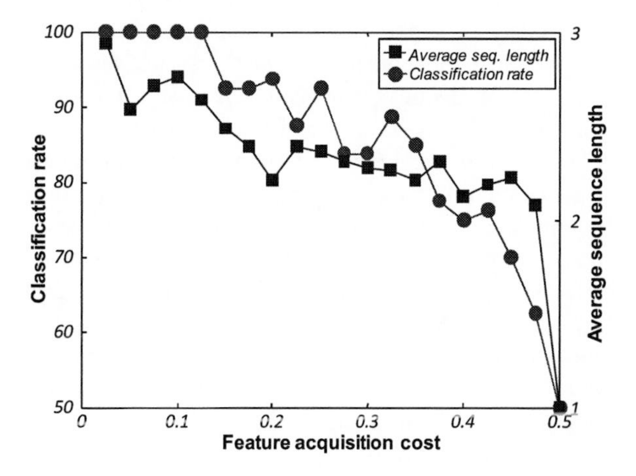

FIGURE 3. Classification performance and average sequence length as a function of feature acquisition costs. Misclassification cost $c_{uv} = 1$ if $u \neq v$; zero otherwise.

Common Principal Component Analysis For Drift Compensation Of Gas Sensor Array Data

A. Ziyatdinov[(1)], A. Chaudry[(2)], K. Persaud[(3)], P. Caminal[(1)] and A. Perera[(1)]

(1) Departament d'Electrònica. Universitat Politècnica de Catalunya, Pau Gargallo 5, 08028 Barcelona, Spain/Centro de Investigación Biomédica en Red en Bioingeniería, Biomateriales y Nanomedicina (CIBER-BBN)
(2) Protea Ltd, 11 Mallard Court, Mallard Way, Crewe Business Park, Crewe, Cheshire, CW1 6ZQ, UK
(3) School of Chemical Engineering and Analytical Science, The University of Manchester, PO Box 88, Sackville St, Manchester, M60 1QD, UK

Abstract. Two Component Correction methods for drift compensation in chemical sensor arrays have been evaluated in terms of classification accuracy. A well established linear method based on computing Principal Components of the reference gas is applied to a dataset, where three different reference gases are examined. The proposed method in this contribution is based on computing a Common Component Principal Analysis, which discovers a variance direction followed by all gasses in feature space. This new method – employing no reference gas – has shown the same performance as the traditional approach with the best fitted reference gas.

Keywords: Gas sensor array, Drift, Common Principal Component Analysis, Component Correction, Electronic nose
PACS: 07.07.Df , 87.85.Ng

1. INTRODUCTION

The technology of chemical sensors can be different: Optical, Piezoelectric, Surface Acoustic Waves (SAW), Metal Oxide (MOX) semiconductors and Conductive Polymeric (CP) sensors, among others. In these devices, the chemical information is transformed into an analytical signal used for further processing. One of the issues in signal processing and data analysis within this technology is related with sensor drift phenomena, defined as an undesired variation of identical samples in sensor signals, due to uncontrolled long-term changes in either environment (temperature, humidity) or experiment equipment (delivery system, sensor aging). The methods on drift counteraction figure prominently in the sensor array technology and assist to improve the time stability of the devices. A variety of approaches for drift compensation have been proposed in the literature that include different techniques on sensor technology and design, and signal processing methods based on baseline-correction, multivariate approaches and adaptive filters. The studies present in the literature can be classified into two categories: linear approaches like Component Correction (CC) [1] based on Principal Component Analysis (PCA) or Partial Least Squares (PLS), and non-linear methods based on Self Organizing Maps (SOM) [8,9] or system identification theory [6,7].

This paper focuses on linear methods, and compares the traditional method by Arthursson in respect to a new method based on Common Principal Component Analysis (CPCA) based on Joint Diagonalization (JD) of covariance matrices of several gas classes.

The rest of the paper is organized as follows. The Section 2 describes the dataset used for data processing and the methods examined in terms of drift counteraction. The section 3 presents the conclusions.

2. EXPERIMENTAL AND METHODS

2.1 The Dataset

The dataset has been obtained at Osmetech plc. facilities at the University of Manchester. Three gases at different concentration level were measured:

CP1137, *Olfaction and Electronic Nose: Proceedings of the 13th International Symposium*, edited by M. Pardo and G. Sberveglieri

ammonia (0.01%, 0.02%, 0.05%), propanoic acid (0.01%, 0.02%, 0.05%), n-buthanol (0.01%, 0.1%). The experiments lasted for 5 months on the array of 17 polymeric sensors, so that 3925 samples labelled to 8 classes have been recorded. The response of the sensors has 328 seconds time-length and consists of 329 points sampled at 1Hz frequency. The compound is induced to the sensor array at instant t=0s, then the clean air enters the chamber at instant t=185s.

In this work, the stable-state point of every transient response is used for data analysis. The operation on removing the outliers has been performed by using the algorithm of Filzmoser et al. [2] with the default parameters. Hence, the number of samples has been reduced to 3484, and the total number of input variables is equal to 17: 1 stable-state from each of 17 sensors.

2.2 Component Correction Through The Reference Gas

The component correction method was proposed by Arthursson in 2000 [1]. This approach assumes that drift is highly correlated along the classes of gases, so that a reference gas can be selected which is representative for all classes. By applying a Principal Component Analysis (PCA) to the entire reference gas, one or several component(s) can be computed which capture the in-class variance for the reference gas and is supposed to explain the drift phenomena for all the classes. This strategy seems to be reasonable, since drift often stands for one preferred direction (according to the definition of drift) in the data space, rather than random distribution. Additionally, the relationship between similar sensors assumed to be linear as well as the relationship between the reference and other gases. The objective of this paper is to study some problems regarding to with the choice of the reference gas and propose an alternative to estimate the drift direction based on computing the common principal components for all classes, rather than only for the reference class.

2.3 Common Component Correction

The Common Principal Component Analysis (CPCA) can be viewed as a generalization of PCA to k groups of datasets. Under the Common Principal Component (CPC) hypothesis there exist an orthogonal matrix V such that k covariance matrices C(i) are diagonal in the data space defined by V: L(i)=V'C(i)V, i=1,2,...k. CPCA was proposed by Flury in 1984 [3], who first derived the normal theory maximum likelihood estimates of V and L(i). When all the matrices commute, this can be achieved exactly. When this is not the case, the approximate Joint Diagonalization (JD) problem is stated and can be solved by optimizing different diagonality criterion [4]. In this work, the algorithm of Cardoso [5] is used which performs the orthogonal diagonalization based on weighted least-squared (WLS) criteria by using the Givens rotations. In terms of the application of CPCA for drift compensation, common components obtained for k classes of gases represent the variation in data caused by drift which affect all gasses in the same manner. Such approach conforms to the definition of drift and captures the information of all data variation, rather than only of the reference gas.

3. RESULTS

To evaluate the algorithms towards drift counteraction, the complete dataset is divided into Training and Validation Sets. All samples are ordered in time. A simple k-NN classifier is used with the parameter of nearest neighbours k = 3. The classification is performed for 3 classes of the given chemical species: ammonia, propanoic acid and n-buthanol.

3.1 Size Of The Training Set

In the experiments, two Training Sets T1 and T2 with different size has been tested: 1000 and 1200 samples (containing all 8 classes) respectively. Consequently, if the complete dataset amounts for 3484 samples, the Validation data V1 and V2 contain 2484 and 2284 samples respectively. Since the measurements of the dataset were obtained continuously for all 8 classes of gases, each gas in Training Set is presented by 115-155 samples. These numbers considered to be statistically significant for computation either PCA of the reference gas or CPCA for k gases. Final Validation Sets will be a sliding windows over Validation data (V1 o V2) in order to check each algorithm against the time distance from the training set.

3.2 Gas Selection for CC-PCA and CC-CPCA

To choose the reference gas, the classes of gases with the highest concentration for each chemical species have been selected, because they have greater signal-to-noise ratio (n-buthanol 0.1 %, propanoic acid 0.05% and ammonia 0.05%). CPCA is performed only for the gases at the highest concentration levels, in order to avoid capturing the gradient direction of concentration for classes of the same chemical species.

3.3 Drift Component

The examined methods on drift counteraction are based on Component Correction (CC) operation, where the drift component is subtracted from the data in multivariate data space [1]. To compute the drift component in CC-PCA method, PCA has to be calculated for the reference gas on samples of Training Set, and the first Principal Component with the largest eigenvalue captures the most variance in data and represents the drift (under the assumption that the reference gas is representative for all classes). Under CC-CPCA approach covariance matrices of k gases (in our case k =3) are jointly diagonalized, and the transformation matrix V gives the basis vectors in the new data space where the covariance matrices are as diagonal as possible. For estimation the quality of performed approximate diagonalization, signal-to-interference (SIR) ratio is used, which comes from signal processing on speech recognition. Since only one component of drift is of the interest, the basis vector with the greatest SIR ratio selected. For both Training Sets T1 and T2, the SIR value is at the acceptable level and equals to 73% and 82% respectively.

Figure 1 presents the drift directions in 2D PCA subspace for both Training Sets T1 and T2 computed by two methods. For both Training Sets, all drift vectors of 3 reference gases are quite different, that underline the weakness of CC-PCA approach when the choice of the reference in not clear. Comparing plots for T1 and T2, the drift direction obtained from propanoic acid 0.05% reference gas appeared to be sensitive to switch from 1000 to 1200 of sample size of Training Set, while the drift vectors from n-buthanol 0.1% and ammonia 0.05% reference preserve the same direction. The drift component calculated by joint diagonalization also reflects the changes in Training Set by aligning towards the propanoic acid 0.05% direction. Following the mathematical intuition applied in CC-CPCA, the drift direction from joint diagonalization is believed to correspond to the actual variation of data caused by drift. Thus, the rest two gases n-buthanol 0.1% and propanoic acid 0.05% are likely to fail on drift compensation for the given Training Set. The numerical results are present further in the next Subsection.

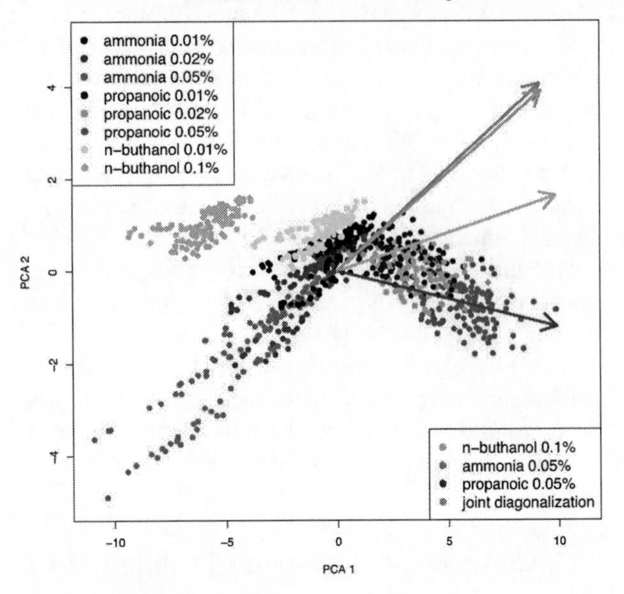

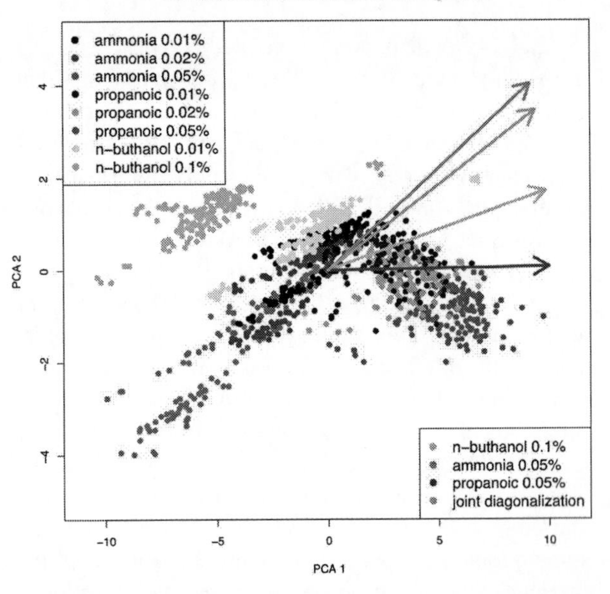

FIGURE 1. PCA plot and principal drift directions for the proposed methods for the different training sets T1(top) and T2 (bottom).

3.4 Evaluation Of CC-PCA And CC-CPCA Methods

The power of the examined methods on drift counteraction is performed on classification problem (k-NN classifier) on the corrected data and compared in respect to the results obtained for non corrected data. The validation process has been accomplished for sliding window of the Validation Set with the window size the same as the size of the Training Set.

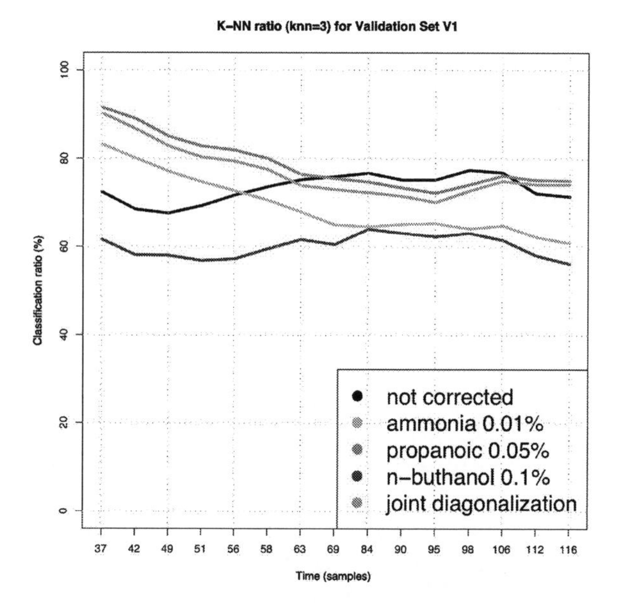

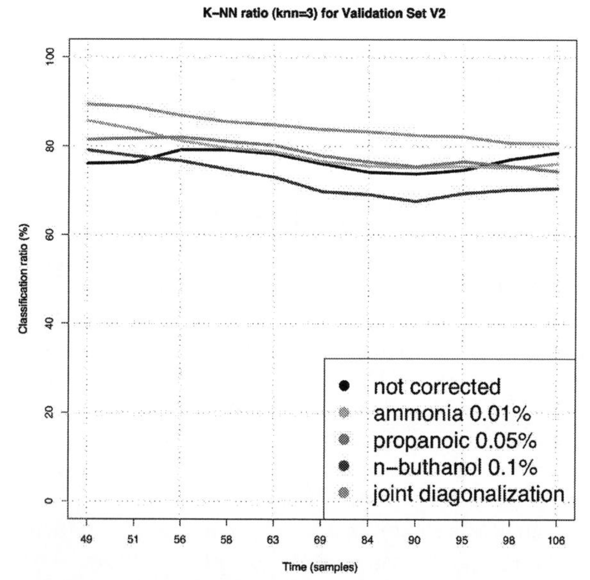

FIGURE 2. K-NN performance as a function of the distance of the validation set from the training set T1/T2(top/bottom)

Figure 2 shows the main results in terms of comparison between the examined methods on drift counteraction. The curve of Classification ratio for non-corrected data is of black colour, the curves of the references ammonia 0.05%, propanoic acid 0.05%, nbuthanol 0.1% and joint diagonalization are marked by grey, green, red and orange respectively. For Training Set T1 the CC-PCA approach with the reference gas ammonia 0.05% performs as well as CC-CPCA methods, as expected from their almost coincident drift directions on the top plot of Figure 1. At the time instant of 63 days, all drift compensation methods fail to improve the classification. For Training Set T2 the CC-CPCA method shows superior efficiency along the others, and the time-stability of the classification is increased being at the level not below than 80%.

4. CONCLUSIONS

In this manuscript we have proposed a new method based on common-class variance CC-CPCA, that has proven to perform at same level of confidence as CC-PCA approach with the best reference gas (ammonia 0.05%). Moreover, the drift direction captured by Joint Diagonalization is able to show better stability to artefacts in the training set for a particular class, as it employs information from all classes, as seen from Training Set T2 experiment. The direction of future work will be related with application of CPCA to the drift compensation problem optimizing the dimension of the subspace obtained by CPCA, using higher order components rather than the first one.

REFERENCES

1. T. Artursson, T. Eklöv, I. Lundström, P. Mårtensson, M. Sjöström, M. Holmberg, "Drift correction for gas sensors using multivariate methods", Journal of Chemometrics, Vol. 14, No. 5-6 (2000), pp. 711-723
2. P. Filzmoser, R. Maronna, M. Werner, "Outlier identification in high dimensions", Computational Statistics and Data Analysis, Vol. 52, No. 3 (2008), pp. 1694-1711
3. Xi-Lin Li , Xian-Da Zhang. "Nonorthogonal Joint Diagonalization Free of Degenerate Solution", IEEE Transactions on Signal Processing, (2007), Vol. 55, No. 5-1, pp. 1803-1814
4. J.-F. Cardoso, A. Souloumiac, "Jacobi angles for simultaneous diagonalization", J. Mat. Anal. Appl., Vol. 17 (1996), pp. 161-164
5. Holmberg M, Winquist F, Lundström I, Davide F, Di Natale C, D'Amico A. "Drift counteraction for an electronic nose". Sens. Actuat. B (1996); 35/36: pp. 528-538.
6. Holmberg M, Davide F, Di Natale C, D'Amico A, Winquist F, Lundström I., "Drift counteraction in odour recognition application: lifelong calibration method". Sens. Actuat. B (1997); 42: pp. 185-194.
7. C. Di Natale, F.A.M. Davide, A. D'Amico, "A self-organizing system for pattern classification: time varying statistics and sensor drift effects", Sens. Actuators B 26–27 (1995) 237–241.
8. S. Marco, A. Ortega, A. Pardo, J. Samitier, "Gas identification with tin oxide sensor array and self-organizing maps: adaptive compensateion of sensor drifts", Proc IEEE Transactions on Instrumentation and Measurement (1998), vol. 47.

A Fuzzy ARTMAP Approach To The Incorporation Of Chromatographic Retention Time Information To An MS Based E-Nose

Cosmin Burian[1], Jesus Brezmes [1,2], Maria Vinaixa [1,2], Eduard Llobet[1], Xavier Vilanova[1], Nicolau Cañellas[1] and Xavier Correig [1,2]

[1]Department of Electronic Engineering, Universitat Rovira i Virgili, Avenida Paisos Catalans 26, 43007 Tarragona, Spain
[2]CIBER de Diabetes y Enfermedades Metabólicas Asociadas (CIBERDEM)
cosmin.burian@urv.cat, jesus.brezmes@urv.cat

Abstract. This paper presents the work done with Fuzzy ARTMAP neural networks in order to improve the performance of mass spectrometry-based electronic noses using the time retention of a chromatographic column as additional information. Solutions of nine isomers of dimethylphenols and ethylphenols were used in this experiment. The gas chromatograph mass spectrometer response was analyzed with an in-house developed Fuzzy ARTMAP neural network, showing that the combined information (GC plus MS) gives better results than MS information alone.

Keywords: gas chromatography mass spectrometry; fuzzy ARTMAP.
PACS: 07.05.Mh; 82.80.Bg; 82.80.Ms

INTRODUCTION

The goal of this work is to improve the performance of an MS based Electronic Nose by using the combination of Gas Chromatography and Mass Spectrometry. This approach adds extra information to the dataset, since both the average mass spectra (AVMS) and Total Ion Chromatogram (TIC) of the GC/MS measurements are used. Fuzzy ARTMAP Neural Networks [1] are used to combine both data dimensions. To generalize conclusions, three different chromatographic retention times were investigated.

Even though in many experiments the MS electronic nose has proven better than gas sensor based multisensor systems [2], some complex mixtures have proven difficult to classify with any E-nose technique. In most of the cases the methods used for the classification of this kind of data are two and three-way methods such as PCA, PARAFAC. By using a fuzzy ARTMAP neural network combining the chromatographic and mass spectra data in two different ways (concatenation and by addition of the neural network results) we want to prove that the extra information added by the chromatographic separation is improving the results in a challenging experiment.

EXPERIMENTAL

Twenty mixtures of nine isomers of dimethylphenol and ethylphenol were measured and analyzed by means of gas chromatography mass spectrometry (GC-MS). The nine isomers were chosen based on their similar theoretical mass spectra.

Based on this information we designed the experiment as shown in Table 1. Benzene acts as an internal standard.

Three chromatographic methods were studied. Method one tried to obtain well-resolved chromatograms; it employed a temperature-programmed separation from 50°C to 180°C, where almost all the isomers were separated. Method two and three were designed to return coeluted peaks. The measurements were conducted through syringe injection of 1 µl per measurement, and ten repetitions for each method and solution were made.

RESULTS

Having the data in a 3D format, we added the axis of the chromatographic separation time obtaining the AVMS, which represents the MS approach. We also

CP1137, *Olfaction and Electronic Nose: Proceedings of the 13th International Symposium*, edited by M. Pardo and G. Sberveglieri

TABLE 1. Experiment design (the numbers represent percentage of substance against methanol)

component	Sol1	Sol2	Sol3	Sol4	Sol5	Sol6	Sol7	Sol8	Sol9	Sol10	Sol11	Sol12	Sol13	Sol14	Sol15	Sol16	Sol17	Sol18	Sol19	Sol20
2,3-Dimethylphenol	-	0.5	0.5	0.5	0.5	0.5	0.5	0.5	0.5	0.25	0.5	0.5	0.5	0.5	0.5	0.5	0.25	0.5	0.25	0.5
2,4-Dimethylphenol	0.5	-	0.5	0.5	0.5	0.5	0.5	0.5	0.5	0.5	0.25	0.5	0.5	0.5	0.5	0.5	0.5	0.25	0.5	0.25
2,5-Dimethylphenol	0.5	0.5	-	0.5	0.5	0.5	0.5	0.5	0.5	0.5	0.5	0.25	0.5	0.5	0.5	0.5	0.5	0.5	0.5	0.5
2,6-Dimethylphenol	0.5	0.5	0.5	-	0.5	0.5	0.5	0.5	0.5	0.5	0.5	0.5	0.5	0.5	0.5	0.5	0.5	0.5	0.5	0.5
3,4-Dimethylphenol	0.5	0.5	0.5	0.5	-	0.5	0.5	0.5	0.5	0.5	0.5	0.5	0.5	0.5	0.5	0.5	0.5	0.5	0.5	0.5
3,5-Dimethylphenol	0.5	0.5	0.5	0.5	0.5	-	0.5	0.5	0.5	0.5	0.5	0.5	0.25	0.5	0.5	0.5	0.5	0.5	0.5	0.5
2-Ethylphenol	0.5	0.5	0.5	0.5	0.5	0.5	-	0.5	0.5	0.5	0.5	0.5	0.5	0.25	0.5	0.5	0.25	0.25	0.5	0.5
3-Ethylphenol	0.5	0.5	0.5	0.5	0.5	0.5	0.5	-	0.5	0.5	0.5	0.5	0.5	0.5	0.25	0.5	0.5	0.5	0.25	0.25
4-Ethylphenol	0.5	0.5	0.5	0.5	0.5	0.5	0.5	0.5	-	0.5	0.5	0.5	0.5	0.5	0.5	0.25	0.5	0.5	0.5	0.5
B/Et-OH/acetone	2%	2%	2%	2%	2%	2%	2%	2%	2%	2%	2%	2%	2%	2%	2%	2%	2%	2%	2%	2%

added the m/z hits at a given scan obtaining the TIC, giving the GC approach. These were the input matrices that were fed into the fuzzy ARTMAP routines.

The two data matrices (MS and GC) were individually pretreated by normalization, and fed into an in-house developed fuzzy ARTMAP algorithm. The network was tested using the leave-one-out cross-validation method. Both data matrices were normalized between 0 and 1.

The results of the fuzzy ARTMAP neural network were displayed as a percentual success rate and in the form of a confusion matrix (figure 1).

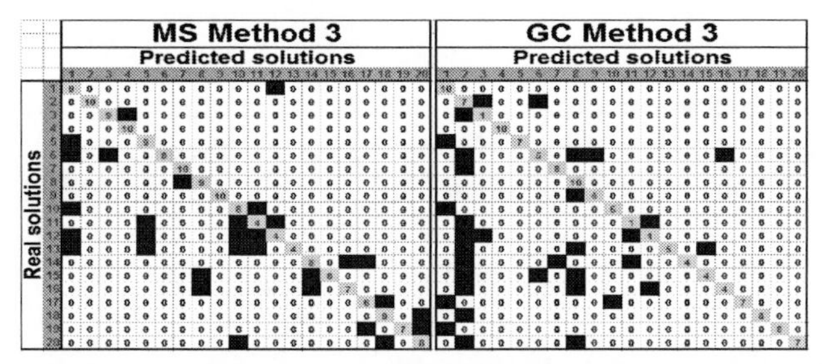

FIGURE 1 Fuzzy ARTMAP results for the Method 3 MS and GC (real (y) vs. predicted (x) solution)

Comparing the fuzzy ARTMAP results we can see (figure 1) that the errors are different for the two approaches (MS and GC). Using both information dimensions (time and m/z) should enhance the classifying ability of the MS-based electronic nose.

To improve the results of the fuzzy ARTMAP neural network a voting strategy was implemented and the winning class was decided by the total number of votes each solution received. Initially, the input data for the fuzzy ARTMAP voting strategy were the AVMS and TIC matrices.

In order to take advantage of both data types we followed 2 approaches: 1) Concatenation of the MS and GC matrices (MSGC) and 2) by summing the MS votes with the GC votes for a given measurement; to combine the MS and GC information through the voting approach, the two matrices were fed into the fuzzy ARTMAP independently and the votes that they had collected were added deciding the winning class.

Possible outputs from the algorithm were correctly classified, wrongly classified or unclassified. Because of the high number of unclassified votes in the combined MS and GC voting method, and because the MS proved to give better results, we decided that in case of an unclassification in the voting strategy of the MS and GC matrices the winner should be the MS winner (MSGC vote MS improved).

The results were monitored by the success rate based on the confusion matrix.

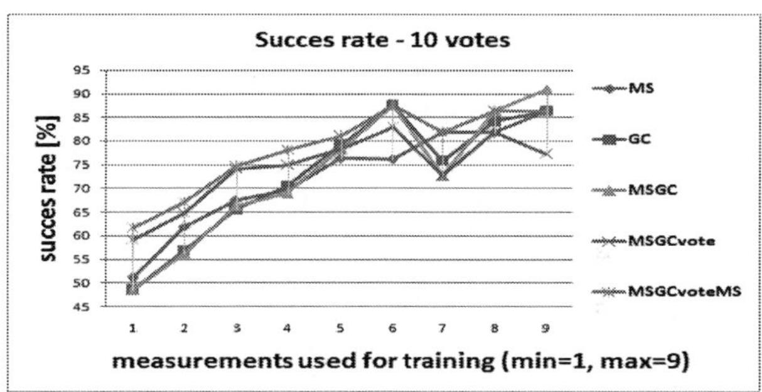

FIGURE 2 Method 1 Voting strategy FuzzyARTMAP success rates

Looking at figure 2 we can see that the MSGCvoteMS results are always on top of the rest of the approaches if 1 to 8 measurements are used for training. Because in practice a high number a measurements is hard to obtain this shows that the MSGCvoteMS approach is the best one, giving the best results when fewer measurements are used for training.

Conclusions

Taking advantage of the differences in classification of the MS and GC signal in 2 distinct ways (concatenations and votes) the new methodology improves the results in all the possible scenarios tested.

REFERENCES

1. Gail A. Carpenter & co. *IEEE Trans. On neural networks* **5** 698-713 (1992)
2. M. Vinaixa & co. *Sensors and actuators B,* **106** 67-75 (2005)

Total solvent amount and human panel test predictions using gas sensor fast chromatography and multivariate linear and non-linear processing

A. Perera[1], F. Röck[2], I. Montoliu[3], U. Weimar[2], S. Marco[3]

[1]*CREB, Universitat Poltècnica de Catalunya, Pau Gargallo 5, 08028 Barcelona Spain*
[1]*Centro de Investigación Biomédica en Red en Bioingeniería, Biomateriales y Nanomedicina (CIBER-BBN)*
[2]*Institute of Physical and Theoretical Chemistry, University of Tübingen, Germany*
[3]*Intitut de Bioenginyeria de Catalunya, Barcelona Spain*

Abstract. Data from a Gas Sensor based Chromatography instrument is used in order to replicate output from a human panel and the estimation of the total solvent amount measured by and FID device in a packaging application. The system is trained on different packaging sample properties and validated with unseen combinations of materials, varnishes and production processes. This contribution will show the difficulties on the prediction of the output of the human panel, and the success on the prediction of the total amount of solvent in the sample

Keywords: Gas sensors, solvent prediction.
PACS: 07.07.Df , 87.85.Ng

INTRODUCTION

The final goal of the described application is to monitor undesired emissions from packaging materials in food industry. In contrast to classical solutions based on HS GC, the idea is to develop an instrument that serves as rapid monitoring tool working at-line in the production chain.

For this purpose easy operability and affordability are requirements that can be achieved with a system based on metal oxide gas sensors. Using a proper chromatography column, it is possible to separate water vapour from organic solvents in two different peaks within a few minutes. The obtained sensor data can be evaluated without humidity interference. Together with an autonomous sampling system the basics are created for a low cost quality control system that doesn't need qualified employees for operation.

Goals reported in this action will cover two aspects: on the one hand, to check the capacity of the provided sensor system to predict the total solvent amount (reference is the FID area) and odour scores (reference is a human test panel), on the other hand, to prove the generalization capability of the instrument and algorithms when testing samples containing unseen production processes or materials. For this objectives, we have exploited the capabilities of Three-way data processing for visualization and regression [1-2], as well as linear Partial Least Squares (PLS), Multilinear Regression (MLR) [3,4] and Principal Components Regression (PCR). A non-linear method based on a Support Vector Machines for regression has been applied as well [5].

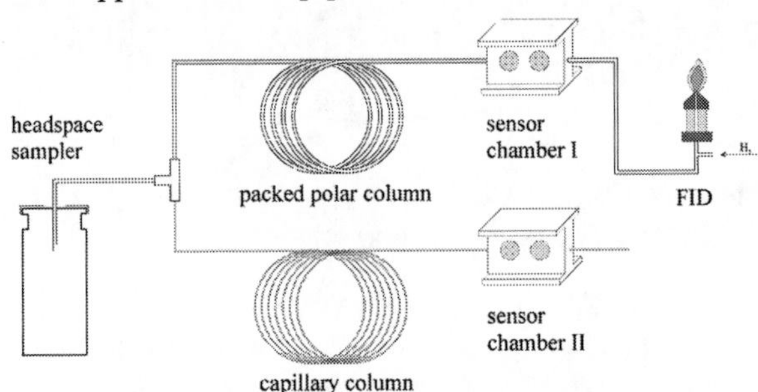

FIGURE 1 Schematic of the experimental setup.

EXPERIMENTAL AND METHODS

The data analyzed in this document was acquired from a prototype built at the IPTC. This prototype is provided with two independent columns, a packed column and a capillary column. Four metal oxide sensors with different doping and different temperature were placed after each column, namely S1 S4 for the packed column and S5 S8 for the capillary column. The packed column (flow rate 20 mL/min; temperature 100° C) only separates water from the organic solvents, whereas the capillary column (flow rate 5 mL/min; temperature 61° C) additionally separates some organic components. A schematic of the experimental setup is shown in Figure 1.

CP1137, *Olfaction and Electronic Nose: Proceedings of the 13th International Symposium*, edited by M. Pardo and G. Sberveglieri

The reference dataset consists of odour scores, Robinson scores, and the total solvent amount for a number of combinations of different production processes, materials, inks and varnishes used in the production of the paper and board packaging material.

Examples for the production process are roto gravure, offset combination or hot foil stamping/reverse side stamping. The density of the paper is in the range from 90 – 315 g/m3. Used inks are water based gravure, UV spot and combination of both. The odour scores obtained by the sniff test, as well as the results of the Robinson test are in the range between 0.0 and 4.0. The FID signal was calibrated using ethanol as calibration standard. Altogether 480 samples were measured. For each paper sample, signals from the four sensors were acquired at a rate of 1Hz during one hour. Therefore, we have a raw input dataset of dimensionality 3600x4 = 14400. If the chromatograms are limited to the first relevant minutes (1500 samples) we may reduce this dimensionality to 1500x4 =6000.

To validate the generalization properties of the instrument and data algorithms, special care has been taken in the selection of the training, calibration and test sets. Also comparison with linear (PCR, MLR, PLS) and non-linear (Support Vector Regression) regression algorithms is reported. Models for each algorithm have been computed for the total solvent amount, the odour scores and the Robinson values. Samples included in the training, calibration and test set are manually selected in order to stress generalization power of the algorithms. In this case, certain combinations of material, ink, varnishes or production process present in the calibration or validation set may not be appear in the test set.

Table 1 Summary of RMSE for test set

	S1—S4 (Packed)			S5—S8 (Capillary)		
	Odour	**Robin**	**FID**	**Odour**	**Robin**	**FID**
MLR	5.26	18.51	2.96	2.69	4.62	1.68
PCR	1.06	0.69	0.12	1.24	0.85	0.34
PLS	1.35	1.44	0.13	1.17	0.69	0.34
SVR	0.61	0.65	0.27	0.71	0.57	0.35

Results

Results cover the output of the regression models for test set (171 samples) given by MLR, PCR, PLS and SVM. A poor generalization of odour scores and results form the Robinson test for unseen data is seen for all algorithms. SVR is able, at most, of reducing data dispersion as seen in the three regressions. The change produced by unseen materials, inks or varnishes invalidate a calibration built with the combinations available in the training set. Different replicas of the combination of the different qualities produce regularly distributed errors in the regression.

On the other hand, the regression of the total solvent amount shows good generalization properties to unseen combinations. In fact, best results are achieved by PCR with the sensors after the packet column. Although residual errors are also clustered with the replicas of the combinations of inks, materials, varnishes and production process, results are with an acceptable error given the stressed validation procedure. As opposite to the regression of the odour scores and the results of the Robinson test, this is a clear case where a properly trained linear method provides better generalization than a non-linear method. These results are summarized in Table 1

Conclusions

From regression models it is clear that panelist values (odour and Robinson scores) will not be correctly predicted in the case that all possible combinations of material, inks, varnish and production process are available within the training set. The variation produced by the change of any of these variables will affect the generalization properties of the regression models. On the other hand, the total solvent amount can be predicted with reasonable generalization to different combinations than those found in the calibration set. Non-linear methods provided improved results in the prediction of odour and Robinson scores.

ACKNOWLEDGMENTS

Authors wish to acknowledge the Ramon y Cajal Program from the Spanish Ministerio de Educación y Ciencia. This work was supported within the framework of the CICYT grant TEC2007-63637 from the Spanish Government and Bio-ICT 216916. CIBER-BBN is an initiative of the Spanish ISCIII.

REFERENCES

1. L. R. Tucker. *Problems in measuring change*. University of Wisconsin Press, 1966.
2. L. R. Tucker. Some methematical notes on three-mode factor analysis. *Psycometrika*, 31:479, 1963.
3. H. Wold. Systems under indirect observation. Part II., chapter Soft Modelling: The basic design and some extensions, page 154. 1982. K. G. Jöreskog and H. Wold (eds).
4. S. Wold, C. Albano, W. D. III, K. Esbensen, S. Hellberg, E. Johansson, and M. Sjöström. Chemometrics: Mathematics and Statistics in chemistry, chapter Multivariate Data Analysis in Chemistry, pages 17– 95. D. Reidel, The Netherlands, 1983a. B.R. Kowaqlski (ed).
5. Harris Drucker, Chris J.C. Burges, Linda Kaufman, Alex Smola and Vladimir Vapnik (1997). "Support Vector Regression Machines". Advances in Neural Information Processing Systems 9, NIPS 1996, 155-161, MIT Press.

Reproducibility and Uniqueness of Information Coding as Key Factors For Array Optimization

Julia Burlachenko[1], Simonetta Capone[2], Pietro Siciliano[2] and Boris Snopok[1]

[1] *V. Lashkaryov Institute of Semiconductor Physics, NAS Ukraine, 03028 Nauki av. 41, Kiev, Ukraine*
[2] *Institute for Microelectronics and Microsystems, CNR, Italy, via Arnesano 1, Lecce, 73100, Italy*

Abstract. An approach for the evaluation of sensor array efficiency is proposed. The reproducibility of sensor responses and the uniqueness of chemical images (CI) formed by the array are considered as dominant factors that determine sensor array functionality. The key feature of the method is representation of array response as a numerical function of sensors responses. This allow making mathematical operations with CI. The dispersion of CIs of given set of analytes can serve as a measure of their uniqueness. The coefficient of covariance in the set of measurements expresses their reproducibility. The rate of influence of both reproducibility and CI uniqueness on the discriminating ability is shown qualitatively on the example of three QCM sensor array with metal phthalocyanines sensitive coatings. The illumination of sensors surfaces in the Q-band region of Pc's adsorption spectra was applied as a factor affecting both the reproducibility and the uniqueness of CIs.

Keywords: Sensor array, chemical image, QCM, reproducibility, optimization
PACS: 07.07.Df

INTRODUCTION

Different approaches for sensor arrays optimization such as generalized mathematical models, statistical methods, estimations of information content through the Fisher information, etc. were proposed. At the same time, for practical optimization of sensor arrays one should know dominant factors that determine their functionality, and have a simple algorithm for their estimation. We consider effects of reproducibility and uniqueness of CI on the array discriminating ability. We propose to use a numerical function of sensors responses as CI and the dispersion of CIs as a measure of their uniqueness.

EXPERIMENTAL AND METHODS

The array of three QCM sensors modified by methal free, copper and lead phthalocyanines was used. The analytes were: Set 1 (water, triethylamine, propilamine) and Set 2 (ethanol, triethylamine, propilamine).

The temporal CIs were calculated as:

$$F_{CI}(f_1, f_2, f_3, t) = \frac{[f_3(t) - f_2(t)] - [f_2(t) - f_1(t)]}{f_3(t) - f_1(t)}, \quad (1)$$

$f_1(t)$, $f_2(t)$ and $f_3(t)$ are sensors responses at the moment t. The dispersion of F_{CI} D(t) were compared with the effectiveness of discrimination expressed by S(t) – silhouette width (cluster analysis in fuzzy logic format [1]). The reproducibilty was calculated as coefficient of covariance of measurements. The illumination was used as factor affecting both CI uniqueness and reproducibility.

RESULTS

The correlation between the effectiveness of discrimination S(t) and the dispersion of CI D(t) was investigated. The dependences S(t) and D(t) for Set 2 under illumination are shown on fig.1. The initial parts of curves are non correllated that can be explained by the lack of reproducibility in kinetic parts of adsorption curves. However after 50s of measurement the correlation is high. The coefficients of correlation calculated for $t>50$ for Set 1 are 0,55 without illumination and 0,98 with illumination; and for Set 2 – 0,83 and 0,92 respectively.

As it was shown previously [2] illumination changes Pc's adsorption properties. The better correlation under illumination proves the suggestion about the uniqueness of CI as one of predominate factors defining the array's discrimination ability

CP1137, *Olfaction and Electronic Nose: Proceedings of the 13th International Symposium*, edited by M. Pardo and G. Sberveglieri

(under illumination the other factor, reproducibility, is high and has less influence).

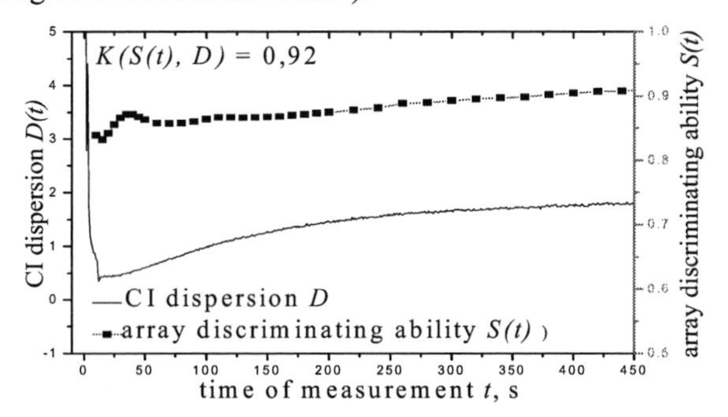

FIGURE 1. S(t) and D(t) for set 2 under illumination.

The influense of the reprodusibility was evaluated when the D(t) has the same value but the reproducibility increases. The discrimination in the second case is better due to reduction of clusters sizes that is related to the improvement of the measurement results reproducibility.

CONCLUSIONS

CI representation in a form of function allows to express numerically the uniqueness of array responses as the dispersion of CIs. The CI dispersion and the reproducibilty of measurements are key factors determinig the array effectiveness.

REFERENCES

1. Struyf A., Hubert M., Rousseeuw P.J. "Integrating robust clustering techniques in S-PLUS" *Computational Statistics and Data Analysis*, 26, (1997), pp. 17-37.
2. Yu. V. Burlachenko, B. A. Snopok "Multisensor Arrays for Gas Analysis Based on Photosensitive Organic Materials: An Increase in the Discriminating Capacity under Selective Illumination Conditions" *Journal of Analytical Chemistry* Vol. 63, No. 6 (2008), pp. 557–565.

Resolution of Ion Mobility Spectra for the Detection of Hazardous Substances in Real Sampling Conditions

I. Montoliu[1], V. Pomareda[1,2], A. Kalms[1], A. Pardo[1], J. Göbel[3], M. Kessler[3], G. Müller[3], S.Marco[1,2]

1 Department of Electronics, Universitat de Barcelona. Martí i Franquès, 1. 08028 Barcelona (Spain)
2 Artificial Olfaction Lab, Institut for BioEngineering of Catalonia, Baldiri I Rexach 13, 08028-Barcelona, (Spain)
3 EADS Innovation Works, Munich, Germany
Email: imontoliu@el.ub.es

Abstract. This work presents the possibilities offered by a blind source separation method such Multivariate Curve Resolution- Alternating Least Squares (MCR-ALS) in the analysis of Ion Mobility Spectra (IMS). Two security applications are analyzed in this context: the detection of TNT both in synthetic and real samples. Results obtained show the possibilities offered by the direct analysis of the drift time spectra when an appropriate resolution method is used.

Keywords: Ion Mobility Spectrometry, Multivariate Curve Resolution, Security, LIMS, MCR-ALS.
PACS: 07.81.+ ; 02.50.Sk

INTRODUCTION

The current challenges in security applications require of reliable instrumentation able to operate with volatile samples, at very low concentration levels, and with minor or inexistent sample preparation. In addition, high sensitivities and selectivities are required to prevent the occurrence of false negatives and false positives. On the other hand, the final applications require of instrumentation capable to operate in uneven conditions and to provide precise identification of threats 'in situ'. Only a selected family of high sensitive detectors is candidates to fulfill the expected requirements. Among the possible options, IMS has generated a great interest on the security industry, mainly due to its sensitivity towards volatiles of very low vapor pressure. In the recent years, this instrumentation has become a standard in the detection of explosives.

In spite of the instrumental advances in IMS, with improved ionization sources and better sensitivities, data processing of the obtained signals is often univariate after time windowing. This fact can reduce the selectivity of the instrument in real scenarios due to sample matrix effects. To overcome somehow these problems, multivariate measurements are a possible alternative. The use of the complete drift spectra becomes an option in this sense and must be analyzed appropriately.

Drift spectra can be considered a complex signal with contributions of different chemical sources. This specially important when dealing with real samples, where vapors from other substances will be also present in the sample. Then aside from the reactant ion peak species, there can be contributions of other environmental volatiles (matrix effect) and the main analyte of interest. Therefore, with the appropriate method, it can be theoretically possible to decompose the global signal.

MCR-ALS is able to provide the simultaneous decomposition of linearly mixed signals following an alternating least squares approach. The method is capable of consider physical characteristics of the measurement and include them in the resolution as constraints.

The focus of the analysis will be put in the possibilities of MCR to isolate the contribution of both the TNT and salicilates to the global signal in two IMS instruments of different kind.

CP1137, *Olfaction and Electronic Nose: Proceedings of the 13th International Symposium,* edited by M. Pardo and G. Sberveglieri

EXPERIMENTAL AND METHODS

Data analyzed contains different IMS recordings of thermal desorption of sampling membranes exposed to different environments and explosives. These recordings have been registered by means of a LIMS prototype, provided by EADS, and a specially designed thermal desorber. This device allows desorption of the contents of the membrane into a carrier gas flow that is later introduced in the reaction chamber.

A first set of samples contained TNT at very low concentration levels (around 5ng TNT). A second set, has been sampled in realistic scenarios from syntheticlly contaminated luggage materials. The preparation of both sets of samples has been done after dilution of the explosive in ketone and posterior deposition onto the material under analysis.

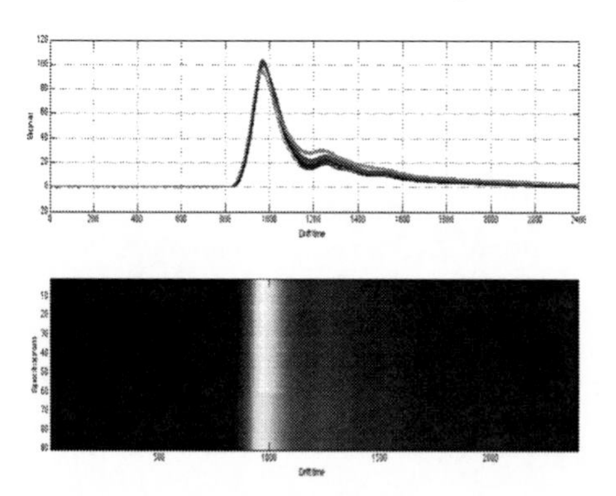

FIGURE 1. Consecutive (blue, red, green) recordings of a synthetic sample of 10ng TNT.

RESULTS AND DISCUSSION

MCR Analysis. Synthetic Samples

Starting up from filtered signals, data matrices were arranged accordingly. The first set of data to be analyzed contained the successive recordings by triplicate of the same sample, with 10ng TNT. Evolving Factor Analysis (EFA) analysis suggested the presence of up to 3 factors. To initialize MCR analysis an estimation of the pure concentration/spectral profiles is necessary. In this case, because the TNT pure spectrogram is unknown, an estimation of the spectral profiles was achieved by SIMPLISMA.

MCR results obtained confirmed the presence of, at least, 3 factors during the measurement. These results were obtained under conditions of nonnegativity for both spectral and concentration modes. Therefore, the behaviour during the measure integration (30 scans per sample) was reflected in the concentration profile. These results obtained agreed with the expected response.

To increase somehow this resolution, to get an isolated response (pure spectrogram) for TNT, it was introduced a constraint to the spectral mode for the first factor. This constraint reflected the 'a priori' knowledge of the system, in which the spectrogram bands ascribable to TNT come slightly after the occurrence of the RIP. The results confirmed the appearance of three peaks around drift time region 1400-1800. Pure concentration and spectral (normalized) profiles are shown in figure 2a,b.

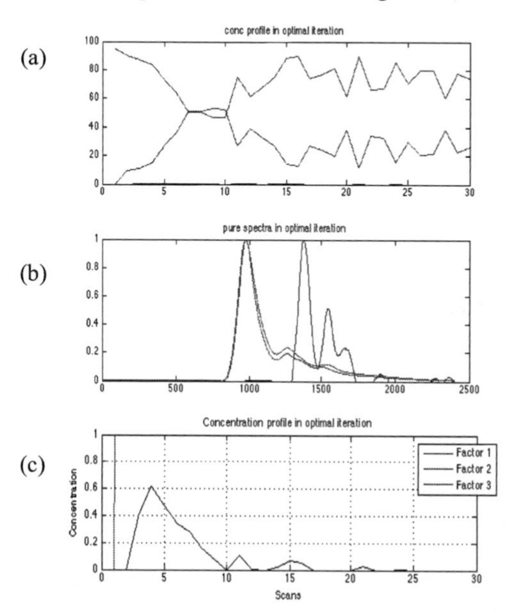

FIGURE 2. Synthetic data. MCR Results after imposition of spectral constraints (a,b). Zooming of concentration mode (c).

Having a look at the pure concentration profiles, was clear that there existed a region in the scan number mode (1-30) in which the concentration / signal ratio was higher. As is seen in figure 2c, this region is located around the fourth sample. Later on, the concentration contribution to the signal decays until scan 30. Hence, it could be recommended the use of the signal in a region around this point, to achieve the maximum sensitivity for TNT.

MCR Analysis. Real Samples

The second set of IMS spectrograms corresponded to the analysis of TNT and salicilate samples in a real environment. For TNT, EFA analysis of the background confirmed the appearance of 4 factors along the measurement time. Results showed almost constant concentration profiles along the entire scan range. Pure spectra did not show any specially differentiated profile, aside from the resolution of overlapped peaks at reactant ion peak area.

A first EFA analysis of a sample doped with TNT was done, detecting up to 5 factors. This number agreed with the increased signal complexity after the introduction of the analyte. Pure spectrogram corresponding to this factor could be also isolated. Inspecting these pure spectral profiles, there appear two peaks in the drift time region comprised between 1250-1600. Due to their position and shape, they almost coincided with the ones expected for TNT. Therefore, these bands were considered as indicators of its presence in the sample.

After this test, a MCR analysis was performed on the spectrogram of a real sample without doping with a solution of TNT. In this case, the IMS signal was more complex. It showed the influence of the different volatiles present in the environment under real measurement conditions, and made it harder for the MCR algorithm to converge to an appropriate solution. However, using a suitable initialization with proper dimensioned EFA, MCR pure concentration and spectral profiles were also obtained.

CONCLUSIONS

The use of Multivariate Curve Resolution offers good possibilities in the direct detection of the presence of TNT in air samples. The method does it following a multivariate approach, instead of other univariate approaches much more common, such areas and peak heights. These last are prone to experiment unexpected variations during the measurement process that can affect its value. By the opposite, MCR can deal with that variability and model these sources. This is achieved by using a blind source separation strategy based in ALS that can be enhanced, if necessary, by using previous knowledge of the system.

MCR also delivers more information about the dynamics of the signal during the accumulation of spectrograms and, therefore, makes possible to determine the most appropriate scan number window that provide the highest sensitivities.

ACKNOWLEDGMENTS

Authors would like to acknowledge the support of the exchange program of GOSPEL FP6-IST 507610.

REFERENCES

1. R. Tauler, D. Barceló, Trends Anal. Chem. **12** 319-327 (1993).
2. R. Tauler, S. Lacorte, D. Barceló, J. Chromatogr. A, **730** 1-2 177-183 (1996).
3. A. De Juan, R. Tauler, Analytica Chimica Acta, **500** 1-2 195-210 (2003).
4. Keller, H.R. and D.L. Massart, Chemom. Intell. Lab. Syst. **12**, 209-224 (1992).
5. W. Windig, J. Guilment, Anal.Chem. **63** 1425-1432 (1991).

Evaluation of taste solutions by sensor fusion

Yohichiro Kojima,* Eriko Sato,** Masahiko Atobe,** Miki Nakashima,** Yukihisa Kato,** Koichi Nonoue,*** and Yoshimasa Yamano***

**Tomakomai National College of Technology, Tomakomai, Hokkaido 059-1275, Japan*
***Taste Res. Dept., Pokka Corporation, Kitanagoya, Aichi 481-8515, Japan*
****Institute of OISHISA science, Takamatsu, Kagawa 761-0301, Japan*

Abstract. In our previous studies, properties of taste solutions were discriminated based on sound velocity and amplitude of ultrasonic waves propagating through the solutions. However, to make this method applicable to beverages which contain many taste substances, further studies are required. In this study, the waveform of an ultrasonic wave with frequency of approximately 5 MHz propagating through a solution was measured and subjected to frequency analysis. Further, taste sensors require various techniques of sensor fusion to effectively obtain chemical and physical parameter of taste solutions. A sensor fusion method of ultrasonic wave sensor and various sensors, such as the surface plasmon resonance (SPR) sensor, to estimate tastes were proposed and examined in this report. As a result, differences among pure water and two basic taste solutions were clearly observed as differences in their properties. Furthermore, a self-organizing neural network was applied to obtained data which were used to clarify the differences among solutions.

Keywords: Taste, Sensor fusion, Ultrasonic, Surface plasmon resonance, Self-organizing neural network
PACS: 43.58.+z

INTRODUCTION

An organism which sensitively perceives the continuously changing outside world accurately detects molecules or light quanta. Therefore, studies on taste and odor sensors, which simulate chemical mechanisms of sensation of organisms, are actively performed [1]. These sensors detect and process much information, such as electrical properties of a lipid membrane, which are subsequently adopted in the simulation of chemical process of sensations. On the other hand, a new sensing method in which biological and related substances are not utilized has recently been reported[2,3]. In this study, a sensor fusion method of ultrasonic wave sensor and various sensors, such as the surface plasmon resonance sensor, to estimate tastes were proposed and examined.

EXPERIMENTAL AND METHODS

A high-voltage pulse of several hundred V transmitted from a pulsar is transformed to an ultrasonic wave pulse using an electroacoustic transducer element and introduced into the solution to be measured. Subsequently, the obtained pulse voltage is sent to an oscilloscope and analyzed using computers. Next, by utilizing the surface plasmon resonance sensor, the change of refractivity was detected. The solution measured was pure water passed through an ion-exchange membrane. As basic tastes, sucrose (sweet), and NaCl and KCl (two types salty taste) were used; their concentrations were constant at 1M. During measurement, the temperature of solutions was held at 30.0°C.

RESULTS AND DISCUSSION

Frequency characteristics of the second reflected wave of KCl and NaCl salty solutions were measured. These characteristics are different from those of high-viscosity sucrose solution. A difference between KCl and NaCl solutions was observed in the peak value at around 5.18 MHz. However, similar peaks were obtained at other frequencies, which we considered was because the characteristics of KCl and NaCl solutions are similar since they are both monovalent cations. The power spectrum of sucrose solution was generally lower than that of salty solutions, which indicated that the ultrasonic wave was greatly attenuated due to the viscosity of sucrose. At the frequency at which marked attenuation of the power

CP1137, *Olfaction and Electronic Nose: Proceedings of the 13th International Symposium,* edited by M. Pardo and G. Sberveglieri

spectrum was observed, the ultrasonic wave was absorbed by the reflecting surface of the transducer.

Next, Changes of refractive index which were obtained from the surface plasmon resonance sensor were measured. For comparison with the characteristics of mixed solution, the changes of refractive index of sweet and salty solutions are included. Depending on the type of solution, the pattern of the changes of refractive index is different. Accordingly, solutions are likely to be roughly classified into three types: mixed solutions, sucrose solutions with high viscosity and salty and electrolyte NaCl solutions.

CONCLUSIONS

In this study, to develop the new taste sensing method, we focused on sensor fusion which are used to study the properties of various substances. To make a taste sensing method applicable to marketed beverages which include many taste substances, it is necessary to extract the maximum amount of information possible using a single device and to increase the number of parameters used for differentiation. As a result, the possibility of taste sensing using sensor fusion was suggested.

REFERENCES

1. K. Toko, *Biomimetic Sensor Technology*, Cambridge University Press (2000).
2. Y. Kojima and H. Kato, *Sensors and Materials*, 13, 145-154 (2001).
3. Y. Kojima and T. Mikami, *IEEJ Trans. SM*, 122, 318-325 (2002) (in Japanese).

Gas chromatography instrumental variation and drift compensation

S. LABRECHE, O.CABROL, F. LOUBET, H. AMINE
ALPHA MOS SA
20 Avenue Didier DAURAT, 31400 Toulouse, France
amine@alpha-mos.com

Abstract. In Electronic Nose domain, Gas Chromatography introduce a new challenge not only in the technology point of view but also in data processing one. Drift and instrument to instrument variations have been observed for different applications. Theses variations concern both retention time and intensity of chromatograms. Statistical and Mathematical models have been adapted in order to compensate the instrumental variations

Keywords: Drift, Compensation, Gas Chromatography.
PACS: 82.80.B

INTRODUCTION

Nowdays, Gas Chromatography is used as an electronic Nose by adapting Multivariate statitics classically used with multisensors systems[1,2]. Drift and instrument variation become an issue as one needs to maintain models over time without retraining.

DRIFT IN GAS CHROMATOGRAPHY

Two elements are concerning by the drift: retention time and peaks intensity.

Retention Time

Tow approches have been used for retention time drift compensation:

1. Kovats Index [3] that allows to have a relative index that depends only on some chemical carateristics of the molecule and the type of the chromatographic column.
2. Correlation Optimisation Warping (COW) method[4] based on correlation between the chromatograms.

Each of these two methods allows to align time retention of the chromatograms. They compensate only the retention times.

Peak Area

An adaptation of the Robust Point Matching (RPM) proposed by Chui and Rangarajan [5] is done . This method, used on image processing, is base on Thin-Plane Spline (TPS). The method is called TPS-RPM. We have used it to compensate both intensities and retention times[5]. The objective of this method is to find mathematical function that matches the two chromatograms.

EXPERIMENTAL AND METHODS

Two GC instruments have been used to analyse complex product.

To compensate the chromatograms generated by the two instruments, we have used COW and TPS-RPM methods.

The objective of each of these methods is to find mathematical function that match the two chromatograms. One of the chromatograms is considered as a "Reference". The second is the chromatogram to be transformed. Mathematically, if we note **A** the Reference, **B** the chromatogram to transform, **f** the function and **C** the result of the transformation, one has

$$C = f(B) \text{ and } C \text{ is very similar to } A$$

Figure 1 Shows the chromatograms generated by the two systems. One can see a difference between the two chromatograms

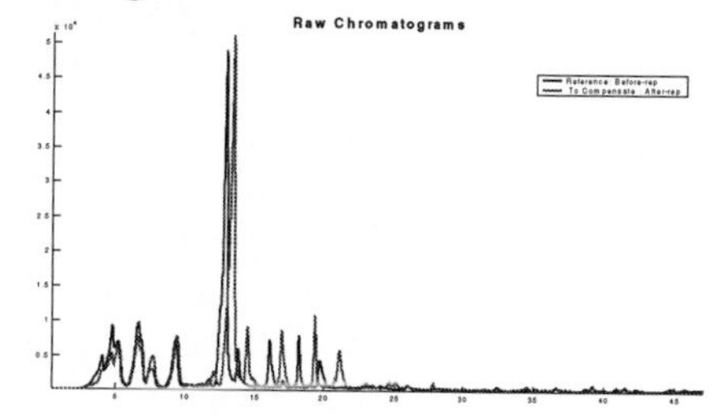

FIGURE 1. Chromatograms before compensation

CP1137, *Olfaction and Electronic Nose: Proceedings of the 13th International Symposium*, edited by M. Pardo and G. Sberveglieri

The results of COW [Respectively. TPS-RPM] method is represented on Figure 2 [Resp. Figure 3]. As we can see on each figure, the two chromatograms are superimposed. One cans deduce that for this example both of two methods compensate the retention times drift.

In order to compare the results of the two methods, we represent on the same figure (Figure 4) three chromatograms

-Chromatogram A (Bleu) (Reference)

-Chromatogram transformed by COW (Magenta)

-Chromatogram transformed by TPS-RPM (Green)

We observe that the two methods well compensate globally the time retention. One cans see that the second method do a good compensation of the intensities.

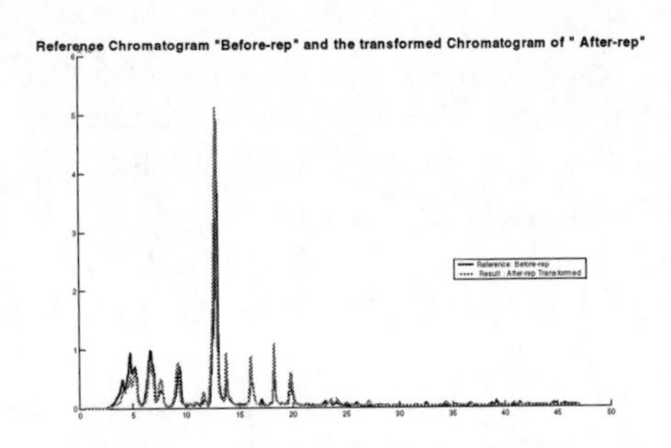

FIGURE 2. Result of COW method

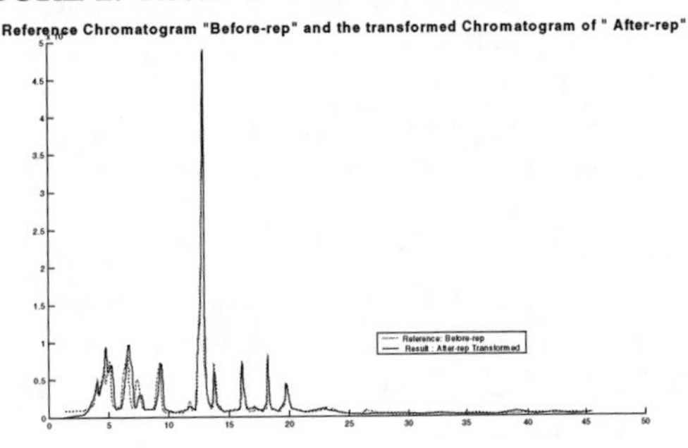

FIGURE 3. Results of TPS-RPM method

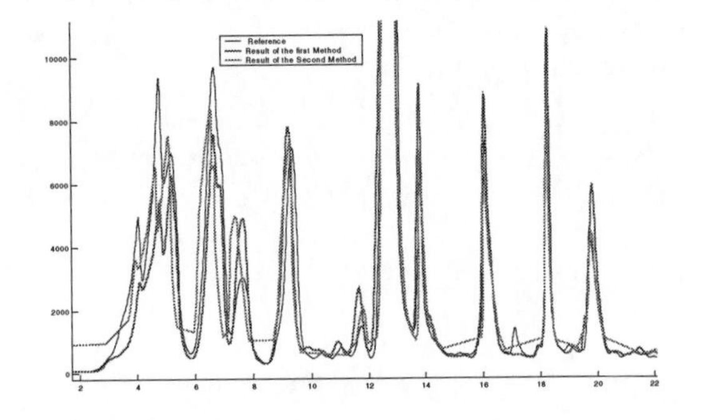

FIGURE 4. Zoom on both results

DISCUSSION

In this paper we have shown the possibility to compensate chromatograms drift by two methods. However, we have faced several problems

1. Calculation time depends on the numbers of acquisition sampling time. When applying directly the algorithms, we were obliged to stop the programs before calculation convergence if we take the whole chromatograms
2. Real sample chromatograms are more complex than some chromatograms area of these samples contains peaks while it does not on the Alcan's samples.

To solve the tow problems encountered we have developed a sampling selection method that takes only some characteristic points of the chromatograms.

CONCLUSIONS

Compensation methods of instrumental variation and drift of chromatograms have been developed. Each of them consists of alignement of two chromatograms of the same sample. In order to facilate the use of Electronic Nose based on Gas Chromatography, we are looking for a possibly to calculate each of the transformation only on chemical stable samples like Alkanes.

REFERENCES

1. Tan, T.T.; Loubet, F.; Labreche, S.; Amine, H. Quality control of coffee using the FOX4000 electronic nose Industrial Electronics, Proceedings of the IEEE International Symposium on ISIE. Volume 1, Issue , 7-11 Jul 1997.
2. H AMINE, S. BAZZO , S. LABRECHE, E. CHANIE Quantitative Applications Stability and transferability using Sensor Array System . ISOEN 2000.
3 A.Wehrli and E.Kovats .Helv. "Gas-chromatographische Charakterisierung organischer Verbindungen. Teil 3: Berechnung der Retentionsindices aliphatischer, alicyclischer und aromatischer Verbindungen. Chim.Acta 42, 2709 (1959)
4. Nielsen N.-P.V.; Carstensen J.M.; Smedsgaard J. Aligning of single and multiple wavelength chromatographic profiles for chemometric data analysis using correlation optimized warping: Journal of Chromatography A, Volume 805, Number 1, 1 May 1998 , pp. 17-35(19).
5. Chui H, Rangarajan A " A new Point Matching Algorithm for Non-rigid Registrations" Technical report, Department of Computer and Information Science & Engineering, University of Florida. 2002

Embedded Electronic Nose for VOC Mixture Analysis

B. Botre, D. Gharpure and A. Shaligram

Department of Electronic Science, University of Pune,
Pune: 411007, India. Phone: +91-0202-25699841.
Email: bab@electronics.unipune.ernet.in

Abstract. This paper details the work done towards a low cost, small size, portable embedded electronic nose (e-nose) and its application for analysis of different VOC mixtures. The sensor array is composed of commercially available metal oxide semiconductor sensors by Figaro. The embedded E-nose consists of an ADuC831 and has an RS 232 interface for Desktop PC for higher level data collection and NN training. The ESP tool with database facility and multilayer perceptron neural network (MLP NN) is employed to interface the embedded hardware and to process the electronic nose signals before being classified. The use of embedded e-nose for the quantification of VOCs in mixtures is investigated.

Keywords: embedded electronic nose, neural network, VOCs mixture analysis.
PACS: 47.66. -P, 87.19.lt

INTRODUCTION

Electronic nose is a emerging field that deals with odor sensing and classification. Research into odor sensing and recognition is widespread because there are potential applications of versatile and a portable electronic nose system in fields like food, medical diagnosis and environment monitoring. [1-2].

The present work describes the application of a portable embedded e-nose for Volatile Organic Compound (VOC) mixture quantification. The traditional techniques used for VOC monitoring are based on gas-chromatography (GC), mass-spectroscopy (MS), FTIR-analysis etc.. These techniques are accurate, but have some drawbacks such as high cost, low portability, prolonged analysis time, enormous weight and volume. On the contrary, the electronic nose provides a low cost solution for the same purpose. In Literature, the nanostructure based sensor array and a sol-gel based electronic nose for qualitative and quantitative analysis of VOCs present in the headspace of food is reported. In another case, the classification of water solution with ethanol and their volatile organic compounds by means of the electronic nose device based on SnO_2 sensor array and the system was employed for wine classification is proposed.

The objective of the present work was to develop a cost effective and a portable embedded e-nose based on array of metal oxide semiconductor sensors to provide efficient, real-time estimation of VOCs mixture. This paper is divided as follows. The first section introduces basic blocks of embedded e-nose and ESP tool to realize the embedded e-nose. The third section deals with the embedded e-nose hardware and its different operating modes. The last section presents an experimental validation of the proposed embedded e-nose as well as an application of the instrument to estimate the concentration of analyte in VOCs mixture.

EXPERIMENTAL AND METHODS

The design of electronic nose starts with acquiring the response of sensor array to specific odors. The next step is obtaining the characteristic features which can be used as the odor signature. Implementing a pattern recognition algorithm and training the MLP NN for identification is the final step in realization. This required a software tool for controlling the data acquisition, storage of sensor array response for characterization and training the neural networks for pattern classification.

A complete library of Virtual Instruments (VI's) for data acquisition, signal preprocessing, feature extraction, normalization and Multilayer Perceptron Neural Network (MLP-NN) for odor classification is developed in the E-nose software package (ESP) tool using LabVIEW programming language [3]. The VI library aids in realization of a portable embedded e-nose, sensor characterization, test, debug and tuning

CP1137, *Olfaction and Electronic Nose: Proceedings of the 13th International Symposium,* edited by M. Pardo and G. Sberveglieri

the system for in-field applications. The training session of MLP NN is shown in figure 1.

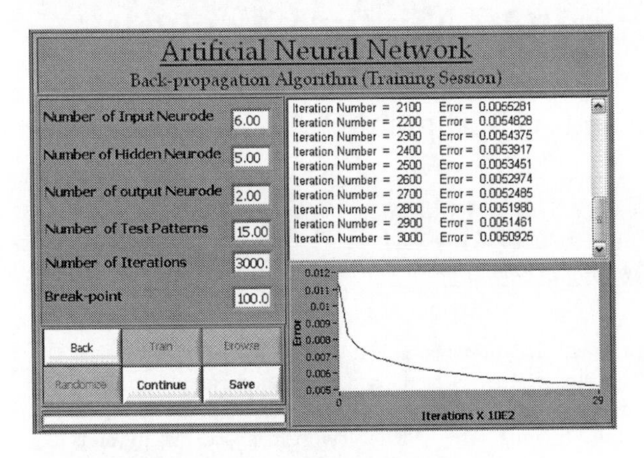

FIGURE 1. Training VI of MLP NN.

Aduc831 Based Embedded E-Nose Hardware

The functional block diagram of the ADuC831 [3] based embedded e-nose system is given in Figure 2. Three main parts composed of this system: sensor module, Odor Delivery System (ODS), ADuC831 embedded hardware platform for measurements.

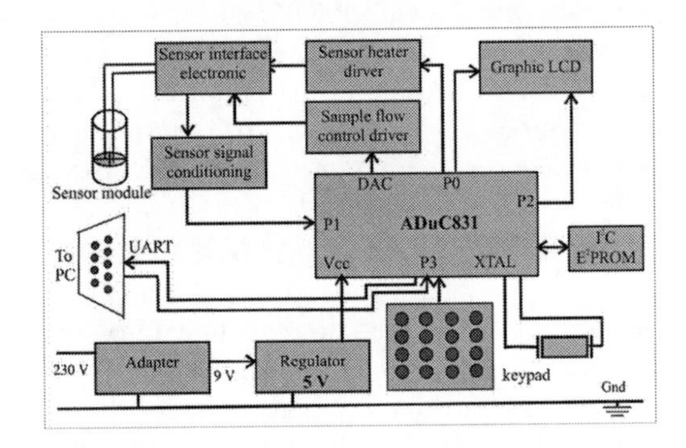

FIGURE 2. Functional block diagram of ADuC831 Embedded E-nose.

The sensor module is composed of three commercially available Figaro gas sensors: 1 TGS2620, 1 TGS2610 and 1 TGS2600. The sensor interface electronic consists of a potential divider circuit followed by an unity gain amplifier, to read the sensor conductance variation. They are supplied with a 5 V circuit voltage and a 5 V sensor heating voltage providing a specific temperature by using sensor heater driver. Two other types of sensors are added to this array to check humidity and temperature.

The odor delivery system (ODS) encompasses the automatic sample flow driver circuit, a pump, rotameter and solenoid valves. The ODS is designed to bring the odorant molecules to the sensing surface of the sensor array in the controlled environment.

The resources used to implement the circuit modules are embedded microcontroller ADuC831, External E^{2}PROM serial memory to record online sensor array data and to store trained network weights, LG128X64 graphic LCD to display waveforms and text messages, 4x4 matrix keypad for user interface and circuitry to control data acquisition and ODS.

The memory utilized on embedded hardware for the storage of program code, sensor array responses and to generate the library of trained network weights for different odor patterns. System power is provided by a switched mode power supply (SMPS) that provides clean and filtered +/- 5 V and +/- 12 V to all the circuit modules.

Embedded E-nose Software For ADuC831

Modular programs are written in cross C compiler for the functioning of ADuC831 embedded e-nose. The main software modules initialize the interfaces to sensor array, heater and sampling system. A menu driven program consisting of three operating modes, namely: Train, Identify and Purge.

Training mode provides interface to the ESP tool on the PC based e-nose using standard serial interface. The ESP tool is used to optimize the parameters required for the functioning of the embedded e-nose. This software module also includes the trained network weight download option to store the optimized parameters in the library generated using the embedded e-nose memory.

Identification mode consists of several sub-modules which facilitates data acquisition, preprocessing and feature extraction on ADuC831. The feed forward calculations are done using floating point variables in C compiler. Sigmoid activation function is implemented using the exponential function. The onchip 1.8 kbyte memory is utilized for calculations of the MLP NN feed forward calculations using floating point variables in C compiler.

Purge mode is developed to drive the pump, sample flow system and the DC fan to clean reagents from the sensor chamber.

All these modes are linked to the main menu program to operate the embedded e-nose for different odor/vapor classification and quantification.

VOCs Mixture Analysis

Experiments were performed using the embedded e-nose platform to estimate the percentage of concentration in the VOCs mixture. The mixtures of VOCs such as (ethanol + water), (methanol + water) and (ethanol + methanol) were prepared as tabulated in Table 1.

TABLE 1. Amount of VOCs mixture.

VOC mixture in %	mixture in ml
20 - 100	2 - 10
40 -100	4 - 10
60 - 100	6 - 10
80 - 100	8 - 10
100 - 100	10 - 10

The sensor heater was kept ON during the whole measurements. The measurement protocol followed was: 20 sec of sensing element preheating time, 20 sec of ambient air, 1min of stabilization time, 20 sec of odor/vapor time and 1 min of odor stabilization time.

The sensor transient was obtained by injecting 250 ppm of alcohol sample and was digitized at 1 samples/second sampling rate using the ESP tool. The measurements comprised four VOC mixtures at 5 percentage levels as given in table 1 with eight replicates of each type of measurements. 250 ppm of each VOC mixture was injected into the glass vial for the measurement. The database of 160 (4 mixture types, 8 replicates, 5 mixture concentrations) measurements and 480 (3 TGS26xx sensors) transient responses were prepared.

The sensor transient responses to every mixture analyte were recorded and processed using ESP tool. Figure 3 shows the sensor transient response to the VOC mixture. The maximum conductance (max_G) and rise time (Tr) features were optimized and normalized between [0 – 1] using the ESP tool [3].

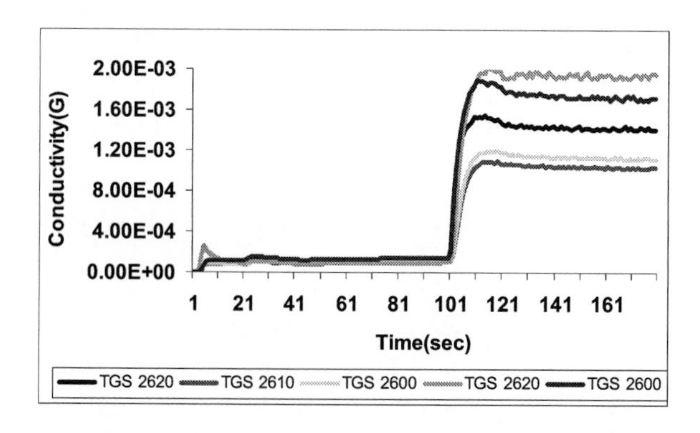

FIGURE 3. Sensor transient response to 60% ethanol and 100% methanol mixture.

RESULTS

Principal Component Analysis

PCA was performed on the optimized feature vectors. Figure 4 depicts the PCA analysis score plot performed on all mixtures, where the % concentration of ethanol and methanol in water, ethanol in methanol and methanol in ethanol formed the separation among them. The PCA plot has shown 58% and 26% variance for the alcohol mixture analysis as clearly shown in Figure 4. PCA analysis showed mixture classification ability based on optimized features (max_G and Tr).

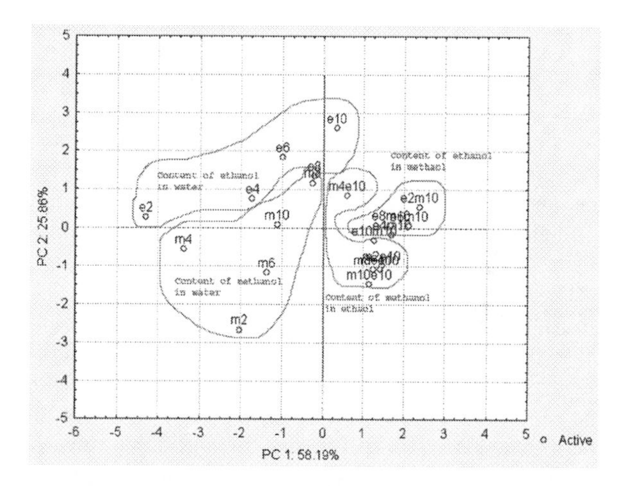

FIGURE 4. PCA of alcohol mixtures, e – ethanol content in water, m – methanol content in water, em – ethanol content in methaol. The numbers are showing the % concentration of alcohol sample in the mixture.

Training MLP NN

The MLP NN structure used in VOC mixture analysis is shown in Figure 5. To train the network, 1920 vectors (for 4 sets of mixture analytes) were used and divided into: 1320 learning vectors and 600 test vectors.

The MLP NN was trained using Back-propagation (BP) algorithm in the ESP tool. The employed network architecture was 6 neuron input layer for the input data (max_G and Tr), one hidden neuron layer with the optimized 5 neurons and finally 3 neuron output layer for the identification of amount of alcohol vapor concentration in the mixture. The MLP NN was trained by optimizing the learning rate 0.3, momentum 0.0075 and 64,000 learning iterations. After training, the connection weights and the configuration of the MLP NN were stored and evaluated with the test feature vectors in the Recall mode.

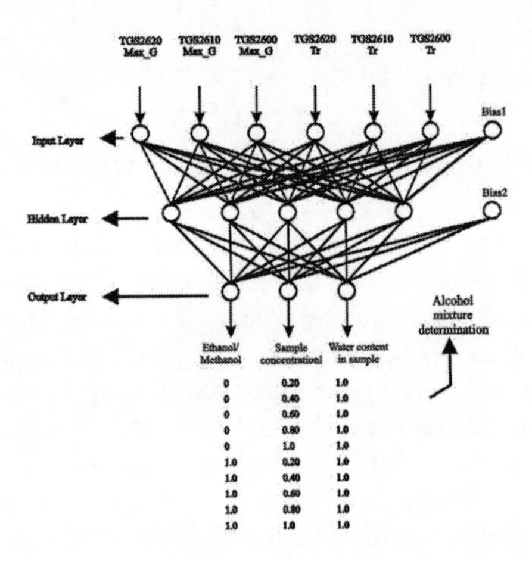

FIGURE 5. MLP NN architecture for mixture determination using the embedded e-nose.

After confirmatory tests, the trained network weights were downloaded on to the embedded e-nose memory. The embedded e-nose has successfully estimated the mixtures of ethanol in water, as shown in Figure 6 A, and methanol in water, as shown in Figure 6 B.

An output neuron in the embedded e-nose yields a 0.20, 0.40, 0.60, 0.80 and 1.0 corresponding to 20%, 40%, 60%, 80% and 100% for the neuron assigned to the vapor and its % concentration of alcohol present in the mixture. A '0' or '1' for the ones that reveal presence of the other analyte in the mixture as shown in the MLP NN structure Figure 5.

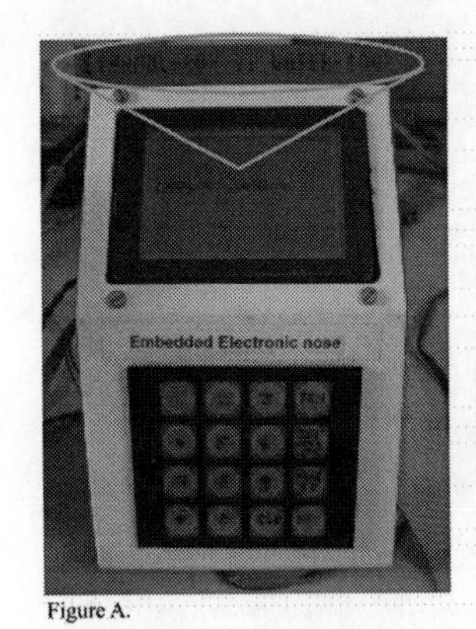

Figure A.

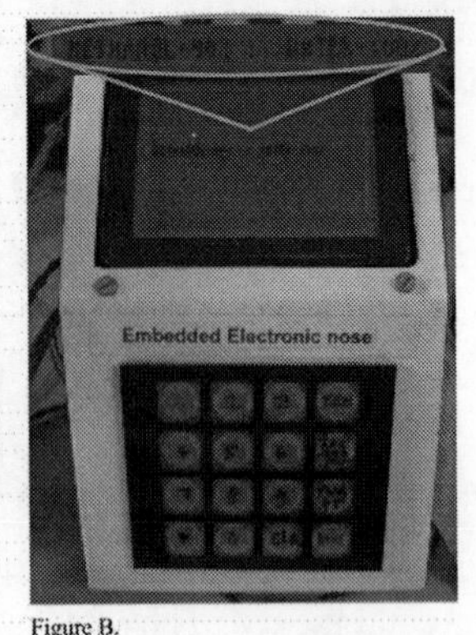

Figure B.

FIGURE 6. Determination of Ethanol and B. Methanol content in water using the embedded e-nose.

DISCUSSION

The estimation of concentration in VOC mixtures is carried out using the embedded e-nose. The concentration of (ethanol + water), (methanol + water), (ethanol + methanol) and (methanol + ethanol) mixtures is analyzed with the help of and PCA. The % amount of VOC in mixtures is determined at the output of the MLP NN. The mean concentration estimation in mixtures along with their errors in RMS are also indicated in table 2. It is evident that the RMS error is quite high for the lower level of concentration mixtures, whereas it is less for the higher level concentration mixtures.

TABLE 2. Concentration determination of VOC mixture.

VOC	Ethanol + Water		Methanol + Water		Ethanol + Methanol	
mixture in %	**Mean**	**RMS (Er +/-)**	**Mean**	**RMS (Er +/-)**	**Mean**	**RMS (Er +/-)**
20 - 100	20.62 –99.90	2.61 – 1.97	21 – 99.67	1.19 – 1.21	20.5 – 99.5	1.77 – 1.64
40 - 100	39.00 –99.87	1.85 – 1.72	40.25 – 100.1	1.03 – 1.14	40.75 – 100.23	1.16 – 1.15
60 - 100	60.50 – 100.00	1.06 1.26	60.5 – 100.00	0.75 – 0.89	59.87 – 99.89	1.24 – 1.10
80 - 100	80.00 –99. 89	1.19 – 1.15	79.62 – 100.00	0.51 – 0.47	79.5 – 100.11	0.53 – 0.64
100 – 100	99.89 – 100.10	0.90 – 0.78	99.79 – 100.12	0.40 - 0.36	99.5 – 100.05	0.37 – 0.28

CONCLUSION

This work detailed a low cost, small size and portable embedded electronic nose with an array of metal oxide semiconductor sensors interfaced to the MLP-NN on embedded processor. To implement the system in embedded processor, parameters like data acquisition, feature extraction and MLP-NN weight values are optimized using the ESP tool. The embedded e-nose is successfully used to estimate concentration in different VOCs mixture. The estimation of different VOC mixtures has been studied and their results are discussed.

ACKNOWLEDGMENTS

The authors thank University of Pune, CSIR and DST for financial support.

REFERENCES

1. Di Natale C et. al, *Biotechnol. Agron. Soc Environ*, **5**, 159-165, (2001).
2. S. Sunshine, *US patent*, No. 6658915, (2003).
3.. B. Botre, "Electronic nose based on embedded technology and neural network", *11th Int. conf. on cognitive and neural network,* (2007), 57.

Semiconductor Sensor Array Based Electronic Nose for Milk, Rancid Milk and Yoghurt Odors Identification

[1]B. Botre, [1]D. Gharpure, [1]A. Shaligram and [2]S. Sadistap

[1]Department of Electronic Science, University of Pune,
Pune: 411007, India. Phone: +91-0202-25699841.
Email: bab@electronics.unipune.ernet.in
[2]CEERI, Pilani, Rajasthan.

Abstract. This paper presents the use semiconductor sensor array based electronic nose for the identification of milk, rancid milk and yoghurt odors. A low cost sensor array, serial data acquisition system and E-nose software package (ESP) tool are used to generate the database, feature extraction and normalization. The MLP NN is trained using the NeuroSolutions for the identification. The network has successfully classified milk, rancid milk and yoghurt odors with 96% success rate. A sensitivity analysis is done to test the performance of the sensor data in the trained network

Keywords: electronic nose, MLP NN, milk, rancid milk, yoghurt.
PACS: 47.66. -P, 87.19.lt

INTRODUCTION

In the recent years, Electronic Noses have been successfully used for different applications particularly for food and beverage industries. Monitoring quality and freshness of the dairy products is one applications for which electronic nose could be used. S. Labreche et al. [1] used Alpha MOS FOX 4000 electronic nose to determine the shelf life of milk. The evaluation of the degradation of yoghurt samples by headspace-gas chromatography-mass spectrometry based electronic nose is reported by C.Carrillo-Carrion et al. [2]. A sensor fusion method for on-line monitoring of yoghurt fermentation is used by C. Cimander et al. [3], wherein an electronic nose, a near-infrared spectrometer (NIRS) and standard bioreactor probes were used.

The electronic noses are also used for classification and quality inspection of sea food like fish, fruits like banana, apple, tomato, meat, eggs, beverages like wine, coffee, juice [4-8] etc. Extending this idea further, this paper reports an attempt of using the semiconductor sensor array and Neurosoluation based electronic nose for analyzing milk, rancid milk and yoghurt odors. The first section deals with the experimentation and measurements using an electronic nose. The MLP NN training and results obtained are discussed along with sensitivity analysis in the second section.

EXPERIMENTAL AND METHODS

The electronic nose system (figure 1) used for the experimentation is composed of sensor module, ADuC831 serial data acquisition (DAQ) system for measurements [9], E-nose software package (ESP) tool and NeuroSolutions 5.0 from NeuroDimension Inc. The sensor array incorporates of five commercially available Figaro gas sensors: 2 TGS2620, 1 TGS2610 and 2 TGS2600.

The sensors are supplied with a 5 V circuit voltage and a 5 V sensor heating voltage providing a specific temperature by using sensor heater driver. The temperature and humidity sensors are also mounted in the sensor chamber to monitor the ambient condition while the measurements.

The Aduc831 based serial DAQ is composed of 12bit, 8-channel analog to digital converters interfaced to the sensor array. A serial communication is achieved by using RS232 communication protocol to the Desktop PC. ESP tool includes data base generation, feature extraction and normalization required before the data to be analyzed using the pattern recognition system. A NeuroSolutions 5.0 is used as the pattern recognition system in electronic nose to classify the odor patterns.

CP1137, *Olfaction and Electronic Nose: Proceedings of the 13th International Symposium,* edited by M. Pardo and G. Sberveglieri

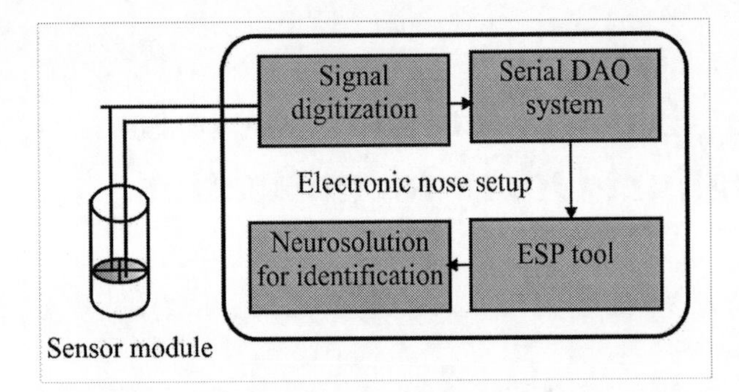

FIGURE 1. Experimental setup of Electronic Nose.

Measurements

Three samples of milk, rancid milk and yoghurt were selected for identification. The response of the sensor array was obtained by injecting 10 gm of sample and keeping the sensor heater ON during the measurement of 80 sec. The data was digitized at 1 samples/second sampling rate using serial DAQ system & ESP tool. Figure 2 A. B. C. shows the sensor array response to milk, rancid milk and yoghurt. The sensor responses clearly illustrate the difference between odor patterns of the three samples which can be identified using the pattern recognition system.

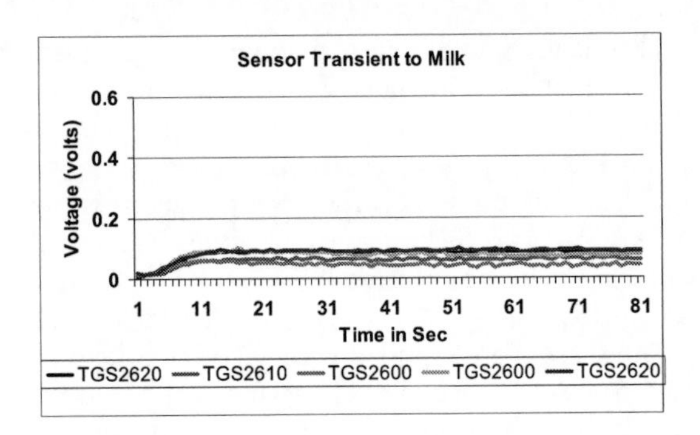

Figure A

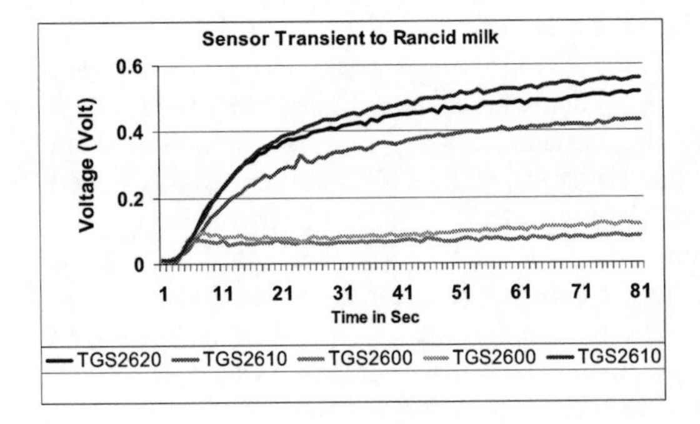

Figure B

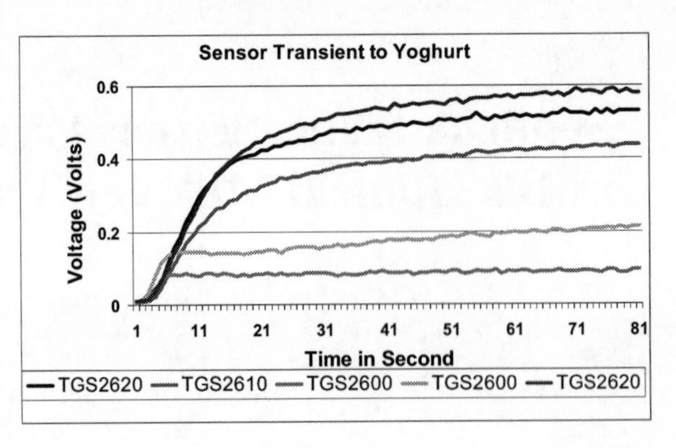

Figure B

FIGURE 2. Sensor response to A. milk, B. rancid milk and C. Yoghurt.

The measurements comprised three samples with eight replicates of each type of measurements. Five such data sets were taken over a period of 3 weeks. The database of 120 (3 samples, 8 replicates, 5 such data sets) measurements were taken.

The feature vector for each odor sample is composed of points in the stable region i.e. voltage values at the 50th second, 60th second and 70th second forming a data base of 360 feature vector. The response patterns of features were then normalized between [0 – 1] using ESP tool and the Multilayer Perceptron Neural Network (MLP NN) was trained for identification.

RESULTS

All the rows were tagged as input and desired output and randomized for better training of the MLP NN. To train the network, 360 vectors were used and divided into: 214 training vectors and 146 test vectors. The architecture of MLP NN is shown in figure 3.

The features consisting of seven neurons for seven sensors were applied as the input to the network. The output layer is composed of three neurons corresponding to milk, rancid milk and yoghurt samples to be identified and the number of neurons in the hidden layer was optimized to 5.

The network was trained using Neurosolutions for 50,000 iterations. On satisfactory performance, the connection weights and the configuration of the MLP NN were stored and evaluated with the 146 test vectors in the test network mode. The output neurons o1, o2 and o3 of the MLP NN, were assigned to milk, rancid milk and yoghurt odors respectively. The confusion matrix in table 1 shows successful identification of milk, rancid milk and yoghurt.

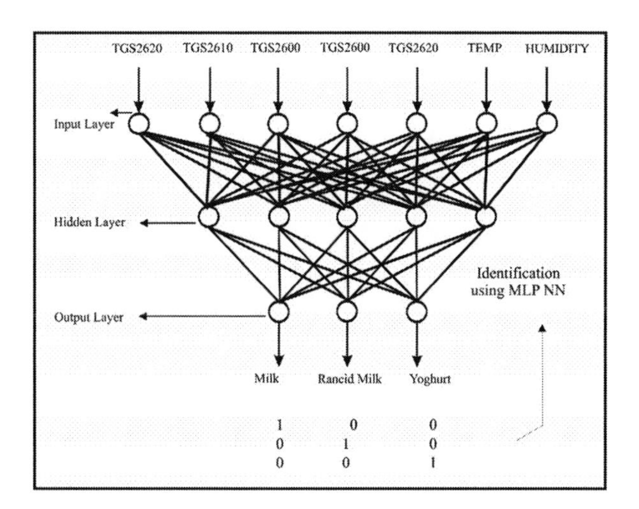

FIGURE 3. Architecture of MLP NN.

TABLE 1. Confusion matrix obtained during the MLP NN test.

Output / Desired	*o1*	*o2*	*o3*
o1	47	0	0
o2	0	41	1
o3	0	5	51

It is possible to identify milk sample up to 100 % success rate, where as rancid milk and yoghurt samples are identified with success rate up to 89% and 98% respectively. The performance of the trained network is given in table 2.

TABLE 2. Performance evaluation of the trained MLP NN for the test feature vector.

Performance	*o1*	*o2*	*o3*
MSE	0.0107	0.0439	0.04366
NMSE	0.0488	0.2027	0.1898
MAE	0.06150	0.0912	0.09056
Min Abs Error	0.02403	0.00321	0.00329
Max Abs Error	1.05237	1.05539	1.0550
r	0.983628	0.90587	0.912923
Percent Correct	100	89.1304	98.07692

DISCUSSION

The electronic nose has been successfully tested for the identification of milk, rancid milk and yoghurt. Three points in the stable region of the sensor response were used as feature vector without any signal pre-processing for the classification. A database of 360 feature vectors was obtained from the 3 points in the stable region of the sensors response. A network of 7 neurons at the input, 5 neurons in the hidden and 3 neurons in the output layer was trained using the Neurosolutions.

The network could able successfully classify theses samples up to 96% overall success rate. The success rate obtained using the trained network is 100%, 89% and 98% for milk, rancid milk and yoghurt samples respectively. The success rate for rancid milk is less as compared to other two sample odors. This could be due to the more or less similar odor patterns obtained for yoghurt and rancid milk odors as can be observed from the sensitivity analysis in figure 4.

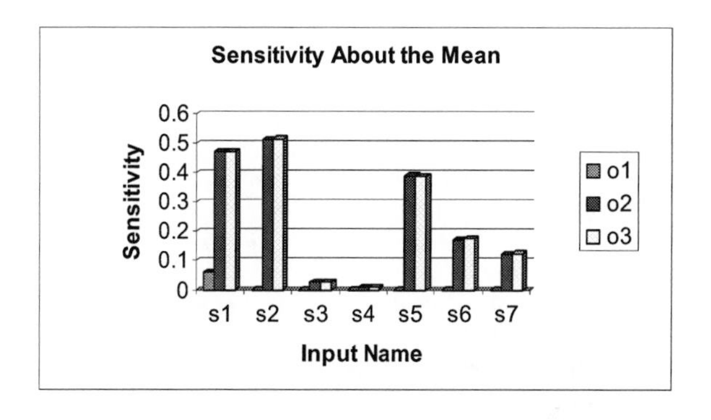

FIGURE 4. Sensitivity test of the sensors in array.

It is also observed from figure 4 that the sensor data obtained from TGS2620 and TGS2610 have shown good sensitivity while training the MLP NN to the odors from the dairy products. On the other side, the data from sensor TGS2600 have list sensitivity to these odors. The temperature and humidity sensors data have equally contributed in the identification as seen from table 3.

TABLE 3. Sensitivities of the sensors obtained for the trained MLP NN.

Sensitivity	*o1*	*o2*	*o3*
s1 – TGS2620	0.05807	0.46875	0.47007
s2 – TGS2610	1.31E-06	0.51167	0.51400
s3 – TGS2600	1.53E-07	0.02685	0.02679
s4 – TGS2600	1.76E-07	0.00967	0.00981
s5 – TGS2620	1.28E-06	0.38907	0.38731
s6 – TEMP	3.77E-07	0.17187	0.17659
s7 – HUMIDITY	3.09E-07	0.12196	0.12421

CONCLUSION

The electronic nose is successfully employed for the identification of milk, rancid milk and yoghurt. The MLP NN in neurosolutions is trained using neurosolution for the classification of these samples. The overall performance of the electronic nose has shown 96% success rate with only 3 points in the stable region as features. The database of sensors

TGS2620 and TGS2610 have contributed more in the classification as compared to the TGS2600 one as observed from the sensitivity analysis.

ACKNOWLEDGMENTS

The authors thank University of Pune, CSIR and DST for financial support.

REFERENCES

1. S Labreche et al, *Sensors and Actuators B,* **106**, 199-206, (2005).
2. E. Llobet te al, *Meas. Sci. Technol*, 10, 538 – 48, (1999).
3. C. Carrillo-Carrion et al, *Journal of Chromatography A*, 1141, 98-105, (2007).
4. Christian Cimander 1 et al., *Journal of Biotechnology*, 99, 237-248, (2002).
5. E. Molto et al, *Journal of Engineering Research*, 72, 311-316, (1999).
6. F. Sinesio et al, *Journal of the Science of food and Agriculture.* 80, 63-71, (2000).
7. R. Dutta et al, *Meas. Sci. Technol,* 14, 190-198, (2003).
8. S. Christophe S et al, *Journal of Agricultural & Food Chemistry*, 49, 3151-3160, (2001).
9. B. Botre et al, "Electronic nose based on embedded technology and neural network*", Proc 11th International conference on cognitive and neural systems*, Boston University, USA, (2007), 57.
10. http://www.neurosolutions.com/

A

B

C

D

E

F

G

H

I

J

K

L

M

N

O

P

R

S

T

U

V

W

X

Y

Z